LEUNG'S ENCYCLOPEDIA OF COMMON NATURAL INGREDIENTS

USED IN FOOD,
DRUGS, AND COSMETICS

Third Edition

LEUNG'S ENCYCLOPEDIA OF COMMON NATURAL INGREDIENTS

USED IN FOOD, DRUGS, AND COSMETICS

Third Edition

IKHLAS A. KHAN
Research Professor and Assistant Director
National Center for Natural Products Research
Professor of Pharmacognosy
University of Mississippi
Oxford, Mississippi

EHAB A. ABOURASHED
Assistant Professor
College of Pharmacy
Chicago State University
Chicago, Illinois
Adj. Res. Associate Professor
National Center for Natural Products Research
University of Mississippi
Oxford, Mississippi

WILEY

A JOHN WILEY & SONS, INC., PUBLICATION

Copyright © 2010 by John Wiley & Sons, Inc. All rights reserved.

Published by John Wiley & Sons, Inc., Hoboken, New Jersey
Published simultaneously in Canada

No part of this publication may be reproduced, stored in a retrieval system, or transmitted in any form or by any means, electronic, mechanical, photocopying, recording, scanning, or otherwise, except as permitted under Section 107 or 108 of the 1976 United States Copyright Act, without either the prior written permission of the Publisher, or authorization through payment of the appropriate per-copy fee to the Copyright Clearance Center, Inc., 222 Rosewood Drive, Danvers, MA 01923, (978) 750-8400, fax (978) 750-4470, or on the web at www.copyright.com. Requests to the Publisher for permission should be addressed to the Permissions Department, John Wiley & Sons, Inc., 111 River Street, Hoboken, NJ 07030, (201) 748-6011, fax (201) 748-6008, or online at http://www.wiley.com/go/permission.

Limit of Liability/Disclaimer of Warranty: While the publisher and author have used their best efforts in preparing this book, they make no representations or warranties with respect to the accuracy or completeness of the contents of this book and specifically disclaim any implied warranties of merchantability or fitness for a particular purpose. No warranty may be created or extended by sales representatives or written sales materials. The advice and strategies contained herein may not be suitable for your situation. You should consult with a professional where appropriate. Neither the publisher nor author shall be liable for any loss of profit or any other commercial damages, including but not limited to special, incidental, consequential, or other damages.

For general information on our other products and services or for technical support, please contact our Customer Care Department within the United States at (800) 762-2974, outside the United States at (317) 572-3993 or fax (317) 572-4002.

Wiley also publishes its books in a variety of electronic formats. Some content that appears in print may not be available in electronic formats. For more information about Wiley products, visit our web site at www.wiley.com.

Library of Congress Cataloging-in-Publication Data:

Khan, I. A. (Ikhlas Ahmad)
  Leung's encyclopedia of common natural ingredients used in food, drugs, and cosmetics / Ikhlas A. Khan, Ehab A. Abourashed. – 3rd ed.
      p. cm.
  Rev. ed. of: Encyclopedia of common natural ingredients used in food, drugs, and cosmetics / Albert Y. Leung, Steven Foster. 2nd ed. c1996.
   Includes bibliographical references and index.
   ISBN 978-0-471-46743-4 (cloth)
  1. Natural products–Encyclopedias. I. Abourashed, Ehab A., 1964- II. Title.
  QD415.A25L48 2010
  660.6'3–dc22

                                                 2009015927

Printed in the United States of America

10 9 8 7 6 5 4 3 2 1

From I. A. Khan

Dedicated to My Parents
and
My Family, Shabana, Farjad, and Sariya
For Their Love and Support

From E. A. Abourashed

To Abir and Our Two Daughters Ayat and Nada
For Their Love, Patience, and Support
And . . . to My Parents

Contents

FOREWORD TO THE FIRST EDITION, BY ARA G. PAUL, PH.D.	ix
FOREWORD TO THE FIRST EDITION, BY DONALD A. DAVIS	xi
PREFACE TO THE THIRD EDITION	xiii
PREFACE TO THE SECOND EDITION	xv
PREFACE TO THE FIRST EDITION	xxi
ACKNOWLEDGMENTS	xxv
INTRODUCTION	xxvii
NATURAL INGREDIENTS	1
INDIAN TRADITIONAL MEDICINE	637
CHINESE COSMETIC INGREDIENTS	657
APPENDIX A–GENERAL REFERENCES	687
APPENDIX B–GLOSSARY/ABBREVIATIONS	695
APPENDIX C–BOTANICAL TERMS	700
APPENDIX D–MORPHOLOGICAL DESCRIPTION OF PLANT ORGANS	706
GENERAL INDEX	711
CHEMICAL INDEX	779

Foreword to the First Edition

The publication of *Encyclopedia of Common Natural Ingredients Used in Food, Drugs, and Cosmetics* is a welcome addition to the libraries of those of us interested in natural products. The reasons for publishing this unique encyclopedia are aptly dealt with by the author in the Preface, and the principal audience has been identified as practicing technologists in the food, drug, and cosmetic industries and their purchasing agents and marketers. But, as well, it should prove to be an important reference for teaching and research in economic botany, food technology, natural products chemistry, and pharmacognosy, for it brings together information about a variety of substances that, for various reasons, are not included in recent compendia dealing with one or another of these disciplines. Yet, as the author points out, these are materials that find significant usage in our society.

Dr. Albert Y. Leung's education as a pharmacist and pharmacognosist, coupled with his extensive experience in natural products industries, provide him with a unique background that accounts for his successful synthesis of this information into a practical compendium. The material is accurately and succinctly presented, the individual monographs are selectively supplemented with a current bibliography that allows for further reading on a particular product, and the selection of products included has been skillful.

Dr. Leung is to be commended for his efforts in bringing us this most worthy publication.

ARA G. PAUL, PH.D.

*Professor of Pharmacognosy
and Dean
College of Pharmacy
The University of Michigan
Ann Arbor, Michigan*

Foreword to the First Edition

By some peculiar irony, the rapid technological advances made by the chemical industry since World War II have worked to obscure the solid basic knowledge the industry once had of some of its natural building blocks, the botanicals that were (and still are) the prime ingredients in so many drugs, cosmetics, flavors, industrial reodorants, and so on. The recently trained chemist, pharmacologist, or food flavorist (or, for that matter, the person involved in sales, marketing, or purchasing of these materials) in all likelihood has missed the fact that these materials have considerable historical significance, that they still have application in so diverse a list of products, and even what specific role they play in familiar products. After all, these older, possibly no-longer-glamorous natural substances may seem unsophisticated and awkward to handle to those trained in the glories of what might be called synthetic chemistry—the molecular juggling of carbons, hydrocarbons, acids, and alcohols to evolve pristine crystals and powders.

Many of the veteran bench chemists with experience in natural materials have retired or passed on to their ultimate reward (hopefully, a golf cart or a fishing boat in some warmer clime), so the time is coming when there will be less use of such fascinating ingredients as bloodroot, horehound, or ylang ylang oil. The veteran chemists used these materials to make cough remedies or perfume oils before there were synthetics, and when they are gone, the individual little pockets of knowledge have been in danger of dying out. They appreciated that these unique materials provide special product attributes, in the same way that classic spices do for a good chef. Then too, much of the chemical and biological information has been buried in foreign scientific literature, thus making it unavailable to the average technologist.

Dr. A. Y. Leung has been observing this widening information gap for several years, perceiving that one logical way to bridge it was to put together a compendium of materials of natural origin. He has gone about the task with logic and a sense of order, selecting the cardinal facts without deluging the reader or peruser of the book with a veritable mountain of biological data. As befits the only reference book that covers food, drug, and cosmetic aspects of common natural ingredients, Dr. Leung has identified each entry according to biological name, its alternative or slang description, a general description of the plant from which it is derived, chemical composition, pharmacological or biological activity, and uses and commercial preparations. And for those needing more information, he has included a comprehensive list of references.

Such handy organization of material makes this book especially useful to the working chemist or technologist, to the purchasing director, and to the person in sales or product development or marketing, for in one fell swoop he or she is given clear, comprehensive information with no unnecessary embellishment. Exotica become less exotic, the strange becomes more familiar.

Because of the ongoing work of the *Cosmetic Ingredient Review* and the Research Institute for Fragrance Materials, it is a safe bet that in the not-so-distant future there will be a demand for a second edition of this monumental work. These efforts will produce much information about the toxicity or safety of these materials, information that will give better clues as to whether it may be opportune or diplomatic or safe from a regulatory standpoint to persist in using a material that may be allergenic or sensitizing. Also, Dr. Leung has tried to emphasize the quality of commercial preparations, with an eye toward the purchasers and end users who will ultimately find the volume so useful.

Anyone with a sense of romance will cherish the names of these materials, and anyone with an appreciation for order and thorough documentation will regard this book as useful and to the point.

DONALD A. DAVIS, EDITOR
Drug and Cosmetic Industry
New York

Preface to the Third Edition

Ten years after the publication of the second edition, we accepted the invitation to complete the third edition of this encyclopedia. As it turned out, it was a massive task that took more than three years to accomplish even though many of the initially planned new ingredients were eventually left behind. The reader may notice that the included ingredients are almost the same as those of the second edition (ca. 300 main entries and about the same number of related species and varieties) and that the new ones are included under a new section entitled "**Indian Traditional Medicine—Ayurveda.**" Nevertheless, the total number of pages of this edition has increased by at least 50% from the previous one. The reason for this is the fact that since the inception of the Dietary Supplement Health and Education Act of 1994 (DSHEA), interest in dietary supplement/herbal medicine research has witnessed an unprecedented revival resulting in an exponential increase in scientific/industrial journals on herbals and alternative medicine and reports published therein over the past decade, not only in the United States but all over the world as well. Faced with such a massive body of data, we had to make a decision to focus our efforts on updating the literature related to the currently included ingredients rather than diverting our attention to the inclusion of new entities. Pursuing this strategy, we ended up searching Medline as our main online source, reviewing thousands of published reports for quality and content, and summarizing the findings of those satisfying our evaluation criteria. This resulted in the inclusion of more than 6500 references in the current edition. A number of the most prominent textbooks published in the field during the same period have also been added to the *general references*. Of these, two textbooks focusing on clinical trials with herbal products (BARRETT; BLUMENTHAL) and a translated atlas of phytopharmaceuticals (WICHTL) are worth mentioning.

The general format of the previous edition has been generally maintained (Source, Description, Chemical Composition, Pharmacology, Uses, Commercial Preparations, Regulatory Status, and References) with minor updated headings for "**Toxicology**" and "**Dietary Supplements/Health Foods**" to reflect the latest trends and regulations. The reader should refer to prefaces to the first and second editions for a thorough description of each heading. The newly added section includes a background on Indian traditional medicine and lists information on nine commonly used herbal ingredients.

The *general references* and *glossary* have been updated and moved to the end of the book and assigned as Appendices A and B, respectively. Two new appendices (Appendices C and D), explaining and illustrating the botanical terminology frequently encountered in the text, have also been added.

It is our hope that this new edition will provide updated information on classical herbs and/or their common ingredients included in the previous edition. Such information should continue to be useful to readers with various backgrounds who share an interest in natural product applications in medicine, nutrition, and cosmetics and who are looking for a comprehensive compilation of the recent literature summarizing the most prominent findings in this field.

IKHLAS A. KHAN AND EHAB A. ABOURASHED

Oxford, Mississippi
March 2009

Preface to the Second Edition

Fifteen years have passed since the publication of the first edition. During this period, basic information on most of the traditional ingredients in the encyclopedia has remained essentially the same. However, usages and use trends of many of these ingredients have changed. A whole new field of food/drug products, loosely categorized as "health foods," has established itself in North America; so has the use of herbal teas. Although a few conventional food and drug companies have tried to capitalize on the market generated by these new fields, most members of the medical, pharmaceutical, and food establishments have so far chosen to ignore them. However, these rapidly expanding fields of health foods and herbal teas have made it very difficult for one to ignore. For this reason, up-to-date and accurate information on their ingredients should be made readily available, not only to provide useful data for technologists and consumers interested in these ingredients, but also to counterbalance the proliferation of promotional literature from marketers that is often grossly inaccurate and misleading, as well as negative information from opposing interest groups that is based on bias and self-interest and not on relevant traditional and scientific data available. Consequently, I have included information to support the "new" uses in this edition. As Chinese herbs constitute a majority of all natural products used in the world, which have increasingly found their way into American cosmetic, health food, and herbal tea products in recent years, new ingredients described in this second edition reflect this trend.

Due to other commitments that limited my availability for this revision, I enlisted the capable assistance of Steven Foster who has earned a reputation for his writings. He has been instrumental in updating much of the information in the original entries of the encyclopedia as well as introducing most of the new non-Chinese ingredients in this revision.

Concomitant with the development of health foods and herbal teas, many books on natural products have appeared since the first edition was published. Some of these books contain well-researched information, while others are simply indiscriminate compilations of data, which only help to perpetuate the confusion relating to information in the herbal products field. In order to help minimize the spread of dubious data, books containing excessive outdated, secondary, and/or misleading data, as well as those consisting primarily of indiscriminate compilations of data, including some English titles on Chinese medicinal plants (even though aggressively marketed), are not included in the *general references*.

An unusual, but positive, alliance emerged during the past decade. Numerous scientists and practitioners from traditional fields (pharmacognosy, pharmacology, chemistry, medicine, botany, etc.) have joined forces with herbalists and manufacturers and developers of health foods and herbal teas to promote research and information dissemination in the field of medicinal plants and herbal products. Thus, the Herb Research Foundation, established in 1983 with an advisory board of respected scientists in various fields, has been engaged in promoting research in the various aspects of herbs. Along with the more recently founded American Botanical Council (1988), it publishes *HerbalGram*, a quarterly journal that provides accurate information on many commonly used natural products. In this new edition, we have selected as **general references** some of the books that we find useful and that we believe had an impact in the industry in recent years. Also, the extensive use of information from the Chinese literature in this new edition reflects the greatly increased availability of data on natural products from China during the past decade. As there is no standard translation of Chinese pharmaceutical and biomedical titles, I have used the transliterated titles of such references whenever there is a possibility of confusion. The *pin-yin* system of transliteration has been selected over the Wade–Giles system because the former is now standard in Chinese literature originating in the People's Republic of China, which is by far the more abundant than that originating elsewhere.

Along with the greatly increased availability of information and books on natural products, I have observed a tendency in both the professional and lay press in the overly simplified interpretation of this information. It is tempting to assign the biological activity of a compound present in a natural product to the product itself prematurely, irrespective of the amount present. For example, taking this approach, the common spice, ginger, could easily be turned into a panacea as it contains dozens of active compounds, each of which by itself has been shown to have various biological activities. These activities include antimicrobial (essential oil components such as linalool, geraniol, chavicol, 1,8-cineole, etc.); narcotic (cumene); spasmolytic (borneol, myrcene); analgesic (borneol, gingerols, shogaols); diuretic (asparagine); antihistaminic (citral); lipotropic (lecithins); anti-inflammatory (α-curcumene, borneol); sedative (gingerols, shogaols); hypotensive (1,8-cineole, gingerols); hypertensive (shogaols); liver protectant (borneol); cardiotonic (gingerols); antipyretic (borneol, gingerols, shogaols); insect repellent (*p*-cymene, geraniol, myrcene); antibronchitic, antitussive, and expectorant (1,8-cineole); nutrient (vitamins, minerals, amino acids); and others.[1] Under certain conditions or in specific dosage forms, ginger could indeed exert some of these effects. But to say that the spice ginger is narcotic or hypotensive is an oversimplification in interpretation.

This brings up one of the major challenges in natural products research, especially in Chinese herbs, which is to make sense out of their myriad of traditional uses. Some of the answers seem to lie in the complex chemical nature of these products. The bioavailability of these chemicals in a herbal formula or in an ingested herb is most likely very selective and dependent upon the physiological state of the individual consumer. This may be one of the major reasons why ginseng and other tonics have been used for so many centuries in China for so many different conditions, and yet despite extensive research over the past 30 years, generating thousands of research publications, ginseng has still not been "proven effective" by modern science. In our current state of specialization and advanced instrumentation and analytical and biological technology, it is very easy for a chemist to discover new chemicals or find known active chemicals in trace amounts in any plant material or for a pharmacologist to test the pharmacological activities of chemicals that are isolated only in traces from plant drugs, which would invariably result in publications that in turn would boost the political and financial status of the researchers involved. There is nothing wrong about such research. However, the challenge is to refrain from over-interpreting the results that are often blown out of context by proponents or opponents of the herbal drug as "preliminary evidence" to promote or restrict use of this particular herb.

There is also a general tendency to consider biomedical publications from Chinese sources as of inferior quality, which consequently should not be taken seriously. However, in my opinion, the most common flaw in publications on natural products, a good amount from "advanced" countries, is the failure of the investigators to identify correctly and quality control the material they are studying.

This leads to results that cannot be duplicated and contributes further to the overflow of misinformation or useless information in this field.

While use of natural ingredients in processed foods and cosmetics was at its peak when the first edition was published, use of these ingredients in drugs was on the decline. Now, the trend is reversed. More and more natural ingredients are being used in "herbal formulations" for the prevention and often the treatment of illnesses, most of which are related to side effects of our modern lifestyle or are common diseases that normally will resolve themselves with adjustment of lifestyle and without drug treatment. The former include obesity, hyperlipemia, and stress-related conditions, while the latter include some digestive problems, minor aches and pains, and the common cold and its related symptoms. Based on traditional consumption patterns and use history, many of these formulations contain ingredients that can truly be considered as food ingredients while others fall under the category of drugs, and still others can be considered as either food or drug, depending on usage. Most of the original entries in the first edition serve as ingredients in both foods and drugs in conventional usage, that is, processed foods and over-the-counter (OTC) drugs. In the second edition, I have added over 70 new entries and included a new category of usage called "**Health Food/Herb Teas.**" All food and drug uses of commercial products that do not fall under conventional processed foods or cosmetic or OTC drug categories are grouped under "**Health Food/ Herb Teas.**" In this section, we simply report on perceived uses of individual ingredients in the health food/herb tea category. This information is not intended to confirm efficacy or safety for a given indication. Rather, it is meant only to indicate for what purposes consumers may be using these products.

The debate whether health foods or herbal teas should be classified legally as genuine foods/teas or as drugs still goes on. While health food and herbal tea companies consider their products as composed of food ingredients, the medical and pharmaceutical industries generally view them or prefer to classify them as drugs requiring strict federal control. Although these opposing views are obviously dictated by economic and political considerations, the truth, in reality, lies somewhere in between. I expect this ongoing debate to continue for a long time. In the meantime, the new "**Health Food/Herb Teas**" category should be adequate in covering reported uses in these areas. However, the information reported here should in no way be construed to be an endorsement of the reported usages.

Also, the "**Folk Medicine**" category has been changed to "**Traditional Medicine**" to accommodate Chinese traditional medical usages of the new Chinese drug and cosmetic entries, as well as to recognize the role that traditional medicine now plays in primary health care delivery, particularly in developing countries. Since 1978, the World Health Organization (WHO) and dozens of collaborating institutions worldwide have sought to assess the value and extent of the use of plants in health care systems. WHO has estimated that as much as 80% of the world's population rely chiefly on traditional medical systems, primarily in the form of plants, plant extracts, and active principles. Observing traditional, historic, folkloric, or ethnobotanical uses of plants is regarded as a useful approach for targeting research leads in the development of new drugs from plants. A recent survey of medicinal plants used in therapy worldwide found that 119 distinct chemical substances derived from 91 species are used as drugs in one or more countries. Of these plant-derived substances, 74% were discovered following chemical studies to determine the active compounds responsible for the use of the plant in traditional medicine.[2] While many traditional uses may not be validated as safe or efficacious by current scientific methodology, they can provide valuable leads for new or expanded utilization in the future.

Under "regulatory status," information has been included on German regulatory monographs. German health authorities have established a separate expert commission ("Commission E") to develop standardized therapeutic monographs on herbal medicines. It has produced nearly 300 "Therapeutic Monographs on Medicinal Products for Human Use." Each monograph, published in the German Federal Gazette (*Bundesanzeiger*), includes details on the name of the drug, constituents, indications (including those for the crude drug or preparations), contraindications (if any), side effects (if known), interactions with other drugs or agents (if known), details on dosage of the crude drug or preparations, the method of administration, and the general properties or therapeutic value of the herb or herb product. The German monograph system is considered to be the best governmental information source on medicinal plant usage produced by a Western industrialized nation.[3] It also serves as the model for the development of a European phytomedicine monograph system produced by the European Scientific Cooperative on Phytotherapy (ESCOP) for use by European Union member countries.

Since the publication of the first edition, use of natural ingredients in cosmetics had been slowly declining until more recently, when a new surge of interest in Chinese cosmetic ingredients prompted the introduction of a number of Chinese natural products into American cosmetics. Although used for centuries in China, these ingredients are new to most American cosmetic formulators. Some of these new ingredients can be found among the more than 24 main entries that I have included in this revision. Others (more than 22) can be found under the new section titled "**Chinese Cosmetic Ingredients.**" This section describes in brief some of the more commonly used natural ingredients in Asia, which may now be found in new cosmetic products on the domestic market.

Despite renewed talks in the herbal/botanical industry to standardize quality and to assure purity of herbal ingredients, trade practices in this industry have not changed significantly during the past decade. And irrespective of claims by individual suppliers, manufacturers, and associated trade groups on quality, no *meaningful* assay standards or quality assurance methods have been introduced to guarantee purity and quality of many natural ingredients. Thus, for example, the most commonly used ingredients, such as aloe vera and ginseng, still lack meaningful assay standards and are frequently adulterated. The practice of this intentional adulteration is implicitly encouraged by manufacturers who purchase only low priced ingredients and who will simply accept dubious "certificates of purity" from suppliers as the sole proof of quality and by the common practice of employing "label claims" in the cosmetic industry. Only a very small number of companies have their own programs to standardize and control the identity and purity of the herbal ingredients used in their products. In addition, due to ignorance, even some well-known herbs, especially in their powdered forms, are misidentified, yet distributed as genuine in the industry. These include echinacea, eleuthero, ginseng, and numerous Chinese herbs such as fo-ti. Thus, it is obvious that much remains to be done in assuring the identity and quality of natural ingredients in the health foods/herb teas field.

This adulteration/misidentification has caused a major problem in the research on commercial natural products. Due to the failure of researchers to recognize the importance of identifying the correct source of test materials, results of studies on unidentifiable commercial herbal products (e.g., "ginseng capsules" or "aloe vera") are irreproducible and mostly worthless. Because of this problem, one should exercise extreme caution when quoting results of these studies. A well-publicized example is an uncontrolled study on ginseng (?) resulting in the so-called "ginseng abuse syndrome," which was published in a reputable journal.[4] This study has been repeatedly quoted worldwide for the past 16 years both in scientific journals and in the lay press. None of the people who quoted this study seemed to have read the original publication, noticed, realized, or cared that the results of that study were based on uncontrolled test materials that included not only Asian ginseng (identity and purity doubtful), but also American ginseng (?), Siberian ginseng (?), desert ginseng (canaigre) (?), caffeinated drinks, other drugs the subjects happened to be taking, as well as other unidentified materials (could be anything) in commercial "ginseng" products! Unfortunately, this is not the only incidence of such publications or research by researchers and editors who lack expertise in the natural products area and who otherwise are eminent in their own fields. If such papers were submitted to journals of natural products such as the *HerbalGram*, *Planta Medica*, and *Journal of Natural Products*, they would be rejected outright. This clearly demonstrates the need for experts of other disciplines to be aware of the intricacies of natural products when investigating, reporting, and evaluating these products.

Another important point to remember when studying natural products is that it is sometimes not enough just to identify correctly the botanical source of the natural product to be studied, especially where Chinese herbs are concerned. While in most cases with Western medicinal plants it is sufficient to simply assure their botanical identity, it is not so with Chinese herbal materials. In addition to their correct botanical sources, Chinese herbal materials require further clarification, including plant parts used and whether or not the materials are simply cleaned and dried or are specially treated with other herbs and/or boiled in water or wine. Thus, simply identifying an herbal drug as *Ephedra sinica* Stapf can mean one of the at least two different drugs with distinctly different medicinal properties: *mahuang* (stem) is diaphoretic, among other properties, while *mahuanggen* (root) is antiperspirant. Another example is *Polygonum multiflorum* Thunb., from which at least three different herbal products are derived, each with distinct medicinal characteristics: stem, raw root tuber (*heshouwu*), and cured root tuber (*zhiheshouwu*). It is obvious the Western term for it, fo-ti, is meaningless. A voucher specimen of *Polygonum multiflorum* to go with fo-ti would further add to the confusion and would not determine whether the fo-ti shipment in question is the laxative (raw root) or the tonic (cured root).

A recent trend in the herbal industry is to market the so-called standardized extracts, such as ginseng extract standardized to "ginsenosides" content or Siberian ginseng extract to "eleutherosides" content. However, as there is normally more than one (or one type of) active component in a natural product, standardization based on one particular type of chemical component is not representative of the total activity of the product. Consequently, these arbitrarily selected components can only be useful as a "marker" of product quality. And these "markers" are only valid for extracts that are total extractions of the herbs concerned. Extraction processes designed to extract these "markers" selectively would produce extracts that are not representative of the original herbs. Thus, a "standardized" ginseng or Siberian ginseng extract may be devoid of polysaccharides that are also biologically active. Also, the ginsenosides in a ginseng extract may not be from ginseng itself but rather from another much cheaper, non-ginseng source (see **ginseng**). To be fair to both traditional and modern science, one should not be overzealous in trying to equate a chemical constituent to a traditional herbal drug.

As more and more biological and toxicological research is performed on commercial natural products, it is increasingly apparent that scientific evaluation of individual purified components from these natural products has rarely produced results that are consistent with the property of the products *in toto*. Consequently, one should not be prematurely alarmed if one of numerous components in a long-used natural product is shown to have toxic effects in the laboratory, unless further research on the product in its complete form produces the same effects. Conversely, one should not be overoptimistic in claiming a particular herb or natural product as "cure" for a certain disease after studies have indicated that one of its numerous chemical components exhibits a positive effect on the disease. This is especially true if this component is present only in minute quantities whose effect may be overshadowed by those of other compounds present, thus making the herb (in its original form) inactive as a cure for the disease. Oleanolic acid is a typical example. It is widely distributed in nature. A recent double-blind study involving 152 cancer patients demonstrated it to have immunomodulating effects (enhanced phagocytosis, E-rosette formation, and delayed hypersensitivity), improving the general condition in two-thirds of the patients. Preliminary studies have also indicated it to be effective against hepatitis and HIV.[5] However, all this does not mean eating cloves and olives, both containing oleanolic acid, will necessarily produce such an effect.

We are currently being literally choked by an overabundance of data on natural products, much of which either has not been evaluated or is of dubious value. In this second edition, as in the first edition, every effort has been made to evaluate all original publications available to assure that the research methods and findings are of decent quality. And I have paid particular attention to the identity and quality control of the test materials. Papers reporting on results of studies based on unidentified or unidentifiable test materials are not cited under the respective entries (e.g., the so-called "ginseng abuse syndrome" not cited under ginseng) for the same reason that results of research on an unidentified "yellow powder" as due to riboflavin or curcumin would not be reported in a medical or pharmaceutical journal.

Appearing in the English literature mostly for the first time, the information on the new Chinese natural products in this second edition has been gathered from dozens of major Chinese classical and modern works and from over 50 Chinese journals on traditional and herbal medicine. I have tried my best to present a balanced view of the traditional and modern aspects of Chinese herb use. The new ingredients selected for this new edition generally reflect the trend in current commercial use of natural products in America. I hope this new edition will provide the readers with an accurate update on the original entries of the first edition as well as an overview of the huge resources in Chinese herbal ingredients.

1. J. A. Duke, *HerbalGram*, **17**, 20 (1988).
2. N. R. Farnsworth et al., *Bull. World Health Organ.*, **63** (6), 965 (1985).
3. V. E. Tyler, *HerbalGram*, **30**, 24 (1994).
4. R. K. Siegel, *JAMA*, **241**, 1614 (1979).
5. Y. Sun et al., *Chin. J. Clin. Pharmacol.*, **6**, 72 (1990).

ALBERT Y. LEUNG

Glen Rock, New Jersey
July 1995

Preface to the First Edition

About 500 natural ingredients are currently used in commercial food, drug, and cosmetic products. These do not include antibiotics, vitamins, and many other natural substances that constitute prescription drugs nor medicinal herbs that are not readily available in commerce. Some of these ingredients are pure chemicals isolated from natural sources while others are extracts of botanicals. Our daily food, drug, and cosmetic items often contain these ingredients. Many of the substances used in foods are also used in drugs and cosmetics, where higher concentrations are involved.

Three major reasons have prompted me to compile this encyclopedia. First, no reference books are presently available that specifically and simultaneously deal with commonly used natural ingredients in processed foods, over-the-counter drugs, and cosmetics. Since many natural flavor ingredients and food additives are also drug and cosmetic ingredients when used in higher concentrations, there has been an acute need for a compact reference book that provides condensed and accurate information on these substances, saving the reader much time and effort that otherwise would have to be spent in consulting various handbooks and journals.

Second, most of the currently available technical reference books in the English language on food, drug, or cosmetic ingredients contain limited and out-of-date information regarding naturally derived substances. Many formerly official botanical drugs that are no longer official in the United States Pharmacopoeia (U.S.P. XIX) or the National Formulary (N.F. XIV) are still widely used in nonprescription pharmaceutical preparations and in food products. Yet they are largely neglected or ignored by editors or authors of readily available handbooks. Presumably, when a botanical drug is deleted from a currently official compendium, there should no longer be any interest in it. Formerly official drugs such as arnica, chamomile, rhubarb, valerian, white pine, and witch hazel are still widely used today in foods, drugs, and cosmetics; so are many plants that have never been admitted as official drugs, examples of which are alfalfa herb, annatto seed, chicory root, fenugreek seed, ginseng root, and rose hips. There is still ongoing, active research on many of these natural products, particularly outside the United States. Since these botanicals are very much a part of our culture and daily life, information on them should be readily available. This encyclopedia is intended to furnish correct, up-to-date information on these materials.

Third, there is a general information gap regarding natural products between technologists of the botanical industry and those of the food, drug, and cosmetic industries, between members of the academic and research communities and those in industry, as well as between the consumer and the industry concerned. Information readily available to one group is often not available to the others. One of the objectives of this book is to try to bridge this gap by supplying information that would make different groups more aware of the practices and happenings outside of their own circle regarding the use of natural ingredients.

In this encyclopedia, each natural product is presented in alphabetical order according to its most common name, with each natural ingredient being cross-referenced with its scientific name (Latin binomial) in the Index. As a natural ingredient often has several common names (synonyms), the reader is advised to use the Index if an ingredient cannot be found in the text under a particular synonym. Data on about 310 natural ingredients are furnished. Information included in each item includes plant or other sources, habitats, parts used or derived from, method of preparation, brief physical description, chemical composition, pharmacology or biological activities, common commercially available forms in the United States, and their qualities, uses, and regulatory status, whenever applicable.

Data on chemical compositions of natural ingredients are constantly increasing as analytical techniques keep improving. Often an ingredient contains hundreds of chemical constituents, yet only a few (occasionally arbitrarily selected) are listed in this encyclopedia. For further information on other compounds, the reader is referred to the original references cited. Incidentally, the absence of a particular compound in a natural ingredient does not necessarily mean that it is actually absent; it may simply mean that nobody has analyzed for it in this particular ingredient. On the other hand, its reported presence in a natural ingredient means only that someone has investigated it in this particular ingredient using a particular analytical technique for whatever reason. Also, the mere presence of a toxic chemical in a natural ingredient does not necessarily make this ingredient toxic. Its concentration and biological availability should be taken into account when the toxicity of the ingredient is considered.

The data on pharmacology or biological activities (be they favorable or unfavorable) reported in this book should be viewed with caution as often they were single reports or reports from a single laboratory or research group that have not been substantiated by other studies. Furthermore, it should be kept in mind that results from animal studies are not necessarily applicable to humans. Purity of the test material (which is often not sufficiently stressed) should also be taken into account when evaluating such data.

Uses are categorized into four major areas: (1) pharmaceutical and/or cosmetic, (2) food, (3) folk medicine, and (4) others. Pharmaceutical and cosmetic uses refer to current uses in commercially available products mainly in the United States. No attempt has been made to identify the function of each ingredient in a product, as often there are over a dozen botanical components present in a single preparation, making it an impossible task. The same situation applies to the food area where the majority (200–250) of the ingredients used in food products are broadly identified only as flavor ingredients. The specific function and use level of a particular ingredient in a flavor formulation are often proprietary information, which is seldom publicly available. Consequently, food uses are reported in this encyclopedia by food categories, as in the report on "Average Maximum Use Levels" published by the Flavor and Extracts Manufacturers' Association of the United States (FEMA). Only in cases where the functions of the ingredients have become widely known in the trade or otherwise in open literature (e.g., fenugreek extract as a major flavoring agent in artificial maple syrup, yucca extracts as foaming agents in root beer, absinthium as a flavor ingredient in vermouths, etc.) are they specifically mentioned in this book. Sometimes an ingredient is reportedly used in various types of food products, yet federal regulations have approved its used in only one particular type of product. This appears to be a typical case of information dissemination lag. Under folk medicinal uses are listed only those traditional uses that are reported in reliable sources available to me, primarily in the English, German, and Chinese languages; they are by no means complete and they should not be regarded as endorsement of such uses. They are included in this volume because of their popular interest. Under the fourth category ("others") are listed potential or unusual uses that do not fall in above categories.

Use levels in foods reported in this encyclopedia are based on the FEMA report; a manufacturer may foreseeably use an ingredient in an amount five times the average maximum use and still be considered within good manufacturing practice. Use levels reported for cosmetics are based on values reported in the Monographs on "Fragrance Raw Materials" prepared by Opdyke of The Research Institute for Fragrance Materials, Inc. and published in *Food and Cosmetic Toxicology*.

Under regulatory status, GRAS means generally recognized as safe as sanctioned by the Food and Drug Administration (FDA); an ingredient described as having been approved for food use is not necessarily GRAS. For more precise and up-to-date information, the reader is referred to §182 and its appropriate sections under Title 21 of the *Code of Federal Regulations* (formerly §121.101), to §172.510 (formerly §121.1163), and to other appropriate sections, to the FDA, and to the latest notices and rulings published in the *Federal Register*.

A glossary of terminology commonly used in the botanical industry is found in the Introduction. Since the primary purpose of this encyclopedia is to serve as a practical reference guide for practicing technologists in the food, drug, and cosmetic industries and their purchasing agents and marketers, theoretical considerations and basic principles in the fields concerned are omitted. For these topics, the reader is referred to standard texts on these subjects such as BALSAM AND SAGARIN, FURIA, HARBORNE, LEWIS AND ELVIN-LEWIS, REMINGTON, and TYLER, listed in the General References.

In the General References are listed textbooks and handbooks from which general and sometimes specific information was obtained. They are identified in the text by the names of the authors in small capital letters, and if there are more than two authors, by the name of the first author. If an author has more than one book, it is identified by a number such as 1 or 2 immediately following the author's name (e.g., BAILEY 2); the number refers to the order of appearance of this author's books in the list.

Specific references are cited under References immediately following each entry, numbered according to their order of citation in the text.

It is hoped that this encyclopedia will serve as a handy and useful reference to technical and nontechnical members of the food, drug, and cosmetic industries, to teachers and students of corresponding sciences and related fields, and to the general public who want to know more about natural ingredients.

ALBERT Y. LEUNG

Glen Rock, New Jersey
January 1980

Acknowledgments

Our thanks to Dr. Vaishali Joshi, National Center for Natural Products Research, School of Pharmacy, University of Mississippi, for botanical description and help with monograph preparation and literature searching. In addition, Dr. Joshi's major contribution to the Indian Traditional Medicine chapter cannot go unnoticed. Thanks are also due to Dr. Shabana Khan, National Center for Natural Products Research, School of Pharmacy, University of Mississippi, for her help with the pharmacology sections and proofreading the manuscript. Without their support the preparation of this book would have been a daunting task.

Introduction

People have been using natural products since the dawn of human history. Only toward the end of the 19th century, however, have we started to know something about the chemistry of some of these products. With our increasing knowledge of chemistry and related sciences, we have begun to duplicate some of the natural chemicals and at the same time make modifications in these compounds, or sometimes produce completely new ones. Consequently, since the advent of the Synthetic Era several decades ago, many natural drugs have been replaced by synthetic ones; natural flavors and fragrances have been duplicated or simulated by manufactured chemicals. However, the number of natural products/natural product-derived drugs used in pharmaceutical products is still sizable, amounting to ca. 25% of the total number of medicines approved by the FDA. This number has not changed appreciably for the last two decades, especially with reference to botanicals. At least 250 plants or their extracts are currently used in commercial food products broadly classified as flavoring ingredients (FEMA). Over the past decade, there has been an increasing interest in the use of natural products, particularly in foods, cosmetics, and complementary medicine, especially after the passage of the Dietary Supplement Health and Education Act (DSHEA) in 1994. The implementation of DSHEA in the United States opened the market to a new class of natural-based products that are collectively known as dietary supplements (more below).

To define a natural product is not a straightforward task, for, strictly speaking, everything is derived from nature. Nevertheless, by natural products it is generally meant that products are not made by chemical synthesis. Theoretically, a natural chemical is the same as its synthetic counterpart in every respect. However, it must be pointed out that unless this chemical is absolutely pure (which it seldom is), it would contain different impurities, depending on its sources. The impurities present in a naturally derived food, drug, or cosmetic ingredient are bound to be different from those of its synthetic counterpart, and if there is more than one way to synthesize this compound, then the impurities would be different from one synthetic process to another. The relative toxicities or merits of these small differences have not been determined. If an impurity, whether it is a natural or synthetic chemical, has unusually high latent biological activity, a minute quantity of it present in a chemical would produce physiological effects besides those elicited by the pure chemical itself. These effects may not be immediately apparent. Most, if not all, of existing standards for food, drug, and cosmetic ingredients do not have provisions for pinpointing small amounts of impurities, as it is impractical to set absolute purity standards for these ingredients. Consequently, in practice, most of these materials are permitted to have a range of errors built into their purity assays. This range of errors can be due either to the assay methods themselves or to actual impurities present in the chemical. In some cases, as analytical methodology advances, this range has become progressively narrower. However, before this range becomes negligible, one should not equate a naturally derived chemical with its synthetic counterpart, and their sources should be indicated, as is the case with certain flavor chemicals.

There are several definitions of a natural product. In the case of flavoring substances, some definitions of a natural product (flavor) limit the product to be one obtained from natural sources by physical processes only. Other definitions allow hydrolysis and fermentation as permissible processes. For all practical purposes in this book, a natural product is defined as a product that is derived from plant, animal, or microbial sources, primarily through physical processing, sometimes facilitated by simple chemical reactions such as acidification, basification, ion exchange, hydrolysis, and salt formation as well as microbial fermentation. These chemical reactions do not drastically alter the chemical structure of the natural product to be isolated.

Ingredients used in foods, drugs, and cosmetics can be divided into two main categories, namely, active and inactive. Active ingredients can be considered as those that supply energy to the body or serve as its nutrients (foods and some food additives), or cause physiological changes in or on the body (drugs and cosmetics) when taken internally or applied externally. Inactive (inert) ingredients are substances that, based on prevalent data, do not exert physiological actions when ingested or applied to the body. Their primary function is to act as diluents (fillers) and/or to facilitate the ultimate intake or utilization of the active ingredients. Among food products, basic foodstuffs such as flour, starch, and milk are not included in this book, although they are considered active ingredients. Only food additives are considered. However, in drug and cosmetic products, both active and inactive substances are included.

Food additives are a large group of substances that are added to foods either directly or indirectly during the growing, storage, or processing of foods for one or more of the following purposes:

1. Improve or maintain nutritional value
2. Enhance quality
3. Reduce wastage
4. Enhance consumer acceptability
5. Improve keeping quality
6. Make the food more readily available
7. Facilitate preparation of the food

There are about 2500 direct food additives currently used by the food industry. Out of this number, perhaps 12–15% are natural products. Many of these food additives are also drugs when used in larger quantities. Some of these are also used in cosmetics. The total number of the more commonly used natural food, drug, and cosmetic ingredients in this encyclopedia is about 310 (first edition).

In spite of the fact that plants have been used for therapeutic purposes for millennia, only a relatively few plants or plant products are currently officially recognized in the United States as effective drugs. This is largely due to the difficulties encountered in plant drug research and the limitations of scientific methodology employed. Quite often, premature publicity on unconfirmed research data has tainted the reputation of many botanical drugs. Since many drug plants have rather complicated chemical compositions and analytical technology has not been adequate in determining their identities and qualities once extracts are made from them, adulteration, sophistication, or substitution has been common. This has led to inconsistencies in drug potency, and many natural drugs have probably been removed from officially recognized status as a result. Many natural drugs formerly recognized by the United States Pharmacopeia (U.S.P.) and National Formulary (N.F.) are no longer official in these compendia, yet many of these continue to be used in pharmaceutical preparations. As mentioned above, the implementation of DSHEA has imparted a new status to the majority of natural formulations currently available in the United States market. As partially defined under DSHEA, the term "dietary supplement" means a product (other than tobacco) intended to supplement the diet that bears or contains one or more of the following dietary ingredients: (a) a vitamin; (b) a mineral; (c) a herb or other botanical; (d) an amino acid; (e) a dietary substance for use by man to supplement the diet by increasing the total dietary intake; or (f) a concentrate, metabolite, constituent, extract, or combination of any ingredient described above. As mandated by the FDA, the label of any dietary supplement marketed in the United States should have the statement:

"These statements have not been evaluated by the Food and Drug Administration. This product is not intended to diagnose, treat, cure, or prevent any disease." The unevaluated "statements" include any structure/function claims introduced by the manufacturer for a particular dietary supplement.

Some of the food, drug, and cosmetic ingredients are pure chemicals isolated from plants, animals, or microbes. However, most are in the form of extracts, oleoresins, fixed oils, and volatile oils, among others. The included appendices contain most of the commonly encountered terms used in the botanical industry.

REFERENCES

See the General References for ARCTANDER; FEMA; FURIA; FURIA AND BELLANCA; REMINGTON; U.S.P. XIX, XXI AND XXXIII.

1. N. R. Farnsworth and R. W. Morris, *Am. J. Pharm.*, **148**, 46 (1976).
2. National Academy of Sciences, *The Use of Chemicals in Food Production, Processing, Storage and Distribution*, Washington, DC, 1973.
3. D. J. Newman and G. M. Cragg, *J. Nat. Prod.*, **70**, 461 (2007)
4. Dietary Supplement Health and Education Act of 1994; Public Law 103–417; 103rd Congress. http://www.fda.gov/opacom/laws/dshea.html#sec3.
5. I. K. Khan, *Life Sci.*, **78**, 2033 (2006)

LEUNG'S ENCYCLOPEDIA OF COMMON NATURAL INGREDIENTS

USED IN FOOD,
DRUGS, AND COSMETICS

Third Edition

Natural Ingredients

ABSINTHIUM

Source: *Artemisia absinthium* L. (Family Compositae or Asteraceae).

Common/vernacular names: Absinthe, absinthe grande, absinthium, armoise, common wormwood, wermut, wermutkraut, and wormwood.

GENERAL DESCRIPTION

Artemisia absinthium is a shrubby perennial herb with grayish white stems covered with fine silky hairs, 30–90 cm high; leaves also silky, hairy, and glandular, 2–3-pinnatisect, petiolate lobes, mostly obtuse; odor aromatic, spicy; taste bitter; native to Europe, northern Africa, and western Asia, naturalized in North America; extensively cultivated. Parts used are the leaves and flowering tops (fresh and dried), harvested just before or during flowering; from these a volatile oil is obtained by steam distillation (EVANS; FERNALD; YOUNGKEN).

CHEMICAL COMPOSITION

Artemisia absinthium contains up to 1.7% volatile oil composed mainly of thujone (α- and β-) and β-caryophyllene.[1] Bitter principles include artabsin (analogous monomer of absinthin), dimeric guaianolides (absinthin and absintholide),[2,3] artabsinolides A–C (EVANS), and artemetin (5-hydroxy-3,6,7,3′,4′-pentamethoxyflavone);[4] other isolated lactones include arabsin,[5] artabin,[6] ketopelenolide a (a germacranolide),[7] artenolide, artemoline, and deacetylglobicin (monomeric guaianolides),[8] anabsin, and isoabsinthin (dimeric guaianolides).[5,7,9,10] Other constituents present in relatively high amounts in the oil include sabinene, *trans*-sabinyl acetate + lavandulyl acetate, (−)-sabinyl acetate, (Z)-epoxy-α-ocimene, chrysanthendiol (+), and chrysanthenyl acetate, among others.[11–13] Cadinene, camphene, bisabolene, thujyl alcohol, myrcene, 1,8-cineole, and azulenes (e.g., chamazulene, 3,6-dihydrochamazulene, and 5,6-dihydrochamazulene) are also found.[11,13] Chamazulene at concentrations of up to 0.29% was detected in the flowers at the beginning of flowering.[14] Also, (−)- and *trans*-epoxyocimenes were isolated from an Italian absinthium oil of which they constituted 16–57%.[15]

Varying geographical origin, altitude, and exposures affect qualitative and quantitative differences in the essential oil. The volatile oil of different chemotypes can contain >40% of either *p*-thujone, chrysanthenyl acetate, *trans*-sabinyl acetate, or (−)-epoxyocimene (EVANS).[11,14] Plant material collected in Argentina (Patagonia) was composed of 59.9% β-thujone (2.34% α-thujone), sabinyl acetate (18.11%), (−)-epoxyocimene (1.48%), caryophyllene (1.92%), linalool (1.15%), and sabinene (1.09%), with trace amounts of α-pinene, α-terpineol, germacrene D, neryl acetate, neryl propionate, nerol, geranyl propionate, and geraniol (<1% each).[13]

Miscellaneous constituents of the plant include inulobiose (an oligofructoside),[16] coumarins (scopoletin, umbelliferone), phenolic acids, flavonoids,[17–19] amino acids,[19] tannins (4.0–7.7%),[20] lignans (3,7-dioxabicyclo[3,3,0]-octanes),[7] pipecolic acid,[21] and sterols, including an antipyretic sterol (24ζ-ethylcholesta-7,22-dien-3β-ol).[22,23]

PHARMACOLOGY AND BIOLOGICAL ACTIVITIES

Despite the postulation that thujone and tetrahydrocannabinol, active principles of absinthe and marijuana, respectively, interact with a common receptor in the central nervous system, no evidence of such activity was found.[24]

Antitumor activity has also been reported[25] and attributed to a flavonoid, artemisetin.[26]

The essential oil and extracts of the plant have shown *in vitro* antimicrobial,[27–31] antifungal,[27,31] nematocidal,[32] acaricidal,[33] and antimalarial activities.[34] *In vitro* antimaliarial activity was found from two homoditerpene peroxides isolated from the aerial parts of the plant.[35] The oil was shown to repel mosquitoes, fleas, and flies.[31]

TOXICOLOGY

The toxicity of *A. absinthium* and thujone remains poorly understood;[36] however, ingestion of large doses of thujone causes convulsions. Habitual use or large doses of alcoholic drinks containing *A. absinthium* have been associated with brain damage, epilepsy, suicide, hallucinations, restlessness, insomnia, nightmares, vomiting, vertigo, tremors, and convulsions.[37,38] With the likely exception of the latter owing to GABA-modulating activity of α- and β-thujones,[39] how many of these observed toxic effects were caused by the plant remains unknown.

Thujone is porphyrogenic and may therefore be hazardous to patients with defective hepatic heme synthesis.[40] Applied externally the essential oil is nontoxic.[41]

USES

Medicinal, Pharmaceutical, and Cosmetic. Oil is used as an ingredient in certain rubefacient preparations; extracts now rarely used internally, except in some bitter tonics for anorexia and dyspeptic symptoms (bitter value of at least 15,000) (BLUMENTHAL 1). The oil has been used as a fragrance component in soaps, detergents, creams, lotions, and perfumes, with maximum use levels of 0.01% in detergents and 0.25% in perfumes.[42]

Food. *Artemisia absinthium* is widely used in flavoring alcoholic bitters and in vermouth formulations; average maximum use level of 0.024% reported. The oil and extracts are also used in alcoholic beverages as well as in other categories of foods such as nonalcoholic beverages, frozen dairy deserts, candy, baked goods, and gelatins and puddings. Reported average maximum use levels for the oil is about 0.006% in the last four food categories.

The popular 19th century alcoholic beverage, absinthe, is made by macerating absinthium and other aromatic herbs in alcohol, distilling the spirit, and then adding flavorings or coloring. Absinthe was banned in many countries soon after the turn of the 19th century (1907 in Switzerland; 1912 in the United States). Its sale persisted in France until 1915[42] where it was popular among writers and artists, including Toulouse-Lautrec and Vincent van Gogh, and for a time was the most heavily consumed alcoholic beverage in the country.[37]

Dietary Supplements/Health Foods. Not commonly used; cut and sifted herb as tea (infusion or decoction) reportedly used as a bitter digestive stimulant (HOFFMAN); also used in the form of an aqueous extract at doses equivalent to 2–3 g of herb for the treatment of anorexia, dyspeptic symptoms, and so on (BLUMENTHAL 1).

Traditional Medicine. Used as an aromatic bitter for promoting appetite, for strengthening the system in colds and flu, and as choleretic for liver and gallbladder disorders, usually in the form of a dilute extract; also as emmenagogue, antianorexic, antidyspetic (EVANS), mental restorative (GRIEVE), febrifuge,[36] and topically for contusions, edema, ulcers, as an antiseptic[18] and a vermifuge;[43] anthelmintic activity is probably the result of lactones related to santonin found in wormseed (*A. cina* Berg.) and other *Artemisia* species.[44] The dried and fragmented leaves are used in the Philippines to treat herpes, purulent scabies, and eczema and to speed the healing of wounds.[45]

COMMERCIAL PREPARATIONS

Crude, dilute extracts (tincture and fluid extract) and the essential oil. *A. absinthium* was formerly official in U.S.P. (1830–1890) (BOYLE).

Regulatory Status. Approved for food use as a natural flavoring substance, provided the finished food is thujone-free (§172.510). *A. absinthium* is found in various European pharmacopoeias, including the British Pharmacopoeia (BP) and EP (EVANS). The herb is subject of a German Commission E monograph (BLUMENTHAL 1).

REFERENCES

See the General References for APhA; APPLEQUIST; BIANCHINI AND CORBETTA; BLUMENTHAL 1; BRUNETON; DER MARDEROSIAN AND BEUTLER; EVANS; FEMA; FERNALD; GRIEVE; GUENTHER; GUPTA; HOFFMAN; KARRER; MASADA; MERCK; SAX; STAHL; TUTIN 5; YOUNGKEN.

1. J. Slepetys, *Polez. Rast. Priblat. Respub. Beloruss., Mater., Nauch, Konf.*, **2**, 289 (1973).
2. T. Nozoe and S. Ito in L. Zechmeister, ed., *Fortschritte der Chemie Organischer Naturstoffe*, Vol. 19, Springer-Verlag, Vienna, Austria, 1959, p. 1.
3. J. Beauhaire et al., *Tetrahedron Lett.*, **21**, 3191 (1980).
4. K. Venkataraman in L. Zechmeister, ed., *Fortschritte der Chemie Organischer Naturstoffe*, Vol. 17, Springer-Verlag, Vienna, Austria, 1959, p. 1.
5. S. K. Zakirov et al., *Khim. Prir. Soedin.*, **4**, 548 (1976).
6. I. S. Akhmedov et al., *Khim. Prir. Soedin.*, **6**, (1970).
7. M. Dermanovic et al., *Glas. Hem. Drus., Beogard*, **41**, 287 (1976).
8. S. Z. Kasymov et al., *Khim. Prir. Soedin.*, **5**, 667 (1987).
9. A. Ovezdurdyev et al., *Khim. Prir. Soedin.*, **5**, 667 (1987).
10. J. Beauhaire et al., *Tetrahedron Lett.*, **25**, 2751 (1984).
11. B. M. Lawrence, *Perf. Flav.* **17**, 39 (1992).
12. D. J. Bertelli and J. H. Crabtree, *Tetrahedron*, **24**, 2079 (1968).
13. T. Sacco and F. Chialva, *Planta Med.*, **54**, 93 (1988).
14. J. Slepetys, *Liet. TSR Mokslu Akad. Darb., Ser. C*, **4**, 29 (1974).
15. F. Chialva et al., *Riv. Ital. Essenze, Profumi, Piante Off, Aromi, Saponi, Cosmet., Aerosol*, **58**, 522 (1976); through *Chem. Abstr.*, **86**, 161105d (1977).
16. M. L. Tourn and A. Lombard, *Atti Acad. Sci. Torino, Cl. Sci. Fis., Mater. Nat.*, **5–6**, 941 (1974).
17. L. Swiatek and E. Dombrowicz, *Farm. Pol.*, **40**, 729 (1984).
18. L. Swiatek et al., *Pharm. Pharmacol. Lett.* **8**, 158 (1998).
19. G. A. Zhukov and V. V. Timofeev, *Khim. Prir. Soedin.*, **3**, 447 (1987).
20. J. Slepetys, *Liet. TSR Mokslu Akad. Darb., Ser. C*, **1**, 43 (1975).
21. V. Rosetti and A. Garrone, *Phytochemistry*, **14**, 1467 (1975).
22. M. Ikram et al., *Planta Med.*, **53**, 389 (1987).
23. M. D. Sayed et al., *Egypt J. Pharm. Sci.*, **19**, 323 (1980).
24. J. P. Meschler and A. C. Howlett, *Pharmacol. Biochem. Behav.*, **62**, 473 (1999).
25. N. V. Gribel and V. G. Pashinskii, *Rastitel'Nye Resursy*, **27**, 65 (1991).
26. I. I. Chemesova et al., *Rastitel'Nye Resursy*, **23**, 100 (1987).

27. V. K. Kaul et al., *Indian J. Pharm.*, **38**, 21 (1975).
28. N. S. Alzoreky and K. Nakahara, *Int. J. Food Micriobiol.*, **80**, 223 (2003).
29. D. Basaran et al., *Turk. J. Biol.*, **23**, 377 (1999).
30. E. O. Turgay, *Pharm. Biol.*, **40**, 269 (2002).
31. F. Juteau et al., *Planta Med.*, **69**, 158 (2003).
32. A. M. Korayem et al., *Anzeiger Schaedlingsk Pflanze Umweltschutz.*, **66**, 32 (1993).
33. H. Chiasson et al., *J. Econ. Entomol.*, **94**, 167 (2001).
34. M. M. Zafar et al., *J. Ethnopharmacol.*, **30**, 223 (1990).
35. G. Ruecker et al., *Phytochemistry*, **31**, 340 (1992).
36. S. G. Khattak et al., *J. Ethnopharmacol.*, **14**, 45 (1985).
37. J. Strang et al., *Br. Med. J.*, **319**, 1590 (1999).
38. C. Gambelunghe and P. Melai, *Forensic Sci. Int.*, **130**, 183 (2002).
39. K. M. Höld et al., *Proc. Natl. Acad. Sci. USA*, **97**, 3826 (2000).
40. H. L. Bonkovsky et al., *Biochem. Pharmacol.*, **43**, 2359 (1992).
41. D. L. J. Opdyke, *Food Cosmet. Toxicol.*, **13**(Suppl.), 721 (1975).
42. D. D. Vogt, *J. Ethnopharmacol.*, **4**, 337 (1981).
43. M. B. Quinlan et al., *J. Ethnopharmacol.*, **80**, 75 (2002).
44. H. J. Woerdenbag et al. in P. A. G. M. De Smet et al., eds., *Adverse Effects of Herbal Drugs*, Vol. 3, Springer-Verlag, Berlin, 1997, p. 15.
45. T. Aburjai and F. M. Natsheh, *Phytother. Res.*, **17**, 987 (2003).

ACACIA (GUM ARABIC)

Source: *Acacia senegal* (L.) Willd. and other *Acacia* spp. (Family Leguminosae or Fabaceae).

Common/vernacular names: Gomme arabique, gomme de Senegal, gum acacia, gum arabic, gum Senegal, gummae mimosae, and kher.

GENERAL DESCRIPTION

The dried gummy exudate from stems and branches of *Acacia senegal* (L.) Willd. (syn. *A. verek* Guill. et Perr.) or other related African *Acacia* species. *Acacia senegal* has triple spines at the base of its branchlets, which distinguishes it from many other *Acacia* spp. in its range. The trees are tapped by making transverse incisions in the bark and peeling off a thin strip of the bark. The gummy exudates form as tears on the surface of the wounds and are collected after they have hardened, usually in 2 or more weeks.

The Republic of Sudan supplies most of the world's gum acacia and produces the best quality product. *A. senegal* ranges from Senegal to northeastern Africa, south to Mozambique. Other suppliers include Senegal, Mauritania, Chad, Nigeria, Tanzania, and Ethiopia.

Gum acacia is one of the most water-soluble plant gums; one part acacia can dissolve in two parts water, forming a weakly acidic solution with pH 4.5–5.5. Its solutions have lower viscosities than those of other natural gums. It is insoluble in alcohol, chloroform, ether, and oils and very slightly soluble in glycerol and propylene glycol. It is almost odorless and has a bland mucilaginous taste.

Gum acacia contains a peroxidase that, unless destroyed by heating briefly at 100°C, forms colored compounds with certain amines and phenols (e.g., aminopyrine, antipyrine, epinephrine, cresol, eugenol, guaiacol, phenol, tannins, thymol, vanillin, etc.). It also causes partial destruction of many alkaloids, including atropine, hyoscyamine, scopolamine, homatropine, morphine, apomorphine, cocaine, and physostigmine.

Gum acacia is incompatible with heavy metals, which destroy the gum by precipitation. Borax and alcohol also precipitate it, but the process can be prevented or reversed.

CHEMICAL COMPOSITION

Crude *A. senegal* gum is a complex polysaccharide that consists of varying numbers of polysaccharide units of molecular weight 200,000 linked to a protein core,[1] forming an arabinogalactan–protein complex.[2]

PHARMACOLOGY AND BIOLOGICAL ACTIVITIES

Ingested orally, gum acacia is nontoxic. Hypersensititivty reactions to the dust or from ingestion of gum acacia are rare and consist of skin lesions and severe asthmatic attacks. Gum acacia can be digested by rats to an extent of 71%; guinea pigs and rabbits also seem to use it for energy, as do humans to a certain extent.[3]

Studies of gum acacia in rats as a potential hypocholesterolemic agent have shown conflicting results.[4–7] In rat models of chronic diarrhea, gum acacia enhanced electrolyte, glucose, and water absorption in rat models of chronic diarrhea and in healthy rats.[8,9] Chronic renal failure patients administered gum acacia have shown increased fecal nitrogen excretion and lowered concentrations of retained metabolites, including urea.[10,11]

USES

Medicinal, Pharmaceutical, and Cosmetic. Mainly in the manufacture of emulsions and in making pills and troches (as an excipient); as a demulcent for inflammations of the throat or stomach and as a masking agent for acrid-tasting substances such as capsicum (MARTINDALE); also as a film-forming agent in peel-off facial masks.

Food. Currently, the major use of gum acacia is in foods, where it performs many functions, for example, as a suspending or emulsifying agent, stabilizer, adhesive, and flavor fixative and to prevent crystallization of sugar, among others. It is used in practically all categories of processed foods, including candy, snack foods, alcoholic and nonalcoholic beverages, baked goods, frozen dairy desserts, gelatins and puddings, imitation dairy products, breakfast cereals, and fats and oils, among others. Its use levels range from <0.004% (40 ppm) in soups and milk products to 0.7–2.9% in nonalcoholic beverages, imitation dairy, and snack foods to as high as 45% in candy products.

COMMERCIAL PREPARATIONS

Available in crude, flake, powdered, granular, and spray-dried forms. It is official in N.F. and F.C.C.

Regulatory Status. Affirmed as GRAS (§184.1330).

USES

Acacia gum has been in use since ancient times.

REFERENCES

See the General References for DAVIDSON; DER MARDEROSIAN AND BEUTLER; FEMA; FURIA; GLICKSMAN; GOSSELIN; GRIEVE; GUPTA; KEAY; KENNEDY; LAWRENCE; MARTINDALE; MCGUFFIN; REMINGTON; SAX; TERRELL; WHISTLER AND BEMILLER; YOUNGKEN.

1. S. Connolly et al., *Carbohydr. Polym.*, **8**, 23 (1988).
2. Y. Akiyama et al., *Agric. Biol. Chem.*, **48**, 235 (1984).
3. A. Jeanes, *ACS Symp. Ser.*, **15**, 336 (1975).
4. S. Kiriyama et al., *J. Nutr.*, **97**, 382 (1969).
5. A. C. Tsai et al., *J. Nutr.*, **106**, 118 (1976).
6. A. A. Al-Othman et al., *Food Chem.*, **62**, 69 (1998).
7. G. Annison et al., *J. Nutr.*, **125**, 283 (1995).
8. K. U. Rehman et al., *Dig. Dis. Sci.*, **48**, 755 (2003).
9. S. Teichberg et al., *J. Pediatr. Gastroenterol. Nutr.*, **29**, 411 (1999).
10. D. Z. Bliss et al., *Am. J. Clin. Nutr.*, **63**, 392 (1996).
11. A. J. Al-Mosawi, *Pediatr. Nephrol.*, **17**, 390 (2002).

ACEROLA

Source: *Malpighia emarginata* DC. (*M. glabra* L., *M. punicifolia* L., *M. berteriana* Spreng., *M. lanceolata* Griseb., *M. retusa* Benth., *M. umbellata* Rose) (Family Malpighiaceae).

Common/vernacular names: Acerola, Barbados, Jamaica, Puerto Rico, and West Indian cherry, and huesito.

GENERAL DESCRIPTION

Acerola is the fruit of a shrub or small tree that grows to a height of 5 m. Fruits (drupes) are globose, ovoid, or subglobose, 1–2 cm in diameter, bright red, slightly resembling cherries. Mature fruits are juicy and soft, with a pleasant tart flavor.[1]

Both *M. glabra* and *M. punicifolia* have been reported in the literature as a source of acerola with high vitamin C content.[1–3] However, *M. punicifolia* or its hybrid with *M. glabra* appears to be the correct source (MORTON2; WATT AND MERRILL).[1] More recently, the plant species that supplies acerola has been renamed *M. emarginata* (USFISB).

Malpighia emarginata is native to the West Indies and is also found in northern South America, Central America, Florida, and Texasand and now increasingly grown worldwide (e.g., USA, Brazil, and Australia) for use in dietary supplements.[1] Its fruit is among the richest known sources of natural vitamin C.

CHEMICAL COMPOSITION

Contains 1–4.5% vitamin C (ascorbic acid) and dehydroascorbic acid, mainly the former, in edible portion of fruit (cf. 0.05% in peeled orange), which makes up about 80% of the fruit. Vitamin C content varies with ripeness of the fruit (highest in green and lowest in fully ripe fruit), season, climate, and locality.[1,3–5]

Other vitamins present include 4300–12,500 IU/100 g vitamin A (cf. 11,000 IU/100 g for raw carrots); thiamine, riboflavin, and niacin in concentrations comparable to those in other fruits.[1]

Furfural was found as the main volatile flavor constituent in the fruit of *M. emarginata* cultivated in Cuba, which was also found to contain over 150 other volatile constituents; among them, aliphatic esters (31%), terpenoids (24%), ketones and aldehydes (15%), and alcohols (13%) made up the major components.[6] Miscellaneous constituents include calcium, iron, and phosphorus in comparable concentrations to those of apple; *l*-malic acid; dextrose, fructose, and sucrose; evidence of a heat-resistant enzyme (not completely deactivated at 103°C) that breaks down ascorbic acid during storage of pasteurized juice, resulting in carbon dioxide buildup, causing swelling of cans or explosion of bottles.[1,2]

PHARMACOLOGY AND BIOLOGICAL ACTIVITIES

Acerola cherry extract inhibits the oxidation of LDL cholesterol *in vitro*.[7] Acetone and hexane fractions of the fresh fruit have shown *in vitro* cytotoxic activity against a human tumor cell lines (oral squamous cells and submandibular gland carcinoma).[8]

USES

Food. As a source of natural vitamin C, in the form of juice, tablet, or capsule; however, most of the vitamin is destroyed during processing (see *rose hips*).

Dietary Supplements/Health Foods. Tablets, capsules, or other products, often combined with other herbs and vitamin C.

Traditional Medicine. The fruits have reportedly been used for the treatment of dysentery, diarrhea, and liver disorders (CSIR VI).

Others. The bark, which contains 20–25% tannin, has been used in the tanning of leathers (MORTON 4).

COMMERCIAL PREPARATIONS

Available as fresh fruit for home consumption in certain East Coast supermarkets and ethnic stores; also in juice and spray-dried form. Canned juice of the fruits has been used to enhance the vitamin C content of other juices, such as pear, apricot, and grape juice.

REFERENCES

See the General References for CSIR VI; DER MARDEROSIAN AND BEUTLER; MCGUFFIN 1 & 2; MORTON 2; MORTON 4; TERRELL; WATT AND MERRILL.

1. C. G. Moscoso, *Econ. Bot.*, **10**, 280 (1956).
2. R. E. Berry et al., *Food Prod. Dev.*, **14**, 109 (1977).
3. H. Y. Nakasone et al., *Proc. Am. Soc. Hort. Sci.*, **89**, 161 (1966).
4. A. Schillinger, *Z. Lebensm, Unters. Forsch.*, **131**, 89 (1966).
5. A. L. Vendramini and L. C. Trugo, *Food Chem.*, **71**, 195 (2000).
6. J. A. Pino and R. Marbot, *J. Agric. Food Chem.*, **49**, 5880 (2001).
7. J. Hwang et al., *J. Agric. Food Chem.*, **49**, 308 (2001).
8. N. Motohashi et al., *Phytother. Res.*, **18**, 212 (2004).

ACONITE

Source: *Aconitum napellus* L. and other *Aconitum* spp. (Family Ranunculaceae).

Common/vernacular names: Aconite, aconitum, monkshood, and wolfsbane.

GENERAL DESCRIPTION

Perennial herbs consisting of many subspecies, varieties, clones, and forms; up to 1.5 m high with tuberous roots that resemble turnips; native to mountainous regions of central Europe; naturalized in Asia, Africa, and North America; cultivated in Russia, Germany, Spain, and France. Part used is the dried tuberous root. Of the 100 northern temperate species in the genus, 35 species in China have been investigated chemically.

CHEMICAL COMPOSITION

Total alkaloids 0.2–2%, consisting mainly of aconitine (acetylbenzylaconine), picraconitine (benzoylaconine), aconine, and napelline (isoaconitine, pseudoaconitine; others include 12-epidehydronapelline, 12-epiacetyldehydronapelline, 1,14-diacetylneoline, *N*-deethylaconitine, aconosine, 14-acetylneoline, hokbusine A, senbusine A, senbusine C, mesaconitine, neoline, and songoramine.[1–3] Alkaloid content decreases with altitude from 0.82% of fresh root of plants grown at 1750 m to 0.29% at 2500 m. Aconitine content is greatest in winter-dormant tubers (FROHNE AND PFANDER).[4]

On hydrolysis, aconitine yields picraconitine, which in turn yields aconine on further hydrolysis.

Compounds identified from raw (dried) *A. carmichaelii* Debx. include aconitines, coryneine chloride, and higenamine, all of which have been implicated in cardioactivity of the tubers.[5]

Other constituents include aconitic acid, itaconic acid, succinic acid, malonic acid, fructose, maltose, melibiose, mannitol, starch, fat, and resin.

PHARMACOLOGY AND BIOLOGICAL ACTIVITIES

Extracts of *A. carmichaelii* have shown cardiotonic activity, including inotropic and chronotropic activities, leading to hypotension and/or hypertension. Analgesic and anesthetic activities have been reported. Hyoscine potentiates the action of aconitine.[5] Aconitine and related compounds exhibit anti-inflammatory and analgesic properties in experimental animals.[6] *dl*-Demethylcoclaurine, a component of prepared (processed) lateral rootlets of *A. carmichaelii* (considered a separate drug in Chinese tradition), has been shown to raise the heart rate in sinus arrhythmia patients.[7]

Certain *Aconitum* species are reported to have antitumor activity in laboratory animals; others show antibacterial, antifungal, and antiviral activities. Extracts of various species also have antipyretic properties (FARNSWORTH 3).[8]

During the past decade, extensive studies have been carried out on the chemistry and pharmacology of aconite in general. While hypaconitine is found to be the active neuromuscular blocking agent in Asian aconite,[9] higenamine (*dl*-demethylcoclaurine) and other chemical components (including a nonalkaloidal fraction) are the cardiotonic principles. These cardiotonic substances are heat resistant, and their activities are realized after prolonged decocting, whereby the deadly aconitine is hydrolyzed to the much less toxic aconine (WANG).

TOXICOLOGY

Aconite is a strong and fast-acting poison, affecting both the heart and the central nervous system. Its active principles are aconitine and its related alkaloids. As little as 2 mg aconitine may cause death from paralysis of the heart or respiratory center. The lethal dose for adults generally ranges only from 3 to 6 mg of aconitine, readily contained in a few grams of plant material (FROHNE AND PFANDER).

When applied to the skin, aconite produces tingling and then numbness; poisoning may result from percutaneous absorption.

USES

Medicinal, Pharmaceutical, and Cosmetic. Now rarely used internally in the United States; its current use is mainly in liniments (rubefacients), often with belladonna, for external applications only.

Traditional Medicine. Used internally as a cardiac depressant and mild diaphoretic; externally as local analgesic in facial neuralgia, rheumatism, and sciatica. Related species such as *A. chinense* Paxt. and *A. kusnezoffii* Reichb. are widely used in Chinese medicines for rheumatoid arthritis, chronic nephritis, sciatica, and other ailments. These and other Asian *Aconitum* spp. have been valued for their analgesic, anti-inflammatory (antirheumatic); antibiotic (antiseptic), antipyretic, and cardiotonic activities. The methods used for treating these crude drugs (roots) are numerous and quite different from that for *A. napellus*. One method involves soaking and washing in clear water for several days and treating with licorice, ginger, black beans, and other drugs, followed by boiling or steaming and then drying. The resulting product therefore cannot be compared directly with the American or European product or any other unprocessed product (JIANGSU; NANJING; WANG).

COMMERCIAL PREPARATIONS

Crude and extracts. Strengths (see *glossary*) of extracts are expressed in weight-to-weight ratios. Crude was formerly official in U.S.P. (1850–1936) and N.F. (1942).

Regulatory Status. Subject of a negative German therapeutic monograph, due to toxicity that can occur within the therapeutic dose (including vomiting, dizziness, muscle spasms, hypothermia, paralysis of respiratory system, and rhythmic heart disorders) (BLUMENTHAL 1).

REFERENCES

See the General References for BLUMENTHAL 1; BRUNETON; CLAUS; DER MARDEROSIAN AND BEUTLER; FARNSWORTH 3; FERNALD; FOGARTY; FROHNE AND PFANDER; GOSSELIN; GRIEVE; GUPTA; JIANGSU; KARRER; MARTINDALE; MCGUFFIN 1 & 2; MERCK; NANJING; SAX; WANG.

1. G. de la Fuente et al., *Heterocycles*, **27**, 1109 (1980).
2. E. Arlandini et al., *J. Nat. Prod.*, **50**, 937 (1987).
3. H. Hikino et al., *J. Nat. Prod.*, **47**, 190 (1984).
4. A. Crema, *Arch. Ital. Sci. Farmacol.*, **7**, 119 (1957).
5. N. G. Bisset, *J. Ethnopharmacol.*, **4**, 247 (1981).
6. M. Murayama et al., *J. Pharm. Pharmacol.*, **35**, 135 (1991).
7. P. G. Xiao and K. J. Chen. *Phytother. Res.*, **1**, 2, 5 (1987).
8. J. L. Hartwell, *Lloydia,* **34**, 103 (1971).
9. M. Kimura et al., *Jpn. J. Pharmacol.*, **48**, 290 (1988).

AGAR

Source: Red algae, including ***Gelidium cartilagineum*** (L.) Gaill., ***Gelidium amansii*** Lamour., ***Gracilaria confervoides*** (L.) Grev., other ***Gelidium*** and ***Gracilaria*** species as well as species of the genera ***Pterocladia***, ***Ahnfeltia***, ***Acanthopeltis***, and ***Suhria***.

Common/vernacular names: Agar-agar, gelosa, gelose, layor carang, and vegetable gelatin; also, Chinese gelatin, colle du Japon, Japanese gelatin, and Japanese isinglass.

GENERAL DESCRIPTION

Agar is the dried hydrophilic, colloidal extract of various red algae (Class Rhodophyceae); the more commonly used red algae are *Gelidium cartilagineum* (L.) Gaill., *Gelidium amansii* Lamour., *Gracilaria confervoides* (L.) Grev., other *Gelidium* and *Gracilaria* species as well as species of the genera *Pterocladia*, *Ahnfeltia*, *Acanthopeltis*, and *Suhria*. Agar is extracted from the algae by boiling them in water at a neutral or slightly acidic pH. The hot liquor is filtered and on cooling forms a gel that is purified by freezing and thawing followed by drying.

The major agar producer has been and still is Japan. Other producing countries include the United States, Spain, Portugal, Chile, Taiwan, Korea, Morocco, New Zealand, Australia, Argentina, and Mexico.

Agar is insoluble in cold water but readily soluble up to 5% in boiling water. The solution (sol) on cooling to 35–40°C forms a firm, resilient gel that does not melt below 85°C. This ability to gel at a much lower temperature than the melting temperature of the gel, commonly called hysteresis lag, is uniquely long in agar, and many of its uses depend on this property. Agar gels also have the property of shrinking and exuding water from their surface (syneresis), particularly when broken. The gel strength of agar can be increased by addition of dextrose, sucrose, and locust bean gum, while it tends to weaken with gelatin, algin, starch, and karaya gum. The colorless, tasteless powder can absorb up to 200 times its volume of water when forming a gel.

Agar solutions have low viscosity; their degree of clarity and color (yellowish to colorless) depend on the quality and source of the agar, as do their gel strength, gelling temperature, and the degree of syneresis. Quality is largely affected by extraction procedures. Physical and rheological properties of agar that provide the greatest determinations of quality are the average molecular weight and molecular weight distribution.[1]

Agar is insoluble in organic solvents and is precipitated from aqueous solution by alcohol and tannin.

CHEMICAL COMPOSITION

The structure of agar is still not fully determined, the problem being complicated by the large number of commercial sources of agar. It is generally believed that all agars consist of two major polysaccharides (neutral agarose and charged agaropectin), although several studies have indicated a much more complicated structure.[2–6] Agarose is the gelling fraction and agaropectin is the nongelling fraction. Both are composed of a linear chain of alternating β-D-galactopyranose and 3,6-anhydro-α-L-galactopyranose residues, with agaropectin having a higher proportion of uronic acid, sulfate, and pyruvic acid residues.[6] Commercial agar may contain free amino acids (arginine, aspartic acid, glutamic acid, and threonine) and free sugars (galactose and gluconic acid).[7] It may also contain other sugar residues including 4-O-methyl-L-galactose, 6-O-methyl-D-galactose,[8–10] D-xylose, and O-methylpentose, as well as boric acid (approximately 0.1%)[11] and various inorganic cations (Na^+, K^+, Ca^{2+}, Mg^{2+}, etc.).[12]

PHARMACOLOGY AND BIOLOGICAL ACTIVITIES

Agar is nontoxic and can be ingested in large doses without much distress. It passes through the intestinal tract mostly unabsorbed.[13] Agar has shown *in vitro* antipeptic activities[14] and the results of a study in rats suggested that it may elevate serum or tissue cholesterol levels.[15]

TOXICOLOGY

Mice fed agar showed significantly more colon tumors per animal (twice as many) as those fed diets without agar. Agar-fed animals also showed decreased levels of fecal neutral sterol and bile acid concentrations.[13] When fed to rats at 5% and 15% levels of the diet, agar impaired protein utilization.[16]

USES

Medicinal, Pharmaceutical, and Cosmetic. As a bulk laxative, particularly in chronic constipation; in the manufacture of emulsions, suspensions, gels, and hydrophilic suppositories; in dentistry as basic constituent of reversible impression and duplicating materials.

Food. Used in canned meat and fish products as gel filler or gel binder; in baked goods (icings and glazes); and in confectionery, dairy products, processed fruits, sweet sauces, and reconstituted vegetables, among others. Highest average maximum usage level usually about 0.4% in baked goods.

Agar has been used as a food in the Far East for centuries.

Others. A major use of agar is in culture media for microorganisms. It is one of the most widely used media for biotechnology purposes.

COMMERCIAL PREPARATIONS

Available in flakes, strips, and powders; grades and quality vary, with the bacteriological grades demanding the most stringent quality. Some high-quality agars from certain commercial sources have higher congealing temperatures than that required by the F.C.C. and N.F., which is due to the source of algae used.

Regulatory Status. Agar is GRAS as a stabilizer (§582.7115) and red algae is affirmed as GRAS (§184.1115).[17]

REFERENCES

See the General References for FEMA; FURIA; GLICKSMAN; GOSSELIN; LAWRENCE; MARTINDALE; PHILLIPS; REMINGTON; TYLER 2; UPHOF; WHISTLER AND BEMILLER; WREN.

1. F. Zanetti et al., *Planta Med.*, **58** (S1), A696 (1992).
2. K. B. Guiseley in A. Standen, ed., *Kirk-Othmer Encyclopedia of Chemical Technology*, Vol. 17, 2nd ed., Wiley–Interscience, New York, 1968, p. 763.
3. M. Duckworth and W. Yaphe, *Carbohydr. Res.*, **16**, 189 (1971).
4. M. Duckworth and W. Yaphe, *Carbohydr. Res.*, **16**, 435 (1971).
5. K. Young et al., *Carbohydr. Res.*, **16**, 446 (1971).
6. K. Izumi, *Carbohydr. Res.*, **17**, 227 (1971).
7. K. Hayashi and K. Nonaka, *Nippon Shokuhin Kogyo Gakkaishi*, **14**, 66 (1967).
8. C. Araki et al., *Bull. Chem. Soc. Jpn.*, **40**, 959 (1967).
9. M. Duckworth et al., *Carbohydr. Res.*, **18**, 1 (1971).
10. Y. Karamanos et al., *Carbohydr. Res.*, **187**, 93 (1989).
11. H. Hayashi et al., *Shokuhin Eiseigaku Zasshi*, **29**, 390 (1988).
12. H. A. Kordan, *Biochem. Physiol. Pflanz.*, **183**, 355 (1988).
13. H. P Glauert et al., *Food Cosmet. Toxicol.*, **19**, 281 (1981).
14. M. W. Gouda and G. S. Godhka, *Can. J. Pharm. Sci.*, **12**, 4 (1977).
15. A. C Tsai et al., *J. Nutr.*, **106**, 118 (1976).
16. S. Y Shiau et al., *Nutr. Rep. Int.*, **39**, 281 (1989).
17. Anon., *Fed. Regist.*, **42** (161), 41876 (1977).

ALETRIS

Source: *Aletris farinosa* L. (Family Liliaceae).

Common/vernacular names: Ague grass, ague root, aletris, blazing star, colic root, stargrass, starwort, true unicorn root, unicorn root, and whitetube stargrass.

GENERAL DESCRIPTION

Perennial herb with grasslike leaves up to 20 cm long, formed as a rosette around a slender, naked flowering stem that grows up to almost 1 m high; flowers white, tubular, mealy at base; native to North America from southern Maine south to Florida and west to

Wisconsin and Texas. Parts used are the dried rhizome and roots, which are collected in the fall (FOSTER AND DUKE; YOUNGKEN).

CHEMICAL COMPOSITION

Diosgenin was detected in the roots,[1] which, along with gentrogenin, was isolated from whole plant samples of two Japanese *Aletris* species: *A. foliata* and *A. formosana*.[2]

Other constituents found in aletris include an amber volatile oil said to be pharmacologically active, a resinous material, and a saponin-like glycoside that yields diosgenin on hydrolysis.

PHARMACOLOGY AND BIOLOGICAL ACTIVITIES

Aletris has shown estrogenic activity.[3]

USES

Medicinal, Pharmaceutical, and Cosmetic. It has been and is still used in proprietary preparations for the treatment of female disorders such as dysmenorrhea and other menstrual discomforts, in laxatives, and also as an antiflatulent.

Dietary Supplements/Health Foods. Crude root, powdered, or cut and sifted used in tablets, capsules, tinctures, teas, often in combination with other herbs for menstrual disorders and as a bitter digestive tonic (CRELLIN AND PHILPOTT).

Traditional Medicine. Tea made from the leaves was used by the Catawba Indians to relieve colic and stomach disorders and to treat dysentery; the Cherokee Indians ingested the leaves to treat rheumatism, flatulent colic, fever in children, coughs, lung diseases, cough, jaundice, and painful urination; Micmac used the root as an emmenagogue and stomachic; Rappahannock Indians used tea made from the plant for female problems.[4]

COMMERCIAL PREPARATIONS

Crude and extracts. Strengths (see *glossary*) of extracts are expressed in weight-to-weight ratios. Crude was formerly official in U.S.P. (1820–1860) and N.F. (1916–1942).

REFERENCES

See the General References for CRELLIN AND PHILPOTT; DER MARDEROSIAN AND BEUTLER; FERNALD; FOSTER AND DUKE; GRIEVE; KROCHMAL AND KROCHMAL; LEWIS AND ELVIN-LEWIS; MCGUFFIN 1 & 2; MERCK; YOUNGKEN.

1. R. E. Marker et al., *J. Am. Chem. Soc.*, **62**, 2620 (1940).
2. T. Okanishi et al., *Chem. Pharm. Bull.*, **23**, 575 (1975).
3. C. L. Butler and C. H. Costello, *J. Am. Pharm. Assoc.*, **33**, 177 (1944).
4. D. E. Moerman, *Native American Ethnobotany*, Timber Press, Portland, OR, 1998, pp. 55–56.

ALFALFA

Source: *Medicago sativa* L. (Family Leguminosae or Fabaceae).

Common/vernacular names: Alfalfa and lucerne.

GENERAL DESCRIPTION

Perennial herb with a deep taproot; leaves resemble those of clover; grows to a height of 1 m with mostly bluish purple flowers in the typical subspecies. Native to the Near East (western Asia and east Mediterranean regions); now cultivated extensively throughout the world. Parts used are the aerial parts. The species has several distinct variants including *M. sativa* (*sensu stricto*) and subsp. *falcata* (L.) Arcangeli (syn. *M. falcata* L.). The former is a purple-flowered form with strongly coiled legumes, originating from an arid continental climate in alkaline soils, principally from Turkey. Wild and cultivated *M. sativa* subsp. *sativa* and their progeny are relatively low in hemolytic saponins. *M. sativa* subsp. *falcata* has yellow flowers and uncoiled fruits, originating from cool, upland, humid climates in acidic soils and is comparatively higher in hemolytic saponins. Both taxa are involved in the parentage of numerous commercial alfalfa cultivars.[1] Modern western European and North American cultivars have intermediate levels of hemolytic alfalfa saponins due to hybridization and introgressions involving *M. sativa* subsp. *falcata*.[2]

CHEMICAL COMPOSITION

Alfalfa has been one of the most studied plants. Its chemical constituents include the following.

Saponins (2–3%) that on hydrolysis yield the aglycones medicagenic acid, soyasapogenols A, B, C, D, and E, and hederagenin and the glycones glucose, arabinose, xylose, rhamnose, galactose, and glucuronic acid;[3–8] sterols (β-sitosterol, α-spinasterol, stigmasterol, cycloartenol, and campesterol, with β-sitosterol as the major component);[9–11] high molecular weight alcohols (octacosanol, triacontanol); and paraffins (nonacosane, triacontane, hentriacontane).[12] β-Sitosterol also occurs as esters with fatty acids (mainly palmitic, lauric, and myristic). Triacontanol has been shown to be a plant growth regulator that increases the growth of rice, corn, and barley as well as the yield of tomato, cucumber, and lettuce.[13]

Flavones and isoflavones (tricin, genistein, daidzein, biochanin A, formononetin, and (−)-5′-methoxysativan); coumarin derivatives (coumestrol, medicagol, sativol, trifoliol, lucernol, and daphnoretin); and pectin methylesterase (an enzyme present in significant quantities believed to be one of the causes for bloating in cattle by releasing pectic acids that combine with calcium in the rumen to form a resinous material, trapping gases produced during digestion).[4,14–16]

Alkaloids (trigonelline, which is in seeds only; stachydrine; and homostachydrine); plant acids (malic, oxalic, malonic, maleic, and quinic, etc.); vitamins and growth factors (vitamins A, B_1, B_6, B_{12}, C, E, and K_1; niacin; pantothenic acid; biotin; folic acid; etc.); amino acids (valine, lysine, arginine, leucine, isoleucine, tryptophan, phenylalanine, methionine, and threonine; asparagine in high concentrations in seeds); sugars (sucrose, fructose, arabinose, xylose, galactose, ribose, mannoheptulose, and D-glycero-D-mannooctulose); plant pigments (chlorophyll, xanthophyll, β-carotene, anthocyanins); crude fibers (17–25%); proteins (15–25% in dehydrated alfalfa meal); minerals; and trace elements (Ca, P, K, Mn, Fe, Zn, Cu) (KARRER; LIST AND HÖRHAMMER).

Medicarpin-β-D-glucoside (in roots); cerebrosides (sphingosines); plastocyanins and ferredoxins; benzoylmesotartaric acid, and benzoyl-(*S*),(−)-malic acid;[17–19] three phytoalexins;[20] medicosides A, C, G, I, J, and L (triterpene glycosides) in roots;[21] and a new amino acid, medicanine, (*S*)-*N*-(3-hydroxypropyl)-azetidine-2-carboxylic acid from seedlings.[22]

In commercial solid extracts of alfalfa and red clover, traces of cannabinol, caffeine, scopolamine, isocoumarin, phenylpentadienal, phenylhexadiene, and nepetalactone have been reported.[23] Whether or not these compounds were results of contamination or adulteration remains to be confirmed.

PHARMACOLOGY AND BIOLOGICAL ACTIVITIES

Coumestrol, genistein, biochanin A, and daidzein have estrogenic activities on ruminants.[14,15] Studies have also demonstrated that coumestrol when fed to pullets increases the age of maturity and depresses egg production.[24]

Alfalfa saponins are hemolytic;[25] they also interfere with vitamin E metabolism and are believed to be one of the causes of ruminant bloat.[4,26] Alfalfa saponins are reported to be fungitoxic, antimicrobial, insecticidal, piscicidal, and taste repellent to rats, swine, and poultry, while attractive to rabbits.

Medicagenic acid, isolated from the roots of alflafa, has shown potent *in vitro* inhibitory activity against medically pathogenic fungi.[27] Medicagenic acid and its glycoside (but not soyasapogenol and glycoside) are toxic to L-cells in culture, lowering mitotic index, viability, and growth of the cells as well as inducing cell death.[28] When administered intramuscularly to Wistar rats, they caused pathological changes in internal organs, especially the kidneys and liver.[29]

Saponins derived from the root of the plant have shown hypocholesterolemic activity in monkeys on a high cholesterol diet (also see **quillaia**).[30] Male rats fed a complex of alfalfa top saponins (1% of diet for 6 months) showed reduced levels of serum cholesterol and triglycerides, with no evidence of toxicity. Alfalfa top saponins have also shown hypocholesterolemic activity and prevention of atherosclerosis.[31]

TOXICOLOGY

Ingesting large amounts of alfalfa seeds can produce reversible pancytopenia with splenomegaly in humans, probably due to the activity of canavanine.[32] The seeds or sprouts may induce systemic lupus erythematosus (SLE).[33–35] Persons with or predisposed to SLE are cautioned to curtail or eliminate alfalfa product intake (TYLER 1). Incorporated in the diets of male rats for up to 6 months, alfalfa saponins produced no evidence of toxicity.[25] Oral toxicity of alfalfa saponins in humans is considered low because they are not absorbed by the gut and then enter the bloodstream. Administered i.v., alfalfa saponins are highly toxic to mammals.[2]

USES

Medicinal, Pharmaceutical, and Cosmetic. The unsaponifiable extract has been claimed to be beneficial in treating skin conditions, including damage caused by radiotherapy and in the healing of gums after orthodontic operations.[9] Alfalfa is also reportedly used in peelable facial masks (DE NAVARRE).

Food. Extract used as a flavor ingredient in most major categories of food products, including nonalcoholic and alcoholic beverages, frozen desserts, candy, baked goods, gelatins and puddings, and meat and meat products, with highest average maximum use level of 0.05% in the last category.

Dietary Supplements/Health Foods. Alfalfa sprouts are a favorite salad ingredient among health food enthusiasts. Dried leaves used in tablets, capsules, teas, tinctures, and so on reported as a source of chlorophyll, vitamins, minerals, and protein, with unsubstantiated benefit in conditions such as rheumatoid arthritis, to prevent absorption of cholesterol, treating diabetes, stimulating appetite, and as a general tonic (TYLER 1).

Traditional Medicine. Reportedly used as a nutrient to increase vitality, appetite, and weight in humans; also as a diuretic, galactogogue, and to increase peristaltic action of the stomach and bowels, resulting in increased appetite; more recently for the treatment of asthma and hay fever (JIANGSU).

Others. Alfalfa meal is used extensively as a poultry and cattle feed and as a source of raw material for the manufacture of leaf protein intended for human consumption. Alfalfa is also a source of chlorophyll manufacture.

COMMERCIAL PREPARATIONS

Crude and extracts.

Regulatory Status. Alfalfa herb and seed are GRAS as natural seasonings and flavorings (§ 582.10); alfalfa essential oil, oleoresin (solvent-free) and natural extractives are GRAS (§182.20).

REFERENCES

See the General References for BAILEY 2; BARNES; DER MARDEROSIAN AND BEUTLER; FEMA; JIANGSU; KARRER; LIST AND HÖRHAMMER; MCGUFFIN 1 & 2; TYLER 1; WILLAMAN AND SCHUBERT.

1. E. Small and B. S. Brookes, *Econ. Bot.*, **38**, 83 (1984).
2. E. Small et al., *Econ. Bot.*, **44**, 226 (1990).
3. M. Jurzysta, *Pamiet. Pulawski*, **62**, 99 (1975).
4. J. F. Morton, *Morris Arboretum Bull.*, **26**, 24 (1975).
5. E. Nowacki et al., *Biochem. Physiol. Pfanz.*, **169**, 183 (1976).
6. B. Gestetner, *Phytochemistry*, **10**, 2221 (1971).
7. B. Berrang et al., *Phytochemistry*, **13**, 2253 (1974).
8. R. L. Baxter, *J. Nat. Prod.*, **53**, 298 (1990).
9. P. de Froment, Fr. Demande 2,187,328 (1974).
10. S. Ito and Y. Fujino, *Nippon Nogei Kagaku Kaisha*, **47**, 229 (1973).
11. S. Ito and Y. Fujino, *Obihiro Chikusan Daigaku Gakujutsu Kenkyu Hokoku, Dai-I-Bu*, **9**, 817 (1976).
12. H. Choichiro et al., *Nippon Kagaku Zasshi*, **77**, 1247 (1956).
13. S. K. Ries et al., *Science*, **195**, 1339 (1977).
14. P. J. Schaible, *Poultry: Feeds and Nutrition*, AVI, Westport, CT, 1970, p. 358.
15. R. F. Keeler, *Lloydia*, **38**, 56 (1975).
16. R. W. Miller et al., *J. Nat. Prod.*, **52**, 634 (1989).
17. Y. Sakagami et al., *Agric. Biol. Chem.*, **38**, 1031 (1974).
18. E. G. Sarukhanyan et al., *Dokl. Akad. Nauk Arm. SSR*, **64**, 112 (1977).
19. T. Yoshihara and S. Sakamura, *Agric. Biol. Chem.*, **41**, 2427 (1977).
20. P. M. Dewick and M. Martin, *J. Chem. Soc., Chem. Comm.*, 637 (1976).
21. A. E. Timbekova and N. K. Abubakirov, *Khim. Prir. Soedin.*, **5**, 607 (1986).
22. S. Fushiya et al., *Heterocycles*, **22**, 1039 (1984).
23. S. R. Srinivas, *Dev. Food Sci.*, **18**, 343 (1988).
24. M. Mohsin and A. K. Pal, *Indian J. Exp. Biol.*, **15**, 76 (1977).
25. B. J. Hudson and S. E. Mahgoub, *J. Sci. Food Agric.*, **31**, 646 (1980).
26. A. J. George, *Food Cosmet. Toxicol.*, **3**, 85 (1965).
27. M. Levy et al., *J. Agric. Food Chem.*, **34**, 960 (1986).
28. M. Slotwinska, *Ann. Univ. Mariae Curie-Sklodowska. Sect. C*, **38**, 177 (1986).
29. M. Gorski et al., *Ann. Univ. Marie Curie-Sklodowska, Sect. C*, **35**, 167 (1980).
30. M. R. Malinow et al., *Steroids*, **29**, 105 (1977).
31. M. R. Malinow et al., *Food Cosmet. Toxicol.*, **19**, 443, (1981).
32. M. R. Malinow et al., *Lancet*, **1**, 615 (1981).
33. M. R. Malinow et al., *Science*, **216**, 415 (1982).
34. J. L. Roberts and J. A. Hayashi, *N. Engl. J. Med.*, **308**, 1361 (1983).
35. P. E. Prete, *Arthritis Rheum.*, **28**, 1198 (1985).

ALGIN

Source: Brown algae, commonly including members of the following genera: ***Macrocystis***, ***Laminaria***, and ***Ascophyllum***.

Common/vernacular names: Algin, salts of alginic acid (alginates), and particularly sodium alginate.

GENERAL DESCRIPTION

Algin is a collective term for the hydrophilic colloidal substance isolated from certain brown algae (class Phaeophyceae). The most commonly used algae include members of the following genera: *Macrocystis*, *Laminaria*, and *Ascophyllum*.

A major source of algin in the United States is *Macrocystis pyrifera* (L.) C. A. Agardh. or "giant kelp" that grows along the West Coast of North America. Other sources include *Ascophyllum nodosum* (L.) LeJolis and *Laminaria digitata* (L.) Edmonson and related species that are used by countries bordering the Atlantic Ocean. *Laminaria* species are also used by Japanese producers.

The process for algin manufacture basically involves a prewash of the seaweed whereby undesirable salts are leached out and removed, followed by extraction with a dilute alkaline solution that solubilizes the alginic acid present in the seaweed. The resulting thick and viscous mass is clarified and the algin is obtained as free alginic acid on treatment with mineral acids. The alginic acid can then be converted to sodium alginate. Sodium alginate is the major form of algin currently in use.[1,2]

The major producing countries include the United States, UK, Norway, France, and Japan.

Alginic acid and its calcium salt are insoluble in water, but its ammonium, sodium, potassium, and magnesium salts as well as its propylene glycol ester are readily soluble in cold and hot water in which they form viscous solutions. The viscosity of algin solutions depends on various factors including concentration, pH, degree of polymerization (DP), temperature, and presence of polyvalent metal ions. Viscosity increases with DP; it decreases with increase in temperature but will regain its original value on cooling to its initial temperature, provided the solutions are not held above 50 °C for long periods. Between pH 4 and 10 the viscosity of algin solutions is generally stable.

Algin solutions form gels with calcium ions due to the formation of insoluble calcium alginate. These gels are not thermally reversible but may be liquefied by calcium sequestrants.

Propylene glycol alginate is more acid-tolerant than the other alginates. Its solutions are stable below pH 4 (down to pH 2.6).[1,2]

CHEMICAL COMPOSITION

Alginic acid is a linear polymer consisting of (1,4)-linked residues of β-D-mannopyranosyluronic acid and α-L-gulopyranosyluronic acid. These D-mannuronic acid and L-guluronic acid residues are arranged in the polymer chain in blocks. Blocks of mannuronic acid are separated from those of guluronic acid by blocks made up of random or alternating units of mannuronic and guluronic acids.[2–4] The homogeneous blocks (those composed of either acid residues alone) are less readily hydrolyzed than the interconnecting heterogeneous blocks.[5] Alginates from different sources vary in their proportions of blocks of mannuronic and guluronic acid residues; for some alginate samples, values of mannuronic acid to guluronic acid ratios range from 0.3 to 2.3.[6–8] These values can be readily determined by infrared spectroscopy.[6]

Molecular weights of alginates range from 10,000 to 1,870,000 depending on algal sources and methods of analysis.[8–11]

PHARMACOLOGY AND BIOLOGICAL ACTIVITIES

Sodium alginate has the ability to reduce strontium absorption, and the sodium alginate with the highest proportion of guluronic acid is the most effective.[12] Sodium alginate can also decrease the retention of other radioactive divalent metallic ions in rats in the following order: Ba > Sr > Sn > Cd > Mn Zn > Hg >, with Ba levels being reduced to 3% of control values and Cd and Mn levels to about 50% in 3 weeks.[13]

Studies have shown that orally fed alginic acid and sodium alginate depress plasma and/or liver cholesterol levels in rats;[14,15] only algin with a high DP is active. Hypocholesterolemic activity was attributed to the inhibition of cholesterol absorption from the gut.[14]

TOXICOLOGY

Animal studies have shown that algin (alginic acid and its sodium and calcium salts, and propylene glycol alginate) is generally nontoxic.[16] Algin is apparently not digested, though this remains to be confirmed.[16,17] Rats fed sodium alginate as 5% of the diet for 2 weeks showed elevated pancreatic-bile secretion and enlarged digestive organs, possibly as the result of interference by algin with absorption and digestion of dietary nutrients.[18]

USES

Algin has been available commercially for several decades and currently is widely used. Its applications generally depend on its thickening, gel-forming, and stabilizing properties.

Medicinal, Pharmaceutical, and Cosmetic. Sodium alginate has many uses: a binding and disintegrating agent in tablets; a binding agent and demulcent in lozenges; a film former in peel-off facial masks; a suspending and thickening agent in water-miscible gels, lotions, and creams; and a stabilizer for oil-in-water emulsions. Calcium alginate is used as absorbable hemostatic; potassium alginate (in conjunction with calcium sulfate and sodium phosphate) is used as an irreversible dental impression material.

Food. Algin is used in virtually every category of food products. Average maximum usage level is about 1% in such products as candy, gelatins, puddings, condiments, relishes, processed vegetables, fish products, and imitation dairy products. Other products in which it is used in lower levels include alcoholic and nonalcoholic beverages, frozen dairy desserts, baked goods, meat and meat products, milk products, fats and oils, cheese, egg products, soups, snack foods, and others.

Others. A 0.2% sodium alginate spray as an effective fungicide against fungal infection of rice by *Pyricularia orysae* Cav. was claimed by a Japanese patent.[19] Alginic acid is used as a sizing agent for textiles and in adhesive formulations.

COMMERCIAL PREPARATIONS

Sodium, potassium, ammonium, calcium salts of alginic acid, and propylene glycol alginate. Alginic acid is official in N.F. and F.C.C.; alginates (potassium, propylene glycol, and sodium) are official in F.C.C.

Regulatory Status. Algin is GRAS for use in foods (§ 182.7133, § 182.7187, § 582.30, § 582.40); alginic acid (§ 184.1011); and brown algae is affirmed GRAS (§ 184.1120).

REFERENCES

See the General References for DER MARDEROSIAN AND LIBERTI; FEMA; FURIA; MARTINDALE; PHILLIPS; UPHOF; WHISTLER AND BEMILLER.

1. K. B. Guiseley in A. Standen, ed., *Kirk-Othmer Encyclopedia of Chemical Technology*, Vol. 17, 2nd ed., Wiley–Interscience, New York, 1968, p. 768.
2. A. Wylie, *R. Soc. Health J.*, **93**, 309 (1973).
3. A. Haug et al., *Carbohydr. Res.*, **32**, 217 (1974).

4. O. Smidsrød, *Carbohydr. Res.*, **27**, 107 (1973).
5. D. A Rees, *Adv. Carbohydr. Chem. Biochem.*, **24**, 296 (1969).
6. W. Mackie, *Carbohydr. Res.*, **20**, 413 (1971).
7. A. Penman and G. R. Sanderson, *Carbohydr. Res.*, **25**, 273 (1972).
8. M. H Ji et al., *Proc. Joint China–U.S. Phycol. Symp.*, 393 (1983).
9. M. Fujihara and T. Nagumo, *J. Chromatogr.*, **465**, 386 (1989).
10. A. Ball et al., *Int. J. Biol. Macromol.*, **10**, 259 (1988).
11. R. S Doubet and R. S Quatrano, *J. Chromatogr.*, **264**, 479 (1983).
12. E. R Humphreys and G. R Howells, *Carbohydr. Res.*, **16**, 65 (1971).
13. A. J. Silva et al., *Health Phys.*, **19**, 245 (1970).
14. K. Ito and Y. Tsuchiya, *Proc. Int. Seaweed Symp.*, **7**, 558 (1972).
15. E. Tsuji et al., *Eiyogaku Zasshi*, **33**, 273 (1975).
16. W. H. McNeely and P. Kovacs, *ACS Symp. Ser.*, **15**, 269 (1975).
17. S. Viola et al., *Nutr. Rep. Int.*, **1**, 367 (1970).
18. S. Ikegami et al., *J. Nutr.*, **120**, 353 (1990).
19. T. Misato et al., *Jpn. Kokai*, **76**, 110,022 (1976).

ALKANET

Source: *Alkanna tinctoria* (L.) Tausch (Family Boraginaceae).

Common/vernacular names: Anchusa, alkanet, alkanna, dyer's alkanet, orcanette, and Spanish bugloss.

GENERAL DESCRIPTION

Biennial or perennial herb about 0.5 m high with hairy leaves and blue or purple trumpet-shaped flowers; indigenous to southeastern Europe, particularly Hungary, Greece, and the Mediterranean region. Part used is the dried root.

CHEMICAL COMPOSITION

Contains polymeric and napthoquinine pigments (isohexenylnaphthazarins),[1] up to 5% alkannin (coloring principle also known as anchusin, anchusic acid, and alkanna red), which occurs mainly in the cortex. Other constituents present include pyrrolizidine alkaloids (7-angeloylretronecine, triangularine, and dihydroxytriangularine),[2] tannin, alkannin isovalerate, alkannin angelate,[3] and alkannan;[4] waxy substances;[5,6] in the root are alcohols, methyl esters, and fatty acids including linoleic, palmitic, 9,2,15-octadecatrienoic, oleic, and γ-linolenic acid.[1]

Alkannin is soluble in organic solvents but almost insoluble in water. Its buffered aqueous solutions are red at pH 6.1, purple at pH 8.8, and blue at pH 10.0.

PHARMACOLOGY AND BIOLOGICAL ACTIVITIES

Fed to mice for 15 weeks at 1% of the diet, alkannin showed no toxic effects.[7] Alkannins have shown antimicrobial and wound-healing properties and have been used externally for the treatment of ulcers.[8] Shikonin, the 1′R-isomer of alkannin, has immunomodulatory effects at low dosage and is immunosuppressive in higher doses (HARBOURNE AND BAXTER). Alkannin angelate and alkannin isovalerate are claimed to have 80% and 85% healing effects, respectively, in patients with leg ulcers.[9]

USES

Medicinal, Pharmaceutical, and Cosmetic. In lipsticks and hair dyes.[10]

Food. Formerly used mainly as dye for sausage casings; oleomargarine, and shortening; also as ink to mark food products.

Traditional Medicine. Used to treat burns, old ulcers,[1] and as an astringent in diarrhea and abscesses.

Others. *A. tinctoria* root was used to stain woods and marble and to impart a red color to salves and port wines (GRIEVE); also used for coloring oils and tars. A tincture of the root is used in microscopy for the detection of fats and oils (EVANS). Alkannin is used as a pH indicator.

COMMERCIAL PREPARATION

Not widely available either in crude or in extract forms; crude was formerly official in U.S.P.

Regulatory Status. Formerly approved by the USDA Meat Inspection Division as a food dye with specific limitations; root and extract thereof not currently approved by the FDA as food colorant.

REFERENCES

See the General References for BISSET; DER MARDEROSIAN AND BEUTLER; FURIA; GLASBY 2; GRIEVE; HARBOURNE AND BAXTER; HOCKING; MARTINDALE; MCGUFFIN 1 & 2; MERCK; POUCHER; TERRELL; UPHOF; USD 23RD; WICHTL; WREN; YOUNGKEN.

1. V. P. Papageorgiou and A. N. Assimopoulou, *Phytochem. Anal.*, **14**, 251 (2003).
2. E. Roeder, *Pharmazie*, **50**, 83 (1995).
3. V. P. Papageorgiou and G. A. Digenis, *Planta Med.*, **39**, 81 (1980).
4. R. D. Gibbs, *Chemotaxonomy of Flowering Plants*, Vol. 2, McGill-Queens University Press, Montreal, 1974, p. 700.
5. A. G. Varvoglis, *Chem. Chron.*, **1**, 156 (1972).
6. B. Papageorgiou, *Chem. Chron.*, **6**, 365 (1977).
7. L. Majlathova, *Nahrung*, **15**, 505 (1971).
8. V. P. Papageorgiou, *Planta Med.*, **39**, 193 (1980).
9. V. P. Papageorgiou, Ger. Offen. 2,829,744 (1979).
10. P. Hatinguais and R. Belle, Fr. Demande FR 2,477,872 (1981).

ALLSPICE

Source: ***Pimenta dioica*** (L.) Merr. (syn. *P. officinalis* Lindl.; *Eugenia Pimenta* DC.) (Family Myrtaceae).

Common/vernacular names: Allspice, Jamaica pepper, pimenta, and pimento.

GENERAL DESCRIPTION

Pimenta dioica, the source of allspice, is a neotropical tree 8–20 m high, with opposite, leathery, oblong leaves 5–15 cm long; fruit globose, about 6 mm in diameter; native to the West Indies, Central America, and Mexico. Part used is the dried, full-grown but unripe fruit; leaves are also used. Major producers include Jamaica and Cuba; also grown in India.[1]

West Indian allspice berries are smaller than Central American and Mexican berries, but they have stronger and smoother flavor. The relatively harsher flavor and aroma of Central American and Mexican berries are due to their relatively high content of monoterpene hydrocarbons, especially myrcene, in their essential oil.[2]

CHEMICAL COMPOSITION

Allspice contains about 4% volatile oil, which is rather stable compared with those of tarragon and black pepper.[3] However, there is evidence that storage of the undried berries under conditions that prevent rapid removal of moisture can increase the volatile oil content by up to 50%; it appears that enzymes released in the fruit after harvest are responsible for producing volatile components from their precursors.[4] The major component of the volatile oil (known as pimenta, pimento, or allspice oil) is eugenol, present at 60–80%. Other constituents include methyleugenol, 1,8-cineole, l-α-phellandrene, caryophyllene, epimeric 10-cadinols (2%),[5] β-phellandrene, camphene, and guaiene.[6] Total identified constituents number more than three dozen.[5,7]

Other constituents of the berries include pimentol, gallic acid, galloylglucosides,[8] phenylpropanoids, vanillin,[9] quercetin glycosides,[10] catechins, proanthocyanidins,[11] protein, lipids, carbohydrates, vitamins (A, C, thiamine, riboflavin, niacin), and minerals.[12]

The leaf oil (pimenta leaf oil) contains more eugenol (up to 96%) than the berry oil and is similar in composition to clove leaf oil.[13,14] Annual leaf oil production exceeds that of the oil of the berries.

PHARMACOLOGY AND BIOLOGICAL ACTIVITIES

Eugenol, the major component of both allspice berry and leaf oils, has local antiseptic and anesthetic properties. It is considered anticonvulsant, antimitotic, antioxidant, and spasmolytic. Eugenol has shown central nervous system a depressant activity and inhibits prostaglandin synthesis in human colonic muscoa (HARBOURNE AND BAXTER). Oral administration of an aqueous suspension of allspice to rats and mice produced anti-inflammatory, antipyretic, antiulcerogenic, analgesic activities and on *ex vivo* gastric mucosa, a protective effect.[15]

When pimento oil and eugenol were applied on intact shaved abdominal skin of the mouse, no percutaneous absorption was observed.[16]

Eugenol, aqueous extracts of allspice, and allspice oil, along with numerous other spices and their volatile oils, have been demonstrated to enhance trypsin activity;[17] they also exhibit larvicidal properties.[18]

USES

Medicinal, Pharmaceutical, and Cosmetic. Allspice oil has been used medicinally as an aromatic carminative at dose of 0.05–0.2 mL. It is also used in cosmetics as an ingredient in fragrance formulations, for spicy, clove-like notes. Eugenol is used as a dental antiseptic and anesthetic.

Food. Allspice, its oil, and its oleoresin (less so) are currently extensively used in food products, including alcoholic and nonalcoholic beverages, frozen dairy desserts, candy, baked goods, gelatins and puddings, meat and meat products, condiments and relishes, and others. The leaf oil is also used for flavoring in food products. The highest average maximum use level of the berry oil is in candy (ca. 0.025%).

Traditional Medicine. Formerly, the berries were used as an appetite stimulant, for stomachache, and for painful menstruation; leaves used for pain, fever, cold remedy, toothache, anodyne, astringent, and carminative. In Jamaica, the fruit is used to treat influenza and stomachache; used in Guatemala to treat rheumatism. In the Dominican Republic, the fruits, decocted with salt, are also used as an antiemetic (WENIGER AND ROBINEAU). Other uses in Middle Eastern, South American, and Asian countries include the treatment of obesity, hyperglycemia, menstrual cramps, abdominal pain, digestive ailments, inflammatory conditions, and high blood pressure.[15]

Others. Solvent extracts of allspice have shown potent *in vitro* antioxidant activity[9,19] and antimutagenic activity.[19] Radical scavenging activity was found from various constituents of the berries including gallic acid, galloylglucosides,[8] phenylpropanoids, eugenol, and vanillin.[9] A fluid extract of the berries has shown *in vitro* antibacterial and antifungal activities.[20]

COMMERCIAL PREPARATIONS

Crude, oleoresin, berry, and leaf oils. Allspice and allspice oil were formerly official in N.F.; allspice oil and pimenta leaf oil are official in F.C.C.

Regulatory Status. Herb as natural flavoring or spice (§182.10) and essential oil, natural extractive, and solvent-free oleoresin are GRAS for use in foods (§182.20).

REFERENCES

See the General References for ARCTANDER; AYENSU; BAILEY 1; BARNES; BAUER, FEMA; FURIA; GOSSELIN; GUENTHER; HARBOURNE AND BAXTER; KARRER; MARSH; MARTINDALE; MCGUFFIN 1 & 2; ROSENGARTEN; TERRELL; WENIGER AND ROBINEAU.

1. M. Ilyas, *Econ. Bot.*, **30**, 273 (1976).
2. C. L. Green and F. Espinosa, *Dev. Food Sci.*, **18**, 3 (1988).
3. E. G. Chinenova et al., *Konserv. Ovoshchesush. Prom.*, **24**, 31 (1969).
4. P. R. Ashurst, *An. Acad. Bras. Cienc.*, **44** (Suppl.), 198 (1972).
5. J. W. Hogg et al., *Am. Perfum. Cosmet.*, **86**, 33 (1971).
6. J. Pino et al., *Nahrung*, **33**, 717 (1989).
7. J. Nabney and F. V. Robinson, *Flav. Ind.*, **3**, 50 (1972).
8. H. Kikuzaki et al., *J. Nat. Prod.*, **63**, 749 (2000).
9. H. Kikuzaki et al., *Phytochemistry*, **52**, 1307 (1999).
10. B. Voesgen et al., *Z. Lebensm. Unters. Forsch.*, **170**, 204 (1980).
11. J. M. Schulz and K. Herrmann, *Z. Lebensm. Unters. Forsch.*, **171**, 278 (1980).
12. M. Teotia et al., *Indian Food Packer*, **41**(5), 49 (1987).
13. E. Calderon Gomez et al., *Rev. Colomb. Cienc. Quim. Farm.*, **2**, 37 (1974).
14. M. E. Veek and G. F. Russell, *J. Food Sci.*, **38**, 1028 (1973).
15. A. J. Al-Rehaily et al., *Pharm. Biol.*, **40**, 200 (2002).
16. F. Meyer and E. Meyer, *Arzneim.-Forsch.*, **9**, 516 (1959).
17. Y. Kato, *Koryo*, **113**, 17, 24 (1975).
18. K. Oishi et al., *Nippon Suisan Gakkaishi*, **40**, 1241 (1974).
19. A. Ramos et al., *J. Ethnopharmacol.*, **87**, 241 (2003).
20. M. Rodriguez et al., *Alimentaria*, **34**, 107 (1996).

ALMONDS

Source: Sweet almond ***Prunus dulcis*** (Mill.) D. A. Webb (syn. *Prunus amygdalus* Batsch var. *dulcis* (DC.) Koehne). Bitter almond ***Prunus dulcis*** (Mill.) D. A. Webb var. ***amara*** (DC.) H. E. Moore (syn. *Prunus amygdalus* Batsch var. *amara* (DC.) Focke) (Family Rosaceae).

GENERAL DESCRIPTION

The almond tree, *Prunus dulcis*, is also known as *P. communis* (L.) Arcang, *Amygdalus dulcis* Mill., and *A. communis* L., in addition to the above synonyms. It grows to a height of about 7 m and has several varieties; two of them, var. *dulcis* and *amara*, yield sweet and bitter almonds, respectively. The tree is native to western Asia and is now extensively cultivated in the Mediterranean countries and in California. The fruit is botanically classified as a drupe (same as peach or plum), except that its outer portion is leathery, not fleshy and edible like the peach; the almond is its seed.

Sweet almonds are used as food, but bitter almonds are not; this is due to the presence of amygdalin in bitter almonds that can be hydrolyzed to yield deadly hydrocyanic acid (HCN).

Two major types of products are derived from the almond, namely a fixed oil and a volatile oil. The fixed oil is commonly called almond oil, expressed almond oil, or sweet almond oil; it is made from both sweet and bitter almonds by pressing the kernels. It does not contain benzaldehyde or HCN.

The volatile oil is called bitter almond oil. It is obtained by water maceration and subsequent steam distillation of the expressed and partially deoleated bitter almonds or kernels of other *Prunus* species that contain amygdalin; these species include apricot (*P. armeniaca* L.), peach (*P. persica* (L.) Batsch.), and plum (*P. domestica* L.). During maceration, the enzyme (emulsin) present hydrolyzes the amygdalin into sugar, benzaldehyde, and HCN, the last two being distilled by steam. Sweet almond does not yield a volatile oil.

CHEMICAL COMPOSITION

Both sweet and bitter almonds have similar chemical composition and contain 35–55% fixed oil (MERCK).[1,2] The only difference appears to be the presence of amygdalin (3–4%) in bitter almond and its absence or presence in trace amounts in sweet almonds.[3,4] Other constituents reported to be present in sweet and/or bitter almonds include protein (18–25%); emulsin; prunasin (0.005% in bitter almond);[5] daucosterol, and other sterols (e.g., sitosterol, citrostadienol, 24-methylenecycloartanol);[2,6] calcium oxalate;[7] tocopherols (mostly α);[8,9] trace amounts of vitamins A, B complex, and E; and amino acids, including glutamic acid, aspartic acid, and arginine.[10]

Expressed almond oil has been reported to contain 53 individual triglycerides of which triolein and dioleolinolein make up 32% and 33%, respectively;[11] fatty acids present include oleic (66–72%), linoleic (18–22%), palmitic (5.7–7.9%), stearic, lauric, myristic, and palmitoleic acids (MERCK).[7,12,13]

Bitter almond oil contains mostly benzaldehyde (95%) and HCN (2–4%). For food and flavor uses the HCN is removed, and the resulting oil is almost pure benzaldehyde.

Almonds also contain varying amounts (3.11–5.25%) of soluble nonreducing sugars (sucrose, raffinose, and stachyose), depending on the variety.[14–16]

PHARMACOLOGY AND BIOLOGICAL ACTIVITIES

The results of a clinical study in men and women suggest that combined with a heart-healthy diet, the addition of 100 g of dry-roasted or raw almonds per day may significantly lower LDL cholesterol.[17]

Expressed almond oil has emollient, demulcent, and mildly laxative properties. It is a weak antibacterial. While easily absorbed and digested orally, it is slowly absorbed through intact skin.[10] Based on prevalent data, almond oil and almond meal are nonirritating and nonsensitizing to the skin and are considered safe for cosmetic use.[10]

Bitter almond oil, containing 2–4% HCN, is poisonous; and fatal poisoning of an adult after taking 7.5 mL has been reported.

Bitter almond oil, FFPA (free from prussic acid, outdated term for HCN), can be regarded as pure benzaldehyde; it has antipeptic, local anesthetic, and antispasmodic properties; it also has narcotic properties at high doses; ingestion of 50–60 mL can be fatal due to central nervous depression with respiratory failure (GOSSELIN).[18]

Low molecular weight peptides (mol. wt. 1000–10,000) with analgesic and anti-inflammatory activities are described in a Japanese patent.[19] Two anti-inflammatory peptides (mol. wt. 257,000 and 19,000) have also been isolated from Chinese almonds (peach kernels, *Prunus persica*).[14]

USES

Medicinal, Pharmaceutical, and Cosmetic. Sweet almond oil is used as a laxative in doses up to 30 mL as well as a solvent for parenterally administered drugs and a solvent for hemorrhoid injectable solutions.[10] It is also used as an emollient and emulsifier for chapped hands, in lotions (both moisturizing and night skin care preparations), suntan gels, blushers, makeup bases, skin cleansing preparations, creams, and as an ointment base. It is used in cosmetic formulations in concentrations up to 50%; 25% in lipstick formulations. Almond meal is used as a skin cleanser and in medicated soaps.[10]

Bitter almond oil was formerly used in the United States as cough sedative and as an antipruritic. It is no longer used commercially for these purposes.

Food. Bitter almond oil (FFPA) is widely used as a flavor ingredient in most categories of food products, including alcoholic and nonalcoholic beverages, frozen dairy desserts, candy, baked goods, gelatins and puddings, and others, generally at use levels below 0.05%. Synthetic benzaldehyde is even more widely used for the same purposes.

Sweet almonds have been used as food for thousands of years. They are a good source of protein, fats, calcium, iron, potassium, phosphorus, and trace minerals such as zinc and copper.

Dietary Supplements/Health Foods. The controversial unofficial anticancer drug laetrile is mainly amygdalin isolated from kernels of apricot, peach, and other related fruits (TYLER 2).[20] Almond oil is often used as a base for massage oil products.

Traditional Medicine. In traditional Chinese medicine, apricot (*P. armeniaca*) kernels are used as an antitussive and antiasthmatic and in treating tumors. Apricot tree inner bark in the form of a decoction is used in treating apricot kernel poisoning, reportedly with great success (JIANGSU). Sweet almond seed or seed oil has been used as a folk cancer remedy for bladder, breast, mouth, spleen, and uterine cancers, among others.

COMMERCIAL PREPARATIONS

Crude and oils (sweet and bitter, FFPA). Sweet almond oil is official in N.F., and bitter almond oil (FFPA) is official in F.C.C.

Regulatory Status. Bitter almond essential oil natural extractive and solvent-free oleoresin (free from prussic acid) are GRAS for use in foods (§182.20).

REFERENCES

See the General References for ARCTANDER; BAILEY 2; BRUNETON; CLAUS; DUKE 3; FEMA; GUENTHER; JIANGSU; MARTINDALE; MCGUFFIN 1 & 2; NANJING; SAX; TERRELL; TYLER 2; WATT AND MERRILL.

1. E. Y. Babekova and V. F. Shcheglova, *Sb. Rab. Aspirantov, Tadzh. Univ. Ser Biol. Nauk*, **3**, 24 (1969).
2. T. M. Jeong et al., *Lipids*, **9**, 921 (1974).
3. I. Karkocha, *Rocz. Panstev. Zakl. Higi.*, **24**, 703 (1973).
4. M. Nishijima et al., *Tokyo Toritsu Eisei Kenkyusho Kenkyu Nempo*, **26**, 183 (1975).
5. U. Schwarzmaier, *Phytochemistry*, **11**, 2358 (1972).
6. T. M. Jeong et al., *Lipids*, **10**, 634 (1975).

7. J. Seidemann, *Seifen, Öle, Fette, Wachse*, **99**, 302 (1973).
8. G. Lambersten et al., *J. Sci. Food Agric.*, **13**, 617 (1962).
9. F. Hotellier and P. Delaveau, *Ann. Pharm. Fr.*, **30**, 495 (1972).
10. K. T. Fisher, *J. Am. Coll. Toxicol.*, **2**, 85 (1983).
11. N. S. Geiko et al., *Khlebopek. Konditer. Prom-st.*, **8**, 25 (1975).
12. K. Aitzetmueller and M. Ihrig, *Fette Wiss. Technol.*, **90**, 464 (1988).
13. M. Farines et al., *Rev. Fr. Corps Gras*, **33**, 115 (1986).
14. F. J. Lopez Andreu et al., *An. Edafol. Agribiol.*, **44**, 207 (1985).
15. X. D. Fang et al., *Zhongyao Tongbao*, **11**, 37 (1986).
16. F. S. Calicto et al., *J. Agric. Food Chem.*, **29**, 509 (1981).
17. G. A. Spiller et al., *J. Am. Coll. Nutr.* **22**, 195 (2003).
18. D. L. J. Opdyke, *Food Cosmet. Toxicol.*, **14**, 693 (1976).
19. M. Kubo, Jpn. Kokai Tokkyo Koho JP 62,198,399 [87,198,399] (1987).
20. J. Jee et al., *J. Pharm. Sci.*, **67**, 438 (1978).

ALOE (AND ALOE VERA)

Source: *Aloe vera* (L.) Burm. f. (syn. *Aloe barbadensis* Mill.); *A. perfoliata* L. var. *vera* L.); *A. arborescens* Miller var. *natalensis* Berger; *A. ferox* Mill. and its hybrids with *A. africana* Mill. and *A. spicata* Baker; *A. perryi* Baker (Family Liliaceae).

GENERAL DESCRIPTION

There are two major products derived from the leaves of *Aloe* spp. The yellow bitter juice present in specialized cells beneath the thick epidermis yields the drug aloe, which is obtained from all above species. The parenchymatous tissue in the center of the leaf contains a mucilaginous gel that yields aloe gel or *Aloe vera* gel, which is currently obtained from *A. vera*.[1]

All the above species are perennial succulents native to Africa that later spread to other parts of the world. They are not cacti and should not be confused with the American aloe or century plant (*Agave* sp.). Considerable confusion has arisen over the nomenclature of *Aloe* species. Currently, there are more than 360 accepted species of *Aloe*. N. L. Burman's *Aloe vera* binomial was published in 1768 and therefore has priority over Miller's *A. barbadensis*.[2] Aloe vera is also a common name.

Aloe vera yields Curaçao aloe or Barbados aloe, which is produced in the West Indies (Curaçao, Aruba, Bonaire). *Aloe ferox* and its hybrids yield Cape aloe that is produced in South Africa. Other *Aloe* species yield aloes of lesser importance.

Aloe is obtained by cutting the leaves at their base and letting the yellow bitter juice drain out. The water is evaporated off from the juice by heat and the resulting light to dark brown mass is the drug aloe.

Commercial aloin is a concentrated form of aloe containing high concentrations of anthraglycosides (mostly barbaloin). This commercial product is not pure aloin.

Aloe vera gel is prepared from the leaves by numerous methods, some patented and others proprietary.[3–6] These methods essentially involve expression and/or solvent extraction, often with harsh physical and chemical treatments; the resulting gel products vary considerably in properties and generally are not representative of the fresh gel.

CHEMICAL COMPOSITION

Aloe contains cathartic anthraglycosides; mostly C-glucosides such as barbaloin, a glucoside of aloe-emodin. Aloes from most species contain cathartic anthraglycosides at concentrations between 10% and 20%, though some contain levels of 30% barbaloin. A single *A. vera* plant from Sri Lanka was found to contain 57% barbaloin in its exudate. Highest concentrations of barbaloin are found in young mature leaf exudates, decreasing in older leaves toward the base of the plant.[7] The concentrations of anthraglycosides vary with the types of aloe; aloin content ranges from 4.5% to 25%. Other constituents include aloesin and its aglycone aloesone (a chromone),[8] free anthraquinones (e.g., aloe-emodin), and resins.

The composition of aloe vera gel is still not clear. Studies to date indicate that the gel consists of more than one type of polysaccharide and that their compositions vary from one season and source to the next. While one study showed it to contain at least four different partially acetylated linear glucomannans with (1,4)-glycosidic linkages,[9] others revealed an acidic galactan, mannan, glucomannan, arabinan, and/or glucogalactomannan. The ratios of hexoses in each polysaccharide differ widely among the studies, as do their molecular weights.[10–13]

The polysaccharides constitute 0.2–0.3% of the fresh gel (0.8–1.2% of dry matter content). Postproduction autodegradation of the glucomannan polysaccharides produces mainly mannans. The gel polysaccharides, consisting mainly of mannose and glucose in a 1 : 3 ratio, can degrade in 48 h at room temperature, with a decrease in glucose content and an increase in mannose:glucose ratio to >10.[14]

Other constituents reported in *A. vera* include other polysaccharides, polypeptides, steroids, chromones, lectins, organic acids, enzymes, amino acids, saponins, and minerals.[1,3–5,13,15–23] Carboxypeptidase, a serine carboxypeptidase enzyme found in *A. arborescens* and other species of *Aloe*, is suggested to be a primary antithermic agent.

PHARMACOLOGY AND BIOLOGICAL ACTIVITIES

Drug aloe and its purified form, aloin, have cathartic properties through action on the colon. They are extremely bitter and are considered the least desirable among the plant purgative drugs (see ***cascara*** and ***senna***) due to their bitterness and their tendency to produce more griping and irritation (GOSSELIN; MARTINDALE). Aloin induces secretion of electrolytes and water in the intestinal lumen, inhibiting reabsorption of electrolytes and water from the large intestine, which thereby increases intestinal volume via increased filling pressure, which in turn stimulates peristalsis (BLUMENTHAL 1).

An alcoholic extract of drug aloe was reported to show antitumor activity (JIANGSU). Aloe-emodin (see ***buckthorn***) administered intraperitoneally (i.p.) has shown antitumor activity in immunodeficient mice and *in vitro* cytoxicity against tumor cells.[24] A phthalate (diethylhexylphthalate or DEHP), isolated from *A. vera*, has shown *in vitro* activity against the growth of several human leukemic cell lines.[25] Daily oral administration of the fresh leaf pulp of *A. vera* produced cancer chemopreventive activity in mice, in part by inducing the phase II enzyme system and by increasing levels of endogenous antioxidants in the liver.[26] Antiviral (herpes simplex types 1 and 2) activities were found from a methanolic extract of the whole leaf of *A. vera*.[17,27,28] In laboratory studies, the leaf gel or extracts of *A. vera* leaf have shown diverse activities, including *in vitro* antimicrobial and *in vivo* hypocholesteremic, dermal protectant, wound healing, burn healing, frostbite healing, anti-ulcerogenic, anti-inflammatory, hypoglycemic, and others.[17]

A number of studies have explored the immunoreactive or immunomodulatory activity of various *Aloe* species.[17] The principle immunoactive constituents identified in *A. vera* are lectins[29–31] and polysaccharides; in particular, acemannan[15,17] and a water-soluble polysaccharide named aloeride.[15]

Chromones isolated from *A. vera* leaves have shown anti-inflammatory[32] and antioxidant activities.[33] From *A. arborescens*, carboxypeptidase showed analgesic activity (comparable with bromelain) and dermoprotectant activity against burns in rats.[34,35]

Male rats fed freeze-dried, skinned (filetted) *A. vera* leaf at 1% of the diet showed significantly fewer instances of death from disease.[36] Oil-in-water aloe extracts significantly increase soluble collagen levels, suggesting a topical antiaging effect.[37] Dermal wound-healing effects of oral and topical *A. vera* gel in rats are partly attributed to their ability to increase the contents of proteoglycans and glycosaminoglycans in the wound matrix.[38]

A systematic review of controlled clinical trials on aloe vera preparations up to 1999 concluded that it is not clear whether *A. vera* promotes wound healing and that overall, either topical or oral effectiveness of its preparations remained insufficiently defined; however, evidence suggested that it might be effective in the treatment of psoriasis, genital herpes, as an adjunctive in the management of diabetes, and in lowering cholesterol levels.[39] A further critical review of clinical trials on *A. vera* preparations in the treatment of various dermatologic conditions up to 2000 concluded that evidence for its efficacy remained unconvincing.[40]

TOXICOLOGY

Overdosage of drug aloe fluid and electrolyte imbalance, abdominal pain, bloody diarrhea, hemorrhagic gastritis, and sometimes nephritis (BLUMENTHAL 1; GOSSELIN; MARTINDALE), the latter also associated in case studies with prolonged use of Cape aloes containing aloesin and aloeresin A.[41]

No severe pathology was evident in mice fed dry aloe extract (50 mg/kg).[42] Aloe is not to be used for longer than 8–14 days; contraindicated for use by children of age 12 years and under and in pregnancy, breast feeding, menstruation, metrorrhagia, menorrhagia, active hemorrhoids, diarrhea, intestinal obstruction or inflammation, inflammatory bowel disease, kidney disease, vomiting, appendicitis, and abdominal pain. Use may temporarily color the urine red. Toxic constituents include aloeresin A and anthraquinone glycosides (aloinosides A and B and aloins A and B) not found in *A. vera* gel (BLUMENTHAL 1; BRINKER).[42] Use with other laxatives, especially those containing anthraquinone glycosides (e.g., cascara, senna) should be avoided owing to potentiating effects.[41,42] Aloe is potassium-depleting. Use with diuretics could or other potassium-depleting substances (e.g., licorice root, thiazide diuretics, corticoadrenal steroids) could result in hypokalemia.

Aloe vera gel (freeze-dried) produced no toxic effects in rats from either acute or sub-chronic oral doses (1–64 mg/kg p.o. twice daily). In mice or rats, the preserved or fresh gel failed to cause any toxicity at doses up to 20 g/kg p.o. or i.p., and no toxic effects were found from a dose of 5 g/kg p.o. per day for 45 days.[43] Life-long dosing of a freeze-dried filet of the leaves at 1% of the diet also failed to produce any deleterious effects.[36]

Reported adverse effects of topical use of *A. vera* gel preparations by humans include allergic contact dermatitis, mild itching, and burning sensation. These effects have been mild, of rare occurrence, and reversible when use was stopped.[17,44,45]

USES

Medicinal, Pharmaceutical, and Cosmetic. Currently the only officially recognized use of aloe is as an ingredient in compound benzoin tincture, presumably for its beneficial properties on the skin.

Aloe and aloin are extensively used as active ingredients in laxative preparations, often with other cathartics such as buckthorn, cascara, and senna; belladonna extracts are often included to lessen griping. Aloin is also used in antiobesity preparations.

In Germany, concentrated dried aloe leaf juice is used for conditions in which ease of defecation and soft stool are desired, for example, anal fissures, hemorrhoids, postanorectal surgery, and refractory constipation (BLUMENTHAL 1).

Aloe gel and sometimes drug aloe are used in various cosmetic and pharmaceutical formulations as moisturizers, emollients, or wound-healing agents. The fatty fraction of the leaf is used in the cosmetics industry as a pigment carrier.[16] Extracts of aloe or aloin are also used in sunscreen and other cosmetic preparations.

Food. Aloe extracts are used as a flavor ingredient primarily in alcoholic and nonalcoholic beverages and in candy to impart a bitter note. Based on the reported average maximum use levels of about 0.02% in alcoholic (186 ppm) and nonalcoholic (190 ppm) beverages and 0.05% in candy, the extracts used must be tinctures or greatly diluted extracts, as standard extracts (e.g., solid extract or fluid extract) would contain too much active anthraglycosides to be safely used. The most preferred commercial aloe exudates used in bitter spirits (Port Elzabeth and Mossel Bay aloes) show a balance of major aromatic constituents not found in many other aloe extracts.[16]

Dietary Supplements/Health Foods. *Aloe vera* gel is used in nonalcoholic beverages that are commonly known as "aloe vera juice." It is normally produced from *A. vera* gel by diluting with water and mixing with citric acid and preservatives. It is also sometimes mixed with fruit juices and/or herbal extracts. Despite label claims, "pure" aloe vera juice is rarely pure; instead, it contains only a minor percentage of *A. vera* gel.[1]

Aloe gel products are available in liquid and solid forms. The most popular liquid products are the 10X, 20X, and 40X concentrates, while spray-dried *Aloe vera* gel extract is the most popular solid product.[1] Commercial liquid concentrates are not always genuine,[46] and the more highly concentrated they are, the more degradation they have undergone, as evidenced by their lack of viscosity.[1]

Despite claims to be 200X concentrated pure aloe gel, solid products normally contain high proportions of carriers such as gums (acacia, guar, locust bean), lactose, mannitol, hydrolyzed starch, and/or others.[1]

Traditional Medicine. Fresh *Aloe vera* gel is a well-known domestic medicine,[47-49] widely used to relieve thermal burn and sunburn and to promote wound healing.[1] It is the most widely used herbal folk remedy among the general population in the United States (LUST).

COMMERCIAL PREPARATIONS

Crude, aloin, and extracts in various forms. Crude Barbados and Cape aloes are official in U.S.P.

Aloin was official in N.F. XI; its current quality is generally governed by standards set forth in this compendium, which do not require specific assays for anthraglycosides.

Aloe gel products are available in liquid and solid forms; qualities vary greatly depending on suppliers.[46] Because the principal component in the gel is a glucomannan similar to guar and locust bean gums, these gums are frequently mixed with aloe gel to increase its viscosity and yield. Dried concentrated aloe gel can be evaluated by infrared spectroscopy.[1]

Regulatory Status. Aloe has been approved for food use as a natural flavoring (§172.510) and is regulated in the United States as a dietary supplement. Dried *Aloe vera* leaf juice (and preparations) calculated to contain at least 28% hydroxyanthracene derivatives (as anhydrous barbaloin), and *A. ferox* dried leaf juice (and preparations) calculated to contain at least 18% hydroxyanthracene derivatives (as anhydrous barbaloin), are the subject of a German BGA monograph (BLUMENTHAL 1).

REFERENCES

See the General References for AHPA; APhA; BLUMENTHAL 1; DUKE 4; FEMA; GUPTA; JIANGSU; LUST; MERCK; TERRELL; USD 26TH; YOUNGKEN.

1. A. Y. Leung, *Drug Cosmet. Ind.*, **120** (6), 34 (1977).
2. D. Grindlay and T. Reynolds, *J. Ethnopharmacol.*, **16**, 117 (1986).
3. A. Farkas, U.S. Pat. 3,103,466 (1963).
4. A. Farkas and R. A. Mayer, U.S. Pat. 3,362,951 (1968).
5. H. H. Cobble, U.S. Pat. 3,892,853 (1975).
6. H. Matsui and T. Matsukura, Jpn. Kokai 75 155,664 (1975).
7. Q. J. Groom and T. Reynolds, *Planta Med.*, **53**, 345 (1987).
8. D. K. Holdsworth, *Planta Med.*, **22**, 54 (1972).
9. D. C. Gowda et al., *Carbohydr. Res.*, **72**, 210 (1979).
10. G. Mandal and A. Das, *Carbohydr. Res.*, **87**, 249 (1980).
11. Q. N. Haq and A. Hannan, *Bangladesh J. Sci. Ind. Res.*, **16**, 68 (1981).
12. G. Mandal et al., *Indian J. Chem., Sect. B*, **22B**, 890 (1983).
13. S. P. Joshi, *J. Med. Arom. Plant Sci.*, **20**, 768 (1998).
14. A. Yaron, *Phytother. Res.*, **7**, S11 (1993).
15. N. Pugh et al., *J. Agric. Food Chem.*, **49**, 1030 (2001).
16. D. Saccù et al., *J. Agric. Food Chem.*, **49**, 4526 (2001).
17. T. Reynolds and A. C. Dweck, *J. Ethnopharmacol.*, **68**, 3 (1999).
18. T. D. Rowe and L. M. Parks, *J. Am. Pharm. Assoc.*, **30**, 262 (1941).
19. E. Roboz and A. J. Haagen-Smit, *J. Am. Chem. Soc.*, **70**, 3248 (1948).
20. L. J. Lorenzetti et al., *J. Pharm. Sci.*, **53**, 1287 (1964).
21. L. B. Fly and I. Kiem, *Econ. Bot.*, **17**, 46 (1963).
22. G. D. Bouchey and G. Gjerstad, *Q. J. Crude Drug Res.*, **9**, 1445 (1969).
23. G. Gjerstad, *Adv. Front. Plant Sci.*, **28**, 311 (1971).
24. T. Pecere et al., *Cancer Res.*, **60**, 2600 (2000).
25. K. H. Lee et al., *J. Pharm. Pharmacol.*, **52**, 1037 (2000).
26. R. P. Singh et al., *Phytomedicine*, **7**, 209 (2000).
27. S. M. Kupchan and A. Karim, *Lloydia*, **39**, 223 (1976).
28. R. J. Sydiskis and D. G. Owen, U.S. Pat. 4,670,265 (1987).
29. L. A. 't Hart et al., *J. Ethnopharmacol.*, **23**, 61 (1988).
30. K. Imanishi, *Phytother. Res.*, **7**, S20 (1993).
31. H. Saito, *Phytother. Res.*, **7**, S14 (1993).
32. J. A. Hutter et al., *J. Nat. Prod.*, **59**, 541 (1996).
33. K. Y. Lee et al., *Free Radic. Biol. Med.*, **28**, 261 (2000).
34. S. Ito et al., *Phytother. Res.*, **7**, S26 (1993).
35. M. Obata et al., *Phytother. Res.*, **7**, S30 (1993).
36. Y. Ikeno et al., *Phytother. Res.*, **16**, 712 (2002).
37. I. E. Danhof et al., *Phytother. Res.*, **7**, S53 (1993).
38. P. Chithra et al., *J. Ethnopharmacol.*, **59**, 179 (1998).
39. B. K. Vogler and E. Ernst, *Br. J. Gen. Pract.*, **49**, 823 (1999).
40. E. Ernst et al., *Am. J. Clin. Dermatol.*, **3**, 341 (2002).
41. V. A. Luyckx et al., *Am. J. Kidney Dis.*, **39**, E13 (2002).

42. A. Fugh-Berman, *The 5-Minute Herb and Dietary Supplement Consult*, Lippincott Williams and Wilkins, Philadelphia, PA, 2003, pp. 8–9.
43. S. Yongchaiyudha et al., *Phytomedicine*, **3**, 241 (1996).
44. E. Ernst, *Br. J. Dermatol.*, **143**, 923 (2000)
45. H. P. Hörmann and H. C. Korting, *Phytomedicine*, **1**, 161 (1994).
46. R. P. Pelley et al., *Subtrop. Plant Sci.*, **50**, 1 (1998).
47. J. F. Morton, *Econ. Bot.*, **15**, 311 (1961).
48. G. Gjerstad and T. D. Riner, *Am. J. Pharm.*, **140**, 58 (1968).
49. J. F. Nieberding, *Am. Bee J.*, **114**, 15 (1974).

ALTHEA ROOT

Source: *Althea officinalis* L. (Family Malvaceae).

Common/vernacular names: Althaea, althea, marshmallow.

GENERAL DESCRIPTION

Perennial herb, velvety hairy, up to 1.5 m high with three-lobed coarsely serrate leaves, pink 3 cm wide flowers in peduncled clusters; grows in marshes and moist places; under cultivation adaptable to drier soils; native to Europe and naturalized in the United States in salt marshes from Massachusetts to Virginia; locally elsewhere. Preferred part is the peeled root collected in the fall; whole dried root and dried leaves enter commerce.

CHEMICAL COMPOSITION

Contains starch, pectin, mucilage, sugar, fats, tannin, asparagine, and calcium oxalate. The mucilage content is generally considered to be 25–35%, but that of the homogeneous mucilaginous polysaccharides is much lower. The mucilage content changes considerably with season (6.2–11.6%) and is highest in winter.[1] A purified, homogeneous mucilage, althaea-mucilage O, is composed of L-rhamnose: D-galactose: D-galacturonic acid: D-glucuronic acid in the molar ratio of 3:2:3:3, with a molecular weight of about 34,000 (as the ammonium salt).[2] The sequence of the component sugars and the configurations of the glycoside linkages have also been examined.[3] Roots also contain scopoletin, asparagines, hypolaetin 8-glucoside, parahydroxybenzoic acid, quercetin, kaempferol, and chlorogenic, syringic, caffeic, and *p*-coumaric acids (BRADLY).[4]

PHARMACOLOGY AND BIOLOGICAL ACTIVITIES

Althaea root extracts reportedly have demulcent, soothing properties on the mucous membranes and an antitussive effect. It is believed that these properties are mainly caused by the mucilaginous substances. Althaea-mucilage O has shown potent hypoglycemic activity.[5] The polysaccharide fraction (50 mg/kg p.o.) isolated from the root (*A. officinalis* L. var. *robusta*) showed significant antitussive activity in a cat model of cough. It was more effective than the crude extract of the root (100 mg/kg p.o.), equally effective compared to a syrup made from the root (1000 mg/kg p.o.), and more effective than prenoxdiazine (30 mg/kg p.o.).[6]

TOXICOLOGY

Crude root may delay absorption of other drugs taken at the same time (BLUMENTHAL 1).

USES

Medicinal, Pharmaceutical, and Cosmetic. Mainly used as a demulcent in various pharmaceutical preparations, particularly cough medicines. In Europe, root and leaf preparations are used to treat irritation of the oral or pharyngeal mucosa and associated dry irritable cough; root also used to treat mild inflammation of the gastric mucosa (BLUMENTHAL 1). Root is used crude or in formulations at a daily dose of 6 g; leaf used in 5 g daily dose or equivalent in formulations (BLUMENTHAL 1).

Food. Used to a limited extent in alcoholic and nonalcoholic beverages, frozen desserts, candy, baked goods, and gelatins and puddings. Extracts of the root are used in confectionaries. Reported use levels are very low; usually below 0.002% (20 ppm).

Traditional Medicine. Used for more than 2000 years in Europe both internally and externally as a wound healer, a remedy for coughs, sore throat, and stomach troubles, among other ailments; in ointments to relieve chapped hands and chilblains (BIANCHINI AND CORBETTA; FOSTER).

COMMERCIAL PREPARATIONS

Crude and extracts. Crude was formerly official in N.F. and U.S.P.

Regulatory Status. Regulated in the United States as a dietary supplement; root and flowers approved for food use as a natural flavoring (§172.510). In Germany, the leaf and root are the subjects of therapeutic monographs; approved for irritation of the oral or pharyngeal mucosa and associated dry irritable cough; root also for mild inflammation of the gastric mucosa (BLUMENTHAL 1; WICHTL).

REFERENCES

See the General References for APPLEQUIST; BARNES; BISSET; BAILEY 2; BIANCHINI AND CORBETTA; BLUMENTHAL 1; DER MARDEROSIAN AND BEUTLER; FEMA; FOSTER; GLEASON AND CRONQUIST; GOSSELIN; GRIEVE; GUPTA; KARRER; MARTINDALE; MCGUFFIN 1 & 2; MERCK; TUTIN 2; USD 23RD; YOUNGKEN.

1. G. Franz, *Planta Med.*, **14**, 90 (1966).
2. M. Tomoda et al., *Chem. Pharm. Bull.*, **25**, 1357 (1977).
3. M. Tomoda et al., *Chem. Pharm. Bull.*, **28**, 824 (1980).
4. St. N. I. Ionkova and D. Kolev, *Fitoterapia*, **63**, 474 (1992).
5. M. Tomoda et al., *Planta Med.*, **53**, 824 (1987).
6. G. Nosal'ova et al., *Pharmazie*, **47**, 224 (1992).

AMBRETTE SEED

Source: *Abelmoschus moschatus* Medik. L. (syn. *Hibiscus abelmoschus* L.) (Family Malvaceae).

Common/vernacular names: Ambrette, musk mallow, and musk seed.

GENERAL DESCRIPTION

An annual or biennial herb with bristly hairs, up to 2 m high; showy flowers yellow, crimson centered, about 10 cm across; seeds musk fragrant, flat, and kidney shaped; indigenous to India; widely cultivated in tropical countries, including the West Indies, Java,

Indonesia, and Africa. Part used is the seed, which are kidney shaped, grayish brown, and about 3 mm in diameter. The seeds are traded as musk grains or musk pods. An aromatic oil is obtained via steam distillation of the crushed, dried seeds (CSIR I).

CHEMICAL COMPOSITION

Compounds isolated from the seed oil include ambrettolide, (Z)-7-hexadecen-16-olide), ambrettolic acid, and farnesol as well as 12,13-epoxyoleic acid, malvalic acid, sterculic acid, and C_{10}–C_{18} acids (oleic, palmitic, C_{10}, C_{12}, C_{14}, C_{16}, C_{18}).[1] (Z)-5-Tetradecen-14-olide, (Z)-5-dodecenyl acetate, and (Z)-5-tetradecenyl acetate were isolated from the absolute in yields of 0.5%, 0.01%, and 0.4%, respectively; in addition, 2-*trans*,6-*trans*-farnesylacetate, 2-*cis*,6-*trans*-farnesylacetate, and oxacyclononadec-10-en-2-one (an ambrettolide homologue) were found in the seed coat. The floral musky odor of the oil is primarily the result of ambrettolide and (Z)-5-tetradecen-14-olide.[2,3]

Other compounds found in the seeds include methionine sulfoxide, phospholipids (α-cephalin, phosphatidylserine, phosphatidylserine plasmalogen, and phosphatidylcholine plasmalogen);[4,5] sterols, including campesterol, sitosterol, stigmasterol, ergosterol, and cholesterol.[6] Sizable amounts of palmitic and myristic acids may also occur in ambrette seed oil or concrete, depending on the method of manufacture (ARCTANDER). The long-chain fatty acids of the seeds result in a crude product of a waxy nature (Ambrette beurre); the fatty acids, removed with alkali, dilute alcohol, calcium, or lithium salts, produce a yellow clear to amber liquid that possesses the musky fragrance of ambrettolide (BAUER). Yield of the oil is from 0.2% to 0.6%.

TOXICOLOGY

Based on available data, ambrette seed oil and its major odor principle, ambrettolide, are nontoxic.[7]

USES

Medicinal, Pharmaceutical, and Cosmetic. Oil and absolute are used in sophisticated types of perfumes and in soaps, detergents, creams, and lotions; maximum use level reported is 0.12% for the oil in perfumes.[7] The oil is valued for its sweet, rich wine or brandy-like, floral, musky scent with a unique bouquet and roundness. While resembling animal-derived musk scents, ambrette seed oil lacks the fecal note sometimes found in the former (CSIR I); one of the most expensive essential oils.

Food. Ambrette seed and its tincture are used in preparing vermouths and bitters. The oil and absolute are also used in vermouths and bitters, but they are more commonly used in flavoring other types of food products, including nonalcoholic beverages, frozen dairy desserts, candy, baked goods, and gelatins and puddings. The use levels are usually very low, <0.001% (10 ppm).

Traditional Medicine. Reportedly used as a stimulant and an antispasmodic; used in Chinese medicine to treat headache; in Western traditions as a folk medicine with stimulant, aromatic, antispasmodic actions and as an insecticide for protecting woolen garments from moths (CSIR I).

Others. The stem bark has been used as a fiber (78% cellulose) and the root mucilage as a paper sizing material (CSIR I).

COMMERCIAL PREPARATIONS

Seeds, oil, absolute, concrete, and tincture; oil is official in F.C.C.

Regulatory Status. Seed GRAS as a natural flavoring or seasoning in foods (§182.10); seed essential oil, extractive, and solvent-free oleoresin GRAS for use in foods (§182.20).

REFERENCES

See the General Reference for ARCTANDER; BAILEY 1; BAUER; CSIR I; DER MARDEROSIAN AND BEUTLER; FEMA; GUENTHER; GUPTA; KARRER; JIANGSU; MCGUFFIN 1 & 2; TERRELL; USD 23RD; WREN.

1. M. Hashmi et al., *J. Oil Technol. Assoc. India*, **14**, 64 (1982).
2. B. Maurer and A. Grieder, *Helv. Chim. Acta*, **60**, 1155 (1977).
3. T. Y. Nee et al., *Phytochemistry*, **25**, 2157 (1986).
4. L. Peyron, *Bull. Soc. Fr. Physiol. Veg.*, **7**, 46 (1961).
5. K. C. Srivastava and S. C. Rastogi, *Planta Med.*, **17**, 189 (1969).
6. U. K. Chauhan, *Proc. Natl. Acad. Sci., India, Sect. B*, **54**, 236 (1984).
7. D. L. J. Opdyke, *Food Cosmet. Toxicol.*, **13**, 705, 707 (1975).

ANGELICA

Source: *Angelica archangelica* L. (syn. *Archangelica archangelica* Hoffm.) (Family Apiaceae or Umbelliferae).

Common/vernacular names: Angelica, archangel, European angelica, and Garden angelica.

GENERAL DESCRIPTION

Stout biennial or perennial herb, up to 2 m high with a large rhizome; fruit with thick corky wings; native to northern and eastern Europe and Iceland, eastward to Siberia; cultivated in Belgium, Hungary, Germany, and other countries; naturalized elsewhere. Parts used are the rhizome and roots, fruits, and stem, with the stem less extensively used; currently the roots and rhizome are the most frequently used.

Other *Angelica* spp. are also used, but infrequently.

CHEMICAL COMPOSITION

Angelica is very rich in coumarins, which occur throughout the plant.

The root (root and rhizome) contains 0.3–1% volatile oil composed mainly of *d*-α-phellandrene, α-pinene, limonene, β-caryophyllene, linalool, borneol, acetaldehyde, and four macrocyclic lactones (ω-tridecanolide, 12-methyl-ω-tridecanolide, ω-pentadecanolide, and ω-heptadecanolide), among others (MASADA);[1,2] coumarins, including osthol, angelicin, osthenol, umbelliferone, archangelicin, bergapten, ostruthol, imperatorin, umbelliprenine, xanthotoxol, xanthotoxin, oxypeucedanin, oreoselone, phellopterin, marmesin, byakangelicol, and 2′-angeloyl-3′-isovaleryl vaginate, with osthol in major concentration (ca. 0.2% of root),[3–7] acids (angelic, aconitic, citric, malic, oxalic, malonic, fumaric, succinic, caffeic, chlorogenic, quinic, lauric, tridecanoic, myristic, pentadecanoic, palmitic, palmitoleic, stearic, oleic, linoleic, linolenic, petroselinic, and behenic acids, etc.).[4,8,9] Other constituents include resin, starch, sugars (sucrose, fructose, glucose, umbelliferose), archangelenone (a flavonone), β-sitosteryl palminate, and arachinate (KARRER).[4,9,10]

The fruits (commonly known as seeds) contain about 1% volatile oil, consisting mainly of β-phellandrene and other terpenes similar to those found in the root oil (MASADA);[11] coumarins, including imperatorin, bergapten, *iso*-imperatorin, *iso*-pimpinellin, 8-hydroxy-5-methoxypsoralen, 4-methoxy-7-hydroxypsoralen, phellopterin, xanthotoxol, and xanthotoxin, with imperatorin and bergapten in larger concentrations (0.5% and 0.1%, respectively) (KARRER).[12,13] The seed oil of a Pakistani variety, *A. archangelica* L. var. *himalaica* (Clarke) Krishna & Badhwar, contains hexylmethylphthalate as its major component (36%).[14]

PHARMACOLOGY AND BIOLOGICAL ACTIVITIES

The root oil has shown antibacterial and antifungal activities.[1,15]

A number of *Angelica* species have shown calcium-antagonist-like effects *in vitro*. As an activity involving relaxation of vascular smooth muscle, calcium-antagonist activity is a topic of interest in cardiovascular disease research. Calcium-antagonist activity was found from coumarin-rich fractions of angelica root extracts (*A. archangelica*).[16] The root extract has shown dose-dependent antiulcerogenic activity in rats with indomethacin-induced gastric ulcers, an effect partly attributed to an increase in mucin secretion, a decrease in leukotrienes, and an increase in the release of prostaglandin E_2.[17] Orally administered, *A. archangelica* also ameliorated ethanol-induced hepatotoxicity in mice and inhibited malondialdehyde formation in mouse livers, *in vitro* and *in vivo*.[18]

TOXICOLOGY

Certain coumarins in the plant (e.g., bergapten, xanthotoxin) are known to be phototoxic (see ***bergamot oil***). Angelica root and seed oils are obtained by steam distillation and are not expected to contain these coumarins; however, extracts (e.g., absolute, solid extract, fluid extract) may contain them. The root oil (but not seed oil) is reported to be phototoxic.[1,11]

USES

Medicinal, Pharmaceutical, and Cosmetic. Now rarely used in pharmaceutical preparations. Its major current use is as a fragrance ingredient in soaps, detergent, creams, lotions, and perfumes. Reported maximum use levels for both root and seed oils are usually very low, the highest for either being about 0.1% in the perfume category.[1,11]

Food. Leaves used as vegetable (MABBERLY); dried seeds and cut and sifted or powdered root occasionally used as tea flavoring (FOSTER); also used as a flavor ingredient in most major categories of food products, including alcoholic (bitters, liqueurs, vermouths) and nonalcoholic beverages, frozen dairy desserts, candy, baked goods, and gelatins and puddings. The seed and root oils and the root extract are more commonly used; average maximum use levels are low, usually below 0.01%, except for the seed extract, which is reported to be 0.2% in alcoholic beverages.

Dietary Supplements/Health Foods. The dried seeds and root powder are used in tinctures or oral formulations, primarily for menstrual regulation and as an expectorant (FOSTER).

Traditional Medicine. Angelica has a long history of use in Europe in the treatment of bronchial ailments, colds, coughs, and stomach troubles caused by indigestion; also used in cosmetics for its allegedly quieting and soothing effect on the nerves of the skin (DE NAVARRE). The roots and seeds have been used in the treatment of arthritic disease, nervous conditions, insomnia, hyperacidity, and intestinal disturbances, as well as for anti-inflammatory, diuretic, and diaphoretic effects.[16]

In Chinese medicine, at least 10 *Angelica* species are used, including *A. dahurica* (Fisch.) Benth. et Hook., *A. anomala* Lalem., *A. formosana* Boiss., and *A. sinensis* (Oliv.) Diels; the latter, known as *danggui* or ***dong quai*** (see following), is widely used in treating female ailments in China (FARNSWORTH 1–4; FOGARTY; JIANGSU; NANJING).

COMMERCIAL PREPARATIONS

Crude, extracts, and oils. Root and seed were both formerly official in U.S.P. and N.F.; both root oil and seed oil are official in F.C.C.

Regulatory Status. Regulated as a dietary supplement in the United States; extract, essential oil, and solvent-free oleoresin of the root, seed, or stem are GRAS for use in foods (§182.20); root and seed are GRAS for use in foods as a spice or natural flavoring (§182.10). In Germany, the fruit and roots are subjects of official monographs. The crude root at a daily dose of 4.5 g and galenical preparations are indicated for internal use for appetite loss and digestive ailments, including mild gastrointestinal tract spasms and flatulence. Crude fruit (seed) and preparations are not recommended for use as diuretics and diaphoretics because efficacy and safety have not been established (BLUMENTHAL 1; WICHTL).

REFERENCES

See the General Reference for BARNES; BARRETT; BIANCHINI AND CORBETTA; BISSET; BLUMENTHAL 1; BRUNETON; DE NAVARRE; DE NAVARRE; DER MARDEROSIAN AND BEUTLER; FEMA; FOSTER; GRIEVE; GUENTHER; GUPTA; JIANGSU; MARTINDALE; MCGUFFIN 1 & 2; TERRELL; TUTIN 2; UPHOF.

1. D. L. J. Opdyke, *Food Cosmet. Toxicol.*, **13**(Suppl.), 713 (1975).
2. J. Taskinen, *Acta Chem. Scand., Ser. B*, **29**, 637 (1975).
3. W. Steck and B. K. Bailey, *Can. J. Chem.*, **47**, 2425 (1969).
4. A. B. Svendsen, *Blyttia*, **11**, 96 (1953).
5. A. Chatterjee and S. Dutta, *Indian J. Chem.*, **6**, 415 (1968).
6. P. Härmälä et al., *Planta Med.*, **58**, 288 (1992).
7. J. Carbonnier and D. Molho, *Planta Med.*, **44**, 162 (1982).
8. G. I. Kas'yanov et al., *Khim. Prir. Soedin.*, **1**, 108 (1977).
9. B. E. Nielsen and H. Kofod, *Acta Chem. Scand.*, **17**, 1161 (1963).
10. S. C. Basa et al., *Chem. Ind. (London)*, **13**, 355 (1971).
11. D. L. J. Opdyke, *Food Cosmet. Toxicol.*, **12**(Suppl.), 821 (1974).
12. A. Patra et al., *Indian J. Chem.*, **14B**, 816 (1976).
13. T. Beyrich, *Arch. Pharm.*, **298**, 672 (1965).
14. M. Ashraf et al., *Pak. J. Sci. Ind. Res.*, **23**, 73 (1980).
15. N. Saksena and H. H. S. Tripathi, *Fitoterapia*, **56**, 243 (1985).
16. P. Härmälä et al., *Planta Med.*, **58**, 176 (1992).
17. M. T. Khayyal et al., *Arzneim.-Forsch.*, **51**, 545 (2001).
18. M. L. Yeh et al., *Pharmacology*, **68**, 70 (2003).

ANGOSTURA BARK

Source: *Angostura trifoliata* (Willd.) T. S. (syn. *Cusparia febrifuga* Humb. ex DC., *Galipea officinalis* Hancock) (Family Rutaceae).

Common/vernacular names: Angustura, carony, and cusparia bark.

GENERAL DESCRIPTION

Angostura trifoliata is a shrubby tree native to northern South America. In Venezuela, it grows in mountainous areas at an altitude of between 200 and 300 m above sea level (USD 23RD).

Formerly, angostura bark was believed to be the bark of *Cusparia trifoliata* Engl. (syn.

C. febrifuga Humb.), a tree related to *Galipea officinalis* reported to grow in Brazil. (Related species found in Brazil include *Galipea dichotoma* Sald. and *G. multiflora* Schult.[1]) Much of the scientific literature before 1960 was based on this species.

CHEMICAL COMPOSITION

The bark contains two unstable bitter principles: angostura bitters 1 and 2 (3,5-dihydroxy-5-ethoxy-2-syringoyl-1-methyl-4-*O*-β-D-glucopyranosylcyclopentane and 3,5-dihydroxy-5-ethoxy-2-vanilloyl-1-methyl-4-*O*-β-D-glucopyranosylcyclopentane);[2] a high content of alkaloids (ca. 40%; largely quinoline type) (cusparine, cuspareine, allocuspareine, galipoline, galipoidine, galipidine, galipinine, galipine, quinaldine, 4-methoxyquinaldine, quinoline, 2-*n*-amyl-quinoline, candicine, etc.) (GLASBY 1);[2–6] and a volatile oil (1–2%) containing some 15 alkaloids (KARRER; WREN) and various sesquiterpenes (e.g., β-bisabolene, cadinol T, germacrene D, δ-curcumene) also found in the trunk bark.[5] The stem and root bark of a related species, *Galipea trifoliata* Aublet, contains phebalosin, ramosin, and galipein (coumarins).[7]

PHARMACOLOGY AND BIOLOGICAL ACTIVITIES

The alkaloids, particularly cusparine and galipine, have shown antispasmodic activity[8] and in dogs, respiration–excitation effects.[9] The alkaloidal fraction has shown *in vitro* activity against the growth of *Mycobacterium tuberculosis*.[6] Galipinine, a tetrahydroquinoline alkaloid, has shown *in vitro* activity against the malarial parasite *Plasmodium falciparum*.[10]

TOXICOLOGY

Large doses may produce nausea.

USES

Medicinal, Pharmaceutical, and Cosmetic. Used in bitter tonics.

Food. Extracts used in most categories of food products such as alcoholic (bitters) and nonalcoholic beverages, frozen dairy desserts, candy, baked goods, gelatins and puddings, and gravies. Average maximum use level in alcoholic beverages is reported to be about 0.3%.

The well-known "angostura bitters" does not contain angostura bark at all and is made from a mixture of gentian root and other botanicals.

Traditional Medicine. Used in treating dyspepsia, chronic diarrhea, and dysentery.[4] Also used as a febrifuge and bitter tonic in doses of 0.3–1 g; large doses cathartic and emetic.

COMMERCIAL PREPARATIONS

Crude and extracts.

Regulatory Status. Bark essential oil, solvent-free oleoresin, and extractive are GRAS for use in foods (§182.20); bark GRAS as natural flavoring or spice in foods (§182.10).

REFERENCES

See the General References for ARCTANDER; FEMA; GLASBY 2; GUENTHER; MARTINDALE; UPHOF; WREN.

1. W. B. Mors et al., *Medicinal Plants of Brazil*, Reference Publications, Algonac, MI, 2000.
2. C. H. Brieskorn and V. Beck, *Phytochemistry*, **10**, 3205 (1971).

3. C. H. Brieskorn and V. Beck, *Präp. Pharm.*, **6**, 177 (1970).
4. I. Jacquemond-Collet et al., *Fitoterapia*, **71**, 605 (2000).
5. I. Jacquemond-Collet et al., *Phytochem. Anal.*, **12**, 312 (2001).
6. P. J. Houghton et al., *Planta Med.*, **65**, 250 (1999).
7. K. R. Wirasutisna et al., *Phytochemistry*, **26**, 3372 (1987).
8. C. E. T. Kraukau, *Kgl. Fysiograf. Saellskap. Lund, Foerh.*, **15**, 289 (1945).
9. L. Binet and M. V. Strumza, *Therapie*, **8**, 669 (1953).
10. I. Jacquemond-Collet et al., *Planta Med.*, **68**, 68 (2002).

ANISE (AND STAR ANISE)

Source: Anise: ***Pimpinella anisum*** L. (syn. *Anisum vulgare* Gaertn.; *A. officinarum* Moench.) (Family Umbelliferae or Apiaceae); Chinese star anise: ***Illicium verum*** Hook. f. (Family Illiciaceae).

Common/vernacular names: Anise seed, aniseed, sweet cumin (***P. anisum***); Chinese star anise, illicium, and star anise (***Illicium verum***).

GENERAL DESCRIPTION

Anise is an annual herb, usually less than 0.6 m high; leaves alternate, below, opposite above; native to Greece and Egypt, now widely cultivated. Part used is the dried ripe fruit; anise oil is obtained from it by steam distillation.

Chinese star anise is an evergreen tree usually 4–6 m high but may reach 12 m; indigenous to southeastern Asia; extensively cultivated in southern China, also in Vietnam, India, Japan. Part used is the dried, ripe fruit that consists of 5–13 (usually 8) seed-bearing woody follicles (one seed per follicle) attached to a central axis in the shape of a star, and therefore, the name star anise. In Chinese, the plant is also called "eight-horned anise" or simply "eight horns," referring to the usually eight-follicled fruit. China is the major producer of star anise. Chinese star anise oil is obtained by steam distillation.

Due to the traditional use of anise oils with licorice in licorice candy, the flavor of anise is often confused with that of licorice, particularly among the public, and is erroneously described as licorice-like.

CHEMICAL COMPOSITION

Anise contains 1–4% volatile oil (FURIA AND BELLANCA);[1] coumarins (bergapten, umbelliprenine, umbelliferone, scopoletin);[2,3] lipids (ca. 16%), including fatty acids (C_{16}, C_{18}, C_{20}, C_{22}, C_{24}, C_{26}, C_{30}, etc.), β-amyrin, and stigmasterol and its salts (palmitate and stearate) (MARSH);[2,4] flavonoid glycosides (quercetin-3-glucuronide, rutin, luteolin-7-glucoside, isoorientin, isovitexin, apigenin-7-glucoside [apigetrin], etc.);[5] phenylpropanoid glucosides;[6–8] myristicin;[9] protein (ca. 18%); carbohydrate (ca. 50%); and others (MARSH).

Anise oil contains *trans*-anethole (75–90%), estragole (methylchavicol) (1%), anise ketone (*p*-methoxyphenylacetone), and β-caryophyllene. Other compounds in minor concentrations include anisaldehyde, anisic acid (oxidation products of anethole), linalool, limonene, α-pinene, acetaldehyde, *p*-cresol, creosol, hydroquinine, β-farnesene, γ-himachalene, neophytadiene, and *ar*-curcumene (KARRER).[1,10–12]

Chinese star anise contains about 5% volatile oil (ca. 10% in follicles and 2.5% in seeds) (NANJING), phenylpropanoid glucosides,[13] lignans,[14] catechins and proanthocyanidins, among others.[15]

Chinese star anise oil contains *trans*-anethole (80–90%) as its major component. Other constituents include estragole, 1,4-cineole, β-bisabolene, β-farnesene, α-copaene, (−)- and *trans*-α-bergamotene, caryophyllene, nerolidol, methylanisoate, *trans*-methylisoeugenol, cadinene, foeniculin, 3-carene, *d*-α-pinene, phellandrene, α-terpineol, hydroquinine, traces of *cis*-anethole, and safrole (KARRER),[16–19] although the presence of safrole is disputed.[17,18]

PHARMACOLOGY AND BIOLOGICAL ACTIVITIES

Oil of anise and Chinese star anise has carminative and expectorant properties.[20] Water and ethanol extracts of the fruits (seeds) of anise have shown greater *in vitro* antioxidant activity than some common antioxidants (α-tocopherol, BHA, and BHT).[21] *In vitro* estrogenic activity was found from the essential oil of anise in human breast cancer cells (MCF-7 cells).[22] A relaxant effect of the essential oil on isolated, precontracted guinea pig tracheal chains suggests bronchodilatory activity and was due to inhibition of muscarinic receptors.[23] An antidiuretic effect was found in rats administered the essential oil (0.05%) in drinking water and was attributed to an increase in Na^+–K^+ ATPase activity.[24] A randomized controlled trial of a preparation containing anise oil and ylang ylang oil found that topical application was equally effective in the treatment of head lice in children as a spray composed of various insecticides.[25]

Anethole, anisaldehyde, and myristicin (in aniseed), along with *d*-carvone (present in *P. anisum* plant), have shown mild insecticidal properties.[26] In houseflies, the major (56% by weight) active insecticide in the oil from anise tops was shown to be *trans*-anethole. Anisaldehyde and anethole increased the toxicity to houseflies when applied at the same as various common insecticides.[27] Anethole also inhibits growth of mycotoxin-producing *Aspergillus* species in culture.[28]

Anethole was formerly considered an active estrogenic agent of the essential oil of anise. However, further research suggests that the active estrogenic compounds are polymers of anethole, such as dianethole and photoanethole.[29]

A methanol extract of the fruits of Chinese star anise (*Illicium verum*) showed *in vitro* growth-inhibitory activity against the periodontopathic bacteria, *Eikenella corrodens*,[30] and against the fungus *Saccharomyces cerevisiae*.[31] *In vitro* antimicrobial and antifungal activities of Chinese star anise are largely attributed to the presence of anethole.[32]

TOXICOLOGY

A double-blind, placebo-controlled allergenicity study using aniseed in spice industry workers found positive results in skin prick, and nasal and oral challenge tests.[33] Other case reports of occupational allergic reactions to anise include a psoriasis-like allergic contact dermatitis from exposure to the seed oil[34] and allergic asthma from exposure to the seed dust.[35]

Anethole, the major component of the oil of both anise and star anise, has been reported as the cause of dermatitis (erythema, scaling, and vesiculation) in some people.[36] Anethole has two isomers (*trans* and *cis*), with the *cis* isomer being 15–38 times more toxic to animals than the *trans* isomer (MERCK).[19] Current U.S.P. and F.C.C. specifications for anethole do not require differentiation between the isomers. Anethole (no isomer given) has shown mutagenic activity in Ames *Salmonella* reversion assay,[36] and estragole is a known genotoxic carcinogen with a recommended limit in food of 0.05 mg/kg.[11]

Japanese star anise (*Illicium lanceolatum* A. C. Smith; formerly identified as *Illicium anisatum* L. or *I. religiosum* Sieb. et Zucc.) should not be confused with true star anise (Chinese star anise) (*I. verum*). Use of the former as a culinary spice could result in fatality.[37] Its mistaken identity in Europe resulted in epidemics of epileptic seizures[38] and poisoning in infants (after being traditionally treated with "star anise" for colicky pain).[39] Symptoms in infants included vomiting, nystagmus, and abnormal movements.[40]

Japanese star anise grows in southern China, Taiwan, and Japan; it looks like a smaller, deformed version of Chinese star anise and is highly poisonous. A 10–15% aqueous extract is used in China as agricultural insecticide. Toxicity is attributed to the sesquiterpene anisatin.

USES

Anise oil and Chinese star anise oil are used interchangeably in the United States, both being officially recognized as anise oil in the U.S.P. and F.C.C.

Medicinal, Pharmaceutical, and Cosmetic. Both anise and Chinese star anise oils are used as carminatives, stimulants, mild spasmolytics, weak antibacterials, and expectorants in cough mixtures and lozenges, among other preparations; internally used for dyspeptic complaints; externally as an inhalant for congestion of the respiratory tract.

Both oils are used to mask undesirable odors in drug and cosmetic products and as fragrance components in toothpastes, perfumes, soaps, detergents, creams, and lotions, with maximum use levels of 0.25% anise oil and 0.4% star anise oil in perfumes.[19,20]

Food. Anise, Chinese star anise (to a lesser extent), anise oil, and star anise oil are widely used as flavoring ingredients in all major categories of foods, including alcoholic (bitters, brandies, and liqueurs, e.g., anisette) and nonalcoholic beverages, frozen dairy desserts, candy (e.g., licorice candies), baked goods, gelatins and puddings, and meat and meat products. Highest average maximum use levels for anise oil are about 0.06% (570 ppm) in alcoholic beverages and 0.07% (681 ppm) in candy.

Both anise and Chinese star anise are widely used as domestic spices; the former is mainly used by Westerners, while the latter is used primarily by Asians, especially in Chinese foods.

Dietary Supplements/Health Foods. Whole, crushed, or ground crude drug is used for infusions and other galenical preparations (BLUMENTHAL 1).

Traditional Medicine. Anise and Chinese star anise have been used as aromatic carminatives, stimulants, and expectorants; also as estrogenic agents to increase milk secretion, promote menstruation, facilitate child birth, increase libido, and alleviate symptoms of male climacteric.[29] Chinese star anise has been used in Chinese medicine for similar purposes for 1300 years (JIANGSU). Anise has also been used as an appetizer, diuretic, tranquillizer,[21] antiseptic, laxative,[24] and as a treatment for epilepsy and seizures.[41]

COMMERCIAL PREPARATIONS

Crude and essential oils are official in N.F. and F.C.C. In Germany, preparations of anise seed containing 5–10% essential oil are used as a respiratory inhalant, whereas Chinese star anise is used internally for peptic discomfort and catarrh (BLUMENTHAL 1).

Regulatory Status. Regulated in the United States as a dietary supplement; *P. anisum* and *I. verum* are GRAS as natural flavoring or spice for use in foods (§182.10); essential oil, extractives, and solvent-free oleoresin of *P. anisum* are GRAS for use in foods (§182.20). Anise seed and Chinese star anise seed are subjects of German official monographs; 3.0 g of seed or 0.3 g essential oil (mean daily dose) allowed as a bronchial expectorant for upper respiratory tract congestion and as gastrointestinal spasmolytic (BLUMENTHAL 1; WICHTL).

REFERENCES

See General References for APPLEQUIST; ARCTANDER; BAILEY 2; BARNES; BISSET; BLUMENTHAL 1; BRUNETON; DER MARDEROSIAN AND BEUTLER; GRIEVE; GUENTHER; GUPTA; JIANGSU; LUST; MARTINDALE; MASADA; NANJING; REMINGTON; TERRELL; UPHOF; WICHTL; YOUNGKEN.

1. M. B. Embong et al., *Can. J. Plant. Sci.*, **57**, 681 (1977).
2. T. Kartnig and G. Scholz, *Fette, Seifen, Anstrichmit.*, **71**, 276 (1969).
3. T. Kartnig et al., *Planta Med.*, **27**, 1 (1975).
4. A. Szegfu et al., *Acta Pharm. Hung.*, **42**, 162 (1972).
5. J. Kunzemann and K. Hermann, *Z. Lebensm. Unters. Forsch.*, **164**, 194 (1977).
6. T. Ishikawa et al., *Chem. Pharm. Bull.*, **50**, 1460 (2002).
7. E. Fujimatu et al., *Phytochemistry*, **63**, 609 (2003).
8. J. Kitajima et al., *Phytochemistry*, **62**, 115 (2003).
9. J. B. Harborne et al., *Phytochemistry*, **8**, 1729 (1969).
10. R. Tabacchi et al., *Helv. Chim. Acta*, **57**, 849 (1974).
11. M. De Vincenzi et al., *Fitoterapia*, **71**, 725 (2000).
12. G. Burkhardt et al., *Pharm. Weekbl. Sci.*, **8**, 190 (1986).
13. S. W. Lee et al., *Arch. Pharm. Res.*, **26**, 591 (2003).
14. L. K. Sy and G. D. Brown, *J. Nat. Prod.*, **61**, 987 (1998).
15. J. M. Schulz and K. Herrmann, *Z. Lebensm. Unters. Forsch.*, **171**, 278 (1980).
16. H. M. Okely and M. F. Grundon, *J. Chem. Soc. D*, **19**, 1157 (1971).
17. R. Kaempf and E. Steinegger, *Pharm. Acta Helv.*, **49**, 87 (1974).
18. J. Bricout, *Bull. Soc. Chim. Fr.*, **9–10**, 1901 (1974).
19. D. L. J. Opdyke, *Food Cosmet. Toxicol.*, **13**(Suppl.), 715 (1975).
20. D. L. J. Opdyke, *Food Cosmet. Toxicol.*, **11**, 865 (1973).
21. I. Gülçin et al., *Food Chem.*, **83**, 271 (2003).
22. M. F. Melzig et al., *Z. Phytother.*, **24**, 112 (2003).
23. M. H. Boskabady et al., *J. Ethnopharmacol.*, **74**, 83 (2001).
24. S. I. Kreydiyyeh et al., *Life Sci.*, **74**, 663 (2003).
25. K. Y. Mumcuoglu et al., *Isr. Med. Assoc. J.*, **4**, 790 (2002).
26. G. T. Carter, *Diss. Abstr. Int. B*, **37**, 766 (1976).
27. G. Marcus and E. P. Lichtenstein, *J. Agric. Food Chem.*, **27**, 1217 (1979).
28. H. Hitokoto et al., *Appl. Environ. Microbiol.*, **39**, 818 (1980).
29. M. Albert-Puleo, *J. Ethnopharmacol.*, **2**, 337 (1980).
30. L. Iauk et al., *Phytother. Res.*, **17**, 599 (2003).
31. G. H. Shahidi et al., *Daru*, **10**, 162 (2002).
32. M. De et al., *Phytother. Res.*, **16**, 94 (2002).
33. J. J. Garcia-Gonzales et al., *Ann. Allergy Asthma Immunol.*, **88**, 518 (2002).
34. D. Assalve et al., *Ann. Ital. Dermatol. Clin. Sper.*, **41**, 411 (1987).
35. J. Fraj et al., *Allergy*, **51**, 337 (1996).
36. J. Sekizawa and T. Shibamoto, *Mutat. Res.*, **101**, 127 (1982).
37. E. Small, *Econ. Bot.*, **50**, 337 (1996).
38. E. S. Johanns et al., *Ned. Tijdschr. Geneeskd.*, **146**, 813 (2002).
39. P. Minodier et al., *Arch. Pediatr.*, **10**, 619 (2003).
40. C. Garzo Fernandez et al., *An. Esp. Pediatr.*, **57**, 290 (2002).
41. M. H. Pourgholami et al., *J. Ethnopharmacol.*, **66**, 211 (1999).

ANNATTO

Source: *Bixa orellana* L. (Family Bixaceae).

Common/vernacular names: Achiote, achiotillo, annatto, arnotta, and lipstick tree.

GENERAL DESCRIPTION

Shrub or small tree, up to 10 m high; native to northern South America; extensively cultivated in tropical areas of Africa, Asia, and in the West Indies. Part used is the seed, which contains the coloring principles in an orange-red waxy covering.

Major producers of annatto seeds are India, Kenya, Ecuador, and Peru; qualities differ considerably depending on sources and seasons.

CHEMICAL COMPOSITION

The coloring principles are carotenoids, mostly bixin (oil-soluble) and norbixin (water-soluble), with bixin (especially the *cis* isomer) in major concentration (2.5% dry wt.). Norbixin is the principle component of the water-soluble dyes of annatto and is formed by the removal of the methyl esters of bixin. Both α- and β-forms of norbixin are found in water-soluble annatto dye. (−)-Bixin (syn. α-bixin) is unstable and during extraction is usually converted to *trans*-bixin (syn. β-bixin), the stable isomer, which is also known as isobixin. Bixin is a monomethyl ester of norbixin (a dicarboxylic acid); it is readily hydrolyzed by alkalis during alkali extraction of annatto to the dicarboxylic acid, norbixin (EVANS; MERCK).[1,2] Annatto also contains β-carotene (6.8–11.3 mg/100 g), apocarotenoids, diapocarotenoids, an essential oil (with the sesquiterpene ishwarane as the major component), pentosans, pectin, protein (13–17%), tannins, and others (EVANS).[3,5,6]

The coloring principle in oil-soluble annatto preparations is free bixin, while that of water-soluble (usually alkaline) annatto extracts is an alkali salt of norbixin, commonly the potassium salt.

Both bixin and norbixin in the free acid state are insoluble in water but soluble in organic solvents (e.g., acetone, alcohols) and aqueous alkaline solutions.

Bixin is one of the more stable natural yellow colors. However, it loses much of its tinctorial power gradually on storage, the process being accelerated by light and heat. Therefore, the fresh seeds are preferred in manufacturing processes.

Tinctorial strength of bixin is comparable to that of β-carotene, but bixin is the more stable.

Annatto is reported to be most stable at pH 8, with decreased stability at pH 4–8.[7]

PHARMACOLOGY AND BIOLOGICAL ACTIVITIES

Broad-spectrum *in vitro* antibacterial activity was found from an ethanolic extract of the dried seeds. An extract of the dried leaves showed greater potency, and both extracts were also active against *Candida albicans*.[8] However, others report a narrow spectrum of antibacterial activity from the leaves, and with activity only found against Gram-positive bacteria.[9] A water extract of the leaves showed platelet antiaggregant activity *in vitro*.[10] Although bixin and norbixin are carotenoids, they do not have vitamin A activity.

Norbixin exhibits antioxidant activity against the deterioration of lipids in bulk olive oil and in olive oil-in-water emulsions.[11]

Administered to rats during and after whole body irradiation, bixin (200 μmol/kg p.o.) significantly reduced levels of lung collagen hydroxyproline and liver and serum lipid peroxidation values.[12] At concentrations of 1–100 μM *in vitro* in rat spleen lymphocytes, bixin inhibited the activity of IgE, suggesting a possible antiallergic effect, and enhanced the production of IgM.[13]

TOXICOLOGY

An acute (3-week) oral toxicity study of an extract of annatto seeds containing 50%

norbixin found that doses of 0.8, 7.5, and 68 mg/kg produced hyperinsulinemia in rats and that doses of 56 and 351 mg/kg produced hypoinsulinemia in mice. In the rats, the extract also caused a significant decrease in globulins and plasma total protein levels without evidence of any adverse effects on the liver or plasma chemistry.[14] Administered to pregnant rats, annatto food color containing 28% bixin caused no adverse effects to the mothers and none to the fetuses. A developmental and maternal No-Observed-Adverse-Effect-Level (NOEL) for the extract was proposed as 500 mg/kg p.o. per day or greater and for bixin, at least 140 mg/kg p.o. per day.[15]

At high oral doses in rats, norbixin (8.5 and 74 mg/kg) produced hyperglycemia.[14] However, in a subchronic (2-week) toxicity test in rats of either sex, the NOEL of norbixin was 0.1% of the diet (76 and 69 mg/kg p.o. for males and females, respectively). At 0.3% and 0.9% of the diet, norbixin produced a pronounced elevation in liver weights, hepatocyte hypertrophy, and in blood work, increases in albumin/globulin, albumin, alkaline phosphatase, total protein, and phospholipid ratios.[16] In the micronucleus test in male mice, annatto food colorant at dietary levels of 1330, 5330, or 10,670 ppm for 7 days failed to produce any mutagenic effects. However, the highest concentration increased the mutagenic effect of a mutagen (cyclophosphamide).[17] Similar results were found from *in vitro* tests in which low concentrations of norbixin (10–50 µM) protected DNA against oxidative damage, while at higher concentrations it enhanced oxidative damage to DNA.[18]

Induction of drug-metabolizing enzymes of the liver (CYP2B and CYP1A) was found from oral dosing of female rats with an extract of the seeds containing 28% bixin; 95% pure bixin showed only weak inducing activity.[19]

Human allergic reactions to annatto food coloring include IgE-mediated anaphylaxis in a man who consumed a breakfast cereal containing the dye. Other clinical reports suggest the possibility that angioedema and urticaria occur as allergic reactions to the dye in some individuals.[20]

USES

The current major commercial uses of annatto colors are in foods, especially dairy;[21] also used in drugs as color coatings for granules, pills, tablets, and herbal preparations, as well as in cosmetics (e.g., lipsticks) and hair dyes.

Aqueous alkaline extracts are extensively used in coloring cheeses (especially cheddar) and to a lesser extent in ice creams and other dairy products. Oil-soluble extracts are used in oily food products such as salad oils, popcorn oil, butter, margarine, and sausage casings. Major food categories in which annatto color is used include alcoholic and nonalcoholic beverages, frozen dairy desserts, baked goods, meat and meat products, condiments and relishes, fats and oils, snack foods, and gravies, with highest average maximum use level reported for the extract in baked goods (ca. 0.24%).

Traditional Medicine. The seeds have been used in treating tumors of the oral cavity, as a purgative, antipruritic,[8] and in the treatment of venereal diseases. The pulp that surrounds the seeds is topically applied as a mosquito repellant.[22] Preparations of the leaves are traditionally used in treating gonorrhea,[23] nausea, as a gargle for sore throat, and for oral hygiene.[8]

COMMERCIAL PREPARATIONS

Crude, oil- and water-soluble extracts and spray-dried powders.

Also available for domestic use as ground or whole seeds in supermarkets or ethnic stores in metropolitan areas in the United States.

Regulatory Status. Approved for use as a colorant in foods (§73.30), drugs (§73.1030), and cosmetics (§73.2030).

REFERENCES

See the General References for EVANS; FEMA; FURIA; GRIEVE; MCGUFFIN 1 & 2; MORTON 2; TERRELL; UPHOF.

1. P. Karrer and E. Jucker, *Carotenoids*, Elsevier Applied Science Publishers, Barking, UK, 1950.
2. F. Mayer and A. H. Cook, *The Chemistry of Natural Coloring Matters*, ACS Monograph Series 89, Rheinhold, New York, 1943.
3. Z. Angelucci et al., *Colet. Inst. Tecnol. Aliment (Campinas Braz.)*, **11**, 89 (1980).
4. R. Bressani et al., *Arch. Latinoam. Nutr.*, **33**, 356 (1983).
5. C. Srinivasulu and S. N. Mahapatra, *Indian Perfum.*, **26**, 132 (1982).
6. R. Bressani et al., *Arch. Latinoam. Nutr.*, **33**, 356 (1983).
7. D. E. Auslander et al., *Drug Cosmet. Ind.*, **121** (6), 55 (1977).
8. T. C. Fleischer et al., *Fitoterapia*, **74**, 136 (2003).
9. O. N. Irobi et al., *Int. J. Pharmacogn.*, **34**, 87 (1996).
10. R. Villar et al., *Phytother. Res.*, **11**, 441 (1997).
11. S. Kiokias and M. H. Gordon, *Food Chem.*, **83**, 523 (2003).
12. K. C. Thresiamma et al., *Indian J. Exp. Biol.*, **34**, 845 (1996).
13. Y. Kuramoto et al., *Biosci. Biotechnol. Biochem.*, **60**, 1712 (1996).
14. A. C. S. Fernandes et al., *J. Nutr. Biochem.*, **13**, 411 (2002).
15. F. J. R. Paumgartten et al., *Food Chem. Toxicol.*, **40**, 1595 (2002).
16. A. Hagiwara et al., *Food Chem. Toxicol.*, **41**, 1157 (2003).
17. R. O. Alves de Lima et al., *Food Chem. Toxicol.*, **41**, 189 (2003).
18. K. Kovary et al., *Br. J. Nutr.*, **85**, 431 (2001).
19. A. C. A. X. De Oliveira et al., *Braz. J. Med. Biol. Res.*, **36**, 113 (2003).
20. W. A. Nish et al., *Ann. Allergy*, **66**, 129 (1991).
21. F. E. Lancaster and J. F. Lawrence, *Food Addit. Contam.*, **12**, 1 (1995).
22. V. C. Gupta et al., *Fitoterapia*, **68**, 45 (1997).
23. A. Caceres et al., *J. Ethnopharmacol.*, **48**, 85 (1995).

ARNICA

Source: *Arnica montana* L. (Family Compositae or Asteraceae).

Common/vernacular names: Arnica, European arnica, leopard's bane, mountain tobacco, and wolf's bane.

GENERAL DESCRIPTION

Perennial herb, up to about 0.6 m high; native to mountainous regions of Europe; cultivated in northern India. Part used is the dried flower head.

Several American *Arnica* species ("American arnica") have also been used (*A. fulgens* Pursh, *A. sororia* Green, and *A. cordifolia* Hook.), all of which are native to the region of the western Rocky Mountains.

CHEMICAL COMPOSITION

Contains up to 1% (normally about 0.3%) of a viscous volatile oil, about half of which is composed of fatty acids, with palmitic, linoleic, myristic, and linolenic acids predominant. Aromatic constituents present include terpenes, thymol, thymol methyl ether, 4-hydroxy-thymol dimethyl ether, and isobutyric acid thymyl ether.[1–5] Other constituents include resins, arnicin (bitter principle), sesquiterpene lactones (e.g., helenalin, 11α,13-dihydrohelenalin, helenalin and 11α,13-dihydrohelenalin esters, 2β-ethoxy-6-O-isobutyryl 1-2,3-dihydrohelenalin and 6-O-isobutyryltetrahydrohelenalin[6–8]); tannin, arnisterin (a sterol), carotenoids (α- and β-carotene, cryptoxanthin, lutein, etc.) (KARRER; MERCK);[9] flavonoids (astragalin, betuletol, 6-methoxykaempferol, hispidulin, isoquercetin, jaceosidin, pectolinarigenin, etc.),[10–12] coumarins (umbelliferone, scopoletin),[12] and phenolic acids (p-hydroxybenzoic, p-coumaric, gentisic, ferulic, caffeic, vanillic, etc.).[13]

Pseudoguaianolides have been reported in leaves of A. montana, including arnifolin, arnicolides A, B, C, and D, and loliolide.[14,15]

PHARMACOLOGY AND BIOLOGICAL ACTIVITIES

An extract of Arnica montana was shown to increase the resistance of animals to bacterial infections by stimulating phagocytosis of the bacteria involved, particularly Listeria monocytogenes and Salmonella typhimurium.[16]

The sesquiterpene lactones helenalin acetate and 11,13-dihydrohelenalin have shown platelet aggregation, 5-hydroxytryptamine secretion, and thromboxane formation-inhibiting,[17] and antibacterial and antifungal activities in vitro.[18] Numerous esters of helenalin have shown anti-inflammatory activity in mice and rats.[19] The anti-inflammatory activity of A. montana flowers is attributed to helenalin,[20] which has the ability to inhibit proinflammatory gene expression[21] by directly modifying the transcription factor NF-κB in vitro.[22] Helenalin has also shown in vitro antitrypanosomal activity (Trypanosoma cruzi and T. brucei rhodesiense),[23] and antitumor activity against human colorectal cancer (COLO 320 cells) and human small cell lung carcinoma (GLC4 cells).[24] It also induces apoptosis in leukemia (Jurkat T cells).[25]

Acidic polysaccharides derived from an extract of A. montana have shown immunostimulating activities in vitro.[18] Helenalin also showed immunostimulating activity in a preliminary screening.[26]

Double-blind, placebo-controlled clinical trials of homeopathic preparations of A. montana have shown questionable or no benefits in the treatment of postoperative pain,[27,28] postoperative bruising,[29,30] postoperative hematomas,[28,31] postoperative swelling,[30] muscle soreness,[32,33] post-laser treatment bruising,[29] wound healing, pain, edema, stroke, and symptoms of acute trauma.[32]

TOXICOLOGY

Arnica montana is not recommended for any internal uses. The plant is an irritant to mucous membranes, and ingestion may cause burning pain in the stomach, diarrhea, vomiting, giddiness, intense muscular weakness, collapse, decrease or increase of the pulse rate, shortness of breath, and death. One ounce (ca. 30 mL) of the tincture (1:0.2 or 20%) has been reported to produce serious but not fatal symptoms (USD 23RD).[34] Evidence suggests that helenalin is responsible for many, if not all, of these toxic effects.[34]

Reported cases of contact allergic reactions to A. montana preparations are numerous and date from as early as 1844. Cross-reactivity with other plants is also reported (e.g., sunflower, Tagetes, Chrysanthemum).[12] Sesquiterpene lactones, especially helenalin and its derivatives, are known sensitizers.[6,34–36] Data on the safety of Arnica Montana extract, which is made from the dried flower heads of A. montana, are lacking. It has an oral LD_{50} of >5 g/kg in rats, whereas in mice the LD_{50} was 123 mg/kg p.o. An extract made from the dried plant showed mutagenic activity in the Ames test, an effect attributed to the flavonol content.[37]

USES

Medicinal, Pharmaceutical, and Cosmetic. No longer (or rarely) used in preparations intended for internal use. Current use is mostly as a local anti-inflammatory in the form of a tincture of the dried flower heads as a component (5–25% v/v) of salves, ointments, compressed, gels, and creams for external application to sprains and bruises[26,38,39] and in various dilutions in homeopathic preparations;[32] also used in hair tonics and antidandruff preparations; oil occasionally used in perfumes and other cosmetic preparations.[37,40]

Food. Used (though not widely) as a flavor ingredient in alcoholic (ca. 0.03%) and nonalcoholic (0.02%) beverages, frozen dairy desserts (0.03%), candy (0.04%), baked goods (0.08%), and gelatins and puddings (0.04%), with reported average maximum use levels in parentheses. These figures apparently cannot apply to the crude flowers as they are too high to be safe in all categories, except perhaps alcoholic beverages because of their limited volume of intake.

Dietary Supplements/Health Foods. Various ointments, salves, lotions, tinctures, and homeopathic products for external use only (FOSTER AND CARAS; LUST; WREN).

Traditional Medicine. Used as a diaphoretic, diuretic, stimulant; externally antiphlogistic, analgesic, antiseptic, vulnerary for hematomas, dislocations, contusions, fracture-induced edema, and insect bites.[26]

COMMERCIAL PREPARATIONS

Crude, tincture, and extracts. Crude, tincture, and fluid extract were formerly official in N.F.; crude flowers and root formerly in U.S.P. In the United States in 1998, Arnica Montana extract was reported to be used in close to 100 body care formulations including bubble baths, hair conditioners, hair dyes, deodorants, skin fresheners, moisturizers, shaving creams, and various others such products.[37]

Regulatory Status. Flowers GRAS as a natural flavoring (§172.510). External applications of preparations containing *A. montana* flower are the subject of a positive German monograph that includes inflammatory conditions (superficial phlebitis, insect bites, furunculosis, oral cavity, and throat conditions), injuries, contusions, hematomas, joint problems, rheumatism, and edema resulting from bone fractures (BLUMENTHAL 1).

REFERENCES

See the General Reference for APPLEQUIST; ARCTANDER; BARNES; BISSET; BLUMENTHAL 1; DER MARDEROSIAN AND BEUTLER; FEMA; FOSTER AND CARAS; GRIEVE; LUST; MARTINDALE; MCGUFFIN 1 & 2; WICHTL; WREN; YOUNGKEN.

1. G. Willuhn, *Planta Med.*, **21**, 221 (1972).
2. H. Kating et al., *Planta Med.*, **18**, 130 (1970).
3. G. Willuhn, *Planta Med.*, **22**, 1 (1972).
4. G. Willuhn, *Planta Med.*, **21**, 329 (1972).
5. R. Schmitz and H. Kating, *Planta Med.*, **31**, 310 (1977).
6. H. D. Herrmann et al., *Planta Med.*, **34**, 299 (1978).
7. G. Willuhn et al., *Planta Med.*, **50**, 35 (1984).
8. G. Willuhn et al., *Planta Med.*, **49**, 226 (1983).
9. M. Vanhaelen, *Planta Med.*, **23**, 308 (1973).
10. I. Merfort, *Planta Med.*, **50**, 107 (1984).
11. I. Merfort and D. Wendisch, *Planta Med.*, **53**, 434 (1987); **54**, 247 (1988).
12. H. P. Hörmann and H. C. Korting, *Phytomedicine*, **1**, 161 (1994).
13. L. Swiatek and J. Gora, *Herba Pol.*, **24**, 187 (1978).

14. V. Herout, in H. Wagner and L. Hörhammer, eds., *Pharmacognosy and Phytochemistry, 1st International Congress*, Munich, 1970, Springer-Verlag, Berlin, Germany, 1971, p. 93.
15. M. Holub et al., *Phytochemistry*, **14**, 1659 (1975).
16. H. Buschmann, *Fortschr. Veterinärmed.*, **20**, 98 (1974).
17. H. Schroder et al., *Thromb. Res.*, **57**, 839 (1990).
18. G. Willuhn et al., *Pharm. Ztg.*, **127**, 2183 (1982).
19. I. H. Hall et al., *Planta Med.*, **53**, 153 (1987).
20. C. A. Klaas et al., *Planta Med.*, **68**, 385 (2002).
21. J. Gertsch et al., *Biochem. Pharmacol.*, **66**, 2141 (2003).
22. G. Lyss et al., *J. Biol. Chem.*, **273**, 33508 (1998).
23. T. J. Schmidt et al., *Planta Med.*, **68**, 750 (2002).
24. H. J. Woerdenbag et al., *Planta Med.*, **60**, 434 (1994).
25. V. M. Dirsch et al., *Cancer Res.*, **61**, 5817 (2001).
26. H. Wagner et al., *Planta Med.*, **50**, 139 (1985).
27. O. Hart et al., *J. R. Soc. Med.*, **90**, 73 (1997).
28. M. Wolf et al., *Forsch. Komplementarmed. Klass. Naturheilkd.*, **10**, 242 (2003).
29. D. Alonso et al., *Dermatol. Surg.*, **28**, 686 (2002).
30. C. Stevinson et al., *J. R. Soc. Med.*, **96**, 60 (2003).
31. A. A. Ramelet et al., *Dermatology*, **201**, 347 (2000).
32. E. Ernst and M. H. Pittler, *Arch. Surg.*, **133**, 1187 (1998).
33. A. J. Vickers et al., *Clin. J. Pain*, **14**, 227 (1998).
34. S. MacKinnon, *Can. Pharm. J.*, **125**, 125 (1992).
35. G. Willuhn, *Dtsch. Apoth. Ztg.*, **126**, 2038 (1986).
36. E. Spettoli et al., *Am. J. Contact Dermat.*, **9**, 49 (1998).
37. M. Z. Fioume, *Int. J. Toxicol.*, **20**(Suppl. 2), 1 (2001).
38. H. Wagner et al., *Arzneim.-Forsch.*, **35**, 1069 (1985).
39. A. R. Bilia et al., *J. Pharm. Biomed. Anal.*, **30**, 321 (2002).
40. H. B. Heath, *Cosmet. Toilet.*, **92**, 19 (1977).

ARTICHOKE

Source: *Cynara scolymus* L. (Family Compositae or Asteraceae).

Common/vernacular names: Globe artichoke, cynara.

GENERAL DESCRIPTION

Large thistle-like perennial herb, up to about 1 m high; native to southern Europe, North Africa, and the Canary Islands; widely cultivated. Parts used are the leaves; the immature flower heads with fleshy bracts are eaten as a vegetable. It should not be confused with Jerusalem artichoke, which is the tuber of *Helianthus tuberosus* L.

CHEMICAL COMPOSITION

The leaves contain cynarin (1,3-dicaffeoylquinic acid), apigenin, cynaroside, chlorogenic acid, rutin, hesperitin, hesperidoside, maritimein, esculetin-6-O-β-glucoside, quercetin, caffeic acid, cosmoside, luteolin,[1] sesquiterpenes (aguerin B, cynaropicrin, grosheimin), sesquiterpene glycosides (cynarascolosides A, B, and C),[2] apigenin-7-rutinoside, narirutin;[3] artichoke also contains up to 2% O-diphenolic derivatives such as caffeic acid, 1-, 3-, 4-, and 5-caffeoylquinic acids, and 1,3-di-O-caffeoylquinic acid; flavonoids (0.1–1.0%), including glycosides luteolin-7-β-rutinoside (scolymoside), luteolin-7-β-D-glucoside, and 4-β-D-glucoside; glycolic and glyceric acids; taraxasterol, Ψ-taraxasterol; inulin; guaianolides (cynaropicrin, 8-epigrosheimin); cynaratriol;[4,5] sugars; enzymes (KARRER);[6–12] and a volatile oil consisting of β-selinene and caryophyllene as its major components, with α-cadrene, oct-1-en-3-one, hex-1-en-3-one, decanal, non-$trans$-2-enal, phenylacetaldehyde, and eugenol as the major aromatic principles.[13,14]

Cynaropicrin, cynarin (1,3-dicaffeoylquinic acid), 3-caffeoylquinic acid (chlorogenic acid), and scolymoside are among various active constituents identified so far.[2,10,15,16] The maximum content of cynarin, the major caffeoylquinic acid derivative in the artichoke heads,[3] is obtained by aqueous ebullition of the drug.[17]

PHARMACOLOGY AND BIOLOGICAL ACTIVITIES

Artichoke leaf extract has shown in $vitro$ antimicrobial[18] and antioxidant activities, the latter attributed to caffeic acid, 1-caffeoylquinic acid, luteolin, apigenin-7-O-glucoside, luteolin-7-O-glucoside, luteolin-7-rutinoside, cynarin, cynaroside, chlorogenic acid, and narirutin.[19,20] Antioxidant activity was also shown in the erythrocytes of rats fed diets containing the edible portions of artichoke, and the effect was accompanied by a decrease in plasma uric acid.[21] Anticlastogenic activity was shown in the in $vivo$ mouse bone marrow micronucleus assay from the leaf homogenate or concentrate,[22] and topical administration of taraxastane-type hydroxy triterpenes isolated from the leaves (faradiol and taraxasterol) inhibited the development of skin tumors in mice.[23] Although more than one study showed no cholerectic activity from either cholorgenic acid or cynarin in rats,[16,24] acute and repeated oral administration of an artichoke leaf extract in rats produced increases in total bile acid levels, however, without any effect on cholesterol or phospholipid levels.[25] Sesquiterpene constituents of the leaves, notably cynaropicrin, lowered serum triglyceride levels in rats when administered orally following olive oil.[2] Hepatoprotective activity has been shown in both animal and in $vitro$ studies with artichoke extracts.[26]

Placebo-controlled clinical trials of artichoke leaf extracts have shown cholersterol-lowering effects[15,26–28] and symptomatic improvement of patients with functional dyspepsia.[29] Cynarin has shown inconsistent hypolipidemic effects in humans (MARTINDALE).[30,31]

TOXICOLOGY

Side effects from artichoke are mild or absent[26–28,32] and drug interactions are unknown (BLUMENTHAL 1). Allergic reactions to artichoke are rare, including allergic contact dermatitis,[33] food allergy,[34] acute edema of the tongue,[35] bronchial asthma, and allergic rhinitis.[36] Potential allergens in artichoke are sesquiterpene lactones, including cynaropicrin.[37,38] In rats, LD_{50} of an extract containing 46% caffeoylquinic acids was 265 mg/kg i.p. and by the oral route the LD_{40} was 2000 mg/kg. The LD_{50} of a hydroalcoholic extract of artichoke containing 19% caffeoylquinic acids was 1000 mg/kg i.p.[39] Use is contraindicated in individuals with known allergies to the Composite plant family, in bile duct obstructions, and in gallstones (except under medical advice) (BLUMENTHAL 1).

USES

Medicinal, Pharmaceutical, and Cosmetic. Artichoke leaf extracts have been widely used in Europe for the treatment of digestive complaints (e.g., nausea, abdominal pains, dyspepsia, loss of appetite), hepatobiliary dysfunction, and to lower cholesterol levels.[32]

Food. Leaves and their extracts used mainly as a flavor ingredient in alcoholic beverages (bitters, liqueurs, etc.). Reported average maximum use level for leaves is 0.0016% (16 ppm).

Cynarin and chlorogenic acid can be used as sweeteners (LEWIS AND ELVIN-LEWIS).

Dietary Supplements/Health Foods. Dried or ground leaves, or expressed juice of fresh plant, in capsules, tablets, tincture, or other products for oral use (WREN).

Traditional Medicine. Has reportedly been used in Europe since Roman times as a choleretic and diuretic, among other uses (BIANCHINI AND CORBETTA).

COMMERCIAL PREPARATIONS

Crude and extracts; there are no uniform standards for extracts.

Regulatory Status. Regulated in the United States as a dietary supplement; leaves are GRAS as a natural flavoring (§172.510). Leaves subject of a German therapeutic monograph indicated as a choleretic for dyspeptic problems (BLUMENTHAL 1).

REFERENCES

See the General References for BAILEY 2; BARNES; BIANCHINI AND CORBETTA; BLUMENTHAL 1; DER MARDEROSIAN AND BEUTLER; FEMA; GRIEVE; LIST AND HÖRHAMMER; MARTINDALE; MCGUFFIN 1 & 2; UPHOF; WICHTL; WREN.

1. J. Hinou et al., *Ann. Pharm. Fr.*, **47**, 95 (1989).
2. H. Shimoda et al., *Bioorg. Med. Chem. Lett.*, **20**, 223 (2003).
3. M. Wang et al., *J. Agric. Food Chem.*, **51**, 601 (2003).
4. H. O. Bernhard et al., *Helv. Chim. Acta*, **62**, 1288 (1979).
5. P. Barbetti et al., *Stud. Carciofo Congr. Int.*, **3** (1979), 77 (1981).
6. M. Jaruzelski et al., *Herba Pol.*, **22**, 144 (1976).
7. L. I. Dranik, *Fenolnye Soedin. Ikh Biol. Funkts., Mater. Vses. Simp.*, **1** (1966), **53** (1968).
8. E. Bombardelli et al., *Fitoterapia*, **48**, 143 (1977).
9. H. Wagner et al., *Chem. Ber.*, **104**, 2118 (1971).
10. J. Michaud, *Bull. Soc. Pharm. Bordeaux*, **106**, 181 (1967).
11. D. G. Constatninescu et al., *Pharmazie*, **22**, 176 (1967).
12. L. Panizzi and M. L. Scarpati, *Nature*, **174**, 1062 (1954).
13. R. G. Buttery et al., paper presented at the ACS 175th National Meeting, Division of Agriculture and Food, Paper no. 35, 1978.
14. A. J. MacLeod et al., *Phytochemistry*, **21**, 1647 (1982).
15. R. Kirchoff, *Phytomedicine*, **1**, 107 (1994).
16. E. Speroni et al., *J. Ethnopharmacol.*, **86**, 203 (2003).
17. J. Ingelsias, et al., *Plant. Med. Phytother.*, **19**, 202 (1985).
18. V. Martino et al., *Acta Hort.*, **501**, 111 (1999).
19. F. Perez-Garcia et al., *Free Radic. Res.*, **33**, 661 (2000).
20. A. Betancor-Fernandez et al., *J. Pharm. Pharmacol.*, **55**, 981 (1003).

21. A. Jimenez-Escrig et al., *J. Agric. Food Chem.*, **51**, 5540 (2003).
22. R. Edenharder et al., *Mutat. Res.*, **537**, 169 (2003).
23. K. Yasukawa et al., *Oncology*, **53**, 341 (1996).
24. P. E. Altman Jr. and I. L. Honigberg, *J. Pharm. Sci.*, **61**, 610 (1972).
25. T. Saenz Rodriguez et al., *Phytomedicine*, **9**, 687 (2002).
26. K. Kraft, *Phytomedicine*, **4**, 369 (1997).
27. W. Englisch et al., *Arzneim.-Forsch.*, **50**, 260 (2000).
28. M. H. Pittler et al., *Cochrane Database Syst. Rev.*, **(3)**, CD003335 (2002).
29. G. Holtmann et al., *Aliment. Pharmacol. Ther.*, **18**, 1099 (2003).
30. W. H. Hammerl et al., *Wien. Med. Wochenschr.*, **123**, 601 (1973).
31. H. Heckers et al., *Atherosclerosis*, **26**, 249 (1977).
32. A. F. Walker et al., *Phytother. Res.*, **15**, 58 (2001).
33. J. C. Mitchell in V. C. Runeckles, ed., *Recent Advances in Phytochemistry*, Vol. 9, Plenum Press, New York, 1975, p. 119.
34. C. Romano et al., *J. Invest. Allergol. Clin. Immunol.*, **10**, 102 (2000).
35. H. Gadban et al., *Ann. Otol. Rhinol. Laryngol.*, **112**, 651 (2003).
36. J. C. Miralles et al., *Ann. Allergy Asthma Immunol.*, **91**, 92 (2003).
37. J. Mitchell and A. Rook, *Botanical Dermatology—Plants and Plant Products Injurious to the Skin*, Greengrass, Vancouver, BC, 1979.
38. B. Meding, *Contact Dermatitis*, **9**, 314 (1983).
39. A. Lietti, *Fitoterapia*, **48**, 153 (1977).

ASAFETIDA

Source: *Ferula assa-foetida* L. (Family Umbelliferae or Apiaceae).

Common/vernacular names: Asafetida, asafoetida, giant fennel; oleogum resin of the plant: devil's dung, food of the gods, and gum asafetida.

GENERAL DESCRIPTION

Asafetida is the oleogum resin obtained by incising or cutting the living rhizomes and roots of *F. assa-foetida*. It is a solid or semisolid with a persistent alliaceous (garlic-like) odor and bitter acrid taste. An essential oil (asafetida oil) is obtained by steam distillation.

F. assa-foetida is a large branching perennial herb, up to 3 m high; native to southwestern Asia (eastern Iran and western Afghanistan).

CHEMICAL COMPOSITION

Contains 40–64% resinous material composed of ferulic acid, asaresinotannols, umbelliferone, and umbelliferone ethers (e.g., farnesiferols A, B, and C, kamolonol) among others; approximately 25% gum composed of glucose, galactose, L-arabinose, rhamnose, and glucuronic acid; volatile oil (3–21%; usual range 7–9%) consisting of disulfides and polysulfides as its major components, notably 2-butyl propenyl disulfide (*E*- and *Z*-isomers), with monoterpenes (α- and β-pinene, etc.), free ferulic acid, valeric acid, and traces of vanillin, are also present.[1–8] A chloroform extract of the resins from the roots yielded sesquiterpene coumarins, assafoetidnols A and B,[9] 3-*O*-acetylepisamarcadin,[10] samarcandin, bradrakemin, galbanic acid, gummosin, neveskone, and polyanthin.[9]

Sulfur-containing compounds are responsible for the characteristic flavor and odor of asafetida (e.g., asadisulfide[11] and other sulfides[12]).

PHARMACOLOGY AND BIOLOGICAL ACTIVITIES

Administered at 1.25% and 2.5% of the diet of female rats subjected to carcinogen-induced mammary carcinogenesis, asafetida increased activity levels of endogenous antioxidants (glutathione-*S*-transferase, reduced glutathione, superoxide dismutase, and catalase), inhibited lipid peroxidation, delayed tumor appearance, and reduced the size of tumors.[13] The essential oil showed a relaxant effect on isolated rat ileum.[14] Asafetida has also shown hypotensive activities in animals.[15] It has also been demonstrated to increase blood coagulation time.[16]

TOXICOLOGY

Available data indicate asafetida to be relatively nontoxic, and ingestion of 0.5 oz (ca. 15 g) has reportedly produced no untoward effects (GOSSELIN). Use of the gum was associated with a case of methemoglobinemia in an infant treated for colic with glycerated asafetida. Tests showed that only the gum portion of the formulation had an oxidative effect on fetal hemoglobin *in vitro* and that adult hemoglobin was unaffected.[17]

USES

Medicinal, Pharmaceutical, and Cosmetic. Now rarely used in pharmaceutical preparations; mainly use in cosmetics is as a fixative or fragrance component in perfumes.

Food. Asafetida is regularly consumed in Nepal and India as part of the daily diet;[18,19] it is reported to be an ingredient in Worcestershire sauce. Together with the oil and fluid extract, it is reportedly used in nonalcoholic beverages, frozen dairy desserts, candy, baked goods, gelatins and puddings, meat and meat products, and condiments and relishes, among others. Use levels are usually very low (<0.004%).

Traditional Medicine. Used in Iranian folk medicine to treat abdominal cramps and diarrhea;[14] in Nepalese folk medicine as an aphrodisiac, diuretic, antispasmodic, emmenagogue, expectorant, anthelmintic, and sedative;[18] in India to treat nervous disorders of children and women, cough, bronchitis, and pneumonia in children, and asthma, bronchitis, or flatulence in adults;[19] used in Chinese medicine (since the 7th century) as a nerve stimulant in treating neurasthenia; in chronic bronchitis; and as an expectorant, antiflatulent, and laxative.

COMMERCIAL PREPARATIONS

Crude, oil, tincture, and extracts. Crude and tincture were formerly official in N.F. Commercial crude asafetida differs widely in quality, depending on sources and is not necessarily derived from *Ferula assa-foetida*.[20]

Regulatory Status. Essential oil, extractive, and solvent-free oleoresin GRAS for use in foods (§182.20).

REFERENCES

See the General References for ARCTANDER; BAILEY 2; BARNES; CLAUS; DER MARDEROSIAN AND BEUTLER; FEMA; GRIEVE; GUENTHER; HUANG; JIANGSU; LUST; MCGUFFIN 1 & 2; MERCK; MORTON 1; NANJING; USD 23RD.

1. G. H. Mahran et al., *Bull. Fac. Pharm., Cairo Univ.*, **12**, 119 (1975).
2. L. Caglioti et al., *Helv. Chim. Acta*, **41**, 2278 (1958).
3. L. Caglioti et al., *Helv. Chim. Acta*, **41**, 2557 (1959).
4. A. Kjaer et al., *Acta Chem. Scand., Ser. B*, **30**, 137 (1976).

5. H. Naimie et al., *Collect. Czech. Chem. Commun.*, **37**, 1166 (1972).
6. B. Rajanikanth et al., *Phytochemistry*, **23**, 899 (1984).
7. O. Hofer et al., *Monatsh. Chem.*, **115**, 1207 (1984).
8. M. Ashraf et al., *Pak. J. Sci. Ind. Res.*, **23**, 68 (1980).
9. M. H. Abd El-Razek et al., *Phytochemistry*, **58**, 1289 (2001).
10. M. I. Nassar and T. K. Mohamed, *Fitoterapia*, **69**, 41 (1998).
11. T. Kajimoto et al., *Phytochemistry*, **28**, 1761 (1989).
12. R. Rajanikanth et al., *Phytochemistry*, **23**, 899 (1984).
13. G. U. Mallikarjuna et al., *Breast Cancer Res. Treat.*, **81**, 1 (2003).
14. H. Sadraei et al., *Saudi Pharm. J.*, **11**, 136 (2003).
15. R. G. Sarkis'yan, *Med. Zh. Uzb.*, **9**, 23 (1969).
16. M. M. Mansurov, *Med. Zh. Uzb.*, **6**, 46 (1967).
17. K. J. Kelly et al., *Pediatrics*, **73**, 717 (1984).
18. D. Eigner and D. Scholz, *J. Ethnopharmacol.*, **67**, 1 (1999).
19. A. K. Nadkarni, *Indian Materia Medica*, Vol. 1, Popular Prakashan, Bombay, India, 1976.
20. M. M. Samimi and W. Unger, *Planta Med.*, **36**, 128 (1979).

ASH, PRICKLY

Source: *Zanthoxylum americanum* Mill. or *Z. clava-herculis* L. (sometimes erroneously spelled *Xanthoxylum*) (Family Rutaceae).

Common/vernacular names: Angelica tree, northern prickly ash, pepper wood, toothache tree, and yellow wood (*Z. americanum*); Hercules' club, prickly yellow wood, sea ash, and southern prickly ash (*Z. clava-herculis*); toothache tree and xanthoxylum.

GENERAL DESCRIPTION

Both species are shrubs or small trees growing to about 3 m high with prickly stems and petioles; native to North America. Northern prickly ash (*Z. americanum*) grows from Quebec south to Mississippi and west to Oklahoma, while southern prickly ash (*Z. clava-herculis*) grows farther to the south, from southern Virginia to Florida, Texas, and perhaps Mexico. Part used is the dried bark.

CHEMICAL COMPOSITION

Northern prickly ash contains coumarins (xanthyletin, xanthoxyletin, alloxanthoxyletin, dipetaline), lignans (asarinin, sesamin),[1] alkaloids (laurifoline, nitidine, chelerythrine, tembetarine, magnoflorine, and candicine, with first two in major amounts),[2,3] resins, tannins, and an acrid volatile oil. Other compounds isolated include 8-(3,3-dimethylallyl)-alloxanthoxyletin (a coumarin) from the root bark and two furoquinoline alkaloids (γ-fagarine and skimmianine) from the leaves.[2] The berries contain furanocoumarins (psoralen, xanthotoxin, cnidilin, imperatorin, and isoimperatoin).[4]

Southern prickly ash contains alkaloids (laurifoline, magnofoline, tembetarine, and candicine in root bark; chelerythrine, nitidine, and tembetarine in stem bark), amides (herculin, neoherculin, and a cinnamamide), pluviatilol γ,γ-dimethylallyl ether, *N*-acetylanonaine, lignans (asarinin and sesamin) (KARRER),[3,5,6] tannins, resins, and an acrid volatile oil.

Previously thought to be present in prickly ash bark, berberine was not detected in later studies,[2,5] although it has been reported as a major alkaloid in another *Zanthoxylum* species, *Z. monophyllum* Lam.[7]

PHARMACOLOGY AND BIOLOGICAL ACTIVITIES

In vitro cytotoxic activity was found against human leukemia (HL-60) cells exposed to coumarins (especially dipetaline) and lignans (asarinin and sesamin) isolated from the root bark and fresh stems of northern prickly ash.[1] Cytotoxic effects on human tumor cell lines were also found from crude extracts of the berries and furanocoumarins isolated therefrom (psoralen, isoimperatoin, and xanthoxtoxin).[4] The bark of southern prickly ash has shown *in vitro* growth inhibition of a methicillin-resistant strain of *Staphylococcus aureus*. Activity was largely attributed to the alkaloid chelerythrine, which showed activity against various other methicillin-resistant strains of the bacteria.[8]

TOXICOLOGY

Southern prickly ash bark has shown toxic effects in cattle.[9]

USES

Food. Extract is reportedly used as a flavor component in major food categories such as alcoholic and nonalcoholic beverages, frozen dairy desserts, candy, baked goods, and gelatins and puddings. Highest average maximum use level is approximately 0.01%.

Dietary Supplements/Health Foods. Used in combination with other herbs in capsules, tablets, tinctures, fluid extract, or decoction for increased circulation, antispasmodic in colds, rheumatism, poor digestion, arthritis, and as "blood purifier" (FOSTER AND DUKE; KROCHMAL AND KROCHMAL).

Traditional Medicine. Used to treat toothache; as a tonic and a stimulant; both internally and externally to treat rheumatism; as a diaphoretic in fever; to treat sores, ulcers, and cancer (as an ingredient in Hoxsey "cure" in the 1950s); and others.[10,11]

COMMERCIAL PREPARATIONS

Crude and extracts; crude was formerly official in N.F. and U.S.P.

Regulatory Status. Essential oil, extractive, and solvent-free oleoresin of bark are GRAS for use in foods (§182.20).

REFERENCES

See the General References for BARNES; CLAUS; FEMA; FERNALD; FOSTER AND DUKE; KROCHMAL AND KROCHMAL; LEWIS AND ELVIN-LEWIS; MCGUFFIN 1 & 2; MERCK; USD 23RD.

1. Y. Ju et al., *Phytother. Res.*, **15**, 441 (2001).
2. G. C. Sun, *Diss. Abstr. Int. B*, **35**, 5826 (1975).
3. F. Fish et al., *Lloydia*, **38**, 268 (1975).
4. Q. N. Saqib et al., *Phytother. Res.*, **4**, 216 (1990).
5. F. Fish and P. G. Waterman, *J. Pharm. Pharmacol.*, **25**(Suppl.), 115P (1973).
6. K. V. Rao and R. Davies, *J. Nat. Prod.*, **49**, 340 (1986).
7. F. R. Stermitz and I. A. Sharifi, *Phytochemistry*, **16**, 2003 (1977).
8. S. Gibbons et al., *Phytother. Res.*, **17**, 274 (2003).
9. J. M. Bowen et al., *Am. J. Vet. Res.*, **57**, 1239 (1996).
10. J. F. Morton, *Bull. Med. Libr. Assoc.*, **56**, 161 (1968).
11. J. L. Hartwell, *Lloydia*, **34**, 103 (1971).

ASPARAGUS

Source: *Asparagus officinalis* L. (Family Liliaceae).

Common/vernacular names: Garden asparagus.

GENERAL DESCRIPTION

A dioecious perennial herb with erect and much branched stem, up to 3 m high; leaves scale-like; native to Europe and western Asia; widely cultivated.

CHEMICAL COMPOSITION

Roots contain inulin and fructo-oligosaccharides;[1] glycosidic bitter principles (officinalisnin-I and officinalisnin-II);[2] β-sitosterol, sarsasapogenin, and steroidal glycosides (named asparagosides A to I, in order of their increasing polarity);[3,4] asparagusic acid;[5] and others.

Shoots (spears and tips) contain sulfur-containing acids, including asparagusic, dihydroasparagusic, and S-acetyldihydro-asparagusic acids;[5,6] α-amino-dimethyl-γ-butyrothetin (an S-methylmethionine derivative), among others;[7] a glycosidic bitter principle that is different from the two found in roots;[8] protodioscin;[9] flavonoids (rutin, hyperoside, isoquercitrin, cosmosiin, kaempferol-3-O-L-rhamno-D-glucoside, kaempferol, quercetin, etc.);[10,11] asparagine, arginine, tyrosine, sarsasapogenin, β-sitosterol, succinic acid, sugars, and others (KARRER; JIANGSU; MERCK).[12]

Asparagusic acid and its derivatives are plant growth inhibitors, inhibiting the growth of lettuce;[6] it also has nematicidal properties, thus being responsible for the resistance of asparagus to several plant parasitic nematodes.[5]

Methylmercaptan (a hydrolysis product of the S-containing compounds) or asparagine–aspartic acid monoamide is believed to be present in urine after eating asparagus, causing its peculiar odor (MERCK).

Seeds contain steroidal saponins (protodioscin, oligofurostanosides);[13] large amounts of NaOH-soluble polysaccharides;[14] carotenoids (mutatoxanthin epimers, antheraxanthin, β-carotene, β-cryptoxanthin, lutein, capsanthin, capsanthin 5,6-epoxide, capsorubin, neoxanthin, violaxanthin, zeaxanthin);[15] and 15.3% oil composed of 43.47% arachidic, 22.16% oleic, 11.52% palmitic, 11.34% linoleic, 5.78% behenic, 3.59% stearic, 2.14% linolenic acids, and 1.43% unsaponifiable matter consisted mostly of β-sitosterol.[16]

PHARMACOLOGY AND BIOLOGICAL ACTIVITIES

The roots are reported to have diuretic and hypotensive properties (JIANGSU). Animal experiments with the roots indicate diuretic activity (BLUMENTHAL 1). Extracts of the spears have shown *in vitro* antioxidant activity.[17] Fibers isolated from the vegetable are claimed to have mutagen-adsorbing (cancer-preventing) properties.[18] An antifungal deoxyribonuclease was isolated from the seeds[19] and the saponin fraction has shown antifungal activity.[20] Among the saponins, protodioscin, found in larger quantities in the bottom of the stalks than the tips,[9] has shown *in vitro* cytotoxic activity against herpes simplex virus type 1,[21] human leukemia HL-60 cells,[13] colon cancer cells, glioma cells, melanoma, renal and CNS cancer cells.[22] Protodioscin has also shown proerectile activity in rabbits.[23]

TOXICOLOGY

Allergic reactions to asparagus have occurred from ingestion or handling of the plant and commonly result in severe asthma or anaphylaxis. Allergens from asparagus have been identified as lipid transfer proteins[24,25] and 1,2,3-trithiane-5-carboxylic acid, a sulfur-containing plant growth inhibitor.[26]

USES

Medicinal, Pharmaceutical, and Cosmetic.
Fleshy fibrous roots and to a lesser extent seeds are used for medicinal purposes. Roots are used in diuretic preparations. Herb is used in diuretic galenical preparations in Germany, although claimed efficacy is not substantially documented (BLUMENTHAL 1).

Food. Aerial stems (asparagus spears) arising from rhizomes are used as vegetable. Seed and root extracts reportedly used in alcoholic beverages at an average maximum use level of 0.0016% (16 ppm). Seeds have been used as coffee substitutes (UPHOF).

Traditional Medicine. Roots are used as diuretic, laxative, and in neuritis and rheumatism; also reportedly used in treating cancer;[27] used in Chinese medicine as an antitussive, tonic, antifebrile, diuretic, and hair growth stimulator.[13]

Fruits (berries) have been reported to be used as contraceptives.[28,29]

Shoots are used in homemade preparations to cleanse face and to dry up pimples and sores (ROSE).

COMMERCIAL PREPARATION

Crude and extracts.

Regulatory Status. Undetermined in the United States. Herb subject of a German therapeutic monograph, though therapeutic use is not recommended (as diuretic) due to insufficient scientific evidence. A positive therapeutic German monograph exists for use of root preparations in irrigation therapy for inflammatory disease of the urinary tract and prevention of renal gravel (contraindicated in inflammatory kidney disease or edema caused by cardiac or renal disease) (BLUMENTHAL 1).

REFERENCES

See the General References for BAILEY 1; BAILEY 2; BIANCHINI AND CORBETTA; BLUMENTHAL 1; DER MARDEROSIAN AND BEUTLER; FEMA; GOSSELIN; GRIEVE; GUPTA; KARRER; JIANGSU; LEWIS AND ELVIN-LEWIS; UPHOF; WREN.

1. N. Shiomi et al., *Agric. Biol. Chem.*, **40**, 567 (1976).
2. K. Kawano et al., *Agric. Biol. Chem.*, **39**, 1999 (1975).
3. G. M. Goryanu et al., *Khim. Prir. Soedin.*, **3**, 400 (1976).
4. G. M. Goryanu and P. K. Kintya, *Him. Prir. Soedin.*, **6**, 762 (1976).
5. M. Takasugi et al., *Chem. Lett.*, **1**, 43 (1975).
6. H. Yanagawa et al., *Tetrahedron Lett.*, **25**, 2549 (1972).
7. R. Tressl et al., *J. Agric. Food Chem.*, **25**, 455 (1977).
8. S. Sakamura et al., *Nippon Shokuhin Kogyo Gakkaishi*, **14**, 491 (1967).
9. M. Wang et al., *J. Agric. Food Chem.*, **51**, 6132 (2003).
10. M. Woeldecke and K. Herrmann, *Z. Lebensm. Unters. Forsch.*, **155**, 151 (1974).
11. T. Kartnig et al., *Planta Med.*, **51**, 288 (1985).
12. R. Tressl et al., *J. Agric. Food Chem.*, **25**, 459 (1977).
13. Y. Shao et al., *Planta Med.*, **63**, 258 (1997).
14. R. Goldberg, *Phytochemistry*, **8**, 1783 (1969).
15. J. Deli et al., *J. Agric. Food Chem.*, **48**, 2793 (2000).
16. Y. R. Prasad and S. S. Nigam, *Proc. Natl. Acad. Sci. India, Sect. A*, **52**, 396 (1982).

17. D. P. Makris et al., *J. Agric. Food Chem.*, **49**, 3216 (2001).
18. Y. Sasaki et al., Jpn. Kokai Tokkyo Koho JP 6140764 (1986).
19. H. Wang and T. B. Ng, *Biochem. Biophys. Res. Commun.*, **289**, 120 (2001).
20. M. Shimoyamada et al., *Agric. Biol. Chem.*, **54**, 2553 (1990).
21. T. Ikeda et al., *Biol. Pharm. Bull.*, **23**, 363 (2000).
22. K. Hu and X. Yao, *Planta Med.*, **68**, 297 (2002).
23. P. G. Adaikan et al., *Ann. Acad. Med. Singapore*, **29**, 22 (2000).
24. A. Tabar et al., *Clin. Exp. Allergy*, **34**, 131 (2004).
25. A. Diaz-Perales et al., *J. Allergy Clin. Immunol.*, **110**, 790 (2002).
26. B. M. Hausen and C. Wolf, *Am. J. Contact Dermat.*, **7**, 41 (1996).
27. J. L. Hartwell, *Lloydia*, **33**, 97 (1970).
28. C. S. Barnes et al., *Lloydia*, **38**, 135 (1975).
29. V. J. Brondegaard, *Planta Med.*, **23**, 167 (1973).

ASPIDIUM

Source: *Dryopteris filix-mas* (L.) Schott or D. *marginalis* (L.) Gray (Family Polypodiaceae).

Common/vernacular names: European aspidium and male fern (*D. filix-mas*); American aspidium and marginal fern (*D. marginalis*).

GENERAL DESCRIPTION

Both are stout perennial ferns with fronds up to about 1 m long; *D. filix-mas* is the larger of the two. Sori in *D. filix-mas* are large and near the midvein, while those of *D. marginalis* are small and near the margin. *Dryopteris filix-mas* is found in Europe, Asia, North America, South America, and South and northern Africa; *D. marginalis* grows in eastern and central North America from Nova Scotia to Georgia and Kansas. Parts used are the dried rhizomes and stipes.

CHEMICAL COMPOSITION

Most studies were done on *D. filix-mas*, which contains 6.5–15% oleoresin. The oleoresin contains not less than 24% of the active principles filicin, which is a collective term for several ether-soluble phloroglucides (filixic acids, flavaspidic acids, paraaspadin, albaspidin, desaspidin, etc.);[1,2] aspidinol; n-alkanes from C_{27} to C_{33}, with C_{29} and C_{31} in major concentrations; triterpene hydrocarbons (12-hopene, 11,13 (18)-hopadiene, 9(11)-fernene);[3] lignins;[4] volatile oil, resins, and others. Presence of aspidinol and desaspidin was not observed in one study.[2]

Dryopteris marginalis contains margaspidin, flavaspidic acids, paraaspidin, phloraspin, and others.[5,6]

PHARMACOLOGY AND BIOLOGICAL ACTIVITIES

Filicin is anthelmintic (especially against tapeworms), but it is a violent poison if absorbed. Castor oil promotes its absorption in the gut and should never be used with it as a laxative. Poisoning symptoms include nausea, vomiting, bloody diarrhea, dizziness, delirium, tremors, convulsions, coma, respiratory or cardiac failure, visual disturbances leading to temporary or permanent blindness, and others.

Margaspidin has been reported to have anti-inflammatory activity in rats.[5]

Extracts of *D. filix-mas* are active against vesicular stomatitis virus in monkey cell cultures.[7]

Extracts containing phloroglucinols from *Dryopteris crassirhizoma* Nakai have antimicrobial activities *in vitro*[8] and antitumor activities in transplanted tumors (ARS, S_{180}, U_{14}, B_{22}, etc.) in rats and mice.[9]

USES

Medicinal, Pharmaceutical, and Cosmetic. Oleoresin is used for expulsion of tapeworms along with a saline laxative such as magnesium or sodium sulfate.

Traditional Medicine. Has been reported to be used in treating tumors.[10]

COMMERCIAL PREPARATIONS

Crude and oleoresin. Was formerly official in U.S.P.

REFERENCES

See the General References for BAILEY 1; CLAUS; DER MARDEROSIAN AND BEUTLER; GOSSELIN; GRIEVE; KARRER; MARTINDALE; MCGUFFIN 1 & 2; MERCK; USD 26TH; YOUNGKEN.

1. C. J. Widen, *Helv. Chim. Acta*, **54**, 2824 (1971).
2. M. Guley and T. Soylemezoglu, *Ankara Univ. Eczacilik Fak. Mecm.*, **6**, 214 (1976).
3. F. Bottari et al., *Phytochemistry*, **11**, 2519 (1972).
4. O. Faix et al., *Holzforschung*, **31**(5), 137 (1977).
5. H. Otsuka et al., *Takeka Kenkyusho Ho*, **30**, 225 (1971).
6. H. S. Puri et al., *Planta Med.*, **33**, 177 (1975).
7. G. P. Husson et al., *Ann. Pharm. Fr.*, **44**, 41 (1986).
8. S. S. A. Vichkanova et al., *Rast. Resur.*, **18**, 93 (1982).
9. D. H. Li et al., *Zhongcaoyao*, **17**(6), 14 (1986).
10. J. L. Hartwell, *Lloydia*, **33**, 288 (1970).

ASTRAGALUS

Source: *Astragalus membranaceus* (Fisch. ex Link) Bunge. (syn. *A. propinguus* B. Schischk.) *A. mongholicus* Bunge. (syn. *A. membranaceus* (L.) (Fish. ex Link) Bunge. var. *mongholicus* (Bunge.) P. K. Hsiao), and other Chinese *Astragalus* spp. (Family Leguminosae or Fabaceae).

Common/vernacular names: Astragalus, huangqi, membranous milk vetch (*A. membranaceus*); astragalus, Mongolian milk vetch (*A. mongholicus*), and milk vetch (*Astragalus* spp.).

GENERAL DESCRIPTION

Astragalus spp. are perennial herbs, up to about 1 m high (normally 0.5–0.8 m). Most are native to northern China and some to high regions such as Sichuan, Yunnan, and Tibet; extensively cultivated. Although several *Astragalus* species serve as source of astragalus root, *A. membranaceus* and *A. mongholicus* yield most of the root in commerce and on which most the chemical and pharmacological research has been performed.[1]

Part used is the dried root from 4- to 7-year-old plants collected in the spring before leaves appear or in autumn after they have fallen. After the root is dug up, the crown and rootlets are removed along with dirt and then usually sun dried. The most commonly used forms are raw astragalus (dried root) and cured (honey-treated) astragalus; the former usually comes in slices (size and shape like tongue depressors), which is produced by thoroughly moistening the raw root, cutting into thick slices, and drying; and the latter is produced by frying the sliced root with honey (from 25–30 parts to 100 parts of root) over medium heat until no longer sticky to touch.

CHEMICAL COMPOSITION

The main constituents of the root are saponins, polysaccharides, isoflavonoids, free amino acids, and trace minerals.[2,3] Other constituents include coumarin, folic acid, nicotinic acid, choline, betaine, phenolic acids (ferulic, isoferulic, caffeic and chlorogenic acids, etc.), sitosterol, sucrose, and linoleic and linolenic acids (HU; JIANGSU).[2,4]

More than 40 triterpene glycosides (saponins) have been isolated from roots of *A. membranaceus*, *A. mongholicus*, and other Chinese *Astragalus* species, including astragalus saponins I, II, and III; astragalosides I, II, III, IV, V, VI, VII, and VIII; acetylastragaloside I; isoastragalosides I and II; astrasieversianin I, II, III, IV (astragaloside I), V, VI, VII (isoastragaloside II), VIII (astragaloside II), and IX–XVI; cyclogaleginosides A and B; astramembrannin I (astragaloside IV, astrasieversianin XIV) and astramembrannin II (cyclogaleginoside B); and soyasaponin I. Soyasapogenol B is the aglycone of soyasaponin I and astragaloside VIII and cycloastragenol (cyclogalegigenin, astramembrangenin), the aglycone of the other saponins.[4–6] Huangqiyenins A and B were isolated from the leaves.[7]

Polysaccharides (from *A. mongholicus* root) include astragalan I (mol. wt. 36,300; D-glucose: D-galactose: L-arabinose = 1.75 : 1.63 : 1; with a trace of pentose), astragalan II (mol. wt. 12,300; α-(1,4)(1,6)-glucan), and astragalan III (mol. wt. 34,600; α-(1,4)(1,6)-glucan);[2,6] and AG-1 (α-glucan with α-(1,4) : α-(1,6) ratio of 5 : 2), AG-2 (water-insoluble α-(1,4)-glucan), AH-1 (acidic; galacturonic acid/glucuronic acid:glucose:rhamnose: arabinose = 1 : 0.04 : 0.02 : 0.01), and AH-2 (glucose:arabinose = 1 : 0.15).[2,8,9]

Flavonoids include kaempferol, quercetin, isorhamnetin, calycosin, formononetin, rhamnocitrin, kumatakenin, (3R)-2′,3′-dihydroxy-7,4-dimethoxy-isoflavone, L-3-hydroxy-9-methoxypterocarpan, (6aR,11aR)-10-hydroxy-3,9-dimethoxypterocarpan, calycosin glucoside, 9,10-dimethoxypterocarpan-3-*O*-β-D-glucoside and 2′-hydroxy-3′,4′-dimethoxy-isoflavone-7-*O*-β-D-glucoside, among others.[2,10–15]

Of over 20 free amino acids identified, asparagine, glutamic acid, canavanine, proline, arginine, β-aminobutyric acid, γ-aminobutyric acid (GABA), aspartic acid, and alanine are present in the highest concentrations; they make up 0.50–1.26% of astragalus root, depending on sources.[2,3,16,17]

Among more than 20 trace minerals found in astragalus are magnesium (1108–1761 ppm), iron (94–694 ppm), manganese (8–52 ppm), zinc (11–23 ppm), copper (5–9 ppm), rubidium (11–13 ppm), molybdenum (0.1–10 ppm) and chromium (0.3–0.8 ppm).[2,18–20]

PHARMACOLOGY AND BIOLOGICAL ACTIVITIES

Astragalus root is a highly valued Chinese herbal tonic with diverse pharmacological properties. Decoctions, alcoholic extracts, and/or powders of the root have shown numerous activities in humans and experimental animals, including immunopotentiating effects;[21–30] tumor-inhibiting,[31] antibacterial,[23] and antiviral;[28] promoting nucleic acid synthesis in liver and spleen; elevating and/or reducing cAMP and cGMP levels in blood, liver, and spleen of mice;[23] cyclooxygenase-2-inhibiting *in vitro*;[32] anti-inflammatory;[33] cardiovascular effects (hypotensive, vasodilating, etc.);[28,34] inhibiting experimentally induced hypoglycemia and hyperglycemia in mice;[35] contracting smooth muscles; prolonging life span of silkworm and of cells *in vitro*;[21,36] larvicidal (*Lymantria dispar* L.);[37] antiprotozoal (*Trypanosoma cruzi*);[38] inducing the release of growth hormone in rat pituitary cells *in vitro*;[39] melanocyte proliferation-stimulating *in vitro*;[40] antioxidant effects[41] (e.g., increasing superoxide dismutase activity); antimutagenic *in vitro*;[28] improving learning and memory; promoting cartilage growth *in vitro*; hepatoprotective against experimental liver toxicity; reducing urinary protein in chronic and in experimental nephritis; diuretic; improving stamina; and others (JIANGSU; WANG; ZHOU AND WANG).[28,42,43] Its effects are not due to a single compound or one single class of compounds but rather to different types of components, with the saponins, polysaccharides,[21–44] and isoflavonoids[3] appearing to play a major role.

A placebo-controlled study on *A. membranaceous* found that blood leucocytes of subjects treated with the root showed a significantly greater level of interferon induction.[45]

TOXICOLOGY

Toxicity of astragalus root is very low: a crude water extract (75 and 100 g/kg p.o.) produced no adverse effects in mice within 48 hours; LD_{50} in mice: 40 g/kg i.p. The former oral doses are 375 and 500 times the usual effective diuretic dose in humans (WANG; ZHOU AND WANG).

USES

Medicinal, Pharmaceutical, and Cosmetic. Extracts of astragalus root are used in skin care cosmetics (e.g., hand and facial creams and lotions) for its traditional healing and nourishing as well as vasodilating properties; also used in hair tonics for similar effects (ZHOU).

Dietary Supplements/Health Foods. Powdered crude and/or extracts are used singly or in combination with other herbs in capsule, tablet, or liquid (syrup or drink) form primarily as a general (*qi*) tonic to improve body resistance (immunity); also used in sliced or tea bag cut form in tea or soup mix packets (FOSTER AND YUE).

Traditional Medicine. One of the major Chinese *qi* (energy) tonics, with a recorded use history of 2000 years. Raw root is traditionally considered to benefit the body's resistance (*yiwei gubiao*), promote diuresis, reduce swelling, promote suppuration (drains pus, *tuo du*), and regenerate tissue or promote muscle growth (*sheng ji*). Cured root is said to reinforce the Middle Burner and replenish the vital energy (*buzhong yiqi*).

Raw astragalus used mainly in spontaneous and night sweating, edema, chronic sores and abscesses, unhealing wounds and ulcers, and painful joints; cured astragalus primarily as an energy (*qi*) tonic to treat general weakness, fatigue, lack of appetite, diarrhea caused by spleen deficiency (*pi xu xie xie*), rectal prolapse, and uterine bleeding. Uses of the two occasionally overlap.

Modern/recent uses include prevention and treatment of the common cold and influenza; stomach ulcer,[46] neurodermatitis, and diabetes, for which high doses (>60 g) are sometimes used (CHP; JIANGSU). Astragalus root has also been studied as one of the *fuzheng guben* (strengthening body defense therapy) herbs used in treating AIDS.[47]

Usual daily oral dose for adults is 9–30 g.

COMMERCIAL PREPARATIONS

Raw astragalus (readily available as sticks or slices in several grades); powdered crude; and extracts (aqueous, hydroalcoholic, glycolic). Most extracts come without standardized strengths, and powdered astragalus may contain adulterants such as starches and pre-extracted plant materials.

Regulatory Status. Regulated in the United States as a dietary supplement.

REFERENCES

See the General References for CHP; DER MARDEROSIAN AND BEUTLER; FOSTER AND YUE; HONGKUI; HU; HUANG; JIANGSU; JIXIAN; WANG; ZHOU; ZHOU AND WANG.

1. S. S. Sun et al., *Zhongguo Yaoxue Zazhi*, **25**, 643 (1990).
2. Z. S. Qi, *Zhongcaoyao*, **18**, 41 (1987).
3. X. Q. Ma et al., *J. Agric. Food Chem.*, **50**, 4861 (2002).
4. K. He and H. K. Wang, *Yaoxue Xuebao*, **23**, 873 (1988).
5. I. Kitagawa et al., *Chem. Pharm. Bull.*, **31**, 716 (1983).
6. I. Kitagawa et al., *Chem. Pharm. Bull.*, **31**, 689 (1983).
7. Y. Ma et al., *Chem. Pharm. Bull.*, **45**, 359 (1997).
8. S. D. Fang et al., *Youji Huaxue*, **1**, 26 (1982).
9. Q. S. Huang et al., *Yaoxue Xuebao*, **17**, 200 (1982).
10. G. B. Lu et al., *Zhongcaoyao*, **15**, 20 (1984).
11. X. Ma et al., *J. Chromatogr. A*, **992**, 193 (2003).
12. X. Ma et al., *J. Chromatogr. A*, **962**, 243 (2002).
13. C. Q. Song et al., *Acta Bot. Sin.*, **39**, 486 (1997).
14. C. Q. Song et al., *Acta Bot. Sin.*, **39**, 764 (1997).
15. L. Z. Lin et al., *J. Chromatogr. A*, **876**, 87 (2000).
16. Y. X. Gong, *Zhongcaoyao*, **18**, 37 (1987).
17. R. Xiao et al., *Zhongyao Tongbao*, **9**, 30 (1984).
18. X. Z. Liu et al., *Chin. J. Integr. Trad. West. Med.*, **5**, 235 (1985).
19. X. Zhao et al., *Chin. J. Integr. Trad. West. Med.*, **8**, 419 (1988).
20. C. L. Ma and S. K. Wang, *Zhongcaoyao*, **16**, 4 (1985).
21. Z. D. Du and Y. M. Lai, *Zhongcaoyao*, **19**, 40 (1988).
22. B. Y. Du and J. Y. Zhang, *Zhongyao Tongbao*, **12**, 53 (1987).
23. C. S. Geng, *Chin. J. Integr. Trad. West. Med.*, **6**, 62 (1986).
24. B. H. S. Lau et al., *Phytother. Res.*, **3**(4), 148 (1989).
25. D. C. Wang et al., *Chin. J. Oncol.*, **11**, 180 (1989).
26. D. T. Chu et al., *J. Clin. Lab. Immunol.*, **25**, 119 (1988).
27. D. T. Chu et al., *J. Clin. Lab. Immunol.*, **25**, 125 (1988).
28. J. L. Ríos and P. G. Waterman, *Phytother. Res.*, **11**, 411 (1997).
29. K. Kajimura et al., *Biol. Pharm. Bull.*, **20**, 1178 (1997).
30. R. T. Wang et al., *Zhongguo Zhong Xi Yi Jie He Za Zhi*, **22**, 453 (2002).
31. R. Cui et al., *Cancer Chemother. Pharmacol.*, **51**, 75 (2003).
32. E. J. Kim et al., *Korean J. Pharmacogn.*, **32**, 311 (2001).
33. Y. H. Shon et al., *Biol. Pharm. Bull.*, **25**, 77 (2002).
34. Y. D. Zhang et al., *Yaoxue Xuebao*, **19**, 333 (1984).
35. X. R. Li et al., *Zhongchengyao*, **11**, 32 (1989).
36. M. She, *Zhongyao Tongbao*, **10**, 41 (1985).
37. S. J. Parks et al., *Appl. Entomol. Zool.*, **32**, 601 (1997).
38. G. R. Schinella et al., *Fitoterapia*, **73**, 569 (2002).
39. C. Kim et al., *Arch. Pharm. Res.*, **26**, 34 (2003).
40. Z. X. Lin et al., *J. Ethnopharmacol.*, **66**, 141 (1999).
41. S. Toda and Y. Shirataki, *Phytother. Res.*, **11**, 603 (1997).
42. Y. Kang et al., *Zhongcaoyao*, **20**(11), 21 (1989).
43. X. R. Li et al., *Zhongchengyao*, **11** (3), 27 (1989).
44. H. Wagner, *Naturheilpraxis*, **3**, 256 (1983).
45. Y. Hou et al., *Chin. Med. J.*, **95**, 35 (1981).
46. J. Grujic-Vasic et al., *Planta Med.*, **55**, 649 (1989).
47. S. M. Bao, *Zhongguo Zhongyao Zazhi*, **14**, 59 (1989).

AVOCADO

Source: *Persea americana* Mill. (syn. *P. gratissima* Gaertn.; *Laurus persea* L.) (Family Lauraceae).

Common/vernacular names: Ahuacate, alligator pear, and avocado.

GENERAL DESCRIPTION

Large evergreen tree up to about 20 m high; fruit large (5–20 cm long) and fleshy, pyriform, ovate, or spherical, with a thin to thick sometimes woody skin; native to tropical America (Mexico, Central America); widely cultivated. Parts used are the fruit and seed.

There are several commercial varieties in the United States, including the Mexican avocado and the West Indian avocado. The Mexican avocado (*P. americana* var. *drymifolia*) is grown in California, while the larger West Indian avocado (*P. americana* var. *americana*) is produced in Florida.[1] Avocado oil (a fixed oil) is produced by expressing the dried pulp of the fruit.

CHEMICAL COMPOSITION

Pulp contains a fatty oil (4–40%, depending on season, location, climate, etc., but usually about 16%); campesterol, high amounts of β-sitosterol (average 76.4 mg/100 g);[2,3] fatty acids (approximately 60% monounsaturated, 20% saturated, and 20% unsaturated);[4] high amounts of glutathione (27.7 mg/100 g);[5] approximately 2% protein; 6–9% carbohydrates and sugars (glucose, fructose, D-mannoheptulose, a taloheptulose, and an alloheptulose); two bitter substances (1-acetoxy-2,4-dihydroxyheptadeca-16-ene and 1,2,4-trihydroxyheptadeca-16-ene); carnitine;[6–14] proanthocyanidins;[15] persenones A and B;[16] magnesium, potassium, vitamin K, vitamin E, folic acid, riboflavin, niacin, thiamin, pantothenic acid, biotin (each in higher quantities than any of the more frequently consumed raw fruits in the United States);[4] and others.[6–14]

The pulp oil (avocado oil) consists mainly of glycerides of oleic acid; the other major fatty acids in the oil are palmitic and linoleic acid. It also contains highly variable amounts of unsaponifiable matter (1.6–11.3%)[7,8] consisting of sterols (β-sitosterol, campesterol, 24-methylenecycloartanol, citrostadienol, etc.), hydrocarbons, volatile acids (propionic, butyric, valeric, etc.), amino acids, and vitamin D (higher than in butter or eggs), among others.[16–18]

Seeds contain polyhydric alcohols (volemitol, perseitol, arabinitol, galactitol, myo-inositol, D-erythro-D-galacto-octitol, glycerol), rare sugars (D-mannoheptulose, etc.);[19] avocatins (avocadene, avocadyne, avocadenone acetate, avocadynone acetate, avocadenofuran, avocadynofuran, avocadienofuran, isoavocadienofuran, etc.);[20] 4,8″-biscatechin (a condensed flavanol);[21] C_{17} oxygenated aliphatic unsaturated compounds (1,2,4-trihydroxyheptadeca-16-ene, etc.),[10,11,22] some of which have an unpleasant bitter flavor,[10,11] and more common constituents, including protein and fats.

Steam distillation of Mexican avocado leaves yielded 3.1% of an essential oil consisting of 95% estragole (see **sweet basil** for toxicity) and 5% anethole.[23] The fruit peel contains *l*-epicatechin.[24]

PHARMACOLOGY AND BIOLOGICAL ACTIVITIES

Antioxidant activity was shown from the fruit pulp against hydroxy radicals *in vitro*.[25] A methanol extract of the pulp was shown to inhibit the *in vitro* generation of nitric oxide (NO) in macrophages.[26] Constituents isolated from the fruit pulp (persenones A and B) inhibited the production of NO, inducible nitric oxide synthase (iNOS), cyclooxygenase-2 (COX-2), and superoxide radical generation *in vitro*. Hydrogen peroxide generation in mouse skin was inhibited by persenone

A.[16,27] Fed to rats with toxin-induced liver injury, the fruit pulp showed hepatoprotective activity; an effect attributed to various fatty acid derivatives (e.g., 2E,5E,12Z,15Z)-1-hydroxyheneicosa-2,5,12,15-tetraen-4-one).[28]

A methanol extract of the fruits showed *in vitro* inhibition of acetyl-CoA carboxylase, an enzyme involved in fatty acid biosynthesis. Activity was attributed to 5E,12Z,15Z)-2-hydroxy-4-oxoheneicosa-5,12,15-trienyl and several other related compounds.[29]

In rabbits fed a high-fat diet, atherogenicity of avocado oil in the diet was not significantly different from that of olive or corn oil.[30] Rats fed a high-cholesterol diet with the addition of defatted avocado fruit pulp showed lower food consumption, body weight gain, and levels of hepatic total fat, compared to rats fed the same diet with added cellulose instead of avocado.[31]

4,8″-Biscatechin, a condensed flavanol isolated from avocado seeds, has shown antitumor activity against Sarcoma 180 in mice and Walker 256 in rats.[21] C_{17} oxygenated unsaturated aliphatics (especially 1,2,4-trihydroxyheptadeca-16-ene) isolated from avocado (pulp and seeds) have shown *in vitro* antimicrobial activity against Gram-positive bacteria, especially *Staphylococcus aureus*.[22,32]

TOXICOLOGY

Avocado is a known cross-reactant in individuals with latex allergy.[33] Severe allergic reactions can occur in these patients after eating avocado.[34]

Poisoning of cattle, horses, goats, rabbits, canaries, and fish by avocado (leaves, fruit, bark, seeds) have been reported (LEWIS AND ELVIN-LEWIS).

USES

Medicinal, Pharmaceutical, and Cosmetic. Avocado oil is believed to have healing and soothing properties to the skin and the pulp oil is used in massage creams, muscle oils, hair products, and others. A pharmaceutical preparation containing the seed oil (unsaponifiable fraction) has been patented for use in the treatment of sclerosis of the skin, pyorrhea, arthritis, and others.[35] The fruit pulp is used in face creams.

The unsaponifiable fraction is combined with those of soy beans for use in the treatment of osteoarthritis.[36,37]

Food. Pulp has been used as a food for thousands of years in tropical America and is today consumed internationally.[4]

Traditional Medicine. Pulp of *Persea planifolia* (American avocado) used by Guatemalan Indians as a hair pomade to stimulate hair growth, to hasten suppuration of wounds, and as an aphrodisiac and emmenagogue; seeds used to treat diarrhea and dysentery; powdered seeds used by American Indians (Mahuna, southwest California) to treat pyorrhea; infusion used to treat toothache.

COMMERCIAL PREPARATIONS

Fruit and pulp oil.

REFERENCES

See the General References for BAILEY 2; BRUNETON; DER MARDEROSIAN AND BEUTLER; MCGUFFIN 1 & 2; MOERMAN; MORTON 2; POUCHER; UPHOF; WATT AND MERRILL.

1. L. O. Williams, *Econ. Bot.*, **31**, 315 (1977).
2. K. C. Duester, *J. Am. Diet. Assoc.*, **101**, 404 (2001).
3. J. L. Weihrauch and J. M. Gardner, *J. Am. Diet. Assoc.*, **78**, 39 (1978).
4. K. C. Duester, *Nutr. Today*, **35**, 151 (2000).

5. D. P. Jones et al., *Nutr. Cancer*, **17**, 57 (1992).
6. T. Itoh et al., *Fruits*, **30**, 687 (1975).
7. M. H. Bertoni et al., *Ann. Asoc. Quim. Argent.*, **55**, 257 (1967).
8. G. G. Slater et al., *J. Agric. Food Chem.*, **23**, 468 (1975).
9. D. Pearson, *J. Sci. Food Agric.*, **26**, 207 (1975).
10. G. Ben-Et et al., *J. Food Sci.*, **38**, 546 (1973).
11. B. I. Brown, *J. Agric. Food Chem.*, **20**, 753 (1972).
12. I. Johansson and N. K. Richtmyer, *Carbohydr. Res.*, **13**, 461 (1970).
13. J. N. Ogata et al., *J. Agric. Food Chem.*, **20**, 113 (1972).
14. M. Tada et al., *Nippon Eiyo Shokuryo Gakkaishi*, **37**, 13 (1984).
15. L. Gu et al., *J. Agric. Food Chem.*, **51**, 7513 (2003).
16. M. A. Joslyn and W. Stepka, *Food Res.*, **14**, 459 (1949).
17. A. Zanobini et al., *Boll. Soc. Ital. Biol. Sper.*, **50**, 887 (1974).
18. A. O. Moreno et al., *J. Agric. Food Chem.*, **51**, 2216 (2003).
19. N. K. Richtmyer, *Carbohydr. Res.*, **12**, 135 (1970).
20. H. M. Alves et al., *An. Acad. Bras. Cienc.*, **42** (Suppl.), 45 (1970).
21. M. M. DeOliveira et al., *An. Acad. Bras. Cienc.*, **44**, 41 (1972).
22. I. Néeman et al., *Appl. Microbiol.*, **19**, 470 (1970).
23. D. I. Acostade Iglesias et al., *Riv. Ital. Essenze, Profumi, Piante Offic., Aromi, Saponi, Cosmet., Aerosol*, **58**, 158 (1976).
24. M. Nose and N. Fujino, *Nippon Shokuhin Kogyo Gakkaishi*, **29**, 507 (1982).
25. M. A. Murcia et al., *J. Food Prot.*, **64**, 2307 (2001).
26. O. K. Kim et al., *Cancer Lett.*, **125**, 199 (1998).
27. O. K. Kim et al., *Biosci. Biotechnol. Biochem.*, **64**, 2504 (2000).
28. H. Kawagishi et al., *J. Agric. Food Chem.*, **49**, 2215 (2001).
29. H. Hashimura et al., *Biosci. Biotechnol. Biochem.*, **65**, 1656 (2001).
30. D. Kritchevsky et al., *J. Am. Coll. Nutr.*, **22**, 52 (2003).
31. E. Naveh et al., *J. Nutr.*, **132**, 2015 (2002).
32. I. Néeman et al., Fr. Demande 2,075,994 (1971).
33. C. Blanco, *Curr. Allergy Asthma Rep.*, **3**, 47 (2003).
34. S. Isola et al., *Allergy Asthma Proc.*, **24**, 193 (2003).
35. H. Thiers, Neth. Appl. 6,601,888 (1966).
36. E. Ernst, *Clin. Rheumatol.*, **22**, 285 (2003).
37. K. L. Soeken, *Clin. J. Pain*, **20**, 13 (2004).

BALM, LEMON

Source: *Melissa officinalis* L. (Family Lamiaceae).

Common/vernacular names: Balm, bee balm, common balm, lemon balm, melissa, and melissa balm.

GENERAL DESCRIPTION

An aromatic perennial herb with yellowish or white flowers, up to approximately 1 m in height, growing in the Mediterranean region, western Asia, southwestern Siberia, and northern Africa; widely cultivated. Parts used are the dried leaves often with flowering tops; an essential oil is obtained from these by steam distillation. Bees are attracted to the plant and bruising the leaves releases a lemony odor.

CHEMICAL COMPOSITION

Contains about 0.1–0.2% volatile oil composed mainly of oxygenated compounds such as citral (a mixture of neral and geranial), caryophyllene oxide, citronellal, eugenol acetate, and geraniol, plus smaller amounts of terpenes, including *trans*- and (−)-β-ocimene, caryophyllene, α-cubebene, copaene, and β-bourbonene.[1–5] Other constituents of lemon balm include polyphenols (caffeic acid, protocatechuic acid, etc.); a tannin composed of caffeic acid units;[6–9] flavonoids (rhamnazin,[10] luteolin, luteolin 7-*O*-beta-D-glucopyranoside and other luteolin glycosides; apigenin 7-*O*-beta-D-glucopyranoside);[11] rosmarinic acid;[10] triterpenoids (ursolitc acid, etc.); and glucosides of geraniol, nerol, eugenol, benzyl alcohol, β-phenylethyl alcohol, neric acid, and geranic acid, among others (KARRER).[6,9,12,13] The main constituents of the essential oil are geranial, neral, citronellal, geranyl acetate, citronellol, and β-caryophyllene.[14]

PHARMACOLOGY AND BIOLOGICAL ACTIVITIES

Extracts of balm have shown *in vitro* antiviral activity against HIV-1,[15] herpes simplex,[16] Newcastle disease virus, paramyxovirus (mumps virus), vaccinia, and other viruses. The active constituents include polyphenols (other than caffeic acid) and tannin.[7,8] Extracts of balm have also shown *in vitro* antioxidant activity against lipid peroxidation;[17] *in vitro* antithyrotropic activity;[18,19] and antiulcerogenic activity in rats against indomethicin-induced ulcer formation.[20]

Balm oil has shown *in vitro* antitrypanosomal[21] and antibacterial activity against *Mycobacterium phlei* and *Streptococcus hemolytica*,[5] as well as antifungal activity,[22] including activity against food spoilage yeasts.[23] *In vitro* antihistaminic and antispasmodic activities from the oil have also been reported. Antispasmodic activity is attributed to the presence of eugenol acetate[5,24] and to citral. Both the essential oil and citral inhibited acetylcholine- and serotonin-induced contractions of rat ileum.[14] *In vitro* CNS receptor binding studies of lemon balm extracts have found nicotinic[25] and muscarinic receptor activities.[26]

A randomized double-blind, placebo-controlled clinical trial of the essential oil as an aromatherapy in the treatment of patients with severe dementia found significant improvements in quality of life and agitation scores from the oil compared to placebo.[27] Another placebo-controlled, randomized double-blind trial examined the benefits of a liquid extract preparation of lemon balm (60 drops/day) in patients with mild to moderate Alzheimer's disease. Patients in the active treatment groups showed significantly less agitation and improved cognitive function compared to placebo.[28] An acute 600 mg dose of an encapsulated extract of lemon balm in healthy young adults in a placebo-controlled, double-blind study found significant improvement in their accuracy of attention and memory functions and increased calmness.[29] In a similar study, an acute 1600 mg dose increased calmness and improved memory scores.[26] Placebo-controlled, double-blind randomized clinical trials of topical lemon balm cream preparations (1% dried extract of the leaves) showed significant benefits in the treatment of herpes simplex.[30,31]

USES

Medicinal, Pharmaceutical, and Cosmetic. Used in numerous European pharmaceutical preparations as a carminative and mild tranquilizer; also used in cough drops oil more often used as a component in perfumes; commonly used in lip balms.

Food. The monoterpene derivative citral, composed of neral and geranial, is widely used in cosmetics and foods to lend a lemon-like aroma and flavor.[1] Balm extract and oil are used in major categories of food products such as alcoholic (bitters, vermouths, etc.) and nonalcoholic beverages, frozen dairy desserts, candy, baked goods, and gelatins and puddings. Highest average maximum use level reported is 0.5% of extract in baked goods.

Dietary Supplements/Health Foods. Cut and sifted herb, powdered herb, liquid and dried extracts, infusions, tinctures, and so on, used as mild sleep aid as well as a stomachic; dried leaves used for tea in doses of 1.5–4.5 g of the herb in infusion; often used in combination with other herbs (BLUMENTHAL 1).

Traditional Medicine. Regarded in European folklore for the treatment of melancholy and for enhancing the memory; Greek physicians used the plant to treat wounds (GRIEVE) and in Iranian folk medicine lemon balm is used in treating gastrointestinal disorders and for analgesic, asntispasmodic, sedative, diuretic, digestive, and carminitive effects.[14]

COMMERCIAL PREPARATIONS

Crude; extracts, and oil; oil is seldom unadulterated, (ARCTANDER). Crude formerly official in U.S.P.

Regulatory Status. Regulated in the United States as a dietary supplement; both lemon balm (§182.10) and the essential oil, extractive, and solvent-free oleoresin are GRAS for use in foods (§182.20). Formerly official in the U.S.P. from 1840 to 1890. The leaves and preparations thereof are the subject of a positive German therapeutic monograph, indicated for difficulty in falling asleep caused by nervous conditions, and functional gastrointestinal symptoms (BLUMENTHAL 1).

REFERENCES

See the General References for APPLEQUIST; ARCTANDER; BARNES; BARRETT; BIANCHINI AND CORBETTA; BLUMENTHAL 1; DER MARDEROSIAN AND BEUTLER; FEMA; GOSSELIN; GRIEVE; GUENTHER; LUST; MARTINDALE; MCGUFFIN 1 & 2; ROSE; UPHOF; WICHTL; YOUNGKEN.

1. F. Enjalbert et al., *Fitoterapia*, **54**, 59 (1983).
2. N. Stankeviciene et al., *Polez. Rast. Priblat. Respub. Beloruss., Mater. Nauch. Konf.*, **2** (1973), 264 (1973).
3. S. Kapetanovic and S. Dugumovic, *Acta Pharm. Jugosl.*, **18**(304), 127 (1968).
4. F. W. Hefendehl, *Arch. Pharm. (Weinheim)*, **303**, 345 (1970).
5. H. Wagner and L. Sprinkmeyer, *Dtsch. Apoth. Ztg.*, **113**, 1159 (1973).
6. H. Thieme and C. Kitze, *Pharmazie*, **28**, 69 (1973).
7. L. S. Kucera and E. C. Herrmann Jr., *Proc. Soc. Exp. Biol. Med.*, **124**, 865 (1967).
8. E. C. Herrmann Jr. and L. S. Kucera, *Proc. Soc. Exp. Biol. Med.*, **124**, 869 (1967).

9. I. Morelli, *Boll. Chim. Farm.*, **116**, 334 (1977).
10. U. Gerhardt and A. Schroeter, *Fleischwirtschaft*, **63**, 1628 (1983).
11. J. Patora et al., *Acta Pol. Pharm.*, **59**, 139 (2002).
12. A. Mulkens et al., *Pharm. Acta Helv.*, **60** (9–10), 276 (1985).
13. M. Burgett, *Bee World*, **61**(2), 44 (1980).
14. H. Sadraei et al., *Fitoterapia*, **74**, 445 (2003).
15. K. Yamasaki et al., *Biol. Pharm. Bull.*, **21**, 829 (1998).
16. Z. Dimitrova et al., *Acta Microbiol. Bulg.*, **29**, 65 (1993).
17. J. Hohmann et al., *Planta Med.*, **65**, 576 (1999).
18. M. Auf'mkolk et al., *Endocrinology (Baltimore)*, **115**, 527 (1984).
19. F. Santini et al., *J. Endocrinol. Invest.*, **26**, 950 (2003).
20. M. T. Khayyal et al., *Arzneim.-Forsch.*, **51**, 545 (2001).
21. J. Mikus et al., *Planta Med.*, **66**, 366 (2000).
22. C. Araujo et al., *J. Food Prot.*, **66**, 625 (2003).
23. A. M. Debelmas and J. Rochat, *Plant Med. Phytother.*, **1**, 23 (1967).
24. G. Wake et al., *J. Ethnopharmacol.*, **69**, 105 (2000).
25. D. O. Kennedy et al., *Neuropsychopharmacology*, **28**, 1871 (2003).
26. C. G. Ballard et al., *J. Clin. Psychiatry*, **63**, 553 (2002).
27. S. Akhondzadeh et al., *J. Neurol. Neurosurg. Psychiatry*, **74**, 863 (2003).
28. D. O. Kennedy et al., *Pharmacol. Biochem. Behav.*, **72**, 953 (2002).
29. R. Koytchev et al., *Phytomedicine*, **6**, 225 (1999).
30. R. H. Wölbling and K. Leonhardt, *Phytomedicine*, **1**, 25 (1994).
31. A. P. Carnat et al., *Pharm. Acta Helv.*, **72**, 301 (1998).

BALM OF GILEAD BUDS

Source: *Populus tacamahacca* Mill. (syn. *P. balsamifera* Du Roi) or *P. candicans* Ait. (Family Salicaceae).

Common/vernacular names: Poplar buds and balsam poplar buds.

GENERAL DESCRIPTION

There has been much confusion regarding the sources of this botanical. In N.F. XI, it was officially described as derived from *P. tacamahacca* Mill. (syn. *P. balsamifera* L.) or from *P. candicans* Ait. However, according to SARGENT, *P. balsamifera* L. is not the same as *P. tacamahacca* Mill. but is a different species; he listed *P. balsamifera* Du Roi as the synonym for *P. tacamahacca*. Balm of Gilead has also been used as a synonym for Canada balsam (*Abies balsamea*) and Mecca balsam (*Balsamodendron opobalsamum* Engl.), among others.

Populus tacamahacca (balsam poplar) is a tree often up to 33 m high, trunk about 3 m in diameter, with stout, erect branches.

Populus candicans (true balm of Gilead) has been considered as a variety of the balsam poplar; only the pistillate tree is known.

Populus balsamifera L. is described as a large tree with massive spreading branches and stout yellow-brown, often angular branchlets.

All above three species are native to North America (SARGENT).

Balm of Gilead buds are the leaf buds collected in the spring before they open. Precise source is not certain; all above species as well as other *Populus* species are probably used.

CHEMICAL COMPOSITION

Contains about 2% volatile oil; resins; C_{25}, C_{27}, and C_{29} *n*-alkanes; salicin and populin; phenolic acids (e.g., caffeic); chalcones; and others. Compounds reportedly present in the volatile oil include *d*-cadinene, cineole, *ar*-curcumene, bisabolene, farnesene, *d*-α-bisabolol, β-phenethyl alcohol, acetophenone (KARRER),[1–3] and humulene (α-caryophyllene) (CLAUS; FURIA AND BELLANCA; YOUNGKEN).

PHARMACOLOGY AND BIOLOGICAL ACTIVITIES

Is considered to have stimulant and expectorant properties.

Salicin (a glucoside of salicyl alcohol) has antipyretic, antirheumatic, analgesic, and other properties as well as toxicity of salicylates (GOODMAN AND GILMAN; MARTINDALE).

USES

Medicinal, Pharmaceutical, and Cosmetic. Its major use is in cough preparations often together with white pine and wild cherry barks, bloodroot, and spikenard root, as in white pine compound or its variations.

Traditional Medicine. Used for relieving minor aches and pains; in colds and coughs; locally for sores, bruises and cuts, and for healing pimples. Bark and leaves are also similarly used.

COMMERCIAL PREPARATIONS

Crude and extracts. Crude was formerly official in N.F. and U.S.P.

Regulatory Status. Has been approved for food use in alcoholic beverages only (§172,510). Subject of a German therapeutic monograph; allowed in external preparations for superficial skin injuries, hemorrhoids; frostbite, and sunburn.[4]

REFERENCES

See the General References for ARCTANDER; CLAUS; GOSSELIN; KROCHMAL AND KROCHMAL; LUST; MCGUFFIN 1 & 2; MERCK; POUCHER; ROSE; SARGENT; UPHOF; YOUNGKEN.

1. E. Wollenweber and W. Weber, *Z. Pflanzenphysiol.*, **69**, 125 (1973).
2. S. Frantisek et al., *Collect. Czech. Chem. Commun.*, **18**, 364 (1953).
3. O. Isaac et al., *Dtsch. Apoth. Ztg.*, **108**, 293 (1968).
4. Monograph *Populi gemma. Bundesanzeiger*, no. 22 (February 1, 1990).

BALSAM CANADA

Source: *Abies balsamea* (L.) Mill. (Family Pinaceae).

Common/vernacular names: Balm of Gilead, balsam fir, balsam of fir, balsam fir Canada, Canada balsam, Canadian balsam, and Canada turpentine.

GENERAL DESCRIPTION

Balsam Canada is an oleoresin occurring normally in the bark of *Abies balsamea*, collected by puncturing the vesicles on the bark. It is a light greenish-yellow, viscous liquid that solidifies on exposure to air. *A. balsamea* is an evergreen tree up to 20 m high with trunk usually 30–45 cm in diameter; native to eastern North America, reaching Minnesota and Wisconsin. Canada balsam is not a true balsam because it does not contain benzoic or cinnamic acid or their esters (see *glossary*).

CHEMICAL COMPOSITION

The bark of *A. balsamea* contains a volatile oil (up to 30%) and an odorless resin. The volatile oil is composed entirely of monoterpenes (37.4% β-phellandrene, 36.4% β-pinene, 23.5% α-pinene, and 2.7% α-phellandrene).[1] The resin makes up the remainder of the oleoresin and contains neutral and acidic materials (e.g., abietic and neoabietic acids; diterpenoids).

PHARMACOLOGY AND BIOLOGICAL ACTIVITIES

In guinea pigs, oral administration of abietic acid (an abietane diterpenoid found in the oleoresin) produced antiallergic activity. *In vitro* inhibition of 5-lipoxygenase by abietic acid may be a contributing factor to the activity.[2]

The essential oil (balsam fir oil) extracted from the foliage has shown *in vitro* cytotoxicity against several tumor cell lines. The main active constituent was identified as α-humulene.[3]

TOXICOLOGY

Oleoresin considered as nontoxic when applied externally.[4] Internal toxicity data not available.

USES

Medicinal, Pharmaceutical, and Cosmetic. Oleoresin used in certain ointments and creams as antiseptic and as treatment for hemorrhoids; as a fixative or fragrance ingredient in soaps, detergents, creams, lotions, and perfumes. Maximum use levels reported are 0.15% in soaps and 0.2% in perfumes.[4] Formerly used in dentistry as an ingredient in root canal sealers and dentifrices (ADA).[5]

Food. Both the oleoresin and oil have had limited use in major categories of foods, including alcoholic and nonalcoholic beverages, frozen dairy desserts, candy, and gelatins and puddings. Use levels have been low, generally below 0.001% (10 ppm).

Traditional Medicine. Bark resin used externally by American Indians for burns, sores, and cuts and to relieve heart and chest pains. Also reportedly used in treating tumors.[6]

Others. Due to its ability to dry to a brittle, clear glass-like residue, Canada balsam (usually freed from volatile oil and dissolved in xylene) was extensively used as a cement for lenses and prepared microscopic slides. Its use in some "balsam" hair grooming products probably takes advantage of this property to stiffen hair and give it "body."

COMMERCIAL PREPARATIONS

Oleoresin and oil. Oleoresin was formerly official in U.S.P.

Regulatory Status. Only needles and twigs of *A. balsamea* and their appropriate preparations are approved for food use as natural flavoring substances (§172.510).

REFERENCES

See the General References for ARCTANDER; CLAUS; FEMA; GUENTHER; KROCHMAL AND KROCHMAL; MERCK; SARGENT; UPHOF.

1. H. J. Petrowitz et al., *Riechst., Aromen, Körperpflegem.*, **12**, 1 (1962).
2. N. N. Ulusu et al., *Phytother. Res.*, **16**, 88 (2003).

3. J. Legault et al., *Planta Med.*, **69**, 402 (2003).
4. D. L. J. Opdyke, *Food Cosmet. Toxicol.*, **13**, 449 (1975).
5. K. M. Kosti, U.S. Pat. 4,348,378 (1982).
6. J. L. Hartwell, *Lloydia*, **33**, 288 (1970).

BALSAM COPAIBA

Source: *Copaifera officinalis* (Jacq.) L. or other South American *Copaifera* species (Family Leguminosae or Fabaceae).

Common/vernacular names: Copaiba, copaiva, and Jesuit's balsam.

GENERAL DESCRIPTION

Copaiba balsam is an oleoresin that accumulates in cavities within the tree trunk of *C. officinalis* and is tapped by drilling holes into the wood of the trunk; it is not a true balsam (see **glossary**). *C. officinalis* is a tree with a height of up to 18 m found in tropical South America (particularly Brazil, Colombia, and Venezuela).

Copaiba oil is obtained by direct vacuum distillation of the oleoresin containing large amounts of the volatile oil (60–90%) (ARCTANDER).

CHEMICAL COMPOSITION

Copaiba balsam contains 30–90% volatile oil; the rest being resins and acids (ARCTANDER; YOUNGKEN).[1] Compounds reported in the oleoresin include caryophyllene (major component of the volatile oil), copaene, β-bisabolene, γ-humulene, caryophyllene oxide, α-ylangene, α-multijugenol;[2,3] and terpenic acids such as copalic, copaiferic, copaiferolic, Hardwick, 7-hydroxy-Hardwick, enantioagathic, and eperu-8(20)-en-15,18-dioic acids, among others.[3–6]

PHARMACOLOGY AND BIOLOGICAL ACTIVITIES

The oil has shown *in vitro* antibacterial activity,[1,7] and the oleoresin from Brazilian *Copaifera* species has shown anti-inflammatory activity following oral administration in rats.[8]

TOXICOLOGY

Available data indicate that the oleoresin is relatively nontoxic;[1,9] the oral LD_{50} in rats is 3.79 mL/kg. At lower dosages in rats (1.92 and 2.86 mL/kg), subacute toxicity was seen in the form of significant reductions in body weight and food and water intake, in addition to diarrhea, gastric irritation, sialorrhea, and symptoms of CNS depression; however, at 30 days these symptoms were no longer significant relative controls.[8] Large doses are reported to cause vomiting and diarrhea as well as measles-like rash (MARTINDALE; SAX).

USES

Medicinal, Pharmaceutical, and Cosmetic. The oleoresin and oil have occasionally been used in pharmaceutical preparations (diuretics, cough medicines, etc.). Both copaiba balsam and oil are widely used in cosmetic preparations (soaps, bubble baths, detergents, creams, lotions, and perfumes), the former primarily as a fixative and the latter as a fragrance component. Maximum use level of the oil or oleoresin in perfumes is 0.8%.[1,9,10]

Food. Occasionally used as a flavoring component in most major categories of foods, including alcoholic and nonalcoholic beverages, frozen dairy desserts, candy, baked goods, gelatins and puddings, and meat and meat products. Average maximum use levels are usually very low, less than 0.002% (16 ppm).

Traditional Medicine. The oleoresin has been used for several centuries in Europe in the treatment of chronic cystitis, bronchitis, hemorrhoids, chronic diarrhea, and other conditions for which it has also been used in Latin America (MORTON 1). Traditional uses in tropical America also include the treatment of hypertension, cancer, wounds, dysentery, skin diseases, vesical catarrh, pneumonia, urinary tract infections, leucorrhea, blenorrhagia, psoriasis, dressing the navel of newborns, and as an anti-inflammatory for sore throat. In large doses, copaiba balsam is believed to provoke diarrhea, nausea, vomiting, and colic.[11]

COMMERCIAL PREPARATIONS

Oleoresin ("balsam") and oil. Copaiba balsam was formerly official in N.F., and oil is official in F.C.C.

Regulatory Status. Approved for food use as a natural flavoring (§172.510).

REFERENCES

See the General References for ARCTANDER; CLAUS; FEMA; GUENTHER; LEWIS AND ELVIN-LEWIS; MARTINDALE; MCGUFFIN 1 & 2; MORTON 1; TERRELL; YOUNGKEN.

1. D. L. J. Opdyke, *Food Cosmet. Toxicol.*, **14**(Suppl.), 687 (1976).
2. G. Delle Monache et al., *Tetrahedron Lett.*, **8**, 659 (1971).
3. M. Ferrari et al., *Phytochemistry*, **10**, 905 (1971).
4. F. Delle Monache et al., *Ann. Chim. (Rome)*, **59**, 539 (1969).
5. F. Delle Monache et al., *Ann. Chim. (Rome)*, **60**, 233 (1970).
6. J. R. Mahajan and G. A. L. Ferreira, *Ann. Acad. Bras. Cienc.*, **43**, 611 (1971).
7. J. C. Maruzzella and N. A. Sicurella, *J. Am. Pharm. Assoc.*, **49**, 692 (1960).
8. A. C. Basile, et al., *J. Ethnopharmacol.*, **22**, 101 (1988).
9. D. L. J. Opdyke, *Food Cosmet. Toxicol.*, **11**, 1075 (1973).
10. M. J. Del Nunzio, *Aerosol Cosmet.*, **7**(41), 7 (1985).
11. W. B. Mors et al. *Medicinal Plants of Brazil*, Reference Publications, Algonac, MI, 2000.

BALSAM FIR OREGON

Source: *Pseudotsuga taxifolia* (Lam.) Britt. (syn. *P. douglasii* Carr.; *P. mucronata* (Raf.) Sudw.) (Family Pinaceae).

Common/vernacular names: Balsam fir, balsam Oregon, Douglas fir, Douglas spruce, Oregon balsam, and red fir.

GENERAL DESCRIPTION

Oregon balsam is an oleoresin that occurs in the tree trunk and is usually collected from felled trees. It is a light amber or yellow, viscous liquid with a piney odor. It is not a true balsam (see *glossary*). The tree varies greatly in size, depending on localities; often up to 60 m high with a trunk of approximately

1 m in diameter; native to western North America, particularly the Pacific Coast (California, Oregon, Washington, and British Columbia); cultivated in Europe.

CHEMICAL COMPOSITION

Although considerable chemical data are available on the needles, wood, and bark (KARRER),[1–6] chemical information on the "balsam" (oleoresin) itself is practically nonexistent, except that it has been reported to be an oleoresin of the turpentine type, yielding a volatile oil on steam distillation and has properties and uses similar to those of Canada balsam (ARCTANDER; CLAUS).

PHARMACOLOGY AND BIOLOGICAL ACTIVITIES

Data on its pharmacological or toxicological properties are not available.

USES

Similar to those of Canada balsam; used as an adulterant of the latter oleoresin; however, Oregon balsam is not suited for use in microscopy because it is less viscous and slower drying than Canada balsam; it does not dry to a glassy and brittle film as Canada balsam.

COMMERCIAL PREPARATION

Oleoresin ("balsam").

REFERENCES

See the General References for ARCTANDER; GOSSELIN; GUENTHER; SARGENT; UPHOF; YOUNGKEN.

1. E. Von Rudloff, *Can. J. Bot.*, **50**, 1025 (1972).
2. E. Von Rudloff, *Pure Appl. Chem.*, **34**, 401 (1973).
3. H. M. Graham and E. F. Kurth, *Ind. Eng. Chem.*, **41**, 409 (1949).
4. E. F. Kurth and H. J. Klefer, *Tappi*, **33**, 183 (1950).
5. H. L. Hergert and E. F. Kurth, *Tappi*, **35**, 59 (1952).
6. R. D. Kolesnikova et al., *Khim. Prir. Soedin.*, **5**, 613 (1976).

BALSAM PERU

Source: *Myroxylon pereirae* Klotzsch (syn. *M. balsamum* var. *pereirae* (Royle) Harms) (Family Leguminosae or Fabaceae).

Common/vernacular names: Balsam of Peru, balsam-of-Peru tree, black balsam, Indian balsam, Peru balsam, and Peruvian balsam.

GENERAL DESCRIPTION

Balsam of Peru is obtained from the exposed wood of *M. pereirae* after strips of bark are removed from the tree trunk. The exudation is soaked up by rags wrapped around the trunk that are then boiled with water. In its commercial crude form, balsam of Peru is a dark-brown, viscous liquid, reddish brown and transparent in thin layers, with an aromatic vanilla-like odor and a bitter acrid taste. It is a true balsam (see ***glossary***). The balsam sinks to the bottom and is separated.

Myroxylon pereirae is a large tree growing up to about 25 m in height and is native to Central America.

CHEMICAL COMPOSITION

Contains not more than 70% w/w and not less than 45% of esters, maily benzylcinnamate and benzyl benzoate, and 50–64% of a high-boiling volatile oil, referred to as cinnamein; and 20–28% resin. The volatile oil consists mainly of benzoic and cinnamic acid esters such as benzyl benzoate, benzyl cinnamate, and cinnamyl cinnamate (styracin), with small amounts of nerolidol, free benzyl alcohol, and free benzoic and cinnamic acids also present. Other constituents include traces of styrene, vanillin, and coumarin (KARRER; MARTINDALE; REMINGTON).

PHARMACOLOGY AND BIOLOGICAL ACTIVITIES

Balsam of Peru has mild antiseptic and antibacterial properties and is believed to promote the growth of epithelial cells; also antiparasitic (especially for scabies) (BLUMENTHAL 1).

TOXICOLOGY

Balsam Peru is one of the most common contact allergens. Dermatitis as a result of contact with this balsam is documented in many countries.[1–4] A double-blind, placebo-controlled study found that taken orally, balsam of Peru caused allergic dermatitis to flare.[5]

USES

Medicinal, Pharmaceutical, and Cosmetic. Balsam of Peru is used extensively in topical preparations for the treatment of wounds, skin graft healing, indolent ulcers, scabies, diaper rash, hemorrhoids, anal pruritus, bedsores, intertrigo, eczema, and others; in hair tonic and antidandruff preparations, feminine hygiene sprays, and as a fixative or fragrance ingredient in soaps, detergents, creams, lotions, and perfumes, with maximum use level up to 0.8% in perfumes. Balsam of Peru oil and resinoid, obtained by high-vacuum distillation and/or solvent extraction of the balsam, are also used in cosmetics.[1,6,7]

Balsam of Peru is also used in dental preparations, especially for the treatment of dry socket (postextraction alveolitis) and as a component in certain dental impression materials, and dentifrices.[8]

Food. Balsam and oil are extensively used as a flavor ingredient in major categories of foods, including alcoholic and nonalcoholic beverages, frozen dairy desserts, candy, baked goods, and gelatins and puddings, with the highest reported average maximum use level of about 0.0015% (15.33 ppm) for the balsam in candy.

Traditional Medicine. Reportedly used in treating cancer.[9]

COMMERCIAL PREPARATIONS

Balsam, resinoid, and oil. Balsam of Peru was formerly official in N.F. XII.

Regulatory Status. Approved in the United States for topical OTC preparations, cosmetics and dietary supplements (WICHTL). Extractive, essential oil, and solvent-free oleoresin are GRAS for use in foods (§182.20). Subject of a German therapeutic monograph; externally preparations allowed for infected or poorly healing wounds, burns, ulcers, frost-bite, ulcus cruris, bruises, hemorrhoids (BLUMENTHAL 1; WICHTL).

REFERENCES

See the General References for ADA; ARCTANDER; BLUMENTHAL 1; DERMARDEROSIAN AND BEUTLER; FEMA; GOSSELIN; GRIEVE; GUENTHER; JIANGSU; MARTINDALE; PHILLIPS; TERRELL; YOUNGKEN.

1. D. L. J. Opdyke, *Food Cosmet. Toxicol.* **12**(Suppl.), 951 (1974).
2. B. M. Hausen, *Am. J. Contact Dermat.*, **12**, 93 (2001).
3. O. Hammershoy, *Contact Dermatitis*, **6**, 263 (1980).
4. Y. Olumide, *Contact Dermatitis*, **17**, 85 (1987).

5. N. K. Veien et al., *Contact Dermatitis*, **12**, 104 (1987).
6. D. L. J. Opdyke, *Food Cosmet. Toxicol.*, **12**(Suppl.), 953 (1974).
7. S. N. Carson et al., *Ostomy Wound Manage.*, **49**, 60 (2003).
8. K. M. Kosti, U.S. Pat. 4,348,378 (1982).
9. J. L. Hartwell, *Lloydia*, **33**, 97 (1970).

BALSAM TOLU

Source: *Myroxylon balsamum* Harms (syn. *M. toluiferum* H. B. and K.) (Family Leguminosae or Fabaceae).

Common/vernacular names: Balsam tolu, opobalsam resin tolu, Thomas balsam, and tolu balsam.

GENERAL DESCRIPTION

Balsam tolu is obtained from the tree trunk of *M. balsamum* by making V-shaped incisions through the bark and sap wood. The liquid balsam is collected in gourds and solidifies on aging. The balsam is a plastic solid with a brown or brownish-yellow color that darkens and hardens on aging. Like balsam Peru, it is a true balsam. Tolu balsam has an aromatic vanilla-like odor and an aromatic, mildly pungent taste; it is insoluble in water but soluble in alcohol, acetone, benzene, and chlorinated hydrocarbons.

Myroxylon balsamum is a tall tree native to northern South America (Colombia, Peru, Venezuela), cultivated in the West Indies.

CHEMICAL COMPOSITION

Contains resin, free cinnamic and benzoic acids, and volatile oil composed mainly of esters of these acids with small amounts of terpenes. Concentrations of these constituents vary greatly with each report, probably due to the great differences in quality of commercial products and the lack of requirements for determining specific components in official compendia (e.g., U.S.P.). Other constituents present include guaiadienes, triterpene acids (oleanolic acid, sumaresinolic acid, etc.), other triterpenoids (e.g., 20R, 24ξ_2-ocotillone), and traces of eugenol and vanillin, among many other compounds (CLAUS; KARRER; MERCK; REMINGTON).[1–5] None of the recent reports identified with certainly the sources of the samples investigated, with the exception of some museum specimens that were shown to contain benzyl cinnamate not found in modern commercial samples.[2,6]

PHARMACOLOGY AND BIOLOGICAL ACTIVITIES

It has mild antiseptic and expectorant properties.

TOXICOLOGY

Allergic reactions to tolu balsam occur in some individuals.[3,7]

USES

Medicinal, Pharmaceutical, and Cosmetic. Balsam tolu is extensively used as a flavor and mild expectorant in cough medicines (e.g., syrups, lozenges, etc.); also used as an ingredient in Compound Benzoin Tincture or similar formulations for treatment of bedsores, cracked nipples, lips, and minor cuts on the skin and for inhalation to treat laryngitis and croup. Both balsam and its oil (obtained by dry or steam distillation) are used as a fixative or fragrance ingredient in cosmetics including soaps, detergents, creams, lotions, and perfumes, with maximum use levels of 0.1% in soaps and 0.2% in perfumes reported.[3]

Food. Used as a flavor ingredient in chewing gum and other major categories of food products, such as alcoholic and nonalcoholic beverages, frozen dairy desserts, candy, baked goods, and gelatins and puddings. Use levels are usually rather low.

Traditional Medicine. Reportedly used in treating cancer.[7] Resin used in Peruvian folk-medicine in treating asthma, bronchitis, catarrh, colds, rheumatism, wounds, sores, sprains, headache, veneral diseases, and fevers.[8,9]

COMMERCIAL PREPARATIONS

Crude, resinoid, extracts, and tincture. Tincture is official in N.F.

Regulatory Status. Approved for food use as a natural flavoring (§172.510). Subject of a German therapeutic monograph; allowed in preparations with a mean daily dose of 0.6 g for catarrhs of the respiratory tract (BLUMENTHAL 1).

REFERENCES

See the General References for ARCTANDER; BLUMENTHAL 1; CLAUS; FEMA; GUENTHER; UPHOF; USD 26th; YOUNGKEN.

1. I. Wahlberg et al., *Acta Chem. Scand.*, **25**, 3285 (1971).
2. K. J. Harkiss and P. A. Linley, *J. Pharm. Pharmacol.*, **25**(Suppl.), 146P (1973).
3. D. L. J. Opdyke, *Food Cosmet. Toxicol.*, **14**(Suppl.), 689 (1976).
4. I. Wahlberg and C. R. Enzell, *Acta Chem. Scand.*, **25**, 70 (1971).
5. H. D. Friedel and R. Matusch, *Helv. Chim. Acta*, **70**, 1616 (1987).
6. K. J. Harkiss, personal communication.
7. T. N. Salam and J. F. Fowler Jr., *J. Am. Acad. Dermatol.*, **45**, 377 (2001).
8. J. L. Hartwell, *Lloydia*, **33**, 97 (1970).
9. J. A. Duke and R. Vasquez, *Amazonian Ethnobotanical Dictionary*, CRC Press, Boca Raton, FL, 1994, p. 121.

BARBERRY

Source: *Berberis vulgaris* L., *Mahonia aquifolium* Nutt. (syn. *B. aquifolium* Pursh), or other *Berberis* species (Family Berberidaceae).

Common/vernacular names: Barberry, berberis, common barberry, and European barberry (*B. Vulgaris*); berberis, Oregon grape, and trailing mahonia (*Mahonia aquifolium*).

GENERAL DESCRIPTION

Berberis vulgaris is a deciduous spiny shrub that may reach 5 m in height; native to Europe, naturalized in eastern North America.

Berberis aquifolium is an evergreen spineless shrub, 1–2 m high; native to the Rocky Mountains, extending to British Columbia and California.

Parts used are the dried rhizome and roots (*B. aquifolium*) and barks of stem and root (*B. vulgaris*).

CHEMICAL COMPOSITION

Berberis spp. are rich in isoquinoline alkaloids; those in *B. vulgaris* include berberine, berbamine, oxyacanthine, jatrorrhizine, columbamine, palmatine, isotetrandine (berbamine methyl ether), bervulcine, and magnoflorine;[1] those in *B. aquifolium* include aromoline, obamegine, oxyberberine, berbamine, and oxyacanthine.[2]

Berberine and other *Berberis* alkaloids were reported to be toxic to seedlings of dog rose and horse chestnut, causing atrophy.[3]

PHARMACOLOGY AND BIOLOGICAL ACTIVITIES

The total ethanol extract of the root of *B. vulgaris* administered i.p. inhibited acute inflammation in rats more potently than berberine, oxyacanthine, or three alkaloidal fractions of the roots. It was also more potent at inhibiting chronic inflammation in a model of adjuvant arthritis in rats.[4] An aqueous extract of the fruits of *B. vulgaris* exhibited *in vitro* anticholinergic and antihistaminergic activity.[5]

Certain *Berberis* alkaloidal salts, particularly berberine, oxyacanthine, and columbamine, have bactericidal activities.[6,7] In one study, berberine chloride showed higher activity than chloramphenicol against *Staphylococcus epidermidis, Neisseria meningitidis, Escherichia coli*, and other bacteria. Oxyacanthine chloride at 0.01% and columbamine chloride at 1.0% killed *Bacillus subtilis* and *Colpidium colpoda*.[7]

Some reports have stated berberine sulfate to be amebicidal, and trypanocidal and berbamine, isotetrandine, and hydroxyacanthine to have hypotensive properties (GLASBY 1; MARTINDALE).[8]

Berberine has antifibrillatory activity, elevating the ventricular fibrillation threshold to electrical stimulation in anesthetized cats.[9] Berberine also has anticonvulsant, sedative, uterine stimulant, and numerous other activities (see *goldenseal*).

A fraction of a root extract of *B. vulgaris* containing 80% berbamine and three unidentified isoquinoline alkaloids has shown spasmolytic effects on smooth muscles.[10]

A study of berbamine in mice infected with influenza viruses yielded results that indicate it to be an immunostimulating agent.[11] It also exhibits various cardiovascular effects, including hypotensive and antiarrhythmic.[12]

USES

Medicinal, Pharmaceutical, and Cosmetic. Berberis is used as an ingredient in certain tonic preparations. Berberine salts have been used in ophthalmic products, usually in eye drops and eyewashes.

Dietary Supplements/Health Foods. Crude root, cut and sifted and powdered used in capsules, teas, and other products, primarily as a bitter tonic (FOSTER AND DUKE).

Traditional Medicine. In Bulgaria, the stem bark and roots of *B. vulgaris* are used to treat arthritis and chronic inflammatory conditions of the kidneys, liver, and gallbaldder.[4] In Europe and the United States, *Berberis* is reportedly used as a bitter tonic, antipyretic, and antihemorrhagic, usually referring to *B. vulgaris*. In China, various *Berberis* species are also used for similar purposes; *B. vulgaris* is listed as a related drug, sometimes used as an adulterant, to *Dichroa febrifuga* Lour., which is widely used in China for its antimalarial and antipyretic activities (NANJING).

COMMERCIAL PREPARATIONS

Barberry (*B. vulgaris*) is available in crude and extract form. Qualities of extracts may vary because the only standards for potency are strengths (see *glossary*) based on a weight-to-weight ratio of extract to crude drug. Crude was formerly official in U.S.P.

Regulatory Status. Root of *B. vulgaris* is the subject of a German therapeutic monograph in which the uses for the claimed therapeutic applications are not recommended owing to lack of documented benefits; such applications including the use of the bark, root bark or fruits for gastrointestinal, kidney, liver, urinary tract, circulatory, spleen, bronchial, and other organ functions (BLUMENTHAL 1).

REFERENCES

See the General References for APPLEQUIST; BAILEY 2; BLUMENTHAL 1; CLAUS; DER MARDEROSIAN AND BEUTLER; FARNSWORTH 3; FOGARTY; FOSTER AND DUKE; GOSSELIN; GRIEVE; JIANGSU; MCGUFFIN 1 & 2; MERCK; NANJING.

1. M. Ikram, *Planta Med.*, **28**, 353 (1975).
2. D. Kostalova et al., *Chem. Pap.*, **40**, 389 (1986).
3. V. M. Oleksevich, *Introd. Eksp. Ekol. Rosl.*, **1**, 224 (1972).
4. N. Ivanovska and S. Philipov, *Int. J. Immunopharmacol.*, **18**, 553 (1996).
5. F. Shamsa et al., *J. Ethnopharmacol.*, **64**, 161 (1999).
6. Z. Kowalewski et al., *Arch. Immunol. Ther. Exp.*, **20**, 353 (1972).
7. E. Andronescu et al., *Clujul Med.*, **46**, 627 (1973).
8. L. P. Naidovich et al., *Farmatsiya (Moscow)*, **25**, 33 (1976).
9. D. C. Fang et al., *Zhongguo Yaoli Xuebao*, **7**, 321 (1986).
10. P. Manalov et al., *Eksp. Med. Morfol.*, **24**, 41 (1985).
11. L. Jin and W. Z. Sui, *Zhongguo Yaoli Xuebao*, **7**, 475 (1986).
12. F. L. Li, *Yaoxue Xuebao*, **20**, 859 (1985).

BASIL, SWEET

Source: *Ocimum basilicum* L. (Family Labiatae or Lamiaceae).

Common/vernacular names: Basil, common basil, and sweet basil.

GENERAL DESCRIPTION

Annual herb, about 0.5 m high, thought to be native to Africa and tropical Asia; cultivated worldwide (e.g., Europe, India, and the United States). There are many varieties, some of which have different compositions and flavoring characteristics. The plant is also strongly affected by environmental factors such as temperature, geographic location, soil, and amount of rainfall.[1,2] Parts used are the dried leaves and flowering tops.

An essential oil is obtained by steam distillation. There are two major types of commercial basil oils, namely, the true sweet basil oil and the so-called exotic or Reunion, basil oil. True sweet basil oil is distilled in Europe and the United States; exotic basil oil is produced in the Comoro Islands, the Seychelles, and the Malagasy Republic. The two differ mainly in their contents of d-camphor, linalool, and estragole (methyl chavicol or 1-allyl-4-methoxybenzene). Generally, the former does not contain camphor and the latter contains little or no linalool; also the former is levorotatory and the latter is dextrorotatory.

CHEMICAL COMPOSITION

The volatile oil (ca. 0.08%) contains d-linalool and estragole as the major components, with the former up to 55% and the latter about 70%, depending on the sources (MASADA).[3,4] Other components include methyl cinnamate (reported to be the major component (ca. 28%) of a variety of sweet basil), 1,8-cineole, eugenol, borneol, ocimene, geraniol, anethole; 10-cadinols, β-caryophyllene, α-terpineol, camphor, 3-octanone, methyleugenol, safrole, sesquithujene, and 1-epibicyclosesquiphellandrene as well as juvocimene 1 and juvocimene 2, which are potent juvenile hormone mimics (JIANGSU).[4–8] There are great variations in concentrations of these components in the volatile oils from different sources.

Other constituents present in sweet basil include protein (14%), carbohydrates (61%), vitamins A and C in relatively high concentrations (MARSH), rosmarinic acid,[9,10] thymol, and xanthomicrol (a flavone).[11]

PHARMACOLOGY AND BIOLOGICAL ACTIVITIES

The volatile oil of a variety of sweet basil was shown to have antiwormal activities.[12] Essential oil has shown antimicrobial and mildly antiseptic activities *in vitro*.[13]

Methyl cinnamate, estragole and, to a lesser extent, ocimene, cineole, and linalool have insecticidal activities.[14]

Xanthomicrol has shown cytotoxic and antineoplastic activities.[11]

TOXICOLOGY

Sweet basil oil is reported to be nontoxic.[15]

Estragole, a major component in some sweet basil oils, has been shown to produce tumors (hepatocellular carcinomas) in mice[16] and genotoxicity. The Council of Europe currently recommends that the level of estragole in food products should not exceed 0.05 mg/kg.[17]

USES

Medicinal, Pharmaceutical, and Cosmetic. Used as a fragrance ingredient in perfumes, soaps, hair dressings, dental creams, and mouth washes.

Food. Used as a spice and in chartreuse liqueur.

The oil and oleoresin are extensively used as a flavor ingredient in all major food products, usually in rather low use levels (mostly below 0.005%).

Traditional Medicine. The herb has been used for head colds and as a cure for warts and worms, as an appetite stimulant, carminative, and diuretic, among other applications (BLUMENTHAL 1).[7]

More widely used as a medicinal herb in the Far East, especially in China and India. It was first described in a major Chinese herbal around AD 1060 and has since been used in China for spasms of the stomach and kidney ailments, among others; it is especially recommended for use before and after parturition to promote blood circulation. The whole herb is also used to treat snakebite and insect bites (JIANGSU; NANJING).

COMMERCIAL PREPARATIONS

Crude, essential oil, and oleoresin.

Regulatory Status. Regulated in the United States as a dietary supplement; in foods, both the use of the herb as a spice, natural flavoring, or seasoning (§182.10), and the use of the essential oil and extractives (§182.20) are GRAS. Subject of a German therapeutic monograph; claimed efficacies not well substantiated; allowed as flavor corrigent at 5% or less (BLUMENTHAL 1; WICHTL).

REFERENCES

See the General References for ARCTANDER; BISSET; BLUMENTHAL 1; BRUNETON; FEMA; FOSTER; GUENTHER; JIANGSU; MASADA; MCGUFFIN 1 & 2; MORTON 1; NANJING; ROSENGARTEN; TERRELL; YOUNGKEN.

1. D. Pogany, *Diss. Abstr. B*, **28**, 1871 (1967).
2. S. N. Sobti et al., *Lloydia*, **41**, 50 (1978).
3. B. C. Gulati et al., *Parfüm. Kosmet.*, **58**, 165 (1977).
4. S. S. Nigam and A. K. Rao, *Riechst., Aromen, Körperpflegem.*, **18**, 169 (1968).
5. S. J. Terhune et al., Paper presented at the 6th International Congress of Essential Oils, 1974, p. 153.

6. M. S. Karawya et al., *J. Agric. Food Chem.*, **22**, 520 (1974).
7. J. W. Hogg et al., *Am. Perfum. Cosmet.*, **86**, 33 (1971).
8. W. S. Bowers and R. Nishida, *Science*, **209** (4460), 1030 (1980).
9. U. Gerhardt and A. Schroeter, *Fleishwirtschaft*, **63**, 1628 (1983).
10. A. Reschke, *Z. Lebensm. Unters. Forsch.*, **176**, 116 (1983).
11. M. O. Fatope and Y. Takeda, *Planta Med.*, **54**, 190 (1988).
12. M. L. Jain and S. R. Jain, *Planta Med.*, **22**, 66 (1972).
13. K. J. Lachowicz et al., *Lett. Appl. Microbiol.*, **26**, 209 (1998).
14. R. S. Deshpande and H. P. Tipnis, *Pesticides*, **11**(5), 11 (1977).
15. D. L. J. Opdyke, *Food Cosmet. Toxicol.*, **11**, 867 (1973).
16. N. R. Drinkwater et al., *J. Natl. Cancer Inst.*, **57**, 1323 (1976).
17. H. Schulz et al., *J. Agric. Food Chem.*, **51**, 2475 (2003).

BAY, SWEET

Source: *Laurus nobilis* L. (Family Lauraceae).

Common/vernacular names: Bay, laurel, bay laurel, Grecian laurel, Mediterranean bay, sweet bay, and true bay.

GENERAL DESCRIPTION

Laurus nobilis is an evergreen tree, up to 20 m high, native to the Mediterranean region; extensively cultivated. Part used is the dried leaf.

An essential oil, commonly known as laurel leaf oil, is produced by steam distillation of the leaves and branchlets.

Its leafy branchlets were used in wreaths by the ancient Greeks and Romans to crown their victors.

There are several botanicals known under the name of *bay*. For example, West Indian bay is *Pimenta racemosa* (Mill.) J. W. Moore, and California bay is *Umbellularia californica* Nutt. The word *bay* in the literature may refer to any one of these botanicals, among others.

CHEMICAL COMPOSITION

Contains 0.3–3.1% volatile oil composed mainly of 1,8-cineole (30–50%), α-pinene (ca. 12%), linalool (ca. 11%), α-terpineol acetate (ca. 10%), α-terpineol, β-pinene, sabinene, limonene, methyleugenol (3,4-dimethoxyallylbenzene), eugenol, *p*-cymene, camphene, and dehydro-1,8-cineole, as well as phenyl-hydrazine, piperidine, and geraniol.[1] The essential oil from a supercritical carbon dioxide extraction contained mainly methyleugenol (8.1%), α-terpinyl acetate (11.4%), linalool (12.5%), and 1,8-cineole (22.8%).[2] Oil content is highest in autumn and lowest in spring, with old leaves containing the most oil.[3] Other constituents reported include costunolide, laurenobiolide (germacranolides), sesquiterpenes (santamarine, dehydrocostus lactone, zaluzanin D, reynosin), catechins, proanthocyanidins, quercetin, isoquercitrin,[4–7] alkaloids (reticuline, boldine, launobine, isodomesticine, neolitsine, nandigerine, etc.),[8] vitamin E,[9] and various acids (e.g., butyric, caproic, enanthic acids, etc.).[10–16]

PHARMACOLOGY AND BIOLOGICAL ACTIVITIES

Methyleugenol, a major constituent of sweet bay and California bay oils (at 4% and 5.4%, respectively), as well as a variety of West Indian bay and other species, has been reported to have sedative and narcotic properties in mice, producing sedation at low doses and reversible narcosis at higher doses; it prevented the death of mice treated with lethal convulsant doses of strychnine.[13]

Bay leaf and some of its volatile compounds (esp. cineole, phenylhydrazine, geraniol, and piperidine) have been shown to repel cockroaches.[17,18] Aqueous extracts of the leaves and flowers have shown toxicity to snails (*Biomphalaria glabrata*).[19]

The essential oil has shown bactericidal and fungicidal properties; it also depressed the heart rate and lowered blood pressure in animals. Formulations containing sweet bay leaf and its volatile oils have been claimed to have antidandruff activities.

In animal studies, the essential oil of the leaf has shown anticonvulsant activity against experimental seizures,[20] analgesic, and anti-inflammatory activities.[21] Aqueous extracts of the seeds and fruits have shown gastroprotective activity against ethanol-induced ulcers.[22,23] Alcohol absorption-inhibiting activity of the leaves in rats is attributed to various sesquiterpenes.[6,24] The bark has shown greater antioxidant activity than the leaves;[25] the alkyl peroxy radical scavenging activity of the leaves, which was higher than that of 120 other extracts of edible plants and herbs, is attributed to quercetin and isoquercitrin.[7] Growth suppression of various human leukemia cell lines was found *in vitro* from 1,8-cineole, the main constituent of the essential oil of the leaves.[26]

TOXICOLOGY

Allergic reactions (contact dermatitis) to sweet bay have been documented.[27–30] Methyleugenol is hepatotoxic.[31]

USES

Medicinal, Pharmaceutical, and Cosmetic. The oil is used mainly as a fragrance ingredient in creams, lotions, perfumes, soaps, and detergents. Maximum use level reported is 0.2% in perfumes.[13]

Food. Sweet bay is a common household spice known as bay leaf.

Both the spice and oil are extensively used in processed foods, including alcoholic (oil only) and nonalcoholic beverages, frozen dairy desserts, baked goods, meat and meat products, condiments and relishes, and others. Use levels are generally low; highest reported are in condiments and relishes, which are 0.1% for the spice and 0.02% for the oil.

Traditional Medicine. Has been used in treating cancer[32] and as a cholagogic, general stimulant, carminative, and diaphoretic; leaves used in Iranian folk medicine as an antiepileptic.[20]

COMMERCIAL PREPARATIONS

Crude and oil.

Regulatory Status. Essential oil, extractive, and solvent-free oleoresin of berries and leaves of *Laurus* species are GRAS for use in foods (§182.20); herb GRAS for use in foods as a natural flavoring or spice (§182.10).

REFERENCES

See the General References for ARCTANDER; BAILEY 1; BIANCHINI AND CORBETTA; BRUNETON; DUKE 4; FEMA; GRIEVE; GUENTHER; JIANGSU; MASADA; MCGUFFIN 1 & 2; ROSENGARTEN; TERRELL; UPHOF; YOUNGKEN.

1. H. Hokwerda et al., *Planta Med.*, **44**, 116 (1982).
2. Caredda et al., *J. Agric. Food Chem.*, **50**, 1492 (2002).
3. Z. Putievsky et al., *Isr. J. Bot.*, **33**, 47 (1984).
4. J. M. Schulz and K. Hermann, *Z. Lebensm. Unters. Forsch.*, **171**, 278 (1980).
5. H. Hibasami et al., *Int. J. Mol. Med.*, **12**, 147 (2003).
6. M. Yoshikawa et al., *Bioorg. Med. Chem.*, **8**, 2071 (2000).

7. H. W. Kang et al., *Biol. Pharm. Bull.*, **25**, 102 (2002).
8. B. Pech and J. Bruneton, *J. Nat. Prod.*, **45**, 560 (1982).
9. D. J. Gomez-Coronado and C. Barbas, *J. Agric. Food Chem.*, **51**, 5196 (2003).
10. U. Asllani, *Bull. Univ. Shteteror Tiranes, Ser. Shkencat Natyr.*, **23**, 93 (1969).
11. N. A. Gugunava, *Subtrop. Kul't.*, **3**, 84 (1971).
12. M. G. Pertoldi and B. Stancher, *Atti Congr. Qual.*, **6**, 303 (1967).
13. J. T. MacGregor et al., *J. Agric. Food Chem.*, **22**, 777 (1974).
14. J. W. Hogg et al., *Phytochemistry*, **13**, 868 (1974).
15. H. Tada and K. Takeda, *Chem. Pharm. Bull.*, **24**, 667 (1976).
16. K. Tori et al., *Tetrahedron Lett.*, **5**, 387 (1976).
17. M. M. Verma, *Diss. Abstr. Int. B*, **41**(12, Pt. 1), 4514 (1981).
18. V. L. L. Machado et al., *Anais Soc. Entomol. Brasil*, **24**, 13 (1995).
19. L. Re and T. Kawano, *Mem. Inst. Oswaldo Cruz*, **82**(Suppl. 4), 315 (1987).
20. M. Sayyah et al., *Phytomedicine*, **9**, 212 (2002).
21. M. Sayyah et al., *Phytother. Res.*, **17**, 733 (2003).
22. F. U. Afifi et al., *J. Ethnopharmacol.*, **58**, 9 (1997).
23. I. Gurbuz et al., *J. Ethnopharmacol.*, **83**, 241 (2002).
24. H. Matsuda et al., *Alcohol Alcohol.*, **37**, 121 (2002).
25. M. Simic et al., *Fitoterapia*, **74**, 613 (2003).
26. H. Moteki et al., *Oncol. Rep.*, **9**, 757 (2002).
27. D. L. J. Opdyke, *Food Cosmet. Toxicol.*, **14**, 337 (1976).
28. J. C. Mitchell in V. C. Runeckles, ed., *Recent Advances in Phytochemistry*, Vol. 9, Plenum Press, New York, 1975, p. 119.
29. M. G. Ozden et al., *Contact Dermatitis*, **45**, 178 (2001).
30. A. Cheminat et al., *Arch. Dermatol. Res.*, **276**, 178 (1984).
31. Council of Europe, *Opinion of the Scientific Committee on Food on Methyleugenol (4-Allyl-1,2-dimethoxybenzene)*, European Commission Health and Consumer Protection Directorate-General, Brussel, Belgium, 2001. Available at: http://europa.eu.int/comm/food/fs/scf/index_en.html.
32. J. L. Hartwell, *Lloydia*, **32**, 247 (1969).

BAY, WEST INDIAN

Source: *Pimenta racemosa* (Mill.) J. W. Moore (syn. *P. acris* Kostel) (Family Myrtaceae).

Common/vernacular names: Bay, bay rum tree, myrcia, and West Indian bay.

GENERAL DESCRIPTION

Tree with leathery leaves, up to about 8 m high; native to the West Indies; cultivated in Venezuela, Puerto Rico, and the Caribbean Islands. Part used is the leaf from which a volatile oil (commonly called bay oil or Myrcia oil) is obtained by steam distillation, with yield of up to 3.9% reported.[1]

Although sometimes also referred to as bay leaf in the literature, the commonly used domestic spice is sweet bay (*Laurus nobilis*), not West Indian bay.

CHEMICAL COMPOSITION

The main components of bay oil are eugenol (up to 56%), chavicol (up to 22%), and myrcene (up to 21%). Those present in lesser amounts include 1,8-cineole, limonene, isoeugenol, linalool, methyleugenol (3,4-dimethoxyallylbenzene), estragole (methyl chavicol), α-terpineol, and others.[1-3]

The leaves of two other general varieties, either anise scented or lemon scented, have been reported to yield volatile oils with quite different proportions of the above components. The anise scented variety contains methyleugenol (43%) and methylchavicol (32%) as the major components, and the lemon scented variety contains mostly citral (>80%).[1] The main constituents of the essential oil of the leaves of *P. racemosa* var. *terebinthina* (Burret) L. R. Landrum are 4-methoxy eugenol (12.6%), α-terpineol (20%), and α-terpineol acetate (27%), while in the essential oil of *P. racemosa* var. *grisea* (Kiaersk.) Fosberg, the major constituents are 4-methoxy eugenol (4.5%) and 4-methoxy-isoeugenol (75.2%).[4] The leaves also contain abietic acid, a diterpene.[5] Lupeol, a triterpene, was isolated from the leaves of *P. racemosa* var. *ozua* (Urban & Ekman) L. R. Landrum.[6]

PHARMACOLOGY AND BIOLOGICAL ACTIVITIES

The volatile oil has antiseptic and astringent properties (GOSSELIN). The essential of bay inhibits the *in vitro* growth of *E. coli* (nontoxigenic strain 0157 : H7).[7] In animal models of inflammation, lupeol, from the leaves of *P. racemosa* var. *ozua*,[6] and abietic acid from the leaves of *P. racemosa* var. *grisea*,[5] have shown anti-inflammatory activity in rats when applied topically or by the oral route.

TOXICOLOGY

The volatile oil is considered moderately toxic on oral administration because of its relatively high content of phenols (GOSSELIN). However, no allergic reactions in humans have been reported.[8] Recently, the Council of Europe recommended that owing to evidence of heptotoxicity, methyleugenol should not be allowed in foods at any level.[9]

USES

Medicinal, Pharmaceutical, and Cosmetic. Volatile oil used extensively as a fragrance ingredient in bay rum; also in creams, lotions (particularly aftershave and hair lotions), soaps, detergents, and perfumes, with maximum use level of 1.5% in certain perfumes.[8]

Food. Volatile oil, oleoresin, and extract (less extensively) all used as a flavor ingredient in major categories of food products, including alcoholic and nonalcoholic beverages, frozen dairy desserts, candy, baked goods, gelatins and puddings, meat and meat products, and condiments and relishes at very low levels, usually below 0.01%.

Traditional Medicine. Reportedly used in cancer therapy.[10] In the Dominican Republic, the crushed leaves are used in herbal mixtures to treat toothache, while in Haiti a decoction of the leaves made with salt is used orally to treat abdominal pain. In Curaçao, the leaves are used to treat toothache and as a stomachic, and in Puerto Rico the leaves are used to treat rheumatism.[11]

COMMERCIAL PREPARATIONS

Volatile oil (regular and terpeneless), oleoresin, and extracts. The volatile oil was formerly official in N.F.; it is currently official in F.C.C.

Regulatory Status. Myrcia oil is GRAS (§182.20).

REFERENCES

See the General References for ARCTANDER; BAILEY 1; FEMA; GUENTHER; MASADA.

1. D. McHale et al., *Food Chem.*, **2**, 19 (1977).
2. R. G. Buttery et al., *J. Agric. Food Chem.*, **22**, 773 (1974).
3. Analytical Methods Committee, *Analyst*, **100**, 593 (1975).
4. D. Garcia et al., *Flav. Fragr. J.*, **10**, 319 (1995).
5. M. A. Fernandez et al., *J. Pharm. Pharmacol.*, **53**, 867 (2001).
6. A. Fernandez et al., *Farmaco*, **56**, 335 (2001).
7. S. A. Burt et al., *Lett. Appl. Microbiol.*, **36**, 162 (2003).
8. D. L. J. Opdyke, *Food Cosmet. Toxicol.*, **11**, 869 (1973).
9. Council of Europe, *Opinion of the Scientific Committee on Food on Methyleugenol (4-Allyl-1,2-dimethoxybenzene)*, European Commission Health and Consumer Protection Directorate-General, Brussel, Belgium, 2001. Available at: http://europa.eu.int/comm/food/fs/scf/index_en.html.
10. J. L. Hartwell, *Lloydia*, **33**, 288 (1970).
11. B. Weniger and L. Robineau, in C. Gyllenhaal and D. D. Soejarto, eds., *Elements for a Caribbean Pharmacopeia*, Enda-Caribe, Ministry of Public Health, Cuba, 1988, p. 207.

BAYBERRY BARK

Source: *Myrica cerifera* L. (Family Myricaceae).

Common/vernacular names: Bayberry, Southern bayberry, southern wax myrtle, and wax myrtle bark.

GENERAL DESCRIPTION

Myrica cerifera is an evergreen shrub or small tree that grows up to 13 m in height; branchlets are waxy; fruits round, grayish green, and coated with bluish wax, which can be removed by boiling in water. *M. cedrifera* is native to eastern United States from New Jersey to South Florida and west to Texas; also grows in the Bahamas, the West Indies, and Bermuda. The part used is the dried root bark.

CHEMICAL COMPOSITION

Contains tannins, triterpenes (myricadiol, taraxerol, and taraxerone), and myricitrin (a flavonoid glycoside).[1] Other constituents reported to be present include an acrid astringent resin, gum, and starch. The twigs contain myricalactone and myrica acid, an oleanane triterpenic acid.[2] The plant also contains the terpenoids myriceric acids A, C, and D.[3]

PHARMACOLOGY AND BIOLOGICAL ACTIVITIES

Dried root bark has astringent, emetic, and antipyretic properties.

In an *in vitro* assay for antithrombin activity, a methylene chloride extract fraction of *M. cerifera* showed greater than 80% activity.[4] Myricitrin has shown choleretic, bactericidal, and paramecicidal activities; myricadiol has shown mineralocorticoid activity.[1] Myriceric acid A is a selective endothelin A receptor antagonist.[3,5]

TOXICOLOGY

Tannins and phenols isolated from bayberry bark administered subcutaneously to rats have shown carcinogenic activity.[6] Myricadiol has shown spermatocidal activity.[1]

USES

Medicinal, Pharmaceutical, and Cosmetic. Root bark has been and probably still is used as astringent, tonic, and stimulant to indolent ulcers and as an ingredient in Composition Powder used for colds and chills.

Dietary Supplements/Health Foods. Powdered root bark still seen as an ingredient in Composition Powders, for colds and fevers (FOSTER AND DUKE).

Traditional Medicine. Used in Puerto Rico to treat stubborn ulcers (MORTON 2). Root bark used historically in the United States as an astringent and in larger doses an emetic, for chronic gastritis, diarrhea, dysentery, leucorrhea, jaundice, fevers; externally for hard-to-heal ulcers (FOSTER AND DUKE). Micmac Indians used the roots to treat headaches and inflammations, and as an analgesic.[7] External inflammations were treated with the crushed roots soaked in water.[8]

Other. Fruit is source of bayberry wax for candles.

COMMERCIAL PREPARATIONS

Available generally as crude botanical.

Regulatory Status. Bayberry root bark was included in N.F. IV–V (1916–1926).

REFERENCES

See the General References for BAILEY 1; DER MARDEROSIAN AND BEUTLER; FOSTER AND DUKE; GRIEVE; KROCHMAL AND KROCHMAL; LUST; MARTINDALE; MERCK; MORTON 2; YOUNGKEN.

1. B. D. Paul et al., *J. Pharm. Sci.*, **63**, 958 (1974).
2. M. Nagai et al., *Chem. Pharm. Bull.*, **48**, 1427 (2000).
3. K. Sakurawi et al., *Chem. Pharm. Bull.*, **44**, 343 (1996).
4. N. Chistokhodova et al., *J. Ethnopharmacol.*, **81**, 277 (2002).
5. S. Mihara and M. Fujimoto, *Eur. J. Pharmacol.*, **12**, 33 (1993).
6. G. J. Kapadia et al., *J. Natl. Cancer Inst.*, **57**, 207 (1976).
7. R. F. Chandler et al., *J. Ethnopharmacol.*, **1**, 49 (1979).
8. W. D. Wallis, *Am. Anthropol.*, **24**, 24 (1922).

BEE POLLEN

Source: Pollen collected by bees or harvested directly from flowers for commercial use.

Common/vernacular names: Buckwheat pollen, maize pollen, pollen, pollen pini, pollen typhae, puhuang, rape pollen, typha pollen, pine pollen, songhuafen, and so on.

GENERAL DESCRIPTION

Pollen is composed of microspores (male reproductive elements) of seed-bearing plants. Bee pollen refers to pollen collected by bees that is in turn harvested for commercial distribution.

The sources and types of bee pollen are extremely variable. Known species that yield commercial bee pollen include buckwheat, rape, maize, and pine, among others. Typha pollen (*puhuang*) is collected from *Typha* species (*T. angustata* Bory et Chaub., *T. angustifolia* L., *T. latifolia* L., etc.) and has probably the longest and most extensively documented use history, dating back to the *Shen Nong Ben Cao Jing* (ca. 200 BC–AD 100). Pine pollen is collected from numerous *Pinus* species (including *P. thunbergii* Parl., *P. massoniana* Lamb., and *P. tabulaeformis* Carr.) and has been in traditional Chinese medical records since the 7th century when it was first described in the *Tang Ben Cao*.

Commercial bee pollen is collected by means of netlike pollen traps, set up next to the beehives that remove some of the pollen from the hind legs of worker bees as they return to their hives. The collected pollen is manually rid of impurities (dirt, floral parts, insect fragments, etc.) and dried. Major bee pollen-producing countries include China and Spain. Bee pollen from China is mostly derived from buckwheat (*Fagopyrum esculentum* Moench) and rape (*Brassica campestris* L).

To collect typha and pine pollen, the male inflorescence or flower head is picked in spring or summer when the flowers just start to bloom. It is sun dried; the pollen is then mechanically separated from the floral parts and other impurities. Major producers are northeastern provinces of China.

CHEMICAL COMPOSITION

Pollen is very rich in nutrients. However, its chemistry varies greatly depending on its botanical source and contains: 3–16% water; 5.9–28.3% crude protein: 14.6–21.9% amino acids, with some in free form; 1–20% lipids; up to 44% carbohydrates; 4–10% simple sugars; 2–2.5% flavonoids; vitamins (A, B_1, B_2, C, D_2, E, K_1, K_3, folic acid, nicotinic acid, etc.); 19–24 trace elements; sterols; and others.[1–5]

The following are some examples of specific chemical constituents reported to be present in certain types of pollen but not necessarily in others: pentacosane, isorhamnetin glycoside, narcissin, free palmitic and stearic acids, 6-aminopurine, turanose and an oligosaccharide, sitosterol, and α-typhasterol in *puhuang* (ZHOU AND WANG);[6] β-sitosterol and cholesterol, ursolic acid, rutin (0–17%), C-3/C-8″-biapigenin, palmitic acid, nonacosane, luteolin, tricetin, kaempferol-3-*O*-sophoroside, and kaempferol-3-*O*-β-D-gluco-7-*O*-β-D-glucoside in buckwheat pollen.[7]

PHARMACOLOGY AND BIOLOGICAL ACTIVITIES

Numerous activity studies have been performed on bee pollen. However, results of these are extremely difficult to evaluate or duplicate due to the highly variable nature of this food/drug. The following are some of the activities of pollen and its extracts: hypolipemic in humans and experimental animals (*puhuang* and pollen mixture);[8,9] antiatherosclerotic in experimental animals (*puhuang* and pollen mixture);[10,11] protecting liver from experimental injury (rape pollen and pollen mixture);[12,13] inhibiting prostatic hypertrophy in aged dogs (rape pollen and pollen mixture);[14–16] immunoregulating and antioxidant (rape pollen and *puhuang*);[11,17,18] antiulcer in humans and experimental animals (*puhuang* and rape pollen;[19,20] antifatigue in mice (maize pollen);[21] laxative in humans (maize pollen and unspecified product);[22,23] anti-inflammatory (*puhuang*); and uterine stimulant (*puhuang*) effects, among others (WANG; ZHOU AND WANG).[24]

TOXICOLOGY

Although rare, bee pollen can cause allergic reactions,[25,26] including anaphylactic reactions.[27] As it is a uterine stimulant, pregnant women are advised not to use it (*puhuang*) (CHP).

USES

Medicinal, Pharmaceutical, and Cosmetic. Extracts (hydroalcoholic and lipoid) of pollen are used in skin care products (facial and hand creams and lotions) for its nutritional and traditional healing and skin-softening properties (ETIC).

Dietary Supplements/Health Foods. Used extensively as a food supplement in tablet, capsule, or liquid (drink or syrup) form (TYLER 1).

Traditional Medicine. Used for centuries by different cultures as a nutrient.

In China, typha pollen (*puhuang*) was first described 2000 years ago as sweet tasting, neutral, and having diuretic, hemostatic and stasis-dispersing properties. It has since been used to treat bleeding of different kinds (nosebleed, vomiting blood, coughing blood, metrorrhagia, bloody diarrhea, traumatic injuries, etc.), amenorrhea, dysmenorrhea, abdominal pain, painful urination, mouth sores, constipation, and externally, eczema.

Pine pollen (*songhuafen*) also has a long use history, dating back to the 7th century AD. Traditionally regarded as sweet tasting, warming, benefiting vital energy, removing wetness (*zao shi*), astringent, and hemostatic; used topically in treating eczema, pustular eruptions, diaper rash, bleeding caused by traumatic injuries, and other skin conditions; also used internally to treat alcohol intoxication, chronic diarrhea, and rheumatism (CHP; JIANGSU).

COMMERCIAL PREPARATIONS

Bee pollen comes in powdered or granular form, usually with color ranging from yellow to orange, depending on sources. Typha pollen comes in two types, one mixed with anthers and filaments, while the other is pure pollen. Extracts (water, hydroalcoholic, and lipoid) are also available.

Regulatory Status. U.S. regulatory status not determined. Subject of a German therapeutic monograph; allowed as an appetite stimulant (BLUMENTHAL 1).

REFERENCES

See General References for BLUMENTHAL 1; CHP; JIANGSU; LU AND LI; NATIONAL; TYLER 1; WANG; ZHOU AND WANG; ZHU.

1. C. Y. Guo et al., *Jilin Zhongyiyao*, (4), 35 (1990).
2. L. Z. Mao et al., *Yingyang Xuebao*, **12**(1), 121 (1990).
3. S. S. Jia and S. M. Yang, *Zhongcaoyao*, **19**(1), 47 (1988).
4. V. E. Tyler, *The New Honest Herbal*, George F. Stickley Co., Philadelphia, PA, 1987, p. 184.
5. T, Seppanen et al., *Phytother. Res.*, **3**(3), 115 (1989).
6. L. F. Chen et al., *Zhongguo Yaoli Xuebao*, **8**(2), 123 (1987).
7. J. X. Wei et al., *Zhongguo Zhongyao Zazhi*, **15**(5), 37 (1990).
8. B. Z. Zhang et al., *Chin. J. Integr. Med.*, **5**, 141 (1985).
9. P. G. Xiao and K. J. Chen, *Phytother. Res.*, **1**(2), 53 (1987).
10. J. Wojcicki et al., *Atherosclerosis*, **62**(1), 39 (1986).
11. Y. L. Yin et al., *Zhongguo Zhongyao Zazhi*, **17**, 374 (1992).
12. J. Wojcicki et al., *Acta Pharmacol. Toxicol. Suppl.*, **59**(7), 233 (1986).
13. M. S. Wang et al., *Zhongcaoyao*, **18**(5), 25 (1987).
14. B. C. Qian et al., *Zhonghua Laonian Yixue Zazhi*, **6**, 177 (1987).

15. X. L. Liu et al., *Zhongcaoyao*, **21**(4), 20 (1990).
16. M. Kimura et al., *Planta Med.*, **52**, 148 (1986).
17. F. Zhao and E. Y. Zhan, *Zhongyao Tongbao*, **11**(8), 45 (1986).
18. B. C. Qian et al., *Zhongguo Zhongyao Zazhi*, **15**(5), 45 (1990).
19. J. Chen et al., *Zhongcaoyao*, **20**(1), 27 (1989).
20. X. H. Yang et al., *Zhongyao Tongbao*, **12**(8), 48 (1987).
21. S. T. Yu et al., *Zhongyao Tongbao*, **13**(12), 44 (1988).
22. S. T. Yu et al., *Zhongcaoyao*, **19**(2), 26 (1988).
23. J. Q. Zhou et al., *Chin. J. Integr. Med.*, **8**, 357 (1988).
24. B. C. Qian, *Chin. J. Integr. Med.*, **9**, 125 (1989).
25. J. P. Geyman, *J. Am. Board Fam. Pract.*, **7**, 250 (1994).
26. S. Puente et al., *Med. Clin. (Barc.)*, **108**, 698 (1997).
27. P. A. Greenberger et al., *Ann. Allergy Asthma Immunol.*, **86**, 239 (2001).

BEESWAX

Source: Honeycomb of the honeybee (*Apis* spp.).

Common/vernacular names: Beeswax, bleached beeswax, white beeswax, white wax, yellow beeswax, and yellow wax.

GENERAL DESCRIPTION

Beeswax is the wax obtained from the honeycomb of the honeybee, *Apis mellifera* L., as well as other *Apis* species, including *A. cerana* Fabricius (Family Apidae).

After the honey is removed from the honeycombs, the combs are washed rapidly and thoroughly with water. They are then melted with hot water or steam, strained, and run into molds to cool and harden.[1]

There are three major beeswax products: yellow beeswax, white beeswax (bleached beeswax), and beeswax absolute (*absolute cire d'abeille*). Yellow beeswax is the crude beeswax first obtained from the honeycombs. White beeswax and beeswax absolute are derived from yellow beeswax, the former from bleaching with the combined action of air, sunlight, and moisture (or with peroxides) and the latter by extraction with alcohol. Beeswax is produced worldwide.

Yellow beeswax is a yellow to brownish yellow or grayish brown solid with an agreeable honey-like odor and faint but characteristic taste; it melts between 62 and 65°C.

White beeswax is a yellowish-white solid with a faint, characteristic odor, less pronounced than yellow beeswax; it is almost tasteless and translucent in thin layers; melts between 62 and 65°C.

Both yellow wax and white wax are insoluble in water, slightly soluble in cold alcohol, partly soluble in cold benzene, and completely soluble in chloroform, ether, and fixed and volatile oils.

Beeswax absolute is a pale yellow solid with a mild, sweet, and oily odor reminiscent of good linseed oil with a trace of honey notes, depending on sources.

CHEMICAL COMPOSITION

Beeswax (yellow and white) contains over 80 different compounds, largely made up of hydrocarbons, alkanes, fatty acids, fatty alcohols, free fatty acids, fatty acid monoesters, fatty acid polyesters, diesters, monoesters, triesters, hydroxypolyesters, 1,2,3-propanetriol monoesters, unsaturated linear fatty acids, and hydroxyacids.[2–4] Oxygenated volatiles in beeswax include octanal, furfural, 1-decanol, and benzaldehyde with decanal making up close to 50% of their content.[4] High molecular weight di-, mono-, and polyesters compounds make up over 60% of beeswax.[2] Beeswax contains about 57–71% esters of fatty acids[3] (mostly palmitic and 15-hydroxypalmitic acid) and C_{24} to C_{34} straight-chained mono- and sometimes di-alcohols, composed of 35% C_{46} to C_{48} monoesters and 12% free acids (cerolein). Up to 23% of the monoesters is myricyl palmitate, which together with myricyl alcohol has been referred to as myricin (ARCTANDER; JIANGSU; REMINGTON).[1,5] The chemical composition of beeswax varies in part according to the bee species that make the wax.[3]

Beeswax absolute contains mostly cerolein; also aromatic volatile compounds.[6] Cerolein is soluble in cold alcohol; myricin is insoluble in cold alcohol but sparingly soluble in boiling alcohol.

PHARMACOLOGY AND BIOLOGICAL ACTIVITIES

In animal studies, a mixture of high molecular weight primary alcohols isolated from beeswax with triacontanol as the main constituent, antioxidant, antiperoxidative,[7] anti-inflammatory,[8] antiulcerogenic, gastroprotective,[9] and anticolitis activities were shown.[10] Randomized, double-blind, placebo-controlled clinical studies of the mixture have also demonstrated antioxidant and antiperoxidative activity.[11,12] Triacontanol, also known as myricyl alcohol, has also shown antiperoxidative activity[13] and is a plant growth regulator that increases yields of tomato, cucumber, and lettuce (see *alfalfa*).

TOXICOLOGY

Although beeswax is generally regarded as inert and nontoxic, allergic reactions have been reported (MARTINDALE).[14–16]

Ethoxylated derivatives of beeswax known as PEG (polyethylene glycol)-6 and PEG-20 Sorbitan Beeswax are currently considered safe for use in cosmetics.[17]

USES

Medicinal, Pharmaceutical, and Cosmetic. Both yellow beeswax and white beeswax are used as thickener, emulsifier, or stiffening agents in ointments, baby products, bath preparations, cold creams, emollient creams, eye and facial makeups, lotions, lipsticks, hair dressings, hair conditioners, shaving products, suntan products, suppositories, and others; also used as a tablet polishing component.[18,19]

Beeswax absolute is used as a fragrance ingredient in soaps, lotions, creams, and perfumes in levels up to 0.4% in perfumes.[14]

Polyethylene glycol-20 (PEG-20 Sorbitan Beeswax), an ethoxylated derivative of beeswax, is currently used as a surfactant in cosmetics at concentrations of up to 11%.[17]

Food. White beeswax and beeswax absolute are used as thickener, emulsifier, or flavor ingredients in all major categories of foods, including nonalcoholic and alcoholic beverages, frozen dairy desserts, baked goods, gelatins and puddings, confectioner's frosting, and sweet sauces. White wax is also used as a candy glaze or polish. Use levels are usually low, the highest being in candy (ca. 0.05%).

Traditional Medicine. In Chinese medicine, beeswax is used to treat diarrhea and hiccups and to relieve pain, among others. For internal use, it is usually dissolved in hot alcohol or wine.

Others. As a source of triacontanol for increasing crop yield.[20,21]

COMMERCIAL PREPARATIONS

Yellow beeswax, white beeswax, and beeswax absolute; first two are official in F.C.C. and N.F.

Regulatory Status. GRAS for use as adjuvants for pesticide chemicals (§582.1972, §582.1975).

REFERENCES

See the General References for ARCTANDER; FEMA; JIANGSU; REMINGTON; SAX; YOUNGKEN.

1. E. S. McLoud in A. Standen, ed., *Kirk-Othmer Encyclopedia of Chemical Technology*, Vol. 22, 2nd ed., Wiley–Interscience, New York, 1970, p. 156.
2. J. J. Jiménez et al., *J. Chromatogr. A*, **1007**, 101 (2003).
3. R. Aichholz et al., *J. Chromatogr. A*, **855**, 601 (1999).
4. M. S. Blum et al., *Comp. Biochem. Physiol. B*, **91**, 581 (1988).
5. A. P. Tulloch, *Lipids*, **5**, 247 (1970).
6. C. E. M. Ferber and H. E. Nursten, *J. Sci. Food Agric.*, **28**, 511 (1977).
7. V. Molina et al., *J. Med. Food*, **5**, 79 (2001).
8. D. Carbajal et al., *Prostaglandins Leukot. Essent. Fatty Acids*, **59**, 235 (1998).
9. D. Carbajal et al., *Pharmacol. Res.*, **42**, 329 (2000).
10. M. Noa et al., *Drugs Exp. Clin. Res.*, **26**, 13 (2000).
11. R. Menendez et al., *J. Med. Food*, **4**, 71 (2001).
12. R. Menendez et al., *Arch. Med. Res.*, **32**, 436 (2001).
13. K. Ramanarayan et al., *Phytochemistry*, **55**, 59 (2000).
14. D. L. J. Opdyke, *Food Cosmet. Toxicol.*, **14**(Suppl.), 691 (1976).
15. M. Garcia et al., *Contact Dermatitis*, **33**, 440 (1995).
16. V. Junghans et al., *Am. J. Contact Dermat.*, **13**, 87 (2002).
17. R. S. Lanigan et al., *Int. J. Toxicol.*, **20**(Suppl. 4), 27 (2001).
18. S. L. Pulco, *Cosmet. Toilet.*, **102**(6), 57 (1987).
19. Anon., *J. Am. Coll. Toxicol.*, **3**(3), 1 (1984).
20. Z. X. Huang et al., *Huaxue Shiji*, **9**, 299 (1987).
21. C. Devakumar et al., *Indian J. Agric. Sci.*, **56**, 744 (1986).

BEET COLOR, RED

Source: *Beta vulgaris* L. (Family Chenopodiaceae).

GENERAL DESCRIPTION

Red beet color is the coloring material derived from the red beet root, *Beta vulgaris* L. *Beta vulgaris* has several varieties with roots ranging in size from small to thick and in color from whitish or yellowish (sugar beets) to deep blood-red (certain garden beets).[1,2]

CHEMICAL COMPOSITION

The coloring principles present in red beet juice are known as betalains (quaternary ammonium amino acids). They consist mostly of betacyanins (red), with a small amount of betaxanthins (yellow). Betanin and to a lesser extent isobetanin account for most of the betacyanins present, while vulgaxanthin-I and vulgaxanthin-II are the major betaxanthins. Betanin is a glucoside of betanidin, and isobetanin is its C_{15} epimer. There is evidence that betanin occurs in red beet root as a sulfate linked through the sugar moiety at the 3- or 6-position.[3,4] Cyclodopa glucoside has recently been found in red beet juice, strengthening its role as intermediate in the biosynthesis of betanin.[5]

Red beet color is most stable at pH 4.5–5.5; it is rather unstable outside this range. There is evidence of an enzyme present in red beet that specifically destroys the betanin chromophore at an optimal pH of 3.35 and a temperature optimum of 42°C.[6] Betacyanin and betaxanthin decolorizing enzyme with an optimal pH of 3.4 have recently been reported in beet root tissue.[7]

Red beet pigments are heat labile, especially in the presence of metals (e.g., Cu, Mn, Fe, Zn); and at temperatures above 121°C, betanin is rapidly destroyed. Copper is the most efficient catalyst for the breakdown. The color can be stabilized by sequestrants and/or antioxidants such as citric acid, sorbic acid, and ascorbic acid;[8,9] though ascorbic acid has also been shown to decrease the color stability of betanin in aqueous solutions.[10]

The tinctorial power of betanin is quite high; however, its concentration in most commercial beet colors is only 1–2%, making it necessary to use these colors at relatively high levels to achieve the desired color effects. At these high levels, the characteristic beet flavor usually is perceptible. According to a patented process, the pigment content can be considerably increased, at the same time eliminating the beet flavor and aroma.[11,12]

PHARMACOLOGY AND BIOLOGICAL ACTIVITIES

Betalains and betanin have shown potent *in vitro* antioxidant activity. Betanin inhibited lipid peroxidation *in vitro* more potently than catechin[13,14] and in humans showed have high bioavailability.[13]

TOXICOLOGY

Red beet color has been reported to have weakly mutagenic activities per Ames test,[15] although these results were not substantiated by others.[16] Red beet color did not initiate or promote hepatocarcinogenesis in rats during a short-term study.[17]

USES

Used in coloring various food products.[2,18]

COMMERCIAL PREPARATIONS

Powdered beet root, juice, concentrated juice, and spray-dried powder are available in different coloring strengths.

Regulatory Status. Beet powder as a color additive is approved for food use (§73.40 and §73.260); exempt from certification.

REFERENCES

See the General References for BAILEY 2; MCGUFFIN 1 & 2.

1. B. V. Ford-Lloyd and J. T. Williams, *Bot. J. Linn. Soc.*, **71**, 89 (1975).
2. F. Delgardo-Vargas et al., *Crit. Rev. Food Sci. Nutr.*, **40**, 173 (2000).
3. T. J. Mabry and A. S. Dreiding in T. J. Mabry et al., eds., *Recent Advances in Phytochemistry*, Vol. 1, Appleton-Century-Crofts, New York, 1968, p. 145.
4. J. B. Harborne in L. Reinhold et al., eds., *Progress in Phytochemistry*, Vol. 4, Pergamon, Oxford, UK, 1977, p. 189.
5. H. Wyler et al., *Helv. Chim. Acta*, **67**, 1348 (1984).
6. M. S. Ul'yanova et al., *Dokl. Akad. Nauk SSSR*, **200**, 990 (1971).
7. C. C. Shih and R. C. Wiley, *J. Food Sci.*, **47**, 164 (1981).
8. A. F. Fang-Yung and A. V. Khotivari, *Isv. Vyssh. Uchebn. Zaved., Pishch. Tekhnol.*, **6**, 152 (1975).
9. N. I. Oragvelidze et al., USSR 565,049 (1977).

10. G. Muschiolik and H. Schmandke, *Nahrung*, **22**, 637 (1978).
11. J. P. Adams et al., *J. Food Sci.*, **41**, 78 (1976).
12. J. Von Elbe and C. H. Ammundson, Ger. Offen. 2,545,975 (1976).
13. J. Kanner et al., *J. Agric. Food Chem.*, **49**, 5178 (2001).
14. L. Tesoriere et al., *Free Radic. Res.*, **37**, 689 (2003).
15. M. Ishidate Jr. et al., *Food Chem. Toxicol.*, **22**, 623 (1984).
16. K. Kawana et al., *Kanagawa-ken Eisei Kenkyusho Kenkyu Hokoku*, **13**, 27 (1983).
17. J. H. Von Elbe and S. J. Schwartz, *Arch. Toxicol.*, **49**, 93 (1981).
18. S. J. Schwartz et al., *Food Chem. Toxicol.*, **21**, 531 (1983).

BELLADONNA

Source: *Atropa belladonna* L. or its variety *acuminata* Royle ex Lindl. (Family Solanaceae).

Common/vernacular names: Belladonna, deadly nightshade.

GENERAL DESCRIPTION

Atropa belladonna is a perennial herb that grows up to 1 m high with black fruit (a berry); native to central and southern Europe and Asia Minor; now cultivated worldwide, including the United States, United Kingdom, China, and India. Parts used are the dried leaves (including flowering and fruiting tops) and roots.

The specific epithet *belladonna* is of Italian origin, meaning "beautiful lady." This refers to the former practice of Italian women in using the juice of the berry on the eyes to dilate the pupils, giving them a striking appearance.

CHEMICAL COMPOSITION

Leaves and roots contain tropane alkaloids (0.3–0.5%) that are composed mainly of *l*-hyoscyamine (95–98%) and traces of *l*-scopolamine (hyoscine) and atropine (*dl*-hyoscyamine); their concentration and proportions vary greatly with age of the plant. On extraction most of the *l*-hyoscyamine is racemized to atropine. Other alkaloids isolated include *l*-hyoscyamine *N*-oxide (equatorial and axial *N*-oxide isomers) and *l*-hyoscine *N*-oxide (equatorial isomer); roots contain cuscohygrine, which is absent in the leaves. A total of at least 14 alkaloids have been found in the root.[1,2] Flavonoids scopolin, scopoletin, 7-methylquercetin, and a methylkaempferol are present in leaves (STAHL).[3,4] The seeds contain spirostane-type steroidal glycosides (atroposides A through H).[5]

PHARMACOLOGY AND BIOLOGICAL ACTIVITIES

Low doses of the plant in mice produced protective effects against the effects of experimental stress, including immunoprotective, gastroprotective, and behavioral (neurotropic) protection.[6] The activity of belladonna is due to its alkaloids, primarily atropine. Atropine is anticholinergic, both central and peripheral. Its effect on the central nervous system is first stimulation and then depression. Its peripheral anticholinergic effects include reducing secretions (e.g., sweat, tears, saliva, nasal, gastric, and intestinal), decreasing gastric and intestinal motility, and increasing heart rate. Other activities include dilatation of the pupil, increase of intraocular pressure, and photophobia. The activities of *l*-hyoscyamine and *l*-scopolamine are essentially the same as those of atropine, except that scopolamine is a powerful hypnotic and usually slows rather than increases the heart rate.

TOXICOLOGY

Symptoms of overdose are typical of anticholinergic syndrome and are known to include acute psychosis, coma, convulsions, difficulty swallowing and walking and impaired articulation of speech, dryness of mouth, intense thirst, difficulty in swallowing, burning pain in the throat, dilatation of the pupils with blurred vision and photophobia, flushing with hot and dry skin, high fever, fast heart rate with palpitations and elevated blood pressure, urge to urinate but inability to, constipation, restlessness, confusion, excitement, hallucinations, and delirium. Death may result from respiratory failure. Above symptoms have been described as "blind as a bat, dry as a bone, red as a beet, hot as a hare, and mad as a hatter" (GOSSELIN).[7] The berries have been mistaken for bilberries with toxic results;[8,9] implicated in the death of children consuming a couple of berries, the chief toxic principle thought to be hyoscine.

USES

Medicinal, Pharmaceutical, and Cosmetic. Its extracts and isolated alkaloids are widely used in both over-the-counter and prescription drugs, including sedatives, antispasmodics in bronchial asthma and whooping cough, cold and hay fever remedies, ophthalmic preparations, laxatives (to lessen griping), suppositories for hemorrhoids, liniments for treatment of muscular rheumatism, sciatica, and neuralgia (often with aconite extract); also in the treatment of Parkinson's disease and intestinal and biliary colic.

Hyoscine is used in preparations to treat motion sickness (MERCK).

Hyoscine-containing plants have been used for centuries in traditional Chinese medicine as anesthetics. Hyoscine from *Flos daturae* (flowers from *Datura* species) has been used as a general anesthetic in China (JIANGSU).[10]

COMMERCIAL PREPARATIONS

Available as crude and as various extracts. Belladonna and its fluid, solid, and powdered extracts are official in U.S.P.

Regulatory Status. Leaves and roots, calculated to specified levels of tropane alkaloids, are the subject of a German therapeutic monograph, indicated for treatment of spasms and colic pains in the gastrointestinal tract and bile ducts (BLUMENTHAL 1).

REFERENCES

See the General References for APhA; BLUMENTHAL 1; BRUNETON; CLAUS; GOODMAN AND GILMAN; GRIEVE; GUPTA; JIANGSU; JIXIAN; MARTINDALE; MCGUFFIN 1 & 2; NANJING; USD 26th.

1. T. Hartmann et al., *Planta Med.*, **52**, 390 (1986).
2. F. Oprach et al., *Planta Med.*, **52**, 513 (1986).
3. J. D. Phillipson and S. S. Handa, *Phytochemistry*, **14**, 999 (1975).
4. G. Clair et al., *C.R. Hebd. Seances Acad. Sci. Ser. D.*, **282**, 53 (1976).
5. S. A. Shvests et al., *Exp. Med. Biol.*, **404**, 475 (1996).
6. D. Bousta et al., *J. Ethnopharmacol.*, **74**, 205 (2001).
7. R. Jaspersen-Schib et al., *Schweiz. Med. Wochenschr.*, **126**, 1085 (1996).
8. K. Jellema et al., *Ned. Tijdschr. Geneeskd.*, **146**, 2173 (2002).
9. H. J. Southgate et al., *J. R. Soc. Health*, **120**, 127 (2000).
10. Anon., *Am. J. Chin. Med.*, **3**, 91 (1975).

BENZOIN

Source: *Styrax* spp. (Family Styraceae).

Common/vernacular names: Gum benjamin, gum benzoin, Siam benzoin, Sumatra benzoin.

GENERAL DESCRIPTION

Benzoin is the balsamic resin obtained from the bark of various *Styrax* spp. *Styrax benzoin* Dryand and *S. paralleloneurum* Perkins yield Sumatra benzoin; *S. tonkinensis* (Pierre) Craib ex Hartwich and other related *Styrax* species yield Siam benzoin.

Benzoin-producing *Styrax* species are mostly small to medium trees (up to 20 m high) growing in tropical Asia. Sumatra benzoin is largely produced from cultivated trees growing in North Sumatra, although the tree occurs natively in Borneo, Java, and the Malay Peninsula; Siam benzoin is produced from trees growing in Thailand in the Province of Luang Probang, although the tree also occurs in Laos, Vietnam, Cambodia, and China.

Benzoin is a pathological product formed when the tree trunk is injured. It is produced by incising the bark; the exuded balsamic resin hardens on exposure to air and sunlight, and is collected. Benzoin resinoid is prepared from crude benzoin by extraction with solvents such as benzene and alcohol, followed by their subsequent removal.

CHEMICAL COMPOSITION

Benzoin contains chiefly esters of cinnamic and benzoic acids together with free acids. Amounts and types of esters and acids vary widely depending on the source.[1,2]

The major constituents of Sumatra benzoin are *p*-coumaryl cinnamate, cinnamic acid, cinnamyl cinnamate (styracin), *p*-coumaryl benzoate, pinoresinol, benzoic acid, coniferyl cinnamate, and coniferyl benzoate. The major constituents of Siam benzoin are *p*-coumaryl benzoate, benzoic acid, pinoresinol, cinnamic acid, benzoic acid ester, and *p*-coumaryl cinnamate. Both Sumatra and Siam benzoin contain small amounts of vanillin and approximately 2–3% triterpenoid compounds[2] (e.g., siaresinolic acid and sumaresinolic acid) (EVANS). The major component of the oils derived from Sumatra and Siam benzoins is benzyl benzoate (76.1–80.1%). Other major constituents in Sumatra benzoin oil are cinnamic acid, benzyl cinnamate, and styrene, whereas those in the oil of Siam benzoin are benzoic acid, allyl benzoate, and methyl benzoate.[3]

PHARMACOLOGY AND BIOLOGICAL ACTIVITIES

Benzoin vapor (with steam) has expectorant properties; its solutions (e.g., tincture) have local antiseptic properties. Following topical application in monkeys, benzoin is partly absorbed through the skin resulting in systemic exposure.[4]

TOXICOLOGY

Benzoin is regarded as moderately toxic, probably due to occasional contact dermatitis developed in some individuals when using Compound Benzoin Tincture, which contains, in addition to benzoin, aloe, storax, balsam tolu, and others (GOSSELIN; MARTINDALE).[5–7] Use of benzoin as a surgical adhesive has resulted postoperative contact dermatitis in some individuals,[8] and produced adverse effects in children when used as a circumcision dressing.[9] The use of tincture of benzoin as a pressure bandage in enucleation of the eye has resulted in necrotizing dermatitis.[10]

USES

Both Siam benzoin and Sumatra benzoin are official in pharmacopoeias of many countries; however, only Sumatra benzoin is found in the BP. Both types are official in U.S.P., but in the United States Sumatra benzoin is more customarily used in pharmaceutical preparations, while Siam benzoin is used in flavors and fragrances. Benzoin is also used in incense (EVANS) and aromatherapy oils.[11]

Medicinal, Pharmaceutical, and Cosmetic. Mainly used in friar's balsam; also as an antiseptic, astringent, and expectorant; in vaporizer fluids for inhalation to relieve respiratory discomforts; in Compound Benzoin Tincture, which is widely used as a skin protectant or topical adhesive agent; and as an antiseptic and styptic on small cuts. Tincture also used in dentistry to treat inflammation of gums and oral herpetic lesions.

Benzoin, especially Siam benzoin, has antioxidative and preservative properties and is used in cosmetics for these properties. The resinoid is extensively used as a fixative in perfumes, soaps, detergents, creams, and lotions, in amounts up to 0.8% in perfumes (ARCTANDER).[5,12]

Food. Classified as a natural flavor; used in most categories of foods, including alcoholic and nonalcoholic beverages, frozen dairy desserts, candy (e.g., chocolate glaze), baked goods, and gelatins and puddings. Use levels usually quite low, with highest average maximum level of 0.014% reported in candy and baked goods.

COMMERCIAL PREPARATIONS

Crude benzoin, benzoin tincture, fluid extract, and resinoid are all readily available. Benzoin is currently official in U.S.P.

Regulatory Status. Resin approved for food use as a natural flavoring substance (§172.510).

REFERENCES

See the General References for ADA; ARCTANDER; CLAUS; EVANS; FEMA; GUENTHER; MARTINDALE; TYLER 3; YOUNGKEN.

1. A. Nitta et al., *Yakugaku Zasshi*, **104**, 592 (1984).
2. I. Pastorova et al., *Phytochem. Anal.*, **8**, 63 (1997).
3. X. Fernandez et al., *Flav. Fragr. J.*, **18**, 328 (2003).
4. R. L. Bronaugh et al., *Food Chem. Toxicol.*, **28**, 369 (1990).
5. D. L. J. Opdyke, *Food Cosmet. Toxicol.*, **11**, 871 (1973).
6. W. D. James et al., *J. Am. Acad. Dermatol.*, **11**, 847 (1984).
7. L. Scardamalgia et al., *Australas. J. Dermatol.*, **44**, 180 (2003).
8. C. B. Lesesne, *J. Dermatol. Surg. Oncol.*, **18**, 990 (1992).
9. D. C. S. Gough and N. Lawton, *Br. J. Urol.*, **65**, 418 (1990).
10. R. C. Tripathi et al., *Lens Eye Toxic Res.*, **7**, 173 (1990).
11. C. Anderson et al., *Phytother. Res.*, **14**, 452 (2000).
12. D. O. Gyane, *Drug Cosmet. Ind.*, **118**, 36 (1976).

BERGAMOT OIL

Source: *Citrus bergamia* Risso & Poit. (syn. *C. aurantium* L. subsp. *bergamia* Wright & Arn. ex Engl.) (Family Rutaceae).

GENERAL DESCRIPTION

Brownish-yellow or greenish oil with an aromatic bitter flavor and fragrant odor; obtained from the peel of the fresh, nearly ripe fruit of

Citrus bergamia, a small tree native to tropical Asia, now extensively cultivated in the Calabrian coast in southern Italy. Bergamot oil is obtained by cold expression of the peel; it is also known as expressed bergamot oil from which rectified or terpeneless bergamot oil is produced by vacuum distillation or by selective solvent extraction, or by chromatography (ARCTANDER).

CHEMICAL COMPOSITION

Approximately 300 compounds have been identified in the expressed oil, including 30–60% linalyl acetate, 11–22% linalool and other alcohols; sesquiterpenes (α-*trans*-bergamotene, caryophyllene, β-farnesene, humulene, β-bisabolene), terpenes (limonene, *p*-cymene, γ-terpinene, phellandrene, α- and β-pinene), C_{20} to C_{33} *n*-alkanes, and furocoumarins (bergaptene, bergamottin, citroptene, 7-methoxy-5-geranoxycoumarin, bergaptol, isopimpinellin, and xanthotoxin, and bergapten at 0.30–0.39%).[1–6]

The distilled oil contains a small concentration of coumarins compared to the cold pressed oil.[7] Rectified (terpeneless) oil contains a lower concentration of terpene components than the expressed oil and no coumarins (ARCTANDER).[8]

PHARMACOLOGY AND BIOLOGICAL ACTIVITIES

5-Methoxypsoralen (5-MOP) appears in blood serum following topical application of bergamot essential oil to human skin[9] and has shown mutagenic effects on mammalian cells *in vitro*.[10]

TOXICOLOGY

Use of bergamot oil is banned or restricted in many countries owing to phototoxic effects.[11,12] Use of the oil in foods is restricted to those with coumarins removed.[13] Topical use of preparations containing bergamot oil has caused photosensitivity reactions owing to the presence of certain furocoumarins (particularly bergapten and xanthotoxin, also known as 5-methoxypsoralen and 8-methoxypsoralen, respectively) in the expressed oil. Due to the photosensitizing activity of these constituents, the use of bergamot oil in cosmetics has caused hyperpigmentation of the face and neck.[5,6,10] When used with long-wave ultraviolet light, however, the same furonocoumarins have been effectively used in the treatment of psoriasis, vitiligo, and mycosis fungoides.[14] Recent cases of phototoxic reactions to the oil have been reported from its use in aromatherapy[11] and from traditional medical colognes known as "Florida Water" and "Kananga."[12]

USES

Medicinal, Pharmaceutical, and Cosmetic. Formerly used extensively used in high-quality perfumes (especially eau de cologne), aromatherapy oils, creams, lotions, suntanning preparations to stimulate melanin production,[10] and in soaps, with use levels up to 0.25% in creams and lotions and 3% in perfumes (EVANS; MARTINDALE).[6]

Food. Allowed for use in foods provided coumarins (e.g., bergapten) are removed;[13] once widely used as an ingredient in flavor formulations with fruity citrus notes in most major food categories, including alcoholic and nonalcoholic beverages, Earl Grey tea, frozen dairy desserts, candy, baked goods, gelatins and puddings, and meat and meat products. Highest average maximum use level was 0.02% in gelatins and puddings.

COMMERCIAL PREPARATIONS

Bergamot oil expressed and bergamot oil rectified; the former is official in F.C.C.; formerly officinal in N.F.'

Regulatory Status. Essential oil, extractive, and solvent-free oleoresin of bergamot orange are GRAS (§182.20).

REFERENCES

See the General References for FEMA; GUENTHER; MASADA; MCGUFFIN 1 & 2.

1. M. Mammi de Leo, *Essenze Deriv. Agrum.*, **46**, 181 (1976).
2. G. Calabro and P. Curro, *Ann. Fac. Econ. Commer., Univ. Studi Messina*, **10**, 67 (1972).
3. A. Liberti and G. Goretti, *Atti Conv. Naz. Olii Essenz. Sui Deriv. Agrum.*, **1–2**, 69 (1974).
4. U. R. Cieri, *J. Assoc. Offic. Anal. Chem.*, **52**, 719 (1969).
5. S. T. Zaynoun et al., *Br. J. Dermatol.*, **96**, 475 (1977).
6. D. L. J. Opdyke, *Food Cosmet. Toxicol.*, **11**, 1031 (1973).
7. F. Buiarelli et al., *Ann. Chim.*, **92**, 363 (2002).
8. D. L. J. Opdyke, *Food Cosmet. Toxicol.*, **11**, 1035 (1973).
9. L. H. Wang and M. Tso, *J. Pharm. Biomed. Anal.*, **30**, 593 (2002).
10. M. J. Ashwood-Smith et al., *Nature*, **285**, 407 (1980).
11. S. Kaddu et al., *J. Am. Acad. Dermatol.*, **45**, 458 (2001).
12. L. Wang et al., *Cutis*, **70**, 29 (2002).
13. B. B. Mandula et al., *Science*, **193**, 1131 (1976).
14. T. Lakshmipathi et al., *Br. J. Dermatol.*, **96**, 587 (1977).

BILBERRY

Source: *Vaccinium myrtillus* L. (Family Ericaceae).

Common/vernacular names: Bilberry, dwarf bilberry, and whortleberry.

GENERAL DESCRIPTION

Bilberry is a deciduous freely branched shrub up to about 35–60 cm high; arising from a creeping rhizome. Found in heaths, moors, and woods in most of Europe (mountains in southern Europe); also found in N. Asia. The parts used are the fruits and leaves.

CHEMICAL COMPOSITION

The fruits contain resveratrol[1] and at least 14 different anthocyanins (malvidin-3-arabinoside, cyanidin-3-xyloside, cyanidin-3-rutinoside, peonidin-3-glucoside, petunidin-3-galactoside, delphinidin-3-glucose, etc.).[2] Fruits or fruit juice contain at least 3% anthocyanosides such as procyanidins B_1, B_2, B_3, B_4; myrtillin; flavonoids including quercitrin, hyperoside, isoquercitrin, astragalin; flavan-3-ols including (+)-catechin and (−)-epicatechin; phenolic acids including caffeic, chlorogenic, *p*-coumaric, ferulic, syringic, gallic, protocatechuic, *p*-hydroxybenzoic, *m*-hydroxybenzoic, vanillic, *m*-coumaric and *o*-coumaric acids, and a hydroxybenzoic acid derivative; vitamin C; quinolizidine alkaloids myrtine and epimyrtine in aerial parts; arbutin and other hydroquinone derivatives, ubiquitous in other *Vaccinum* spp., are absent in bilberry.[3–6]

PHARMACOLOGY AND BIOLOGICAL ACTIVITIES

Bilberry fruits and leaves exhibit astringent and diuretic activity. Clinical use of anthocyanoside-rich extracts of the fruit is largely found in for degenerative retinal conditions.[7,8] Anthocyanosides extracted from the fruit have shown diverse activities in animal studies: a protective effect on the liberation of lactate dehydrogenase in heart and plasma and cardiac isoenzymes, indicating an angina-protecting effect.[9] The anthocyanosides have also shown retinal phosphoglucomutase and glucose-6-phosphatase inhibiting activity;[10] a hypoglycemic effect (due to neomyrtillin content in leaves);[11] vasoprotective activity (twofold as active in protecting capillary permeability as rutin); and antiedema activity.[12] A long-lasting increase in capillary resistance produced by the anthocyanins of the fruit is believed to result from their greater affinity for the skin and kidney tissue rather than plasma.[13] The anthocyanins decrease collagen hydrolysis, significantly reducing permeability of the blood–brain barrier.[14] Their vasodilating effect stimulates local synthesis of vasodilator prostaglandins;[15,16] they also inhibit platelet aggregation and thrombus formation via stimulation of PGI_2-like substances in vascular tissue.[17] Orally administered to rats with experimental type 1 diabetes, leaf extracts lowered plasma, glucose, cholesterol, and triglyceride levels.[18]

TOXICOLOGY

In mice and rats, the acute oral LD_{50} of a bilberry extract (standardized to contain 36% anthocyanins) was greater than the equivalent anthocyanins at 720 mg/kg. Long-term (6 months) oral administration of bilberry extract equivalent to up to anthocyanins at 180 mg/kg/day failed to produce toxic effects; no teratogenic or mutagenic effects were found.[19] Prolonged use of the leaves may result in chronic intoxication. At a chronic dose of 1.5 g/kg/day, bilberry leaf caused toxicity and eventual death of animals. Side effects of the berries are unknown (BLUMENTHAL 1). However, inhibition of platelet aggregation was found after 30–60 days in human volunteers taking an extract that provided 480 mg of bilberry anthocyanosides per day.[20]

USES

Medicinal, Pharmaceutical, and Cosmetic. Fruit preparations, calculated at a daily dose of 20–60 g, are used for the treatment of acute diarrhea and for localized mild inflammation of the mucous membranes of the mouth and throat. In Europe, leaf preparations are used for the supportive treatment of diabetes mellitus; prevention and treatment of gastrointestinal, kidney, and urinary tract disorders, as well as arthritis, dermatitis, functional heart problems, gout, hemorrhoids, poor circulation, and for metabolic stimulation of circulation. Efficacy of the leaves is not established and their therapeutic use is not recommended (BLUMENTHAL 1).

Food. The fruit is best known for its food value; used in alcoholic and nonalcoholic beverages; conserves, pastries, compote, syrups, or eaten raw; fruit extracts also used as red coloring in wine.

Dietary Supplements/Health Foods. Dried fruits in encapsulated products; teas; primarily for improved vision (BLUMENTHAL 1).

Traditional Medicine. Historically, in Europe, the fruits and to a lesser extent the leaves, have been used for astringent and antiseptic activities in diarrhea, dysentery, dyspepsia, intestinal dyspepsia in infants; leaf tea as antidiabetic.

COMMERCIAL PREPARATIONS

Crude berries or extract.

Regulatory Status. Regulated in the United States as a dietary supplement. Both the leaves and the fruits are subjects of German therapeutic monographs. Efficacy of leaves is not documented; fruits allowed for acute diarrhea and mild inflammation of mouth and throat (BLUMENTHAL 1; WICHTL).

REFERENCES

See the General References for AHPA; APPLEQUIST; BARNES; BLUMENTHAL 1 & 2; DER MARDEROSIAN AND BEUTLER; DER MARDEROSIAN AND LIBERTI; HORTUS 3rd; MCGUFFIN 1 & 2; STEINMETZ; TUTIN 2; UPHOF; WEISS; WREN.

1. R. B. van Breemen et al., *J. Agric. Food Chem.*, **51**, 5867 (2003).
2. P. Dugo et al., *J. Agric. Food Chem.*, **49**, 3987 (2001).
3. P. Slosse and C. Hootelé, *Tetrahedron Lett.*, **34**, 397 (1978).
4. P. Slosse and C. Hootelé, *Tetrahedron Lett.*, **37**, 4287 (1981).
5. M. Azar et al., *J. Food Sci.*, **52**, 1255 (1987).
6. H. Friedrich and J. Schöert, *Planta Med.*, **24**, 90 (1973).
7. Y. Levy and Y. Glovinsky, *Eye*, **12**, 967 (1998).
8. E. R. Muth et al., *Altern. Med. Rev.*, **5**, 164 (2000).
9. M. Marcollet et al., *C. R. Soc. Biol.*, **163**, 8 (1969).
10. C. Cluzel et al., *C. R. Soc. Biol.*, **163**, 147 (1969).
11. R. N. Zozulya et al., *Rast. Resur.*, **11**, 87 (1975).
12. A. Lietti et al., *Arzneim.-Forsch.*, **26**, 829 (1976).
13. A. Lietti and G. Forni, *Arzneim.-Forsch.*, **26**, 832 (1976).
14. A. M. Robert et al., *J. Med.*, **8**, 321 (1977).
15. V. Bettini et al., *Fitoterapia*, **55**, 265 (1984).
16. V. Bettini et al., *Fitoterapia*, **56**, 3 (1985).
17. P. Morazzoni and M. J. Magistretti, *Fitoterapia*, **57**, 11 (1986).
18. A. Cignarella et al., *Thromb. Res.*, **84**, 311 (1996).
19. S. Mills and K. Bone, *Principles and Practice of Phytotherapy*, Chruchill Livingstone, Edinburgh, 2000, p. 301.
20. G. Pulleirio et al., *Fitoterapia*, **60**, 69 (1989).

BIRCH OIL, SWEET

Source: ***Betula lenta*** L. (syn. *B. carpinefolia* Ehrh.) (Family Betulaceae).

Common/vernacular names: Birch, black birch, cherry birch oil, and sweet birch oil.

GENERAL DESCRIPTION

Tree up to about 25 m high with dark reddish brown bark, which is not peeling but broken into plates; native to southern Canada and northern United States, from Maine to Ohio and south to Florida and Alabama.

There are numerous species of birch with habitats spanning several continents.

Sweet birch oil is produced by steam distillation of the warm water-macerated bark. During maceration the enzyme system present hydrolyzes gaultherin, setting free methyl salicylate, which is the major component of the oil. The yield is about 0.6%.[1]

Sweet birch oil should not be confused with other birch oils such as birch bud and birch tar oils, which are produced from different species of birch and have different physical and chemical characteristics; they are used for quite different purposes. For example, birch tar oil is obtained by destructive distillation of the wood and bark of the European white birch (*Betula pendula* Roth; syn. *B. alba* L.) and is used in psoriasis, eczema, and other chronic skin diseases.

CHEMICAL COMPOSITION

Birch oil is almost entirely composed of methyl salicylate (98%).

PHARMACOLOGY AND BIOLOGICAL ACTIVITIES

Like salicylates in general, methyl salicylate has antipyretic, anti-inflammatory, and analgesic properties (GOODMAN AND GILMAN).

TOXICOLOGY

Methyl salicylate is much more toxic than salicylates. It can be absorbed through the skin, and fatal poisoning via this route has been reported. As little as 4 mL (4.7 g) methyl salicylate may be fatal in children (GOODMAN AND GILMAN). Side effects of the leaves are unknown (BLUMENTHAL 1).

USES

Medicinal, Pharmaceutical, and Cosmetic. Methyl salicylate has limited use as a counterirritant in antiarthritic and antineuralgic preparations such as ointments, liniments, and analgesic balms; as an antiseptic; and as a fragrance ingredient in perfumes and other cosmetic preparations. Presently, synthetic methyl salicylate is mostly used.

Betula pendula leaves are reportedly diuretic, used in irrigation therapy for bacterial and inflammatory disease of the urinary tract (BLUMENTHAL 1).

Food. Extensively used for its wintergreen (or root beer) flavor in most major categories of foods, especially nonalcoholic and alcoholic beverages, frozen dairy desserts, candy, chewing gum, gelatins and puddings, and baked goods. It is a common flavor ingredient in root beers. The highest average maximum use level reported is in candy (ca. 0.1%), but most of it is probably lost during processing.

Traditional uses. American Indians used the bark tea for milky urine and stomachache, the bark decoction for diarrhea, pneumonia, and pulmonary problems, and the leaf tea for dysentery and colds (MOERMAN). Essential oil reportedly used for rheumatism, gout, scrofula, bladder infection, and neuralgia (FOSTER AND DUKE).

COMMERCIAL PREPARATIONS

Volatile oil; it was formerly official in U.S.P. and is currently official in F.C.C. monographed under methyl salicylate together with wintergreen oil. *Betula pendula* leaves are the subject of a German therapeutic monograph, used in irrigation therapy as a diuretic (BLUMENTHAL 1).

REFERENCE

See the General References for ARCTANDER; BAILEY 2; BLUMENTHAL 1; BRUNETON; FEMA; FOSTER AND DUKE; GUENTHER; MCGUFFIN 1 & 2; MERCK; UPHOF; USD 26th.

1. G. A. Nowak, *Am. Perfum. Cosmet.*, **81**, 37 (1966).

BLACKBERRY BARK

Source: *Rubus fructicosus* L. (Family Rosaceae).

Common/vernacular names: Blackberry, bramble.

GENERAL DESCRIPTION

A spiny shrub with an edible black berry; extensively hybridized; native to temperate Europe and adjacent countries. Part used is the dried bark of the rhizome and roots collected in the spring and fall.

CHEMICAL COMPOSITION

Active constituents believe to be tannins; other constituents present include gallic acid, villosin, and calcium oxalate.

USES

Food. In flavor formulations in all major categories of foods, including alcoholic and nonalcoholic beverages, frozen dairy desserts, candy, baked goods, gelatins and puddings, and sweet sauces, with highest average maximum use level of 0.08% in frozen dairy desserts.

Dietary Supplements/Health Foods. Blackberry bark in capsules, tablets, teas, and other preparations; primarily as an astringent (FOSTER AND DUKE).

Traditional Medicine. The root is used by the Micmac Indians as an astringent and antidiarrheal (MOERMAN); root and root bark used as topical treatment for sore throat, inflammation of the gums, and mouth ulcers (BOWN). In Morocco, the leaves are used to treat diabetes.[1]

COMMERCIAL PREPARATIONS

Crude and extracts; formerly official in U.S.P. and N.F.

Regulatory Status. Approved for food use as a natural flavoring substance (§172.510).

REFERENCE

See the General References for BAILEY 2; FEMA; FERNALD; FOSTER AND DUKE; KROCHMAL AND KROCHMAL; MERCK; YOUNGKEN.

1. H. Jouad et al., *J. Ethnopharmacol.*, **81**, 351 (2002).

BLACK COHOSH

Source: *Actaea racemosa* L. (syn. *Cimicifuga racemosa* (L.) Nutt.) (Family Ranunculaceae).

Common/vernacular names: Black cohosh, black snakeroot, cimicifuga, cohosh bugbane, rattleroot, rattleweed, rattle top, squaw root.

GENERAL DESCRIPTION

Perennial herb, up to 3 m high, with knotted rhizome; leaves are three divided; terminal leaflet three lobed; middle lobe is largest; flowers, white, in tall raceme; native to rich woods of eastern North America, from Maine west to Ontario and Wisconsin, and south to Georgia. Parts used are the dried rhizome and roots.

CHEMICAL COMPOSITION

Roots contain triterpene glycosides: 26-deoxyactein, 23-*epi*-26-deoxyactein (27-deoxyactein or actein),[1] cimicifugoside (cimigoside), cimifugoside M,[2] cimiracemosides,[3] and others;[4] organic acids and esters (2-hexylcyclopropaneoctanoic acid,[5] caffeic, fukinolic acid cimicifugic acids, ferulic acid, isoferulic acid, and others).[6] Other constituents reported to be present include salicylic acid,[7] cimigonite, tannin, and volatile oils (DUKE 2; WREN).

PHARMACOLOGY AND BIOLOGICAL ACTIVITIES

Extracts of the rhizome have failed to show either estrogenic or antiestrogenic activity in either animal or *in vitro* studies.[8–10] However, selective estrogen receptor modulating (SERM) activity has been demonstrated following oral administration of black cohosh extracts (e.g., inhibition of pituitary luteinizing hormone secretion[11] and estrogenic-like effects in fat tissue and osteoblasts in the bone of rats).[10] *In vitro* studies have shown that the triterpene glycoside fraction inhibits the growth of human breast cancer cells,[12] that extracts of the rhizome show serotonin receptor-binding[8] and dopaminergic activity,[9] and that cimicifugoside exhibits nicotinic acetylcholine receptor (nAChR) agonist activity.[13] In a rat model of hot flashes, a standardized ethanolic-aqueous extract of the rhizome administered orally reduced the symptoms and in a behavioral test in rats, showed antidepressant activity.[14] The extract also reduced the loss of bone mineral density in ovariectomized rats.[10]

Clinical trials of standardized isopropanolic extracts of the rhizome (randomized, double-blind, placebo-controlled) have found significant benefits in the treatment of menopause symptoms (MCKENNA), including an improvement in bone metabolism.[15]

TOXICOLOGY

A critical review on clinical studies of black cohosh in the treatment of menopausal symptoms concluded that specific extracts of the rhizome are safe alternatives to estrogen therapy. In trials of black cohosh preparations involving over 2800 patients, the incidence of adverse effects was 5.4%; the majority (97%) were minor and none attributed to black cohosh were serious.[16] An isopropanolic extract failed to increase estrogen-dependent mammary tumors in rats[17] and in a rat model of endometrial cancer, failed to increase growth or metastasis of the primary tumor.[18] No toxic effects were found in humans administered a fluid extract of the rhizome at doses of up to 890 mg/day. Rats administered an isopropanolic extract (up to 5 g/kg) for 26 weeks showed no organ toxicity. The minimum acute lethal oral dose of a tincture of black cohosh in rats was reported to be >1 g/kg. No mutagenicity was found from an isopropanolic extract in the Ames test (MCKENNA).

USES

Medicinal, Pharmaceutical, and Cosmetic. Used in certain analgesic and tonic preparations, among others. In European phytomedicine, isopropanol or ethanol extracts of the dried rhizome standardized to triterpene glycoside contents are used in treating menopausal symptoms, menstrual disorders, including premenstrual discomfort, dysmenorrhea, and postoperatively in patients after hysterectomy or ovariectomy in the treatment of functional deficits (BLUMENTHAL 1; MCKENNA).

Dietary Supplements/Health Foods. Used alone and in herbal formulas to relieve problems related to menopause; capsules, tablets, tinctures, fluid extract, crude root in infusion or decoction (BRADLY; MCKENNA; WREN).

Traditional Medicine. Used in treating amenorrhea, postpartum and labor pains, uterine disorders, support for natural uterine contractions during labor, cough, dropsy, fever, nervous disorders, smallpox, yellow fever, lumbago, pain of acute rheumatism, headache, hysteria, nervous system disorders, influenza, and itch.[16] Used by American Indian tribes as a galactogogue, diuretic, to stimulate menstruation, and in the treatment of colds, rheumatic pains, constipation, coughs, and kidney problems (MOERMAN).

COMMERCIAL PREPARATIONS

Available as crude and extracts (fluid, solid, and powdered). Formerly official in N.F. and U.S.P. Potencies of extracts are expressed only in strength (see *glossary*) based on weight-to-weight ratio of crude and extracts.

Regulatory Status. Regulated in the U.S. as a dietary supplement. The root is the subject of a German therapeutic monograph, indicated for premenstrual discomfort and dysmenorrhea (BLUMENTHAL 1).

REFERENCES

See the General References for BAILEY1; BARNES ET AL.; BLUMENTHAL 1 & 2; BRADLY; DER MARDEROSIAN AND BEUTLER; FOGARTY; FOSTER; FOSTER AND DUKE; JIANGSU; KROCHMAL AND KROCHMAL; GRIEVE; MCKENNA; MERCK; UPHOF; WREN; YOUNGKEN.

1. S. Chen et al., *J. Nat. Prod.*, **65**, 601 (2002).
2. K. He et al., *Planta Med.*, **66**, 635 (2000).
3. E. Bedir and L. A. Khan, *Chem. Pharm. Bull.*, **48**, 425 (2000).
4. K. Wende et al., *J. Nat. Prod.*, **64**, 986 (2001).
5. A. Panossian et al., *Phytochem. Anal.*, **15**, 100 (2004).
6. S. O. Kruse, *Planta Med.*, **65**, 763 (1999).
7. H. Jarry et al., *Planta Med.*, **51**, 316 (1985).
8. J. E. Burdette et al., *J. Agric. Food Chem.*, **51**, 5661 (2003).
9. H. Jarry et al., *Maturitas*, **44**(Suppl. 1), S31 (2003).
10. D. Seidlova-Wuttke et al., *Maturitas*, **44**(Suppl. 1), S39 (2003).
11. E. M. Düker et al., *Planta Med.*, **57**, 420 (1991).
12. L. S. Einbond et al., *Breast Cancer Res. Treat.*, **83**, 221 (2004).
13. K. Woo et al., *J. Pharmacol. Exp. Ther.*, **309**, 641 (2004).
14. H. Winterhoff et al., *Maturitas*, **44**(Suppl. 1), S51 (2003).
15. W. Wuttke et al., *Maturitas*, **44**(Suppl. 1), S67 (2003).
16. T. Low Dog et al., *Menopause*, **10**, 299 (2003).
17. J. Freudenstein et al., *Cancer Res.*, **62**, 3448 (2002).
18. T. Nilein and J. Freudenstein, *Toxicol. Lett.*, **150**, 271 (2004).

BLACK HAW BARK

Source: *Viburnum prunifolium* L. (Family Caprifoliaceae).

Common/vernacular names: Black haw, nanny bush, stag bush, viburnum.

GENERAL DESCRIPTION

Viburnum prunifolium is a spreading deciduous shrub or small tree, up to 5 m high, with blue-black fruits and white flowers native to North America, from Connecticut to Florida and west to Michigan and Texas.

Parts used are the root and stem barks.

CHEMICAL COMPOSITION

Viburnum prunifolium contains coumarins (scopoletin,[1] scopolin, aesculetin), alkaloids, β-sitosterin, arbutin, triterpenes (α- and β-amyrin, ursolic acid, oleanolic acid), amentoflavone,[2,3] and iridoid glucosides (*Valeriana* type).[4,5] Other constituents include malic, citric, oxalic, and valeric acids); tannin; bitter resin; and others (MERCK). 1-Methyl-2,3-dibutyl hemimellitate was isolated from the stem bark.[6]

PHARMACOLOGY AND BIOLOGICAL ACTIVITIES

Black haw has shown uterine antispasmodic properties *in vitro*.[7,8]

USES

Medicinal, Pharmaceutical, and Cosmetic. Root bark and its extracts are used as tonics and in uterine-relaxant, antidiarrheal, diuretic, and general antispasmodic preparations.

Food. Stem bark extract is used as a flavor ingredient primarily in alcoholic and nonalcoholic beverages in very low concentrations, with maximum use level at less than 0.001%.

Dietary Supplements/Health Foods. Root bark used in capsules, tablets, tinctures, and infusion, primarily for uterine-relaxant and antidiarrheal activity (WREN).

Traditional Medicine. American Indians used the root and/or stem bark for the treatment of painful menses, to prevent miscarriage, as a postpartum antispasmodic, and for asthma (FOSTER AND DUKE). The root bark was also used as a tonic for the female reproductive organs and as a diaphoretic (MOERMAN) and the bark was used to treat dysentery.

COMMERCIAL PREPARATIONS

Crude, fluid extract, solid extract, and powdered extract are readily available; crude and fluid extract were formerly official in N.F. and U.S.P. Strengths (see *glossary*) of extracts are expressed in weight-to-weight ratios between crude and extracts.

Regulatory Status. Regulated in the U.S. as a dietary supplement; approved for food use in food as a natural flavoring substance (§172.510).

REFERENCES

See the General References for BAILEY 1; FEMA; FOSTER AND DUKE; KROCHMAL AND KROCHMAL; MCGUFFIN 1 & 2; uphof; APPLEQUIST; WICHTL; WREN; YOUNGKEN.

1. C. H. Jarboe et al., *J. Med. Chem.*, **10**, 488 (1967).
2. R. Upton, ed., *Cramp bark (Viburnum prunifolium)*, American Herbal Pharmacopeia, Santa Cruz, CA, 2000.
3. R. Upton, ed., *Cramp bark (Viburnum opulus)*, American Herbal Pharmacopeia, Santa Cruz, CA, 2000.
4. L. Tomassini et al., *Planta Med.*, **65**, 195 (1999).
5. L. Tomassini et al., *Phytochemistry*, **44**, 751 (1997).
6. C. H. Jarboe et al., *J. Org. Chem.*, **34**, 4202 (1969).
7. C. H. Jarboe et al., *Nature*, **212**, 837 (1966).
8. G. Balansard et al., *Plantes Méd. Phytothér.*, **17**, 123 (1983).

BLESSED THISTLE

Source: *Cnicus benedictus* L. (Family Compositae or Asteraceae).

Common/vernacular names: Carbenia benedicta, carduus benedictus, cnicus, Holy thistle.

GENERAL DESCRIPTION

Thistle-like, highly branched, reddish annual up to 60 cm in height; leaves are long and narrow with prominent white veins beneath; flowers pale yellow in prickly green heads; whole plant covered in thin down; indigenous to waste lands and fields of the Near East and Mediterranean region; naturalized in the United States; cultivated in Europe (BAILEY 1 & 2; GRIEVE). Part used is the herb (flowering tops).

CHEMICAL COMPOSITION

Chloroform and light petrol extracts of the herb yielded sitosterol-3β-D-glucoside, oleanolic acid, multiflorenol, multiflorenol acetate, α-amyrine, and α-amyrenone.[1] Leaves contain lignans, 2-acetylnortracheloside, arctigenin, nortracheloside, salonitenolide, trachelogenin;[2] arctiin in fruit; sesquiterpene lactone cnicin (bitter index = 1 : 1800);[3] lithospermic acid; minute amounts of volatile oil, small amounts of polyacetylenes, approximately 8% tannins, and high amounts of potassium, calcium, and manganese (BRADLY; WREN).

PHARMACOLOGY AND BIOLOGICAL ACTIVITIES

Essential oil bacteriostatic against *Staphylococcus aureus*; inactive in *Bacillus coli*.[4] The extremely bitter sesquiterpene lactone cnicin has shown *in vitro* antibacterial and *in vivo* antitumor activities; also, antifeedant activity against certain insects (HARBOURNE AND BAXTER). Following oral ingestion, the lignans arctiin and tracheloside are metabolized in the intestinal tract to their genins, arctigenin, and trachelogenin, which have inhibitory effects on cyclic-AMP phosphodiesterase and on histamine release in rat mast cells. They also show Ca^{2+} and platelet-activating factor antagonist activities (see also **burdock** and **safflower**).[5]

TOXICOLOGY

An ether extract of blessed thistle showed a strong sensitizing effect in guinea pigs, suggesting that individuals who experience allergic contact dermatitis from exposure to the Compositae family should avoid the plant.[6] Use of the herb is contraindicated in pregnancy (BRADLY).

The approximate acute oral LD_{50} in mice of cnicin is 1.6–3.2 μM/kg.[7]

USES

Food. Extract an ingredient in alcoholic beverages (Benedictine); also in bitters.

Dietary Supplements/Health Foods. Leaf capsules, tablets; tea, extract, tincture, primarily as bitter digestive (BLUMENTHAL 1), antiflatulent and in gallbladder disease.

Traditional Medicine. Used as an emmenagogue, galactogogue, emetic, appetite stimulant, diaphoretic, and in the treatment of intestinal worms, fevers, headaches, migraines, body aches, memory, hearing loss, and sores (GRIEVE; WREN).

COMMERCIAL PREPARATIONS

Crude leaves, leaf extract.

Regulatory Status. Regulated in the United States as a dietary supplement; approved for use as a natural flavoring in alcoholic beverages only (§172.510); subject of German therapeutic monograph; allowed as a bitter digestive to treat loss of appetite and dyspeptic discomfort (BLUMENTHAL 1; WICHTL).

REFERENCES

See the General References for AHPA; BISSET; BLUMENTHAL 1; BRUNETON; DUKE 2; GLEASON AND CRONQUIST; HARBOURNE AND BAXTER; MCGUFFIN 1 & 2; NIKITAKIS; STEINMETZ; TUTIN 4; TYLER 1; UPHOF; WREN.

1. A. Ulubelen and T. Berkan, *Planta Med.*, **31**, 375 (1977).
2. M. Vanhaelen and R. Vanhaelen-Fastré, *Phytochemistry*, **14**, 2709 (1975).

3. R. Vanhaelen-Fastré and M. Vanhaelen, *Planta Med.*, **29**, 179 (1976).
4. R. Vanhaelen-Fastré, *Planta Med.*, **24**, 165 (1973).
5. M. Nose et al., *Planta Med.*, **59**, 131 (1993).
6. W. Zeller et al., *Arch. Dermatol. Res.*, **277**, 28 (1985).
7. G. Schneider and I. Lachner, *Planta Med.*, **53**, 247 (1987).

BLOODROOT

Source: *Sanguinaria canadensis* L. (Family Papaveraceae).

Common/vernacular names: Bloodroot, Indian red paint, red puccoon, red root, sanguinaria.

GENERAL DESCRIPTION

A low perennial herb with horizontal, branching rhizome bearing slender roots; up to about 35 cm high; both rhizome and roots contain an orange-red latex. The plant grows in Quebec and in the United States from New England south to Florida and west to Wisconsin and Texas. Part used is the dried rhizome, sometimes referred to as "root" in the literature.

CHEMICAL COMPOSITION

Both the rhizome and root contain benzo[c]phenanthridine alkaloids (approximately 4–7% and 1.8%, respectively).[1] The major alkaloid of the rhizome is sanguinarine, followed in order of concentration by chelerythrine, sanguilutine, chelilutine, chelirubine, and sanguirubine.[2] Other alkaloids identified in the plant include protopine, oxysanguinarine, α- and β-allocryptopine, dihydrosanguilutine, berberine, coptisine, and homochelidonine. Also contains a reddish resin and citric and malic acids.[3–5]

PHARMACOLOGY AND BIOLOGICAL ACTIVITIES

Sanguinarine has broad *in vitro* antibacterial activity and also displays antifungal and antitrichomonas activities[2] and local anesthetic property.[6,7] It has also shown inhibition of NF-κB[3], anti-inflammatory and antioxidant activities,[8] and antitumor activity in animals.[9] At micromolar concentrations *in vitro*, sanguinarine inhibited the growth of human epidermoid carcinoma cells through the induction of apoptosis.[10] Protopine has shown platelet aggregation-inhibitory effects in animals.[11]

The majority of double-blind, placebo-controlled, and other controlled and open label trials of sanguinaria oral rinses with zinc and sanguinaria tooth pastes (alone and in combination) have shown that the preparations are effective at reducing gingival inflammation and plaque and in decreasing symptoms of periodontitis.[2,12–20]

TOXICOLOGY

Sanguinarine and dehydrosanguinarine were implicated in the pathogenesis of glaucoma in epidemic dropsy in India in the years 1880–1930 when it was found as a constituent (approximately 0.5%) of Mexican prickly-poppy seed oil (*Argemone mexicana* L.).[1,21] Although the contribution of sanguinarine was later disputed,[22] further studies found the alkaloid inhibited Na^+/K^+-ATPase activity of the heart, which suggested that it could be responsible for degenerative changes seen in cardiac heart muscle of rats fed Mexican prickly-poppy seed oil and consequently for cardiac failure and tachycardia in epidemic dropsy patients. Other studies have shown that the alkaloid interferes with pyruvic acid oxidation that leads to pyruvic acid accumulation in the blood.[21] Long-term studies in rats administered the alkaloid orally failed to show teratogenic effects.[1] Oral administration of a sanguinaria extract containing approximately 68% total benzo[c]phenanthridine alkaloids and approximately 33% sanguinarine chloride to rabbits and rats failed to show any effects on fetal and neonatal development, fertility, or reproduction at doses below 60 mg/kg/day and 25 mg/kg/day, respectively. The acute oral LD_{50} of sanguinarine and extracts of the rhizome in rats ranges between 1250 and 1658 mg/kg.[23] A recent study of oral leukoplakia in 10 patients who had used a rinse and/or dentifrice containing sanguinaria for 6 months called for avoiding such preparations until the risk of malignancy could be determined.[24]

USES

Medicinal, Pharmaceutical, and Cosmetic. Formerly used in treating carcinoma of the nose following surgical treatment[25] and in cough remedies, almost always in combination with other herbal ingredients (e.g., spikenard root, balm of Gilead bud, white pine, and wild cherry barks, as in Compound White Pine Syrup and other formulations). Also used in cosmetics for its alleged healing properties. Sanguinarine has also been used as an antiplaque agent in toothpaste and mouthwash preparations.

Traditional Medicine. Used by various eastern North American Indians for face painting; by Canadian Indians in Quebec as a tonic; root chewed to treat heart troubles. Used by American Indians as a blood purifier and to treat burns, cuts, sores, ulcers, debility, pain, hemorrhages, fevers, asthma, coughs, colds, sore throat, tuberculosis, gonorrhea, rheumatism, stomach cramps, and other health problems (MOERMAN).

Regulatory Status. Classified by the U.S. FDA as being unsafe for use in drugs, foods, or beverages (WICHTL).E Crude and fluid extracts were formerly official in N.F. and U.S.P. Strengths (see *glossary*) of extracts are expressed in weight-to-weight ratios between crude and extracts.

REFERENCES

See the General References for APPLEQUIST; BARNES; CLAUS; DE NAVARRE; DER MARDEROSIAN AND BEUTLER; DER MARDEROSIAN AND LIBERTI; KROCHMAL AND KROCHMAL; MARTINDALE; MERCK; YOUNGKEN.

1. P. Kosina et al., *Food Chem. Toxicol.*, **42**, 85 (2004).
2. K. C. Godowski, *J. Clin. Dent.*, **1**, 96 (1989).
3. F. Santavy in R. H. F. Manske, ed., *The Alkaloids*, Vol. 12, Academic Press, New York, 1970, p. 333.
4. M. Tin-Wa et al., *J. Pharm. Sci.*, **61**, 1846 (1972).
5. D. K. Kim and F. R Stermitz, *Phytochemistry*, **14**, 834 (1975).
6. V. Preininger in R. H. F. Manske, ed., *The Alkaloids*, Vol. 15, Academic Press, New York, 1975, p. 207.
7. I. S. Shenolikar et al., *Food Cosmet. Toxicol.*, **12**, 699 (1974).
8. M. M. Chaturvedi et al., *J. Biol. Chem.*, **272**, 30129 (1997).
9. M. Tin-Wa et al., *Lloydia*, **33**, 267 (1970).
10. N. Ahmad et al., *Clin. Cancer Res.*, **6**, 1524 (2000).
11. H. Matsuda, *Planta Med.*, **54**, 498 (1988).
12. R. T. Boulware et al., *J. Soc. Cosmet. Chem.*, **36**, 297 (1985).
13. H. Tenebaum et al., *J. Periodontol.*, **70**, 307 (1999).

14. R. A. Kopczyk et al., *J. Periodontol.*, **62**, 617 (1991).
15. M. P. Cullinan et al., *Aust. Dent. J.*, **42**, 47 (1997).
16. A. M. Polson et al., *J. Periodontol.*, **68**, 119 (1997).
17. A. M. Polson et al., *J. Clin. Periodontol.*, **23**, 782 (1996).
18. D. S. Harper et al., *J. Clin. Periodontol.*, **16**, 352 (1990).
19. E. Grossman et al., *J. Periodontol.*, **60**, 435 (1989).
20. H. Etemadzadeh et al., *J. Clin. Periodontol.*, **14**, 176 (1987).
21. M. Das and S. K. Khanna, *Crit. Rev. Toxicol.*, **27**, 273 (1997).
22. G. Lord et al., *J. Clin. Dent.*, **1**, 110 (1989).
23. P. J. Becci, *J. Toxicol. Environ. Health*, **20**, 199 (1987).
24. L. R. Eversole et al., *Oral Surg. Oral Med. Oral Pathol. Oral Radiol. Endod.*, **89**, 455 (2000).
25. J. T. Phelan and J. Juardo, *Surgery*, **53**, 310 (1963).

BLUE COHOSH

Source: *Caulophyllum thalictroides* (L.) Michx. (Family Berberidaceae).

Common/vernacular names: Blue cohosh, caulophyllum, papoose, squaw root.

GENERAL DESCRIPTION

Perennial herb with a thick, crooked, and horizontal rhizome, up to 1 m high; grows in eastern North America. Parts used are the dried rhizome and roots.

CHEMICAL COMPOSITION

The rhizomes and roots contain triterpene saponins (e.g., caulophyllogenin 3-O-α-L-arabinopyranoside, hederagenin 3-O-α-L-arabinopyranoside);[1] alkaloids, including N-methylcytisine (caulophylline), baptifoline, anagyrine, lupanine, α-isolupanine, 5,6-dehydro-α-lupanine, magnoflorine, sparteine, and taspine.[2,3] Other constituents present include caulosaponin and resins.[4–6]

The rhizomes and roots of an eastern Asian species of *Caulophyllum* (*C. robustum* Maxim.) have been extensively studied in the former USSR and found to be rich in triterpene glycosides (caulosides A, B, C, D, and G), most of which have hederagenin as their aglycone; they possess fungicidal activities.[7–10]

PHARMACOLOGY AND BIOLOGICAL ACTIVITIES

An alcoholic extract of the aerial parts after treatment with petroleum ether showed anti-inflammatory activity in rats.[11] A hot water extract and the saponin-containing fraction of the roots and rhizomes exhibited a uterine stimulant effect in isolated rat uterine muscle preparations.[6] An extract of blue cohosh produced tonic contraction of isolated uteri of guinea pigs or rats.[12]

Methylcytisine stimulates motility of the small intestine, stimulates respiration, and increases blood pressure (TYLER 1).

TOXICOLOGY

In vitro teratogenic activity was found in the rat embryo culture system from the alkaloids taspine and N-methylcytisine.[3] Whereas the latter is found in varying amounts in dietary supplements containing blue cohosh

(8–850 ppm),[13] very little taspine was found in the dried rhizome (0.00013%).[3] Reduced coronary flow was observed in rat heart preparations infused with *N*-methylcytisine and a vasoconstrictive effect was found in hog and cattle carotid artery preparations exposed to the compound.[14]

Case reports of adverse effects attributed to blue cohosh include contact dermatitis from handling the root;[12] severe congestive heart failure in a neonate after the mother ingested 3 times the recommended dose of blue cohosh tablets for several weeks prior to parturition;[15] poisoning in children from ingestion of the seeds.[12]

USES

Medicinal, Pharmaceutical, and Cosmetic. Used in diuretic, uterine, antispasmodic, and emmenagogue as well as laxative preparations.

Dietary Supplements/Health Foods. Tea, tincture, capsules, tablets, and other products used primarily for menstrual difficulties (amenorrhea and dysmenorrhea) (FOSTER).

Traditional Medicine. A decoction or syrup of the root was used by American Indians as a sedative to treat "hysterics" and "fits." Often in the form of a decoction, the root was also used to treat rheumatism, fevers, stomach cramps (alone or in conjunction with painful menstruation), genitourinary problems in men, profuse menstruation, to aid childbirth, and in treatments of lung problems, including hemorrhages (MOERMAN).

COMMERCIAL PREPARATIONS

Available as crude and extracts; crude was formerly official in N.F. (1916–1942); U.S.P (1880–1890). Strengths (see ***glossary***) of extracts are expressed as weight-to-weight ratios between crude and extract.

Regulatory Status. Class 2b (not to be used in pregnancy).

REFERENCES

See the General References for APPLEQUIST; BARNES; DER MARDEROSIAN AND BEUTLER; FOSTER; GRIEVE; KROCHMAL AND KROCHMAL; MCGUFFIN 1 & 2; MERCK; YOUNGKEN.

1. J. W. Jhoo et al., *J. Agric. Food Chem.*, **49**, 5969 (2001).
2. M. Ganzera et al., *Phytochem. Anal.*, **14**, 1 (2003).
3. E. J. Kennelly et al., *J. Nat. Prod.*, **62**, 1385 (1999).
4. M. S. Flom et al., *J. Pharm. Sci.*, **56**, 1515 (1967).
5. M. S. Flom, *Diss. Abstr. Int. B*, **32**, 2312 (1971).
6. C. T. Che, *Diss. Abstr. Int. B*, **43**, 1049 (1982).
7. L. I. Strigina et al., *Khim. Prir. Soedin.*, **6**, 552 (1970).
8. L. I. Strigina et al., *Phytochemistry*, **14**, 1583 (1975).
9. M. M. Anisimov, *Antibiotiki (Moscow)*, **17**, 834 (1972).
10. L. I. Strigina et al., *Khim. Prir. Soedin.*, **5**, 619 (1976).
11. P. S. Benoit et al., *Lloydia*, **39**, 160 (1976).
12. P. A. G. M. De Smet in P. A. G. M. De Smet, et al., eds., *Adverse Effects of Herbal Drugs*, Vol. 2, Springer-Verlag, Berlin, Germany, 1993, p. 153.
13. J. Betz et al., *Phytochem. Anal.*, **9**, 232 (1998).
14. H. C. Ferguson and L. D. Edwards, *J. Am. Pharm. Assoc.*, **43**, 16 (1954).
15. T. K. Jones and B. M. Lawson, *J. Pediatr.*, **132**, 550 (1998).

BOIS DE ROSE OIL

Source: *Aniba rosaeodora* Ducke (Family Lauraceae).

Common/vernacular names: Rosewood oil, cayenne rosewood oil.

GENERAL DESCRIPTION

Medium-size evergreen tree, growing wild in the Amazon region, notably Brazil, French Guiana, Peru, and East Surinam. The essential oil is obtained from chipped wood by steam distillation and occasionally water distillation. Brazil and Peru are the major producers of the oil; French Guiana is also a producer, producing cayenne bois de rose oil that is considered the best quality among the bois de rose oils (ARCTANDER; MASADA).

CHEMICAL COMPOSITION

The major component is linalool, which is present in 90–97% in cayenne bois de rose oil and 80–90% in Brazilian Oil (MASADA).[1,2] The balance is made up of cineole (up to 10%), α-terpineol, geraniol, citronellal, limonene, α-pinene, β-pinene, β-elemene, *cis*- and *trans*-linalool oxides, sesquiterpenes, and others (MASADA).[3,4]

PHARMACOLOGY AND BIOLOGICAL ACTIVITIES

Available data indicate acetylated bois de rose oil to be nontoxic when applied externally.[5]

Linalool, the major component of bois de rose oil, has weak tumor-promoting properties in mice.[6,7] It also has been reported to have anticonvulsant activity in mice and rats, spasmolytic activity on isolated guinea pig ileum, antimicrobial properties, and others.[7]

USES

Medicinal, Pharmaceutical, and Cosmetic. Used as a source of natural linalool or linalool acetate, which are extensively used in perfumery. Acetylated bois de rose oil is reportedly used in soaps, detergents, creams, lotions, and perfumes, with maximum use level of 1.2% in perfumes.[5]

Food. Used extensively as a flavor ingredient in most major categories of foods, ages, frozen dairy desserts, candy, baked goods, gelatins and puddings, meat and meat products, and gravies. Average maximum use levels are generally below 0.003% (24.9 ppm).

COMMERCIAL PREPARATIONS

Oil and acetylated oil. Bois de rose is official in F.C.C.]

Regulatory Status. GRAS (§182.20).

REFERENCES

See the General References for ARCTANDER; FEMA; GUENTHER; LIST AND HÖRHAMMER; MA; MCGUFFIN 1 & 2; UPHOF.

1. A. Alpande de Morais et al., *Acta Amazon.*, **2**, 41 (1972).
2. A. Alpande de Morais et al., *An. Acad. Bras. Cienc.*, **44**(Suppl.) 303 (1972).
3. G. Chiurdoglu et al., *Ind. Chim. Belge.*, **28**, 636 (1963).
4. I. C. Nigam and L. Levi, *Perfum. Essent. Oil Rec.*, **54**, 814 (1963).
5. D. L. J. Opdyke, *Food Cosmet. Toxicol.*, **11**, 1039 (1973).
6. F. Homburger and E. Boger, *Cancer Res.*, **28**, 2372 (1968).
7. D. L. J. Opdyke, *Food Cosmet. Toxicol.*, **13**, 827 (1975).

BOLDO LEAVES

Source: ***Peumus boldus*** Molina (syn. ***Boldu boldus*** (Mol.) Lyons) (Family Monimiaceae).

Common/vernacular names: Boldo, boldo leaves, boldus.

GENERAL DESCRIPTION

Dioecious evergreen shrub or small tree, up to about 6 m high; native to mountainous regions of Chile and naturalized in Europe (Mediterranean region). Parts used are the dried leaves.

CHEMICAL COMPOSITION

The dried leaves contain 0.25–0.5% aporphine alkaloids, including laurotetanine, N-methyllaurotetanine, boldine (0.06%), isoboldine, laurolitsine, norisocorydine, isocorydine, isocorydine-N-oxide, and reticuline;[1–5] approximately 2.5% volatile oil composed mainly of ascaridole (45%),[1] p-cymene (28.6%), ascaridole (16.1%), 1,8-cineole (16.0%), and linalool (9.1%), among 38 identified compounds;[5,6] flavonol glycosides (e.g., isorhamnetin-3-α-L-arabinopyranoside-7-α-L-rhamnopyranoside,[7] boldoside, fragroside, and pneumoside),[1] resin, and tannins.

PHARMACOLOGY AND BIOLOGICAL ACTIVITIES

The leaves have shown choleretic, diuretic, stomachic, cholagogic, and other activities.[5,7] One study indicated that in rats, the total alcoholic extract of the leaves, as opposed to partial extracts, had the highest choleretic activity.[8] Boldine has shown anti-inflammatory activity in guinea pigs in the carrageenan-induced paw edema test (ED_{50}, 34 mg/kg p.o.), and in rabbits (60 mg/kg p.o.) reduced hyperthermia induced by bacterial pyrogen. *In vitro* tests showed that boldine inhibits prostaglandin biosynthesis.[9] In streptozotocin-induced diabetic rats, the addition of boldine in the drinking water resulted in less weight loss and hyperglycemia, restoration of aberrant enzyme activities in the pancreas and liver, and attenuation of various oxidative processes.[10]

Ethanolic and ether extracts of the leaves exhibit strong *in vitro* antioxidant activity,[11] as does boldine in a variety of assays.[1,12] *In vitro* free radical scavenging activity of extracts of the dried leaves is mainly attributable to catechin.[13] Boldine has also shown *in vitro* chemoprotectant activity, in part by decreasing metabolic activation of chemical mutagens and stimulating glutathione S-transferase.[14] Antihilminthic activity of the leaves is attributed to ascaridole.[1]

In a placebo-controlled study, volunteers administered a powder extract of the leaves (2.5 g p.o.) showed a significant increase in oro-cecal transit time, thereby confirming results found in animal studies, which showed that boldo could prolong the intestinal transit time and relax smooth muscles.[15]

TOXICOLOGY

No toxicity was observed in rats that administered a hydro-alcoholic extract of the leaves containing boldine at doses of up to 3 g/kg p.o., which was three times the lethal oral dose of boldine in mice. Boldine is nonmutagenic and showed no genotoxicity in the mouse micronucleus test following oral doses of up to 900 mg/kg.[16,17] In chronic toxicity tests in rats, both a hydro-alcoholic extract of the dried leaves and boldine produced a low level of fetal toxicity from oral doses of 800 mg/kg and none from 500 mg/kg p.o. Similarly, alterations in cholesterol, bilirubin, urea, AST, and ALT were found from the 800 mg doses of the extract or boldine over 90 days, whereas the 500 mg dose was without significant effects. The authors of the study advised that boldo should not be used during pregnancy.[18] The German Commission E advises against the use of the essential or distillates of the leaves because of the content of ascaridole, a toxic, antihilminthic principle (BLUMENTHAL 1; WICHTL).

USES

Medicinal, Pharmaceutical, and Cosmetic. In tonic and diuretic preparations; in Europe for gastrointestinal spasms (BLUMENTHAL 1); rarely, if at all, used in cosmetics.

Food. Only used in alcoholic beverages (liqueurs, bitters, etc.). Average maximum use level reported is about 0.002% (16 ppm).

Dietary Supplements/Health Foods. Leaves used in combination products, capsules, tablets, and infusions as antioxidant (WREN) and for liver and gallbladder (WICHTL).

Traditional Medicine. Reportedly used as a diuretic and biliary stimulant in hepatic illnesses and cholelithiasis (gallstones) (MARTINDALE); used by the Mapuche Indians of Chile in the treatment of rheumatism. Other traditional uses of boldo include nervous weakness, headache, nasal congestion, menstrual pain, earache, gonorrhea, syphilis, flatulence, and dyspepsia.[1]

COMMERCIAL PREPARATIONS

Crude and extracts. Crude and fluid extract were formerly official in N.F. Strengths (see *glossary*) of extracts are expressed in weight-to-weight ratios.

Regulatory Status. Regulated in the United States as a dietary supplement; leaves permitted for use as a natural flavoring substance in alcoholic beverages only (§172.510). Leaf preparations that are practically free of the toxic principle ascaridole are the subject of a German therapeutic monograph that allows use for mild gastrointestinal spasms and dyspeptic disorders. It advises against any use in cases of biliary obstruction and severe liver diseases, and in cases of gallstones without the advice of a physician (BLUMENTHAL 1; WICHTL).

REFERENCES

See the General References for BIANCHINI AND CORBETTA; BLUMENTHAL 1; DER MARDEROSIAN AND BEUTLER; FEMA; GOSSELIN; GRIEVE; MARTINDALE; RAFFAUF; TERRELL; WILLAMAN AND SCHUBERT; WREN; YOUNGKEN.

1. H. Speisky and B. K. Cassels, *Pharmacol. Res.*, **29**, 1 (1994).
2. M. Sobiczewska and B. Borkowski, *Acta Pol. Pharm.*, **29**, 271 (1972).
3. M. Vanhaelen, *J. Pharm. Belg.*, **28**, 291 (1973).
4. N. Didry, *Bull. Soc. Pharm. Lille*, **31**, 51 (1977).
5. H. Schindler, *Arzneim.-Forsch.*, **7**, 747 (1957).
6. K. Bruns and M. Köhler, *Parfüm. Kosmet.*, **55**(8), 225 (1974).
7. E. Bombardelli et al., *Fitoterapia*, **47**, 3 (1976).
8. M. C. Levy-Appert-Collin and J. Levy, *J. Pharm. Belg.*, **32**, 13 (1977).
9. N. Backhouse et al., *Agents Actions*, **42**, 114 (1994).
10. Y. Y. Jang et al., *Pharmacol. Res.*, **42**, 361 (2000).
11. T. Hirosue et al., *Nippon Shokuhin Kohyo Gakkaishi*, **35**, 630 (1988).
12. I. Jimenez et al., *Phytother. Res.*, **14**, 339 (2000).

13. G. Schmeda-Hirschmann et al., *Free Radic. Res.*, **37**, 447 (2003).
14. R. Kubinova et al., *Pharmazie*, **56**, 242 (2001).
15. M. Gotteland et al., *Rev. Med. Chil.*, **123**, 955 (1995).
16. P. R. Moreno et al., *Mutat. Res.*, **260**, 145 (1991).
17. D. C. Tavares and C. S. Takahashi, *Mutat. Res.*, **321**, 139 (1994).
18. E. R. de Almeida et al., *Phytother. Res.*, **14**, 99 (2000).

BONESET

Source: *Eupatorium perfoliatum* L. (Family Compositae or Asteraceae).

Common/vernacular names: Agueweed, bonest, common bonest, Eupatorium, feverwort, thoroughwort.

GENERAL DESCRIPTION

Perennial herb with opposite sessile clasping leaves; up to about 1.5 m high; native to eastern and central North America, from Quebec south to Florida, Alabama, and Louisiana, and west to Texas and the Dakotas. Parts used are the dried leaves and flowering tops collected in late summer.

CHEMICAL COMPOSITION

The chemical information on various *Eupatorium* species is considerable (KARRER).[1–8] *E. perfoliatum* is reported to contain flavonoids, including quercetin, kaempferol, quercetin-3-β-galactoside (hyperoside), kaempferol-3-β-glucoside (astragalin), quercetin-3-rutinoside (rutin), kaempferol-3-rutinoside, and eupatorin (MERCK);[1] terpenoids, including chromenes, sesquiterpenes, sesquiterpene lactones (euperfolin, euperfolitin, eufoliatin, and euperfolide, among others), and diterpenes (dendroidinic acid and hebeclinolide);[2–4] triterpenes (e.g., α-amyrin); sterols (sitosterol and stigmasterol); dotriacontane;[5] a volatile oil; a polysaccharide (4-*O*-methylglucuronoxylan);[9,10] and resin, among others.

PHARMACOLOGY AND BIOLOGICAL ACTIVITIES

Boneset is reported to have stimulant and diaphoretic properties; in large doses it is both emetic and cathartic (CLAUS).

An ethanol extract of the leaves showed cytotoxicity to mammalian cells *in vitro* and weak *in vitro* antibacterial activity against Gram-positive and Gram-negative organisms.[11] Numerous sesquiterpene lactones (e.g., eupatilin, eupafolin, eupatorin acetate, and eupaformosanin) and flavones (especially eupatorin) isolated from *Eupatorium* species have shown cytotoxic and/or antineoplastic activities.[6–8,12]

An ethanol extract of the whole plant after treatment with petroleum ether exhibited weak anti-inflammatory activity in rats.[13]

The polysaccharide has shown immunostimulating activities following i.v. administration in mice and *in vitro* (chemoluminescence, carbon clearance, and granulocyte tests).[9,10] Preliminary screening also indicates that eufoliatin has immunostimulating activity.[14]

TOXICOLOGY

Use discouraged due to potential presence of hepatotoxic pyrrolizidine alkaloids ubiquitous in *Eupatorium* species (NEWALL).[15] Abortion in cattle grazing on the plant is attributed to a high content of nitrate.[16] Toxic principles are believed to include eupatorin (cytotoxic, emetic), bitter components (gastrointestinal irritants), and lactones (skin sensitizers) (BRINKER).

USES

Medicinal, Pharmaceutical, and Cosmetic. Used in certain antipyretic and urinary antiseptic preparations, among others.

Dietary Supplements/Health Foods. Crude herb in infusion, also tinctures and extracts as immunostimulant and for fevers; use relatively uncommon (FOSTER AND DUKE).

Traditional Medicine. Used in infusion as tonic, febrifuge, diaphoretic, emetic, and cathartic; also used to treat skin rashes (KROCHMAL AND KROCHMAL). Used by North American Indians to break fevers and induce sweating; adopted by settlers to treat colds, influenza, typhoid, malaria, intestinal worms, and rheumatism.[15]

COMMERCIAL PREPARATIONS

Crude; extracts not readily available. Crude was formerly official in N.F. and U.S.P. (1820–1900).

Regulatory Status. Classified by the U.S. FDA as a herb of unknown safety (NEWALL).

REFERENCES

See the General References for BARNES; CLAUS; DE NAVARRE; DER MARDEROSIAN AND BEUTLER; FERNALD; FOSTER AND DUKE; GOSSELIN; GRIEVE; KROCHMAL AND KROCHMAL; LIST AND HÖRHAMMER; LUST; MCGUFFIN 1 & 2; TYLER 1; YOUNGKEN.

1. H. Wagner et al., *Phytochemistry*, **11**, 1504 (1972).
2. F. Bohlmann and M. Grenz, *Chem. Ber.*, **110**, 1321 (1977).
3. W. Herz et al., *J. Org. Chem.*, **42**, 2264 (1977).
4. F. Bohlmann et al., *Phytochemistry*, **16**, 1973 (1977).
5. X. A. Dominguez et al., *Phytochemistry*, **13**, 673 (1974).
6. M. D. Midge and A. V. R. Rao, *Indian J. Chem.*, **13**, 541 (1975).
7. K. H. Lee et al., *Phytochemistry*, **16**, 1068 (1977).
8. E. O. Arene et al., *Lloydia*, **41**, 186 (1978).
9. H. Wagner et al., *Arzneim.-Forsch.*, **35**, 1069 (1985).
10. A. Vollmar et al., *Phytochemistry*, **25**, 377 (1986).
11. S. Habtemariam and M. Macpherson, *Phytother. Res.*, **14**, 575 (2000).
12. E. Rodriguez et al., *Phytochemistry*, **15**, 1573 (1976).
13. P. S. Benoit et al., *Lloydia*, **39**, 160 (1976).
14. H. Wagner et al., *Planta Med.*, **51**, 139 (1985).
15. R. A Locock, *Can. Pharm. J.*, **123**, 229 (1990).
16. J. M. Sund et al., *Agron. J.*, **49**, 278 (1957).

BORAGE

Source: *Borago officinalis* L. (Family Boraginaceae).

Common/vernacular names: Borage.

GENERAL DESCRIPTION

Coarse, hispid annual, 15–100 cm high; leaves rough, wrinkled; flowers blue, star shaped with protruding cone; indigenous to dry, waste places of south Europe; grown as an ornamental or potherb; naturalized in central, eastern, and western Europe; established as a casual weed in the eastern United States. The parts used are the nutlets (seeds), leaves, and flowers.

CHEMICAL COMPOSITION

Seeds, leaves, and flowers contain pyrrolizidine alkaloids (PA);[1] total alkaloid content of the plant is less than 0.001%;[2] mature seeds contain about 0.03% crude alkaloids.[3] Leaves contain small amounts, including lycopsamine, intermedine, their acetyl derivatives, supinine, supinidine, and amabiline; also choline (WREN); 9.1% fatty acids, including α-linolenic acid (55%) and γ-linolenic acid (>4%); silicic acid (1.5%–2.2%); potassium, calcium, potassium nitrate (3%), acetic, lactic, and malic acids; δ-bornesitol, cyanogens; fresh leaves contain up to 30% mucilage hydrolyzing to glucose, galactose, arabinose, and allantoin (especially in seedlings). Seed oil (28–38% lipids) is of recent interest as a rich source of γ-linolenic acid (GLA; 17–25%).[2] Other seed fatty acids include linoleic (38%), oleic (14.5–23%), palmitic (11%), and stearic (4.7%).[3,4] Immature seeds and flowers contain amabiline and the rare nontoxic, saturated PA, thesinine;[3] in mature seeds contain a glucoside of thesinine (thesinine-4′-O-β-D-glucoside[5]); thesinine and toxic PA are absent from seed oil;[6,7] roots contain minor amounts of PA (supinine, intermedine, lycopsamine, amabiline, 7-acetyllycopsamine, and 7-acetylintermedine).[7]

PHARMACOLOGY AND BIOLOGICAL ACTIVITIES

Seed oil of interest for GLA content as a prostaglandin precursor, especially for PGE_1; prostaglandins help regulate metabolic functions. Normal synthesis of GLA from linoleic acid via δ-6-desaturase may be blocked or diminished in mammalian systems as the result of aging, diabetes, excessive carbohydrate intake, or fasting. Seeds of *Oenothera biennis* (see ***evening primrose***), various *Ribes* species, and borage serve as GLA sources for dietary supplements, borage seed oil having the highest concentration.[4] GLA is of potential therapeutic interest in treatments of atopic eczema, premenstrual syndrome, diabetes, alcoholism, inflammation, and prevention of heart disease and stroke.[8]

A methanol extract of the leaves exhibited antioxidant activity using the DPPH free radical method. The major antioxidant was identified as rosmarinic acid.[9] Solvent extracts of the defatted seeds have shown *in vitro* antioxidant activity in a meat model system[10] and the DPPH free radical method.[11]

A double-blind, placebo-controlled trial of borage seed oil (providing 1.4 g GLA/day) in patients with active synovitis and rheumatoid arthritis found significant symptomatic improvements in both tender and swollen joints.[12] Compared to placebo, borage seed oil also significantly attenuated heart and blood pressure rates, decreased task performance and the increase in skin temperature in humans in response to acute stress.[13]

TOXICOLOGY

The dried herb, tincture, and decoction fed to guinea pigs for 5 weeks produced no adverse effects except for fatty liver. The seed oil administered orally to mice (0.1 mL) was also without toxic effects and only produced a mild laxative effect.[6] Toxic PA in borage are lycopsamine and intermedine and their 7-acetyl derivatives and the only slightly toxic alkaloid amabiline.[2]

USES

Medicinal, Pharmaceutical, and Cosmetic. Borage extract used in skin care products (NIKITAKIS).

Food. Seldom used in flavoring; fresh flowers with saline, cucumber-like flavor (free of toxic PA) used in salads.

Dietary Supplements/Health Foods. Capsulated seed oil products available as dietary supplement; dried tops sometimes used in teas.[4]

Traditional Medicine. Leaves reportedly used as diuretic, demulcent, emollient, expectorant, and refrigerant; in fevers, lung diseases, colds; externally a poultice; a folk cancer remedy in breast or facial cancers (DUKE 2; FOSTER; WREN).

COMMERCIAL PREPARATIONS

Crude leaves, leaf extract, and seed oil.

Regulatory Status. Regulated in the United States as a dietary supplement. Borage leaf and flower subject of a German therapeutic monograph primarily concerning inflammatory conditions; use suspended due to PA content (BLUMENTHAL 1).

REFERENCES

See the General References for AHPA; BARNES; BLUMENTHAL 1; DER MARDEROSIAN AND BEUTLER; DUKE 2; FOSTER; GLEASON AND CRONQUIST; GRIEVE; HARBOURNE AND BAXTER; MCGUFFIN 1 & 2; NIKITAKIS; STEINMETZ; TUTIN 3; TYLER 1; UPHOF; WREN.

1. K. M. Larson et al., *J. Nat. Prod.* **47**, 747 (1984).
2. E. Roeder, *Pharmazie*, **50**, 83 (1995).
3. C. D. Dodson and F. R. Stermitz, *J. Nat. Prod.*, **49**, 727 (1986).
4. D. V. C. Awang, *Can. Pharm. J.*, **121** March (1990).
5. M. Hermann et al., *Phytochemistry*, **60**, 399 (2002).
6. P. A. G. M. De Smet in P. A. G. M. De Smet et al., eds., *Adverse Effects of Herbal Drugs*, Vol. 2, Springer-Verlag, Berlin, 1993, pp. 147–152.
7. T. Langer and C. Franz, *Sci. Pharm.*, **65**, 321 (1997).
8. J. Janick et al. in L. E. Craker and J. E. Simon, eds., *Herbs, Spices, and Medicinal Plants: Recent Advances in Botany, Horticulture, and Pharmacology*, Vol. 4, Oryx Press, Phoenix, AZ, 1989, p. 145.
9. D. Bandoniene and M. Murkovic, *Biochem. Biophys. Methods*, **53**, 45 (2002).
10. M. Wettasinghe and F. Shahidi, *Food Chem.*, **67**, 399 (1999).
11. M. Wettasinghe and F. Shahidi, *Food Chem.*, **70**, 17 (2000).
12. L. J. Leventhal et al., *Ann. Int. Med.*, **119**, 867 (1993).
13. D. E. Mills et al., *J. Hum. Hypertens.*, **3**, 111 (1989).

BORONIA ABSOLUTE

Source: *Boronia megastigma* Nees ex. Bartl. (Family Rutaceae).

GENERAL DESCRIPTION

Shrub about 3 m high; native to southwestern Australia. Part used is the flower, from which a concrete is produced in 0.4–0.8% yield by extraction with petroleum ether; the absolute is obtained by alcohol washing of the concrete in a yield of about 60%.[1]

CHEMICAL COMPOSITION

Contains ionones as principal constituents, which are composed mainly of β-ionone

(95%), with the rest being *d*-α-ionone (KARRER),[2] eugenol, hydrocarbons (mostly heptacosane), and others.[3]

TOXICOLOGY

Its major constituents, ionones, are reportedly allergenic (MERCK).

USES

Medicinal, Pharmaceutical, and Cosmetic. Used mainly in expensive perfumes.

Food. Quite extensively used in fruit-type flavors (e.g., strawberry, raspberry, plum, and peach) in most major categories of food products, including alcoholic and nonalcoholic beverages, frozen dairy desserts, candy, baked goods, gelatins and puddings, meat and meat products, and condiments and relishes. Use levels are very low, with highest average maximum of about 0.001% (11.6 ppm) reported in baked goods.

COMMERCIAL PREPARATION

Absolute.

Regulatory Status. Has been approved for food use (§172.510).

REFERENCES

See the General References for ARCTANDER; FEMA; GUENTHER; TERRELL.

1. A. R. Penfold and J. L. Willis, *Econ. Bot.*, **8**, 316 (1954).
2. F. V. Wells, *Indian Perfum.*, **2** (Pt. 2), 27 (1958).
3. Y. R. Naves and G. R. Parry, *Helv. Chim. Acta*, **30**, 419 (1947).

BROMELAIN

Source: *Ananas comosus* (L.) Merr. (syn. *A. sativus* Schult. f.) and varieties (Family Bromeliaceae).

Common/vernacular names: Bromelain, bromelains, bromelin (fruit-bromelin), plant protease concentrate.

GENERAL DESCRIPTION

Bromelains are sulfhydryl proteolytic enzymes obtained from the pineapple plant, a perennial herb with many varieties native to tropical America. Two kinds of bromelain are known: stem bromelain and fruit bromelain, which are crude aqueous extracts of the stem and immature fruit of pineapple (*A. cosmosus* Merr., mainly from var. *Cayenne*), respectively.[1]

Commercial bromelain is currently prepared by centrifugation, ultrafiltration and freeze-drying of cooled pineapple juice, which produces a yellowish powder. Stem bromelain has been generally prepared from the juice of pineapple wastes (mainly stems) by precipitation with organic solvents (e.g., acetone and methanol) or by ultrafiltration.[2]

CHEMICAL COMPOSITION

Crude bromelain contains proteinases mainly consisting of glycosylated multiple enzymes having molecular masses from 20 to 31 kDa. Pineapple fruit contains fruit bromelain and pineapple stem contains stem bromelain, ananain, and comosain. It also contains various incompletely characterized constituents, including glycoproteins, carbohydrates, peroxidases, phosphatases, cellulases, and others.[1]

Like papain, stem bromelain has broad specificity, hydrolyzing various proteinaceous substrates (e.g., proteins, amides, esters, and small peptides).

PHARMACOLOGY AND BIOLOGICAL ACTIVITIES

Bromelain has shown a wide variety of pharmacological effects in clinical, *in vitro* and *in vivo* studies. These effects include burn debridement, anti-inflammatory activity, prevention of epinephrine-induced pulmonary edema, smooth muscle relaxation, stimulation of muscle contractions, enhanced antibiotic absorption, immunomodulation, cancer prevention and remission, antitumor activity, ulcer prevention, sinusitis relief, appetite inhibition, shortening of labor, and enhanced excretion of fat.[1,2–10] The precise nature of these effects (some of which are not produced by other proteases such as ficin, papain, and trypsin) is not clear.[1,4,5]

In double-blind, placebo-controlled clinical trials, bromelain has shown superior results versus placebo in the treatment of sinusitis,[11] inflammation, edema and pain from mediolateral episiotomy,[12] postoperative tumifications of the feet,[13] arthritis,[14] and amelioration of ecchymoses and edema in patients with head and face trauma.[15] Oral bioavailability has also been demonstrated (randomized, double-blind, crossover design trial).[1]

TOXICOLOGY

Acute oral doses of up to 10 g/kg in mice failed to produce lethality, and chronic oral administration of bromelain (500 mg/kg) in rats produced no adverse effects.[16]

From placebo-controlled studies, side effects of bromelain appear to be few and of low incidence (e.g., 1.8% incidence of allergic reactions, occasional gastric complaints, diarrhea, and nausea).[1]

Bromelain may increase plasma levels of tetracyclines when taken concomitantly (BLUMENTHAL 1).

USES

Medicinal, Pharmaceutical, and Cosmetic. Used primarily in preparations to treat inflammation and edema associated with surgical or accidental trauma, infections, and allergies.

Bromelain is used in certain cosmetics, such as facial cleansers and bath preparations.

Food. Due to the high cost of papain, bromelain is increasingly used to replace or supplement papain usage. Current major uses of bromelain are in meat tenderizing, manufacturing precooked cereals, modifying dough (bread, wafers, pizza, etc.), and in chill-proofing beer.

Other uses or potential uses include preparation of protein hydrolyzates, liquefying fish protein to facilitate fish oil extraction, clarifying fruit juices, and manufacture of sausage casings as well as their removal from sausages.

Dietary Supplements/Health Foods. Used in some vitamin and herbal formulations, mainly as a digestive aid (MARTINDALE).

Others. Used in bating hide and in desizing fabrics.

COMMERCIAL PREPARATIONS

Available in numerous grades with different activities that are expressed in different enzyme units, depending on the suppliers. Bromelain is official in F.C.C.

Regulatory Status. A German therapeutic monographic allows use for treating acute post-traumatic and postoperative edemas, notably those of the sinuses (BLUMENTHAL 1). Affirmed as GRAS (§184.1024).

REFERENCES

See the General References for BAILEY 1; BRUNETON; DER MARDEROSIAN AND BEUTLER; GUPTA; MARTINDALE; MCGUFFIN 1 & 2; MERCK; TERRELL; USD 26th.

1. H. R. Maurer, *Cell. Mol. Life Sci.*, **58**, 1234 (2001).
2. R. M. Heinicke, U.S. Pat. 3,002,891 (1961).
3. T. Enomoto et al., *Jpn. J. Pharmacol.*, **17**, 331 (1967).
4. S. Mineshita and Y. Nagai, *Jpn. J. Pharmacol.*, **27**, 170 (1977).
5. W. M. Cooreman et al., *Pharm. Acta Helv.*, **51**, 73 (1976).
6. G. E. Felton, *Hawaii Med. J.*, **36** (2), 39 (1977).
7. International Commission on Pharmaceutical Enzymes, *Farm. Tijdschr. Belg.*, **54**, 85 (1977).
8. S. J. Taussig and S. Batkin, *J. Ethnopharmacol.*, **22**, 191 (1988).
9. S. Kumakura et al., *Eur. J. Pharmacol.*, **150**, 295 (1988).
10. L. L. Bolton and B. E. Constantine, Eur. Pat. Appl. 194,647 (1986); through *Chem. Abstr.*, **106**, 201776 (1987).
11. H. R. Maurer et al., *Planta Med.*, **54**, 377 (1988).
12. R. E. Ryan, *Headache*, **4**, 13 (1967).
13. G. J. Zatucchini and D. Colombi, *Obst. Gynecol.*, **29**, 275 (1967).
14. E. Nitzschke and R. Leonhardt, *Orthopäd. Praxis*, **25**, 111 (1989).
15. W. Vogler, *Natur-Ganzheits-Med.*, **1**, 27 (1988).
16. A. P. Seltzer, *Eye, Ear, Nose, Throat Mon.*, **43**, 54 (1964).
17. I. N. Moss et al., *Arch. Int. Pharmacodyn.*, **145**, 166 (1963).

BROOM TOPS

Source: ***Cytisus scoparius*** (L.) Link (syn. *Sarothamnus scoparius* (L.) Wimm.; *S. vulgaris* Wimm.; *Spartium scoparium* L.) (Family Leguminosae or Fabaceae).

Common/vernacular names: Banal, broom tops, hogweed, Irish broom, scoparius, Scotch broom.

GENERAL DESCRIPTION

Deciduous shrub with erect slender branches, up to about 3 m high; native to central and southern Europe; naturalized in North America (invasive along the west coast from North California to British Columbia); also grows in Asia and South Africa; widely grown in Japan as an ornamental. Parts used are the dried flowering tops collected just before blooming.

Scotch broom should not be confused with its relative Spanish broom, which is *Spartium junceum* L. (see **genet**).

CHEMICAL COMPOSITION

Contains alkaloids, including the major alkaloids, sparteine (ca. 0.3%) and lupanine, and genisteine, sarothamnine, and others; simple amines (tyramine, hydroxytyramine, epinine, salsolidine, etc.)[1,2]; flavonoids, including scoparin (scoparoside), spiraeoside, and others; flavones (vitexin and others); isoflavones (genistein, orobol, sarothamnoside); pigments (e.g., taraxanthin and flavoxanthin); amino acids; volatile oil (containing 4-mercapto-4-methylpentane-2-one); coumarins, cafeic acid derivatives; tannin; wax; fat; and sugars (KARRER; WICHTL).[3] Essential oil of the fresh flowers contains alcohols, coumarins, eugenol, benzoic acid, guaiacol, fatty acids, and others.[4]

PHARMACOLOGY AND BIOLOGICAL ACTIVITIES

The cardiac activity of broom tops is the result of its alkaloids (principally sparteine), which have cardiac-depressant and curare-like properties and are highly toxic (MARTINDALE, SAX). Sparteine has shown antiarhythmic activity (WICHTL).

TOXICOLOGY

Sparteine was withdrawn by the U.S. FDA as an injectable drug because of its ability to produce titanic uterine contractions and its unpredictability.[5] Due to the tyramine content of the herb, use may cause blood pressure crisis with simultaneous administrations of MAO inhibitors (BLUMENTHAL 1); due to the ability of the herb to increase the tonus of the gravid uterus, use is contraindicated in pregnancy; sparteine is a potent oxytocic agent; broom tops is also contraindicated in patients with high blood pressure (NEWALL; WICHTL). The oral LD_{50} of sparteine in mice is 220 mg/kg.[6]

USES

Medicinal, Pharmaceutical, and Cosmetic. Broom tops is regarded as having diuretic, emetic, and cathartic properties and is used in certain laxative, diuretic, and tonic preparations; also used in treating circulatory disorders (WICHTL).

Dietary Supplements/Health Foods. Teas, capsules, and so on, primarily as a diuretic.

Traditional Medicine. Reportedly used internally as a diuretic and externally to treat sore muscles, abscesses, and swellings; flowers used in hair rinses for their lightening and brightening effects (LUST, ROSE, UPHOF).

Broom flowers, seeds, and root as well as the whole herb have reportedly been used in treating tumors.[7]

COMMERCIAL PREPARATIONS

Crude and extracts. Crude was formerly official in N.F., and U.S.P. Strengths (see *glossary*) of extracts are based on weight-to-weight ratios.

Regulatory Status. Regulated in the United States as a dietary supplement. Subject of a German therapeutic monograph, with preparations not to contain more than 1 mg/mL of sparteine for treatment of functional heart and circulatory disorders (BLUMENTHAL 1; WICHTL).

REFERENCES

See the General References for ARCTANDER; BAILEY 1; BIANCHINI AND CORBETTA; BLUMENTHAL 1; DER MARDEROSIAN AND BEUTLER; GOSSELIN ET AL.; LIST AND HÖRHAMMER; LUST; GRIEVE; MCGUFFIN 1 & 2; MERCK; ROSE; TERRELL; TYLER 1; APPLEQUIST; UPHOF.

1. G. Gresser et al., *Z. Naturforsch., C-A*, **51**, 791 (1996).
2. I. Murakoshi et al., *Phytochemistry*, **25**, 521 (1986).
3. K. Egger, *Planta*, **80**, 65 (1968).
4. T. Kurihara and M. Kikuchi, *Yakugaku Zasshi*, **100**, 1054 (1980).
5. E. Yarnell and K. Abascal, *Altern. Compl. Ther.*, **125** June, (2003)
6. K. Yovo et al., *Planta Med.*, **50**, 420 (1984).
7. J. L. Hartwell, *Lloydia*, **33**, 97 (1970).

BUCHU

Source: *Agathosma betulina* (P. J. Bergius) Pillans (syn. *Barosma betulina* (Berg.) Bartl. et Wendl.) or *A. crenulata* (L.) Pillans (syn. *B. crenulata* (L.) Hook. and *B. serratifolia* (Curt.) Willd.) (Family Rutaceae).

Common/vernacular names: Bookoo, buchu, bucku, bucco, buku, diosma; round buchu, short buchu (*A. betulina*); buchu long buchu (*A. crenulata*).

GENERAL DESCRIPTION

Low, aromatic shrubs, usually less than 2 m high with opposite and/or alternate leaves that are finely toothed at the margin and bear oil glands beneath and at the base of the teeth; native to South Africa. Parts used are the dried leaves, from which an essential oil is obtained by steam distillation. There are two types of buchu leaves. Round or short buchu is from *A. betulina*, and long buchu is from *A. crenulata*; they differ in the relative compositions of their volatile oils. Major supplies of the leaves (from both wild and cultivated plants) come from the Cape Province of South Africa, which also produces some of the world's supply of the oil; the rest of the oil is distilled in Europe and the United States. Buchu is currently a threatened species.

CHEMICAL COMPOSITION

Contains 1.0–3.5% (usually 1.5–2.5%) volatile oil composed mainly of *l*-pulegone, isopulegone, diosphenol (buchu camphor), ψ-diosphenol, *l*-isomenthone, *d*-menthone, *d*-limonene, 8-mercapto-*p*-menthan-3-one, 8--acetylthio-*p*-menthan-3-one, and piperitone epoxide, among more than 100 other identified minor compounds;[1–6] presence of piperitone epoxide disputed.[6] Relative proportions of pulegone and diosphenols vary considerably in commercial oils. In general, round buchu (*A. betulina*) leaf yields oils with high proportions of diosphenols, while long buchu (*A. crenulata*) yields oils containing high proportions of pulegone with little or no diosphenols present.[1,4,6] In addition, round buchu leaf contains higher concentrations of volatile oil than long buchu leaf. Other constituents present in buchu leaf include flavonoids (diosmin, rutin, quercetin-3,7-glucoside, etc.), resin, mucilage, and others (KARRER; LIST AND HÖRHAMMER; MERCK).

PHARMACOLOGY AND BIOLOGICAL ACTIVITIES

Buchu leaves are reported to have urinary antiseptic, diuretic, and carminative properties. The essential oils of *Agathosma betulina* and *A. crenulata* leaves have shown spasmolytic activity on the smooth muscle of isolated guinea pig ileum.[7]

A formulation containing 90% diosmin and 10% hesperidin has been used in the treatment of acute hemorrhoids[8–10] and chronic venous insufficiency.[11–13]

USES

Medicinal, Pharmaceutical, and Cosmetic. Extensively used in diuretic preparations; also in laxative, stomachic, and carminative formulas.

Food. Oil is used as a component in artificial fruit flavors, especially black currant flavor. Round buchu oil is preferred because of its higher contents of diosphenols and 8-mercapto-*p*-menthan-3-one, which are considered to be the more desirable flavor components.[6] Major categories of food products in which the oil is used include alcoholic and nonalcoholic beverages, frozen dairy desserts, candy, baked goods, gelatins and puddings, and condiments and relishes. Use levels reported are rather low, with average maximum highest in gelatins and puddings at about 0.002% (15.4 ppm).

Dietary Supplements/Health Foods. Capsules, tablets, crude herb in teas reportedly as a diuretic and urinary antiseptic (WREN).

Traditional Medicine. Common household medicine used in South Africa for the treatment of urinary tract and kidney diseases, symptoms of rheumatism, and externally on wounds and bruises.[14] Also used to treat cystitis, urethritis, and others; also as a diuretic, tonic, and stimulant (BLUMENTHAL 1; WICHTL), and for the treatment of coughs and colds.[15]

COMMERCIAL PREPARATIONS

Crude, extracts, and oil. Crude and fluid extracts were formerly official in N.F. Strengths (see *glossary*) of extracts are expressed in weight-to-weight ratios. Buchu leaf is the subject of a German therapeutic monograph; however, use as a urinary tract anti-inflammatory and diuretic is not recommended since effectiveness is not well documented. Leaf is allowed as an aroma or flavor corrigent in teas.[7]

Regulatory Status. Regulated in the United States as a dietary supplement; GRAS as a natural flavoring in foods (§172.510).

REFERENCES

See the General References for APhA; ARCTANDER; BARNES; BLUMENTHAL 1; CLAUS; DER MARDEROSIAN AND BEUTLER; GUENTHER; LIST AND HÖRHAMMER; LUST; MARTINDALE; TERRELL; WICHTL; WREN; YOUNGKEN.

1. A. A. J. Fluck et al., *J. Sci. Food Agric.*, **12**, 290 (1961).
2. E. Klein and W. Rojahn, *Dragoco Rep.*, **14**, 183 (1967).
3. E. Sundt et al., *Helv. Chim. Acta*, **54**, 1801 (1971).
4. K. L. J. Blommaert and E. Bartel, *J. S. Afr. Bot.*, **42**, 121 (1976); through *Chem. Abstr.*, **87**, 58401u (1977).
5. D. Lamparsky and P. Schudel, *Tetrahedron Lett.*, **36**, 3323 (1971).
6. R. Kaiser et al., *J. Agric. Food Chem.*, **23**, 943 (1975).
7. M. Lis-Balchin et al., *J. Pharm. Pharmacol.*, **53**, 579 (2001).
8. K. Buckshee et al., *Int. J. Gynecol. Obstet.*, **57**, 145 (1997).
9. M. Sarabia et al., *Curr. Ther. Res. Clin. Exp.*, **62**, 524 (2001).
10. A. N. Nicolaides, *Angiology*, **54**(Suppl. 1), S33 (2003).
11. G. Jantet, *Angiology*, **53**, 245 (2002).
12. A. A. A. Ramelet, *Angiology*, **52** (Suppl. 1), S49 (2001).
13. J. R. Struckmann, *J. Vasc. Res.*, **36** (Suppl. 1), 37 (1999).
14. B. van Wyk et al., *Medicinal Plants of South Africa*, Briza Publications, Pretoria, 1997, p. 34.
15. D. Simpson, *Scott. Med. J.*, **43**, 189 (1998).

BUCKTHORN, ALDER

Source: *Frangula alnus* Mill. (syn. *Rhamnus frangula* L.) (Family Rhamnaceae).

Common/vernacular names: Alder buckthorn, arrow wood, black dogwood, frangula, glossy buckthorn.

GENERAL DESCRIPTION

Shrub or small tree with shiny, dark green, short oblong to obovate leaves 3–7 cm long; up to 6 m high; native to Europe, western Asia, and northern Africa; naturalized in North America. Part used is the dried bark aged for 1 year to get rid of an emetic principle (see *cascara*).

CHEMICAL COMPOSITION

Contains 3–7% anthraquinone glycosides as active principles, which include glycofrangulin A and B, frangulin A and B,[1] emodin-1-glucoside, emodin-8-glucoside, emodin-8-O-β-gentiobioside, and others.[2–10] Other constituents include chrysophanol, physcion (WICHTL), and an alkaloid, armepavine, which is present in fresh bark but not in dried bark;[11] tannins; flavonoids;[12] and free anthraquinones, among others (STAHL). The fresh bark also contains anthrones and anthrone glycosides that are believed to constitute the emetic principle;[13] they are oxidized to anthraquinones or their glycosides on storage.

PHARMACOLOGY AND BIOLOGICAL ACTIVITIES

The active principles of buckthorn bark (anthraglycosides) are cathartic; they act on the large intestine (colon), with the diglycosides being more active than the monoglycosides (see also *cascara* and *senna*).[13,14]

Aloe emodin isolated from the seed of buckthorn has been reported to have significant inhibitory activity against P-388 leukemia in mice; it exhibited such activity only when administered as a suspension in acetone–Tween 80.[15]

TOXICOLOGY

Safety data are lacking; use is contraindicated in pregnancy and lactation, Crohn's disease, appendicitis, ileus, colitis ulcerosa, abdominal pain of undetermined origin, severe dehydration, and children 10 years old and younger (WICHTL).

USES

Medicinal, Pharmaceutical, and Cosmetic. Used in certain laxative preparations, more commonly in Europe. Extracts have been used in sunscreen preparations.[16,17]

Traditional Medicine. Used as a laxative and tonic; also reportedly used in treating cancers and as a component in Hoxsey cancer "cure."[18]

COMMERCIAL PREPARATIONS

Crude and extracts: crude and fluid extract were formerly official in N.F. and U.S.P. Strengths (see *glossary*) of extracts are expressed in weight-to-weight ratios.

Regulatory Status. Regulated in the United States as a dietary supplement; not recognized as safe and effective as OTC laxative (§310.545(12)(iv)). Allowed for sale in Germany as a botanical stimulant laxative for the short-term treatment of constipation; only sold with a license; restricted to pharmacies (WICHTL).

REFERENCES

See the General References for APhA; BAILEY 2; BIANCHINI AND CORBETTA; BRUNETON; FERNALD; LUST; MCGUFFIN 1 & 2; MERCK; STAHL; TERRELL; WICHTL; YOUNGKEN.

1. G. Matysik and E. Wojtasik, *J. Planar Chromatogr.*, **7**, 34 (1994).
2. B. Kaminski and W. Grzesiuk, *Farm. Pol.*, **33**, 157 (1977).
3. A. V. Gotsiridze and E. P. Kemertelidze, *Rast. Resur.*, **13**, 64 (1977).
4. H. Auterhoff and E. Eujen, *Dtsch. Apoth. Ztg.*, **112**, 1533 (1972).
5. H. Wagner and G. Demuth, *Tetrahedron Lett.*, **49**, 5013 (1972).
6. M. Rosca and V. Cucu, *Planta Med.*, **28**, 343 (1975).

7. K. Savonius, *Farm. Aikak.*, **82**(9–10), 136 (1973).
8. M. Rosca and V. Cucu, *Planta Med.*, **28**, 178 (1975).
9. A. Bonati and G. Forni, *Fitoterapia*, **48**, 159 (1977).
10. G. Dermuth et al., *Planta Med.*, **33**, 53 (1978).
11. M. Pailer and E. Haslinger, *Monatsh. Chem.*, **103**, 1399 (1972).
12. J. C. Dauguet and R. R. Paris, *C. R. Acad. Sci., Ser. D*, **285**, 519 (1977).
13. F. H. L. van Os, *Pharmacology*, **14** (Suppl. 1), 7, 18 (1976).
14. F. A. Nelemans, *Pharmacology*, **14** (Suppl. 1), 73 (1976).
15. S. M. Kupchan and A. Karim, *Lloydia*, **39**, 223 (1976).
16. G. Prosperio, *Cosmet. Toilet.*, **91**, 34 (1976).
17. S. Bader et al., *Cosmet. Toilet.*, **96**(10), 67 (1981).
18. J. L. Hartwell, *Lloydia*, **34**, 103 (1971).

BURDOCK

Source: *Arctium lappa* L. (syn. *A. majus* Bernh.); *Arctium minus* Bernh. (Family Compositae or Asteraceae).

Common/vernacular names: Bardana, beggar's buttons, burdock, edible burdock, great bur, great burdock, clotbur, lappa.

GENERAL DESCRIPTION

A. lappa is a biennial or perennial herb up to about 3 m high in its second year of growth; native to Asia and Europe; naturalized in North America. Part used is the dried first-year root collected in the fall; fruits and leaves are also used.

Common burdock (*A. minus*) is also used; it resembles *A. lappa* but is smaller.

CHEMICAL COMPOSITION

Root contains inulin (up to approx. 45%); polyacetylenes[1] (0.001–0.002%, dry-weight basis) consisting mainly of 1,11-tridecadiene-3,5,7,9-tetrayne and 1,3,11-tridecatriene-5,7,9-triyne;[2] arctic acid and lappaphens a and b (acetylenic acids containing S);[3,4] volatile acids (acetic, propionic, butyric, isovaleric, 3-hexenoic, 3-octenoic, costic, etc.) and nonhydroxy acids (lauric, myristic, stearic, palmitic, etc.);[5,6] a crystalline plant hormone;[7] γ-guanidino-*n*-butyric acid; dehydrocostuslactone; tannin; polyphenolic acids (e.g., caffeic and chlorogenic), and others (KARRER; JIANGSU; WICHTL).[8]

Fruit contains arctiin (a bitter glucoside) as a major constituent;[9] seeds contain 15–30% fixed oils (oleic and octadecanoic acids major constituents), arctiin, arctigenin, chlorogenic acid, lignans (lappaols A, and diarctigenin, neoarctin, and others), a germacranolide, and others (KARRER).[8–12]

Leaves contain arctiol (8-α-hydroxyeudesmol), $\Delta^9(10)$-fukinone (dehydrofukinone), fukinone, fukinanolide, β-eudesmol, petasitolone, eremophilene, and taraxasterol, among others.[13]

PHARMACOLOGY AND BIOLOGICAL ACTIVITIES

Burdock (fruit and root) has shown diuretic, diaphoretic, antipyretic, antimicrobial, and antitumor activities (FARNSWORTH 1–4; JIANGSU; LIST AND HÖRHAMMER). Antimicrobial activity is attributed to polyacetylenes.[2] The fresh root juice has shown *in vitro* antimutagenic activity.[14] The active constituent was reported to be polyanionic with a molecular weight of >300,000; data suggest it to be a lignin-like compound containing 10% sugar.[15,16] Water extracts of the dried root have shown greater *in vitro* antiperoxidative activity than solvent extracts.[17] Antioxidant activity of the roots has been attributed to caffeoylquinic acid derivates.[18] Hepatoprotective activity against acetaminophen- and carbon tetrachloride-induced toxicity in mice was found from oral administration of a decoction of the root.[19] Others have reported that root ameliorated ethanol induced liver toxicity in rats.[20]

A methanol extract of the fruits inhibited the *in vitro* formation of a marker of oxidative stress in carcinogenesis (8-OH-dG).[21] An ethanol extract of the fruits has shown *in vitro* cytotoxicity against human hepatoma (HepG2) cells and activity against sarcoma 180 cells in mice. Acrtiin and arctigenin showed the most activity against Hep2G cells *in vitro*.[22] In vitro differentiation-inducing, antiproliferative activity and phagocytic cell-increasing activity in mouse myeloid leukemia cells were found from various lignans isolated from the fruit. Arctigenin was the most active.[23] Chemopreventive activity against heterocyclic amine-induced mammary carcinogenesis in female rats was evident from the addition of arctiin to the diet at a concentration of as little as 0.02%.[24]

In addition, hypoglycemic activity in rats was reported from administration of burdock fruit extracts,[25] and a water extract of the seeds inhibited the *in vitro* binding of platelet activating factor to rabbit platelets.[26]

TOXICOLOGY

A decoction of the leaves as 6.25% of the daily diet of mice for 28 days before a streptozotocin-induced diabetes was found to aggravate the condition.[27] A methanol extract of the fruits exhibited mutagenic activity in the rec-assay and in the Ames test only after enzymatic activation.[28]

USES

Medicinal, Pharmaceutical, and Cosmetic. Used in some diuretic, laxative, and other preparations. Also reportedly useful in cosmetic and toiletry preparations for its alleged skin-cleansing properties (DE NAVARRE);[29] used in hair tonic and antidandruff preparations.

Food. Root is used in Asia as a food; Iroquois Indians used the dried roots to make soup and the young leaves were eaten cooked.

Dietary Supplements/Health Foods. Seeds used in cold remedies; leaves used in teas, combination products, primarily as "blood purifier" for skin ailments (acne, psoriasis, etc.); root used as nutritive food (FOSTER AND DUKE).

Traditional Medicine. Root, leaves, and seeds (fruits) of both species have been used in treating cancers;[30] decoctions or tea of the root have been used in treating rheumatism, catarrh, gout, and stomach ailments. The root is used as a diuretic, diaphoretic, and mild laxative, among other uses. Decoctions and teas of roots and leaves have been used both externally and internally for skin problems (e.g., eczema and scaly skin) (FOSTER AND DUKE; LUST; TYLER 1). The Micmac Indians used the roots and buds to treat sores, and a tea made from the seeds or roots was used by the Cherokee as a blood cleanser (MOERMAN).

In Chinese medicine, roots collected are from plants that are at least 2 years old.

Dried and dried and roasted fruits of *A. lappa* are widely used in Chinese medicine to treat colds, sore throat, tonsillitis, coughs, measles, sores, and abscesses, among other applications, usually in combination with other drugs (JIANGSU; TU).

COMMERCIAL PREPARATIONS

Crude and extracts. Crude root was formerly official in N.F. and U.S.P. Strengths (see *glossary*) of extracts are expressed in weight-to-weight ratios.

Regulatory Status. Regulated in the United States as a dietary supplement and food. A German therapeutic monograph on burdock root does not recommend use since efficacy is not confirmed (BLUMENTHAL 1; WICHTL).

REFERENCES

See the General References for BAILEY 2; BISSET; BRUNETON; DER MARDEROSIAN AND BEUTLER; FARNSWORTH 1–4; FERNALD; FOGARTY; FOSTER AND DUKE; GUPTA; GRIEVE;HONGKUI; JIANGSU; JIXIAN; KROCHMAL AND KROCHMAL; LUST; MOERMAN; NANJING; ROSE; TERRELL; TYLER 1; YOUNGKEN.

1. T. Washino et al., *Nippon Nogeikagaku Kaishi*, **60**, 377 (1986)
2. K. E. Schulte et al., *Arzneim.-Forsch.*, **17**, 829 (1967).
3. T. Washino et al., *Agric. Biol. Chem.*, **50**, 263 (1986).
4. T. Washino et al., *Agric. Biol. Chem.*, **51**, 1475 (1987).
5. S. Obata et al., *Nippon Nogeikagaku Kaishi*, **44**, 437 (1970).
6. T. Washino et al., *Nippon Nogeikagaku Kaishi*, **59**, 389 (1985).
7. T. Kimura, Jpn. Kokai 76115,914 (1976).
8. Y. Yamada et al., *Phytochemistry*, **14**, 582 (1975).
9. A. Ichihara et al., *Tetrahedron Lett.*, **44**, 3961 (1976).
10. B. H. Han et al., *Phytochemistry*, **37**, 1161 (1994).
11. H. Y. Wang and J. S. Yang, *Yaoxue Xuebao*, **28**, 911 (1993).
12. C. Wang et al., *J. Plant Res. Environ.*, **11**, 58 (2002).
13. K. Naya et al., *Chem. Lett.*, **3**, 235 (1972).
14. K. Morita et al., *Agric. Biol. Chem.*, **42**, 1235 (1978).
15. K. Morita et al., *Agric. Biol. Chem.*, **49**, 925 (1985).
16. K. Morita et al., *Mutat. Res.*, **129**, 25 (1984).
17. P. Duh, *J.A.O.C.S.*, **7**, 455 (1998).
18. M. Yoshihiko et al., *J. Agric. Food Chem.*, **43**, 2592 (1995).
19. S. Lin et al., *Am. J. Chin. Med.*, **28**, 163 (2000).
20. S. Lin et al., *J. Biomed. Sci.*, **9**, 401 (2002).
21. H. Kasai et al., *Food Chem. Toxicol.*, **38**, 467 (2000).
22. S. Moritani et al., *Biol. Pharm. Bull.*, **19**, 1515 (1996).
23. K. Umehara et al., *Chem. Pharm. Bull.*, **41**, 1774 (1993).
24. M. Hirose et al., *Environ. Mol. Mutagen.*, **39**, 271 (2002).
25. L. O. Lapinina and T. F. Sisoeva, *Farm. Zh. (Kiev)*, **19**, 52 (1964).
26. S. Iwakami et al., *Chem. Pharm. Bull.*, **40**, 1196 (1992).
27. S. K. Swanston-Flatt et al., *Diabetes Res.*, **10**, 69 (1989).
28. I. Morimoto et al., *Mutat. Res.*, **129**, 25 (1984).
29. H. B. Heath, *Cosmet. Toilet.*, **92**, 19 (1977).
30. J. L. Hartwell, *Lloydia*, **31**, 71 (1968).

CADE OIL

Source: *Juniperus oxycedrus* L. (Family Pinaceae).

Common/vernacular names: Cade oil, juniper tar, oil of cade, oil of juniper tar.

GENERAL DESCRIPTION

The source of cade oil is *Juniperus oxycedrus* or "prickly cedar," a shrub or small tree native to the Mediterranean region, which grows up to about 4 m in height. The volatile oil (cade oil) is obtained by destructive distillation of the branches and wood, usually in the form of shavings or chips. The resultant distillate separates into three layers of which the uppermost dark brown viscous layer is cade oil. Rectified cade oil (the vapor of juniper tar) is obtained by steam or vacuum distillation of crude cade oil.

CHEMICAL COMPOSITION

The volatile oil contains the sesquiterpene δ-cadinene as the major component (27.3%).[1,2] Other major components include *p*-cresol and guaiacol (LIST AND HÖRHAMMER; MARTINDALE).

PHARMACOLOGY AND BIOLOGICAL ACTIVITIES

Reported to have keratolytic and antipruritic properties and *in vitro* antimicrobial and antifungal activities.[3–5]

TOXICOLOGY

Has been considered as an allergen (SAX); and although other available data show it to be relatively nontoxic,[6] recent *in vivo* studies have shown that cade oil produces DNA adduct formation in mouse and human skin and mouse lung tissue.[5] Applied topically to psoriasis patients, cade oil produced sufficient damage to DNA to be potentially carcinogenic.[7] The Cosmetic Ingredient Review Expert Panel concluded that the data on the safety of cade oil are not sufficient to support its use in cosmetic products.[5]

Acute oral LD_{50} in rats: 8014 mg/kg. Toxic effects noted include GI irritation and signs of depression. Acute dermal toxicity in rabbits: >5 g/kg. No skin irritation or phototoxicity found from topical application of cade oil in humans, mice, and swine.[3]

USES

Medicinal, Pharmaceutical, and Cosmetic. Cade oil is widely used in topical preparations for the treatment of parasitic skin diseases and eczema; in antiseptic wound dressings, analgesic and antipruritic preparations, and dermatologic creams and ointments; also in antidandruff shampoos, among others. Rectified cade oil is used as a fragrance component in soaps, detergents, creams, lotions, and perfumes. Maximum use level reported is 0.2% in perfumes.[6] Maximum use concentration of cade oil in cosmetics ranges from 1% to 5%.[5]

Traditional Medicine. Used in treating various skin disorders and problems of the scalp and hair loss; also used in cancers.[8]

COMMERCIAL PREPARATIONS

Crude and rectified.

Regulatory Status. Regulated in the United States as a dietary supplement. Cade oil (juniper tar oil) is official in U.S.P.

REFERENCES

See the General References for ARCTANDER; BAILEY 2; CLAUS; DUKE 4; GUENTHER; LUST; MARTINDALE; MCGUFFIN 1 & 2; ROSE.

1. J. C. Chalchat et al., *Flav. Frag. J.*, **3**, 19 (1988).
2. B. M. Lawrence, *Perfum. Flav.*, **14**(3), 71 (1989).
3. J. C. Maruzzella and L. Liguori, *J. Am. Pharm. Assoc.*, **47**, 250 (1958).
4. J. C. Maruzzella and P. A. Henry, *J. Am. Pharm. Assoc.*, **47**, 294 (1958).
5. W. Johnson Jr., *Int. J. Toxicol.*, **20**, 41 (2001).
6. D. L. J. Opdyke, *Food Cosmet. Toxicol.*, **13** (Suppl.), 733 (1975).
7. B. Schoket et al., *J. Invest. Dermatol.*, **94**, 241 (1990).
8. J. L. Hartwell, *Lloydia*, **33**, 288 (1970).

CAJEPUT OIL

Source: *Melaleuca cajuputi* Powel (syn. *M. leucadendron*, *M. minor*); *M. quinquenervia* (Cav.) S. T. Blake; *M. alternifolia* (Maiden & Betche) Cheel (syn. *M. larinariifolia* var. *alternifolia*); and other *Melaleuca* species and subspecies (Family Myrtaceae).

Common/vernacular names: Cajuput, punk tree, and paperbark tree oils; tea tree oil (*M. alternifolia*), oil of Melaleuca, terpinen-4-ol-type tea tree oil.

GENERAL DESCRIPTION

Cajeput are large evergreen trees with whitish, paper-like, spongy bark; up to about 30 m high; native to Australia and southeastern Asia.

CHEMICAL COMPOSITION

Cajeput oil is obtained by steam distillation of the fresh leaves of various species of *Melaleuca*. The main sources of commercial oil are Australian plantations of *M. alternifolia* (Australian tea tree oil) and *M. cajuputi* subsp. *cajuputi* from Indonesia and Vietnam. The composition of the oil of *M. cajuputi* subsp. *cajuputi* is highly variable, partly due to natural variations and to industry practice of blending oils from different species of *Melaleuca* and even with added individual compounds. Otherwise, the oil from this species and its subspecies most frequently, but not always, contain large amounts of 1,8-cineole (3–60%), sesquiterpene alcohols (spathulenol, up to 30%; viridiflorol, up to 16%; and globulol, up to 9%), in addition to usually significant amounts of ledene (viridiflorene) (up to 9%), terpinen-4-ol (up to 5.6%), α-terpineol (1–8%), limonene (up to 5%), β-caryophyllene (up to 4%), caryophyllene oxide (up to 7%), and α- and β-selinene (up to 3%).[1] The oil from *M. alternifolia* contains close to 100 different constituents. Examples of major constituents of the leaf oil and their quantifications include terpinen-4-ol (37%), γ-terpinene (21.2%), α-terpinene (9.9%), and terpinolene (3.2%).[2]

In 1985 a quality standard for *Melaleuca* oils was established in Australia[3] (Australian Standard, *Oil of Melaleuca*, "Terpinen-4-ol Type," AS 2782), which in 1996 was revised and adopted as the International Standard (ISO 4730). The standard calls for the oil to contain 30% or more of terpinen-4-ol and a maximum of 15% cineole; the terpinen-4-ol type oil is found in various species of *Melaleuca* (*M. alternifolia*, *M. dissitiflora*, and

M. linariifolia); however, other species may qualify.[2]

PHARMACOLOGY AND BIOLOGICAL ACTIVITIES

Australian tea tree oil (*M. alternifolia*) exhibits a broad spectrum of *in vitro* antibacterial activity that includes laboratory strains of oral bacteria,[4] methicillin-resistant *Staphylococcus aureus*,[5] *Propionbacterium acnes*,[6] and *Escherichia coli*.[7] *In vitro* antifungal activity from the oil is found against *Malassezia furfur*, *Candida albicans*, and various other species of *Candida*.[8–10] Among eight components of the essential oil, greatest *in vitro* antibacterial (11 species) and antifungal activities (*C. albicans*) were found from linalool, followed by terpinen-4-ol, α-terpineol, α-terpinene, and terpinolene.[7] (Linalool occurs in only trace amounts in the oil.[2]) The oil has also shown *in vitro* activity against herpes simplex virus.[11]

Essential oils from the leaves of *Melaleuca* (*M. armillaris*, *M. ericifolia*, and *M. leucadendron* = *M. cajuputi*) have shown *in vitro* antibacterial, antifungal (*C. albicans*), and antiviral activities (HIV-1); *in vitro* potentiation of catalase, superoxide dismutase, and glutathione in erythrocytes; and *in vitro* inhibition of lipid peroxidation.[12] Both *M. alternifolia* leaf oil and terpinen-4-ol induced differentiation of white blood cells *in vitro* in human myelocytic (HL-60) cells.[13]

Contact hypersensitivity response and skin swelling in mice were inhibited by topical application of tea tree oil, α-terpineol, or terpinen-4-ol; however, ultraviolet B- or irritant-induced swelling was not suppressed.[14]

Antitussive activity against capsaicin-induced cough in guinea pigs was shown from oral administration of tea tree oil and from some of its main constituents (cineole, terpinen-4-ol, α-pinene, and γ-terpinene).[15]

A clinical trial in 124 patients provided evidence of the effectiveness of topical *M. alternifolia* oil in the treatment of moderate (noninflamed) acne vulgaris. A 5% tea tree oil in a water-based gel was less effective than a 5% benzoyl peroxide in a water-based lotion because of slower onset of action. However, clinical assessment and self-reporting of side effects indicated that the tea tree oil preparation was better tolerated by facial skin, producing less skin scaling, dryness, pruritus, and irritation than a benzoyl peroxide preparation (TYLER 1–3).[16] A study in 27 volunteers found that topical application of tea tree oil inhibited histamine-induced weal volume on the arm, but not flare volume.[17] In placebo-controlled clinical studies, symptomatic improvement of tinea pedis was found from topical use of a 10% tea tree oil cream;[18] reduced time before re-epithelization was reported in patients using a 6% tea tree oil gel topically on herpes labialis sores;[19] and reduced itchiness and greasiness but not scaliness of dandruff after using a 5% tea tree oil shampoo.[20]

TOXICOLOGY

Use of tea tree oil for the topical treatment of burn wounds is not recommended due to *in vitro* inhibition of human epithelial cell and fibroblast viability from 24 h exposure to the oil.[21,22] Application of the pure oil to the abraded skin of rabbits (Draize acute dermal irritation test) resulted in increased skin irritation, suggesting that the oil should not be applied in cases of dermatitis.[23]

There are a number of case reports of contact dermatitis from topical use of tea tree oil, including allergic contact dermatitis.[23,24]

No irritant effects were found in the skin sensitization test in guinea pigs from tea tree oil. No skin irritation was visible from topical application of tea tree oil in a 30-day irritation test in rabbits; no apparent signs of dermal toxicity were found after 24 h exposure of the normal skin to a high concentration of the undiluted oil (2 g/kg). The healing of superficial dermal wounds in rabbits was neither inhibited nor prolonged from topical application of tea tree oil. No *in vitro* mutagenic activity was produced by tea tree oil in the Ames test.[23]

The acute oral LD_{50} of tea tree oil in rats is 1.9–2.6 mL/kg; similar to eucalyptus and other essential oils.[23]

Case reports of overdosing from ingestion of large amounts of undiluted tea tree oil noted CNS toxicity. Similar effects have been reported in veterinary medicine from topical applications of large amounts of tea tree oil.[25]

USES

Medicinal, Pharmaceutical, and Cosmetic. Cajeput oil has been used in expectorant formulations and antiseptic and pain-relieving liniments, mouthwashes, shampoos, body washes, deodorants, acne products, barrier creams, lip balms, sunscreens, toothpastes, insect repellents, veterinary products, and others.[1,25,26] The oil has been used in dentistry for relieving discomfort due to dry sockets; also used as a fragrance component in soaps, detergents, creams, lotions, and perfumes, with maximum use level of 0.4% in the last category.[27]

Food. Cajeput oil is used as a flavor component in nonalcoholic beverages, frozen dairy desserts, candy, baked goods, meat and meat products, condiments, and relishes. Average maximum use level reported is very low (less than 0.001% or 9.9 ppm).

Dietary Supplements/Health Foods. Tea tree oil is found in numerous product forms, either singly or with other ingredients, with a broad range of health claims, including treatments for fungal infections, burns, sunburn, pimples, boils, stings, ringworm, sore throat, oral infections, bronchial congestion, lice, scabies, cuts, abrasions, and vaginal infections (TYLER 1–3).[26]

Traditional Medicine. Australian aborigines used cajeput oil from *M. cajuputi* subsp. *cajuputi* in treating pain. Vapors from the crushed leaves were inhaled to treat bronchial and nasal congestion. The same species has been a common household remedy in Vietnam, India, Malaysia, and Indonesia for generations.[1] They used an infusion of the macerated leaves of *Melaleuca* to treat the common cold, sore throat, insect bites, fungal infections, skin wounds, and for delousing.[23] Cajeput oil has also been used to treat indolent tumors.[28]

Others. In natural disinfectants.

COMMERCIAL PREPARATION

Oil formerly official in U.S.P.

Regulatory Status. Approved for food use as a natural flavoring (§172.510). The essential oil of *M. viridiflora* is the subject of a German therapeutic monograph allowing a dose of 0.2 g or less in the treatment of catarrhs of the upper respiratory tract (BLUMENTHAL 1).

REFERENCES

See the General References for ADA; ARCTANDER; FEMA; GRIEVE; GUENTHER; JIANGSU; LEWIS AND ELVIN-LEWIS; LIST AND HÖRHAMMER; MCGUFFIN 1 & 2; MORTON 2; ROSE; TYLER 1; UPHOF; YOUNGKEN.

1. J. C. Doran, *Med. Arom. Plants Ind. Profiles*, **9**, 221 (1999).
2. I. Soutwell, *Med. Arom. Plants Ind. Profiles*, **9**, 29 (1999).
3. J. J. Brophy et al., *J. Agric. Food Chem.*, **37**, 1330 (1989).
4. S. Shapiro et al., *Oral Microbiol. Immunol.*, **9**, 202 (1996).
5. C. F. Carson et al., *J. Antimicrob. Chemother.*, **35**, 421 (1995).
6. C. F. Carson and T. V. Riley, *Lett. Appl. Microbiol.*, **19**, 24 (1994).

7. C. F. Carson and T. V. Riley, *J. Appl. Bacteriol.*, **78**, 264 (1995).
8. A. Hammer et al., *J. Antimicrob. Chemother.*, **42**, 591 (1998).
9. P. Nenoff et al., *Skin Pharmacol.*, **9**, 388 (1996).
10. S. D. Cox et al., *J. Appl. Microbiol.*, **88**, 170 (2000).
11. P. Schnitzler et al., *Pharmazie*, **56**, 343 (2001).
12. R. S. Farag et al., *Phytother. Res.*, **18**, 30 (2004).
13. S. Budhiraja et al., *J. Manipulative Physiol. Ther.*, **22**, 447 (1999).
14. C. Brand et al., *Inflamm. Res.*, **51**, 236 (2002).
15. A. Saitoh et al., *Nippon Nogeikagaku Kaishi*, **77**, 1242 (2003).
16. I. B. Bassett et al., *Med. J. Aust.*, **153**, 455 (1990).
17. K. J. Koh et al., *Br. J. Dermatol.*, **147**, 1212 (2002).
18. M. M. Tong et al., *Aust. J. Dermatol.*, **33**, 145 (1992).
19. C. F. Carson et al., *J. Antimicrob. Chemother.*, **48**, 450 (2001).
20. A. C. Satchell et al., *J. Am. Acad. Dermatol.*, **47**, 852 (2002).
21. J. Faoagali et al., *Burns*, **23**, 349 (1997).
22. T. A. Soderberg et al., *Toxicology*, **107**, 99 (1996).
23. R. Saller et al., *Phytomedicine*, **5**, 489 (1998).
24. T. M. Fritz et al., *Ann. Dermatol. Venereol.*, **128**, 123 (2001).
25. F. Osborne et al., *Can. Pharm. J.*, 42 (1998).
26. D. Priest, *Med. Arom. Plants Ind. Profiles*, **9**, 203 (1999).
27. D. L. J. Opdyke, *Food Chem. Toxicol.*, **14**, 701 (1976).
28. J. L. Hartwell, *Lloydia*, **33**, 288 (1970).

CALAMUS

Source: *Acorus calamus* L. (Family Araceae).

Common/vernacular names: Calamus, sweet cinnamon, sweet flag, sweet myrtle, sweet root, sweet sedge.

GENERAL DESCRIPTION

Perennial herb growing in wet or swampy areas with stiff, sword-shaped leaves; up to about 2 m high; native to Northern Hemisphere (North America, Europe, and Asia). Part used is the stout aromatic rhizome after it is peeled and dried. The essential oil is obtained by steam distillation of both the fresh and the dried unpeeled rhizome. Roots are also used.

CHEMICAL COMPOSITION

The rhizome contains highly variable amounts of volatile oil (up to 9%, but usually 2–6%), depending on sources. Asian (Pakistani, Japanese, Indian) plants yield more oil than the European plants, but the European oil is considered superior in flavor and fragrance qualities (JIANGSU).[1–4] Constituents present in the oil include β-asarone (*cis*-isoasarone), (−)-methyl isoeugenol, asarone, asaryladehyde, calamene, linalool, calamol, calameone, eugenol, methyl eugenol, azulene, pinene, cineole, camphor, and others, with β-asarone being the major component (up to 76% of the oil; 85% in Chinese oil) and the European type containing larger numbers of aromatic compounds (JIANGSU; MASADA).[4,5] The essential oil from the rhizome of the North American variety (*A. calamus* L. var. *americanus*

Wolff) does not contain the carcinogenic β-asarone.[6] The essential oil of the eastern European variety (*A. calamus* var. *vulgaris*) contains 13% β-asarone while the essential oil of an Indian chemotype (var. *augustata* Engler) contains up to 80% or more (WICHTL).

Other constituents present in the rhizome include acoragermacrone, acolamone, and isoacolamone (all sesquiterpenes),[7-10] acoradin, 2,4,5-trimethoxy benzaldehyde, 2,5-dimethoxybenzoquinone, galangin, sitosterol,[11] acoric acid, tannin, resin, and others (WICHTL).[12]

PHARMACOLOGY AND BIOLOGICAL ACTIVITIES

The oil and extracts have shown numerous pharmacological activities, including spasmolytic activities on isolated animal organs (smooth muscle), hypotensive activities in cats and rabbits, anticonvulsant and CNS-depressant activities (FARNSWORTH 3; JIANGSU),[13-16] hypolipidemic activity in rats,[17] *in vitro* immunosuppressive activities,[18] antineurotoxic activity,[19] *in vitro* antimicrobial activity,[20] and others.[21]

TOXICOLOGY

Calamus oil is toxic and the Jammu (Indian) variety has been reported to be carcinogenic in rats.[16,22,23]

β-Asarone has shown antigonadal activity in insects.[24] It is also mutagenic in the Ames test, as are commercial samples of the drug containing various concentrations of β-asarone.[25]

Indian calamus root oil repels houseflies.[26]

USES

Medicinal, Pharmaceutical, and Cosmetic. Oil used as a fragrance component in soaps, detergents, creams, lotions, and perfumes. Maximum use level reported is 0.4% in perfumes.[16] The root extract has been used in hair tonic and antidandruff preparations.

Traditional Medicine. Used for more than 2000 years in China to treat numerous disorders, including rheumatoid arthritis, strokes, epilepsy, gastritis, and lack of appetite; also externally in skin diseases (JIANGSU); used in India in treatments of debility, mental disorders including depression, convulsions, cough, inflammation, tumors, skin diseases, and other conditions;[17] used in Tibetan medicine in treating diptheria, indigestion, and sudden coma related to heart disease (WICHTL); used in Western cultures for centuries as stomachic, carminative, sedative, febrifuge, and others.[27] When chewed it is said to kill the taste for tobacco and to clear phlegm from the throat; the roots of *A. calamus* L. var. *americanus* were used by the Indians of eastern Canada to treat symptoms of diabetes[28] and menopause; also used by many Indian tribes in Canada and the United States as a remedy for colds (MOERMAN).

COMMERCIAL PREPARATION

Oil; crude formerly official in U.S.P. and N.F.

Regulatory Status. Calamus and its derivatives (oil, extracts, etc.) are prohibited from use in human food (§189.110). The North American variety is also prohibited, despite containing none or negligible amounts of β-asarone (WICHTL).

REFERENCES

See the General References for ARCTANDER; BAILEY 2; BARNES; BISSET; CLAUS; DER MARDEROSIAN AND BEUTLER; FARNSWORTH 3; FOGARTY; FOSTER; GRIEVE; GUENTHER; GUPTA; JIANGSU; KROCHMAL AND KROCHMAL; LUST; MARTINDALE; MCGUFFIN 1 & 2; MERCK; MORTON 1; NANJING; ROSE; UPHOF; WICHTL.

1. G. Jukneviciene, *Liet. TSR Mokslu Akad. Darb., Ser. C*, **4**, 69 (1972).
2. G. Pamakstyte-Jukneviciene, *Bot. Sady Pribaltiki*, **445** (1971).
3. M. Raquibuddowla et al., *Sci. Res. (Dacca)*, **4**, 234 (1967).
4. G. Cavazza, *Ann. Falsif. Expert. Chim.*, **69**, 833 (1976).
5. M. Jacobson et al., *Lloydia*, **39**, 412 (1976).
6. K. Keller and E. Stahl, *Planta Med.*, **47**, 71 (1983).
7. S. Yamamura et al., *Tetrahedron*, **27**, 5419 (1971).
8. M. Niwa et al., *Chem. Lett.*, **9**, 823 (1972).
9. M. Iguchi et al., *Tetrahedron Lett.*, **29**, 2759 (1973).
10. M. Niwa et al., *Bull. Chem. Soc. Jpn.*, **48**, 2930 (1975).
11. A. Patra and A. K. Mitra, *Indian J. Chem., Sect. B*, **17B**, 412 (1979).
12. E. G. El'yashevich et al., *Khim. Prir. Soedin.*, **10**, 94 (1974).
13. T. Shipochliev, *Vet. Med. Nauki*, **5**(6), 63 (1968).
14. J. Maj et al., *Acta Pol. Pharm.*, **23**, 477 (1966).
15. N. S. Dhalla and I. C. Bhattacharya, *Arch. Int. Pharmacodyn. Ther.*, **172**, 356 (1968).
16. D. L. J. Opdyke, *Food Cosmet. Toxicol.*, **15**, 623 (1977).
17. R. S. Parab and S. A. Mengi, *Fitoterapia*, **73**, 451 (2002).
18. S. Mehotra et al., *Int. Immunopharmacol.*, **3**, 53 (2003).
19. P. K. Shukla et al., *Phytother. Res.*, **16**, 256 (2002).
20. I. Ahmad and A. Z. Beg, *J. Ethnopharmacol.*, **74**, 113 (2001).
21. V. Derle Deelip and S. A. Mengi, *Indian Drugs*, **38**, 444 (2001).
22. Anon., *Fed. Reg.*, **33**, 6967 (1968).
23. P. M. Jenner et al., *Food Cosmet. Toxicol.*, **2**, 327 (1964).
24. B. P. Saxena et al., *Nature*, **270**, 512 (1977).
25. W. Gogglemann and O. Schimmer, *Mutat. Res.*, **121**, 191 (1983).
26. V. E. Adler and M. Jacobson, *J. Environ. Sci. Health*, **A17**, 667 (1982).
27. T. J. Motley, *Econ. Bot.*, **48**, 397 (1994).
28. L. M. McCune and T. Johns, *Pharm. Biol.*, **41**, 363 (2003).

CALENDULA

Source: *Calendula officinalis* L. (Family Compositae or Asteraceae).

Common/vernacular names: Calendula, goldbloom, holligold, marigold, Marybud, pot marigold.

GENERAL DESCRIPTION

Hairy annual to perennial (absent freezing), 20–50 cm high; leaves oblanceolate to oblong, 7–17 × 1–4 cm; flowers yellow or orange; 4–7 cm in diameter; widely cultivated ornamental in Europe and North America; naturalized in south and west Europe; origin undetermined. Parts used are the flower and herb; not to be confused with *Tagetes* species, also commonly known as marigolds (FOSTER).

CHEMICAL COMPOSITION

Saponins (glycosides A–D, D_2, and F);[1] triterpenoids (helinatriol C and F, ursadiol, 12-ursene-3,16,21-triol;[2] faradiol, brein, arnidiol, erthrodiol, calenduladiol, longispoinogenine; calendulosides, α- and β-amyrin, taraxasterol, τ-taraxasterol and lupeol); triterpenoidal monoesters;[3] flavonoids (narcissin, rutin, and others); trace amounts of essential oil; chlorogenic acid (GLASBY 2; WREN); polysaccharides including a rhamnoarabinogalactan and arabinogalactans;[4] flower yellow pigment is a mixture of β-carotene, lycopene, violaxanthin, and other xanthophylls (CSIR II). Carotenoids in the flower petals are mainly flavoxanthin, luteoxanthin, and auroxanthin and while those in the leaves and stems are mainly β-carotene and lutein. However, the major carotenoid in the dried flowers used to make herbal teas is lutein.[5]

PHARMACOLOGY AND BIOLOGICAL ACTIVITIES

Flower, flower/herb preparations, and extracts have shown anti-inflammatory, immunomodulating, and wound-healing activities; stimulation of granulation at wound site; also increasing glycoprotein, nucleoprotein, and collagen metabolism at site; antibacterial; antifungal, antiviral; antiparasitic (trichomonacidal); stimulation of phagocytosis *in vitro* and in the carbon clearance test in mice; choleretic activity; isolated polysaccharides have shown *in vitro* and *in vivo* tumor-inhibiting activity (ESCOP 3) and immunostimulating activity in the carbon clearance and granulocytes tests.[6] An ethanol extract of the flowers enhanced the *in vitro* proliferation of lymphocytes.[7]

Wound-healing effects of the flowers have been demonstrated in various animal tests.[8] From *in vitro* and animal tests, topical anti-inflammatory activity of the flowers is attributed to Ψ-taraxasterol,[10–12] isorhamnetic glycosides,[13] and triterpenoidal fatty acid esters;[9] notably, faradiol monoester,[9] which showed the same topical antiedematous activity in the Croton oil-induced edema model as indomethacin and at the same dose.[9,11]

A butanolic extract of the flowers showed potent *in vitro* antioxidant and radical scavenging activity.[14]

In a phase III clinical trial, a cream containing an extract of calendula was found to provide good protection against acute dermatitis in women undergoing radiation therapy for postoperative breast cancer.[15]

TOXICOLOGY

No dermal irritation was produced by a 10% aqueous extract of the flowers in rabbits. Ocular irritation from a 10% aqueous extract in the eyes of rabbits was minimal. No genotoxic effects were found a "herbal" tea of the flowers and none was found from six saponins isolated from the flowers.[16] A fluid extract of the flowers was non-mutagenic in the mouse bone micronucleus and in the Ames test (*Salmonella*/microsome assays). However, *in vitro* genotoxic effects were found in *Aspergillus nidulans*.[17] In rat liver cell cultures, unscheduled DNA synthesis was inhibited by nanogram concentrations of various solvent extracts of the flowers; genotoxic effects were only found from high (g/mL) concentrations.[18] No sensitizing effects were found in humans from occlusive patches supplying cosmetics containing 1% calendula extract; however, safety data to support the use of the flowers in cosmetic preparations are considered insufficient. The acute oral LD_{50} of calendula flower extract in rats: >4640 mg/kg.[16]

USES

Medicinal, Pharmaceutical, and Cosmetic. Preparations of calendula flowers are used externally to treat dermal and mucous membrane inflammations, hard-to-heal wounds, leg ulcers, dermatitis, mild burns, and sunburn; internally for inflammatory lesions of the oral and pharyngeal mucosa; also used as an immunostimulant in treating skin inflammations and herpes zoster infections (ESCOP 3; WICHTL).[19]

Used in diverse body care products, including face, body, and hand creams, lotions, night creams, ointments, shampoos, lipsticks, deodorants, shaving creams, suntan products, baby products, eye makeup, and others (NIKITAKIS).[16] Concentrations of calendula and extracts thereof in cosmetic products are from 0.1% to 5%.[16]

Food. Flowers primarily used as mildly saline flavoring and coloring; saffron substitute.

Dietary Supplements/Health Foods. Flowers in tincture ("lotion") for external/internal use; teas (FOSTER).

Traditional Medicine. Flower historically considered vulnerary, antiseptic, styptic; externally used as lotion or ointment for burns and scalds (1st degree), bruises, cuts, rashes, sore nipples; internally for stomach ailments; gastric and duodenal ulcers, and jaundice (FOSTER; WREN). Herb and its preparations reportedly used to stimulate circulation, promote healing; for gastric hemorrhage, ulcers, spasms, glandular swelling, jaundice, anemia; externally for abscesses, wounds, bleeding, and eczema (BLUMENTHAL 1).

COMMERCIAL PREPARATIONS

Crude ligulate florets; flower heads; flower and herb; tincture; ointment, and so on; crude formerly official in U.S.P. and N.F.

Regulatory Status. GRAS as spice, natural flavoring, and seasoning (§182.10). Flowers are subject of a positive German therapeutic monograph. Herb subject of a German therapeutic monograph; however, therapeutic use is not recommended since claimed effectiveness has not been demonstrated (BLUMENTHAL 1).

REFERENCES

See the General References for APPLEQUIST; BARNES; BISSET; BLUMENTHAL 1; BRUNETON; CSIR II; DER MARDEROSIAN AND BEUTLER; ESCOP 3; FOSTER; GRIEVE; HARBOURNE AND BAXTER; MCGUFFIN 1 & 2; NIKITAKIS; STEINMETZ; TUTIN 4; TYLER 1; UPHOF; WICHTL; WREN.

1. E. Vidal-Olliveier and G. Balansard, *J. Nat. Prod.*, **52**, 1156 (1989).
2. J. Pyrek et al., *Pol. J. Chem.*, **53**, 1071 (1979).
3. H. Neukirch et al., *Phytochem. Anal.*, **15**, 30 (2004).
4. J. Varljen et al., *Phytochemistry*, **28**, 2379 (1989).
5. E. Bako et al., *J. Biochem. Biophys. Methods*, **53**, 241 (2002).
6. H. Wagner et al., *Arneim.-Forsch.*, **35**, 1069 (1985).
7. Z. Amirghofran et al., *J. Ethnopharmacol.*, **72**, 167 (2000).
8. S. C. Rao et al., *Fitoterapia*, **62**, 508 (1991).
9. R. Della Loggia et al., *Planta Med.*, **60**, 516 (1994).
10. T. Akihisa et al., *Phytochemistry*, **43**, 1255 (1996).
11. K. Zitterl-Eglseer et al., *J. Ethnopharmacol.*, **57**, 139 (1997).
12. M. Hamburger et al., *Fitoterapia*, **74**, 328 (2003).
13. L. Bezakowa et al., *Pharmazie*, **51**, 126 (1996).
14. C. A. Cordova et al., *Redox Rep.*, **7**, 95 (2002).
15. P. Pommier et al., *J. Clin. Oncol.*, **22**, 1447 (2004).
16. Cosmetic Ingredient Review Expert Panel, *Int. J. Toxicol.*, **20**(Suppl. 2), 13 (2001).
17. A. Ramos et al., *J. Ethnopharmacol.*, **61**, 49 (1998).
18. J. I. Perez-Carreon et al., *Toxicol. In Vitro*, **16**, 253 (2003).
19. ESCOP, Vol. 3. Proposals for European Monographs on Calendulae flos/Flos cum Herba.

CANANGA OIL

Source: *Cananga odorata* J. D. Hook. & Thompson (syn. *Canangium odoratum* Baill. forma *macrophylla*) (Family Annonaceae).

GENERAL DESCRIPTION

Large tree with fragrant flowers; native to islands of tropical Asia (Java, Malaysia, the Philippines, the Moluccas, etc.). The oil is obtained by water distillation of the flowers. A similar essential oil, *ylang ylang oil* (see *ylang ylang oil*), is obtained in a similar manner from *Canangium odoratum* Baill. forma *genuina*.

CHEMICAL COMPOSITION

Contains mainly β-caryophyllene, benzyl acetate, benzyl alcohol, farnesol, α-terpineol, borneol, geranyl acetate, methyl salicylate, benzaldehyde, safrole, linalool, eugenol, isoeugenol, limonene, and other minor components totaling over 100 compounds (MASADA).[1]

PHARMACOLOGY AND BIOLOGICAL ACTIVITIES

Cananga oil is nontoxic, except for causing irritation when applied full strength to rabbit skin.[2]

USES

Medicinal, Pharmaceutical, and Cosmetic. Used as a fragrance component in soaps, detergents, creams, lotions, and perfumes (especially men's fragrances). Maximum use level reported is 0.8% in perfumes.[2]

Food. Used as a flavor ingredient in alcoholic and nonalcoholic beverages, frozen dairy desserts, candy, baked goods, and gelatins and puddings, with highest average maximum use level of about 0.003% (32.3 ppm) in the last category.

COMMERCIAL PREPARATION

Oil official in F.C.C.

Regulatory Status. GRAS (§ 182.20).

REFERENCES

See the General References for ARCTANDER; BAILEY 2; FEMA; FURIA AND BELLANCA; GUENTHER; MASADA.

1. R. N. Duve et al., *Int. Flav. Food Addit.*, **6**, 341 (1975).
2. D. L. J. Opdyke, *Food Cosmet. Toxicol.*, **11**, 1049 (1973).

CAPSICUM

Source: *Capsicum frutescens* L.; *C. annuum* L. and its varieties; *C. chinense* Jacq. (syn. *C. angulosum* Mill.); *C. baccatum* L. var. *pendulum* (Willd.) Eshbaugh (syn. *C. pendulum* Willd.); *C. pubescens* Ruiz & Pavon. (Family Solanaceae).

Common/vernacular names: Capsicum, cayenne pepper, paprika, red pepper, Tabasco pepper, hot pepper, chili pepper.

GENERAL DESCRIPTION

There has been much dispute and confusion regarding the classification of *Capsicum*. All peppers, hot and mild (not to be confused with black and white pepper), have been at one time or another considered as fruits of a single species, *C. annuum* and its varieties, or of two species, *C. annuum* and *C. frutescens*, and their varieties (ARCTANDER; BAILEY 2; UPHOF). Currently, five major *Capsicum* species and their varieties are recognized: *C. frutescens, C. chinense, C. baccatum, C. pubescens*, and *C. annuum* (DE SMET; ROSENGARTEN; TERRELL).

Capsicum annuum is an annual herb (from 1–5 m in height), while the other species are usually perennial woody shrubs, all native to tropical America and now widely cultivated. *C. frutescens* is readily distinguished from *C. annuum* in that its stem is shrubby, its flowers are borne in groups,[1] it grows up to 2 m in height, and it is a perennial (DE SMET; ROSENGARTEN; TERRELL). All five species yield pungent fruits commonly called red pepper or simply capsicum. Mild fruits commonly known as paprika, bell pepper, sweet pepper, or green pepper are usually produced by varieties of *C. annuum*.

Capsicum oleoresin is obtained by extracting red pepper with a suitable organic solvent; extraction of sweet pepper (paprika) with similar solvents yields paprika oleoresin, which contains high concentrations of carotenoids but little or no pungent principles, depending on the process.

CHEMICAL COMPOSITION

Capsicum contains up to 1.5% (usually 0.1–1.0%) pungent principles, composed mainly of capsaicin; other pungent alkaloid principles (capsaicinoids) include dihydrocapsaicin, nordihydrocapsaicin, homocapsaicin, and homodihydrocapsaicin, with the last two in minor concentrations.[2–4] Other constituents present include carotenoids (capsanthin, capsorubin, β-carotene, lutein, zeaxanthin, etc.);[5–10] fats (9–17%), proteins (12–15%), vitamins A, C, and others; and a small amount of a volatile oil made up of more than 125 components of which 24 were identified, including 4-methyl-1-pentyl-2-methyl butyrate, 3-methyl-1-pentyl-3-methyl butyrate, and isohexyl isocaproate (JIANGSU; MARSH).[11]

Mild peppers (e.g., paprika and bell pepper) contain similar constituents as hot peppers but with little or no pungent principles.[12]

PHARMACOLOGY AND BIOLOGICAL ACTIVITIES

While a single dose of capsaicin activates pain, inflammation and hypersensitivity, repeated (long-term) application in appropriately formulated product forms leads to desensitization, analgesic, and anti-inflammatory activity. Capsiacin-induced analgesia and desensitization has been explained on the basis of neuropeptide release and depletion, selective targeting of C fibers in the pain pathway, and activation of the vanilloid receptor type 1.[13,14]

Extracts of five species of hot peppers showed *in vitro* antimicrobial activity.[15] Lipid peroxidation and bacterial counts were inhibited by the addition of hot or sweet peppers to beef patties.[16] Antioxidant activity of capsicum is attributed to capsaicin.[17,18] *In vitro* inhibition of bacteria and platelet aggregation by capsaicin has been associated with *in vitro* fluidization of lipid membranes.[19] Gerbils fed a high-cholesterol diet containing capsicum oleoresin showed reduced serum levels of cholesterol and triglycerides.[20] High oral doses of capsicum in rats lowered serum glucose levels.[21] Administered intragastrically to rats, capsaicin inhibited the formation of hydrochloric acid-induced ulcers,[22] damage to the gastric mucosa, myeloperoxidase activity, lipid peroxidation, and hemorrhagic erosion. Capsaicin also inhibits constitutive activation of NF-κB in malignant melanoma cells and when topically applied to the skin of mice.[17]

Placebo-controlled studies of topical preparations containing capsaicin have found benefits in the treatment of lower back pain,[23] cluster headache, postmastectomy pain syndrome, pruritis, psoriasis, fibromyalgia, arthritis (FUGH-BERMAN; MCKENNA), and osteoarthritis.[24] Oral administration of red pepper powder (providing 1.75 mg capsaicin and hydrocapsaicin/day) in gelatin capsules was reported to be more effective than placebo in reducing the intensity of dyspepsia symptoms,[25] as was intranasal capsaicin spray in decreasing idiopathic rhinitis[26] and intranasal capsaicin in postsurgical recurrence of nasal obstruction and nasal polyposis.[27] Other clinical studies have shown benefits from capsaicin in the treatment of neurogenic incontinence (intravesically applied),[28–30] hypersensitive and overactive bladder,[31] diabetic neuropathy,[32] and postherpetic neuralgia.[33]

In patients with heartburn, capsaicin (5 mg in gelatin capsules) taken 30 min before meals was no different than placebo in effects on dyspepsia, heartburn scores, gastric pH, and gastric emptying, yet it enhanced heartburn by shortening the time to peak heartburn following a meal.[34]

TOXICOLOGY

Capsicum is a powerful local stimulant; its oleoresin or active principles (capsaicin) are strongly irritant to the eyes, tender skin, and mucous membranes, producing an intense burning sensation (MARTINDALE).

Cardio-respiratory arrests, seizures, and subsequent death of an 8-month-old infant was associated with the administration of a tea prepared from powdered red pepper.[35]

The safety of pepper sprays that contain high amounts of capsaicinoids for use in riot control and self-defense products is controversial[36] and is associated with death and respiratory failure in animals and people. Inhalation exposure of rats to capsaicinoids resulted in acute respiratory inflammation and dose-related damage to alveolar, bronchial, nasal and tracheal cells, and death of respiratory epithelial cells.[37]

Following a body of conflicting results, the carcinogenity and genotoxicity of capsicum and capsaicin are controversial (DE SMET; ROSENGARTEN; TERRELL).[38–41] However, the evidence suggests that whereas capsaicin in large amounts taken over a long time may be carcinogenic, in low amounts it appears to act as an anticarcinogen.[42]

The oral LD_{50} of capsaicin mice is 190 mg/kg,[43] which is 190 times the human consumption in tropical countries.[44]

The no-observed-adverse-effect level (NOAEL) of paprika color in rats of either sex is 5% of the diet.[45]

USES

Medicinal, Pharmaceutical, and Cosmetic. Capsicum tincture and oleoresin are used in topical counterirritant preparations to treat arthritis, rheumatism, neuralgia, lumbago, and chilblains; also used in certain preparations for stopping thumb sucking or nail biting in children.

Food. Capsicum, in whole and ground forms, is widely used as a spice.

Capsicum and its extracts and oleoresin are widely used in food products, including alcoholic and nonalcoholic beverages, frozen dairy desserts, candy, baked goods, gelatins and pudding, meat and meat products, and condiments and relishes, among others. Highest average maximum use levels are reported in alcoholic beverages for the oleoresin and extract, 0.09% and 0.12%, respectively.

Paprika and its oleoresin are primarily used as a colorant in all the above food categories to impart a yellow to orange color.

Dietary Supplements/Health Foods. Used as a synergistic ingredient in various herbal formulas, including general tonics, laxatives, sedatives, and hay fever remedies (FOSTER; LUST).

Traditional Medicine. Capsicum has been used internally to treat diarrhea, cramps, colic, toothache, sore throat, laryngitis, asthma, pneumonia, flatulence, poor appetite, and other ailments; externally as a counterirritant in rheumatism, arthritis, lumbago, neuralgia, cold injuries (chilbains), and others (DE SMET; ROSENGARTEN; TERRELL NADKARNI; NEWALL).

Others. The oleoresin of the fruit is used in spray-delivered riot control and self-defense products (pepper sprays).[46]

COMMERCIAL PREPARATIONS

Crude, capsicum oleoresin, and extracts (e.g., tincture), and paprika oleoresin. Both capsicum oleoresin and tincture were formerly official in N.F. and U.S.P.; pungency is determined by a taste test and is generally expressed in Scoville units. Paprika oleoresin comes in various color strengths.

Regulatory Status. Regulated in the United States as a dietary supplement. Capsicum (red pepper, cayenne pepper) and paprika are GRAS as natural seasonings and flavorings (§182.10). Their essential oils, solvent-free oleoresins and natural extractives are also GRAS (§182.20); paprika and paprika oleoresin are also approved as color additives for food use exempt from certification (§73.340 and §73.345). Capsaicin-containing topical products are approved in over-the-counter and prescription drug form in the United States.

Standardized capsaicin products are approved in Germany for topical therapeutic use in painful muscle spasms in the shoulders, arms, and spine. Low capsaicin-containing *Capsicum* products are the subject of a negative German monograph; efficacy for digestive disturbances, and supportive treatment of heart and circulatory functions has not been scientifically established (BLUMENTHAL 1).

REFERENCES

See the General References for BLUMENTHAL 1 & 2; DER MARDEROSIAN AND BEUTLER; FEMA; FOSTER; FUGH-BERMAN; JIANGSU; LIST AND HÖRHAMMER; LUST; MARSH; MCKENNA; NANJING; ROSENGARTEN; STAHL; USD 26th.

1. J. G. Vaughan and C. Geissler, *The New Oxford Book of Plants*, Oxford University Press, New York, 1997, pp. 138–139.
2. J. A. Maga, *CRC Crit. Rev. Food Sci. Nutr.*, **6**, 177 (1975).
3. S. I. Balbaa et al., *Lloydia*, **31**, 272 (1968).
4. J. Jurenitsch and W. Kubelka, *Planta Med.*, **33**, 285 (1978).
5. C. E. C. Lord and A. S. L. Tirimanna, *Mikrochim. Acta*, **1**, 469 (1976).
6. B. Camara and R. Monéger, *Phytochemistry*, **17**, 91 (1978).
7. D. Hornero-Méndez and M. I. Mínguez-Mosquera, *J. Agric. Food Chem.*, **46**, 4087 (1998).
8. J. Deli and P. Molnar, *Curr. Org. Chem.*, **6**, 1197 (2002).
9. T. Maoka et al., *J. Agric. Food Chem.*, **49**, 1601 (2001).
10. M. Materska et al., *Phytochemistry*, **63**, 893 (2003).
11. L. W. Haymon and L. W. Aurand, *J. Agric. Food Chem.*, **19**, 1131 (1971).
12. K. Iwai et al., *Agric. Biol. Chem.*, **41** (1873), 1877 (1977).
13. W. Robbins, *Clin. J. Pain*, **16**(Suppl. 2), S86 (2000).
14. M. J. Caterina and D. Julius, *Annu. Rev. Neurosci.*, **2**, 487 (2001).
15. R. H. Chichewicz and P. A. Thorpe, *J. Ethnopharmacol.*, **52**, 61 (1996).
16. A. Sánchez-Escalante et al., *J. Sci. Food Agric.*, **83**, 187 (2003).
17. Y. J. Suhr, *Food Chem. Toxicol.*, **40**, 1091 (2002).
18. T. Ochi et al., *J. Nat. Prod.*, **66**, 1094 (2003).
19. H. Tsuchiya, *J. Ethnopharmacol.*, **75**, 295 (2001).
20. R. S. Gupta et al., *Phytother. Res.*, **16**, 273 (2002).
21. Y. Monsereenusorn, *Quart. J. Crude Drug Res.*, **18**, 1 (1980).
22. S. Horie et al., *Scand. J. Gastroenterol.*, **39**, 303 (2004).
23. H. Frerick et al., *Pain*, **106**, 59 (2003).
24. G. McCleane, *Eur. J. Pain*, **4**, 355 (2000).
25. M. Bortolotti et al., *Aliment. Pharmacol. Ther.*, **16**, 1075 (2002).
26. J. B. Van Rijswijk et al., *Allergy*, **58**, 754 (2003).
27. C. Zheng et al., *Acta Oto-Laryngol.*, **120**, 62 (2000).
28. M. A. Cerruto et al., *Urodinamica*, **12**, 29 (2002).

29. M. Lazzeri et al., *Urol. Int.*, **72**, 145 (2004).
30. M. De Seze et al., *J. Urol.*, **171**, 251 (2004).
31. S. Soontrapa et al., *J. Med. Assoc. Thailand*, **86**, 861 (2003).
32. T. Forst et al., *Acta Diabetol.*, **39**, 1 (2002).
33. M. Pappagallo and E. J. Haldey, *CNS Drugs*, **17**, 771 (2003).
34. S. Rodriguez-Stanley et al., *Aliment. Pharmacol. Ther.*, **14**, 129 (2000).
35. T. Snyman et al., *Forensic Sci. Int.*, **124**, 43 (2001).
36. E. J. Olajos and H. Salem, *J. Appl. Toxicol.*, **21**, 355 (2001).
37. C. A. Reilly et al., *Toxicol. Sci.*, **73**, 170 (2003).
38. Y. J. Surh and S. S. Lee, *Food Chem. Toxicol.*, **34**, 313 (1996).
39. S. Marques et al., *Mutat. Res.*, **517**, 39 (2002).
40. V. E. Archer and D. W. Jones, *Med. Hypoth.*, **59**, 450 (2002).
41. S. Chanda et al., *Mutat. Res. Genet. Toxicol. Environ. Mutagen.*, **557**, 85 (2004).
42. E. Ernst and J. Barnes, *Side Effect Drugs Ann.*, **21**, 489 (1998).
43. T. Glinsukon et al., *Toxicon*, **18**, 215 (1980).
44. A. Szallasi and P. M. Blumberg, *Pharmacol. Rev.*, **51**, 159 (1999).
45. K. Kanki et al., *Food Chem. Toxicol.*, **41**, 1337 (2003).
46. C. A. Reilly et al., *J. Forensic Sci.*, **46**, 502 (2001).

CARAMEL COLOR

Common/vernacular names: Burnt sugar coloring, caramel, caramel color.

GENERAL DESCRIPTION

Caramel color was initially produced by heating sugar in an open pan. Later, various sugar sources were used (corn syrup, malt syrup, molasses, invert sugar, etc.) with small amounts of ammonia or ammonium salts under controlled temperature and pressure until the sweet taste was destroyed and the desired color was obtained. Small quantities of mineral acids, bases, or salts were also added during heating. Currently, the use of sugars has largely been replaced with starch hydrosylates reacted with ammonia and/or sulfite compounds in pressurized reaction vessels raised from an initial 50–70 to about 0, until the desired color is achieved. After cooling, the resulting product is usually a thick dark reddish-brown to brown-black viscous liquid or hygroscopic powder.[1,2]

Caramel color has four classes, known as caramel colors I, II, III, and IV. End use, manufacturing process, physical and chemical properties differ among them, and the classes are primarily based on the use of ammonium compounds, sulfite compound or both types of reactants in their manufacture.[3] However, the preparation of caramel colors I and II also requires the use of salts, alkalis, and food-grade acids.[1] Having different colloidal characteristics, tinctorial strengths, and varying acidic pHs, each color has preferred uses as colorant that is reflected in their synonyms: Caramel color I (spirit caramel); caramel color II (process caramel); caramel color III (beer caramel); and caramel color IV (soft-drink caramel).[2] About 70% of all caramel color used worldwide is caramel color IV that is made using both ammonium and sulfite compounds as reactants.[4]

Caramel for flavoring is prepared by heating milk and sugar; it is heated to a much lesser degree than caramel color and has a characteristic pleasant flavor but little tinctorial power.

CHEMICAL COMPOSITION

Many of the higher molecular weight constituents in caramel colors are still unknown. Among the low molecular substances identified are pyrroles, pyrazines, pyridines, and imidazoles. Other types of compounds reported to be present in certain caramel colors include anhydrosugars, oligosaccharides, furanoid compounds,[1] about 50% digestible carbohydrate, 25% nondigestible carbohydrate, and 25% melanoidins. Minor components include ammonia (0.2–2.0%), iron, and copper.[5,6] Available data also show the presence of substituted imidazoles such as 4-methylimidazole (0.005–0.1%) and other nitrogenous compounds.[6–9]

TOXICOLOGY

The four caramel colors have undergone extensive testing for toxic effects in animal and *in vitro* studies, including genotoxicity studies, and were found safe at dosage levels far in excess of those used in foods.[10–20]

USES

Pharmaceutical, Cosmetic, and Food. Typical uses of the four caramel colors include desserts, distilled spirits, spice blends (I),[1] ice creams, brandies, liqueurs, vermouths (II),[12] beer, gravies, baked goods (III), pet foods, soups, and soft drinks (IV).[1]

Caramel color is one of the most commonly used food colors and produces pale yellow to dark brown colors in products intended for drug, cosmetic, and food uses. Formerly, extracts produced by the botanical industry were mostly dark due to the antiquated technology used; for a time, it became the tradition since to add caramel color to lighter extracts that were produced by modern methods to obtain the same color effects, as it was generally believed that a light extract means low strength. Furthermore, caramel color is used directly in practically every category of food product, including baked products, cola beverages, root beers, wines, gravies, jams, prepared meats, frozen desserts, and confections. Its ubiquitous presence in foods is matched perhaps only by salt or sugar. Highest average maximum use levels of caramel color are reported in gravies (5.4%) and reconstituted vegetables (4.8%).

COMMERCIAL PREPARATIONS

Various types (acid proof, beer, spirit, bakers, and confectioners, etc.); also powdered form.[21] Official in N.F. and F.C.C.

Regulatory Status. GRAS as a multiple purpose food substance (§182.1235); also approved as a color additive exempt from certification, to be used in foods and drugs (§73.85 and §73.1085, respectively).

REFERENCES

See the General References for FEMA; FURIA; MARTINDALE; MERCK.

1. D. V. Myers and J. C. Howell, *Food Chem. Toxicol.*, **30**, 359 (1992).
2. C. I. Chappel and J. C. Howell, *Food Chem. Toxicol.*, **30**, 351 (1992).
3. B. H. Licht et al., *Food Chem. Toxicol.*, **30**, 375 (1992).
4. B. H. Licht et al., *Food Chem. Toxicol.*, **30**, 365 (1992).
5. W. R. Fetzer in A. Standen, ed., *Kirk-Othmer Encyclopedia of Chemical Technology*, Vol. 4, 2nd ed., Wiley–Interscience, New York, 1964, p. 63.
6. I. F. Gaunt et al., *Food Cosmet. Toxicol.*, **15**, 509 (1977).
7. G. Fuchs and S. Sundell, *J. Agric. Food Chem.*, **23**, 120 (1975).

8. R. A. Wilks et al., *J. Chromatogr.*, **87**, 411 (1973).
9. M. Komoto et al., *Seito Gijutsu Kenkyukaishi*, **25**, 25 (1975).
10. J. A. Allen et al., *Food Chem. Toxicol.*, **30**, 389 (1992).
11. K. Adams et al., *Food Chem. Toxicol.*, **30**, 397 (1992).
12. K. M. MacKenzie et al., *Food Chem. Toxicol.*, **30**, 397 (1992).
13. D. J. Brusick et al., *Food Chem. Toxicol.*, **30**, 403 (1992).
14. K. M. MacKenzie et al., *Food Chem. Toxicol.*, **30**, 431 (1992).
15. K. M. MacKenzie et al., *Food Chem. Toxicol.*, **30**, 417 (1992).
16. G. F. Houben and A. H. Penninks, *Toxicology*, **91**, 289 (1994).
17. G. F. Houben et al., *Fundam. Appl. Toxicol.*, **20**, 30 (1993).
18. A. Thuvander and A. Oskarsson, *Food Chem. Toxicol.*, **32**, 7 (1994).
19. G. F. Houben et al., *Food Chem. Toxicol.*, **30**, 749 (1992).
20. G. F. Houben et al., *Food Chem. Toxicol.*, **30**, 427 (1992).
21. M. Berdick, *Cosmet. Toilet.*, **92**, 26 (1977).

CARAWAY

Source: *Carum carvi* L. (syn. *Apium carvi* Crantz) (Family Umbelliferae or Apiaceae).

Common/vernacular names: Caraway, caraway fruit, caraway seed, carum.

GENERAL DESCRIPTION

Biennial herb with second-year stem up to about 0.75 m high, widely branching; native to Europe and western Asia and naturalized in North America; widely cultivated. Part used is the dried ripe fruit. An essential oil is obtained from the fruit by steam distillation (GUENTHER).

CHEMICAL COMPOSITION

The seeds contain glucides (coniferin, syringin, benzyl-β-D-glucopyranoside and others), monoterpenoid alcohols,[1] 2–8% (usually 3–7%) volatile oil, about 15% lipids, 20% protein, a β(1 → 4) mannan, and flavonoids (quercetin-3-glucuronide, isoquercitrin, etc.), furano- and hydroxycoumarins, among others (JIANGSU; MARSH; WICHTL).[2–6]

The volatile oil is composed mainly of a ketone, carvone (50–65%), and a terpene, limonene (35–45%) (a terpene), with minor amounts of carveol, dihydrocarveol, dihydrocarvone, thujone, pinene, phellandrene, α-thujene, β-fenchene, and others (MASADA; WICHTL).[3,6–9] Concentrations of the components vary, depending on the degree of ripeness of the fruit; contents of carvone, and other oxygenated components increase as fruit ripens.[8] Therefore, oils obtained from fully mature seeds contain more carvone and less limonene (and other terpenes) and are considered to be of a better quality (ARCTANDER).[3]

PHARMACOLOGY AND BIOLOGICAL ACTIVITIES

Caraway is generally considered to have carminative and stomachic properties. Caraway oil has been reported to exhibit antibacterial activities in vitro as well as larvicidal properties.[10,11] It also has antispasmodic and antihistaminic activities on isolated animal organs (JIANGSU).[12–14]

Powdered caraway inhibited lipid peroxidation in stored chicken meat. An ethanol extract of the seeds showed greater potency.[15] As a supplement in the diet of mice, the oil inhibited carcinogen-induced skin tumors; however, best results were found from topical application of the oil.[16] Carvone and limonene have shown *in vitro* chemopreventive activity by inducing the detoxifying enzyme glutathione S-transferase (GST) in several mouse target tissues. Compounds that induce an increase in the activity of GST detoxification are considered potential inhibitors of carcinogenesis.[17]

TOXICOLOGY

Caraway seeds showed no mutagenic activity in the Ames test.[18] The acute oral LD_{50} of caraway oil in rats is 6.68 g/kg and 3.5 mL/kg. The acute dermal LD_{50} of the oil in rabbits is 1.78 mL/kg.[19] Carvone showed no mutagenic activity in the Ames test, although it did produce chromosomal aberrations in CHO cells and induced sister chromatid changes; however, no carcinogenic activity was found in mice of either sex administered carvone by gavage daily (375 or 750 mg/kg/day, 5 days/week) for 2 years.[20] Other toxicity studies have found that at 1% of the diet of rats for 16 weeks, carvone produced testicular atrophy and retarded growth; at 0.1% and 0.25% of the diet of rats for 28 weeks it had no deleterious effects; and that based on a 12-week study, the maximum acceptable daily intake of carvone in rats was 1 mg/kg.[19]

USES

Medicinal, Pharmaceutical, and Cosmetic. Used in some carminative, stomachic, and laxative preparations. Caraway oil is widely used in Europe in combination with peppermint oil (see ***peppermint***) and other oils in the treatment of dyspepsia;[6,20–26] also used as a flavor in pharmaceuticals and as a fragrance component in cosmetic preparations including toothpaste, mouthwash, soaps, creams, lotions, and perfumes, with maximum use level of 0.4% reported in perfumes.

Food. Caraway is widely used as a domestic spice. It is also extensively used in commercial food products particularly baked goods (rye bread, etc.) and meat and meat products, among others. Caraway oil is used in all major categories of foods, including alcoholic and nonalcoholic beverages, frozen dairy desserts, candy, baked goods, gelatins and puddings, meat and meat products, condiments and relishes, and others. Highest average maximum use level is reported to be about 0.02% (225 ppm) in baked goods.

Dietary Supplements/Health Foods. Oil and extract are used as ingredients in some products as spasmolytic aids to digestion (FOSTER; WREN).

Traditional Medicine. Used as an antispasmodic, antiflatulent, carminative, expectorant, and stomachic for dyspepsia; also used to treat incontinence and indigestion;[6,27] relieving menstrual discomforts, promoting milk secretion, and others (WICHTL; WREN).

COMMERCIAL PREPARATIONS

Fruit and oil; official in N.F. and F.C.C.; formerly official in U.S.P.

Regulatory Status. GRAS as a spice, natural flavoring, or seasoning (§182.10); essential oil, oleoresin, and natural extractives are also GRAS (§182.20). Both essential oil and fruits are subjects of German therapeutic monographs. Essential oil (in daily dose of 3–6 drops); spasmolytic, antimicrobial; for dyspeptic complaints such as mild gastrointestinal spasm, bloating, and fullness. Seed used similarly, though therapeutic use not recommended since efficacy is not well documented (BLUMENTHAL 1).

REFERENCES

See the General References for BAILEY 1; BARRETT; BISSET; BLUMENTHAL 1; CLAUS; FEMA; FOSTER; GRIEVE; GUENTHER; JIANGSU; LIST AND HÖRHAMMER; LUST; MCGUFFIN 1 & 2; USD 26th; WREN.

1. T. Matsumura et al., *Phytochemistry*, **61**, 455 (2002).
2. I. Bochenska and J. Kozlowski, *Herba Pol.*, **15**, 251 (1969).
3. M. B. Embong et al., *Can. J. Plant Sci.*, **57**, 543 (1977).
4. H. Hopf and O. Kandler, *Phytochemistry*, **16**, 1715 (1977).
5. J. Kunzemann and K. Herrmann, *Z. Lebensm. Unters. Forsch.*, **164**, 194 (1977).
6. E. Nemeth and R. Hardman, eds., *Med. Arom. Plants Ind. Profiles*, **7** (1998).
7. H. Rothbaecher and F. Suteu, *Planta Med.*, **28**, 112 (1975).
8. D. Razinskaite, Nauji Laimejimai Biol. Biochem., Liet. TSR Jaunuju Mokslininku Biol. Biochem. Moksline Konf., 35 (1967).
9. A. Salveson and A. B. Svendsen, *Planta Med.*, **30**, 93 (1976).
10. F. M. Ramadan et al., *Chem. Microbiol. Technol. Lebensm.*, **2**, 51 (1972).
11. K. Oishi et al., *Nippon Suisan Gakkaishi*, **40**, 1241 (1974).
12. A. M. Debelmas and J. Rochat, *Plant. Med. Phytother.*, **1**, 23 (1967).
13. M. Reiter and W. Brandt, *Arzneim.-Forsch.*, **35**, 408 (1985).
14. K. J. Goerg and T. Spilker, *Aliment. Pharmacol. Ther.*, **7**, 445 (2003).
15. S. S. L. A. El-Amin et al., *J. Sci. Food Agric.*, **79**, 277 (1999).
16. M. H. Shwaird, *Nutr. Cancer*, **19**, 321 (1993).
17. G. Q. Zheng et al., *Planta Med*, **58**, 338 (1992).
18. B. A. Al-Bataina et al., *J. Trace Elem. Med. Biol.*, **17**, 85 (2003).
19. European Scientific Cooperative on Phytotherapy (ESCOP), *Monographs on the Medicinal Use of Plant Drugs*, Carvi Fructus (Caraway), Exeter, UK, 1997.
20. National Toxicology Program, *Natl. Toxicol. Program Tech. Rep. Ser.*, **381**, 1 (1990).
21. A. Madisch et al., *Digestion*, **69**, 45 (2004).
22. G. Holtman et al., *Phytomedicine*, **20** (Suppl. 4), 56 (2003).
23. B. May et al., *Aliment. Pharmacol. Ther.*, **14**, 1671 (2000).
24. A. Madisch et al., *Arzneim.-Forsch.*, **49**, 925 (1999).
25. B. May et al., *Arzneim.-Forsch.*, **46**, 1149 (1996).
26. J. Westphal et al., *Phytomedicine*, **2**, 285 (1996).
27. J. A. Duke and C. F. Reed., *Q. J. Crude Drug Res.*, **16**, **3**, 116 (1978).

CARDAMOM

Source: *Elettaria cardamomum* (L.) Maton var. *cardamomum* (syn. *E. cardamomum* var. *miniscula* Burkill) (Family Zingiberaceae).

Common/vernacular names: Cardamom, cardamom seed.

GENERAL DESCRIPTION

Perennial reed-like plant with lance-shaped leaves borne on long sheathing stems, up to about 4 m high; native to tropical Asia; now cultivated extensively in tropical regions, particularly India (Malabar coast), Sri Lanka (Ceylon), Laos, Guatemala, and El Salvador. Parts used are the dried, nearly ripe fruits with seeds from which an essential oil is obtained by steam distillation. The long wild native cardamon of Sri Lanka is obtained from *E. cardamomum* var. *major* Thwaites (syn. *E. cardamomum* var. *miniscula* Burkill), which has comparatively more elongated fruits (up to approximately 4 cm) than var. *cardamomum*, and dark brown pericarps with coarse striations, the oil of the which is used as a natural flavoring in liqueurs.

CHEMICAL COMPOSITION

Contains 2.8–6.2% volatile oil, approximately 10% protein, 1–10% fixed oil, up to 50% starch, manganese, and iron, among others (LIST AND HÖRHAMMER; MARSH; WICHTL). The volatile oil is composed mainly of α-terpinyl acetate and 1,8-cineole, each of which may be present at concentrations of up to 50% or more; lesser components include limonene, sabinene, linalool, linalyl acetate, α-pinene, α-terpineol, camphene, myrcene, 1,4-cineole, borneol, and others (MASADA).[1–4] Acid constituents of the oil include acetic, butyric, decanoic, dodecanoic, citronellic, geranic, hexanoic, heptanoic, nerylic, and perillic acids.[3,4] The fixed oil mainly consists of waxes containing *n*-alkanes and sterols, including β-sitostenone, stigmasterol, and β-sitosterol.[5]

Compositions of oils vary, depending on types (e.g., Mysore and Malabar). Oils containing a low content of cineole but high content of terpinyl acetate are considered to be of superior quality for flavor applications.[6]

PHARMACOLOGY AND BIOLOGICAL ACTIVITIES

Cardamom is considered to have carminative, stimulant, and stomachic properties.

Cardamom oil has shown *in vitro* antispasmodic activity on isolated mouse[7] and rabbit intestine.[8] Various constituents of the essential oil show antimicrobial activity *in vitro*. Against 14 different species, 1,8-cineole was only active against *Propionibacterium acnes*.[9]

Alcohol and aqueous extracts of various plant parts of *E. cardamomum* inhibited the *in vitro* growth of a human pathogenic strain of *Salmonella typhi*.[10]

An aqueous extract of the seeds increases trypsin activity in buffer solution.[11]

TOXICOLOGY

Available data indicate cardamom oil to be nontoxic.[12] No mutagenic activity was found from cardamom in the Ames test.[13]

USES

Medicinal, Pharmaceutical, and Cosmetic. Cardamom is used in some carminative, stomachic, and laxative preparations. The seed oil is mainly used as a flavor ingredient in Compound Cardamom Spirit to flavor pharmaceuticals; also used as a fragrance component in soaps, detergents, creams, lotions, and perfumes, with maximum use level of 0.4% reported in perfumes.[12]

Food. Cardamom is used extensively as a domestic spice in curries, breads, and cakes; also in coffee, especially in India, Britain, Germany, Scandinavia, the Middle East, and Latin America. Both cardamom seed and its oil are widely used as flavor components in most categories of food products, including alcoholic and nonalcoholic beverages, frozen desserts, candy, baked goods, gelatins and puddings, meat and meat products, condiments and relishes, and gravies, among others. Highest average maximum use level reported for the seed is 0.5% in gravies and about 0.01% (117 ppm) for the oil in alcoholic beverages.

Dietary Supplements/Health Foods. Whole or ground cardamom used as a flavoring ingredient in India-inspired popular tea known as chai.

Traditional Medicine. Cardamom has been used in medicine for centuries in India and

China as a carminative, stimulant, and to treat urinary problems, among other conditions.

The cardamom used in China for these purposes is the fruit of *Amomum cardamomum* L., which is considered in Chinese medicine to be superior to that of *Elettaria cardamomum* (JIANGSU). *Amomum* cardamoms are from Java and Siam (WICHTL).

Regulatory Status. Use of the seed as a spice, natural flavoring, and natural seasoning (§182.10), and the essential, natural extractive, and solvent-free oleoresins of the seed are GRAS (§182.20). Fruits subject of a German therapeutic monograph in medium daily dose of 1.5 g for treatment of dyspeptic disorders (BLUMENTHAL 1).

COMMERCIAL PREPARATIONS

Seed and oil; official in N.F. and F.C.C.

REFERENCES

See the General References for ARCTANDER; BARRETT; BIANCHINI AND CORBETTA; FEMA; GREIVE; GUENTHER; GUPTA; JIANGSU; LUST; MASADA; MCGUFFIN 1 & 2; NANJING; ROSENGARTEN; STAHL; TERRELL; UPHOF; WICHTL.

1. M. Miyazawa and H. Kameoka, *Yukagaku*, **24**, 22 (1975).
2. J. S. T. Chou, *Koryo*, **106**, 55 (1974).
3. B. M. Lawrence, *Perfum. Flav.*, **14**, 87 (1989).
4. B. M. Lawrence, *Perfum. Flav.*, **16**, 39 (1991).
5. M. Gopalakishnan et al., *J. Agric. Food Chem.*, **38**, 2133 (1990).
6. Y. S. Lewis et al., 6th International Congress of Essential Oils (Pap.), 1974, p. 65.
7. J. Haginiwa et al., *Yakugaku Zasshi*, **83**, 624 (1963).
8. H. al-Zuhair et al., *Pharmacol. Res.*, **34**, 79 (1996).
9. I. Kubo et al., *J. Agric. Food Chem.*, **39**, 1984 (1991).
10. L. Daswani and A. Bohra, *Adv. Plant Sci.*, **16**, 87 (2003).
11. Y. Kato, *Koryo*, **113**, 17 (1975).
12. D. L. J. Opdyke, *Food Cosmet. Toxicol.*, **12**(Suppl.), 837 (1974).
13. B. A. Al-Bataina et al., *J. Trace Elem. Med. Biol.*, **17**, 85 (2003).

CAROB

Source: *Ceratonia siliqua* L. (Family Leguminosae or Fabaceae).

Common/vernacular names: Carob, carob bean, locust bean, St. John's bread.

GENERAL DESCRIPTION

Dome-shaped evergreen tree with dark green compound leaves, consisting of two to five pairs of large, rounded glossy leaflets; fruits (pods) up to 30 cm long, indehiscent, and sometimes borne on the tree trunk; tree up to 15 m high; native to southeastern Europe and western Asia; widely cultivated in the Mediterranean region. Part used is the dried ripe fruit, from which three major commercial products are obtained: carob extract of the dried pod, either roasted or unroasted; carob flour from the pulp or the whole pod; and carob bean gum or *locust bean gum* from the endosperm of the seed. Spain, Italy, and Portugal are the major producers of carob.[1]

Carob pods are believed to be the locusts consumed by St. John the Baptist, hence the name St. John's bread. Seeds were used in ancient times as weight units for gold from which the term *carat* is reportedly derived.

CHEMICAL COMPOSITION

Pod pulp contains 30–40% total fiber,[1] 30–60% (usually 40–50%) sugars mainly composed of sucrose (up to 26% in pulp), fructose (13%), xylose, maltose, dextrose, inositols, among others; proteins; amino acids (alanine, proline, valine, etc.); gallic acid; fats; starch; abscisic acid (a plant growth inhibitor); and others.[1–5] Carob pods also contain high quantities of dietary fiber[1] and polyphenols, including proanthocyanidins,[6] catechin, (−)-epicatechin gallate, (−)-epigallocatecin gallate, gallic acid, quercetin, and ellagic acid. The fiber component (derived from the water-insoluble fraction of the pods) contains 3.94% polphenols (dry weight), including gallotannins, cinnamic, ferulic, gallic (1.65%), *p*-coumaric, and syringic, acids; flavones (luteolin, apigenin, and chrysoeriol), flavanones (genistein, isoflavone, and naringenin), flavonols (isorhamnetin, kaempferol, myricetin, and quercetin), and flavonol glycosides (quercetin arabinoside and others).[7]

The seeds contain protein, a high content of essential fatty acids (mostly oleic, linoleic, and palmitic acids), tannins, gum (a galactomannan), and others.[3,8–10] The protein is localized in the embryo and cotyledons while the gum is present mainly in the endosperm (LIST AND HÖRHAMMER).

PHARMACOLOGY AND BIOLOGICAL ACTIVITIES

Rats fed preparations of carob pods rich in dietary fiber along with a high-fat diet showed significantly lower serum cholesterol levels and greater fecal mass compared to controls.[11] A tannin-rich carob pod preparation also lowered cholesterol levels in rats.[12] Rats fed a diet containing 15% carob gum lost weight compared with control animals and showed decreases in blood glucose, plasma cholesterol, and insulin levels, along with an increase in glucose tolerance.[13]

An infusion of the pods inhibited the *in vitro* proliferation of mouse hepatocellular tumor (T1) cells and induced apoptosis in the cells.[14] The crude polyphenol fraction of carob pods exhibits *in vitro* lipid peroxidation-inhibiting, antioxidant, and free radical scavenging activities.[15]

Selective *in vitro* binding to peripheral benzodiazepine receptors was found from a methanol extract of the pods, as well as an extract of the leaves.[16]

In a double-blind, placebo-controlled trial in patients diagnosed with hypercholesterolemia, a preparation of carob pulp rich in insoluble fiber (15 g/day) significantly lowered total and LDL-cholesterol levels. However, triglyceride levels were only lowered in the female patients and total cholesterol levels decreased by 4% in the females versus only 1% in the male patients.[17]

Infants aged 3–21 months diagnosed with acute diarrhea treated with a carob pod powder containing 40% tannins (1.5 g/kg/day) showed a significant increase in normal defecation and a faster return to normal body temperature and weight and the cessation of vomiting compared to placebo.[18]

TOXICOLOGY

The addition of carob bean gum (9.5 g) to the normal daily diet of healthy humans (ages 19–25 years) was found to significantly reduce the absorption of iron, calcium, and zinc, but not copper.[19]

USES

Food. Carob has served as an emergency food and as a sweet for children[7] and was used by the ancient Egyptians to make beer (MANNICHE). Carob flour and carob extracts (carob syrup, etc.) have been used as food for centuries. Currently, the flour is popular in health foods and as a cocoa substitute, for which roasted kibbles are used, while carob extracts are widely used as flavor ingredients (e.g., butterscotch, imitation chocolate, and vanilla) in all kinds of food products, including alcoholic and nonalcoholic beverages, frozen dairy desserts, candy, baked goods, gelatins and puddings, meat and meat products, condiments and relishes, fruit and ices, sweet sauces, gravies, imitation dairy, and many others. Highest average maximum use levels reported are in imitation dairy (0.50%), fruit and ices (0.50%), gravies (0.46%), sweet sauces (0.46%), and condiments and relishes (0.42%). Carob seed is also the source of *locust bean gum* (galactomannan) that is widely used in foods to increase viscosity.

Dietary Supplements/Health Foods. Carob flour is widely used in health food products, including weight-loss formulations, "energy" bars, tea formulations, and other products, primarily as a chocolate substitute.

Use of carob in the United States is largely as a chocolate substitute; imitation chocolate products containing carob include brownies, carob chip cookies, candy bars, bits, creams, fudge, carob-flavored milk, and so on.[20]

Traditional Medicine. The dried seed kernels (GHAZANFAR) and carob flour has long been used as an antidiarrheal by people of the Mediterranean and Aegean regions.[21] The ancient Egyptians used the pods in topical treatments of wounds and eye conditions, and internally in other conditions (MANNICHE). A decoction of the pods has been used for catarrhal infections (UPHOF).

COMMERCIAL PREPARATIONS

Roasted and unroasted crude (kibbles), syrup, and extracts; extracts usually come in specific flavor strengths, depending on users' requirements.

Regulatory Status. Essential oil, solvent-free oleoresin, and natural extractives of carob bean/St. John's bread are GRAS (§182.20).

REFERENCES

See the General References for ARCTANDER; BIANCHINI AND CORBETTA; FEMA; UPHOF.

1. B. Haber, *Cereal Foods World*, **47**, 365 (2002).
2. T. G. Loo, *Public R. Trop. Inst. Amsterdam*, **288** (1969).
3. Y. Vardar et al., *Qual. Plant. Mater. Veg.*, **21**, 367 (1972).
4. M. A. Joslyn et al., *J. Sci. Food Agric.*, **19**, 543 (1968).
5. B. H. Most et al., *Planta*, **92**, 41 (1970).
6. R. Avallone et al., *Food Comp. Anal.*, **10**, 166 (1997).
7. R. W. Owen et al., *Food Chem. Toxicol.*, **41**, 1727 (2003).
8. J. Artaud et al., *Ann. Falsif. Expert. Chim.*, **70**(749), 39 (1977).
9. J. Artaud et al., *Ann. Falsif. Expert. Chim.*, **69**(737), 23 (1976).
10. I. Orhan and B. Sener, *J. Herbal Pharmacother.*, **2**, 29 (2002).
11. K. Perez-Olleros et al., *J. Sci. Food Agric.*, **79**, 173 (1999).
12. P. Wuersch, *J. Nutr.*, **109**, 685 (1970).
13. A. M. Forestieri, *Phytother. Res.*, **3**, 1, (1989).
14. L. Corsi et al., *Fitoterapia*, **73**, 674 (2002).
15. S. Kumazawa et al., *J. Agric. Food Chem.*, **50**, 373 (2002).
16. R. Avallone et al., *Fitoterapia*, **73**, 390 (2002).

17. H. J. F. Zunft et al., *Eur. J. Nutr.*, **42**, 235 (2003).
18. H. Loeb et al., *J. Pediatr. Gastroenterol. Nutr.*, **8**, 480 (1989).
19. A. E. Harmuth-Hoene et al., *Z. Ernahrungswiss.*, **21**, 202 (1982).
20. J. Ott, *The Cacahuatl Eater: Ruminations of an Unabashed Chocolate Addict*, Natural Products, Co., Vashon, WA, 1985.
21. S. Aksit et al., *Paediatr. Perinat. Epidemiol.*, **12**, 176 (1998).

CARRAGEENAN

Source: *Chondrus crispus* (L.) Stackh., *Eucheuma*, and *Gigartina* species or related red algae (seaweeds) of the class *Rhodophyceae*.

Common/vernacular names: Carrageenan, carrageenin, carragheenan, chondrus extract, Irish moss extract.

GENERAL DESCRIPTION

Carrageenan is a seaweed gum (hydrocolloid) obtained from various red algae growing along the Atlantic coast of Europe and North America, with *C. crispus* (Irish moss) as its major source. It occurs in the intercellular matrix and cell walls of the algae and constitutes 60–80% of their salt-free dry weight.[1–3] Its primary use is as a food stabilizer.

In the manufacture of carrageenan, the dried seaweed is first cleaned with cold water and mechanical devices to remove salt and other extraneous materials. It is then extracted with hot water containing calcium or sodium hydroxide. The extract is clarified by filtration, its pH adjusted to slightly basic, and carrageenan is obtained either by direct drum or roll drying of the filtrate or by precipitation with alcohol (e.g., ethyl or isopropyl), depending on the type or purity desired.[4,5] The United States and European countries (e.g., Denmark, France, and Spain) are the major carrageenan producers, with the United States being by far the largest.

Carrageenan comes in many types with different solubilities and gel characteristics, depending on the process and types of algae used for its manufacture. It readily dissolves in water to form viscous solutions or gels, depending on its chemical composition. Its reaction to heating and shearing forces also depends on its chemical nature.

Carrageenan has high reactivity with certain proteins, particularly milk protein, to form weak to strong gels.

CHEMICAL COMPOSITION

Carrageenan is a sulfated, straight-chain polygalactan composed of residues of D-galactose and 3,6-anhydro-D-galactose with a molecular weight usually of 100,000–500,000. It contains a high content of sulfate (20–40% dry weight basis). The number and position of the sulfate groups and the ratio of galactose to 3,6-anhydrogalactose vary greatly. Carrageenan generally contains two major fractions: a gelling fraction called κ-carrageenan and a nongelling fraction called λ-carrageenan; κ-carrageenan contains D-galactose, 3,6-anhydro-D-galactose and ester sulfate groups, while λ-carrageenan contains D-galactose and its monosulfate and disulfate esters. Other types of carrageenan include ι-carrageenan, which is composed mainly of monosulfates of D-galactose and 3,6-anhydro-D-galactose (FURIA).[2,6,7]

λ-Carrageenan is readily soluble in cold water to form a viscous solution regardless of the cations present, whereas κ-carrageenan is precipitated by potassium ions. The potassium salt of carrageenan, however, is soluble in water on heating and forms an elastic gel on cooling; the elasticity or rigidity of the gel depends on the amounts of potassium ions present. ι-Carrageenan (obtained mostly from *Eucheuma spinosum*) forms thermally reversible elastic gelatin-like gels with calcium ions.

Solutions and gels of carrageenan are degraded rapidly by low pH and high temperatures. Degraded carrageenans (molecular weight approximately 15,000) do not have the viscosity or gelling properties of food-grade carrageenans (molecular weight >100,000).

PHARMACOLOGY AND BIOLOGICAL ACTIVITIES

Carrageenan has been reported to exhibit many pharmacological activities in animals, including lowering of blood cholesterol level, reducing gastric secretions and food absorption, and increasing water content of the gut when large doses are ingested. However, carrageenan is not absorbed following oral administration and does not undergo metabolization to lower molecular weight substances. When administered parenterally, it has shown anticoagulant, hypotensive, and immunosuppressive activities. Furthermore, when injected into a rodent's paw, carrageenan produces a reproducible inflammatory condition, which is extensively used as a model for screening potential anti-inflammatory drugs.[2,8,9]

Carrageenan, both in the degraded (poligeenan; molecular weight ≤20,000) and undegraded forms, has been reported to alleviate peptic and duodenal ulcers in humans (MARTINDALE). Carrageenan has inhibitory effects on pepsin activity *in vitro*. Its degraded form (no viscosity) and forms with low and high viscosities, all exhibit antiproteolytic activities *in vitro* against papain.[10]

TOXICOLOGY

At high doses the degraded form of carrageenan (poligeenan) shows various toxic effects, especially when administered by injection. Food-grade carrageenan (molecular weight >100,000) has undergone extensive toxicological testing, which demonstrated that they are not absorbed through the gutand are nontoxic,[2] noncarcinogenic, nongenotoxic, and not tumor promoting. Studies that have shown any toxicity from carrageenan involved injecting the substance into animals or used doses far in excess of amounts that pertain to human exposure.[7]

USES

Medicinal, Pharmaceutical, and Cosmetic. Used extensively as binder, emulsifier, or stabilizer in toothpastes; also in hand lotions, creams, tablets, and others.

The degraded form has been used in preparations for treating peptic ulcers, primarily in Europe (France).

Food. Carrageenan (or its salts) is extensively used in food products for its gelling, stabilizing, and thickening properties. Carrangeenans have been used by the meat processing industry to improve the texture of produced meat products with reduced fat.[11] Other products that incorporate carrageenans include milk products such as chocolate milk, ice cream, sherberts, cottage cheese, cream cheese, evaporated milk, milk desserts, puddings, yogurts, infant formulas, and others. Carrageenan is also used in gravies, thickening sauces, bread doughs, jams, and jellies, among others. Its major functions are as thickening, gelling, emulsifying, stabilizing, and suspending agents, preventing the settling of solids, and in the case of ice cream, to prevent ice crystal formation, among others. For use in gel products such as jams and jellies, its use level is usually 0.5–1.1%; ***locust bean gum*** is often used with carrageenan to improve its gel strength and elasticity.

Dietary Supplements/Health Foods. Carrageenan is used in various weight-loss formulations; also in drinks, especially aloe vera, fruit juice, and herbal drinks.

COMMERCIAL PREPARATIONS

Carrageenan and its ammonium, calcium, potassium, and sodium salts come in many grades and types to meet specific end use requirements. Official in F.C.C. and N.F.

Regulatory Status. Carrageenan and its salts are GRAS (§172.620, §172.623, §172.626), §182.7255, §582.7115, §582.7255).

REFERENCES

See the General References for FEMA; FURIA; GLICKSMAN; GRIEVE; LAWRENCE; LEWIS AND ELVIN-LEWIS; LUST; MARTINDALE; MCGUFFIN 1 & 2; UPHOF; WHISTLER AND BEMILLER; WREN; YOUNGKEN.

1. E. L. McCandless et al., *Can. J. Bot.*, **55**, 2053 (1977).
2. D. J. Stancioff and D. W. Renn, *ACS Symp. Ser.*, **15**, 282 (1975).
3. R. L. Whistler and C. L. Smart, *Polysaccharide Chemistry*, Academic Press, New York, 1953, p. 218.
4. C. T. Blood in L. W. Codd et al., eds., *Chemical Technology: An Encyclopedic Treatment*, Vol. **5**, Barnes & Noble, New York, 1972, p. 27.
5. K. B. Guiseley in A. Standen, ed., *Kirk-Othmer Encyclopedia of Chemical Technology*, Vol. 17, 2nd ed., Wiley–Interscience, New York, 1968, p. 763.
6. R. L. Whistler in W. H. Schultz et al., eds., *Symposium on Foods: Carbohydrates and Their Roles*, AVI, Westport, CT, 1969, p. 73.
7. S. M. Cohen and N. Ito, *Crit. Rev. Toxicol.*, **32**, 413 (2002).
8. J. E. Sawicki and P. J. Catanzaro, *Int. Arch. Allergy Appl. Immunol.*, **49**, 709 (1975).
9. C. J. Morris, *Methods Mol. Biol.*, **225**, 115 (2003).
10. E. E. Deschner et al., *Clin. Gastroenterol.*, **10**, 755 (1981).
11. F. Arnal-Peyrot and J. Adrian, *Med. Nutr.*, **13**, 49 (1977).
12. A. Trius and J. G. Sebranek, *Food Sci. Nutr.*, **36**, 69 (1996).

CARROT OILS

Source: *Daucus carota* L. (Family Umbelliferae or Apiaceae).

Common/vernacular names: Carrot oil, oil of carrot, wild carrot oil, Queen Anne's lace.

GENERAL DESCRIPTION

Annual or biennial herb with erect, much branched stem; up to about 1.5 m high. The common cultivated carrot, *D. carota* L. subsp. *sativus* (Hoffm.) Arcang., has an edible fleshy, orange-red taproot, while the wild carrot, or Queen Anne's lace, *D. carota* L. subsp. *carota*, has an inedible, tough whitish root; wild carrot is native to Europe, Asia, and North America; naturalized in North America. Part used is the dried fruit, from which carrot seed oil is obtained by steam distillation. Carrot root oil is obtained by solvent extraction of the red carrot (root); it contains high concentrations of carotenes (α, β, etc.).

CHEMICAL COMPOSITION

Carrot seed oil contains α-pinene (up to 13.3%), β-pinene, carotol (up to 18.29%), daucol, limonene, β-bisabolene, β-elemene, (−)-β-bergamotene, γ-decalactone, β-farnesene, geraniol, geranyl acetate (up to 10.39%), caryophyllene, caryophyllene oxide, methyl eugenol, nerolidol, eugenol, *trans*-asarone, vanillin, asarone, α-terpineol, terpinen-4-ol, γ-decanolactone, coumarin, β-selinene, 2,4,5-trimethoxybenzaldehyde, among others. Other constituents present include oleic acid, palmitic acid, butyric acid, 4-hydroxybenzyl alcohol, and others (JIANGSU).[1–13] Oil content is highly variable, from 0.05% to 7.15%.[10]

PHARMACOLOGY AND BIOLOGICAL ACTIVITIES

COX-I and -II-inhibiting activity of the oil is largely attributed to *trans*-asaraone, 2,4,5-trimethoxybenzaldehyde, geraniol, and oleic acid.[14] *trans*-Asaraone is toxic to mosquito larvae *in vitro* and some species of nematodes and caterpillars *in vivo*. It also inhibits the *in vitro* growth of *Candida albicans*, *C. kruseii*, and *C. parapsilasis*.[11] The seed oil inhibits the *in vitro* growth of *Campylobacter jejuni*[15] and *Helicobacter pylori*. In mice infected with *H. pylori*, oral administration of the oil cleared the bacteria in 20–30% of cases.[16] *In vitro* activity of the oil was also shown against *Aspergillus parasiticus* and its production of aflatoxin. Active constituents were identified as limonene and terpinene.[17] Carrot seed oil has also shown *in vitro* vasodilatory and smooth-muscle relaxant activities on isolated animal organs. It also depressed cardiac action in isolated frog and dog hearts.

TOXICOLOGY

Data indicate carrot seed oil to be nontoxic.[5] However, the chloroform/methanol fraction and petroleum ether extracts of the seeds have shown antifertility activity in female rats (20 mg/kg p.o.).[18]

USES

Medicinal, Pharmaceutical, and Cosmetic. Carrot seed oil is used primarily as a fragrance component in soaps, detergents, creams, lotions, and perfumes. Highest use level reported is 0.4% in perfumes.[5]

Carrot root oil is used in certain sunscreen preparations and as a source of β-carotene and vitamin A.

Food. Carrot seed oil is used as a flavor ingredient in most major categories of food products, including alcoholic (particularly liqueurs) and nonalcoholic beverages, frozen dairy desserts, candy, baked goods, gelatins and puddings, meat and meat products; condiments and relishes, and soups, usually in rather low use levels (<0.003%).

Carrot root oil is used mainly as a yellow food color because of its carotene content.

Traditional Medicine. Seeds used as a diuretic and emmenagogue and for flatulence in the form of a decoction or infusion; in Chinese medicine to treat chronic dysentery and as an anthelmintic.

Others. Carrot seed oil can serve as sources of carotol and daucol, which have potential as starting materials for the synthesis of new fragrance compounds.[4,19]

COMMERCIAL PREPARATIONS

Available as carrot oil (that can be either seed oil or root oil); seed oil is official in F.C.C.

Regulatory Status. Carrot (seed) oil is GRAS (§182.20), and carrot (root) oil has been approved for use as a food color (§73.300); no distinction between root oil and seed oil is given in §182.20.

REFERENCES

See the General References for ARCTANDER; BAILEY 2; BARNES; CLAUS; DER MARDEROSIAN AND BEUTLER; FEMA; GRIEVE; GUENTHER; HUANG; JIANGSU; JIXIAN; LUST; MARTINDALE; MCGUFFIN 1 & 2; NANJING; TERRELL.

1. R. M. Siefert et al., *J. Sci. Food Agric.*, **19**, 383 (1968).
2. L. N. Chelovskaya and A. G. Nikolaev, *Mezhdunar. Kongr. Efirnym Maslam (Mater.)*, 4th (Pub. 1972), **2** 227, (1968).
3. H. Strzelecka and T. Soroczynska, *Farm. Pol.*, **30**, 13 (1974).
4. J. Kuleska et al., *Riechst., Aromen, Körperpflegem.*, **23**, 34 (1973).
5. D. L. J. Opdyke, *Food Cosmet. Toxicol.*, **14**, 705 (1976).
6. J. B. Harborne et al., *Phytochemistry*, **8**, 1729 (1969).
7. G. V. Pigulevskii and V. I. Kovaleva, *Rast. Resur.*, **2**, 527 (1966).
8. A. S. Cheema et al., *Riechst., Aromen, Körperpflegem.*, **25**, 138 (1875).
9. J. W. Hogg et al., *Cosmet. Perfum.*, **89**, 64 (1974).
10. B. M. Lawrence, *Perfum. Flav.*, **15**, 4, 63 (1990).
11. R. A. Momin and M. G. Nair, *J. Agric. Food Chem.*, **50**, 4475 (2002)
12. T. Kobayashi et al., *J. Plant Physiol.*, **160**, 713 (2003).
13. V. Mazzoni et al., *Flav. Fragr. J.*, **14**, 268 (1999).
14. R. A. Momin et al., *Phytother. Res.*, **17**, 976 (2003).
15. M. Friedman et al., *J. Food Prot.*, **65**, 1545 (2002).
16. G. E. Bergonzelli et al., *Antimicrob. Agents Chemother.*, **47**, 3240 (2003).
17. C. Batt et al., *J. Food Sci.*, **48**, 762 (1983).
18. S. K. Garg and V. S. Mathur, *J. Reprod. Fertil.*, **31**, 143 (1972).
19. J. Kulesza et al., *An. Acad. Bras. Cienc.*, **44**(Suppl.), 412 (1972).

CASCARA SAGRADA

Source: *Frangula purshiana* (DC.) J. G. Cooper (syn. *Rhamnus purshiana* DC.) (Family Rhamnaceae).

Common/vernacular names: Cascara, cascara sagrada, chittem bark, sacred bark.

GENERAL DESCRIPTION

Small- to medium-size, deciduous tree with reddish brown bark and hairy twigs; up to about 13 m high; native to the Pacific Coast of North America (northern California, Oregon, Washington, British Columbia, Idaho, and Montana). Part used is the bark, which is removed from trees with trunk diameter of about 10 cm or more; it is then allowed to dry and age for 1 year before use, as the fresh bark has an emetic principle that is destroyed on prolonged storage or by heating. This emetic principle is now generally considered to be composed of monoanthrones and their *O*-glycosides that are oxidized to nonemetic anthraquinones or anthrone *C*-glycosides on storage or heat treatment.[1]

CHEMICAL COMPOSITION

Contains 6–10% (usually ca. 8%) anthraglycosides as its active principles that consist primarily (60% or more) of *C*-glucosides (cascarosides) A, B, C, D, E, and F and barbaloin and chrysaloin), with minor concentrations of *O*-glycosides (e.g., frangulin) also present. Other constituents include free anthraquinones (e.g., emodin, aloe-emodin, isoemodin, and chrysophanol, also known as chrysophanic acid), resins, tannins, and lipids, among others (STAHL).[1–6]

Cascarosides account for 60–70% of the total anthraglycosides (anthracene derivatives) present; the remainder (10–20%) consists mostly of anthraquinone-O-glucosides (e.g., frangulaemodin- and aloe-emodin-8-O-glucoside), aglycones of chrysophanol, frangulaemodin, aloe-emodin, and physcion), and small amounts of hetero- and isodianthrones (WICHTL). Cascarosides A and B are O-glucosides of barbaloin, while cascarosides C and D are O-glucosides of chrysaloin. Mild acid hydrolysis of cascarosides yields aloins, which, however, can be broken down to their aglycones only by strong oxidative hydrolysis.

PHARMACOLOGY AND BIOLOGICAL ACTIVITIES

Cascarosides A and B are responsible for most of the cathartic properties in cascara and act on the large intestine by inducing increased peristalsis (see **senna**).[7–9] Cascarosides are more active than their hydrolyzed products (aloins and free anthraquinones).[10–12]

Laxatives can potentially interfere with orally administered drugs taken at the same time through a decrease in intestinal transit time. Long-term use of anthranoid-containing laxatives can result in potassium depletion; for those taking cardiac glycosides potassium deficiency may result in cardiac arrhythmias (FUGH-BERMAN; WICHTL).

USES

Medicinal, Pharmaceutical, and Cosmetic. Used extensively in laxative preparations; also used in sunscreens.

Food. Only the bitterless extract is reportedly used as a flavor component in foods, including nonalcoholic beverages, frozen dairy desserts, and candy and baked goods. Average maximum use level is below 0.008% (75 ppm).

Dietary Supplements/Health Foods. Crude aged bark used in laxative and detoxicant teas; extracts in herbal formulas (capsules, tablets, drinks, etc.) for the laxative and alleged detoxicant effects.

Traditional Medicine. Bark infusion used by various North American Indian tribes as a laxative and purgative; also taken for arthritis and rheumatism, as a treatment for worms in children, and used topically for sores (MOERMAN); reportedly used in treatments of cancer.[13]

COMMERCIAL PREPARATIONS

Crude and fluid extract, aromatic fluid extract, solid extract (bitter, bitterless, aromatic), powder extract, granular extract, and cascaroside concentrates; crude, aromatic fluid extract, and bitter fluid and powder extracts are official in U.S.P. Strengths of extracts (see **glossary**) are expressed in weight-to-weight ratios as well as total cascaroside content.

Regulatory Status. Allowed as a natural flavoring substance at subtherapeutic levels (§172.510). In 2002, the U.S. FDA. ruled that in over-the-counter products cascara sagrada is not GRAS and effective.[14] Bark subject of a German therapeutic monograph; allowed for short-term use in laxative formulations; only sold in pharmacies (BLUMENTHAL 1; WICHTL).

REFERENCES

See the General References for APhA; BAILEY 1; BARNES; BLUMENTHAL 1; BRUNETON; DER MARDEROSIAN AND BEUTLER; ESCOP 1; FEMA; FUGH-BERMAN; KROCHMAL AND KROCHMAL; LUST; MARTINDALE; STAHL; WICHTL; YOUNGKEN.

1. F. H. L. van Os, *Pharmacology*, **14**(Suppl. 1), 7 (1976).
2. F. J. Evans et al., *J. Pharm. Pharmacol.*, **27**(Suppl.), 91P (1975).
3. J. W. Fairbairn et al., *J. Pharm. Sci.*, **66**, 1300 (1977).
4. H. Wagner and G. Demuth, *Z. Naturforsch. C.*, **29**, 444 (1974).
5. H. Wagner and G. Demuth, *Z. Naturforsch. B.*, **31b**, 267 (1976).
6. A. Y. Leung, *Drug Cosmemet. Ind.*, **121**, 42 (1977).
7. F. A. Nelemans, *Pharmacology*, **14**(Suppl. 1), 73 (1976).
8. E. W. Godding, *Pharmacology*, **14**(Suppl. 1), 78 (1976).
9. F. H. L. van Os, *Pharmacology*, **14**(Suppl. 1), 18 (1976).
10. P. de Witte and L. Lemli, *Hepatogastroenterology*, **37**, 601 (1990).
11. J. W. Fairbairn and G. E. D. H. Mahran, *J. Pharm. Pharmacol.*, **5**, 827 (1953).
12. J. W. Fairbairn and S. Simic, *J. Pharm. Pharmacol.*, **16**, 450 (1964).
13. J. L. Hartwell, *Lloydia*, **34**, 103 (1971).
14. Food and Drug Administration, HHS, *Fed. Regist.*, **67**, 31125 (2002).

CASCARILLA BARK

Source: *Croton eluteria* (L.) Sw. (Family Euphorbiaceae).

Common/vernacular names: Cascarilla, sweet bark, and sweetwood bark.

GENERAL DESCRIPTION

Large shrub to small tree, flowering and fruiting year round; up to 12 m high; native to the West Indies (Bahamas, Jamaica, Cuba, etc.); also grows in tropical America (Mexico, Colombia, and Ecuador).[1] Part used is the dried bark from which an essential oil is obtained by steam distillation; mainly from the Bahamas.

CHEMICAL COMPOSITION

Contains cascarillan as the major constituent and other diterpenes (cascarillans A–D, cascallin, cascarillone, cascarilladone,[2] eleuterins A–K, pseudoeleuterin B[3,4]); lupeol (triterpene);[3] 1.5–3.0% volatile oil, a bitter principle (cascarillin A), resins, tannin, starch, and lipids, among others (LIST AND HÖRHAMMER).

The volatile oil consists primarily of *p*-cymene, camphene, dipentene, *d*-limonene, β-caryophyllene, α-terpineol, α- and β-pinene, α-thujene, borneol, terpinen-4-ol, eugenol, and others.[5,6] Additional constituents include cineole, methylthymol, cuparophenol, and cascarilladiene (MASADA).[7–10]

PHARMACOLOGY AND BIOLOGICAL ACTIVITIES

An acetone extract of the bark and carscarillin potentiated histamine-stimulated gastric acid secretion in a mouse stomach preparation.[3]

The essential oil has shown antimicrobial activities.[1,12]

TOXICOLOGY

Data indicate that cascarilla bark is not toxic.[13]

USES

Medicinal, Pharmaceutical, and Cosmetic. Tinctures and extracts are used in certain bitter tonic preparations; essential oil as a fragrance component in soaps, detergents, creams, lotions, and perfumes (particularly oriental types and men's fragrances), with maximum use level of 0.4% reported in perfumes.[13] The oil is used by compounders, especially for men's fragrances, because of its power and tenacity.[5]

Food. Extract is mainly used as a tobacco additive[4] and in alcoholic (bitters and liqueurs) and nonalcoholic beverages, with average maximum use levels of 0.01% and about 0.08% (775 ppm), respectively. Essential oil is used as a flavor ingredient in most major categories of food products, including alcoholic and nonalcoholic beverages, frozen dairy desserts, candy, baked goods, and condiments and relishes. Highest average maximum use level is 0.007% (72.1 ppm) reported for the last category.

Traditional Medicine. Used in the treatment of fevers,[14] malaria, dysentery, dyspepsia (DUKE 2), and as digestive,[3] aromatic bitter, aromatic stimulant, and body tonic (AYENSU).

Others. Used in flavoring smoking tobacco.

COMMERCIAL PREPARATIONS

Crude, extracts, and essential oil; oil is official in F.C.C.

Regulatory Status. Essential oil, solvent-free oleoresin, and natural extractives are GRAS (§182.20).

REFERENCES

See the General References for ARCTANDER; AYENSU; BAILEY 2; FEMA; GUENTHER; GRIEVE; MCGUFFIN 1 & 2; MERCK; TERRELL; UPHOF; YOUNGKEN.

1. C. D. Adams, *Flowering Plants of Jamaica*, University of the West Indies, Mona, Jamaica, 1972, p. 414.
2. C. Vigor et al., *Phytochemistry*, **57**, 1209 (2001).
3. G. Appendino et al., *J. Agric. Food Chem.*, **51**, 6970 (2003).
4. E. Fattorusso et al., *J. Agric. Food Chem.*, **50**, 5131 (2002).
5. B. M. Lawrence, *Perfum. Flavor.*, **2**(1), 3 (1977).
6. M. L. Hagedorn et al., *Flav. Fragr. J.*, **6**, 193 (1991).
7. A. Claude-LaFontaine et al., *Bull. Soc. Chim. Fr.*, **9–10**, 2866 (1973).
8. A. Claude-LaFontaine et al., *Bull. Soc. Chim. Fr.*, **1–2**, 88 (1976).
9. O. Motl and A. Trka, *Parfüm. Kosmet.*, **54**, 5 (1973).
10. O. Motl et al., *Phytochemistry*, **11**, 407 (1972).
11. J. C. Maruzzella and N. A. Sicurella, *J. Am. Pharm. Ass.*, **49**, 692 (1960).
12. J. C. Maruzzella and L. Liguori, *J. Am. Pharm. Ass.*, **47**, 250 (1958).
13. D. L. J. Opdyke, *Food Cosmet. Toxicol.*, **14**, 707 (1976).
14. W. B. Mors et al., *Medicinal Plants of Brazil*, Reference Publications, Algonac, MI, 2000.

CASSIE ABSOLUTE

Source: *Acacia farnesiana* (L.) Willd. (Family Leguminosae or Fabaceae).

Common/vernacular names: Cassie absolute, huisache, popinac absolute, and sweet acacia.

Cassie absolute

GENERAL DESCRIPTION

Thorny shrub to small tree, 3–9 m high, with very fragrant flowers; believed to be native of the Old World, now widespread and cultivated in subtropical and tropical regions of the world.[1] Parts used are the flowers from which a concréte is first prepared by extraction with petroleum ether; the absolute is then obtained by alcohol extraction of the concréte. Produced primarily in Cannes, France; also in India.

CHEMICAL COMPOSITION

The absolute contains approximately 25% of volatile constituents, which are composed mainly of benzyl alcohol, methyl salicylate, farnesol, and geraniol, with more than 40 other minor compounds, including α-ionone, geranyl acetate, linalyl acetate, nerolidol, dihydroactinidiolide, (−)-3-methyl-dec-3-en-1-ol, (−)-3-methyl-dec-3-enoic acid, and *trans*-3-methyl-dec-4-enoic acid. The last three compounds are responsible for much of the characteristic fragrance of cassie oil.[2–5]

The nonfragrant material present accounts for about 75% of the absolute and consists mostly of high molecular weight lipids (e.g., hydrocarbons and waxes).[2]

USES

Medicinal, Pharmaceutical, and Cosmetic. The absolute is used as a fragrance component in some high-cost perfumes.

Food. The absolute is used as a flavor ingredient (fruit flavors) in most major categories of food products, including alcoholic and nonalcoholic beverages, frozen dairy desserts, candy, baked goods, and gelatins and puddings, with average maximum use levels generally below 0.002%.

Traditional Medicine. Flowers used topically to relieve headache; also used as antispasmodic, aphrodisiac, antidiarrheal, febrifuge, antirheumatic, stimulant, and insecticide in the form of an infusion; used in baths for dry skins. In India, leaves are used to treat inflammatory conditions[6] and gonorrhea; root chewed for sore throat; dried gum ground to powder for diarrhea.[7] Root has been used in treating stomach cancer in Venezuela;[8] also used in China to treat rheumatoid arthritis and pulmonary tuberculosis.

Others. Bark and pods (23% tannin) used for tanning; gum marketed with other acacia gums; used in confectionery.

COMMERCIAL PREPARATIONS

Absolute and oil.

Regulatory Status. Approved for food use as a natural flavoring (§172.510).

REFERENCES

See the General References for ARCTANDER; BAILEY 1; CSIR I; DUKE 1; DUKE 2; GUENTHER; GUPTA; JIANGSU; LIST AND HÖRHAMMER; MCGUFFIN 1 & 2; MORTON 2; ROSE; UPHOF.

1. C. D. Adams, *Flowering Plants of Jamaica*, University of the West Indies, Mona, Jamaica, 1972, p. 336.
2. E. Demole et al., *Helv. Chim. Acta*, **52**, 24 (1969).
3. M. S. Karawaya et al., *Bull. Fac. Pharm. Cairo Univ.*, **13**, 183 (1974).
4. A. El-Hamidi and I. Sidrak, *Planta Med.*, **18**, 98 (1970).

5. A. M. El-Gamassy and I. S. Rofaeel, *Egypt J. Hortic.*, **2**, 39, 53 (1975).
6. V. I. Hukkeri et al., *Indian Drugs*, **39**, 664 (2002).
7. M. B. Siddiqui and W. Husain, *Fitoterapia*, **62**(4), 325 (1991).
8. J. L. Hartwell, *Lloydia*, **33**, 97 (1970).

CASTOR OIL

Source: *Ricinus communis* L. (Family Euphorbiaceae).

Common/vernacular names: Castor bean, palma christi, and ricinus.

GENERAL DESCRIPTION

An annual herb (up to 5 m high) when grown in temperate zones and a perennial shrub or tree (up to about 15 m high) in warmer climates. Castor bean is composed of many varieties. It is generally believed to be a native of Africa or India and is extensively cultivated worldwide. Parts used are the ripe seeds from which a colorless to pale oil is obtained by cold pressing; yield, 25–35%; hot pressing and solvent extraction yield darker grades of oil, which are of lower quality. Castor oil is remarkably stable and does not easily turn rancid. Major castor oil-producing countries include Brazil, China, and India.[1,2]

CHEMICAL COMPOSITION

Castor oil contains fatty acid glycerides of linoleic, oleic, dihydrostearic, and stearic acids, with ricinoleic acid comprising 80–90% of the total fatty acid glyceride content. Ricinoleic acid is a hydroxy acid, and as a result of hydrogen bonding of its hydroxyl groups, castor oil has a characteristically high viscosity.[1–3]

Castor bean (seed) contains a highly poisonous protein (ricin), which remains in the seed cake (pomace) after the expression of castor oil. Ricin is reported to contain 18 different amino acids and to have a molecular weight of 53,000–54,000. Steam or moist cooking of the pomace destroys the ricin. The seed also contains ricinine (an alkaloid), lectins, and a very powerful heat-stable allergen (MARTINDALE).[1,4]

PHARMACOLOGY AND BIOLOGICAL ACTIVITIES

Due to the content of ricinoleic acid, castor oil has cathartic properties, acting on the small intestine and producing purgation 2–8 h after ingestion; its usual dose is about 15 mL. Large doses may produce, besides purgation, nausea, vomiting, and colic. It also has emollient properties on the skin and is soothing to the eyes.

TOXICOLOGY

Castor oil facilitates the absorption of oil-soluble anthelmintics and should not be used with them.

Castor seed is extremely toxic due to its content of ricin, which is not present in the oil; chewing a single seed may be fatal to a child (MARTINDALE), and whereas seven or eight seeds are believed to be sufficient to kill an adult, as few as three seeds can cause fatal gastroenteritis.[5]

Glycoprotein allergens present in the seed pulp can cause serious symptoms (asthma, eye irritations, hay fever, skin rashes, etc.) in certain individuals (MARTINDALE).[5] Allergic reactions are also reported from wearing ornamental necklaces made of the seeds.[6]

A feeding study in chicks found that at concentrations of 0.5–5% of the diet, castor seed produced toxicity characterized by locomotor disturbances, impaired vision, abnormal posture, growth depression, anemia with significant increases in serum sorbitol dehydrogenase, glutamic dehydrogenase, glutamic oxaloacetic transaminase, potassium, and total hepatic and cardiac lipids. Decreases in hepatic vitamin A, serum protein totals, and manganese were also observed.[7] Other animal studies on the seeds have reported fever, perspiration, vomiting, diarrhea, and eventual death.[5]

USES

Medicinal, Pharmaceutical, and Cosmetic. Used as a cathartic, particularly in the treatment of food poisoning and in evacuation of the bowel before X-ray examination; as a solvent or vehicle in some parenteral and ophthalmic preparations; as an ingredient in lipsticks, hair-grooming products, ointments, creams, lotions, transparent soaps, suppository bases, and others.

Food. Castor oil is used as an antisticking and release agent in hard candy production and as a component of protective coatings in tablets (vitamins, minerals, etc.); as a flavor component (e.g., butter and nut flavors) in major categories of foods such as nonalcoholic beverages, frozen dairy desserts, candy, baked goods, and meat and meat products, with highest average maximum use level of 0.055% reported for frozen dairy desserts.

Traditional Medicine. Castor oil has been used for centuries in India, Egypt, and China as a cathartic and externally for sores and abscesses, among others; seeds also reportedly used as an oral contraceptive in Algiers[8] and India.[5] A teaspoon of the oil with parsley has been used to treat asthma (Haiti). Other uses of the oil include intestinal colic (Iran), vomiting, prolapse of the rectum, hemorrhoids (Brazil), diarrhea (South Africa), and as a purgative (Italy, Indonesia, Philippines, Mexico, Argentina, Columbia, and Brazil). Topical uses include bone deformities, limb paralysis (Algeria), bedsores (Nigeria), bronchial catarrh (Guatemala), flatulence in children, mastitis during breastfeeding (Indian and Pakistan), tinea or seborrhea of the scalp (Ethiopia), scabies (Angola), sedative (East and South Africa), warts, old age spots, ulcerated feet (Columbia), scalds (Brazil), burns, eczema (Russia), conjunctivitis (India), sties, and reddening and irritation of the eyes (Columbia). The oil has also been instilled into the ear to treat otitis (Sri Lanka) and earache (Italy).[5]

Others. The seed oil was used by the ancient Egyptians for embalming in mummification;[9] however, by far the largest use of castor oil is as its dehydrated or partially dehydrated form in industrial lubricants, coatings, paints, varnishes, and others; also in synthesis of urethanes, foams, plastics, and certain perfume chemicals, among others.[1,2,10] Sulfonated castor oil (Turkey red oil) is widely used in the textile and printing industry as a surfactant.[11]

COMMERCIAL PREPARATIONS

No. 1 and no. 3 quality oils. Castor oil is official in U.S.P. and F.C.C.

Regulatory Status. Castor oil is approved for use in foods as a natural flavoring substance (§172.510) and as a diluent in color additive mixtures for finished foods in a concentration of not more than 500 ppm (§73.1).

REFERENCES

See the General References for BAILEY 2; BIANCHINI AND CORBETTA; DER MARDEROSIAN AND BEUTLER; DUKE 4; FEMA; GRIEVE; JIXIAN; BRUNETON; JIANGSU; HUANG; LEWIS AND ELVIN-LEWIS; GRIEVE; MARTINDALE; MCGUFFIN 1 & 2; MERCK; NANJING; UPHOF; USD 26th; YOUNGKEN.

1. G. J. Hutzler in A. Standen, ed., *Kirk-Othmer Encyclopedia of Chemical Technology*, Vol. 4, 2nd ed., Wiley–Interscience, New York, 1967, p. 524.
2. L. A. O'Neill in L. W. Codd et al., eds., *Chemical Technology: An Encyclopedic Treatment*, Vol. 5, Barnes & Noble, New York, 1972, p. 187.
3. T. Khadzhiiski and M. Kalichkov, *Nauch. Tr., Vissh Inst. Khranit. Vkusova Prom. Plovdiv*, **21**, 61 (1974).
4. N. Koja and K. Mochida, *Eisei Kagaku*, **20**, 204 (1974).
5. A. Scarpa and A. Guerci, *J. Ethnopharmacol.*, **5**, 117 (1982).
6. R. Torricelli and B. Wuthrich, *Allergologie*, **20**, 34 (1997).
7. S. M. A. El Badwi et al., *Phytother. Res.*, **6**, 205 (1992).
8. V. J. Brondegaard, *Planta Med.*, **23**, 167 (1973).
9. A. Tchapla et al., *J. Sep. Sci.*, **27**, 217 (2004).
10. C. N. Subramanian, *Indian Chem. J. Ann.*, **107** (1972).
11. A. V. Nawaby et al., *J. High Resolut. Chromatogr.*, **21**, 401 (1998).

CASTOREUM

Source: *Castor fiber* L. or *C. canadensis* Kuhl (Family Castoridae).

Common/vernacular names: Secretion of Canadian beaver (*C. canadensis*) and Siberian or European beaver (*C. fiber*).

GENERAL DESCRIPTION

Beavers are large pale brown to chestnut-brown rodents. The Canadian beaver inhabits lakes and rivers of Canada and northern United States, while the Siberian beaver is found in Europe and Siberia. Castoreum is the secretion accumulated in glands located near the pubis (between anus and sex organs) of these animals. These scent glands with their secretion (castoreum) are collected and dried, from which extracts (absolute, tincture; etc.) are prepared by solvent extraction. Canadian castoreum is considered superior in quality to the Siberian castoreum.[1]

CHEMICAL COMPOSITION

Contains 1–2% volatile oil; 0.33–2.5% castorin (a waxy crystalline substance separated from the hot alcoholic extract on cooling); up to 80% of an alcohol-soluble resinoid material; acids (benzoic, salicylic, cinnamic acids, etc.); phenols (phenol, *o*-ethylphenol, *p*-ethylphenol, *p*-propylphenol, chavicol, betuligenol, etc.); ketones (acetophenone and its derivatives, an ionone derivative, etc.); castoramine; cholesterol and other alcohols (benzylalcohol, *cis*-1,2-cyclohexanediol, etc.); 1.4% calcium phosphate; and others (ARCTANDER; LIST AND HÖRHAMMER; POUCHER).[2] Canadian castoreum and Siberian castoreum differ considerably in their relative concentrations of certain of these constituents, with the Siberian material generally higher in volatile oil, castorin, and resinoid matter (LIST AND HÖRHAMMER).

PHARMACOLOGY AND BIOLOGICAL ACTIVITIES

Believed to have sedative, nervine, and other properties. No pharmacological data are available.

TOXICOLOGY

Few tests (primarily dermatological) using castoreum tincture have indicated it to be nontoxic.[3]

USES

Medicinal, Pharmaceutical, and Cosmetic. Rarely used in pharmaceuticals. Main use (generally as a tincture) is in cosmetics as a fragrance component or fixative in perfumes (particularly men's fragrances and oriental types), soaps, creams, and lotions, with maximum use level of 0.4% reported in perfumes.[3]

Food. Extracts used as flavor components (particularly in vanilla flavors) in most major categories of foods such as alcoholic and nonalcoholic beverages, frozen dairy desserts, candy, baked goods, gelatins and puddings, meat and meat products, and gravies. Average maximum use levels reported are usually below 0.009% (93.7 ppm).

Traditional Medicine. Used in amenorrhea, dysmenorrhea, hysteria, restless sleep, and as analeptic and nervine, among others.

COMMERCIAL PREPARATIONS

Mainly crude.

REFERENCES

See the General References for ARCTANDER; FEMA; LIST AND HÖRHAMMER; MARTINDALE; POUCHER.

1. E. Shiftan in A. Standen, ed., *Kirk-Othmer Encyclopedia of Chemical Technology*, Vol. 14, 2nd ed., Wiley–Interscience, New York, 1967, p. 717.
2. Z. Valenta et al., *Experientia*, **17**, 130 (1961).
3. D. L. J. Opdyke, *Food Cosmet. Toxicol.*, **11**, 1061 (1973).

CATECHU (BLACK AND PALE)

Source: *Black catechu*: *Acacia catechu* (L.f.) Willd. (Family Leguminosae or Fabaceae). *Pale catechu*: *Uncaria gambir* (W. Hunter) Roxb. (Family Rubiaceae).

Common/vernacular names: Catechu, dark catechu, black cutch, cutch, cachou, pegu catechu, cashou (*A. catechu*); catechu, brown cutch, white cutch, gambir, gambier, gambir catechu, pale catechu, terra japonica (*U. gambir*).

GENERAL DESCRIPTION

Acacia catechu is a spiny, deciduous medium-size tree, up to 13 m high, native to India and Myanmar. Part used is the heartwood, which is extracted with boiling water; the aqueous extract after filtration, evaporation, and drying yields black catechu, a shiny black mass.

Uncaria gambir is an evergreen woody vine, native to southeastern Asia (Malaysia, Indonesia, etc.). Parts used are the leaves and twigs, which are extracted with boiling water to yield pale catechu after filtration and evaporation of the extract to dryness; it is a pale brown to dark mass occurring in cubes (EVANS).

Both black catechu and pale catechu are incompatible with alkaloids, proteins (e.g., gelatin), and metallic salts (e.g., iron).

The terms *catechu* and *cutch* can also mean products other than black catechu and pale catechu; examples include Bombay catechu and Borneo cutch, which are derived from *Areca catechu* (betel nut) and a mangrove species, respectively.[1]

CHEMICAL COMPOSITION

Black catechu contains 2–12% *l*- and *dl*-catechin, 22–50% catechutannic acid, *l*- and *dl*-epicatechin, quercitin, quercitrin, 25–33% phlobatannin (EVANS), fisetin, red pigments, and others (JIANGSU).

Pale catechu contains *d*- and *dl*-catechin (7.33%) and the condensation product catechutannic acid (22–50%), quercitin, gambirfluorescein, catechu red (EVANS), gallic acid, ellagic acid, catechol, pigments, and others. In addition, it contains several indole alkaloids, including gambirtannine, dihydrogambirtannine, and oxogambirtannine. Gambirine, gambirdine, and others are also found in leaves and/or stems (JIANGSU).[2–5]

PHARMACOLOGY AND BIOLOGICAL ACTIVITIES

Because of their high tannin content, black catechu and pale catechu have astringent, antibacterial, and other pharmacological properties, as well as toxicities of tannins (see *tannic acid*).

d-Catechin has been reported to cause constriction of isolated rabbit ear blood vessels and suppression followed by enhancement of the amplitude of the isolated toad heart (JIANGSU).

Gambirine has been reported to have hypotensive properties (GLASBY 1).

An aqueous extract of *A. catechu* (small branches) has shown hypotensive effects in anesthetized dogs and rats following i.v. administration.[6]

TOXICOLOGY

Fed to rats as part of the normal diet (0.1%), Indian catechu ("katha") was found to decrease liver and blood levels of niacin by 43% and 48%, respectively.[7] Also, at 0.5% of the diet, the minimum daily requirement of niacin was increased and niacin content of liver, blood, and muscle was decreased.[8]

USES

Medicinal, Pharmaceutical, and Cosmetic. Both black and pale catechus are used primarily as an astringent in certain antidiarrheal preparations and in mouthwashes.

Food. Both are used as flavor components in major categories of food products, including alcoholic and nonalcoholic beverages, frozen dairy desserts, candy, baked goods, and gelatins and puddings. Black catechu extract (type of extract not specified) has been reported as the more commonly used, with highest average maximum use levels of 0.01% and 0.016% reported in candy and alcoholic beverages, respectively.

Traditional Medicine. Both black and pale catechus are used in stopping nose bleeding and in treating boils, sores, ulcers, hemorrhoids, and others; black catechu is also reported to be used in cancers.[9] In India, an extract of the bark of black catechu has been orally administered in the treatment of leprosy.[10] Various parts of the plant have also been used topically for bathing leprous sores (NADKARNI). Other Indian uses of black catechu include abortifacient,[11] antipyretic, anti-inflammatory, anthelminitic, ulcers, anemia, psoriasis, bronchitis (bark extract), expectorant, anodyne, and others (WILLIAMSON). In China, one of the major uses of black catechu is in treating indigestion in children.

Pale catechu has been used to treat aphthous ulcers of the mouth, and, diluted with water, used as a gargle to treat sore throat (NADKARNI).

Others. As a source of tannic acid used for tanning and dyeing.

COMMERCIAL PREPARATIONS

Crude and extracts (e.g., tincture). Black catechu was formerly official in U.S.P., while pale catechu was official in N.F.

Regulatory Status. Black catechu is approved for food use as a natural flavoring (§172.510).

REFERENCES

See the General References for CLAUS; EVANS; FEMA; GRIEVE; KARRER; JIANGSU; MARTINDALE; MERCK; NANJING; TERRELL.

1. W. Gardner, *Chemical Synonyms and Trade Names*, 6th ed., The Technical Press, London, 1968, p. 182.
2. L. Merlini et al., *Tetrahedron*, **23**, 3129 (1967).
3. C. Cardani, *Corsi Semin. Chim.*, **11**, 131 (1968).
4. L. Merlini et al., *Phytochemistry*, **11**, 1525 (1972).
5. K. C. Chan, *Tetrahedron Lett.*, **30**, 3403 (1968).
6. J. S. K. Sham et al., *Planta Med.*, **50**, 177 (1984).
7. P. N. Chaudhari and V. G. Hatwalne, *J. Vitaminol.*, **17**, 105 (1971).
8. P. N. Chaudhari and V. G. Hatwalne, *J. Vitaminol.*, **17**, 125 (1971).
9. J. L. Hartwell, *Lloydia*, **33**, 97 (1970).
10. D. Ojha et al., *Int. J. Lepr. Other Mycobact. Dis.*, **37**, 302 (1969).
11. A. Jain et al., *J. Ethnopharmacol.*, **90**, 171 (2004).

CATNIP

Source: *Nepeta cataria* L. (Family Lamiaceae).

Common/vernacular names: Catnep, catnip, and catmint.

GENERAL DESCRIPTION

Gray, hairy, erect, branched perennial, 40–100 cm high; leaves ovate, crenate, base cordate; 2–8 cm long; flowering in spike, white, tinged with purple; native to southern and eastern Europe; widely naturalized elsewhere in Europe and North America, Central Asia, and the Iranian plateaus; commercially harvested from naturalized populations in Virginia, North Carolina; cultivated in Washington, Europe, and Argentina. Part used is the flowering tops and the essential oil obtained from steam distillation.

CHEMICAL COMPOSITION

Contains 0.3–1% essential oil consisting mainly of terpenoids, nepetalic acid, β-caryophyllene, nepetalic anhydride, plus high amounts of nepetalactone (GUENTHER) and its two isomers, (*E,Z*- and *Z,E*-nepetalactone), plus 5,9-dehydronepetalactone, dihydronepetalactone, isodihydronepetalactone, and neonepetalactone;[1,2] high amounts of geranyl acetate, citronellyl acetate, citronellol, geraniol,[3] geranial (citral a), and neral (citral b); also β-caryophyllene, nerol, humuline, limonene, β-pinene, myrcene, β-ocimene,[4] carvacrol, pulegone, thymol, and others; plant also contains tannins (LIST AND HÖRHAMMER); iridoids, including 1,5,9-epideoxyloganic acid[5] and 7-deoxyloganic acid.[6]

PHARMACOLOGY AND BIOLOGICAL ACTIVITIES

Best known for its ability to elicit behavioral responses in cats, including sniffing, licking, and chewing with head shaking, chin and cheek rubbing, sexual stimulation, and head-over rolling and body rubbing; known as "the catnip response."[2] The response is observed in domestic and large cats, such as lions, jaguars, tigers, leopards, and others following exposure to the odor of the plant; however, not all cats, domestic or large, respond; outgoing cats respond well, whereas withdrawn cats respond poorly. Nepetalactone or catnip oil elicit the response when applied as an odor,[2] but not when administered orally or by i.p. injection.[7] Acute doses of catnip in mice (10% of diet) increased locomotion frequencies, rearing, and susceptibility to induced seizures and decreased sodium pentobarbital-induced sleeping time. Short-term effects were "amphetamine-like," whereas long-term administration produced tolerance with adaptative changes.[8]

Diethyl ether extract of plant and nepetalactones have shown *in vitro* antibacterial and antifungal activities.[9,10] Vapors of nepetalactone have shown repellent activity in 13 families of insects.[11] The essential oil and the two isomers of nepetalactone have shown insect repellent activity to subterranean termites (*Reticulitermes* spp)[12] and to male German cockroaches (*Blattella germanica*); E,Z-nepetalactone showed greater repellent activity than DEET (*N,N*-diethyl-3-methylbenzamide).[13]

TOXICOLOGY

As 10% of the diet of pregnant mice, the dried leaves decreased maternal body weight and reduced fetal, placental, and offspring weights; some organ development delayed in both sexes.[14] Others have reported increased food consumption in mice administered catnip.[8] A hypotonic episode (CNS depression lasting approximately 60 h) was reported in a 19-month-old male who ingested raisons soaked in a commercial catnip tea.[15]

USES

Medicinal, Pharmaceutical, and Cosmetic. Essential oil used in cosmetics and perfumes.[4]

Food. The leaves and flowering tops have been used as a flavoring in sauces and cooked foods; dried in mixtures for soups, stews, and so on (DUKE 2).

Dietary Supplements/Health Foods. Tops in teas; pleasant-tasting, mint-like characteristic (FOSTER).

Traditional Medicine. American Indian uses include colds, fever, colic, sedative, sleep aid, headaches, constipation, diarrhea, rheumatism and pains in babies, and tea; also used as a diaphoretic; majority of uses in infants (MOERMAN). Used in Europe in the treatment of colds, fever, headaches, insanity, restlessness, nervousness, flatulence; bruised leaves in ointment for hemorrhoids; also diaphoretic, antispasmodic, and mild stimulant; children's remedy (GRIEVE).

Others. Dried, loosely powdered leaves alone and as stuffing in cat toys; nepetalactones commercially derived from catnip used in the production of aphid sex pheromones (insect attractants).[16] Oil formerly used as an attractant in wild cat traps (DUKE 2).

COMMERCIAL PREPARATIONS

Crude herb; extracts, essential oil; formerly official in both N.F. and U.S.P.

Regulatory Status. Regulated in the United States as a dietary supplement. Formerly included in U.S.P. (1840–1870) and N.F. (IV–VII).

REFERENCES

See the General References for CSIR VII; DER MARDEROSIAN AND BEUTLER; DUKE 2; GUENTHER; HARBOURNE AND BAXTER; LIST AND HÖRHAMMER; GRIEVE; MCGUFFIN 1 & 2; SIMON, STEINMETZ; TUTIN 3; TYLER 1–3; UPHOF; WREN.

1. S. D. Sastry et al., *Phytochemistry*, **11**, 453 (1972).
2. A. O. Tucker and S. S. Tucker, *Econ. Bot.*, **42**, 214 (1988).
3. R. Baranauskiene et al., *J. Agric Food Chem.*, **51**, 3840 (2003).
4. M. E. Ibrahim and A. A. Ezz El-Din, *Egypt. J. Hortic.*, **26**, 281 (1999).
5. F. Murai et al., *Chem. Pharm. Bull.*, **32**, 2809 (1984).
6. M. Tagawa and F. Murai, *Planta Med.*, **47**, 109 (1983).
7. G. R. Waller et al., *Science*, **164**, 1281 (1969).
8. C. O. Mossoco et al., *Vet. Hum. Toxicol.*, **37**, 530 (1995).
9. A. Nostro et al., *Int. J. Antimicrob. Agents*, **18**, 583 (2001).
10. C. Bourrel et al., *J. Essent. Oil Res.*, **5**, 159 (1993).
11. T. Eisner, *Science*, **146**, 1318 (1964).
12. C. J. Peterson and J. Ems-Wilson, *J. Econ. Entomol.*, **96**, 1275 (2003).
13. C. J. Peterson et al., *J. Econ. Entomol.*, **95**, 377 (2002).
14. M. M. Bernardi et al., *Toxicon*, **36**, 1261 (1998).
15. K. C. Osterhoudt et al., *Vet. Hum. Toxicol.*, **39**, 373 (1997).
16. M. A. Birkett and J. A. Pickett, *Phytochemistry*, **62**, 651 (2003).

CEDAR LEAF OIL

Source: *Thuja occidentalis* L. (Family Cupressaceae).

Common/vernacular names: American arborvitae, cedar leaf, eastern white cedar, northern white cedar, thuja, and white cedar oils.

GENERAL DESCRIPTION

Thuja occidentalis is a small- to medium-size tree belonging to the cypress family and grows up to about 20 m high; native to northeastern North America (Nova Scotia south to North Carolina and west to Illinois). There are many cultivated varieties. Parts used are the fresh leaves and twigs, from which cedar leaf oil is obtained by steam distillation. Major producers of the oil are Canada and the United States.

CHEMICAL COMPOSITION

Contains mainly thujone, isothujone, *l*-fenchone, borneol, *l*-bornyl acetate, *dl*-limonene, *d*-sabinene, *d*-terpinen-4-ol, pinene, camphor, myrcene, and *l*-α-thujene, among others, with thujone in major concentration and accounting for up to 65% (w/w) of the oil (KARRER).[1–5]

PHARMACOLOGY AND BIOLOGICAL ACTIVITIES

Oil is believed to have expectorant, uterine stimulant, emmenagogue, anthelmintic, and counterirritant properties.

Antiviral activity has been demonstrated *in vitro*.[6] Leaf extract stimulates phagocytosis (erythrocytes) through Kupffer's cells in isolated rat liver.[7]

TOXICOLOGY

The oil is reported to be nontoxic when applied externally.[4] Due to its high thujone content, the oil is poisonous when ingested in large quantities, producing symptoms such as hypotension and convulsions and eventually death (see ***absinthium***) (MERCK).

USES

Medicinal, Pharmaceutical, and Cosmetic. Used primarily as a counterirritant in certain analgesic ointments and liniments. In Europe, tincture used externally for its antifungal and antiviral activities in treating warts (WREN). Used in German phytomedicine for nonspecific immunostimulant therapy.[8] Principal use in the United States is as a fragrance ingredient in soaps, detergents, creams, lotions, and perfumes, with maximum use level of 0.4% reported in perfumes.[4]

Food. Oil is used as a flavor ingredient in most categories of foods, including alcoholic and nonalcoholic beverages, frozen dairy desserts, candy, baked goods, gelatins and puddings, meat and meat products, condiments and relishes, and others. Reported average maximum use levels are quite low, with the highest being 0.002% in condiments and relishes.

Traditional Medicine. Ointment or decoction of fresh leaves is used to treat rheumatism, coughs, fever, gout, and other ailments. Oil is used internally as an expectorant, antirheumatic, diuretic, and emmenagogue and externally to treat skin diseases and as insect repellent; also used in treating condyloma and cancers.[9]

COMMERCIAL PREPARATION

Essential oil; official in F.C.C.

Regulatory Status. Approved for food use as a natural flavoring, provided that the finished food is thujone-free (§172.510).

REFERENCES

See the General References for ARCTANDER; BAILEY 1; FEMA; GUENTHER; KROCHMAL AND KROCHMAL; LUST; GRIEVE; MCGUFFIN 1 & 2; merck; terrell; uphof; wren; youngken.

1. A. C. Shaw, *Can. J. Chem.*, **31**, 277 (1953).
2. R. M. Ideda et al., *J. Food Sci.*, **27**, 455 (1962).
3. D. V. Banthorpe et al., *Planta Med.*, **23**, 64 (1973).
4. D. L. J. Opdyke, *Food Cosmet. Toxicol.*, **12** (Suppl.), 843 (1974).
5. S. Simard et al., *J. Wood Chem. Technol.*, **8**, 561 (1988).
6. N. Beuscher and L. Kopanski, *Planta Med.*, **52**, 111P (1986).
7. T. Vomel, *Arzneim.-Forsch.*, **35II**, 1437 (1985).
8. H. Wagner and A. Proksch, in N. Farnsworth, N. H. Hikino, and H. Wagner, eds., *Economic and Medicinal Plant Research*, Vol. 1, Academic Press, New York, 1985, p. 113.
9. J. L. Hartwell, *Lloydia*, **33**, 288 (1970).

CEDARWOOD OIL

Source: *Cedarwood oil Virginia: Juniperus virginiana* L. (Family Cupressaceae); *cedarwood oil Texas*: *Juniperus mexicana* Spreng. (Family Cupressaceae); *cedarwood oil Atlas*: *Cedrus atlantica* Manetti (Family Pinaceae).

Common/vernacular names: Cedar oil, cedarwood oil, red cedarwood oil (cedarwood oil Virginia); cedarwood oil Moroccan (cedarwood oil Atlas).

GENERAL DESCRIPTION

There are several cedarwood oils with different physical and chemical properties. They are often referred to in the literature simply as cedarwood oil. The most common ones, cedarwood oil Virginia, cedarwood oil Texas, and cedarwood oil Atlas, are derived from *J. virginiana*, *J. mexicana*, and *C. atlantica*, respectively. Others such as cedarwood oil Himalaya, cedarwood oil East Africa, and cedarwood oil Japanese are obtained from other conifers (ARCTANDER). *Juniperus ashei* Buchh. is also used as a source of cedarwood oil; other species (e.g., *J. erythrocarpa* Cory and *J. scopulorum* Sarg.) containing high oil content are potential sources.[1]

Cedarwood oil Virginia is obtained by steam distillation of the wood (sawdust, shavings, and other lumber wastes) of *J. virginiana*, commonly known as red cedar, eastern red cedar, and savin, which is a tree up to about 33 m high growing in North America east of the Rocky Mountains. The tree has many cultivated varieties. This oil is primarily produced in the United States and is most commonly referred to as cedarwood oil or cedar oil; it has a sweet "pencil wood" and balsamic odor.

Cedarwood oil Texas is prepared by steam distillation of the wood (shavings, etc.) of *J. mexicana*, which is a small tree up to about 6 m high growing in mountains of southwestern United States, Mexico, and Central America. The oil is produced in Texas; it has an odor similar to that of Virginia cedarwood oil.

Cedarwood oil Atlas is obtained by steam distillation of the wood of *C. atlantica*, which is a pyramidal tree closely related to the pines up to about 40 m high and growing in the Atlas Mountains of Algeria. This oil is produced primarily in Morocco; it has different odor characteristics than the Virginian and Texan oils.

CHEMICAL COMPOSITION

Cedarwood oil Virginia contains mainly α- and β-cedrene (ca. 80%), cedrol (3–14%), and cedrenol. Other sesquiterpenes present include thujopsene, β-elemene, caryophyllene, cuparene, α-acoradiene ("acorene"), and others. Monoterpenes are also present (mostly sabinene and sabinyl acetate) (MASADA).[2–8]

Cedarwood oil Texas contains similar major constituents as cedarwood oil Virginia (ARCTANDER; KARRER).[3,5,9]

Cedarwood oil Atlas contains as its major odoriferous components α- and γ-atlantone. Other constituents include acetone, α-ionone, and α-caryophyllene, among others.[10–12]

PHARMACOLOGY AND BIOLOGICAL ACTIVITIES

Oil of *Cedrus atlantica* inhibited the growth of *Candida albicans in vitro*[13] and has shown molluscicidal activity.[14]

Exposure to cedrol evaporated in the air produced sedative effects in mice and rats. In rats, cedrol exposure also caused a significant prolongation of sleeping time.[15]

Humans exposed to cedrol fumes showed significant decreases in heart rate, diastolic and systolic blood pressure, and respiratory rate. Tests indicted that cedrol inhalation also caused a reduction in sympathetic and an increase in parasympathetic nervous system activity, results supporting the alleged relaxant effect of cedar oil.[16] In addition, a decrease in nonrapid eye movement sleep latency was found in humans exposed to cedar essence.[17]

TOXICOLOGY

Cedarwood oil, most likely Virginia, showed tumor-inducing properties on mouse skin.[18] Cedarwood oil (Virginia and/or Texas) is reported to have slight local allergenic (acute and chronic) and acute local irritant properties (SAX).[19] However, other dermatological data indicate that cedarwood oils (Virginia, Texas, and Atlas) are generally nontoxic.[6,9,12]

USES

Medicinal, Pharmaceutical, and Cosmetic. All three types of cedarwood oils (Virginia, Texas, and Atlas) are primarily used as fragrance components or fixatives in cosmetic and household products, particularly soaps and detergents; others include creams, lotions, and perfumes. The maximum use level reported is 0.8% for all three in perfumes.[6,9,12]

Traditional Medicine. Cedarwood oil Virginia has been used as an insect repellent. Decoctions of the leaves, bark, twigs, and seeds of *J. virginiana* are used to treat various illnesses including coughs, bronchitis, rheumatism, venereal warts, and skin rash, among others.[20] Cedar wood oils (*J. ashei*, *J. virginiana*, and *Cedrus atlantica*) are also used in aromatherapy in the treatment of stress-related disorders (EVANS) and topically (*Cedrus altantica*) in treating alopecia areata.[21]

Others. Cedarwood oil Virginia is used in microscopy as a clearing agent and, thickened together with resins, as an immersion oil. It can also serve as source of cedrene, a starting material for fragrance chemicals.

COMMERCIAL PREPARATIONS

The essential oils.

REFERENCES

See the General References for ARCTANDER; BAILEY 1; FERNALD; GUENTHER; KROCHMAL AND KROCHMAL; MARTINDALE; POUCHER; SAX.

1. R. P. Adams, *Econ. Bot.*, **41**, 48 (1987).
2. J. A. Wenninger et al., *J. Assoc. Offic. Anal. Chem.*, **50**, 1304 (1967).
3. W. D. Fordham in L. W. Codd et al., eds., *Chemical Technology: An Encyclopedic Treatment*, Vol. 5, Barnes & Noble, New York, 1972, p. 1.
4. D. V. Banthorpe et al., *Planta Med.*, **23**, 64 (1973).
5. G. C. Kitchens et al., *Givaudanian*, **1**, 3 (1971).
6. D. L. J. Opdyke, *Food Cosmet. Toxicol.*, **12**(Suppl.), 845 (1974).
7. J. A. Marshall et al. in W. Herz et al., eds., *Progress in the Chemistry of Organic Natural Products*, Vol. 31, Springer-Verlag, Vienna, 1974, p. 283.
8. G. C. Kitchens et al., Ger. Offen. 2,202,249 (1972).
9. D. L. J. Opdyke, *Food Cosmet. Toxicol.*, **14**, 711 (1976).
10. A. S. Pfau and P. Plattner, *Helv. Chim. Acta*, **17**, 129 (1934).
11. D. R. Adams et al., *Tetrahedron Lett.*, **44**, 3903 (1974).
12. D. L. J. Opdyke, *Food Cosmet. Toxicol.*, **14**, 709 (1976).
13. S. Abe et al., *Jpn. J. Med. Mycol.*, **44**, 285 (2003).
14. M. Lahlou, *Pharm. Biol.*, **41**, 207 (2003).
15. D. Kagawa et al., *Planta Med.*, **69**, 637 (2003).
16. S. Dayawansa et al., *Auton. Neurosci. Basic Clin.*, **108**, 79 (2003).
17. A. Sano et al., *Psychiatry Clin. Neurosci.*, **52**, 133 (1998).
18. F. J. C. Roe and W. E. H. Field, *Food Cosmet. Toxicol.*, **3**, 311 (1965).
19. H. Franz et al., *Contact Dermatitis*, **38**, 182 (1998).
20. J. L. Hartwell, *Lloydia*, **33**, 288 (1970).
21. I. C. Hay et al., *Arch. Dermatol.*, **134**, 1349 (1998).

CELERY SEED

Source: *Apium graveolens* L. (Family Umbelliferae or Apiaceae).

Common name: Celery fruit and celery seed.

GENERAL DESCRIPTION

An erect biennial herb, up to about 1 m high; native to southern Europe; extensively cultivated. There are many varieties. *Apium graveolens* var. *dulce* (Mill.) Pers. yields the celery vegetable that is its leafstalk (petiole), and *A. graveolens* var. *rapaceum* (Mill.) Gaudich., the turniprooted celery, yields celeriac. The seeds (dried ripe fruits) used for oil production or as spices are produced from other varieties. Major seed-producing countries are France and India. Celery seed oil is obtained by steam distillation of the whole or crushed seeds in about 2–3% yield. An oleoresin and extracts are also prepared by extracting the seeds with solvents.

CHEMICAL COMPOSITION

Celery seed contains coumarins (aprigravin, celerin, osthenol) and coumarin glycosides, including bergapten, apiumoside, apiumetin, vellein, celereoin, nodakenin, and celereoside (WICHTL);[1] also, pthalide glycosides (celephtalides A–C), sesquiterpenoid glucosides (celeriosides A–E),[2] phthalides (senkyunolide-J and -N), 3′-methoxy apiin, tryptophan, and others.[3]

Major components of the oil are *d*-limonene (ca. 60%) and other limonene-type terpenes4selinene (ca. 10%), and about 3% phthalides. Other constituents include santalol, α- and β-eudesmol, dihydrocarvone, and fatty acids (linoleic, palmitic, petroselinic, stearic, oleic acids, etc.), among others (JIANGSU; LIST AND HÖRHAMMER).[5–11]

The phthalides are the odoriferous principles and consist mostly of 3-*n*-butyl phthalide, sedanenolide (3-*n*-butyl-4,5-dihydrophthalide), sedanolide, and sedanonic anhydride, with several others in minor amounts.[7–9]

Celery seed oleoresin contains more odoriferous principles and less terpenes; it also contains apiin and other flavonoids (JIANGSU).

PHARMACOLOGY AND BIOLOGICAL ACTIVITIES

In vitro inhibition of proinflammatory chemical messengers (COX-1 and -2 and of topoisomerase-I and -II) was found from various constituents of the seeds (sedanolide, senkyunolide-J, senkyunolide-N, 3′-methoxy apiin, and tryptophan), the latter two also showing *in vitro* antioxidant activity.[3]

In rat models of acute inflammation and chronic arthritic inflammation, oral administration of extracts (alcohol and supercritical fluid) of wild, green celery seed suppressed NSAID- and ethanol-induced gastric injury, whereas against ibuprofen-induced gastrotoxicity, most commercial flavorant celery seed oils (derived from aged, brown seeds) were inactive.[12] A similar outcome was found from celery seed products in a rat model of polyarthritis, some being effective anti-inflammatories (notably one made from wild, green seeds), whereas the majority were ineffective.[13]

Hepatoprotective activity in rats was found from oral administration of methanol extracts of celery seeds against chemically induced liver toxicity.[14,15]

Phthalides (sedanolide and 3-*n*-butyl phthalide) present in celery seed oil have shown tumor-inhibiting activity and glutathione-inducing activity from oral administration in mice.[4] Sedative activities in mice from phthalides in the seed oil have also been reported.[7] 3-*n*-Butyl phthalide has shown anticonvulsant effects in experimental chronic epilepsy induced by coriaria lactone in rats. Its anticonvulsant effects were weaker than those of diazepam, but its ability to counteract the learning and memory impairment caused by coriaria lactone was greater than that of diazepam, causing no damage to brain cells.[16,17] It also has low acute and chronic toxicities and showed no teratogenic activity in animal experiments.[18]

The petroleum ether-soluble fraction of celery seed exhibited antioxidative properties on lard.[19]

TOXICOLOGY

Celery seed oil is reported to be generally nonirritating, nonsensitizing, and nonphototoxic, though cases of mild to severe dermatitis resulting from contact with celery plants are well documented[20,21] and owe to the presence of phototoxic furanocoumarins also found in the seeds (WICHTL). The seeds are contraindicated in pregnancy (NEWALL).

USES

Medicinal, Pharmaceutical, and Cosmetic. Oil is used in certain tonic, sedative, and carminative preparations and as a fragrance component in soaps, detergents, creams, lotions, and perfumes. Use levels in cosmetics range from a low of 0.0003% (3 ppm) in detergents to a maximum of 0.4% in perfumes.[20]

Food. Celery seed oil, celery seed, and celery seed extracts are all extensively used as flavoring ingredients in all major food products, including alcoholic and nonalcoholic beverages, frozen dairy desserts, candy, baked goods, gelatins and puddings, meat and meat products, condiments and relishes, soups, gravies, snack foods, and others. Use levels reported for the oil are usually very low, with the highest average maximum of about 0.005% (46.6 ppm) in condiments and relishes.

Dietary Supplements/Health Foods. Celery seed or celery seed extracts are used as flavoring or for anti-inflammatory, sedative, urinary antiseptic, and mild diuretic effects in herbal dietary supplements; also in antirheumatic formulations (NEWALL; WREN).

Traditional Medicine. The seeds are used in India as an antispasmodic to treat asthma and bronchitis and diseases of the liver and spleen (NADKARNI); also in kidney failure, bladder and kidney calculi, edema, gout, pleurisy, flatulence, and others (WICHTL). Oil reportedly used as diuretic in dropsy and bladder ailments, as a nervine and antispasmodic, and in rheumatoid arthritis. In the European tradition, the seeds have been used as carminative, stomachic, emmenagogue, diuretic, and laxative; also for glandular stimulation, gout, kidney stones, rheumatic complaints, nervous unrest, loss of appetite, and exhaustion (BLUMENTHAL 1; WICHTL). Leaves and petioles are used for skin problems in addition to above uses.

COMMERCIAL PREPARATIONS

Oil and oleoresin (extracts); seed was formerly official in N.F., and oil is official in F.C.C. Extracts come in various forms with strengths (see *glossary*) expressed in weight-to-weight ratios or in flavor intensities. Celery (including stems, roots, herb, and seed) is the subject of a German therapeutic monograph. Effectiveness of traditional claims is not documented; therefore, use is not recommended (BLUMENTHAL 1).

Regulatory Status. The seed is regulated as a dietary supplement and is GRAS as a natural flavoring or seasoning (§182.10). The solvent-free oleoresin, essential oil, and natural extractives of the seed are also GRAS (§182.20).

REFERENCES

See the General References for APPLEQUIST; ARCTANDER; BAILEY 2; BARNES; BISSET; DER MARDEROSIAN AND BEUTLER; FEMA; GRIEVE; GUENTHER; GUPTA; HUANG; JIANGSU; LIST AND HÖRHAMMER; LUST; MARTINDALE; NEWALL: ROSENGARTEN; TERRELL; UPHOF; WICHTL; WREN.

1. A. K. Jain et al., *Planta Med.*, **52**, 246 (1986).
2. J. Kitajima et al., *J. Agric. Food Chem.*, **64**, 1003 (2003).
3. R. A. Momin and M. G. Nair, *Phytomedicine*, **9**, 312 (2002).
4. G. Zheng et al., *Nutr. Cancer*, **19**, 77 (1993).
5. M. M. Ahuja and S. S. Nigam, *Riechst., Aromen, Körperpflegem.*, **21**, 281 (1971).
6. S. I. Balbaa et al., *Egypt. J. Pharm. Sci.*, **16**, 383 (1976).
7. L. F. Bjeldanes and I. Kim, *J. Org. Chem.*, **42**, 2333 (1977).
8. D. H. R. Barton and J. X. De Vries, *J. Chem. Soc. (London)*, 1916 (1963).
9. H. J. Gold and C. W. Wilson III, *J. Org. Chem.*, **28**, 985 (1963).
10. R. M. Ideda et al., *J. Food Sci.*, **27**, 455 (1962).
11. F. Destaillats and P. Angers, *Lipids*, **37**, 527 (2002).
12. M. W. Whitehouse et al., *Inflammopharmacology*, **9**, 201 (2001).
13. M. W. Whitehouse et al., *Inflammopharmacology*, **7**, 89 (1999).
14. A. Singh and S. S. Handa, *J. Ethnopharmacol.*, **49**, 119 (1995).
15. B. Ahmed et al., *J. Ethnopharmacol.*, **79**, 313 (2002).
16. S. R. Yu et al., *Yaoxue Xuebao*, **23**, 656 (1988).
17. S. R. Yu et al., *Acta Pharmacol. Sin.*, **9**, 385 (1988).
18. S. R. Yu et al., *Yaoxue Tongbao*, **20**, 187 (1985).
19. Y. Saito et al., *Eiyo To Shokuryo*, **29**, 505 (1976).
20. D. L. J. Opdyke, *Food Cosmet. Toxicol.*, **12**(Suppl.), 849 (1974).
21. R. M. Adams, in J. B. Lippincott, ed., *Occupational Contact Dermatitis*, Philadelphia, PA, 1969, p. 190.

CENTAURY

Source: *Centaurium erythraea* Rafn. (syn. *C. umbellatum* Gilib., *C. minus* Moench and *Erythraea centaurium* Pers.) (Family Gentianaceae).

Common/vernacular names: Bitter herb, common centaury, drug centaurium, European centaury, lesser centaury, minor centaury, red cantarone.

GENERAL DESCRIPTION

Annual, mostly biennial herb with upright stem branching near the top; up to 0.5 m high; leaves opposite; flowers in corymbiform cymes, sessile, rose-purple; native to Europe, western Asia, and northern Africa and naturalized in North America in dry grassland, scrub, and mountain slopes. Variable (stem branching, leaf shape and size, flower size, etc.), separated into six subspecies in Europe, more or less restricted by geographic region (TUTIN ET AL. 3). Part used is the dried flowering herb; not to be confused with the genus *Centaurea* (Compositae). Major commercial sources are Morocco and eastern Europe (WICHTL).

CHEMICAL COMPOSITION

Contains several bitter glucosides (gentiopicrin, centapicrin, swertiamarin, gentioflavoside, and sweroside), alkaloids, (gentianine, gentianidine, gentioflavine, etc.), phenolic acids (protocatechuic, *m*- and *p*-hydroxybenzoic, vanillic, syringic, *p*-coumaric, ferulic, 3,4-dihydroxyphenylacetic, sinapic, caffeic acids, etc.), triterpenes (α- and β-amyrin, erythrodiol, crataegolic acid, oleanolic acid, oleanolic lactone, etc.), xanthones (decussatin, eustomin, desmethyleustomin, methylbellidifolin, etc.), sterols (sitosterol, campesterol, etc.), fatty acids (palmitic and stearic acids, etc.), *n*-alkanes (nonacosane and heptacosane, etc.), wax, amino acids, and others (WICHTL).[1–13]

PHARMACOLOGY AND BIOLOGICAL ACTIVITIES

An infusion of the dried flowering tops showed *in vitro* hydroxyl radical and hypochlorus acid scavenging activity.[14] Orally administered, a filtered water suspension of the dried herb showed anti-inflammatory activity in a rat model of polyarthritis and when topically applied in a cream (2.5–10%) in the air-pouch granuloma bioassay. Antipyretic activity was found in rats from oral administration of the suspension against 2,4-dinitrophenol- and amphetamine-induced hyperthermia.[15] Antipyretic activity of the herb is reported to be due to phenolic acids.[16] Hepatoprotective activity against acetaminophen-induced toxicity was shown from oral administration of a methanol extract of the leaves.[17]

Gentiopicrin is reported to have antimalarial properties (MERCK).

USES

Medicinal, Pharmaceutical, and Cosmetic. Used in some bitter tonic preparations in Europe to increase gastric secretions for dyspeptic discomfort and loss of appetite (BLUMENTHAL 1; WICHTL). Also reportedly used in some cosmetic and toiletry preparations for its alleged soothing and astringent properties.[18]

Food. Used in bitters and vermouth formulations; average maximum use level reported is very low, about 0.0002% (2.29 ppm). Also reportedly used in nonalcoholic beverages at an average maximum use level of 0.0008%.

Traditional Medicine. Used since ancient times in Egypt to treat hypertension and to eliminate kidney stones, and in Europe as a tonic, stomachic, febrifuge, and sedative. Used in lotions to remove freckles, spots, and other skin blemishes and in treating cancers.[19]

COMMERCIAL PREPARATIONS

Mainly the crude herb.

Regulatory Status. Regulated in the United States as a dietary supplement. Approved for use as a natural flavoring in alcoholic beverages only (§172.510). Herb subject of a German therapeutic monograph (in daily dose of 1–2 g); use not recommended since effectiveness is not verified (BLUMENTHAL 1).

REFERENCES

See the General References for BARNES; BIANCHINI AND CORBETTA; BISSET; BLUMENTHAL 1; DER MARDEROSIAN AND BEUTLER; FEMA; LIST AND HÖRHAMMER; LUST; MCGUFFIN 1 & 2; STAHL; TUTIN ET AL. 3; UPHOF.

1. K. Sakina and K. Aota, *Yakugaku Zasshi*, **96**, 683 (1976).
2. S. S. Popov et al., *Dokl. Bulg. Akad. Nauk*, **25**, 1225 (1972).
3. D. W. Bishay et al., *Planta Med.*, **33**, 422 (1978).
4. F. Rulko and K. Witkiewicz, *Diss. Pharm. Pharmacol.*, **24**, 73 (1972).
5. F. Rulko, *Pr. Nauk. Akad. Med. Wroclawiu*, **8**, 3 (1976).
6. M. Hatjimanoli and A. M. Debelmas, *Ann. Pharm. Fr.*, **35**, 107 (1977).
7. V. Bellavita et al., *Phytochemistry*, **13**, 289 (1974).
8. J. H. Zwaving, *Pharm. Weekbl.*, **101**, 605 (1966).

9. N. Marekov and S. Popov, *C. R. Acad. Bulg. Sci.*, **20**, 441 (1967).
10. R. Lacroix et al., *Fitoterapia*, **5**, 213 (1983).
11. P. Valentão et al., *J. Agric. Food Chem.*, **50**, 460 (2002).
12. B. Nikolova-Damyanova and N. Handjieva, *Phytochem. Anal.*, **7**, 140 (1996).
13. R. Petlevski et al., *Farm. Glasnik*, **58**, 119 (2002).
14. P. Valentão et al., *Phytomedicine*, **10**, 517 (2003).
15. T. Berkan et al., *Planta Med.*, **57**, 34 (1991).
16. R. Lacroix et al., *Tunisie Med.*, **51**, 327 (1973).
17. M. Mroueh et al., *Phytother. Res.*, **18**, 431 (2004).
18. H. B. Heath, *Cosmet. Toilet.*, **92**, 19 (1977).
19. J. L. Hartwell, *Lloydia*, **32**, 153 (1969).

CHAMOMILE (GERMAN AND ROMAN)

Source: *German chamomile* Matricaria recutita L. (syn. *Matricaria chamomilla* L., *Chamomilla recutita* (L.) Rauschert) (Family Compositae or Asteraceae); ***Roman chamomile*** *Chamaemelum nobile* (L.) All. (syn. *Anthemis nobilis* L.) (Family Compositae or Asteraceae).

Common/vernacular names: Chamomile, German chamomile, Hungarian chamomile, manzanilla, matricaria, sweet false chamomile, wild chamomile (*M. recutita*); English chamomile, garden chamomile, Roman chamomile (*C. nobile*); camomile.

GENERAL DESCRIPTION

German chamomile is a fragrant, low annual herb, with ligulate flowerheads about 2 cm broad; up to about 0.6 m high; native to Europe and northern and western Asia; naturalized in North America; extensively cultivated, particularly in Hungary, Romania, Bulgaria, the Czech Republic, Slovakia, Germany, Greece, Argentina, and Egypt. Parts used are the dried flowerheads. An essential oil obtained from the flowers by steam distillation is blue when fresh.

Frequent changes in interpretation of the scientific name of German chamomile have led to great confusion over the past two decades. The currently accepted scientific name is *Matricaria recutita*, though *Chamomilla recutita* and, to a lesser extent, *Matricaria chamomilla* are still commonly seen in the literature.[1]

Roman chamomile is a strongly fragrant, hairy, half-spreading, and much-branched perennial, with mostly white ligulate flowerheads up to 3 cm across; up to about 0.3 m high; native to southern and western Europe; naturalized in North America; cultivated in England, Belgium, the United States, Argentina, and other countries. Parts used are the dried expanded flowerheads.

CHEMICAL COMPOSITION

Four chemotypes, characterized as A, B, C, and D, dominate *M. recutita*, although as many as 11 different German chamomile chemotypes have been reported, which show considerable variation in flavonoid concentration and bisaboloid profile, but with only slight differences in coumarin and phenolic acid content.[1] The four main chemotypes are dominated, respectively, by bisabolol, bisabolol oxide A, bisabolol oxide B, and bisabolone oxide A.[2]

German chamomile contains variable amounts of volatile oil (0.24–1.9%); flavonoids including apigenin, apigetrin (apigenin-7-D-glucoside), apigenin-7-*O*-glucoside, and acetylated derivatives thereof, apiin (apigenin-7-apiosylglucoside), quercetin, rutin (quercetin-3-rutinoside), luteolin, patuletin, and quercimeritrin (quercetin-7-D-glucoside), among others; coumarins, including aesculetin, scopoletin, scopoletin-7-β-glucoside, umbelliferone (7-hydroxycoumarin) and its methyl ether (herniarin); proazulenes (matricin, matricarin, etc.); triterpene alcohols (α- and β-amyrin, lupeol, taraxerol); sterols (campesterol, cholesterol, sitostanol, sitosterol, stigmasterol, taraxasterol, etc.); sesquiterpenes ((−)-α-bisabolol oxides A, B, and C, cadinene, calamene, chamazulene, matricin, spathulenol, and others); plant acids (anisic, caffeic, syringic, and vanillic) and fatty acids (linoleic, oleic, and others; tannins (catechin and gallic); water-soluble polysaccharides; choline; amino acids; and others (JIANGSU; LIST AND HÖRHAMMER; MCKENNA; STAHL; WICHTL).[1–7]

The volatile oil contains sesquiterpenoids, including matricarin and matricin and the fragrant constituent farnesol. Matricin and its degradation product, chamazulene, are mainly responsible for the characteristic blue coloration; chamazulene, farnesene, (−)-α-bisabolol oxides A, B, and C, spathulenol, *E*- and *Z*-en-yn-dicycloether are the major constituents, with their relative concentrations varying considerably, depending on the sources (JIANGSU; LIST AND HÖRHAMMER; MCKENNA: STAHL; WICHTL).[8–16]

Roman chamomile contains 0.6–2.4% volatile oil that mainly consists of angeloyl, isobutyryl, methacryl, and tigloyl esters of aliphatic alcohols (isobutanol, amylalcohol, 3-phenylpropanol, and others); also, bisabolol, chamzulene, pinocarvone, and others. Other constituents include about 0.6% of bitter sesquiterpene lactones (germacranolides and guaianolides), which include nobilin, 3-epinobilin, 1,10-epoxynobilin, 3-dehydronobilin, 4α-hydroperoxyromanolide, and others; flavonoids (apigenin, apigenin-7-*O*-glucoside, apigetrin, apiin, anthemoside, chamaemeloside, quercitrin, luteolin-7-*O*-glucoside, etc.); coumarins, phenolic acids; acetylenes; choline; fatty acids; and others (LIST AND HÖRHAMMER; WICHTL).[17,18]

The volatile oil contains mainly (ca. 85%) esters of angelic and tiglic acids (e.g., butyl, amyl, isoamyl, and hexyl angelates or tiglates). The major constituents are 2-methylbutyl angelate and isobutyl angelate. Other constituents reported to be present include amyl angelate, α-pinene, farnesol, nerolidol, chamazulene, myrcene, *l-trans*-pinocarveol, *l-trans*-pinocarvone, and 1,8-cineole, among others; the last two terpenoids, reported to be present in greatest quantity in Roman chamomile oil, which is held to have the highest content in esters of any known essential oil.[1] Relative concentrations of the constituents vary, depending on sources of the oil (ARCTANDER; MARTINDALE; WICHTL).[18–22]

The dark blue to blue-green color of chamomile oils is due to azulenes, formed by steam distillation or exposure to acidic pH, from proazulenes (azulene from matricarin, chamazulene from matricin). Guaiazulene (1,4-dimethyl-7-isopropylazulene), as blue as chamazulene (1,4-dimethyl-7-ethylazulene) but less biologically active, is synthetically available and has been used as an adulterant to increase the blueness of commercial chamomile extracts. Chamazulene and biabolol are very unstable, resulting in discharge of blue color on aging of chamomile oil.[23]

Adulteration of commercial chamomile oil and chamomile preparations with essential oil of the wood of the Brazilian tree *Vanillosmopsis erythropappa* Schultz Bip. was noted to be widespread in the last decade. The cheaper Brazilian oil contains almost exclusively (−)-α-bisabolol, which can only be detected by Isotope Ratio Mass Spectrometry (IRMS).[24]

PHARMACOLOGY AND BIOLOGICAL ACTIVITIES

An extract of German chamomile inhibited the formation indomethacin-induced ulcers in rats.[25] Addition of German chamomile to the diet of rats suppressed chemically induced itching.[26] The oil has bactericidal and fungicidal activities, particularly against Gram-positive bacteria (e.g., *Staphylococcus aureus*) and *Candida albicans*. It also reduced blood urea concentration in rabbits to a normal level.[27,28]

Umbelliferone has fungistatic properties.[25] Chamazulene, a major component of the oil, has shown pain-relieving, wound-healing, antispasmodic, anti-inflammatory, and antimicrobial properties; α-bisabolol, another constituent of the oil, has shown anti-inflammatory, antiedemic, antimicrobial, and *in vitro* antipeptic activities; the cyclic spiro ethers (e.g., *cis/trans*-en-yn-dicyloethers) have shown antimicrobial, anti-inflammatory, antianaphylactic, and antispasmodic activities (JIANGSU; LIST AND HÖRHAMMER; MARTINDALE; MCKENNA).[15,29–33] Anti-inflammatory and spasmolytic activities are also attributed to apigenin that also exhibits growth-inhibitory activity against human cancer lines, including breast and prostate tumor cell lines and melanoma cells and cancer chemopreventive activity against chemical- and ultraviolet B-induced skin tumors in mice. Apigenin and (−)-α-bisabolol exhibited *in vitro* antispasmodic activity in smooth muscles, and apigenin has shown anxiolytic activity in rats. Both the essential oil and en-yn-dicycloether inhibited histamine release from rat mast cells (MCKENNA; WICHTL). *In vitro* histamine release from rat mast cells is moderately stimulated by en-yn-dicyclother at low concentrations and strongly inhibited at higher concentrations. Chamazulene and (−)-α-bisabolol showed little effect at the lower concentrations and stimulated histamine release at the higher concentrations.[34] Whereas (−)-α-bisabolol is the most active antispasmodic constituent of the essential oil,[2] topical anti-inflammatory (antiphlogistic) activity of chamomile is largely attributed to apigenin, followed by matricin, (−)-α-bisabolol, and chamazulene. Teas contain little or no apigenin and an extract prepared from the fresh flowers was devoid of apigenin. Matricin is preserved in alcoholic tinctures.[23]

A double-blind clinical trial of standardized extract of German chamomile applied in dressings to dermabrasion wounds following tattoo removal found more rapid would healing compared to placebo.[35] In a randomized placebo-controlled trial, patients with the common cold who inhaled the steam from hot water to which an extract of German chamomile was added experienced greater relief of upper and middle respiratory tract symptoms compared to those treated with a placebo inhalant.[36] In a placebo-controlled double-blind study, an extract of German chamomile combined with an extract of *Angelica sinensis* was found to significantly reduce hot flashes in postmenopausal women compared to placebo.[37]

Three germacranolide sesquiterpene lactones (nobilin, 1,10-epoxynobilin, and 3-dehydronobilin) isolated from Roman chamomile are reported to exhibit antitumor activities *in vitro* against human tumor cells.[17] Chamaemeloside has shown hypoglycemic activity in animals.[38] Topical application of an extract of the dried flowers is reported to be preventative against sunburn and to facilitate more rapid healing of sunburned skin (WICHTL).

TOXICOLOGY

Allergic contact dermatitis to German (rare) or Roman chamomile may occur in people sensitized to certain sesquiterpene lactones or who are already allergic to ragweed (BRADLY; FUGH-BERMAN; MCKENNA; WICHTL).[39] Allergic conjunctivitis of the eyelids was reported in sensitive patients who applied chamomile tea as an eyewash.[40] Anaphylaxis attributed to chamomile has been shown to be almost certainly due to *Anthemis cotula* L. (dog or stinking chamomile).[41] German Kamillosan®, a total extract of the bisabolol chemotype, has been shown to possess a low sensitizing capacity while the bisabolol oxide B chemotype of *M. recutita* has an evident moderate allergenic potential. Allergenicity is apparently due to low variable levels of the highly allergic sequiterpene lactone anthecotulid, dominant in *A. cotula*.[42] British Kamillosan®, an ointment containing Roman chamomile, has been implicated in a case of allergic contact dermatitis of the nipple.[43]

The acute oral LD_{50} of German or Roman chamomile in rats is >5 g/kg; oral LD_{50} of German chamomile oil in mice is 2.5 mL/kg. The oil of Roman chamomile lacks phototoxic effects in animals, and neither oil showed irritating nor sensitizing effects on human skin (NEWALL).[44,45] Long-term oral toxicity studies of German chamomile in dogs and rats found no toxicity and no changes in pups from prenatal dosing; also, no teratogenic effects in rats after long-term administration; no toxicity from 3-week topical application on rabbits; no toxic effects from the oil applied to the backs of hairless mice and limited irritation applied for 24 h to the skin of rabbits.[1]

USES

Medicinal, Pharmaceutical, and Cosmetic. Both German chamomile and Roman chamomile extracts are used in pharmaceutical preparations, with the former more frequently used; they are used in antiseptic ointments, creams, and gels to treat cracked nipples, sore gums, inflammations, irritation of the skin and mucosa, respiratory tract inflammation, and for wound healing. The volatile oils are used in carminative, antispasmodic, and tonic preparations, among others. An infusion or tincture of the flowerheads is used for gastrointestinal spasms, inflammatory conditions of the gastrointestinal tract, and peptic ulcers, menstrual disorders, in addition to mild sleep disorders, especially in children (BRADLY; ESCOP 1; WICHTL).[1]

Extracts of both German and Roman chamomiles are used in cosmetics and body care products including bath preparations, hair dye formulas (for blond hair), shampoos, sunscreen preparations, mouthwashes, and others. The oils are used as fragrance components or active ingredients in soaps, detergents, creams, lotions, and perfumes. Use levels reported range from a low of 0.0005% in detergents to a maximum of 0.4% in perfumes; considered deodorant and stimulative to skin metabolism (BLUMENTHAL 1).[1]

Food. The essential oils and extracts of both German and Roman chamomiles are used as flavor components in most major food categories, including alcoholic (bitters, vermouths, Benedictine liqueurs, etc.) and nonalcoholic beverages, frozen dairy desserts, candy, baked goods, and gelatins and puddings. Average maximum use levels reported are usually less than 0.002% for the oils.

Dietary Supplements/Health Foods. German chamomile and to a lesser extent Roman chamomile crude flower heads or extracts are one of the most widely used herb tea ingredients, singly or in combination with other ingredients. Topical products are used in cosmetics against inflammation. Tinctures and extracts are used as mild sleep aids, antispasmodics, and digestive aids (FOSTER).

Traditional Medicine. An infusion of German chamomile is used in Turkish folk medicine to treat bronchitis and as a laxative and digestive. A compress containing the infusion is applied to treat eye strain and to clean the eyes and face of babies.[37] Since ancient times, German chamomile has been used in treating colic, diarrhea, malarial symptoms, depression, indigestion, flatulence, insomnia, infantile convulsions in teething, mouth sores, toothache, bleeding and swollen gums, sore throat, and other ailments, usually in the form of an infusion, decoction, or tincture. Also used for sciatica, gout, lumbago, skin problems, and inflammation in the form of compresses.[1]

Roman chamomile is used essentially for the same purposes (WICHTL). Both German and Roman chamomiles have reportedly been used against cancers.[46]

COMMERCIAL PREPARATIONS

Crude, extracts, and volatile oils. German chamomile was formerly official in N.F., and

English chamomile (i.e., Roman) in U.S.P.; both oils are prone to insect infestation on storage. Strengths (see *glossary*) of extracts are expressed in weight-to-weight ratios.

Regulatory Status. Both German and Roman chamomiles are regulated as dietary supplements and are GRAS as natural seasonings and flavorings (§182.10) Essential oils, solvent-free oleoresins, and natural extractives are also GRAS (§182.20). German chamomile flowers subject of a positive German therapeutic monograph; preparations allowed; internally for gastrointestinal spasms and inflammatory diseases of the gastrointestinal tract; externally for skin and mucous membrane inflammation, bacterial skin disease or the oral cavity and gums; inflammatory disease of the respiratory tract (as inhalations); bath and irritation or inflammation of the genital and anal areas. Owing to a lack of documented effectiveness, Roman chamomile flowers are the subject of a negative German therapeutic monograph (BLUMENTHAL 1).

REFERENCES

See the General References for ARCTANDER; BAILEY 2; BIANCHINI AND CORBETTA; BLUMENTHAL 1 & 2; ESCOP 1; FEMA; FOSTER; GUENTHER; JIANGSU; LEWIS AND ELVIN-LEWIS; LUST; MASADA; MCKENNA; STAHL; TERRELL; TYLER 1–3; UPHOF; WICHTL.

1. C. Mann and E. J. Staba in L. E. Craker and J. E. Simon, eds., *Herbs, Spices, and Medicinal Plants: Recent Advances in Botany, Horticulture, and Pharmacology*, Vol. 1, Oryx Press, Phoenix, AZ, 1986, p. 235.
2. R. Carle and K. Gomma, *Br. J. Phytother.*, **2**, 147 (1991–1992).
3. A. G. Gorin and A. I. Yakovlev, *Sb. Nauchn. Tr., Ryazan Med. Inst.*, **50**, 9 (1975); through Chem. Abstr., **84**, 79626z (1976).
4. J. Hölzl and G. Demuth, *Planta Med.*, **27**, 37 (1975).
5. C. Redaelli et al., *Planta Med.*, **42**, 288 (1981).
6. Y. Ganeva et al., *Pharmacia*, **50**, 3 (2003).
7. V. Svehlikova et al., *Phytochemistry*, **65**, 2323 (2004).
8. J. Hölzl et al., *Z. Naturforsch.*, **30**, 853 (1975).
9. J. Reichling and H. Becker, *Dtsch. Apoth. Ztg.*, **117**, 275 (1977).
10. O. Motl et al., *Arch. Pharm. (Weinheim)*, **310**, 210 (1977).
11. H. Schilcher, *Planta Med.*, **23**, 132 (1973).
12. G. Verzar-Petri et al., *Herba Hung.*, **15**, 69 (1976).
13. G. Verzar-Petri et al., *Herba Hung.*, **12**, 119 (1973).
14. L. Z. Padula et al., *Planta Med.*, **30**, 273 (1976).
15. O. Isaac et al., *Dtsch. Apoth. Ztg.*, **108**, 293 (1968).
16. O. Motl et al., *Arch. Pharm. (Weinheim)*, **311**, 75 (1978).
17. M. Holub and Z. Samek, *Collect. Czech. Chem. Commun.*, **24**, 1053 (1977).
18. A. Carnat et al., *Fitoterapia*, **75**, 32 (2004).
19. S. I. Balbaa et al., *Egypt J. Pharm. Sci.*, **16**, 161 (1975).
20. Y. Chretien-Bessiere et al., *Riv. Ital. Essenze, Profumi, Piante Office, Aromi, Saponi, Cosmet., Aerosol*, **52**, 211 (1970).
21. G. M. Nano et al., 6th International Congress of Essential Oils (Pap.), 1974, p. 114
22. G. M. Nano et al., *Essenze Deriv. Agrum.*, **46**, 171 (1976).
23. R. Bauer, *Q. Am. Herb Assoc.*, **9**, 4 (1992).

24. R. Carle et al., *Planta Med.*, **56**, 456 (1990).
25. M. T. Kayyal et al., *Arzneim.-Forsch.*, **51**, 545 (2001).
26. Y. Kobayashi et al., *Phytomedicine*, **10**, 657 (2003).
27. M. E. Aggag and R. T. Yousef, *Planta Med.*, **22**, 140 (1972).
28. A. Grochulski and B. Borkowski, *Planta Med.*, **21**, 289 (1972).
29. M. Szalontai et al., *Parfüm Kosmet.*, **58**, 121 (1977).
30. O. Isaac and K. Thiemer, *Arzneim.-Forsch.*, **25**, 1352 (1975).
31. H. Wirth, *Am. Perfum. Cosmet.*, **82**, 81 (1967).
32. N. R. Farnsworth and B. M. Morgan, *J. Am. Med. Assoc.*, **221**, 410 (1972).
33. A. Tubaro et al., *Planta Med.*, **50**, 359 (1984).
34. T. Miller et al., *Planta Med.*, **62**, 60–61 (1996).
35. H. J. Glowania et al., *Z. Hautkrank.*, **62**, 1262 (1987).
36. R. Saller et al., *Eur. J. Pharmacol.*, **183**, 728 (1990).
37. C. Kupfersztain et al., *Clin. Exp. Obstet. Gynecol.*, **30**, 203 (2003).
38. G. M. Konig et al., *Planta Med.*, **64**, 612 (1998).
39. B. M. Hausen in P. A. G. M. De Smet et al., eds., *Adverse Effects of Herbal Drugs*, Vol. 1, Springer-Verlag, Berlin, 1992, p. 243.
40. E. Tuzlaci and P. E. Aymaz, *Fitoterapia*, **72**, 323 (2001).
41. D. V. C. Awang, *Leung's (Chinese) Herb News*, **40**, 2 (2003).
42. B. M. Hausen et al., *Planta Med.*, **50**, 229 (1984).
43. C. Bruce et al., *Contact Dermatitis*, **24**, 139 (1991).
44. D. L. J. Opdyke, *Food Cosmet. Toxicol.*, **12**(Suppl.), 853 (1974).
45. D. L. J. Opdyke, *Food Cosmet. Toxicol.*, **12**(Suppl.), 851 (1974).
46. J. L. Hartwell, *Lloydia*, **31**, 71 (1968).

CHAPARRAL

Source: *Larrea tridentata* (Sessé & Moç. ex DC) Coville (Family Zygophyllaceae).

Common/vernacular names: Chaparral, creosote bush, gobernadora, greasewood, hediondilla.

GENERAL DESCRIPTION

Erect to prostrate evergreen shrub, 1–3 m high; resinous, distinctively aromatic leaves, lanceolate to curved, 18 mm long, 8.5 mm wide; flowers solitary in axils, yellow, five-petaled, twisted or propeller-like; 2.5 cm wide; dominant shrub of desert scrub in much of the arid western United States (southwest Utah to California and Texas); also central Mexico. Clones known to live 10,000+ years; longer than any other plants (HICKMAN). *Larrea* is represented by five species (one North American; four South American); cytologically distinct *L. tridentata* populations in the Sonoran, Chihuahuan, and Mojave deserts often considered conspecific with South American *L. divaricata* Cav.[1] Parts used are leaves and stems.

CHEMICAL COMPOSITION

Lignans dominate chemistry of resin, stems, and leaves, especially nordihydroguaiaretic acid (NDGA), at 1.6–6.55%[2] and 10–15% of the dried leaves;[3] plus dihydroguaiaretic acid, mesodihydroguaiaretic acid, 3′-methoxyisoguaiacin, 3′-demethoxyisoguaiacin,[4,5] 6,3′-di-O-demethylisoguaiacin (previously designated 3′-hydroxynorisoguaiacin), 6-O-demethylisoguaiacin (norisoguaiacin), didehydro-3′-demethoxy-6-O-demethylguaiacin, 3′-demethoxy-6-O-demethylguaiacin,[6] 4-epi-larreatricin, larreatricin, 3′,3″-dimethoxylarreatricin, 3,4-dehydrolarreatricin, larreatridenticin, and others;[7] flavonoids include 2,6-di-C-glucopyranosylapigenin, 6,8-di-C-glucopyranosylchrysoeriol, gossypetin 3,7-dimethyl ether, 5,8,4′-trihydroxy-3,7,3′-trimethoxyflavone,[8] quercetin, kaempferol, rhamnetin, rutin;[9] triterpenes, including larreagenin A, larreic acid, erythrodiol-3-β-(4-hydroxy-cinnamdyl), erythrodiol-3-β-(4-dihydroxy cinnamdyl); essential oil contains α-pinene, Δ-3-carene, limonene, camphene, linalool, borneol, camphor, bornyl acetate, etc.[9–11]

PHARMACOLOGY AND BIOLOGICAL ACTIVITIES

Uterine relaxation activity *in vitro* led to the isolation of an antiimplantation agent from the leaves and twigs (3′-demethoxyisoguaiacin).[6,12] *Larrea tridentata* extracts have also shown *in vitro* antimicrobial,[13] antioxidant,[14] and hepatic enzyme-inhibiting activity.[11]

In vitro studies have shown that NDGA induces apoptosis in various human and animal cancer cell lines[15,16] and cell growth of estrogen-positive human breast cancer cells (MCF-7).[17] Among other activities, it also inhibits platelet aggregation,[18] testosterone release,[19] insulin secretion,[20] human lipoxygenases,[21] and from topical application, skin tumor formation in rodents.[22]

A clinical trial of chaparral tea (16–24 oz./day) in patients with advanced malignancies found a significant number appeared to show tumor progression.[23]

TOXICOLOGY

At high doses (580 mg/kg p.o.), solvent extracts of the leaves and twigs and of the stems have shown antifertility effects in pregnant rats. Behavioral and reproductive organ toxicity was shown in male hamsters (water/ethanol extract of dried leaves at 4% of diet). Fed to chicks (3 weeks old) at 5% of the diet, the ground leaves resulted in body weight-gain inhibition, but not at 2.5% of the feed. *In vitro* tests found cytotoxicity in human and rat kidney and liver slices (ethanol/water tincture) and in primary rat hepatocytes (water, ethanol/water tincture, and various solvent extracts).[11] Chaparral tea was only weakly cytotoxic to rat hepatocytes *in vitro*.[24]

Contact dermatitis from *Larrea* species is common (DE SMET ET AL). Case reports of liver toxicity in subjects largely taking chaparral orally in tablets or capsules (alone or in combination with other substances)[25–28] are confounded by a lack of hepatotoxicity from chaparral or NDGA in animal studies, poor characterization of the "chaparral" taken, possible pre-existing liver disease, and other factors.[11] In a retrospective clinical study, small amounts of a tincture of the leaves and flowers (water/ethanol extract) as part of a more complex herbal preparation (8–10 other herbs) taken by patients over 2 years appeared to be safe. Topical use of chaparral in *Ricinus* oil also appeared to be safe.[29]

NDGA was extensively used as a food antioxidant from 1943 to 1968. It was removed from GRAS status after a chronic feeding study in rats (0.5–1% NDGA for 74 weeks) found lymph node and kidney lesions.[11]

USES

Medicinal, Pharmaceutical, and Cosmetic. NDGA is approved for use in topical form in the U.S. in the treatment of skin cancer, psoriasis, and actinic keratosis.[11]

Food. NDGA was formerly used as an antioxidant in numerous food products (0.02%); currently, use is permitted only in animal fat products (lard, animal shortenings at 0.01%) under USDA authority (TYLER 3).[11]

Dietary Supplements/Health Foods. Dried leaves, stems and twigs in loose, powder, and extract forms in capsules, tablets, teas, and other products (MOORE 1).[11]

Traditional Medicine. A decoction of chaparral leaves has been taken internally by southwest American Indian tribes in the treatment of rheumatism, sores, gonorrhea, tuberculosis, bowel complaints, cramps, stomachaches, sore gums (gargle), and as an emetic to bring down high fevers; externally for rheumatism, arthritis, sore body parts,, sores on animals; leaf infusion taken internally to treat colds, dysuria, asthma, and congestion and used externally against arthritis, rheumatism, sprains, aching bones, infected skin, dandruff, and impetigo sores; stems also used in various preparations, alone and combined with leaves (MOERMAN); widely used popular folk remedy for cancer (in the United States) for facial, stomach, liver, lung, kidney, skin cancers, melanoma, and leukemia.[30]

COMMERCIAL PREPARATIONS

Crude herb; extracts.

Regulatory Status. NDGA is prohibited from use in human foods (§189.165); herb considered unsafe by the F.D.A;[27] oral forms voluntarily removed from U.S. market in 1992, but continue to be sold.[11] Class 2b (not to be used during pregnancy).

REFERENCES

See the General References for BARNES; DER MARDEROSIAN AND LIBERTI; DE SMET ET AL. ; DUKE 2; GLASBY 1; HARBOURNE AND BAXTER; HICKMAN; HUANG; MCGUFFIN 1 & 2 ; MOORE 1; NIKITAKIS; STEINMETZ; TYLER 1; TYLER 3; UPHOF.

1. D. M. Porter, *Taxon*, **23**, 339 (1974).
2. J. L. Valentine et al., *Anal Lett.*, **17**, 1617 (1984).
3. W. R. Obermeyer et al., *Proc. Soc. Exp. Biol. Med.*, **208**, 6 (1995).
4. F. R. Fronczek et al., *J. Nat. Prod.*, **40**, 497 (1987).
5. O. Gisvold and E. Thaker, *J. Pharm. Sci.*, **63**, 1905 (1974).
6. C. Konno et al., *J. Nat. Prod.*, **52**, 1113 (1989).
7. C. Konno et al., *J. Nat. Prod.*, **53**, 396 (1990).
8. M. Sakakibrara et al., *Phytochemistry*, **15**, 727 (1976).
9. J. T. Mabry et al., eds., *Creosote Bush, Biology and Chemistry of Larrea in New World Deserts*, Dowden, Hutchinson and Ross, Stroudsburg, PA, 1977.
10. C. F. M. T. J. Bohnsted, *Rev. Latinoamer. Quim.*, **10**, 128 (1979).
11. Institute of Medicine and National Research Council, National Academies, *Prototype Monograph on Chaparral. Dietary Supplement Ingredient Prototype Monographs*, The National Academies Press, Washington, DC, 2004.
12. C. Konno et al., *Proceedings of International Congress on Natural Products*, **2**, 328 (1987).
13. M. Angeles Verastegui et al., *J. Ethnopharmacol.*, **52**, 175 (1996).
14. L. Y. Zang et al., *Mol. Cell. Biochem.*, **196**, 157 (1999).
15. D. G. Tang and K. V. Honn, *J. Cell. Physiol.*, **172**, 155 (1997).
16. T. Seufferlein et al., *Br. J. Cancer*, **86**, 1188 (2002).
17. N. Sathyamoorthy et al., *Cancer Res.*, **54**, 957 (1994).
18. J. Chung et al., *J. Biol. Chem.*, **272**, 14740 (1997).

19. F. Romanelli et al., *Life Sci.*, **63**, 255 (1998).
20. B. R. Hsu et al., *Cell Transplant.*, **10**, 255 (2001).
21. S. Whitman et al., *J. Med. Chem.*, **45**, 2659 (2002).
22. Z. Y. Wang et al., *Mutat. Res.*, **261**, 153 (1991).
23. C. R. Smart et al., *Rocky Mt. Med. J.*, **67**, 39 (1970).
24. D. Pritchard et al., *In Vitro Cell. Dev. Biol. Anim.*, **30a**, 91 (1994).
25. M. Katz Miriam and F. Saibil, *J. Clin. Gastroenterol.*, **12**, 203 (1990).
26. M. Blumenthal, *HerbalGram*, **28**, 38 (1993).
27. FDA, HHS News (December 10, 1992); FDA Talk Paper (December 11, 1992); *Food Drug Cosmet. Law Rep.*, 1577; §43,127 (December 28, 1992).
28. Centers for Disease Control and Prevention, *Morb. Mort. Weekly*, **41**, 812 (1992).
29. S. Heron and E. Yarnell, *J. Altern. Complement. Med.*, **7**, 175 (2001).
30. F. Brinker in E. K. Alstat, ed., *Eclectic Dispensatory of Natural Therapeutics*, Vol. 1, Eclectic Medical Publications, Portland, OR, 1989.

CHASTE-TREE

Source: *Vitex agnus-castus* L. (Family Verbenaceae).

Common names: Agnus castus, chasteberry, chaste-tree, monk's pepper, and vitex.

GENERAL DESCRIPTION

Deciduous shrub growing to 6 m in height; leaves palmately compound; leaflets linear lanceolate, entire, white tomentose beneath, glabrous above, to 10 cm long; flowers blue, purple, or pink in spike-like panicles; indigenous to the Mediterranean region, Crimea, Central Asia and western Asia to northwest India; naturalized in north America (Florida, Georgia, Alabama, Mississippi, Louisiana, Arkansas, Texas, southeast Oklahoma, north to Maryland), Nigeria and northern Brazil. Part used is the dried fruit (a small, brownish- to olive-black drupe) with four seeds.

CHEMICAL COMPOSITION

Fruits contain the fatty acid linoleic acid[1] and the flavone apigenin;[2] other flavonoids in the fruits and leaves include casticin, vitexin, isovitexin, orientin, isoorientin, xyloside; others include penduletin, 6-hydroxykaempferol-3,6,7,4′-tetramethylether, chrysosplenol-D, luteolin-7-glucoside, homoorientin, and so on;[3–6] iridoid glycosides in the leaf, fruit, and flower include aucubin and agnuside and eurostoside in the leaves;[7,8] other iridoids in the leaves, flowers and twigs include agnucastosides A, B, and C and mussaenosidic acid;[9] essential oil of the fruits, leaves, and flowers contains monoterpenes (limonene, α-pinene, 1,8-cineole);[10] in essential oil of leaves also contains β-pinene, citronellol, myrcene, linalool, sabinene, and others; sesquiterpenes in the essential oil of the flowers, fruit, and leaves include β-caryophylline oxide, caryophylline, farnesene, and others; diterpenes in the fruit include vitexilactone, rotundifuran (MCKENNA; WICHTL), and others;[11] triterpenoids include 3β-acetoxyolean-12-en-27-oic acid; 2α,3α-dihydroxyolean-5,12-dien-28-oic acid; 2β,3α-diacetoxyolean-5,12-dien-28-oic acid; and 2α,3β-diacetoxy-18-hydroxyolean-5,12-dien-28-oic acid;[12] ketosteroids detected in flower extracts include progesterone, 17α-hydroxyprogesterone, testosterone, and epitestosterone; leaf extracts yield andro-stenedione.[13]

PHARMACOLOGY AND BIOLOGICAL ACTIVITIES

Ethanol extracts of the leaves inhibit the *in vitro* growth of various bacteria and fungi (*E. coli, Staphylococcus aureus, Streptococcus faecalis, Candida krusei, Epidermophyton floccosum, Trichophyton mentagrophytes*, and others). Ethanol extracts of the fruits, flowers, and leaves inhibited the *in vitro* growth of *Bacillus subtilis, E. coli, Candida albicans*, and *Shigella sonei*. Flavonoids (casticin, orientin, vitexin, xyloside, and others) and iridoids (aucubin and agnoside) of the fruits and leaves also showed *in vitro* antibacterial activity (*S. aureus, Bacillus cereus*, and *B. megaterium*) (MCKENNA).

Extracts of the berries have shown *in vitro* cycotoxic activity against human carcinoma cells (breast, gastric, small lung, and colon carcinoma);[14] *in vitro* growth inhibition of an estrogen- and progesterone-positive human mammary cancer cell line (T-47D) (MCKENNA); *in vitro* estrogenic activity with binding of both α- and β-estrogen receptors;[1,15] and estrogenic activity in female rats administered an extract in feed (MCKENNA). Oral administration of a standardized extract of the berries with β-estrogen receptor-specific activity to female rats produced estrogenic activity without effects on uterine genes or uterine weight.[11] Estrogenic activity of the berries is attributed to linoleic acid,[1] apigenin, penduletin, and vitexin.[2]

Dopaminergic compounds in the fruits with *in vitro* prolactin-suppressive activity are diterpenes, including rotundifuran[16] and others (clerodadienols).[11]

Clinical trials of standardized extracts of the berries (double-blind, placebo-controlled) have shown significant benefits in the treatment of women with latent hyperprolactinemia, abnormal menstrual cycles, and corpus luteum insufficiency, cyclical mastalgia, premenstrual syndrome, and premenstrual tension syndrome (MCKENNA).[17,18]

TOXICOLOGY

Generally considered safe; rare occurrences of itching and urticaria have been reported (BLUMENTHAL 1). No toxicity was found in mice or rats administered a standardized extract of the berries at a single dose of 7000 mg. Doses of up to 3500 mg in rats for over 5 weeks also failed to produce toxic effects. Decreased milk production in rats was attributed to the prolatic-depressing activity of a vitex berry extract (MCKENNA).

USES

Medicinal, Pharmaceutical, and Cosmetic. Used in modern Europe in the treatment of premenstrual and menopausal disorders including premenstrual mastodynia (mastalgia) and other symptoms of PMS;[11] also in treatments of premenstrual tension syndrome, Parkinson's disease, acne, uterine myomas, as a galactogogue, as an adjunctive therapy in endometriosis (MCKENNA), and upon termination of use of birth control pills; also used to help re-establish normal menstruation and ovulation, to increase or stimulate milk flow; reduce water retention during menstruation; allay effusions in the knee joints associated with premenstrual syndrome.[10,11]

Dietary Supplements/Health Foods. Dried fruits and extracts thereof in capsules, tea, and so on (FOSTER).

Food. Seeds used as a spice and substitute for pepper (GRIEVE; KIPLE AND ORNELAS).

Traditional Medicine. Syrup made from the seeds was used in medieval Europe to suppress the libido. Berries and leaves were used in 16th century Europe as an emmenagogue. Eclectic medical practitioners (19th century) used a tincture of the fresh fruit as galactagogue and emmenagogue (FELTER AND LLOYD). Plant used in Anatolia to treat anxiety, stomachache, to prevent early birth, and as a diuretic, digestive, and antifungal.[9] Seed decoctions and fruits used in ancient Greece to treat uterine inflammations (MCKENNA) and in Arabian medicine in baths to treat uterine tumors and pain;[19] plant also used in Arabian medicine to treat insanity, hysteria, epilepsy;

leaves and root bark used in Nigeria to treat depression. Seeds used in Ayurvedic medicine to induce abortion, as a diuretic, and for stomachache and ocular conditions; seeds used in Unani medicine as a contraceptive and purify the liver and brain; plant used in Amazonia Brazil as an emmenagogue, anaphrodisiac, diuretic, and for stomachache, headache, flu, and other conditions (MCKENNA).

COMMERCIAL PREPARATIONS

Crude herb, alcoholic and aqueous extracts of pulverized fruits and their formulations.

Regulatory Status. Class 2b dietary supplement (not to be used during pregnancy). In Germany, formulation indicated for menstrual disorders due to primary or secondary corpus luteum insufficiency; premenstrual syndrome, mastodynia, inadequate lactation, menopausal symptoms (BLUMENTHAL 1).

REFERENCES

See the General References for APPLEQUIST; BARNES; BLUMENTHAL 1 & 2; DER MARDEROSIAN AND BEUTLER; FELTER AND LLOYD; FOSTER; MCGUFFIN 1 & 2; MCKENNA; STEINMETZ; TUTIN 3; UPHOF; WEISS; WICHTL.

1. J. Liu et al., *Phytomedicine*, **11**, 18 (2004).
2. H. Jarry et al., *Planta Med.*, **69**, 945 (2003).
3. G. Hahn et al., *Notabene Med.*, **16**, 233, 297 (1986).
4. E. Wollenweber and K. Mann, *Planta Med.*, **48**, 126 (1983).
5. R. Hansel and H. Rimpler, *Arch. Pharm. (Wienhiem)*, **296**, 598 (1963).
6. C. S. Gomaa et al., *Planta Med.*, **52**, 277 (1978).
7. R. Hansel and E. Winder, *Arzneim.-Forsch*, **9**, 180 (1959).
8. K. Gorler et al., *Planta Med.*, **51**, 530 (1985).
9. A. Kuruuzum-Uz et al., *Phytochemistry*, **63**, 959 (2003).
10. E. Winder and R. Hansel, *Arch. Pharm. (Weinheim)*, **293**, 556 (1960).
11. W. Wuttke et al., *Phytomedicine*, **10**, 348 (2003).
12. A. S. Chawala et al., *J. Nat. Prod.*, **55**, 163 (1992).
13. M. Saden-Krehula et al., Short Reports of Short Lectures and Poster Presentations, Bonn BACANS Symposium, P_1 77, July 17–22, 1990, p. 59.
14. K. Ohyama et al., *Biol. Pharm. Bull.*, **26**, 10 (2003).
15. J Liu et al., *J. Agric. Food Chem.*, **49**, 2472 (2001).
16. A. Antolic and Z. Males, *Acta Pharm. (Zagreb)*, **47**, 207 (1997).
17. D. Propping, *Therapiewoche*, **38**, 2992 (1988).
18. D. Propping and T. Katzorke, *Z. Allg.*, **63**, 932 (1987).
19. F. Alakbarov, *HerbalGram*, **57**, 40 (2003).

CHENOPODIUM OIL

Source: *Chenopodium ambrosioides* L. var. *anthelminticum* (L.) A. Gray or *C. ambrosioides* L. (Family Chenopodiaceae).

Common/vernacular names: Oils of American wormseed (C. ambrosioides var. anthelminticum), Mexican tea, and epazote (C. ambrosioides).

GENERAL DESCRIPTION

Strongly aromatic, hairy annual or perennial herb up to about 1.5 m high; native to tropical

America; naturalized and cultivated worldwide. Part used is the fresh aboveground flowering and fruiting plant, from which the volatile oil (with a disagreeable odor and bitter taste) is obtained by steam distillation. Major producing countries include India, China, Brazil, and the United States. Due to its high ascaridole (a peroxide) content, chenopodium oil may explode when heated or treated with acids and should be handled with caution.

CHEMICAL COMPOSITION

Contains variable amounts of ascaridole (17–90%, usually 60–80%), *l*-limonene, myrcene, *p*-cymene, α-terpinene, saturated hydrocarbons (C_{21} to C_{31} with C_{29} predominant), triacontyl alcohol, α-spinasterol, and others (JIANGSU; LIST AND HÖRHAMMER).[1–4]

PHARMACOLOGY AND BIOLOGICAL ACTIVITIES

Ascaridole, the active principle of the oil, has anthelmintic properties, particularly against roundworms (*Ascaris*); it is also effective against hookworms and dwarf tapeworms but not large tapeworms.

TOXICOLOGY

The oil is considered as very toxic. Toxic effects include irritation of skin and mucous membranes, vomiting, headache, vertigo, kidney and liver damage, temporary deafness, and circulatory collapse, among others. Effects may be cumulative. Cases of death have also been reported. (GOSSELIN; MARTINDALE).[4]

USES

Medicinal, Pharmaceutical, and Cosmetic. Now seldom (if at all) used in pharmaceutical preparations as it is largely replaced by synthetic anthelmintics such as piperazine and other compounds. Major use is as a fragrance component in soaps, detergents, creams, lotions, and perfumes, with maximum use level of 0.4% reportedly used in perfumes.[4]

Food. The leaves and seeds of *C. ambrosioides* are used in Mexican cooking as a carminative flavoring with bean dishes (MOORE 1).

Traditional Medicine. Used as an anthelmintic for roundworms, hookworms, and dwarf tapeworms, among others. Leaf, root, and plant of *C. ambrosioides* have been used in tumors.[5] In China, the fresh root is used to treat articular rheumatism.

COMMERCIAL PREPARATION

Essential oil. It was formerly official in N.F. and U.S.P.

Regulatory Status. Not permitted in foods.

REFERENCES

See the General References for FERNALD; GRIEVE; GUENTHER; JIANGSU; MOORE 1; NANJING; TYLER 3; USD 26th; YOUNGKEN.

1. L. Bauer et al., *Rev. Bras. Farm.*, **54**, 240 (1973).
2. G. S. Gupta and M. Behari, *J. Indian Chem. Soc.*, **49**, 317 (1972).
3. G. S. Gupta and M. Behari, *Indian Perfum.*, **18**(Pt. 2), 40 (1975).
4. D. L. J. Opdyke, *Food Cosmet. Toxicol.*, **14**, 713 (1976).
5. J. L. Hartwell, *Lloydia*, **31**, 71 (1968).

CHEROKEE ROSEHIP

Source: *Rosa laevigata* Michx. (syn. *R. sinica* Murr.; *R. cherokensis* Donn.; *R. ternata* Poir.; *R. nivea* DC.; *R. Camellia* Hort.) (Family Rosaceae).

Common/vernacular names: Chinese rosehip, Fructus Rosae Laevigatae, and *jinyingzi*.

GENERAL DESCRIPTION

High-climbing shrub (up to 5 m), with slender green prickly branches; flowering in May and fruiting in September to October. Native to China and Japan; naturalized in the southern United States; now widely distributed throughout China.

Part used is the dried ripe prickly fruit (hip) collected when it turns red in autumn. After partially dried under the sun, the fruits are placed in a barrel and their prickles removed by stirring with a wooden bat. They are then further dried to yield whole *jinyingzi*. The whole hips are normally further processed to yield *jinyingzi rou* (meat) by soaking in water until soft, slicing in half, removing the seeds and again sun drying. Commercial Cherokee rosehips are either whole or sliced, the latter yielding more extractives.

CHEMICAL COMPOSITION

Fruits contain hydrolysable tannins (laevigatins A to G,[1,2] agrimoniin, agrimonic acids A and B,[1] pedunculagin, and sanguiin H-4[2]) euscaphic, oleanolic, and ursolic acid derivatives, sterol glucosides,[3] laevigatanoside A ($2\alpha,3\beta,19\alpha,23$-trihydroxy-12-ursorlic-28-glucopyester),[4] saponin glycosides; vitamin C; sugars; plant acids (citric, malic, etc.); amino acids,[5] starch; pigments; resin; and others (IMM-3; JIANGSU).

PHARMACOLOGY AND BIOLOGICAL ACTIVITIES

A hot water extract of Cherokee rosehip showed potent antimutagenic activities in the Ames test.[6]

TOXICOLOGY

Polyhydroxy pigments: LD_{50} (white mice) 519 ± 105 mg/kg i.p.; and s.c. injection of 500 mg/kg and 1100 mg/kg in white rats (observed 1–2 weeks) retarded weight gain and caused an increase in white blood cells and decrease in red cells but no pathological changes in heart, liver, kidney, spleen, intestine, and adrenal tissues.[7]

USES

Dietary Supplements/Health Foods. Used in tonic (especially male tonic) preparations in various forms (drinks, soup packets, tablets, and capsules).

Traditional Medicine. Considered one of the most important Chinese health tonics. First described around AD 500, traditional Chinese medicine considers it to have *gu jing* (strengthens male essence), *suo niao* (antidiuretic), *yi shen* (tonic; invigorates urinary and reproductive functions), and *se chang* (intestinal astringent) properties. Traditionally used to treat male sexual inadequacies, including nocturnal emission and spermatorrhea; female problems (e.g., uterine bleeding and leukorrhea); chronic diarrhea and enteritis; sweating and night sweating; polyuria and enuresis; also used in sexual neurasthenia, hypertension, and chronic cough (LU AND LI).

COMMERCIAL PREPARATIONS

Mainly crude (both whole and sliced).

Regulatory Status. Although essential oil solvent-free oleoresins and natural extractives of other *Rosa* spp. are GRAS (§182.20), the U.S. regulatory status of *Rosa laevigata* is not established.

REFERENCES

See General References for BAILEY 2; CHEUNG AND LI; CHP; FOSTER AND DUKE; IMM-3; JIANGSU; JIXIAN; LU AND LI; MCGUFFIN 1 & 2.; NATIONAL.

1. T. Yoshida et al., *Chem. Pharm. Bull.*, **37**, 920 (1989).
2. T. Yoshida et al., *Phytochemistry*, **28**, 2451 (1989).
3. J. Fang et al., *Phytochemistry*, **30**, 3383 (1991).
4. X. Li, *Zhongguo Zhong Yao Za Zhi/Zhongguo Zhongyao Zazhi*, **22**, 298 (1997).
5. X. She et al., *Acta Hort. Sin.*, **15**, 240 (1988).
6. X. B. Ni, *Zhongcaoyao*, **22**, 429 (1991).
7. L. Sun et al., *Jiangxi Yixueyuan Xuebao*, **30**(3), 5 (1990).

CHERRY BARK, WILD

Source: *Prunus serotina* Ehrh. (Family Rosaceae).

Common/vernacular names: Black cherry, capulin, rum cherry bark, Wild black cherry, and wild cherry bark.

GENERAL DESCRIPTION

Large tree with rough dark trunk and reddish brown branches; up to about 1.5 m in diameter and 30 m in height; native to North America (Nova Scotia to Florida and west to Nebraska and Texas.) Part used is the dried stem bark, free of the rough outer bark, preferably collected in the fall.

CHEMICAL COMPOSITION

Bark contains condensed tannins,[1] prunasin (*d*-mandelonitrile glucoside); emulsin; eudesmic acid (3,4,5-trimethoxybenzoic acid); *p*-coumaric acid; scopoletin; sugars; and others (KARRER). Fruit skin contains capulin anthocyanins.[2]

Prunasin is a cyanogenic glucoside that is hydrolyzed by the enzyme prunase into hydrocyanic acid (HCN, prussic acid), glucose, and benzaldehyde. The yield of HCN by the bark varies with the times of collection and the thickness and types of bark. Bark collected in the fall has the highest HCN yield (ca. 0.15%), while that collected in the spring has only about 0.05% yield of HCN. In contrast, leaves have been reported to yield the highest amounts of HCN in the spring, up to about 0.25% of potential HCN yield in fresh leaves (LIST AND HÖRHAMMER).[3] Leaves also contain amygdalin.[4]

PHARMACOLOGY AND BIOLOGICAL ACTIVITIES

Believed to have astringent, sedative, and antitussive properties.

TOXICOLOGY

Hydrocyanic acid, present in bark extracts (MARTINDALE), is a lethal poison.

USES

Medicinal, Pharmaceutical, and Cosmetic. Extracts used extensively in cold and cough preparations in the form of syrups, particularly in formulations based on White Pine Compound.

Food. Extracts used in most major food products as a flavoring substance, including alcoholic and nonalcoholic beverages, frozen dairy desserts, candy, baked goods, gelatins and puddings, processed fruits, and others. Highest average maximum use level of the extract (strength and type unspecified) reported is 0.06% in alcoholic beverages.

Dietary Supplements/Health Foods. Used in cough syrups or bronchial formulations; tea ingredient (FOSTER AND DUKE).

Traditional Medicine. Bark used by a number of American Indian tribes for colds and coughs; by the Cherokee Indian to treat laryngitis and as a wash for sores and ague; by the Delaware to treat diarrhea and women's diseases; by the Iroquois in a poultice for headaches (MOERMAN). Also reportedly used against cancers.[5] Used by Canadian Indians (Delaware, Iroquois, Malecite, and Ojibwa) in treatments of diabetes and its complications.[6]

In China, the stem bark and root of a related *Prunus* species (*P. armeniaca* L., the apricot) has been used for centuries in treating apricot kernel poisoning. Recent clinical reports have substantiated this usage. Decoctions of the fresh bark were used to treat 80 cases of apricot kernel poisoning; all patients were reported to recover completely within 4 h (JIANGSU).

COMMERCIAL PREPARATIONS

Crude and extracts. Crude and syrup were formerly official in U.S.P., and fluid extract was official in N.F. The crude comes in two types: thin and thick with the former being considered superior in quality. Strengths (see ***glossary***) of extracts are expressed in weight-to-weight ratios; extracts may contain detectable amounts of prunasin or its hydrolysis product, HCN.

Regulatory Status. Essential oil, solvent-free oleoresin and natural extractives GRAS (§182.20); no HCN limits are specified.

REFERENCES

See the General References for BAILEY 2; CLAUS; FEMA; FERNALD; FOSTER AND DUKE; GRIEVE; KARRER; MARTINDALE; UPHOF; USD 26th; YOUNGKEN.

1. L. Buchalter, *J. Pharm, Sci.*, **58**, 1272 (1968).
2. A. Ordaz-Galindo et al., *Food Chem.*, **65**, 201 (1999).
3. D. M. Smeathers et al., *Agron. J.*, **65**, 775 (1973).
4. E. F. Santamour, *Phytochemistry*, **47**, 1537 (1998).
5. J. L. Hartwell, *Lloydia*, **34**, 103 (1971).
6. L. M. McCune and T. Johns, *J. Ethnopharmacol.*, **82**, 197 (2002).

CHERRY LAUREL LEAVES

Source: ***Prunus laurocerasus*** L. (syn. *Laurocerasus officinalis* M. Roem.) (Family Rosaceae).

Common/vernacular names: Cherry laurel leaves, common cherry laurel, and laurocerasus leaves.

GENERAL DESCRIPTION

An evergreen bush to small tree with oblong leathery leaves 7.5–15 cm long; up to about 6 m high; native to western Asia; widely cultivated. Parts used are the fresh leaves. The oil is obtained by steam distillation of the warm-water macerated leaves. During maceration the enzyme prunase (or emulsin) hydrolyzes the cyanogenic glucoside present to yield benzaldehyde and hydrocyanic acid (HCN), which are volatile and distilled with steam. Most of the HCN is removed by neutralization and washing of the oil. Cherry laurel water is the water distillate adjusted to contain 0.1% HCN.[1]

CHEMICAL COMPOSITION

The leaves contain variable amounts (usually ca. 1.5%) of prunasin (*d*-mandelonitrile glucoside), with the young and small leaves containing the highest concentrations. During isolation, prunasin is partially converted to its isomer sambunigrin (*l*-mandelonitrile glucoside), resulting in a racemic mixture of the two isomers known as prulaurasin (*dl*-mandelonitrile glucoside). Other constituents present in the leaves include 1% ursolic acid, wax, tannin, emulsin, and others (KARRER).[2–4]

The oil, like bitter almond oil, is composed almost entirely of benzaldehyde and HCN, with small amounts of benzyl alcohol.

Cherry laurel oil (FFPA) for food use should not contain HCN.

PHARMACOLOGY AND BIOLOGICAL ACTIVITIES

Cherry laurel oil (FFPA) is practically equivalent to pure benzaldehyde and has the pharmacological and toxicological properties of benzaldehyde (see *bitter almond*).

TOXICOLOGY

Hydrocyanic acid is a deadly poison.

USES

Medicinal, Pharmaceutical, and Cosmetic. Not used in the United States, but cherry laurel water is used in Europe as sedative, anodyne, and antispasmodic and in eye lotions.

Food. Cherry laurel oil (FFPA) is used as a flavor component in numerous food products, including alcoholic (liqueurs such as cordials, etc.) and nonalcoholic beverages, frozen dairy desserts, candy, and baked goods. Highest average maximum use level reported is 0.014% in candy.

Traditional Medicine. Leaves used in treating coughs, insomnia, stomach and intestinal spasms, vomiting, and other ailments; also reportedly used in cancers.[5]

COMMERCIAL PREPARATION

Cherry laurel oil (FFPA).

Regulatory Status. Leaves approved for food as natural flavoring substance provided prussic acid (HCN) does not exceed 25 ppm (§172.510).

REFERENCES

See the General References for ARCTANDER; BAILEY 1; BIANCHINI AND CORBETTA; FEMA; GRIEVE; LIST AND HÖRHAMMER; MCGUFFIN 1 & 2; MARTINDALE; MERCK; UPHOF.

1. A. Puech et al., *Trav. Soc. Pharm. Montp.*, **36**, 101 (1976).
2. M. Henriet et al., *J. Pharm. Belg.*, **29**, 437 (1974).

3. L. P. Miller in L. P. Miller, ed., *Phytochemistry*, Vol. **1**, Van Nostrand Reinhold, New York, 1973, p. 297.

4. L. Stammitti et al., *Phytochemistry*, **43**, 45 (1996).

5. J. L. Hartwell *Lloydia*, **34**, 103 (1971).

CHERVIL

Source: *Anthriscus cerefolium* (L.) Hoffm. (syn. *A. longirostris* Bertol.) (Family Umbelliferae or Apiaceae).

Common/vernacular names: Chervil, garden chervil, and salad chervil.

GENERAL DESCRIPTION

Slender annual with small leaves and erect branching stem, hairy near the nodes; up to about 0.8 m high; native to Europe (the Caucasus and south Russia) and western Asia; naturalized in North America (Quebec to Pennsylvania), Australia, and New Zealand; widely cultivated. Parts used are the leaves (fresh or dried) and the dried flowering herb.

CHEMICAL COMPOSITION

Contains a volatile oil (ca. 0.03% in the herb and 0.9% in the fruits), apiin (apigenin-7-apiosylglucoside), bitter principles, and high concentrations of potassium (ca. 4.7%), calcium (ca. 1.3%), magnesium (130 mg/100 g), phosphorus (450 mg/100 g), and others. Fruits (seeds) contain luteolin-7-glucoside and about 13% fixed oils, which are composed of petroselinic acid and linoleic acid as the main components, with minor concentrations of palmitic acid and short-chain hydrocarbons (C_{23} or less, mainly branched-chain C_{17}). The volatile oil contains estragole (methyl chavicol) and 1-allyl-2,4-dimethoxybenzene as major constituents, with anethole also reported to be present in the oil of Indian origin (KARRER; LIST AND HÖRHAMMER; MARSH).[1–4]

PHARMACOLOGY AND BIOLOGICAL ACTIVITIES

In vitro antioxidant, lipid peroxidation-inhibiting, and copper chelating activities were found from aqueous extracts of the leaves and roots.[5]

TOXICOLOGY

Estragole, the major component of the volatile oil, has been reported to produce tumors in mice (see *sweet basil*).

USES

Food. Used as a flavor ingredient in food products, including nonalcoholic beverages, frozen dairy desserts, candy, baked goods, meat and meat products, and condiments and relishes. Highest average maximum use level reported is 0.114% of the herb in meats and meat products.

The leaves (particularly when fresh) are used as a domestic spice in soups, salads, vinegar for salad dressings, omelets, and other dishes.

Traditional Medicine. Used as diuretic, expectorant, and digestive; also to lower blood pressure, in the form of an infusion. The juice from the fresh herb is used to treat eczema, gout stones, and abscesses, among others.

COMMERCIAL PREPARATION

Available mainly as the crude.

Regulatory Status. GRAS as a natural flavoring or seasoning (§182.10). Essential oil, natural extractives, and solvent-free oleoresin also GRAS §182.20).

REFERENCES

See the General References for BAILEY 2; FEMA; LUST; ROSENGARTEN; TERRELL; UPHOF.

1. J. H. Zwaving et al., *Pharm. Weekbl.*, **106**, 182 (1971).
2. J. Van Loon, *Z. Lebensm. Unters. Forsch.*, **153**, 289 (1973).
3. S. O. Brown et al., *Phytochemistry*, **14**, 2726 (1975).
4. J. B. Harborne and C. A. Williams, *Phytochemistry*, **11**, 1741 (1972).
5. S. Fejes et al., *J. Ethnopharmacol.*, **69**, 259 (2000).

CHESTNUT LEAVES

Source: *Castanea dentata* (Marsh.) Borkh. (syn. *C. americana* (Michx.) Raf.) (Family Fagaceae).

Common/vernacular names: American chestnut leaves, chestnut leaves.

GENERAL DESCRIPTION

Deciduous tree with rough bark and glabrous mature leaves that reach about 25 cm in length and 5 cm in width; up to about 30 m high; native to North America. Parts used are the dried leaves.

Once a dominant hardwood species in eastern North America, the American chestnut has been extensively destroyed by a fungal disease during recent years caused by *Cryphonectria* (*Endothia*) *parasitica* (Murrill) Barr (chestnut disease fungus).[1] Consequently, the leaves used in commerce are mostly derived from Spanish chestnut leaves (*C. sativa* Mill.), a native of the Mediterranean region, or from other *Castanea* species.

CHEMICAL COMPOSITION

Contains 8–9% tannins, mucilage, resins, and others.

PHARMACOLOGY AND BIOLOGICAL ACTIVITIES

Considered to have tonic and astringent properties; also antitussive, antirheumatic. Offers no advantage over other antitussives (WEISS; WREN).

USES

Food. Used in alcoholic and nonalcoholic beverages in the form of an extract; reported use levels are low, less than 0.0075%.

Traditional Medicine. Decoction of leaves used by the Cherokee Indians in a compound formula to treat coughs; leaves used by the Mohegans to treat colds and rheumatism; leaf infusion used to treat whopping cough (MOERMAN); infusion of a gargle in pharyngitis; also as sedative, tonic, and astringent.

COMMERCIAL PREPARATIONS

Crude and extracts. Crude was formerly official in N.F. and U.S.P. Strengths (see *glossary*) of extracts are expressed either in weight-to-weight ratios or, when intended for food use, in flavor intensities.

Regulatory Status. Leaves approved for food use as natural flavoring substance (§172.510); only *C. dentata* is listed. *Castanea sativa* leaves are the subject of a German therapeutic monograph; however, use is not recommended since efficacy is not well documented (BLUMENTHAL 1).

REFERENCE

See the General References for BAILEY 2; BIANCHINI AND CORBETTA; BLUMENTHAL 1; FEMA; FERNALD; KROCHMAL AND KROCHMAL; MCGUFFIN 1 & 2; STAHL; UPHOF; WEISS; WREN.

1. L. K. Rieske et al., *Environ. Entomol.*, **32**, 359 (2003).

CHICKWEED

Source: *Stellaria media* (L.) Villars (Family Caryophyllaceae).

Common/vernacular names: Chickweed, starweed, star chickweed, and hakobe (Japan).

GENERAL DESCRIPTION

Prostrate to decumbent annual herb to 40 cm; leaves ovate, sessile, glabrous; flower white, star-shaped, petals two lobed, to 3 mm long, sepals to 5 mm; mainly a weed of cultivated ground around human dwellings; throughout Europe, North America; cosmopolitan elsewhere; part used is the herb.

CHEMICAL COMPOSITION

Carboxylic acids, oxalic acid, coumarins, hydroxycoumarins; glycosides; flavonoids, including rutin, hexosylapigenins, pentosylapigenins,[1] apigenin *C*-glycosylflavones (e.g., 7,2″-di-*O*-β-glucopyranosyl vitexin),[2] luteolin, vicenin-2, and the isoflavone genistein;[3] saponins, steroids, triterpene glycosides,[1,4] hydroxybenzoic acids (e.g., *p*-hydroxybenzoic acid, vanillic acid), hydroxycinnamic acids (caffeic, chlorogenic, ferulic,[3] and *trans*-ferulic acid); galactolipids;[5] vitamin C, dehydroascorbic acid, thiamine, riboflavin, niacin, carotenoids; linolenic acids[4,6] (DUKE 2); aminoadipic acid, and saccharopine;[7] octadecatetraenoic acid in leaf lipids concentrated in monogalactoxyl diglyceride fraction; γ-linolenic acid in polar fraction.[8]

Stellaria dichotoma var. *lanceolata* contains wogonin (flavone derivative), α-spinasterol, stimast-7-enol, palmitate and furan 3-carboxylic acid, and *C*-glycosyl-flavonoids.[9]

PHARMACOLOGY AND BIOLOGICAL ACTIVITIES

Extracts of the aerial parts have shown *in vitro* antioxidant activity (xanthane oxidase-inhibiting).[10]

USES

Food. Aerial part used in salads in Italy;[11] forage plant of chickens living among the Iroquois Indians (MOERMAN); formerly a source of vitamin C; traditionally a pot herb, emergency food; seeds once a commercial bird seed source.

Dietary Supplements/Health Foods. Leaves in capsules, teas; widely used emollient in salves and ointments (WREN); used in Japan in a herbal toothpaste ("Hakobe-salt").[12]

Traditional Medicine. Chippewa Indians used a strained decoction of the leaves used to wash sore eyes; used by Iroquois in a compound poultice to treat rheumatism (MOERMAN); also antirheumatic,[11] galactogogue, gastroenteric diseases, toothache, swellings, antipuritic,[12] demulcent, emollient, and vulnerary activity reported; externally applied in poultices for boils, eczema, inflammation, psoriasis, sores, ulcers; ointment also used to allay itching. WEISS reports negative results as antirheumatic.

COMMERCIAL PREPARATIONS

Crude herb; extracts.

Regulatory Status. Class 1 dietary supplement (herbs can be safely consumed when used appropriately).

REFERENCES

See the General References for APPLEQUIST; DER MARDEROSIAN AND BEUTLER; DUKE 2; FOSTER AND DUKE; GRIEVE; MCGUFFIN 1 & 2; STEINMETZ; TUTIN 1; TYLER 1; UPHOF; WEISS; WREN.

1. G. G. Tsotsoriua et al., *Kromatogr. Metody. Farm.*, 172 (1977).
2. J. Budzianowski and G. Pakulski, *Planta Med.*, **57**, 290 (1991).
3. G. M. Kitanov, *Pharmazie*, **47**, 470 (1992).
4. J. L. Guil et al., *Plant Foods Hum. Nutr.*, **51**, 99 (1997).
5. J. Hohmann et al., *Fitoterapia*, **67**, 381 (1996).
6. M. L. Salo and T. Makinen, *Maataloust. Aikakousk*, **37**, 127 (1965).
7. R. Nawaz and H. Sorenson, *Phytochemistry*, **16**, 599 (1977).
8. G. R. Jamieson and E. H. Reid, *Phytochemistry*, **10**, 1575 (1971).
9. A. M. Rizk, *The Phytochemistry of the Flora of Qatar*, Scientific and Applied Research Centre, University of Qatar, Doha, 1987, p. 26.
10. A. Pieroni et al., *Phytother. Res.*, **16**, 467 (2002).
11. P. M. Guarrera, *Fitoterapia*, **74**, 515 (2003).
12. F. Kuginuki and T. Nanba, *Jpn. J. Pharmacol.*, **47**, 182 (1993).

CHICLE

Source: *Manilkara zapota* (L.) P. Royen (syn. *M. zapotilla* (Jacq.) Gilly; *M. achras* (Mill.) Fosb.; *Sapota achras* Mill.; *Achras sapota* L.; *A. zapotilla* (Jacq.) Nutt.) (Family Sapotaceae).

Common name: Chicle.

GENERAL DESCRIPTION

An evergreen tree with shiny leaves, up to about 33 m high; native to Central America and the Yucutan Peninsula of Mexico; now extensively cultivated in the tropics for its edible fruit (*naseberry* or *sapodilla*). Part used is the latex present in the bark, pith, and leaves; it is collected by making multiple incisions in the trunk and dried by careful boiling to remove excess water. Crude chicle is purified by repeatedly washing with strong alkali and neutralizing with sodium acid phosphate, followed by drying and powdering. The resulting product is a water-insoluble, amorphous powder that softens on heating.[1,2]

CHEMICAL COMPOSITION

Crude chicle contains 15–20% hydrocarbons that are polyisoprenes (mixture of low molecular weight *cis*-1,4 and *trans*-1,4 units in an approximately 2:7 ratio); up to 55% of a yellow resin, consisting primarily of lupeol acetate with minor amounts of β-amyrin and α-spinasterol acetates; also, taraxasterol and other triterpene alcohol acetates; a gum composed of a (1 → 4)-linked xylan backbone highly substituted with oligosaccharide chains; sugar; inorganic salts; and others (KARRER).[1–9]

Refined chicle for use in chewing gums does not contain the water-soluble constituents present in crude chicle. However, data on its precise chemical composition are limited.

TOXICOLOGY

Limited available data indicate chicle to be nontoxic.[10]

USES

Medicinal, Pharmaceutical, and Cosmetic. Claimed to be a useful ingredient in hair preparations (dressings and pomades).[11,12]

Food. The primary use of chicle is as the "gum" base in chewing gum; its use level in chewing gum is about 20%. The rest of the chewing gum is sugar and corn syrup, with small amounts of flavorings. This "gum" is not a true gum (see *glossary*) but is close in chemical and physical nature to natural rubber and resins; hence it is soft and plastic when chewed and is reportedly not soluble in saliva.[13]

COMMERCIAL PREPARATIONS

Mainly crude.

Regulatory Status. Approved for foods use as a chewing gum base (§172.615).

REFERENCES

See the General References for HORTUS 3rd; LIST AND HÖRHAMMER; MCGUFFIN 1 & 2; TERRELL; UPHOF; YOUNGKEN.

1. B. L. Archer and B. G. Audley in L. P. Miller, ed., *Phytochemistry*, Vol. 2, Van Nostrand Reinhold, New York, 1973, p. 310.
2. P. D. Strausbaugh and E. L. Core in D. N. Lapedes et al., eds., *McGraw-Hill Encyclopedia of Science and Technology*, Vol. 3, McGraw-Hill, New York, 1977, p. 63.
3. F. W. Stavely et al., *Rubber Chem. Technol.*, **34**, 423 (1961).
4. E. Azpeitia et al., *Can. J. Chem.*, **39**, 2321 (1961).
5. Y. Tanaka and H. Sato, *Polymer*, **17**, 113 (1976).
6. G. G. S. Dutton and S. Kabir, *Carbohydr. Res.*, **28**, 187 (1973).
7. E. Anderson and H. D. Ledbetter, *J. Am. Pharm. Assoc.*, **40**, 623 (1951).
8. Y. Tanaka et al., *J. Nat. Rubber Res.*, **3**, 177 (1988).
9. Y. Sato et al., *J. Jpn. Soc. Food Sci. Technol.*, **38**, 595 (1991).
10. T. Shoji et al., *Shokuhin Eiseigaku Zasshi*, **6**, 27 (1965).
11. E. M. Mendez, U.S. Pat. 3,453,361 (1969).
12. M. Fujiwara, Jpn. Kokai 72 47,665 (1972).
13. H. W. Conner in D. N. Lapedes et al., eds., *McGraw-Hill Encyclopedia of Science and Technology*, Vol. 5, McGraw-Hill, New York, 1977, p. 447.

CHICORY ROOT

Source: *Cichorium intybus* L. (Family Compositae or Asteraceae).

Common/vernacular names: Blue sailors, chicory, chicory root, common chicory root, wild chicory, and succory.

GENERAL DESCRIPTION

Biennial or perennial herb with spindle-shaped taproot, bright blue flowers, and cauline hairy leaves (borne on stem) resembling those of dandelion; up to 2 m high; believed to be native to Europe and Asia; naturalized and weedy in North America (FOSTER). Parts used are the dried root and the dried aboveground parts, collected in autumn.

CHEMICAL COMPOSITION

The root contains a high concentration (up to 58% in fresh cultivated root) of inulin (a mixture of linked fructans), which yields on hydrolysis mostly fructose, with glucose in minor amounts. The root also contains fatty acids (mostly palmitic and linoleic), the bitter principles lactucin and lactucopicrin (intybin), cichoriin (esculetin-7-glucoside), α-lactucerol (taraxasterol), tannins, sugars (fructose, mannose, etc.), pectin; fixed oils, choline, and others (JIANGSU; LIST AND HÖRHAMMER).[1–5]

The roasted root contains a steam-distillable fraction (aroma), which is composed of pyrazines, benzothiazoles, aldehydes, aromatic hydrocarbons, furans, phenols, organic acids, and others, totaling 33 identified compounds, among which acetophenone is a characteristic component of roasted chicory not previously reported as a component of aroma of any heated food products such as coffee.[6] Other constituents of the roasted root include 2-acetylpyrrole, furfural, phenylacetaldehyde, phenylacetic acid, and vanillin.[5] Small amounts of two indole alkaloids (β-carbolines), harman and norharman, have also been isolated from the roasted root.[7]

The herb (leaves, flowers, shoots, etc.) contains inulin, fructose, choline, resin, chicoric acid (dicaffeoyl tartaric acid), esculetin, esculin (esculetin-6-glucoside), cichoriin, and others (LIST AND HÖRHAMMER).[8–10]

PHARMACOLOGY AND BIOLOGICAL PROPERTIES

Addition of inulin to ground beef before frying inhibited the formation of mutagenic compounds (heterocyclic aromatic hydrocarbons) after frying.[11]

As part of the diet of rats, chicory inulin inhibited tumor formation (colon,[12] mammary, and lung) and potentiated the cytotoxic effects of various common anticancer drugs at subtherapeutic dosages (cyclophosphamide, cytarabine, doxorubicine, 5-fluorouracil, methotrexate, and vincristine).[13]

Chicory inulin stimulates the growth of bifidobacteria in the human colon and is classified as a prebiotic.[14] In a placebo-controlled study, dietary supplementation with chicory inulin increased bifidobacteria counts and reduced clostridia, fusobacteria, and bacteroides.[15] In a randomized, double-blind study in hypercholesterolemic men on a controlled diet, a supplement of ice cream containing chicory inulin produced a significant decrease in serum triglyceride levels.[16]

TOXICOLOGY

The herb (vegetable) has been reported to cause contact dermatitis in humans.[17,18] An aqueous suspension of the root at a sublethal dosage (8.7 g/kg p.o. per day for 10 days) decreased body weight and impaired spermatogenesis in mice, whereas the threshold dosage (4.5 g/kg) had neither effect.[19]

USES

Medicinal, Pharmaceutical, and Cosmetic. In Germany, cut herb and root used in infusion or extract for loss of appetite and dyspeptic disorders. Contraindicated in allergies to chicory or other Compositae; in gallstones, only after consultation with physician (BLUMENTHAL 1).

Food. Extracts are used extensively as a flavor ingredient in major food products, including alcoholic (primarily bitter formulations) and nonalcoholic beverages (e.g., instant coffee substitutes), frozen dairy desserts, candy, baked goods, and gelatins and puddings, among others. Highest average maximum use level reported for the extract (type unspecified) is about 0.61% (6116 ppm) in frozen dairy desserts, though use level in instant coffee substitutes could be much higher.

Ground roasted root is increasingly more used admixed with coffee to impart "richer" flavor and to decrease the caffeine content of the resulting coffee formulation. This use is very common in Europe.

Chicory leaf buds known as "chicons," usually obtained from the Witloof variety, are used as a vegetable and salad.[20]

Chicory inulin has 10% of the sweetness of sugar and is used as a fat and sugar replacement, fiber, and prebiotic in dairy products, frozen desserts, fruit preparations, breads and baked goods, and dietetic products; also as a fat replacement in table spreads, salad dressings, meat products, and fillings; as a fiber and prebiotic in breakfast cereals; sugar replacement and fiber in chocolate; also used to provide form stability, moisture retention, texture improvement, texture, crispness, and mouthfeel in diverse foods.[21]

Dietary Supplements/Health Foods. The root and leaves reportedly used as a flavor component in herb teas; also in diuretic and digestive formulations (FOSTER). Inulin is taken in tablets as a sugar replacement, dietary fiber, and prebiotic.[22]

Traditional Medicine. Both root and herb reportedly used as bitter tonics to increase appetite and to treat digestive problems, usually in the form of tea or as juice. Also used as diuretics and in treating gallstones, liver ailments (e.g., hepatitis), and cancers, among others (JIANGSU).[23]

COMMERCIAL PREPARATIONS

Crudes (both roasted and unroasted) and their extracts; strengths (see *glossary*) of extracts are either expressed in weight-to-weight ratios or in terms of flavor potencies.

Regulatory Status. Essential oil, natural extractive, and solvent-free oleoresin are GRAS (§182.20). Chicory inulin is confirmed GRAS.[22] Root and herb subject of a combined positive German therapeutic monograph for mild dyspeptic disorders and loss of appetite (BLUMENTHAL 1).

REFERENCES

See the General References for BAILEY 2; BIANCHINI AND CORBETTA; BLUMENTHAL 1; FEMA; FERNALD; FOSTER; GRIEVE; JIANGSU; LIST AND HÖRHAMMER; LUST.

1. M. P. J. Kierstan, *Biotechnol. Bioeng.*, **20**, 447 (1978).
2. S. I. Balbaa et al., *Planta Med.*, **24**, 133 (1973).
3. J. Promayon et al., *Cafe, Cacao, The*, **20**, 209 (1976).
4. M. Blanc, *Lebensm. Wiss. Technol.*, **11**, 19 (1978).
5. A. Sannai et al., *Agric. Biol. Chem.*, **46**, 429 (1982).
6. S. Kawabata and M. Deki, *Kanzei Chuo Bunsekishoho*, **17**, 63 (1977).
7. A. Proliac and M. Blanc, *Helv. Chim. Acta*, **59**, 2503 (1976).
8. G. F. Fedorin et al., *Rast. Resur.*, **10**, 573 (1974).
9. V. G. Dem'yanenko et al., *USSR*, **577**, 033 (1977).
10. H. Schmidtlein and K. Herrmann, *Z. Lebesm. Unters. Forsch.*, **159**, 255 (1975).
11. H. Shin et al., *J. Agric. Food Chem.*, **51**, 6726 (2003).
12. B. Pool-Zobel et al., *Br. J. Nutr.*, **87**(Suppl. 2), S273 (2002).
13. H. S. Taper and M. B. Roberfroid, *Br. J. Nutr.*, **87**(Suppl. 2), S283 (2002).

14. M. B. Roberfroid et al., *J. Nutr.*, **128**, 11 (1998).
15. G. R. Gibson et al., *Gastroenterol.*, **108**, 975 (1995).
16. J. L. Causey et al., *Nutr. Res.*, **20**, 191 (2000).
17. J. C. Mitchell in V. C. Runeckles, ed., *Recent Advances in Phytochemistry*, Vol. 9, Plenum Press, New York, 1975, p. 119.
18. R. M. Adams, *Occupational Contact Dermatitis*, J. B. Lippincott, Philadelphia, 1969, p. 188.
19. C. A. Roy, *Indian J. Ind. Med.*, **42**, 217 (1996).
20. J. Swabey in L. W. Codd et al., eds., *Chemical Technology: An Encyclopedic Treatment*, Vol. 7, Barnes & Noble, New York, 1975, p. 287.
21. A. Franck, *Br. J. Nutr.*, **87**(Suppl. 2), S287 (2002).
22. R. A. A. Coussement, *J. Nutr.*, **129**(Suppl.), 7S, 1412S (1999).
23. J. L. Hartwell, *Lloydia*, **31**, 71 (1968).

CHIRATA

Source: *Swertia chirata* (Roxb. ex Fleming) (syn. *Swertia chirata* (Wall.) C.B. Clarke) (Family Gentianaceae).

Common/vernacular names: Bitter stick, chirata, chirayta, chiretta, and East Indian balmony.

GENERAL DESCRIPTION

Annual herb with opposite leaves and branching, four-angled stem, with large continuous pith; about 1 m high; leaves broadly lanceolate, subsessile; flowers in large panicles, greenish yellow, purple-tinged; native to northern India, Nepal, and Pakistan in temperate altitudes of the Himalayas from 1200 to 3000 m. Part used is the whole dried herb. Other *Swertia* species appear as adulterants (especially *S. angustifolia*; also *Andrographis paniculata* and roots of *Rubia cordifolia*); *S. chirata* is distinguished by the large dark stem pith and intensely bitter flavor.

CHEMICAL COMPOSITION

Contains a bitter glucoside amarogentin (chiratin), bitter glycosides (amarogentin, amaroswerin, sweroside); (−)-syringaresinol (lignan);[1] numerous tetraoxygenated xanthones, including chiratol (1,5-dihydroxy-3,8-dimethoxy xanthone),[2] swertinin (7,8-dihydroxy-1,3-dimethoxyxanthone), swertianin (1,7,8-trihydroxy-3-methoxyxanthone), swerchirin (1,8-dihydroxy-3,5-dimethoxyxanthone), decussatin (1-hydroxy-2,6,8-trimethoxyxanthone), isobellidifolin (1,6,8-trihydroxy-4-methoxyxanthone), 1,3,7,8-tetrahydroxyxanthone, 1,8-dihydroxy-3,7-dimethoxyxanthone, 1-hydroxy-3,5,8-trimethoxyxanthone, mangiferin, and others; also triterpenes (β-amyrin, chiratenol, lupeol, oleanolic acid, pichierenol, swertenol, swertanone, taraxerol, ursolic acid, etc.); and monoterpene alkaloids, among others (KARRER).[1,3–7]

PHARMACOLOGY AND BIOLOGICAL ACTIVITIES

Chirata xanthones (swertianin, 1,3,7,8-tetrahydroxyxanthone, and 1,8-dihydroxy-3,7-dimethoxyxanthone) are claimed to have antituberculous activities.[4] Amarogentin has shown *in vitro* hepatoprotective activity against carbon tetrachloride toxicity.[8] Swerchirin has shown antimalarial activity *in vivo*.[9]

Oral administration of a total benzene extract of the aerial parts of chirata inhibited hind paw edema induced in rats by bradykinin, carrageenin, prostaglandin, and serotonin.[10] In a mouse model of arthritis, an aqueous extract of the stems administered orally produced a dose-dependent increase in the anti-inflammatory cytokine interleukin-10 and reduced levels of proinflammatory cytokines (tumor necrosis factor-α, interferon-γ, interluekin-6, and interleukin-1β).[11]

Oral feeding of an ethanolic extract of the whole plant lowered blood glucose levels in alloxan diabetic[12] and tolbutamide- and glucose-loaded rats.[13] Hypoglycemic activity is attributed to the xanthone swerchirin.[14–16]

USES

Medicinal, Pharmaceutical, and Cosmetic. Used in certain bitter tonic preparations.

Food. Reportedly used in alcoholic (bitters) and nonalcoholic beverages. Average maximum use levels reported are 0.0016% and 0.0008%, respectively.

Dietary Supplements/Health Foods. Sometimes used in bitter tonic formulations; tea (GRIEVE).

Traditional Medicine. Chirata is used in India as a bitter tonic, febrifuge, anti-inflammatory, and in the treatment of thirst, biliousness,[17] diarrhea, skin diseases, sciatica, depression, cough (NADKARNI), asthma, anemia, liver disorders;[11] also taken to prevent epidemic malaria, cholera, and gastroenteritis during the rainy season;[18] used against cancer.[19]

COMMERCIAL PREPARATIONS

Limited availability as crude

Regulatory Status. Approved for use as a natural flavoring substance in alcoholic beverages only (§172.510).

REFERENCES

See the General References for CSIR 1; FEMA; GRIEVE; GUPTA; MCGUFFIN 1 & 2; UPHOF; YOUNGKEN.

1. A. K. Chakravarty et al., *Indian J. Chem.*, **33B**, 405 (1994).
2. R. K. Asthana et al., *Phytochemistry*, **30**, 1037 (1991).
3. K. K. Purushothaman et al., *Leather Sci. (Madras)*, **20**, 132 (1973).
4. M. Komatsu et al., Jpn. Kokai, 7127,558 (1971).
5. S. Ghosal et al., *J. Pharm. Sci.*, **62**, 926 (1973).
6. L. Bennaroche et al., *C. R. Acad. Sci. Ser. D*, **280**, 2493 (1975).
7. L. Bennaroche et al., *Plant. Med. Phytother.*, **8**, 15 (1974).
8. H. Goyal et al., *J. Res. Ayur. Siddha*, **2**, 286 (1981).
9. H. Hikino et al., *Shoyakugky Zasshri*, **38**, 359 (1984).
10. S. Mandal et al., *Fitoterapia*, **63**, 122 (1992).
11. I. V. M. L. R. Sirish Kumar et al., *Immunopharmacol. Immunotoxicol.*, **25**, 573 (2003).
12. A. Kar et al., *J. Ethnopharmacol.*, **84**, 105 (2003).
13. B. C. Sekar et al., *J. Ethnopharmacol.*, **21**, 175 (1987).
14. M. B. Bajpai et al., *Planta Med.*, **57**, 102 (1991).
15. A. M. Saxena et al., *Indian J. Exp. Biol.*, **29**, 674 (1991).
16. A. M. Saxena et al., *Indian J. Exp. Biol.*, **31**, 178 (1993).
17. C. L. Malhotra and B. Balodi, *J. Econ. Taxon. Bot.*, **5**, 841 (1984).
18. A. K. Goel and B. S. Aswal, *J. Econ. Taxon. Bot.*, **14**, 185 (1990).
19. J. L. Hartwell, *Lloydia*, **32**, 153 (1969).

CINCHONA (RED AND YELLOW)

Source: *Red cinchona*: *Cinchona officinalis* L., *C. pubescens* M. Vahl. (syn. *C. succirubra* Pavón ex Klotzsch) and its hybrids; *Yellow cinchona*: *Cinchona calisaya* Weddell, *C. ledgeriana* Moens ex Trimen, and their hybrids with other *Cinchona* species (Family Rubiaceae).

Common/vernacular names: Red bark, red Peruvian bark, cinchona rubra (*C. pubescens*); yellow bark, calisaya bark, ledger bark, brown bark, cinchona flava (*C. calisaya* and *C. ledgeriana*); Jesuit's bark, Peruvian bark, China bark, cortex chinae, and fever tree.

GENERAL DESCRIPTION

Evergreen shrubs or trees, up to about 30 m high; *C. calisaya* being the tallest, while *C. pubescens* reaching about 24 m and *C. ledgeriana* only up to 6 m; native to mountains of tropical America (Bolivia, Costa Rica, Ecuador, Guatemala, Peru, etc.) between altitudes of about 900 and 3400 m; extensively cultivated in Central and South America, Southeast Asia (India, Java, Sumatra, China, etc.), and Africa. Part used is the dried bark.

CHEMICAL COMPOSITION

Contains up to about 16% (average 6–10%) total quinoline alkaloids that consist mainly of quinine, quinidine, cinchonine, and cinchonidine, with quinine usually in major concentration. Other alkaloids in minor amounts include quinamine, epiquinamine, epiquinine, hydroquinine, hydroquinidine, and many others, totaling over three dozen. Contents of the total alkaloids vary, depending on the sources, with *C. ledgeriana* generally containing a higher amount than *C. pubescens*. Other constituents present include norsolorinic acid (an anthraquinine), quinovic acid, quinovin A, B, and C (bitter glycosides), quinic acid, β-sitosterol, tannins, starch, resin, wax, and others (JIANGSU; LIST AND HÖRHAMMER; MORTON 3; NANJING; STAHL; USD 26th).[1–4]

PHARMACOLOGY AND BIOLOGICAL ACTIVITIES

Cinchona has astringent and bitter tonic properties; also reportedly has analgesic and local anesthetic properties, among others (GOODMAN AND GILMAN; JIANGSU; NANJING).

Quinine, quinidine, cinchonine, cinchonidine, and cinchonicinol have shown *in vitro* inhibitory activity against the cytotoxicity of polyorphonuclear leucocytes.[5] Potent monoamine oxidase inhibiting activity *in vitro* was found from quinine, cinchonicinol, and cinchonaminone derived from *C. succirubra*.[6]

The alkaloids of cinchona have antimalarial and antipyretic activities, with quinine being the most potent. Certain strains of malarial parasites, particularly those of Vietnamese origin that have become resistant to synthetic antimalarials, are still susceptible to quinine treatment.[7]

A double-blind, placebo-controlled trial of quinine in the treatment of nocturnal cramps found significant reductions in the pain, intensity, and frequency of cramps.[8]

TOXICOLOGY

Use of the bark is contraindicated in pregnancy and ulcers, intestinal or gastric, and if taken concomitantly with anticoagulants can increase their effects (WICHTL). For some individuals, even low doses of quinine, such as those found in tonic and gin drinks, can elicit thrombocytopenia with purpura ("cocktail purpura"); usually a self-limited and benign syndrome provided immediate cessation of quinine.[8] Quinidine and quinine have cardiac-depressant properties, with quinidine being twice as active as quinine (GOODMAN AND GILMAN; MARTINDALE; USD 26th). High plasma levels of quinine (16 mg/L) are strongly associated with cardiac arrhythmias.[9]

Quinine has been reported to cause a hypersensitivity reaction known as "black water fever," which consists of hemolysis, hemoglobinuria, and hemoglobinemia and can advance to renal failure. The syndrome is more commonly reported in patients with G6PD deficiency, pregnant women, and those with malaria.[9] Ground cinchona bark and quinine have been reported to cause urticaria, contact dermatitis, and other hypersensitivity reactions in some individuals (GOODMAN AND GILMAN; MARTINDALE; USD 26th; WICHTL).

Cinchona alkaloids are toxic. Poisoning (cinchonism) is usually due to overdosage or hypersensitivity, with symptoms including blindness, deafness, severe headache, tinnitus, delirium, vasodilation, abdominal pain, diarrhea, convulsions, paralysis, and collapse. Cinchonism has resulted from as little as a single dose of 4 g quinine. A single oral dose of 8 g quinine may be fatal to an adult (GOODMAN AND GILMAN; MARTINDALE; USD 26th).[9]

USES

Medicinal, Pharmaceutical, and Cosmetic. In European phytomedicine, the dried bark is seldom used alone. Extracts are ingredients of herbal formulas (WICHTL) used to stimulate saliva and gastric secretions in the treatment of loss of appetite and dyspeptic discomfort (BLUMENTHAL 1).

The current use of quinine, apart from treating malaria, is primarily in the form of the sulfate salt in preparations for treating cold and flu and nocturnal leg cramps, mostly as prescription drugs (MARTINDALE). Besides being used as an antimalarial, quinine has been used for treating various conditions, including hemorrhoids and varicose veins (as hardening agent), and in eye lotions for its astringent, bactericidal, and anesthetic effects.

Quinidine is used in prescription preparations mainly for treating cardiac arrhythmias.

In cosmetics, extracts of cinchona are primarily used in hair tonics, reportedly for stimulating hair growth and controlling oiliness (DE NAVARRE).

Food. Quinine and extracts of cinchona (mostly red cinchona) are extensively used as a bitter in tonic water,[9] alcoholic bitters, liqueurs, and soft drinks (bitter lemon drinks); amounts in commercial soft drinks are approximately 61–67 mg/L.[10]

Other food products in which red cinchona extract has been reportedly used include frozen dairy desserts, candy, baked goods, and condiments and relishes. Use levels reported are lower than those reported in beverages.

Traditional Medicine. Cinchona is used in treating malaria, fevers, indigestion, and for mouth and throat problems, usually in the form of an infusion; has been used in China to treat hangovers; also reportedly used in cancers.[11] An infusion of yellow cinchona was used by the Cherokee Indians as a tonic and treatment for impotence (MOERMAN).

COMMERCIAL PREPARATIONS

Available as crude and extracts (fluid extract, solid extract, etc.). Both red and yellow cinchonas were formerly official in U.S.P. and N.F.; quinine sulfate and quinidine sulfate and gluconate are official in U.S.P.; quinine hydrochloride and sulfate are also official in F.C.C.

Regulatory Status. Both red and yellow cinchona barks have been approved for use in beverages only, with the limitation that the total cinchona alkaloids do not exceed 83 ppm (0.0083%) in the finished beverage (§172.510 and §172.575). Concentration must be declared on the label; in China, quinine is not permitted for use in beverages.[10] Crude drug subject of a positive. German therapeutic monograph (BLUMENTHAL 1).

REFERENCES

See the General References for BLUMENTHAL 1; FEMA; GOSSELIN; GUPTA; JIANGSU; LUST; MCGUFFIN 1 & 2; MERCK; NANJING; TERRELL; UPHOF; YOUNGKEN.

1. A. Haznagy, *Acta Pharm. Hung.*, **47**, 249 (1977); through *Chem. Abstr.*, **88**, 158517a (1978).
2. P. Niaussat et al., *P. V. Seances Soc. Sci. Phys. Nat. Bordeaux*, **35** (1974).
3. R. Adamski and J. Bitner, *Farm. Pol.*, **32**, 661 (1976).
4. S. Dorairaj et al., *Planta Med.*, **54**, 469 (1988).
5. K. Kinoshita et al., *Planta Med.*, **58**, 137 (1991).
6. N. Mitsui et al., *Chem. Pharm. Bull.*, **37**, 363 (1989).
7. N. R. Farnsworth in L. P. Miller, ed., *Phytochemistry*, Vol. 3, Van Nostrand Reinhold, New York, 1973, p. 351.
8. H. C. Diener et al., *Int. J. Clin. Pract.*, **56**, 243 (2002).
9. L. D. Morrison et al., *Vet. Hum. Toxicol.*, **45**, 303 (2003).
10. V. F. Samanidou et al., *J. Liq. Chromatogr. Rel. Technol.*, **27**, 2397 (2004).
11. J. L. Hartwell, *Lloydia*, **34**, 103 (1971).

CINNAMON (AND CASSIA)

Source: *Cinnamon Cinnamomum verum* Berchthold and Presl (syn. *C. zeylanicum* Nees), and *C. burmanii* (C.G. Th. Nees) Blume; *Chinese cassia Cinnamomum cassia* (L.) Berchthold & Presl. (syn. *C. aromaticum* Nees, *C. loureirii* Nees) (Family Lauraceae).

Common/vernacular names: Sri Lankan cinnamon and true cinnamon (*C. verum*); Batavia cassia, Fagot cassia, Indonesian cassia, Indonesian cinnamon, Java cassia, Korintji cassia, and Padang cinnamon (*C. burmanii*); Chinese cassia, Chinese cinnamon, false cinnamon, cassia lignea, and Vietnam cassia (*C. cassia*). In the United States, the common name cinnamon applies to *C. verum*, *C. burmanii*, and *C. cassia*, whereas in the United Kingdom and continental Europe, the name applies only to *C. verum*.

GENERAL DESCRIPTION

Sri Lankan or true cinnamon (*C. verum*) is a medium-size evergreen tree up to 16 m in height and up to 60 cm in diameter at breast height; bark is thin, smooth, light pinkish-brown; leaves opposite, elliptic or oval to lanceolate–oval; flowers pale yellowish-green; native to south India and Sri Lanka. Indonesian cinnamon (*C. burmanii*) reaches 15 m in height; leaves are sub-opposite; native to Sumatra–Java. Chinese or Vietnam cassia (*C. cassia*) is native to China and Vietnam reaching heights of 18–20 m with a diameter of 40–60 cm; bark is gray brown; trunk is cylindrical and straight; leaves simple or sub-opposite; flowers white. Parts used are the dried bark, leaves, and twigs. The trees are mostly cultivated for commercial production of cinnamon and are usually cut back (coppiced) to form bushes or shrubs. The essential oils are obtained by steam distillation. Cassia oil (Chinese cinnamon oil) is obtained from leaves, bark, and twigs of *C. cassia* and is mainly produced in China; cinnamon bark oil is derived from the dried inner bark of *C. verum* and cinnamon leaf oil from the leaves and twigs of the same species (RAVINDRAN).

In the United States, cinnamon bark (particularly from *C. verum*) and its oils are generally considered superior in flavor characteristics to Chinese cassia bark and cassia oil and are also more valued than oils from Indonesian cinnamon (RAVINDRAN).

CHEMICAL COMPOSITION

C. verum bark yields 0.4–0.8% oil; tannins, consisting of polymeric 5,7,3′,4′-tetrahydroxyflavan-3,4-diol units;[1] large amounts of catechins and proanthocyanidins (condensed tannins)[2] and procyanidins; resins; mucilage; gum; sugars; calcium oxalate; two insecticidal compounds (cinnzelanin and cinnzelanol);[3] coumarin (lowest concentration in Ceylon cinnamon); and others (LIST AND HÖRHAMMER; RAVINDRAN).[1,3–5]

C. verum cinnamon bark oil contains as its major component cinnamic aldehyde (usually 60–80%); other major constituents include sesquiterpenoids (4–5%) (e.g., α-humulene and β-caryophyllene that make up 3–4% of the total, limonene, and others), eugenol, eugenol acetate, cinnamyl acetate, cinnamyl alcohol, methyl eugenol, benzaldehyde, cuminaldehyde, benzyl benzoate, monoterpenes (e.g., linalool, pinene, phellandrene, and cymene), carophyllene, safrole, and others (LIST AND HÖRHAMMER; MASADA; RAVINDRAN).[6–10]

C. verum leaf oil contains high concentrations of eugenol (Ceylon type 80–88%; Seychelles type 87–96%); it also contains many of the major constituents present in cinnamon bark oil (e.g., benzyl benzoate (6%), cinnamaldehyde, cinnamyl acetate, eugenol acetate, benzaldehyde, linalool, α-terpinene, and others), as well as other minor compounds, including α-humulene, β-caryophyllene, α-ylangene, methyl cinnamate, and cinnamyl acetate (MASADA; RAVINDRAN).[5,6,10]

The bark of *C. burmanii* yields 1.32% oil containing 1,8-cineole (51.4%), α-terpineol (12.5%), camphor (9.0%), terpinen-4-ol (8.5%), borneol (1.8%), α-pinene (1.6%), β-caryophyllene (1.6%), *p*-cymene (1.0%), and lesser amounts of myristicin, α-humulene, β-eudesmol, and others. The leaf oil contains mostly 1,8-cineole (28.5%) and borneol (16.5%) with lesser amounts of α-terpineol (6.4%), *p*-cymene (6.1%), spathulenol (5.8%), terpinen-4-ol (4.1%), β-caryophyllene (2.9%), and others. Eugenol is absent in both the bark and leaf oils (RAVINDRAN).

Cassia bark (*C. cassia*) contains 1–2% of volatile oil and other constituents, including glycosides (cassioside, cinnamoside),[11] diterpenes (cinnacassiol B and D_1),[12,13] 2′-hydoxycinnamaldehyde,[14] cinnamic acid, vanillic acid, syringic acid, choline, protocatechuic acid, condensed tannins (proanthocyanidins), procyanidins, resins, sugars, calcium oxalate, coumarin, mucilage, and minerals, notably manganese (LIST AND HÖRHAMMER; NANJING; RAVINDRAN).[15]

Cassia bark oil grown in Australia contains mainly cinnamic aldehyde (87%), whereas the bark oil from trees grown in China contains mostly (*E*)-cinnamic aldehyde (65.5%). Other major constituents present in Australian bark oil are benzaldehyde (4.7%), 2-phenylethanol (2.5%), and 3-phenylpropanol (2.0%); others found in lesser amounts include cuminaldehyde, coumarin, eugenol, linalool, ethyl cinnamate, chavicol, and others. In Chinese cassia bark oil, the other major constituents are coumarin (8.7%), cinnamyl acetate (3.6%), and 2-methoxycinnamaldehyde (2.7%); others found in lesser amounts include benzyl benzoate, cinnamyl alcohol, 2-phenylethyl acetate, eugenyl acetate, (*Z*)-cinnamic aldehyde, and others. Eugenol occurs in trace amounts (JIANGSU; LIST AND HÖRHAMMER; NANJING; RAVINDRAN).[9]

The leaf oil of *C. cassia* grown in Australia contains mostly cinnamic aldehyde (77.2%), coumarin (15.3%), cinnamyl acetate (3.6%), benzaldehyde (1.2%), and in lesser amounts, 4-ethylguaiacol, ethyl cinnamate, 2-phenylethyl acetate, α-terpineol, terpinen-4-ol, and others. The leaf oil of China-grown trees contains mostly cinnamic aldehyde (74.1%), 2-methoxycinnamaldehyde (10.3%), cinnamyl acetate (6.6%), coumarin (1.2%), and lesser amounts of benzaldehyde (1.1%), salicyaldehyde, cinnamyl alcohol, 2-phenylethyl acetate, α-pinene, and others (RAVINDRAN).

PHARMACOLOGY AND BIOLOGICAL ACTIVITIES

Extracts of the dried bark of *C. verum* have shown *in vitro* inhibitory activity against *Candida albicans*.[16] The essential oil of the bark inhibits the growth of human pathogenic fungi (*Aspergillus niger*, *Candida albicans*, *Rhizopus oligosporus*) and various bacteria (*Escherichia coli*, *Enterobacter cloacae*,

Micrococcus luteus, Staphylococcus aureus, Streptococcus faecalis, and others).[17] A methanol extract exhibited significant *in vitro* nematicidal activity against *Toxocara canis*.[18] Extracts of the bark have also shown significant insulin-like activity *in vitro* (increased glucose oxidation and uptake,[19,20] and insulin receptor kinase activation and inhibition of insulin receptor dephosphorylation activity);[21,22] insulin activity potentiating activities are attributed to a methylhydroxychalcone polymer[23] and polyphenol type A polymers isolated from korintje cinnamon from Sumatra (*C. burmannii*) consisting of oligomeric procyanidins.[20]

Rats fed a high-fat diet containing 10% *C. verum* bark powder showed significant decreases in heart and liver levels of hydroperoxides and significantly increased levels of antioxidant enzymes (catalase, glutathione, glutathione *S*-transferase, and superoxide dismutase) in the same organs.[24] Anti-inflammatory and pain-inhibiting activities were found in mice orally administered an ethanolic extract of the bark.[25]

Extracts of the dried stem bark of *C. cassia* have shown *in vitro* growth inhibition of human intestinal bacteria (*Bacteriodes fragilis* and *Clostridium perfringens*), an effect attributed to cinnamaldehyde.[26] Extracts of the bark have also shown *in vitro* insulin-like activity,[19] protective activity against glutamate-induced toxicity to rat cerebellar granule cells,[27] antioxidant,[28,29] free radical scavenging activity, and inhibition of HMG-CoA reductase,[30,31] cyclooxygenase-2, and inducible nitric oxide synthase.[32] An aqueous extract[33] and diterpenes from the bark (cinnacassiol B and D_1) have shown anticomplement activity.[12,13] *In vitro* stimulation of human lymphocyte proliferation, interleukin-1, and immunoglobulin G production by an infusion of the bark was attributed to glycoproteins.[34] Cinnamaldehydes (2'-benzoxycinnamaldehyde and 2'-hydroxycinnamaldehyde) derived from the bark inhibited *in vitro* proliferation of lymphocytes and induced *in vitro* T-cell differentiation.[35] 2'-Hydroxycinnamaldehyde (HCA) inhibited the *in vitro* growth of 29 different human cancer cell lines.[36] Cinnamaldehyde induced apoptotic cell death in human leukemia cells *in vitro*.[37] A methanol extract of the bark inhibited the *in vitro* growth of human hepatocellular carcinoma HepG2 cells.[38] Glycosides (cassioside, cinnamoside, and a β-D-glucopyranoside) isolated from an aqueous extract of the dried stem bark are attributed to antiulcerogenic activity.[11]

In mice with influenza-induced fever, an aqueous extract of the dried stem bark (p.o.) showed antipyretic activity and suppressed the production of interleukin-1α. Solvent fractions were also active. Active constituents were identified as acetic acid cinnamylester, cinnamic acid ethylester, 7-hydroxycoumarin, and 4-allylanisole, which were active by the oral route.[39]

A randomized placebo-controlled clinical study of powdered *C. cassia* bark in type 2 diabetics taking sulfonylurea drugs and maintaining their usual diets found that daily supplementation with the bark immediately after each of three daily meals produced significant decreases in triglyceride, LDL, and total serum cholesterol levels and serum glucose levels.[40]

TOXICOLOGY

No significant chronic (90 days) or acute (500 mg, 1 g, or 3 g/kg p.o.) mortality was found in mice administered an ethanol extract of *C. verum* bark. No change in body weight was found from chronic dosing; liver weight and hemoglobin levels were reduced, and reproductive organ weights, sperm counts, and sperm motility were increased without spermatotoxic effects.[41] Allergic skin reactions to *C. verum* are common (WICHTL).

Cinnamaldehyde can cause dermatitis in humans and allergic reactions have occurred from contact with products (foods, toothpastes, ointments, mouthwashes) containing either cinnamaldehyde or cinnamon oil (DE SMET ET AL.). Cinnamon may cause allergic reactions in some people who are allergic to balsam of Peru.[42] Cassia oil causes mucous membrane and dermal irritation, both effects being attributed to cinnamaldehyde (DE SMET ET AL; RAVINDRAN).

An alcoholic extract of cinnamon, cinnamon oil, and cassia oil have shown *in vitro* mutagenic activity (DE SMET ET AL.). However, a recent test of the essential of *C. cassia* found no mutagenic activity in the Ames test.[43] Cinnamaldehyde is also reported to have mutagenic[44] and both *in vitro*[45,46] (Ames test, micronucleus test, and bone marrow chromosomal aberration assay) and *in vivo* antimutagenic activities.[47] Microencapsulated *trans*-cinnamaldehyde failed to produce neoplasms in rats after 2 years of exposure to 1000, 2100, and 4100 ppm in their feed. Reductions in body weights were seen in rats exposed to 4100 ppm and in mice from 2100 ppm.[48] The oral LD_{50} of cinnamaldehyde in mice is 2225 mg/kg.[49]

USES

Medicinal, Pharmaceutical, and Cosmetic. Cassia, cinnamon, and their bark oils have been used either as flavors or as carminative, stomachic, tonic, or counterirritants in pharmaceutical and cosmetic preparations, including liniments, suntan lotions, nasal sprays, mouthwashes or gargles, and toothpaste, among others.

In European phytomedicine, cassia and cinnamon bark (2.0–4.0 g daily) or the essential oils (0.05–0.2 g daily) are used in teas and other galenicals as antibacterial, carminative, fungistatic; also as gastrointestinal remedies for loss of appetite and dyspeptic disturbances (BLUMENTHAL 1; WICHTL).

Sri Lankan cinnamon leaf oil is used as a fragrance component in soaps, detergents, creams, lotions, and perfumes, with highest reported maximum use level of 0.8% in perfumes.[50] Cinnamon bark oil and Chinese cassia oil find limited use in perfume industries owing to their skin sensitizing properties (RAVINDRAN).

Food. The dried inner bark of cinnamon (*C. burmannii* and *C. verum*), widely used as a spice in domestic cooking and for flavoring processed foods, is tan in color; used as an ingredient of curry powders, mulled wines, baked products, candies, desserts, beverages, chewing gum, sauces, soups, pickles, canned fruits; added to chocolate in Mexico and Spain. The bark oil is more commonly used in the food industry than the bark powder owing to the more uniform flavor it imparts; less expensive leaf oil also used in flavor industry; eugenol derived from the leaf oil used to prepare synthetic vanillin (RAVINDRAN).

Chinese cassia has a more powerful aroma than cinnamon and is reddish-brown; used as an ingredient of Chinese five-spice powder and in flavoring beverages, confectioneries, meat dishes, bakery products, sauces, and pickles. The cassia oil of commerce (made from the leaves, stalks, and twigs) is widely used for the same purposes as the powdered bark and is also widely used for flavoring soft drinks and liqueurs (RAVINDRAN).

Dietary Supplements/Health Foods. Ground bark widely used as flavor ingredient in numerous herbal tea formulations and herbal tonics; also in digestive and stimulant in capsulated, tableted products, tinctures, and so on. (DUKE 2).

Traditional Medicine. Both cassia and cinnamon bark have been used for several thousand years in Eastern and Western cultures in treating chronic diarrhea, flatulence, dyspepsia, vomiting, rheumatism, colds, abdominal and heart pains, kidney troubles, hypertension, female disorders (amenorrhea, cramps, menorrhagia, etc.), and cancer, among others (BIANCHINI AND CORBETTA; FARNSWORTH 1; FOGARTY; JIANGSU; NADKARNI; NANJING; RAVINDRAN).[51]

Others. A major use of cinnamon leaf oil is for the isolation of eugenol.

COMMERCIAL PREPARATIONS

Bark, extracts, and oils; oils are of various types and qualities. Cassia oil (Chinese cinnamon oil) is official in N.F., where it is simply monographed as cinnamon oil; it is also official in F.C.C. Sri Lankan cinnamon bark oil and cinnamon leaf oil (both Sri Lankan and Seychelles types) are official in F.C.C.

Regulatory Status. Indonesia cassia or cinnamon (*C. burmanni*), Sri Lankan (*C. verum*), and Chinese or Vietnam (*C. cassia*) are GRAS as spices, natural flavorings, and seasonings (§182.10). Essential oils, solvent-free oleoresins, and natural extractives of the barks of the same species are GRAS, while only those of the leaves of Sri Lankan, Chinese, or Vietnam cinnamon are GRAS (§182.20). Both cassia and cinnamon bark are subjects of positive German therapeutic monographs indicated for treatment of loss of appetite and dyspeptic discomfort such as mild gastrointestinal spasms. Cinnamon flowers are the subject of a neutral monograph, since efficacy is not established, and risks (allergic skin reactions and mucosal irritation) outweigh benefits (BLUMENTHAL 1).

REFERENCES

See the General References for ARCTANDER; BLUMENTHAL 1; DER MARDEROSIAN AND BEUTLER; DUKE 2; FURIA; GOSSELIN; GUENTHER; JIANGSU; KARRER; LIST AND HÖRHAMMER; MARTINDALE; RAVINDRAN; ROSENGARTEN; TERRELL; UPHOF.

1. L. Buchalter, *J. Pharm. Sci.*, **60**, 144 (1971).
2. J. M. Schulz and K. Herrmann, *Z. Lebensm. Unters. Forsch.*, **171**, 278 (1980).
3. A. Isogai et al., *Agric. Biol. Chem.*, **41**, 1779 (1977).
4. Y. S. Lewis et al., *Curr. Sci.*, **46**, 832 (1977).
5. J. E. Angmor et al., *Planta Med.*, **21**, 416 (1972).
6. R. O. B. Wijeskera et al., *J. Sci. Food Agric.*, **25**, 1211 (1974).
7. M. S. F. Ross, *J. Chromatogr.*, **118**, 273 (1976).
8. J. S. Chou, *T' ai-wan Ko Hsueh*, **31**(2), 8 (1977).
9. A. Herisset et al., *Plant. Med. Phytother.*, **6**, 11 (1972).
10. A. Bhramaramba and G. S. Sidhu, *Perfum. Essent. Oil Records*, **54**, 732 (1963).
11. Y. Shiraga et al., *Tetrahedron*, **44**, 4703 (1988).
12. T. Nohara et al., *Tetrahedron Lett.*, **21**, 2647 (1980).
13. T. Nohara et al., *Chem. Pharm. Bull.*, **28**, 2682 (1980).
14. B. Kwon et al., *Planta Med.*, **62**, 183 (1996).
15. Y. Mino and N. Ota, *Chem. Pharm. Bull.*, **38**, 709 (1990).
16. J. M. Quale et al., *Am. J. Chin. Med.*, **24**, 103 (1996).
17. S. C. Chao et al., *J. Essent. Oil Res.*, **12**, 639 (2000).
18. F. Kiuchi et al., *Jpn. J. Pharmacogn.*, **43**, 353 (1989).
19. C. L. Broadhurst et al., *J. Agric. Food Chem.*, **48**, 849 (2000).
20. R. Anderson et al., *J. Agric. Food Chem.*, **52**, 65 (2004).
21. L. F. Berrio et al., *Horm. Res.*, **37**, 225 (1992).
22. J. Imparl-Radosevich et al., *Horm. Res.*, **50**, 177 (1998).
23. K. J. Jarvill-Taylor et al., *J. Am. Coll. Nutr.*, **20**, 327 (2001).
24. J. N. Dhuley, *Indian J. Exp. Biol.*, **37**, 238 (1999).
25. A. H. Atta and A. Alkofahi, *J. Ethnopharmacol.*, **60**, 117 (1998).

26. H. S. Lee and Y. J. Ahn, *J. Agric. Food Chem.*, **46**, 8 (1998).
27. Y. Shimada et al., *Phytother. Res.*, **14**, 466 (2000).
28. H. Lee et al., *J. Agric. Food Chem.*, **50**, 7700 (2002).
29. C. Lin et al., *Phytother. Res.*, **17**, 726 (2003).
30. N. Kim et al., *Korean J. Pharmacogn.*, **31**, 190 (2000).
31. R. Kim et al., *Nat. Prod. Sci.*, **6**, 49 (2000).
32. C. H. Hong et al., *J. Ethnopharmacol.*, **83**, 153 (2002).
33. H. Nagai et al., *Jpn. J. Pharmacol.*, **32**, 813 (1982).
34. B. E. Shan et al., *Int. J. Immunopharmacol.*, **21**, 149 (1999).
35. W. S. Koh et al., *Int. J. Immunopharmacol.*, **20**, 643 (1998).
36. W. L. Lee et al., *Planta Med.*, **65**, 263 (1999).
37. H. Ka et al., *Cancer Lett.*, **196**, 143 (2003).
38. K. J. Park et al., *Pharm. Biol.*, **40**, 189 (2002).
39. M. Kurokawa et al., *Eur. J. Pharmacol.*, **348**, 45 (1998).
40. A. Khan et al., *Diabetes Care*, **26**, 3215 (2003).
41. A. H. Shah et al., *Plant Foods Hum. Nutr.*, **52**, 231 (1998).
42. A. Niinimaki, *Contact Dermatitis*, **33**, 78 (1995).
43. H. Park, *Korean J. Pharmacogn.*, **33**, 372 (2002).
44. M. Ishidate Jr. et al., *Food Chem. Toxicol.*, **22**, 623 (1984).
45. T. Ohta et al., *Mutat. Res.*, **107**, 219 (1983).
46. T. Ohta et al., *Mutat. Res.*, **117**, 135 (1983).
47. N. Sharma et al., *Mutat. Res., Fund. Mol. Mech. Mutagen*, **480–481**, 179 (2001).
48. M. J. Hooth et al., *Food Chem. Toxicol.*, **42**, 1757 (2004).
49. M. Harada and Y. Ozaki, *Yakugaku Zasshi*, **92**, 135 (1972).
50. D. L. J. Opdyke, *Food Cosmet. Toxicol.*, **13**, 545 (1975).
51. J. L. Hartwell, *Lloydia*, **32**, 247 (1969).

CITRONELLA OIL (CEYLON AND JAVA)

Source: *Cymbopogon nardus* (L.) Rendle (syn. *Andropogon nardus* L.); *C. winterianus* Jowitt and their varieties (Family Poaceae).

Common/vernacular names: Ceylon Lenabatu citronella oil (*C. nardus*); Java or Maha Pengiri citronella oil (*C. winterianus*).

GENERAL DESCRIPTION

Cymbopogon nardus (citronella) and *C. winterianus* (Java citronella) are both perennial grasses. The former is extensively cultivated in southern Sri Lanka, while the latter is widely cultivated in many parts of the tropical world (e.g., Java, Taiwan, Hainan Island, Indonesia, India, Nepal, Africa, Vietnam, Guatemala, Brazil, Paraguay, and Argentina).

The essential oils are obtained by steam distillation of the fresh, partly dried, or dried grass. The Java-type oil is generally considered to be of superior quality to the Ceylon oil.[1]

CHEMICAL COMPOSITION

Both Ceylon and Java citronella oils contain citronellal, citronellol, and geraniol and as the major components, with the Java type having a higher concentration of these constituents than the Ceylon type; the relative proportions of these components vary greatly, depending on the sources (ARCTANDER; LIST AND HÖRHAMMER; MARTINDALE; MASADA; YOUNGKEN).[1–9]

Other constituents include esters (acetates, propionates, etc.) of geraniol, citronellol, and linalool; monoterpenes (limonene, pinene, camphene, etc.); sesquiterpenes and alcohols (bourbonene, caryophyllene, elemol, farnesol, etc.); phenols (eugenol, methyl eugenol, etc.); and free acids, among others. Java citronella oil contains higher amounts of sesquiterpenes, while the Ceylon type contains much larger amounts of monoterpenes (ARCTANDER; LIST AND HÖRHAMMER).[1,2,4–6,8] Other major constituent monoterpenes of Ceylon citronella oil are *cis*-sabinene hydrate, and γ-terpineol, and among sesquiterpenes, nerolidol, β-caryophyllene, and germacrene-ol.[10]

PHARMACOLOGY AND BIOLOGICAL ACTIVITIES

Citronella oil has shown antibacterial and antifungal activities *in vitro*,[11–13] the Ceylon oil being as active as penicillin against certain Gram-positive bacteria.[14,15] The most active volatile constituents of Ceylon oil against the growth of *Aspergillus*, *Eurotium*, and *Penicillium* species were citronellal and linalool.[16] Java citronella oil also displays *in vitro* nematicidal activity.[17] Both types of citronella oil have mosquito-repellent activity.[18,19] Ceylon oil has shown mosquito larvicidal activity against *Culex quinquefasciatus* larvae. The most active constituent in the monoterpene fraction was myrcene.[20]

TOXICOLOGY

Citronella oil is mildly irritant to the skin (TISSERAND AND BALAZS) and is reported to cause contact dermatitis in humans (DE SMET ET AL.; LEWIS AND ELVIN-LEWIS).[21] However, at 8% concentration no sensitization was found in a 2-day closed patch test. The LD_{50} of citronella oil in rats was over 5 g/kg p.o.[21]

USES

Medicinal, Pharmaceutical, and Cosmetic. Both oils are used as a component in certain insect repellent formulations. Major current use is as a fragrance component in soaps, brilliantines, disinfectants, and perfumes, among others. Maximum use levels reported for the Ceylon oil were 0.6% in soaps and 0.8% in perfumes.[21]

Food. The Ceylon oil is reportedly used as flavor ingredient in numerous food products, including alcoholic and nonalcoholic beverages, frozen dairy desserts, candy, baked goods, gelatins and puddings, and breakfast cereals. Highest average maximum use level reported is about 0.005% in candy and baked goods (45.9 and 47.6 ppm, respectively).

Dietary Supplements/Health Foods. Essential oil of citronella widely available in health food stores, primarily used as an insect repellent for humans and pets (ROSE).

Traditional Medicine. Leaves of Ceylon citronella are used in medicinal and aromatic teas, as vermifuge, febrifuge, stomachic, diaphoretic, diuretic, emmenagogue, antispasmodic, and stimulant in various cultures (LIST AND HÖRHAMMER; ROSE). The oil is used as a rubifacient and is reputed to be carminitive, diaphoretic, stimulant, and antispasmodic (KIRTIKAR).

COMMERCIAL PREPARATIONS

Both Oils are Available

Regulatory Status. Essential oil, natural extractive, and solvent-free oleoresin GRAS, with only *C. nardus* listed (§182.20). Subject of a German therapeutic monograph as a mild astringent and stomachic; efficacy not documented (BLUMENTHAL 1).

REFERENCES

See the General References for ARCTANDER; DER MARDEROSIAN AND BEUTLER; FEMA; GUENTHER; MASADA; MCGUFFIN 1 & 2; ROSE; TERRELL; UPHOF.

1. W. D. Fordham in L. W. Codd et al., eds., *Chemical Technology: An Encyclopedic Treatment*, Vol. 5, Barnes & Noble, New York, 1972, p. 1.
2. R. O. B. Wijesekera et al., *Phytochemistry*, **12**, 2697 (1973).
3. E. Guenther, *Am. Perfum. Cosmet.*, **83**, 57 (1968).
4. B. C. Gulati and Sadgopal, *Indian Oil Soap J.*, **37**, 305 (1972).
5. T. K. Razdan and G. L. Koul, *Indian Chem. J.*, **8**, 27 (1973).
6. H. S. Singh et al., *Indian Perfum.*, **20**(1B), 77 (1976).
7. B. L. Kaul et al., *Indian J. Pharm.*, **39**, 42 (1977).
8. B. M. Lawrence, *Perfum. Flav.*, **2**(2), 3 (1977).
9. R. M. Ideda et al., *J. Food Sci.*, **27**, 455 (1962).
10. V. S. Mahalwal and M. Ali, *Flav. Fragr. J.*, **18**, 73 (2003)
11. V. G. de Billerbeck et al., *Can. J. Microbiol.*, **47**, 9 (2001).
12. K. K. Aggarwal et al., *J. Med. Aromatic Plant Sci.*, **22**(1B), 544 (2000).
13. A. Dikshit et al., *Fitoterapia*, **55**, 171 (1984).
14. C. K. Kokate and K. C. Varma, *Sci. Cult.*, **37**, 196 (1971).
15. B. G. V. N. Rao and P. L. Joseph, *Riechst., Aromen, Körperpflegem.*, **21**, 405 (1971).
16. K. Nakahara et al., *JARQ*, **37**, 249 (2003).
17. N. K. Sangwan et al., *Nematologica*, **31**, 93 (1985).
18. M. A. Ansari and R. K. Razdan, *Indian J. Malariol.*, **32**, 104 (1995).
19. A. Tawatsin et al., *J. Vector Ecol.*, **26**, 76 (2001).
20. S. S. Ranaweera, *J. Natl. Sci. Council Sri Lanka*, **24**, 63 (1996).
21. D. L. J. Opdyke, *Food Cosmet. Toxicol.*, **11**, 1067 (1973).

CIVET

Source: *Viverra civetta* Schreber (syn. *Civettictis civetta* Schreber), *V. zibetha* L., and other related species (Family Viverridae).

Common/vernacular names: African civet (*V. civetta*), large Indian civet (*V. zibetha*), and zibeth.

GENERAL DESCRIPTION

Civets, also known as civet cats, are not related to cats and have shorter legs and longer muzzles than do cats. Both the African civet and the large Indian civet have gray coats with black markings, erectile manes, and short tails; their overall length is about 1.2 m.[1–3] Part used is the secretion from their anal glands, which is called civet and is collected by curetting (scraping) the glands with a wood or horn spatula at regular intervals (about once a week for the African civet and two to three times a week for the Indian civet). The civet cats are raised in captivity for this purpose (JIANGSU).[4] A concrète is prepared from crude civet by extracting with hydrocarbons; from this, the absolute is obtained by alcohol extraction. Major suppliers of crude civet are African countries (primarily Ethiopia but also Belgian Congo, Kenya, etc.); minor suppliers include India, Indonesia, Malaya, and China (ARCTANDER; JIANGSU).[4]

CHEMICAL COMPOSITION

Contains civetone (9-*cis*-cycloheptadecenone) as its major aromatic principle; others include butyric acid, skatole, cycloheptadecanone, cyclononadecanone, and various saturated and unsaturated cyclic ketones and alcohols.[1,5,6]

TOXICOLOGY

Limited available data indicate civet (absolute) to be nontoxic.[7]

USES

Medicinal, Pharmaceutical, and Cosmetic. Civet absolute and tincture are extensively used as fixatives and fragrance components in perfumes (especially Oriental and rose types), with maximum use level of 0.4% reported for the absolute. Other cosmetic products in which the absolute is reported to be used include soaps, detergents, creams, and lotions.[7]

Food. Civet absolute has been reportedly used as a flavor component in most major food products, which include alcoholic and nonalcoholic beverages, frozen dairy desserts, candy, baked goods, and gelatins and puddings. Average maximum use levels reported are very low, less than 0.0014% (14.2 ppm).

Traditional Medicine. Used in Chinese medicine for centuries to relieve pain and as cardiac and neural; sedatives, among others (JIANGSU).

COMMERCIAL PREPARATIONS

Crude and extracts; crude has been reported to be frequently adulterated (ARCTANDER).[4]

Regulatory Status. GRAS (§182.50).

REFERENCES

See the General References for ARCTANDER; FEMA; GUENTHER; JIANGSU; YOUNGKEN.

1. E. Shiftan in A. Standen, ed., *Kirk-Othmer Encyclopedia of Chemical Technology*, Vol. 14, 2nd ed., Wiley–Interscience, New York, 1976, p. 717.
2. *The Larousse Encyclopedia of Animal Life*, McGraw-Hill, New York, 1967, p. 564.
3. R. F. Ewer, *The Carnivores*, Cornell University Press, Ithaca, NY, 1972, p. 400.
4. C. L. Fischbeck, *Am. Perfum. Cosmet.*, **82**(12), 45 (1967).
5. Y. Ohno and S. Tanaka, *Bunseki Kagaku*, **26**, 232 (1977).
6. D. A. Van Drop et al., *Recl. Trav. Chim. Pays-Bas*, **92**, 915 (1973).
7. D. L. J. Opdyke, *Food Cosmet. Toxicol.*, **12**(Suppl.), 863 (1974).

CLARY SAGE

Source: *Salvia sclarea* L. (Family Labiatae or Lamiaceae).

Common/vernacular names: Clary, clary wort, muscatel sage, clear eye, see bright, and eyebright.

GENERAL DESCRIPTION

Erect biennial or perennial aromatic herb with large hairy leaves and stout hairy stem; up to about 1 m high; native to southern Europe; cultivated worldwide (e.g., Mediterranean region, central Europe, Russia, the United Kingdom, and the United States). Parts used are the flowering tops and leaves, from which an essential oil is obtained by steam distillation and an absolute obtained by solvent extraction (ARCTANDER; POUCHER).[1]

CHEMICAL COMPOSITION

Essential oil, which largely contains α-terpineol, α-terpinyl, linalool, linalyl acetate, geranyl acetate, germacrene D, β-caryopyllene, nerol, neryl acetate, myrcene, and the diterpene alcohols, manool, and sclareol, is obtained from the flowering tops and leaves.[2–5] The oil of the leaves contains high amounts of germacrene D.[4] The essential oil and flowers contain 1-methoxyhexane-3-thiol, one of the most potent odorants known, having a unique odor profile described as alliaceous, herbaceous-green, and perspirant.[6] Some commercial and genuine samples of the essential oil contain *trans*-anethole.[7] The yield of the volatile oil and the relative concentrations of its components vary with the sources (MASADA; POUCHER).[8–11] An acetone extract of the whole plant collected in Turkey contained various diterpenes (candidissiol, ferruginol, manool, microstegiol, 7-oxoroyleanone, sclareol, 7-oxoferruginol-18-al, and 2,3-dehydrosalvipisone), sesquiterpenes (caryophyllene oxide and spathulenol), and flavonoids (apigenin, 4′-methylapigenin, 6-hydroxy apigenin-7,4′-dimethyl ether, luteolin, and 6-hydroxyluteolin-6).[12]

PHARMACOLOGY AND BIOLOGICAL ACTIVITIES

Clary sage oil has shown *in vitro* antimicrobial (*Escherichia coli*, *Staphylococcus aureus*, and *S. epidermis*) and antifungal activities (*Candida albicans*). Active constituents were identified as linalool and α-terpineol.[3] Other antimicrobial compounds obtained from the whole plant include caryophyllene oxide, 2,3-dehydrosalvipisone, and 7-oxoroyleanone.[12] A methanolic extract of the dried leaves and stems showed high antioxidant activity in the DPPH radical scavenging assay.[13] The oil has also shown activity against the growth of *Trichomonas vaginalis*[14] and anticonvulsive activity in animals; it also potentiated the narcotic effects of Evipam and chloral hydrate.[15]

TOXICOLOGY

Except for being moderately irritating to rabbit skin, available data indicate clary sage oil to be generally nontoxic.[16] The essential oil failed to show *in vitro* DNA-damaging activity.[17] The oral LD_{50} of the essential in rodents was over 5 g/kg (TISSERAND AND BALAZS).

USES

Medicinal, Pharmaceutical, and Cosmetic. Both oil and absolute are used as fragrance components in soaps, detergents, creams, lotions, and perfumes (e.g., eau de cologne). Maximum use level reported for the oil is 0.8% in perfumes.[16]

Food. While clary sage is reportedly used only in beverages (e.g., wines and liqueurs with muscatel flavor), clary sage oil is used rather extensively in major food products such as alcoholic (vermouths, etc.) and nonalcoholic beverages, frozen dairy desserts, candy, baked goods, gelatins and puddings, and condiments and relishes. Highest average maximum use level reported is about 0.016% (155 ppm) for the oil in alcoholic beverages.

Traditional Medicine. The flowering tops and leaves have been used to treat catarrh and as an antiseptic and emmenagogue (BOULOS); herb also used as a stomachic in digestive disorders, as an antispasmodic, and powdered to serve as a snuff to treat headache (UPHOF) and in kidney diseases. Mucilage of seeds is used in tumors and in removing dust particles from the eyes, among others.[18]

Others. After the essential oil is removed by distillation, the crude material is used as a source of sclareol, which can be solvent extracted from the plant and converted to sclareolide; both are used in flavoring tobaccos. Sclareolide is also used in the production of an ambergris substitute.[1]

COMMERCIAL PREPARATIONS

Crude and oil; oil is official in F.C.C.

Regulatory Status. GRAS as a natural seasoning or flavoring (§182.10). Essential oil, solvent-free oleoresin, and natural extractive also GRAS (§182.20).

REFERENCES

See the General References for ARCTANDER; BAILEY 1; BIANCHINI AND CORBETTA; FEMA; GRIEVE; GUENTHER; MCGUFFIN 1 & 2; POUCHER; TERRELL.

1. S. E. Allured, *Cosmet. Perfum.*, **90**(4), 69 (1975).
2. C. Souleles and N. Argyriadou, *Int. J. Pharmacogn.*, **35**, 218 (1997).
3. A. T. Peana et al., *Planta Med.*, **65**, 752 (1999).
4. A. Carrubba et al., *Flav. Fragr. J.*, **17**, 191 (2002).
5. D. Lorenzo et al., *Flav. Fragr. J.*, **19**, 303 (2004).
6. M. Van de Waal et al., *Helv. Chim. Acta*, **85**, 1246 (2002).
7. J. Fiori et al., *J. Sep. Sci.*, **25**, 703 (2002).
8. H. B. Heath, *Cosmet. Toilet.*, **92**(1), 19 (1977).
9. S. Chorbadzhiev et al., paper given at the 4th Mezhdunar. Kongr. Efirnym Maslam (Mater.), 1968.
10. G. Petri Verzar and M. Then, *Herba Hung.*, **13**, 51 (1974).
11. A. I. Karetnikova et al., *Maslo Zhir. Prom.*, **7**, 29 (1974).
12. A. Ulubelen et al., *Phytochemistry*, **36**, 971 (1994).
13. G. Miliauskas et al., *Food Chem.*, **85**, 231 (2004).
14. C. Viollon et al., *Fitoterapia*, **68**, 279 (1996).
15. S. Atanasova-Shopova and K. S. Rusinov, *Izv. Acad. Nauk. Inst. Fiziol., Bulg.*, **13**, 89 (1970); through *Chem. Abstr.*, **74**, 123533m (1971).
16. D. L. J. Opdyke, *Food Cosmet. Toxicol.*, **12**(Suppl.), 865 (1974).
17. F. Zani et al., *Planta Med.*, **57**, 237 (1991).
18. J. L. Hartwell, *Lloydia*, **32**, 247 (1969).

CLOVER TOPS, RED

Source: *Trifolium pratense* L. (Family Leguminosae).

Common/vernacular names: Cow clover, meadow clover, purple clover, red clover, and trifolium.

GENERAL DESCRIPTION

A biennial or perennial herb with rose-purple flowers and leaves consisting of three, often hairy, white-blotched leaflets; up to 0.8 m high; native to Europe and naturalized in North America. Parts used are the flowering tops (inflorescence).

CHEMICAL COMPOSITION

Contains the isoflavones biochanin A, formononetin, genistein, and daidzein as major isoflavone constituents, in addition to pratensein, pseudobaptigenin, glycitein, calycosin, prunetin,[1] and trifoside (5-hydroxy-7-methoxy-isoflavone-4'-O-β-D-glucopyranoside); flavones (e.g., pectolinarin); coumarins (coumarin, medicagol, and coumestrol, etc.),[2-5] with the presence of coumestrol disputed (JIANGSU).[6] Other constituents include procyanidin polymers,[7] isorhamnetin glucosides, trans- and cis-clovamide (L-dopa conjugated with trans- and cis-caffeic acids); trifoliin (isoquercitrin); phaselic acid; a galactoglucomannan composed of a backbone of β-(1 → 4)-linked D-glucose and D-mannose units with α-(1 → 6)-linked D-galactose side chains; sugars; protein; a volatile oil containing furfural; resins, fat, and minerals (particularly rich in magnesium, copper, and calcium); phosphorus, vitamins, and others (DUKE 1; JIANGSU; KARRER).[8-11] Red clover produces phytoalexins (pterocarpan types) in response to viral or fungal infections.[5,12-14]

Some commercial solid extracts of red clover have been reported to contain traces of cannabinol, caffeine, scopolamine, isocoumarin, phenylpentadienal, phenylhexadiene, and nepetalactone (see *alfalfa*).[15] Contents of the major isoflavone constituents show wide variability.[16]

PHARMACOLOGY AND BIOLOGICAL ACTIVITIES

Topical application of genistein to the skin of hairless mice after they were exposed to ultraviolet radiation ameliorated the inflammatory edema reaction.[17] A red clover isoflavone-rich diet in mice with enlarged prostate glands reduced the enlargement and serum androgen levels.[18]

Red clover has shown estrogenic properties in animals that are due to its isoflavones (see *alfalfa*).[2,5,19-23] Oral administration of a red clover tops extract (standardized to contain 15% isoflavones) showed evidence of weak estrogenic activity in ovariectomized rats.[24] *In vitro* transactivation of human estrogen receptors α and β was greater from biochanin A, followed by daidzein, formononetin, and genistein. However, the orders of potency were 2–5 orders lower than that of 17β-estradiol. Biochanin A also exhibited a higher affinity to progesterone and to androgen receptor *in vitro* and was followed in potency by genistein, daidzein, and formononetin.[16] Oral administration of a red clover tops extract (standardized to contain 15% isoflavones) showed evidence of weak estrogenic activity in ovariectomized rats.[24] *In vitro*, biochanin A inhibited DNA damage in human mammary gland cells exposed to a polycyclic aromatic hydrocarbon carcinogen (DMBA).[25]

In a randomized, double-blind, placebo-controlled (RDPC) crossover trial, a biochanin-rich extract of red clover (but not a formononetin-rich extract) significantly lowered LDL cholesterol in men, whereas neither extract was effective in postmenopausal women, nor had any effect on levels of HDL-c or triglycerides in the men or women.[26] In premenopausal women, a RDPC parallel study found that subjects taking an isoflavone-rich extract of red clover showed no significant change in cholesterol, triaglycerol, lipoprotein (a), insulin, or glucose levels.[27] Double-blind, placebo-controlled trials have found that red clover extracts caused HDL-cholesterol levels to significantly increase in postmenopausal women.[28,29] Although they have failed to effect serum total cholesterol, LDL-c, or triglycerides, bone mineral density (BMD) was significantly increased.[28,30] A RDPC in postmenopausal women who experienced at least five hot flushes/day found that daily supplementation with an isoflavone-rich extract of red clover (standardized to contain largely biochanin A and formononetin) reduced both the severity and the frequency of hot flushes.[31] However, a study of the same length and design using the same extract and dosage failed to find a significant reduction in hot flush frequency or any significant improvement in quality-of-life scores.[32] A larger year-long study of the same design using the same extract and dosage also failed to find significant effects on menopausal symptoms and frequency of hot flushes. The study also found that in contrast to standard hormone replacement therapies, the red clover extract did not increase mammographic breast density.[33] In postmenopausal women and men with normal blood pressure, a RDPC trial of an isoflavone-rich extract of red clover was found to significantly improve total vascular resistance and arterial stiffness.[34] In postmenopausal women with type 2 diabetics, a randomized double-blind crossover trial of a red clover extract significantly lowered diastolic and systolic blood pressure and improved endothelial function.[35]

TOXICOLOGY

Mice fed diets containing 20% and 40% red clover showed no significant changes in spermatocytes or the diameter of seminiferous tubules. The LD_{50} was 4237 mg/kg i.p.[36]

USES

Food. The solid extract is reportedly used as a flavor ingredient in many food products, including nonalcoholic beverages, frozen dairy desserts, candy, baked goods, gravies, and jams and jellies. Average maximum use levels reported are usually below 0.002%, except in jams and jellies, where it is about 0.053% (525 ppm).

Dietary Supplements/Health Foods. Whole or ground flowering tops used as an herb tea ingredient; also in capsules, tablets, tinctures, and so on, primarily as an "alterative" (blood purifier) used for skin ailments, such as psoriasis, eczema; ingredient in unconventional anticancer formulas including the Hoxsey formula (FOSTER AND DUKE).

Traditional Medicine. Dried inflorescence and whole herb are used in both Eastern and Western cultures as diuretic, sedative, and antitussive and in treating whooping cough, asthma, bronchitis, eczema, psoriasis, and burns, among others, usually in the form of a tea, infusion, or salve (BARNES; DUKE 1); flowers used in the form of a cold tea by Iroquois women for "the change of life" (MOERMAN). The whole plant or its various parts (leaves, flowers, roots, etc.) have also been extensively used in treating cancers.[37]

COMMERCIAL PREPARATIONS

Crude and extracts (solid, fluid, etc.); crude and fluid extract were formerly official in N.F. Strengths (see glossary) of extracts are expressed in weight-to-weight ratios.

Regulatory Status. *Trifolium* species are GRAS as natural flavorings and seasonings (§182.10). Essential oils, solvent-free oleoresins and natural extractives of *Trifolium* species are also GRAS (§182.20).

REFERENCES

See the General References for APPLEQUIST; BARNES; BARRETT; DER MARDEROSIAN AND BEUTLER; DUKE 1; FEMA; FOSTER AND DUKE; GRIEVE; HORTUS 3rd; HUANG; JIANGSU; KROCHMAL AND KROCHMAL; LUST; MCGUFFIN 1 & 2; ROSE; UPHOF; WREN; YOUNGKEN.

1. Q. Wu et al., *J. Chromatogr. A*, **1016**, 195 (2003).
2. Z. Rolinski, *Ann. Univ. Mariae Curie Sklodowska, Sect. DD*, **24**, 165 (1970).
3. G. Schultz, *Deut. Tieraerztl. Wochenschr.*, **74**, 118 (1967).
4. N. S. Kattaev et al., *Khim. Prir. Soedin.*, **6**, 806 (1972).

5. P. M. Dewick, *Phytochemistry*, **16**, 93 (1977).
6. W. Dedio, *Diss. Abstr. Int. B*, **34**, 5281 (1974).
7. S. Sivakumaran et al., *J. Agric. Food Chem.*, **52**, 1581 (2004).
8. T. Yoshihara et al., *Agric. Biol. Chem.*, **38**, 1107 (1974).
9. A. J. Buchala and H. Meier, *Carbohydr. Res.*, **31**, 87 (1973).
10. D. Smith et al., *Agron. J.*, **66**, 817 (1974).
11. B. D. E. Guillard and R. W. Bailey, *Phytochemistry*, **7**, 2037 (1968).
12. P. M. Dewick, *Phytochemistry*, **14**, 979 (1975).
13. J. L. Ingham, *Phytochemistry*, **15**, 1489 (1976).
14. J. N. Bitton et al., *Phytochemistry*, **15**, 1411 (1976).
15. S. R. Srinivas, *Dev. Food Sci.*, **18**, 343 (1988).
16. V. Beck et al., *J. Steroid Biochem. Mol. Biol.*, **84**, 259 (2003).
17. S. Widyarini et al., *Photochem. Photobiol.*, **74**, 465 (2001).
18. R. A. Jarred et al., *Prostate*, **56**, 54 (2003).
19. A. K. Tuskaev, *Rast. Resur.*, **7**, 295 (1971).
20. E. Krause, *Acta Vet. (Brno)*, **39**, 279 (1970).
21. Z. Rolinski, *Ann. Univ. Mariae Curie Sklodowska, Sect. DD*, **24**, 187 (1970).
22. F. Garcia and P. H. Reinshagen, *Nutr. Bromatol. Toxicol.*, **5**, 67 (1966).
23. J. M. Bergeron and M. Goulet, *Can. J. Zool.*, **58**, 1575 (1980).
24. J. E. Burdette et al., *J. Nutr.*, **132**, 27 (2002).
25. H. Y. Chan et al., *Br. J. Nutr.*, **90**, 87 (2003).
26. P. Nestel et al., *Eur. J. Clin. Nutr.*, **58**, 403 (2004).
27. S. J. Blakesmithh et al., *Br. J. Nutr.*, **89**, 467 (2003).
28. P. B. Clifton-Bligh et al., *Menopause*, **8**, 259 (2001).
29. M. J. Campbell et al., *Eur. J. Clin. Nutr.*, **58**, 173 (2004).
30. C. Atkinson et al., *Am. J. Clin. Nutr.*, **79**, 326 (2004).
31. P. H. M. van de Weijer et al., *Maturitas*, **42**, 187 (2002).
32. J. A. Tice et al., *JAMA*, **290**, 207 (2003).
33. C. Atkinson et al., *Breast Cancer Res.*, **6**, R170 (2004).
34. H. J. Teede et al., *Arterioscler. Thromb. Vasc. Biol.*, **23**, 1066 (2003).
35. J. B. Howes et al., *Diabetes Obes. Metab.*, **5**, 325 (2003).
36. T. Bakirel et al., *Turk. Vet. Hayvan. Dergisi*, **26**, 555 (2002).
37. J. L. Hartwell, *Lloydia*, **33**, 97 (1970).

CLOVES

Source: *Syzygium aromaticum* (L.) Merr. et Perry (syn. *Eugenia aromatica* (L.) Baill., *E. caryophyllata* Thunb., and *E. caryophyllus* (Spreng.) Bull. et Harr.) (Family Myrtaceae).

GENERAL DESCRIPTION

The clove is an evergreen tree with narrowly elliptic, pinkish (young) to dark green (mature) leaves; up to about 12 m high; believed to be native of Southeast Asia (eastern Indonesia); now cultivated worldwide (tropical Asia, Africa, tropical America, etc.). Parts used are the buds (cloves), stems, and leaves from which their respective essential oils are produced; bud and leaf oils by water distillation and stem oil by steam distillation (ARCTANDER). The major clove-producing country is Tanzania; other producers include Madagascar, Indonesia, Malaysia, and Sri Lanka. Clove bud oil is considered more valuable than stem and leaf oils in flavor applications.

CHEMICAL COMPOSITION

Clove buds yield 15–18% volatile oil; clove stems yield 4–6%; and clove leaves yield 2–3%.

Other constituents present in clove buds include glucosides of sterols (sitosterol, stigmasterol, and campesterol), crategolic acid methyl ester, oleanolic acid, quercetin, eugeniin, kaempferol, rhamnetin, about 6% protein, 20% lipids, 61% carbohydrates, vitamins, and others (JIANGSU; MARSH).[1–3]

Clove bud oil contains 60–90% eugenol, 2–27% eugenol acetate, and 5–12% β-caryophyllene, with minor constituents such as methyl salicylate, methyl eugenol, benzaldehyde, methyl amyl ketone, α-ylangene, and chavicol also present (ARCTANDER; MASADA).[4,5]

Clove stem oil usually contains 90–95% and clove leaf oil 82–88% eugenol (ARCTANDER; MASADA); they contain little or no eugenyl acetate. Naphthalene (not present in the bud oil) is reportedly present in both oils in trace amounts; also many of the minor constituents in the bud oil are absent or present in much smaller concentrations in the leaf and stem oils.[5]

PHARMACOLOGY AND BIOLOGICAL ACTIVITIES

A tincture of cloves (15% in 70% alcohol) has been reported to be effective in treating ringworms such as athlete's foot (JIANGSU). Clove oil has antihistaminic and spasmolytic (musculotropic) properties, the latter probably due to its content of eugenyl acetate (see *balm*).[6–8]

Clove oil (due to its eugenol) has anodyne and mildly antiseptic properties, exhibiting broad antimicrobial activities (against Gram-positive, Gram-negative, and acid-fast bacteria, and fungi),[5,7–10] as well as anthelmintic and larvicidal properties.[11] No data are available that correlate the pharmacological properties of noneugenol clove constituents such as eugenyl acetate, methyl eugenol (see *sweet bay*), and caryophyllene, which are often present in relatively large amounts in cloves and clove derivatives. Aqueous extracts of cloves, clove oil, eugenol, eugenyl acetate, and methyl eugenol all have trypsin-potentiating activity (see *cinnamon*).

Eugeniin exhibited strong antiviral activity against herpes simplex virus.[3]

TOXICOLOGY

Clove oil is reported to cause skin irritation and sensitization in humans.[7,8] Despite its possible toxicity in high dosage levels, eugenol (and presumably cloves and clove derivatives) is considered nontoxic at normal use levels.[5]

USES

Medicinal, Pharmaceutical, and Cosmetic. Clove bud oil (or eugenol) is used for the symptomatic relief of toothache; the oil is applied directly without pressure on the carious tooth with a small piece of cotton. It is also extensively used as a major component in preparations for the treatment of postextraction alveolitis (dry socket) and in dental cements and fillings, among others.

Clove bud and stem oils are used extensively as fragrance components in dentifrices, soaps, detergents, creams, lotions, and perfumes. Maximum use levels reported for the bud and stem oils are, respectively, 0.15% and 0.25% in soaps and 0.7% and 1.0%, respectively, in perfumes.[7,8] Clove leaf oil is primarily used in soaps and low-cost perfumes, and to a much lesser extent than the other oils (ARCTANDER).

Food. Cloves, clove bud oil, clove stem oil, clove leaf oil, and eugenol are widely used in flavoring many food products, with cloves and clove bud oil by far the most used. Clove bud extract and oleoresin are also used, though to a lesser scale. Major food products in which cloves and their derivatives are used include alcoholic (bitters, vermouths, etc.) and nonalcoholic beverages, frozen dairy desserts, candy, baked goods, gelatins and puddings, meat and meat products, condiments and relishes, and gravies, among others. Highest average maximum use level reported for cloves is 0.236% in condiments and relishes, that for the oils is 0.06% of clove stem oil in alcoholic beverages, and that for clove bud oleoresin is about 0.078% (775 ppm) in alcoholic beverages.

Dietary Supplements/Health Foods. Powdered cloves are used as a flavoring ingredient in Oriental-type herb teas (DUKE 2).

Traditional Medicine. Cloves are used as a carminative, antiemetic, and counterirritant. Clove tea is used to relieve nausea. Clove oil is also used as an antiemetic as well as in relieving toothache.

In Chinese medicine, clove oil is used in diarrhea, hernia, and bad breath, in addition to the above and other uses (JIANGSU; NANJING).

Others. Clove leaf oil is used as a source for the isolation of eugenol.

A large portion of the world's clove production goes to Indonesia for use in Kretak cigarettes, which consist of a mixture of two parts tobacco and one part ground cloves and when smoked produce a crackling noise (ROSENGARTEN).

Clove extracts and oil have been demonstrated to have strong antioxidative properties.[12–14] Clove oil (also eugenol) and clove aqueous extract also markedly increase trypsin activity.[15] These properties could be useful in food and drug applications.

COMMERCIAL PREPARATIONS

Cloves, extracts (e.g., oleoresin), and oils (bud, stem, and leaf). Clove bud oil is official in N.F. and all three oils are official in F.C.C.

Regulatory Status. Cloves and their derivatives (oils, extracts, etc.) have been affirmed as GRAS (§184.1257).[5] Subject of a German therapeutic monograph indicated for inflamed oral and pharyngeal mucosa; topical anesthesia in dentistry.[16]

REFERENCES

See the General References for ADA; ARCTANDER; BARNES; BARRETT; BISSET; BRUNETON; DUKE 2; FEMA; FURIA AND BELLANCA; GUENTHER; JIANGSU; LUST; MARTINDALE; MASADA; MCGUFFIN; ROSENGARTEN; USD 26th.

1. C. H. Brieskorn et al., *Phytochemistry*, **14**, 2308 (1975).
2. B. Voesgen and K. Herrmann, *Z. Lebensm. Unters. Forsch.*, **170**, 204 (1980).
3. M. Takechi and Y. Tanaka, *Planta Med.*, **42**, 69 (1981).
4. E. Cerma and B. Stancher, Paper given at the 4th Atti Conv. Reg. Aliment., 1st Conv. Naz. Qual., Trieste, 1965.
5. Anon., *Fed. Regist.*, **42**(146), 38613 (1977).
6. A. M. Debelmas and J. Rochat, *Plant. Med. Phytother.* **1**, 23 (1967).
7. D. L. J. Opdyke, *Food Cosmet. Toxicol.*, **13**, 761 (1975).
8. D. L. J. Opdyke, *Food Cosmet. Toxicol.*, **13**, 765 (1975).
9. N. G. Martinez Nadal et al., *Cosmet. Perfum.*, **88**(10), 37 (1973).
10. F. M. Ramadan et al., *Chem. Mikrobiol. Technol. Lebensm.*, **1**, 96 (1972).
11. K. Oishi et al., *Nippon Suisan Gakkaishi*, **40**, 1241 (1974).
12. Y. Saito et al., *Eiyo To Shokuryo*, **29**, 505 (1976).
13. H. Fujio et al., *Nippon Shokuhin Kogyo Gakkaishi*, **16**, 241 (1969).

14. F. Hirahara et al., *Eiyogaku Zasshi*, **32**, 1 (1974).
15. Y. Kato, *Koryo*, **113**, 17, 24 (1975).
16. Monograph *Caryophylli flos*, *Bundesanzeiger*, no. 223 (November 30, 1985).

COCA

Source: *Erythroxylum coca* Lamarck var. ***coca*** (syn. *E. chilpei* E. Machado), ***E. coca*** var. ***ipadu*** Plowman; *E. novogranatense* var. *novogranatense* Rusby var. *novogranatense*, *E. novogranatense* var. *truxillense* (Rusby) Plowman (syn. *E. hardinii* E. Machado) (Family Erythroxylaceae).

Common/vernacular names: Bolivian or Huánuco coca (*E. coca* var. *coca*), Truxillo coca (*E. novogranatense* var. *truxillense*), Columbian or Java coca (*E. novogranatense* var. *novogranatense*)), Amazonian coca (*E. coca* var. *ipadu*), cocaine plant, and spadic.

GENERAL DESCRIPTION

Leafy evergreen shrubs to small trees, with slender branches; up to about 5 m high at lower altitudes and 2 m at higher altitudes; native to the South American Andes; primarily cultivated at altitudes between 500 and 2000 m. Bolivian or Huánuco coca (*Erythroxylum coca* var. *coca*) is cultivated in moist, tropical mountain forests of the eastern Andes (Ecuador, Peru, Bolivia, and extreme NW Argentina) and wet inter-Andean valleys. *E. novogranatense* var. *novogranatense* (Columbian coca) is grown in moist regions of the Columbian Andes, along the coast of the Caribbean, and in arid inter-Andean valleys of Colombia. *E. novogranatense* var. *truxillense* (Trujillo coca) is grown in river valleys in northern Peru and in the arid upper Marañón river valley of Peru. Amazonian coca (*E. coca* var. *ipadu*) is cultivated in the upper Amazon in regions of Brazil, Columbia, and Peru.[1] Coca has also been grown in Asia (China, India, Indonesia, Taiwan, etc.) and Australia.

Part used is the leaf. Major producers are Bolivia, Colombia, and Peru (LIST AND HÖRHAMMER; MORTON 3; NANJING).

CHEMICAL COMPOSITION

Coca leaves contain alkaloids, including benzoylecgonine, ecgonine, ecgonine methyl ester, cocaine (methyl ester of benzoylecgonine), hydroxycocaines, $3',4',5'$-trimethoxycocaine, *cis*- and *trans*-cinnamoylcocaine, $3',4',5'$-trimethoxy-*cis*- cinnamoylcocaine, $3',4',5'$-trimethoxy-*trans*-cinnamoylcocaine, α- and β-truxilline, tropine, tropacocaine, $3',4',5'$-trimethoxy tropacocaine, valerine, hygrine, hygroline, and cuscohygrine.[2,3] The concentrations of these alkaloids and their relative proportions vary widely, depending on the sources of the leaves, their age when harvested, storage age, among other factors (LIST AND HÖRHAMMER; MARTINDALE).[4-7] Nicotine was found in coca growing in Peru,[8] whereas it was not found in leaves of Bolivian coca.[9] Mean amounts of cocaine in the dried leaves vary from 0.25% in Amazonian coca, 0.77% in Columbian coca, 0.72% in Trujillo coca, and 0.63% in Bolivian coca.[10] Leaves of the latter variety also contain pseudococaine.[11]

Other constituents present in the leaves of *E. coca* var. *coca* include 0.02–0.13% of a volatile oil composed mainly of methyl salicylate, two dihydrobenzaldehydes (tentatively identified), *cis*-3-hexen-1-ol, *trans*-2-hexenal, 1-hexanol, and *n*-methylpyrrole;[12] α- and β-truxillic acids; rutin and isoquercitrin (LIST AND HÖRHAMMER); about 19% protein and 44% carbohydrates; high contents of calcium, iron, vitamin A, riboflavin, and phosphorus;[13] amino acids (e.g., arginine and phenylalanine).[14] Flavonoid constituents of the leaves appear to show considerable interspecies variation and although they have yet to be characterized for all four varieties.[15] For example, ombuin-3-O-rutinoside was found in the leaves of both varieties of *E. novogranatense*, but not in either variety of *E. coca*. Flavonoids found in the leaves of all four species are kaempferol 3-O-glucoside and quercetin 3-O-glucoside.[16] *E. novogranatense* (variety not stated) also contains procyanidins B_1 and B_3, a catechin rhamnopyranoside,[17] and a procyanidin glycoside (catechin 3-O-rhamnosyl-(4α → 8)-catechin).[18]

PHARMACOLOGY AND BIOLOGICAL ACTIVITIES

Oral administration of whole coca leaf extract (*E. coca* from Peru) to rats resulted in reduced food consumption, which could not be entirely attributed to the content of cocaine.[19] A cocaine-free extract of the leaves administered intraperitoneally also reduced food consumption in rats.[20]

Preliminary studies indicate that after chewing coca leaves, chronic coca leaf chewers of Bolivia show increased plasma levels of cocaine and benzoylecgonine during exercise;[21] from chewing approximately 15 g of leaves, no significant alteration in blood pressure or heart rate during maximal[22] or prolonged submaximal exercise;[23] no change in maximal exercise capacity; and compared to nonchewers without coca, greater ventilatory output and free fatty acid availability during incremental exercise.[22] At rest, after chewing approximately 50 g of leaves for 1 h, chronic coca chewers showed a significant increase in plasma norepinephrine, heart rate, hematocrit, and hemoglobin concentration compared to nonchewers without coca. During submaximal exercise, the same subjects showed a significantly higher mean arterial blood pressure and a higher heart rate compared to the nonchewers, suggesting a compromised circulatory adjustment during exercise.[24] After chewing 15 g of leaves for 1 h, nonhabitual coca leaf chewers showed a significant decrease in plasma insulin levels. During steady-state exercise, they displayed significant increases in heart rate, oxygen uptake, and respiratory exchange, whereas the decrease in plasma insulin and glucose induced by exercise appeared to be prevented by the coca chewing. No evidence was found of acute coca leaf use increasing tolerance to exercise.[25]

The pharmacological activity and toxicity of coca is generally attributed to cocaine. In addition to its local anesthetic, central nervous system stimulant, and addictive (similar to amphetamines) properties, cocaine has many other activities (GOODMAN AND GILMAN; LIST AND HÖRHAMMER; MARTINDALE; MORTON 3). Little is known of the effects of the other alkaloids present in coca leaf.[26]

TOXICOLOGY

In a chronic feeding in rabbits, decocainized leaves of Trujillo coca leaves from which the majority of cocaine-like alkaloids were removed, doses of 21 or 210 mg produced no significant signs of toxicity. Rats fed the same leaf extract at 150 mg showed a significant decrease in weight gain, but not from doses of 1.5 and 15 mg. At all doses, the rats showed pyometra often with endometrial metaplasia, renal tubular calcification, and portal triaditis in the liver.[27] The LD_{50} of a coca leaf (*E. coca*) extract in male mice was 3450 mg/kg i.p.[28]

Biopsies of the buccal mucosa of chronic coca leaf chewers of Bolivia showed no evidence of carcinoma or chronic ulceration. Leukoedema was present in 76% of samples taken from the side of the mouth that the subjects placed their coca leaf quid and may have resulted from the known irritant property of the lime traditionally added to the leaves in the course of chewing.[29]

The fatal dose of cocaine in humans is reported to be about 1.2 g, but a dose as low as 20 mg (0.02 g) has been reported to cause severe toxic effects (GOODMAN AND GILMAN). Cocaine abusers show cerebral perfusion defects, cerebral thrombosis, and high incidences of neuropsychological impairment. (HIGGINS AND KATZ;).[30]

USES

Medicinal, Pharmaceutical, and Cosmetic.
Cocaine (free base or salt form) is used as a local anesthetic, mainly for eye (cornea), nose, and throat during surgery. Owing to ocular toxicity, its use in ophthalmology is limited and it has been replaced by other agents (MARTINDALE). Coca leaf and its extracts are not used in pharmaceutical preparations in the United States.

Food. Coca extract, from which cocaine is removed, is used together with extracts of kola (*Cola nitida*), cinnamon, ginger, lime, orange peel, and others as a flavor component in cola drinks; average maximum use level is reported to be 0.02%. Other food products in which the decocainized extract is used include alcoholic beverages, frozen dairy desserts, and candy. The highest average maximum use level is 0.055% in frozen dairy desserts.

Dietary Supplements/Health Foods. Leaf sold as a herbal tea in the United States.[31]

Traditional Medicine. Used in South America by natives to relieve hunger, fatigue, rheumatic pains, hangover, sore throat, asthma, stomach pain, constipation, nausea, cramps, diarrhea, general malaise, altitude sickness, mental disturbances, hemorrhage, amenorrhea, and other conditions; the leaves are usually chewed with alkaline substances such as plant ashes, lime from burnt seashells, bone, or limestone, often carried in a gourd. During this process, absorption of cocaine is increased.[1]

COMMERCIAL PREPARATIONS

Coca leaf formerly used in diverse products in the United States, including soft drinks, wine, cigarettes, cigars, syrups, cordials, chocolate tablets, and preparations for catarrh, hay fever, opium and morphine addiction, and timidity in young persons.[32] Coca was formerly official in U.S.P. Cocaine and cocaine hydrochloride are official in N.F. and U.S.P., respectively. Coca and cocaine are controlled as narcotic agents in the United States.

Regulatory Status. Decocainized essential oils, natural extractives, and solvent-free oleoresins of *E. coca* and other *Erythroxylum* species are GRAS (§182.20).

REFERENCES

See the General References for FEMA; GOODMAN AND GILMAN; GRIEVE; HORTUS 3rd; HUANG; MORTON 3; NANJING; TERRELL; UPHOF; USD 26th.

1. T. Plowman, in G. T. Prance and J. A. Kallunki, eds, *Ethnobotany in the Neotropics*, New York Botanical Garden, New York, NY, 1984, p. 62.
2. J. M. Moore and J. F. Casale, *J. Chromatogr. A*, **674**, 165 (1994).
3. J. M. Moore et al., *J. Chromatogr. A*, **659**, 163 (1994).
4. J. R. Ehleringer et al., *Nature*, **408**, 311 (2000).
5. B. Holmstedt et al., *Phytochemistry*, **16**, 1753 (1977).
6. G. H. Anilian et al., *J. Pharm. Sci.*, **63**, 1938 (1974).
7. G. Espinel Ovalle and I. Guzman Parra, *Rev. Colomb. Cienc. Quim. Farm.*, **1**, 95 (1971); through *Chem. Abstr.*, **76**, 89994s (1972).
8. E. Machado, *Raymondiana*, **5**, 5 (1972).
9. M. Sauvain et al., *J. Ethnopharmacol.*, **56**, 179 (1997).
10. T. Plowman and L. Rivier, *Ann. Bot. (London)*, **51**, 641 (1983).

11. J. F. Casale and J. M. Moore, *J. Forensic Sci.*, **39**, 1537 (1994).
12. M. Novak and C. A. Salemink, *Planta Med.*, **53**, 113 (1987).
13. J. A. Duke et al., *Bot. Mus. Leafl. Harv. Univ.*, **24**, 113 (1975).
14. E. L. Johnson, *Z. Naturforsch. C*, **48**, 863 (1993).
15. E. L. Johnson et al., *Biochem. Syst. Ecol.*, **26**, 743 (1998).
16. B. A. Bohm et al., *Syst. Bot.*, **7**, 121 (1982).
17. M. Bonefeld et al., *Phytochemistry*, **25**, 1205 (1986).
18. H. Kolodziej et al., *Phytochemistry*, **30**, 1255 (1991).
19. G. L. Vee et al., *Pharmacol. Biochem. Behav.*, **18**, 515 (1983).
20. J. A. Bedford et al., *Pharmacol. Biochem. Behav.*, **14**, 725 (1981).
21. C. Rerat et al., *J. Ethnopharmacol.*, **56**, 173 (1997).
22. H. Spielvogel et al., *J. Appl. Physiol.*, **80**, 643 (1996).
23. R. Favier et al., *J. Appl. Physiol.*, **80**, 650 (1996).
24. H. Spielvogel et al., *Eur. J. Appl. Physiol. Occup. Physiol.*, **75**, 400 (1997).
25. R. Favier et al., *J. Appl. Physiol.*, **81**, 1901 (1996).
26. M. Novak et al., *J. Ethnopharmacol.*, **10**, 261 (1984).
27. J. L. Valentine et al., *Vet. Hum. Toxicol.*, **7**, 21 (1988).
28. J. A. Bedford et al., *Pharmacol. Biochem. Behav.*, **17**, 1087 (1982).
29. J. E. Hamner Jr. and O. L. Villegas, *Oral Surg. Oral Med. Oral Pathol.*, **28**, 287 (1969).
30. T. R. Kosten, *Drug Alcohol Depend.*, **49**, 133 (1998).
31. R. K. Siegel et al., *JAMA*, **255**, 40 (1986).
32. L. Grinspoon and J. B. Bakalar, *J. Ethnopharmacol.*, **3**, 149 (1981).

COCILLANA BARK

Source: *Guarea rusbyi* (Britt.) Rusby (syn. *Sycocarpus rusbyi* Britt.) and closely related species (Family Meliaceae).

Common/vernacular names: Cocillana bark, grape bark, guapi, trompillo, and upas.

GENERAL DESCRIPTION

Trees native to the South American Andes. Part used is the dried bark. The original cocillana bark is believed to be derived from *G. rusbyi*. However, there is evidence that the current drug is obtained from other closely related *Guarea* species and is not the same as the cocillana first introduced into modern medicine (USD 23rd).[1] The bark is collected in Haiti and Bolivia (EVANS).

CHEMICAL COMPOSITION

Chemical studies on cocillana are limited. Cocillana bark is reported to contain small amounts (0.003–0.023%) of alkaloid(s) (rusbyine) of which the chemical structure(s) has not been determined. It also contains β-sitosterol, a volatile oil, tannin, anthraquinones, flavonols, terpenoids, and others (LIST AND HÖRHAMMER).[1]

PHARMACOLOGY AND BIOLOGICAL ACTIVITIES

Cocillana is reported to have expectorant and, in higher doses (1.3–3.0 g), emetic properties similar to those of ipecac (BRADLY; LIST AND HÖRHAMMER).[1] These properties are based on findings reported at the end of the last century. No recent pharmacological or toxicological data on cocillana are available.

USES

Medicinal, Pharmaceutical, and Cosmetic. Used in cough syrups and similar preparations as an alternative to ipecac, being particularly popular in the British Commonwealth countries.

Dietary Supplements/Health Foods. Seldom seen on the American market; common cough syrup ingredient in the United Kingdom (WREN).

Traditional Medicine. Used by natives in South America as an expectorant for alleviating coughs.[1] Among related species, the bark of *G. trichilioides* L. is used in Amazonian Brazil as an abortifacient, purgative, febrifuge, astringent, and anthelmintic. The bark of *G. spiciflora* Juss. is used in Brazilian folk medicine to treat dermatoses, hydropsy, and syphilis).

COMMERCIAL PREPARATIONS

Crude and extracts; the bark and fluid extract were formerly official in N. F. Strengths (see *glossary*) of extracts are expressed in weight-to-weight ratios.

REFERENCE

See the General References for EVANS; GOSSELIN; GRIEVE; MARTINDALE; MCGUFFIN 1 & 2; YOUNGKEN.
1. E. B. Ritchie and J. W. Steel, *Planta Med.*, **14**, 247 (1966).

COCOA (CACAO)

Source: *Theobroma cacao* L. subsp. *cacao* (Family Sterculiaceae or Byttneriaceae).

Common/vernacular names: Theobroma.

GENERAL DESCRIPTION

Evergreen tree with leathery oblong leaves, about 8 m high; fruits are berries borne directly on trunk and branches, with seeds within a mucilaginous pulp. Parts used are the seeds, which are commonly called cacao or cocoa beans. Cacao is generally used to describe the crude materials (e.g., cacao tree and cacao beans), while cocoa is used to describe the processed products. However, it is increasingly common to use the term cocoa for both crude and processed products, and thus cocoa tree, cocoa beans, cocoa powder, cocoa butter, and so on.[1,2]

There are three varieties of cacao: forastero, criollo, and trinitario. Forastero accounts for more than 90% of the world's usage and is produced primarily in West African countries (e.g., Ghana, Nigeria, Cameroon, and the Ivory Coast), while the criollo variety is produced in Venezuela and Central America, as well as Papua New Guinea, Java, and Samoa. Trinitario is believed to be a hybrid of the other two varieties and is produced in Venezuela, Trinidad, Sri Lanka, and other countries. Both criollo and trinitario cacao are considered to have better flavor qualities than forastero cacao. Thus, criollo is blended with forastero to improve the flavor of forastero in the manufacture of cocoa and, like trinitario, is also used in certain high-quality eating chocolates.[1]

Three main types of ingredients are produced from cacao seeds: cocoa powder, cocoa butter, and cocoa extracts. For the manufacture of these products, the cacao beans are first cured by fermentation and drying, during which time the pulp surrounding the seeds is decomposed and removed and flavor precursors develop in the seeds. The dried beans, now called raw cocoa containing about 6–8% moisture, are roasted to produce the required flavor, aroma, and color and to facilitate the removal of the seed coat (shell); temperatures vary from 100 to 150°C, depending on the types of beans and the products to be made. Beans for manufacturing cocoa butter or chocolate are roasted at lower temperatures, while those for cocoa powder production are roasted at higher temperatures. After roasting, the shell and hypocotyl are separated from the cotyledons (called nibs).

The nib, containing about 55% cocoa butter, is ground while hot to a liquid mass called cocoa or chocolate liquor, from which variable amounts of the cocoa butter is removed by hydraulic pressing. The cocoa cake left on the filter is cooled and then ground to a fine powder under controlled cool temperatures to yield cocoa powder that has cocoa fat contents of up to 22% or more.

Currently, most cocoa powders are produced by the so-called Dutch or alkalized process, in which the nib is treated with a warm aqueous solution of up to three parts of anhydrous potassium carbonate to 100 parts of nib (or equivalent amounts of other alkalis such as potassium bicarbonate and hydroxide; carbonates, bicarbonates, and hydroxides of sodium, magnesium, and ammonium; or their combinations). After the alkali is completely absorbed, the nib is processed as in the above method to yield alkalized cocoa powder. Alkalized cocoa is considered to have improved dispersibility, color, and flavor over unalkalized cocoa.

Cacao nibs, cocoa powder, and certain other cocoa products (e.g., chocolates) are governed by standards of identity set forth in the Code of Federal Regulations (21 CFR §§163.110–163.155). For example, cacao nibs used for cocoa manufacture are required to have no more than 1.75% cacao shell. Breakfast cocoa (or high-fat cocoa) must have at least 22% cacao fat (cocoa butter), cocoa (or medium-fat cocoa) must contain less than 22% but not less than 10%, and low-fat cocoa must contain less than 10% cacao fat. Alkalized cocoa must be labeled "processed with alkali," and so on.

Cocoa butter (also called cacao butter and theobroma oil) is produced commonly by three methods: hydraulic pressing, extrusion or expeller pressing, and solvent extraction. Cocoa butter produced by the first two methods has a faint chocolate flavor and aroma that can be removed by steam distillation under vacuum; it is brittle at temperatures below 25°C and melts at 34–35°C.[1,2]

Cocoa extracts are generally prepared by extraction of the roasted seeds (nibs) with hydroalcoholic solvents; an essential oil is also produced by steam distillation.

CHEMICAL COMPOSITION

Cocoa contains more than 300 volatile compounds, including hydrocarbons, monocarbonyls, pyrroles, pyrazines, esters, lactones, and others.

The important flavor components are reported to be aliphatic esters, polyphenols, unsaturated aromatic carbonyls, pyrazines, diketopiperazines, and theobromine.[3–5] Cocoa also contains about 18% proteins (ca. 8% digestible);[2,6,7] fats (cocoa butter); amines and alkaloids, including theobromine (0.5–2.7%), caffeine (ca. 0.25% in cocoa; 0.07–1.70% in fat-free beans, with forasteros containing less than 0.1% and criollos containing 1.43–1.70%),[8–10] tyramine, dopamine, salsolinol,[11,12] trigonelline, nicotinic acid, and free amino acids;[9] tannins; phospholipids;[13] starch and sugars;[2] minerals (particularly high in sodium or potassium in alkalized cocoa); and others (MARTINDALE; MORTON 3; WATT AND MERRILL).[14]

The characteristic bitter taste of cocoa is reported to be due to the diketopiperazines (especially those containing phenylalanine) reacting with the theobromine present during roasting.[5]

Cocoa butter contains mainly triglycerides of fatty acids that consist primarily of oleic

($C_{18:1}$), stearic ($C_{18:0}$), and palmitic ($C_{16:0}$) in decreasing concentrations, with small amounts of linoleic ($C_{18:2}$) and arachidic ($C_{20:2}$) acids. Over 73% of the glycerides are present as monounsaturated forms (oleopalmitostearin and oleodistearin), the remaining being mostly diunsaturated glycerides (palmitodiolein and stearodiolein) with lesser amounts of fully saturated and triunsaturated (triolein) glycerides. Linoleic acid levels have been reported to be up to 4.1% (MERCK; MORTON 3).[2,15,16] Also present in cocoa butter are small amounts of sterols and methylsterols; sterols consist mainly of β-sitosterol, stigmasterol, and campesterol, with a small quantity of cholesterol (0–0.28%) present.[17–19]

In addition to alkaloids (mainly theobromine), tannins, and other constituents, cocoa husk contains a pigment that is a polyflavone glucoside with a molecular weight of over 1500. This pigment is claimed to be heat and light resistant, highly stable at pH 3–11, and useful as a food colorant; it was isolated at a 7.9% yield.[20]

PHARMACOLOGY AND BIOLOGICAL ACTIVITIES

Theobromine, the major alkaloid in cocoa, has similar pharmacological activities as caffeine. However, its stimulant activities on the central nervous system, respiration, and skeletal muscles are much weaker than those of caffeine, but its cardiac stimulant, coronary dilating, smooth muscle relaxant, and diuretic properties are stronger (GOODMAN AND GILMAN).

TOXICOLOGY

Cocoa butter has been reported to have skin allergenic and comedogenic (forming blackheads) properties in animals.[21,22]

Depending on the alkali used, cocoa powder produced by the Dutch process may contain relatively high concentrations of sodium, which may cause problems in persons who are on a low-sodium diet.

USES

Medicinal, Pharmaceutical, and Cosmetic. Cocoa powder (or cocoa syrup) is used in flavoring pharmaceutical preparations.

Cocoa butter is used extensively as a suppository and ointment base; also used as emollient, skin softener, and skin protectant in creams (e.g., massage), lotions, lipsticks, and soaps, among others.

Food. Beverages made from cacao flavored with vanilla and other spices have been used by native Mexicans (Aztecs) for centuries.

Cocoa powder is used extensively as a flavor or nutrient component in nonalcoholic beverages, ice cream, cakes, biscuits, and others.

Cocoa butter is extensively used in chocolate manufacture, where it is mixed with cocoa liquor (ground cacao nibs), sugar, milk, and other ingredients such as flavors. Dark chocolate does not contain milk.

Cocoa extract is used in both alcoholic (liqueurs such as creme de cacao) and nonalcoholic beverages, frozen dairy desserts, candies, baked goods, and others.

Dietary Supplements/Health Foods. Cocoa butter is used in creams, massage oils, and other cosmetic preparations sold in health food stores (ROSE).

Traditional Medicine. Cocoa butter is used to treat neck wrinkles on neck (turkey neck), around the eyes, and at the corners of the mouth (ROSE). Reportedly used in European tradition in combination with other ingredients for infectious intestinal disease, diarrhea; bronchial expectorant in asthma, bronchitis, irritating cough, and lung congestion; to regulate function of endocrine glands, especially the thyroid.[23]

Others. Cocoa and cocoa butter have been reported to contain fat-soluble antioxidants and could be a source of such substances.[24]

COMMERCIAL PREPARATIONS

Cocoa powders, cocoa butter, cocoa syrup, and cocoa extracts. Cocoa (10–22% fat), cocoa butter, and cocoa syrup are official in N.F.

Regulatory Status. Standards of identity for cocoa products apply (§§163.110–163.155). Cocoa extracts are GRAS (§182.20). Cocoa seed is the subject of a German therapeutic monograph; not recommended as claimed efficacy is unsubstantiated; allowed as flavoring agent.[23]

REFERENCES

See the General References for BAILEY 2; BLUMENTHAL 1; GOSSELIN; GRIEVE; HORTUS 3rd; KARRER; MARTINDALE; TERRELL; USDA.

1. R. J. Clarke and J. W. Drummond in L. W. Codd et al., eds., *Chemical Technology: An Encyclopedic Treatment*, Vol. 7, Barnes & Noble, New York, 1975, p. 645.
2. B. D. Powell and T. L. Harris in A. Standen, ed., *Kirk-Othmer Encyclopedia of Chemical Technology*, Vol. **5**, Wiley–Interscience, New York, 1964, p. 363.
3. D. Reymond, *Chemtech*, **7**, 664 (1977).
4. P. G. Keeney, *J. Am. Oil Chem. Soc.*, **49**, 567 (1972).
5. W. Pickenhagen et al., *Helv. Chim. Acta*, **58**, 1078 (1975).
6. D. J. Timbie and P. G. Keeney, *J. Agric. Food Chem.*, **25**, 424 (1977).
7. D. L. Zak and P. G. Keeney, *J. Agric. Food Chem.*, **24**, 483 (1976).
8. Y. Asamoa and J. Wurziger, *Gordian*, **76**, 138 (1976).
9. G. Barbiroli, *Atti Cong. Qual.*, **6** 149 (1968).
10. M. Mironescu, *Rev. Fiz. Chim. Ser. A*, **11**, 218 (1974).
11. T. M. Kenyhercz and P. T. Kissinger, *Phytochemistry*, **16**, 1602 (1977).
12. R. M. Riggin and P. T. Kissinger, *J. Agric. Food Chem.*, **24**, 900 (1976).
13. L. Biino and E. Clabot, *Atti Soc. Peloritana Sci. Fis. Mat. Natur.*, **16**, 257 (1970).
14. T. M. Kenyhercz and P. T. Kissinger, *Lloydia*, **41**, 130 (1978).
15. R. F. Looney in L. W. Codd et al., eds., *Chemical Technology: An Encyclopedic Treatment*, Vol. 8, Barnes & Noble, New York, 1975, p. 1.
16. Y. Asamoa and J. Wurziger, *Gordian*, **74**, 280 (1974).
17. H. Chaveron, *Choc. Confiserie Fr.*, **273**, 12 (1971).
18. T. Itoh et al., *J. Am. Oil Chem. Soc.*, **50**, 300 (1973).
19. T. Itoh et al., *Oleagineux*, **29**, 253 (1974).
20. K. Kimura et al., Jpn. Kokai 73 17, 825 (1973).
21. V. V. Ivanov, *Vestn. Dermatol. Venerol.*, **3**, 57 (1976).
22. O. H. Mills et al., *Br. J. Dermatol.*, **98**, 145 (1978).
23. Monograph *Cacao semen*, Bundesanzeiger, no. 40 (February 27, 1991).
24. P. A. Dewdney and M. L. Meara, *Sci. Tech. Surv. Br. Food Manuf. Ind. Res. Assoc.*, 96 (1977).

CODONOPSIS

Source: *Codonopsis pilosula* (Franch.) Nannf., *C. pilosula* Nannf. var. *modesta* (Nannf.) L. T. Shen, *C. tangshen* Oliv., *C. tubulosa* Kom., and many other *Codonopsis* species (Family Campanulaceae).[1]

Common/vernacular names: Radix codonopsis, bonnet bellflower, bastard ginseng, and *dangshen*.

GENERAL DESCRIPTION

Mostly small herbaceous perennials, strongly scented, with thick fleshy cylindrical to slightly spindle-shaped roots; native to Asia; distributed throughout China, including the provinces of Shanxi, Shaanxi, Sichuan, Yunnan, Xinjiang, Gansu, Jilin, and Liaoning; now extensively cultivated, also as ornamental in the United States. Part used is the root, collected in autumn from wild or cultivated plants at least 3 years old, washed clean. It is sorted and strung out to sun dry to about half dry, then massaging or rubbing by hand or between two wood boards to bring internal tissues together, followed by further drying and rubbing until completely dry; this process eliminates or minimizes the presence of air space, cracks, or holes in the dried herb, which would not keep well (CMH). There are five major types each with different grades: *xidang* (western codonopsis) from Gansu; *dongdang* (eastern codonopsis) from Jilin, Liaoning, and Heilongjiang; *ludang* (Shanxi codonopsis) from Shanxi; *chuandang* or *tiaodang* (Sichuan codonopsis) from Sichuan, Hubei, and Shaanxi; and *baidang* or *guanhua dangshen* (white or tubular-flowered codonopsis) from Guizhou and Yunnan (CMH; ZHU). *Ludang* is the most often encountered in the United States.

Considered as the poor man's ginseng, it is frequently used as a substitute for ginseng.

CHEMICAL COMPOSITION

Most of the chemical studies on codonopsis were performed during the 1980s. Compounds identified include polysaccharides and sugars (e.g., inulin, starch, glucose, sucrose, fructose);[2,3] 1.24–10.38% saponins (tangshenosides I, II, III, and IV);[3–6] amino acids (1.47–5.33%), with ca. 0.1% as free amino acids;[4,5,7] a β-carboline alkaloid (perlolyrine) and other nitrogen compounds (choline, nicotinic acid, *n*-butylallophanate, etc.);[8] triterpenes (e.g., taraxeryl acetate, taraxerol, friedelin) and sterols (stigmasterol and spinasterol) and their glucosides;[9–11] atractylenolides II and III (see ***baizhu***);[3] scutellarein glucoside; oroxylin A;[12] volatile oil (ca. 0.12%) composed of 50% acidic compounds, predominantly methyl palmitate;[13] and trace minerals, among others.[4,5]

Four polysaccharides (CP-1, CP-2, CP-3, and CP-4) have been isolated with molecular weights of 10,500, 12,000, 14,000, and 79,000, respectively; the first two have β-glycosidic linkages, while the latter two have α-linkages, involving an unusually large number of sugars (glucose, fructose, galactose, arabinaose, mannose, xylose, and rhamnose).[2]

No ginseng saponins have been found in codonopsis.

PHARMACOLOGY AND BIOLOGICAL ACTIVITIES

As a frequent substitute for ginseng, *dangshen* has many of the properties of ginseng. Its biological activities include central stimulation in mice, weight gain in rabbits, increased swimming time in mice, prevention of leukocytosis induced by turpentine oil in experimental animals, increased tolerance to anoxia and elevated temperatures in mice, and prolonged survival, radioprotective, enhanced phagocytosis of macrophages, immunoregulating, improved blood picture in mice (red and white cells and hemoglobin all increased), increased serum corticosterone in mice, hypotensive and peripheral vasodilatory as well as adrenolytic, and stimulation as well as relaxation of isolated guinea pig ileum (IMM-2; WANG).[5,14,15]

Dangshen polysaccharides have exhibited immunomodulating effects in guinea pigs, inhibited experimental ulcers (stress, indomethacin, acetic acid, and pyloric ligation models) in rats, and had antistress effects (prolonging swimming time and increasing tolerance to anoxia and to elevated temperatures) in mice, and so on.[16–18]

Oroxylin A was shown to have antihistaminic effects on isolated guinea pig ileum.[12]

USES

Dietary Supplements/Health Foods. Powdered herb and extracts are used in tonic formulas (in tablet, capsule, or liquid form), often as oriental ginseng substitute, for boosting immune system and replenishing *qi* (vital energy); cut or tea bag cut herb is used in tea or soup mixes (FOSTER AND YUE).

Traditional Medicine. A relatively recent addition to Chinese materia medica, *dangshen* was first described in the *Ben Jing Feng Yuan* (AD 1695) as a lung-clearing (*qing fei*) drug with a sweet taste and neutral nature. Its tonic properties were later described in the *Ben Cao Cong Xin* (1757) and has since become a highly valued *qi* tonic of equal status as some ancient ones such as astragalus, ginseng, and common jujube (see those entries). It is used to treat many of the same conditions as these tonics, including general weakness, lack of appetite, chronic diarrhea, shortness of breath, palpitations, asthma, cough, thirst and diabetes, conditions due to spleen and blood deficiencies, and damaged *qi*.

COMMERCIAL PREPARATIONS

Crude (whole or powdered) and extracts, with no uniform standards.

Regulatory Status. Class 1 dietary supplement (herbs can be safely consumed when used appropriately).

REFERENCES

See the General References for CHP; CMH; FOSTER AND YUE; IMM-2; JIXIAN; JIANGSU; LU AND LI; NATIONAL; WANG; ZHU.

1. Z. T. Wang and G. J. Xu, *Zhongcaoyao*, **23**, 144 (1992).
2. S. J. Zhang and S. Y. Zhang, *Zhongcaoyao*, **18**(3), 2 (1987).
3. D. G. Cai, *Zhongguo Zhongyao Zazhi*, **16**, 376 (1991).
4. S. M. Wang et at., *Shanxi Zhongyi*, **6**(3), 35 (1990).
5. S. M. Wang and Y. Yang, *Shanxi Zhongyi*, **5**(1), 37 (1989).
6. G. R. Han et al., *Zhongguo Zhongyao Zazhi*, **15**(2), 41 (1990).
7. Y. X. Gong, *Zhongcaoyao*, **18**(11), 37 (1987).
8. T. Liu et al., *Planta Med.*, **54**, 472 (1988).
9. M. P. Wong et al., *Planta Med.*, **49**, 60 (1983).
10. G. R. Han et al., *Zhongcaoyao*, **22**, 422 (1991).
11. Y. Z. Wang et al., *Zhongcaoyao*, **17**(5), 41 (1986).
12. D. X. Zhou et al., *Zhongguo Zhongyao Zazhi*, **16**, 564 (1991).
13. J. Liao and Y. Q. Lu, *Zhongcaovao*, **18**(9), 2 (1987).
14. X. L. Mao et al., *Chin. J. Integr. Trad. West. Med.*, **5**, 739 (1985).
15. C. O. Ling, *Henan Zhongyi*, **13**, 94 (1993).
16. M. X. Zhuang et al., *Zhongguo Yaoxue Zazhi*, **27**, 653 (1992).
17. J. C. Cui et al., *Zhongcaoyao*, **19**(8), 21 (1988).
18. W. Li et al., *Jilin Zhongyiyao*, (6), 33 (1990).

COFFEE

Source: *Coffea arabica* L., *C. canephora* Pierre ex Froehner (syn. *C. robusta* Linden ex De Wild.), and other *Coffea* species, varieties or hybrids (Family Rubiaceae).

Common/vernacular names: Arabica, Arabian, Colombian, or Santos coffee (*C. arabica*); and robusta coffee (*C. robusta*).

GENERAL DESCRIPTION

Evergreen shrubs to small trees with two-seeded, deeply crimson fruits (berries) that are commonly called "cherries"; *C. arabica* up to 6 m and *C. canephora* to 8 m high; believed to be native to Ethiopia; now extensively cultivated in tropical and subtropical countries. Parts used are the roasted seeds, commonly called "beans".

During the production of coffee beans the freshly picked ripe cherries are either sun dried (requiring 2–3 weeks), followed by mechanical removal of the dried husk (pulp, skin, etc.) and seed coat, or they are placed in water and subjected to pulping machines to remove most of the pulp, followed by fermentation (requiring up to several days), drying, and mechanical removal of the silver skin. The former method is called the dry process, and the latter is called the wet process, producing respectively the so-called natural and washed coffees. The dried beans at this stage are known as green coffee and are exported.[1,2]

Arabica coffee is produced mostly in South and Central America, particularly Brazil, Colombia, Mexico, and Guatemala, while robusta coffee is produced mainly by African countries (Ivory Coast, Uganda, Angola, etc.).[1,2]

In the United States, Colombian and Central American coffees are preferred over Brazilian and African coffees.

To develop the characteristic coffee aroma and taste, the green coffee is roasted to the required time at temperatures up to about 220°C, depending on the types of coffee beans to be produced. During roasting, the beans acquire their typical flavor, at the same time turning dark and slightly increasing in size (swelling) as well as losing weight (due to loss of moisture, carbon dioxide, and other volatile compounds from pyrolysis). Coffee beans are often blended before or after roasting to produce various commercial grades or brands.[1,2]

Decaffeinated coffee is produced by removing most of its caffeine content while at the green coffee stage, generally by extraction of the whole beans with organic (e.g., chlorinated) solvents. The beans are then rid of solvent and roasted.[1,2]

Instant coffee is produced by extracting ground-roasted coffee with hot water, often under pressure. The extract is concentrated and freeze-dried or spray-dried to produce a granular or powdered product. For instant coffee manufacture, robusta coffees are more commonly used due probably to their higher contents of soluble materials and thus giving higher yields.[1,2]

Coffee extracts for flavoring purposes are prepared by extracting roasted coffee with water or water–alcohol mixtures.[1,2]

CHEMICAL COMPOSITION

Green coffee contains 0.6–3.2% (usually 1.5–2.5%) caffeine; 0.3–1.3% trigonelline; 5–10% chlorogenic acid (robusta more than arabica);[3] 7.4–17% oil called coffee oil (arabica more than robusta);[1,2,4] up to 60% carbohydrates (mostly a galactomannan);[5–7] protein (ca. 12%); about 2% free amino acids consisting mainly of glutamic and aspartic acids and asparagine;[8] polyamines (putrescine, spermine, and spermidine);[9] tannins (ca. 9%); B vitamins and trace of niacin; and others (LIST AND HÖRHAMMER; MERCK; MORTON 3; WATT AND MERRILL).[1,2,10]

Coffee oil contains mainly glycerides of fatty acids (e.g., linoleic, palmitic, oleic, and stearic acids, with the first two in predominant concentrations) and 5–8% of unsaponifiable matter, which consists of squalene, *n*-nonacosane, lanosterol, cafestol, cahweol, sitosterol, stigmasterol, methylsterols, tocopherols (α, β, and γ, with the last two being predominant), and others (LIST AND HÖRHAMMER).[4,11–13]

Roasted coffee contains slightly less caffeine than green coffee, but contains much lower concentrations of trigonelline, chlorogenic acid, tannins, polyamines, proteins, and sugars, which are degraded and involved in flavor formation during roasting. The most important flavor precursors are reported to be trigonelline, sugars, free amino acids, and peptides.[1,9,10,14]

More than 100 aroma compounds have been identified in roasted coffee, including such important flavor contributors as furan derivatives, pyrazines, pyrroles, oxazoles, and acids (LIST AND HÖRHAMMER).[1,10]

Roasted coffee contains a relatively high content of niacin, and coffee has been suggested as a source of niacin, for treating pellagra or niacin deficiency.[10,15] A cup of coffee contains about 100 mg caffeine, which is within the therapeutic dose range.

PHARMACOLOGY AND BIOLOGICAL ACTIVITIES

The physiological activities of coffee are generally attributed to its caffeine. Caffeine is a powerful stimulant of the central nervous system, respiration, and skeletal muscles; other activities include cardiac stimulation, coronary dilation, smooth muscle relaxation, and diuresis.

Apart from the biological activities of caffeine, those of chlorogenic acid (which is present in substantial quantities in coffee) should not be ignored, as chlorogenic acid is reported to have stimulant, diuretic, and choleretic properties (see *artichoke* and *honeysuckle*); it also has allergenic properties (MORTON 3).

TOXICOLOGY

The fatal dose of caffeine in humans is reported to be 10 g. A dose of 1 g or more would produce toxic effects, including headache, nausea, insomnia, restlessness, excitement, mild delirium, muscle tremor, tachycardia, and extrasystoles.

In addition to the above well-known activities, caffeine has been reported to have many other activities, including mutagenic, teratogenic, and carcinogenic activities; it is also reported to cause temporary increase in intraocular pressure, to have calming effects on hyperkinetic children (effect similar to methyl phenidate or dextroamphetamine), and to cause chronic recurrent headache, among others.[16,17] More than 50% of the total mutagenic activity of coffee can be attributed to the activity of methylglyoxal.[18] Coffee drinking has also been linked to myocardial infarction (a kind of blood clot in blood vessels that supply blood to heart muscles), cancer of the lower urinary tract (e.g., bladder), ovaries, prostate, and others.[19,20] However, most of these findings were disputed by later reports (GOODMAN AND GILMAN; MARTINDALE; USD 23rd).[15,21,22]

Mutagenic activity of coffee is inactivated by sodium sulfite (completely suppressing the mutagenicities of the 1,2 dicarbonyls, diacetyl and glyoxal) as well as by sodium bisulfite and metabisulfite. Sodium sulfite also inactivates the phage-inducing activity of coffee. It has been suggested that sulfites should be added to coffee to reduce mutagenicity.[23]

Coffee (even decaffeinated) is reported to stimulate gastric secretion and should be taken only with proper precautions (e.g., with cream or during meals) by individuals with peptic ulcer (GOODMAN AND GILMAN; MARTINDALE).

USES

Medicinal, Pharmaceutical, and Cosmetic. In addition to its use as a central and respiratory stimulant, usually as caffeine, U.S.P., and caffeine and sodium benzoate injection, U.S.P., caffeine is extensively used as an ingredient in many types of pharmaceutical preparations, particularly internal analgesics, cold and allergy products, weight-control formulations (appetite depressants), and others.

Food. Coffee has been used for centuries by various cultures as a beverage to stay alert and to improve work efficiency.

Coffee extract (type not specified) is widely used as a flavor ingredient in many food products, including alcoholic (e.g., liqueurs) and nonalcoholic beverages, frozen dairy desserts, candy, baked goods, gelatins and puddings, sweet sauces, and milk products. Highest average maximum use level reported is about 2.8% (28,216 ppm) in baked goods.

Caffeine is extensively used in nonalcoholic beverages (particularly colas), with reported average maximum use level of about 0.014% (141 ppm). It is also used in frozen dairy desserts, candy, gelatins and puddings, and baked goods. Average maximum use level reported is 0.04% in all except the last category that is about 0.007% (68 ppm).

COMMERCIAL PREPARATIONS

Extracts (e.g., fluid, solid, and tincture) and natural caffeine. Caffeine is official in U.S.P. and F.C.C.

Regulatory Status. Coffee extracts (§182.20) and caffeine (§182.1180) are GRAS.[17] Coffee charcoal, consisting of the milled, roasted to blackened, carbonized outer parts of green dried fruits is the subject of a German therapeutic monograph for treatment of nonspecific acute diarrhea.[24]

REFERENCES

See the General References for APhA; BAILEY 1; BLUMENTHAL 1; BRUNETON; FEMA; GOSSELIN; GRIEVE; LIST AND HÖRHAMMER; MARTINDALE; TERRELL; USDA; YOUNGKEN.

1. R. G. Moores and A. Stefanucci in A. Standen, ed., *Kirk-Othmer Encyclopedia of Chemical Technology*, Vol. 5 Wiley–Interscience, New York, 1964, p. 748.
2. R. J. Clarke and J. W. Drummond in L. W. Codd et al., eds., *Chemical Technology: An Encyclopedic Treatment*, Vol. 7 Barnes & Noble, New York, 1975, p. 645.
3. H. Vilar and L. A. B. Ferreira, *Coll. Int. Chim. Cafes (C. R.)*, **6**, 135 (1974).
4. P. Folstar et al., *J. Agric. Food Chem.*, **25**, 283 (1977).
5. V. Ara and H. Thaler, *Z. Lebensm. Unters. Forsch.*, **161**, 143 (1976).
6. M. Asante and H. Thaler, *Chem. Mikrobiol. Technol. Lebensm.*, **4**, 110 (1975).
7. V. Ara and H. Thaler, *Z. Lebensm. Unters. Forsch.*, **164**, 8 (1977).
8. W. Walter et al., *Naturwissenschaften*, **57**, 246 (1970).
9. H. V. Amorim et al., *J. Agric. Food Chem.*, **25**, 957 (1977).
10. D. Reymond, *Chemtech*, **7**, 664 (1977).
11. E. Cerma and P. Baradel, *Atti Cong. Qual.*, **6**, 321 (1967).
12. J. Wurziger, *Fette, Seifen, Anstrichmit.*, **79**, 334 (1977).
13. B. A. Nagasampagi et al., *Phytochemistry*, **10**, 1101 (1971).
14. R. Viani and L Horman, *J. Food Sci.*, **39**, 1216 (1974).
15. G. Czok, *Z. Ernaehrungswiss.*, **16**, 248 (1977).
16. N. Loprieno et al., *Mutat. Res.*, **21**, 275 (1973).
17. M. T. O'Brien, *Food Prod. Dev.*, **12** (9), **86** (1978).
18. H. Kasai et al., *Gann*, **73**, 381 (1982).
19. P. Stocks, *Br. J. Cancer*, **24**, 215 (1970).
20. D. H. Shennon, *Br. J. Cancer*, **28**, 473 (1973).
21. D. Simon et al., *J. Natl. Cancer Inst.*, **54**, 587 (1975).
22. H.-P. Wuerzner et al., *Food Cosmet. Toxicol.*, **15**, 289 (1977).
23. Y. Suwa et al., *Mutat. Res.*, **102**, 383 (1982).
24. Monograph *Coffea carbo*, Bundesanzeiger, no. 85 (May 5, 1988).

COMFREY

Source: *Symphytum officinale* L.; *Symphytum* × *uplandicum* Nym.; *S. asperum* Lepechin. (Family Boraginaceae).

Common/vernacular names: Common comfrey, Russian comfrey (*S.* × *uplandicum*), prickly comfrey (*S. asperum*), and blackwort.

GENERAL DESCRIPTION

Perennial herbs with branching stems and thick root. Parts used are the dried rhizome and root; also the leaves.

Common comfrey (*S. officinale*) is an erect, stout, often branched perennial, to 1 m; leaves broadly lance shaped; middle and upper ones sessile, but at point of insertion extend downward on stalk; stalk distinctly winged; flowers variable from white or cream, yellowish to rose, pink, or light to dark violet; anther about as wide as filament; petal lobes recurved; calyx segments distinctly lanceolate; nutlets smooth; occurs in moist grasslands and riverbanks most of Europe, is rare in extreme south, naturalized alien in northern Europe; eastern North America.

Russian comfrey (*S.* × *uplandicum*), a hybrid of *S. officinale* and *S. asperum*, robust perennial, 1–2 m. Intermediate between parents. Leaves narrow winged on the main stalk, ending between internodes; flowers dark violet to blue, or pinkish to pink-blue. Native to northern Europe, mostly persisted after cultivation.

Prickly comfrey, (*S. asperum*), stems not winged, upper leaves on short stalks; flowers rose to bluish; anthers significantly shorter than filaments; from Southwest Asia; naturalized in Europe, and the eastern United States (rare).

CHEMICAL COMPOSITION

S. officinale root contains 0.75–2.55% allantoin;[1–3] about 0.3% alkaloids, including the pyrrolizidine alkaloids symphytine, echimidine, heliosupine, viridiflorine, echinatine, 7-acetyllycopsamine, 7-angelylretronecine viridiflorate, lasiocarpine, and acetylechimidine;[3–5] the presence of lasiocarpine is questioned;[6] lithospermic acid;[7] 29% mucopolysaccharide that is composed of glucose and fructose;[8] a gum consisting of L(−)-xylose, L-rhamnose, L-arabinose, D-mannose, and D-glucuronic acid;[9] pyrocatechol tannins (2.4%); 0.63% carotene; glycosides, sugars; isobauerenol, β-sitosterol, and stigmasterol; steroidal saponins; triterpenoids; rosmarinic acid, and others.[1,4,10,11]

S. asperum contains the pyrrolizidine alkaloids asperumine, echinatine, heliosupine, 7-acetyllycopsamine, and acetylechimidine.[5] *S.* × *uplandicum* contains symphytine, symlandine, echimidine; 7-acetyllycopsamine, 7-angelylintermidine, uplandicine, lycopsamine, and intermedine.[5]

Leaves also contain substantial quantities of allantoin,[2,3] alkaloids (ca. 0.15%),[12] and possibly other similar constituents as the root.

PHARMACOLOGY AND BIOLOGICAL ACTIVITIES

Allantoin is reported to have healing properties (MARTINDALE).

An aqueous extract of comfrey containing lithospermic acid and other common plant acids has been reported to exhibit antigonadotropic activity in mice,[13] though lithospermic acid itself has no such activity unless oxidized by a plant phenol oxidase preparation.[7]

Comfrey is reported to have anti-inflammatory properties.[14] It is also considered to have many beneficial properties, including astringent, demulcent, emollient, hemostatic, and expectorant properties, among others.

Anti-inflammatory efficacy has been confirmed, correlating clinical and analytic data of topical comfrey products by measuring redness and pain sensitivity *in vivo*, to the allantoin and rosmarinic acid fractions of preparations.[10]

TOXICOLOGY

The root and leaves of *S. officinale* have been found to be carcinogenic in rats.[15]

Pyrrolizidine alkaloids from *S.* × *uplandicum* have also been found to cause chronic hepatotoxicity.[15] Veno-occlusive disease from ingestion of various comfrey species, including leaves and roots, has been clearly documented in humans.[6,16–18]

USES

Medicinal, Pharmaceutical, and Cosmetic. Comfrey root and leaves and their extracts are used as ingredients in various types of cosmetic preparations such as lotions, creams, ointments, eyedrops, hair products, and others.

Food. Young shoot and leaves have been used as vegetables, though that use is currently discouraged because of pyrrolizidine alkaloid toxicity.

Dietary Supplements/Health Foods. The root and leaves of various *Symphytum* species have been sold and labeled as "*Symphytum officinale*."[19] Formerly widely available in teas, capsules, tablets, tinctures, extracts, and so on. Many product manufacturers have withdrawn comfrey products from sale because of toxicity. Topical products, including salves, ointments, and balms, are still widely available (FOSTER).

Traditional Medicine. A root decoction is reportedly used as a gargle or mouthwash for throat inflammations, hoarseness, and bleeding gums. The root, in one form or another, is also used to treat a wide variety of ailments such as gastrointestinal problems (e.g., ulcers), excessive menstrual flow, diarrhea, dysentery, bloody urine, persistent cough, bronchitis, cancers, and others.[20]

Externally, the powdered root is used as a hemostatic and in poulticing wounds, bruises, sores, and insect bites. The mucilage is believed to help soften the skin when used in baths.[21,22]

COMMERCIAL PREPARATIONS

Mainly as crude; no uniform standards in extracts (see "Strength" in ***glossary***).

Regulatory Status. The United Kingdom, Australia, Canada, and Germany have restricted the availability of products containing comfrey. In July 2001, the FDA advised dietary supplement manufactures not to use comfrey due to safety concerns. Class 2a (for external use only).

REFERENCES

See the General References for BAILEY 1; BLUMENTHAL 1; FOSTER; GOSSELIN; LUST; MARTINDALE; ROSE; UPHOF.

1. G. V. Makarova et al., *Farm. Zh. (Kiev)*, **21**(5), 41 (1966).
2. D. Fijalkowski and M. Seroczynska, *Herba Pol.*, **23**, 47 (1977).
3. T. Furuya and K. Araki, *Chem. Pharm. Bull.*, **16**, 2512 (1968).
4. T. Furuya and M. Hikichi, *Phytochemistry*, **10**, 2217 (1971).
5. L. W. Smith and C. C. J. Culvenor, *J. Nat. Prod.*, **44**(2), 29 (1981).
6. D. V. C. Awang, *Can. Pharm. J.*, 101 (1987).
7. H. Wagner et al., *Arzneim.-Forsch.*, **20**, 705 (1970).
8. G. Franz, *Planta Med.*, **17**, 217 (1969).
9. Z. Michalska and T. Jakimowicz, *Farm. Pol.*, **25**, 185 (1969).
10. R. Andres et al., *Planta Med.*, **55**, 643 (1989).
11. V. U. Ahmad et al., *J. Nat. Prod.*, **56**(3), 329 (1993).
12. I. V. I. Man'ko et al., *Rast. Resur.*, **5**, 508 (1969).

13. I. S. Kozhina et al., *Rast. Resur.*, **6**, 345 (1970).
14. G. Furnadzhiev et al., *Stomatologiya (Sofia)*, **58**, 37 (1976).
15. 1. Hirono et al., *J. Natl. Cancer Inst.*, **61**(3), 865 (1978).
16. P. M. Ridker et al., *Gastroenterology*, **88**, 1050 (1985).
17. R. Huxtable et al., *N. Engl. J. Med.*, **315**, 1095 (1986).
18. P. M. Ridker, *Lancet*, **1**, 657 (1989).
19. R. Huxtable, *Am. J. Med.*, **89**, 548 (1990).
20. J. L. Hartwell, *Lloydia*, **31**, 71 (1968).
21. Monograph *Symphyti herba-folium*, *Bundesanzeiger*, no. 138 (July 27) 1990.
22. Monograph *Symphyti radix*, *Bundesanzeiger*, no. 138 (July 27) 1990.

CORIANDER

Source: *Coriandrum sativum* L. (Family Umbelliferae or Apiaceae).

Common/vernacular names: Cilantro, Chinese parsley.

GENERAL DESCRIPTION

Strong-smelling annual herb with erect hollow stem, up to about 1 m high; native to Europe and Western Asia; naturalized in North America; widely cultivated. Parts used are the dried ripe fruits (commonly called coriander seeds) and leaves (both fresh and dried). An essential oil is obtained by steam distillation of the crushed fruits; it is mainly produced in Europe.

CHEMICAL COMPOSITION

Fruits contain 0.2–2.6% (usually 0.4–1.0%) volatile oil. The major component of the oil is *d*-linalool (coriandrol), which is present in 55–74%, depending on the ripeness of the fruits, geographical locations, and other factors.[1–4] Other compounds present in the oil include decyl aldehyde, *trans*-tridecene-(2)-al-(1), borneol, geraniol, geranyl acetate, camphor, carvone, anethole, caryophyllene oxide, elemol, and monoterpene hydrocarbons (mainly γ-terpinene, and α- and β-pinene, *d*-limonene, *p*-cymene, β-phellandrene, and camphene, with relative proportions varying considerably with sources).[2,3,5–11]

Other constituents present in fruits include up to 26% fats made up of glycerides (primarily of oleic, petroselinic and linolenic acids), a small amount of unsaponifiable matter (containing β-sitosterol, δ-sitosterol, triacontane, triacontanol, tricosanol, etc.), and $\Delta^{5,6}$-octadecenoic acid; proteins (11–17%); about 1.0% starch and 20% sugars; coumarins (psoralen, angelicin, scopoletin, umbelliferone, etc.); flavonoid glycosides, including quercetin-3-glucuronide, isoquercitrin, coriandrinol (β-sitosterol-D-glucoside), and rutin; tannins; chlorogenic and caffeic acids; and others (JIANGSU; LIST AND HÖRHAMMER; WATT AND MERRILL).[4,8,12–16]

Leaves contain less volatile oil than fruits; about 5% fats; about 22% proteins; sugars; coumarins and flavonoid glycosides similar to those in fruits; chlorogenic and caffeic acids; vitamin C; and others. The volatile oil contains mainly decyl and nonyl aldehydes, and linalool, among others (JIANGSU; WATT AND MERRILL).[15,16]

PHARMACOLOGY AND BIOLOGICAL ACTIVITIES

Coriander has been reported to have strong lipolytic activity.[17] Its petroleum ether-soluble fraction is reported to have antioxidative activity when mixed with lard.[18]

Coriander possesses hypoglycemic activities in experimental animals.[19]

Coriander oil is reported to have larvicidal properties (see *allspice* and *clove*) as well as bactericidal and weakly cytotoxic activities.[20,21]

A liquid carbon dioxide extract of coriander seeds has been reported to exhibit antibacterial and antifungal activities.[22]

Coriander oil when tested at a concentration of 6% in petrolatum on human subjects (25 per test) did not produce skin irritation or sensitization reactions.[23]

An aqueous extract of fresh coriander seeds produced a dose-dependent significant antiimplantation effect in rats (related to a significant decrease in serum progesterone levels after 5 days).[24]

USES

Medicinal, Pharmaceutical, and Cosmetic. Oil is used mainly as a flavoring agent in pharmaceutical preparations (e.g., Aromatic Cascara Sagrada Fluid extract); fruits are used as aromatic and carminative and in preparations to prevent griping (MARTINDALE).

In cosmetics, oil is used as a fragrance component in soaps, creams, lotions, and perfumes, with maximum use level of 0.6% in perfumes.[23]

Oil is also used in flavoring tobacco.

Food. The young leaves are widely used as a garnish in cooking (e.g., Chinese, Armenian, Spanish, etc.); they are known as Chinese parsley in Chinese cuisine and cilantro in Spanish cooking.

The seeds (fruits) and oil are extensively used as flavor ingredients in all types of food products, including alcoholic (vermouths, bitters, gin, etc.) and nonalcoholic beverages, frozen dairy desserts, candy, baked goods, gelatins and puddings, meat and meat products, condiments and relishes, and others. Highest average maximum use levels reported for seeds and oil were 0.52% and 0.012%, respectively, in meat and meat products and in alcoholic beverages.

Dietary Supplements/Health Foods. Fruits sometimes used in carminative and digestive products (FOSTER).

Traditional Medicine. Fruits are used as an aromatic carminative, stomachic, and antispasmodic, usually in the form of an infusion.

In Chinese medicine, in addition to being used as a stomachic, they are used in measles, dysentery, hemorrhoids, and other ailments; a decoction is also used as a gargle to relieve toothache. The whole herb is also used in stomachache, nausea, measles, and painful hernia (JIANGSU).

COMMERCIAL PREPARATIONS

Crude and oil. Crude was formerly official in N.F., and oil is official in N.F. and F.C.C.

Regulatory Status. GRAS (§182.10 and §182.20). Fruits subject of a German therapeutic monograph indicated for dyspeptic complaints and loss of appetite.[25]

REFERENCES

See the General References for ARCTANDER; BLUMENTHAL 1; BRUNETON; DUKE 4; FEMA; FOGARTY; FOSTER; GRIEVE; JIANGSU; LUST; MASADA; NANJING; ROSENGARTEN.

1. E. Gliozheni, *Bul. Shkencave Nat., Univ. Shteteror Tiranes*, **28**(3), 41 (1974).
2. G. Jukneviciene et al., *Liet. TSR Mokslu Akad. Darb., Ser. C*, **3**, 9 (1977).
3. N. N. Glushchenko et al., *Maslo-Zhir. Prom.*, **6**, 28 (1977).
4. H. Karow, *Riechst., Aromen, Körperpflegem.*, **19**(2), 60 (1969).
5. S. K. Chogovadze and D. M. Bakhtadze, *Lebensm. Ind.*, **24**, 513 (1977).
6. E. Schratz and S. M. J. S. Qadry, *Planta Med.*, **14**, 310 (1966).

7. S. Rasmussen et al., *Medd. Nor. Farm. Selsk.*, **34**(3–4), 33 (1972).
8. G. K. Gupta et al., *Indian Perfum.*, **21**, 86 (1977).
9. R. M. Ikeda et al., *J. Food Sci.*, **27**, 455 (1962).
10. J. S. T. Chou, *Koryo*, **106**, 55 (1974).
11. J. Taskinen and L. Nykanen, *Acta Chem. Scand., Ser. B*, **29**, 425 (1975).
12. G. A. Stepanenko et al., *Khim. Prir. Soedin.*, **10**, 37 (1974).
13. L. T. Lee and M. W. Wah, *Hua Hsueh*, **2**, 52 (1973).
14. A. R. S. Kartha and Y. Selvaraj, *Chem. Ind. (London)*, **25**, 831 (1970).
15. N. V. Sergeeva, *Khim. Prir. Soedin.*, **10**, 94 (1974).
16. J. Kunzemann and K. Herrmann, *Z. Lebensm. Unters. Forsch.*, **164**, 194 (1977).
17. G. Paulet et al., *Rev. Fr. Corps Gras.*, **21**, 415 (1974).
18. Y. Saito et al., *Eiyo To Shokuryo*, **29**, 505 (1976).
19. N. R. Farnsworth and A. B. Segelman, *Tile Till*, **57**, 52 (1971).
20. K. K. Abdullin, *Uch. Zap. Kazansk. Vet. Inst.*, **84**, 75 (1962).
21. K. Silyanovska et al., *Parfum. Kosmet.*, **50**, 293 (1969).
22. M. L. Khanin et al., *Khim. Farm. Zh.*, **2**, 40 (1968).
23. D. L. J. Opdyke, *Food Cosmet. Toxicol.*, **11**, 1077 (1973).
24. M. S. Al-Said et al., *J. Ethnopharmacol.*, **21**, 165 (1987).
25. Monograph *Coriandri fructus, Bundesanzeiger*, no. 173 (September 18) 1986.

CORN SILK

Source: *Zea mays* L. subsp. *mays* (Family Poaceae or Gramineae).

Common/vernacular names: Stigmata maydis and zea.

GENERAL DESCRIPTION

Coarse erect annual with prop (adventitious) roots near the ground and long sword-shaped leaves, one at each node; up to 4 m high; generally thought to be a native of tropical America. Parts used are the long styles and stigmata of the pistils called corn silk; the dried product is normally used.

CHEMICAL COMPOSITION

Corn silk is reported to contain 2.5% fats, 0.12% volatile oil, 3.8% gums, 2.7% resin, 1.15% bitter glucosidic substances, 3.18% saponins, 0.05% alkaloids, cryptoxanthin, vitamins C and K, sitosterol, stigmasterol, plant acids (malic, tartaric, etc.), anthocyanins, and others (JIANGSU).[1,2]

PHARMACOLOGY AND BIOLOGICAL ACTIVITIES

Corn silk has diuretic, hypoglycemic, and hypotensive activities in experimental animals. A dialyzed methanol-insoluble fraction of its aqueous extract has been demonstrated to be strongly diuretic in humans and rabbits; the toxicity of this fraction was low when compared with its effective dose. Lethal intravenous dose in rabbits was 250 mg/kg, while effective dose was 1.5 mg/kg (JIANGSU).

A crystalline constituent from an aqueous extract of corn silk has also been reported to be hypotensive and to stimulate uterine contraction in rabbits.[3]

Corn silk and its aqueous extracts are also reported to be effective in kidney and other diseases during clinical trials (JIANGSU).

USES

Medicinal, Pharmaceutical, and Cosmetic. Both crude and extracts are used as an ingredient in certain diuretic preparations; crude is also used in face powders, among others.

Food. Extracts are used as flavor components in major food products such as alcoholic and nonalcoholic beverages, frozen dairy desserts, candy, baked goods, and others. Use levels are generally lower than 0.002%.

Traditional Medicine. Used as a diuretic in urinary problems (cystitis, pyelitis, etc.); also as a demulcent.

In Chinese medicine, in addition to being used as a diuretic in dropsy, corn silk is used to treat sugar diabetes (diabetes mellitus) in the form of a decoction and to treat hypertension when decocted with watermelon peel and banana, as well as other ailments.

COMMERCIAL PREPARATIONS

Crude and extracts. Extracts come in vaning strengths (see *glossary*), with those for food use expressed in flavor intensities and those for pharmaceutical applications expressed in weight-to-weight ratios. Crude and fluid extract were formerly official in N.F.

Regulatory Status. GRAS (§182.20).

REFERENCES

See the General References for APhA; BARRETT; FEMA; GOSSELIN; GRIEVE; JIANGSU; LUST; ROSE; TERRELL; UPHOF.

1. N. E. Bobryshev, *Kukuruza*, **9**, 59 (1962).
2. E. D. Styles and O. Ceska, *Phytochemistry*, **14**, 413 (1975).
3. S. J. Hahn, *K'at'ollik Taehak Uihakpu Nonmunjip*, **25**, 127 (1973).

COSTUS OIL

Source: *Saussurea lappa* Clarke (syn. *Aucklandia costus* Falc.) (Family Compositae or Asteraceae).

GENERAL DESCRIPTION

Large erect perennial herb with a thick taproot; up to about 2 m high; native to the mountains of northern India (the Himalayas); cultivated in India and southwestern China. Part used is the dried root, from which a volatile oil is obtained by steam distillation followed by solvent extraction of the distilled water. India is the major producer of the oil.

CHEMICAL COMPOSITION

Root contains 0.3–3% volatile oil; saussurine (an alkaloid); betulin; stigamsterol; about 18% inulin; and resins (JIANGSU; NANJING; WILLAMAN AND SCHUBERT).

The major components in the oil are sesquiterpene lactones, including the crystalline dehydrocostus lactone and costunolide, which together make up about 50% of the oil; α- and β-cyclocostunolide; alantolactone; isoalantolactone; dihydrodehydrocostus lactone; cynaropicrin, and others.[1–6] Also present are other sesquiterpenes such as β-costol, elema-1,3,11(13)trien-l2-ol, α-costol, γ-costol, β-selinene, β-elemene, elemol, caryophyllene, caryophyllene oxide, *ar*-curcumene, aselinene, α-costal, β-costal, γ-costal, and others, with their concentrations in decreasing order and the first four accounting for about 18% of commercial root oil;[7,8] (Z,Z,Z)-1,8,11,14-heptadecatetraene;[9,10] an unusual terpenoid C_{14}-ketone (E)-9-isopropyl-6-methyl-5,9-decadien-2-one);[11] α- and β-ionones; dihydro-α-ionone; (E)-geranylacetone; aplotaxene and dihydroaplotaxene; 3,9,11-guaiatriene-12-carboxylic acid;[12] costic acid; palmitic, linoleic, and oleic acids; friedelin; β-sitosterol; and others (JIANGSU; MASADA).[2–4,7]

The sesquiterpene lactones (especially alantolactone, dehydrocostus lactone, and costunolide) have plant growth-regulating activities.[1,13]

PHARMACOLOGY AND BIOLOGICAL ACTIVITIES

Various fractions of costus oil have been reported to have hypotensive activities in anesthetized dogs, with 12-methoxydihydrocostunolide and the delactonized oil being the most potent, acting through direct peripheral vasodilation and cardiac depression. Most fractions were also effective in relieving bronchial spasm induced by histamine and acetylcholine in guinea pigs, but none had antitussive activity.[5]

A decoction of costus root is reported to exhibit weak inhibitory activities on paratyphoid A bacterium and certain other pathogenic bacteria (JIANGSU).

TOXICOLOGY

Costus and its derivatives (e.g., absolute used in perfumes) are known to cause allergic reactions (e.g., contact dermatitis) in humans.[13–15]

USES

Medicinal, Pharmaceutical, and Cosmetic. Costus and its derivatives (essential oil, absolute, and concrete) are used as fixatives and fragrance components in creams, lotions, and perfumes (e.g., Oriental types); reported maximum use level is 0.4% (no specific product form given) in perfumes.[14]

Food. Oil is used as a flavor component in most major food products, including alcoholic and nonalcoholic beverages, frozen dairy desserts, candy, baked goods, gelatins and puddings, and confectioner's frosting. Use levels are low, with highest average maximum of about 0.0004% (4.04 and 4.16 ppm) reported for alcoholic beverages and baked goods.

Traditional Medicine. Root has been used for millennia in China and India as a tonic, stomachic, carminative, and stimulant and in treating asthma, cough, dysentery, and cholera, among others; also used in incense.

COMMERCIAL PREPARATIONS

Mainly oil; it is official in F.C.C.

Regulatory Status. Has been approved for food use (§1721.510.)

REFERENCES

See the General References for ARCTANDER; BARRETT; FEMA; FOGARTY; GUENTHER; GUPTA; JIANGSU; NANJING.

1. P. S. Kalsi et al., *Phytochemistry*, **16**, 784 (1977).
2. S. B. Mathur, *Phytochemistry*, **11**, 449 (1972).
3. S. V. Govindan and S.C. Bhattacharyya, *Indian J. Chem., Sect. B*, **15**, 956 (1977).
4. S. B. Mathur and S. C. Bhattacharyya, *Int. Cong. Essent. Oils (Pap.)*, **6**, 126 (1974).

5. O. P. Gupta and B. J. R. Ghatak, *Indian J. Med. Res.*, **55**, 1078 (1967).
6. F. Bohlmann et al., *Planta Med.*, **51**, 74 (1985).
7. B. Maurer and A. Grieder, *Helv. Chim. Acta*, **60**, 2177 (1977).
8. A. S. Bawdekar et al., *Tetrahedron*, **23**, 1993 (1967).
9. R. G. Binder et al., *Phytochemistry*, **14**, 2085 (1975).
10. M. Romanuk et al., *Collect. Czech. Chem. Commun.*, **24**, 2018 (1959).
11. B. Maurer and G. Ohloff, *Helv. Chim. Acta*, **60**, 2191 (1977).
12. E. Klein and F. Thoemel, *Tetrahedron*, **32**, 163 (1976).
13. E. Rodriguez et al., *Phytochemistry*, **15**, 1573 (1976).
14. D. L. J. Opdyke, *Food Cosmet. Toxicol.*, **12**(Suppl.) 867 (1974).
15. J. C. Mitchell, *Arch. Dermatol.*, **109**, 572 (1974).

CRANBERRY

Source: *Vacciniummacrocarpon* Aiton (Family Ericaceae).

Common/vernacular names: Low-bush cranberry.

GENERAL DESCRIPTION

Trailing, evergreen, slender-stemmed shrub; leaves leathery, flower white to pink, 1 cm wide; fruit glossy red, 1–1.5 cm wide, tart flavored; occurs in bogs from Newfoundland to Manitoba south to Virginia, Ohio, and northern Illinois; also grown in northern and central Europe; locally naturalized. Part used is the fruit from which a juice (cranberry juice) is produced and extensively marketed in the United States; Massachusetts is the major producer. *Vaccinium oxycoccus* L., found in peat bogs in northern and central Europe used as a cranberry source in Europe. *V. macrocarpon* should not be confused with *Viburnum opulus* L., sometimes known as "high-bush cranberry."

CHEMICAL COMPOSITION

Fruits contain anthocyanins; flavonol glycosides (leptosine); catechin; triterpenoids; citric, malic, and quinic acids; smaller levels of benzoic and glucuronic acids; trace of alkaloids; carbohydrates (10%), protein, and vitamin C.[1] Quinic acid 0.5–0.9%, conjugated with glycine to produce hippuric acid, to which biological activity was formerly attributed.

PHARMACOLOGY AND BIOLOGICAL ACTIVITIES

Fruit widely regarded as possessing bacteriostatic activity for urinary tract infections. Antibacterial activity has been variously, though not conclusively attributed to anthocyanins; flavonol glycosides; catechin; volatile components; and benzoic, quinic, malic, and citric acids (CRELLIN AND PHILPOTT). It has been suggested and disputed that the urinary antiseptic effect is due to the action of hippuric acid.[2,3] Other studies suggest cranberry juice possesses antiadhesion activity to mucous membrane surfaces.[4–6] *Escherichia coli*, the most common causative bacteria for urinary tract infections adheres to cells of the urinary and alimentary tracts, enhancing their capacity to withstand nutrient deprivation and cleansing mechanisms, enhancing both toxicity and colonization. Blocking the adhesion of *E. coli* to urinary bladder mucosal cells has been shown to prevent development of urinary tract infections in mice. Cranberry juice fructose and an uncharacterized high molecular weight polymeric compound inhibit cellular adhesion of uropathogenic strains of *E. coli*. The antiadhesive agent of cranberry juice may prevent *E. coli* colonization in the gut, in the bladder, or in both.[7] Dosage of juice ranges from 5 to 20 oz daily (6 oz are equivalent to 90 g of fresh fruit). One study found that drinking 4–6 oz of cranberry juice daily for 7 weeks appeared to prevent urinary tract infections in 19 of 28 nursing home patients; a preventative rather than curative effect was suggested.[8]

USES

Food. Fruit juice, jelly, sauce: commonly eaten with poultry. Leaves a folk tea substitute. Cranberry juice cocktail is a 33% dilution of pure juice with added fructose. Approximately 1500 g of fresh fruit produce 1 L of juice. Anthocyanin pigment from fruit pulp used as commercial food coloring.

Dietary Supplements/Health Foods. Fruit juice concentrates or dried fruit in capsules and tablets, intended for relief of urinary tract infections (LUST).

Traditional Medicine. Fruit juice traditionally considered diuretic, antiseptic, febrifuge, refrigerant; a home remedy for treatment of urinary tract infections; folk cancer remedy in eastern Europe (STEINMETZ).

COMMERCIAL PREPARATIONS

Fruit juice, juice concentrates, dried fruit, juice concentrate in capsules.

REFERENCES

See the General References for BLUMENTHAL 2; CRELLIN AND PHILPOTT; GLEASON AND CRONQUIST; LUST; STEINMETZ; TUTIN 3; UPHOF.

1. L. Liberti, *Lawrence Rev.* (1987).
2. P. T. Bodel et al., *J. Lab. Clin. Med.*, **54**, 881 (1959).
3. P. Sternlieb, *N. Engl. J. Med.*, **268**, 57 (1963).
4. A. E. Sobata, *J. Urol.*, **131**, 1013 (1984).
5. D. R. Schmidt and A. E. Sobata, *Microbios*, **55**, 173 (1988).
6. D. Zafriri et al., *Antimicrob. Agents Chemother.*, **33**, 92 (1989).
7. I. Ofek et al., *N. Eng. J. Med.*, **324**(22), 1599 (1991).
8. L. Gibson et al., *J. Naturopathic Med.*, **2**(1), 45 (1991).

CUBEBS

Source: *Piper cubeba* L. f. (syn. *Cubeba officinalis* Miq.) (Family Piperaceae).

Common/vernacular names: Cubeba, cubeb berries, and tailed pepper.

GENERAL DESCRIPTION

Evergreen shrub or climbing vine, up to about 6 m high, often grown with other economic plants such as coffee; native to Indonesia and cultivated throughout Southeast Asia. Part used is the dried, fully grown but unripe fruit, commonly called cubeb berry, with its attached peduncle or stem (hence also called tailed pepper); cubeb oil is produced by steam distillation of the crushed fruits, usually in Europe and in the United States.

Cubeb oil has been reported to exhibit antiviral activities in rats as well as weak to strong antibacterial activities *in vitro*; it also has certain urinary antiseptic properties (JIANGSU; MERCK).[5]

CHEMICAL COMPOSITION

Fruits contain 10–20% volatile oil; about 2.5% cubebin;[1] 1–1.7% amorphous cubebic acid of undetermined structure; an acidic resin and 3–3.7% neutral resins, 8% gum; fats; and others (JIANGSU; LIST AND HÖRHAMMER).

The essential oil contains mainly sesquiterpenes and monoterpenes and their alcohols, with their relative concentrations varying considerably according to different reports. Sesquiterpene hydrocarbons present include caryophyllene, cadinene, α- and β-cubebene,[2] copaene, and 1-isopropyl-4-methylene-7-methyl-1,2,3,6,7,8,9-heptahydronaphthalein; monoterpene hydrocarbons including sabinene, α-thujene, β-phellandrene, α-pinene, myrcene, β-pinene, α-phellandrene, γ- and α-terpinene, limomene, and ocimene, among others;[3] oxygenated terpenes include 1,4-cineole, α-terpineol, cadinol, and cubebol; and others (JIANGSU; LIST AND HÖRHAMMER; STAHL).[1,4]

PHARMACOLOGY AND BIOLOGICAL ACTIVITIES

Fruit is reported to have local stimulant effect on mucous membranes (urinary and respiratory tract). Cubebic acid (not cubebin) is mainly responsible for its physiological properties (JIANGSU).

Ground cubeb was found 90% clinically effective in treating amebic dysentery (JIANGSU).

USES

Medicinal, Pharmaceutical, and Cosmetic. Berry and oil have been used in diuretic and urinary antiseptic preparations. Oil is used as a fragrance component in soaps, detergents, creams, lotions, and perfumes, with highest maximum use level of 0.8% reported in perfumes.[5] Also used in flavoring tobacco.

Food. Oil is used as a flavor ingredient in most major categories of food products, including alcoholic (liqueurs) and nonalcoholic beverages, frozen dairy desserts, candy, baked goods, gelatins and puddings, meat and meat products, condiments and relishes, and others. Highest average maximum use level reported is about 0.004% (38.2 ppm) in condiments and relishes. Fruits are reportedly used only in nonalcoholic beverages at an average maximum use level of 0.085%.

Traditional Medicine. Used as a diuretic, urinary antiseptic, carminative, and stimulating expectorant, among others. Also used in treating gonorrhea and cancer.[6]

COMMERCIAL PREPARATIONS

Crude and oil. Crude was formerly official in N.F., and oil is official in F.C.C.

Regulatory Status. Has been approved for food use (§172.510).

REFERENCES

See the General References for ARCTANDER; BRUNETON; FEMA; GRIEVE; GUENTHER; JIANGSU; LUST; MASADA; TERRELL.

1. C. K. Atal et al., *Lloydia*, **38**, 256 (1975).
2. Y. Ohta et al., *Tetrahedron Lett.*, **51**, 6365 (1966).
3. R. M. Ikeda, *J. Food Sci.*, **27**, 455 (1962).
4. S. J. Terhune et al., *Int. Congr. Essent. Oils (Pap.)*, **6**, 153 (1974).
5. D. L. J. Opdyke, *Food Cosmet. Toxicol.*, **14**, 729 (1976).
6. J. L. Hartwell, *Lloydia*, **33**, 288 (1970).

CUMIN

Source: *Cuminum cyminum* L. (syn. *C. odorum* Salisb.) (Family Umbelliferae or Apiaceae).

Common/vernacular names: Cummin and cumin seed.

GENERAL DESCRIPTION

Small annual with a slender stem, much branched above; up to about 0.6 m high; native to the Mediterranean region, now extensively cultivated there (Morocco, Turkey, Greece, Egypt, etc.) and in Iran, India, and other countries. Part used is the dried ripe fruit, commonly called "seed." An essential oil (cumin oil) is obtained by steam distillation of the crushed fruit. Major cumin seed producers include Egypt, Iran, India, Morocco, Turkey, and the former U.S.S.R.; major oil producers include India and the United States.

CHEMICAL COMPOSITION

Contains 2–5% volatile oil;[1,2] up to about 22% fats with a small amount of $\Delta^{5,6}$-octadecenoic acid;[3] 14 free amino acids, including five essential ones; about 18% protein;[4] flavonoid glycosides, including apigenin-7-glucoside (apigetrin), apigenin-7-glucuronosyl glucoside, luteolin-7-glucoside, and luteolin-7-glucuronosyl glucoside;[5] tannin; resin; gum; and others (LIST AND HÖRHAMMER; MARSH).

The volatile oil contains aldehydes (up to 60%) as its major components, which consist mainly of cuminaldehyde, 1,3-*p*-menthadien-7-al, 1,4-*p*-menthadien-7-al, and 3-*p*-menthen-7-al; commercial volatile oil and the volatile oil from previously ground commercial cumin contain more cuminaldehyde than the other aldehydes, with the absence of 1,4-*p*-menthadien-7-al, while the essential oil from freshly ground cumin contains primarily 1,4-*p*-menthadien-7-al, with cuminaldehyde in a much smaller amount and the other two aldehydes only in traces. Other major components of the oil are monoterpene hydrocarbons (up to 52%) composed mainly of β-pinene, γ-terpinene, and *p*-cymene, with α- and β-phellandrene, myrcene, α-terpinene, and limonene also present. Minor constituents include sesquiterpene hydrocarbons (β-farnesene, β-caryophyllene, β-bisabolene, etc.); cuminyl alcohol (believed to be an artifact, as it is present only in trace quantities in the volatile oil from freshly ground cumin); perillaldehyde; phellandral; *cis*- and *trans*-sabinene hydrate; cryptone; and others (LIST AND HÖRHAMMER; MASADA).[1,6,7]

Fine milling of cumin is reportedly responsible for up to 50% loss of its essential oil content, with the greatest loss occurring during the first hour of storage after milling.[8]

Cuminaldehyde, 1,4-*p*-menthadien-7-al, and 1,3-*p*-menthadien-7-al have indistinguishable odors; they appear to be mostly responsible for the characteristic aroma of unheated whole cumin seeds. The chief odor characteristics of heated cumin are due to 3-*p*-menthen-7-al in combination with the other three aldehydes.[2]

The petroleum ether-soluble fraction of cumin reportedly has antioxidative activity when mixed in lard.[9]

PHARMACOLOGY AND BIOLOGICAL ACTIVITIES

Cumin oil (especially cuminaldehyde) has been reported to exhibit strong larvicidal activities (see *cinnamon*, *clove*, and *coriander*);[10] it also has antibacterial properties.[11] It is rapidly absorbed through the shaved intact abdominal skin of mice.[12]

TOXICOLOGY

Undiluted cumin oil has been demonstrated to have distinct phototoxic effects that were not due to cuminaldehyde, its principal component.[13]

USES

Medicinal, Pharmaceutical, and Cosmetic. Oil is used as a fragrance component in creams, lotions, and perfumes, with a maximum use level of 0.4% reported in perfumes.[13]

Food. Cumin is a major flavor component of curry and chili powders. It is also used in other food products, including baked goods, meat and meat products, condiments and relishes, processed vegetables, soups, gravies, snack foods, with the highest maximum use level of about 0.4% (4308 ppm) reported in soups.

The oil is used as a flavor component in alcoholic and nonalcoholic beverages, frozen dairy desserts, candy, baked goods, gelatins and desserts, meat and meat products, condiments and relishes, gravies, snack foods, and others. Highest average maximum use level is reported to be about 0.025% (247 ppm) in condiments and relishes.

Dietary Supplements/Health Foods. Commonly used in specialty curry products, also as carminative and tea ingredient.

Traditional Medicine. Used as a stimulant, antispasmodic, carminative, diuretic, aphrodisiac, and emmenagogue, among others.

COMMERCIAL PREPARATIONS

Crude and oil. Oil is official in F.C.C.

Regulatory Status. GRAS (§182.10 and §182.20).

REFERENCES

See the General References for ARCTANDER; BAILEY 1; BRUNETON; DUKE 4; FEMA; GRIEVE; GUENTHER; GUPTA; LIST AND HÖRHAMMER; MARSH; ROSENGARTEN; TERRELL.

1. H. Karow, *Riechst., Aromen, Körperpflegem.*, **19**(2), 60 (1969).
2. C. G. Tassan and G. F. Russell, *J. Food Sci.*, **40**, 1185 (1975).
3. A. R. S. Kartha and Y. Selvaraj, *Chem. Ind. (London)*, **25**, 831 (1970).
4. F. Toghrol and H. Daneshpejouh, *J. Trop. Pediatr. Environ. Child Health*, **20**, 109 (1974).
5. J. B. Harborne and C. A. Williams, *Phytochemistry*, **11**, 1741 (1972).
6. P. T. Varo and D. E. Heinz, *J. Agric. Food Chem.*, **18**, 234 (1970).
7. P. T. Varo and D. E. Heinz, *J. Agric. Food Chem.*, **18**, 239 (1970).
8. E. Georgiev and Van Hong Tam, *Nauch. Tr., Vissh Inst. Khranit. Vkusova Prom. Plovdiv*, **20**, 99 (1973).
9. Y. Saito et al., *Eiyo To Shokuryo*, **29**, 505 (1976).
10. K. Oishi et al., *Nippon Suisan Gakkaishi*, **40**, 1241 (1974).
11. F. M. Ramadan et al., *Chem. Mikrobiol. Technol. Lebensm.*, **2**, 51 (1972).
12. F. Meyer and E. Meyer, *Arzneim.-Forsch.*, **9**, 516 (1959).
13. D. L. J. Opdyke, *Food Cosmet. Toxicol.*, **12**(Suppl.) 869 (1974).

DAMIANA

Source: *Turnera diffusa* Willd. (syn. *T. aphrodisiaca* L.F. Ward and *T. microphylla* Desv.) (Family Turneraceae).

GENERAL DESCRIPTION

A shrub with small aromatic leaves (mostly 1–2 cm long), up to 2 m high; native to tropical America (Mexico, Texas, Central America, South America, and the West Indies).[1] Part used is the dried leaf.

CHEMICAL COMPOSITION

Contains 0.5–1% volatile oil;[2,3] triacontane; hexacosanol-1; β-sitosterol; gonzalitosin I (5-hydroxy-7,3′,4′-trimethoxyflavone)[2] and other flavonoids,[4] a cyanogenic glycoside;[5] arbutin;[4,6] resin; tannin; a bitter substance (called damianin) of undetermined structure.[3]

The volatile oil is composed of two main fractions. A low-boiling fraction contains mainly 1,8-cineole, α- and β-pinenes, and *p*-cymene, while the higher boiling fraction consists primarily of thymol and sesquiterpenes (α-copaene, δ-cadinene, and calamenene).[6,7] The essential oil of micropropagated damiana contains caryophyllene, caryophyllene oxide, δ-cadinene, elemene, and 1,8-cineole.[8]

PHARMACOLOGY AND BIOLOGICAL ACTIVITIES

Damiana is widely believed to have aphrodisiac properties and has also been reported to have diuretic, laxative, and stimulant properties, among others (LEWIS AND ELVIN-LEWIS; TYLER 1; UPHOF).[3]

USES

Food. Used as a flavor ingredient in major food products, including alcoholic and nonalcoholic beverages, frozen dairy desserts, candy, baked goods, and gelatins and puddings. Highest maximum use level reported to be 0.125% for the crude in baked goods.

Dietary Supplements/Health Foods. Various products, including capsules, tablets, and drinks; tea ingredient; primarily for alleged aphrodisiac reputation (TYLER 1).[9]

Traditional Medicine. Reportedly used as a laxative, nervous stimulant, tonic, and aphrodisiac; also used in coughing, nephritis, menstrual disorders, and other ailments.[3]

COMMERCIAL PREPARATIONS

Crude and extracts. Crude was formerly official in N.F. Extracts do not have uniform standards.

Regulatory Status. Has been approved for food use (§172.510). Subject of a German therapeutic monograph; not recommended as claimed efficacy is unproven.[10]

REFERENCES

See the General References for APPLEQUIST; BARNES; BLUMENTHAL 1; CLAUS; DER MARDEROSIAN AND BEUTLER; LUST; MCGUFFIN 1 & 2; TERRELL; TYLER 1; UPHOF.

1. P. C. Standley, Vol. 23 Part 3 of *Contributions From the United States* NATIONAL *Herbarium, Trees and Shrubs of Mexico*, Smithsonian Press, Washington, DC, 1923 p. 848.
2. X. A. Dominguez and M. Hinojosa, *Planta Med.*, **30**, 68 (1976).
3. E. F. Steinmetz, *Acta Phytother.*, **7**, 1 (1960).
4. S. Piacente et al., *Z. Naturforsch. C*, **57**, 983 (2002).

5. B. Tantisewie, et al., *Pharm. Weekbl.*, **104**, 1341 (1969).
6. H. Auterhoff and H. P. Haufel. *Arch. Pharm. (Weinheim)*, **301**, 537 (1968).
7. H. Auterhoff and H. Momberger, *Arch. Pharm. (Weinheim)*, **305**, 455 (1972).
8. L. caraz-Melendez et al., *Fitoterapia*, **75**, 696 (2004).
9. E. Tyler. *Pharmacy in History*, **25**, 55, (1983).
10. Monograph *Turnerae diffusae folium et herba*, *Bundesanzeiger*, no. 43 (March 2, 1989).

DANDELION ROOT

Source: ***Taraxacum officinale*** Wiggers (syn. *T. vulgare* (Lam.) Schrank) and other ***Taraxacum*** species (Family Compositae or Asteraceae).

Common/vernacular names: Common dandelion, lion's tooth, taraxacum.

GENERAL DESCRIPTION

Taraxacum officinale is a perennial herb with deeply cut leaves forming a basal rosette in the spring and flower heads borne on long stalks, up to about 45 cm high; native to Europe and naturalized in North America, occurring as a common weed on lawns. Related species are found worldwide. Parts used are the dried rhizome and root; leaves and flowers are sometimes used for direct domestic consumption.

CHEMICAL COMPOSITION

Root contains several triterpenes, including taraxol, taraxerol, taraxasterol, ψ-taraxasterol, and β-amyrin; sterols (stigmasterol, β-sitosterol); inulin (ca. 25%); sugars (fructose, glucose, sucrose, etc.); pectin; glucosides; choline; phenolic acids (e.g., caffeic and *p*-hydroxyphenylacetic acids; gum; resins and vitamins (CLAUS; JIANGSU; KARRER).[1] Sesquiterpene lactones (free, e.g., dihydrolactucin, ixerin D and ainslioside and glycoside, e.g., glucosyltaraxinic acid) have been reported in addition to benzyl glucoside, syringin, dihydrosyringin, and dihydroconiferin.[2,3]

Flowers contain carotenoids (e.g., lutein, lutein epoxide, cryptoxanthin, cryptoxanthin epoxide, flavoxanthin, chrysanthemaxanthin, and violaxanthin) and their monoesters and diesters with fatty acids (mainly myristic; also lauric, palmitic and stearic acids); 2–4 arnidiol (JIANGSU); flavonoids (free, e.g., luteolin and chrysoeriol; and glycosides, e.g., luteolin-7-glucoside) and coumaric acids.[4,5]

Leaves contain lutein, violaxanthin, among other carotenoids; bitter substances; vitamins A, B, C, and D; and others. Vitamin A content (14,000 IU/100 g) is higher than that in carrots (11,000 IU/100 g) (JIANGSU; WATT AND MERRILL).

The herb and root extracts contain hydroxycinnamic acids, for example, chicoric, caffeoylquinic and caffeoyl tartaric acids, and others.[5,6]

Coumesterol (see ***alfalfa*** and ***red clover***, has been reported present in dandelion with plant part not specified (JIANGSU).[7]

PHARMACOLOGY AND BIOLOGICAL ACTIVITIES

Dandelion (plant part not specified) has been reported to exhibit hypoglycemic effects in experimental animals and to cause contact dermatitis in humans.[8,9]

Dandelion root is generally considered to have diuretic, choleretic, tonic and laxative properties, among others (ESCOP 3; MARTINDALE)[10]

Rodent experiments have confirmed diuretic activity of a fluid extract of the herb

(8 g dried/kg body weight); greater activity than that of the root; comparable with furosemid (80 mg/kg body weight) high potassium content (4% in dried leaves) replaces that eliminated in urine (ESCOP 3).

Bitter sesquiterpene lactones in the root increase bile secretion in rats by more than 40%; and increases gastric secretion (ESCOP 3).

The aqueous and ethyl acetate flower extracts rich in luteolin and luteolin-7-glucoside exhibited antioxidant activity in the 2,2-iphenyl-1-picrylhydrazyl (DPPH) free radical scavenging assay as well as in the phosphatidylcholine liposome assay. Lower concentrations, however, had a pro-oxidant effect.[11]

Traditional use for rheumatic conditions could be related to observed anti-inflammatory activity.[12] The anti-inflammatory activity has further been demonstrated by the protective effect of the extract against cholecystokinin-induced acute pancreatitis in rats. A concomitant reduction in interleukin-6 (IL-6) and tumor necrosis factor-α (TNF-α) was also observed.[13]

Numerous clinical studies using Chinese *Taraxacum* species have been reported. Root, leaves, juice, and extracts were effective in treating infections of various kinds (e.g., upper respiratory infections, pneumonia, chronic bronchitis, hepatitis, etc.) with few side effects (JIANGSU).

USES

Medicinal, Pharmaceutical, and Cosmetic. Extracts are quite extensively used in tonics (especially those for female ailments). Also used in diuretic, laxative, and antismoking preparations as well as in cosmetic and toiletry formulations, presumably for their tonic properties.[14] In Germany, the root is used for disturbances in bile flow, as diuretic, and as an appetite stimulant. The herb is used for appetite and dyspeptic disorders, such as abdominal fullness and flatulence. Products are contraindicated in obstruction of bile ducts, gallbladder empyema, and in gallstones, only under advice of physician.[15,16]

Food. Extracts are used as flavor components in various food products, including alcoholic (e.g., bitters) and nonalcoholic beverages, frozen dairy desserts, candy, baked goods, gelatins and puddings, and cheese. Highest average maximum use levels reported are about 0.014% (143 ppm) for the fluid extract in cheese and 0.003% (33.3 ppm) for the solid extract in baked goods.

The roasted root and its extract are used as coffee substitutes or in instant coffee substitute preparations.

Young leaves, particularly those of cultivated forms, are used as salad or vegetables.

Flowers are used in home wine making.

Dietary Supplements/Health Foods. Root used as a flavoring ingredient in tea formulations; capsules, tincture, tablet formulations for choleretic activity; leaf in tea as flavoring, bulk filler; capsules, tablets, tinctures, and so on, primarily as diuretics (FOSTER).

Traditional Medicine. The root is reportedly used as a laxative, tonic, and diuretic and to treat various liver and spleen ailments. Root and leaves are also used for heartburn and bruises and in treating chronic rheumatism, gout, and stiff joints as well as eczema, other skin problems, and cancers.[17]

In China, *Taraxacum mongolicum* Hand.-Mazz. and other *Taraxacum* species have been used for more than 1100 years in treating breast cancer and other breast problems (inflammation of the mammary glands, lack of milk flow, etc.), liver diseases (e.g., hepatitis), stomach problems, and others (JIANGSU; NANJING).

Others. Due to its high content of inulin, juice of root can serve as source of a special high fructose syrup; a very light-colored syrup containing 71% total sugars of which 77% was fructose has been produced from dandelion root by hydrolysis and other treatment.[18]

COMMERCIAL PREPARATIONS

Crude and extracts (solid, fluid, tincture, etc.). Crude and fluid extract were formerly official in N.F. Strengths (see **glossary**) of extracts are expressed in weight-to-weight ratios or flavor intensities. In Europe the drug consists of the root, with herb, gathered while blooming. Dandelion herb consists of the fresh or dried aboveground portions of the plant.[15,16]

Regulatory Status. GRAS (§182.20). Dandelion root (with herb) and dandelion leaves are subjects of positive German therapeutic monographs, allowed as a diuretic, for loss of appetite, dyspeptic problems, and disturbances in bile flow.[15,16]

REFERENCES

See the General References for APPLEQUIST; BAILEY 1; BARNES; BLUMENTHAL 1; BRUNETON; DER MARDEROSIAN AND BEUTLER; DUKE 4; ESCOP 3; FEMA; FOSTER; FOSTER AND DUKE; GOSSELIN; GRIEVE; JIANGSU; KROCHMAL AND KROCHMAL; LUST; MARTINDALE; MCGUFFIN 1 & 2; NANJING; ROSE; TERRELL.

1. B. Proda and E. Andrzejewska, *Farm. Polska*, **22**, 181 (1966); through Chem. Abstr., **65**, 9341c (1966).
2. Y. Kashiwada et al., *J. Asian Nat. Prod. Res.*, **3**, 191 (2001).
3. W. Kisiel and B. Barszcz, *Fitoterapia*, **71**, 269 (2000).
4. C. Hu and D. D. Kitts, *Phytomedicine*, **12**, 588 (2005).
5. C. A. Williams et al., *Phytochemistry*, **42**, 121 (1996).
6. K. Schutz et al., *Rapid Commun. Mass Spectrom.*, **19**, 179 (2005).
7. J. Chury and F. Prosek, *Vet. Med. (Prague)*, **13**, 305 (1968).
8. N. R. Farnsworth and A. B. Segelman, *Tile Till*, **57**, 52 (1971).
9. J. C. Mitchell, *Recent Advances in Phytochemistry*, Plenum Press, New York, 1975, p. 119.
10. K. Faber, *Pharmazie*, **13**, 423 (1958).
11. C. Hu and D. D. Kitts, *J. Agric. Food Chem.*, **51**, 301 (2003).
12. N. Mascolo, *Phytother. Res.*, **1**, 28 (1987).
13. S. W. Seo et al., *World J. Gastroenterol.*, **11**, 597 (2005).
14. H. B. Heath, *Cosmet. Toilet.*, **92**, 19 (1977).
15. Monograph *Taraxaci radix cum herba. Bundesanzeiger*, no. 228 (December 5, 1984); corrected (September 1, 1990).
16. Monograph *Taraxaci radix cum herba. Bundesanzeiger*, no. 228 (August 29, 1992).
17. J. L. Hartwell, *Lloydia*, **31**, 71 (1968).
18. V. F. Belyaev and P. V. Golovin, *Uch. Zap., Beloruss. Gos Univ. V. I. Lenina, Ser. Khim.*, **20**, 220 (1954).

DEERTONGUE

Source: *Trilisa odoratissima* (J. F. Gmel.) Cass. (syn. *Carphephorus odoratissimus* (J. F. Gmel.) Hebert; *Liatris odoratissima* Michx.) (Family Compositae or Asteraceae).

Common/vernacular names: Deer's tongue, Carolina vanilla, wild vanilla, liatris, vanilla trilisa, vanilla plant, vanilla leaf, and hound's tongue.

GENERAL DESCRIPTION

A perennial herb with large, thick leaves and a stem that is branched near the top; up to about 1.2 m high; native to eastern United

States. Part used is the dried leaf which has a characteristic coumarin-like (new-mown hay) fragrance.

CHEMICAL COMPOSITION

Contains about 1.6% coumarin as the major aromatic constituent.[1] Other volatile components include dihydrocoumarin, 2,3-benzofuran, terpenes, and straight-chain aldehydes and ketones, among others, totaling more than 90 identified compounds.[2]

Nonvolatile compounds reported present include triterpenes (e.g., β-amyrin, lupeol, lupenone, 11-oxo-β-amyrin, 11-oxo-α-amyrin, and others); sesquiterpenes (e.g., (±)-eudesmin and (±)-epieudesmin); and others.[1,3]

Fresh leaves contain high concentrations of cis- and trans-O-hydroxycinnamic acids, present mostly as glucosides. It is postulated that during drying (curing) cis-O-hydroxycinnamic acid glucoside is converted to coumarin by hydrolysis followed by cyclization.[4]

PHARMACOLOGY AND BIOLOGICAL ACTIVITIES

Coumarin has toxic properties, including liver injury and hemorrhages (GOSSELIN; LEWIS AND ELVIN-LEWIS).[5]

Coumarin and related compounds have been reported to be effective in reducing high-protein edemas, especially lymphedema.[6]

Some studies have shown coumarin to be nonteratogenic in animals.[7]

TOXICOLOGY

Reported to have diaphoretic, demulcent, and febrifuge activity (WREN). The charcoal-decolorized extract (deertongue incolore) has been reported to be nonirritating, nonsensitizing, nonphototoxic, and nonphotoallergenic when applied to the skin of animals and/or humans.[8]

USES

Medicinal, Pharmaceutical, and Cosmetic. Extracts are used as fixatives or fragrance components in perfumes and other cosmetic products (e.g., soap, detergent, creams, and lotions), with maximum use level of 0.5% of the incolore being reported in perfumes.[8]

Traditional Medicine. Used as a tonic in treating malaria.

Others. Extracts are used extensively in flavoring various types of tobaccos.

COMMERCIAL PREPARATIONS

Leaves and extracts (oleoresin, solid extract, etc.).

Regulatory Status. Not permitted in foods.

REFERENCES

See the General References for BAILEY 2; BRUNETON; KROCHMAL AND KROCHMAL; MARTINDALE; WREN.

1. R. A. Appleton and C. R. Enzell, *Phytochemistry*, **10**, 447 (1971).
2. K. Karlsson et al., *Acta Chem. Scand.*, **26**, 2837 (1972).
3. I. Wahlberg et al., *Acta Chem. Scand.*, **26**, 1383 (1972).
4. F. A. Haskins et al., *Econ. Bot.*, **26**, 44 (1972).
5. D. L. J. Opdyke, *Food Cosmet. Toxicol.*, **12**, 385, (1974).
6. T. Bolton and J. R. Casley-Smith, *Experientia*, **31**, 271 (1975).

7. W. Grote and I. Weinmann, *Arzneim.-Forsch.*, **23**, 1319 (1973).

8. D. L. J. Opdyke, *Food Cosmet. Toxicol.*, **14**(Suppl.), 743 (1976).

DEVIL'S CLAW

Source: *Harpagophytum procumbens* DC (Family Pedaliaceae)

Common/vernacular names: Grapple plant, wood spider.

GENERAL DESCRIPTION

Herbaceous trailing perennial; flowers red in axils; fruit with pointed and barbed woody grapples to 2.5 cm long; occurring in steppes on red sand in south tropical Africa, especially in the Kalahari desert and in the Namibian steppes plus Madagascar. The secondary tuber, about 6 cm in diameter; 20 cm long, is the part used.

CHEMICAL COMPOSITION

Iridoid glycosides, including harpagoside (0.1–3%), found at twice the concentration in secondary tubers than primary root, in trace amounts in leaves; also harpagide and procumbide; phytosterols (β-sitosterol and stigmasterol);[1–3] tricyclic diterpenes (stable abietane- and totarane-type; and unstable chinane-type);[4,5] flavonoids, including kaempferol and luteolin glycosides; cinnamic, chlorogenic, oleanolic, and ursolic acids; harpagoquinone; and others (WEISS; WREN). The new glycosides 8-cinnamoylmyoporoside, pagoside, and 6'-O-acetylacteoside have recently been reported alongside of 8-feruloylharpagide and caffeic acid reported in devil's claw for the first time.[6]

PHARMACOLOGY AND BIOLOGICAL ACTIVITIES

Anti-inflammatory, analgesic, antirheumatic, and reputed sedative activity.

Anti-inflammatory activity evaluated in the carrageenan edema assay in rats showed no significant activity (6% inhibition) compared with indomethacin (63% inhibition) in oral administration of an aqueous extract of the roots; positive results were reported, however, in acute inflammatory reactions with intravenous administration.[7] A lack of significant experimental anti-inflammatory activity in orally administered extracts in the carrageenan-induced edema test has been reported by several research groups.[8–11] Oral administration of dried aqueous extract in 13 arthritic patients showed no significant improvement after 6 weeks.[11] Experimental anti-inflammatory activity, however, was later demonstrated in dose-dependent i.p. administration of a dried aqueous extract, at a dose corresponding to 100 mg of the dried secondary root/kg, comparable to the efficacy of 2.5 mg/kg of indomethacin. The same tuber extract at a dose corresponding to 400 mg of the dried secondary root/kg proved more efficient than indomethacin at 10 mg/kg. Harpagoside administered i.p. at 10 mg/kg (5 mg harpagoside equals the content in 400 mg of dried secondary tubers) did not have significant anti-inflammatory effects. It has been suggested that simultaneous action of various principles other than iridoid glycosides is responsible for positive experimental anti-inflammatory activity.[12]

Peripheral analgesic activity of an aqueous extract of the tuber against a chemical stimulus has been observed. Other compounds besides harpagoside are responsible for analgesic activity. Conflicting results in various studies have been attributed to nonefficacy of oral versus i.p. administration. Analgesic activity, with a reduction of high uric acid and cholesterol levels, has also been observed.[13] The analgesic and anti-inflammatory activities of devil's claw aqueous root extract have been tested in mice using the hot-plate/acetic acid writhing and rat paw edema methods, respectively. The extract produced significant effects in all experiments as compared to diclofenac.[14] Similar results were obtained with an ethanolic extract containing 1.5% harpagoside.[15] The use of devil's claw in chronic back pain and osteoarthritis of the hip and knee has been the subject of recent resear h and reviews that further supports the analgesic/anti-inflammatory activity of devil's claw.[16–20] Involved mechanisms of actions are believed to involve inhibition of COX-2 and leukocyte elastase,[6,21] suppression of iNOS expression and NO production,[21–23] and reduced induction of NF-κB and TNF-α.[21,24]

Other activities reported for devil's claw include

Anticonvulsant activity. The aqueous extract of the root administered at 100–800 mg/kg i.p. to mice significantly decreased the onset and severity of seizures induced by pentylenetetrazole and picrotoxin. The displayed anticonvulsant effect was similar to that of phenobarbitone and diazepam. The authors suggested that an enhancing effect on GABA neurotransmission and/or GABAergic action was the mechanism underlying the anticonvulsant activity of devil's claw.[25]

Antidiabetic activity. i.p. administration of 50–800 mg/kg of the aqueous root extract to streptozotocin-treated mice resulted in a significant reduction in blood glucose levels comparable to that produced by the reference hypoglycemic agent chlorpropamide.[14]

Antiplasmodial activity. The petroleum ether extract of the root displayed selective *in vitro* antiplasmodial activity against two strains of *Plasmodium falciparum* (chloroquine-resistant and -sensitive). Bioassay-guided fractionation resulted in the isolation of two diterpenes to which the activity was attributed.[5]

Cardiovascular effects. Extracts have exhibited an arterial blood pressure reduction in rats, a decrease in heart rate in rabbits, and a protective effect against arrythmias.[26]

USES

Dietary Supplements/Health Foods. Oral dosage forms, including tablets, capsules, and extracts (e.g., tinctures) are sold in natural food or health food stores; primarily for relief of arthritic symptoms (WEISS; WREN).

Traditional Medicine. Used in Africa and since the early 20th century in Europe for indigestion (bitter tonic), blood diseases, headache, allergies, rheumatism, arthritis, lumbago, neuralgia; also as febrifuge, purgative; externally for sores, ulcer, boils, and skin lesions; folk cancer remedy.

COMMERCIAL PREPARATIONS

Crude and extracts. Extracts do not have uniform standards. In Germany, ampoules are available for i.v. and i.m. administration.

Regulatory Status. Undetermined in the United States. Root the subject of an official monograph in Germany, indicated for loss of appetite, dyspeptic discomfort, and others.[27]

REFERENCES

See the General References for BARNES; BLUMENTHAL 1; BRUNETON; MCGUFFIN 1 & 2; STEINMETZ; TYLER 1; UPHOF; WEISS; WREN.

1. H. Litichi and A. Wartburg, *Tetrahedron Lett.*, **15**, 835 (1964).
2. O. Sticher, *Dtsch. Apoth. Ztg.*, **32**, 1279 (1977).
3. F. C. Cyzgan and A. Kreuger, *Planta Med.*, **31**, 305 (1977).
4. C. Clarkson et al., *J. Nat. Prod.*, **69**, 527 (2006).

5. C. Clarkson et al., *Planta Med.*, **69**, 720 (2003).
6. K. Boje et al., *Planta Med.*, **69**, 820 (2003).
7. A. Erdos et al., *Planta Med.*, **34**, 97 (1978).
8. O. Eichler and C. Koch, *Arzneim.-Forsch*, **20**, 107 (1970).
9. D. McLeod et al., *Br. J. Pharmacol.*, **6**, 140 (1979).
10. I. Whitehouse et al., *Can. Med. Assoc.*, **129**, 249 (1983).
11. R. Graham and B. V. Robinson, *Ann. Rheum. Dis.*, **40**, 632 (1981).
12. M. C. Lanhers et al., *Planta Med.*, **58**, 117 (1992).
13. R. Kampf, *Schweiz. Apothek. Zeitung*, **114**, 337 (1976).
14. I. M. Mahomed and J. A. Ojewole, *Phytother. Res.*, **18**, 982 (2004).
15. M. L. Andersen et al., *J. Ethnopharmacol.*, **91**, 325 (2004).
16. P. Chantre et al., *Phytomedicine*, **7**, 177 (2000).
17. S. Chrubasik et al., *Phytother. Res.*, **18**, 187 (2004).
18. J. J. Gagnier et al., *BMC Complement Altern. Med.*, **4**, 13 (2004).
19. D. Laudahn and A. Walper, *Phytother. Res.*, **15**, 621 (2001).
20. T. Wegener and N. P. Lupke, *Phytother. Res.*, **17**, 1165 (2003).
21. T. H. Huang et al., *J. Ethnopharmacol.*, **104**, 149 (2006).
22. M. H. Jang et al., *J. Pharmacol. Sci.*, **93**, 367 (2003).
23. M. Kaszkin et al., *Phytomedicine*, **11**, 585 (2004).
24. B. L. Fiebich et al., *Phytomedicine*, **8**, 28 (2001).
25. I. M. Mahomed and J. A. Ojewole, *Brain Res. Bull.*, **69**, 57 (2006).
26. C. Circost et al., *J. Ethnopharmacol.*, **11**, 259 (1984).
27. Monograph *Harpagophyti radix*. *Bundesangeizer*, no. 43 (March 2, 1989).

DILL AND INDIAN DILL

Source: *Dill Anethum graveolens* L.; *Indiandill Anethum sowa* Roxb. (Family Umbelliferae or Apiaceae).

Common/vernacular names: European dill and American dill (*A. graveolens*); East Indian dill (*A. sowa*).

GENERAL DESCRIPTION

Dill (*A. graveolens*) is an annual or biennial herb with a smooth and erect stem; up to 1 m high; native to Mediterranean region and Asia (southern Russia); now cultivated worldwide (Germany, the Netherlands, the United Kingdom, Italy, the United States, India, China, etc.). Parts used are the dried ripe fruit (commonly called "seed") and the whole aboveground herb (dill herb) harvested immediately before the fruits mature or, for best quality, before flowering. Dill seed oil is obtained by steam distillation of the crushed dried fruits, and dillweed oil (dill oil or dill herb oil) is obtained by steam distillation of the freshly harvested herb. Dill seed oil is produced mainly in Europe, while dillweed oil is produced primarily in the United States.

Indian dill (*A. sowa*) is a smooth perennial herb, up to about 1 m high; native to tropical Asia and widely cultivated in India and Japan. Part used is the dried ripe fruit (commonly called "seed"); Indian dill seed oil (Indian dill oil or East Indian dill seed oil) is obtained by steam distillation of the crushed fruit.

CHEMICAL COMPOSITION

Dill seeds contain 1.2–7.7% (usually 2.5–4%) volatile oil, with concentrations varying according to geographical origin and seasons;[1-3] dillanoside (a xanthone glucoside);[4] coumarins (scopoletin, esculetin, bergapten, umbelliferone, umbelliprenine, etc.); kaempferol and its 3-glucuronide;[5,6] vicenin (6,8-di-C-glucosyl-5,7,3′-trihydroxyflavone) and other flavonoids;[7] petroselinic acid triglyceride and β-sitosterol glucoside;[8] phenolic acids (caffeic, ferulic, and chlorogenic); protein (ca. 16%); fats (ca. 15%); and others (JIANGSU; LIST AND HÖRHAMMER; MARSH).

Dill seed oil contains mainly carvone (35–60%), d-limonene, and α-phellandrene, which together can account for 90% of the oil.[1,2,9] Other components present include dihydrocarvone, eugenol, β-phellandrene, α-pinene, anethole, dillapiole, myristicin, carveol, β-caryophyllene, and others.[1,2,9,10]

Dillweed oil contains α-phellandrene, limonene, and carvone as its major components, usually with carvone in lesser amount than that in dill seed oil.[1,3,11,12] Other constituents include terpinene, apinene, dillapiole, myristicin, and two coumarans, among others (LIST AND HÖRHAMMER).[1,12-14]

Thirty-three compounds have been identified in the water-soluble portion of the methanol extract of dill seed. Of these compounds, nine were reported for the first time and included a monoterpene, six monoterpene glycosides, an aromatic, and an alkyl glucoside.[15]

Indian dill seeds contain 2–6% volatile oil;[6] kaempferol, isorhamnetin, quercetin, and their 3-glucuronides;[5,6] petroselinic acid triglyceride and β-sitosterol;[8] fats; proteins; and others (LIST AND HÖRHAMMER).[16]

Indian dill seed oil consists mainly of carvone, dihydrocarvone, and limonene in varying amounts;[6] dillapiole (often in significant quantities) and apiole;[6,17-20] myristicin;[19] and others (LIST AND HÖRHAMMER).

Two minor aromatic compounds (a propiophenone and a biphenyl derivative) were isolated from 3 kg of dried Indian dell seeds. Their yields were ca. 0.001% and 0.002%, respectively, and were reported for the first time.[21]

Egyptian dill seed was found to contain limonene (30.3%), dillapiole (26.8%), carvone (22%), piperitone (8.2%); reported in the oil for the first time were Δ^8-dehydro-p-cymene, camphor, and linalylacetate.

The whole herb of *A. graveolens* was reported to contain the new furanocoumarin, 5-[4″-hydroxy-3″-methyl-2″-butenyloxy]-6,7-furocoumarin, in addition to oxypeucedanin, oxypeucedanin hydrate, and falcarindiol.[22] The whole herb also contains 9-hydroxypiperitone glucosides, p-menth-2-ene-1,6-diol, and 8-hydroxygeraniol.[23]

PHARMACOLOGY AND BIOLOGICAL ACTIVITIES

Dill seed oil has been reported to have spasmolytic effects on smooth muscles, especially those of the GIT.[24] A 5% emulsion in physiological saline administered intravenously to cats (5–10 mg/kg) increased respiratory volume and lowered blood pressure, while guinea pigs receiving a higher dose (35 mg/kg) via i.p. injection went into anaphylactic shock.[25]

Dill seed oil has antibacterial properties, and Indian dill seed oil has antifungal activities.[26,27] The oxypeucedanins and falcarindiol isolated from the whole herb were reported to exhibit antimycobacterial activity with MIC values between 2 and 128 µg/mL.[22]

An infusion of the young dill herb when administered intravenously to animals is reported to lower the blood pressure, dilate blood vessels, stimulate respiration, slow the heart rate, and other activities (JIANGSU).

Ethanolic extracts of dill seeds as well as the volatile oil have been shown to produce diuresis in dogs, while significantly increasing Na^+ and Cl^- excretion.[28]

Oral administration of an aqueous extract of dill leaves to hyperlipidemic rats for 14 days reduced their blood cholesterol and triacylgleceride levels by almost 20% and 50%, respectively. Oral administration of the essential oil had a comparable effect on triacylglycerides but had no effect on cholesterol level.[29]

Administration of aqueous and ethanolic extracts of dill seed to mice models of gastric irritation/hyperacidity resulted in a dose-dependent decrease in total gastric acid and a rise in pH comparable to that of cimetidine. Gastric lesions induced by HCl or ethanol were also reduced after oral administration of the extracts (dose 0.45 g/kg vs. 0.1 g/kg of sucralfate). The reduction of gastric lesions was attributed not only to the antacid effect of the extracts but also to a possible cytoprotective effect of their terpene and flavonoid constituents.[30]

TOXICOLOGY

At ordinary use levels, dill and Indian dill oils are considered nontoxic.[18] However, a number of case reports and studies have been published about allergic reactions to dill seed and related herbs.[31–34]

USES

Medicinal, Pharmaceutical, and Cosmetic. Dill seed and dill seed oil are occasionally used in digestive preparations. Dill weed oil is used as a fragrance component in cosmetics, including soaps, detergents, creams, lotions, and perfumes. Maximum use level reported is 0.4% in perfumes.

Food. Dill, Indian dill, and their oils are extensively used in many food products. Dill and Indian dill are reportedly used in baked goods, meat and meat products, condiments and relishes, fats and oils, and others; highest average maximum use level is about 2.9% (28,976 ppm) for Indian dill in condiments and relishes. Dill oil is used in all above foods as well as in alcoholic and nonalcoholic beverages, frozen dairy desserts, cheese, snack foods, gravies, and others, with highest average maximum use level of 0.075% reported in snack foods.

Dietary Supplements/Health Foods. Bulk dillweed and dill seeds are available at natural food stores valued as a carminative (FOSTER).

Traditional Medicine. Both in domestic Western and Chinese medicine, dill seed and dill seed oil are used as aromatic carminative and stimulant in the treatment of flatulence, especially in children. Reported to have carminative, antispasmodic, sedative, lactagogue, and diuretic properties. Used in India, Africa, and elsewhere for hemorrhoids, bronchial asthma, neuralgias, renal colic, dysuria, genital ulcers, dysmenorrhea, and others.[35] In European tradition, dill herb is reportedly used as an antispasmodic for conditions of the gastrointestinal tract, kidney and urinary tract; also for sleep disorders.[36]

COMMERCIAL PREPARATIONS

Crude and oils. Dill seed oil, dillweed oil, and Indian dill seed oil are official in F. C. C., with ketone (as carvone) contents highest in dill seed oil and lowest in Indian dill seed oil.

Regulatory Status. GRAS status affirmed (§184.1282). Dill seed is the subject of a positive German therapeutic monograph, indicated as a spasmolytic and bacteriostatic for dyspeptic disorders, with a mean daily dosage of 3 g (or 0.1–0.3 g of the essential oil).[37] Dillweed is not allowed to carry therapeutic claims, since efficacy for reported uses has not been documented.

REFERENCES

See the General References for ARCTANDER; BLUMENTHAL 1; FOSTER; GUENTHER; JIANGSU; LIST AND HÖRHAMMER; MARTINDALE; ROSENGARTEN; UPHOF.

1. H. Karow, *Riechst, Aromen, Korperpflegem,* **19**, 60 (1969).
2. M. B. Embong et al., *Can. Inst. Food Sci. Technol. J.,* **10**, 208 (1977).
3. R.Gupta, *Cultiv. Util. Med Aromat. Plants,* 337 (1977).
4. M. Kozawa et al., *Chem. Pharm. Bull.,* **24**, 220 (1976).

5. J. B. Harborne and C. A. Williams, *Phytochemistry*, **11**, 1741 (1972).
6. C. S. Shah, *Cultiv. Util. Med. Aromat. Plants*, 335 (1977).
7. L. I. Dranik, *Khim. Prir. Soedin.*, **6**, 268 (1970).
8. M. Bandopadhyay et al., *Curr. Sci.*, **41**, 50 (1972).
9. R. K. Baslas et al., *Flav. Ind.*, **2**, 241 (1971).
10. M. Miyazawa and H. Kameoka, *Yukagaku*, **23**, 746 (1974).
11. S. Zlatev, *Riv. Ital. Essenze, Profumi, Piante offic., Aromi, Saponi, Cosmet., Aerosol*, **58**, 553 (1976).
12. K. Belafi-Rethy et al., *Acta Chim. Acad. Sci. Hung.*, **83**, 1 (1974).
13. K. Belafi-Rethy and E. Kerenyi, *Acta Chim. Acad. Sci. Hung.*, **94**, 1 (1977).
14. E. P. Lichtenstein et al., *J. Agric. Food Chem.*, **22**, 658 (1974).
15. T. Ishikawa et al., *Chem. Pharm. Bull. (Tokyo)*, **50**, 501 (2002).
16. A. R. S. Kartha and Y. Selvaraj, *Chem. Ind. (London)*, **25**, 831 (1970).
17. B. C. Gulati et al., *Perfum. Essent. Oil Record*, **60**, 277 (1969).
18. Anon, *Fed. Regist.*, **39**, 34211 (1974).
19. J. B. Harborne et al., *Phytochemistry*, **8**, 1729 (1969).
20. R. K. Sahdev et al., *J. Inst. Chem., Calcutta*, **47**, 234 (1975).
21. S. S. Tomar and P. Dureja, *Fitoterapia*, **72**, 76 (2001).
22. M. Stavri and S. Gibbons, *Phytother. Res.*, **19**, 938 (2005).
23. B. Bonnlander and P. Winterhalter, *J. Agric. Food Chem.*, **48**, 4821 (2000).
24. T. Fleming, *PDR for Herbal Medicines*, Medical Economics Company, New Jersey, 2000, pp. 252–253.
25. T. Shipochiliev, *Vet. Med. Nauki*, **5**, 63 (1968).
26. F. M. Ramadan et al., *Chem. Mikrobiol. Technol. Lebensm*, **1**, 96 (1972).
27. B. Dayal and R. M. Purohit, *Flav. Ind.*, **2**, 484 (1971).
28. G. H. Mahran et al., *Phytother. Res.*, **5**, 169 (1992).
29. R. Yazdanparast and M. Alavi, *Cytobios*, **105**, 185 (2001).
30. H. Hosseinzadeh et al., *BMC Pharmacol.*, **2**, 21 (2002).
31. J. J. Garcia-Gonzalez et al., *Ann. Allergy Asthma Immunol.*, **88**, 518 (2002).
32. A. W. van Toorenenbergen et al., *Int. Arch. Allergy Appl. Immunol.*, **86**, 117 (1988).
33. J. Monteseirin et al., *Allergy*, **57**, 866 (2002).
34. G. L. Freeman, *Allergy*, **54**, 531 (1999).
35. G. H. Mahran et al., *Int. J. Pharmacognosy*, **30**, 139 (1992).
36. Monograph *Anethi Herba*, Bundesanzeiger, no. 193 (October 15, 1987).
37. Monograph *Anethi Herba*, Bundesanzeiger, no. 193 (October 15, 1987); revised (March 13, 1990).

DOGGRASS

Source: *Elytrigia repens* (L.) Desv, ex B. D. Jackson (syn. *Agropyron repens* (L.) Beauv.; *Triticum repens* L.) (Family: Gramineae or Poaceae).

Common/vernacular names: Quack grass, couch grass, witchgrass, quick grass, quitch grass, triticum, and agropyron.

GENERAL DESCRIPTION

A perennial grass, up to 1.5 m high; native to Eurasia, naturalized in North America; widely distributed as a weed. Parts used are the dried rhizomes and roots collected in the fall or early spring.

CHEMICAL COMPOSITION

Contains up to about 8% triticin (a fructosan); 2–3% inositol and mannitol; 1.5% fixed oil; vitamins A and B; glycosides (e.g., vanillin glucoside); mucilaginous substances; ash with high silicon and iron contents; and small amount of volatile oil (up to ca. 0.05%); among others (LIST AND HÖRHAMMER).[1,2]

The volatile oil is composed of up to 95% agropyrene (1-phenyl-2,4-hexadiyne) and monoterpenes, including carvacrol (10.81%), *trans*-anethole (6.80%), carvone (5.5%), thymol (4.30%), menthol (3.5%), menthone, (1.40%), and *p*-cymene (1.10%), among others (LIST AND HÖRHAMMER).[3]

The leaves contain a lectin that is a dimer of two identical subunits. The lectin binds to *N*-acetylgalactosamine and agglutinates erythrocytes of blood group A.[4]

PHARMACOLOGY AND BIOLOGICAL ACTIVITIES

Agropyrene and its oxidation product, 1phenylhexa-2,4-diyne-l-one, have been reported to have broad antibiotic activities (LIST AND HÖRHAMMER)[3,5]

A doggrass infusion has been demonstrated to have pronounced sedative effects in mice.[6]

Both the aqueous and hydroalcoholic extracts of doggrass have diuretic activities in rats.[7]

USES

Medicinal, Pharmaceutical, and Cosmetic. Extracts used in certain diuretic preparations. Used in European phytomedicine for irrigation therapy in the treatment of inflammatory diseases of the urinary tract and the prevention of kidney stones.

Food. Extracts are used as flavor components in nonalcoholic beverages, frozen dairy desserts, candy, baked goods, and gelatins and puddings. Use levels reported are low, with the highest average maximum of about 0.003% (32.4 ppm) being reported (type of extract not specified) in baked goods.

Traditional Medicine. Used as a diuretic and expectorant, to reduce blood cholesterol, and in treating nephrolithiasis (kidney stones), diabetes, chronic skin diseases, and liven ailments, among others.[1]

COMMERCIAL PREPARATIONS

Crude and extracts (fluid, solid, powdered, etc.). Crude and fluid extract were formerly official in N.F. and U.S.P. Strengths (see *glossary*) of extracts are expressed in weight-to-weight ratios.

Regulatory Status. GRAS (§182.20). Root subject of a positive German therapeutic monograph for inflammatory diseases of the kidney tract.[8]

REFERENCES

See the General References for APhA; BLUMENTHAL 1; FEMA; GOSSELIN; HORTUS 3RD; LIST AND HÖRHAMMER; WREN; YOUNGKEN.

1. S. Paslawska and R. Piekos, *Planta Med.*, **30**, 216 (1976).
2. D. Smith and R. D. Grotelschen, *Crop Sci.*, **6**, 263 (1966).
3. R. Soiesel and H. Schilcher, *Planta Med.*, **55**, 399 (1989).
4. B. Cammue et al., *Eur. J. Biochem.*, **148**, 315 (1985).
5. M. Hejmanek and V. Dadak, *Cesk. Mykol.*, **13**, 183 (1959).
6. R. Kiesewetter and M. Müller, *Pharmazie*, **13**, 777 (1958).
7. E. Racz-Kotilla and E. Mozes, *Rev. Med. (Tirgu-Mures)*, **17**, 82 (1971).
8. Monograph Graminis rhizoma, Bundesanzeiger, no. 22 (February 1, 1990).

DOGWOOD, JAMAICAN

Source: *Piscidia piscipula* (L.) Sarg. (syn. *P. communis* Harms, *P. erythrina* L., and *Ichthyomethia piscipula* (L.) A. S. Hitchc. ex. Sarg.) (Family Leguminosae or Fabaceae)

Common/vernacular names: West Indian dogwood, fishfuddle, fish-poison tree.

GENERAL DESCRIPTION

Evergreen tree with scaly gray bark, up to about 15 m high; native to tropical America (West Indies, south Florida, Mexico, etc.). Part used is the dried root bark.

Jamaican dogwood bark is different from the bark of flowering dogwood (also known as common dogwood, boxwood, and simply, dogwood). The latter is *Cornus florida* L. (family Cornaceae), which is widely distributed in eastern United States; its dried bark is usually used in domestic medicine as a febrifuge, tonic, and astringent (KROCHMAL AND KROCHMAL; LUST).

CHEMICAL COMPOSITION

Contains ichthynone and jamaicin (isoflavones);[1-3] rotenone and related compounds, including milletone, isomilletone, and dehydromilletone;[1,2,4] piscidic acid (*p*-hydroxybenzyltartaric acid) and its monoethyl and diethyl esters, fukiic acid, and 3′-*O*-methylfukiic acid;[5,6] β-sitosterol;[1,4] simple plant acids (e.g., malic, succinic, and tartaric acids); a saponin glycoside; tannin; and others (LIST AND HÖRHAMMER).

PHARMACOLOGY AND BIOLOGICAL ACTIVITIES

An extract of Jamaican dogwood has been reported to exhibit a sedative effect in cats and guinea pigs as well as marked antitussive and antipyretic activities; it also has anti-inflammatory properties and antispasmodic action on smooth muscles. Its toxicity was reported to be very low in most of the animal species tested.[7]

The aqueous extract of Jamaican dogwood was among 22 extracts of Guatemalan plants that inhibited *in vitro* one or more of the most common dermatophytes, such as *Epidermophyton floccosum* and *Trichophyton mentagrophytes*.[8]

USES

Medicinal, Pharmaceutical and Cosmetic. Extracts are widely used in certain female tonic preparations.

Traditional Medicine. It is used as analgesic, narcotic, and antispasmodic; in promoting sleep; and in treating whooping cough, toothache, asthma, and other ailments.

COMMERCIAL PREPARATIONS

Crude and extracts (solid, powdered, etc.). There are no uniform standards.

REFERENCES

See the General References for HORTUS 3RD; LIST AND HÖRHAMMER; MCGUFFIN 1 & 2; MORTON 2; SARGENT; YOUNGKEN.

1. J. Buechi et al., *Arch. Pharm. Chem.*, **68**, 183 (1961).
2. J. S. P. Schwarz et al., *Tetrahedron*, **20**, 1317 (1964).
3. O. A. Stamm et al., *Helv. Chim. Acta*, **41**, 2006 (1958).
4. A. L. Kapoor et al., *Helv. Chim. Acta*, **40**, 1574 (1957).

5. W. Heller and C. Tamm, *Helv. Chim. Acta*, **58**, 974 (1975).
6. A. Nordal et al., *Acta Chem. Scand.*, **20**, 1431 (1966).
7. M. Aurousseau et al., *Ann. Pharm. Fr.*, **23**, 251 (1965).
8. A. Caceres et al., *J. Ethnopharmacol.*, **31**, 263 (1991).

ECHINACEA

Source: *Echinacea angustifolia* DC, *E. pallida* (Nutt.) Nutt. and *E. purpurea* (L.) Moench. (Family Compositae or Asteraceae).

Common/vernacular names: Purple coneflower, common purple coneflower (*E. purpurea*), pale purple coneflower (*E. pallida*); mixed lots of *E. angustifolia* and *E. pallida* ambiguously traded as "Kansas snakeroot."

GENERAL DESCRIPTION

The genus *Echinacea* has nine indigenous North American herbaceous perennial species. *E. angustifolia*, to 60 cm high; leaves lanceolate; flowers violet, ray florets as long as or less than the width of receptacle; found in barrens, dry prairies; Minnesota to Texas, west to eastern Colorado and Montana; taproot wild dug; little cultivation. *E. pallida*, to 120 cm, leaves lanceolate; purple ray flowers to 9 cm; in glades, prairies; Wisconsin to Arkansas, eastern Texas to Iowa; taproot wild dug; cultivation in the United States and Europe. *E. purpurea*, to 90 cm, leaves ovate, coarsely toothed, basal ones often cordate; root fibrous; widely distributed in Midwestern United States; entire market supply cultivated in Europe, North America, and Australia.[1]

Commercial supplies involve the roots of the above three species, dried tops or fresh flowering herbage of *E. purpurea*, and to a lesser extent, dried tops of *E. angustifolia* and *E. pallida*. Endemic or rare species, including *E. atrorubens, E. paradoxa,* and *E. simulata*, documented in commercial supplies, vicariously harvested; misidentified or intentionally substituted. Root supply of *E. purpurea*, also *E. angustifolia*, historically and persistently adulterated with *Parthenium integrifolium*, traded as the ambiguous "Missouri snakeroot."[1]

CHEMICAL COMPOSITION

Misidentification of source plants involved in chemical analysis before 1986, except for authenticated cultivated *E. purpurea*, renders earlier chemical studies unreliable. Components attributed to *E. angustifolia*, which may instead have involved *E. pallida*, include flavonoid components of the leaves; essential oil constituents, including echinolone, humulene, caryophyllene epoxide, and various polyacetylene components.[2] Four sesquiterpene esters (cinnamates of echinadiol, epoxyechinadiol, echinaxanthol, and dihydroxynardol) attributed to *E. purpurea* in fact do not occur in *Echinacea* but are constituents of *Parthenium integrifolium*.[2] Chemical work by R. Bauer and coworkers at Munich and Düsseldorf now makes distinction of source species in commercial supplies possible.

Echinacoside, once believed a marker compound for *E. angustifolia* (0.3–1.3% roots dry weight, flowers 0.1–1%), has also been found in *E. atrorubens, E. pallida* (0.4–1.7% roots), *E. paradoxa, E. simulata*, and tissue cultures of *E. purpurea*.[2]

Essential oil components common to the aerial parts of *E. pallida, E. purpurea*, and *E. angustifolia* include borneol, bornylacetate, pentadeca-8-en-2-one, germacrene D, caryophyllene, caryophyllene epoxide, and palmitic acid.[2] According to another report, the aerial parts contain β-myrcene, α- and β-pinene, limonene, camphene, *trans*-ocimene, 3-hexen-1-ol, and 2-methyl-4-pentenal.[3] The same report lists dimethyl sulfide, 2- and 3-methylbutanal, 2-propanal, 2-methylpropanal, acetaldehyde, camphene, and limonene as the main volatile constituents of the roots, in addition to α-phellandrene that is present only in *E. purpurea* and *E. angustifolia* roots.[3]

E. angustifolia root contains cynarin (1,5-di-*O*-caffeoylquinic acid); cichoric acid (trace amounts); essential oil (less than 0.1%), containing dodeca-2,4-dien-l-yl-isovalerate, penta-(1,8Z)-diene, 1-pentadecene, and palmitic and linolenic acids; alkylamides, including dodeca-(2E,6Z,8E,10E)-tetraenoic acid isobutylamide (echinacein, at 0.01%

dried roots);[4] and 14 additional isobutylamides from the dried roots of *E. angustifolia* at 0.009–0.151%.[5] Dodeca-2Z,4E,10Z-trien-8-ynoic acid isobutylamide was recently isolated from the root by Chen et al.[6] *E. angustifolia* polysaccharides include inulin (5.9%) and fructans. Other constituents include a resin (yielding oleic, linoleic, cerotic, and palmitic acids on hydrolysis), myristic and linolenic acids, n-triacontanol, β-sitosterol, stigmasterol, sitosterol-3-β-O-glucoside (leaves and stems), behenic acid ethyl ester (roots), and three glycoproteins. Chlorogenic acid and isochlorogenic acids are found in the leaves and stems of *E. angustifolia* and *E. pallida*.[2]

Rutoside is the major leaf flavonoid of *E. angustifolia*, *E. pallida*, and *E. purpurea*. Flavonoids from *E. angustifolia* (possibly involving *E. pallida*) leaves include luteolin, kaempferol, quercetin, quercetagetin-7-glucoside, luteolin-7-glucoside, kaempferol-3-glucoside, apigenin, and isorhamnetin. *E. purpurea* leaves contain quercetin, quercetin-7-glucoside, kaempferol-3-rutinoside, rutin, and others.

E. pallida root essential oil (0.2–2.0%) contains a series of ketoalkynes and ketoalkenes, including tetradeca-8Z-en-11,13-diyn-2-one; pentadeca-8Z-en,11,13-diyn-2-one; pentadeca-8Z,13Z-dien-11-yn-2-one; pentadeca-8Z,11Z,13E-trien-2-one; pentadeca-8Z,11E,13Z-trien-2-one; pentadeca-8Z,11Z-dien-2-one; pentadeca-8Z-en-2-one; and heptadeca-8Z,11Z-dien-2-one. In stored roots (commercial dried root), these components are oxidized by atmospheric oxygen to the hydroxylated derivatives 8-hydroxytetradeca-9E-en-11,13-diyn-2-one; 8-hydroxypentadeca-9E,13Z-dien-11-yn-2-one; and 8-hydroxypentadeca-9E-en-11,13-diyn-2-one. These compounds differentiate *E. pallida* extracts from *E. angustifolia* (absent).[7]

E. purpurea, the best studied species, contains 1.2–3.1% (flowers) and 0.6–2.1% (roots) cichoric acid, and other caffeic acid derivatives; essential oil (in addition to above reported essential oil components) contains vanillin, p-hydroxycinnamic acid methyl ester, and germacrene alcohol (characteristic of fresh plant extracts). Alkylamides include isomeric dodeca-(2E,4E,8Z,10E/Z)-tetraenoic acid isobutylamides, dodeca-2Z,4E-diene-8,10-diynoic acid isobutylamide (also in *E. pallida*),[6] along with 10 additional alkylamides possessing a 2,4-diene structure (compared with one double bond in conjugation with the carbonyl group in *E. angustifolia* alkylamides). The aerial parts of *E. pallida*, *E. angustifolia*, and *E. purpurea* have a comparable alkylamide spectrum, mainly of the 2,4-diene type.[2]

Various polysaccharides from *E. purpurea* dried root include a heteroxylan (mean mol. wt. 35,000), rhamnoarabinogalactan (mean mol. wt. 450,000), and a xyloglucan.[8] Hot water extracts of *E. purpurea* roots contain water-soluble inulin-type fructans mainly composed of linear glucose/fructose polymers with limited branch points.[9] An acidic arabinogalactan and two neutral fucogalactoxyloglucans are found in *E. purpurea* cell tissue cultures.[2] From the pressed juice of *E. purpurea*, Classen et al. identified hydroxyproline-rich glycoproteins with arabinogalactan constituting the polysaccharide portion of the molecule.[10,11] Similar glycoproteins (sugar 96.1%, protein 3.6% w/w), in addition to an arabinan polysaccharide, were also identified in the root of *E. pallida*.[12]

Alkaloids include glycine betaine in *E. purpurea*.[2] The pyrrolizidine alkaloids tussilagine (0.006%) and isotussilagine have been identified in *E. angustifolia* and *E. purpurea* dried roots, neither of which possesses a 1,2-unsaturated necine ring structure associated with hepatotoxicity; therefore, they are not considered to be problematic.[13]

A summary of the chemical composition of *E. angustifolia*, *E. pallida*, and *E. purpurea* can be found in the recent review by Barnes et al. and elsewhere.[14]

PHARMACOLOGY AND BIOLOGICAL ACTIVITIES

Numerous studies report on the wound-healing mechanism of a preparation of the expressed juice of fresh flowering *E. purpurea* on local tissues: inhibits hyaluronidase (both direct and indirect via formation of a complex with hyaluronic acid, causing depolymerization of hyaluronidase), promotes the formation of mesenchymal mucopolysaccharides, stimulates histogenic and haematogenic phagocytes, promotes differentiation of fibrocytes from fibroblasts, stimulates the anterior pituitary–adrenal cortex, and has anti-inflammatory activity.[1,2,15,16] The antihyaluronidase activity was also exhibited by *E. angustifolia* root extracts and was mainly due to the four caffeoyl acid esters (cynarine, chicoric, caftaric and chlorogenic acids) identified by mass spectrometry.[17] Topical application of the root extract of *E. pallida* as well as echinacoside resulted in a marked reduction of local inflammation and improved wound healing in treated rats in comparison to the control group. The authors attributed the observed effect to the antihyaluronidase activity of echinacoside.[18]

Bacteriostatic and fungistatic activities of isolated compounds and/or plant extracts are reported against *Staphylococcus aureus* (weak *in vitro*), *Escherichia coli*, *Pseudomonas aeruginosa*, *Trichomonas vaginalis* (weak *in vitro*), and *Epidermophyton interdigitale*. Yeasts, such as *Saccharomyces cerevisiae* and different species of *Candida*, are susceptible to the light-mediated activity of *E. purpurea* root extracts rich in polyacetylenes and alkamides. Conventional antifungal activity, in absence of UV irradiation, was not as marked as when combined with UV irradiation (enhanced phototoxicity).[19] Different alkamides of *E. purpurea* produced 100% mortality in *Aedes aegyptii* mosquito larvae when tested *in vitro* at 100 μg/mL.[20]

E. purpurea extracts have shown indirect antiviral activity against encephalomyocarditis, vesicular stomatitis, influenza, herpes, and poliovirus; described as interferon-like. The 70% ethanolic extract of *E. pallida* and the *n*-hexane extract of *E. purpurea* inhibited HSV-1 *in vitro* at MIC 0.026 mg/mL and 0.12 mg/mL, respectively.[21] The alkamide constituents of the extracts may be behind the reported anti-HIV activity. Chicoric acid inhibits HIV-1 integrase *in vitro* at concentration as low as 500 nM and inhibits viral integration *in vivo* at concentrations above 1 μM.[22] On the other hand, a 1-year clinical study to investigate the effect of an *E. purpurea* product on recurrent genital herpes in 50 patients failed to demonstrate any significant benefit of the product versus placebo.[23]

Weak oncolytic activity has been reported from pentane extracts against Walker carcinoma 256 and P-388 lymphocytic leukemia, but not lymphoid leukemia.[15] The acidic arabinogalactan isolated from *E. purpurea* tissue culture activated macrophages to cytotoxicity against tumor cells and *Leishmania enriettii*, in addition to stimulating macrophages to produce TNF, IL-1, and IF-β_2, and a slight increase in T-cell proliferation.[24]

The major activity for *Echinacea* species and chemical fractions thereof is nonspecific stimulation of the immune system. Immunostimulant activity involves an overall increase in phagocytosis by macrophages and granulocytes. Oral dosage is as effective as parenteral dosage forms, though acts more slowly.[1,2,15,16]

Immunostimulatory principles have been demonstrated both in lipophilic and polar fractions of extracts of various species and plant parts. Active components of lipophilic (chloroform) extracts may include polyacetylenes, alkylamides, and essential oils. Enriched alkylamide fractions of *E. purpurea*, *E. angustifolia*, and *E. pallida* root ethanol extracts eliminate carbon particles (carbon clearance test) by a factor of 1.5–1.7. Cichoric acid from the polar water-soluble fraction of *E. purpurea* increased carbon elimination by a factor of 2.1. In the granulocyte smear test, all ethanolic root extracts of the three species increased *in vitro* phagocytosis by 20–30% (*E. purpurea* most active). Lipophilic (chloroform) fractions of *E. angustifolia* and *E. pallida* were more active than hydrophilic fractions, while hydrophilic fractions of *E. purpurea* stimulated phagocytosis by 40%.[15,25]

Two high molecular weight polysaccharides from an aqueous extract of *E. purpurea* stimulated T-lymphocyte activity 20–30% more than a potent T-cell stimulator, with a concurrent enhancement of phagocytosis in the carbon clearance test.[26] These compounds are found only in low concentrations in the expressed juice of the herb and precipitate out of ethanol extracts.[2]

Bauer and Wagner concluded that the immunostimulatory activity of alcoholic and aqueous extracts depend on the combined action of several constituents. In lipophilic fractions, alkylamides and the polar caffeic acid derivative cichoric acid contribute to activity of alcoholic extracts. Polysaccharides are implicated in the expressed juice of *E. purpurea* and aqueous extracts, as well as orally administered powdered whole drug.[2,27]

A recent double-blind, placebo-controlled study indicates that a dose of 450 mg/day of *E. purpurea* root extract (1:5 in 55% ethanol) significantly relieved the severity and duration of flu symptoms.[28] A double-blind, monocentric, placebo-controlled clinical trial examined the immunostimulating influence of an expressed fresh juice *E. purpurea* preparation on the course and severity of colds and flu-like symptoms with patients deemed to have greater susceptibility to infections. At a dose of 2–4 mL/day, patients with diminished immune response (expressed by a low T4:T8 cell ratio) were found to benefit significantly from preventative treatment with the Echinacea preparation.[29]

The caffeic acid glycoside echinacoside, once regarded as a significant active principle (based on one report of weak antibacterial activity) and inactive in the carbon clearance test and granulocyte smear test, does not appear to possess immunostimulatory activity.[2,16]

Further studies have suggested that echinacea could play a therapeutic role in a wide range of clinical disciplines, including dermatology, pediatrics, gynecology, surgery, urology, and the treatment of allergies.[30] More than 10 clinical studies have been conducted over the past 10 years. These studies investigated the efficacy of echinacea in immunostimulation,[31–33] upper respiratory tract infections, common cold, and flu. Numerous *in vivo* and *in vitro* studies have been reported about new activities of echinacea. The most important of these are

1. *Antioxidant activity.* Caffeoyl derivatives from echinacea, for example, echinacoside and cynarine, are capable of reducing collagen damage produced by reactive oxygen species and/or other radicals (IC_{50} 15–90 μM).[34]

   Alcoholic extracts of the roots and leaves of *E. angustifolia*, *E. pallida*, and *E. purpurea* exhibited comparable *in vitro* antioxidant activities in ABTS free radical scavenging assay, and the root extracts were also active in the lipid peroxidation inhibition assay.[35] Methanolic extracts of freeze-dried roots of the three species act as *in vitro* antioxidants through the scavenging of DPPH and ABTS radicals, as well as by chelation of the copper ion (Cu^{2+}).[36] Copper-catalyzed oxidation of human LDL was further demonstrated *in vitro* by different preparations from *E. purpurea* root containing alkamides, caffeoyl derivatives, and polysaccharides, as well as by pure caffeoyl derivatives. The antioxidant effect was synergistic and dose dependent.[37]

   The dried juice of *E. purpurea* (i.p. 360 mg/kg every other day, 3-week pre- and 4-week post-irradiation) significantly increased serum SOD activity resulting in an enhanced blood antioxidant activity in response to oxidative stress caused by exposure to a high dose of X-rays.[38]

2. *Anti-inflammatory activity.* Alkamides from *E. angustifolia*, and other medicinal herbs, exhibited *in vitro* inhibition of sheep seminal microsome cyclooxygenase and porcine leukocyte 5-lipooxygenase.[39]

   The extract of *E. purpurea* significantly inhibited the formation of rat paw edema in mice (oral administration, 100 mg/kg, twice daily). The involved mechanism was down-regulation of COX-2 expression as shown by Western blotting.[40] Several alkamides from *E. purpurea* showed ca. 48% and 30% inhibition of COX-1 and -2, respectively, at 100 μg/mL.[20]

Two extracts of *E. purpurea* were capable of *in vitro* reversing the release of more than 30 cytokine-related mediators of inflammation in human bronchial cells infected with rhinovirus 14. The fact that the same extracts stimulated cytokine production in uninfected cells further substantiated the reputed immunostimulant effects of Echinacea.[41]

3. *Cannabinomimetic activity.* The echinacea alkamides dodeca-2*E*,4*E*,8*Z*,10*Z*-tetraenoic acid isobutylamide and dodeca-2*E*,4*E*-dienoic acid isobutylamide bind more strongly to the human CB2 cannabinoid receptor (K_i 60 nM) than the endogenous cannabinoid anandamide ($K_i > 200$ nm). The CB2 binding resulted in inhibition of LPS-induced inflammation in human whole blood.[42] An earlier study with rodent cannabinoid receptors (CB1 and CB2) also showed that alkamides from *E. angustifolia* roots displayed selective CB2 binding.[43]

4. *Antiandrogenic activity.* In two separate studies, *E. purpurea* extract added to the usual diet of Wistar rats (dose: 50 mg/Kg) resulted in a significant reduction in the weights of the testicles and prostate gland after 8 weeks of treatment in comparison to the control group.[44,45]

The importance of echinacea in health and medicine has been the subject of numerous review articles that are either broad or specific in nature. Also, detailed monographs have been published that describe in detail the botany, chemistry, and pharmacology of the different species of Echinacea.

TOXICOLOGY

Some adverse effects have been reported for echinacea. For example, more than 50 cases of allergy were detected in Australia and were linked to the use of echinacea preparations. Symptoms included acute asthma, maculopapular rash/urticaria, and anaphylaxis. There was evidence of the involvement of an echinacea-binding IgE in ca. 50% of the cases and the possibility of cross-reactivity between echinacea and other environmental allergens was proposed.[46,47]

USES

Medicinal, Pharmaceutical, and Cosmetic. Echinacea preparations, especially oral (liquid extract), topical (ointment), and parenteral products of the fresh aboveground preparations of *E. purpurea*, and, to a lesser extent, the roots of *E. angustifolia* and *E. pallida*, are used in Germany for the external treatment of hard-to-heal wounds, eczema, burns, psoriasis, herpes simplex, and so on. As immunostimulants; internally a prophylactic at the onset of cold and flu symptoms and for treatment of *Candida albicans* infections, chronic respiratory infections, prostatitis, polyarthritis (rheumatoid arthritis), and so on.[27]

Echinacea extracts, increasingly seen in cosmetics, are used in lip balms, shampoos, toothpaste, and other product categories.

Dietary Supplements/Health Foods. Numerous oral dosage products, vicariously positioned as cold and flu preventatives.[1]

Traditional Medicine. "This plant (*E. angustifolia*) was universally used as an antidote for snakebite and other venomous bites and stings and poisonous conditions. Echinacea seems to have been used as a remedy for more ailments than any other plant."[48] Diseases and conditions for which echinacea was employed by physicians (1887–1939) included old sores, wounds, snakebite, gangrene, and as a local antiseptic; internally for diphtheria, typhoid conditions, cholera infantum, syphilis, and blood poisoning.[1]

COMMERCIAL PREPARATIONS

Ointment, oral liquid, intravenous and intramuscular ampoules from expressed juice of fresh flowering *E. purpurea*; tinctures, extracts, capsules, tablets, and so on of *E. angustifolia*, *E. pallida*, and *E. purpurea*. More than 280 echinacea pharmaceutical products are available in Europe.[2] Echinacea extracts "standardized" to echinacoside persist in the marketplace, despite the fact that the compound has insignificant biological activity. *E. angustifolia* and *E. pallida* crude were formerly official in N.F.

Regulatory Status. Class 1 dietary supplement (can be safely consumed when used appropriately); formerly *E. angustifolia/E. pallida* official in N.F. (1916–1950). *E. purpurea* fresh aboveground parts,[49] and *E. pallida* root[50] subjects of positive therapeutic monographs for human use in Germany.

REFERENCES

See the General References for APPLEQUIST; BARNES; BISSET; BLUMENTHAL 2; DER MARDEROSIAN AND BEUTLER; FOSTER; FOSTER AND DUKE; GRIEVE; MCGUFFIN 1 & 2 ; STEINMETZ; TYLER 1; UPHOF; WEISS; WREN.

1. S.Foster, *Echinaceae—Nature's Immune Enhancer*, Healing Arts Press, Rochester, VT, 1991.
2. R. Bauer and H. Wagner, *Economic and Medicinal Plant Research*, Academic Press, New York, 1991, p. 253.
3. G. Mazza and T. Cottrell, *J. Agric. Food Chem.*, **47**, 3081 (1999).
4. M. Jacobson, *J. Org. Chem.*, **32**, 1646 (1967).
5. R. Bauer et al., *Phytochemistry*, **28**, 505 (1989).
6. Y. Chen et al., *J. Nat. Prod.*, **68**, 773 (2005).
7. R. Bauer et al., *Planta Med.*, **54**, 426 (1988).
8. H. Wagner and A. Proksch, *Economic and Medicinal Plant Research*, Academic Press, New York, 1985, p. 113.
9. M. Wack and W. Blaschek, *Carbohydr. Res.* (2006).
10. B. Classen et al., *Planta Med.*, **71**, 59 (2005).
11. B. Classen et al., *Carbohydr. Res.*, **327**, 497 (2000).
12. S. Thude and B. Classen, *Phytochemistry*, **66**, 1026 (2005).
13. E. Roder et al., *Dtsch. Apoth. Ztg.*, **124**, 2316 (1984).
14. J. Barnes et al., *J. Pharm. Pharmacol.*, **57**, 929 (2005).
15. C. Hobbs, *The Echinacea Handbook*, Eclectic Medical Publications, Portland, OR, 1989.
16. D. V. C. Awang and D. G. Kindack, *Can. Pharm. J.*, **124**, 512 (1991).
17. R. M. Facino et al., *Farmaco*, **48**, 1447 (1993).
18. E. Speroni et al., *J. Ethnopharmacol.*, **79**, 265 (2002).
19. S. E. Binns et al., *Planta Med.*, **66**, 241 (2000).
20. L. J. Clifford et al., *Phytomedicine*, **9**, 249 (2002).
21. S. E. Binns et al., *Planta Med.*, **68**, 780 (2002).
22. R. A. Reinke et al., *Virology*, **326**, 203 (2004).
23. B. Vonau et al., *Int. J. STD AIDS*, **12**, 154 (2001).
24. B. Luetting et al., *J. Natl. Cancer Inst.*, **81**, 669 (1989).
25. R. Bauer et al., *Z. Phytother.*, **10**, 43 (1989).
26. H. Wagner and A.A. Proksch, *Angew Phytother.*, **2**, 166 (1981).
27. S. Foster, *Echinacea—The Purple Coneflowers*, The American Botanical Council, Austin, TX, 1991.
28. B. Braunig et al., *Z. Phytother.*, **13**, 7 (1992).
29. D. Schoneberger, *Forum Immunol.*, **8**, 2 (1992).
30. G. Hahn and A. Mayer, *Oster. Apoth. Ztg.*, **38**, 1040 (1984).
31. L. L. Agnew et al., *J. Clin. Pharm. Ther.*, **30**, 363 (2005).
32. L. S. Kim et al., *Altern. Med. Rev.*, **7**, 138 (2002).

33. E. Schwarz et al., *Phytomedicine*, **12**, 625 (2005).
34. R. M. Facino et al., *Planta Med.*, **61**, 510 (1995).
35. B. D. Sloley et al., *J. Pharm. Pharmacol.*, **53**, 849 (2001).
36. C. Hu and D. D. Kitts, *J. Agric. Food Chem.*, **48**, 1466 (2000).
37. L. Dalby-Brown et al., *J. Agric. Food Chem.*, **53**, 9413 (2005).
38. S. Mishima et al., *Biol. Pharm. Bull.*, **27**, 1004 (2004).
39. B. Muller-Jakic et al., *Planta Med.*, **60**, 37 (1994).
40. G. M. Raso et al., *J. Pharm. Pharmacol.*, **54**, 1379 (2002).
41. M. Sharma et al., *Phytother. Res.*, **20**, 147 (2006).
42. S. Raduner et al., *J. Biol. Chem.* (2006).
43. K. Woelkart et al., *Planta Med.*, **71**, 701 (2005).
44. D. Skaudickas et al., *Medicina (Kaunas.)*, **39**, 761 (2003).
45. D. Skaudickas et al., *Medicina (Kaunas.)*, **40**, 1211 (2004).
46. R. J. Mullins, *Med. J. Aust.*, **168**, 170 (1998).
47. R. J. Mullins and R. Heddle, *Ann. Allergy Asthma Immunol.*, **88**, 42 (2002).
48. M. R. Gilmore in *Thirty-Third Annual Report of the Bureau of the American Ethnology* U.S. Government Printing Office, Washington, DC, 1919, p. 145.
49. Monograph *Echinaceae Purpureae herba*, *Bundesanzeiger*, no. 43 (March 2, 1989).
50. Monograph *Echinaceae Pallide radix*, *Bundesangeizer*, no. 44 (August 29, 1992).

ELDER FLOWERS (AMERICAN AND EUROPEAN)

Source: *American elder Sambucus canadensis* L.; *European elder Sambucus nigra* L. (Family Caprifoliaceae).

Common/vernacular names: Sweet elder, common elder, and American elderberry (*S. canadensis*); sambucus.

GENERAL DESCRIPTION

American elder (*S. canadensis*) is a tall shrub with white pith, spreading by suckers; up to about 4 m high; native to eastern North America. Part used is the flower.

European elder (*S. nigra*) is a tall shrub or small tree; up to about 10 m high; native to Europe and naturalized in the United States. Parts used are the dried flower and leaf.

CHEMICAL COMPOSITION

European elder flowers contain a small amount of an essential oil (ca. 0.3%) that is composed primarily of free fatty acids (66%) and alkanes (ca. 7%), with palmitic and linolenic acids being the major acids and C_{19}, C_{21}, C_{23} and C_{25} alkanes the major alkanes;[1] triterpenes, including α-amyrin and β-amyrin (mainly as fatty acid esters), ursolic acid, 30β-hydroxyursolic acid, and oleanolic acid;[2,3] sterols (as free sterols, esters, and glycosides);[3] flavonoids, flavone glycosides, and phenolic acids, including quercetin, kaempferol, isoquercitrin, rutin (up to 1.9%), and chlorogenic acid;[4-6] pectin; sugar; and others.[6]

Elder leaf contains sambunigrin (a cyanogenic glucoside) at 0.042% concentration, according to one source;[7] choline; rutin and quercetin;[7,8] sterols (sitosterol, stigmasterol, and campesterol); triterpenes (α- and β-amyrin palmitates, oleanolic acid, and ursolic acid); alkanes (mainly *n*-nonacosane and *n*-hentriacontane); fatty acids (stearic, oleic, linoleic, etc.);[8,9] tannins; resins; fats; sugars; vitamin C; and others.[7,10]

American elder fruits contain acylated anthocyanin glycosides, such as cyanidin 3-*O*-(6-*O*-*E*-*p*-coumaroyl-2-*O*-β-D-xylopyranosyl)-β-D-glucopyranoside, and other related glycosides.[11,12]

PHARMACOLOGY AND BIOLOGICAL ACTIVITIES

Elder flowers are generally considered to have diuretic, diaphoretic, and laxative properties.

One study found that sambuculin A and α- and β-amyrin palmitate from *S. formosana* have antihepatotoxic activity against carbon tetrachloride-induced liver damage.[13]

Compounds from *Sambucus* species contain a number of plant lectins with hemagglutinin characteristics, which could prove useful in blood typing and hematological tests.[14,15]

USES

Medicinal, Pharmaceutical, and Cosmetic. European elder flower water (water phase from steam distillate) has been used as a vehicle for eye and skin lotions.

Extracts (e.g., absolute) of the flowers are used in perfumes.

Food. Flowers are reported to be used as flavor components in numerous food products, including alcoholic (bitters and vermouths) and nonalcoholic beverages, frozen dairy desserts, candy, baked goods, and gelatins and puddings. Highest average maximum use level reported is 0.049% in nonalcoholic beverages.

Dietary Supplements/Health Foods. The flowers of *S. canadensis* are used as a tea ingredient, also in capsules, tablets, and so on; primarily as diaphoretic for colds and flu (FOSTER AND DUKE).

Traditional Medicine. Flowers are used as diuretic, laxative, and diaphoretic, as well as a gentle astringent for the skin and in treating rheumatism, usually in the form of a tea, infusion, poultice, or water distillate.

Bark of American elder and bark, leaf, flower, root, and fruit of European elder have been reported to be used in cancers.[16]

In Chinese medicine, a related species, *Sambucus williamsii* Hance, is widely used. Flowers are used as diaphoretic and diuretic; twigs, leaves, and roots in treating rheumatoid arthritis, among other conditions, usually in the form of a decoction taken internally (JIANGSU).

COMMERCIAL PREPARATIONS

Crude and extracts; flowers were formerly official in N.F. Strengths (see *glossary*) of extracts are expressed in weight-to-weight ratios or in flavor intensities.

Regulatory Status. Flowers are GRAS (American, §182.10 and §182.20; European, §182.20); leaves have been approved for use in alcoholic beverages only, with the provision that hydrocyanic acid (HCN) should not exceed 25 ppm (0.0025%) in the flavor (§172.510). *S. nigra* flowers are subject of a German therapeutic monograph; allowed as a diaphoretic and to increase bronchial secretion in the treatment of colds.[17]

REFERENCES

See the General References for ARCTANDER; BAILEY 1; BLUMENTHAL 1; FEMA; FERNALD; FOSTER AND DUKE; FURIA AND BELLANCA; KROCHMAL AND KROCHMAL; MARTINDALE; ROSE; TERRELL; UPHOF.

1. W. Ritchter and G. Willuhn, *Dtsch. Apoth. Ztg.*, **114**, 947 (1974).

2. R. Hänsel and M. Kussmaul, *Arch. Pharm. (Weinheim)*, **308**, 790 (1975).

3. W. Ritchter and G. Willuhn, *Pharm. Ztg.*, **122**, 1567 (1977).
4. A. Radu et al., *Farmacia (Bucharest)*, **24**, 9 (1976).
5. I. Leifertova et al., *Acta Fac. Pharm., Univ. Comeniana*, **20**, 57 (1971).
6. I. I. Hajkova and V. Brazdova, *Farm. Obzor.*, **32**, 343 (1963).
7. Z. N. Guseinova, *Azerb. Med. Zh.*, **42**, 29 (1965).
8. T. Inoue and K. Sato, *Phytochemistry*, **14**, 1871 (1975).
9. A. M. Al-Moghazy Shoaib, *Egypt J. Pharm. Sci.*, **13**, 255 (1972).
10. S. R. Jensen and B. J. Nielsen, *Acta Chem. Scand.*, **27**, 2661 (1973).
11. N. Nakatani et al., *Phytochemistry*, **38**, 755 (1995).
12. O. P. Johansen et al., *Phytochemistry*, **30**, 4137 (1991).
13. C. N. Lin and W. P. Tome, *Planta Med.*, **54**, 223 (1988).
14. L. Marklc et al., *Biochem. J.*, **278**, 667 (1991).
15. H. Kaku et al., *Arch. Biochem. Biophys.*, **277**, 255 (1990).
16. J. L. Hartwell, *Lloydia*, **31**, 71 (1968).
17. Monograph *Sambuci flos*, *Bundesangeizer*, no. **50** (March 13, 1986); revised (March 13, 1990).

ELECAMPANE

Source: *Inula helenium* L. (syn. *Helenium grandiflorum* Gilib.; *Aster officinalis* All.; *A. helenium* (L.) Scop.) (Family Compositae or Asteraceae).

Common/vernacular names: Scabwort, inula, alant, horseheal, and yellow starwort.

GENERAL DESCRIPTION

A perennial herb, covered with soft short hairs; up to about 1.8 m high; native to Europe and Asia; naturalized in North America; cultivated in Europe (Belgium, France, Germany, etc.) and Asia (e.g., China). Parts used are the dried roots and rhizomes collected in late fall or early winter. An essential oil is obtained by steam distillation.

CHEMICAL COMPOSITION

Contains 1–4% volatile oil, which is composed primarily of sesquiterpene lactones, including alantolactone (also called helenin, elecampane camphor, and alant camphor), isoalantolactone, dihydroisoalantolactone, and dihydroalantolactone; alantic acid; and azulene (JIANGSU; LIST AND HÖRHAMMER).[1–5] Other sesquiterpene lactones present in the root include elemane, isocostunolide, 4β,5α-epoxy-1(10),11(13)-germacradiene-8,12-olide in addition to the alantolactones.[6,7]

Other constituents reported to be present include up to about 44% inulin; sterols (stigmasterol, β- and γ-sitosterols, damaradienol, etc.);[8] friedelin; resin; pectic substances; and others (LIST AND HÖRHAMMER).

Thymol isobutyrate derivatives have been isolated from the root cultures.[9]

Alantolactone was reported to have strong inhibitory effects on seed germination and seedling growth.[10]

PHARMACOLOGY AND BIOLOGICAL ACTIVITIES

An infusion of elecampane has been reported to have a pronounced sedative effect on mice.[11] A tincture preparation was recently shown to alleviate the symptoms of experimentally induced stress in cellular models.[12]

Much of the pharmacological properties of elecampane is due to alantolactone, which has been reported to have anthelmintic activities *in vitro*[13] and in humans (similar to santonin but better and less toxic) and hypotensive, hyperglycemic (in large doses), and hypoglycemic (smaller doses) effects in experimental animals. It also has antibacterial and antifungal properties, among others (JIANGSU).

Among 105 plant lactones studied, alantolactone and isoalantolactone are among the few that have been reported to exhibit the highest bactericidal and fungicidal properties *in vitro*.[14] The recently isolated epoxythymol isobutyrate also exhibits moderate activity against Gram-negative, Gram-positive, *Pseudomonas*, and *Candida*.[9] Root extracts rich in sesquiterpene lactones were significantly active against *Mycobacterium tuberculosis in vitro*.[15]

Alantolactone was reported as an immunostimulant.[16]

Root extracts and their major sesquiterpene constituents have been shown to possess selective cytotoxic activity against different tumor cell lines, such as HeLa, HepG2, HT-29, MCF-7, and Capan-2.[6,7,17] Caspase-dependent apoptosis may be one of the mechanisms involved in the displayed antitumor activity.[6]

TOXICOLOGY

Some individuals are extremely sensitive to the oil when applied to the skin.[18] Contact dermatitis and allergic cross-reactions with other plants belonging to the same family, for example, Echinacea and Chamomile, are not uncommon.[19,20]

USES

Medicinal, Pharmaceutical, and Cosmetic. Alantolactone is used as an anthelmintic, primarily in the United Kingdom and Europe.

The oil is used as a fragrance component in cosmetic products, including soaps, detergents, creams, lotions, and perfumes (e.g., oriental types). Highest maximum use level reported is 0.4% in perfumes.[18]

Food. Reportedly used as a flavor ingredient in major food products, including alcoholic (aromatic bitters, vermouths, etc.) and nonalcoholic beverages, frozen dairy desserts, candy, baked goods, and gelatins and puddings. Highest average maximum use level reported is 0.08% for the crude in baked goods.

Dietary Supplements/Health Foods. Powdered root used in tea formulations; sometimes in other product forms for lung conditions (FOSTER).

Traditional Medicine. Reportedly used in treating asthma, bronchitis, whooping cough, nausea, diarrhea, and other ailments. Also used as a diuretic, stomachic, and anthelmintic both in Western domestic medicine and in Chinese medicine, usually in the form of a decoction or tea. It has reportedly been used in treating cancers.[21]

COMMERCIAL PREPARATIONS

Crude, oil, and extracts; crude was formerly official in N.F.

Regulatory Status. Has been approved for use in alcoholic beverages only (§172.510). Root subject of a German therapeutic monograph; not recommended due to lack of evidence of efficacy and risk for allergic reactions.[22]

REFERENCES

See the General References for ARCTANDER; BAILEY 1; BARNES; BIANCHINI AND CORBETTA; BLUMENTHAL 1; BRUNETON; FEMA; FOSTER; FURIA AND BELLANCA; GRIEVE; GUENTHER; JIXIAN; JIANGSU; LUST; MARTINDALE; MCGUFFIN 1 & 2; NANJING; TERRELL; UPHOF; WREN.

1. A. Boeva et al., *Farmatsiya (Sofia)*, **21**, (1971).
2. S. S. Kerimov and O. S. Chizhov, *Khim. Prir. Soedin.*, **10**, 254 (1974).
3. P. P. Khvorost and N. F. Komissarenko, *Khim. Prir. Soedin.*, **6**, 820 (1976).
4. V. G. Sinitsina and Z. I. Boshko, *Nek. Probl. Farm. Nauki Prakt., Mater S'ezda Farm. Kaz.*, **1**, 87 (1975).
5. Y. Kashman et al., *Isr. J. Chem.*, **5**, 23 (1967).
6. C. N. Chen et al., *Cancer Lett.*, **246**, 237 (2007).
7. T. Konishi et al., *Biol. Pharm. Bull.*, **25**, 1370 (2002).
8. W. Olechnowicz-Stepien et al., *Rocz. Chem.*, **49**, 849 (1975).
9. A. Stojakowska et al., *Z. Naturforsch. C*, **59**, 606 (2004).
10. E. Rodriguez et al., *Phytochemistry*, **15**, 1573 (1976).
11. R. Kiesewetter and M. Muller, *Pharmazie*, **13**, 777 (1958).
12. K. L. Zelenskaya et al., *Bull. Exp. Biol. Med.*, **139**, 414 (2005).
13. M. F. El Garhy and L. H. Mahmoud, *J. Egypt. Soc. Parasitol.*, **32**, 893 (2002).
14. S. A. Vichkanova et al., *Rastit. Resur.*, **13**, 428 (1977).
15. C. L. Cantrell et al., *Planta Med.*, **65**, 351 (1999).
16. H. Wagner and A. Proksch, *Economic and Medicinal Plant Research*, Academic Press, New York, 1985, p. 113.
17. D. C. Dorn et al., *Phytother. Res.*, **20**, 970 (2006).
18. D. L. J. Opdyke, *Food Cosmet. Toxicol.*, **14**, 307 (1976).
19. E. Paulsen et al., *Contact Dermatitis*, **45**, 197 (2001).
20. E. Paulsen, *Contact Dermatitis*, **47**, 189 (2002).
21. J. L. Hartwell, *Lloydia*, **31**, 71 (1968).
22. Monograph *Helenii radix*, *Bundesangeizer*, no. 85 (May 5, 1988).

ELEMI GUM

Source: *Canarium commune* L. and *C. luzonicum* Miq. (Family Burseraceae).

Common/vernacular names: Manila elemi, elemi oleoresin, and elemi resin.

GENERAL DESCRIPTION

Trees up to about 30 m high; native to the Philippines and Moluccas. Part used is their resinous pathological exudation from which an essential oil (elemi oil) is obtained by steam distillation.

CHEMICAL COMPOSITION

Elemi gum (oleoresin) contains 65–75% triterpenoid resinous compounds, which include amyrin, brein, maniladiol, elemadienolic acid, and elemadienonic acid, and 10–25% volatile oil, consisting mainly of phellandrene, dipentene, elemol, elemicin, terpineol, carvone, and terpinolene (KARRER; POUCHER).[1–4]

TOXICOLOGY

Available data indicate elemi and elemi oil to be relatively nontoxic.[5]

USES

Medicinal, Pharmaceutical, and Cosmetic. Elemi resinoid and elemi oil are used as fixatives and fragrance components in soaps, detergents, creams, lotions, and perfumes. Maximum use level of the oil reported is 0.6% in perfumes.

Food. Elemi oil is reported to be used as a flavor component in major categories of food products, including alcoholic and nonalcoholic beverages, frozen dairy desserts, candy, baked goods, gelatins and puddings, meat and meat products, and condiments and relishes. Average maximum use levels are low, with highest reported being about 0.002% (17.2 ppm) in baked goods.

Elemi gum is also used but much less frequently.

Traditional Medicine. Resin is reportedly used as a stomachic and as an expectorant; also used externally as a local stimulant (MARTINDALE).[2]

COMMERCIAL PREPARATIONS

Gum and oil.

Regulatory Status. Has been approved for food use (§172.510).

REFERENCES

See the General References for ARCTANDER; FEMA; GUENTHER; POUCHER; UPHOF.

1. A. F. Summa, *Diss. Abstr.*, **21**, 1772 (1961).
2. R. Pernet, *Lloydia*, **35**, 280 (1972).
3. G. D. Manalo and A. P. West, *Philippine J. Sci.*, **78**, 111 (1949); through *Chem. Abstr.*, **44**, 7564f (1950).
4. M. Mladenović and D. Fodor-Mandušić, *Acta Pharm. Jugoslav.*, **8**, 59 (1958); through *Chem. Abstr.*, **52**, 17614i (1958).
5. D. L. J. Opdyke, *Food Cosmet. Toxicol.*, **14**(Suppl.) 755 (1976).

ELEUTHERO

Source: *Eleutherococcus senticosus* (Rupr. and Maxim.) Maxim. (syn. *Acanthopanax senticosus* Harms) (Family Araliaceae).

Common/vernacular names: Siberian ginseng, eleutheroco, eleuthero ginseng, and Ussurian thorny pepperbush.

GENERAL DESCRIPTION

Deciduous shrub, 1–3 m high, branches beset with numerous small sharp spines; leaves palmate; in northeast Asia, including much of far southeastern Russia (middle Amur region in the north to Sakhalin), northeast China (Heilongjiang, Jilin, Liaoning, Nei Monggol, Hebei, Shanxi); abundant in the Xiaoxinganling Mountains of Heilongjiang, adjacent Korea, and Japan (Hokkaido). In traditional Chinese medicine, the bark of the root historically used; currently root, rhizome, stems, and leaves enter commerce.[1]

CHEMICAL COMPOSITION

Russian workers initially isolated seven compounds from a methanol extract of roots, deemed eleutherosides A–G (in a ratio of 8:30:10:12:4:2:1), ranging from 0.6% to 0.9% (roots); 0.6% to 1.5% (stems); seven additional eleutherosides have been identified.[2] Eleutheroside A is the sterol daucosterol; eleutheroside B is the phenylpropanoid syringin; eleutheroside B_1 is isofraxidin-7-O-α-L-glucoside (β-calycanthoside); eleutheroside B_2 and B_3. Eleutheroside B_4 ((−)-sesamin), eleutheroside D ((−)-syringaresinol-di-O-β-D-glucoside), and eleutheroside E (acanthoside D) are lignans. A new lignan, termed eleutheroside E_2, together with isomaltol glucoside and thymidine were later isolated by Li et al.[3] Triterpenes include eleutherosides I–M, (eleutheroside I is mussenin B; eleutheroside M is hederasaponin), senticosides A–F (incompletely characterized oleanolic acid glycosides), and oleanolic acid. Additional phenylpropanoids include caffeic acid, caffeic acid ethyl ester, dicaffeoylquinic acids, coniferyl aldehyde, and sinapyl alcohol; other components include β-sitosterol, galactose, α- and β-glucose, α- and β-maltose, sucrose, vitamin E, β-carotene,[2,4] polysaccharides (eleutherans A–G),[5] and two glucose-, galactose-, and arabinose-containing polysaccharides.[6] Saponin glycosides, well known in various *Panax* species, are absent in *E. senticosus*.

PHARMACOLOGY AND BIOLOGICAL ACTIVITIES

In vivo animal studies of root ethanol extracts have been evaluated for adaptogenic activity in hyperthermia, electroshock-induced convulsions, toxic cardioglycoside dose resistance, cortisone-induced lymphatic stress, gastric ulcers, X-ray irradiation, increased metabolic efficiency in swimming-induced stress, increased conditioned response to stimuli, inhibition of conditioned avoidance response, NK cell activity, and corticosterone level.[2,7,8] Clinical trials have also been conducted in humans to substantiate the adaptogenic effect of eleuthero. Some observed end points included quality of life in geriatrics, steroidal hormone indices and lymphocyte subset numbers, endurance, cellular stress, and physical fitness. The collective results of these trials indicate an inconsistent pattern of effectiveness for Siberian ginseng.[9–12] A number of reviews about the adaptogenic activity of eleuthero have been published.[13–15]

Antioxidant (free radical scavenging) activity has been demonstrated in animals.[2] The hepatoprotective effect of the aqueous extract and polysaccharide fraction of the root was demonstrated in significant attenuation of hepatic failure in mice (300 mg/kg, oral and i.p.) induced by D-galactosamine/lipopolysaccharide injection.[16]

A root extract had a marked antiviral effect against the RNA viruses human rhinovirus, respiratory syncytial virus, and influenza A virus *in vitro*. DNA viruses were not affected by the same extract.[17]

A hypoglycemic effect has been demonstrated in animals, though the mechanism of action is unclear.[2]

Antiedema, anti-inflammatory, diuretic, gonadotropic activity, estrogenic activity, and antihypertensive activity were also reported *in vivo*.[2] The antihypertensive activity may be explained in view of a recent report by Kwan et al. in which Siberian ginseng aqueous extract resulted in a concentration-dependent relaxation in different contracted vascular preparations (dog carotid arterial rings, rat aorta, and rat artery) at 0.04–2.0 mg/mL. The observed relaxation was attributed to NO and/or EDHF, in addition to other possible mechanisms.[18]

Polysaccharides are responsible for immunostimulatory activity (carbon clearance and granulocyte tests); lessened thioacetamide, phytohemagglutin, and X-ray toxicity; antitumor activity; and hypoglycemic activity.[5,6] Recent studies demonstrated that eleutherosides B and E may also contribute to the immunostimulant activity. For example, the ethanolic extract of eleuthero stimulated the *in vitro* production of IL-1 and IL-6 but not IL-2.[19] Two oral preparations containing the root extracts displayed varying levels of enhanced humoral responses in mice as evidenced by the increased levels of immunogobulins in mouse serum.[20]

A carcinostatic effect, slowing the spread of metastases, is suggested by Russian research (DUKE 2). A comprehensive review of Russian studies is available.[2]

USES

Medicinal, Pharmaceutical, and Cosmetic. Used to increase general resistance as an adaptogenic and immunostimulant.[2] Hydroalcoholic and glycolic extracts increasingly used in skin care products, including creams and lotions.

Dietary Supplements/Health Foods. Numerous oral dosage products, including liquid extract, tinctures, capsules, tablets, crude herb, and so on as tonic tea (FOSTER AND YUE).

Traditional Medicine. In China, the whole root and rhizome is known as *ciwujia*; reportedly used as stimulant, tonic, adaptogenic, diuretic; for sleeplessness, lower back or kidney pain, lack of appetite, rheumatoid arthritis; to enhance overall resistance to disease or adverse physical influences or stress (FOSTER AND YUE).

Bark of *E. senticosus* (*jiapi* or *ciwujiapi*) is a substitute to the bark of *E. gracilistylus* (*Acanthopanax gracilistylus*), source of *wujiapi* (also *jiapi*). Bark of *Periploca sepium* (Asclepidaceae), also known as "*wujia*" (*xiangjiapi, gangliupi*, and *beiwujiapi*), has prompted market confusion, and consequently entered trade as an adulterant to *E. senticosus*.[1] There is a report of a purported case of neonatal androgenization associated with maternal "ginseng" use; attributed to "pure Siberian ginseng."[21] Follow-up revealed that the product in question did not contain eleuthero but *Periploca sepium* instead, prompting a pharmacological study in which no androgenicity was observed.[22–24]

COMMERCIAL PREPARATIONS

Crude (bark of stem, whole root and rhizome, rarely bark of root) and extracts (liquid, solid, powder, etc.).

Regulatory Status. Class 1 dietary supplement (herb that can be safely consumed when used appropriately). Root subject of a German regulatory monograph indicated as a tonic for invigoration during fatigue, debility, declining work capacity and concentration, and during convalescence.[25]

REFERENCES

See the General References for APPLEQUIST; BARNES; BLUMENTHAL 1 & 2; BRUNETON; DER MARDEROSIAN AND BEUTLER; DUKE 2; FOSTER AND YUE; MCGUFFIN 1 & 2; TYLER 1; WEISS.

1. S. Foster, *Eleutherococcus senticosus*, American Botanical Council, Austin, TX, 1991.
2. N. R. Farnsworth et al., *Economic and Medical Plant Research*, Academic Press, New York, 1985, p. 155.
3. X. C. Li et al., *Planta Med.*, **67**, 776 (2001).
4. A. Tolonen et al., *Phytochem. Anal.*, **13**, 316 (2002).
5. N. H. Hikino et al., *J. Nat. Prod.*, **49**, 293 (1986).
6. H. Wagner and A. Proksch, *Economic and Medical Plant Research*, Academic Press, New York, 1985, p. 113.
7. Y. Kimura and M. Sumiyoshi, *J. Ethnopharmacol.*, **95**, 447 (2004).
8. T. Deyama et al., *Acta Pharmacol. Sin.*, **22**, 1057 (2001).
9. A. F. Cicero et al., *Arch. Gerontol. Geriatr. Suppl.*, 69 (2004).
10. B. T. Gaffney et al., *Life Sci.*, **70**, 431 (2001).
11. J. Szolomicki et al., *Phytother. Res.*, **14**, 30 (2000).
12. L. F. Eschbach et al., *Int. J. Sport Nutr. Exerc. Metab.*, **10**, 444 (2000).
13. A. Panossian and H. Wagner, *Phytother. Res.*, **19**, 819 (2005).
14. E. D. Goulet and I. J. Dionne, *Int. J. Sport Nutr. Exerc. Metab.*, **15**, 75 (2005).
15. M. Davydov and A. D. Krikorian, *J. Ethnopharmacol.*, **72**, 345 (2000).
16. E. J. Park et al., *Basic Clin. Pharmacol. Toxicol.*, **94**, 298 (2004).
17. B. Glatthaar-Saalmuller et al., *Antiviral Res.*, **50**, 223 (2001).

18. C. Y. Kwan et al., *Naunyn Schmiedebergs Arch. Pharmacol.*, **369**, 473 (2004).
19. G. G. Steinmann et al., *Arzneim.-Forsch.*, **51**, 76 (2001).
20. J. Drozd et al., *Acta Pol. Pharm.*, **59**, 395 (2002).
21. G. Koren et al., *J. Am. Med. Assoc.*, **264**, 2866 (1990).
22. D. V. C. Awang, *J. Am. Med. Assoc.*, **265**, 1828 (1991).
23. D. V. C. Awang, *J. Am. Med. Assoc.*, **266**, 363 (1991).
24. D. P. Waller et al., *J. Am. Med. Assoc.*, **267**, 2329 (1992).
25. Monograph *Eleutherococci radix*, *Bundesanzeiger*, no. 11 (January 17, 1991).

EPHEDRA

Source: *Chinese ephedra Ephedra sinica* Stapf.; *Intermediate ephedra Ephedra intermedia* Shrenk et C. A. Mey.; *Mongolian ephedra Ephedra equisetina* Bge.; *Other Ephedra* spp. (Family Ephedraceae).

Common/vernacular names: MaHUANG, Herba Ephedrae, cao maHUANG (Chinese ephedra), zhong maHUANG (intermediate ephedra), and muzei maHUANG (Mongolian ephedra); maHUANGgen (root).

GENERAL DESCRIPTION

Low shrubs with scale-like leaves, 1.5–3.3 m high; stems herbaceous above and woody below; herbaceous stems greenish, with those of *E. intermedia* and *E. equisetina* often covered with a white powder. *E. equisetina* is the largest among the three, while *E. sinica* is the smallest, with herbaceous features; all flowering in spring and fruiting in late summer. Native to central Asia, now distributed throughout northern China from Xinjiang to Inner Mongolia and Jilin; also cultivated.

Parts used are the herbaceous green stems (maHUANG) and the root (maHUANGgen, ephedra root). Stems are collected in autumn either by cutting the green parts aboveground or the whole plant is pulled out and rid of dirt, and then the stems and roots separated and sun dried.

CHEMICAL COMPOSITION

Stem (maHUANG) contains 1–2% alkaloids composed mainly of *l*-ephedrine and *d*-pseudoephedrine, with ephedrine ranging from 30% to 90%, depending on the source. Thus, *E. sinica* contains ca. 1.3% alkaloids with more than 60% ephedrine; *E. intermedia* contains ca. 1.1% alkaloids with 30–40% ephedrine; and *E. equisetina* contains ca. 1.7% alkaloids with 85–90% ephedrine. Other alkaloids include *l*-N-methylephedrine, *d*-N-methylpseudoephedrine, *l*-norephedrine, *d*-norpseudoephedrine (cathine), ephedine, ephedroxane, and pseudoephedroxane (IMM-4).[1,2] The alkaloids are concentrated in the internodes, with lesser amount (ca. 50%) in the nodes and none in the root (ZHOU).

Other compounds present include glycans (ephedrans A, B, C, D, and E with mol. wt. of 1.2×10^6, 1.5×10^6, 1.9×10^4, 6.6×10^3, and 3.4×10^4, respectively);[3] a volatile oil;[4,5] catechin, gallic acid, and condensed tannin; flavonoid glycosides; inulin, dextrin, starch, and pectin; and other common plant constituents, including plant acids (citric, malic, oxalic, etc.), sugars, and trace minerals.

Active components in the volatile oil include limonene, caryophyllene, phellandrene, linalool, *l*-α-terpineol, and 2,3,5,6-tetramethylpyrazine. Concentration and composition vary considerably, depending on botanical sources (e.g., 0.250% volatile oil in Chinese ephedra vs. 0.124% in Mongolian ephedra)[4] and type of processing. For example, honey-cured and stir-fried maHUANG contains higher concentration of antiasthmatic (*l*-α-terpineol, caryophyllene, and tetramethylpyrazine) and antitussive, expectorant, antibacterial, and antiviral components (limonene and linalool) but at the same time is devoid of other compounds (nerolidol, farnesol, selinene, nonadecane, eicosane, octadecane, dodecanoic acid, tetradecanoic acid, 1,2-benzenedicarboxylic acid dibutyl ester, etc.) that are present in raw *ma*HUANG.[5]

Root (*ma*HUANG*gen*) contains macrocyclic spermine alkaloids (ephedradines A, B, C, and D); an imidazole alkaloid (feruloylhistamine); *l*-tyrosine betaine (maokonine); and diflavonols (mahuannins A and B) (IMM-4; WANG).[6–10]

Various compounds isolated from other ephedra species include cyclopropyl-α-amino acids (*E. altissima* and *E. foeminea*);[11] 6-methoxykynurenic acid (*E. pachyclada*);[12] lariciresinol, isolariciresinol, 9-acetoxylariciresinol, and (+)-9-acetoxyisolariciresinol (*E. viridis*).[13]

PHARMACOLOGY AND BIOLOGICAL ACTIVITIES

Comprehensive reviews about the chemistry and pharmacology of ephedra and its use in sports and weight loss products have recently been published.[14–17] The pharmacological effects of *ma*HUANG are generally attributed to ephedrine, which include central nervous system (CNS) stimulation, peripheral vasoconstriction, elevation of blood pressure, bronchodilatation, cardiac stimulation, decrease of intestinal tone and motility, mydriasis, and tachycardia, among others (ZHOU). The central stimulant action of ephedrine appears to be mediated by *l*-adrenoceptors and not by dopamine receptors.[18]

d-Pseudoephedrine has similar activities as ephedrine except that its pressor (hypertensive) and CNS effects are weaker; it also has strong diuretic action in animals (dog and rabbit) (MARTINDALE; ZHOU).

d-Norpseudoephedrine (major active chemical present also in the African stimulant khat) is also a CNS stimulant (MARTINDALE; TYLER 3).

Ephedrine, pseudoephedrine, ephedroxane, and pseudoephedroxane have been shown to have anti-inflammatory effects on experimental edema in animals.[2]

The glycans (ephedrans A, B, C, D, and E) exhibited marked hypoglycemic effects in normal and alloxan-induced hyperglycemic mice.[3]

*Ma*HUANG decoction and volatile oil have diaphoretic action in humans, and its decoction and alcoholic extract have antiallergic effects *in vitro*. When administered subcutaneously, the volatile oil was effective in treating mice infected with the Asian influenza virus strain AR_8 (ZHOU).

*Ma*HUANG*gen* extract when injected intravenously into cats and rabbits caused vasodilation and lowered blood pressure, as well as stimulated respiration and inhibited perspiration; it also contracted isolated guinea pig and rabbit smooth muscle preparations (e.g., uterus and intestine) (HU). The ephedradines, tyrosine betaine, and feruloylhistamine are the hypotensive principles (IMM-4).[7,8]

Apart from the sympathomimetic effects of ephedra, a few studies probed its other possible effects. In a herb-acupuncture study, the water distillate of *E. sinica* was effective in reducing the inflammatory responses of arthritis, both *in vitro* and in a rat model.[19] An aqueous fraction of *E. sinica*, at 30 μg/kg/day, exhibited anti-invasive and antiangiogenic activities in murine melanoma cells and in human umbilical venous endothelial cells, respectively. The same fraction also inhibited the growth of a melanoma cell mass in mice in comparison to adrimaycin.[20]

In a study to investigate the anorexigenic effect of ephedra, the expression of neuropeptide Y in rats was suppressed by an aqueous extract of *ma*HUANG. Since this peptide is an endogenous stimulant of feeding desire, its inhibition may be one of the mechanisms by which ephedra causes weight loss.[21]

TOXICOLOGY

Reported side effects of abuse include insomnia, motor disturbances, high blood pressure, glaucoma, impaired cerebral circulation, urinary disturbances, and others.[22] Case reports of severe to fatal conditions associated with ephedra, together with collective reviews, are continuously being published even after the banning of its sales in the United States.[23–28]

USES

Dietary Supplements/Health Foods. Was used in diet formulas for its appetite suppressive effect and in "energy" formulas for its central nervous system stimulant action; also used in cold and flu remedies, usually in tablet, capsule, and tea forms.

Traditional Medicine. *Ma*HUANG and *ma*HUANG*gen* (root) are traditionally used for different purposes: the former as diaphoretic and the latter as antisudorific and antiperspirant.

*Ma*HUANG is traditionally considered to have diaphoretic, diuretic, antiasthmatic, cold-dispersing, lung-soothing (*xuan fei*), and antiswelling (*xiao zhong*) properties (CHP); used for more than 2000 years to treat bronchial asthma, cold and flu, fever, chills, lack of perspiration, headache, nasal congestion, aching joints and bones, cough and wheezing, and edema, among others.

*Ma*HUANG*gen* is traditionally used to treat spontaneous and night sweating due to body deficiency (*ti xu*); now also used externally to treat excessive perspiration (e.g., foot), where the powder is topically applied (NATIONAL).

Others. *Ma*HUANG serves as raw material for the extraction and production of natural ephedrine and pseudoephedrine. *Ma*HUANG*gen*, due to its antiperspirant properties, is a potential ingredient in antiperspirant preparations.

COMMERCIAL PREPARATIONS

Crude (mainly raw, cut, and powdered) and extracts; extracts normally come in 5–9% total alkaloid content. Powdered crude *ma*HUANG claimed to contain 6–8% ephedrine has been offered, which should be considered adulterated, as commercial crude *ma*HUANG normally contains only about 1% ephedrine alkaloids. Ephedrine, ephedrine hydrochloride, and ephedrine sulfate formerly official in U.S.P.

Regulatory Status. Herb subject of German therapeutic monograph for treatment of diseases of the respiratory tract with mild bronchospasms in adults.[22] The United States FDA has long regarded dietary supplements containing ephedra. These effects are linked to adverse health effects like heart attacks and stroke. In February 2004, the FDA issued a rule prohibiting the sale of dietary supplements containing ephedrine alkaloids because they present an unreasonable risk of illness or injury.

REFERENCES

See the General References for BLUMENTHAL 1 & 2; CHP; HU; IMM-4; JIANGSU; NATIONAL; TYLER 2 & 3; WANG; ZHOU AND WANG.

1. C. Konno et al., *Phytochemistry*, **18**, 697 (1979).
2. Y. Kasahara et al., *Planta Med.*, **325**, (1985).
3. C. Konno et al., *Planta Med.*, **162**, (1985).
4. Y. Y. Jia et al., *Zhongguo Yaoxue Zazhi*, **24**, 402 (1989).
5. Q. Zeng et al., *Zhongguo Yaoxue Zazhi*, **17**, 83 (1992).
6. N. H. Hikino et al., *Planta Med.*, **478**, (1984).

7. N. H. Hikino et al., *Planta Med.*, **48**, 108 (1983).
8. N. H. Hikino et al., *Planta Med.*, **48**, 290 (1983).
9. N. H. Hikino et al., *Heterocycles*, **17**, 155 (1982).
10. N. H. Hikino et al., *Heterocycles*, **19**, 1381 (1982).
11. A. N. Starrat and S. Caveney, *Phytochemistry*, **40**, 479 (1995).
12. A. N. Starrat and S. Caveney, *Phytochemistry*, **42**, 1477 (1996).
13. S. V. Pullela et al., *Planta Med.*, **71**, 789 (2005).
14. P. A. Sharpe et al., *J. Am. Diet. Assoc.*, **106**, 2045 (2006).
15. R. Andraws et al., *Prog. Cardiovasc. Dis.*, **47**, 217 (2005).
16. M. H. Pittler and E. Ernst, *Int. J. Obes. (London)*, **29**, 1030 (2005).
17. E. A. Abourashed et al., *Phytother. Res.*, **17**, 703 (2003).
18. A. Li and B. H. Li, *Zhongguo Yaoli Xuebao*, **12**, 468 (1991).
19. M. J. Yeom et al., *J. Pharmacol. Sci.*, **100**, 41 (2006).
20. N. H. Nam et al., *Phytother. Res.*, **17**, 70 (2003).
21. E. H. Kim et al., *Am. J. Chin Med.*, **32**, 659 (2004).
22. Monograph *Ephedrae herba*, Bundesanzeiger, no. 11 (January 17, 1991).
23. R. M. Fleming, *Angiology*, **58**, 102 (2007).
24. C. E. Stahl et al., *Med. Sci. Monit.*, **12**, CS81 (2006).
25. C. Chen-Scarabelli et al., *Eur. J. Heart Fail.*, **7**, 927 (2005).
26. B. D. Keisler and R. G. Hosey, *Curr. Sports Med. Rep.*, **4**, 231 (2005).
27. C. A. Haller et al., *Clin. Pharmacol. Ther.*, **77**, 560 (2005).
28. M. Maglione et al., *Am. J. Psychiatry*, **162**, 189 (2005).

EPIMEDIUM

Source: ***Epimedium brevicornum*** Maxim., ***E. pubescens*** Maxim., ***E. koreanum*** Nakai, ***E. wushanense*** T. S. Ying, ***E. acuminatum*** Franch., ***E. sagittatum*** (Sieb. et Zucc.) Maxim., and nine other ***Epimedium*** species (Family Berberidaceae).[1]

Common/vernacular names: Herba epimedii, yinyanghuo, and xian ling pi.

GENERAL DESCRIPTION

Perennial herbs, mostly under 50 cm high, with compound leaves, leaflets thin leathery; native to China and Korea; now widely distributed in China; can be differentiated microscopically by the characteristic features of their hairs, especially nonglandular hairs.[2] The species of most commercial importance are *E. brevicornum*, *E. pubescens*, *E. koreanum*, *E. wushanense*, and *E. acuminatum* in decreasing order.[1] Although *E. sagittatum* is often described as a major source, it is in fact a minor source. Also, although *E. grandiflorum* Morr. (syn. *E. macranthum* Komarov.) has been listed as a major source, it is a Japanese species not commercially available in China.[1]

Part used is the aerial portion (mostly leaves) collected from wild plants in summer or autumn when leaves are bright green; then rid of thick petioles and impurities and sun dried or dried in the shade. Normally exported in neatly tied rectangular bundles; sometimes also in loose form, with stems and petioles.

CHEMICAL COMPOSITION

Leaves and stems from different species (including *E. brevicornum*, *E. pubescens*, *E. davidii*, *E. wushanense*, *E. acuminatum*, and *E. sagittatum*) contain flavonoid glycosides, mostly rhamnosides (ca. 4.5% in *E. brevicornum*):[3] icariin and epimedosides, wushanicariin (3′,5′,7-trihydroxy-4′-methoxy-6-(3,3-dimethylallyl)-flavone-7-β-D-glucopyranoside), hyperin (hyperoside, quercetin galactoside), quercitrin (quercetin rhamnoside), kaempferol-3-*O*-α-L-rhamnopyranoside, acuminatin (6″,6″-dimethylpyrano-(2″,3″,7,8)-4′-methyl, kaempferol-3-*O*-α-L-rhamnopyranoside), kaempferol-3-dirhamnoside, baohuosides, sagittatins A and B, and others.[3–9] Different prenylflavonol glycosides, such as acetylicariin, breviflavone B, ikarisoside A, icarisid II, and others, have recently been isolated from *E. brevicornum*, *E. koreanum*, and *E. sagittatum*.[10–12] More flavonoids (icaritin, baohuosu, tricin, desmethylicaritin, etc.) isolated from different *Epimedium* species have also been reported.[5,7,13–18]

Traditional curing of epimedium herb by stir frying with lamb fat (20% w/w) followed by drying only slightly reduces total flavonoids content but significantly improves their water extractability (MA).[19]

Other chemical constituents present include polysaccharides;[20] volatile oil, phytosterols (daucosterol), tannin, fatty acids, and others (JIANGSU).[9]

PHARMACOLOGY AND BIOLOGICAL ACTIVITIES

Epimedium is a highly valued tonic herb in traditional Chinese medical practice and has been shown to have numerous pharmacological effects in humans and experimental animals, including

1. Stimulation or improvement of male sexual function in experimental animals, with cured (fried with 20% w/w lamb fat) but not raw epimedium (*E. brevicornum*) being the active herb;[19] The aqueous extract of *E. brevicornum* was recently found to relax rabbit corpus cavernosum strips, through an NO-mediated mechanism, and may thus have potential for the management of erectile dysfunction in men.[21] A follow-up study further demonstrated that intracavernous administration of *E. brevicornum* elicited penile erection in rats (300–1000 μg/mL) and that NO may be involved.[22]

2. Broad cardiovascular effects (hypotensive, peripheral vasodilatory, increasing peripheral and coronary blood flow volumes, stimulating ADP-induced platelet aggregation, etc.);[9] and catecholamine inhibition.[23]

3. Promotion of growth of chick embryonic femur and its protein and polysaccharide synthesis *in vitro*.[24]

4. Immunomodulating; regulating nucleic acid metabolism; antiviral and antibacterial; anti-inflammatory; antitussive and expectorant; possible antitumor activity[25] and others (JIANGSU; WANG).[26]

The last decade has witnessed a shift in focus toward the following new activities of epimedium and/or its major constituents:

1. *Reduction of bone resorption*. Zhang et al. investigated this activity in more than one study. The first two studies investigated the mechanism of reduction of bone loss in animal models and found that epimedium-derived phytoestrogenic flavones prevented estrogen deficiency-induced osteoporosis and steroid-associated osteonecrosis in rats and rabbits, respectively.[27,28] The final study, a double-blind placebo-controlled clinical trial conducted on postmenopausal women, showed that the same preparation had a beneficial effect toward reducing bone resorption after 24 months of use.[29]

2. *Estrogenic activity*. A polyphenolic extract from the leaves was equally active in the *in vitro* yeast and Ishikawa assays for estrogenic activity.[30] The prenylflavone breviflavone B, isolated from the leaves of *E. brevicornum*, displayed a biphasic effect of stimulating the estrogen receptor at low concentration (EC_{50} 200 nM) and inhibiting it at a higher one (>2 μM).[11] Icaritin and desmethylicaritin resulted in an estrogen-like dose-dependent effect (1 nM–10 μM) when tested in the MCF-7 cell proliferation assay. The

phytoestrogenic effect was blocked by a specific estrogen receptor antagonist.[31]

3. *Antidepressant activity*. In two different studies, Pan et al. demonstrated that icariin, isolated from *E. brevicornum*, had an antidepressant-like effect in the forced swimming test and tail suspension test in mice after oral administration for 21 and 7 days, respectively. The 21-day treatment also resulted in a decrease in the levels of MAO and serum CRF.[32] Further investigation of its mechanism of action in rats showed that icariin reduced the serum levels of CRF, IL-6, and TNF-α resulting in the chronic mild stress model utilized in the study.[33]

4. *Antioxidant activity*. Icariin, isolated from the aerial parts of *E. koreanum*, displayed *in vitro* antihepatotoxic activity in cultured rat heptocytes exposed to carbon tetrachloride. Cytotoxicity, GPT, and sorbitol dehydrogenase were all reduced at 1–20 μM.[17] Pretretment icariin pretreatment, 0.1–50 μM) was also active in protecting human umbilical vein endothelial cells from oxidative injury by hydrogen peroxide. Reduction of caspase expression was also observed, which may be one of the underlying mechanisms of action of icariin.[34]

Thus far, the total flavonoids, icariin, and polysaccharides have been shown to be the active constituents of epimedium. A comprehensive review about the chemistry and pharmacology of epimedium has recently been published.[35]

Acute toxicity of edimedium is low: the LD_{50} of its total flavonoids in white mice was 2.99 ± 0.14 g/kg after i.p. administration.[26]

USES

Medicinal, Pharmaceutical, and Cosmetic. Extracts used in personal care products (e. g., disinfectant sprays) for its antimicrobial effects.

Dietary Supplements/Health Foods/Herb Teas. Powdered herb and extracts used in tonic formulas and teas for its traditional male tonic (aphrodisiac) properties (JIANGSU).

Traditional Medicine. First recorded use dates back to the *Shen Nong Ben Cao Jing* (ca. 200 BC–AD 100). Traditionally regarded as acrid and sweet tasting; warming; invigorating kidney *yang* (*bu shen yang*); benefiting *jing* (semen, life essence); strengthening bones, tendons, and muscles (*qiang jin gu*); and antirheumatic and antiarthritic (*qu feng shi*). Used in impotence, spermatorrhea, weak back and knees, rheumatism and arthritic pain, mental fatigue and poor memory, and postmenopausal hypertension, among others.

In recent years, extensively used in China in the treatment of coronary heart disease, hypertension (including postmenopausal), bronchitis, and neurasthenia;[36,37] also used in chronic hepatitis, poliomyelitis, chronic leukopenia, and others (JIANGSU; WANG).[36,38]

COMMERCIAL PREPARATIONS

Crude and extracts. Crude (raw herb) comes in whole bundled form composed of mostly leaflets with little or no petioles or in loose form containing leaves with stems and petioles, also in powdered form. Cured form not available in bulk quantities. Extracts occasionally come with assay for flavonoids.

Regulatory Status. Class 2d dietary supplement (not for long-term use).

REFERENCES

See the General References for CHP; IMM-4; JIANGSU; LU AND LI; MA; NATIONAL; WANG; ZHU.

1. H. R. Liang et al., *Zhongyao Tongbao*, **13**, 7 (1988).
2. H. R. Liang et al., *Beijing Zhongyi Xueyuan Xuebao*, **13**, 42 (1990).

3. L. X. Xu and X. Q. Zhang, *Yaoxue Xuebao*, **24**, 606 (1989).
4. H. R. Liang et al., *Yaoxue Xuebao*, **23**, 34 (1988).
5. F. Li and Y. L. Liu, *Yaoxue Xuebao*, **23**, 739 (1988).
6. Y. S. Li and Y. L. Liu, *Zhongcaoyao*, **23**, 8 (1992).
7. F. Li and Y. L. Liu, *Yaoxue Xuebao*, **23**, 672 (1988).
8. Y. Oshima et al., *Planta Med.*, **55**, 309 (1989).
9. B. H. Hu et al., *Yaoxue Xuebao*, **27**, 397 (1992).
10. J. Dou et al., *Anal. Sci.*, **22**, 449 (2006).
11. S. P. Yap et al., *Planta Med.*, **71**, 114 (2005).
12. M. Kuroda et al., *Planta Med.*, **66**, 575 (2000).
13. C. M. Liu et al., *Zhongguo Zhong Yao Za Zhi*, **30**, 1511 (2005).
14. H. R. Liang et al., *Planta Med.*, **63**, 316 (1997).
15. W. K. Li et al., *Phytochemistry*, **43**, 527 (1996).
16. W. K. Li et al., *Phytochemistry*, **42**, 213 (1996).
17. M. K. Lee et al., *Planta Med.*, **61**, 523 (1995).
18. W. K. Li et al., *Phytochemistry*, **38**, 263 (1995).
19. R. Niu et al., *Zhongchengyao*, **13**, 18 (1991).
20. R. S. Li et al., *Zhongyao Tongbao*, **12**, 40 (1987).
21. J. H. Chiu et al., *Int. J. Impot. Res.*, **18**, 335 (2006).
22. K. K. Chen and J. H. Chiu, *Urology*, **67**, 631 (2006).
23. H. C. Li and G. H. Huang, *Zhongcaoyao*, **15**, 26 (1984).
24. Z. F. Gao et al., *Chin. J. Integr. Trad. Western Med.*, **5**, 172 (1985).
25. C. C. Lin et al., *Clin. Exp. Pharmacol. Physiol.*, **31**, 65 (2004).
26. J. H. Liu and L. S. Shen, *Beijing Zhongyi Xueyuan Xuebao*, **16**, 29 (1993).
27. G. Zhang et al., *Bone*, **40**, 685 (2007).
28. G. Zhang et al., *Bone*, **38**, 818 (2006).
29. G. Zhang et al., *J. Bone Miner. Res.* (2007).
30. N. A. De et al., *Fitoterapia*, **76**, 35 (2005).
31. Z. Q. Wang and Y. J. Lou, *Eur. J. Pharmacol.*, **504**, 147 (2004).
32. Y. Pan et al., *Pharmacol. Biochem. Behav.*, **82**, 686 (2005).
33. Y. Pan et al., *Biol. Pharm. Bull.*, **29**, 2399 (2006).
34. Y. K. Wang and Z. Q. Huang, *Pharmacol. Res.*, **52**, 174 (2005).
35. H. Wu et al., *Prog. Drug Res.*, **60**, 1 (2003).
36. F. C. Liu, *Zhongcaoyao*, **16**, 44 (1985).
37. L. Yu et al., *Zhongyi zazhi*, **36** (1990).
38. P. G. Xiao and K. J. Chen, *Phytother. Res.*, **2**, 55 (1988).

EUCALYPTUS

Source: *Eucalyptus globulus* Labill. (Family Myrtaceae).

Common/vernacular names: Blue gum, Tasmanian blue gum, fever tree, and gum tree.

GENERAL DESCRIPTION

Evergreen tree with bluish green leaves often covered with a white powder; up to about 90 m high; native to Australia; extensively cultivated worldwide (e.g., Europe, United States, China, Africa, and South America).

Part used is the fresh or partially dried leaf from which the essential oil is produced by steam distillation. Major oil-producing countries include Spain, Portugal, and Brazil.[1]

Essential oils from other Eucalyptus species are also used, some of which may have quite different chemical compositions.

CHEMICAL COMPOSITION

Eucalyptus leaves contain 0.5–3.5% volatile oil, tannins, polyphenolic acids (gallic, caffeic, ferulic, gentisic, protocatechuic acids, etc.), flavonoids (quercetin, quercitrin, rutin, hyperoside, eucalyptin, etc.), wax, and others (JIANGSU; LIST AND HÖRHAMMER).[2–6] A newly reported ellagic acid was isolated from eucalyptus fruits and was identified as 3-O-methylellagic acid 4'-O-α-L-2''-O-acetylrhamnopyranoside. Two related glycosides and ellagic acid were also isolated.[7] Eucalyptone, another newly reported compound, was also isolated from the leaves.[8]

Eucalyptus oil contains usually 70–85% of eucalyptol (1,8-cineole);[1,2,9] other constituents present are mostly monoterpene hydrocarbons (α-pinene, δ-limonene, p-cymene, β-pinene, α-phellandrene, camphene, γ-terpinene, etc., with the first three in major amounts), with lesser amounts of sesquiterpenes (e.g., aromadendrene, allo-aromadendrene, globulol, epiglobulol, ledol, and viridiflorol), aldehydes (e.g., myrtenal), ketones (e.g., carvone and pinocarvone), and others (JIANGSU).[1,10,11] The eucalyptol content is 61.2% and 83.9% in Brazilian and Chinese eucalyptus, respectively.[12,13]

The rectified oil contains little or no unpleasant smelling lower aliphatic aldehydes.

The epicuticular wax contains esters of triterpene and fatty acids, such as ursolic acid, hexadecanoic acid, sesamin, and others.[14]

PHARMACOLOGY AND BIOLOGICAL ACTIVITIES

Eucalyptus oil and eucalyptol reportedly have antiseptic (antibacterial) and expectorant properties;[15–17] strongly antibacterial against several strains of *Streptococcus*.[18] The oil also has antiparasitic activity against a number of organisms, such as head lice, scabies, mites, that affect humans living in poor hygienic conditions.[19–21] The leaf extract was shown to be active against *Staphylococcus aureus*, *Streptococcus pyogenes*, *S. pneumoniae*, and *Haemophilus influenza* isolated from patients with respiratory tract infections ($MIC_{50} < 100$ μg/mL).[22] The flavonoids quercitrin and hyperoside have been reported to have eliminated influenza type A viral infections in mouse tissue and in chick embryos.[23]

A crude extract of *E. globulus* leaves rich in phenolic glycoside(s) has been reported to have antihyperglycemic activity in rabbits; a loss of this activity resulted upon purification of this material.[5] Oral administration of an aqueous extract of eucalyptus to streptozotocin-treated mice resulted in a reduction of hyperglycemia and weight loss. The same extract enhanced *in vitro* insulin secretion from a pancreatic beta-cell line after 20 min incubation at a concentration of 0.25–0.5 mg/mL.[24]

The *in vitro* anti-inflammatory effect of eucalyptus leaf extract was demonstrated due to its ability to scavenge and reduce NO production in a murine macrophage cell line.[25] The essential oil was also tested in a number of *in vivo* models of pain (acid-induced writhing and hot plate) and inflammation (paw edema) in rats and was found to be active in reducing both effects.[26]

When tested among other Sardinian plant extracts/volatile oils, eucalyptus oil inhibited the oxidation of linoleic acid as compared to the BHT and α-tocopherol antioxidant controls.[27]

TOXICOLOGY

Eucalyptus oil has been reported to be rapidly absorbed through the intact, shaved abdominal skin of the mouse and to promote the formation of tumors (papillomas) by 9,10-dimethyl-l,2-benzanthracene.[28,29] Later studies demonstrated that eucalyptus oil and eucalyptol were generally nonirritating, nonsensitizing, and nonphototoxic to the skin.[30,31]

When taken internally, eucalyptus oil is toxic, and ingestion of as little as 3.5 mL has been reported to be fatal (JIANGSU).

Rare instances of nausea, vomiting, and diarrhea have been reported after ingestion of nonfatal doses of leaf preparations or essential oil.[32,33]

USES

Medicinal, Pharmaceutical, and Cosmetic. Both eucalyptus oil and eucalyptol are extensively used as expectorants and/or flavoring agents in cold and cough medicines (e.g., cough drops and syrups), vaporizer fluids, antiseptic liniments, ointments, toothpastes, and mouthwashes. Also widely used as fragrance components in soaps, detergents, creams, lotions, and perfumes, with maximum use levels of 1.0 and 1.6% in perfumes reported for eucalyptus oil and eucalyptol, respectively.[30,31] Eucalyptus oil and eucalyptol are used in dentistry as components of certain root canal sealers; also used as solvents for root canal fillings.

Food. Both eucalyptus oil and eucalyptol are used as flavor ingredients in most food products, including alcoholic and nonalcoholic beverages, frozen dairy desserts, candy; baked goods, gelatins and puddings, meat and meat products, and others. Average maximum use levels reported are generally low, with the highest being about 0.002% (19.5 ppm) for eucalyptol in candy.

Dietary Supplements/Health Foods. Leaves used in tea; oil a fragrance ingredient in topical balms and massage oils (ROSE).

Traditional Medicine. Leaves and oil are reportedly used as antiseptic and febrifuge, and as expectorant and stimulant in respiratory ailments; also used for wounds, burns, ulcers, and cancers.[34] In Chinese medicine, leaves and oil are used for similar purposes. In addition, aqueous extracts and decoctions of the leaves are used to treat aching joints, bacterial dysentery, ringworms, pulmonary tuberculosis, and others; successful clinical studies on some of these uses have been reported (JIANGSU).

COMMERCIAL PREPARATIONS

Leaves, oil, and eucalyptol. The leaves were formerly official in U.S.P., while eucalyptol was official in N.F. Eucalyptus oil and eucalyptol are currently official in F.C.C., and eucalyptus oil is official in N.F.

Regulatory Status. Has been approved for food use (§172.510); eucalyptol is listed as a synthetic flavoring agent (§172.515). Leaves and essential oil subjects of German therapeutic monographs, indicated for catarrhs of the upper respiratory tract;[33] oil topically for rheumatic complaints.[32]

REFERENCES

See the General References for ADA; ARCTANDER; BAILEY 1; BARNES; BIANCHINI AND CORBETTA; BISSET; BLUMENTHAL 1; FEMA; GUENTHER; JIANGSU; JIXIAN; LIST AND HÖRHAMMER; LUST; GRIEVE; MCGUFFIN 1 & 2 ; ROSE; UPHOF.

1. W. D. Fordham, *Chemical Technology: An Encyclopedia Treatment*, Barnes and Noble, New York, 1972, p. 1.
2. R. K. Baslas, *Indian Oil Soap J.*, **35**, 136 (1969).
3. K. Boukef et al., *Plant. Med. Phytother.*, **10**, 24 (1976).
4. K. Boukef et al., *Plant. Med. Phytother.*, **10**, 30 (1976).
5. K. Boukef et al., *Plant. Med. Phytother.*, **10**, 119 (1976).
6. M. A. Elkeiy et al., *Bull. Fac. Pharm.*, **31**, 83 (1964).

7. Q. M. Guo and X. W. Yang, *Pharmazie*, **60**, 708 (2005).
8. K. Osawa et al., *Phytochemistry*, **40**, 183 (1995).
9. C. H. Brieskorn and W. Schlicht, *Pharm. Acta Helv.*, **51**, 133 (1976).
10. R. M. Ikeda et al., *J. Food Sci.*, **27**, 455 (1962).
11. J. De Pascual Teresa et al., *An. Quim.*, **73**, 751 (1977).
12. T. Nakashima et al., *Trib. Farm.*, **53**, 29 (1985).
13. B. M. Lawrence, *Perfum. Flavor*, **15**, 45 (1990).
14. S. I. Pereira et al., *Phytochem. Anal.*, **16**, 364 (2005).
15. A. C. Pizsolitto et al., *Rev. Fac. Farm. Odontol. Araraquara*, **9**, 55 (1975).
16. S. Prakash et al., *Indian Oil Soap J.*, **37**, 230 (1972).
17. J. C. Maruzella and P. A. Henry, *J. Am. Pharm. Assoc.*, **47**, 294 (1958).
18. A. Benouda et al., *Fitotherapia*, **59**, 115 (1988).
19. Y. C. Yang et al., *J. Agric. Food Chem.*, **52**, 2507 (2004).
20. T. A. Morsy et al., *J. Egypt. Soc. Parasitol.*, **33**, 47 (2003).
21. T. A. Morsy et al., *J. Egypt. Soc. Parasitol.*, **32**, 797 (2002).
22. M. H. Salari et al., *Clin. Microbiol. Infect.*, **12**, 194 (2006).
23. S. A. Vichkanova and L. V. Goryunova, *Tr. Vses. Nauch. Issled. Inst. Lek. Rast.*, **14**, 212 (1971).
24. A. M. Gray and P. R. Flatt, *J. Nutr.*, **128**, 2319 (1998).
25. E. Vigo et al., *J. Pharm. Pharmacol.*, **56**, 257 (2004).
26. J. Silva et al., *J. Ethnopharmacol.*, **89**, 277 (2003).
27. M. A. Dessi et al., *Phytother. Res.*, **15**, 511 (2001).
28. F. Meyer and E. Meyer, *Arzeim.-Forsch.*, **9**, 516 (1959).
29. F. J. C. Roe and W. E. H. Field, *Food Cosmet. Toxicol.*, **3**, 311 (1965).
30. D. L. J. Opdyke, *Food Cosmet. Toxicol.*, **13**, 107 (1975).
31. D. L. J. Opdyke. *Food Cosmet. Toxicol.*, **13**, 105 (1975).
32. Monograph *Eucalypti aetheroleum*, *Bundesanzeiger*, 177a (September 24, 1986); revised (March 6, 1990).
33. Monograph *Eucalypti folium*, *Bundesanzeiger*, 177a (September 24, 1986); revised (March 6, 1990).
34. J. L. Hartwell, *Lloydia*, **33**, 288 (1970).

EUPHORBIA

Source: *Euphorbia pilulifera* L. (syn. *E. hirta* L.; *E. capitata* Lam.) (Family Euphorbiaceae).

Common/vernacular names: Snakeweed and pill-bearing spurge.

GENERAL DESCRIPTION

An upright hairy annual; up to 0.5 m high; native to India. Part used is the whole flowering or fruiting plant.

CHEMICAL COMPOSITION

Contains choline and shikimic acid as active constituents.[1] Other compounds present include triterpenes (e.g., free taraxerol and α-amyrin; esters of taraxerone, α-amyrin, and β-amyrin; friedelin); sterols (campesterol, sitosterol, stigmasterol, etc.);[2,3] flavonoids (quercitrin, quercetin, leucocyanidin, xanthorhamnin, etc.);[4] *n*-alkanes (e.g., hentriacontane);[3] phenolic acids (e.g., gallic and ellagic), *l*-inositol, sugars (glucose, fructose, and sucrose), and resins; and others (LIST AND HÖRHAMMER).[1]

PHARMACOLOGY AND BIOLOGICAL ACTIVITIES

Euphorbia has been mentioned in earlier reports to have antispasmodic and histamine-potentiating properties,[5] as well as antitumor activities in laboratory animals.[6] Choline produces contraction of isolated guinea pig ileum, while shikimic acid produces a relaxation of guinea pig ileum.[1] Shikimic acid, a ubiquitous constituent of higher plants, has been reported to have carcinogenic properties in mice,[7,8] though no mutagenic activities have been observed using the Ames assay.[9]

More recent literature supports some of the earlier reports and introduce new findings about the activities of euphorbia. The majority of recent reports utilize the synonym *E. hirta* as an identifier of the plant source.

The antispasmodic effect of *E. hirta* was reported by Tona et al., whereby a polyphenolic extract inhibited the contractions of isolated guinea pig ileum at 80 μg/mL.[10] The gastric motility of normal rats was also decreased by an aqueous leaf extract of *E. hirta*.[11]

Further effects on the GIT include antiamoebic and antidiarrheal activities as reported by Tona et al. and by Galvez et al., whereby euphorbia polyphenolic extract inhibited the growth of *Entamoeba histolytica* at a concentration below 10 μg/mL,[10,12] and the lyophilized decoction and one of its constituents, quercitrin, displayed antidiarrheal effect in experimental animal models.[13]

Euphorbia ethanolic extracts displayed moderate antibacterial activity against *E. coli*, *Proteus vulgaris*, *Pseudomonas aeruginosa* and *Staph. aureus* in the agar well diffusion and tube dilution assays.[14] A methanolic extract of the flower inhibited the cytotoxic effect of *Shigella* spp. on Vero cells at 1.56 mg/mL for up to 48 h.[15] Aqueous bark and leaf extracts displayed molluscicidal activity on the vector snail *Lymnaea acuminata*.[16] The ethanolic and dichloromethane extracts of the aerial parts exhibited *in vitro* antiplasmodial activity against *Plasmodium falciparum* at IC_{50} 0.3–6 μg/mL. Oral administration of the same extracts (100–400 mg/kg/d) resulted in a significant suppression of parasitaemia in mice infected with *P. berghei*.[17,18]

An ethanolic extract of *E. hirta* was recently reported to display an antiallergic effect by inhibiting rat peritoneal mast cell degranulation and by relieving the symptoms in a model of mild asthma.[19] This finding is contradictory to the earlier one reported by Hellerman and Hezelton in 1950.

The aqueous and ethanolic leaf extracts (50 and 100 mg/kg, respectively) produced a diuretic effect in mice similar to that produced by furosemide and acetazolamide. Both urine output and electrolyte excretion were significantly increased.[20]

Lanhers et al. investigated other effects of the lyophilized aqueous extract of euphorbia in two separate studies. In the first study, the extract showed sedative and anxiolytic effects at a wide dose range of 12.5–100 mg/kg when administered orally and i.p. in mice.[21,22] In the second study, the extract was shown to possess analgesic, antipyretic, and anti-inflammatory properties in mice and rat models. The anti-inflammatory effect was displayed at 100 mg/kg in a rat model of acute inflammation. The analgesic effect was believed to be central in nature at a dose of 20–25 mg/kg. The antipyretic effect was displayed at higher dose of 100–400 mg/kg and was accompanied by a strong sedative effect, in agreement with the first study.[23]

USES

Dietary Supplements/Health Foods. Used primarily in certain cough preparations.

Traditional Medicine. Reportedly used mainly in treating respiratory ailments (e.g., asthma, bronchitis, coughs, and hay fever); also in tumors.[24]

In India, it is used in treating worms in children and for dysentery, gonorrhea, digestive problems, and others.

In China, numerous *Euphorbia* species (e.g., *E. humifusa* Willd., *E. pekinensis* Rupr., *E. lunulata* Bunge, *E. lathyris* L., and *E. sieboldiana* Morr. et Decne.) are traditionally used in treating conditions that include dysentery, enteritis, ascites, bleeding, dropsical nephritis, and chronic bronchitis. Some of these uses have been clinically substantiated (JIANGSU).

COMMERCIAL PREPARATIONS

Crude and extracts (fluid, solid, etc.); crude and fluid extract were formerly official in N.F. Strengths of extracts (see *glossary*) are expressed in weight-to-weight ratios.

REFERENCES

See the General References for FARNSWORTH 3; FOGARTY; FOSTER AND DUKE; GOSSELIN; JIANGSU; LIST AND HÖRHAMMER; YOUNGKEN.

1. L. El-Naggar et al., *Lloydia*, **41**, 73 (1978).
2. A. Atallah and H. Nicholas, *Phytochemistry*, **11**, 1860 (1972).
3. R.Gupta and S. Garg, *Bull. Chem. Soc. Jpn.*, **39**, 2532 (1966).
4. P. Blanc and G. Sannes, *Plant. Med. Phytother.*, **6**, 106 (1972).
5. R. C. Hellerman and L. Hazelton, *J. Am. Pharm. Assoc.*, **39**, 142 (1950).
6. M. Belkin and D. B. Fitzgerald, *J. Natl. Cancer Inst.*, **13**, 139 (1952).
7. I. A.Evans and M. A. Osman, *Nature (London)*, **250**, 348 (1974).
8. B. Stavric and D. R. Stoltz, *Food Cosmet. Toxicol.*, **14**, 141 (1976).
9. L. B. Jacobsen et al., *Lloydia*, **41**, 450 (1978).
10. L. Tona et al., *Phytomedicine*, **7**, 31 (2000).
11. S. K. Hore et al., *Fitoterapia*, **77**, 35 (2006).
12. L. Tona et al., *Phytomedicine*, **6**, 59 (1999).
13. J. Galvez et al., *Planta Med.*, **59**, 333 (1993).
14. M. Sudhakar et al., *Fitoterapia*, **77**, 378 (2006).
15. K. Vijaya et al., *J. Ethnopharmacol.*, **49**, 115 (1995).
16. S. K. Singh et al., *Chemosphere*, **59**, 263 (2005).
17. L. Tona et al., *J. Ethnopharmacol.*, **93**, 27 (2004).
18. L. Tona et al., *J. Ethnopharmacol.*, **68**, 193 (1999).
19. G. D. Singh et al., *Phytother. Res.*, **20**, 316 (2006).
20. P. B. Johnson et al., *J. Ethnopharmacol.*, **65**, 63 (1999).
21. J. Galvez et al., *J. Pharm. Pharmacol.*, **45**, 157 (1993).
22. M. C. Lanhers et al., *J. Ethnopharmacol.*, **29**, 189 (1990).
23. M. C. Lanhers et al., *Planta Med.*, **57**, 225 (1991).
24. J. L. Hartwell, *Lloydia*, **32**, 153 (1969).

EVENING PRIMROSE

Source: *Oenothera biennis* L. (Family Onagraceae).

GENERAL DESCRIPTION

Annual or biennial, 1–3 m. Leaves in basal rosette before anthesis, lanceolate, 10–22 cm long, 1 cm wide, margins undulate or minutely toothed; flowers four-merous, yellow; fruit a dry pod to 4 cm, with numerous minute seeds; throughout North America, pastures, old fields, roadsides; cultivated in Europe, North America, and elsewhere for seed oil.

CHEMICAL COMPOSITION

Seed contains about 14% fixed oil (evening primrose oil, EPO), with about 50–70% *cis*-linolenic acid and 7–10% *cis*-γ-linolenic acid (GLA); *cis*-6,9,12-octadecatrienoic acid, plus small amounts of oleic, palmitic, and stearic acid; steroids campesterol and β-sitosterol.[1] 3-*O-trans*-Caffeoyl esters of triterpene acids (betulinic, morolic, and oleanolic acids) are also present in the seed oil.[2,3]

Evening primrose meal contains about 1.7% of low molecular weight phenolic compounds (catechin, epicatechin, and gallic acid).[4]

PHARMACOLOGY AND BIOLOGICAL ACTIVITIES

GLA inhibits platelet aggregation, reduces blood pressure, and restores motility of red blood cells in multiple sclerosis. Exaggerated claims to efficacy in obesity are unsubstantiated (WREN). EPO of interest for GLA content as a prostaglandin precursor, especially for PGE_1; prostaglandins help regulate metabolic functions. Normal synthesis of GLA from linoleic acid via δ-6-desaturase may be blocked or diminished in mammalian systems as a result of aging, diabetes, excessive carbohydrate intake, or fasting. GLA supplement is valid for increased demand for GLA in alcoholism.[5]

Therapeutic use for atopic eczema initially produced modest, but significant, improvement (20–25% over controls)[6]. More recent clinical trials, however, indicate that evening primrose may have a marked beneficial effect in atopic eczema and dermatitis.[7–9]

EPO is used in premenstrual syndrome (PMS), diabetes, alcoholism, inflammation, and multiple sclerosis; and as a preventive in heart disease and stroke.[10] A prospective randomized, double-blind, placebo-controlled, crossover trial to evaluate efficacy in relief of PMS symptoms showed improvement, though statistically insignificant.[11] However, other double-blind, placebo-controlled studies showed that evening primrose oil significantly reduced irritability, breast pain and tenderness, and mood changes associated with PMS.[12] A study comparing the effectiveness of topical EPO and a nonsteroidal anti-inflammatory drug (NSAID) in breast pain (mastalgia) showed that EPO was ca. 70% as effective as the NSAID.[13]

Two separate studies in rats demonstrated that EPO prevented some of the side effects associated with diabetes, such as prostacyclin release, impairment of blood flow, and endoneurial oxygenation.[14,15]

Other reported pharmacological activities of evening primrose include

1. *Anti-inflammatory / antioxidant activity.* Seed extracts have antioxidant activity comparable to that of BHT,[16] while the triterpene acid esters have radical scavenging, COX, and neutrophil elastase inhibitory activity *in vitro*.[2] EPO administration reduced tissue oxidative stress in hyperlipemic rabbits.[17]

2. *Antitumor activity.* Caspase-dependent apoptosis in Ehrlich ascites carcinoma and other tumor cell lines has been demonstrated by evening primrose extract.[18,19] Other mechanisms of apoptosis may also include a rapid increase of intracellular peroxide levels in tumor cells.[20]

3. *Antithrombotic activity.* EPO supplemented in hyperlipemic diets and administered to rabbits over 6 weeks resulted in reduced vascular thrombogenesis and platelet aggregation. Hypercholesterolemia and endothelial lesions were also significantly reduced after administration.[21,22]

4. *Antiulcer effect.* EPO, administered at 5 and 10 mg/kg via gastric intubation, was effective

in inhibiting gastric mucosal damage in different rat models. The author proposed that the antisecretory and antiulcerogenic effects were attributed to the linolenic acid content of the oil.[23]

Gallic acid isolated from evening primrose root was reported to be phytotoxic causing ca. 85% inhibition of root growth in germinating wheat seeds at 250 ppm. A fourfold dose resulted in complete inhibition of root growth.[24]

USES

Medicinal, Pharmaceutical, and Cosmetic. Evening primrose oil used clinically in the United Kingdom for the treatment of atopic eczema, mastalgia, and premenstrual syndrome; increasingly seen in cosmetic products, including hand lotions, soaps, shampoos, and so on.

Dietary Supplements/Health Foods. Capsulated seed oil products widely available; dietary supplement for addition of essential fatty acids to diet (FOSTER).

Traditional Medicine. Whole plant infusion as astringent, sedative, antispasmodic in asthmatic coughs, gastrointestinal disorders, whooping cough; poulticed to enhance wound healing, anodyne; root rubbed on muscles to give athletes strength. Leaves, shoots, root, and seeds also used as food by American Indians.[5]

COMMERCIAL PREPARATIONS

Seed oil; seed oil capsules with GLA and vitamin E. Products also combined with linseed oil and safflower oil.

Regulatory Status. In the United Kingdom, approved therapeutic agent for treatment of atopic eczema. In Canada, a dietary supplement for increased essential fatty acid intake. Undetermined in the United States; the FDA has treated EPO as a both "misbranded drug" and "unsafe food additive," seizing product by treating EPO as a "food additive," rather than a food. However, a recent decision of the U. S. Court of Appeals for the 7th Circuit (decision rendered Jan. 27, 1993), involving seizure of black currant oil (also a GLA source), found in favor of the defendant (Traco Labs, Inc.). Writing for the court, Judge Cudahy stated that "the [Food, Drug, and Cosmetic] Act distinguishes between food additives and food in the generic sense, and this distinction is critical in allocating the burden of proof. The FDA's food additive definition is so broad, however, that it would blur this distinction. It would classify every component of food—even single, active ingredients—as food additives. Thus, it would seem that even the addition of water to food would make the food a food additive. The only justification for this Alice-in-Wonderland approach is to allow the FDA to make an end-run around that statutory scheme and shift to the processors the burden of proving the safety of a substance in all circumstances."[25]

The court's decision positively affects the dietary supplement status of EPO.

REFERENCES

See the General References for BLUMENTHAL 2; BRUNETON; DER MARDEROSIAN AND BEUTLER; DUKE 2; FOSTER; FOSTER AND DUKE; GLASBY 1 & 2; MARTINDALE; TYLER 1–3; WEISS; WREN.

1. Fedeli et al., *Riv. Ital. Sostanze Grasse*, **53**, 23 (1976).
2. M. Hamburger et al., *J. Agric. Food Chem.*, **50**, 5533 (2002).
3. J. Zaugg et al., *J. Agric. Food Chem.*, **54**, 6623 (2006).
4. M. Wettasinghe et al., *J. Agric. Food Chem.*, **50**, 1267 (2002).

5. C. J. Briggs, *Can. Pharm. J.*, **250** (1986).
6. C. R. Lovell et al., *Lancet*, **1**, 278 (1981).
7. C. A. Hederos and A. Berg, *Arch. Dis. Child*, **75**, 494 (1996).
8. S. Yoon et al., *Skin Pharmacol. Appl. Skin Physiol.*, **15**, 20 (2002).
9. N. L. Morse and P. M. Clough, *Curr. Pharm. Biotechnol.*, **7**, 503 (2006).
10. J. Janick et al., *Herbs, Spices, and Medicinal Plants: Recent Advances in Botany, Horticulture, and Pharmacology*, Oryx Press, Phoenix, AZ, 1989, p. 145.
11. S. K. Khoo et al., *Med. J. Aust.*, **152**, 189 (1990).
12. J. K. Pyke et al., *Lancet*, **2**, 373 (1985).
13. S. Qureshi and N. Sultan, *Surgeon*, **3**, 7 (2005).
14. E. J. Stevens et al., *Prostaglandins Leukot. Essent. Fatty Acids*, **49**, 699 (1993).
15. N. E. Cameron and M. A. Cotter, *Acta Diabetol.*, **31**, 220 (1994).
16. A. E. Birch et al., *J. Agric. Food Chem.*, **49**, 4502 (2001).
17. J. P. De La Cruz et al., *Life Sci.*, **65**, 543 (1999).
18. T. Arimura et al., *Amino Acids*, **28**, 21 (2005).
19. C. D. Pellegrina et al., *Cancer Lett.*, **226**, 17 (2005).
20. T. Arimura et al., *Chem. Biol. Interact.*, **145**, 337 (2003).
21. J. P. De La Cruz et al., *Thromb. Res.*, **87**, 141 (1997).
22. M. A. Villalobos et al., *Thromb. Haemost.*, **80**, 696 (1998).
23. O. A. al-Shabanah, *Food Chem. Toxicol.*, **35**, 769 (1997).
24. Y. N. Shukla et al., *J. Ethnopharmacol.*, **67**, 241 (1999).
25. M. Blumenthal 1, *HerbalGram*, **29**, 38 (1993).

EYEBRIGHT

Source: *Euphrasia rostkoviana* F. Hayne and other *Euphrasia* spp. (*E. officinalis* L.) (Family Scrophulariaceae).

GENERAL DESCRIPTION

Euphrasia officinalis has been used by modern authors to refer collectively to the genus; hence is a "collective species" and, as such, a *nomen ambiguum*. Close to 450 species described; many into hardly distinguishable microspecies; *Euphrasia officinalis* probably the most useful designation for commercial supplies.

Small, hemiparasitic, herbaceous, mostly annual herbs to 4 dm, simple or freely branched; leaves opposite sessile, ovate to rotund, palmately veined, coarsclv toothed above; bracteal leaves tend to alternate; flowers small, four lobed, deeply cleft above; corolla bilabiate, upper lip concave, two lobed or notched. Cold temperate regions, Northern and Southern Hemisphere; subarctic, alpine areas of tropical mountains (GLEASON AND CRONQUIST; TYLER 1; TUCKER 2). Part used is the whole herb. Most commercial supply from Europe and limited wild harvest in North America.

CHEMICAL COMPOSITION

E. rostkoviana contains iridoid glycosides including aucubin, catapol, and erostosidc; eukovoside; geniposide and luproside; gallotannins; caffeic and ferulic acids; trace amounts of an essential oil at 0.017%; choline; β-sitosterol; oleic, linoleic, linolenic, palmitic, and stearic acids.[1,2] *E. pectinata* contains the iridoid glucosides hydroxy- and dihydroxyboschnaloside, aucubin, euphroside, plantarenaloside, and geniposidic acid, together with the phenylethyl glycosides verbascoside (acetoside) and leucosceptoside A.[3]

PHARMACOLOGY AND BIOLOGICAL ACTIVITIES

Astringent ophthalmic; infusion internally and externally (as eye wash) recommended historically and in modern literature for eye irritations, particularly conjunctivitis, without scientific substantiation (TYLER 1).

A report of clinical success with compresses of an eyebright decoction to provide surprisingly rapid relief of redness, swelling, and visual disturbances in acute and subacute eye inflammations, particularly conjunctivitis, blepharitis, and recent eye injuries with risk of serpiginous corneal ulcers developing; used internally at the same time (WEISS). In a prospective cohort trial conducted on 65 conjunctivitis patients, a single drop of a *E. rostkoviana* preparation was administered 1–5 times daily for up to 14 days. The observed symptoms included conjunctival reddening and burning, foreign body sensation, and blurred vision. This treatment resulted in a complete recovery from symptoms of conjunctivitis in 53 patients and a marked improvement in 11 patients.[4]

The aqueous extract of *E. officinalis* leaves reduced the blood glucose levels in a diabetes model of alloxanized rats, while it had no hypoglycemic effect in normal rats.[5]

USES

Medicinal, Pharmaceutical, and Cosmetic. Used mainly in Europe as rinse, compress, or eye bath for eye-related inflammatory and vascular conditions, including eye lid inflammation, conjunctivitis, secreting and inflamed eyes, catarrh of eyes, and prevention of mucous secretion from eyes.[6]

Dietary Supplements/Health Foods. Tea, capsules, tablets, tincture, and so on, presumably for traditional applications.

Traditional Medicine. Use as ophthalmic for eye inflammations with mucous discharge; a folk remedy for allergy, cancer, coughs, earache, headache with congestion, hoarseness, inflammation, jaundice, rhinitis, and sore throat (DUKE 2; FOSTER AND DUKE). In European tradition also reported as stomachic and for skin diseases.[6]

COMMERCIAL PREPARATIONS

Crude herb, extracts, and so on.

Regulatory Status. Undetermined in the United States. Subject of a German therapeutic monograph; use not recommended for eye conditions because of hygienic concerns and nondocumented efficacy.[6]

REFERENCES

See the General References for BARNES; BLUMENTHAL 1; BRUNETON; DER MARDEROSIAN AND BEUTLER; DUKE 3; FOSTER AND DUKE; GLASBY 1 & 2; GLEASON AND CRONQUIST; LUST; MARTINDALE; MCGUFFIN 1 & 2; STEINMETZ; TYLER 1; TUCKER 2; WEISS; WREN.

1. O. Salama and O. Sticher, *Planta Med.*, **47**, 90 (1983).
2. O. Sticher et al., *Helv. Chim. Acta*, **65**, 1538 (1982).
3. T. Ersoz et al., *J. Nat. Prod.*, **63**, 1449 (2000).
4. M. Stoss et al., *J. Altern. Complement Med.*, **6**, 499 (2000).
5. E. Porchezhian et al., *Fitoterapia*, **71**, 522 (2000).
6. Monograph *Euphrasia*, *Bundesanzeiger* (Aug. 29, 1992).

FANGFENG

Source: *Saposhnikovia divaricata* (Turcz.) Schischk. (syn. *Ledebouriella divaricata* (Turcz.) Hiroe; *L. seseloides* (Hoffm.) Wolff; *Siler divaricatum* (Turcz.) Benth. et Hook. f.) (Family Umbelliferae or Apiaceae).

Common/vernacular names: Guan fangfeng, saposhnikovia, siler, ledebouriella, radix saposhnikoviae, radix ledebouriellae, and radix sileris.

GENERAL DESCRIPTION

Herbaceous glabrous perennial, 30–80 cm high, with thick root; base of stem covered with brown fibrous remains of petioles; basal leaves numerous; petioles flattened, with ovate sheaths; leaf blades oblong–ovate to broad–ovate, 14–35 × 6–8(–18) cm, 2-pinnate; pinnae 3–4 pairs, petiolulate; ultimate segments linear–lanceolate or cuneate–obovate, three lobed at apex, 2–5 × 0.5–2.5 cm. Leaves reduced upward. Umbels numerous, ca. 6 cm across; petals ca. 1.5 mm; fruit 4–5 × 2–3 mm, tuberculate when young, becoming smooth when mature. Plant native to China and is distributed throughout northern and northeastern provinces, with some fields still in their virgin primitive state existent in Heilongjiang and Inner Mongolia (HU). Part used is the root collected in spring or fall (when it reaches about 30 cm long and 1.2 cm thick) from cultivated (2–3 years old) or wild plants, rid of rootlets and dirt, and sun dried. Heilongjiang is the largest producer.

Substitutes produced in other regions of China are not saposhnikovia but are from other species; they include chuan fangfeng from Sichuan (*Peucedanum dielsianum* Fedde ex Wolff) and *yun fangfeng* from Yunnan (*Seseli mairei* Wolff, *Pimpinella candolleana* Wight et Arn. or *Seseli yunnanense* Franch.) (CMH; ZHU).[1–3]

CHEMICAL COMPOSITION

Contains coumarins (anomalin, bergapten, imperatorin, phellopterin, xanthotoxin, psoralen, scopoletin, etc.); chromones (5-*O*-methylvisamminol, 4-*O*-β-glucopyranosyl-5-*O*-methylvisamminol, hamaudol, sec-*O*-glucosylhamaudol, 3′-*O*-acetylhamaudol, 3′-*O*-angeloylhamaudol, cimifugin, prim-*O*-glucosylcimifugin, ledebouriellol, and divaricatol); lignoceric acid; polyacetylenes (falcarinone, falcarindiol); acylglycerols; a volatile oil; mannitol and sucrose; β-sitosterol and its glucoside; and others.[2,4–7] The roots and rhizomes also contain two pectin-like acidic arabinogalactan polysaccharides (saposhinkovan A and C).[8,9]

PHARMACOLOGY AND BIOLOGICAL ACTIVITIES

Both aqueous and alcoholic extracts have marked antipyretic, analgesic, and anti-inflammatory effects in mice and rats (WANG).[10] The chromones seem to display the most potent analgesic activity (inhibition of writhing) when orally administered to mice at a dose of 1 mg/kg. The analgesic activity is also displayed by coumarin, polyacetylene, and acylglycerol constituents but at higher doses.[4] Polyacetylenes isolated from fangfeng were found to inhibit iNOS (IC$_{50}$ of falcrindiol = 1.98 μM).[11]

The arabinogalactan saposnikovan C was reported to possess a significant potentiating effect on the reticuloendothelial system in animals.[9]

Alcohol extractives had mild antihistaminic activity on isolated guinea pig trachea;[12] they also had tyrosinase inhibitory effects.[13,14] Aqueous extract also exhibited immunopotentiating effects in mice, markedly increasing phagocytosis by macrophages.[15]

TOXICOLOGY

Certain coumarins (e.g., bergapten and xanthotoxin) are phototoxic (see ***angelica*** and ***bergamot oil***).

USES

Medicinal, Pharmaceutical, and Cosmetic. Listed in Li Shi-Zhen's *Ben Cao Gang Mu* (ca. 1590) as one of the herbs for removing facial dark spots as well as having antipruritic properties, *fangfeng* is used in skin care products (creams, lotions, bath preparations, and antiallergic ointments) and in hair tonics for these properties and for its anti-inflammatory and whitening (tyrosinase inhibitory) effects.

Dietary Supplements/Health Foods. Powdered herb and extracts used extensively as ingredients in formulas for treating the common cold, influenza, arthritis, rheumatism, pruritus, and urticaria, usually in capsule or tablet form (JIANGSU).

Traditional Medicine. First described in the *Shen Nong Ben Cao Jing* (ca. 200 BC–AD 100), *fangfeng* is one of the major wind- and dampness-dispelling drugs (*qu feng chu shi*; anti-inflammatory) and has since been used in countless prepared formulas for relieving pain due to arthritis, rheumatism, common cold, and influenza. Traditionally considered pungent and sweet tasting, warming, and also to have diaphoretic (*jie biao*) properties, it is commonly used in treating cold and flu and their associated headaches, migraine, rheumatism and arthritis, urticaria, pruritus, hard-to-heal carbuncles, and tetanus. It is one of the three ingredients (with astragalus and *baizhu*) of the famous 15th century formula, *yu ping fan san* ("jade screen powder") for fortifying body defense against outside pathogens, whose efficacy in enhancing immune functions and preventing colds and flu has been well documented (DENG).

Commercial Preparations. Crude comes in sticks (20–30 cm long), slices, or as powder. Extracts (water or hydroalcoholic) do not have uniform strength nor assays of chemical components.

Regulatory Status. U.S. regulatory status not determined.

REFERENCES

See the General References for CHP; CMH; HONGKUI; HU; JIXIAN; JIANGSU; MCGUFFIN 1 & 2; NATIONAL; WANG; ZHU.

1. J. H. Wang and Z. C. Lou, *Zhongyao Tongbao*, **13**, 9 (1988).
2. J. H. Wang and Z. C. Lou, *Zhongguo Yaoxue Zazhi*, **27**, 323 (1992).
3. J. H. Wang and Z. C. Lou, *Zhongyao Tongbao*, **13**, 5 (1988).
4. E. Okuyama et al., *Chem. Pharm. Bull.*, **49**, 154 (2001).
5. A. R. Ding et al., *Zhongcaoyao*, **18**, 7 (1987).
6. G. Z. Jin et al., *Zhongguo Zhongyao Zazhi*, **17**, 38 (1992).
7. H. Sasaki et al., *Chem. Pharm. Bull.*, **30**, 3555 (1982).
8. N. Shimizu et al., *Chem. Pharm. Bull.*, **37**, 1329 (1989).
9. N. Shimizu et al., *Chem. Pharm. Bull.*, **37**, 3054 (1989).
10. J. H. Wang et al., *Acta Med. Sinica*, **4**, 20 (1989).
11. C. N. Wang et al., *Planta Med.*, **66**, 644 (2000).
12. R. D. Xiang et al., *Zhongcaoyao*, **15**, 22 (1985).
13. X. T. Liu, *Zhongchengyao*, **13**, 9 (1991).
14. Y. Masamoto et al., *Planta Med.*, **40**, 361 (1980).
15. S. Y. Zhang et al., *Zhongcaoyao*, **18**, 9 (1987).

FENNEL

Source: *Foeniculum vulgare* Mill. (syn. *F. officinale* All.; *F. capillaceum* Gilib.; *Anethum foeniculum* L.) (Family Umbelliferae or Apiaceae).

Common/vernacular names: Florence fennel and finocchio.

GENERAL DESCRIPTION

Perennial herb with erect stem; up to 1.5 m high; lower petioles 5–15 cm; blade broadly triangular in outline, 4–30 × 5–40 cm, 4–5-pinnatisect; ultimate segments linear, 1–6 × ca. 0.1 mm. Umbels 5–9 cm across; peduncles 2–25 cm; rays 6–29, unequal, 1.5–10 cm; umbellules 14–39-flowered; pedicels thin, 2–10 mm, unequal. Fruit 4–6(–10) × 1.5–2.2 (–2.5) mm; generally considered to be native of the Mediterranean region; cultivated as an annual or a perennial worldwide (Argentina, Hungary, Bulgaria, Germany, France, Italy, Greece, China, India, etc.). Part used is the dried ripe fruit (commonly called seed) from which an essential oil is obtained by steam distillation.

There are two commonly used varieties of fennel: common fennel (or bitter fennel) and sweet fennel, with the latter occurring only in the cultivated form.[1,2] Common fennel appears to be the more commonly used fennel whenever the spice is called for. However, although fennel oils are official in the N.F. and F.C.C. (without specific distinctions between them), sweet fennel oil is reported to be the one generally used; bitter fennel oil is used only to a limited extent, mainly in cosmetics (ARCTANDER; FEMA; FURIA AND BELLANCA; ROSENGARTEN).[3,4]

CHEMICAL COMPOSITION

Fruits contain 1.5–8.6% (usually 2–6%) volatile oil;[1,5,6] 9–28% (usually 17–20%) fixed oil composed primarily of petroselinic acid (60–75%), oleic acid, and linoleic acid with a relatively high concentration of tocopherols (mostly γ-tocotrienol);[6–8] flavonoids (mainly quercetin-3-glucuronide, rutin, isoquercitrin, and quercetin-3-arabinoside, with minor amounts of kaempferol-3-arabinoside and kaempferol-3-glucuronide);[9,10] umbelliferone (7-hydroxycoumarin); hydroxycinnamic acid derivatives; stigmasterol; protein (16–20%); sugars; vitamins; minerals (relatively high in calcium and potassium); and others (JIANGSU; LIST AND HÖRHAMMER; MARSH).[11,12] Low concentrations of polyacetylenes were recently detected in the root.[13] An antimicrobial phenylpropanoid (dillapional) was also isolated from the stem.[14]

The volatile oil contains mostly *trans*-anethole (72–74%), with lesser amounts of fenchone (11–16%), estragole (methyl chavicol, 3–5%), limonene, camphene, and α-pinene. Other compounds present include more monoterpene hydrocarbons (β-pinene; α-thujene, α-fenchene, 3-carene, sabinene, α-phellandrene, myrcene, α- and β-terpinene, *cis*- and *trans*-ocimenes, terpinolene, and *p*-cymene), fenchyl alcohol, anisaldehyde, p-anisic acid, *trans*-1,8-terpin, myristicin, and apiole, the last two reportedly only present in the cultivated sweet variety.[1,2,7,15–20]

The concentrations of *trans*-anethole in the oil vary widely, with reported values ranging from 50% to 90%, depending on the varieties, sources, ripeness of fruits, and other factors.[1,2,7,16,18] There are also considerable variations in the amounts of fenchone (0–22%) and estragole.[2,7,16,18,21] Common fennel is reported to contain usually lower amounts of anethole but higher amounts of fenchone than sweet fennel (JIANGSU; LIST AND HÖRHAMMER).

PHARMACOLOGY AND BIOLOGICAL ACTIVITIES

Fennel and its volatile oil have carminative and stimulant properties.

Fennel oil has been reported to have spasmolytic effects on smooth muscles of experimental animals.[22] The essential oil and the ethanolic extract have a relaxant effect on tracheal chain of guinea pigs resulting in bronchodilation. The effect is suggested to be through the opening of potassium channels.[23] A recent clinical study showed that fennel oil is effective in the treatment of infantile colic, which supports a common traditional use of the oil.[24] The oil has a relaxant effect on isolated rat uterus pre-exposed to oxytocin and PGE_2 as a model of primary dysmenorrhea.[25] The same effect was observed in a clinical study conducted in young females with dysmenorrhea whereby the effect of the oil was comparable to that of mefenamic acid.[26]

Fennel oil exhibited antimicrobial activities *in vitro* against a number of bacteria and fungi, such as *Helicobacter pylori*, *Bacillus subtilis* and *Aspergillus niger*.[14,27,28] *p*-Anisaldehyde and (+)-fenchone were found to have an acaricidal activity against two *Dermatophagoides* species that is stronger than that of benzyl benzoate.[29]

The volatile oil and extracts have a cytotoxic effect on the larvae of *Culex pipiens* mosquito and a repellent effect against *Aedes aegypti* malaria mosquito. Moderate repellent activity was exhibited by fenchone and *E*-9-octadecenoic acid of the oil.[30,31]

Fennel oil displayed an anti-inflammatory, central analgesic, and antioxidant effect in experimental animals.[32] The oil also had a hepatoprotective effect against carbon tetrachloride-induced toxicity in a liver injury model in rats.[33]

Fennel oil has antiplatelet activity and ability to inhibit clot retraction in experimental animals. The activity has been correlated with the phenylpropanoid content of the oil.[34]

A terpene fraction of fennel oil has shown strong cytotoxic properties.[35]

Aqueous extracts of fennel experimentally increase ciliary action of ciliary epithelium in frogs.[36]

A cream containing 2% of the ethanolic extract of fennel was effective in reducing excessive hair growth in women diagnosed with idiopathic hirsutism.[37]

Anethole is reported to have allergenic, weakly insecticidal, and toxic properties (see *anise*). It also stimulates secretions of the upper respiratory tract, stimulating ciliary action, and ciliary epithelium in frogs.[36] Recent research shows that anethole is one of many phytochemicals that interrupt NFκB involved in the etiology of many diseases.[38] It is suggested that polymers of anethole, such as dianethole and photoanethole, are active estrogenic compounds.[39] Anethole and fenchone experimentally reduce secretions of upper respiratory tract.[36]

TOXICOLOGY

Estragole has been reported to cause tumors in animals (see *sweet basil*).

USES

Medicinal, Pharmaceutical, and Cosmetic. Fennel and sweet fennel oil are used as a carminative or flavoring agent in certain laxative preparations.

In Germany, the fruits used in phytomedicines for dyspeptic disorders; mild gastrointestinal antispasmodic, also upper respiratory tract conditions (expectorant); in syrup for children's coughs.[36,40]

Bitter (common) fennel and sweet fennel oils are used as fragrance components in cosmetics, including soaps, detergents, creams, lotions, and perfumes, with highest maximum use levels of 0.4% reported for both oils in perfumes.

Food. Common fennel is used as a flavor component in alcoholic beverages (especially liqueurs), baked goods, meat and meat products, fats and oils, snack foods, and gravies, with highest average maximum use level of about 0.119% (1186 ppm) reported in meat and meat products.

Sweet fennel is reportedly used in nonalcoholic beverages, candy, baked goods, meat and meat products, condiments and relishes, gravies, and processed vegetables. Highest average maximum use level reported is about 0.305% (3049 ppm) in meat and meat products.

Sweet fennel oil is widely used in most major food products; including alcoholic (e.g., liqueurs) and nonalcoholic beverages, frozen dairy desserts, candy, baked goods, gelatins and puddings, meat and meat products, and condiments and relishes, among others. Highest average maximum use level reported is about 0.023% (234 ppm) in alcoholic beverages.

Dietary Supplement/Health Food. Crushed or ground fruit in teas, tincture, or honey syrup (FOSTER).[40]

Traditional Medicine. Fennel fruits and oil are reportedly used as a stomachic and as a carminative in treating flatulence and other stomach troubles, as well as for catarrhs of the upper respiratory tract.[36,40]

Traditionally, fennel fruits have been used for many similar purposes as dill fruits, both in Western traditional medicine and in Chinese medicine.

In Chinese medicine, fennel has also been used for centuries in treating hard-to-heal snakebites, for which the powdered drug is used as a poultice, and for cholera, backache, and bedwetting, usually decocted with other drugs (JIANGSU).

COMMERCIAL PREPARATIONS

Crude and oil. Crude was formerly official in N.F. and U.S.P. Oil official in N.F. and F.C.C.

Regulatory Status. GRAS: common fennel and sweet fennel (§182.10); sweet fennel (§182.20). Seeds and essential oil subjects of German therapeutic monographs.[36,40] The oil (0.1–0.6 mL daily dose) and fruits (5–7 g daily dose) allowed for stimulation of gastrointestinal motility (or spasmolytic effect at high end of dosage) for dyspeptic discomfort, gastrointestinal spasms, and congestion of upper respiratory tract.[36,40]

REFERENCES

See The General References For APPLEQUIST;BAILEY 1; BIANCHINI AND CORBETTA; BISSET; BLUMENTHAL 1; DER MARDEROSIAN AND BEUTLER; FEMA; FOSTER; GRIEVE; GUENTHER; GUPTA; JIXIAN; JIANGSU; HUANG; LIST AND HÖRHAMMER; MARTINDALE; MCGUFFIN 1 & 2; ROSENGARTEN; TERRELL; UPHOF.

1. M. B. Embong et al., *Can. J. Plant Sci.*, **57**, 829 (1977).
2. J. Karlsen et al., *Planta Med.*, **17**, 281 (1969).
3. D. L. J. Opdyke, *Food Cosmet. Toxicol.*, **12**(Suppl.), 879 (1974).
4. D. L. J. Opdyke, *Food Cosmet. Toxicol.*, **14**, 309 (1976).
5. M. A. Wahid and M. Ikram, *Pak. J. Sci. Ind. Res.*, **4**, 40 (1961).
6. A. Seher and S. Ivanov, *Fette, Seifen, Anstrichmit*, **78**, 224 (1976).
7. M. R. I. Saleh et al., *J. Pharm. Sci. U. Arab. Rep.*, **5**, 55 (1964).
8. J. P. Moreau et al., *J. Am. Oil Chem. Soc.*, **43**, 352 (1966).
9. J. B. Harborne and C. A. Williams, *Phytochemistry*, **11**, 1741 (1972).
10. J. Kunzemann and K. Herrmann, *Z. Lebensm. Unters. Forsch.*, **164**, 194 (1977).
11. I. Parejo et al., *J. Agric. Food Chem.*, **52**, 3679 (2004).
12. A. Z. Abyshev et al., *Farmatsiya (Moscow)*, **26**, 42 (1977).
13. C. Zidorn et al., *J. Agric. Food Chem.*, **53**, 2518 (2005).
14. Y. S. Kwon et al., *Arch. Pharm. Res.*, **25**, 154 (2002).
15. N. Mimica-Dukic et al., *Phytother. Res.*, **17**, 368 (2003).
16. M. Ashraf and M. K. Bhatty, *Pak. J. Sci. Ind. Res.*, **18**, 236 (1975).

17. H. Rothbacher and A. Kraus, *Pharmazie*, **25**, 566 (1970).
18. G. A. de A. Brasil e Silva and L. Bauer, *Rev. Bras. Farm.*, **54**, 143 (1973).
19. J. B. Harborne et al., *Phytochemistry*, **8**, 1729 (1969).
20. L. Peyron et al., *Bull. Soc. Chim. Fr.*, **1**, 339 (1969).
21. C. S. Shah et al., *Planta Med.*, **18**, 285 (1970).
22. T. Shipochliev, *Vet. Med. Nauki*, **5**, 63 (1968).
23. M. H. Boskabady et al., *Pharmazie*, **59**, 561 (2004).
24. I. Alexandrovich et al., *Altern. Ther. Health Med.*, **9**, 58 (2003).
25. S. N. Ostad et al., *J. Ethnopharmacol.*, **76**, 299 (2001).
26. J. B. Namavar et al., *Int. J. Gynaecol. Obstet.*, **80**, 153 (2003).
27. G. B. Mahady et al., *Phytother. Res.*, **19**, 988 (2005).
28. F. M. Ramadan et al., *Chem. Mikrobiol. Technol. Lebensm.*, **2**, 51 (1972).
29. H. S. Lee, *J. Agric. Food Chem.*, **52**, 2887 (2004).
30. A. F. Traboulsi et al., *Pest. Manag. Sci.*, **61**, 597 (2005).
31. D. H. Kim et al., *J. Agric. Food Chem.*, **50**, 6993 (2002).
32. E. M. Choi and J. K. Hwang, *Fitoterapia*, **75**, 557 (2004).
33. H. Ozbek et al., *Fitoterapia*, **74**, 317 (2003).
34. M. Tognolini et al., *Life Sci.*, **78**, 1419 (2006).
35. K. Silyanovska et al., *Parfüm. Kosmet.*, **50**, 293 (1969).
36. Monograph *Foeniculi fructus*, *Bundesanzeiger*, no. 74 (April 19, 1991).
37. K. Javidnia et al., *Phytomedicine*, **10**, 455 (2003).
38. B. B. Aggarwal and S. Shishodia, *Ann. N. Y. Acad. Sci.*, **1030**, 434 (2004).
39. M. Albert-Puleo, *J. Ethnopharmacol.*, **2**, 337 (1980).
40. Monograph *Foeniculi aetheroleum*, *Bundesanzeiger*, no. 74 (April 19, 1991).

FENUGREEK

Source: *Trigonella foenum-graecum* L. (Family Leguminosae or Fabaceae).

Common/vernacular names: Foenugreek, Greek hay.

GENERAL DESCRIPTION

An annual herb; erect up to 0.6 m high, nearly smooth herb with alternate leaves; leaves are trifoliate, leaflets oblong–lanceolate, to 5 cm; flowers yellowish ca. 12–18 mm long in leaf axils; fruits are almost straight and flattened with a pronounced beak; they are 50–110 mm long excluding the beak of 10–35 mm; seeds brownish, about 1/8 inch long, oblong, rhomboidal, with a deep furrow dividing them into two unequal lobes; contained, 10–20 together, in long, narrow, sickle-like pods; native to western Asia and southeastern Europe; cultivated worldwide (e.g., Mediterranean region, northern Africa, South America, China, and India). Part used is the dried ripe seed; it is hard, smooth, and oblong, somewhat flattened, resembling a triangle.

CHEMICAL COMPOSITION

Contains simple alkaloids consisting mainly of trigonelline (up to 0.13%), choline (0.05%), gentianine, and carpaine; much of the trigonelline is degraded during roasting to nicotinic acid and other pyridines and pyrroles, which probably account for much of the flavor of roasted fenugreek (JIANGSU; MARTINDALE).[1]

Other constituents include (1) saponins that yield on hydrolysis 0.6–1.7% steroid sapogenins consisting mainly of diosgenin and its isomer yamogenin usually in a 3:2 ratio, with tigogenin and neotigogenin also present;[2–7] treatment of the seeds with enzymes before acid hydrolysis has increased the yield of diosgenin and yamogenin by 10–90%;[5,8,9] yamogenin tetrosides B and C have been reported to be two of the glycosides (saponins) present.[10] (2) Flavonoids, including vitexin, vitexin-7-glucoside, orientin arabinoside, homoorientin, saponaretin (isovitexin), vicenin-1, vicenin-2, quercetin, luteolin, and vitexin cinnamate.[11–13] (3) Fixed oils (5–8%), which on extraction with fat solvents yield an extract with a strong odor; varying from fishy to nutty, depending on age of the extract (MARSH; ROSENGARTEN).[3,14] (4) Considerable amount of a mucilage, which appears to be mostly a galactomannan and is probably responsible for swelling of the seed in water. (5) Protein (23–25%), which is low in S-amino acids but high in lysine and tryptophan; it has been suggested as a supplement of cereal proteins.[15] (6) Free amino acids, including ($2S,3R,4R$)-4-hydroxyisoleucine, histidine, lysine, and arginine, with the first one isolated at 0.09% yield as the major component.[16] (7) Vitamins, especially A, B_1, and C.[17] (8) Minerals (especially calcium and iron). (9) Volatile components (more than 50), which include n-alkanes, sesquiterpenes, and oxygenated compounds (undecane to hexadecane, elemenes, muurolenes, γ-nonalactone, 5-methyl-δ-caprolactone, etc.); and others (JIANGSU; NANJING).[18–21]

PHARMACOLOGY AND BIOLOGICAL ACTIVITIES

Both water and alcoholic extracts have been reported to have a stimulating effect on the isolated guinea pig uterus, especially during the last period of pregnancy, indicating that these extracts may have a highly oxytocic activity; they were suggested as possible replacements for oxytocin. The water extract has also been reported to have accelerating effects on the heartbeats of the isolated mammalian heart.[22]

Trigonelline and fenugreek infusion have been shown to have hypoglycemic effects in animals. However, the effects of trigonelline in diabetics have been inconclusive (MARTINDALE).[23,24]

When fed both before and after experimental diabetes induction, fenugreek has antidiabetic activities in rats.[25]

TOXICOLOGY

A report has indicated that fenugreek absolute is nonirritating, nonsensitizing, and nonphototoxic to human skin.[24]

USES

Medicinal, Pharmaceutical, and Cosmetic. Fenugreek extracts are used in certain perfume bases as well as in soaps, detergents, creams, and lotions, with maximum use level of 0.2% reported in perfumes.[24]

Food. Used as an ingredient of curry powder and many spice blends. Its major use in the United States is in imitation maple syrups for which solid extracts are mostly employed; flavor of the extracts varies with the extent of roasting and the solvents used. Other food products in which it is used include alcoholic and nonalcoholic beverages, frozen dairy desserts, candy, baked goods, gelatins and puddings, meat and meat products, and others. Use levels for extracts are usually below 0.05%.

Traditional Medicine. Has been used for millennia as a drug and a food or spice in Egypt, India, and the Middle East. Its medicinal uses include fever reducing and treating mouth ulcers, bronchitis, chronic coughs, and chapped lips, for milk promotion, as digestive aid, for cancers, and others;[26,27] also reported to be used in Java in hair tonics and to cure baldness (ROSENGARTEN).

Fenugreek was first introduced into Chinese medicine in the Sung Dynasty (ca. 1057) and has since been used as a nutrient and in treating kidney ailments, beriberi, hernia, impotence, other male problems, and others. Both unroasted and roasted (fried and sprayed with salt water) seeds are used.

Others. Extracts used in flavoring tobacco. Used extensively in foreign countries as a feed for livestock.

Due to its content of sapogenins, particularly diosgenin, fenugreek seed is a potential source of sapogenins for the manufacture of steroid hormones and related drugs. Because it is an annual herb, the time required for its planting to seed harvesting is much shorter than that for *Dioscorea* species and may prove to have a distinct advantage.

COMMERCIAL PREPARATIONS

Crude and extracts in liquid and spray-dried forms. Strengths (see *glossary*) of extracts are expressed in flavor intensities.

Regulatory Status. GRAS (§182.10 and §182.20). Seeds subject of a German therapeutic monograph; allowed internally for loss of appetite; externally as poultice for local inflammation.[28]

REFERENCES

See the General References for APPLEQUIST; ARCTANDER; BARNES; BISSET; BLUMENTHAL 1; DER MARDEROSIAN AND BEUTLER; DUKE 4; FEMA; HORTUS 3rd; JIXIAN; BRUNETON; JIANGSU; HUANG; LUST; GRIEVE; MCGUFFIN 1 & 2; NANJING; POUCHER; ROSENGARTEN; UPHOF; YOUNGKEN.

1. D. Reymond, *Chemtech*, **7**, 664 (1977).
2. H. S. Puri et al., *Planta Med.*, **30**, 118 (1976).
3. T. M. Jefferies and R. Hardman, *Analyst*, **101**, 122 (1976).
4. M. B. Bohannon et al., *Phytochemistry*, **13**, 1513 (1974).
5. R. Hardman and F. R. Y. Fazli, *Planta Med.*, **21**, 322 (1972).
6. R. Hardman and K. R. Brain, *Planta Med.*, **21**, 426 (1972).
7. J. C. Knight, *J. Chromatogr.*, **133**, 222 (1977).
8. D. A. Voloshina et al., *Prikl. Biokhim. Mikrobiol.*, **11**, 896 (1975).
9. A. A. Elujoba and R. Hardman, *Planta Med.*, **51**, 113 (1985).
10. N. G. Bogacheva et al., *Khim. Farm. Zh.*, **11**, 65 (1977).
11. H. Wagner et al., *Phytochemistry*, **12**, 2548 (1973).
12. M. Adamksa and J. Lutosmski, *Planta Med.*, **20**, 224 (1971).
13. A. R. Sood et al., *Phytochemistry*, **15**, 351 (1976).
14. F. R. Y. Fazli and R. Hardman, *Phytochemistry*, **10**, 2497 (1971).
15. I. Elmadfa, *Nahrung*, **19**, 683 (1975).
16. L. Fowden et al., *Phytochemistry*, **12**, 1707 (1973).
17. N. Saleh et al., *Z. Ernahrungswiss.*, **16**, 158 (1977).
18. S. Ghosal et al., *Phytochemistry*, **13**, 2247 (1974).
19. R. C. Badami and G. S. Kalburgi, *J. Karnatak Univ.*, **14**, 16 (1969).
20. D. K. Bhardwaj et al., *Indian J. Chem. Sect., B*, **15**, 94 (1977).
21. P. Girardon et al., *Planta Med.*, **51**, 533 (1985).
22. M. S. Abdo and A. A. Al-Kafawi, *Planta Med.*, **17**, 14 (1969).

23. N. R. Farnsworth and A. B. Segelman, *Tile Till*, **57**, 52 (1971).
24. D. L. J. Opdyke, *Food Cosmet. Toxicol.*, **16**(Suppl. 1), **755** (1978).
25. M. A. Riyad et al., *Planta Med.*, **54**, 286 (1988).
26. J. L. Hartwell, *Lloydia*, **33**, 97 (1970).
27. S. B. Vohora et al., *Planta Med.*, **23**, 381 (1973).
28. Monograph *Foenugraeci semen*, *Bundesanzeiger*, no. 22 (February 1, 1990).

FEVERFEW

Source: *Tanacetum parthenium* (L.) Schulz. Bip. (syn. *Chrysanthemum parthenium* (L.) Bernh., *Leucanthemum parthenium* (L.) Gren. et Godron, and *Pyrethrum parthenium* L. Sm.) (Family Compositae or Asteraceae).

GENERAL DESCRIPTION

Strongly aromatic perennial, stems ridged, up to 1 m; leaves yellowish green, pinnatisect into 3–7 oblong to ovate segments, divided into crenate-toothed to entire lobes; Flower heads 5–30, in dense corymb, ligules white, disk yellow to white; single, double (both ligulate and disk) forms common in horticulture; indigenous to rocky mountain scrub of the Balkan peninsula; cultivated for many centuries, naturalized throughout Europe, occasionally escaped in eastern North America, Central and South America, and elsewhere (GLEASON AND CRONQUIST; TUTIN 4). Parts used are the leaves and/or stems.

CHEMICAL COMPOSITION

Leaves contain sesquiterpenoids: artecanin, canin, chrysanthemolide, chrysanthemonin, 10-epi-canin, 1β-hydroxyarbusculin, 8β-hydroxyreynosin, 3β-hydroxyparthenolide, magnoliolide, parthenolide (up to 85% of sesquiterpene content), reynosin, santamarin, *seco*-tanapartholide A, tanaparthin, tanaparthin-1α,4α-epoxide, and tanaparthin-1β,4β-epoxide (GLASBY 2).[1]

Parthenolide, deemed the active sesquiterpene lactone, is highly variable in quantity or absent, depending on chemotype and geographic location of originating material; the sesquiterpene fraction dominated by parthenolide contains smaller levels of other germacranolides and guaianolides. A Mexican chemotype is dominated by the eudesmanolides reynosin and santamarin, as well as the guaianolides canin and artecanin; parthenolide is absent. Other chemotypes (Balkan) also dominated by eudesmanolides and guaianolides canin and artecanin; parthenolide is absent.[2-6] A cultivar *T. parthenium* f. *flosculosum* (DC) Beck. was found to contain as much as 1.27% parthenolide in leaves.[7]

Lipophilic methyl ethers of the flavonols 6-hydroxykaemferol and quercetagetin are present in the leaves and flowers, in addition to apigenin and the 7-glucuronide and 7-glucoside of luteolin.[8] The three flavonols centaureidin, jaceidin and santin were also isolated from *T. parthenium*.[9]

PHARMACOLOGY AND BIOLOGICAL ACTIVITIES

Crude extracts inhibit both human blood platelet aggregation and secretory activity in platelets and neutrophils (polymorphonuclear leucocytes). *In vitro*, extracts protect endothelial cell wall of rabbit aortas from perfusion-induced injury and inhibit deposition of human platelets on collagen.[3]

The ability of feverfew extracts to inhibit release of serotonin (5-hydroxytryptamine) from blood platelets has been suggested as mechanism of action in treatment of migraines.[10–13] Inhibition of serotonin release *in vitro* correlates well with parthenolide levels.[6] A minimum level of 0.2% parthenolide has been proposed by the Health Protection Branch of Health and Welfare Canada (roughly half the parthenolide content of feverfew samples used in clinical trials).[7]

A 1985 double-blind, placebo-controlled trial on the use of feverfew as a prophylactic treatment for migraine assessed a 25 mg dose of freeze-dried leaf capsules on 17 patients, concluding that when taken prophylactically, feverfew reduces frequency and severity of migraine symptoms.[14] A 1988 randomized, double-blind, placebo-controlled trial involving 72 volunteers clearly associated feverfew treatment with a reduction in the frequency of and vomiting associated with migraine attacks, as well as a reduction in their severity. However, duration of individual attacks was unaltered.[15]

Based on previous reports of anti-inflammatory activity, a 1989 double-blind, placebo-controlled, randomized study evaluated the effect of dried leaves (70–86 mg) in the treatment of rheumatoid arthritis. Over the 6-week trial, 41 female patients with symptomatic rheumatoid arthritis received feverfew or placebo. More than 13 laboratory and/or clinical parameters were assessed. The authors concluded that there were no important differences between the control group and those receiving feverfew. Participating patients, however, had not previously responded to conventional therapies. The results do not preclude possible benefits for the use of feverfew in osteoarthritis and soft tissue lesions.[16] Later *in vivo* studies showed that both feverfew extract and parthenolide possessed anti-inflammatory and antinociceptive activities in mice and rats and that such activities were dose dependent.[17] The extract, parthenolide, and some of the constituent flavonoids were also shown to inhibit the arachidonic acid pathways in leukocytes.[8,18] Another possible mechanism for the anti-inflammatory effect of feverfew is the inhibition of the expression of intercellular adhesion molecule-1 (ICAM-1).[19]

Feverfew ethanolic extract demonstrated a significant dose-dependent binding affinity to the GABA(A)–benzodiazepine receptor.[20]

Parthenolide and feverfew extract have *in vitro* antiproliferative activities against a number of human tumor cell lines of lymphoma, breast, and cervical cancers.[21,22]

USES

Medicinal, Pharmaceutical, and Cosmetic. Feverfew extracts, standardized to contain 0.1% parthenolide (leaf and stem; France) or 0.2% parthenolide (leaf only; Canada), are used for the prophylactic treatment of migraines.

Dietary Supplements/Health Foods. Tea; capsules, tablets, tincture, and so on.[3]

Traditional Medicine. In the United Kingdom, lay use of fresh or dried leaf primarily for allaying rheumatic and arthritic joint inflammation is more widespread than use in migraine prophylaxis.[16]

Anti-inflammatory, anodyne, antipyretic, antispasmodic, carminative, emmenagogue, febrifuge, stimulant, tonic, and vermifuge activity; used in migraine headache, asthma, rheumatism, gynecological problems, and so on. In Latin America used to promote functional activity of digestion, for colic, earache, stomachache, morning sickness, kidney pains, and so on.[23]

In Danish folk medicine as one of many herbs used for the treatment of epilepsy and convulsions.[20]

COMMERCIAL PREPARATIONS

Crude herb, extracts, freeze-dried leaf capsules, dried leaf capsules, tinctures, and so on.

Regulatory Status. Class 1 dietary supplement (herb that can be safely consumed when used appropriately). Health and Welfare Canada issued a Drug Identification Number

(D.I.N. 01958712) to a feverfew leaf capsules product, standardized to 0.2% parthenolide. Product allowed to carry the claim "used as a prophylactic against migraines."[23]

REFERENCES

See the General References for APPLEQUIST; BARNES; BLUMENTHAL 2; BRUNETON; DER MARDEROSIAN AND BEUTLER; DUKE 2; FOSTER AND DUKE; GLASBY 1 & 2; GLEASON AND CRONQUIST; MARTINDALE; MCGUFFIN 1 & 2; STEINMETZ; TUCKER 2; TUTIN 4; TYLER 1; WEISS; WREN.

1. C. Hobbs, *HerbalGram*, **20**, 26 (1989).
2. D. V. C. Awang, *Can. Pharm. J.*, **239**, 487 (1987).
3. D. V. C. Awang, *Can. Pharm. J.*, **122**, 266 (1989).
4. D. V. C. Awang, paper presented to the 57 Congrès de l' Association Canadienne Francaise pour l' Advancement des Sciences, Montréal, Québec, May 15–19, 1989.
5. D. V. C. Awang, *HerbalGram*, **22**, 34, 42 (1990).
6. S. Heptinstall et al., *J. Pharm. Pharmacol.*, **44**, 391 (1992).
7. D. V. C. Awang et al., *J. Nat. Prod.*, **54**, 1516 (1991).
8. C. A. Williams et al., *Phytochemistry*, **51**, 417 (1999).
9. C. Long et al., *Phytochemistry*, **64**, 567 (2003).
10. H. C. Diener et al., *Cephalalgia*, **25**, 1031 (2005).
11. E. Ernst and M. H. Pittler, *Public Health Nutr.*, **3**, 509 (2000).
12. S. Mittra et al., *Acta Pharmacol. Sin.*, **21**, 1106 (2000).
13. R. J. Marles et al., *J. Nat. Prod.*, **55**, 1044 (1992).
14. E. S. Johnson et al., *Br. Med. J.*, **291**, 589 (1985).
15. J. J. Murphy et al., *Lancet*, **22**, 189 (1988).
16. M. Pattrick et al., *Ann. Rheumatic Dis.*, **48**, 547 (1989).
17. N. K. Jain and S. K. Kulkarni, *J. Ethnopharmacol.*, **68**, 251 (1999).
18. H. Sumner et al., *Biochem. Pharmacol.*, **43**, 2313 (1992).
19. T. H. Piela-Smith and X. Liu, *Cell Immunol.*, **209**, 89 (2001).
20. A. K. Jager et al., *J. Ethnopharmacol.*, **105**, 294 (2006).
21. C. Wu et al., *J. Med. Food*, **9**, 55 (2006).
22. J. J. Ross et al., *Planta Med.*, **65**, 126 (1999).
23. S. Foster, *Feverfew-Tanacetum parthenium*, Botanical series, no. 310, American Botanical Council, Austin, TX, 1991.

FICIN

Source: *Ficus insipida* Willd. (syn. *F. glabrata* H.B.K., *F. anthelmintica* Mart., and *F. laurifolia* Hort. ex. Lam.) (Family Moraceae).

Common/vernacular names: Leche de oje, leche de higueron.

GENERAL DESCRIPTION

A tree with tall, erect cylindrical trunk over 1 m in diameter; up to about 45 m high, trunk smooth grayish brown; leaves bright, shiny green, with yellow veins; tip of a branch has a long, yellow, pointed stipule, which falls as the branch grows leaving a circular scar on branch at the base of each leaf; broken leaves drip white latex rapidly; native to tropical South America; growing in Peru, Colombia, Venezuela, and Central America. Part used is the latex from the tree. It is generally collected by felling the tree, making incisions all over the surface, and allowing the latex to drip into wooden or non-iron containers placed beneath the incised areas. To prevent coagulation, acetic acid is generally added; sodium benzoate (1%) is also added as a preservative. This is the usual form of crude ficin imported into the United States. The major producing countries are Peru and Colombia; the Peruvian material has a higher solids content.[1]

Crude ficin (latex) is cream to pinkish in color and has an acidic pH (usually 3–4). It is usually purified by filtration, followed by spray-drying. Commercial purified ficin is not pure ficin but is a mixture of several proteases and small amounts of other enzymes (e.g., peroxidases) in addition to diluents (e.g., lactose, dextrose, or starch) and other constituents.[1–3]

Commercial purified ficin is a beige to light-brown powder, odorless or with a putrid odor, depending on the quality. It is relatively soluble in water but insoluble in most organic solvents. It is active over a pH range of 4–9. It can withstand an acidic pH of 2, and its solutions are reported to be stable over a pH range of 3.5–9.5, with maximum stability at pH 5–8.[1,4,5] It is inactivated at 80.6°C.[6]

The optimal pH for ficin activity varies with the substrate, ranging from 5 (for gelatin and elastin) to 9.5 (for casein). With gelatin there is another optimum at pH 7.5, and casein has another optimum at pH 6.7.[4,6]

The optimal temperature range for ficin activity is generally considered to be between 50 and 65°C, depending on the substrates and the pH as well as the purity of the commercial ficin preparations. Thus with gelatin at pH 7.5, it is reported to be 62.6°C.[1,6]

CHEMICAL COMPOSITION AND PROPERTIES

Ficin is a sulfhydryl proteinase containing a carbohydrate moeity as bromelain.[7] It has been reported to consist of three major components and 248 amino acid residues, with a molecular weight of 23,800–26,000 and an isoelectric point at pH 9 or 10.[1,3,7,8] It is similar to papain in its chemical properties, being affected similarly by the usual papain activators (e.g., cysteine, sulfide, bisulfite, and cyanide) and inhibitors (e.g., methyl bromide, iodoacetate, and hydrogen peroxide); its activity has been reported to be reduced by ethanol and propanol.[9] It is also easily inactivated by metals (e.g., iron, copper, aluminum, and mercury) and sorbic and maleic acids.[1] Like bromelain and papain, the lost activity due to inactivation by certain metals can be restored by EDTA and a reducing agent such as cysteine, mercaptoethanol, or 1,2-dimercaptopropanol.

Like papain, ficin has broad specificity in hydrolyzing proteins, amides, esters, and small peptides.

PHARMACOLOGY AND BIOLOGICAL ACTIVITIES

Ficin (also bromelain but not papain) when administered orally to rats has been demonstrated to have anti-inflammatory activity against paw edema induced by serotonin, egg white, dextran, brewer's yeast, and carrageenan.[10]

Ficin is well known for its ability to digest intestinal worms *in vitro*.

Crude ficin (latex) is corrosive to the skin and may cause bleeding on prolonged contact.

Like other proteases, ficin may cause contact allergies in certain individuals, and large oral doses are reported to cause catharsis (MERCK).

USES

Medicinal, Pharmaceutical, and Cosmetic. Used in anti-inflammatory preparations, primarily in Europe; also used as digestive aid.

Food. Used mainly in meat tenderizers (usually in combination with papain and/or bromelain) and in the preparation of protein hydrolysates, edible collagen films, and sausage casings. Also used in cheese making (curdling milk) and in chillproofing beer.

Traditional Medicine. The latex is used in South America by natives as a vermifuge.

Others. Due to its increasingly high cost and limited availability, ficin is used in special applications where the other proteases are less suitable. Such uses include cleaning and preparation of intestinal submucosa in the manufacture of sutures; cleaning and preparation of animal arteries for implantation in humans; and in serology (e.g., determination of the Rh factor).[1]

COMMERCIAL PREPARATIONS

Crude (latex) and purified; purified ficin is official in F.C.C. Available grades and activities vary, depending on suppliers (see ***bromelain***).

REFERENCES

See the General References for MERCK; TYLER 3.

1. E. R. L. Gaughran, *Q. J. Crude Drug Res.*, **14**, 1 (1976).
2. J. P. G. Malthouse and K. Brocklehurst, *Biochem. J.*, **159**, 221 (1976).
3. I. K. Jones and A. N. Glazer, *J. Biol. Chem.*, **245**, 2765 (1970).
4. A. Yamamoto in G. Reed, ed., *Enzymes in Food Processing*, 2nd ed., Academic Press, New York, 1975, p. 123.
5. I. E. Liener and B. Friedenson in G. E. Perlmann and L. Lorand, eds., *Methods in Enzymology*, Academic Press, New York, 1970, p. 261.
6. J. R. Whitaker, *Food Res.*, **22**, 468 (1957).
7. I. E. Liener, *Adv. Chem. Ser.*, **136**, 202 (1974).
8. T. Murachi and N. Takahashi in P. Desnuelle et al., eds., *Structure–Function Relationships of Proteolytic Enzymes*, Academic Press, New York, 1970, p. 298.
9. E. Van den Eeckhout and R. Ruyssen, *Farm. Tijdschr. Belg.*, **47**, 73 (1970).
10. C. Netti et al., *Farmaco, Ed. Prat.*, **27**, 453 (1972).

FO-TI (RAW AND CURED)

Source: ***Polygonum multiflorum*** Thunb. (Family Polygonaceae).

Common/Vernacular names: Heshouwu, shouwu, polygonum, Radix Polygoni Multiflori; *zhiheshouwu, zhishouwu*, Radix Polygoni Multiflora Preparata (cured fo-ti); also *hoshou-wu, ho-show-wu, ho-shau-wu, ho-shao-wu*.

GENERAL DESCRIPTION

A climbing perennial herb with thick rhizomes, reaching 3–4 m long, hollow stems, somewhat woody at the base and mostly branched at the top; branches herbaceous; roots slender, bearing reddish brown to dark brown thick tubers near the tip; leaves alternate, simple, ovate-cordate, 3–4.5 cm long, 2.5–4.5 cm wide, short-acuminate, glabrous; petioles 1–5 cm; sheaths rather short. Spikes branched, terminal, paniculate; flowers greenish white, small in slender panicles, 2 mm long; fruit an achene, 3 angular, with 3 wings, entirely enclosed by sepals, 7–8 mm long; native to China; now distributed mainly along coastal provinces, from Guangxi to Hebei, and extending inland to Sichuan and Yunnan. Part used is the tuberous root collected from autumn through spring (preferably autumn and winter) and processed into different types of fo-ti. Since the early 1970s, fo-ti has been increasingly produced from cultivated plants, for which root tubers of 3- to 4-year-old plants are used (IMM-1; IMM-CAMS; JIANGSU).

Although easy to pronounce, the term fo-ti does not mean anything and was originally coined in America for marketing purposes only. Depending on methods of processing, there are at least four types of fo-ti: raw fo-ti, wine fo-ti, steamed fo-ti, and prepared or cured fo-ti. Raw and cured fo-ti are the most commonly used and are the ones imported into the United States.

To prepare raw fo-ti, freshly collected tubers are washed with water. After having both ends removed, the larger tubers are normally cut in half or sliced and sun or oven dried (CHP; IMM-1; JIANGSU).

To prepare cured fo-ti, raw fo-ti slices are stirred into black soybean broth in a noniron container, sealed and cooked in a water or steam bath until all liquid is absorbed and the fo-ti slices turn dark brown to reddish brown. This heating may take many hours; the optimal heating time for yielding fo-ti with the best immuno-modulating effects (tonic properties) being 32 h.[1] For every 100 kg of raw fo-ti slices, broth from 10 kg of black soybean is used. The slices are then dried (CHP).

Much confusion exists in the Western literature regarding fo-ti, with some major or popular works making no mention of the existence of different types (BENSKY AND GAMBLE; CHEUNG AND LI; LIST AND HÖRHAMMER; TYLER 1).[2-5] For instance, in one bilingual reference, even though raw fo-ti is mentioned in the Chinese version as having lubricating effects on the intestines as well as antitoxic and nodule-dispersing (antiswelling) properties, these properties are simply stated in the English version, without giving any hints as to the existence of the two types of fo-ti with distinct differences in properties and uses (CHEUNG AND LI). This ambiguity has caused problems in fo-ti products as manufacturers not familiar with Chinese herbs have been using the cheaper raw fo-ti in place of cured fo-ti in their tonic formulas. Adding to this confusion is the lack of specificity in reporting findings on fo-ti, particularly in the already meager English literature, where authors frequently do not specify type of fo-ti used in their research or report, thus making the reported information of dubious value (LIST AND HÖRHAMMER).[6]

Thus the two types of fo-ti have distinctly different chemical compositions, traditional properties, and modern pharmacologic effects and uses; they should be sourced carefully and used specifically.

CHEMICAL COMPOSITION

Many chemical studies have been performed on the genus *Polygonum* but few on *Polygonum multiflorum* itself (LIST AND HÖRHAMMER). Results obtained from earlier studies were somewhat confusing as most did not specify the types of fo-ti used, especially in the non-Chinese literature. Thus one report (consisting of only three lines) mentions the presence of anthraquinones in the rhizome, rootstock, and stem (LIST AND HÖRHAMMER), which has been the only chemical study quoted in a well-known herbal, attributing cathartic activity as solely responsible for the action of fo-ti (TYLER 1). Two other reports simply mention the dried roots or *heshouwu* as being used,[7,8] with no indication as to what type of fo-ti; the only clue to their using raw fo-ti was the use of description "for the treatment of suppurative dermatitis, gonorrhea, favus athlete's foot, inflammation," for which raw fo-ti is traditionally used. Nevertheless, from these few reports and from the Chinese literature that does describe raw and cured fo-ti, in addition to a few recently published Western reports, the following chemical profile of fo-ti can be described.

Both raw and cured fo-ti contain anthraquinones, including chrysophanol, chrysophanic acid anthrone, emodin, 6-hydroxyemodin, rhein, physcion, and digitolutein, which exist both free and as glucosides (mainly glucosides, such as emodin-8-β-D-glucoside) (LIST AND HÖRHAMMER);[9–12] the concentration can reach 1.1% in raw fo-ti.[9] Curing by successively steaming for 12 h, standing overnight and sun drying for 8 h (repeated nine times), or according to the *Chinese Pharmacopeia* for 32 h, reduced the concentrations of both the free and conjugated anthraquinones by 42–96%.[10,12] In addition, the proportion of free to conjugated anthraquinones is also greatly increased in cured fo-ti, thus further reducing the laxative effects.[9,13]

Fo-ti (type not specified) has been reported to contain up to 3.7% lecithins (JIANGSU).[9] Curing was found to increase the phosphorus (presumably lecithins) content by 36.9%;[14] it also increases the sugar content.[13] However, another study found that curing reduced the amount of phospholipids in fo-ti.[15]

Other chemical constituents reported to be present in fo-ti (mainly raw fo-ti) include rhapontin, β-sitosterol, catechins, cyanidins, stilbene glycosides, gallic acid/gallates, 2,3,4′,5-tetrahydroxystilbene-2-O-(3-D-glucoside (>1.2%),[7,11,16–18] *cis*- and *trans*-E-3-butylidene-4,5,6,7-tetrahydro-6,7-dihydroxy-1(3H)-isobenzofuranone[19] and relatively high concentrations of calcium (2225 ppm), iron (350 ppm), zinc (24.5 ppm), manganese (18.5 ppm), and other trace minerals[20] as well as copious amounts of starch and others (WANG).[21]

PHARMACOLOGY AND BIOLOGICAL ACTIVITIES

Fo-ti has exhibited numerous pharmacological effects, including the following:

1. *Antiaging effects.* Cured fo-ti was found to influence favorably various biological processes related to aging. Thus, its decoction (p.o.) significantly increased the levels of superoxide dismutase (SOD), biogenic amines (5-hydroxytryptamine, norepinephrine, and dopamine) and proteins but decreased the levels of monoamine oxidase-B (MAO-B), lipid peroxide, and malonyl dialdehyde (MDA; a product of destructive lipid peroxidation) in key organs of aging mice; extracts of fo-ti (type not specified) showed strong stimulating effects on SOD activity in red blood cells and strong inhibition of thiobarbituric acid reactive material formation (indicator of lipid peroxidation) in the liver.[22] Decoction of cured fo-ti (p.o.) significantly inhibited the increase of serum ceruloplasmin level in aging mice; it also significantly inhibited atrophy of the thymus gland of mice due to natural aging or induced by hydrocortisone as well as inhibited atrophy of their, adrenal glands.[9,23]

2. *Immunologic effects.* Fo-ti, cured per method of *Chinese Pharmacopeia* for 32 h, was shown to enhance nonspecific immunity and cellular immunity in mice; along with fo-ti cured by pressure cooking for 6 h at 120°C, it also had antagonistic activity against the immunosuppressive effects of prednisolone or hydrocortisone. In contrast, raw fo-ti had little or no such activities.[1,24]

3. *Hypolipemic and antiatherosclerotic effects.* A water-soluble fraction of fo-ti rich in stilbene glucosides was tested for reduction of the severity of atherosclerosis in rabbits fed a high-cholesterol diet for 12 weeks. Concomitant administration of the fraction for the whole period resulted in a decrease in plasma cholesterol, low-density lipoprotein (LDL), and very low-density lipoprotein (VLDL) cholesterol as well as plasma triglycerides. The atherosclerotic lesioned area was also reduced by 40–60%.[25] In a study using Japanese quails, an alcoholic extract of cured fo-ti raised blood plasma high-density lipoprotein (HDL) cholesterol: total cholesterol ratio; markedly reduced plasma total cholesterol, free cholesterol, and triglyceride levels; inhibited hyperlipemia; and retarded the development of atherosclerosis in the animals.[24] Earlier studies in other

animal species (e.g., rabbits, pigeons, and rats) also showed that fo-ti (type not reported) or its formulations exhibited similar effects (WANG).[9,26]

4. *Hepatoprotective effects.* Stilbene glucosides (e.g., 2,3,5,4'-tetrahydroxystilbene-2-O-D-glucoside) isolated from fo-ti were found to partially inhibit the deposition of lipid peroxides in the liver of rats fed peroxidized corn oil. They inhibited the elevation of GOT and GPT levels in the serum of the rats. They also inhibited lipid peroxidation induced by ADP and NADPH in rat liver microsomes.[8]

5. *Resistance to cold.* Daily intragastric administration of fo-ti extract (type not specified) at a dose of 0.5 mL, (equivalent to 0.2 g crude drug) for 14 days markedly decreased the mortality rate in mice induced by refrigeration at −5°C (WANG).

6. *Antimicrobial activities.* Among 80 herbs commonly used in treating hepatitis tested against hepatitis B virus *in vitro*, water extract of cured fo-ti was found to be one of the eight most actives in inhibiting HBV DNA replication.[27] Raw fo-ti and four different types of cured fo-ti all exhibited varying degrees of inhibition against nine species of bacteria, but there was no predictable pattern in their activities.[28]

7. *Neuroprotection and enhancement of cognitive performance.* In passive avoidance and water-maze tests, a water extract of fo-ti significantly reduced the cognitive deficit induced in mice by intracerebroventricular injection of amyloid β-peptide 25–35 (Abeta25–35). Abeta25–35 accumulates in Alzheimer's disease and causes accelerated lipid peroxidation and increased levels of thiobarbituric acid in the brain.[29] Daily administration of ethanolic extracts of fo-ti resulted in a significant reduction in the symptoms of Parkinsonism induced in mice by paraquat and maneb over a 6-week period.[30] The learning ability and memory of senescence accelerated mice (SAMP8) were significantly improved, and brain pathological changes were also reduced when the mice were fed for 18 weeks with casein diets supplemented with aqueous and ethanolic extracts of fo-ti.[31] The same effects were later confirmed by the same authors who attributed them to a probable involvement of the antioxidant constituents of fo-ti.[32]

8. *Others.* Fo-ti (type not specified) is reported to be nonmutagenic per the Ames *Salmonella*/microsome assay;[6] the hydxoxyanthraquinones in its aqueous extract (decoction) had antimutagenic effects.[33] The same effect was also demonstrated by the root extract in a Tradescantia micronucleus assay. The authors attributed the antimutagenic effect to a possible combination of antioxidant, radical scavenging,[34] and DNA repair activity of fo-ti.[35] Raw fo-ti has cathartic activities.

TOXICOLOGY

Compared to cured fo-ti, raw fo-ti is relatively toxic. Thus, the LD_{50} of an alcoholic percolate of raw fo-ti per i.p. administration was 2.7 g/kg, while the LD_{50} of an alcoholic percolate of cured fo-ti was 169.4 g/kg (animal species not reported).[9]

One case of allergic reaction (cystitis, patient recovered) due to cured fo-ti has been reported.[36] Gastrointestinal disturbances (diarrhea, abdominal pain, nausea, and vomiting) reported for fo-ti were most likely due to raw or improperly cured fo-ti (WANG). Other reported adverse effects due to fo-ti (or could be due to other ingredients present in the formulas) include numbness of the extremities and skin rashes (a few cases) (WANG).

USES

Medicinal, Pharmaceutical, and Cosmetic. Extract of cured fo-ti is a popular ingredient in hair care products (e.g., shampoos and tonics) for its alleged hair-darkening and growth-promoting properties, especially in China and Hong Kong (ZHOU); also used in skin care products (e.g., creams and lotions) for its traditional detoxicant (antiallergic) and nourishing properties.

Dietary Supplements/Health Foods. Powder and occasionally extracts are used in tonic formulas in America, though often with no distinction between the raw and cured forms; also used in sliced form in soup mix packets.

Traditional Medicine. Fo-ti was first described in *He Shou Wu Lu* (before 10th century) and later in *Kai Bao Ben Cao* (10th century, in the Song Dynasty). Raw fo-ti is traditionally considered a detoxicant and laxative and is used to treat scrofula (lymph node tuberculosis), sores, carbuncles, skin eruptions (*feng zhen*), pruritis, and constipation, among other conditions. On the other hand, cured fo-ti is traditionally considered to have slightly warming properties and a liver and kidney tonic; it is believed to tone up the vital essence and blood and to fortify the muscles, tendons, and bones. Cured fo-ti is traditionally used to treat dizziness with tinnitus, insomnia, premature graying, soreness and weakness of lower back and knees, and numbness of limbs and others. In recent years, both raw and cured fo-ti are also used in treating hyperlipemia (CHP; JIANGSU).

COMMERCIAL PREPARATIONS

Crude, powder, and extracts. Raw fo-ti comes in whole, halved, or thick slices, light brown to brown in color; cured fo-ti in very dark brown to dark reddish brown slices. Due to the much lower price of raw fo-ti, this is normally the powder available in America. Also, there is no simple practical assay method to determine whether a given extract from a supplier is genuine fo-ti and which type it is.

Regulatory Status. Class 2d dietary supplement (contraindicated with diarrhea, prepared root and stem may cause gastric distress, raw root causes catharsis).

REFERENCES

See the General References for CHEUNG AND LI; CHP; DER MARDEROSIAN AND BEUTLER; FOSTER AND YUE; HONGKUI; HUANG; IMM-1; IMM-CAMS, JIANGSU; JIXIAN; LIST AND HÖRHAMMER; WANG.

1. D. J. Ye et al., *Zhongyao Tongbao*, **12**, 21 (1987).
2. *Herbal Pharmacology in the People's Republic of China*, National Academy of Sciences, Washington, DC, 1975, p. 186.
3. *A Barefoot Doctor's Manual*, Running Press, Philadelphia, 1977, p. 743.
4. J. D. Keys, *Chinese Herbs*, Charles E. Tuttle Co., Rutland, VT, 1976, p. 151.
5. M. Tierra, *The Way of Herbs*, Pocket Books, New York, 1989, p. 183.
6. J. K. Kam, *Am. J. Chin. Med.*, **9**, 213 (1981).
7. K. Hata et al., *Yakugaku Zasshi*, **95** (1975).
8. Y. Kimura et al., *Planta Med.*, **49**, 51 (1983).
9. W. L. Deng and S. R. Gong, *Zhongcaoyao*, **18**, 42 (1987).
10. G. G. Yao et al., *Zhongcaoyao*, **14**, 15 (1983).
11. S. Yao, *J. Chromatogr. A*, **1115**, 64 (2006).
12. D. J. Ye et al., *Zhongyao Tongbao*, **11**, 23 (1986).
13. L. Q. Ling et al., *Shanghai Zhongyiyao Zazhi*, **78** (1966).
14. M. X. Chang et al., *Zhongcaoyao*, **19**, 17 (1988).
15. C. H. Ma and J. S. Wang, *Zhongguo Zhongyao Zazhi*, **16**, 662 (1991).
16. Y. Chen, *J. Agric. Food Chem.*, **47**, 2226 (1999).
17. X. Z. Yan, *Shanghai Ti I I Hsueh Pao*, **8**, 123 (1981).

18. M. Yoshizaki et al., *Planta Med.*, **53**, 273 (1987).
19. J. N. Grech, *J. Nat. Prod.*, **57**, 1682 (1994).
20. Z. X. Li et al., *Zhongcaoyao*, **16**, 15 (1985).
21. J. B. Li and M. Lin, *Zhongcaoyao*, **24** (1993).
22. Y. R. Dai et al., *Planta Med.*, **53**, 309 (1987).
23. M. C. Yao et al., *Yaoxue Tongbao*, **19**, 28 (1984).
24. J. H. Ying et al., *Zhongguo Zhongyao Zazhi*, **17**, 722 (1992).
25. P. Y. Yang, *J. Pharmcol. Sci.*, **99**, 294 (2005).
26. M. Z. Mei, *Yao Hsueh Hsueh Pao*, **14**, 8 (1979).
27. J. Y. Yang et al., *Chin. J. Integr. Trad. Western Med.*, **9**, 494 (1989).
28. H. S. Zhen et al., *Zhongyao Tongbao*, **11**, 53 (1986).
29. M. Y. Um, *J. Ethnopharmacol.*, **104**, 144 (2006).
30. X. Li, *Pharmacol. Biochem. Behav.*, **82**, 345 (2005).
31. Y. C. Chan, *J. Nutr. Sci. Vitaminol. (Tokyo)*, **48**, 491 (2002).
32. Y. C. Chan, *Am. J. Chin Med.*, **31**, 171 (2003).
33. Z. H. Xie et al., *J. Zhejiang Coll. Trad. Chin. Med.*, **14**, 22 (1990).
34. G. Ryu, *Arch. Pharm. Res.*, **25**, 636 (2002).
35. H. Zhang, *J. Environ. Pathol. Toxicol. Oncol.*, **18**, 127 (1999).
36. K. G. Ren, *Zhongcaoyao*, **16**, 40 (1985).

GALBANUM

Source: *Ferula gummosa* Boiss. (syn. *F. galbaniflua* Boiss. et Buhse) and other *Ferula* species (Family Umbelliferae or Apiaceae).

Common/vernacular names: Galbanum resin, galbanum gum, galbanum oleoresin, galbanum gum resin, and galbanum oleogum resin.

GENERAL DESCRIPTION

Resinous perennial herbs up to 1 m high, spreading, stems solid, with resin ducts throughout the plants; leaves gray-green hairy; flowers in umbels, yellow; seeds flat; native to the Middle East and western Asia (e.g., Iran, Turkey, and Afghanistan). Part used is the dried resinous exudation obtained by incising the stems near the ground. There are two types of galbanum: soft galbanum (or Levant galbanum) is a viscous liquid, while hard galbanum (or Persian galbanum) is a solid. The former contains more volatile oil and is used for the production of galbanum oil. A resinoid is prepared by solvent extraction; it often contains a high-boiling odorless solvent such as diethyl phthalate, diethyl sebacate, or propylene glycol used as diluent (ARCTANDER; GUENTHER).

CHEMICAL COMPOSITION

Galbanum contains from 5% (Persian) to more than 26% (Levant) volatile oil; about 60% resin consisting mainly of resinic acids; 30–40% gummy substances containing galactose, arabinose, galacturonic acid, and 4-methylglucuronic acid residues; and umbelliferone and its esters, among others (ARCTANDER; CLAUS; GUENTHER; LIST AND HÖRHAMMER; KARRER; MARTINDALE).[1–3]

The volatile oil contains 63–75% monoterpene hydrocarbons (mostly 3-pinene, 3-carene, and α-pinene, with small amounts of *d*-limonene, terpinolene, etc.);[4,5] monoterpene alcohols (linalool, terpineol, borneol, fenchol, etc.) and their acetates;[1] sesquiterpenes (e.g., cadinene, guaiol, bulnesol, galbanol, and 10-epijunenol);[1,6,7] azulenes (e.g., guaiazulene and isoguaiazulene); thiol esters (e.g., *S*-isopropyl-3-methylbutanethioate and *S-sec*-butyl-3methylbutanethioate);[8] polysulfanes;[9] pyrazines (e.g., tetramethylpyrazine, 2,6-diethyl-3-methylpyrazine, and 2-methoxy-3-*sec*butylpyrazine);[10,11] and (*E,Z*)-l,3,5-undecatriene and (*E,E*)-1,3,5-undecatriene (LIST AND HÖRHAMMER).[12–14]

The (*E,Z*)-isomer of *n*-1,3,5-undecatriene is reported to be the major odor principle of galbanum.[12]

PHARMACOLOGY AND BIOLOGICAL ACTIVITIES

Extracts (aqueous, hydroalcoholic, and chloroform) of galbanum gum have been reported to have antimicrobial properties (especially against *Staphylococcus aureus*) *in vitro*; they were also effective in preserving emulsions for up to 6 months without change in physical or organoleptic characteristics.[15,16]

The methanol–chloroform extract has been reported to alleviate morphine withdrawal symptoms induced in mice by naloxone.[17]

The essential oil, the hydroalcoholic and the methanolic extracts displayed a concentration-dependent spasmolytic activity on isolated rat ileum. Such activity was partially attributed to α- and β-pinene.[18]

The acetone extract of the seed exhibited anticonvulsant activity at sub-toxic levels against experimental seizures in mice.[19]

TOXICOLOGY

One report indicated galbanum oil to be non-irritating and nonsensitizing to human skin.[20]

USES

Medicinal, Pharmaceutical, and Cosmetic. Now rarely used in pharmaceuticals. Galbanum oil and resinoid are used as fragrance components and fixatives in cosmetics, including soaps, detergents, creams, lotions, and perfumes, with a maximum use level of 0.7% reported for the oil in perfumes.[20]

Food. Galbanum resin and galbanum oil are used as flavor components in most food products, including nonalcoholic beverages, frozen dairy desserts, candy, baked goods, gelatins and puddings, and condiments and relishes. Galbanum oil is also used in alcoholic beverages, meat and meat products, snack foods, and gravies. Highest average maximum use level reported about 0.003% (33 ppm) for the resin in candy and gelatins and puddings.

Traditional Medicine. Used as a carminative, stimulant, expectorant, and antispamodic for purposes similar to those of *asafetida*; also used in treating wounds.

COMMERCIAL PREPARATIONS

Crude, oil, and resinoid.

Regulatory Status. Has been approved for food use (§172.510).

REFERENCES

See the General References for ARCTANDER; CLAUS; DUKE 4; FEMA; GRIEVE; GUENTHER; LEWIS AND ELVIN-LEWIS; LIST AND HÖRHAMMER; MARTINDALE; MCGUFFIN 1 & 2; TERRELL; UPHOF.

1. P. Teisseire, *Recherches (Paris)*, **14**, 81 (1964).
2. M. G. Jessenne et al., *Plant. Med. Phytother.*, **8**, 241 (1974).
3. E. Graf and M. Alexa, *Planta Med.*, **51**, 428 (1985).
4. R. M. Ikeda et al., *J. Food Sci.*, **27**, 455 (1962).
5. Y. R. Naves, *Parfum. Cosmet. Savons* (1969).
6. M. Wichtl, *Planta Med.*, **11**, 53 (1963).
7. A. G. Thomas et al., *Tetrahedron*, **32**, 2261 (1976).
8. J. W. K. Burrel et al., *Tetrahedron Lett.*, **30**, 2837 (1971).
9. Z. D. Min et al., *Planta Med.*, **53**, 300 (1987).
10. J. W. K. Burell et al., *Chem. Ind.*, **44**, 1409 (1970).
11. F. Bramwell et al., *Tetrahedron Lett.*, **37**, 3215 (1969).
12. F. Naef et al., *Helv. Chim. Acta*, **58**, 1016 (1975).
13. Y. Chretien-Bessiere et al., *Bull. Soc. Chim. Fr.*, **1**, 97 (1967).
14. Y. R. Naves, *Bull. Soc. Chim. Fr.*, **9**, 3152 (1967).
15. A. Vaziri, *Planta Med.*, **28**, 370 (1975).
16. F. Eftekhar et al., *Fitoterapia*, **75**, 758 (2004).
17. M. Ramezani et al., *J. Ethnopharmacol.*, **77**, 71 (2001).
18. H. Sadraei et al., *Phytomedicine*, **8**, 370 (2001).
19. M. Sayyah et al., *J. Ethnopharmacol.*, **82**, 105 (2002).
20. D. L. J. Opdyke, *Food Cosmet. Toxicol.*, **16**(Suppl. 1), 765 (1978).

GANODERMA

Source: *Ganoderma lucidum* (Leyss. ex Fr.) Karst. (syn. *Polyporus japonicus* Fr.) and *G. japonicum* (Fr.) Lloyd. (syn *G. sinense* Zhao, Xu et Zhang) (Family Polyporaceae).

Common/Vernacular names: Reishi, lingzhi, *ling zhi cao*, *ling chi*, mannentake, holy mushroom, *chizhi* (red *lingzhi* or *G. lucidum*), *zizhi* (purple *lingzhi* or *C. japonicum*), and so on.

GENERAL DESCRIPTION

Fungi of the polypore family. Part used is the fruiting body.

The pileus (cap) of G. *lucidum* is corky, kidney shaped to semicircular, with a hard upper surface, yellow at first but gradually changing to reddish brown, reddish purple, or dull purple; shiny, with annular grooves or ridges and radial wrinkles; edges thin, often curved downward. Cap sizes vary considerably, ranging from 4 × 3 cm to 20 × 10 cm in area and 0.5–2 cm in thickness. Woody stalk (stipe) is mostly lateral, 0.5–2.5 cm thick and up to 19 cm long, purplish brown to black and shellacked.[1,2]

The fruiting body of *G. japonicum* (purple *lingzhi*) resembles closely that of *G. lucidum* (red *lingzhi*), with cap sizes ranging from 2 × 1.4 cm to 20 × 20 cm; stalk up to 15 cm long and 0.9 cm thick. The only major difference is the dark purple to black colored cap and stalk of *G. japonicum*. However, some old specimens of red *lingzhi* also have dark purple caps and stalks and thus cannot be readily distinguished from purple *lingzhi* (JIANGSU).[1–3]

Both ganodermas are widely distributed in China, especially along coastal provinces, growing at stumps and decaying logs of oak and other broad-leaved trees as well as on decaying conifers, especially *Tsuga chinensis* (Franch.) Pritz., which is parasitized by *G. lucidum*. The latter can also be found on hardwoods in North America as well as in Japan and Korea.[4,5]

The mushrooms are collected in autumn, washed to rid of dirt, and dried under the sun.[1] They are not processed further.

Although now commercially cultivated in China, much of the ganoderma is still gathered wild. The type imported into the United States is mainly red *lingzhi (G. lucidum)*; this species is now mostly cultivated in America as well as in China, Taiwan, Japan, and Korea (WANG).

In addition to above two *Ganoderma* species, other species of polypores are occasionally used as substitutes of *lingzhi*, including *Ganoderma applanatum* (Pers. ex Gray), *G. lobatum* (Schw.) Atk., *G. capense* (Lloyd) Teng, *Fomes pinicola* (Swartz ex Fr.) Cke., *Trametes dickinsii* Berk., *Polyporus montanus* (Quel.) Freey., *P. grammocephalus* Berk., and *Polysticus vernicipes* (Berk.) Cke.[2,6]

Ganoderma lucidum, G. japonicum, G. capense, and *G. applanatum* are used as fungal sources for the fermentative production of *lingzhi* biomass (WANG).

CHEMICAL COMPOSITION

Most chemical studies have been performed on red *lingzhi* (*G. lucidum*), including its spores and cultivated biomass (mycelium). Hence, unless otherwise stated, the data reported here are from this fungus and its strains, either wild crafted or cultivated.

Chemical constituents present include sterols, mainly ergosterol (0.3–0.4%) and ergosterol peroxide, β-sitosterol, 24-methylcholesta-7,22-dien-3-β-ol, and other sterol esters, fungal lysozyme, acid protease, and other enzymes (laccase, endopolygalacturonase, cellulase, amylase, etc.); water-soluble protein, polypeptides; amino acids; trehalose and other sugars; mannitol; betaine; adenosine; alkanes (tetracosane, hentriacontane); and fatty acids (tetracosanoic, stearic, palmitic, nonadecanoic, and behenic acids) (JIANGSU; WANG).[7–19]

Triterpenes (mainly lanostane type) including ganoderic acids A, B, C, D, E, F, G, H, I, J, K, L, M, N, O, P, Q, R, S, T, U, V, W, X, Y, Z, DM, LM2, SP1, beta, gamma, delta, epsilon, zeta, eta, theta, and others;[20–43] lucidenic acids A, B, C, D, E, F, G, O, P, Q, and others;[22,23,27,31,32,44–47] lucidones A, B, and C;[31,44] lucialdehydes;[48] ganolucidic acids A, B, C, D, and E;[26,30,31,49] lucidumol A;[43] ganoderal A;[21] ganoderiols

A, B, C, D, E, F, G, H, and I;[49,50] ganoderols A (ganodermanonol) and B (ganodermadiol);[21,51] ganodermanontriol and ganodermatriol, among others.[50–53]

Polysaccharides present include a water-soluble branched arabinoxyloglucan (polysaccharide GL-1) of mol. wt. 40,000;[54] an alkali-extracted, water-soluble heteroglycan of mol. wt. 38,000;[55] several water-insoluble heteroglucans of mol. wt.10,000–100,000;[56] ganoderans A and B of mol. wt. 7400–23,000;[57] and others.[5,58–66] Lectins and glycopeptides with carbohydrate-to-peptide ratio of ca. 10:1 are also present.[67–69]

The essential oil hydrodistillate contains $trans$-anethol, R-($-$)-linalool, S-($+$)-carvone, and α-bisabolol as major constituents.[19]

Major inorganic elements present include Ca, Mg, Na, Mn, Fe, Zn, Cu, and Ge.[70]

Purple $lingzhi$ ($G. japonicum$) contains ergosterol in lesser amount (0.03%), ergosta-7,22-dien-3-β-ol, trehalose, alkaloids (betaine, γ-butyrobetaine, etc.), 14 free amino acids, glucosamine, resin, and polysaccharides, including a water-insoluble alkali-soluble glucan (WANG).[71,72]

The mycelium of cultivated $G. capense$ contains adenine, adenosine, uridine, and uracil; D-mannitol; ergosterol, β-sitosterol, soyasapogenol B, and other steroids; stearic, docosanoic, tricosanoic, and tetracosanoic acids; nicotinic acid, and furans, among others (WANG).[71,72]

PHARMACOLOGY AND BIOLOGICAL ACTIVITIES

Most of the pharmacological studies have been performed on various extracts of red $lingzhi$ ($G. lucidum$). Hence, unless otherwise stated, the data reported here are for this species.

Bioactivities exhibited by ganoderma in humans and/or experimental animals include (1) antitumor or cytotoxic activities against experimental tumors (leukemia, lymphoma, solid sarcoma 180, hepatoma cells, Ehrlich ascites carcinoma; bladder, breast, colorectal, prostate cancer, and others). Involved antitumor mechanisms include apoptosis, induction of T-lymphocytes and NK cell activity, cell cycle arrest, antiangiogenesis, inhibition of NF-κB signaling and others with some of the polysaccharides and triterpenoids being the active principles (e.g., GL-1, ganoderans A and B, ganoderic acids T and Z, etc.).[5,54,56,59,60,62,73–95] Cancer chemoprevention mediated through antioxidant and antimutagenic activities have also been reported for $G. lucidum$.[96–101] Recent reviews on the antitumor effects of $G. lucidum$ have also been published.[102,103] (2) Central effects (sedative, analgesic, and anticonvulsive), which are due to adenosine.[104,105] (3) Cardiovascular effects (both hypertensive and hypotensive as well as increase of coronary blood flow); hypotensive effects due to triterpenoids, including ganoderols A and B, and ganoderic acids B, D, F, H, K, S, and Y, with ganoderic acid F being the most active (WANG).[21] (4) Antiallergic effects (inhibition of passive cutaneous anaphylaxis reaction, inhibition of histamine release and prevention of experimental asthma and contact dermatitis); antihistaminic effects due to ganoderic acids C and D.[25,106,107] (5) Antioxidant effects with protective activities against carbon tetrachloride-induced and BCG-induced liver damage for extractives of both red and purple $lingzhi$ as well as $G. capense$.[25,108–113] The antioxidant activity is mediated through free radical scavenging as the major mechanism involved in the protective effect against organ damage.[98,114–121] (6) Antiviral activities against hepatitis B, HSV-1 & 2, HIV-1, EBV, Flu A, and VSV.[45,122–130] The extract was found to be active alone against the bacteria $Micrococcus luteus$ and in combination with cefazolin against $Bacillus subtilis$ and $Klebsiella oxytoca$.[131] An antifungal protein, ganodermin, isolated from $G. lucidum$ was active against $Botrytis cinerea$, $Fusarium oxysporum$, and $Physalospora piricola$.[132] (7) Immunomodulating effects involving different cellular and molecular mechanisms of the extracts,[63–66,69,70,86,133–142] (8) hypoglycemic activities, with polysaccharides (especially ganoderans A and B) being the active principles.[57,143,144] Involved mechanisms include α-glucosidase inhibition as well as enhanced release of insulin from pancreatic beta cells.[145,146] (9) Platelet aggregation inhibitory effects, with adenosine being the active agent.[12,147,148] (10) Hypolipemic/hypocholesterolemic effects through the inhibition of cholesterol synthesis by 26-oxysterols and ganoderic acid derivatives[7,109,149,150]. Other reported activities include anticholinergic;[58] antipandrogenic;[151] anti-inflammatory;[98,152,153] antiulcer;[154,155] smooth muscle relaxant, antitussive, vasodilative, diuretic, anabolic, antifatigue, and others (JIANGSU; WANG).[104,156–158] Recent reviews on the biological activities of *G. lucidum* and its constituents are available.[159,160]

TOXICOLOGY

Toxicities of ganoderma are very low, with LD_{50} (i.p.) in mice varying considerably among different ganoderma preparations, ranging from 3.42 ± 0.11 to 38.3 ± 1.04 g/kg.

Oral doses of up to an equivalent of 1.88 g/kg of crude material could be tolerated by mice for 20 days with no toxic effects. The sensitivity to ganoderma toxicity varies with the animal species, with rabbits being the least sensitive and mice, the most sensitive (JIANGSU; WANG; ZHOU AND WANG).

Human toxicities are rare and not serious, consisting of dizziness, dry mouth and nasal passage, dry throat, nosebleeds, pruritus, stomach upset and bloody stools, which have been observed after long-term (3–6 month) continuous oral use. Clinical examinations revealed no toxic effects on vital organs such as heart, liver, and kidney (ZHOU AND WANG). Recent documented adverse side effects include a case of skin rash after drinking 200 mL of a *lingzhi* wine and a case of allergic shock due to the use of a *lingzhi* injection (i.m.).[161,162]

USES

Medicinal, Pharmaceutical, and Cosmetic. Extracts are used in skin-care products, especially creams and lotions for their traditional moisturizing, nourishing, and whitening properties (ETIC). Known for at least 2000 years for its complexion-benefiting properties, *lingzhi* has been demonstrated to contain a wide spectrum of bioactive chemical constituents.

Dietary Supplements/Health Foods. Powdered crude and/or extracts are used singly or in combination with other herbs in capsule, tablet, or liquid (syrup or drink) form as a general (*qi*) tonic to improve energy, stamina, and resistance to stress and diseases; hydroalcoholic extracts also used to flavor instant soup mixes and herbal drinks because of their mushroom aroma (similar to shiitake).

Traditional Medicine. *Lingzhi* was first described in the *Shen Nong Ben Cao Jing* (ca. 200 BC–AD 100) under the superior category of drugs as good for deafness, beneficial to the joints, calming, benefiting vital energy (*jing qi*), strengthening tendons and bones, and good for one's complexion. It is generally considered sweet and slightly bitter tasting and neutral or warm, and nontoxic. One of the major Chinese tonics, it was once considered the "elixir of life" and was for centuries reserved for emperors and glorified in Chinese literary classics. Although six types of *lingzhi* have been recorded and used since ancient times, only red *lingzhi* and purple *lingzhi* are currently used, and interchangeably.

Major traditional uses include treatment of general weakness (*xu lao*), cough, asthma, insomnia, and indigestion. Modern uses also include the treatment of excessive dreams (nightmares), lack of appetite, neurasthenia, chronic hepatitis, pyelonephritis, mushroom poisoning (large single dose of 120 g used),[1] coronary heart disease, arrhythmia, hyperlipemia, hypertension, chronic bronchitis, rhinitis (topical treatment), acute altitude sickness, Keshan disease, leukocytopenia, and others (JIANGSU; NATIONAL; WANG).

COMMERCIAL PREPARATIONS

Crude (mainly G. *lucidum*, both cultivated and wildcrafted) and extracts. Extracts come in various types with different physicochemical and biological properties, depending on extraction menstruums used.

Regulatory Status. Ganoderma is an ethnic food/herb; its U.S. regulatory status is not clear.

REFERENCES

See the General References for CHEUNG AND LI; DER MARDEROSIAN AND BEUTLER; HUANG; JIANGSU; LU AND LI; MCGUFFIN 1 & 2; NATIONAL; WANG; ZHOU AND WANG.

1. B. Liu, *Zhongguo Yao Yong Zhen Jun (Chinese Medicinal Fungi)*, Shanxi People's Press, Taiyuan, 1984, p. 70.
2. R. A. Ren, ed., *Zhong Yao Jian Ding Xue (Identification of Chinese Drugs)*, Shanghai Scientific and Technical Publications, Shanghai, 1986, p. 517.
3. W. S. Mao et al., eds., *Zhong Yao Zhen Wei Jian Bie (Differentiation of Genuine and Adulterated Chinese Drugs*, Shaanxi Scientific and Technical Publications, Xian, 1987, p. 540.
4. O. K. Miller, *Mushrooms of North America*, E. P. Dutton & Co., New York, 1973, p. 178.
5. Y. Sone et al., *Agric. Biol. Chem.*, **49**, 2641 (1985).
6. Institute for Assay of Drugs and Biologicals, Ministry of Health, and Institute of Botany, Academia Sinica, eds., Zhong Yao Jian Bie Shou Ce (Manual of Chinese Drug Identification), Science Publishers, Beijing, 1981, p. 174.
7. H. Hajjaj et al., *Appl. Environ. Microbiol.*, **71**, 3653 (2005).
8. Y. Mizushina et al., *Biol. Pharm. Bull.*, **21**, 444 (1998).
9. D. Kac et al., *Phytochemistry*, **23**, 2686 (1984).
10. C. Y. Hou et al., *Zhiwu Xuebao*, **30**, 66 (1988).
11. S. S. Subramanian and M. N. Swamy, *J. Sci. Ind. Res. (India)*, **20B**, 39 (1961).
12. A. Shimizu et al., *Chem. Pharm. Bull.*, **33**, 3012 (1985).
13. L. Lukacs and J. Zellner, *Monatsh. Chem.*, **62**, 214 (1933).
14. H. L. Kumari and M. Sirsi, *Arch. Mikrobiol.*, **84**, 350 (1972).
15. H. L. Kumari and M. Sirsi, *J. Gen. Mikrobiol.*, **65**, 285 (1971).
16. T. Terashita et al., *Agric. Biol. Chem.*, **48**, 1029 (1984).
17. J. H. Do and S. D. Kim, *Han'guk Kyunhakhoechi*, **14**, 79 (1986).
18. J. H. Do and S. D. Kim, *Sanop Misaengmul Hakhoechi*, **13**, 173 (1985).
19. Z. F. Campos et al., *Phytochemistry*, **67**, 202 (2006).
20. T. Kubota et al., *Helv. Chim. Acta*, **65**, 611 (1982).
21. A. Morigiwa et al., *Chem. Pharm. Bull.*, **34**, 3025 (1986).
22. T. Nishitoba et al., *Agric. Biol. Chem.*, **48**, 2905 (1984).
23. T. Nishitoba et al., *Agric. Biol. Chem.*, **49**, 1793 (1985).
24. M. Hirotani et al., *Phytochemistry*, **24**, 2055 (1985).
25. H. Kohda et al., *Chem. Pharm. Bull.*, **33**, 1367 (1985).
26. T. Nishitoba et al., *Agric. Biol. Chem.*, **49**, 3637 (1985).
27. T. Kikuchi et al., *Chem. Pharm. Bull.*, **33**, 2624 (1985).
28. Y. Komoda et al., *Chem. Pharm. Bull.*, **33**, 4829 (1985).
29. M. Hirotani and T. Furuya, *Phytochemistry*, **25**, 1189 (1986).

30. T. Kikuchi et al., *Chem. Pharm. Bull.*, **33**, 2628 (1985).
31. T. Nishitoba et al., *Agric. Biol. Chem.*, **50**, 809 (1986).
32. T. Nishitoba et al., *Phytochemistry*, **26**, 1777 (1987).
33. M. Hirotani et al., *Phytochemistry*, **26**, 2797 (1987).
34. M. Hirotani et al., *Chem. Pharm. Bull.*, **34**, 2282 (1986).
35. J. Toth et al., *J. Chem. Res., Synop.*, **12**, 299 (1983).
36. J. Toth et al., *Tetrahedron Lett.*, **24**, 1081 (1983).
37. J. Liu et al., *Biol. Pharm. Bull.*, **29**, 392 (2006).
38. C. Li et al., *Nat. Prod. Res.*, **19**, 461 (2005).
39. J. Luo et al., *J. Asian Nat. Prod. Res.*, **4**, 129 (2002).
40. J. Ma et al., *J. Nat. Prod.*, **65**, 72 (2002).
41. B. S. Min et al., *Planta Med.*, **67**, 811 (2001).
42. B. S. Min et al., *Chem. Pharm. Bull. (Tokyo)*, **48**, 1026 (2000).
43. B. S. Min et al., *Chem. Pharm. Bull. (Tokyo)*, **46**, 1607 (1998).
44. T. Nishitoba et al., *Agric. Biol. Chem.*, **49**, 1547 (1985).
45. K. Iwatsuki et al., *J. Nat. Prod.*, **66**, 1582 (2003).
46. T. S. Wu et al., *J. Nat. Prod.*, **64**, 1121 (2001).
47. Y. Mizushina et al., *Bioorg. Med. Chem.*, **7**, 2047 (1999).
48. J. J. Gao et al., *Chem. Pharm. Bull. (Tokyo)*, **50**, 837 (2002).
49. T. Nishitoba et al., *Agric. Biol. Chem.*, **52**, 367 (1988).
50. H. Sato et al., *Agric. Biol. Chem.*, **50**, 2887 (1986).
51. M. Arisawa et al., *J. Nat. Prod.*, **49**, 621 (1986).
52. L. J. Lin et al., *Phytochemistry*, **27**, 2269 (1988).
53. T. B. Ha et al., *Planta Med.*, **66**, 681 (2000).
54. T. Miyazaki and M. Nishijima, *Chem. Pharm. Bull.*, **29**, 3611 (1981).
55. T. Miyazaki and M. Nishijima, *Carbohydr. Res.*, **109**, 290 (1982).
56. T. Mizuno et al., *Nippon Nogei Kagaku Kaishi*, **59**, 1143 (1985).
57. H. Hikino et al., *Planta Med.*, 339 (1985).
58. Y. L. Lin et al., *Mol. Pharmacol.*, (2006).
59. T. Miyazaki, *Shinkin Shinkinsho*, **24**, 95 (1983).
60. T. Mozuno et al., *Nippon Nogei Kagaku Kaishi*, **58**, 871 (1984).
61. C. H. Wang et al., *Linchan Huaxue Yu Gongye*, **4**, 42 (1984).
62. T. Mizuno and S. Sakamura, *Kagaku to Seibutsu*, **23**, 797 (1985).
63. L. S. Lei and Z. B. Lin, *Yaoxue Xuebao*, **27**, 331 (1992).
64. Y. Q. He et al., *Zhongguo Zhongyao Zazhi*, **17**, 226 (1992).
65. X. Bao et al., *Carbohydr. Res.*, **332**, 67 (2001).
66. X. Bao et al., *Biosci. Biotechnol. Biochem.*, **65**, 2384 (2001).
67. Q. Z. Cao and Z. B. Lin, *Life Sci.*, **78**, 1457 (2006).
68. H. Kawagishi et al., *Phytochemistry*, **44**, 7 (1997).
69. J. Zhang et al., *Life Sci.*, **71**, 623 (2002).
70. H. W. Shin et al., *Saengyak Hakhoechi*, **16**, 181 (1986).
71. S. Ukai et al., *Chem Pharm. Bull.*, **31**, 741 (1983).
72. Z. P. Fang et al., *Zhongchengyao*, **11**, 36 (1989).
73. H. Ito et al., *Mie Med. J.*, **26**, 147 (1977).
74. T. Kosuge et al., *Yakugaku Zasshi*, **105**, 791 (1985).

75. K. C. Kim et al., *Cancer Lett.* (2006).
76. C. I. Muller et al., *Leuk. Res.* (2006).
77. J. T. Xie et al., *Exp. Oncol.*, **28**, 25 (2006).
78. Y. Gao et al., *Immunol. Invest*, **34**, 171 (2005).
79. Y. H. Gu and M. A. Belury, *Cancer Lett.*, **220**, 21 (2005).
80. G. Stanley et al., *Biochem. Biophys. Res. Commun.*, **330**, 46 (2005).
81. H. L. Yang, *Biotechnol. Lett.*, **27**, 835 (2005).
82. Q. Z. Cao and Z. B. Lin, *Acta Pharmacol. Sin.*, **25**, 833 (2004).
83. K. J. Hong et al., *Phytother. Res.*, **18**, 768 (2004).
84. J. Jiang et al., *Int. J. Oncol.*, **24**, 1093 (2004).
85. J. Jiang et al., *Nutr. Cancer*, **49**, 209 (2004).
86. Z. B. Lin and H. N. Zhang, *Acta Pharmacol. Sin.*, **25**, 1387 (2004).
87. Q. Y. Lu et al., *Cancer Lett.*, **216**, 9 (2004).
88. Q. Y. Lu et al., *Oncol. Rep.*, **12**, 659 (2004).
89. Y. S. Song et al., *J. Ethnopharmacol.*, **90**, 17 (2004).
90. H. Lu et al., *Oncol. Rep.*, **10**, 375 (2003).
91. D. Sliva et al., *J. Altern. Complement Med.*, **9**, 491 (2003).
92. H. Hu et al., *Int. J. Cancer*, **102**, 250 (2002).
93. Y. Kimura et al., *Anticancer Res.*, **22**, 3309 (2002).
94. X. Liu et al., *Cancer Lett.*, **182**, 155 (2002).
95. D. Sliva et al., *Biochem. Biophys. Res. Commun.*, **298**, 603 (2002).
96. K. N. Lai et al., *Nephrol. Dial. Transplant.*, **21**, 1188 (2006).
97. B. Lakshmi et al., *J. Ethnopharmacol.* (2006).
98. B. Lakshmi et al., *Teratog. Carcinog. Mutagen.* (Suppl 1), 85 (2003).
99. H. Lu et al., *Int. J. Mol. Med.*, **9**, 113 (2002).
100. H. Lu et al., *Oncol. Rep.*, **8**, 1341 (2001).
101. K. C. Kim and I. G. Kim, *Int. J. Mol. Med.*, **4**, 273 (1999).
102. J. W. Yuen and M. D. Gohel, *Nutr. Cancer*, **53**, 11 (2005).
103. D. Sliva, *Integr. Cancer Ther.*, **2**, 358 (2003).
104. Y. Kasahara and H. Hikino, *Phytother. Res.*, **1**, 17 (1987).
105. Y. Kasahara and H. Hikino, *Phytother. Res.*, **1**, 173 (1987).
106. M. Nogami et al., *Yakugaku Zasshi*, **106**, 594 (1986).
107. M. Nogami et al., *Yakugaku Zasshi*, **106**, 600 (1986).
108. G. T. Liu et al., *Chin. Med. J.*, **92**, 1979 (1979).
109. M. J. Lee and M. H. Chung, *Korean J. Pharmacogn.*, **18**, 254 (1987).
110. J. F. Wang et al., *J. Trad. Chin. Med.*, **5**, 55 (1985).
111. W. C. Lin and W. L. Lin, *World J. Gastroenterol.*, **12**, 265 (2006).
112. X. J. Yang et al., *World J. Gastroenterol.*, **12**, 1379 (2006).
113. G. L. Zhang et al., *World J. Gastroenterol.*, **8**, 728 (2002).
114. J. Sun et al., *J. Agric. Food Chem.*, **52**, 6646 (2004).
115. K. L. Wong et al., *Phytother. Res.*, **18**, 1024 (2004).
116. H. B. Zhao et al., *J. Pharmacol. Sci.*, **95**, 294 (2004).
117. H. N. Zhang et al., *Life Sci.*, **73**, 2307 (2003).
118. Y. H. You and Z. B. Lin, *Acta Pharmacol. Sin.*, **23**, 787 (2002).
119. J. M. Lee et al., *Phytother. Res.*, **15**, 245 (2001).

120. Y. H. Shieh et al., *Am. J. Chin Med.*, **29**, 501 (2001).
121. M. Zhu et al., *Phytother. Res.*, **13**, 529 (1999).
122. Y. Q. Li and S. F. Wang, *Biotechnol. Lett.* (2006).
123. Z. Li et al., *J. Biochem. Mol. Biol.*, **38**, 34 (2005).
124. J. Liu et al., *J. Ethnopharmacol.*, **95**, 265 (2004).
125. S. K. Eo et al., *J. Ethnopharmacol.*, **72**, 475 (2000).
126. Y. S. Kim et al., *J. Ethnopharmacol.*, **72**, 451 (2000).
127. K. W. Oh et al., *J. Ethnopharmacol.*, **72**, 221 (2000).
128. S. K. Eo et al., *J. Ethnopharmacol.*, **68**, 175 (1999).
129. S. K. Eo et al., *J. Ethnopharmacol.*, **68**, 129 (1999).
130. S. el-Mekkawy et al., *Phytochemistry*, **49**, 1651 (1998).
131. S. Y. Yoon et al., *Arch. Pharm. Res.*, **17**, 438 (1994).
132. H. Wang and T. B. Ng, *Peptides*, **27**, 27 (2006).
133. M. C. Kuo et al., *J. Ethnopharmacol.*, **103**, 217 (2006).
134. W. K. Chan et al., *J. Altern. Complement Med.*, **11**, 1047 (2005).
135. Y. Gao et al., *J. Med. Food*, **8**, 159 (2005).
136. Z. B. Lin, *J. Pharmacol. Sci.*, **99**, 144 (2005).
137. X. L. Zhu and Z. B. Lin, *Acta Pharmacol. Sin.*, **26**, 1130 (2005).
138. H. S. Chen et al., *Bioorg. Med. Chem.*, **12**, 5595 (2004).
139. H. Y. Hsu et al., *J. Immunol.*, **173**, 5989 (2004).
140. B. M. Shao et al., *Biochem. Biophys. Res. Commun.*, **323**, 133 (2004).
141. Y. Gao et al., *Immunol. Invest.*, **32**, 201 (2003).
142. X. F. Bao et al., *Chem. Pharm. Bull. (Tokyo)*, **50**, 623 (2002).
143. H. Hikino and T. Mizuno, *Planta Med.*, **55**, 385 (1989).
144. Y. Kimura et al., *Planta Med.*, **54**, 290 (1988).
145. S. D. Kim and H. J. Nho, *J. Microbiol.*, **42**, 223 (2004).
146. H. N. Zhang and Z. B. Lin, *Acta Pharmacol. Sin.*, **25**, 191 (2004).
147. M. Kubo et al., *Yakugaku Zasshi*, **103**, 871 (1983).
148. J. Tao and K. Y. Feng, *Chin. J. Integr. Trad. West. Med.*, **9**, 733 (1989).
149. Y. Komoda et al., *Chem. Pharm. Bull.*, **37**, 531 (1989).
150. A. Berger et al., *Lipids Health Dis.*, **3**, 2 (2004).
151. R. Fujita et al., *J. Ethnopharmacol.*, **102**, 107 (2005).
152. C. W. Woo et al., *Mol. Cell. Biochem.*, **275**, 165 (2005).
153. K. Koyama et al., *Planta Med.*, **63**, 224 (1997).
154. Y. Gao et al., *J. Med. Food*, **7**, 417 (2004).
155. Y. Gao et al., *Life Sci.*, **72**, 731 (2002).
156. F. C. Wan and D. Z. Huang, *Zhongguo Zhongyao Zazhi*, **17**, 619 (1992).
157. C. Lin et al., *Zhongchengyao*, **14**, 31 (1992).
158. M. H. Jiang et al., *Zhongchengyao*, **13**, 24 (1991).
159. D. Sliva, *Mini Rev. Med. Chem.*, **4**, 873 (2004).
160. M. S. Shiao, *Chem. Rec.*, **3**, 172 (2003).
161. Z. M. Zeng, *Chin. J. Integr. Trad. West. Med.*, **6**, 494 (1986).
162. Z. T. Yan, *Zhongguo Yaoxue Zazhi*, **24**, 166 (1989).

GARLIC

Source: *Allium sativum* L. (Family Amaryllidaceae or Liliaceae).

Common/vernacular names: Common garlic, allium.

GENERAL DESCRIPTION

A strong scented perennial herb with long, flat, and firm leaves, 0.5–1.5 cm wide; flowering stem up to 1.2 m high; bulbs with several bulblets (cloves), all enclosed in membranous skins; origin unclear; a variable cultigen (found only in cultivation), garlic's wild progenitor, *Allium longicuspis*, is thought to have originated in the high plains of west-central Asia perhaps in the Kirgiz Desert; spread east and west with nomadic tribes; known to be cultivated in the Middle East more than 5000 years ago; naturalized in North America; cultivated worldwide. Part used is the fresh or dehydrated bulb. Garlic oil is obtained by steam distillation of the crushed fresh bulbs; powdered garlic is derived from the dried bulbs.

CHEMICAL COMPOSITION

Contains 0.1–0.36% (usually ca. 0.2%) volatile oil, alliin (S-allyl-L-cysteine sulfoxide), S-methyl-L-cysteine sulfoxide, enzymes (e.g., alliinase, peroxidase, and myrosinase), ajoenes (E,Z-ajoene, E,Z-methylajoene, and dimethylajoene), protein (16.8%, dry weight basis), minerals, vitamins (thiamine, riboflavin, niacin, etc.), lipids, amino acids, and others (JIANGSU; KARRER; LIST AND HÖRHAMMER; MARSH; MARTINDALE).[1–3]

The volatile oil contains allicin (diallyldisulfide-S-oxide; diallyl thiosulfinate), allylpropyl disulfide, diallyl disulfide, and diallyl trisulfide as the major components, with lesser amounts of dimethyl sulfide, dimethyl disulfide; dimethyl trisulfide, allylmethyl sulfide, 2,3,4-trithiapentane, bis-2-propenyl tri-, tetra-, and pentasulfides, and other related sulfur compounds.[2,4] Boiled garlic also contains sodium 2-propenyl thiosulfate.[5] Most published data since 1892 have indicated diallyl disulfide to be the main compound in garlic oil (60%). However, one study indicated that diallyl trisulfide dominated in freshly distilled oils.[6] Other volatile compounds present include citral, geraniol, linalool, and α- and β-phellandrene (JIANGSU; KARRER; LIST AND HÖRHAMMER).[7]

Prostaglandins A_2 and F_{1a} were isolated from a homogenized garlic extract.[3] High-molecular weight fructans and agglutinins (homo- and heterodimeric mannose-binding lectins) were also isolated from garlic.[8–11]

Allicin is the major odor principle that is produced by the enzymatic action of alliinase on alliin; it is decomposed by heat and alkali but is unaffected by dilute acids in solution (JIANGSU; MERCK).

Composition of garlic products depends on product form. Thiosulfinates (e.g., allicin) were found to be released only from garlic cloves and garlic powder products; vinyl dithiins and ajoenes were found only in products containing garlic macerated in vegetable oil; diallyl, methylallyl, and dimethyl sulfide series components were exclusive to products containing the oil of steam distilled garlic.[12]

PHARMACOLOGY AND BIOLOGICAL ACTIVITIES

Garlic and garlic oil have been reported to exhibit numerous pharmacological properties.[13–17] The ones most frequently reported include (i) hypolipidemic/lowering of serum cholesterol (or lipids) in rabbits and humans (including lowering triglycerides and total low-density lipoprotein cholesterol), while raising levels of high-density lipoprotein cholesterol;[18–28] (ii) cardiovascular enhancing and hypotensive properties in humans and animals (MARTINDALE). Possible mechanisms include sodium pump modulation and/or beta-receptor blockade;[28–35] (iii) antibacterial, antifungal, and ativiral properties (*in vitro* and *in vivo* against *H. pylori*, pneumococci, *Candida*, *Aspergillus*, and dermatophytes) with the hydroalcoholic extract reportedly being much more potent than the essential oil;[2,15,28,36–48] a pilot study in 10 AIDS patients found that a garlic extract produced an improved helper: suppresser ratio in natural killer cell activity, with a concurrent improvement in AIDS-related conditions, including diarrhea, genital herpes, candidiasis, and pansinusitis with recurrent fever;[49,50] garlic also has larvicidal, insecticidal, and amebicidal activities (e.g., antigiardial and antitrypanosomal);[48,51–53] (iv) anti-inflammatory (inhibition of COX and prostaglandin synthesis);[54–57] (v) antitumor activities against various tumor models in mice and rats;[58–61] (vi) hypoglycemic activities in rabbits and rats;[21,62–64] (vii) antioxidant/antihepatotoxic/radioprotective activity in rats (mainly due to *S*-allyl cysteine);[65–67] garlic was recently reported to be effective in the treatment of hepatopulmonary syndrome.[68] Garlic also lowers blood viscosity; improves microcirculation; with expectorant, diaphoretic, and diuretic properties, among others (ESCOP 2; JIANGSU; LIST AND HÖRHAMMER; MARTINDALE).[69] It has recently been reported to reduce osteoporosis and to possess immunomodulatory and spermicidal activities.[70–72]

Antithrombotic activity has been attributed to various fractions of garlic. Ajoene has antithrombotic activity triggered by inhibiting exposure of fibrinogen receptors on platelet membranes. Diallyl sulfide and methyl allyl sulfide, claimed responsible for the antithrombotic activity, were found to be inactive in inhibiting platelet aggregation.[73] However, a recent study attributes higher thrombocyte aggregation-inhibiting potential to chloroform fractions of thiosulfinate than to the ajoenes.[3] The platelet aggregation-inhibiting activity, as well as possible antiplaque effect,[74] may be responsible for the potential utility of garlic as a useful protective agent in atherosclerosis, coronary thrombosis, and stroke (TYLER 1).[49]

The volatile sulfur compounds (especially allicin, diallyl disulfide, and diallyl trisulfide) are generally considered to be responsible for much of garlic's pharmacological activities (e.g., hypoglycemic, hypocholesterolemic, antimicrobial, insecticidal, and larvicidal).[23,45,48,53,63] Allicin, believed to be the most important biologically active compound in garlic, is primarily responsible for its antibiotic and antimutagenic effects.[75,76]

One study found that thiosulfinates (e.g., allicin) were not formed below pH 3.6, thus alliinase is completely and irreversibly inhibited by stomach acid. A second (unidentified) enzyme, in addition to alliinase is involved in thiosulfinate formation. A stomach acid-resistant coating on garlic powder tablets is necessary for thiosulfinate release, which if prepared carefully can release amounts of total thiosulfinates similar to whole garlic cloves.[77]

Garlic oil acts as a gastrointestinal smooth muscle relaxant, suggested for further research in patients with hypermotile intestinal disorders.[78]

TOXICOLOGY

Allergic contact dermatitis due to garlic has been reported (MARTINDALE).

USES

Medicinal, Pharmaceutical, and Cosmetic. Strong popular interest has made garlic preparations the best-selling over-the-counter drugs in Germany, with sales topping $250 million per year in Europe.[79]

Food. Fresh garlic and powdered garlic are widely used as domestic spices. Garlic oil is extensively used as a flavor ingredient in most food products, including nonalcoholic beverages, frozen dairy desserts, candy, baked goods, gelatins and puddings, condiments and relishes, meat and meat products, fats and oils, snack foods, and gravies, with highest average maximum use levels generally much below 0.003% (34.4 ppm).

Dietary Supplements/Health Foods. Use of various dosage forms and formulations, including fresh, dried, powdered, or freeze-dried garlic, essential oil (garlic oil), and various proprietary preparations of powdered, aged, or extracted fresh or dried garlic (FOSTER).

Traditional Medicine. Has been used for thousands of years in treating coughs, colds, chronic bronchitis, toothache, earache, dandruff, high blood pressure, arteriosclerosis, hysteria; also used extensively in cancers; generally as the juice, cold infusion, or tincture.[80] In addition, the fresh cloves, garlic tea, syrup, tincture, and other preparations have been used as an aphrodisiac; to treat fever, flu symptoms, shortness of breath, sinus congestion, headache, stomach ache, hypertension, gout, rheumatism, pinworms, old ulcers, and snakebites; and for numerous other ailments, conditions, and applications (FOSTER AND DUKE).[49]

In Chinese medicine, in addition to above uses, garlic is used for diarrhea, dysentery (amebic and bacterial), pulmonary tuberculosis, bloody urine, diphtheria, whooping cough, typhoid, hepatitis, trachoma, scalp ringworm, hypersensitive dentin, vaginal trichomoniasis, and others, some of which have been reported successful in clinical trials (JIANGSU).

COMMERCIAL PREPARATIONS

Crude, powder, and oil. Crude was formerly official in N.F. and U.S.P., and oil is official in F.C.C. Composition of organosulfur compounds varies greatly with the method of processing, and changes occur in their chemistry when garlic is crushed, cooked, ingested, metabolized, or commercially processed. Analysis of garlic compounds is necessary for standardization of these products, especially commercial ones and for biological studies.[12,77,81]

Regulatory Status. Has been affirmed as GRAS (§184.1317), but only as oil, extract, or oleoresin.[69]

A German therapeutic monograph allows use of the minced bulb and preparations calculated to an average daily dose of 4 g (fresh garlic) or 8 mg (essential oil) for supportive dietary measures to reduce blood lipids and as a preventative for age-dependent vascular changes.[82]

A proposed European monograph indicates use for prophylaxis of atheroselerosis; treatment of elevated blood lipid levels influenced by diet; improvement of arterial vascular disease blood flow; and use for relief of coughs, colds, catarrh, and rhinitis (ESCOP 2).

REFERENCES

See the General References for ARCTANDER; BAILEY 1; BARNES; BLUMENTHAL 1 & 2; DER MARDEROSIAN AND BEUTLER; DUKE 4; ESCOP 2; FEMA; FERNALD; FOSTER; FOSTER AND DUKE; GRIEVE; GUENTHER; GUPTA; HUANG; JIANGSU; LEWIS AND ELVIN-LEWIS; LIST AND HÖRHAMMER; LUST; MABBERLY; MCGUFFIN 1 & 2; NANJING; ROSE; ROSENGARTEN; TERRELL; TYLER 1; UPHOF; YOUNGKEN.

1. J. Lee and J. M. Harnly, *J. Agric. Food Chem.*, **53**, 9100 (2005).
2. J. R. Whitaker et al., *Adv. Food Res.*, **22**, 73 (1976).
3. S. A. Al-Nagdy et al., *Phytother. Res.*, **2**, 196 (1988).
4. Q. Hu et al., *J. Agric. Food Chem.*, **50**, 1059 (2002).
5. O. Yamato et al., *Biosci. Biotechnol. Biochem.*, **67**, 1594 (2003).
6. H. Miething, *Phytother. Res.*, **2**, 149 (1988).
7. O. E. Schultz and H. L. Mohrmann, *Pharmazie*, **20**, 441 (1965).
8. S. Baumgartner et al., *Carbohydr. Res.*, **328**, 177 (2000).

9. A. Gupta and R. S. Sandhu, *Mol. Cell. Biochem.*, **166**, 1 (1997).
10. K. Smeets et al., *Plant Mol. Biol.*, **33**, 223 (1997).
11. H. Kaku et al., *Carbohydr. Res.*, **229**, 347 (1992).
12. L. D. Lawson et al., *Planta Med.*, **57**, 363 (1991).
13. L. D. Brace, *J. Cardiovasc. Nurs.*, **16**, 33 (2002).
14. M. Thomson and M. Ali, *Curr. Cancer Drug Targets*, **3**, 67 (2003).
15. J. C. Harris et al., *Appl. Microbiol. Biotechnol.*, **57**, 282 (2001).
16. U. P. Singh et al., *Indian J. Exp. Biol.*, **39**, 310 (2001).
17. K. T. Augusti, *Indian J. Exp. Biol.*, **34**, 634 (1996).
18. S. Gorinstein et al., *J. Agric. Food Chem.*, **54**, 4022 (2006).
19. B. Turner et al., *Br. J. Nutr.*, **92**, 701 (2004).
20. K. T. Augusti et al., *Indian J. Exp. Biol.*, **39**, 760 (2001).
21. R. C. Jain, *Lancet*, **1**, 1240 (1975).
22. A. Bordia et al., *Atherosclerosis*, **21**, 15 (1975).
23. K. T. Augusti, *Indian J. Exp. Biol.*, **15**, 489 (1977).
24. R. C. Jain and C. R. Vyas, *Medikon*, **6**, 12 (1977).
25. A. Bordia et al., *Atherosclerosis*, **26**, 379 (1977).
26. K. K. Sharma et al., *Indian J. Nutr. Diet.*, **13**, 7 (1976).
27. D. Kritchevsky, *Artery*, **1**, 319 (1975).
28. V. Petkov, *Dtsch. Apoth. Ztg.*, **106**, 1861 (1966).
29. S. K. Verma et al., *Indian J. Physiol. Pharmacol.*, **49**, 115 (2005).
30. R. K. Yadav and N. S. Verma, *Indian J. Exp. Biol.*, **42**, 628 (2004).
31. K. K. Al-Qattan et al., *Prostaglandins Leukot. Essent. Fatty Acids*, **69**, 217 (2003).
32. W. Qidwai et al., *J. Pak. Med. Assoc.*, **50**, 204 (2000).
33. A. Bordia et al., *Prostaglandins Leukot. Essent. Fatty Acids*, **58**, 257 (1998).
34. H. Isensee et al., *Arzneim.-Forsch.*, **43**, 94 (1993).
35. N. Martin et al., *J. Ethnopharmacol.*, **37**, 145 (1992).
36. M. Shams-Ghahfarokhi et al., *Fitoterapia* (2006).
37. D. Dikasso et al., *Ethiop. Med. J.*, **40**, 241 (2002).
38. P. Canizares et al., *Biotechnol. Prog.*, **18**, 1227 (2002).
39. K. M. Lemar et al., *J. Appl. Microbiol.*, **93**, 398 (2002).
40. G. P. Sivam et al., *Nutr. Cancer*, **27**, 118 (1997).
41. L. Cellini et al., *FEMS Immunol. Med. Microbiol.*, **13**, 273 (1996).
42. P. V. Venugopal and T. V. Venugopal, *Int. J. Dermatol.*, **34**, 278 (1995).
43. S. T. Pai and M. W. Platt, *Lett. Appl. Microbiol.*, **20**, 14 (1995).
44. L. E. Davis et al., *Planta Med.*, **60**, 546 (1994).
45. F. E. Barone and M. R. Tansey, *Mycologia*, **69**, 793 (1977).
46. A. V. Kolodin, *Sovrem. Metody. Issled.*, **1**, 101 (1968).
47. Y. Kominato et al., *Oyo Yakuri*, **11**, 941 (1976).
48. N. B. K. Murthy and S. V. Amonkar, *Indian J. Exp. Biol.*, **12**, 208 (1974).
49. S. Foster, *Garlic-Allium sativum*, Botanical Series, no. 311, American Botanical Council, Austin, TX, 1991.
50. T. H. Abdullah et al., *Dtsch. Z. Onkologie*, **21**, 52 (1989).

51. J. C. Harris et al., *Microbiology*, **146** (Pt 12), 3119 (2000).
52. A. J. Nok et al., *Parasitol. Res.*, **82**, 634 (1996).
53. S. V. Amonkar and A. Banerji, *Science*, **174**, 1343 (1971).
54. H. P. Keiss et al., *J. Nutr.*, **133**, 2171 (2003).
55. G. Hodge et al., *Cytometry*, **48**, 209 (2002).
56. M. Ali, *Prostaglandins Leukot. Essent. Fatty Acids*, **53**, 397 (1995).
57. M. Ali et al., *Prostaglandins Leukot. Essent. Fatty Acids*, **49**, 855 (1993).
58. M. D. Samaranayake et al., *Phytother. Res.*, **14**, 564 (2000).
59. D. L. Lamm and D. R. Riggs, *Urol. Clin. North Am.*, **27**, 157 (2000).
60. D. R. Riggs et al., *Cancer*, **79**, 1987 (1997).
61. Y. Kimura and K. Yamamoto, *Gann*, **55**, 325 (1964).
62. C. G. Sheela and K. T. Augusti, *Indian J. Exp. Biol.*, **30**, 523 (1992).
63. P. T. Mathew and K. T. Augusti, *Indian J. Biochem. Biophys.*, **10**, 209 (1973).
64. R. C. Jain and C. R. Vyas, *Am. J. Clin. Nut.*, **28**, 684 (1975).
65. G. Saravanan and J. Prakash, *J. Ethnopharmacol.*, **94**, 155 (2004).
66. S. K. Jaiswal and A. Bordia, *Indian J. Med. Sci.*, **50**, 231 (1996).
67. H. Hikino et al., *Planta Med.*, **52**, 163 (1986).
68. G. A. Abrams and M. B. Fallon, *J. Clin. Gastroenterol.*, **27**, 232 (1998).
69. Anon., *Fed. Regist.*, **39**, 34213 (1974).
70. T. Ghazanfari et al., *J. Ethnopharmacol.*, **103**, 333 (2006).
71. M. Mukherjee et al., *Phytother. Res.*, **20**, 408 (2006).
72. K. Chakrabarti et al., *Asian J. Androl*, **5**, 131 (2003).
73. E. Block in R. P. Steiner, ed., *Folk Medicine the Art and the Science*, American Chemical Society, New York, 1986, p. 125.
74. J. Koscielny et al., *Atherosclerosis*, **144**, 237 (1999).
75. L. D. Lawson et al., *Planta Med.*, **57**, 263 (1991).
76. H. Koch and G. Hahn, *Garlic—Fundamentals of the Therapeutic Application of Allium Sativum*, Urban & Schwarzenberg, Munich, 1988.
77. L. D. Lawson and B. G. Hughes, *Planta Med.*, **58**, 347 (1992).
78. D. J. Joshi et al., *Phytother. Res.*, **1**, 140 (1987).
79. P. Mansell and J. P. D. Reckless, *Br. Med. J.*, **303**, 379 (1991).
80. J. L. Hartwell, *Lloydia*, **33**, 97 (1970).
81. L. D. Lawson in A. D. Kinghorn and M. F. Balandrin, eds., *Human and Medicinal Agents from Plants*, American Chemical Society, Washington, DC, 1994.
82. Monograph *Allii sativi bulbus*, *Bundesanzeiger*, no. 122 (July 6, 1988).

GELSEMIUM

Source: *Gelsemium sempervirens* (L.) Ait. f. (syn. *G. nitidum* Michx. and *Bignonia sempervirens* L.) (Family Loganiaceae or Spigeliaceae)

Common/vernacular names: Yellow jasmine, wild jessamine, woodbine, Carolina yellow jessamine, and evening trumpet flower.

GENERAL DESCRIPTION

Evergreen woody vine sometimes climbing to about 6 m; leaves opposite, lance-shaped to narrowly oval; flowers very fragrant, bright yellow, occur in clusters; native to southwestern United States; also grows in Mexico and Guatemala; widely cultivated as an ornamental plant. Parts used are the dried rhizome and roots.

CHEMICAL COMPOSITION

Contains as active constituents about 0.5% oxindole alkaloids, consisting mainly of gelsemine, with lesser amounts of gelsemicine, gelsedine, gelsevirine, sempervirine, gelsemidine, 1-methoxygelsemine, 21-oxo-gelsemine, and 14-hydroxygelsemicine (GLASBY 2; LIST AND HÖRHAMMER).[1–5]

Other constituents present include the coumarin scopoletin (also called gelsemic acid and β-methylesculetin), an iridoid compound,[1] volatile oil (0.5%), pregnane-type steroids (12-β-hydroxypregna-4,16-diene-3,20-dione and 12-β-hydroxy-5-α-pregn-16-ene-3,20-dione),[6] fatty acids (palmitic, stearic, oleic, and linoleic acids), n-pentatriacontane, and tannins (KARRER; LIST AND HÖRHAMMER).

PHARMACOLOGY AND BIOLOGICAL ACTIVITIES

Gelsemium and gelsemine, its major alkaloid, have been reported to have central stimulant and analgesic properties; they also potentiated the analgesic effects of aspirin and phenacetin (JIANGSU; LIST AND HÖRHAMMER).

Oral administration of a dilute solution of gelsemium extract for 30 days totally inhibited the development of seizures in rats with status epilepticus.[7] An earlier study also demonstrated that low dose of gelsemium extract had beneficial neurotropic, immunological and gastric effects in a stressed mice model.[8] This indicates that low doses of gelsemium may have potential in management of certain nervous disorders.

TOXICOLOGY

Gelsemium alkaloids are very toxic; and gelsemicine is reportedly more toxic than glesemine.[9] Ingestion of as little as 4 mL of a fluid extract (1:1) or a tea made from as few as three leaves has been reported as fatal. Children have been severely poisoned by chewing leaves or sucking flower nectar. Toxic symptoms include giddiness, weakness, ptosis, double vision, dilated pupils, and respiratory depression (GOSSELIN; HARDIN AND ARENA; MARTINDALE). A case of lethal goat intoxication has been reported in which 3 goats died after accidental feeding on gelsemium leaves that resulted in diffuse neuronal degeneration and myofiber atrophy leading to death within 24 h.[10]

USES

Medicinal, Phamaceutical, and Cosmetic. Used as an ingredient in some analgesic, sedative, and antispasmodic preparations.

Traditional Medicine. Due to its high toxicity, gelsemium is rarely used as a domestic medicine in the United States. However when it is used, it is usually for treating nervous heart conditions, migraine, neuralgia, and sciatica; also used in cancer.[11]

Widely used in the 19th century, especially by eclectic practitioners; root tincture used for fevers, inflammations of the spinal column, and neuralgia; believed to diminish blood to the cerebrospinal centers, reducing spasmodic action.[12]

A related species (*G. elegans* Benth.) is used in China to treat conditions, including neuralgia, rheumatic pain, various kinds of sores, scrofula, and wounds. It is used only externally; all parts of the plant are employed (JIANGSU).

A rare species *G. rankinii* Small occurs from Alabama and Florida to North Carolina and may have been involved in traditional usage in the southern United States.[12]

COMMERCIAL PREPARATIONS

Crude and extracts (fluid, solid, etc.). Strengths (see *glossary*) of extracts are expressed in weight-to-weight ratios. Crude and fluid extract were formerly official in N.F.

REFERENCES

See the General References for BAILEY 1; DER MARDEROSIAN AND BEUTLER; GOSSELIN; HARDIN AND ARENA; HUANG; LIST AND HÖRHAMMER; LUST; MARTINDALE; MCGUFFIN 1 & 2; RAFFAUF; UPHOF; WILLAMAN AND SCHUBERT.

1. M. Kitajima et al., *Chem. Pharm. Bull.*, **51**, 1211 (2003).
2. A. Nikiforov et al., *Monatsh. Chem.*, **105**, 1292 (1974).
3. E. Wenkert et al., *Experimentia*, **28**, 377 (1971).
4. M. Wichtl et al., *Monatsh. Chem.*, **104**, 87 (1973).
5. M. Wichtl et al., *Monatsh. Chem.*, **104**, 99 (1973).
6. Y. Schun and G. A. Cordell, *J. Nat. Prod.*, **50**, 195 (1987).
7. O. Peredery and M. A. Persinger, *Phytother. Res.*, **18**, 700 (2004).
8. D. Bousta et al., *J. Ethnopharmacol.*, **74**, 205 (2001).
9. B. R. Olin et al., *Lawrence Rev. Nat. Prod.*, (1993).
10. L. J. Thompson et al., *Vet. Hum. Toxicol.*, **44**, 272 (2002).
11. J. L. Hartwell, *Lloydia*, **33**, 97 (1970).
12. S. Foster, *East–West Botanicals: Comparisons of Medicinal Plants Disjunct between Eastern Asia and Eastern North America*, Ozark Beneficial Plant Project, Brixey, MO, 1986.

GENET

Source: *Spartium junceum* **L.** (syn. *Genista juncea* Lam. (Family Leguminosae or Fabaceae)

Common/vernacular names: Spanish broom, weavers broom.

GENERAL DESCRIPTION

A monotypic genus, upright perennial, evergreen shrub that can reach six to ten feet tall, shrub with green branches, up to about 3 m high; stems bright green rounded (rush-like) and mainly leafless; leaves simple, alternate, linear to lanceolate, short-lived, less than one inch long; flowers grow in clusters at the branches fragrant, bright yellow, and pea-shaped, approximately 1 in. long; fruits hairy seed pods, flat and linear, up to 2.5 cm long; native to the Mediterranean region, southwest Europe, and the Canary Islands; extensively cultivated, especially as an ornamental plant in western United States. Part used is the dried flower from which genet absolute and other extracts are produced by solvent extraction.

CHEMICAL COMPOSITION

Contains alkaloids (mainly anagyrine, cytisine, thermopsine, N-methylcytisine, and sparteine), *n*-nonacosane, 1,18-octadecanediol, 1,26-hexacosanediol, lupeol, sitosterol, phenols, acids and trace of volatile oil (KARRER; WILLAMAN AND SCHUBERT).[1–3] An isoflavone and different glycosides of four flavonoids (azaleatin, carthamidine, luteolin, and quercetin) have been isolated.[4,5] Triterpene saponine glycosides (junceoside and spartitrioside) and seed lectins have also been reported.[4,6,7]

Major constituents of the absolute include ethyl palmitate (14.56%), linalool (10.91%), methyl linoleate (7.13%), ethyl stearate (3.67%), ethyl oleate (4.86%), methyl linolenate (1.53%), phenethyl alcohol (1.30%), and at least 50 additional, mostly ubiquitous, volatile components; plus caproic, caprylic, capric, lauric, myristic, palmitic, stearic, oleic, linoleic, and linolenic acids. Occurrence of pentanal diethyl acetal in the absolute has been attributed to formation during ethanol back extraction of the concrete or added as part of a coupage. β-terpineol and diethyl phthalate occurrence in the absolute may result from addition of synthetic additives.[8]

Due to the extraction methods used, genet absolute (see *glossary*), the more commonly used form, should not contain any alkaloids. However, genet extracts obtained by other methods may contain alkaloids.

PHARMACOLOGY AND BIOLOGICAL ACTIVITIES

Sparteine has oxytocic properties (see *broom tops*).

A saponin-rich extract was reported to reduce gastric secretion/acidity and gastric lesions in rat models. This antiulcerogenic effect was attributed to spartitrioside.[9]

Flower extracts possess strong *in vitro* antioxidant activity resembling that of superoxide dismutase (SOD) with luteolin and azaleatin glycosides possessing the highest activity.[5]

Reversible antifirtility effect targeting the acrosomal protease system of adult male rats and rabbits has been reported.[10,11]

TOXICOLOGY

Available dermatological data indicate genet absolute to be nonphototoxic to mice and swine and nonirritating and nonsensitizing to humans.[12]

USES

Medicinal, Pharmaceutical, and Cosmetic. Genet absolute is used as a fragrance ingredient in soaps, detergents, creams, lotions, and perfumes, with maximum use level of 0.2% reported in perfumes.[12]

Food. Genet absolute and genet extract (type not specified) are used as flavor components in major categories of food products, including alcoholic and nonalcoholic beverages, frozen dairy desserts, candy, baked goods, and gelatins and puddings. Reported average maximum use levels are usually below 0.001%.

Traditional Medicine. Reportedly, laxative, slightly narcotic, diuretic. Stems have appeared as an adulterant to *Cytisus scoparius* (STEINMETZ) (see *broom tops*).

Others. Fiber formerly used for ropes, mattress filler, pillows, paper making, rubberized belts for mine conveyors; maintains strength under high humidity; high-grade cellulose extract from dry stems (CSIR X).

COMMERCIAL PREPARATIONS

Crude and extracts.

Regulatory Status. Has been approved for food use (§172.510).

REFERENCES

See the General References for BAILEY 2; CSIR X:FEMA; GUENTHER; HAY AND SYNGE; STEINMETZ; TUTIN 2; UPHOF.

1. A. D. Dauksha and E. K. Denisova, *Nauch. Dokl. Vysshei Shkoly, Biol. Nauki*, **3**, 178 (1966).
2. N. L. Gurvich and Z. I. Abasova, *Maslo. Zhir. Promy.*, **33**, 27 (1967).
3. G. Faugergas et al., *Plant. Med. Phytother.*, **7**, 68 (1973).
4. A. R. Bilia et al., *Phytochemistry*, **34**, 847 (1993).
5. E. Yesilada et al., *J. Ethnopharmacol.*, **73**, 471 (2000).
6. E. Yesilada and Y. A. Takahishi, *Phytochemistry*, **51**, 903 (1999).

7. C. N. Hankins et al., *Plant Physiol.*, **96**, 98 (1991).
8. B. M. Lawrence, *Perfurn. Flav.*, **6** (2), 59 (1981).
9. E. Yesilada et al., *J. Ethnopharmacol.*, **70**, 219 (2000).
10. B. Baccetti et al., *Zygote*, **1**, 71 (1993).
11. J. S. Chen et al., *Zygote*, **1**, 309 (1993).
12. D. L. J. Opdyke, *Food Cosmet. Toxicol.*, **14**(Suppl.), 779 (1976).

GENTIAN

Source: *Gentian* Gentiana lutea L.; ***Stemless gentian*** Gentiana acaulis L. (Family Gentianaceae).

Common/vernacular names: Yellow gentian, bitter root (*G. lutea*).

GENERAL DESCRIPTION

Gentian (*G. lutea*) is a perennial herb with large tap root simple erect stems up to 1 m tall; leaves large and deeply veined, in basal rosette until flowering; flowers in terminal and axillary clusters yellow, each cluster with a large leaf below it; corolla deeply divided; native to mountains of central and southern Europe and western Asia. Parts used are the dried rhizome and roots.

Stem less gentian (*G. acaulis*) is a small variable taxonomic complex of perennial herbs with basal rosettes; leaves lance shaped; up to about 10 cm high; native to Europe (Alps and Pyrenees). Part used is the whole herb.

CHEMICAL COMPOSITION

Gentian contains iridoid bitter principles (mainly amarogentin, also gentiopicroside or gentiopicrin, loganic acid, trifloroside, sweroside, and swertiamarin);[1–4] alkaloids (mainly gentianine and gentialutine);[5,6] xanthones (gentisein, gentisin, isogentisin, 1,3,7-trimethoxyxanthone, 1-hydroxy-3,7-dimethoxyxanthone, 2,3′,4,6tetrahydroxybenzophenone, etc.);[7,8] triterpenes (lupeol, β-amyrin, ursenediol, seco-oxoursenoic acid, seco-oxooleanenoic acid, and betulin);[9–11] gentianose (glucosyl gluconyl fructose), gentiobiose, and common sugars;[3,12] and traces of volatile oil (KARRER; LIST AND HÖRHAMMER).[3]

It has been reported that content of gentiopicroside does not vary significantly on drying or at various stages of vegetative growth.[13] However, another study found that amarogentin and gentiopicroside content were highest in spring, with concurrent decreased levels of sugar accumulation in the roots.[14]

It has been observed that cultivated plants are much richer in amarogentin than wild ones.[13]

G. lutea ecotypes have been found to retain morphological and chemical characteristics, with amarogentin content ranging from 0.05% to 0.33% in 3-year-old ecotypes, with varying differences in total sugar and essential oil composition.[3] Wild plants harvested at lower altitudes in Italy were found to have higher levels of bitter principles.[15]

Stemless gentian contains gentiacauloside (a xanthone glycoside) and gentisin. It probably also contains similar constituents as gentian (LIST AND HÖRHAMMER).[16]

Amarogentin is one of the most bitter glycosides known (STAHL). Amarogentin and gentiopicrin are the main compounds responsible for the bitter taste (TYLER 1).

PHARMACOLOGY AND BIOLOGICAL ACTIVITIES

Gentian extracts have been reported to exhibit choleretic activities in animals. Gentianine also exhibited strong anti-inflammatory properties in laboratory animals (FARNSWORTH 1).[5,17,18]

Recent reports suggest that *G. lutea* has an effect on the CNS. A parenterally administered methanolic extract exhibited a central effect in mice that resulted in a significant increase in their swimming endurance and a slight analgesic activity.[19] The methanolic extract of the dried bark potently inhibited rat brain monoamine oxidase (MAO). The active MAO inhibitors were determined to be a chalcone dimer, a chromanone, and 5-hydroxyflavanone.[20]

The root extract and two of gentian's constituents (sweroside and swertiamarine) possessed wound-healing properties in a chicken embryonic fibroblasts *in vitro* model. Stimulation of mitotic activity and collagen production were speculated to mediate the activity.[21]

Gentiopicrin has been reported to be lethal to mosquito larvae (JIANGSU).

TOXICOLOGY

The root may not be well tolerated by patients with extreme high blood pressure or by pregnant women (TYLER 1).

Highly toxic *Veratrum album* L., growing wild in proximity to gentian has resulted in several cases of vicarious accidental poisoning in Europe among persons making gentian preparations for home use.[22]

USES

Medicinal, Pharmaceutical, and Cosmetic. Gentian extracts are used in tonic preparations; also used in certain antismoking formulas. They are also used in cosmetics.[23]

Food. Gentian and stemless gentian are widely used in alcoholic bitters and vermouth formulations; average maximum use levels reported are about 0.02% (199 ppm) for gentian extract (type not specified) and 0.001% (12.9 ppm) for stemless gentian. Other food products in which both are used include nonalcoholic beverages, frozen dairy desserts, candy, baked goods, and gelatins and puddings, with highest average maximum use level of about 0.015% (153 ppm) reported for gentian extract (type not specified) in baked goods.

Gentian is also a major constituent of "angostura bitters" (see ***angostura***).

Dietary Supplements/Health Foods. Dried root and extract used in various digestive formulations; tea flavoring; also in nervine formulations. In England, the most popular of gastric stimulants (WREN).

Traditional Medicine. Yellow gentian as well as related gentians reportedly used to stimulate the appetite, improve digestion, and to treat numerous gastrointestinal problems (stomachache, heartburn, gastritis, diarrhea, vomiting, etc.) as well as externally for wounds; also used in cancers.[24]

In Chinese medicine several related *Gentiana* species (e.g., G. *scabra* Bunge, G. *triftora* Pall., and G. *rigescens* Franch.) are used for similar purposes. In addition, they are used for treating jaundice, sore throat, headache, sores, inflammations, and rheumatoid arthritis, among others, both internally and externally (JIANGSU; NANJING).

COMMERCIAL PREPARATIONS

Crude and extracts. Gentian and its fluid extract, compound tincture, and glycerinated elixir were formerly official in N.F. Strengths (see ***glossary***) of extracts are expressed in weight-to-weight ratios.

Regulatory Status. Gentian and stemless gentian both have been approved for food use, the latter in alcoholic beverages only (§172.510). The root is the subject of a German therapeutic monograph, approved for digestive disorders, such as appetite loss and flatulence; use contraindicated in gastric and duodenal peptic ulcers.[25]

REFERENCES

See the General References for BAILEY 1; BIANCHINI AND CORBETTA; BLUMENTHAL 1; FARNSWORTH 3; FEMA; GOSSELIN; JIANGSU; LUST; NANJING; POLUNIN AND SMYTHIES; RAFFAUF; TERRELL; TUTIN 3; TYLER 1; WREN.

1. J. Bricout, *Phytochemistry*, **13**, 2819 (1974).
2. Y. Hatakeyama et al., *Shoyakugaku Zasshi*, **26**, 75 (1972).
3. C. Franz and D. Fritz, *Planta Med.*, **28**, 289 (1975).
4. A. Bianco et al., *Nat. Prod. Res.*, **17**, 221 (2003).
5. F. Rulkco, *Pr. Nauk. Akad. Med. Wroclawiu*, **8**, 3 (1976).
6. F. Sadritdinov, *Farmakol. Alkaloidov Serdechnykh Glikozidov*, **146** (1971).
7. J. R. Lewis and P. Gupta, *J. Chem. Soc. C*, (4), 629 (1971).
8. J. E. Atkinson et al., *Tetrahedron*, **25**, 1507 (1969).
9. L. Benarroche et al., *Plant. Med. Phytother.*, S, 15 (1974).
10. Y. Toriumi et al., *Chem. Pharm. Bull.*, **51**, 89 (2003).
11. R. Kakuda et al., *Chem. Pharm. Bull.*, **51**, 885 (2003).
12. P. Rivaille and D. Raulais, *C. R. Acad. Sci., Ser. D*, **269**, 1121 (1969).
13. V. Rossetti et al., *Q. J. Crude Drug Res.*, **19**(1), 37 (1981).
14. Ch. Franz, et al., *Sci. Pharm.*, **53**, 31 (1985).
15. V. Rossetti et al., *Plant. Med. Phytother.*, **18**(1), 15 (1984).
16. V. Plouvier et al., *C. R. Acad. Sci., Ser. D*, **264**, 1219 (1967).
17. H. C. Chi et al., *Sheng Li Hsueh Pao*, **23**, 151 (1959).
18. B. R. Olin, ed., *Lawrence Rev. Nat. Prod.* 1993.
19. N. Ozturk et al., *Phytother. Res.*, **16**, 627 (2002).
20. H. Haraguchi et al., *Phytochemistry*, **65**, 2255 (2004).
21. N. Ozturk et al., *Planta Med.*, **72**, 289 (2006).
22. R. Garnier et al., *Ann. Med. Interne (Paris)*, **136**, 125 (1985).
23. H. B. Heath, *Cosmet. Toilet.*, **92**(1), 19 (1977).
24. J. L. Hartwell, *Lloydia*, **32**, 153 (1969).
25. Monograph *Gentianae radix*, *Bundesanzeiger*, no. 223 (November 30, 1985); revised (March 13, 1990).

GERANIUM OIL, ROSE

Source: *Pelargonium graveolens* (L.) L'Her. ex Ait. (Family Geraniaceae).

Common/vernacular names: Algerian geranium oil, Bourbon geranium oil, and Moroccan geranium oil.

GENERAL DESCRIPTION

A perennial erect shrubby, hairy, and glandular plant up to 1 m high, becoming woody with age, with fragrant, deeply incised leaves; flowers pinkish in umbel-like inflorescence; native to South Africa; widely cultivated in Africa (Algeria, Morocco, etc.) and Europe (Spain, Italy, France, etc.). Parts used are the fresh leaves and stems, from which geranium oil is obtained by steam distillation, generally in a 0.08–0.4% yield.[1,2]

There are several types of geranium oils produced from cultivated forms, varieties, and hybrids of *P. graveolens*, and other *Pelargonium* species such as *P. odoratissimum* Ait., *P. capitatum* Ait., *P. crispum* (L.) L'Her., and *P. radula* (Cav.) L'Her. ex Ait. (syn. *P. roseum*

Willd.). The more commonly used ones are Algerian or African geranium oil, Reunion or Bourbon geranium oil, and Moroccan geranium oil. Despite mention of several commercial source species in the literature, *P. graveolens* appears to be the only one commercially cultivated (TUCKER AND LAWRENCE).

Geranium oils should not be confused with East Indian or Turkish geranium oil; the latter is palmarosa oil and is derived from a completely different plant source.

CHEMICAL COMPOSITION

All three major types of geranium oil (Algerian, Bourbon, and Moroccan) contain large amounts of alcohols (primarily *l*-citronellol and geraniol, with linalool and phenethyl alcohol in lesser amounts);[1–7] esters (e.g., geranyl tiglate, geranyl acetate, citronellyl formate, and citronellyl acetate);[1–6] and aldehydes and ketones (*l*-isomenthone, citronellal, citral, decyl aldehyde, etc.).[4–7] The total alcohol content is usually 60–70%, and the ester content is normally 20–30% but they vary considerably, depending on the type of oil.[1,4–6] The proportions of alcohols (especially *l*-citronellol, geraniol, and phenethyl alcohol) also vary with the type of oil.[2,3,5,6]

Other compounds reported in geranium oils include sesquiterpene hydrocarbons (α- and β-bourbonene, α-santalene, β-elemene, caryophyllene, γ-muurolene, humulene, δ-cadinene, *ar*-curcumene, selina-3,7(11)-diene, selina-4(14),7(11)-diene, germacratriene, 6,9-guaiadiene, etc.);[8–10] sesquiterpene alcohols (10-*epi*-γ-eudesmol, junenol, *l*-selin-11-en-4-ol, etc.);[10,11] acids (formic, acetic, propionic, caprylic, tiglic acids, etc.);[6,7] dimethyl sulfide, *cis*- and *trans*-dehydrocitronellol, menthol, citronellyl-diethylamine, and others.[6,12,13]

PHARMACOLOGY AND BIOLOGICAL ACTIVITIES

A geranium oil (from leaves of *P. roseum*) with its alcohol component consisting almost entirely of *l*-citronellol has been reported to exhibit *in vitro* inhibitory activities against several fungi that are pathogenic to humans.[2] Other rose geranium oils have also been reported to have antibacterial and antifungal activities *in vitro*.[14–16]

Antioxidant free radical scavenging activity of geranium oil has recently been reported.[17]

TOXICOLOGY

Cases of dermatitis in hypersensitive individuals caused by geranium oil present in cosmetics have been well documented, though some data indicate geranium oil to be nonsensitizing, nonirritating, and nonphototoxic to human skin.[14,18,19]

USES

Medicinal, Pharmaceutical, and Cosmetic. Rose geranium oil (especially Bourbon) is widely used as a fragrance component in all kinds of cosmetic products, including soaps, detergents, creams, lotions, and perfumes, with maximum use level of 1.0% reported in perfumes.[14,18,19]

Food. Rose geranium oil is extensively used in most major food products, including alcoholic and nonalcoholic beverages, frozen dairy desserts, candy, baked goods, and gelatins and puddings. Use levels are generally below 0.001%.

Dietary Supplements/Health Foods. Rose geranium oil is commonly available; primarily used as an inexpensive substitute for rose oil in massage oil formulations, and so on (ROSE).

Traditional Medicine. In Africa the roots of various *Pelargonium* spp. are reportedly used as astringent; in diarrhea, dysentery, and so on. (WATT AND BREYER-BRANDWIJK).

COMMERCIAL PREPARATIONS

Most common types (e.g., Algerian, Bourbon, Moroccan, and Spanish). Algerian geranium oil is official in F.C.C.

Regulatory Status. GRAS: rose geranium (§182.20); other types (§182.10 and §182.20).

REFERENCES

See the General References for ARCTANDER; BAILEY 2; FEMA; FURIA AND BELLANCA; GUENTHER; LIST AND HÖRHAMMER; MARTINDALE; MCGUFFIN 1 & 2; MERCK; ROSE; TERRELL; TUCKER AND LAWRENCE; UPHOF; WATT AND BREYER-BRANDWIJK.

1. S. Angjeli, *Bull. Univ. Sheteror Tiranes, Ser. Shkencat Natyrore,* **1**, 64 (1964).
2. H. Wollman et al., *Pharmazie,* **28**, 56 (1973).
3. E. Gliozheni and H. Ahmataj, *Bull. Univ. Shteteror Tiranes, Ser. Shkencat Natyrore,* **21**, 129 (1967).
4. L. Peyron, *Compt. Rend.,* **255**, 2981 (1962).
5. J. Y. Conan et al., *Riv. Ital. Essenze Profumi, Piante Off, Aromi, Saponi, Cosmet., Aerosol,* **58**, 556 (1976).
6. V. T. Gogiya and L. I. Ivanova, *Mezhdunar. Kongr. Efirnym Maslam. (Mater.),* **1**, 71 (1968).
7. C. De La Torre et al., *An. Acad. Bras. Cienc.,* **44**(Suppl.), 168 (1972).
8. J. Krepinsky et al., *Tetrahedron Lett.,* **3**, 359 (1966).
9. E. Tsankova and I. Ognyanov, *Dokl. Bolg. Akad. Nauk,* **25**, 1229 (1972).
10. P. Pesnelle et al., *Riv. Ital. Essenze Profumi, Piante Off, Aromi, Saponi, Cosmet, Aerosol,* **54**, 92 (1972).
11. P. Pesnelle et al., *Recherches,* **18**, 45 (1971).
12. B. H. Kingston, *Manuf. Chem.,* **10**, 463 (1961).
13. W. Rojahn and E. Klein, *Dragoco Rep. (Ger. Ed.),* **24**, 150 (1977).
14. D. L. J. Opdyke, *Food Cosmet. Toxicol.,* **14**(Suppl.), 781 (1976).
15. J. C. Maruzella and P. A. Henry, *J. Am. Pharm. Assoc.,* **47**, 294 (1958).
16. J. C. Maruzella and L. Liquori, *J. Am. Pharm. Assoc.,* **47**, 250 (1958).
17. W. Sun et al., *Zhong. Yao Cai.,* **28**, 87 (2005).
18. D. L. J. Opdyke, *Food Cosmet. Toxicol.,* **12**(Suppl.), 883 (1974).
19. D. L. J. Opdyke, *Food Cosmet. Toxicol.,* **13**, 451 (1975).

GINGER

Source: *Zingiber officinale* Roscoe (Family Zingiberaceae).

Common/vernacular names: Common ginger, ganzabeel, and zangabeel.

GENERAL DESCRIPTION

An erect perennial herb with thick tuberous rhizomes (underground stems) from which the aerial stem grows up to about 1 m high; rarely flowers and produces seeds; native to southern Asia; extensively cultivated in the tropics (e.g., India, China, Jamaica, Haiti, and Nigeria). Part used is the pungent rhizome commonly called "root," both in fresh and dried forms. Ginger oil is usually produced from freshly ground, unpeeled dried ginger by steam distillation. Extracts and oleoresin are produced from dried unpeeled ginger, as peeled ginger loses much of its essential oil content.

CHEMICAL COMPOSITION

Ginger has been reported to contain 0.25–3.3% (usually 1–3%) volatile oil; pungent principles (gingerols and shogaols);[1–5] glycosides of geraniol and gingerdiol;[6,7] about 6–8% lipids composed of triglycerides, phosphatidic acid, lecithins, free fatty acids (lauric, palmitic, stearic, oleic, linoleic, etc.), gingerglycolipids,[8] and others;[9] protein (ca. 9%); starch (up to 50%); vitamins (especially niacin and A); minerals; amino acids; gingesulfonic acid,[8] resins; and others (JIANGSU; MARSH; ROSENGARTEN; STAHL).[10] New diarylheptanoids were recently isolated from Chinese ginger.[11] Newly reported 1-dehydrogingerdione was also isolated from ginger.[12]

Ginger oil contains as its major components the sesquiterpene hydrocarbons zingiberene and bisabolene.[1–3] Other sesquiterpene hydrocarbons and alcohols present include *ar*-curcumene, β-sesquiphellandrene, sesquithujene, zingiberol, zingiberenol, *cis*-sesquisabinene hydrate, and *cis*- and *trans*-β-sesquiphellandrol.[1,3,13–15] It also contains monoterpene hydrocarbons, alcohols, and aldehydes (e.g., phellandrene, camphene, geranial, neral, linalool, and *d*-borneol),[3,10,13] methylheptenone, nonyl aldehyde; gingediacetate and gingerol;[16] and others[17] (GUENTHER; JIANGSU).

Ginger oleoresin contains mainly the pungent principles gingerols and shogaols as well as zingerone.[1,2,5] Shogaols and zingerones are dehydration and degradation products, respectively, of gingerols; the latter have been reported to constitute about 33% of a freshly prepared oleoresin.[3] Shogaols have recently been found to be twice as pungent as gingerols.[4] The pungency of ginger oleoresin is lost on prolonged contact with alkalis (ARCTANDER).

A protease has been isolated from fresh ginger; yield of the crude enzyme was 2.26% from the fresh rhizome.[18]

PHARMACOLOGY AND BIOLOGICAL ACTIVITIES

Ginger extracts are reported to exhibit numerous pharmacological properties, including stimulating the vasomotor and respiratory centers of anesthetized cats as well as direct heart stimulation;[7] lowering of serum and hepatic cholesterol in rats previously fed cholesterol;[19] and killing vaginal trichomonads *in vitro* (JIANGSU). Ginger also has carminative properties.

Dried ethanol and acetone extracts of fresh rhizome are given orally significantly inhibited gastric secretion of induced acute stress in rats, producing an antiulcer effect.[20] The antiulcer effect may also be mediated through the inhibitory effect of ginger on *Helicobacter pylori*.[21]

Ginger (dried) has been the subject of several clinical trials to assess efficacy in motion sickness, early pregnancy, and postoperative nausea and vomiting with positive results.[22–26] Its antiemetic effect is attributed to an effect on gastric activity rather than a CNS mechanism, characteristic of conventional antimotion sickness drugs.[27] The involvement of 5-HT$_3$ receptor blocking activity has been recently reported.[25,28,29] Powdered ginger root (250 mg q.i.d.) over 4 days was better than placebo in diminishing or eliminating symptoms of hyperemesis gravidarum.[30] The clinical efficacy of ginger as an antiemetic has been reviewed.[31] *In vivo* studies in animal models have also been published.[32,33]

Fresh ginger (juice, aqueous extract, in poultice or sliced form) has been reported highly effective in China in the clinical treatment of rheumatism, acute bacterial dysentery, malaria, and orchitis (inflammation of the testicles) (JIANGSU). Numerous studies have been published on the anti-inflammatory effects of ginger in human patients and animal models.[34–36] Neuroprotective effects involving anti-inflammatory mechanisms have also been reported for ginger.[37,38] The anti-inflammatory effect appears to be mediated by mechanisms involving the inhibition of different mediators of inflammation, for example, TNF-α, COX-2, and leukotriene inhibition.[39–46] The anti-inflammatory effects of ginger have been reviewed.[47]

In vitro and *in vivo* antioxidant, cancer chemoprevention, and radiprotective activities have been reported for ginger extracts.[48–51] Gingerol, paradol, and dehydrozingerone were among the constituents showing highest activity.[52–54] 6-Gingerol was also shown to possess antitumor-promoting activity through the inhibition of angiogenesis.[55]

Ginger extracts, gingerols, and gingerdiol were found to exhibit antiviral, antibacterial, and antifungal activities against *Helicobacter pylori*, influenza and rhinovirus, and 13 fungal pathogens, respectively.[21,56–58] A synergistic antimicrobial effect was recently reported between 10-gingerol and aminoglycoside antibiotics against vancomycin-resistant enterococci.[59]

Other activities reported for ginger include hypoglycemic,[60,61] antihyperlipidemic,[62] immunomodulatory,[63] antiplatelet aggregation,[64] androgenic,[65] and antiwrinkling effects.[66]

TOXICOLOGY

Ginger oil is reported to be nonirritating and nonsensitizing in humans, and its low phototoxicity is not considered significant.[67]

USES

Medicinal, Pharmaceutical, and Cosmetic. Ginger (mainly as the oleoresin) is used as an ingredient in certain digestive, laxative, antitussive, carminative, and antacid preparations. Ginger oil is used as a fragrance component in cosmetic products, including soaps, detergents, creams, lotions, and perfumes (especially oriental and men's fragrances). Maximum use level is 0.4% reported in perfumes.[67]

Food. Ginger is widely used as a domestic spice, especially in Oriental cooking. It is used commercially in many foods, including nonalcoholic beverages, baked goods (e.g., cookies), gelatins and puddings, meat and meat products, and condiments and relishes, with the highest average maximum use level of about 0.525% (5248 ppm) reported in baked goods.

Ginger oil, oleoresin, and extract are widely used in soft drinks (e.g., gingerales and colas). In addition to the above uses, they are used in alcoholic beverages (liqueurs, bitters, etc.), frozen dairy desserts, and candy. Highest average maximum use levels reported are about 0.004% for the oil in baked goods (36.9 ppm) and gelatins and puddings (37.9 ppm); about 0.01% for the oleoresin in alcoholic beverages (99.8 ppm), baked goods (104 ppm), and condiments and relishes (108 ppm); and about 0.023% (233 ppm) for the extract (type not given) in baked goods.

Dietary Supplements/Health Foods. Dried powdered root and extracts used in teas, capsules, and drinks, tablets, singly or in combination as a digestive aid, or antinauseant; also for colds, flu, and as general stimulant (FOSTER AND YUE).

Traditional Medicine. Ginger is usually used as a carminative and diaphoretic and to stimulate the appetite. Dried ginger has been used for thousands of years in China to treat numerous ailments, including stomachache, diarrhea, nausea, cholera, and bleeding. In addition to the above uses, fresh ginger is used to treat rheumatism, poisonous snakebite, baldness, toothache, and other conditions (JIANGSU).

Recently, fresh ginger juice has been reported to be used in treating thermal burns with considerable success.[68]

Others. Ginger and its extracts have strong antioxidative activities on various foods (e.g., lard, cookies, potato chips, oils, and fats) and could serve as potential source of food antioxidants.[69–72]

Due to its reportedly high content of a protease, fresh ginger could serve as a source of this enzyme, which could have applications similar to those of papain or other plant proteases (see ***bromelain*** and ***ficin***).

COMMERCIAL PREPARATIONS

Dried crude in peeled (e.g., Jamaican), rough-peeled (e.g., Cochin), and unpeeled (e.g., Nigerian) forms; extracts; oils; and oleoresin. Strengths (see *glossary*) of extracts are expressed either in weight-to-weight ratios or in flavor intensities. Crude, fluid extract, and oleoresin were formerly official in N. F.; oil is official in F.C.C.

Regulatory Status. GRAS (§182.10 and §182.20). Subject of a German therapeutic monograph; allowed for dyspeptic complaints and prevention of motion sickness.[73]

REFERENCES

See the General References for APPLEQUIST; ARCTANDER; BAILEY 1; BARNES; BARRETT; BLUMENTHAL 1 & 2; BRUNETON; DER MARDEROSIAN AND BEUTLER; FEMA; FOSTER AND YUE; GOSSELIN; GRIEVE; GUENTHER; JIXIAN; JIANGSU; LUST; MARTINDALE; MCGUFFIN 1 & 2; NANJING; ROSENGARTEN; TERRELL.

1. C. R. Mitra, *Riechst., Aromen, Korperpflegem.*, **25**, 170 (1975).
2. S. M. Annathakrishna and V. S. Govindarajan, *Lebensm. Wiss. Technol.*, **7**, 220 (1974).
3. D. W. Connell, *Aust. Chem. Process. Eng.*, **24**, 27 (1971).
4. S. Narasimhan and V. S. Govindarajan, *J. Food Technol.*, **13**, 31 (1978).
5. D. W. Connell and R. McLachlan, *J. Chromatogr.*, **67**, 29 (1972).
6. Y. Sekiwa et al., *Nat. Prod. Lett.*, **15**, 267 (2001).
7. Y. Sekiwa et al., *J. Agric. Food Chem.*, **48**, 373 (2000).
8. M. Yoshikawa et al., *Chem. Pharm. Bull. (Tokyo)*, **42**, 1226 (1994).
9. I. P. Singh et al., *Indian J. Agric. Sci.*, **45**, 545 (1975).
10. M. M. Ally, *Proc. Pan Indian Ocean Sci. Congr.*, **4**, 11 (1960).
11. J. Ma et al., *Phytochemistry*, **65**, 1137 (2004).
12. R. Charles et al., *Fitoterapia*, **71**, 716 (2000).
13. D. W. Connell and R. A. Jordan, *J. Sci. Food Agric.*, **22**, 93 (1971).
14. S. J. Terhune et al., *Can. J. Chem.*, **53**, 3285 (1975).
15. A. A. Bednarczyk et al., *J. Agric. Food Chem.*, **23**, 499 (1975).
16. Y. Masada et al., *Int. Congr. Essent. Oils (Pap)*, **6**, 97 (1974).
17. S. D. Jolad et al., *Phytochemistry*, **66**, 1614 (2005).
18. E. H. Thompson et al., *J. Food Sci.*, **38**, 652 (1973).
19. S. Gujral et al., *Nutr. Rep. Int.*, **17**, 183 (1978).
20. J. A. A. Sertie et al., *Fitoterapia*, **63**, 55 (1992).
21. G. B. Mahady et al., *Anticancer Res.*, **23**, 3699 (2003).
22. D. B. Mowrey and D. E. Clayson, *Lancet*, 655 (1982).
23. A. Grøntved et al., *Acta Otolaryngol. (Stockh.)*, **105**, 45 (1988).
24. F. Borrelli et al., *Obstet. Gynecol.*, **105**, 849 (2005).
25. A. Keating and R. A. Chez, *Altern. Ther. Health Med.*, **8**, 89 (2002).
26. S. Phillips et al., *Anaesthesia*, **48**, 715 (1993).
27. S. Holtmann et al., *Acta Otolaryngol. (Stockh.)*, **108**, 168 (1989).
28. H. bdel-Aziz et al., *Eur. J. Pharmacol.*, **530**, 136 (2006).
29. H. bdel-Aziz et al., *Planta Med.*, **71**, 609 (2005).
30. W. Fischer-Rasmussen et al., *J. Obstet. Gynecol. Reprod. Biol.*, **38**, 19 (1990).

31. E. Ernst and M. H. Pittler, *Br. J. Anaesth.*, **84**, 367 (2000).
32. S. S. Sharma et al., *J. Ethnopharmacol.*, **57**, 93 (1997).
33. T. Kawai et al., *Planta Med.*, **60**, 17 (1994).
34. S. K. Verma et al., *Indian J. Exp. Biol.*, **42**, 736 (2004).
35. S. C. Penna et al., *Phytomedicine*, **10**, 381 (2003).
36. R. D. Altman and K. C. Marcussen, *Arthritis Rheum.*, **44**, 2531 (2001).
37. R. Grzanna et al., *J. Altern. Complement Med.*, **10**, 1009 (2004).
38. D. S. Kim et al., *Planta Med.*, **68**, 375 (2002).
39. R. C. Lantz et al., *Phytomedicine* (2006).
40. C. L. Shen et al., *J. Med. Food*, **8**, 149 (2005).
41. S. O. Kim et al., *Oncogene*, **24**, 2558 (2005).
42. S. O. Kim et al., *Biofactors*, **21**, 27 (2004).
43. S. D. Jolad et al., *Phytochemistry*, **65**, 1937 (2004).
44. E. Nurtjahja-Tjendraputra et al., *Thromb. Res.*, **111**, 259 (2003).
45. F. Kiuchi et al., *Chem. Pharm. Bull. (Tokyo)*, **40**, 387 (1992).
46. M. Thomson et al., *Prostaglandins Leukot. Essent. Fatty Acids*, **67**, 475 (2002).
47. R. Grzanna et al., *J. Med. Food*, **8**, 125 (2005).
48. V. Manju and N. Nalini, *Clin. Chim. Acta*, **358**, 60 (2005).
49. G. Jagetia et al., *Cancer Biother. Radiopharm.*, **19**, 422 (2004).
50. S. K. Katiyar et al., *Cancer Res.*, **56**, 1023 (1996).
51. G. C. Jagetia et al., *Radiat. Res.*, **160**, 584 (2003).
52. P. C. Kuo et al., *Arch. Pharm. Res.*, **28**, 518 (2005).
53. Y. Masuda et al., *Biofactors*, **21**, 293 (2004).
54. Y. J. Surh et al., *J. Environ. Pathol. Toxicol. Oncol.*, **18**, 131 (1999).
55. E. C. Kim et al., *Biochem. Biophys. Res. Commun.*, **335**, 300 (2005).
56. N. Imanishi et al., *Am. J. Chin. Med.*, **34**, 157 (2006).
57. C. Ficker et al., *Phytother. Res.*, **17**, 897 (2003).
58. C. V. Denyer et al., *J. Nat. Prod.*, **57**, 658 (1994).
59. C. Nagoshi et al., *Biol. Pharm. Bull.*, **29**, 443 (2006).
60. U. Bhandari et al., *J. Ethnopharmacol.*, **97**, 227 (2005).
61. S. P. Akhani et al., *J. Pharm. Pharmacol.*, **56**, 101 (2004).
62. U. Bhandari et al., *J. Ethnopharmacol.*, **61**, 167 (1998).
63. H. L. Zhou et al., *J. Ethnopharmacol.*, **105**, 301 (2006).
64. J. H. Guh et al., *J. Pharm. Pharmacol.*, **47**, 329 (1995).
65. P. Kamtchouing et al., *Asian J. Androl*, **4**, 299 (2002).
66. K. Tsukahara et al., *Int. J. Dermatol.*, **45**, 460 (2006).
67. D. L. J. Opdyke, *Food Cosmet. Toxicol.*, **12**(Suppl.), 901 (1974).
68. N. X. Cui, *Zhejiang Zhongyi Zazhi*, 451 (1990).
69. Y. Saito et al., *Eiyo To Shokuryo*, **29**, 505 (1976).
70. Y. Kihara and T. Inoue, *Nippon Shokuhin Kogyo Gakkaishi*, **9**, 290 (1962).
71. F. Hirahara et al., *Eiyogaku Zasshi*, **32**, 1 (1974).
72. H. Fujio, *Nippon Shokuhin Kogyo Gakkaishi*, **16**, 241 (1969).
73. Monograph *Zingiberis rhizoma*, *Bundesanzeiger*, no. 85 (May 5, 1988).

GINKGO

Source: *Ginkgo biloba* L. (Family Ginkgoaceae)

Common/vernacular names: Maidenhair Tree.

GENERAL DESCRIPTION

Monotypic deciduous tree to 40 m; leaves alternate or borne on spurs in clusters of 3–5; parallel veined, broad fan shaped, up to 12 cm, with notch at apex, forming two distinct lobes, hence the species name "*biloba*". Flowers dioecious; male flowers on pendulous catkins with numerous, loosely arranged anthers in stalked pairs on a slender axis; female flowers are in pairs on long foot-stalks. The drupe-like fruits have an acrid, foul-smelling pulp (likened to dog droppings) surrounding a single smooth, oval, thin-shelled, semi-edible nut (seed).

Ginkgo is known only from cultivation; widely planted ornamental tree worldwide; limited occurrence in undisturbed forests in Zhejiang province disputed as natural spontaneous specimens or progeny of planted specimens.

The leaves are used in Western pharmaceuticals; seeds and leaves traditionally used in China. Leaves are grown on a commercial scale in China, South Carolina and Maryland in the United States, and in the Bordeaux region of France (FOSTER AND YUE).

CHEMICAL COMPOSITION

Leaves and root bark contain terpenoids, including the monomethyl-mononorditerpenes: ginkgolide A (3α-OH, 10β-OH), ginkgolide B ($1,3\alpha$-OH, 10β-OH), ginkgolide C ($1,3\alpha$-OH, $7,10\beta$-OH) in the root bark and leaves; ginkgolide M (1α-OH, $7,10\beta$-OH) in the root bark; ginkgolide J (3α-OH, $7,10\beta$-OH) in the leaves; and the sesquiterpene bilobalide (10α-OH, 8β-OH) in the leaves. The ginkgolides differ in the number and position of hydroxyl groups present on C1, C3, and/or C7 of the spirononane framework.[1–3]

Flavonoids from the leaves include the flavones luteolin and tricetin (dalphidenon). Biflavones including amentoflavone, bilobetin, ginkgetin; isoginkgetin, sciadopitysin, 5'-methoxybilobetin, ginkgetin, and isoginkgetin glucosides; flavonols including kaempferol, kaempferol-3-rutinoside, kaempferol-3-O-α-($6'''$-p-coumaroyl-glucosyl-β-1,4-rhamno side), quercetin, quercetin-3-rutinoside (rutin), quercetin-3-glucoside (isoquercitrin), quercetin 3-O-α-($6'''$-p-coumaroyl-glucosyl-β-1,4-rhamnoside), isorhamnetin, myricetin rhamnoglucoside, and 3-O-methylmyricetin-3-rutinoside. Catechins include (+)-catechin, (−)-epicatechin, (+)-gallocatechin, and (−)-epigallocatechin. Proanthocyanidins include gallocatechin-4,8''-catechin (procyanidin) and gallocatechin-4,8''-gallocatechin (prodelphinidin).[1,2,4–6]

The bioactive ginkgolic acids, 6-(Z-8'-heptadecenyl) salicylic and 6-(Z-10'-heptadecenyl) salicylic acids have recently been isolated from the leaves and seed coat.[7,8]

Other leaf components include the lignin Z,Z,-4,4'-(1,4-pentadien-1,5-diyl)diphenol; steroids including megastigmenone, sitosterol and its glucoside (ipuranol); starch; oil, balsam, benzoic acid, and chromenyl methoxybenzoic acid; calcium oxalate raphides; wax (0.7–1% in dry leaves), consisting of alkanes and alcohols (75%), esters (15%), and free acids (10%); pinitol, sequoyitol, D-glucaric acid, plus shikimic and succinic acids, and others.[1,2]

The nut (seed) contains 67.9% starch, 13.1% proteins (globulins, glutelin, and albumins), 2.9% lipids (composed of linoleic, oleic, palmitic, stearic, linolenic, and α-hydroxypalmitic acids), 1.6% pentosans; plus sucrose, glucose, fructose, citric and quinic acids, fiber, and others.[1]

Total synthesis of ginkgolide B has been achieved.[9]

PHARMACOLOGY AND BIOLOGICAL ACTIVITIES

Ginkgo leaf extract has vascular tone-regulating properties, an antihypoxic effect, experimental inhibitory activity in cerebral edema, and neurotoxicity; modulates cerebral energy metabolism;[10] free radical scavenging properties;[11] inhibits lipid peroxidation of membranes, helping to maintain integrity and permeability of cell walls;[12] antiasthmatic, bronchodilator, and so on. (WREN).

Perhaps the most extensively studied activity of ginkgo is its antioxidant activity and it is probably the key to its significant protective effects. Numerous reports have appeared in recent literature that link such activity to neuroprotection,[13–19] hepatoprotection,[20–23] radioprotection,[21] cancer chemoprevention and apoptosis,[24–29] Alzheimer's disease (amyloid-beta protein inhibition),[30–32] reduced cerebral ischemia/stroke,[19,33–36] gastroprotection (antiulcer),[26,37,38] nephroprotection,[39] reduced mitochondrial stress,[40,41] and even skin disorders (vitiligo).[42]

Ginkgo's inhibitory effect on the production of amyloid-beta protein has also been mediated by other mechanisms, such as acetylcholinesterase inhibition, modulation of amyloid protein oligomeric species, and lowering of free cholesterol levels.[43–45]

Numerous pharmacological and clinical studies of ginkgo leaf extract have demonstrated a positive effect in increasing vasodilation and peripheral blood flow rate in capillary vessels and end arteries in various circulatory disorders, Reynaud's disease, varicose conditions, post-thrombotic syndrome (WEISS),[46] chronic cerebral vascular insufficiency,[47] short-term memory improvement,[48] cognitive disorders secondary to depression, dementia,[49] tinnitus,[50] vertigo, obliterative arterial disease of the lower limbs,[51] and other conditions.[46] A retrospective critical review of the quality and methodology of 40 trials (published since 1975) on the use of ginkgo extracts in cerebral insufficiency has found eight such trials to be of high quality.[52] Vasodilatation and antihypertension have been the subject of more research supporting earlier findings. Some of the involved mechanisms include inhibition of cGMP-phosphodiesterase-5, induction of NO release, and Ca^{2+} channel modulation.[4,53–58] Behavioral effects of ginkgo as well as its effects on memory and cognition have also been under investigation in various animal and human trials. Significant memory and behavioral enhancements were observed in rats and dogs under different stressors.[59–66] Results of recent human trials further supported earlier results.[67–69]

Ginkgolide B is a selective antagonist of platelet aggregation induced by platelet-activating factor (PAF). PAF, an inflammatory autacoid, is involved in various inflammatory, cardiovascular, and respiratory disorders.[70–72] Ginkgo extract has also displayed anti-inflammatroy activity in a number of conditions and disease models such as colitis, arthritis, skin inflammation, and rat paw edema.[73–78] Anti-inflammatory mediators have been suppressed by the extract.[73,79]

Ginkgo leaf extracts, bilobalide and ginkgolide A, have shown antidepressant/antistress effects in different animal models, which appears to be mediated by antagonism of the GABA receptor and by elevating brain catecholamines and plasma corticosterone levels.[3,80–84]

Other recently reported activities of ginkgo include estrogenic/antiestrogenic,[85,86] immunostimulatory,[87] larvicidal,[88] and modulation of certain cytochrome enzyme subtypes.[89–91] Many reviews about the pharmacology and therapeutic potential of ginkgo have been published.[32,92–100]

TOXICOLOGY

A number of isolated incidents attributed to the use of ginkgo have recently appeared in the literature, including accidental intoxication of 2-year-old male and female,[101,102] exanthematous pustulosis,[103] spontaneous bleeding,[104] and ventricular arrhythmia.[105]

USES

Medicinal, Pharmaceutical, and Cosmetic. A standardized extract of the dried leaves is among the best-selling phytomedicines in Europe; used clinically for heart disease, eye ailments, tinnitus, cerebral and peripheral vascular insufficiency, injuries involving brain trauma, dementias, short-term memory improvement, cognitive disorders secondary to depression, vertigo, and various conditions associated with senility.[46]

Leaf extracts used in cosmetics include shampoos, creams, and lotions.

Food. Seeds considered a delicacy in Japan and China; edible after acrid, foul-smelling pulp is removed; seeds are then boiled or roasted and eaten sparingly (no more than 8–10 per day). Fresh seeds are toxic and have reportedly caused death in children; pulp may cause contact dermatitis similar to poison ivy rash; handled with rubber gloves. Ingestion of fresh seeds may cause stomachache, nausea, diarrhea, convulsions, weak pulse, restlessness, difficult breathing, and shock. Dyed-red, the nuts were traditionally eaten at weddings (FOSTER AND YUE).

Dietary Supplements/Health Foods. In the United States, various ginkgo leaf preparations or crude leaf are sold as dietary supplements in the form of tablets, capsules, tincture, standardized extracts, tea, and so on.

Traditional Medicine. In China, the dried, processed seed (*baiguo*) is used in prescriptions for asthma, coughs with phlegm, enuresis, mucous vaginal discharges, bronchitis with asthma, chronic bronchitis, tuberculosis, frequent urination, seminal emissions, turbid urine, and so on. Externally, seed are poulticed for scabies and sores.

The leaves (*bai guo ye*) used in prescriptions for arteriosclerosis, angina pectoris, high serum cholesterol levels, dysentery, and filariasis. An infusion of the boiled leaves used as wash for chilblains (FOSTER AND YUE).

Other. The root and inner bark are Chinese folk medicines; bark used in prescriptions for mucous vaginal discharges, seminal emission, or weak, convalescing patients; dried bark, burned to ash, mixed with vegetable oil as poultice for neurodermatitis.

COMMERCIAL PREPARATIONS

Crude herb, extracts, dried leaf in capsules, tinctures, and so on. Purified ginkgolide B is known commercially as BN 52021. A complex standardized extract of the dried leaves (EGb 761), produced by a German/French consortium; standardized to 24% flavone glycosides and ginkgolide B is widely sold in Europe (especially in Germany and France).

Regulatory Status. Class 1 dietary supplement (can be safely consumed when used appropriately).

REFERENCES

See the General References for BARNES; BLUMENTHAL 2; DER MARDEROSIAN AND BEUTLER; EVANS; FOSTER AND YUE; GLASBY 2; HUANG; TYLER 1; WEISS; WREN.

1. N. Boralle et al. in P. Braquet, ed., *Ginkgolides—Chemistry, Biology, Pharmacology, and Clinical Perspectives*, Vol. 1, J. R. Prous Science Publishers, Barcelona, 1988, p. 9.
2. H. Huh and E. J. Staba, *J. Herbs Spices Med. Plants*, **1**, 91 (1992).
3. S. H. Huang et al., *Eur. J. Pharmacol.*, **494**, 131 (2004).
4. M. Dell'Agli et al., *Planta Med.*, **72**, 468 (2006).
5. S. K. Hyun et al., *Chem. Pharm. Bull. (Tokyo)*, **53**, 1200 (2005).
6. E. Bedir et al., *J. Agric. Food Chem.*, **50**, 3150 (2002).
7. W. Pan et al., *Pest. Manag. Sci.*, **62**, 283 (2006).

8. J. H. Lee et al., *Planta Med.*, **70**, 1228 (2004).
9. E. Cory et al., *J. Am. Chem. Soc.*, **110**, 649 (1988).
10. S. S. Chatterjee, in A. Agnoli et al., eds., *Effects of Ginkgo biloba Extract on Organic Cerebral Impairment*, John Libbey Eurotext Ltd., London, 1985.
11. M. Ellnain-Wojtaszek et al., *Fitoterapia*, **74**, 1 (2003).
12. J. Pincemail and C. Deby in E.W. Funfgeld, ed., *Rökan (Ginkgo biloba), Recent Results in Pharmacology, and Clinic*, Springer-Verlag, Berlin, 1988, p. 5.
13. A. Ilhan et al., *Prog. Neuropsychopharmacol. Biol. Psychiatry*, **30**, 1504 (2006).
14. Q. Ao et al., *Spinal Cord.*, **44**, 662 (2006).
15. Y. S. Wang et al., *Chin Med. J. (Engl.)*, **118**, 948 (2005).
16. M. S. Kim et al., *Phytother. Res.*, **18**, 663 (2004).
17. G. Ozturk et al., *Toxicol. Appl. Pharmacol.*, **196**, 169 (2004).
18. K. Chandrasekaran et al., *Pharmacopsychiatry*, **36**(Suppl. 1), S89 (2003).
19. K. Chandrasekaran et al., *Cell Mol. Biol. (Noisy.-le-grand)*, **48**, 663 (2002).
20. S. X. He et al., *World J. Gastroenterol.*, **12**, 3924 (2006).
21. G. Sener et al., *Mol. Cell Biochem.*, **283**, 39 (2006).
22. S. Q. Liu et al., *Am. J. Chin Med.*, **34**, 99 (2006).
23. J. Ding et al., *Liver Int.*, **25**, 1224 (2005).
24. N. Altiok et al., *Neurotoxicology*, **27**, 158 (2006).
25. K. S. Kim et al., *Oral Oncol.*, **41**, 383 (2005).
26. J. C. Chao et al., *World J. Gastroenterol.*, **10**, 560 (2004).
27. A. H. Xu et al., *World J. Gastroenterol.*, **9**, 2424 (2003).
28. F. V. DeFeudis et al., *Fundam. Clin. Pharmacol.*, **17**, 405 (2003).
29. Q. Chen et al., *World J. Gastroenterol.*, **8**, 832 (2002).
30. L. Brunetti et al., *Planta Med.*, **72**, 1296 (2006).
31. J. V. Smith and Y. Luo, *J. Alzheimers Dis.*, **5**, 287 (2003).
32. B. Ahlemeyer and J. Krieglstein, *Cell Mol. Life Sci.*, **60**, 1779 (2003).
33. L. H. Fan et al., *Chin J. Traumatol.*, **9**, 77 (2006).
34. K. P. Loh et al., *Neurosci. Lett.*, **398**, 28 (2006).
35. B. Hu et al., *Chin Med. J. (Engl.)*, **115**, 1316 (2002).
36. H. Peng et al., *Acta Pharmacol. Sin.*, **24**, 467 (2003).
37. A. Mustafa et al., *Pharmacol. Res.*, **53**, 324 (2006).
38. S. H. Chen et al., *World J. Gastroenterol.*, **11**, 3746 (2005).
39. M. Gulec et al., *Toxicol. Ind. Health*, **22**, 125 (2006).
40. A. Eckert et al., *Pharmacopsychiatry*, **36**(Suppl. 1), S15 (2003).
41. A. Eckert et al., *Ann. N. Y. Acad. Sci.*, **1056**, 474 (2005).
42. D. Parsad et al., *Clin. Exp. Dermatol.*, **28**, 285 (2003).
43. Y. Wu et al., *J. Neurosci.*, **26**, 13102 (2006).
44. M. Mazza et al., *Eur. J. Neurol.*, **13**, 981 (2006).
45. Z. X. Yao et al., *J. Nutr. Biochem.*, **15**, 749 (2004).
46. S. Foster, *Ginkgo biloba*, Botanical Series, no. 304, American Botanical Council, Austin, TX, 1991.
47. G. Vorberg, *Clin. Trails J.*, **22**, 149 (1985).
48. I. Hindmarch, in E. W. Fünfgeld, ed., *Rökan (Ginkgo biloba), Recent Results in Pharmacology, and Clinic*, Springer-Verlag, Berlin, 1988, p. 321.

49. D. M. Warburton, in E. W. Fünfgeld, ed., *Rökan (Ginkgo biloba), Recent Results in Pharmacology, and Clinic*, Springer-Verlag, Berlin, 1988, p. 327.
50. B. Meyer, in E. W. Fünfgeld, ed., *Rökan (Ginkgo biloba), Recent Results in Pharmacology, and Clinic*, Springer Verlag, Berlin, 1988, p. 245.
51. U. Bauer, *Arzeim.-Forsch.*, **34**, 716 (1984).
52. J. Kleijnen and P. Knipschild, *Br. J. Clin. Pharmacol.*, **34**, 352 (1992).
53. Y. Kubota et al., *Biol. Pharm. Bull.*, **29**, 266 (2006).
54. Y. Kubota et al., *J. Pharm. Pharmacol.*, **58**, 243 (2006).
55. H. Satoh and S. Nishida, *Clin. Chim. Acta*, **342**, 13 (2004).
56. S. Nishida and H. Satoh, *Clin. Chim. Acta*, **339**, 129 (2004).
57. S. Nishida and H. Satoh, *Life Sci.*, **72**, 2659 (2003).
58. D. Jezova et al., *J. Physiol. Pharmacol.*, **53**, 337 (2002).
59. J. Reichling et al., *Schweiz. Arch. Tierheilkd.*, **148**, 257 (2006).
60. Q. H. Gong et al., *Chin J. Integr. Med.*, **12**, 37 (2006).
61. M. Zhang and J. Cai, *Behav. Pharmacol.*, **16**, 651 (2005).
62. Q. H. Gong et al., *Life Sci.*, **77**, 140 (2005).
63. M. Ahmad et al., *J. Neurochem.*, **93**, 94 (2005).
64. J. R. Hoffman et al., *Pharmacol. Biochem. Behav.*, **77**, 533 (2004).
65. C. C. Lin et al., *Neuropsychobiology*, **47**, 47 (2003).
66. A. Das et al., *Pharmacol. Biochem. Behav.*, **73**, 893 (2002).
67. S. Elsabagh et al., *Psychopharmacology (Berl)*, **179**, 437 (2005).
68. R. F. Santos et al., *Pharmacopsychiatry*, **36**, 127 (2003).
69. D. O. Kennedy et al., *Physiol Behav.*, **75**, 739 (2002).
70. P. Braquet, ed., *Ginkolides—Chemistry, Biology, Pharmacology and Clinical Perspectives*, Vol. 1, J. R. Prous Science Publishers, Barcelona, 1988.
71. P. Braquet, ed., *Ginkgolides—Chemistry, Biology, Pharmacology and Clinical Perspectives*, Vol. 2, J. R. Prous Science Publishers, Barcelona, 1989.
72. F. V. DeFeudis, ed., *Ginkgo biloba Extract (EGb 761): Pharmacological Activities and Clinical Applications*, Elsevier, New York, 1991.
73. Y. H. Zhou et al., *Mediators Inflamm.*, **2006**, 92642 (2006).
74. I. Hedayat et al., *Pharmazie*, **60**, 614 (2005).
75. Y. Han, *Int. Immunopharmacol.*, **5**, 1049 (2005).
76. I. Ilieva et al., *Exp. Eye Res.*, **79**, 181 (2004).
77. O. M. bdel-Salam et al., *Pharmacol. Res.*, **49**, 133 (2004).
78. W. J. Kwak et al., *Planta Med.*, **68**, 316 (2002).
79. Y. B. Jiao et al., *Acta Pharmacol. Sin.*, **26**, 835 (2005).
80. H. Sakakibara et al., *Biol. Pharm. Bull.*, **29**, 1767 (2006).
81. D. Rai et al., *J. Pharmacol. Sci.*, **93**, 458 (2003).
82. H. Kuribara et al., *J. Nat. Prod.*, **66**, 1333 (2003).
83. L. Ivic et al., *J. Biol. Chem.*, **278**, 49279 (2003).
84. Z. A. Shah et al., *Eur. Neuropsychopharmacol.*, **13**, 321 (2003).
85. S. M. Oh and K. H. Chung, *J. Steroid Biochem. Mol. Biol.*, **100**, 167 (2006).
86. S. M. Oh and K. H. Chung, *Life Sci.*, **74**, 1325 (2004).
87. M. M. Villasenor-Garcia et al., *Int. Immunopharmacol.*, **4**, 1217 (2004).

88. L. Sun et al., *J. Med. Entomol.*, **43**, 258 (2006).
89. T. Kupiec and V. Raj, *J. Anal. Toxicol.*, **29**, 755 (2005).
90. L. L. von Moltke et al., *J. Pharm. Pharmacol.*, **56**, 1039 (2004).
91. T. Sugiyama et al., *Food Chem. Toxicol.*, **42**, 953 (2004).
92. G. X. Wang et al., *Chin J. Integr. Med.*, **12**, 234 (2006).
93. W. Zhou et al., *Cardiovasc. Drug Rev.*, **22**, 309 (2004).
94. Y. Christen, *Front. Biosci.*, **9**, 3091 (2004).
95. K. Stromgaard and K. Nakanishi, *Angew. Chem. Int. Ed. Engl.*, **43**, 1640 (2004).
96. H. J. Gertz and M. Kiefer, *Curr. Pharm. Des.*, **10**, 261 (2004).
97. V. S. Sierpina et al., *Am. Fam. Physician*, **68**, 923 (2003).
98. L. L. Ponto and S. K. Schultz, *Ann. Clin. Psychiatry*, **15**, 109 (2003).
99. J. Birks et al., *Cochrane Database Syst. Rev.*, CD003120 (2002).
100. P. H. Canter and E. Ernst, *Psychopharmacol. Bull.*, **36**, 108 (2002).
101. S. Hasegawa et al., *Pediatr. Neurol.*, **35**, 275 (2006).
102. Y. Kajiyama et al., *Pediatrics*, **109**, 325 (2002).
103. R. S. Pennisi, *Med. J. Aust.*, **184**, 583 (2006).
104. S. Bent et al., *J. Gen. Intern. Med.*, **20**, 657 (2005).
105. C. Cianfrocca et al., *Ital. Heart J.*, **3**, 689 (2002).

GINSENG (ASIAN AND AMERICAN)

Source: *Asian ginseng Panax ginseng* C. A. Mey. (syn. *P. schinseng* Nees); **American ginseng** *Panax quinquefolius* L. (Family Araliaceae).

Common/vernacular names: Chinese ginseng, Korean ginseng, Japanese ginseng (*P. ginseng*); Western ginseng (*P. quinquefolius*); seng and sang.

GENERAL DESCRIPTION

Both are perennial herbs with simple single stems bearing at flowering a whorl of three to six long-petioled compound leaves at the top.

Asian ginseng bears only a single leaf with three leaflets in the first year. In the second year, it bears a single leaf with five leaflets, and in its third year two leaves with five leaflets. It usually starts flowering at its fourth year when bearing three leaves (JIANGSU).[1-5] Asian ginseng is native to northeastern China and extensively cultivated there and in nearby Russia, as well as Korea and Japan.

American ginseng is native to eastern North America, from Quebec to Manitoba, south to northern Florida, Alabama, and Oklahoma. Once considered abundant in eastern North America, it is now considered a threatened, rare, or endangered species in many areas due to overzealous harvest of the root for commercial purposes. It is cultivated in Canada (Quebec, Ontario, British Columbia) and in the United States (New England, Wisconsin, Illinois, North Carolina, Tennessee, Georgia, Missouri, etc.), as well as in China; Chinese material enters commerce in Hong Kong under the ambiguous name "China White."[6,7]

Parts used are the dried, often specially treated (cured) roots; normally roots of plants about 6 years old are used.

There are many types and grades of Asian ginseng, depending on the sources, ages, and parts of the roots, as well as methods of preparation. Old, wild, well-farmed roots are most valued, while rootlets of cultivated plants are considered the lowest grade. Powdered ginseng currently imported from Korea for use in cosmetics and health foods is probably from the latter.

American ginseng does not undergo special curing as Asian ginseng, and there are considerably fewer grades, mostly separated on the basis of "wild" or "cultivated." The international trade of American ginseng is regulated under the provisions of the Convention on International Trade in Endangered Species (CITES), which regulates trade through permit requirements for imports, exports, and reexports of listed species.[6,7]

Siberian ginseng is the dried root of *Eleutherococcus senticosus* (Rupr. and Maxim.) Maxim. It does not have a long history of usage as Asian ginseng (or even American ginseng), but it is reported to have similar properties as Asian ginseng and is consequently gaining popularity in the United States and Canada (see *eleuthero*).[8,9]

CHEMICAL COMPOSITION

Asian ginseng contains numerous saponins, which are one of the major groups of active constituents;[8–16] a trace of volatile oil consisting mainly of the sesquiterpene hydrocarbons panacene, panaxene, panaginsene, and ginsinsene;[17] sterols (e.g., β-sitosterol and its β-glucoside); 8–32% starch;[18] 7–9% ginseng polysaccharides (panaxans A–U and pectin-like polysaccharide PG-F$_2$) and pectin;[19–23] free sugars (e.g., glucose, fructose, sucrose, maltose, trisaccharides, etc.); pectin; vitamins (e.g., vitamins B$_1$, B$_2$, and B$_{12}$, nicotinic acid, pantothenic acid, and biotin); 0.1–0.2% choline; fats; minerals (Zn, Cu, Mn, Ca, Fe, etc.); polyacetylenes (e.g., panaxynol, panaxydol, panaxydiol, and panaxytriol);[24,25] oligo- and polypeptides;[26,27] and others (JIANGSU; LIST AND HÖRHAMMER).[8,28]

The saponins are called ginsenosides by Japanese and panaxosides by Russian researchers. There are at least 18 saponins found in Asian ginseng, including ginsenosides R$_0$, R$_{b2}$, R$_{b3}$, R$_c$, R$_d$, R$_e$, R$_f$, R$_{f2}$, R$_{b3}$, R$_{20\text{-gluco-f}}$, R$_{g1}$, and R$_{g2}$, which are all triterpenoids. Ginsenoside R$_0$ is an oleanane type; the rest are all dammarane type. The sapogenin of ginsenoside R$_0$, is oleanolic acid, that of ginsenosides R$_{b1}$ to R$_d$ is 20-*S*-protopanaxadiol, and that of ginsenosides R$_e$ to R$_{g2}$, is 20-*S*-protopanaxatriol. Ginsenosides R$_{b1}$, R$_{b2}$, R$_c$, R$_e$, and R$_{g1}$ are present in major concentrations in Asian ginseng.[6,7] Minor dammarane saponines also present include koryoginsenoside R$_1$ and R$_2$ in addition to ginsenoside R$_{f2}$.[29,30]

American ginseng contains primarily ginsenosides R$_{b1}$ and R$_e$; it does not contain ginsenosides R$_{b2}$, R$_f$, and R$_{g2}$, and in some instances R$_{g1}$.[11,12] The new ginsenoside R$_{g8}$ and quinquenosides I–V as well as some of the known ones (ginsenosides R$_{f1}$, R$_{f4}$, R$_{h1}$, R$_{g2}$, and R$_{h1}$) have also been reported in American ginseng.[24,31] Prized American ginseng contains a high ratio of ginsenosides R$_{b1}$ to R$_{g1}$. Other dammarane-type triterpenes present include chikusetsusaponin IVa, pseudoginsenoside R$_{c1}$, and notoginsenosides A–C.[24]

Six panaxosides (A, B, C, D, E, and F) have been reported, with panaxosides A, B, and C having panaxatriol as their sapogenin and the sapogenin of panaxosides D, E, and F being panaxadiol.[8,9] Since the genuine aglycones reported for ginsenosides are 20-*S*-protopanaxadiol and 20-*S*-protopanaxatriol, they are most likely the sapogenins of panaxosides also.[15,16] Panaxoside A is reported to be the same as ginsenoside R$_{g-1}$ (LIST AND HÖRHAMMER).

Ginseng herb oil (*P. ginseng*) has been reported to contain sesquiterpenes, including bicyclogermacrene, α- and β-panasinsenes, caryophyllene, α- and β-humulenes, α- and β-neoclovenes, β-farnesene, and α-, β-, and γ-selinenes.[32]

A study showed that highest yields of ginsenosides were obtained at the end of summer of the fifth year; the root doubles in weight between the fourth and fifth years.[33]

At least 56 closely related saponins, called gynosaponins, have been isolated from *Gymnostemma pentaphyllum* (Thunb.) Makino of the gourd family, four of which (gynosaponins 3, 4, 8, and 12) are identical to ginsenosides R_{b1}, R_{b3}, R_d, and R_{f2}.[34–36]

New polyacetylenic compounds isolated from Asian ginseng include (9R,10S)-epoxy-16-heptadecen-4,6-diyn-3-one, (9R,10S)-epoxy-16-heptadecan-4,6-diyn-3-one, and 1-methoxy-(9R,10S)-epoxy-16-heptadecan-4,6-diyn-3-one.

PHARMACOLOGY AND BIOLOGICAL ACTIVITIES

The pharmacological properties of Asian ginseng are multiple, which are due to its various components, notably saponins and polysaccharides.

Asian ginseng has been reported to be an adaptogen with numerous pharmacological activities in humans and in laboratory animals, including (i) *qualtiy of life enhancement*: general stimulatory effect, decreasing sensitivity to stress, enhancing cognitive functions, and raising mental and physical capacity for work;[37–42] (ii) *protective effect*: radioprotection and remedial effect on radiation sickness,[43–48] neuroprotective,[49,50] musculoprotective,[51–53] and cancer chemopreventive;[51,54–56] (iii) *immunomodulatory/stimulant*: mainly due to the polysaccharides;[57–70] (iv) *antioxidant*: in various *in vitro* and *in vivo* models of oxidative stress and hepatotoxicity;[71–77] (v) *antidiabetic/hypoglycemic*: ginseng polysaccharides (especially panaxan A) and the polypeptide lowered blood sugar in experimental animals;[22,26] studies suggest that hypoglycemic mechanisms include reducing insulin resistance, enhancing glucose tolerance, lowering blood sugar level and liver glycogen content, and altering carbohydrate and albumin metabolism;[9,37,38,78–81] (vi) *primary metabolism and tissue regeneration*: by promoting the biosynthesis of cholesterol, lipid, RNA, DNA, and protein[17,32,33] and by stimulating angiogenesis (ginsenosides R_{b2} and R_e);[82,83] ginsenosides R_{b1} and R_d also potentiate nerve growth factor (JIANGSU); (vii) *aphrodisiac*: ginsenosides can stimulate the sexual response in both male and female animals, but there are conflicting results on their ability to exhibit sex hormonal effects, for example, estrogenic effects;[82,84,85] the ginsenosides were also reported to increase male copulatory behavior in rats and to enhance fertility in human males;[86,87] (viii) *other effects*: include antagonizing the effects of depressants such as alcohol, chloral hydrate, opiates, and barbiturates;[67,88–90] antiepileptic;[91] anti-inflammatory;[92] antitumor;[69,93,94] antihypetensive (ACE inhibition);[95,96] and antimicrobial.[23,97,98]

Some members of the ginseng saponins produce effects directly opposed to those produced by others, and under certain conditions, ginseng acts in opposite directions (JIANGSU).[12,41,90] Ginsenoside R_{b1} reportedly is CNS-tranquilizing, hypotensive, antipyretic, antipsychotic, and ulcer protective;[99] inhibits conditioned avoidance response; is weakly anti-inflammatory; antihemolytic; increases gastrointestinal motility; accelerates glycolysis; and accelerates serum and liver cholesterol, nuclear RNA, and serum protein synthesis.[6,100,101] Ginsenoside R_{g1} reportedly has weak CNS-stimulant, hypertensive, and antifatigue activity; aggravates stress ulcer; and increases motor activity. In behavioral tests, it accelerates discrimination behavior in pole-climbing tests and Y-maze tests, a reversal learning response in the Y-maze test, and one-trial passive avoidance learning using the step-down method.[6,100,101]

Conflicting results of various studies are attributed to type of preparation, route of administration, dosage, and presence or absence of biologically active compounds, among other factors.[6,7,100–104]

Oleanolic acid has antiallergenic activities in experimental animals (see also **ligustrum**).[105]

The chemistry, pharmacology, toxicology, and therapeutics of American and Asian ginseng have been extensively reviewed.[56,107–110]

TOXICOLOGY

American ginseng was found to be non-mutagenic.[106]

USES

Medicinal, Pharmaceutical, and Cosmetic. Used in all kinds of cosmetic products such as lotions, creams, soaps, bath preparations, and perfumes. Ginseng oil and extracts (probably both American and Asian) are used.

Food. Used in soft drinks; those manufactured in America are mostly made from American ginseng, while those manufactured overseas are from Asian ginseng, although type of ginseng is not always labeled.

Dietary Supplements/Health Foods. Both American and Asian ginseng are available in a wide variety of product forms, including powdered root (or leaf) as single or combination teas, capsules, tablets, liquid extracts, chewing gum, extract, and instant tea (up to 5% extract on fructose carrier). Some products are standardized to 4–7% ginsenoside content (FOSTER).

Traditional Medicine. In Chinese medicine, Asian ginseng is considered to have warming properties, while American ginseng is said to have cooling properties. They are generally used for different purposes. Thus, American ginseng is normally used for its cooling and thirst-quenching effects in summer and as a febrifuge. This is also true with Asian ginseng leaf, which is considered to have similar properties as American ginseng (cooling and thirst quenching) and is similarly used; both are also used to treat hangovers (JIANGSU; LEUNG).

Asian ginseng is used generally as a tonic, for its revitalizing properties, especially after a long illness. Either alone or in combination with other drugs, it is used to treat a wide variety of conditions, including amnesia, dizziness, headache, tiredness, convulsions, impotence, vomiting, rheumatism, dysentery, lack of appetite, difficulties in pregnancy and childbirth, internal hemorrhage, nosebleed, and cancers, among others (JIANGSU; LEUNG; NANJING).[2]

Primarily consumed by Asians, American ginseng has always been regarded as an export commodity. The root was official in the *United States Pharmacopoeia* from 1842 to 1882; primarily used as a stimulant and a stomachic.[7]

COMMERCIAL PREPARATIONS

Crude, extracts, and oils. Currently there are no standards for ginseng. Powdered ginseng and ginseng extracts should be tested for ginsenosides and ginseng polysaccharides, as well as diluents such as dextrose, lactose, corn syrup, and caramel. Chromatographic methods are available and can be used.[11–13]

Regulatory Status. Not yet determined. Asian ginseng is the subject of a German therapeutic monograph. The root is used as a tonic for invigoration for fatigue and reduced work capacity and concentration and during convalescence. Daily dosage is 1–2 g of root in appropriate formulations.

REFERENCES

See the General References for BAILEY 1; BLUMENTHAL 1 & 2; FOSTER; FOSTER AND YUE; JIANGSU; LEUNG; LIST AND HÖRHAMMER; NANJING; TYLER 1.

1. S. Y. Hu, *Econ. Bot.*, **30**, 11 (1976).
2. S. Y. Hu, *Am. J. Chin. Med.*, **5**, 1 (1977).
3. B. Goldstein, *Am. J. Chin. Med.*, **3**, 223 (1975).

4. T. E. Hemmerly, *Econ. Bot.*, **31**, 160 (1977).
5. L. Veninga, *The Ginseng Book*, Big Trees Press, Santa Cruz, CA, 1973.
6. S. Foster, *Asian Ginseng-Panax ginseng*, Botanical Series, no. 303, American Botanical Council, Austin, TX, 1991.
7. S. Foster, *American Ginseng-Panax ginseng*, Botanical Series, no. 308, American Botanical Council, Austin, TX, 1991.
8. I. I. Brekham and I. V. Dardymov, *Lloydia*, **32**, 46 (1969).
9. F. Sandberg, *Planta Med.*, **24**, 392 (1973).
10. G. I. Shaposhnikova et al., *Carbohydr. Res.*, **15**, 319 (1970).
11. E. Bombardelli et al., *Fitoterapia*, **3**, 99 (1976).
12. H. Otsuka et al., *Planta Med.*, **32**, 9 (1977).
13. H. Wagner and A. Wurmböck, *Dtsch. Apoth. Ztg.*, **117**, 743 (1977).
14. T. Komori et al., *Org. Mass Spectrom.*, **9**, 744 (1974).
15. S. Hiai et al., *Planta Med.*, **28**, 131 (1975).
16. S. Hiai et al., *Planta Med.*, **28**, 363 (1975).
17. R. Richter et al., *Phytochemistry*, **66**, 2708 (2005).
18. K. Y. Yim, *Hanguk Saenghwa Hakhoe Chi*, **10**, 26 (1977).
19. C. Konno et al., *Planta Med.*, **50**, 434 (1984).
20. Y. Oshima et al., *J. Ethnopharmacol.*, **14**, 255 (1985).
21. C. Konno et al., *Int. J. Crude Drug Res.*, **25**, 53 (1987).
22. S. J. Wu and D. Y. Li, *Zhongcaoyao*, **23**, 549 (1987).
23. J. H. Lee et al., *Carbohydr. Res.*, **341**, 1154 (2006).
24. M. Yoshikawa et al., *Chem. Pharm. Bull. (Tokyo)*, **46**, 647 (1998).
25. B. M. Kwon et al., *Planta Med.*, **63**, 552 (1997).
26. B. X. Wang et al., *Yaoxue Xuebao*, **25**, 401 (1990).
27. Z. K. Chen et al., *J. Pept. Res.*, **52**, 137 (1998).
28. H. O. Cho et al., *Hanguk Sikp'um Kwahakhoe Chi*, **8**, 95 (1976).
29. J. D. Park et al., *Arch. Pharm. Res.*, **21**, 615 (1998).
30. D. S. Kim et al., *Phytochemistry*, **40**, 1493 (1995).
31. D. Dou et al., *Chem. Pharm. Bull. (Tokyo)*, **54**, 751 (2006).
32. K. Yoshihara and Y. Hirose, *Bull. Chem. Soc. Jpn.*, **48**, 2078 (1975).
33. F. Soldati and O. Tanaka, *Planta Med.*, **50**, 351 (1984).
34. G. Q. Liu et al., *Zhongcaoyao*, **18**, 47 (1987).
35. S. Z. Guo et al., *Zhongcaoyao*, **18**, 37 (1987).
36. S. L. Ding and Z. L. Zhu, *Zhongcaoyao*, **23**, 627 (1992).
37. J. L. Reay et al., *J. Psychopharmacol.*, (2006).
38. J. L. Reay et al., *J. Psychopharmacol.*, **19**, 357 (2005).
39. H. Nitta et al., *Biol. Pharm. Bull.*, **18**, 1439 (1995).
40. H. Nitta et al., *Biol. Pharm. Bull.*, **18**, 1286 (1995).
41. K. Takagi, *Yakhak Hoeji*, **17**, 1 (1973).
42. S. L. Friedman and A. N. Khlebnikov, *Biol. Akt. Veshchestva (Mikroelem., Vitam. Drugie) Rastenievod., Zhivotnovod. Med.*, **113** (1975).
43. H. J. Lee et al., *Phytother. Res.*, **20**, 392 (2006).
44. T. Ivanova et al., *Food Chem. Toxicol.*, **44**, 517 (2006).

45. S. R. Kim et al., *In Vivo*, **17**, 77 (2003).
46. M. Kumar et al., *Biol. Pharm. Bull.*, **26**, 308 (2003).
47. S. H. Kim et al., *In Vivo*, **15**, 407 (2001).
48. S. H. Kim et al., *In Vivo*, **12**, 219 (1998).
49. F. Mannaa et al., *J. Appl. Toxicol.*, **26**, 198 (2006).
50. J. H. Lee et al., *Neurosci. Lett.*, **325**, 129 (2002).
51. A. C. Cabral de Oliveira et al., *J. Ethnopharmacol.*, **97**, 211 (2005).
52. J. S. You et al., *Phytother. Res.*, **19**, 1018 (2005).
53. A. C. Cabral de Oliveira et al., *Comp. Biochem. Physiol C Toxicol. Pharmacol.*, **130**, 369 (2001).
54. M. Panwar et al., *Biol. Pharm. Bull.*, **28**, 2063 (2005).
55. S. Y. Hwang et al., *BJU Int.*, **94**, 663 (2004).
56. T. K. Yun et al., *J. Korean Med. Sci.*, **16** (Suppl.), S6 (2001).
57. M. X. Zhuang et al., *Zhongguo Yaoxue Zazhi*, **27**, 653 (1992).
58. T. S. Lim et al., *J. Med. Food*, **7**, 1 (2004).
59. T. A. Nakaya et al., *J. Interferon Cytokine Res.*, **24**, 93 (2004).
60. J. Y. Song et al., *Arch. Pharm. Res.*, **27**, 531 (2004).
61. M. Wang et al., *Int. Immunopharmacol.*, **4**, 311 (2004).
62. V. A. Assinewe et al., *Phytomedicine*, **9**, 398 (2002).
63. D. S. Lim et al., *J. Infect.*, **45**, 32 (2002).
64. J. Y. Shin et al., *Immunopharmacol. Immunotoxicol.*, **24**, 469 (2002).
65. R. Friedl et al., *Br. J. Pharmacol.*, **134**, 1663 (2001).
66. K. M. Park et al., *Planta Med.*, **67**, 122 (2001).
67. Y. R. Kim et al., *Gen. Pharmacol.*, **32**, 647 (1999).
68. Y. Sonoda et al., *Immunopharmacology*, **38**, 287 (1998).
69. Y. S. Lee et al., *Anticancer Res.*, **17**, 323 (1997).
70. J. Liu et al., *Mech. Ageing Dev.*, **83**, 43 (1995).
71. L. M. Feng et al., *Yaoxue Xuebao*, **25**, 401 (1990).
72. C. H. Jung et al., *J. Ethnopharmacol.*, **98**, 245 (2005).
73. S. H. Kim et al., *J. Sports Med. Phys. Fitness*, **45**, 178 (2005).
74. S. H. Kim and K. S. Park, *Pharmacol. Res.*, **48**, 511 (2003).
75. J. Li et al., *Life Sci.*, **64**, 53 (1999).
76. F. R. Maffei et al., *Planta Med.*, **65**, 614 (1999).
77. H. Hikino et al., *Planta Med.*, **51**, 62 (1985).
78. J. T. Xie et al., *Acta Pharmacol. Sin.*, **26**, 1104 (2005).
79. T. P. Liu et al., *Horm. Metab. Res.*, **37**, 146 (2005).
80. V. Vuksan et al., *Arch. Intern. Med.*, **160**, 1009 (2000).
81. T. Yokozawa et al., *Chem. Pharm. Bull.*, **23**, 3095 (1975).
82. Y. C. Huang et al., *Pharm. Res.*, **22**, 636 (2005).
83. S. Choi, *Arch. Pharm. Res.*, **25**, 71 (2002).
84. J. Cho et al., *J. Clin. Endocrinol. Metab.*, **89**, 3510 (2004).
85. B. X. Wong, *Yaoxue Tongbao*, **19**, 41 (1984).
86. L. L. Murphy et al., *Physiol. Behav.*, **64**, 445 (1998).
87. G. Salvati et al., *Panminerva Med.*, **38**, 249 (1996).
88. H. C. Kim et al., *Arch. Pharm. Res.*, **28**, 995 (2005).
89. H. S. Kang et al., *Hanguk Saenghwa Hakhoe Chi*, **8**, 225 (1975).

90. S. Fulder, *New Sci.*, **1**, 138 (1977).
91. Y. K. Gupta et al., *Indian J. Physiol. Pharmacol.*, **45**, 502 (2001).
92. J. C. Yang et al., *Am. J. Chin. Med.*, **29**, 149 (2001).
93. J. Moon et al., *Biochem. Pharmacol.*, **59**, 1109 (2000).
94. J. Sohn et al., *Exp. Mol. Med.*, **30**, 47 (1998).
95. I. A. Persson et al., *J. Ethnopharmacol.*, **105**, 321 (2006).
96. M. F. Caron et al., *Ann. Pharmacother.*, **36**, 758 (2002).
97. J. H. Lee et al., *Planta Med.*, **70**, 615 (2004).
98. N. I. Belogortseva et al., *Planta Med.*, **66**, 217 (2000).
99. C. S. Jeong et al., *Arch. Pharm. Res.*, **26**, 906 (2003).
100. S. O. Shibata et al., in H. Wagner, H. Hikino, and N. R. Farnsworth, eds., *Economic and Medicinal Plant Research*, Vol. 1, Academic Press, Orlando, FL, 1985, p. 218.
101. T. B. Ng and H. W. Yeung, in R. P. Steiner, ed., *Folk Medicine, The Art and the Science*, American Chemical Society, Washington, DC, 1986, p. 139.
102. J. A. Duke, *Ginseng—A Concise Handbook*, Reference Publications, Algonac, MI, 1989.
103. H. Saito in Proceedings of the 3rd International Ginseng Symposium, Korean Ginseng Research Institute, Seoul, Korea, 1980, p. 181.
104. W. H. Lewis in N. L. Etkin, ed., *Plants in Indigenous Medicine and Diet: Biobehavioral Approaches*, Red grave Publishing Co., Bedford Hills, NY, 1986, p. 290.
105. Y. Dai et al., *Zhongguo Yaoli Xuebao*, **9**, 562 (1988).
106. Y. S. Chang et al., *Planta Med.*, **52**, 338 (1986).
107. S. Helms, *Altern. Med. Rev.*, **9**, 259 (2004).
108. V. A. Assinewe et al., *J. Agric. Food Chem.*, **51**, 4549 (2003).
109. Y. S. Chang et al., *Integr. Cancer Ther.*, **2**, 13 (2003).
110. J. T. Coon and E. Ernst, *Drug Saf.*, **25**, 323 (2002).

GOLDENSEAL

Source: *Hydrastis canadensis L.* (Family Ranunculaceae).

Common/vernacular names: Orange root, yellow root, jaundice root, Indian turmeric, eye root, and eye balm.

GENERAL DESCRIPTION

A perennial herb with a knotty yellow rhizome (rootstock) from which arise a single leaf (radical leaf) and an erect hairy stem in early spring bearing two five- to nine-lobed rounded leaves near the top, terminated by a single greenish white flower; up to about 30 cm high; native to rich, moist, deciduous forests, Vermont to Georgia, west to Alabama and Arkansas, north to eastern Iowa and Minnesota; formerly cultivated in Oregon and Washington. Various botanical writers note rarity where it once flourished due to overcollection of the root.[1] Parts used are the dried rhizome and roots.

CHEMICAL COMPOSITION

Contains as active principles isoquinoline alkaloids consisting mainly of hydrastine (1.5–4%) and berberine (0.5–6%), with lesser amounts of canadine (tetrahydroberberine), canadaline, 1-α-hydrastine, 5-hydroxytetrahydroberberine, and other related alkaloids.[2–7] The roots also contain *C*-methyl flavonoids (methylluteolin methyl ethers) and feruloyl quinic acid glucoside esters that were reported for the first time in nature and in goldenseal.[2,8,9]

Other constituents include meconin, chlorogenic acid, lipids with 75% unsaturated and 25% saturated fatty acids, β-sitosterol glucoside,[2] resin, starch, sugar, and a small amount of volatile oil (LIST AND HÖRHAMMER).

PHARMACOLOGY AND BIOLOGICAL ACTIVITIES

The alkaloids present in goldenseal have earlier been reported to have anticonvulsive activity on mouse intestine and uterus.[10] The extract also has a relaxant effect on guinea pig isolated trachea, mainly due to canadine and candaline.[11]

In vitro antibacterial activity of the extract and its major alkaloids has been demonstrated against oral pathogens, for example, *Streptococcus mutans*, peptic ulcer bacteria *Helicobacter pylori*, other Gram-positive and Gram-negative bacteria, as well as *Mycobacterium tuberculosis*.[2,9,12,13] Goldenseal extract possesses antioxidant activity,[14] and it has an immunostimulant effect on antigen-induced IgM production *in vivo*.[15]

Berberine is reported to have pharmacological properties resembling those of hydrastine.[16] Its many activities include stimulating secretion of the bile in humans, sedative effect on cats and mice upon intraperitoneal administration, lowering the blood pressure of laboratory animals, antiarrhythmic effects and strong antibacterial effects (as mentioned above), among many others (see also **barberry**).[6,16,17]

TOXICOLOGY

Numerous modern secondary reports of contact ulceration and inflammation in topical use can be traced to an ointment of goldenseal containing zinc chloride and *Datura stramonium*.[1] Skin inflammation may also result from alkaloid-induced phototoxicity.[18]

USES

Medicinal, Pharmaceutical, and Cosmetic. Hydrastis extracts and hydrastine hydrochloride have been used for stopping uterine hemorrhage and in relieving menstrual pain. Current use is mainly as components in eyewashes and in certain bitter tonic preparations.

Dietary Supplements/Health Foods. One of the most popular indigenous North American botanicals in health and natural food markets in the United States. Root in capsules, tablets, tinctures, extracts, teas, and various other product forms, single or in combination (often with Echinacea). Presence of green mass in powdered root may indicate presence of goldenseal leaf as filler (FOSTER).

Uses are numerous including, but not limited to, that as antiseptic, hemostatic, diuretic, laxative, and tonic; anti-inflammatory for inflammations of the mucous membranes.[1]

Goldenseal consumption has increased due to word-of-mouth circulation of the belief that root products may be used to mask urine tests for illicit drugs. This grows out of the fictional plot of *Stringtown on the Pike* (1900), a novel by pharmacist John Uri Lloyd (1849–1936). This use has persisted throughout the 20th century; especially in attempts to mask morphine detection in racehorses. There is no scientific evidence to support this use. In fact, it may instead promote false-positive readings (TYLER 2).[19]

Despite goldenseal's continued popularity, it has been poorly researched.[1]

Traditional Medicine. Reportedly used as antiperiodic, antiseptic, hemostatic, diuretic, laxative, and tonic. Conditions for which it is used include inflammation of mucous membranes (vaginal and uteral), hemorrhoids, nasal congestion, sore gums, sore eyes, wounds, sores, acne, dandruff, and ringworm, among others. It has also been used in cancers.[20]

American Indian tribes used the roots as a wash for local inflammations; decoction for general debility, dyspepsia, whooping cough, diarrhea, jaundice, fever, sour stomach, flatulence, pneumonia, and with whisky for heart disease (FOSTER AND DUKE; MOERMAN).[1]

COMMERCIAL PREPARATIONS

Crude, extracts, and hydrastine salts. Crude, fluid extract, and tincture were formerly official in N.F. Strengths (see *glossary*) of extracts are expressed in weight-to-weight ratios or in total alkaloids content.

Regulatory Status. Class 2b dietary supplement (not used during pregnancy). Official in the first revision (1830) of New York edition of U.S.P., but absent in the Philadelphia 1830 U.S.P. Dropped in 1840, and then official in U.S.P. from 1860 to 1926. Included in N.F. 1888; also 1936–1955.[21]

In 1982 the alkaloid, hydrastine, was still official in the pharmacopoeias of nine countries.[21]

REFERENCES

See the General References for APPLEQUIST; BAILEY 1; BARNES; BLUMENTHAL 2; DER MARDEROSIAN AND BEUTLER; FERNALD; FOSTER; FOSTER AND DUKE; GOSSELIN; KROCHMAL AND KROCHMAL; LUST; MARTINDALE; MCGUFFIN 1 & 2; MERCK; MOERMAN; ROSE; TERRELL; TYLER 1; UPHOF.

1. S. Foster, *Goldenseal—Hydrastis canadensis*, Botanical Series, no. 309, American Botanical Council, Austin, TX, 1991.
2. B. Y. Hwang et al., *Planta Med.*, **69**, 623 (2003).
3. J. Gleye and E. Stanislas, *Plant. Med. Phytother.*, **6**, 306 (1972).
4. G. Caille et al., *Can. J. Pharm. Sci.*, **5**, 55 (1970).
5. J. Gleye et al., *Phytochemistry*, **13**, 675 (1974).
6. K. Genest and D. W. Hughes, *Can. J. Pharm. Sci.*, **4**, 41 (1969).
7. S. El-Masry et al., *J. Pharm. Sci.*, **69**, 5597 (1980).
8. C. E. McNamara et al., *J. Nat. Prod.*, **67**, 1818 (2004).
9. E. J. Gentry et al., *J. Nat. Prod.*, **61**, 1187 (1998).
10. J. Haginiwa and M. Harada, *Yakugaku Zasshi*, **82**, 726 (1962).
11. H. bdel-Haq et al., *Pharmacol. Toxicol.*, **87**, 218 (2000).
12. G. B. Mahady et al., *Phytother. Res.*, **17**, 217 (2003).
13. F. Scazzocchio et al., *Planta Med.*, **67**, 561 (2001).
14. S. A. Periera da et al., *Phytother. Res.*, **14**, 612 (2000).
15. J. Rehman et al., *Immunol. Lett.*, **68**, 391 (1999).
16. V. Preininger in R. H. F. Manske, ed., *The Alkaloids*, Vol. 15, Academic Press, New York, 1975, p. 207.
17. C. W. Lau et al., *Cardiovasc. Drug Rev.*, **19**, 234 (2001).
18. J. J. Inbaraj et al., *Chem. Res. Toxicol.*, **14**, 1529 (2001).
19. S. Foster, *HerbalGram*, **21** (1989).
20. J. L. Hartwell, *Lloydia*, **34**, 103 (1971).
21. C. Hobbs, *Pharm. Hist.*, **32**, 79 (1990).

GOTU KOLA

Source: *Centella asiatica* (L.) Urban (syn. *Hydrocotyle asiatica* L. and *Centella coriacea* Nannfd.) (Family Umbelliferae or Apiaceae).

Common/vernacular names: Indian pennywort.

GENERAL DESCRIPTION

Slender herbaceous creeping weakly aromatic prostrate perennial; stems long, prostrate, filiform, often orange with long internodes, rooting at nodes; leaves orbicular, reniform up to 4 cm in diameter; petiole 5–10 cm long; flowers in fascicled umbel, growing near water or marshy places, moist rocky outcrops to 700 m, in India, China, Indonesia, Sri Lanka, western South Sea Islands, Australia, Madagascar, southern Africa, Hawaii, and so on.[1] Parts used are the fresh or dried leaves, aboveground herb, or whole herb with root.

CHEMICAL COMPOSITION

Leaves contain triterpenoid saponins at highly variable levels (1.1–8.0%), including asiaticoside, oxyasiaticoside, and madecassoside (Madagascar chemotype), centelloside and centellasaponins (Sri Lanka chemotype), brahmoside, brahminoside, thankunoside, isothankunoside (India chemotype); sapogenins from various chemotypes including asiatic, madecassic, centellic, indocentoic, brahmic, thankunic, and iosthankunic acids.[1–4]

Volatile oil with *trans*-β-farnesene, germacrene D, β-caryophyllene, camphor, cineole, *n*-dodecane, *p*-cymol, α-pinene, methanol, allyl mustard, and an unidentified terpene acetate (36% of oil content).[1]

Other components include rhamnose, arabinose, glucose, fructose, sucrose, and raffinose; a pectin composed of the aforementioned sugars; an oligosaccharide, centellose; a fatty oil containing glycerides of oleic, linoleic, lignoceric, palmitic, stearic, linolenic, and elaidic acids; steroids, including β-sitosterol, stigmasterol, campesterol, and sitosterol; amino acids, including glutamic acid, serine, and alanine; flavonols, including kaempferol, quercetin, 3-glycosylkaempferol, and 3-glycosylquercetin; polyphenols, tannins, carotenoids, villarin, and ascorbic acid (13.8 mg/100 dry weight).[1,5,6]

PHARMACOLOGY AND BIOLOGICAL ACTIVITIES

Topically, asiaticoside or leaf extracts standardized to asiaticoside have been shown to accelerate the wound-healing process and significantly improved tensile strength of tissues, promoting keratization and stimulating rapid and healthy growth of the reticuloendothelium. It is suggested that inhibition of the biosynthesis of collagen and acidic mucopolysaccharides is involved in the mechanism of action.[1] A leaf extract (standardized to asiaticoside) was evaluated in clinical patients with soiled wounds and chronic atony, resistant to treatment; results showed complete healing in 64% and improvement in 16% of 20 patients.[7] Cellular proliferation, collagen formation, epithelization, and wound healing were further demonstrated in a number of *in vitro* and *in vivo* experiments and in an antipsoriasis model utilizing asiaticoside-rich preparations.[8–13]

The healing effect of *C. asiatica* was also demonstrated in a number of studies on induced gastric ulcers whereby the extract and asiaticoside enhanced ulcer healing in all experiments. The healing effect was mediated through suppression of iNOS, free radical scavenging, and strengthening of the mucosal barrier.[14–17]

Based on traditional use in India claiming the herb improves intelligence, various studies (both animal and human) have suggested the herb to be beneficial in improving memory, adaptogenic in fatigue and stress, tranquilizing in rats (alcoholic extract), to increase general mental ability, behavioral patterns, and to increase I.Q. in mentally retarded children. A two-compartment passive avoidance task test (with rats) showed an improvement in

24 h retention. Assessment of turnover of biogenic amines (norepinephrine, dopamine, and serotonin) showed significant reductions of these amines and their metabolites in the brain following oral administration of a fresh juice (1 mL = 0.38 g fresh leaves), at a dose of 0.18 g/kg for 15 days. The decrease of amine levels was correlated to improved learning and memory in rats.[18] The nootropic effect (enhancement of cognition, learning, and memory) of *C. asiatica* aqueous extract was further demonstrated in two *in vivo* experiments in rats and mice.[19,20]

C. asiatica has an antioxidant effect that directly impacts its usefulness as a neuroprotective nerve tonic, nootropic agent, and in wound healing (mentioned above) and cardiomyopathy.[11,21–24]

A water-soluble fraction was shown to have an antianxiety effect in animals comparable to diazepam and to inhibit hepatic enzymes responsible for barbiturate metabolism, and hence may prolong pentobarbitone-induced sleep.[25] The same effect was exhibited by the methanolic and ethyl acetate extracts as well as by asiaticoside in a number of rat behavioral models.[26]

Other reported activities include anti-inflammatory, CNS-depressant, anticonvulsant, antidepressant, and analgesic;[4] antitumor;[2,27] immunomodulatory;[28] radioprotective;[29] and to improve circulation in microangiopathy.[30–32] The chemistry, pharmacology, and clinical applications of *C. asiatica* have been reviewed.[33]

TOXICOLOGY

Three cases of hepatotoxicity and jaundice due to *C. asiatica* have been recently reported. Treatment with ursodeoxycholic acid and discontinuation of *C. asiatica* resulted in a marked improvement in all patients.[34]

USES

Medicinal, Pharmaceutical, and Cosmetic. Oral extracts, injectable extracts, and ointments standardized to asiaticoside have been used successfully in India for the treatment of mal perforant lesions in leprosy patients; in Europe used clinically for leg ulcers in postphlebitic patients (injectable extract); accelerating healing of superficial postsurgical wounds; inhibiting hypertrophic formation of scar tissue in the treatment of second- and third-degree burns; and general use in wounds, ulcer, and scleroderma.[1]

Leaf extracts used in cosmetics, including hand creams and lotion, hair conditioners, and shampoos.

Food. In Bangladesh, Thailand, and Sri Lanka, the leaves are sold as a leafy vegetable, rich in digestible protein, carotene, and vitamin C.[35]

Dietary Supplements/Health Foods. In the United States, various gotu kola leaf preparations or crude leaves are sold as dietary supplements in the form of tablets, capsules, tincture, standardized extracts, tea, and so on, used for memory improvement; topically for the treatment of wounds (WEISS; WREN).

Traditional Medicine. In Chinese folk medicine, a decoction of the whole aboveground herb is used for treatment of colds, sunstroke, tonsillitis, pleurisy, urinary tract infections, infectious hepatitis, jaundice, and dysentery; as an antidote for arsenic poisoning, poisoning by *Gelsemium elegans*, and toxic mushrooms; external poultice for snakebites, scabies, traumatic injuries, and herpes zoster (JIANGSU).

In India, used as a folk remedy for leprosy, lupus, syphilis, tuberculosis, improving mental function, and others (CSIR II).

In Uttar Pradesh, fresh leaf juice used externally for elephantiasis, inflammations, swelling; whole plant decoction used for skin diseases, such as itching and fungal infections, chronic rheumatism, amenorrhea, and as "blood purifier."[36]

In East Africa, leaf used for fevers, bowel complaints, and syphilitic and scrofulous conditions; in Central Africa, widely used as a folk medicine, particularly for leprosy (WATT AND BREYER-BRANDWIJK).

Other. Insecticidal properties have been reported from leaf extracts.

COMMERCIAL PREPARATIONS

Crude (leaf, whole herb, etc.), tincture, extracts (calculated to contain 70% total triterpenes), and so on.

Regulatory Status. Undetermined in the United States; official in the French pharmacopoeia as early as 1884; official in Indian pharmacopoeia; also once official in the Dutch, Mexican, Spanish, and Venezuelan pharmacopoeias.

REFERENCES

See the General References for APPLEQUIST; BARNES; BARRETT; BRUNETON; CHP; CSIR II; DER MARDEROSIAN AND BEUTLER; GLASBY 2; GUPTA; JIXIAN; HUANG; MCGUFFIN 1 & 2; TYLER 1; WATT AND BREYER-BRANDWIJK; WEISS; WREN.

1. T. Kartnig in L. E. Craker and J. E. Simons, eds., *Herbs, Spices, and Medicinal Plants—Recent Advances in Botany, Horticulture and Pharmacology*, Vol. 3, Oryx Press, Phoenix, 1988, p. 145.
2. M. Yoshida et al., *Biol. Pharm. Bull.*, **28**, 173 (2005).
3. H. Matsuda et al., *Chem. Pharm. Bull. (Tokyo)*, **49**, 1368 (2001).
4. M. R. Sakin and P. C. Dandiya, *Fitoterapia*, **61**, 291 (1990).
5. P. S. Rao and T. R. Seshardi, *Curr. Sci.*, **38**, 77 (1969).
6. X. S. Wang et al., *Carbohydr. Res.*, **338**, 2393 (2003).
7. R. Morisset et al., *Phytother. Res.*, **1**, 117 (1987).
8. J. H. Sampson et al., *Phytomedicine*, **8**, 230 (2001).
9. F. X. Maquart et al., *Eur. J. Dermatol.*, **9**, 289 (1999).
10. A. Shukla et al., *J. Ethnopharmacol.*, **65**, 1 (1999).
11. A. Shukla et al., *Phytother. Res.*, **13**, 50 (1999).
12. Sunilkumar et al., *Indian J. Exp. Biol.*, **36**, 569 (1998).
13. L. Suguna et al., *Indian J. Exp. Biol.*, **34**, 1208 (1996).
14. J. S. Guo et al., *Planta Med.*, **70**, 1150 (2004).
15. C. L. Cheng et al., *Life Sci.*, **74**, 2237 (2004).
16. K. Sairam et al., *Indian J. Exp. Biol.*, **39**, 137 (2001).
17. C. L. Cheng and M. W. Koo, *Life Sci.*, **67**, 2647 (2000).
18. K. Nalin et al., *Fitoterapia*, **63**, 232 (1992).
19. S. B. Rao et al., *Physiol. Behav.*, **86**, 449 (2005).
20. Y. K. Gupta et al., *Pharmacol. Biochem. Behav.*, **74**, 579 (2003).
21. A. Soumyanath et al., *J. Pharm. Pharmacol.*, **57**, 1221 (2005).
22. M. Subathra et al., *Exp. Gerontol.*, **40**, 707 (2005).
23. A. Gnanapragasam et al., *Life Sci.*, **76**, 585 (2004).
24. G. Jayashree et al., *Fitoterapia*, **74**, 431 (2003).
25. P. V. Diwa et al., *Fitoterapia*, **62**, 253 (1991).
26. P. Wijeweera et al., *Phytomedicine* (2006).
27. P. Bunpo et al., *Food Chem. Toxicol.*, **42**, 1987 (2004).
28. K. Punturee et al., *Asian Pac. J. Cancer Prev.*, **6**, 396 (2005).

29. J. Sharma and R. Sharma, *Phytother. Res.*, **16**, 785 (2002).
30. M. R. Cesarone et al., *Angiology*, 52(Suppl. 2), S49 (2001).
31. M. R. Cesarone et al., *Angiology*, 52(Suppl. 2), S19 (2001).
32. M. R. Cesarone et al., *Angiology*, 52(Suppl. 2), S15 (2001).
33. B. Brinkhaus et al., *Phytomedicine*, **7**, 427 (2000).
34. O. A. Jorge and A. D. Jorge, *Rev. Esp. Enferm. Dig.*, **97**, 115 (2005).
35. H. S. Puri and D. C. Bagchi, *Am. Herb Assn. Nltr.*, **8**, 4 (1991).
36. M. Badruzzaman et al., *Fitoterapia*, **63**, 245 (1992).

GRAPE SKIN EXTRACT (ENOCIANINA)

Source: Varieties of *Vitis vinifera* L. (Family Vitaceae).

GENERAL DESCRIPTION

Grape skin extract or enocianina is the coloring matter derived from the skin of certain varieties of the wine grape. It is commonly obtained by acidic aqueous extraction of fermented grape skin (or marc) after the juice has been expressed from it.

The major producer of enocianina is Italy; United States is a potential producer.

CHEMICAL COMPOSITION

The pigments present in enocianina are anthocyanins. They are glycosides of polyhydroxy derivatives of 2-phenylbenzopyrylium salts. Their aglycones are generally called anthocyanidins.[1,2] The most common anthocyanidins in grape skin extract are peonidin, malvidin, delphinidin, and petunidin. The grape anthocyanins are usually either monoglycosides or diglycosides.[3,4] Oligomeric anthocyanins have also been detected by MS analysis.[5]

In addition to anthocyanins, grape skin extract contains plant acids (mainly tartaric acid), tannins, sugars, amino acids, minerals, and other constituents present in grapes. It may also contain diluents used in processing.

Grape skin anthocyanins are soluble in water and water–alcohol mixtures to yield red to magenta solutions with an acidic pH (ca. 3). The color varies with the pH, from red to purple, and finally to almost blue as the pH is changed from 1 to 8. The pigments are most stable below pH 3 and are not stable at higher pH values: they are also sensitive to heat, oxygen, and light.[6,7]

PHARMACOLOGY AND BIOLOGICAL ACTIVITIES

Grape skin extract obtained from *V. labrusca* has antihypertensive vasodilator effect.[8,9] The same extract and others obtained from *V. vinifera* are strong antioxidants that may be beneficial as cardioprotective and neuroprotective (in Alzheimer's disease) agents.[9–11]

Anthocyanins from grapes have been reported to have antifungal activities (e.g., against *Penicillium notatum* and *Aspergillus* species)[12] and strong "vitamin P" (bioflavonoids) activity, strengthening blood capillaries of guinea pigs.[13] A grape anthocyanin (delphinidin-3-monoglucoside) has also been reported to inhibit the growth of *Lactobacillus acidophilus*.[14]

TOXICOLOGY

A tannic substance (leucocyanidin) present in some grape juices has been reported to be toxic to laboratory animals; symptoms included cardiac failure and hepatic lesions.[15] Recent studies, however, demonstrated that grape skin extract lacked any significant *in vivo* toxicity in rats after 3 months of administration.[16] Moreover, acute DNA toxicity could not be detected in an *in vivo* assay for clastogenic activity in mice.[17]

USES

Food. Used in coloring alcoholic and nonalcoholic beverages. The powder is also used in drink mixes.

COMMERCIAL PREPARATIONS

Liquid and powder (vacuum dried or spray dried); the powder is more stable than the liquid. Strengths are expressed in color intensities (absorbance values).

Regulatory Status. Has been approved for food use in beverages only, with specific restrictions (§73.170); exempt from certification.

REFERENCES

See the General References for APPLEQUIST; BARRETT; BRUNETON; DER MARDEROSIAN AND BEUTLER; DUKE 4; GRIEVE; MCGUFFIN 1 & 2.

1. H. B. Hass, *Chemtech*, **7**(9), 525 (1977).
2. J. B Harborne, in L. Zechmeister, ed., *Fortschritte der Chemie Organischer Naturstoffe*, Vol. 20, Springer-Verlag, Vienna, Austria, 1962, p. 165.
3. V. M. Malikov, *Vinogradarstvo*, 158 (1973).
4. L. S. Diaz et al., *Rev. Agroquim. Tecnol. Aliment.*, **16**, 509 (1976).
5. S. Vidal et al., *J. Agric. Food Chem.*, **52**, 7144 (2004).
6. C. Skalski and W. A. Sistrunk, *J. Food Sci.*, **38**, 1060 (1974).
7. L. Szechenyi, *Ind. Aliment Agr. (Paris)*, **80**, 521 (1963).
8. S. V. Madeira et al., *Pharmacol. Res.*, **52**, 321 (2005).
9. M. R. de Soares et al., *J. Pharm. Pharmacol.*, **54**, 1515 (2002).
10. A. Russo et al., *Life Sci.*, **72**, 2369 (2003).
11. J. F. Young et al., *Br. J. Nutr.*, **84**, 505 (2000).
12. K. Rizvanov and B. Karadimcheva, *Lozar. Vinar.*, **21**, 26 (1973).
13. A. S. Sturua et al., *Prikl. Biokhim. Mikrobiol.*, **7**, 606 (1971).
14. D. E. Pratt et al., *Food Res.*, **25**, 26 (1960).
15. R. Patay et al., *C. R. Congr. Natl. Soc. Savantes, Paris Dept. Sect. Sci.*, **86**, 69 (1961).
16. S. S. Bentivegna and K. M. Whitney, *Food Chem. Toxicol.*, **40**, 1731 (2002).
17. G. L. Erexson, *Food Chem. Toxicol.*, **41**, 347 (2003).

GRAPEFRUIT OIL

Source: *Citrus ×paradisi* Macf. (syn. *C. racemosa* (Risso et Poit.) Marcov. ex Tanaka; *C. decumana* var. *racemosa* (Risso et Poit.) Roem.; *C. maxima* (L.) Osbeck var. *racemosa* (Roem.) Stone) (Family Rutaceae).

Common/vernacular names: Expressed grapefruit oil, cold-pressed grapefruit oil, and shaddock oil.

GENERAL DESCRIPTION

A cultivated tree with large fruits, often over 10 m high; is considered to be a relatively recent hybrid of *C. maxima* and *C. sinensis*. Numerous cultivars are grown commercially; "Duncan" is the standard type grown in Florida (TUCKER AND LAWRENCE); cultivated in the United States (especially California, Florida, and Texas), the West Indies (e.g., Jamaica and the Dominican Republic), Nigeria, Brazil, and Europe (e.g., Israel and Portugal). Part used is the fresh peel of the fruit from which grapefruit oil is produced by cold expression. Naringin extract is a bitter flavoring material prepared by extraction of the expressed peel; it is not pure naringin.

CHEMICAL COMPOSITION

Grapefruit oil contains mostly the monoterpene hydrocarbon, limonene (ca. 90%). Other volatile constituents include sesquiterpenes (e.g., cadinene and paradisiol or intermedeol);[1,2] aldehydes (C_7 to C_{12} aldehydes, neral, geranial, perillaldehyde, citronellal, α-sinensal, and β-sinensal); esters (e.g., geranyl acetate, neryl acetate, perillyl acetate, octyl acetate, decyl acetate, citronellyl acetate, *trans*-carvyl acetate, 1,8-*p*-menthadien-2-yl acetate, and 1,8-*p*-menthadien-9-yl acetate); and nootkatone (a bicylcic sesquiterpene ketone); among others (GUENTHER; LIST AND HÖRHAMMER).[1–6]

The oil also contains sizable amounts (ca. 1.4%) of coumarins and furocoumarins (bergaptens) composed mainly of 7-geranoxycoumarin, with marmin, osthol, limettin, 7-methoxy-8-(2-formyl-2-methylpropyl) coumarin, bergapten, bergamottin, dihydroxybergamottin, bergaptol, byakangelicin, 5-[(3,6-dimethyl-6-formyl-2-heptenyl)-oxy] psoralen, and a chromen-7-one also present.[7–10]

The characteristic grapefruit aroma and flavor of grapefruit oil is reported to be due primarily to nootkatone and other carbonyls present (e.g., geranyl acetate, neryl acetate, octyl acetate, 1,8-*p*-menthadien-2-yl acetate, and others).[1,11–13]

The characteristic bitter taste of grapefruit is due to naringin present mainly in the peel.[14,15]

PHARMACOLOGY AND BIOLOGICAL PROPERTIES

Grapefruit oil is reported to have antibacterial activities.[16] Its chromenone constituent has been shown to enhance antibiotic effect against MSSA and MRSA by acting as an inhibitor of the bacterial efflux pump mechanism.[7]

The olfactory stimulation resulting from the scent of grapefruit oil has a sympathomimetic stimulating effect that results in an increase in blood pressure, increased lipolysis, and reduced body weight gain in experimental animals and in humans.[17–20]

TOXICOLOGY

Grapefruit oil has been reported to promote tumor formation on mouse skin by the primary carcinogen, 9,10-dimethyl-1,2-benzanthracene.[21]

Certain bergaptens are known to be phototoxic and allergenic to humans (see **bergamot oil**).

Dermatological studies have indicated grapefruit oil to be nonirritating, nonsensitizing, and nonphototoxic to humans.[16]

USES

Medicinal, Pharmaceutical, and Cosmetic. Grapefruit oil is used as a fragrance component in soaps, detergents, creams, lotions, and perfumes, with maximum use level of 1.0% reported in perfumes.[16]

Food. Grapefruit oil is extensively used as a flavor ingredient in alcoholic and nonalcoholic beverages (especially soft drinks), frozen dairy desserts, candy, baked goods, gelatins and puddings, and milk products, with highest maximum average use level of about 0.108% (1084 ppm) reported in candy.

Naringin extract is used mainly in soft drinks. Other foods in which it is also used include alcoholic beverages, frozen dairy desserts, candy, baked goods, and gelatins and puddings. Highest average maximum use level is about 0.018 (175 ppm) reported in alcoholic beverages; average maximum use level reported in soft drinks (nonalcoholic beverages) is about 0.004% (38.4 ppm).

COMMERCIAL PREPARATION

Grapefruit oil (also partially deterpenized) and naringin extract. Oil is official in F.C.C.

Regulatory Status. GRAS with both grapefruit and naringin listed (§182.20).

REFERENCES

See the General References for ARCTANDER; BAILEY 1; FEMA; GUENTHER; JIANGSU; KARRER; POLUNIN AND SMYTHIES; TERRELL; TUCKER AND LAWRENCE; UPHOF.

1. H. Sulser et al., *J. Org. Chem.*, **36**, 2422 (1971).
2. J. W. Huffman and L. H. Zalkow, *Tetrahedron Lett.*, **10**, 751 (1973).
3. M. Koketsu, et al., *Bol. Pesqui EMBRAPA Cent. Technol. Agr. Ailment*, **7**, 21 (1983).
4. J. A. Remar in J. Erghese, ed., *On Essential Oils*, Synthie Industrial Pte Ltd., Lolenchery, India, 1986, p. 123.
5. G. Dugo et al., *Flav. Frag. J.*, **5**, 205 (1990).
6. W. A. König et al., *J. High Res. Chromatogr.*, **13**, 328 (1990).
7. A. N. Abulrob et al., *Phytochemistry*, **65**, 3021 (2004).
8. P. Schmiedlin-Ren et al., *Drug Metab. Dispos.*, **25**, 1228 (1997).
9. W. L. Stanley and L. Jurd, *J. Agric. Food Chem.*, **19**, 1106 (1971).
10. J. F. Fisher and H. E. Nordby, *J. Food Sci.*, **30**, 869 (1965).
11. M. G. Moshonas, *J. Agric. Food Chem.*, **19**, 769 (1971).
12. J. L. K. Hunter and W. B. Brogden, *J. Food Sci.*, **30**, 383 (1965).
13. J. L. K. Hunter and M. G. Moshonas, *J. Food Sci.*, **31**, 167 (1966).
14. G. M. Fishman and M. N. Gumanitskaya, *U. S. S. R.*, **261**, 166 (1970).
15. V. A. Bandyukova and G. M. Fishman, *Subtrop. Kul't.*, **5–6**, 137 (1976).
16. D. L. J. Opdyke, *Food Cosmet. Toxicol.*, **12**, 723 (1974).
17. M. Tanida et al., *Brain Res.*, **1058**, 44 (2005).
18. J. Shen et al., *Neurosci. Lett.*, **383**, 188 (2005).
19. A. Niijima and K. Nagai, *Exp. Biol. Med. (Maywood.)*, **228**, 1190 (2003).
20. S. Haze et al., *Jpn. J. Pharmacol.*, **90**, 247 (2002).
21. F. J. C. Rose and W. E. H. Field, *Food Cosmet. Toxicol.*, **3**, 311 (1965).

GUAIAC WOOD OIL

Source: *Bulnesia sarmienti* Lorentz ex Griseb. (Family Zygophyllaceae).

Common/vernacular names: Champaca wood oil.

GENERAL DESCRIPTION

A tree native to South America, growing in Brazil, Paraguay, and Argentina. Guaiac wood oil is obtained by steam distillation of the comminuted wood and sawdust; it is a thick semisolid mass with an odor resembling tea roses and may sometimes have an undesirable "smoked ham" note.

Guaiac wood oil is different from **guaiac resin** or **guaiac gum** products. The latter are obtained from the wood of other trees (*Guaiacum officinale* L. and *G. santum* L. (Family Zygophyllaceae)). This guaiac contains a small amount of α-guaiaconic acid mixed with large amounts of other phenolic lignans (formerly collectively called guaiaconic acid) and other substances. It was formerly used in treating rheumatism and gout but currently is mainly used as a diagnostic reagent (e.g., in testing for occult blood).[1] Nevertheless, it has been approved for food use as an equivalent of guaiac wood oil or products derived from *B. sarmienti*; it is used in foods mainly as an antioxidant and is official in F.C.C.

CHEMICAL COMPOSITION

Guaiac wood oil contains 42–72% guaiol, bulnesol, δ-bulnesene (δ-guaiene), β-bulnesene, α-guaiene, β-patchoulene, and guaioxide (LIST AND HÖRHAMMER).

PHARMACOLOGY AND BIOLOGICAL ACTIVITIES

One available report indicates guaiac wood oil to be nonirritating, nonsensitizing, and nonphototoxic to human skin.[2] Another report indicates guaiac wood oil to be nontoxic to rats on short-term feeding.[3]

Guaiazulene, obtained from guaiac wood by dehydration of guaiol, has been demonstrated to have anti-inflammatory activity in animal studies. It has been used in Germany as an ingredient in combination bath products, claimed for efficacy in a wide variety of skin conditions and inflammation, but without clear demonstration of pharmacological or clinical efficacy.[4]

USES

Medicinal, Pharmaceutical, and Cosmetic. Guaiac wood oil is used as a fixative, modifier, or fragrance component in soaps, detergents, creams, lotions, and perfumes, with maximum use level of 0.8% reported in perfumes.

Food. Guaiac wood oil is used as a flavor component in most categories of food products, including alcoholic and nonalcoholic beverages, frozen dairy desserts, candy, baked goods, gelatins and puddings, and meat and meat products, with highest average maximum use level of about 0.002% (22 ppm) reported in meat and meat products.

Dietary Supplements/Health Foods. Guaiac wood preparations are occasionally used in formulation for anti-inflammatory activity, mostly in Europe, including gout and rheumatism formulations (WREN).

COMMERCIAL PREPARATIONS

Oil.

Regulatory Status. Has been approved for food use (§172.510); gum guaiac is GRAS when used in edible oils or fats in accordance with good manufacturing practice (§182.3336).

Guaiac wood is the subject of a German therapeutic monograph, used for supportive therapy for rheumatic complaints.[4]

REFERENCES

See the General References for ARCTANDER; BLUMENTHAL 1; FEMA; GUENTHER; MERCK; TERRELL; UPHOF; WREN.

1. J. F. Kratochvil et al., *Phytochemistry*, **10**, 2529 (1971).
2. D. L. J. Opdyke, *Food Cosmet. Toxicol.*, **12**(Suppl.), 905 (1974).

3. B. L. Oser et al., *Food Cosmet. Toxicol.*, **3**, 563 (1965).

4. Monograph *Guajaci lignum Bundesanzeiger*, no. 76 (April 23, 1987).

GUAR GUM

Source: *Cyamopsis tetragonoloba* (L.) Taub. (syn. *C. psoralioides* DC.) (Family Leguminosae or Fabaceae).

Common/vernacular names: Guar flour, jaguar gum.

GENERAL DESCRIPTION

Guar gum is derived from the seed of the guar plant. The guar plant is a small nitrogen-fixing annual that bears fruits known as legumes (pods) containing five to nine seeds per pod; up to about 1.8 m high. It is believed to be native to tropical Asia and has been grown in India and Pakistan for centuries as food for both humans and animals. It was introduced into the United States in the early 1900s and is now grown in Texas and Oklahoma. Major guar producers are India, Pakistan, and the United States.

Part used is the endosperm of the seed. The endosperm constitutes 35–42% of the seed; it is separated from the outer components of the seed (seed coat or hull and embryo or germ) during processing. Processing involves hull removal by water or acid soaking and grinding, followed by preferential grinding to remove the embryo. The endosperm left is then ground to a fine powder, which is commercial guar gum.

There are different grades of guar gum with varying amounts of the hull and germ present as the main impurities. Food- and pharmaceutical-grade guar gum is a white to yellowish white, nearly odorless powder. It is easily dispersed in cold or hot water (see ***locust bean gum***) to form solutions (sols) with a slightly acidic to almost neutral pH (5.4–6.4). The rate of hydration (dispersion) is dependent on its particle size, water temperature, and the rate of agitation. Its optimal rate of hydration occurs between pH 7.5 and 9. Even at low concentrations (1–2%), guar sols have high viscosity and form gels with borate ions at alkaline pH values as locust bean gum and aloe vera gel; these gels can be liquefied by lowering the pH below 7, by heating, or by adding simple polyols (glycerol, mannitol, etc.) that can react with the borate ions. The borate gels are not edible. Guar gum is not soluble or dispersible in organic solvents.

The viscosity of guar gum sols is unaffected by pH changes between 4 and 10.5.

Guar gum is reported to be compatible with gelatin, starch, and most water-soluble gums (e.g. acacia, agar, algin, karaya gum, locust bean gum, pectin, and tragacanth).

CHEMICAL COMPOSITION

Commercial food-grade guar gum is reported to contain usually about 80% guaran (a galactomannan), 5–6% crude protein, 8–15% moisture, 2.5% crude fiber, 0.5–0.8% ash, and small amounts of lipids composed mainly of free and esterified fatty acids.[1]

Guaran (the pure galactomannan from guar gum) is a polysaccharide with a primary structure consisting of regular repeating units of the trisaccharide 4-*O*-(6-*O*-α-D-galactopyranosyl-β-D-mannopyranosyl)-β-D-mannopyranose. Its molecular weight has been reported to be around 220,000 (WHISTLER AND BEMILLER).[2]

PHARMACOLOGY AND BIOLOGICAL ACTIVITIES

Guar gum has been reported to lower the serum and liver cholesterol levels in chickens and rats, as well as the serum cholesterol and postprandial (after meal) blood glucose in humans (MARTINDALE).[3–5] When included at different levels in the diets of chickens, guar gum has been demonstrated to cause growth depression, though with inconsistent results. It also reduced the metabolizable energy of the diets in which it was included.[3,6,7]

Guar gum does not seem to be digested by animals.[3]

In women, one study found that ingestion of guar gum led to permanent weight loss, but did not influence serum lipids in hypercholesterolemia.[8] Similar results were noted in male patients and elderly patients.[9–11] However, positive results are reported in the use of guar gum as a long-term dietary supplement in control of hypercholesterolemia in diabetics.[10,12–14] Long-term administration (21 g/day) produced a sustained improvement in control of type 2 diabetes, with significantly lower serum total and LDL cholesterol concentrations.[15] An average reduction of 14% total cholesterol levels was observed in doses of 10 g b.i.d. immediately before meals as well as a reduction in postprandial glucose levels.[16] Other studies have produced similar positive results including a recent one in alloxan-induced diabetic rats.[10,12,17,18]

A blood pressure lowering effect (8% systolic; 7% diastolic) has been observed in overweight men with mild hypertension.[19]

TOXICOLOGY

Flatulence has frequently been reported as a side effect to guar gum dietary supplementation.[20] Occupational asthma has been reported in subjects working with industrial production of guar gum.[21] A severe case of contact urticaria has recently been linked to guar gum present in a local anesthetic gel.[22]

Guar gum did not produce teratogenic effects in rats.[23] A human study found that guar gum consumption in diabetic mellitus patients did not adversely affect mineral balance.[24]

USES

Medicinal, Pharmaceutical, and Cosmetic. It is used as a binding and disintegrating agent in tablets and as a thickener in lotions and creams;[25] also used as an appetite depressant and in certain antihypercholesterolemic preparations.

Food. Used extensively as a thickener, stabilizer, suspending agent, and binder of free water in many food products, including nonalcoholic beverages (e.g., fruit drinks), frozen dairy desserts (especially ice cream and sherbets where it binds free water to prevent ice crystals formation), baked goods, gelatins and puddings, meat and meat products, condiments and relishes, breakfast cereals, cheeses (especially soft cheeses and spreads), milk products, soups, sweet sauces, gravies, snack foods, and processed vegetables, among others.[26,27] Highest average maximum use level reported is about 1% in breakfast cereals (11,260 ppm), sweet sauces (9000 ppm), and processed vegetables (10,747 ppm).

Dietary Supplements/Health Foods. Capsules, tablets, powder, and other product forms have, until recently, been widely used in weight loss formulations. Guar gum was blamed for causing esophageal obstruction. A death has been attributed to the use of one guar gum tablet product, which apparently swelled in the esophagus, indirectly resulting in complications that caused the fatality.[28] Major adverse reactions appear to result from product formulations (tablets) that dissolve in the mouth or esophagus before they reach the stomach. The FDA issued regulatory letters to manufacturers of guar gum capsules or tablets. Products that included claims implying use in weight loss, appetite suppression, or cholesterol or blood glucose lowering effects were deemed misbranded drugs.[28]

Others. Technical guar gum is extensively used in other industries (especially paper, oil drilling, and textile).

COMMERCIAL PREPARATIONS

Various grades with different particle sizes and viscosities. It is official in N.F. and F.C.C.

Regulatory Status. Has been affirmed as GRAS (§184.1339), as a thickener, stabilizer, suspending agent, or binder in food products.

Based on "numerous adverse reaction reports of esophageal, gastric, or intestinal obstruction associated with the use of guar gum in a weight control drug," the FDA now considers guar gum, when labeled as a drug, to be a hazardous ingredient.[28]

REFERENCES

See the General References for BRUNETON; DER MARDEROSIAN AND BEUTLER; FEMA; FURIA; GLICKSMAN; LAWRENCE; MARTINDALE; MCGUFFIN 1 & 2; MERCK; TERRELL; UPHOF; WHISTLER AND BEMILLER.

1. J. Gynther et al., *Planta Med.*, **46**, 60 (1982).
2. C. W. Baker and R. L. Whistler, *Carbohydr. Res.*, **45**, 237 (1975).
3. S. E. Davis and B. A. Lewis, *ACS Symp. Ser.*, **15**, 296 (1975).
4. D. J. A. Jenkins et al., *Lancet*, 1351 (1976).
5. D. J. A. Jenkins et al., *Ann. Intern. Med.*, **86**, 20 (1977).
6. F. H. Kratzer et al., *Poultry Sci.*, **46**, 1489 (1967).
7. R. S. Thakur and K. Pradham, *Indian J. Anim. Sci.*, **45**, 880 (1975).
8. J. Tuomilehto et al., *Acta Med. Scand.*, **208**, 45 (1980).
9. A. Aro et al., *Am. J. Clin. Nutr.*, **39**, 911 (1984).
10. A. Lakdhar et al., *Br. Med. J.*, **296**, 1471 (1988).
11. S. A. Rajala et al., *Compr. Gerantol. A*, **2**, 83 (1988).
12. J. Tuomilehto et al., *Atherosclerosis*, **76**, 71 (1989).
13. M. Uusitupa et al., *Int. J. Clin. Pharmcol. Ther. Toxicol.*, **28**, 153 (1990).
14. P. R. Turner et al., *Atherosclerosis*, **81**, 145 (1990).
15. A. Aro et al., *Diabetologia*, **21**, 29 (1981).
16. U. Smith and G. Holm, *Atherosclerosis*, **45**, 1 (1982).
17. H. M. Mukhtar et al., *Indian J. Exp. Biol.*, **42**, 1212 (2004).
18. A. C. Frias and V. C. Sgarbieri, *Plant Foods Hum. Nutr.*, **53**, 15 (1998).
19. M. Krotkiewski, *Acta Med. Scand.*, **222**, 43 (1987).
20. P. A. Todd et al., *Drugs*, **39**, 917 (1990).
21. F. Lagier et al., *J. Allergy Clin. Immunol.*, **85**, 785 (1990).
22. A. Roesch et al., *Contact Dermatitis*, **52**, 307 (2005).
23. T. F. X. Collins et al., *Food Chem. Toxicol.*, **25**, 807 (1987).
24. K. M. Behall, *Diabetes Care*, **12**, 357 (1989).
25. B. N. Patel, *Drug Cosmet Ind.*, **95**, 337 (1964).
26. G. Meer et al., *Food Technol*, **29**, 22 (1975).
27. P. Kovacs and R. S. Igoe, *Food Prod. Dev.*, **10**, 32 (1976).
28. Anon., *Aust. Adv. Drug React. Bull.* (Aug. 1989).

GUARANA

Source: *Paullinia cupana* Kunth ex H.B.K. (syn. *P. sorbilis* (L.) Mart.) (Family Sapindaceae).

Common/vernacular names: Guarana paste, guarana gum.

GENERAL DESCRIPTION

A climbing evergreen liana native to South America in the Amazon region (e.g., Brazil and Venezuela); under cultivation it becomes a shrub up to 2 m high. Leaves alternate with five folioles; tendrils, if present, are axillary; inflorescences axillary racemes or originate on tendrils; fruits small, round, septicidal capsules, bright red to orangey-red in color, and grow in clusters; as it ripens the fruit splits and a black seed emerges, giving it the appearance of an eye. Guarana paste (also called gum) is prepared from the pulverized and roasted seeds by mixing with water to form a paste, which is then molded into bars and dried. Commercial production is in the middle Amazon in northern Brazil, with the city and county of Maués accounting for 80% of the world's supply. It is cultivated there for its seeds.[1]

CHEMICAL COMPOSITION

Contains usually 2.6–7% caffeine as its active constituent, together with traces of related alkaloids (theophylline, theobromine, xanthine, adenine, guanine, hypoxanthine, etc.). Other constituents include tannins (ca. 12%), *d*-catechin, starch (5–6%), fats (ca. 3%), resin (ca. 7%), saponins, mucilage, red pigment, and choline (KARRER; LIST AND HÖRHAMMER; MERCK).

The fixed oil of the seed contains cyanolipids (ca. 3%) and acylglycerols (ca. 28%),[2] while the essential oil contains at least nine identified components of which estragole and anethole are prominent.[3]

PHARMACOLOGY AND BIOLOGICAL ACTIVITIES

Guarana has stimulant and astringent properties due to its caffeine and tannins contents (also see *coffee* and *cocoa*).

The ethanolic extract of *P. cupana* has antimicrobial effects against Gram-positive and Gram-negative bacteria, as well as antioxidant activity at µg/mL levels. The antioxidant activity correlates with the levels of phenolic compounds and catechols in the extract.[4,5] The extract also displayed hepato- and gastroprotective effects when administered to mice and rats at doses below 100 mg/kg.[6–8]

Enhanced cognitive functions were observed in young human subjects after the administration of 75 mg of dried ethanolic extract of guarana. The authors suggested that the enhanced cognition was not due to the caffeine content of the extract.[9]

An aqueous extract of guarana decreased the aggregation of rabbit platelets, which may be due to its inhibition of thromboxane formation observed in the same experiment.[10]

TOXICOLOGY

Aqueous extracts of guarana were found to be genotoxic and mutagenic when tested in *Escherichia coli* and *Salmonella typhimurium*. The effect was attributed to the formation of a toxic caffeine/catechin complex.[11]

USES

Food. Guarana extract (especially fluid extract) is widely used as a flavor ingredient of cola drinks. It is also reportedly used in alcoholic beverages (e.g., liqueurs and cordials) and in candy. Use levels are generally below 0.002%, reported for guarana gum (paste). In Brazil, a carbonated soft drink made from the seeds is considered a NATIONAL beverage, offered commercially since 1909.

Dietary Supplements/Health Foods. Powdered seeds are used in tablets, capsules, combinations, chewing gum, and tea, primarily as a stimulant; also in weight loss formulations (DUKE 3; TYLER 1).

Traditional Medicine. Guarana paste is used by South American natives mainly as a stimulant, astringent, in treating chronic diarrhea. The seeds, grated into water are also used for fevers, heart problems, headache (associated with menstrual or rheumatic conditions), rheumatism, lumbago, migraine, and reduction of heat stress; diuretic (DUKE 2; DUKE 3).[1]

COMMERCIAL PREPARATION

Guarana seeds, guarana paste, and extracts.

Regulatory Status. Has been approved for food use (§172.510).

REFERENCES

See the General References for ARCTANDER; BRUNETON; DUKE 2; DUKE 3; FEMA; FURIA AND BELLANCA; MCGUFFIN 1 & 2; STAHL; TERRELL; TYLER 1; UPHOF.

1. H. T. Erickson et al., *Econ. Bot.*, **38**, 273 (1984).
2. P. Avato et al., *Lipids*, **38**, 773 (2003).
3. H. Benoni et al., *Z. Lebensm. Unters. Forsch.*, **203**, 95 (1996).
4. A. Basile et al., *J. Ethnopharmacol.*, **102**, 32 (2005).
5. R. Mattei et al., *J. Ethnopharmacol.*, **60**, 111 (1998).
6. H. Fukumasu et al., *Food Chem. Toxicol.*, **44**, 862 (2006).
7. H. Fukumasu et al., *Cancer Lett.*, **233**, 158 (2006).
8. A. R. Campos et al., *Phytother. Res.*, **17**, 1199 (2003).
9. D. O. Kennedy et al., *Pharmacol. Biochem. Behav.*, **79**, 401 (2004).
10. S. P. Bydlowski et al., *Braz. J. Med. Biol. Res.*, **24**, 421 (1991).
11. C. A. da Fonseca et al., *Mutat. Res.*, **321**, 165 (1994).

HAWTHORN

Source: *Crataegus laevigata* (Poir) DC (syn. *C. oxyacantha* L.), *C. monogyna* Jacq., *C. pinnatifida* Bge., *C. pinnattifida* var. major N.E. Br., *C. cuneata* Seib. Et Zucc. and other *Crataegus spp.* (Family Rosaceae).

Common/vernacular names: Bei shanzha or northern Chinese hawthorn (*C. pinnatifida*), *nan shanzha* or southern Chinese hawthorn (*C. cuneata*).

GENERAL DESCRIPTION

The genus *Crataegus* includes approximately 280 species primarily from northern temperate zones in East Asia (20), Europe (20) and Eastern North America (200+).

C. laevigata is a spiny shrub; leaves mostly smooth obovate, three to five lobed, serrulate; flowers white to pink with red anthers; fruit globose or ellipsoid, deep red; found in woods from northwest and central Europe, from England to Latvia, west to the Pyrenees and northern Italy; naturalized in eastern North America and India. Many smooth-leaved or slightly pubescent species in Europe have been treated as *C. oxyacantha*, a synonym for the official European source, resulting in much confusion in the botanical and pharmacy literature.

C. pinnatifida, *C. pinnatifida* var. *major*, and *C. cuneata* are source plants for the Chinese drug shanzha (dried hawthorn fruit). *Crataegus pinnatifida* is distributed in northeast China, Shanxi, and JIANGSU provinces. *Crataegus pinnatifida* var. *major* ("northern *shanzha*") is distributed in north and northeast China; supply mostly cultivated, in Shangdong, Henan, and Hebei. *Crataegus cuneata* "Southern *shanzha*," is grown the Changjian Valley in Guangdong and Guangxi (TUCKER AND LAWRENCE; JIANGSU).[1]

The part used is the dried fruits (China); or flowering tops, leaves, or fruits (Europe).

CHEMICAL COMPOSITION

C. laevigata and *C. monogyna* contain flavonoids, including hyperoside (hyperin), quercetin, vitexin, vitexin-4′-L-rhamno-D-glucoside, vitexin-4′-L-rhamnoside, vitexin-4′-7-di-D-glucoside, rutin, quercetin-3-rhamno-galactoside, 8-methoxykaempferol malonylglucoside, C-glycosides of acetylapigenin and acetyl vitexin, and others;[1–8] pentacyclic triterpenes (0.5–1.4% in fruits), including oleanolic acid, ursolic acid, acantolic acid, neotegolic acid, 2-α-hydroxyoleanolic acid (crataegolic acid);[9] xanthine derivatives such as adenosine, adenine, guanine, and uric acid;[1] amines, including ethylamine, dimethylamine, trimethylamine, isobutylamine, isoamylamine, ethanolamine, β-phenylethylamine, choline, acetylcholine, O-methoxyphenethylamine, tyramine, and others;[1,10] proanthocyanidins;[11] plus β-sitosterol, chlorogenic acid, caffeic acid, (+)-catechin, (−)-epicatechin, vitamins B_1, B_2, and C; calcium, iron, phosphorus, fructose, traces of an essential oil; and others (LIST AND HÖRHAMMER).[1]

Crataequinones A(1) and B(2) are two new bioactive furo-1,2-naphthoquinones that have recently been isolated from *C. pinnatifida*.[12]

PHARMACOLOGY AND BIOLOGICAL ACTIVITIES

The main effect of hawthorn is on the cardiovascular system. Pharmacological studies report enhanced coronary blood flow and myocardial perfusion;[13,14] improvement of cardiac muscle contractility;[14] increased left ventrical output velocity; lowering of blood pressure;[15–18] an antiarrhythmic effect;[19,20] increased myocardium tolerance to oxygen deprivation under hypoxic conditions; cardioprotective effect against myocardial infarction;[21,22] and stimulation of revascularization after myocardial ischemia (ESCOP 2).[23,24] Various clinical studies reveal efficacy in congestive heart failure;[25,26] increased cardiac performance; decrease in peripheral vascular resistance; decrease in pulmonary arterial and capillary pressures; reductions in blood pressure at rest and during exercise; and improved metabolic parameters (ESCOP 2).[27]

Compounds associated with cardiotonic activity include hyperoside, vitexin, vitexin-2′-rhamnoside, oligomeric procyanidins, and (−)-epicatechin. The flavonoids and oligomeric procyanidins have a tonic effect on the cardiac muscles, are negatively chronotropic and dromotropic, and also show the bradycardiac effect commonly noted for *Crataegus* (LIST AND HÖRHAMMER).[1] Flavonoids may also be the major constituents responsible for the hypolipidemic effect of hawthorn by regulating lipoprotein lipases expression.[28,29]

Similar activity has been reported for *C. pinnatifida*; also, oral administration of the fruit extract facilitates the clearance of serum cholesterol but does not prevent absorption (WANG).

Hawthorn extract or decoction is antibacterial against *Shigella flexneri*, *S. sonneni*, *Proteus vulgaris*, and *Escherichia coli* (WANG).

Antioxidant and free radical scavenging activities of the extracts of *C. pinnatifida* and *C. monogyna* have been reported.[5,30–33] This effect may play a role in the cardioprotective and hepatoprotective effects of hawthorn.[32–35]

Other reported activities of hawthorn and its preparations include stomachic, hypoglycemic, and analgesic/anti-inflammatory.[36–39] Many reviews on the pharmacology and therapeutic applications of hawthorn have been published recently.[40–43]

TOXICOLOGY

No toxic effects, contraindications, or drug interactions are known (ESCOP 2). A single case of immediate-type hypersensitivity to *C. monogyna* has been reported.[44]

USES

Medicinal, Pharmaceutical, and Cosmetic. Various drug preparations (oral or parenteral as i.m. or i.v. injections) are used in Europe for declining cardiac performance, corresponding to stages I and II of the New York Heart Association (NYHA) classification, senile heart conditions not requiring digitalis, and mild stable forms of angina pectoris, and mild forms of dysrythmia. Flowering tops are used in sleep-inducing preparations (ESCOP 2).

Food. Fruits of various hawthorn species have served as food in Europe, Asia, and at least a dozen species were used by American Indian tribes. Candied fruit slices, jam, jelly, and wine available in major American Chinatowns.

Dietary Supplements/Health Foods. Fruits (or flowers) used in tea, tablets, capsules, tinctures, and so on.[1]

Traditional Medicine. In traditional Chinese, medicine hawthorn fruits (dried, stir-fried, or charred) are used to stimulate digestion, promote function of the stomach, and stimulate blood circulation in epigastric distension, diarrhea, abdominal pain, amenorrhea, abdominal colic, indigestion, enteritis, acute bacillus dysentery, hypertension, hyperlipemia, and coronary heart disease. Charred fruits are used to promote digestion in stagnation of undigested meat, diarrhea, and with inadequate discharge from the bowels (TU). Up to 500 g of the fruits are eaten to treat tapeworm infections; externally as a wash for lacquer sores, itching, and frost bite.

In European tradition, the fruits, flowers, leaves, or a combination thereof reportedly used as astringent, antispasmodic, cardiotonic, diuretic, hypotensive, and antisclerotic (STEINMETZ).

American Indian groups reportedly used a poultice of the leaves for boils, sores, ulcers; root decoction a gastrointestinal aid, diuretic, and to increase circulation (MOERMAN).

COMMERCIAL PREPARATIONS

Crude, and extracts (e.g., powdered, solid, and liquid). Tablets and parenteral dosage forms with at least 5 mg of flavones (calculated as hyperoside), or 10 mg total flavonoid fraction (total phenols, calculated as hyperoside), or 5 mg oligomeric procyanidins (calculated as epicatechol).

Regulatory Status. Undetermined in the United States. Subject of a positive German therapeutic monograph,[9] and a proposed European Union monograph (ESCOP 2).

REFERENCES

See the General References for APPLEQUIST; BARNES; BARRETT; BLUMENTHAL 1 & 2; BISSET; CSIR II; ESCOP 2; HSU; JIANGSU; LIST AND HÖRHAMMER; MABBERLY; MARTINDALE; MCGUFFIN 1 & 2; MOERMAN; STEINMETZ; TU; TUCKER AND LAWRENCE; TYLER 1; WANG; WEISS; WREN.

1. C. Hobbs and S. Foster, *HerbalGram*, **22**, 19 (1990).
2. S. Rayyan et al., *Phytochem. Anal.*, **16**, 334 (2005).
3. P. C. Zhang and S. X. Xu, *J. Asian Nat. Prod. Res.*, **5**, 131 (2003).
4. P. C. Zhang et al., *J. Asian Nat. Prod. Res.*, **3**, 77 (2001).
5. Z. Zhang et al., *J. Nutr. Biochem.*, **12**, 144 (2001).
6. J. Fisel, *Arzniem.-Forsch.*, **15**, 1417 (1965).
7. P. Ficcara et al., *Farmaco Ed. Prat.*, **39**, 148 (1984).
8. M. Simova and T. Pangarova, *Pharmazie*, **38**, 791 (1983).
9. Monograph *Crataegus*, *Bundesanzeiger*, no. **85** (December 22, 1983); revised (May 5, 1988).
10. T. Hockerts and G. Mülke, *Arzniem.-Forsch.*, **5**, 755 (1955).
11. U. Svedstrom et al., *Phytochemistry*, **60**, 821 (2002).
12. B. Min and C. M. White, *Conn. Med.*, **68**, 161 (2004).
13. C. Roddewig and H. Hensel, *Arzniem.-Forsch.*, **27**, 1407 (1977).
14. F. Occhiuto et al., *Plant. Med. Phytother.*, **20**, 52 (1986).
15. Z. C. Kocyildiz et al., *Phytother. Res.*, **20**, 66 (2006).
16. S. H. Kim et al., *Life Sci.*, **67**, 121 (2000).
17. W. Stepka and A. D. Winters, *Lloydia*, **36**, 146 (1973).
18. H. P. T. Ammon and M. Handel, *Planta Med.*, **43**, 105, **209**, 313 (1981).
19. S. R. Long et al., *Phytomedicine* (2006).
20. S. Popping et al., *Arnziem.-Forsch.*, **45**, 1157 (1995).
21. R. Jayalakshmi and D. S. Niranjali, *J. Pharm. Pharmacol.*, **56**, 921 (2004).
22. M. Veveris et al., *Life Sci.*, **74**, 1945 (2004).
23. B. Gabhard and E. Schuler in N. Rietbrock et al., eds., *Wandlungen in der Therapie der Herzinsuffizienz*, Friedr. Vieweg & Soyn, Wiesbaden, 1983, p. 43.
24. J. Guendjev, *Arnziem.-Forsch.*, **27**, 1576 (1977).
25. F. H. Degenring et al., *Phytomedicine*, **10**, 363 (2003).
26. J. G. Zapfe, *Phytomedicine*, **8**, 262 (2001).
27. A. F. Walker et al., *Phytother. Res.*, **16**, 48 (2002).
28. C. Fan et al., *J. Pharmacol. Sci.*, **100**, 51 (2006).
29. Z. Zhang et al., *J. Nutr.*, **132**, 5 (2002).
30. C. Y. Chu et al., *J. Agric. Food Chem.*, **51**, 7583 (2003).
31. C. Quettier-Deleu et al., *Pharmazie*, **58**, 577 (2003).
32. T. Bahorun et al., *Arnziem.-Forsch.*, **46**, 1086 (1996).

33. Y. R. Dai et al., *Planta Med.*, **53**, 309 (1987).
34. C. J. Thirupurasundari et al., *J. Med. Food*, **8**, 400 (2005).
35. D. L. Zhang et al., *J. Neurochem.*, **90**, 211 (2004).
36. M. Fujisawa et al., *Am. J. Chin. Med.*, **33**, 167 (2005).
37. E. S. Kao et al., *J. Agric. Food Chem.*, **53**, 430 (2005).
38. H. Jouad et al., *J. Herb. Pharmacother.*, **3**, 19 (2003).
39. J. Vibes et al., *Prostaglandins Leukot. Essent. Fatty Acids*, **50**, 173 (1994).
40. W. T. Chang et al., *Am. J. Chin. Med.*, **33**, 1 (2005).
41. M. H. Pittler et al., *Am. J. Med.*, **114**, 665 (2003).
42. H. H. Fong and J. L. Bauman, *J. Cardiovasc. Nurs.*, **16**, 1 (2002).
43. J. M. Rigelsky and B. V. Sweet, *Am. J. Health Syst. Pharm.*, **59**, 417 (2002).
44. H. K. Steinman et al., *Contact Dermatitis*, **5**, 321 (1984).

HENNA

Source: *Lawsonia inermis* L. (syn. *L. alba* Lam.) (Family Lythraceae).

Common/vernacular names: Egyptian privet.

GENERAL DESCRIPTION

A glabrous much branched shrub, up to about 6 m high, lateral branches four gonus, often ending in a spinous point with opposite leaves, elliptic or broadly lanceolate; flowers many in paniculate cymes fragrant white or rose colored; generally considered a native of Africa and Asia; widely cultivated in tropical regions of the world (e.g., Egypt, Sudan, China, India, Florida, and the West Indies). Part used is the dried leaf. Major producing countries include Sudan, Egypt, and India.

CHEMICAL COMPOSITION

Contains 0.55–1.0% lawsone (2-hydroxy-1,4-naphthoquinone);[1,2] 1,4-naphthoquinone;[3] 2-methoxy-3-methyl-1,4-naphthoquinone;[4] flavonoids, coumarins, and phenolic acids;[4] 5–10% gallic acid and tannin; about 11% sugars; resin; and others (LIST AND HÖRHAMMER).[2,3]

Two xanthones (laxanthone-I and laxanthone-II) and a substituted coumarin named lacoumarin (5-allyloxy-7-hydroxycoumarin) have been isolated from the whole plant and are probably present in the leaves.[5,6] The aerial parts also contain the triterpenes lawsonin (3-α-*E*-ferulyloxy-urs-11-en-13-β-ol) and lawsonic acid (3-α-*E*-ferulyloxy-lup-20(29)-en-28-oic acid).[7]

Lawsone is the major active principle (coloring and pharmacological) in henna. It is not present in the bark, stem, or root of the henna plant. Its concentrations in the leaves vary with climatic conditions; hot localities yield henna with higher lawsone content than temperate areas.[1] There has been a report indicating lawsone to be a degradation and autoxidation product of primary glycosides called hennosides A, B, and C.[8]

PHARMACOLOGY AND BIOLOGICAL ACTIVITIES

Lawsone has various biological activities, including antifungal (fungicidal and fungistatic) activities toward *Alternaria*, *Aspergillus*, *Absidia*, *Penicillium*, and other species, being effective at 0.1% (1000 ppm) concentrations;[9–11] antibacterial activities toward *Brucella*, *Neisseria*, *Staphylococcus*, *Salmonella*, *Streptococcus*, and others, with a concentration of 0.005–0.02% (50–200 μg/mL) being effective against the first two groups;[1,3,12] antitumor activities (e.g., against sarcoma 180 in mice, Walker 256 carcinosarcoma in rats, and human hepatoma cell lines);[12–14] and antispasmodic properties as well as weak vitamin K activity.[1,2]

In addition to the antibacterial properties of lawsone, crude henna extracts and fractions containing gallic acid, and 1,4-naphthoquinone have also exhibited antibacterial activities.[3,9]

Ethanolic extracts containing luteolin, β-sitosterol, and lawsone displayed anti-inflammatory, antihyaluronidase, analgesic, and antipyretic activities.[9,15–17]

Antioxidant, hepatoprotective, and immunomodulatory activities have also been reported for henna extracts.[4,15,18,19]

Henna leaves (but not the seeds) have been reported to exhibit antifertility activity in female rats.[20] At least two reports, however, indicated that henna was nongenotoxic to mouse bone marrow and did not produce DNA damage in Chinese hamster ovary cells.[21,22] The major adverse reactions reported for henna after topical application are complicated skin allergy (severe dermatitis, eczema, etc.) and less frequent cases of hemolytic anemia in children, but this has been attributed to the synthetic *p*-phenylenediamine (PPD) added to many henna tattoo/dye products to impart a darker shade.[23–37]

USES

Medicinal, Pharmaceutical, and Cosmetic. Henna is used in numerous hair care products (e.g., dyes, conditioners, rinses, etc.). Prolonged use of henna on the hair would turn the hair orange red, unless henna is mixed with other dyes such as indigo and logwood to obtain different shades. To obtain a long-lasting color, the henna preparation must be rendered slightly acid (ca., pH 5.5) by adding a weak acid (e.g., citric, boric, or adipic).

Traditional Medicine. Leaves have been extensively used for centuries in the Middle East, the Far East, and northern Africa as a dye for nails, hands, hair, clothing, and others; they are also used in treating skin problems, headache, jaundice, amebiasis, enlargement of the spleen, and cancers, among others.[1,38]

Others. Lawsone can be used as an acid–base indicator for the titration of strong acids with weak bases.[39]

COMMERCIAL PREPARATIONS

Mainly the crude.

Regulatory Status. Has been approved for use as a color additive exempt from certification, to be used in cosmetics (hair) only (§73.2190).

REFERENCES

See the General References for GUPTA; BALSAM AND SAGARIN; DUKE 4; BRUNETON; JIANGSU; LUST; GRIEVE; MARTINDALE; MCGUFFIN 1 & 2; MERCK; MORTON 2; ROSE; TERRELL; UPHOF; WREN.

1. M. S. Karawaya et al., *Lloydia*, **32**, 76 (1969).
2. A. Latif, *Indian J. Agric. Sci.*, **29**, 147 (1959).
3. Y. Abd-el-Malek et al., *Zentralbl. Bakteriol., Parasitenk., Infektionskr. Hyg., Abt. 2*, **128**, 61 (1973).
4. B. R. Mikhaeil et al., *Z. Naturforsch. C*, **59**, 468 (2004).
5. D. K. Bhardwaj et al., *Phytochemistry*, **16**, 1616 (1977).
6. D. K. Bhardwaj et al., *Phytochemistry*, **15**, 1789 (1976).

7. B. S. Siddiqui and M. N. Kardar, *Phytochemistry*, **58**, 1195 (2001).
8. G. J. Kapadia et al., *Lloydia*, **32**, 523 (1969).
9. O. A. Habbal et al., *Saudi. Med. J.*, **26**, 69 (2005).
10. R. D. Tripathi et al., *Experientia*, **34**, 51 (1978).
11. N. R. Farnsworth and G. A. Cordell, *Lloydia*, **39**, 420 (1976).
12. H. Kamei et al., *Cancer Biother. Radiopharm.*, **13**, 185 (1998).
13. H. Babich and A. Stern, *J. Appl. Toxicol.*, **13**, 353 (1993).
14. O. Goncalves de Lima et al., *Rev. Inst. Antibiot., Univ. Fed. Pernambuco, Recife*, **11**, 21 (1971).
15. B. H. Ali et al., *Pharmacology*, **51**, 356 (1995).
16. A. Gupta et al., *Indian J. Pharmacol.*, **18**, 113 (1986).
17. S. B. Vohara and P. C. Dandiya, *Fitoterapia*, **63**, 195 (1992).
18. M. Bhandarkar and A. Khan, *Indian J. Exp. Biol.*, **41**, 85 (2003).
19. T. Das Gupta et al., *Mol. Cell. Biochem.*, **245**, 11 (2003).
20. S. R. Munshi, *Planta Med.*, **31**, 73 (1977).
21. D. Kirkland and D. Marzin, *Mutat. Res.*, **537**, 183 (2003).
22. D. Marzin and D. Kirkland, *Mutat. Res.*, **560**, 41 (2004).
23. R. Wolf et al., *Skinmed*, **5**, 39 (2006).
24. A. N. Kok et al., *J. Emerg. Med.*, **29**, 343 (2005).
25. R. P. Eager, *Eur. J. Emerg. Med.*, **12**, 189 (2005).
26. J. Matulich and J. Sullivan, *Contact Dermatitis*, **53**, 33 (2005).
27. H. Prosen et al., *Arh. Hig. Rada Toksikol.*, **56**, 1 (2005).
28. I. A. Bukhari, *Saudi. Med. J.*, **26**, 142 (2005).
29. A. M. Nawaf et al., *J. Dermatol.*, **30**, 797 (2003).
30. J. R. Pegas et al., *J. Investig. Allergol. Clin. Immunol.*, **12**, 62 (2002).
31. W. H. Chung et al., *Int. J. Dermatol.*, **40**, 754 (2001).
32. R. R. Brancaccio et al., *Am. J. Contact Dermat.*, **13**, 15 (2002).
33. P. Raupp et al., *Arch. Dis. Child*, **85**, 411 (2001).
34. S. Lauchli and S. Lautenschlager, *Swiss. Med. Wkly.*, **131**, 199 (2001).
35. J. C. Garcia Ortiz et al., *Int. Arch. Allergy Immunol.*, **114**, 298 (1997).
36. H. H. Kandil et al., *Ann. Trop. Paediatr.*, **16**, 287 (1996).
37. I. M. Majoie and D. P. Bruynzeel, *Am. J. Contact Dermat.*, **7**, 38 (1996).
38. J. L. Hartwell, *Lloydia*, **33**, 97 (1970).
39. K. C. Joshi et al., *Z. Naturforsch. B*, **32B**, 890 (1977).

HONEY

Source: Sugar secretions collected in honeycomb by honeybees (*Apis* spp.).

Common/vernacular names: Purified honey, mel depuratum, clarified honey, and strained honey.

GENERAL DESCRIPTION

Sugar secretions collected and stored in honeycomb by Apis *mellifera L.*, and other Apis species (Family Apidae). A thick, syrupy, transparent liquid, honey is extracted from bee hives, then strained through a sieve and allowed to sit in settling tanks for 24 h to allow air bubbles to rise to the surface. Depending on pasturage source, color varies from amber to reddish brown to black.

CHEMICAL COMPOSITION

Honey consists chiefly of dextrose and levulose (70–80%) with smaller amount of water, sucrose (2–10%), dextrin, wax, proteins, volatile oil, minerals, acids, and coloring and flavoring components, based on derivative plant source; contains vitamin B_1, vitamin B_2, vitamin C, nicotinic acid, and formic acid (CSIR I; MARTINDALE).

PHARMACOLOGY AND BIOLOGICAL ACTIVITIES

Antibacterial, generally attributed to production of locally high osmolality due to water activity of honey; antimicrobial activity may be produced enzymatically by liberating hydrogen peroxide. Individual source plants may also contribute to antimicrobial activity. An earlier assessment of 26 honeys in New Zealand found that of *Leptospermacm scoparium* to have high antibacterial activity against *Escherichia coli* and *Staphylococcus aureus*; weaker against *Streptococcus pyogenes*, and *Salmonella typhimurium*.[1] A recent study on stored Egyptian honey also showed that it was more effective against *E. coli* than *S. typhimurium* both *in vitro* and *in vivo* and that the antibacterial activity decreased with longer storage times.[2]

TOXICOLOGY

Spores of *Clostridium botulinum*, responsible for infant botulism, are often contained in honey, which may germinate in adults without adverse effects, but may cause serious illness in infants. In 1976, of 43 cases of infant botulism in California, 13 involved honey (*C. botulinum* found in 13% of 60 tested samples). It has been recommended that honey not be given to infants under 1 year old.[3]

USES

Medicinal, Pharmaceutical, and Cosmetic. Demulcent and sweetener in cough mixtures; in China as pill binder.

Honey is used as a fragrance ingredient and humectant (in skin conditioners), also a biological additive in shampoos; face, body, and hand creams and lotions; bath products, hair conditioners; cleansing products; moisturizing creams and lotions; and paste masks (mud packs) (NIKITAKIS).

Food. Raw honey or honey in honeycomb sections (in beeswax) used as a sweetener.

Dietary Supplements/Health Foods. A Vermont physician, D. C. Jarvis, catapulted honey into the health food realm by claiming that it improved digestion; facilitated wound healing; and had sedative, antiarthritic, and antibacterial effects (TYLER 1). The primary use, however, is as a sweetener for herb teas.

Traditional Medicine. In Indian folk medicine, honey is considered demulcent; used for the treatment of eye ailments; sore throat, coughs, colds, and constipation. Heated or mixed with hot liquids, it is used in conjunction with emetic and enema formulations.

In Chinese medicine, it is used as a nutritive, demulcent, emollient laxative, in cough due to lung dryness, constipation, stomachache, sinusitis, oral ulcerations, scalds, and as a detoxicant for aconitine (HSU).

Traditionally considered useful topically for indolent skin ulcers where antibiotics fail to achieve results. Has also been used to soothe dermal lesions and necrotic malignant breast ulcers.

COMMERCIAL PREPARATIONS

Raw honey, honey in beeswax comb.

Regulatory Status. A common food; GRAS.

REFERENCES

See the General References for CSIR I; HSU; MARTINDALE; NIKITAKIS; TYLER 1.

1. D. Greenwood, *Lancet*, **341**, 191 (1993).
2. O. F. Badawy et al., *Rev. Sci. Tech.*, **23**, 1011 (2004).
3. R. A. Mangione, *Am. Pharm.*, **NS23**, 5 (1983).

HOPS

Source: *Humulus lupulus* L. (Family Moraceae or Cannabaceae).

Common/vernacular names: European hops, common hops.

GENERAL DESCRIPTION

A twinning perennial herb with male and female flowers on separate plants (dioecious); up to about 8 m high; native to Eurasia and North America; extensively cultivated worldwide (e.g., the United States, Germany, and the Czech Republic). Part used is the female membranous cone-like inflorescence (strobile) with its glandular hairs, collected in the fall and carefully dried, often bleached with sulfur dioxide from burning sulfur; an essential oil is obtained by steam distillation of the freshly dried cones. The glandular hairs, separated from the strobiles, compose lupulin, which contains more resins and volatile oil than hops and is also used like hops.

Major producers of hops include the United States, Germany, and the Czech Republic.

CHEMICAL COMPOSITION

Contains 0.3–1% volatile oil; 3–12% resinous bitter principles composed of α-bitter acids (humulone, isohumulone, cohumulone, isocohumulone, adhumulone, prehumulone, posthumulone, and so on, with first three in predominance) and β-bitter acids (lupulone, colupulone, adlupulone, etc., in decreasing concentration); other resins, some of which are oxidation products of the α- and β-acids; chalcones (xanthohumol and desmethylxanthohumol);[1] prenylflavones and flavanones;[1–3] flavonoid glycosides (astragalin, quercitrin, isoquercitrin, rutin, kaempferol-3-rutinoside, etc.); phenolic acids; acylphloroglucinols;[4] tannins; sitosterol; lipids; amino acids; estrogenic substances; and many other compounds (KARRER; JIANGSU; LIST AND HÖRHAMMER).[3,5–8] Proanthocyanidins containing catechin and epicatechin monomers were identified in the extract.[9] Also, (*p*-methoxyphenyl) diphenylmethanol and tribenzylamine have been identified from hops as a natural source for the first time.[10]

The volatile oil is made up mostly of humulene (α-caryophyllene), myrcene, β-caryophyllene, and farnesene, which together may account for more than 90% of the oil.[5] There are more than 100 other compounds present, including germacratriene, α- and β-selinenes, selina-3,7(11)-diene, selina-4(14),7(11)-diene, α-copaene, α- and β-pinenes, limonene, *p*-cymene, linalool, nerol, geraniol, nerolidol, citral, methylnonyl ketone, other oxygenated compounds, 2,3,4-trithiapentane (present only in oil of unsulfured hops in ca. 0.01%), *S*-methylthio-2-methylbutanoate, *S*-methylthio-4-methylpentanoate, and 4,5-epithiocaryophyllene.[5,11–16]

PHARMACOLOGY AND BIOLOGICAL ACTIVITIES

Hops extracts have been reported to have various biological activities, including antimicrobial activities, which are due to the bitter acids (especially lupulone and humulone), the more hydrophobic ones being the more active.[17–20] The essential oils and extracts of different cultivars and varieties were active against Gram-positive bacteria and the fungus *Trichophyton mentagrophytes*.[21] Xanthohumol was also found to be active against Gram-positive bacteria, *T. mentagrophytes*, HSV-1 and -2, HIV-1 and the malaria parasite *Plasmodium falciparum*.[22–24] Alcoholic extracts of hops in various dosage forms have been used clinically in treating numerous forms of leprosy, pulmonary tuberculosis, and acute bacterial dysentery, with varying degrees of success (JIANGSU).

Hops has strong spasmolytic effects on isolated smooth muscle preparations;[25] hypnotic, sedative, and antidepressant effects,[26] supported by the finding that 2-methyl-3-butene-2-ol (present in hops up to 0.15%) has sedative effects in rats;[27,28] estrogenic[29–32] and recently reported antiestrogenic properties due to the observed *in vitro* inhibitory effect of hops prenylflavonoids on estrogen synthase.[33] Allergenic activity has been observed in humans, causing contact dermatitis due to the pollen (JIANGSU; LIST AND HÖRHAMMER; MERCK). A stimulant effect on gastric secretion has been demonstrated for hops in laboratory animals.[34]

The prenylflavonoids of hops, including xanthohumol, have displayed antiproliferative antitumor activities against a number of human cancer cell lines, such as prostate, breast, ovarian, and colon cancers.[1,35–37] Induction of apoptosis and inhibition of NF-κB may be involved in the antitumor activity.[35] Cancer chemopreventive activities of the prenylflavonoids have also been reported.[38,39]

Other recently reported activities of hops include anti-inflammatory (inhibition of COX-1 and COX-2 and iNOS expression);[4,40–42] and hypoglycemic (improved insulin sensitivity) activities.[8]

USES

Medicinal, Pharmaceutical, and Cosmetic. Extracts are used in certain skin creams and lotions, especially in Europe, for their alleged skin-softening properties.

Food. Major use is in beer, with the bitter taste derived primarily from oxidation products of humulone. Extracts and oil are also used as flavor components in nonalcoholic beverages, frozen dairy desserts, candy, baked goods, and gelatins and puddings, with the highest average maximum use level of 0.072% reported for an extract (type not indicated) in baked goods.

Dietary Supplements/Health Foods. Used in sleeping preparations. Cut strobiles, powdered, or dried extract powder for tea, tincture, capsules, tablets, and so on. Also used in "dream pillows," to promote sleep; and bath preparations (FOSTER). Use in medicinal bath preparations in Germany has been disallowed.[43]

Traditional Medicine. Used as a diuretic and anodyne and in treating nervous diarrhea, insomnia, restlessness, and other nervous conditions as well as intestinal cramps and lack of appetite, among others, usually in the form of a tea; also used in Chinese medicine for pulmonary tuberculosis and cystitis (JIANGSU).

It has reportedly been used in cancers.[44]

COMMERCIAL PREPARATIONS

Crude, extracts, and oil. Crude was formerly official in N.F., and oil is official in F.C.C. Strengths (see *glossary*) are either expressed in flavor intensities or in weight-to-weight ratios.

Regulatory Status. GRAS with both hops and lupulin listed (§182.20). The strobiles are the subject of a German therapeutic monograph, with use approved for mood disturbance (unrest, anxiety) and sleep disturbances. The proposed ESCOP monograph indicated usage for nervous tension, excitability, restlessness, sleep disturbances, and lack of appetite (ESCOP 2).

REFERENCES

See the General References for ARCTANDER; BAILEY 2; BARNES; BIANCHINI AND CORBETTA; BLUMENTHAL 1; DER MARDEROSIAN AND BEUTLER; ESCOP 2; FEMA; FOSTER; FOSTER AND DUKE; GRIEVE; GUENTHER; HUANG; JIANGSU; LEWIS AND ELVIN-LEWIS; LIST AND HÖRHAMMER; LUST; MCGUFFIN 1 & 2; TERRELL; UPHOF; WREN; YOUNGKEN.

1. L. Delmulle et al., *Phytomedicine* (2006).
2. F. Zhao et al., *J. Nat. Prod.*, **68**, 43 (2005).
3. L. R. Chadwick et al., *J. Nat. Prod.*, **67**, 2024 (2004).
4. G. Bohr et al., *J. Nat. Prod.*, **68**, 1545 (2005).
5. A. Mijavec and P. Spevok, *Bilt. Hmelj. Sirak.*, **7**, 23 (1975).
6. M. Anguelakova et al., *Riv. Ital. Essenze, Profumi, Piante Offic., Aromi, Saponi, Cosmet., Aerosol*, **53**, 275 (1971).
7. A. Strenkovskaya, *Mezhdunar. Kongr. Efirnym Maslam (Mater.)*, **4**, 325 (1971).
8. H. Yajima et al., *J. Biol. Chem.*, **279**, 33456 (2004).
9. A. W. Taylor et al., *J. Agric. Food Chem.*, **51**, 4101 (2003).
10. Y. Qu et al., *Z. Naturforsch. C*, **58**, 640 (2003).
11. R. G. Buttery and L. C. Ling, *Brewers Dig.*, **8**, 71 (1966).
12. R. G. Buttery and L. C. Ling, *J. Agric. Food Chem.*, **15**, 531 (1967).
13. R. D. Hartley and C. H. Fawcett, *Phytochemistry*, **8**, 1793 (1969).
14. R. G. Buttery et al., *Chem. Ind. (London)*, **28**, 1225 (1966).
15. A. S. Morieson, *Inst. Brewing (Austral. Sect). Proc. Conv.*, **6**, 102 (1960).
16. T. L. Peppard and F. R. Sharpe, *Phytochemistry*, **16**, 2020 (1977).
17. M. Teuber and A. F. Schmalreck, *Arch. Mikrobiol.*, **94**, 159 (1973).
18. Y. Kuroiwa et al., *Jpn. Kokai.*, 73 58, 114 (1973).
19. A. F. Schmalreck et al., *Can. J. Microbiol.*, **21**, 205 (1975).
20. J. Boatwright, *J. Inst. Brew.*, **82**, 334 (1976).
21. C. R. Langezaal et al., *Pharm. Weekbl. Sci.*, **14**, 353 (1992).
22. C. Gerhauser, *Mol. Nutr. Food Res.*, **49**, 827 (2005).
23. S. Frolich et al., *J. Antimicrob. Chemother.*, **55**, 883 (2005).
24. Q. Wang et al., *Antiviral Res.*, **64**, 189 (2004).
25. F. Caujolle et al., *Agressologie*, **10**, 405 (1969).
26. P. Zanoli et al., *J. Ethnopharmacol.*, **102**, 102 (2005).
27. R. Hänsel et al., *Planta Med.*, **45**, 224 (1982).
28. R. Wohlfart et al., *Planta Med*, **48**, 120 (1983).
29. C. R. Overk et al., *J. Agric. Food Chem.*, **53**, 6246 (2005).
30. H. Rong et al., *Eur. J. Cell. Biol.*, **80**, 580 (2001).
31. S. R. Milligan et al., *J. Clin. Endocrinol. Metab.*, **85**, 4912 (2000).
32. S. R. Milligan et al., *J. Clin. Endocrinol. Metab*, **84**, 2249 (1999).
33. R. Monteiro et al., *J. Agric. Food Chem.*, **54**, 2938 (2006).
34. St. Tamasdan, *Farmacia (Bucharest)*, **29**, 71 (1985).
35. E. C. Colgate et al., *Cancer Lett.* (2006).
36. B. Vanhoecke et al., *Int. J. Cancer*, **117**, 889 (2005).
37. C. L. Miranda et al., *Food Chem. Toxicol.*, **37**, 271 (1999).
38. B. M. Dietz et al., *Chem. Res. Toxicol.*, **18**, 1296 (2005).

39. C. L. Miranda et al., *Drug Metab Dispos.*, **28**, 1297 (2000).
40. S. Hougee et al., *Planta Med.*, **72**, 228 (2006).
41. M. Lemay et al., *Asia Pac. J. Clin. Nutr.*, **13**, S110 (2004).
42. F. Zhao et al., *Biol. Pharm. Bull.*, **26**, 61 (2003).
43. Monograph *Lupuli strobulus*, Bundesanzeiger, no. 228 (December 5, 1984); revised (March 13, 1990).
44. J. L. Hartwell, *Lloydia*, **33**, 97 (1970).

HOREHOUND

Source: *Marrubium vulgare* L. (Family Labiatae or Lamiaceae).

Common/vernacular names: Marrubium, hoarhound, common hoarhound, and white horehound.

GENERAL DESCRIPTION

A perennial aromatic shrub with hairy stems; leaves round-ovate gray wooly below crenate dentate, rough and veiny above, 1 or 2 in. in diameter; flowers small, white, in sessile, axillary, hairy dense whorls; up to about 1 m high; four seeded, in the base of the calyx; native to Europe and Asia; naturalized in North America. Parts used are the dried leaves and flowering tops.

Ballota nigra L. and its six subspecies, traded (primarily in Europe) as black horehound, have similar indications to horehound, though use is limited because of extreme bitter taste (WREN).

CHEMICAL COMPOSITION

Contains 0.3–1% of a bitter principle called marrubiin (a diterpene lactone);[1,2] several diterpene alcohols (e.g., marrubiol, marrubenol, peregrinol, and vulgarol);[1,3,4] small amounts of alkaloids (ca. 0.3% betonicine, its stereoisomer turicine, and ca. 0.2% choline);[5] trace of a volatile oil containing monoterpenes (α-pinene, camphene, limonene, sabinene, *p*-cymene, etc.) and a sesquiterpene;[6] C_{27} to C_{34} alkanes (normal and branched);[7] free phytol;[4] and tannin, pectic substances, saponin, resin, β-sitosterol, and others (LIST AND HÖRHAMMER).[8] Phenylpropanoid esters (e.g., caffeoyl malic acid) and glycosides (e.g., acteoside, forsythoside, arenarioside, and ballotetroside), as well as a phenylethanoid glycoside (marrubosied) have been isolated from the aerial parts.[9,10]

PHARMACOLOGY AND BIOLOGICAL ACTIVITIES

Marrubiin is reported to have expectorant properties, and when its lactone ring is opened, the resulting acid (marrubic acid) has strong choleretic activity (TYLER 1). It also has a normalizing effect on extrasystolic arrythmias but in large doses disturbs the heart rhythm, which can be counteracted by atropine (LIST AND HÖRHAMMER). Orally administered aqueous extract has antihypertensive effect in rats.[11,12] This effect appears to be linked to marrubenol and to be mediated through inhibition of aortic smooth muscle relaxation resulting from calcium channel blocking.[13,14] Another mechanism by which the extract, and the phenylpropanoids therein, can lower blood pressure is through the inhibition of endothelin-1 secretion leading to reduced vasoconstriction and retarded development of atherosclerosis.[15]

The phenylpropanoid glycosides isolated from horehound have COX inhibitory activity that is more selective to COX-2.[10] Marrubiin was also found to exhibit a dose-dependent antinociceptive effect in mice.[16]

Horehound leaf extract has an antiproliferative and apoptosis-inducing effect on human colorectal cancer cells.[17]

Brazilian horehound extract was reported to possess a hypoglycemic activity in diabetic rats.[18] However, the same effect was not significantly displayed in humans.[19]

The volatile oil of horehound has also been reported to have vasodilatative and expectorant properties.[8]

An aqueous extract of horehound has been reported to be antagonistic toward serotonin *in vitro* and was anti-inflammatory in the rat paw edema assay (WREN).[20]

Weak antioxidant activity is linked to total hydroxycinnamic acid derivatives (0.6%).[21]

USES

Medicinal, Pharmaceutical, and Cosmetic. Extracts used as an expectorant in cold and cough medicines, especially in Europe.

Food. Extracts are used in flavoring numerous food products, especially candy and alcoholic beverages (e.g., bitters and liqueurs). Other foods in which horehound extracts are used include nonalcoholic beverages, frozen dairy desserts, baked goods, and gelatins and puddings. Highest average maximum use level is 0.073% reported in candy.

Dietary Supplements/Health Foods. Herb used as a minor (bitter) flavoring component in some tea formulations; also in confectionaries intended to have a soothing effect on coughs and colds (FOSTER).

Traditional Medicine. It is used in treating sore throat, colds, coughs, and other respiratory ailments; also used as a diuretic, bitter tonic, and diaphoretic as well as in treating cancers.[22]

COMMERCIAL PREPARATIONS

Crude and extracts. Crude was formerly official in U.S.P. Strengths (see *glossary*) of extracts are expressed in flavor intensities or weight-to-weight ratios.

Regulatory Status. GRAS (§182.10 and §182.20). The leaves are the subject of a German therapeutic monograph, with approval of use of herb and extracts for coughs, colds, and as digestive aid and appetite stimulant.[23] Declared ineffective as an OTC cough suppressant and expectorant by the FDA (TYLER 1).

REFERENCES

See the General References for APPLEQUIST; BAILEY 1; BARNES; BIANCHINI AND CORBETTA; BLUMENTHAL 1; DER MARDEROSIAN AND BEUTLER; FEMA; FOSTER; GOSSELIN; KROCHMAL AND KROCHMAL; LEWIS AND ELVIN-LEWIS; LUST; MCGUFFIN 1 & 2; TERRELL; TYLER 1; UPHOF; WREN; YOUNGKEN.

1. D. P. Popa et al., *Khim. Prir. Soedin.*, **4**, 345 (1968).
2. H. J. Nicholas, *J. Pharm. Sci.*, **53**, 895 (1964).
3. D. P. Popa and G. S. Pasechnik, *Khim. Prir. Soedin.*, **11**, 722 (1975).
4. D. P. Popa and L. A. Salei, *Rast. Resur.*, **9**, 384 (1973).
5. W. W. Paulder and S. Wagner, *Chem. Ind. (London)*, **42**, 1693 (1963).
6. M. O. Karryev et al., *Izv. Akad. Nauk Turkm. SSR, Ser. Biol.*, **3**, 86 (1976).
7. C. H. Brieskorn and K. Feilner, *Phytochemistry*, **7**, 485 (1968).
8. I. M. Baratarelli, *Boll. Chim. Farm.*, **105**, 787 (1966).
9. S. Sahpaz et al., *Nat. Prod. Lett.*, **16**, 195 (2002).
10. S. Sahpaz et al., *J. Ethnopharmacol.*, **79**, 389 (2002).

11. S. El-Bardai et al., *Clin. Exp. Hypertens.*, **26**, 465 (2004).
12. S. El-Bardai et al., *Clin. Exp. Hypertens.*, **23**, 329 (2001).
13. S. El-Bardai et al., *Br. J. Pharmacol.*, **140**, 1211 (2003).
14. S. El-Bardai et al., *Planta Med.*, **69**, 75 (2003).
15. F. Martin-Nizard et al., *J. Pharm. Pharmacol.*, **56**, 1607 (2004).
16. R. A. De Jesus et al., *Phytomedicine*, **7**, 111 (2000).
17. K. Yamaguchi et al., *Oncol. Rep.*, **15**, 275 (2006).
18. A. P. Novaes et al., *Therapie*, **56**, 427 (2001).
19. A. Herrera-Arellano et al., *Phytomedicine*, **11**, 561 (2004).
20. R. Cahen, *C. R. Soc. Biol.*, **164**, 1467 (1970).
21. J. L. Lamaison et al., *Fitoterapia*, **62**, 166 (1991).
22. J. L. Hartwell, *Lloydia*, **32**, 247 (1969).
23. Monograph *Marrubii herba*, Bundesanzeiger, no. 22 (February 1, 1990).

HORSE CHESTNUT

Source: *Aesculus hippocastanum* L. (Family Hippocastanaceae).

GENERAL DESCRIPTION

Deciduous tree up to 25 m; leaves opposite, digitate with five to seven obovate, irregularly crenate–serrate leaflets; glabrous above, tomentose beneath. Flowers white, with yellow to pink spot at base, in large cylindrical panicle. Spiny globose fruits 2–6 cm in diameter with large brown smooth seed, 2–4 cm in diameter; found in mountain woods, indigenous to central Balkan peninsula, widely planted and established throughout the northern hemisphere as a shade and ornamental tree (TUTIN 2). The parts used are the seed, branch bark, and leaves.

CHEMICAL COMPOSITION

The seeds and bark contain a mixture of triterpene saponins known as aescin (escins I–VI and isoescins I–V),[1,2] composed of acylated glycosides of protoeasigenin and barringtogenol-C, hippoaesculin and others;[3] quinones, including plastoquinone 8; flavones, including 3,5-dihydroxy-3′,4′,7-trimethoxyflavone; myricetin 3′,4′,7-trimethyl ether; a catechin dimer (proanthocyanidin-A2);[4] sterols, including stigmasterol, α-spinasterol, and β-sitosterol; linolenic, palmitic, and stearic acids; and others. The glycoside aesculin (esculin) (7-hydroxycoumarin 6-β-glucoside) is considered the most toxic component of the seed (GLASBY 2; WREN).[5]

PHARMACOLOGY AND BIOLOGICAL ACTIVITIES

Horse chestnut extracts, notably aescin, have anti-inflammatory, antiedematous, antiexudative, and venotonic activities. The clinical pharmacology of horse chestnut and aescin has been the subject of many recent reviews.[6–10]

Aescin was found to be responsible for the antiexudative and edema-protective activity.[11] Aescin acts on the capillary membrane, normalizing vascular permeability, enhancing capillary resistance, and reducing the outflow of fluid into the extracapillary space.[12] Aescin has a "sealing" venotonic effect on the capillaries and reduces the number and diameter of the small pores of the capillary wall by which exchange of water occurs. *In vitro*, aescin has been found to stimulate an increase in venous tone, with a decrease in the volume of venous district of the saphenous vein and its collaterals, facilitating return blood flow to the heart.[5,13] Efficacy of a 2%-aescin-containing gel in reducing tenderness of experimentally induced hematoma has recently been confirmed.[14] Anti-inflammatory activity at the initial exudation phase of inflammation has been confirmed in various *in vitro* and *in vivo* models, such as the perfused canine saphenous vein and the rat paw edema test.[15–18] The efficacy of aescin-rich horse chestnut extract in the treatment of leg ulceration and chronic venous insufficiency in humans has repeatedly been demonstrated.[19–22] Such efficacy was shown to be comparable to the use of leg compression stocking.[23] The efficacy of horse chestnut is probably due to a combination of anti-inflammatory, antioxidant, antielastase, and antihyaluronidase activities.[18,24]

Escins Ia, Ib, IIa, IIb, and IIIa have demonstrated a hypoglycemic effect in rats;[25] while escin IIb was shown to inhibit gastric emptying in mice.[26]

In vitro antitumor activity has been observed from hippoaesculin and barringtogenol-C-2l-angelate.[5,27]

TOXICOLOGY

Horse chestnut seeds are considered inedible and poisonous. The bitter flavor prevents consumption of large amounts. The leaves, flowers, young sprouts, and seeds are toxic. Symptoms of poisoning include nervous muscle twitching, weakness, dilated pupils, vomiting, diarrhea, depression, paralysis, and stupor (HARDIN AND ARENA). Incidents of anaphylactic shock after i.v. injections of horse chestnut have been reported, along with renal toxicity or failure (FROHNE AND PFANDER).[5] Horse chestnut pollens are also a common cause of allergic condition in urban children.[28]

USES

Medicinal, Pharmaceutical, and Cosmetic. Horse chestnut extract or aescin (0.25–0.5%) has reportedly been used in shampoos, shower foams, foam baths, skin care products, body and hand creams, lotions, and toothpastes. Cosmetic use in Europe has been based on its clearing and redness reducing properties, and its effectiveness in preventing cellulitis.[29]

Numerous clinical studies and published case reports confirm the efficacy of aescin-containing topical products, especially in the treatment of sport injuries, including blunt trauma of the lower limbs,[30] joint sprains, tendonitis, hematomas, muscle strain, traumatic edema,[31] Achilles' tendonitis; surgical outpatient trauma, including fractures, sprains, crush injuries, and contusions;[32] post-operative or postpartum edema in obstetrics and gynecology;[12] and others.

Intravenous (never extravenous) administration of aescin in ampoules is used clinically by physicians in Germany and other European countries for treatment of post-traumatic, intraoperative or postoperative conditions of cerebral edema, and other surgical specialties.[33]

Traditional Medicine. Fruits, bark, or seed has reportedly been used externally for ulcers; a folk cancer remedy. Seeds used for gastritis, enteritis, and hemorrhoids (DUKE 2). Bark tea astringent, used in malaria, dysentery; externally for lupus and skin ulcers (FOSTER AND DUKE).

Leaf preparations used in European traditions for eczema, varicose veins, supportive treatment of varicose ulcers, phlebitis, thrombophlebitis, hemorrhoids, menstrual spastic pain, soft tissue swelling from bone fracture and sprains, and other uses. Effectiveness of leaf preparation claims is unsubstantiated.[34]

COMMERCIAL PREPARATIONS

In Germany and other countries topical gels contain 1% aescin; ampoules containing 5.1 mg sodium aescinate (equivalent to 5 mg aescin); sugar-coated tablets; and liquid oral preparations are available. Topical products are available in Canada.[5] No horse chestnut or aescin-containing drug formulations are available in the United States, though aescin is available in bulk.

Regulatory Status. In Germany, horse chestnut seeds are the subject of a positive therapeutic monograph, indicated for chronic venous insufficiency, including edema, cramps in the calves, itching, pain, and sensations of heaviness in the legs, varicose veins, post-thrombotic syndrome; plus post-traumatic and postoperative swelling of soft tissue, in average daily doses equivalent to 30–150 mg of aescin in liquid or solid preparations for oral administration.[35] Horse chestnut leaf preparation claims are not substantiated, therefore, therapeutic use is not recommended.[34]

REFERENCES

See the General References for APPLEQUIST; BARNES; BARRETT; BLUMENTHAL 1 & 2; BRUNETON; DER MARDEROSIAN AND BEUTLER; DUKE 2; FOSTER AND DUKE; FROHNE AND PFANDER; GLASBY 2; GUPTA; HARDIN AND ARENA; HUANG; LIST AND HÖRHAMMER; MABBERLY; MARTINDALE; MCGUFFIN 1 & 2; STEINMETZ; TUTIN 2; WEISS; WREN.

1. Y. Li et al., *Bioorg. Med. Chem. Lett.*, **9**, 2473 (1999).
2. M. Yoshikawa et al., *Chem. Pharm. Bull. (Tokyo)*, **44**, 1454 (1996).
3. T. Konoshima and H. K. Lee, *J. Nat. Prod.*, **49**, 650 (1986).
4. P. Ambrogini et al., *Boll. Soc. Ital. Biol. Sper.*, **71**, 227 (1995).
5. R. F. Chandler, *Can. Pharm. J.*, **297** (1993).
6. M. H. Pittler and E. Ernst, *Cochrane Database Syst. Rev.*, CD003230 (2006).
7. N. Tiffany et al., *J. Herb. Pharmacother.*, **2**, 71 (2002).
8. M. J. Leach, *Complement. Ther. Nurs. Midwifery*, **10**, 97 (2004).
9. U. Siebert et al., *Int. Angiol.*, **21**, 305 (2002).
10. C. R. Sirtori, *Pharmacol. Res.*, **44**, 183 (2001).
11. M. D. Lorenz and M. L. Marek, *Arzneim.-Forsch.*, **10**, 263 (1960).
12. D. Tenhaeff, *Arztliche Praxis*, **24**, 559 (1972).
13. F. Annoni et al., *Arzneim.-Forsch.*, **29**, 672 (1979).
14. C. C. Calabrese and P. Preston, *Planta Med.*, **59**, 394 (1993).
15. G. Vogel et al., *Arzneim.-Forsch.*, **20**, 699 (1970).
16. M. I. Rothkopf and G. Vogel, *Arzneim.-Forsch.*, **26**, 225 (1976).
17. H. Matsuda et al., *Biol. Pharm. Bull.*, **20**, 1092 (1997).
18. M. Guillaume and F. Padioleau, *Arzneim.-Forsch.*, **44**, 25 (1994).
19. A. Suter et al., *Adv. Ther.*, **23**, 179 (2006).
20. M. J. Leach et al., *Ostomy Wound Manage.*, **52**, 68 (2006).
21. M. J. Leach et al., *J. Wound Care*, **15**, 159 (2006).
22. S. Dickson et al., *J. Herb. Pharmacother.*, **4**, 19 (2004).
23. C. Diehm et al., *Lancet*, **347**, 292 (1996).
24. R. M. Facino et al., *Arch. Pharm. (Weinheim)*, **328**, 720 (1995).
25. M. Yoshikawa et al., *Chem. Pharm. Bull. (Tokyo)*, **42**, 1357 (1994).
26. H. Matsuda et al., *Life Sci.*, **67**, 2921 (2000).
27. J. De Meirsman and N. Rosselle, *Ars. Med.*, **9**, 247 (1980).
28. W. Popp et al., *Allergy*, **47**, 380 (1992).
29. G. Properio et al., *Fitoterapia*, **2**, 113 (1980).
30. J. Rothhaar and W. Theil, *Med. Welt.*, **33**, 1006 (1982).
31. C. Zuinen, *Rev. Med. Liege*, **31**, 169 (1976).

32. F. Jakob and B. Fassbender, *Fortschr. Med.*, **87**, 893 (1969).
33. F. Heppner et al., *Wien. Med. Wschr.*, **117**, 706 (1967).
34. Monograph *Hippocastani folium*, *Bundesanzeiger* (July 14, 1993).
35. Monograph *Hippocastani semen*, *Bundesanzeiger*, no. 228 (December 5, 1984).

HORSETAIL

Source: *Equisetum arvense* L.; *E. hymale* L. (Family Equisetaceae).

Common/vernacular names: Common horsetail, field horsetail, running clubmoss, *shenjincao* (*E. arvense*); rough horsetail, common scouring rush, and *muzei* (*E. hymale*).

GENERAL DESCRIPTION

Rhizomatous cryptogam, with hollow stems, jointed, with 6–16 grooves and ridges up to 5–50 cm tall with internodes; leaves minute, in whorls united into a sheath at the base. Spore-bearing cones on separate stalks; *E. hymale* stems are unbranched, with 10–30 grooves, 30–100 cm high; both species occur in moist soils; common in much of temperate Northern Hemisphere (Asia, Europe, North America).[1] The part used is the dried stems.

CHEMICAL COMPOSITION

E. arvense contains 0.3–1% flavonoids including quercetin-3-glucoside, luteolin-5-glucoside, protogenkwanin-4″-O-glucoside kaempferol, apigenin, isoquercitrin, and others (depending on chemo-type).[2–4] Two distinct chemo-types are recognized, one from Europe, characterized by quercetin-3-O-sophoroside, protogenkwanin-4′-O-β-D-glucopyranoside, and genkwanin-4′-O-β-D-glucopyranoside. North American and Asian materials are characterized by the presence of flavone-5-glucosides and their 6″-malonyl esters, especially luteolin. Both chemo-types contain quercetin-3-O-β-D-glucopyranoside and its malonyl ester as the major flavonoids.[2,3]

Equisetum hybrids are extremely variable in morphological features. It has been suggested that hybrid parent species can be verified with analysis of flavonoid patterns.[5]

Phenolic acids of *E. arvense* include di-*E*-caffeoyl-*meso*-tartaric acid; methy esters of protocatechuic and caffeic acids aconitic, oxalic, malic, tannic, arabinoic, and threonic acids;[6,7] minerals, including silicic acid and silicates (5–8%) water soluble up to 80%,[8] potassium, aluminum, and manganese; sterols, including campesterol isocuosterol, and brassinosteroids;[7] and others. Trace amounts of nicotine have been found in *E. arvense*.[9]

The stems of *E. arvense* contain volatile oil with 25 identified compounds. The major constituents include hexahydrofarnesyl acetone, geranyl acetone, thymol, and *trans*-phytol.[10]

PHARMACOLOGY AND BIOLOGICAL ACTIVITIES

Horsetail is considered mildly diuretic hemostyptic (astringent), and vulnerary.

A study of four Mexican *Equisetum* species revealed that chloroform extracts of *E. hymale* var. *affine* had the greatest diuretic activity, more effective than spironolactone, furosemide, and hydrochlorothiazide. An increase in excretion of sodium, chloride, and potassium, with a rise in urine pH was also observed.[7,11]

The essential oil is strongly antimicrobial against Gram-positive and Gram-negative bacteria and fungi.[10]

The hydroalcoholic extract showed significant sedative and anticonvulsant effects when tested in mice at 200 and 400 mg/kg.[12]

Chronic i.p. administration of 50 mg/kg of the hydroalcoholic extract of *E. arvense* significantly improved cognitive functions in aged rats. The observed effect may be attributed to the antioxidant activity of the extract.[13] The antioxidant/hepatoprotective activity of a number of phenolic constituents of the extract (petrosins and flavonoids) has also been demonstrated *in vitro*.[14]

The hydroalcoholic extract of the stems also exhibited dose-dependent antinociceptive and anti-inflammatory properties when administered i.p. in mice and tested in a panel of established experiments for these two activities. The most effective doses were 50 and 100 mg/kg.[15]

Other activities include an astringent effect observed in animals, and a strengthening and regenerating effect on connective tissue (BRADLY; WEISS).

Silica, necessary for the formation of articular cartilage and connective tissue, is taken up by the plant in the form of a bioavailable monosilicic acid,[7] perhaps accounting for traditional uses of the plant.

TOXICOLOGY

E. arvense and other species are known to cause toxicity in livestock, including horses, sheep, and rarely cattle. Toxicity, similar to nicotine poisoning has been reported in children who have chewed the stems. Horsetail may also cause seborrheic dermatitis,[7] especially when taken with cholesterol-rich diets, as shown in a feeding experiment conducted on rats.[16] An antithiamine action has been recognized in *Equisetum* ingestion in horses, destroying thiamine (vitamin B_1) in the stomach of monogastric animals, including humans.[7,17]

E. palustre L., which contains the alkaloid palustrine, is also toxic to livestock. The German Pharmacopoeia requires examination of *E. arvense* for adulteration with other *Equisetum* species, especially *E. palustre* (FROHNE AND PFANDER). A critical review of the German Pharmacopeial methods, along with new analytical procedures, investigations, and improvements, has recently been proposed.[18]

USES

Medicinal, Pharmaceutical, and Cosmetic. Horsetail extract is used as a biological additive in shampoos, skin care products, and so on.

Preparations of the herb are used in German phytotherapy for post-traumatic and static edema and in irrigation therapy for bacterial and inflammatory conditions of the lower urinary and renal tract (contraindicated in cases of impaired heart or kidney function). Externally, compresses or poultices are used for the supportive treatment of poorly healing wounds.[19]

Dietary Supplements/Health Foods. Horsetail is primarily used as a dietary supplement for mineral content, also in diuretic formulations, including teas, tinctures, capsules, tablets, and so on (FOSTER AND DUKE).

Traditional Medicine. In Europe, the herb has reportedly been used to promote renal function; digestive elimination; an adjuvant in weight loss products (BRADLY). American Indian groups used plant tea for kidney and bladder ailments and constipation. In India, horsetail is used as a diuretic, hemostatic; root as an analgesic for teething babies. A folk remedy for bloody urine, gout, gonorrhea, stomach disorders; poulticed for wounds (FOSTER AND DUKE).

A French patent exists for the use of isolated silica compounds from *E. arvense* for the treatment of bone fractures, osteoporosis, connective tissue, and tooth and nail injuries.[7]

In traditional Chinese medicine, *E. hymale* and *E. debile* Roxb. (substitute in Yunnan), have been used to treat bloody stools, dysentery with blood, anal prolapse, malaria, sore throat; externally poulticed for sores; an injectable drug has been used in China to treat neurodermatitis (JIANGSU).

COMMERCIAL PREPARATIONS

Crude dried herb, fresh stems, and extracts.

Regulatory Status. Undetermined in the United States. *E. arvense* is the subject of a positive German monograph, as a mild diuretic; externally for supportive treatment of poorly healing wounds.[19] Health and Welfare Canada requires manufacturers to prove *E. arvense* products are free of thiaminase-like activity (though the compound responsible for antithiamine action has not been identified). The action is based on the concern that irreversible brain damage may occur in thiamine-deficient individuals.[7]

REFERENCES

See the General References for APPLEQUIST; BLUMENTHAL 1; BRADLY; BRUNETON; BISSET; DER MARDEROSIAN AND BEUTLER; DUKE 2; FOSTER and DUKE; FROHNE AND PFANDER; GLASBY 2; GRIEVE; HARDIN AND ARENA; JIANGSU; LIST AND HÖRHAMMER; MABBERLY; MARTINDALE; MCGUFFIN 1 & 2; STEINMETZ; TUTIN 1; WEISS; WREN.

1. R. L. Hauke, *New Bot.*, **1**, 89 (1974).
2. M. Veit et al., *Planta Med.*, **55**, 214 (1989).
3. M. Veit et al., *Phytochemistry*, **29**, 2555 (1990).
4. N. A. M. Saleh et al., *Phytochemistry* **11**, 1095 (1972).
5. M. Veit et al., paper presented at the Bonn BACANS Symposium, July 17–22, 1990.
6. I. L. F. Bakke et al., *Acta Pharm. Seuc.*, **15**, 141 (1978).
7. N. W. Hamon and D. V. C. Awang, *Can. Pharm. J.*, 399 (1992).
8. R. Piekos et al., *Plant Med.*, **27**, 145 (1975).
9. J. D. Phillipson and C. Mellville, *J. Pharm. Pharmacol.*, **12**, 506 (1960).
10. N. Radulovic et al., *Phytother. Res.*, **20**, 85 (2006).
11. R. M. Perez-Guitierrez, *J. Enthnopharmacol.*, **14**, 269 (1985).
12. S. J. Dos Jr. et al., *Fitoterapia*, **76**, 508 (2005).
13. S. J. Guilherme dos Jr. et al., *Pharmacol. Biochem. Behav.*, **81**, 593 (2005).
14. H. Oh et al., *J. Ethnopharmacol.*, **95**, 421 (2004).
15. F. H. Do Monte et al., *Pharmacol. Res.*, **49**, 239 (2004).
16. H. Maeda et al., *J. Nutr. Sci. Vitaminol. (Tokyo)*, **43**, 553 (1997).
17. J. A. Henderson et al., *J. Am. Vet. Med. Assoc.*, 375 (1952).
18. M. Viet et al., *Dtsch. Apoth. Ztg.*, **129**, 1591 (1989).
19. Monograph *Equisete herba*, *Bundesanzeiger*, no. 173, (September 18, 1986).

HYDRANGEA

Source: *Hydrangea arborescens* L. (Family Saxifragaceae or Hydrangeaceae).

Common/vernacular names: Smooth hydrangea, mountain hydrangea, wild hydrangea, and seven barks.

GENERAL DESCRIPTION

An erect shrub with large ovate leaves (up to 20 cm long); up to about 3 m high; native to the eastern United States, growing from New York south to Florida and west to Iowa and Oklahoma. Part used is the dried root, collected in the fall.

CHEMICAL COMPOSITION

Although much work has been done on other *Hydrangea* species, especially those of Asian origin such as *H. macrophylla* (Thunb.) Ser. and *H. paniculata* Sieb., chemical data relating to *H. arborescens* have been very limited. Constituents reported present in *H. arborescens* have included hydrangin, saponin, resin, rutin, starch, a fixed oil, and a volatile oil. The identity of hydrangin is unclear; it has been reported in the literature as a glucoside, an alkaloid, as synonymous with umbelliferone, and as a glycoside with formula of $C_{34}H_{25}O_{11}$ (GRIEVE; LIST AND HÖRHAMMER; MERCK; UPHOF).

Compounds found in other *Hydrangea* species (*H. macrophylla*, *H. paniculata*, *H. chinensis*, etc.) include hydrangenol, hydrangeic acid, phyllodulcin, lunularic acid, their glycosides, lunularin, and 3,4'-dihydroxystilbene;[1] paniculatan (a mucous polysaccharide);[2–4] and hydrangetin (7-hydroxy-8-methoxycoumarin), rutin, umbelliferone, febrifugine, isofebrifugine, and neodichroine (alkaloids), hydrachoside A and hydrangenoside E (secoiridoid glycosides),[5] gum, resin, and others (JIANGSU; LIST AND HÖRHAMMER). A series of antiallergic, antimicrobial dihydroisocoumarins (thunberginols A–F) were isolated from the fermented dried leaves of *H. macrophylla*.[6,7]

Of the leaves of numerous *Hydrangea* and *Viburnum* species tested for tannic substances and flavonoids (quercetin, cyanidin, kaempferol, and luteolin), only those of *H. arborescens* and a subspecies of *H. macrophylla* were found to lack tannins and cyanidin but to contain comparable quantities of quercetin and kaempferol.[8] This perhaps can give some indication as to the possible similarity in other constituents.

PHARMACOLOGY AND BIOLOGICAL ACTIVITIES

Hydrangea is believed to have diuretic properties. Hydrangenol and its derivatives have antiallergic activities (antihistaminic and hyaluronidase inhibition).[9]

The hot water extract of *H. macrophylla* leaves displayed marked antimalarial activity against mice infected with *Plasmodium yoelii*. The activity of the aqueous extract was higher than that of the alkaloidal fraction of the same plant (1 mg/kg, b.i.d, 5 days).[10]

USES

Medicinal, Pharmaceutical, and Cosmetic. Used in certain diuretic preparations, often in combination with corn silk, doggrass, and others.

Traditional Medicine. Used as a tonic, diaphoretic, and diuretic as well as in removing and preventing kidney and bladder stones. Used in American Indian tradition as diuretic, cathartic, emetic; externally bark for wounds, burns, sore muscles, sprains; folk cancer remedy (FOSTER AND DUKE).

In Chinese medicine, the roots, leaves, and flowers of certain *Hydrangea* species (e.g., *H. macrophylla*, *H. strigosa* Rehd., *H. umbellata* Rehd., and *H. paniculata*) are used to treat malaria and as diuretics and antitussives (JIANGSU).

Dietary Supplements/Health Foods. Rarely used, tincture available from small manufacturers (LUST).

Others. The leaves of *H. macrophylla* var. *thunbergii* Makino, *H. strigosa*, and *H. umbellata* are sweet; the sweet principle from the first species has been isolated and found to be phyllodulcin (an isocoumarin). These species are a potential source of natural sweeteners (JIANGSU).[11]

COMMERCIAL PREPARATIONS

Crude and extracts. Crude was formerly official in N.F. Strengths (see *glossary*) of extracts are expressed in weight-to-weight ratios.

REFERENCES

See the General References for APhA; BAILEY 1; BARNES; FOSTER AND DUKE; GOSSELIN; GRIEVE; JIANGSU; KROCHMAL AND KROCHMAL; LIST AND HÖRHAMMER; LUST; MCGUFFIN 1 & 2; UPHOF.

1. J. Gorham, *Phytochemistry*, **16**, 249 (1977).
2. A. T. Khalil et al., *Arch. Pharm. Res.*, **26**, 15 (2003).
3. M. Tomoda and N. Satoh, *Chem. Pharm. Bull.*, **24**, 230 (1976).
4. M. Tomoda and N. Satoh, *Chem. Pharm. Bull.*, **25**, 2910 (1977).
5. F. R. Chang et al., *J. Nat. Prod.*, **66**, 1245 (2003).
6. M. Yoshikawa et al., *Chem. Pharm. Bull.*, **40**, 3352 (1992).
7. M. Yoshikawa et al., *Chem. Pharm. Bull.*, **40**, 3121 (1992).
8. E. C. Bate-Smith, *Phytochemistry*, **17**, 267 (1978).
9. H. Kakegawa et al., *Planta Med.*, **54**, 385 (1988).
10. A. Ishih et al., *Phytother. Res.*, **17**, 633 (2003).
11. G. E. Inglett, personal communication.

HYPERICUM

Source: *Hypericum perforatum* L. (Family Hypericaceae or Guttiferae).

Common/vernacular names: St. Johns Wort, Klamath weed, Rose of Sharon, Tutsan, and goatweed.

GENERAL DESCRIPTION

The species vary from small annual or perennial herbaceous plants 5–10 cm tall, to shrubs and small trees up to 12 m tall; stems ridged or two sided, smooth, erect, branching toward top; dark glands along ridges, with distinct dark rings at the lower nodes; leaves, oblong to ovate entire, opposite, sessile, 1–8 cm long, either deciduous or evergreen; flowers with five (rarely four) petals, pale to dark yellow, 0.5–6 cm wide, petal margins glandular, in terminal corymbs; June through August; fruit is usually a dry capsule that splits to release numerous small seeds; indigenous to Europe; naturalized in waste places and along roadsides in Asia, Africa, North America, and Australia; serious weed in range lands of western North America and Australia.

All members of the genus may be referred to as "St. John's worts", though they are also commonly called Hypericums. Some species are used as ornamentals and have large, showy flowers. Two introduced leaf-feeding beetles *Chrysolina quadrigemina* and *C. hyperici* have been used successfully in Canada as a biological control.[1]

The part used is the fresh or dried flowering tops or herb.

CHEMICAL COMPOSITION

The herb contains various phenolic compounds,[2] mainly naphthodianthrones, hypericin (0.5–0.7%), and pseudohypericin; flavonols, including 0.5–2% of the glucoside hyperin (hyperoside), quercetin, isoquercetin, quercitrin, isoquercitrin, rutin, kaempferol, luteolin, and amantoflavone; flavanols, including (+)-catechin, leucocyanidin, (−)-epicatechin; procyanidins; acylphloroglucinols, hyperforin, adhyperforin, furohyperforin, and other hyperforin analogs.[3–6] Four bisanthraquinone glycosides of skyrin have also been isolated.[7]

Volatile oil (0.05–0.9%) has monoterpenes, including α-pinene, β-pinene, myrcene, and limonene, and the sesquiterpenes, caryophyllene, and humulene; xanthones including kielcorin, magniferin, 1,3,6,7-tetrahydroxy-xanthone; I3-II8-biapigenin; tannins (3.8–10%, up to 16% in flowers); nicotinic acid; vitamins C and A; β-sitosterol; choline; pectin; phlobaphene; rhodan; and others.[8–11]

Recently identified compounds include a novel diterpene, 5-methyl-5-(4,8,12-tri-methyl-tridecyl)-dihydro-furan-2-one, a new chromone, 5-hydroxy-7-methoxy-3-methyl-chromen-4-one, and phytol.[12] An anti-HIV-1 protein has also been isolated from a callus culture of *H. perforatum*.[13]

PHARMACOLOGY AND BIOLOGICAL ACTIVITIES

The antidepressant activity of hypericum has been extensively investigated over the last two decades in animal models (forced-swimming and tail-suspension tests) as well as in humans. Clinical trials have demonstrated an improvement in symptoms of anxiety, dysphoric mood, hypersomnia, anorexia, depression, insomnia, psychomotor retardation, and other subjective indicators.[14,15] Potential for the treatment of premenstrual syndrome (PMS) also exists.[16] Earlier studies also showed that hypericum enhanced mice exploratory activity in a foreign environment, extended narcotic sleeping time (dose dependent), is a reserpine antagonist, and decreased aggression in socially isolated male mice.[15,17] Hypericin has been found to inhibit *in vitro* almost irreversibly both type A and B monoamine oxidase (MAO) in rat brain mitochondria. Type A MAO (serotonin) inhibition was more pronounced,[18] but with long-term use (8 weeks of daily treatment).[19] Other mechanisms of action, such as serotonin transport and uptake,[20–23] interaction with brain adenosine, GABA and sigma receptors,[22,24–27] and the involvement of IL-6 and brain cortecosteroids,[25,28] and others,[29,30] have been reported. There are indications that hypericum has a nootropic effect when tested in animal models of cognitive dysfunction and passive avoidance and may thus have potential in the treatment of old-age dementia associated with depression.[31,32] Hypericum extracts have also been tested in alcohol-preferring rats whereby they significantly reduced alcohol intake and could thus have potential in the treatment of alcoholism.[33–35]

Potent antiretroviral activity has been demonstrated for hypericin and pseudohypericin *in vivo* and *in vitro*. It is postulated that the compounds interfere with viral infections and/or spread by direct inactivation of the virus or prevent virus shedding, budding, or assembly at the cell membrane.[36] The National Institutes of Health was involved in human clinical trials with oral doses of hypericin in HIV positive subjects. The compound was also being developed as an antiretroviral agent for the transfusion blood supply. Hypericin achieves complete inactivation of more than five logs (more than 100,000 HIV particles/mL of blood) of infectious HIV The hypericin dose used for inactivation has been shown to be safe to normal blood cells and does not interfere with standard blood tests.[37] Transcription of HIV-1 in different human cells associated with brain HIV-1 infection was suppressed by a novel protein, p27(SJ) isolated from callus cultures of *H. perforatum*.[13]

Hyperforin is antibacterial against staphylococci (normal and MRSA) and streptococci. A butanol fraction was also active against *Helicobacter pylori*.[6,38]

Hepatoprotective activity of hypericin has been demonstrated in mice and rats.[39] A flavonoid-rich extract had a hypocholesterolemic effect in rats fed a cholesterol-rich diet for 16 weeks.[40]

Antioxidant neuroprotective effect on brain cells, spinal cord injury, β-amyloid neurodegradation and against lipid peroxidation has been reported for extracts and flavonoids of St. John's wort. Different mechanisms seem to be involved, such as free radical scavenging, NOS inhibition and iron chelation, among others.[41–49]

Hypericum has anti-inflammatory activity *in vitro* and in different animal models of pancreatitis, pleuritis, and edema possibly due to inhibition of iNOS and COX-2 expression.[50–54] Analgesic and antinociceptive effects were also demonstrated in some studies.[52,54,55]

Other recently reported activities of St. John's wort include *in vitro* and *in vivo* antispasmodic, smooth muscle relaxing, and bronchodilating effects (rabbit jejunum, guinea pig tracheal preparations, rat bladder, and whole mice);[56,57] anticonvulsant effects in epileptic mice;[58] as well as inhibition of prolactin secretion in mice (100 mg/kg, 15 days).[59]

Different extracts (methanolic, petroleum ether, chloroform, etc.) possess photosensitizing and/or proapoptotic antitumor activity against a number of human cancer cells of the bladder, prostate, blood, and others.[60–66] Constituents other than hyperforin and flavonoids are believed to be responsible for the antitumor activity.

Extensive reviews about the chemistry, general pharmacology, and antidepressant activity of St. John's wort have been published.[67–69]

TOXICOLOGY

Hypericin is absorbed in the intestine and concentrates near the skin. Photosensitization has been described for cattle, sheep, and other livestock. Moderate doses, up to 4 g crude, 30 mL of 1 : 5 tincture (40% EtOH), or 240 mg of 1 : 5 powdered extract (calculated at 0.125% hypericin), have been reported not to cause photodermatitis in humans; however, light-skinned persons should be cautioned to avoid direct sunlight after ingestion.[15]

USES

Medicinal, Pharmaceutical, and Cosmetic. Hypericum extract is reportedly used in various types of skin care products, including face, body, and hand creams and lotions, night creams and lotions, skin fresheners, and skin cleansers.

In European phytomedicine, a dose of 2–4 g of herb (0.2–1.0 mg hypericin) is used for mild antidepressant action (MAO inhibitor) or nervous disturbances; externally, oil is used for the treatment of wounds, abrasions, myalgias, and first-degree burns; antiphlogistic activity for topical preparations.[70]

Synthetic hypericin is currently under development as an antiretroviral agent for the transfusion blood supply.

Food. Hypericin-free extract used in alcoholic beverages.

Dietary Supplements/Health Foods. Hypericum is used in herbal teas, tinctures, capsules, tablets, and other product forms; fresh flowers soaked in mineral or olive oil used topically (FOSTER).

Traditional Medicine. Fresh flowers in tea, tincture, or olive oil, reportedly a popular domestic medicine for treatment of external ulcers, wounds (especially with severed nerve tissue), sores, cuts, bruises, and so on. Tea, a folk remedy for bladder ailments, depression, dysentery, diarrhea, and worms; folk cancer remedy (FOSTER AND DUKE).

Modern applications of hypericum (except antiviral use) date back 2000 years.[71]

COMMERCIAL PREPARATIONS

Crude dried or fresh herb, fresh or dried flowers in olive oil, extracts, tinctures, synthetic hypericin, and so on. Hypericum oil, prepared by maceration of fresh flowers in sunlight for several weeks (usually in olive oil) acquires a red hue usually ascribed to hypericin or lipophilic substituted compounds with a hypericin-like color and fluorescence.[72] Hyperforin is found in freshly macerated oil but is unstable, breaking down within 30–90 days; stability can be increased to 6 months in storage by excluding air and using alternate preparation methods.[72]

Regulatory Status. Hypericin-free extract has been affirmed GRAS for use in alcoholic beverages (§121.1163). Herb is subject of a positive German therapeutic monograph allowed as a mild antidepressant for depression, anxiety, or nervous unrest. Oil allowed internally for dyspeptic complaints; externally for treatment of injuries, myalgia, and first degree burns.[70]

REFERENCES

See the General References for APPLEQUIST; BARNES; BARRETT; BLUMENTHAL 1 & 2; DER MARDEROSIAN AND BEUTLER; DUKE 2; FOSTER; FOSTER AND DUKE; LIST AND HÖRHAMMER; MARTINDALE; MCGUFFIN 1 & 2; NIKITAKIS; STEINMETZ; WEISS; WREN.

1. C. W. Crompton et al., *Can. J. Plant Sci.*, **68**, (1988).
2. G. Jurgenliemk and A. Nahrstedt, *Planta Med.*, **68**, 88 (2002).
3. O. Ploss et al., *Pharmazie*, **56**, 509 (2001).
4. M. D. Shan et al., *J. Nat. Prod.*, **64**, 127 (2001).
5. L. Verotta et al., *J. Nat. Prod.*, **62**, 770 (1999).
6. P. Maisenbacher and K.-A. Kovar, *Planta Med.*, **58**, 291 (1992).
7. A. Wirz et al., *Phytochemistry*, **55**, 941 (2000).
8. I. Schwob et al., *C. R. Biol.*, **325**, 781 (2002).
9. G. Hahn, *J. Naturopathic Med.*, **3**(1), 94 (1992).
10. D. V. C. Awang, *Can. Pharm. J.*, **124**, 33 (1991).
11. E. V. Rao and Y. R. Prasad, *Fitoterapia*, **63**(5), 473 (1992).
12. M. D. Shan et al., *Nat. Prod. Res.*, **18**, 15 (2004).
13. N. Darbinian-Sarkissian et al., *Gene Ther.*, **13**, 288 (2006).
14. H. Muldner and M. Zoller, *Arzneim.-Forsch.*, **34**, 918 (1984).
15. C. Hobbs, *HerbalGram*, **18/19**, 24 (1989).
16. C. Stevinson and E. Ernst, *BJOG*, **107**, 870 (2000).
17. S. N. Okpanyi, *Arzzneim.-Forsch.*, **43**, 10 (1987).
18. O. Suzuki et al., *Planta Med.*, **50**, 272 (1984).
19. V. Butterweck et al., *Brain Res.*, **930**, 21 (2002).
20. K. Hirano et al., *J. Pharm. Pharmacol.*, **56**, 1589 (2004).
21. S. Schulte-Lobbert et al., *J. Pharm. Pharmacol.*, **56**, 813 (2004).
22. V. Butterweck, *CNS Drugs*, **17**, 539 (2003).
23. G. Calapai et al., *Pharmacopsychiatry*, **34**, 45 (2001).
24. T. Mennini and M. Gobbi, *Life Sci.*, **75**, 1021 (2004).
25. M. Franklin et al., *Eur. Neuropsychopharmacol.*, **14**, 7 (2004).
26. M. Gobbi et al., *Pharmacopsychiatry*, **34** (Suppl. 1), S45 (2001).
27. C. E. Muller, *Farmaco*, **56**, 77 (2001).
28. G. Calapai et al., *Pharmacopsychiatry*, **34** (Suppl 1), P S8 (2001).
29. U. Simmen et al., *J. Recept. Signal. Transduct. Res.*, **19**, 59 (1999).
30. J. M. Cott, *Pharmacopsychiatry*, **30** (Suppl. 2), 108 (1997).
31. A. E. Khalifa, *J. Ethnopharmacol.*, **76**, 49 (2001).
32. V. Kumar et al., *J. Ethnopharmacol.*, **72**, 119 (2000).
33. I. Panocka et al., *Pharmacol. Biochem. Behav.*, **66**, 105 (2000).
34. V. J. De et al., *Eur. Neuropsychopharmacol.*, **9**, 461 (1999).
35. M. Perfumi et al., *Alcohol Alcohol*, **34**, 690 (1999).
36. D. Meruelo et al., *Proc. Natl. Acad. Sci.*, **85**, 5320 (1988).

37. America On-Line Business News Wire, VIMRx Pharmaceuticals, Inc., Stamford, CT (November 23, 1993).
38. J. Reichling et al., *Pharmacopsychiatry*, **34**(Suppl. 1), S116 (2001).
39. Y. Özturk et al., *Phytother. Res.*, **6**, 44 (1992).
40. Y. Zou et al., *J. Agric. Food Chem.*, **53**, 2462 (2005).
41. Y. Zou et al., *J. Agric. Food Chem.*, **52**, 5032 (2004).
42. L. Luo et al., *J. Ethnopharmacol.*, **93**, 221 (2004).
43. J. Benedi et al., *Life Sci.*, **75**, 1263 (2004).
44. D. A. El-Sherbiny et al., *Pharmacol. Biochem. Behav.*, **76**, 525 (2003).
45. M. H. Jang et al., *Neurosci. Lett.*, **329**, 177 (2002).
46. Y. B. Tripathi and E. Pandey, *Indian J. Exp. Biol.*, **37**, 567 (1999).
47. T. Genovese et al., *Shock*, **25**, 608 (2006).
48. Y. H. Lu et al., *Am. J. Chin. Med.*, **32**, 397 (2004).
49. B. A. Silva et al., *Neurotox. Res.*, **6**, 119 (2004).
50. T. Genovese et al., *Shock*, **25**, 161 (2006).
51. M. Menegazzi et al., *Free Radic. Biol. Med.*, **40**, 740 (2006).
52. O. M. bdel-Salam, *Scientific World Journal*, **5**, 586 (2005).
53. G. M. Raso et al., *J. Pharm. Pharmacol.*, **54**, 1379 (2002).
54. V. Kumar et al., *Indian J. Exp. Biol.*, **39**, 339 (2001).
55. I. A. Bukhari et al., *Pak. J. Pharm. Sci.*, **17**, 13 (2004).
56. A. H. Gilani et al., *Fundam. Clin. Pharmacol.*, **19**, 695 (2005).
57. R. Capasso et al., *Urology*, **64**, 168 (2004).
58. H. Hosseinzadeh et al., *J. Ethnopharmacol.*, **98**, 207 (2005).
59. C. G. Di et al., *Phytomedicine*, **12**, 644 (2005).
60. L. A. Schmitt et al., *J. Agric. Food Chem.*, **54**, 2881 (2006).
61. N. E. Stavropoulos et al., *J. Photochem. Photobiol. B* (2006).
62. C. Quiney et al., *Leukemia*, **20**, 491 (2006).
63. D. Skalkos et al., *Planta Med.*, **71**, 1030 (2005).
64. D. Martarelli et al., *Cancer Lett.*, **210**, 27 (2004).
65. G. Roscetti et al., *Phytother. Res.*, **18**, 66 (2004).
66. K. Hostanska et al., *J. Pharm. Pharmacol.*, **55**, 973 (2003).
67. K. Linde et al., *Cochrane. Database. Syst. Rev.*, CD000448 (2005).
68. A. R. Bilia et al., *Life Sci.*, **70**, 3077 (2002).
69. J. Barnes et al., *J. Pharm. Pharmacol.*, **53**, 583 (2001).
70. Monograph *Hyperici herba, Bundesanzeiger*, no. 228 (December 5, 1984); revised (March 2, 1989).
71. C. Hobbs, *Pharm. Hist.*, **32**(4), 166 (1990).
72. P. Maisenbacher and K.-A. Kovar, *Planta Med.*, **58**, 351 (1992).

HYSSOP

Source: *Hyssopus officinalis* L. (Family Labiatae or Lamiaceae).

GENERAL DESCRIPTION

A perennial aromatic subshrub with slender herbaceous stems arising from a woody base;

up to about 0.5 m high; leaves linear; flowers generally blue, rarely pink or white, native to southern Europe and temperate Asia; naturalized in the United States. Parts used are the leaves and flowering tops from which hyssop oil is obtained by steam distillation. Major producing countries include France, Hungary, and Holland.

CHEMICAL COMPOSITION

Contains 0.3–2% volatile oil;[1] hyssopin (a glucoside); syringoylglycerol glucosides;[2] 5–8% tannin; flavonoid glycosides (5–6% hesperidin and 3–6% diosmin); sterols and triterpenes: ursolic acid, oleanolic acid, stigmasterol, β-sitosterol;[3] marrubiin (a bitter substance, see *horehound*); a polysaccharide (MAR-10);[4] resin, gum, and others (KARRER; LIST AND HÖRHAMMER).

The essential oil contains mainly pinocamphone, isopinocamphone, α- and β-pinenes, camphene, and α-terpinene, which together constitute about 70% of the oil.[5] Other constituents present include pinocampheol, cineole, linalool, terpineol, terpinyl acetate, bornyl acetate, *cis*-pinic acid, *cis*-pinonic acid, myrtenic acid, myrtenol methyl ether, *d*-2-hydroxy-isopinocamphone, methyl myrtenate, cadinene, and other unidentified compounds totaling more than 50 (KARRER; LIST AND HÖRHAMMER).[1,5–7]

Crude also contains 0.5% rosmarinic acid and total hydroxycinnamic derivatives at 2.2%.[8]

PHARMACOLOGY AND BIOLOGICAL ACTIVITIES

Hyssop extracts have been reported to exhibit antiviral activities (against herpes simplex virus), which may be due to the presence of tannins (see *balm*).[9] The extract and its MAR-10 polysaccharide were found to inhibit the replication of HIV-1 *in vitro*.[4,10]

Hyssop has mild antioxidant activity, due probably to its rosmarinic acid.[8]

Hyssop extracts and the syringoylglycerol glucosides were recently found to inhibit α-glucosidase that might be useful in the management of hyperglycemia.[2,11] The essential oil was also reported to have a muscle relaxing effect on the isolated guinea pig ileum.[12]

TOXICOLOGY

One report indicated that hyssop oil was nonirritating and nonsensitizing to human skin as well as nonphototoxic to mice and swine skin.[13]

USES

Medicinal, Pharmaceutical, and Cosmetic. Hyssop oil is used as a fragrance component in soaps, creams, lotions, and perfumes, with a maximum use level of 0.4% in perfumes.[13]

Food. Hyssop, its extracts, and its oil are used in the formulation of bitters and liqueurs as well as in pickles and meat sauces. The extract is also used as a flavor component in candy. The oil is also used in nonalcoholic beverages, candy, baked goods, and gelatins and puddings. Highest average maximum use level reported for hyssop herb used in alcoholic beverages is 0.06%, that reported for hyssop extract (type not indicated) is 0.03% both in alcoholic beverages and in candy, and that reported for the oil is about 0.004% (36.9 ppm) in alcoholic beverages.

Dietary Supplements/Health Foods. Rarely used in tea formulations (FOSTER).

Traditional Medicine. Reportedly used in treating sore throat, coughs, colds, breast and lung problems, digestive disorders, intestinal ailments, menstrual complaints, and others, usually in the form of a tea or gargle; externally as a diaphoretic (in baths) and in treating skin irritations, burns, bruises, and frostbite.[14] It has also been used in tumors.[15]

COMMERCIAL PREPARATIONS

Mainly crude and oil.

Regulatory Status. GRAS (§182.10 and §182.20). Subject of a German therapeutic monograph; not recommended since claimed efficacy is not documented; below 5% in tea mixtures allowed as a flavor corrigent.[14]

REFERENCES

See the General References for APPLEQUIST; ARCTANDER; BAILEY 2; BIANCHINI AND CORBETTA; BLUMENTHAL 1; DER MARDEROSIAN AND BEUTLER; FEMA; FOSTER; GUENTHER; LIST AND HÖRHAMMER; LUST; MCGUFFIN 1 & 2; ROSE; TERRELL; UPHOF.

1. K. K. Khodzhimatov and N. Ramazanova, *Rastit. Resur.*, **11**, 238 (1975).
2. H. Matsuura et al., *Phytochemistry*, **65**, 91 (2004).
3. Z. Skrzypek and H. Wysokinska, *Z. Naturforsch. C*, **58**, 308 (2003).
4. S. Gollapudi et al., *Biochem. Biophys. Res. Commun.*, **210**, 145 (1995).
5. D. Joulain, *Riv. Ital. Essenze Profumi, Piante Off., Aromi, Saponi, Cosmet., Aerosol*, **58**, 479 (1976).
6. D. Joulain and M. Ragault, *Riv. Ital. Essenze, Profumi, Piante Off., Aromi, Saponi, Cosmet., Aerosol*, **58**, 129 (1976).
7. L. N. Misra et al., *Planta Med.*, **54**, 165 (1988).
8. J. L. Lamaison et al., *Fitoterapia*, **62**, 166 (1991).
9. E. C. Herrmann Jr. and L. S. Kucera, *Proc. Soc. Exp. Biol. Med.*, **124**, (1967).
10. W. Kreis et al., *Antiviral Res.*, **14**, 323 (1990).
11. H. Miyazaki et al., *J. Nutr. Sci. Vitaminol. (Tokyo)*, **49**, 346 (2003).
12. M. Lu et al., *Planta Med.*, **68**, 213 (2002).
13. D. L. J. Opdyke, *Food Cosmet. Toxicol.*, **16**, 783 (1978).
14. Monograph *Hyssopi herba/Hyssopi aetheroleum, Bundesanzeiger* (August 29, 1992).
15. J. L. Hartwell, *Lloydia*, **32**, 247 (1969).

IMMORTELLE

Source: *Helichrysum angustifolium* DC. (syn. *H. italicum* G. Don; *H. italicum* (Roth) Guss.) (Family Compositae or Asteraceae).

Common/vernacular names: Helichrysum, everlasting.

GENERAL DESCRIPTION

A strongly aromatic subshrub with much branched stems that are woody at the base; up to about 0.6 m high; native to the Mediterranean region, especially the eastern part, and growing in dry sandy soil. Parts used are the fresh flowers or flowering tops. Extracts (e.g., concretes and subsequently absolute) are prepared from the flowers by solvent extraction, and the essential oil is prepared from the flowering tops by steam distillation.

Other *Helichrysum* species (e.g., *H. orientale* Gaertn., *H. arenarium* (L.) Moench, and *H. stoechas* (L.) DC.) are also used: *H. orientale* in the production of the essential oil and *H. stoechas* the absolute.

Major producing countries include Italy, Spain, and France.

Should not be confused with the southwestern United States native *Asclepias asperula* (Dcne.) Woods, also known as immortal.

CHEMICAL COMPOSITION

Immortelle (*H. angustifalium*) contains a small amount (0.075–0.2%) of volatile oil, two phthalides (5-methoxy-7-hydroxyphthalide and 5,7-dimethoxyphthalide),[1] helipyrone,[2] triterpenes (α-amyrin, uvaol, and ursolic acid) and a triterpene lactone (ursolic acid lactone),[3] wax consisting mostly of C_{31} and C_{29} *n*-alkanes (ca. 2 : 1 ratio),[4] flavonoids, β-sitosterol, caffeic acid, and others (LIST AND HÖRHAMMER).[5]

The volatile oil contains 30–50% of nerol and neryl acetate (the major components); α- and β-pinenes, geraniol, isovaleric aldehyde, myrcene, limonene, 1,8-cineole, borneol, linalool (not present in a stem oil according to one report),[5] 4,7-dimethyl-6-octen-3-one, several β-diketones (with 3,5-dimethyloctane-4,6-dione and 2,4-dimethylheptane-3,5-dione reported to be the odor principles), sesquiterpenes, furfurol, and eugenol (LIST AND HÖRHAMMER).[5–10]

Numerous flavonoids including naringenin, apigenin, luteolin, kaempferol, helichrysin, isohelichrysin, naringenin-5-glucoside, luteolin-7-glucoside, and quercitrin have been isolated from various *Helichrysum* species.[10–13]

PHARMACOLOGY AND BIOLOGICAL ACTIVITIES

The volatile oil of *H. italicum* flowers has been reported to exhibit antimicrobial properties *in vitro* against *Staphylococcus aureus*, *Escherichia coli*, a *Mycobacterium* species, and *Candida albicans*.[5,6] High activities were observed in oil samples containing higher concentrations of nerol, geraniol, eugenol, β-pinene, and furfural.[6]

Certain flavonoids isolated from *Helichrysum* flowers (especially quercitrin, kaempferol, naringenin, and isohelichrysin) have been reported to increase bile secretion in experimental animals. Quercitrin also increased the detoxifying function of the liver and exhibited an anti-inflammatory activity.[12]

In *H. picardii* Boiss et Reuter phenolic fractions are responsible for antimicrobial activity.[13]

Lyophilized aqueous extracts of the inflorescence of *H. arenarium* have antioxidant properties that are mainly attributed to their phenolic compounds and flavonoid content.[14]

USES

Medicinal, Pharmaceutical, and Cosmetic. The absolute is used as a fixative and fragrance component in perfumes. The essential oil is also used as a fragrance ingredient in perfumes. Extracts have also been reported useful in before- and after-sun products making use of the UV absorption properties of the flavonoids present.[11]

Food. Extracts are used as flavor components (fruit types) in major food products, including alcoholic and nonalcoholic beverages, frozen dairy desserts, candy, baked goods, and gelatins and puddings. Use levels are generally below 0.003% reported for immortelle extract (type not given).

Traditional Medicine. Immortelle is used as an expectorant, antitussive, choleretic, diuretic, anti-inflammatory, and antiallergic agent in Europe. Conditions for which it is used include chronic bronchitis, asthma, whooping cough, psoriasis, burns, rheumatism, headache, migraine, allergies, and liver ailments, among others; usually in the form of a decoction or infusion.[5]

Others. The absolute is used in flavoring certain tobaccos.

COMMERCIAL PREPARATIONS

Extracts and oil.

Regulatory Status. GRAS (§182.20).

REFERENCES

See the General References for ARCTANDER; BIANCHINI AND CORBETTA; FEMA; GUENTHER; KARRER; LIST AND HÖRHAMMER; POLUNIN AND SMYTHIES.

1. L. Optiz and R. Hänsel, *Arch. Pharm. (Weinhem)*, **304**, 228 (1971).
2. L. Optiz and R. Hänsel, *Tetrahedron Lett.*, **38**, 3369 (1970).
3. T. Mezzitti et al., *Planta Med.*, **18**, 326 (1970).
4. C. Bicchi et al., *Planta Med.*, **28**, 389 (1975).
5. P. Manitto et al., *Phytochemistry*, **11**, 2112 (1972).
6. N. N. Chirkina and E. A. Osipova, *Biol. Nauki*, **17**, 86 (1974).
7. S. Tira et al., *Tetrahedron Lett.*, **2**, 143 (1967).
8. L. Trabaud, *Fr. Ses Parfums*, **12**, 215 (1969).
9. L. Peyron and M. Roubaud, *Soap, Perfum. Cosmet.*, **43**, 726 (1970).
10. L. Peyron et al., *Perfum. Flav.*, **3**, 25 (1978).
11. G. Prosperio, *Cosmet. Toilet.*, **91**, 34 (1976).
12. O. P. Prokopenko et al., *Farm. Zh. (Kiev)*, **27**, 3 (1972).
13. F. Tomas-Lorente et al., *Fitoterapia*, **62**, 521 (1991).
14. E. Czinner et al., *J. Ethnopharmacol.*, **73**, 437 (2000).

IPECAC

Source: *Cephaelis ipecacuanha* (Brot.) A. Rich. (syn. *Uragoga ipecacuanha* Baill.; *Psychotria ipecacuanha* Stokes) and *C. acuminata* Karsten (syn. *Uragoga granatensis* Baill.) (Family Rubiaceae).

Common/vernacular names: Rio ipecac, Brazilian ipecac, Brazil root (*C. ipecacuanha*); Cartagena, Nicaragua, Panama ipecac (*C. acuminata*).

GENERAL DESCRIPTION

Small evergreen shrubs (or herbs according to certain authors), up to about 0.5 m high. *Cephaelis ipecacuanha* is native to the tropical forests of South America (Brazil and Bolivia), and *C. acuminata* is native to tropical Central

America (Nicaragua, Panama, etc.) and Colombia. *Cephaelis ipecacuanha* is also cultivated in southern Asia (e.g., India, Burma, and Malaysia). Parts used are the dried roots and rhizome collected during the dry season; roots are generally annulated. Cartagena ipecac roots are thicker than those of Rio ipecac.

CHEMICAL COMPOSITION

Contains 1.8–4% (usually 2–3%) alkaloids that are mainly emetine and cephaeline with minor amounts of *O*-methylpsychotrine, psychotrine, emetamine, protoemetine, and others; most of the alkaloids are present in the cortex below the cork. Emetine constitutes 60–75% of the total alkaloids present in *C. ipecacuanha* and only 30–50% of the total alkaloids in *C. acuminata*, the remaining being mostly cephaeline. *Cephaelis acuminata* generally contains higher amounts of total alkaloids than *C. ipecacuanha*. More cephaeline analogs (e.g., neocephaeline free and glycoside, and cephaeline glycoside) were also isolated from the dried roots of *C. acuminata*.[1]

Other constituents include starch (30–40%), tannins (e.g., ipecacuanhin), choline, alkaloidal tetrhaydroisoquinoline monoterpene glycosides[2] (e.g., alangiside, isoalangiside, and ipecoside), resins, an allergen composed of a mixture of glycoproteins with average molecular weight of about 35,000–40,000,[3,4] and others (LIST AND HÖRHAMMER; MORTON 3; STAHL).

PHARMACOLOGY AND BIOLOGICAL ACTIVITIES

Ipecac has emetic properties, acting both centrally (possibly through the involvement of a 5-HT$_4$ receptor)[5] and locally to cause vomiting.[6] Emetic action starts to fade after 30 min from administration.[7] In small doses it has diaphoretic, expectorant, and stimulant properties (USD 26th). Ipecac also reduces absorption of ingested substances in a time-dependent fashion. This effect becomes insignificant after ca. 90 min from administration of ipecac syrup.[8]

It also has amebicidal activities that are due to its contained alkaloids. Emetine is much more active than cephaeline and less toxic, causing less nausea and vomiting than cephaeline.

An ipecac extract has been reported to exhibit anti-inflammatory effects in Wistar rats.[9]

TOXICOLOGY

Toxic effects of ipecac and its alkaloids include gastrointestinal irritation (nausea, vomiting, diarrhea, abdominal pain, etc.), dizziness, hypotension, dyspnea, and tachycardia, among others (GOODMAN AND GILMAN; USD 26th).[6] Side effects can result from abuse by individuals with eating disorders. Cardiac side effects can be severe, for example, myocarditis, cardiomyopathies, and arrhythmias, or even lethal.[10,11]

Severe allergenic reactions (e.g., asthmatic attacks) have been reported in people who have inhaled ipecac powder (dust); the allergen has been identified as a mixture of glycoproteins.[3]

USES

Medicinal, Pharmaceutical, and Cosmetic. Ipecac Syrup U.S.P. is widely used as a domestic emetic for children, but it should not be used if the swallowed poisons are alkalis, strong acids, strychnine, petroleum distillates, and cleaning fluids.

Crude and extracts are used as components in cold and cough medicines as well as in diaphoretic, analgesic, and antipyretic preparations.

Emetine hydrochloride is used to treat amebic infections; however, the clinical use of etnetine and related compounds is limited because of severe side effects.[12]

Traditional Medicine. Used in treating amebic dysentery and, in small doses, to stimulate the appetite. Also has been reported used in cancers (LEWIS AND ELVIN-LEWIS).[13]

COMMERCIAL PREPARATIONS

Crude and extracts. Crude and syrup as well as emetine hydrochloride are official in U.S.P.; fluid extract was formerly official in U.S.P., and tincture was formerly official in N.F. The tincture is reported to be unstable.[14,15]

REFERENCES

See the General References for CLAUS; DER MARDEROSIAN AND BEUTLER; GOODMAN AND GILMAN; GOSSELIN; LIST AND HÖRHAMMER; MARTINDALE; MORTON 3; NANJING; UPHOF; YOUNGKEN.

1. A. Itoh et al., *Phytochemistry*, **52**, 1169 (1999).
2. A. Itoh et al., *Phytochemistry*, **59**, 91 (2002).
3. L. Berrens and E. Young, *Int. Arch. Allergy Appl. Immunol.*, **21**, (1962).
4. L. Berrens and E. Young, *Int. Arch. Allergy Appl. Immunol.*, **22**, (1963).
5. M. Hasegawa et al., *Jpn. J. Pharmacol.*, **89**, 113 (2002).
6. B. R. Mano and E. J. Manno, *Clin. Toxicol.*, **10**, (1977).
7. A. Saincher et al., *J. Toxicol. Clin. Toxicol.*, **35**, 609 (1997).
8. A. S. Manoguerr and D. J. Cobaugh, *Clin. Toxicol. (Phila)*, **43**, 1 (2005).
9. M. Kroutil and J. Kroutilova, *Acta Univ. Palacki. Olomu. Fac. Med.*, **48**, 55 (1968).
10. T. J. Silber, *J. Adolesc. Health*, **37**, 256 (2005).
11. P. C. Ho et al., *Clin. Cardiol.*, **21**, 780 (1998).
12. M. H. Ansari and S. Ahmad, *Fitoterapia*, **62**, (1991).
13. L. J. Hartwell, *Lloydia*, **34**, (1971).
14. P. Rozsa, *Acta Pharm. Hung.*, **31**, (1961).
15. I. P. Ionescu-Stoain et al., *Farmacia (Bucharest)*, **15**, 333 (1967).

JASMINE

Source: *Jasminum officinale* L., *J. grandiflorum* L. (syn. *J. officinale* L. var. *grandiflorum* Bailey), and other *Jasminum* species (Family Oleaceae).

Common/vernacular names: Royal jasmine, Italian jasmine, Catalonian jasmine (*J. grandiflorum*); poet's Jessamine, and common jasmine (*J. officinale*).

GENERAL DESCRIPTION

Evergreen or deciduous shrubs or shrubby vines puberulous when young; leaves opposite, imparipinnate 5–10 cm long; leaflets 3–7, ovate or lanceolate, acuminate; flowers in terminal corymbs or cymes with very fragrant flowers; most currently used species are native to China, India, or western Asia. *Jasminum officinale* and *J. grandiflorum* are cultivated extensively in Mediterranean countries. *J. officinale* L. var. *grandiflorum* (L.) Kobuski, *J. sambac* (L.) Ait., and other *Jasminum* species are cultivated in China and India. Part used is the flower.

The concrete (see **glossary**) is obtained from the flowers in about 3% yield, and the absolute (see **glossary**) is obtained from the concrete in approximately 50–60% yield. The essential oil is obtained from the absolute by steam distillation.

Concrete-producing countries include Italy, France, Morocco, Egypt, China, and Japan. The major absolute producer is France.

CHEMICAL COMPOSITION

The aroma chemicals of jasmine essence (concrete, absolute, and oil) have been reported to number more than 100.[1,2] The compound present in the highest concentration is benzyl acetate.[3–5] Other compounds include phenylacetic acid, linalool, benzyl alcohol, methyl anthranilate, methyl heptenone, farnesol, *cis*-3-hexenyl acetate, *cis*-linalool oxide, *trans*-linalool oxide, nerolidol, *cis*-jasmone, *trans*-methyl jasmonate, *cis*- and *trans*-ethyl jasmonates, jasmolactone, δ-jasmonic acid lactone, and methyl dehydrojasmonate, among others.[1,2,4,6–9]

The absolute contains about 2% acids including benzoic, myristic, palmitic, stearic, oleic, linolenic, arachidic, pelargonic, lauric, tetradecanoic, ursolic acids, and others.[10]

Among the carbonyl compounds, jasmone, jasmolactone, and methyl jasmonate predominate and have been reported to account for up to 94% of the total carbonyls present in jasmine absolute.[8]

PHARMACOLOGY AND BIOLOGICAL ACTIVITIES

A clinical trial in breastfeeding women showed that the effect of local application of jasmine flowers was comparable to bromocriptine in reducing serum prolactin and breast engorgement and in suppressing milk production.[11]

The i.p. injection of an ethanolic extract of jasmine was shown to possess an anti-inflammatory effect in an acute inflammation model in mice. The same extract was also orally active when administered to rats with chronic inflammation.[12]

An *in vitro* experiment utilizing guinea pig ileum and rat uterus demonstrated that jasmine absolute had a spasmolytic activity on both organs. The spasmolytic activity was suggested to be mediated through cAMP.[13]

TOXICOLOGY

Available data indicate jasmine absolute to be nonirritating, nonsensitizing, and nonphototoxic.[5,14] However, allergenic components of the absolute have been characterized as coniferyl acetate and coniferyl benzoate.[15]

USES

Medicinal, Pharmaceutical, and Cosmetic. Due to its strong power in rounding off undesirable rough notes, jasmine has been widely used in fragrance formulations. Jasmine absolute is reported to be used in creams, lotions, and perfumes, with maximum use level of 0.3% in perfumes.[5]

Food. Jasmine absolute and oil are widely used as flavor ingredients in most major food products, including alcoholic (e.g., liqueurs) and nonalcoholic beverages, frozen dairy desserts, candy, baked goods, and gelatins and puddings. The concrete is also used in all the above foods, but less extensively. Average maximum use levels are generally much lower than 0.001% (usually 1–3 ppm).

Dried jasmine flowers (*J. sambac*) are used as an ingredient of Chinese jasmine tea.

Traditional Medicine. The flowers and volatile oil of several *Jasminum* species are used in Western culture mainly as a calmative as well as an aphrodisiac.

In China, numerous *Jasminum* species are also used in medicine. Different parts of the plants (flowers, leaves, roots, etc.) are used and often for different purposes. Thus, flowers of *J. officinale* var. *grandiflorum* (L.) Kobuski are used mainly to treat hepatitis, pain due to liver cirrhosis, and abdominal pain due to dysentery, while flowers of *J. sambac* are used to treat conjunctivitis, skin ulcers and tumors, as well as abdominal pain due to dysentery. The root of *J. sambac* is used to treat headaches, insomnia, and pain due to dislocated joints and broken bones; it is reported to have anesthetic properties. The root and stem of *J. lanceolarium* Roxb. are also used to alleviate pain around the waist and near the joints due to rheumatism and to treat headaches and the other conditions as *J. sambac*.

Several *Jasminum* species have been used in cancers.[16]

COMMERCIAL PREPARATIONS

Concrete, absolute, and oil.

Regulatory Status. GRAS (§182.20).

REFERENCES

See the General References for ARCTANDER; BAILEY 1; BAUER; FEMA; GUENTHER; JIANGSU; KIRTIKAR; LUST; MCGUFFIN 1 & 2; MORTON 2; NANJING; ROSE; TERRELL; UPHOF.

1. R. Kaiser and D. Lamparsky, *Tetrahedron Lett.* (38), 3413 (1974).
2. E. H. Polak, *Cosmet. Perfum.*, **88**(6), 46 (1973).
3. L. Peyron and J. Acchiardi, *Riv. Ital. Essenze, Profumi, Piante Offic., Aromi, Saponi, Cosmet., Aerosol*, **58**, 2 (1976).
4. M. S. Karawya et al., *Bull. Fac. Pharm., Cairo Univ.*, **13**, 183 (1974).
5. D. L. J. Opdyke, *Food Cosmet. Toxicol.*, **14**, 337 (1976).
6. S. Lemberg, *Int. Congr. Essent. Oils (Pap.)*, **6**, 91 (1974).
7. M. Calvarano, *Essenze Deriv. Agrum.*, **36** 237 (1966).
8. B. D. Mookherjee et al., *Int. Congr. Essent. Oils (Pap.)*, **6**, 150 (1974).
9. E. P. Demole in E. T. Theimer, ed., *Fragrance Chemistry—The Science of the Sense of Smell*, Academic Press, New York, 1982, p. 349.
10. J. Garnero et al., *Riv. Ital.*, **62**, 8 (1980).
11. P. Shrivastav et al., *Aust. N. Z. J. Obstet. Gynaecol.*, **28**, 68 (1988).
12. A. H. Atta and A. Alkofahi, *J. Ethnopharmacol.*, **60**, 117 (1998).
13. M. Lis-Balchin et al., *Phytother. Res.*, **16**, 437 (2002).
14. J. P. Guillot et al., *Perfums. Cosmet., Aromes*, **18**, 61 (1977).
15. S. Kato, *Perfum. Flavor.*, **9**(2), 137 (1984).
16. J. L. Hartwell, *Lloydia*, **33**, 288 (1970).

JOB'S TEARS

Source: *Coix lachryma-jobi* L. and *C. lachryma-jobi* L. var. *ma-yuen* (Roman.) Stapf (syn. *C. ma-yuen* Roman.; *C. lachryma-jobi* L. var. *frumentacea* Makino) (Family Gramineae).

Common/vernacular names: Coix seeds, Chinese pearl barley, pearl barley, semen coicis, adlay, *yokuinin*, *yiyiren*, and *yimi*.

GENERAL DESCRIPTION

Annual or perennial herbs, 1–1.5 m high, flowering from July to September and fruiting from September to October; native to East Asia, but now distributed worldwide, especially in tropical regions; cultivated in China, India, and other Asian countries (BAILEY 1).[1]

Part used is the seed, collected when it ripens in late summer to autumn (depending on latitude), and sun dried. After removing husks, seed coats, and impurities, the resulting raw kernels constitute Job's tears. Two types are commercially available in the United States: raw and roasted (fried), usually labeled as pearl barley. However, occasionally it is substituted with barley (*Hordeum* spp.), especially in certain packaged tonic tea formulas.

CHEMICAL COMPOSITION

Seed contains 50–79% starch, 16–19% proteins, 2–7% fixed oil, lipids (5.67% glycolipids, 1.83% phospholipids, sterols, fatty acids,[2] such as palmitic, stearic, oleic, and linoleic acids, etc.), thiamine, amino acids, adenosine and trace minerals, among others. Coixenolide (ca. 0.25%) and coixol have been isolated from the fixed oil; coixans A, B, and C (glycans with 2.4%, 33.5%, and 3.1% peptides, respectively) from the aqueous extract; and feruloyl stigmasterol and feruloyl campesterol are two of the sterols isolated.[3,4]

Coniferyl alcohol, syringic acid, ferulic acid, syringaresinol, 4-ketopinoresinol, and the lignan mayuenolide were recently isolated from the 1-butanol fraction of the total methanolic extract of Coix hulls.[5]

Root contains bioactive benzoxazinoid compounds.[6]

PHARMACOLOGY AND BIOLOGICAL ACTIVITIES

Both alcoholic and acetone extractives of Coix seed inhibited growth of Ehrlich ascites carcinoma (ECA) in mice; acetone extractives also had marked inhibitory effects against murine uterine cervix carcinoma-14 (U-14) and hepatoma (HCA) (activity due to coixenolide), as well as antimutagenic effects against aflatoxin B_1 mutagenesis (WANG).[4,7] The neutral lipid component of the seed inhibited the growth of two human pancreatic cancer cell lines (PaTu-8988 and SW1990) via two possible mechanisms: apoptosis induction and/or gene expression regulation.[8] The free fatty acids were also found to possess *in vivo* antitumor activity against a transplantable mouse tumor.[2]

Coixol has anti-inflammatory, antihistaminic,[9] and numerous other activities, including inhibiting frog muscle contraction, tranquilizing in mice and rabbits, and lowering normal body temperature and experimental fever in rats. Its toxicity is very low: daily oral administration of 500 mg/kg to mice for one month did not produce any toxic reaction; a single i.p. dose of 500 mg/kg in mice only elicited temporary tranquilizing effects but no deaths and a single i.v. dose of 100 mg/kg also did not result in any fatalities or obvious abnormalities (IMM-3; JIANGSU; WANG). A recent report, however, indicated that the aqueous seed extract (1 g/kg, oral) induced embryotoxicity and uterine contractility that lead to an abortifacient effect in pregnant rats.[10]

Coixans A, B, and C exhibited marked hypoglycemic effects in normal and hyperglycemic mice treated with alloxan; coixan A showed the strongest activity.[4]

The methanolic extract of adlay decreased progesterone production in rat granulose cells via at least four different mechanisms involving signal transduction, gene expression, and enzyme inhibition.[11]

Alcohol extractives retarded the oxidation of linoleic acid *in vitro*.[12] The six compounds isolated from Coix hull possessed free radical scavenging activity in the DPPH assay.[5] The methanolic extract of the seed also inhibited the production of nitric oxide and superoxide ions in activated macrophages *in vitro*, which supports the anti-inflammatory effect of job's tears.[13]

USES

Medicinal, Pharmaceutical, and Cosmetic. Mainly used in Asia: fixed oil primarily in hair care products; aqueous and alcoholic extracts in facial creams (especially for acne and freckles), body lotions and bath preparations for their traditional/alleged skin softening, soothing, whitening, moisturizing, and sun screening properties (ETIC; ZHOU).

Dietary Supplements/Health Foods. Powdered roasted seed and extracts of both raw and roasted seed are used in tonic, antioxidant, and pain relief formulas in liquid (e.g., soy milk and herbal drinks), dried tea, tablet, or capsule form.

Traditional Medicine. Traditional Chinese medicine considers Job's tears to have spleen-invigorating, diuretic, heat-dissipating, and pus-expelling (*pai nong*) properties. It has been on Chinese medical records for more than 2000 years for treating stiff and painful joints, rheumatism, and edema; also used in chronic enteritis, diarrhea (*pi xu xie xie*), lung abscesses, and acute appendicitis (*chang yong*), as well as flat wart and eczema (CHEUNG AND LI; CHP). It is often cooked and eaten with or without other herbs or meat for its nutritional qualities as well as to help ease movement of stiff joints.

Others. Concentrated boiling water extract has metal cleansing (e.g., stainless steel) properties that could be utilized for producing a biodegradable natural cleansing product.

COMMERCIAL PREPARATIONS

Crude: both raw and roasted. Care should be taken to ensure that what is offered is not regular barley (*Hordeum* spp.).

Regulatory Status. Class 2b dietary supplement (not used during pregnancy). It is an ethnic food.

REFERENCES

See the General References for BAILEY 1; CHEUNG AND LI; CHP; JIXIAN; IMM-3; JIANGSU; MCGUFFIN 1 & 2; WANG; ZHOU.

1. R. K. Arora, *Econ. Bot.*, **31**, 358 (1977).
2. M. Numata et al., *Planta Med.*, **60**, 356 (1994).
3. Y. G. Gu, *Zhongchengyao*, **12**, 38 (1990).
4. M. Takahasi et al., *Planat Med.*, **50** (1986).
5. C. C. Kuo et al., *J. Agric. Food Chem.*, **50**, 5850 (2002).
6. H. Otsuka et al., *J. Nat. Prod.*, **51**, 74 (1988).
7. C. C. Ruan et al., *Chin. J. Cancer*, **8**, 29 (1989).
8. Y. Bao et al., *J. Gastroenterol. Hepatol.*, **20**, 1046 (2005).
9. H. Ostuka et al., *J. Nat. Prod.*, **51**, 74 (1988).
10. H. P. Tzeng et al., *J. Toxicol. Environ. Health A*, **68**, 1557 (2005).
11. S. M. Hsia et al., *Int. J. Impot. Res.*, **18**, 264 (2006).
12. Y. L. Zhou and X. R. Xu, *Zhongguo Zhongyao Zazhi*, **17**, 368 (1992).
13. W. G. Seo et al., *Immunopharmacol. Immunotoxicol.*, **22**, 545 (2000).

JOJOBA

Source: *Simmondsia chinensis* (Link) C. Schneid. (syn. *Buxus chinensis* Link; *Simmondsia californica* Nutt.) (Family Simmondsiaceae).

Common/vernacular names: Goatnut, pignut, and deernut.

GENERAL DESCRIPTION

Evergreen, much branched shrub, 1–2 m high; leaves opposite, entire, oblong–ovate, 2–4 cm long, dull green, subsessile, and erect; flowers unisexual, small in peduncles; fruit nut-like, ovoid, leathery, three-angled, 2.5 cm long; seed contains waxy liquid; occurs in arid habitats, dry slopes, along washes, below 1500 m, often abundant; southern Arizona, southern California, Sonora, and Baja California. Now grown in different parts of the world; the largest grower is Catamarca, Argentina.

The part used is the seed and its liquid wax, traded as jojoba oil, obtained by expression or solvent extraction.

CHEMICAL COMPOSITION

Upon expression the seeds yield about 50% (by weight) of a liquid wax composed of high molecular weight C_{20} and C_{22} esters of straight long-chain, monounsaturated fatty acids and alcohols (up to 85% of the oil), including monoethylenic acids composed mainly of eicosenoic acid (34%) and docosenoic acid (14%); alcohols, including eicosenol (22%) and docosenol (21%); trace amounts of oleic and palmitoleic acids.[1–5]

The seed meal contains simple cynaophoric glucosides (simmondsins) and inositol galactosides (pinitols).[6,7] Other constituents of the seed meal include phosphatidylcholine fatty acid esters, proteins rich in albumin and globulin (4 : 1), and polysaccharides that yield xylogalactan oligosaccharides on enzymatic hydrolysis.[8–10]

Physical properties of jojoba oil resemble those of sperm oil; hydrogenated to produce a solid wax very similar to spermaceti. Unlike other vegetable oils, it is not a triglyceride fat.[5]

The oil is highly stable, does not oxidize, volatilize, or become rancid after standing for long periods of time. Repeated heating to temperatures above 285°C for 4 days and exposure to high pressure does not alter its properties. Oil, isomerized at the double bonds, produces a cream (jojoba butter). The hardness of the hydrogenated oil is second only to carnauba wax.[5]

PHARMACOLOGY AND BIOLOGICAL ACTIVITIES

Oil is considered emollient, skin, and hair conditioning.

Orally administered wax in mice has been reported to be mostly excreted; only a small amount absorbed by internal organs and epididymal fat, which diminished with time.[3] However, the wax has also been reported to exhibit anti-inflammatory activity in a number of *in vivo* and *in vitro* experimental models of inflammation.[11]

The seed oil is considered nontoxic, though teratogenic toxicity[12] and appetite-depressant activity has been reported for simmondsin in the seed and seed meal (DUKE 3).

The appetite-depressant activity has been the subject of many reports describing the effect of jojoba seed/simmondsin on reducing food intake and body weight.[13,14] Such an anorexigenic effect is believed to be mediated by cholecystokinin and/or vagal effect.[15,16]

USES

Medicinal, Pharmaceutical, and Cosmetic. Jojoba oil is used as a skin- and hair-conditioning agent and occlusive in shampoos, lipsticks, makeup products, cleansing products, face, body, and hand creams and lotions, and moisturizing creams and lotions (NIKITAKIS). Jojoba wax beads are used as an exfoliating agent in facial scrubs, body polishing preparations, soaps, and shower gels. Hydrogenated jojoba wax is used in lipsticks, lotions, creams, ointments, hair styling gels, and other products; provides lubricity and emolliency in creams and lotions; forms matrix with other waxes for holding pigments and oils in lipsticks; increases viscosity; as hair- and skin-conditioning agent; replaces petroleum waxes.[4]

The oil has been suggested as an antifoaming agent in penicillin and cephalosporin fermentations.[1]

Foods. Indian groups and early settlers ate seeds as a survival food; seeds roasted as a coffee substitute.

Dietary Supplements/Health Foods. Used primarily in cosmetic products available through health and natural food outlets; crude oil topically as a skin moisturizer and emollient (ROSE).

Traditional Medicine. Indian groups in Mexico use the oil as a hair restorer, also a folk cancer remedy; externally applied to head sores (DUKE 3).

Others. The oil is sulfurized to produce a lubricating oil and a factice for use in print ink and linoleum. Sulfurized oil is stable, remains liquid, and darkens only slightly. Incompletely hydrogenated oil is used for polishes, coatings, impregnation, and carbon paper.[1,5]

COMMERCIAL PREPARATIONS

Crude wax (as "jojoba oil"), hydrogenated wax, jojoba beads, jojoba butter (isomerized oil).

REFERENCES

See the General References for BRUNETON; DER MARDEROSIAN AND BEUTLER; DUKE 2; DUKE 3; HICKMAN; MARTINDALE; MCGUFFIN 1 & 2; NIKITAKIS; TYLER 1.

1. U.S. NATIONAL Research Council, *Jojoba: New Crop for Arid Lands, New Material for Industry*, NATIONAL Academy Press, Washington, DC, 1985.
2. H. S. Gentry, *Econ. Bot.*, **12**, 261 (1958).
3. A. Yaron et al., *Lipids*, **17**(3), 169 (1982).
4. R. Wilson, *Drug Cosmetic Ind.*, **43**, 1992.
5. A. Benzioni and M. Forti in G. Robbelen et al., eds., *Oil Crops of the World*, McGraw-Hill, New York, 1989, p. 448.
6. C. A. Elliger et al., *J. Org. Chem.*, **39**, 2930 (1974).
7. M. Van Boven et al., *J. Agric. Food Chem.*, **49**, 4278 (2001).
8. F. Leon et al., *J. Agric. Food Chem.*, **52**, 1207 (2004).
9. M. K. Shrestha et al., *J. Agric. Food Chem.*, **50**, 5670 (2002).
10. S. Hantus et al., *Carbohydr. Res.*, **304**, 11 (1997).
11. R. R. Habashy et al., *Pharmacol. Res.*, **51**, 95 (2005).
12. M. Cokelaere et al., *Food Chem. Toxicol.*, **39**, 247 (2001).
13. C. N. Boozer and A. J. Herron, *Int. J. Obes. (Lond)*, **30**, 1143 (2006).
14. S. Lievens et al., *Physiol. Behav.*, **78**, 669 (2003).
15. G. Flo et al., *Horm. Metab. Res.*, **30**, 504 (1998).
16. G. Flo et al., *Appetite*, **34**, 147 (2000).

JUJUBE COMMON

Source: *Ziziphus jujuba* Mill. (syn. *Zizyphus jujuba*; *Z. sativa* Gaertn.) and ***Z. jujuba*** Mill. var *inermis* (Bge.) Rehd. (syn. *Z. vulgaris* Lam. var. *inermis* Bge.) (Family Rhamnaceae).

Common/vernacular names: Chinese jujube, red date, black date, jujube plum, *da zao*, *hong zao*, *hei zao*, and *zao*.

GENERAL DESCRIPTION

Deciduous shrub or small tree, up to 10 m high, with short spines, straight or curved; stems glabrous; fruit (drupe) oval, 1.5–5 cm long, deep red when ripe, sweet and edible; native to southern Europe and Asia; escaped in Gulf States; now mostly cultivated. Part used is the ripe fruit collected in autumn, rid of impurities, and sun dried or first briefly boiled in water and then sun dried (CMH; JIANGSU). Produced throughout China, with major production in the northern and central provinces. Several types of jujube are produced, depending on the process used; the red type (red date) is the most commonly available in the United States.

CHEMICAL COMPOSITION

Jujube fruit is rich in nutrients, including common ones such as vitamins (A, B_2, C, etc.), free amino acids, sugars, proteins, trace elements, fats, and β-carotene. Other constituents include plant acids (malic, tartaric), sterols, coumarins, flavonoids (kaempferol, myricetin), triterpenes and triterpenic glycosides (oleanolic acid; oleanonic acid; ursolic acid; maslinic acid; betulin; betulinic acid; betulonic acid; ziziphin, zizyphus saponins; jujubasaponins, jujuboside B, and so on), isoquinoline alkaloids (stepharine, asimilobine, *N*-nor-nuciferine, etc.), and cyclic AMP (adenosine-3′,5′-monophosphate) (in the largest amount ever reported in plants or animals), among others.[1–8]

PHARMACOLOGY AND BIOLOGICAL ACTIVITIES

Decoction (per os) has been shown to increase body weight and prolong swimming time in mice as well as protect liver damage from carbon tetrachloride in rabbits (JIANGSU).

Ethanolic extract has exhibited anti-inflammatory (vs. carrageenan-induced paw edema and cotton pellet granuloma in rats) and analgesic effects; it also inhibited growth of *Bacillus subtilis*.[4]

Methanolic extract markedly inhibited water-insoluble glucan synthesis (an initial step in tooth decay) by the cariogenic bacterium *Streptococcus mutans*; the active principles were found to be oleanolic acid (see also **ligustrum**) and ursolic acid. Common jujube has no significant toxic effects on experimental animals.[4]

Ethanolic extract exhibited an anxiolytic effect (black and white test and elevated plus maze) when orally administered to mice at a dose of 0.5–2.0 g/kg.[9]

At least 10 of jujube's saponins (e.g. ziziphin, jujubasaponins, and jujuboside B) are sweetness inhibitors that reduce the sensation of sweet taste (elevate the sweetness threshold) of glucose, fructose, aspartame, and other sweetening agents. No taste modifying effect was observed with bitter, salty, or sour flavors.[10]

USES

Medicinal, Pharmaceutical, and Cosmetic. Water and hydroalcoholic extracts of common jujube are used in skin care products for its anti-inflammatory and traditional moisturizing, soothing, antiwrinkle, and sunburn-relieving properties (ETIC).

Dietary Supplements/Health Foods. Extracts used as an ingredient in herbal tonic formulas in capsule, tablet, or liquid form; crude used in soup mixes.

Traditional Medicine. *Da zao* is one of the major Chinese *qi* (*chi*, energy) tonics whose recorded use dates back to the *Shan Hai Jing* (ca. 800 BC). It is a common food, normally eaten in candied form or in soups, often in winter, to normalize dry skin and to relieve itching. Traditionally regarded as sweet tasting and warming, invigorating vital energy (*bu qi*), promoting the secretion of body fluids (*sheng jin*), regulating body nutritional balance and defense (*tiao he ying wei*), tonifying blood and tranquilizing, and neutralizing drug toxicities. Traditionally used in treating lack of appetite, fatigue, and diarrhea due to spleen deficiency; hysteria; also more recently in treating anemia, hypertension, and purpura.

In Arab system of medicine, common jujube is used in treating fever, wounds and ulcers, inflammatory conditions, asthma, and eye diseases, and as a blood purifier.[4]

COMMERCIAL PREPARATIONS

Mainly crude, whole. Due to its high content of sugar and other nutrients, common jujube is prone to mold and insect attack if not treated and stored properly. One effective method is to store it in a wood container in an airy place mixed with black ashes of rice husks.[7]

Regulatory Status. It is an ethnic food; U.S. regulatory status not clear.

REFERENCES

See the General References for BAILEY 1; CHP; CMH; FOSTER AND YUE; HONGKUI; IMM-3; JIANGSU; JIXIAN; LU AND LI; MCGUFFIN 1 & 2; PETRIDES.

1. S. Z. Li and B. Zhang, *Zhongcaoyao*, **14**(10), 39 (1983).
2. N. Okamura et al., *Chem. Pharm. Bull.*, **29**, 676 (1984).
3. R. Q. Zhang and Q. Z. Yang, *Zhongcaoyao*, **23**, 609 (1992).
4. A. H. Shah et al., *Phytother. Res.*, **3**, 232 (1989).
5. K. Hanabusa et al., *Planta Med.*, **42**, 380 (1981).
6. H. Kohda et al., *Planta Med.*, **52**, 119 (1986).
7. H. P. Huang, *Zhongguo Yaoxue Zazhi*, **25**, 142 (1990).
8. K. Yoshikawa et al., *Chem. Pharm. Bull.*, **40**, 2275 (1992).
9. W. H. Peng et al., *J. Ethnopharmacol.*, **72**, 435 (2000).
10. R. Suttisri et al., *J. Ethnopharmacol.*, **47**, 9 (1995).

JUNIPER BERRIES

Source: *Juniperus communis* L. and its varieties (Family Cupressaceae).

Common/vernacular names: Common juniper berries.

GENERAL DESCRIPTION

Evergreen shrub or tree; some varieties up to 6 m or more high; native to the temperate regions of the northern hemisphere; leaves linear–lanceolate (sword-like), about 1/3 to 1/2 inch long, ternate (arranged in whorls of three); sessile (no petiole); flower mostly dioecious; rarely monoecious; male cones small, yellow, and solitary; female cones small, round, and solitary; fruit cones are small (about 1/4 inch diameter) and round with smooth, leathery scales; green when young and bluish black when mature, but always covered with white bloom. Part used is the dried, mature female cone, which is generally called "berry" because of its berry-like appearance. An essential oil is obtained by steam distillation of the crushed, dried, partially dried, or fermented berries; the essential oil produced from unfermented berries is considered to be superior in flavor qualities.

Major producing countries include Italy, Hungary, France, Austria, Czech Republic, Slovakia, Germany, Poland, Russia, and Spain. Berries collected in northern Italy, Hungary, France, Austria, and the Czech Republic are considered superior in quality than those collected in other regions.

CHEMICAL COMPOSITION

Berries contain 0.2–3.42% (usually 1–2%) volatile oil, depending on the geographic location, altitude, degree of ripeness, and other factors;[1–4] sugars (glucose and fructose); glucuronic acid; L-ascorbic acid;[5] resin (ca. 10%); catechins, proanthocyanidins;[6] fatty acids (lauric, palmitic, oleic, linoleic, etc.), sterols (β-sitosterol, campesterol, cholesterol, etc.);[7] gallotannins;[8] geijerone (a C_{12} terpenoid);[9] 1,4-dimethyl-3-cyclohexen-1-yl methyl ketone;[10] diterpene acids (myrceocommunic, cis- and trans-communic, sandaracopimaric, isopimaric, torulosic, labdatrienoic acids, etc.);[11,12] β-elemen-7α-ol;[13] flavonoid (rutin, isoquercitrin, etc.), neolignan, monoterpene, and megastigmane glycosides;[14–16] and others (LIST AND HÖRHAMMER; MARTINDALE).

The volatile oil is composed mainly of monoterpenes (ca. 58%), which include α-pinene, myrcene, and sabinene as the major components, with limonene, p-cymene, γ-terpinene, β-pinene, α-thujene, camphene, and others also present in minor amounts;[17] small amounts of sesquiterpenes (caryophyllene, cadinene, elemene); 1,4-cineole; terpinen-4-ol; esters; and others (LIST AND HÖRHAMMER).[17–20]

PHARMACOLOGY AND BIOLOGICAL ACTIVITIES

Juniper berry oil is generally considered to have diuretic properties; it also has gastrointestinal antiseptic and irritant properties.

Antimicrobial activity of the essential oil shas been reported against many fungal (yeasts and dermatophytes) and bacterial strains.[21–23]

Moderate anti-inflammatory activity may be present in different juniper extracts as shown by their ability to inhibit human 12(S)-lipoxygenase (12(S)-LOX) *in vitro*. The activity was attributed to cryptojaponol, β-sitosterol, and unsaturated fatty acids.[24]

Juniper oil inhibited caspase-3 activation resulting from heat shock in human astrocytes, thus preventing the onset of apoptosis in these cells.[25]

An aqueous decoction of the berries reduced blood glucose levels in normal and hyperglycemic rats when administered at 250 and 125 mg/kg, respectively, for 24 days.[26]

The oil has an antispasmodic effect on smooth muscles.[20]

TOXICOLOGY

Juniper berry oil has been reported to be generally nonsensitizing, nonphototoxic, and only slightly irritating when applied externally to human and animal skins.[27]

Phytomedicine preparations in Europe are contraindicated in pregnancy and inflammatory kidney diseases. Prolonged use or overdose may cause kidney damage.[27] Studies using albino rats showed that juniper extract reduced fertility and had an abortifacient effect.[28]

USES

Medicinal, Pharmaceutical, and Cosmetic. Berries and extracts are used as components in certain diuretic and laxative preparations. Liquid and solid product forms for oral use.[27] Oil is used as a fragrance component in soaps, detergents, creams, lotions, and perfumes, with maximum use level of 0.8% reported in perfumes.

Food. Berries are widely used as a flavor component in gin and also in alcoholic bitters. Extracts and oils are used in most major food categories, including alcoholic and nonalcoholic beverages, frozen dairy desserts, candy, baked goods, gelatins and puddings, and meat and meat products. Highest average maximum use level reported for the oils is 0.006% in alcoholic beverages and 0.01% for the extract in alcoholic and nonalcoholic beverages.

Dietary Supplements/Health Foods. Dried fruits (whole, crushed, and powdered) used as tea flavoring; also in capsules, tablets, tincture, and so on, presumed for diuretic and digestive activities, and in combination with other botanicals for bladder and kidney preparations (FOSTER AND DUKE).

Traditional Medicine. Used as a carminative and diuretic; to treat flatulence, colic, snakebite, intestinal worms, and gastrointestinal infections; vapor (with steam) used in bronchitis. Also reportedly used in cancers.[29]

COMMERCIAL PREPARATIONS

Crude, extracts (e.g., fluid and solid), and oil; crude and fluid extract were formerly official in N.F., and oil is official in F.C.C. Strengths (see *glossary*) of extracts are expressed in flavor intensities or in weight-to-weight ratios.

Regulatory Status. GRAS (§182.20). Subject of a German therapeutic monograph; dried berries at a daily dosage of 2–10 g, calculated to 20–100 mg of essential oil, allowed for dyspeptic complaints.[27]

REFERENCES

See the General References for APhA; APPLEQUIST; ARCTANDER; BARNES; BIANCHINI AND CORBETTA; BLUMENTHAL 1; DER MARDEROSIAN AND BEUTLER; FEMA; FOSTER AND DUKE; GOSSELIN; GRIEVE; GUENTHER; HORTUS 3RD; KROCHMAL AND KROCHMAL; LIST AND HÖRHAMMER; LUST; MCGUFFIN 1 & 2; MERCK; TERRELL; UPHOF.

1. M. Mihajlov and J. Tucakov, *Bull. Acad. Serbe Sci. Arts, Cl. Sci. Med.*, **44**, 19 (1969).
2. M. V. Staicov et al., *Riv. Ital. Essenze, Profumi, Piane Offic., Oli Vegetali, Saponi*, **39**, 559 (1957).
3. M. Mihajlov, *Lek. Sirovine*, **6**, 75 (1968).
4. H. Hoerster, *Planta Med.*, **26**, 45 (1974).
5. M. I. Panaitov, *Bulgar. Akad. Nauk, Izvest. Khim. Inst.*, **6**, 113 (1958).
6. M. J. Schulz and K. Herrmann, *Z. Lebensm. Unters. Forsch.*, **171**, 278 (1980).
7. G. J. Guerra Hernandez et al., *Cienc. Ind. Farm.*, **7**, 8 (1988).
8. A. Baytop and N. Tanker, *Bull. Fac. Med. Istanbul*, **23**, 113 (1960).
9. A. F. Thomas, *Helv. Chim. Acta*, **55**, 2429 (1972).
10. A. F. Thomas, *Helv. Chim. Acta*, **56**, 1800 (1973).
11. A. M. Martin et al., *Phytochem. Anal.*, **17**, 32 (2006).
12. J. de Pascual Teresa et al., *An. Quim.*, **69**, 1065 (1973).
13. J. de Pascual Teresa et al., *An. Quim.*, **73**, 463 (1977).
14. T. Nakanishi et al., *Chem. Pharm. Bull. (Tokyo)*, **53**, 783 (2005).
15. T. Nakanishi et al., *Phytochemistry*, **65**, 207 (2004).
16. E. Lamer-Zarawska, *Pol. J. Chem.*, **54**, 213 (1980).
17. E. Klien and H. Farnow, *Dragoco Rep.*, **11**, 223 (1964).

18. J. Taskinen and L. Nykanen, *Int. Flav. Food Addit.*, **7**, 228 (1976).
19. H. Hoerster et al., *Rev. Med. (Tirgu-Mures, Rom.)*, **20**, 215 (1974).
20. Monograph *Juniperi fructus*, *Bundesanzeiger*, no. 228 (Dec. 5, 1984).
21. C. Cavaleiro et al., *J. Appl. Microbiol.*, **100**, 1333 (2006).
22. S. Pepeljnjak et al., *Acta Pharm.*, **55**, 417 (2005).
23. N. Filipowicz et al., *Phytother. Res.*, **17**, 227 (2003).
24. I. Schneider et al., *Planta Med.*, **70**, 471 (2004).
25. H. J. Na et al., *Clin. Chim. Acta*, **314**, 215 (2001).
26. M. F. de Sanchez et al., *Planta Med.*, **60**, 197 (1994).
27. D. L. J. Opdyke, *Food Cosmet. Toxicol.*, **14**, 307 (1976).
28. Anon., *Int. J. Toxicol.*, **20**(Suppl. 2), 41 (2001).
29. J. L. Hartwell, *Lloydia*, **33**, 288 (1970).

KARAYA GUM

Source: *Sterculia urens* Roxb. (Family Sterculiaceae).

Common/vernacular names: Sterculia gum, Indian tragacanth, kadaya, and mucara.

GENERAL DESCRIPTION

Gum karaya is the dried exudation from the trunk of *Sterculia urens,* which is a softwooded tree with an erect trunk and broad top, up to 9 m high. Leaves are ca. 20–30 cm diameter, crowded at the end of branches, palmately lobed, glabrous above, velvety beneath, and flowers, numerous, yellow, small ca. 6–9 mm in diameter.[1] It is native to India and is cultivated there for gum karaya production.

Karaya gum is collected by blazing or charring the tree trunk and removing a piece of bark or by drilling a hole into the trunk. The gum exudes from the wound and solidifies to form large tears or worm-like strips. After being collected, the tears and strips are broken up and the fragments are graded based on color and amount of adhering bark. India is currently the only producer of karaya gum.

Food-grade karaya gum is produced from the crude gum fragments by a series of physical processes whereby most of the impurities (especially bark, wood, and soil) are removed and the gum is ground, sized, and blended to yield uniform grades of gum containing no more than 3% of water-insoluble impurities. Technical grades contain larger amounts of these impurities.

Food-grade karaya gum is usually a white to pinkish gray powder with a slightly vinegary odor and taste. The better grades are white and contain less insoluble impurities than the lower grades.

Gum karaya is the least soluble of the commercial plant exudates, but it absorbs water rapidly and swells to form viscous colloidal solutions (sols) or dispersions at low concentrations (e.g., 1%). Higher concentrations (up to 4%) when hydrated in cold water will produce a viscous gel-like paste. A 1% karaya gum dispersion has an acidic pH (usually 4.5–4.7); its viscosity is comparable to those of tragacanth, guar, and locust bean gums at the same concentration. On aging, sols of karaya gum lose viscosity and tend to develop a vinegary taste. The powdered gum behaves similarly on storage, especially under hot and humid conditions. Heat (e.g., boiling), excess acid, or electrolytes can lower the viscosity of its sols; alkalis can turn the sols stringy.

Unlike other plant gums, karaya gum swells in 60% alcohol, but it is insoluble in organic solvents as the other gums.

Gum karaya is generally compatible with proteins, carbohydrates, and other plant gums.

CHEMICAL COMPOSITION

Commercial karaya gum contains moisture (12–14%) and impurities in addition to the polysaccharide itself.

The polysaccharide of karaya gum has been reported to have a high molecular weight (9,500,000). Its structure is complicated and has not been determined. Available data have indicated that the complex polysaccharide contains at least three different types of chains. It has been postulated that one chain (constituting 50% of the total polysaccharide) contains repeating units of four galacturonic acid residues containing β-D-galactose branches and an L-rhamnose residue at the reducing end of the unit. A second chain (17% of the polysaccharide) contains an oligorhamnan having D-galacturonic acid branch residues and interrupted occasionally by a D-galactose residue; galacturonic acid was present in 50%, rhamnose 40%, and galactose 10% by weight. A third chain (33% of the polysaccharide) contains D-glucuronic acid residues.[2]

Karaya gum polysaccharide has also been reported to be partially acetylated and has the tendency of splitting off free acetic acid on storage (FURIA; WHISTLER AND BEMILLER).[3–5]

PHARMACOLOGY AND BIOLOGICAL ACTIVITIES

Karaya gum is not digested or absorbed by humans. It has laxative activities, due to the ability of its granules to absorb water and swell up to 100 times their original volume, forming a discontinuous type of mucilage.

Use of the gum as an aphrodisiac in Arabic tradition lacks experimental conformation.[6]

USES

Most of the uses of karaya gum are based on its ability to swell in cold water, its water-absorbing and water-binding properties, and its thickening and suspending powers.

Medicinal, Pharmaceutical, and Cosmetic. In its larger particle size (8–30 mesh), karaya gum is used extensively as a bulk laxative in laxative preparations. Its fine powder is used in dentistry as a dental adhesive and in related preparations. Also used as a thickener and suspending agent in lotions, creams, and hair-setting preparations.

In the past decade, gum karaya has been investigated as (i) excipient for sustained release tablets;[7,8] (ii) dissolution enhancer for poorly water-soluble drugs;[9] (iii) transdermal delivery enhancer for AZT;[10] (iv) denture adhesive;[11] and (v) gel component for skin contact electrodes.[12]

Food. It is used extensively as a water binder to prevent water separation or formation of ice crystals in sherbets, ice pops, and cheese spreads; as a stabilizer in French dressing, meringues, whipped cream, and toppings; and as a binder in meat products (e.g., bologna). Also used in nonalcoholic beverages and candy. Highest average maximum use level is about 0.805% (8045 ppm) reported in candy.

Dietary Supplements/Health Foods. Used as a filler and ingredient in some weight loss formulations.

Traditional Medicine. Gum traditionally used in aphrodisiac formulations.

Others. Used extensively in other industries (e.g., paper and textile), as a thickener in printing inks.

COMMERCIAL PREPARATIONS

Various grades with different particle sizes (granules to fine powders). It was formerly official in N.F. and is currently official in F.C.C.

Regulatory Status. Has been affirmed as GRAS (§184.1349).[13,14]

REFERENCES

See the General References for ADA; APhA; CSIR X; DER MARDEROSIAN AND BEUTLER; FEMA; FURIA; GLICKSMAN; GOSSELIN; LAWRENCE; MARTINDALE; MORTON 3; TERRELL; WHISTLER AND BEMILLER.

1. K. R. Kirtikar and B. D. Basu, *Indian Medicinal Plants*, International Book Distributors, Dehradun, India, 1999, p. 366.
2. R. W. Raymond and W. C. Nagel, *Carbohydr. Res.*, **30**, 293 (1973).
3. R. C. Ordonez et al., *Rev. Farm. (Buenos Aires)*, **110**, 112 (1968).
4. G. O. Aspinall and G. R. Sanderson, *J. Chem. Soc. C*, **16**, 2256 (1970).
5. G. O. Aspinall and G. R. Sanderson, *J. Chem. Soc. C*, **16**, 2259 (1970).

6. M. Tariq et al., Short Reports of Short Lectures and Poster Presentations, Bonn BACANS Symposium, P$_3$72, July 17–22, 1990, p.199.
7. C. R. Park and D. L. Munday, *Drug Dev. Ind. Pharm.*, **30**, 609 (2004).
8. N. K. Jain et al., *Pharmazie*, **47**, 277 (1992).
9. G. V. Murali Mohan Babu et al., *Int. J. Pharm.*, **234**, 1 (2002).
10. S. Y. Oh et al., *J. Control Release*, **51**, 161 (1998).
11. K. D. Collys et al., *Eur. J. Prosthodont. Restor. Dent.*, **5**, 63 (1997).
12. B. R. Eggins, *Analyst*, **118**, 439 (1993).
13. D. M. Anderson, *Food Addit. Contam.*, **6**, 189 (1989).
14. Annon., *Fed. Regist.*, **39**(185), 34209 (1974).

KAVA

Source: *Piper methysticum* Forst. f. (Family Piperaceae).

Common/vernacular names: Kava kava, ava, and awa.

GENERAL DESCRIPTION

Highly variable perennial deciduous shrub up to 3 m high; leaves sparse, thin, single, alternate, cordate, petiolate, 8–25 cm wide (wider than long), petioles to 6 cm long; flowers in irregular spadices; rootstock is knotty, thick, sometimes tuberous, with lateral roots up to 3 m long; under cultivation rootstocks become voluminous; average weight at 10 months is 1 kg; occurs throughout the South Pacific from Hawaii to New Guinea; exact origin unclear, present distribution result of cultivation by Polynesians. The highest degree of diversification is the Vanuatu archipelago. *P. methysticum* is a cultigen of relatively recent development (2500–3000 years); the binomial refers to sterile cultivars of *P. wichmannii* C. D.C.[1]

The part used is the rootstock.

CHEMICAL COMPOSITION

Kava rootstock contains 43.0% starch, 20.0% fiber, 12.0% water, 3.2% sugars, 3.6% proteins, 3.2% minerals, and 3–20% kavalactones (depending on plant age and cultivar).

Kavalactones contained in the resin are α-pyrones bearing a methoxyl group at C-4 and an aromatic styryl moiety at C-6. They include kavain, 7,8-dihydrokavain, 5,6-dehydrokavain, yangonin, 5,6,7,8-tetrahydroyangonin, methysticin, dihydromethysticin, 5,6-dehydromethysticin, 5,6-dihydroyangonin, 7,8-dihydroyangonin, 10-methoxy-yangonin, 11-methoxy-yangonin, 11-hydroxy-yangonin, hydroxykavain, and 11-methoxy-l2-hydroxy-dehydrokavain, and 11-methoxy-5,6-dihydroyangonin.[1–3]

The rootstock also contains flavokavins A–C; piperidine alkaloids pipermethystin, cepharadione A, and awaine;[2,4,5] ketones including cinnamalaketone and methylenedioxy-3,4-cinnamalaketone; an alcohol, dihydrokavain-5-ol; minerals including potassium, calcium, magnesium, sodium, aluminum, iron, and silica; sugars include saccharose, maltose, fructose, and glucose; over 15 amino acids; and others.[1]

PHARMACOLOGY AND BIOLOGICAL ACTIVITIES

Anodyne, anesthetic, analgesic, antimycotic, antiseptic, antispasmodic, diuretic, expectorant, sedative, stimulant, and tonic activities have been attributed to kava, which was found to be sedative, anticonvulsive, and spasmolytic in animal experiments.[6]

The analgesic effect of dihydrokavain and dihydromethysticin (120 mg/kg) is comparable to 200 mg/kg of aspirin. Local anesthesia is produced in the mouth in mastication of fresh kava, especially by kavain. However, subcutaneous injections of an alcoholic extract of kavain produces anesthesia for several hours (or days) but can cause paralysis of peripheral nerves; therefore, it is an unsuitable local anesthetic drug.[1] Kavain and flavokavains A and B have anti-inflammatory activity mediated through activation of NF-κB and/or inhibition of COX-2.[7,8]

Among the kavalactones, dihydromethysticin was earlier reported to have the highest potentiating effect on barbituric narcosis.[1] More recent research further demonstrated that kava extracts possess anxiolytic and antidepressant activities.[9,10] Possible mechanisms include serotonin reuptake inhibition, MAO inhibition, and GABA binding modulation.[9,11–15] Further experimentally supported CNS activities include enhancement of cognitive performance and quality of sleep in animals and humans.[16–18]

Dihydrokavain and dihydromethysticin have been described as muscle relaxants superior to propanediol, benzazoles, and benzodiazepines. The muscular relaxant activity is the result of a direct effect on muscular contractility rather than inhibition of neuromuscular transmission.[1]

Kavalactones inhibit the growth of *Aspergillus niger*, *Fusarium oxysporum*, *F. solani*, and *Trichoderma viride*. They also have herbicidal activity.[19]

Kawain has antithrombotic effect on human platelets.[20] Flavokavain A has antiproliferative activity against human bladder cancer cells.[21]

Kava has been classified as a narcotic and hypnotic; however, it is neither hallucinogenic nor stupefying, is nonaddictive, and does not cause dependency.[22]

TOXICOLOGY

Excessive consumption may result in photophobia and diplopia or rarely temporary oculomotor paralysis. Heavy kava consumption may also result in skin lesions and drying up of the epidermis, producing advance exanthema characterized by urticarial patches with pronounced itching.[22] A case of altered mental status accompanied with ataxia has recently been associated with acute kava overdosing.[23]

A number of recent reports about the possible involvement of kava in liver toxicity, possibly due to inhibition of CYP450 and/or reduction of glutathione levels, resulted in its withdrawal from the U.S. market even though kava's role in such an adverse effect may be remote.[24–29]

USES

Medicinal, Pharmaceutical, and Cosmetic. In German phytomedicine, the dried rhizome and its preparations were used for conditions of nervous anxiety, stress, and unrest. Combined with pumpkin seed oil, kava extracts have been used in the treatment of irritable bladder syndrome. Use is contraindicated during pregnancy, lactation, and depression. The German monograph on the plant also notes that continuous use can cause temporary yellow discoloration of the skin, hair, and nails, or rare allergic skin reactions. May interact with alcohol, barbiturates, and other psychopharmaceuticals or interfere with operation of machinery or vehicles (WEISS).[6,29–31]

Food. Melanesians, Micronesians, and Polynesians grind the fresh or dry roots to prepare a traditional beverage, often imbibed in social or ceremonial settings. Its cultural role in the Pacific societies is compared with the role of wine in southern Europe.[22]

Dietary Supplements/Health Foods. Dried root used in teas, capsules, tablets, tinctures, primarily as a mild relaxant or weak euphoric (DUKE 2).

Traditional Medicine. In Pacific islands, a decoction of the rootstock has been used for the treatment of gonorrhea, chronic cystitis, urogenital infections, menstrual problems, migraine headache, vaginal prolapse, sleeping problems, respiratory tract infections, tuberculosis; externally juice applied to skin diseases including leprosy; fresh leaves or juice poulticed for intestinal problems, otitis, and abscesses.[22]

Kava has also been reportedly valued as an anesthetic, galactagogue, diaphoretic, diuretic, and expectorant for backache, bronchitis, chills, colds, coughs, gonorrhea, myalgia, gout, rheumatism, and others (DUKE 2; STEINMETZ).

COMMERCIAL PREPARATIONS

Crude and extracts (e.g, ethanolic).

Regulatory Status. Undetermined in the United States. Kava is the subject of a positive German therapeutic monograph; allowed for conditions of nervous anxiety, stress, and unrest.[6]

REFERENCES

See the General References for BARRETT; BLUMENTHAL 1 & 2; DER MARDEROSIAN AND BEUTLER; DUKE 2; GRIEVE; MARTINDALE; STEINMETZ; WEISS.

1. V. Lebot et al., *Kava, the Pacific Drug*, Yale University Press, New Haven, CT, 1992.
2. O. Meissner and H. Haberlein, *J. Chromatogr. B. Analyt. Technol. Biomed. Life Sci.*, **826**, 46 (2005).
3. H. R. Dharmaratne et al., *Phyto-chemistry*, **59**, 429 (2002).
4. K. Dragull et al., *Phytochemistry*, **63**, 193 (2003).
5. H. Jaggy and H. Achenbach, *Planta Med.*, **58**, 111 (1992).
6. Monograph *Piperis methystici rhizoma*, *Bundesanzeiger*, no. 101 (June 1, 1990).
7. F. Folmer et al., *Biochem. Pharmacol.*, **71**, 1206 (2006).
8. D. Wu et al., *J. Agric. Food Chem.*, **50**, 701 (2002).
9. K. K. Smith et al., *Psychopharmacology (Berl)*, **155**, 86 (2001).
10. H. P. Volz and M. Kieser, *Pharmacopsychiatry*, **30**, 1 (1997).
11. S. Witte et al., *Phytother. Res.*, **19**, 183 (2005).
12. R. Uebelhack et al., *Pharmacopsychiatry*, **31**, 187 (1998).
13. S. S. Baum et al., *Prog. Neuropsychopharmacol. Biol. Psychiatry*, **22**, 1105 (1998).
14. G. Boonen and H. Haberlein, *Planta Med.*, **64**, 504 (1998).
15. A. Jussofie et al., *Psychopharmacology (Berl)*, **116**, 469 (1994).
16. K. Shinomiya et al., *Psychopharmacology (Berl)*, **180**, 564 (2005).
17. R. Thompson et al., *Hum. Psychopharmacol.*, **19**, 243 (2004).
18. S. Lehrl, *J. Affect. Disord.*, **78**, 101 (2004).
19. T. D. Xuan et al., *J. Agric. Food Chem.*, **54**, 720 (2006).
20. J. Gleitz et al., *Planta Med.*, **63**, 27 (1997).
21. X. Zi and A. R. Simoneau, *Cancer Res.*, **65**, 3479 (2005).
22. V. Lebot in P. A. Cox and S. A. Banack, eds., *Islands, Plants and Polynesians*, Dioscorides Press, Portland, OR, 1991, p. 169.
23. J. Perez and J. F. Holmes, *J. Emerg. Med.*, **28**, 49 (2005).
24. J. Anke and I. Ramzan, *Planta Med.*, **70**, 193 (2004).
25. D. L. Clouatre, *Toxicol. Lett.*, **150**, 85 (2004).
26. P. V. Nerurkar et al., *Toxicol. Sci.*, **79**, 106 (2004).

27. R. Teschke et al., *Phytomedicine*, **10**, 440 (2003).
28. Y. N. Singh and A. K. Devkota, *Planta Med.*, **69**, 496 (2003).
29. L. Zou et al., *Phytomedicine*, **11**, 285 (2004).
30. R. Bressler, *Geriatrics*, **60**, 24 (2005).
31. J. M. Mathews et al., *Drug Metab. Dispos.*, **33**, 1555 (2005).

KOLA NUT

Source: *Cola acuminata* (Beauv.) Schott et Endl. (syn. *Sterculia acuminata* Beauv.), *C. nitida* (Vent.) Schott et Endl., and other *Cola* species (Family Sterculiaceae).

Common/vernacular names: Guru nut, cola nut, and cola seed.

GENERAL DESCRIPTION

Evergreen trees with long leathery leaves up to about 20 m high; native to western Africa. Fruit consists of four to five leathery or woody follicles (pods), each containing one to several seeds. Part used is the dried seed from which the seed coat has been removed; it is commonly called "nut" because of its hard consistency when dried; resembling a nut.

Cola nitida is cultivated extensively in the tropics (e.g., Jamaica, Brazil, Nigeria, Sri Lanka, and Indonesia) and is the major source of commercial kola nuts; its nuts are larger than those from *C. acuminata.*

Kola nuts are prone to infestation by insects.

CHEMICAL COMPOSITION

Contains 1–2.5% (usually 1.5–2%) caffeine and small amounts of theobromine (up to 0.1%) as the active principles.[1]

Other constituents include *d*-catechin, *dl*-catechin, *l*-epicatechin, betaine, a red pigment, a glucoside, a proanthocyanidin, tannin, protein (ca. 6.7%), starch (ca. 34%), fats, sugar, and cellulose, and others (KARRER; LIST AND HÖRHAMMER).[2–4]

Part of the caffeine is linked to catechin and/or tannins.[5,6]

PHARMACOLOGY AND BIOLOGICAL ACTIVITIES

The major active principle of kola is caffeine, which has central stimulant and other properties (see *coffee*). Kola extract affects locomotor activity in mice in a biphasic mode whereby a higher dose (10 mg/kg) had a depressive effect while a lower one (5 mg/kg) significantly increased activity.[7]

A proanthocyanidin isolated from kola nut displayed a selective, dose-dependent lethal antitrypanosomal activity *in vitro* and a trypanostatic effect in mice.[2]

The extract of *C. nitida* inhibits LH release from pituitary cells both *in vitro* and *in vivo*. The mechanism of this antigonadotropic activity is believed to be through binding of the extract with basic glycoproteins and preventing their cellular uptake.[8]

A caffeine-free *C. nitida* nuts extract has a protective effect against tissue damage caused by the release of ROS and elastase from polymorphonuclear neutrophils in response to infection. The protective effect occurs through scavenging of peroxides and inhibition of elastase release and activity.[9]

TOXICOLOGY

Habitual use (mastication) of kola nuts can lead to development of intraoral carcinomas, as was recently observed in northeast Nigeria.[10]

USES

Medicinal, Pharmaceutical, and Cosmetic. Was formerly used in central stimulant preparations and in treating migraine, neuralgia, diarrhea, and others. Now rarely used.

Food. Extracts are widely used as a flavor ingredient in cola drinks. Other food products in which kola extracts are used include alcoholic beverages, frozen dairy desserts, candy, baked goods, and gelatins and puddings. Highest average maximum use level is about 0.045% (446 ppm) reported in frozen dairy desserts.

Dietary Supplements/Health Foods. Used in diet and "energy" formulas, especially in combination with ephedra (before ephedra was banned).[11]

Traditional Medicine. Used by natives as a stimulant; the fresh nuts are chewed. Also used as a tonic and astringent.

COMMERCIAL PREPARATIONS

Crude and extracts (alcoholic and aqueous); crude and fluid extract were formerly official in N.F. Strengths (see *glossary*) of extracts are expressed in flavor intensities.

Regulatory Status. GRAS; both cola nut and kola nut are listed (§182.20). Subject of a German therapeutic monograph as an analeptic for use in mental and physical fatigue.[12]

REFERENCES

See the General References for ARCTANDER; BAILEY 1; CLAUS; LIST AND HÖRHAMMER; MARTINDALE; MCGUFFIN 1 & 2; UPHOF.

1. G. J. Woolley and A. V. Woolley in L. W. Codd et al., eds., *Chemical Technology: An Encyclopedia Treatment*, Vol. 5, Barnes & Noble, New York, 1972, p. 707.
2. B. K. Kubata et al., *Int. J. Parasitol.*, **35**, 91 (2005).
3. K. Freudenberg and K. Weinges, *Bull. Natl. Inst. Sci., India*, **31** (1965).
4. R. Paris and H. Moyse-Mignon, *Ann. Pharm. Fr.*, **14** (1956).
5. C. Maillard et al., *Planta Med.*, **51**, 515 (1985).
6. O. A. M. Oladokon, *Econ. Bot.*, **43**(1), 17 (1989).
7. J. S. Ajarem, *Acta Physiol. Pharmacol. Bulg.*, **16**, 10 (1990).
8. T. Benie and M. L. Thieulant, *Phytomedicine*, **11**, 157 (2004).
9. D. A. els-Rakotoarison et al., *J. Ethnopharmacol.*, **89**, 143 (2003).
10. E. C. Otoh et al., *Oral Dis.*, **11**, 379 (2005).
11. C. N. Boozer et al., *Int. J. Obes. Relat. Metab. Disord.*, **26**, 593 (2002).
12. Monograph *Colae Semen*, *Bundesanzeiger*, no. 127 (July 12, 1991).

KUDZU ROOT

Source: *Pueraria lobata* (Willd.) Ohwi. (syn. *P. montana* (Lour.) Merr. var. *lobata* (Willd.) Maesen et S. Almeida, *P. thunbergiana* (Sieb. et Zucc.) Benth., *P. pseudohirsuta* Tang et Wang, and *Dolichos lobatus* Willd.) and ***P. thomsonii*** Benth. (Family Leguminosae or Fabaceae).

Common/vernacular names: Yege (*P. lobata*); gange (*P. thomsonii*); fenge (*P. thomsonii*, also occasionally *P. lobata*); pueraria root, radix puerariae, and gegen.

GENERAL DESCRIPTION

Pueraria lobata is a fast-growing, high-climbing, and twining hairy perennial vine, reaching about 10 m long; mature stems woody; roots tuberous, up to 60 cm long and 45 cm in diameter;[1] native to eastern Asia (e.g., China and Japan) and widely distributed throughout China (except the provinces of Xinjiang, Xizang, and Qinghai); now growing worldwide; introduced into the United States a century ago and has since run wild in the Southeast (FOSTER AND YUE).

Pueraria thomsonii is also a vine similar to *P. lobata;* native to China; distributed mainly in southern China, including the provinces of Guangdong, Guangxi, Hainan, Yunnan, and Sichuan.[2]

Part used is the tuberous root, collected from fall to early spring. Root of *P. lobata* is mostly wildcrafted, while that of *P. thomsonii* is mainly collected from cultivated plants; the former produced throughout China (especially Hunan, Henan, Guangdong, Zhejiang, and Sichuan) while the latter mainly in the south (especially Guangxi and Guangdong) (IMM-1). After washing, the outer bark is removed (normally for *P. thomsonii*), and the root is sliced or cut into cylindrical pieces before being oven or sun dried (CHP; IMM-1).

CHEMICAL COMPOSITION

Most chemical studies on kudzu root have been performed on *P. lobata*. Hence, unless otherwise stated, the chemical information reported here is for this species.

Contains isoflavones (daidzein or 4′,7-dihydroxyisoflavone and formononetin) and isoflavone glycosides (daidzin, daidzein-4,7-diglucoside, puerarin, puerarin-7-xyloside, pueraria glycosides (PG)-1–6, etc.), flavonoids, coumarins (6,7-dimethoxycoumarin and puerarol or 6-geranyl-7,4′-dihydroxycoumestan), allantoin, sterols (spinasterol,[3] β-sitosterol, and β-sitosterol-β-D-glucoside), 5-methylhydantoin, lupenone, glycerol-1-monotetracosanoate, arachidic acid, large amounts of starch (up to 27% in fresh root), and others (JIANGSU; JILIN)[4–9] Puerarin is its major active principle whose water solubility is greatly enhanced by certain basic amino acids (e.g., arginine, lysine and histidine).[10] More than 10 oleanane-type triterpene saponin glycosides have also been reported in kudzu root.[11–13]

Depending on geographic location, the total isoflavones contents vary from 1.77% to 12.0% (average 7.64%), with puerarin in highest concentration followed by daidzin, daidzein-4′,7-diglucoside, and daidzein; the amount of total isoflavones is much less in root of *P. thomsonii*, ranging from traces to 2.22% (IMM-1).[2,6]

PHARMACOLOGY AND BIOLOGICAL ACTIVITIES

The cardiovascular effects of kudzu root extracts have been well documented in experimental animals, with the isoflavonoids being the active principles. These effects include (1) dilating coronary and cerebral vessels, increasing coronary and cerebral blood flow, decreasing vascular resistance, decreasing the oxygen consumption of the myocardium, increasing the blood oxygen supply, and depressing the production of lactic acid in oxygen-deficient heart muscles.[14,15] (2) Puerarin exhibited protective effects on adrenaline-induced microcirculatory disturbance in mice; these effects were stronger than those of papaverine and with much lower toxicity (i.v. LD_{50} 1000 mg/kg vs. 33 mg/kg of papaverine).[16] (3) Extracts and total flavonoids have hypotensive effects (WANG). Specifically, puerarin decreased blood pressure (by 15%), heart rate (by 19%), and plasma renin activity (by 67%) in spontaneously hypertensive rats; these effects may be due to its adrenergic receptor blocking action.[17,18] (4) Puerarin antagonized the cardiac arrhythmia induced by chloroform–epinephrine in rabbits but not aconitine-induced arrhythmia in rats.[19] Total flavonoids, daidzein, and alcoholic extracts also have antiarrhythmic activities.[15] (5) Puerarin inhibited ADP-induced platelet aggregation and the release of 5-hydroxytryptamine (serotonin) from blood platelets induced by thrombin.[14] (6) Puerarin exhibited hypoglycemic effects in alloxan/streptozotocin diabetic rats and mice;[20,21] it also decreased serum cholesterol levels and reduced capillary hyperpermeability.[22]

Isoflavone-rich kudzu extracts have estrogenic effects in ovariectomized rats and bind to estrogen beta receptor.[23,24] The root extract also prevents bone loss in ovariectomized rats.[25]

In a number of recent studies, both daidzin and daidzein, as well as kudzu extracts, were found to suppress the free choice of ethanol in experimental animals (rats, mice) and in humans. Kudzu extract and/or its active isoflavones may offer therapeutic choices as antidipsotropic agents in the treatment of alcohol abuse. These studies provide a scientific basis for traditional Chinese use of the root (as well as the flowers) for treatment of patients under influence of alcohol.[26–29]

Other bioactivities include smooth muscle relaxation and contraction, the former due to alcohol-soluble extractives and flavonoids, and the latter due to certain water-soluble extractives; anti-inflammatory and antipyretic activities (hydroalcoholic extract much stronger than decoction);[30] antioxidant activity (etahnolic extract, pueraria glycoside (PG)-1); and others (JILIN; WANG).[14,15,31–33] Due to its antioxidant activity, kudzu root has displayed neuroprotective, hepatoprotective, and anticarcinogenic activities, as well as a protective effect on islet cells against pancreatic toxins.[30,34–36]

TOXICOLOGY

Toxicities of kudzu root are very low: daily oral doses of 50–100 g in humans have not produced any adverse effects;[15] LD_{50} in mice is 1.6–2.1 g/kg (i.v.) for total kudzu flavonoids and 1 g/kg (i.v.) for puerarin; daidzein 0.8 g/kg administered i.p. or 1 and 5 g/kg p.o. to mice produced no toxic symptoms.[14] Puerarin has been shown to be nonmutagenic and nonteratogenic.[37]

USES

Food. Kudzu root (*fenge*) is often used by Asian-Americans in soups, for which it is cut into slices and slowly cooked for hours, sometimes together with tangerine peel, meat, and other ingredients. In Hong Kong and southern China, fresh root tubers are eaten in the form of stews. The starch (kudzu flour or *gefen*) is used in pastries and puddings, and as a thickener for sauces.[1,9]

Dietary Supplements/Health Foods. Extracts used in formulas for cold, flu, and allergies and their associated symptoms; in formulas for drunkenness (tuber) and hangover (flower); also in formulas for cardiovascular problems (especially headache).[38]

Traditional Medicine. Plant was first mentioned in the *Shi Jing* (ca. 5th century BC), and the medicinal use of its root (*gegen*) first described in the *Shen Nong Ben Cao Jing* (ca. 200 BC–AD 100). *Gegen* is traditionally regarded as cool natured; tastes sweet and pungent; diaphoretic and reduces fever; promotes production of body fluids; speeds up eruption of measles (*tou zhen*); stops diarrhea; and quenches thirst (CHP). Used in treating various conditions, including cold and flu and associated fever and headache, stiffness and soreness in the neck, thirst, inadequate eruption of measles, diarrhea and dysentery, drunkenness, and others (CHP; JIANGSU). More recent clinical uses include the treatment of hypertension, angina pectoris, migraine, sudden deafness, diabetes, traumatic injuries, nasal sinusitis, urticaria, psoriasis, and pruritus, among others (JIANGSU).[15,39,40]

Root and starch as well as flower are also used to treat alcohol overdose (e.g., unconsciousness and hangover) (JIANGSU).[41]

COMMERCIAL PREPARATIONS

Crude (in small blocks and slices), extracts, total flavonoids, and kudzu starch *(gefen* or kudzu flour).

Regulatory Status. Class 1 dietary supplement (can be safely consumed when used appropriately). Kudzu root is an ethnic food.

REFERENCES

See the General References for BRUNETON; CHEUNG AND LI; CHP; CMH; DER MARDEROSIAN AND BEUTLER; FOSTER AND YUE; HUANG; IMM-1; JIANGSU; JILIN; JIXIAN; NATIONAL; WANG; XIAO.

1. R. D. Tanner et al., *Econ. Bot.*, **33**, 400 (1979).
2. R. Z. Feng et al., *Zhongguo Yaoxue Zazhi*, **28**, 273 (1993).
3. G. C. Jeon et al., *Exp. Mol. Med.*, **37**, 111 (2005).
4. M. H. Chen and S. J. Zhang, *Zhongyao Tongbao*, **10**, 34 (1985).
5. C. X. Zuo et al., *Zhongcaoyao*, **18**, 10 (1987).
6. S. P. Zhao and Y. Z. Zhang, *Yaoxue Xuebao*, **20**, 203 (1985).
7. Y. Ohshima et al., *Planta Med.*, **54**, 250 (1988).
8. L. X. Xu et al., *Yaoxue Xuebao*, **22**, 208 (1987).
9. S. Y. Zee and L. H. Hui, *Hong Kong Food Plants*, The Urban Council, Hong Kong, 1981, p. 64.
10. C. Wang, et al., *Zhongguo Yaoxue Zazhi*, **14**, 52 (1989).
11. T. Arao et al., *Biol. Pharm. Bull.*, **20**, 988 (1997).
12. T. Arao et al., *Chem. Pharm. Bull. (Tokyo)*, **45**, 362 (1997).
13. T. Arao et al., *Chem. Pharm. Bull. (Tokyo)*, **43**, 1176 (1995).
14. Y. P. Zhou, *Chin. J. Integr. Trad. West. Med.*, **11**, 699 (1984).
15. X. L. Lai and B. Tang, *Zhongguo Zhongyao Zazhi*, **14**, 52 (1989).
16. X. L. Jiang and L. N. Xu, *Yaoxue Xuebao*, **24**, 251 (1989).
17. X. P. Song et al., *Zhongguo Yaoli Xuebao*, **9**, 55 (1988).
18. X. R. Lu et al., *Zhongguo Yaoli Xuebao*, **7**, 537 (1986).
19. X. S. Chai et al., *Zhongguo Yaoli Xuebao*, **6**, 166 (1985).
20. W. C. Chen et al., *Planta Med.*, **70**, 113 (2004).
21. F. L. Hsu et al., *J. Nat. Prod.*, **66**, 788 (2003).
22. Z. F. Shen and M. Z. Xie, *Yaoxue Xuebao*, **20**, 863 (1985).
23. S. Malaivijitnond et al., *J. Ethnopharmacol.* (2006).
24. S. M. Boue et al., *J. Agric. Food Chem.*, **51**, 2193 (2003).
25. X. Wang et al., *J. Bone Miner. Metab.*, **21**, 268 (2003).
26. S. E. Lukas et al., *Alcohol Clin. Exp. Res.*, **29**, 756 (2005).
27. E. Benlhabib et al., *J. Med. Food*, **7**, 168 (2004).
28. W. M. Keung and B. L. Vallee, *Phytochemistry*, **47**, 499 (1998).
29. W. M. Keung and B. L. Vallee, *Proc. Natl. Acad. Sci. USA*, **90**, 10008 (1993).
30. M. Jun et al., *Mol. Nutr. Food Res.*, **49**, 1154 (2005).
31. K. A. Kang et al., *Biol. Pharm. Bull.*, **28**, 1154 (2005).
32. A. Sato et al., *Chem. Pharm. Bull.*, **40**, 721 (1992).
33. P. G. Xiao and K. J. Chen, *Phytother. Res.*, **1**, 53 (1987).
34. S. Zhang et al., *J. Nutr. Biochem.* (2005).

35. F. L. Xiong et al., *Eur. J. Pharmacol.*, **529**, 1 (2006).
36. X. Xu et al., *Planta Med.*, **71**, 585 (2005).
37. J. L. Shi et al., *Chin. J. Pharmacol. Toxicol.*, **6**, 223 (1992).
38. S. Dharmananda, *Bestways*, 52 (1988).
39. C. L. Zhang et al., *Yaoyong Shucai (Medicinal Vegetables)*, Guangxi People's Publications, Nanning, 1985, p. 5.
40. Y. Z. Dai and Y. M. Wang, *Fujian Zhongyiyao*, **23**, 12 (1992).
41. S. Meng and D. Zhang, Shi Liao Ben Cao (Diet Therapy Herbal), 7th Century A.D. (reprinted), Beijing, 1984, p.6.

LABDANUM

Source: *Cistus ladanifer* L. (syn. *C. ladaniferus* L.) and other *Cistus* species, including *C. incanus* L. and its subspecies (syn. *C. villosus* auct. vix L., including *C. polymorphus* Willk) (Family Cistaceae).

Common/vernacular names: Ambreine, rockrose, gum cistus, ciste, and cyste.

GENERAL DESCRIPTION

Shrubs native to the Mediterranean region. *Cistus ladanifer* is very sticky and fragrant, with white flowers and linear, lance-shaped leaves that are viscid above and densely white woolly beneath; up to 3 m high. Parts used are the leaves and twigs.

Labdanum gum is the oleoresin obtained by boiling the plant material in water; the top and bottom layers containing the oleoresin are then separated from the water later and combined. Labdanum oil is obtained from labdanum gum by distillation. The concrete is produced by hydrocarbon solvent extraction of the dried plant materials (which usually contain flowering tops). Labdanum absolute (also called cyste absolute) is produced from the concrete by the usual method (see *glossary*). Spain is the major producer of labdanum gum, while France is the major producer of the absolute.

Material previously identified as myrrh in the Bible (Genesis 37:25) is now realized to be correctly translated as labdanum.[1]

CHEMICAL COMPOSITION

Labdanum gum and concrete contain volatile oil, paraffins, and resins. The resins consist of acidic constituents (e.g., labdanolic acid, 6-oxo-cativic acid, cinnamic acids, anisic acid, arachic acid, behenic acid; butyric, malonic, and oxalic acids) and neutral labdane compounds (e.g., labdane-8α,15-diol, labdane 8α,15,19α-triol, 15-nor-8-labdanol, labd- 8(17)-en-l5-ol, and 15-hydroxy-7-labden-6-one), as well as viridiflorol and azulene (KARRER; LIST AND HÖRHAMMER).[2–6]

The volatile oil has been reported to contain at least 170 compounds, including α- and β-pinenes, camphene, sabinene, myrcene, α-phellandrene, α- and β-terpinenes, limonene, *p*-cymene, 1,8-cineole; borneol, nerol, linalool, geraniol, *cis*-3-hexen-1-ol, *trans*-2-hexen-1-ol, terpinen-4-ol, eugenol, 2,2,6-trimethylcyclohexanone, fenchone, α-thujone, isomenthone, acetophenone, ledol, diacetyl, benzaldehyde, *cis*- and *trans*-citral, bornyl acetate, geranyl acetate, and fatty acids (KARRER; LIST AND HÖRHAMMER).[7–10]

α-Pinene has generally been reported to be the major component in the volatile oil (LIST AND HÖRHAMMER),[8] though a study of the oil found that no single component predominated.[9] It has also been reported that the yield of oil and its composition vary considerably with cultivation conditions, seasons, growth stages, and other factors. Field-grown plants yielded twice the amount of essential oil as greenhouse plants, with alcohol content lower in the fall than in the spring (56% versus 70%), while the content of hydrocarbons and carbonyls (e.g., pinene, camphene, and bornyl acetate) was highest in the fall.[7–9]

Comprehensive reviews of the chemistry have been published.[10,11]

PHARMACOLOGY AND BIOLOGICAL ACTIVITIES

The essential oil and resin extracted from *C. ladanifer* have been reported to have antimicrobial activities against *Staphylococcus aureus*, *Escherichia coli*, *Candida albicans*, and other microbes, with β-pinene, eugenol, eucalyptol (cineole), and benzaldehyde being the most active components.[12] Leaf extracts of *C. incanus* and *C. monspeliensis* were active against bacteria and fungi.[13] Labdanes from the aerial parts of *C. incanus* displayed antimicrobial activity against a panel of Gram-positive and Gram-negative bacteria and fungi, as well as cytotoxic activity against KB, P-388, and NCCLC-N6 tumor cell lines.[3]

Aqueous extracts of *C. incanus* and *C. monspeliensis* displayed significant antioxidant activity in three *in vitro* assays for free radical scavenging, inhibition of lipid peroxidation, and protection against DNA cleavage. The *C. monspeliensis* was more active in all assays.[14]

A study on Moroccan medicinal plants showed that the extract of *C. ladanifer* dose dependently inhibited thrombin and ADP-induced platelet aggregation *in vitro*. The authors attributed the observed effects to a possible contribution of the polyphenolic compounds in the extract.[15]

The antispasmodic effect of *C. incanus* were recently demonstrated whereby the aqueous extract of the leaves and stems resulted in reversible dose-dependent inhibition of the spontaneous motility of the rabbit jejunum in one study,[16] and dose dependently inhibited the contractile response to acetylcholine, phenylephrine, and potassium chloride in the isolated smooth muscle of rat ileum and aorta.[17] On the other hand, the flavonoids of *C. ladanifer* inhibited the rabbit skeletal muscle Ca^{2+}-ATPase pump *in vitro*, which was suggested by the authors to be a defensive mechanism against herbivores.[18] The short-boiled aqueous extract of *C. ladanifer* displayed a gastroprotective antiulcer activity when orally administered to rats at 0.25–0.50 g/kg dose. The protective effect was dose dependent in four models of induction of gastric lesions by necrotizing agents and was suggested to be due to enhancement of gastric mucosal microcirculation resulting from the flavonoid constituents of the extract.[19]

TOXICOLOGY

One study has indicated the presence in the plant (*C. ladanifer*) of an alcohol-extractable but not water-extractable nonalkaloidal substance that had toxic effects (hepatic changes, etc.) on experimental animals.[20]

Both labdanum oil and cyste absolute have been reported to be nonirritating, nonsensitizing, and nonphototoxic to human skin.[21,22]

USES

Medicinal, Pharmaceutical, and Cosmetic. Labdanum absolute and oil are used as a fixative and/or fragrance component in soaps, detergents, creams, lotions, and perfumes, with maximum use levels of 0.4% and 0.8% reported for the absolute and oil, respectively, in perfumes.[21,22]

Food. The absolute, oil, and oleoresin are all used as flavor ingredients in major food products, including alcoholic and nonalcoholic beverages, frozen dairy desserts, candy, baked goods, and gelatins and puddings. The oil is also used in meat and meat products, condiments and relishes, and sweet sauces. Average maximum use levels reported are mostly below 0.001%, with the exception of the absolute at about 0.002% (20.3 ppm) used in candy.

Traditional Medicine. Used as an expectorant and in catarrh of the respiratory tract, and diarrhea; also as astringent, nervine, stimulant, and hemostatic; gum used as a fumigant in Turkey (STEINMETZ).

COMMERCIAL PREPARATIONS

Oleoresin, absolute, and oil. Labdanum oil is official in F.C.C.

Regulatory Status. Has been approved for food use (§172.510).

REFERENCES

See the General References for ARCTANDER; BAILEY 1; FEMA; GUENTHER; LIST AND HÖRHAMMER; POLUNIN AND SMYTHIES; STEINMETZ; TUCKER AND LAWRENCE; TUTIN 2.

1. A. O. Tucker, *Econ. Bot.*, **40**, 425 (1986).
2. N. Chaves et al., *J. Chem. Ecol.*, **27**, 611 (2001).
3. I. Chinou et al., *Planta Med.*, **60**, 34 (1994).
4. J. De Pascual Teresa et al., *An. Quim.*, **73**, 1024 (1977).
5. C. Tabacik and M. Bard, *Phytochemistry*, **10**, 3093 (1971).
6. J. De Pascual Teresa et al., *Phytochemistry*, **21**, 899 (1982).
7. P. G. Gülz, *Parfüm. Kosmet.*, **56**, 344 (1975).
8. P. G. Gülz, *Int. Congr. Essent. Oils (Pap)*, **6**, 125 (1974).
9. R. Königs and P. G. Gülz, *Z. Pflanzenphysiol.*, **72**, 237 (1974).
10. B. M. Lawrence, *Perfum. Flav.*, **6**, 43 (1981).
11. B. M. Lawrence, *Perfum. Flav.*, **9**, 49 (1984).
12. N. N. Chirkina and A. V. Patudin, *Biol. Nauki*, **14**, 100 (1971).
13. H. Bouamama et al., *Therapie*, **54**, 731 (1999).
14. G. Attaguile et al., *Cell Biol. Toxicol.*, **16**, 83 (2000).
15. H. Mekhfi et al., *J. Ethnopharmacol.*, **94**, 317 (2004).
16. M. Aziz et al., *Fitoterapia*, **77**, 425 (2006).
17. G. Attaguile et al., *J. Ethnopharmacol.*, **92**, 245 (2004).
18. T. Sosa et al., *J. Chem. Ecol.*, **30**, 1087 (2004).
19. G. Attaguile et al., *Pharmacol. Res.*, **31**, 29 (1995).
20. E. Ballesteros Moreno, *An. Inst. Invest. Vet. (Madrid)*, **14–15**, 77 (1964).
21. D. L. J. Opdyke, *Food Cosmet. Toxicol.*, **14**, 335 (1976).
22. D. L. J. Opdyke, *Food Cosmet. Toxicol.*, **12**, 403 (1974).

LAVENDER

Source: ***Lavender*** *Lavandula angustifolia* Mill. (syn. *L. spica* L.; *L. officinalis* Chaix.; *L. Vera* DC.); ***Spike lavender*** *Lavandula latifolia* Medic. or Vill. (syn. *L. spica* Cav. or DC.); ***Lavandin*** hybrid of *L. angustifolia* and *L. latifolia* (Family Labiatae or Lamiaceae).

Common/vernacular names: True lavender, garden lavender (*L. angustifolia*); broad-leaved lavender, and aspic (*L. latifolia*).

GENERAL DESCRIPTION

Aromatic evergreen subshrubs with linear or lance-shaped leaves; leaves of spike lavender broader than those of true lavender; up to about 0.9 m high; native to the Mediterranean region; cultivated elsewhere. Lavandin, being a hybrid of true lavender and spike lavender, has several forms with varying degrees of resemblance to its parents. Parts used are the fresh flowering tops from which the essential oils are obtained by steam distillation and extracts (concrete and absolute) by solvent extraction (see *glossary*). For spike lavender oil production, the sun-dried flowers are used. Taxonomy of lavender cultivars in the United States, the United Kingdom, and the Netherlands has been reviewed.[1,2]

France is the major producer of all three types of lavender products. Tasmania is a major producer of true lavender oil.

The flowers of *L. stoechas* L. ("Spanish lavender") also enter commerce.[3]

CHEMICAL COMPOSITION

Lavender contains 0.5–1.5% volatile oil,[4] tannin, coumarins (coumarin, umbelliferone, and herniarin),[5] flavonoids (e.g., luteolin), triterpenoids (e.g., ursolic acid), and others (KARRER; LIST AND HÖRHAMMER).

Spike lavender contains 0.5–1% volatile oil and triterpenoids (e.g., ursolic, and oleanolic acids) (LIST AND HÖRHAMMER).

Lavandin is reported to have a higher volatile oil content than lavender and spike lavender. Due to its hybrid nature, the composition of its essential oil is much more variable than either of its parents.

Lavender oil has been reported to contain more than 100 components, including linalool, linalyl acetate, lavandulyl acetate, terpinen-1-ol-4, 1,8-cineole, camphor, β-phellandrene, terpinolene, α-thujene, n-hexanal, n-heptanal, methyl amyl ketone, ethyl amyl ketone, perillaldehyde, perillyl alcohol, d-borneol, α-terpineol, α-pinene, limonene, lactones (4-butanolide, dihydrocoumarin, 4-methyl-4-vinyl-4-butanolide, 5-pentyl-5-pentanolide, 4,4-dimethyl-2-buten-4-olide, etc.), sesquiterpenes (caryophyllene, cadinene, etc.), fatty acids (propionic acid, isobutyric acid, caproic acid, p-coumaric acid, etc.), and others (LIST AND HÖRHAMMER).[4,6–17]

Spike lavender oil and lavandin oil contain many of the constituents present in lavender oil. The important components include linalool, 1,8-cineole, camphor, and linalyl acetate.

The major distinction among the three oils is in their relative contents of linalyl acetate, linalool, 1,8-cineole, and camphor. Lavender oil contains high concentrations of linalyl acetate (ca. 40%) but only traces of 1,8-cineole and camphor (ca. 1%), while spike lavender oil contains large amounts of 1,8-cineole and camphor (40–60%) with only small amounts of linalyl acetate (ca.1%).[9,14,18] The linalool concentration is usually higher in spike lavender oil than in lavender oil.[9,12,18] The amounts of linalool, linalyl acetate, cineole, and camphor in lavandin oil are between those of the other two oils.[9,14] Presence of lower priced, lower quality lavandin oils has reduced the demand and use of higher quality oils.[19] Wide variation of constituents in various cultivars has also been determined.[20]

The volatile oil of Spanish lavender (L. stoechas) contains pulegone (40%), menthol (18%), and menthone (12%) as the main constituents.[21] At least 13 triterpenes, including two new oleanane-type acids, have been isolated from the root.[22]

PHARMACOLOGY AND BIOLOGICAL ACTIVITIES

Lavender oil has been demonstrated to exhibit CNS-depressive activities on experimental animals (e.g., mice).[23] Such activities include anticonvulsive effects, inhibition of the spontaneous motor activity, and potentiation of the narcotic effects of chloral hydrate.[24,25] The hydroalcoholic extract of the flowers of L. stoechas has similar activities.[26] Spike lavender oil has been reported to have spasmolytic effects on smooth muscles of laboratory animals.[27] The mode of antispasmodic action is believed to occur postsynaptically.[28] Lavender has recently been included in aromatherapy preparations that were shown to be effective in dysmenorrhea, in reducing anger, pain and anxiety, and in producing a state of calmness and enhanced well-being.[29–34] The tincture of L. angustifolia was also found to be effective as an adjuvant to imipramine in the treatment of mild to moderate depression.[35]

Lavender oil has also been reported to have antimicrobial (antibacterial, antifungal) and antiparasitic activities.[36–38]

Other activities reported for lavender oil and lavender extracts include antimutagenic, neuroprotective, local anesthetic, analgesic, and anti-inflammatory.[7,39–41] Such activities and others have been reviewed.[42]

Available data from one source indicate spike lavender oil, lavandin oil, and lavender absolute to be nonirritating and nonsensitizing to human skin, though lavender absolute has been reported elsewhere as a sensitizer. No human phototoxicity data were reported.[43–45]

TOXICOLOGY

When tested by subcutaneous administration in mice, lavender oil exhibited low toxicity.[46]

Large doses of lavender oil, however, are considered to be a narcotic poison; one source reports that the oil can cause dermatitis and that more toxicological studies are needed.[3] This statement supported by the recent finding that lavender oil, rich in linalyl acetate and linalool, was cytotoxic to human skin cells *in vitro*.[6]

USES

Medicinal, Pharmaceutical, and Cosmetic. Among the several lavender products, lavender oil is usually the one used in pharmaceuticals; it is used as a fragrance component in products such as antiseptic ointments, creams, lotions, and jellies, among others.

All types of lavender products (especially essential oils) are used as fragrance ingredients (some extensively) in cosmetic products, including soaps, detergents, creams, lotions, and perfumes (e.g., lavender waters and other colognes), spike lavender oil being more extensively used in soaps and detergents. Maximum use levels are 1.2, 1.0, and 0.8% reported for lavandin oil, lavender absolute, and spike lavender oil, respectively, in perfumes.[43–45]

In Europe used in phytomedicine preparations for digestive and mild nervous disorders as an antispasmodic, carminative, and mild tranquilizer.[47]

Food. Lavender oil and, to a lesser extent, other lavender products (e.g., lavender absolute and concrete, lavandin oil, and spike lavender oil) are used as flavor components (e.g., fruit types) in food products, including alcoholic and nonalcoholic beverages, frozen dairy desserts, candy, baked goods, gelatins, puddings, and aromatic vinegars. Average maximum use levels reported are generally below 0.002%, except for spike lavender oil whose highest maximum use level is about 0.004% in frozen dairy desserts (35.6 ppm), candy (35.2 ppm), baked goods (43.5 ppm), and gelatins and puddings (35.0 ppm).

Dietary Supplements/Health Foods. Flowers and oil used as flavoring in tea formulations; oil also used in aromatherapy (FOSTER).

Traditional Medicine. Lavender is reportedly used as an antispasmodic, carminative, stimulant, diuretic, sedative, tonic, and stomachic. Conditions for which it is used include flatulence, spasms, colic, giddiness, nervous headache, migraine, toothache, sprains, neuralgia, rheumatism, acne, pimples, sores, nausea, vomiting, and others, usually in the form of an infusion, a decoction, or the oil, both internally and externally.

Spike lavender has been used in Europe to promote menstruation and to treat cancers.[48]

Other. Lavandin oil is sometimes used as a source of linalool and linalyl acetate. Lavender oil is sometimes used externally as an insect repellent (MARTINDALE).

COMMERCIAL PREPARATIONS

Crudes (lavender and lavandin) and oils are readily available; concrete and absolute of lavender also available. All three oils are official in F.C.C.; lavender oil is also official in N.F.

Regulatory Status. GRAS (lavender, §182.10 and §182.20; lavandin, §182.20; spike lavender, §182.20). *L. angustifolia* flowers subject of a German therapeutic monograph, indicated for sleep disorders, restlessness, and functional abdominal pains (e.g., nervous irritation of the stomach, Roehmheld syndrome, and nervous intestinal syndromes).[47]

REFERENCES

See the General References for ARCTANDER; BAILEY 1; BAILEY 2; BIANCHINI AND CORBETTA; BLUMENTHAL 1; FEMA; FOSTER; FURIA AND BELLANCA; LEWIS AND ELVIN-LEWIS; LUST; ROSE; STAHL; TERRELL; UPHOF.

1. A. O. Tucker, *Baileya*, **21**, 131 (1981).
2. A. O. Tucker and K. J. W. Hensen, *Baileya*, **22**, 168 (1985).
3. M. De Vincenzi and M. R. Dessi, *Fitoterapia*, **62**, 39 (1991).
4. C. De La Torre and P. Carmen, *Int. Congr. Essent. Oils (Pap.)*, **6**, 81 (1974).
5. L. G. Ianova et al., *Khim. Prir. Soedin.*, **1**, 111 (1977).
6. A. Prashar et al., *Cell Prolif.*, **37**, 221 (2004).
7. C. Ghelardini et al., *Planta Med.*, **65**, 700 (1999).
8. I. Ognyanov and L. Panaiotova, *Riv. Ital. Essenze, Profumi, Piante Offic., Aromi, Saponi, Cosmet., Aerosol*, **55** (1973).
9. R. Ter Heide et al., *J. Chromatogr.*, **50**, 127 (1970).
10. R. Timmer et al., *J. Agric. Food Chem.*, **23**, 53 (1975).
11. L. Peyron, *C. R. Seances Acad. Agric. Fr.*, **57**, 1368 (1971).
12. A. I. Karetnikova et al., *Maslo Zhir. Prom.*, **35**, 23 (1969).
13. R. Vlakhov et al., *Riechst., Aromen, Körperpflegem.*, **19**, 293 (1969).
14. A. Herisset et al., *Plant. Med. Phytother.*, **5**, 305 (1971).
15. R. Kaiser and D. Lamparsky, *Tetrahedron Lett.*, **7**, 665 (1977).
16. Y. R. Naves et al., *Helv. Chim. Acta*, **44**, 316 (1961).
17. B. D. Mookherjee and R. W. Trenkle, *J. Agric. Food Chem.*, **21**, 298 (1973).
18. G. Franchi, *Riv. Ital. Essenze, Profumi, Piante Offic., Aromi, Saponi, Cosmet., Aerosol*, **53**, 245 (1971).
19. E. F. K. Denny et al., *Perfum. Flav.*, **6**, 23 (1981).
20. A. O. Tucker et al., *Perfum. Flav.*, **9**, 49 (1984).
21. A. Goren et al., *Z. Naturforsch. C*, **57**, 797 (2002).
22. G. Topcu et al., *Pharmazie*, **56**, 892 (2001).
23. G. Buchbauer et al., *Z. Naturforsch. C*, **46**, 1067 (1991).
24. S. Atanasova-Shopova et al., *Izv. Inst. Fiziol. Bulg. Akad. Nauk*, **15**, 149 (1973).
25. S. Atanasova-Shopova and K. S. Rusinov, *Izv. Inst. Fiziol. Bulg. Akad. Nauk*, **13**, 69 (1970).
26. A. H. Gilani et al., *J. Ethnopharmacol.*, **71**, 161 (2000).
27. T. Shipochliev, *Vet. Med. Nauki*, **5**, 63 (1968).
28. M. Lis-Balchin and S. Hart, *Phytother. Res.*, **13**, 540 (1999).
29. S. H. Han et al., *J. Altern. Complement Med.*, **12**, 535 (2006).
30. M. Imura et al., *J. Midwifery Womens Health*, **51**, e21 (2006).
31. J. Lehrner et al., *Physiol. Behav.*, **86**, 92 (2005).
32. T. Field et al., *Int. J. Neurosci.*, **115**, 207 (2005).
33. N. Morris, *Complement. Ther. Med.*, **10**, 223 (2002).
34. M. Louis and S. D. Kowalski, *Am. J. Hosp. Palliat. Care*, **19**, 381 (2002).
35. S. Akhondzadeh et al., *Prog. Neuropsychopharmacol. Biol. Psychiatry*, **27**, 123 (2003).
36. T. Moon et al., *Parasitol. Res.*, **99**, 722 (2006).
37. F. D. D'Auria et al., *Med. Mycol.*, **43**, 391 (2005).
38. B. N. Uzdennikov, *Nauch. Tr. Tyumen. Sel. Khoz. Inst.*, **7**, 116 (1970).
39. M. G. Evandri et al., *Food Chem. Toxicol.*, **43**, 1381 (2005).
40. V. Hajhashemi et al., *J. Ethnopharmacol.*, **89**, 67 (2003).
41. M. E. Buyukokuroglu et al., *J. Ethnopharmacol.*, **84**, 91 (2003).

42. H. M. Cavanagh and J. M. Wilkinson, *Phytother. Res.*, **16**, 301 (2002).
43. D. L. J. Opdyke, *Food Cosmet. Toxicol.*, **14**, 453 (1976).
44. D. L. J. Opdyke, *Food Cosmet. Toxicol.*, **14**, 449 (1976).
45. D. L. J. Opdyke, *Food Cosmet. Toxicol.*, **14**, 447 (1976).
46. G. M. Zavarzin and I. M. Chudnova, *Vop. Med. Teor., Klin. Prakt. Kurortnogo Lech.*, **4**, 300 (1971).
47. Monograph *Lavandulae flos*, *Bundesanzeiger*, no. 228 (December 5, 1984); corrected (March 13, 1990).
48. J. L. Hartwell, *Lloydia*, **32**, 247 (1969).

LEMON OIL (AND LEMON PETITGRAIN OIL)

Source: *Citrus limon* (L.) Burm. f. (syn. *C. limonum* Risso) (Family Rutaceae).

Common/vernacular names: Expressed lemon oil, cedro oil (terpeneless).

GENERAL DESCRIPTION

A small evergreen tree with very fragrant flowers and stiff thorns; up to about 6 m high; native to Asia; now cultivated worldwide, especially in the United States (e.g., California and Florida), Italy, Cyprus, and Guinea. Parts used are the peel as well as the leaves and twigs together with undeveloped fruits. Lemon oil is obtained from the peel by cold expression, while lemon petitgrain oil is produced from the leaves and twigs, sometimes including undeveloped small fruits by steam distillation. Major lemon oil producers include the United States, Italy, Guinea, and Cyprus. Major producers of lemon petitgrain oil include Guinea and Italy.

CHEMICAL COMPOSITION

Lemon oil contains about 90% monoterpene hydrocarbons, composed mainly of limonene (ca. 70%), with lesser amounts of γ-terpinene, β-pinene, sabinene, α-pinene, and myrcene;[1] 2–6% aldehydes (mainly citral, neral and geranial); alcohols and esters (linalool, octanol, nonanol, decanol, terpinen-4-ol, α-terpineol, geraniol, neryl acetate, geranyl acetate, etc.);[2,3] small amounts of sesquiterpenes (bisabolene, α-bergamotene, and caryophyllene);[2,4] waxes; and 0.41–0.87% coumarins, consisting primarily of bergamottin, 7-methoxy-5-geranoxy coumarin, and limettin (citropten), with imperatorin, isoimperatorin, phellopterin, 8-geranoxypsoralen, and others also present.[5–8]

A study to examine storing comminuted lemon before oil production showed that α-terpineol rose from 0.21% to 10% and that the chemical and organoleptic qualities of the oil changed completely.[9]

Some components in the waxes have been reported to have antioxidant properties.[7]

Lemon petitgrain oil contains large amounts of citral (up to 50%); other components present include limonene, α-pinene, linalool, and nerol.[10]

PHARMACOLOGY AND BIOLOGICAL ACTIVITIES

Lemon oil has been reported to promote tumor formation on the skin of mice by the primary carcinogen 9,10-dimethyl-1,2-benzanthracene.[11] Coumarins from the fruit, however, have been reported to inhibit TPA-induced tumor promotion, as well as superoxide and nitric oxide generation *in vitro*.[12]

Lemon oil has a cidal effect on the larvae of *Culex quinquefasciatus* mosquito.[13] It has also exhibited antimicrobial activities.[11,14–16]

Certain coumarin derivatives are known to be phototoxic and allergenic to humans as well as effective in treating psoriasis (see **bergamot oil**). Lemon oil has been reported to have phototoxic effects, most likely due to its coumarins.[16]

Available data from one source indicate lemon oil to be nonirritating and nonsensitizing to human skin, though some samples have been demonstrated to be irritating to the backs of hairless mice.[16]

The flavonoid diosmin extracted from lemon is used clinically for treatment of venous insufficiency. Increased vascular tone has been observed after *in vivo* oral administration.[17] Anti-inflammatory, antihistamine, and diuretic activities are also reported (WREN).

TOXICOLOGY

Lemon petitgrain oil has been reported to be nonirritating, nonsensitizing, and nonphototoxic to human skin.[10]

USES

Medicinal, Pharmaceutical, and Cosmetic. Lemon oil is used in pharmaceuticals mainly as a flavoring agent.

Lemon oil is used as a fragrance ingredient in soaps, detergents, creams, lotions, and perfumes (e.g., colognes). Lemon petitgrain oil is used in creams, lotions, and perfumes. Maximum use levels reported are 1.0% and 0.3% for lemon oil and lemon petitgrain oil, respectively, in perfumes.[10,16]

Food. Lemon oil and terpeneless lemon oil are extensively used as flavor ingredients in most food products, including alcoholic (bitters, vermouths, sweet liqueurs, etc.) and nonalcoholic beverages (soft drinks, drink mixes, etc.), frozen dairy desserts, candy, baked goods, gelatins and puddings, meat and meat products, breakfast cereals, and fats and oils, among others. Lemon extract and lemon petitgrain oil are also used in many of above food categories but to a much lesser extent. Highest average maximum used levels are 0.968% and 1.208% reported for the extract (type not specified) in candy and baked goods, respectively. Average maximum use levels for the other ingredients are much lower, with the highest reported being about 0.046% (457 ppm) for lemon oil in candy.

Dietary Supplements/Health Foods. Essential oil used as a flavoring ingredient in herb tea formulations.

COMMERCIAL PREPARATION

Lemon oil (California, Italian, etc.), lemon petitgrain oil, and lemon peel extracts. Lemon oil is official in N.F. and F.C.C.

Regulatory Status. GRAS; listed under lemon, lemon peel, and petitgrain lemon (§182.20)

REFERENCES

See the General References for ARCTANDER; BAILEY 1; BARRETT; BISSET; BRUNETON; CLAUS; FEMA; FURIA AND BELLANCA; GUENTHER; JIANGSU; MCGUFFIN 1 & 2; TERRELL; WREN.

1. L. Pennisi and A. Di Giacomo, *Riv. Ital. Essenze, Profumi, Piante Offic., Aromi, Saponi, Cosmet., Aerosol*, **47**, 370 (1965).
2. S. A. Vekiari et al., *J. Agric. Food Chem.*, **50**, 147 (2002).
3. G. L. K. Hunter and M. G. Moshonas, *J. Food Sci.*, **31**, 167 (1966).
4. G. Rispoli and A. Di Giocomo, *Riv. Ital. Essenze, Profumi, Piante Offic., Aromi, Saponi, Cosmet.*, **47**, 650 (1965).
5. G. Calabro and P. Curro, *Essenze Deriv. Agrum.*, 46, 215 (1976).
6. W. L. Stanley and L. Jurd, *J. Agric. Food Chem.*, **19**, 1106 (1971).

7. A. Di Giacomo et al., *Essenze Deriv. Agrum.*, **40**, 143 (1970).
8. B. M. Lawrence, *Perfum. Flav.*, **14**, 41 (1989).
9. I. Calvarava and G. Di Giacomo, *Essenze Deriv. Agrum.*, **54**, 200 (1984).
10. D. L. J. Opdyke, *Food Cosmet. Toxicol.*, **16**(Suppl. 1), **807**(1978).
11. F. J. C. Rose and W. E. H. Field, *Food Cosmet. Toxicol.*, **3**, 311 (1965).
12. Y. Miyake et al., *J. Agric. Food Chem.*, **47**, 3151 (1999).
13. G. L. Mwaiko, *East Afr. Med. J.*, **69**, 223 (1992).
14. M. S. Subba et al., *J. Food Sci.*, **32**, 225 (1967).
15. A. Poretta and A. Casolari, *Ind. Conserve (Parma)*, **41**, (1966).
16. D. L. J. Opdyke, *Food Cosmet. Toxicol.*, **12**, 725 (1974).
17. A. Codignola et al., *Planta Med.*, **58**(S1), **A628**(1992).

LEMONGRASS

Source: *Cymbopogon citratus* (DC.) Stapf (syn. *Andropogon citratus* DC.) and *C. flexuosus* (Nees ex Steud.) W. Wats. (syn. *A. flexuosus* Nees) (Family Gramineae or Poaceae).

Common/vernacular names: West Indian lemongrass, Madagascar lemongrass, Guatemala lemongrass (*C. citratus*); East Indian lemongrass, Cochin lemongrass, native lemongrass, and British Indian lemongrass (*C. flexuosus*).

GENERAL DESCRIPTION

Perennial grasses both native to tropical Asia. East Indian lemongrass (*C. flexuosus*) is cultivated mainly in western India and nearby countries, while West Indian lemongrass (*C. citratus*) is cultivated in the tropics worldwide (West Indies, Central and South America, Africa, and tropical Asia). Leaf blades are linear, long attenuate toward the base tapering upward to a long setaceous point ca. 90×0.6 cm, glaceous green more or less smooth, less rough upward.[1] Parts used are the freshly cut and partially dried leaves of cultivated plants from which the essential oils are obtained by steam distillation. India is the major producer of East Indian lemongrass oil, while major producers of the West Indian oil include Guatemala, Madagascar, the Comoro Islands, Brazil, Malaysia, and Vietnam.

CHEMICAL COMPOSITION

West Indian lemongrass (*C. citratus*) contains a volatile oil (usually 0.2–0.4% yield from fresh grass); an unknown alkaloid; a saponin; β-sitosterol; hexacosanol and triacontanol;[2] cymbopogonol (a triterpenoid);[3] flavonoids (orientin, isoorientin, isoscoparin, swertiajaponin) and phenolic acids (chlorogenic and caffeic);[4] and others (KARRER; LIST AND HÖRHAMMER).

West Indian lemongrass oil contains citral (65–85%) as its major component.[5,6] Cameroonian *C. citratus* contains geranial (33%) as the major component.[7] Other compounds present include myrcene (12–20%), dipentene, methylheptenone, β-dihydropseudoionone, neral, β-pinene, alcohols (linalool, methylheptenol, α-terpineol, geraniol, nerol, farnesol, citronellol, etc.), volatile acids (isovaleric, geranic, caprylic, citronellic, etc.), and others (GUENTHER; JIANGSU; LIST AND HÖRHAMMER).[7]

East Indian lemongrass contains a volatile oil in about 0.5% yield from fresh grass. It probably also contains many of the other constituents present in West Indian lemongrass.

East Indian lemongrass oil contains normally citral as its major component in a 70–85% concentration.[8] Other components include geraniol and methyleugenol as well as many of the compounds present in the West Indian oil (e.g., dipentene, myrcene, methylheptenol, farnesol, n-decanal, guanic acid, and others).

Geraniol-rich strains of East Indian lemongrass have been reported to yield oils that contain citral only as a minor component (10–20%), with their major components being gerianol (35–50%) and methyl eugenol (ca. 20%).[9,10] Another type is reported to contain no citral at all but has borneol (ca. 30%) as one of its major components (LIST AND HÖRHAMMER).

East Indian lemongrass oil usually contains a slightly higher content of citral than West Indian lemongrass oil; it is also more soluble in 70% alcohol than the West Indian oil (GUENTHER).[11]

PHARMACOLOGY AND BIOLOGICAL ACTIVITIES

West Indian lemongrass oil has antimicrobial properties, especially against Gram-positive bacteria and fungi.[6,12,13] The oil was shown to possess a fungicidal effect on Saccharomyces cerevisiae at 4 µmol/L.[14] The powdered dry leaves and the oil prevented the growth and aflatoxin production of four moulds (*Aspergillus* and *penicillium*) at 0.1 mg/mL.[15,16] Lemongrass oil had a bactericidal effect on Helicobacter pylori both *in vitro* and *in vivo* without development of resistance.[17] The essential oil completely inhibited *in vitro* HSV-1 viral replication at 0.1% concentration.[18] Lemongrass oil also displayed a significant synergistic effect with antimicrobial agents against *Staphylococcus aureus* in the agar diffusion assay.[19]

The oil has *in vivo* antimalarial activity in mice infected with *Plasmodium berghi*.[7] It also has an extended insect repellent effect against mosquitoes, lice, and other insects.[20–22]

Lemongrass oil has been reported to have CNS-depressant effects as well as analgesic, antinociceptive, antipyretic, and antioxidant properties.[12,23,24] The antioxidant activity was displayed by some oil components such as citral and phenolic constituents.[4,25]

The oil has antimutagenic and antihepatocarcinogenic effects both *in vivo* and *in vitro*.[26–28] Citral is also an inducer of caspase-3 and apoptosis in hematopoietic and cancer cell lines at 44.5 µM.[29]

Available data from one source indicate the oil to be mildly to moderately irritating to the skin of experimental animals but nonirritating and nonsensitizing to human skin. Its phototoxicity on human skin has not been determined.[30]

The aqueous extract of the leaves lowered the rate of isolated rat hearts without affecting their contractile force.[31]

East Indian lemongrass oil has been reported to exhibit antifungal properties.[32,33] Available data from one source indicate it to be mildly to moderately irritating to the skin of laboratory animals but nonirritating and nonsensitizing to human skin. Its phototoxicity on human skin is not known.[34]

TOXICOLOGY

Citral has been reported to produce sensitization reactions in humans when applied alone but to produce no such reactions when applied as a mixture with other compounds.[35]

A recent study found that lemongrass oil has no adverse effects on the blood, liver function, kidney function, protein, carbohydrate, and lipid metabolism of rats;[36] and other studies have failed to detect mutagenic or toxicologic reactions in humans.[37–39]

USES

Medicinal, Pharmaceutical, and Cosmetic. Lemongrass oil (especially the West Indian type) is used extensively as a fragrance component in soaps arid detergents. Also used in creams, lotions, and perfumes, with maximum use level of 0.7% reported for both types of oil in perfumes.[30,34]

Food. Lemongrass oil is used in most major categories of foods, including alcoholic and nonalcoholic beverages, frozen dairy desserts, candy, baked goods, gelatins and puddings, meat and meat products, and fats and oils. Highest average maximum use levels reported are about 0.003% and 0.004%, respectively, in candy (33.3 ppm) and baked goods (36.3 ppm).

Dietary Supplements/Health Foods. Dried leaves widely used as a "lemon" flavor ingredient in herb teas, and other formulations.

Traditional Medicine. West Indian lemongrass is used in Chinese medicine to treat colds, headache, stomachache, abdominal pain, rheumatic pain, and others (JIANGSU).

Others. Lemongrass oils (both types) are used as starting materials for the synthesis of ionones and vitamin A as well as the production (isolation) of natural citral. The oil possesses biological activity against storage pests and has been used as a postharvest pesticide for some food commodities.[36]

COMMERCIAL PREPARATIONS

Both East and West Indian lemongrass oils; they are official in F.C.C. Also crude.

Regulatory Status. GRAS (§182.20). Subject of a German therapeutic monograph as a mild astringent and stomachic; efficacy not documented.[39]

REFERENCES

See the General References for ARCTANDER; BAILEY 1; DER MARDEROSIAN AND BEUTLER; DER MARDEROSIAN AND LIBERTI; FEMA; GUENTHER; LIST AND HÖRHAMMER; MARTINDALE; MCGUFFIN 1 & 2.

1. K. R. Kirtikar and B. D. Basu, *Indian Medicinal Plants*, International Book Distributors, Dehradun, India, 1999, p. 2681.
2. A. A. Olaniyi et al., *Planta Med.*, **28**, 186 (1975).
3. S. W. Hanson et al., *Phytochemistry*, **15**, 1074 (1976).
4. J. Cheel et al., *J. Agric. Food Chem.*, **53**, 2511 (2005).
5. M. Saleem et al., *Nat. Prod. Res.*, **17**, 159 (2003).
6. C. K. Kokate and K. C. Varma, *Sci. Cult.*, **37**, 196 (1971).
7. F. Tchoumbougnang et al., *Planta Med.*, **71**, 20 (2005).
8. K. S. Ayyar et al., *Perfum. Essent. Oil Records*, **59**, 669 (1968).
9. R. K. Thappa et al., *Cultiv. Util. Med. Aromat. Plants*, 227 (1977).
10. C. K. Atal and B. L. Bradu, *Indian J. Pharm.*, **38**, 63 (1976).
11. W. D. Fordham, *Chemical Technology: An Encyclopedia Treatment*, Barnes and Noble, New York, 1972, p. 1.
12. D. O. Gyane, *Drug Cosmet. Ind.*, **118**, 36 (1976).
13. F. M. Ramadan et al., *Chem. Mikrobiol. Technol. Lebensm.*, **1**, 96 (1972).
14. G. A. Helal et al., *J. Basic Microbiol.*, **46**, 375 (2006).
15. S. A. Bankole et al., *J. Basic Microbiol.*, **45**, 20 (2005).
16. P. A. Paranagama et al., *Lett. Appl. Microbiol.*, **37**, 86 (2003).
17. T. Ohno et al., *Helicobacter*, **8**, 207 (2003).
18. M. Minami et al., *Microbiol. Immunol.*, **47**, 681 (2003).
19. J. E. Betoni et al., *Mem. Inst. Oswaldo Cruz*, **101**, 387 (2006).
20. K. K. Wong et al., *J. Agric. Food Chem.*, **53**, 4633 (2005).

21. K. Y. Mumcuoglu et al., *Isr. Med. Assoc. J.*, **6**, 756 (2004).
22. A. O. Oyedele et al., *Phytomedicine*, **9**, 259 (2002).
23. G. Seth et al., *Indian J. Exp. Biol.*, **14**, 370 (1976).
24. G. S. Viana et al., *J. Ethnopharmacol.*, **70**, 323 (2000).
25. S. I. Rabbani et al., *Pak. J. Pharm. Sci.*, **19**, 108 (2006).
26. R. Puatanachokchai et al., *Cancer Lett.*, **183**, 9 (2002).
27. R. Suaeyun et al., *Carcinogenesis*, **18**, 949 (1997).
28. U. Vinitketkumnuen et al., *Mutat. Res.*, **341**, 71 (1994).
29. N. Dudai et al., *Planta Med.*, **71**, 484 (2005).
30. D. L. J. Opdyke, *Food Cosmet. Toxicol.*, **14**, 457 (1976).
31. R. Gazola et al., *Pharmacol. Res.*, **50**, 477 (2004).
32. B. G. V. N. Rao and P. L. Joseph, *Riechst., Aromen, Korperpflegem.*, **21**, 405 (1971).
33. A. Dikshit and A. Husain, *Fitoterapia*, **55**, 171 (1984).
34. D. L. J. Opdyke, *Food Cosmet. Toxicol.*, **14**, 455 (1976).
35. D. L. J. Opdyke, *Food Cosmet. Toxicol.*, **14**, 197 (1976).
36. A. K. Mishra et al., *Phytother. Res.*, **6**, 279 (1992).
37. E. A. Carlinin et al., *J. Ethnopharmacol.*, **17**, 37 (1986).
38. J. R. Leite et al., *J. Ethnopharmacol.*, **17**, 65 (1986).
39. Monograph, *Cympogon species*, *Bundesanzeiger*, no. 22 (February 1, 1990).

LICORICE ROOT

Source: ***Spanish licorice*** *Glycyrrhiza glabra* L. var. *typica* Reg. et Herd.; ***Persian licorice*** *Glycyrrhiza glabra* L. var. *violacea* Boiss.; ***Russian licorice*** *Glycyrrhiza glabra* L. var. *glandulifera* Waldst. et Kit.; ***Chinese licorice*** *Glycyrrhiza uralensis* Fisch. (Family Leguminosae or Fabaceae).

Common/vernacular names: Italian licorice (*G. glabra* var. *typica*), Turkish licorice (*G. glabra* var. *violacea*), glycyrrhiza, and sweet wood.

GENERAL DESCRIPTION

Perennial herbs or subshrubs generally with horizontal underground stems (stolons or runners); up to 1–2 m high, with Chinese licorice being the smallest plant among the four; native to Eurasia and cultivated in Europe (Spain, Italy, France, etc.), Middle East (Syria, Iran, Turkey, Iraq, etc.), and Asia (e.g., China). Parts used are the dried runners and roots collected in the fall; some types of licorice are peeled. Much of the licorice used in the United States is imported in an extract form, usually in sticks or solid blocks. These extracts are usually prepared by hot water extraction followed by evaporating off the water and drying the extracts to form sticks or blocks.

A commonly used form of licorice extract is a concentrated extract containing large amounts of glycyrrhizin in its ammonium salt form; it is not pure ammonium glycyrrhizin, though the pure form is also used. Glycyrrhizin is 50 times sweeter than sucrose and has synergistic effects with sucrose and other sugars; its sweet taste is lost in an acidic medium.

Major producers of licorice include Spain, Iraq, Iran, Turkey, Russia, and China.

CHEMICAL COMPOSITION

Licorice is one of the most extensively investigated economic plant products. It contains as its major active principle the triterpene glycoside glycyrrhizin (also known as glycyrrhizic or glycyrrhizinic acid) in concentrations ranging from 1% to 24%, depending on sources and methods of assay; a 10-fold difference in glycyrrhizin values due solely to different assay methods has been reported.[1] Glycyrrhizin on hydrolysis yields glycyrrhetinic (or glycyrrhetic) acid and two molecules of glucuronic acid.

Other constituents of licorice include flavonoids, isoflavonoids and pterocarpans (licoflavonol, kumatakenin, licoricone, glabrol, glabrone, glyzarin, licoisoflavones A and B, licoisoflavanone, glycyrol, formononetin, liquiritigenin, liquiritin, neoliquiritin, rhamnoliquiritin, glyzaglabrin, 7-hydroxy-2-methylisoflavone, 4′,7-dihydroxyflavone, glabranine, glabridin, etc.),[2,2–19] chalcones (liquiritigenin, isoliquiritigenin, isoliquiritin, neoisoliquiritin, licuraside, rhamnoisoliquiritin, echinatin, licochalcones A and B, 4-hydroxychalcone, etc.),[2,4,7,13,20] coumarins (umbelliferone, herniarin, liqcoumarin, glycyrin, etc.),[3,21] triterpenoids (liquiritic acid, glycyrrhetol, glabrolide, isoglabrolide, licoric acid, β-amyrin, 18-β-glycyrrhetinic acid, etc.),[22] triterpene saponin glycosides (licorice saponins A–K),[23–25] sterols (β-sitosterol, stigmasterol, 22,23-dihydrostigmasterol etc.),[22] 2–20% starch, 3–14% sugars (glucose and sucrose), lignin, amino acids (proline, serine, aspartic acid, etc.),[26] amines (asparagine, betaine, choline), gums, wax, a volatile oil consisting of many aroma chemicals (including acetol, 2-acetylfuran, propionic acid, 2-acetylpyrrole, furfuryl alcohol, benzaldehyde, pentanol, hexanol, *trans*-hex-3-en-1-ol, oct-l-en-3-ol, linalool, linalyl oxide, α-terpineol, butyrolactone, thujane, and fenchone, among others, none of which alone can account for the licorice flavor),[27–29] and others (JIANGSU; LIST AND HÖRHAMMER). Two new prenylflavones, licoflavones B and C, and a new dibenzoylmathane, glycyridione, were isolated from *G. inflata*, along with the known flavones licoflavone A and 4′,7-dihydroxyflavone.[30]

Glycyrrhizin has been found to concentrate in the woody parts of the thickened root and stolon of *G. glabra*. Betulinic acid and soyasaponins have been previously reported from *G. glabra* cell cultures. Recently, soyasaponins (oleanane-type triterpene glycosides) have been reported in all parts of the intact plant, especially the seeds and young roots, particularly rootlets.[31]

The number of constituents and their relative concentration vary with the sources and types of licorice (JIANGSU; LIST AND HÖRHAMMER; MORTON 3; NANJING).

PHARMACOLOGY AND BIOLOGICAL ACTIVITIES

Licorice is known to exhibit many pharmacological activities, including estrogenic *in vitro* and in laboratory animals,[22,32–40] antiulcer,[41–45] cancer chemoprevention and inhibition of tumor growth (extract, flavonoids glycyrrhetinic acid salt, and a derivative on various cell lines),[15,35,46–49] antitrichomonas,[50] antibacterial, antiviral,[51–53] anti-inflammatory,[45,54–56] antiallergic,[57] antitoxic, antitussive (comparable to codeine, due to a derivative of 18-β-glycyrrhetinic acid) and anticonvulsive,[58] among others (FARNSWORTH 1–4; JIANGSU; LIST AND HÖRHAMMER).

Liquiritigenin and isoliquiritigenin have MAO-inhibitory activities.[59]

Glycyrrhetinic acid inhibits 11-β-hydroxysteroid dehydrogenase in rats, and potentiates the action of hydrocortisone in humans.[60]

In screening for novel anxiolytics using a benzodiazepine-binding assay, a dichloromethane extract of *G. uralensis* was found to stimulate the binding of benzodiazepine to rat synaptosomes.[61]

Licochalcones A and B from *G. inflata* strongly inhibited lysosomal enzyme release and degranulation from human polymorphonuclear neutrophils.[62]

The flavonoids have been shown to have strong antioxidant and antihepatotoxic activities.[63,64] Similarly, licorice extract and glycyrrhizic acid have hepatoprotective effect.[65] Roasted licorice has recently demonstrated neuroprotective effects against neuronal damage in gerbils.[66]

Licorice, glycyrrhetinic acid, and glabridin modulate calcium metabolism, enhance osteoblast functions, and reduce osteoporosis and bone diseases.[2,67]

Based on recent trials, licorice may have potential in the treatment of atopic dermatitis and as a hypocholesterolemic dietary supplement.[68,69]

Licorice extracts have been used in China in the clinical treatment of numerous illnesses (gastric and duodenal ulcers, bronchial asthma, infectious hepatitis, malaria, diabetes insipidus, contact dermatitis, etc.) with considerable success (JIANGSU).

TOXICOLOGY

Chronic ingestion of licorice produces mineralocorticoid effect with sodium retention and potassium loss leading to hypertension (due to glycyrrhizin);[35,70–74] The safety, pharmacology, and chemistry of licorice have been reviewed.[75–77]

USES

Medicinal, Pharmaceutical, and Cosmetic. Licorice extracts are used extensively as ingredients in cough drops and syrups, tonics, laxatives, antismoking lozenges (see *lobelia*), and other preparations. They are also used as flavoring agents to mask bitter, nauseous, or other undesirable tastes in certain medicines (e.g., cascara, ammonium chloride, and quinine preparations).

Average daily doses of 5–15 g root (calculated to 200–600 mg of glycyrrhizin) or root juice (0.5–1 g for respiratory tract catarrhs, or 1.5–3 g for gastric duodenal ulcers) are used in European phytomedicine. Duration is limited to 4–6 weeks, because of potential adverse side effects (described above). Use is contraindicated for cholestatic liver disorders, cirrhosis, hypertonia, pregnancy, and others.[58] Known drug interactions include potassium loss due to thiazine diuretics, as well as increased sensitivity to digitalis glycosides.[58]

Food. Licorice is widely used in flavoring foods.[78,79]

The most well-known use of licorice and its extracts as well as ammoniated glycyrrhizin is in licorice candy where they are mixed with anise oil (see *anise*), with average maximum use levels of about 3.279% (32,792 ppm) and 0.151% (1512 ppm) reported for the powdered extract and ammoniated glycyrrhizin, respectively. Licorice, its extracts, and ammoniated glycyrrhizin are also used in many other food products, including alcoholic (certain kinds of beer) and nonalcoholic (e.g., root beer) beverages, frozen dairy desserts, baked goods, gelatins and puddings, and meat and meat products. Average maximum use levels reported are below 0.25% for licorice and licorice extracts; the use levels reported for ammoniated glycyrrhizin are usually below 0.01%.

Dietary Supplements/Health Foods. Root, powdered or cut and sifted, is widely used as tea ingredient; and in capsules, tablets, tinctures, and other dietary supplement formulations for flavoring and traditional indications; extracts also used in capsules, tablets, and drinks.

Traditional Medicine. The medicinal use of licorice in both Western and Eastern cultures dates back several thousand years.[80] Many of its uses are the same in both cultures. It is commonly used as a demulcent, expectorant, antitussive, and mild laxative.

Licorice is one of the most popular drugs in Chinese medicine and is often an ingredient in Chinese prescriptions. Its traditional uses include treating ulcers (gastric and duodenal), sore throat, malaria, abdominal pain, insomnia, tuberculosis (consumption), sores, abscesses, and food poisoning, among others. For food poisoning and poisoning due to swallowing of poison of undetermined nature, equal amounts of licorice and black beans or mung beans are decocted together and the liquid taken orally.

Licorice has been reported used in many countries to treat cancer.[81]

Others. Licorice extracts are used in flavoring tobaccos.

COMMERCIAL PREPARATIONS

Crude, extracts, and ammoniated glycyrrhizin. Licorice syrup was formerly official in U.S.P. Licorice, licorice extract (solid), and licorice fluid extract are official in N.F. The strength (see *glossary*) of the fluid extract is 1:1, but that of the solid extract is not precisely specified in N.F.

Regulatory Status. GRAS (§182.10 and §182.20) with both licorice and glycyrrhiza listed; ammoniated glycyrrhizin is also GRAS (§182.20). Licorice root is the subject of a German therapeutic monograph, indicated for catarrhs of the upper respiratory tract and gastric/duodenal ulcers.[58]

REFERENCES

See the General References for ARCTANDER; BAILEY 1; BLUMENTHAL 1 & 2; FEMA; FOSTER AND YUE; JIANGSU; MARTINDALE; MORTON 3; NANJING.

1. J. Killacky et al., *Planta Med.*, **30**, 310 (1976).
2. E. M. Choi, *Biochem. Pharmacol.*, **70**, 363 (2005).
3. D. K. Bhardwaj et al., *Phytochemistry*, **16**, 402 (1977).
4. M. Hoton-Dorge, *J. Pharm. Belg.*, **29**, 560 (1974).
5. T. Saitoh et al., *Chem. Pharm. Bull.*, **24**, 752 (1976).
6. T. Kinoshita et al., *Chem. Pharm. Bull.*, **26**, 141 (1978).
7. T. Saitoh et al., *Chem. Pharm. Bull.*, **26**, 144 (1978).
8. D. K. Bhardwaj and R. Singh, *Curr. Sci.*, **46**, 753 (1977).
9. D. K. Bhardwaj et al., *Phytochemistry*, **15**, 352 (1976).
10. N. S. Kattaev and G. K. Nikonov, *Khim. Prir. Soedin.*, **6**, 805 (1972).
11. T. Saitoh et al., *Chem. Pharm. Bull.*, **24**, 1242 (1976).
12. T. Kinoshita et al., *Chem. Pharm. Bull.*, **24**, 991 (1976).
13. T. Saitoh and S. Shibata, *Tetrahedron Lett.*, **50**, 4461 (1975).
14. C. Van Hulle, *Pharm. Tijdschr. Belg.*, **45**, 137 (1968).
15. B. Fu et al., *J. Agric. Food Chem.*, **53**, 7408 (2005).
16. T. Hatano et al., *Chem. Pharm. Bull. (Tokyo)*, **48**, 1286 (2000).
17. I. Kitagawa et al., *Chem. Pharm. Bull. (Tokyo)*, **42**, 1056 (1994).
18. T. Kinoshita et al., *Chem. Pharm. Bull. (Tokyo)*, **53**, 847 (2005).
19. S. Tsukamoto et al., *Biol. Pharm. Bull.*, **28**, 2000 (2005).
20. Y. Fu et al., *Biochem. Biophys. Res. Commun.*, **322**, 263 (2004).
21. T. Kinoshita et al., *Chem. Pharm. Bull.*, **26**, 135 (1978).
22. C. Van Hulle, *Pharmazie*, **25**, 620 (1970).
23. I. Kitagawa et al., *Chem. Pharm. Bull. (Tokyo)*, **41**, 1567 (1993).
24. I. Kitagawa et al., *Chem. Pharm. Bull. (Tokyo)*, **41**, 1337 (1993).
25. I. Kitagawa et al., *Chem. Pharm. Bull. (Tokyo)*, **41**, 43 (1993).
26. H. Nishi and I. Morishita, *Nippon Nogei Kagaku Kaishi*, **45**, 507 (1971).

27. B. Toulemonde et al., *Ind. Aliment Agric.*, **94**, 1179 (1977).
28. C. Frattini et al., *Chim. Ind. (Milan)*, **59**, 522 (1977).
29. C. Frattini et al., *J. Agric. Food Chem.*, **25**, 1238 (1977).
30. K. Kajiyama et al., *J. Nat. Prod.*, **55**, 1197 (1992).
31. H. Hayashi et al., *Planta Med.*, **59**, 351 (1993).
32. D. Somjen et al., *J. Steroid Biochem. Mol. Biol.*, **91**, 241 (2004).
33. D. Somjen et al., *J. Steroid Biochem. Mol. Biol.*, **91**, 147 (2004).
34. R. Ofir et al., *J. Mol. Neurosci.*, **20**, 135 (2003).
35. K. Y. Yen, *Pei I Hseuh Pao*, **5**, 23 (1973).
36. C. H. Costello and E. V. Lynn, *J. Am. Pharm. Assoc.*, **39**, 177 (1950).
37. I. A. Murav'ev and N. F. Kononikhina, *Rast. Resur.*, **8**, 490 (1972).
38. A. Sharaf and N. Goma, *J. Endocrinol.*, **31**, 289 (1965).
39. M. I. Elghamry et al., *Zentr. Veterinarmed.*, **11**, 70 (1964).
40. I. M. Shihata and M. I. Elghamry, *Zentr. Veterinarmed. Ser. A.*, **10**, 155 (1963).
41. Y. T. WANG et al., *Tai-Wan I Hseuh Hui Tsa Chih*, **71**, 256 (1972).
42. K. Takagi et al., *Jpn. J. Pharmacol.*, **21**, 832 (1971).
43. S. Y. Chow et al., *Chung-Hua I Hseuh Tsa Chih (Taipei)*, **23**, 217 (1976).
44. Y. Ishii and N. Sugawara, *Oyo Yakuri*, **7**, 871 (1973).
45. A. M. Aly et al., *AAPS PharmSciTech.*, **6**, E74 (2005).
46. E. H. Jo et al., *Cancer Lett.*, **230**, 239 (2005).
47. T. Takahashi et al., *Cancer Sci.*, **95**, 448 (2004).
48. M. M. Rafi et al., *J. Agric. Food Chem.*, **50**, 677 (2002).
49. I. F. Shvarev et al., *Vop. Izuch. Ispol'z. Solodki SSSR, Akad. Nauk SSSR*, **167** (1966).
50. S. A. Vichkanova and M. A. Rubinchik, *Vop. Izuch. Ispol'z. Solodki SSSR, Akad. Nauk SSSR*, **176**(1966).
51. J. Cinatl et al., *Lancet*, **361**, 2045 (2003).
52. L. Badam, *J. Commun. Dis.*, **29**, 91 (1997).
53. T. Utsunomiya et al., *Antimicrob. Agents Chemother.*, **41**, 551 (1997).
54. J. K. Kim et al., *Biochem. Biophys. Res. Commun.*, **345**, 1215 (2006).
55. I. Furuhashi et al., *J. Pharm. Pharmacol.*, **57**, 1661 (2005).
56. J. S. Kang et al., *J. Pharmacol. Exp. Ther.*, **312**, 1187 (2005).
57. H. Q. Zhang et al., *Zhongguo Yaoli Xuebao*, **7**, 175 (1986).
58. Monograph *Liquiritiae radix*, Bundesanzeiger, no. 90 (May 15, 1985); corrected (March 13, 1990; April 4, 1991).
59. S. Tanaka et al., *Planta Med.*, **53**, 5 (1987).
60. S. Teelucksingh et al., *Lancet*, **335**, 1060 (1990).
61. Y. K. Tony Lam, *Planta Med.*, **58**, 221 (1991).
62. Y. Kimura et al., *Phytother. Res.*, **7**, 335 (1993).
63. H. S. Ju et al., *Yaoxue Xuebao*, **24**, 807 (1989).
64. G. S. WANG and Z. W. Han, *Yaoxue Xuebao*, **28**, 572 (1993).
65. Y. T. Wu et al., *Phytother. Res.*, **20**, 640 (2006).
66. I. K. Hwang et al., *Acta Pharmacol. Sin.*, **27**, 959 (2006).
67. M. J. Mattarello et al., *Steroids*, **71**, 403 (2006).
68. M. Saeedi et al., *J. Dermatolog. Treat.*, **14**, 153 (2003).
69. B. Fuhrman et al., *Nutrition*, **18**, 268 (2002).
70. A. Janse et al., *Neth. J. Med.*, **63**, 149 (2005).

71. A. E. van den Bosch et al., *Neth. J. Med.*, **63**, 146 (2005).
72. S. H. van Uum, *Neth. J. Med.*, **63**, 119 (2005).
73. R. M. Salassa et al., *J. Clin. Endocrinol. Metab.*, **22**, 1156 (1962).
74. M. T. Epstein et al., *Br. Med. J.*, **19**, 488 (1977).
75. R. A. Isbrucker and G. A. Burdock, *Regul. Toxicol. Pharmacol.*, (2006).
76. Z. Y. Wang and D. W. Nixon, *Nutr. Cancer*, **39**, 1 (2001).
77. S. Shibata, *Yakugaku Zasshi*, **120**, 849 (2000).
78. M. K. Cook, *Flav. Ind.*, **2**, 155 (1971).
79. M. K. Cook, *Food Eng.*, **145**(1973).
80. M. R. Gibson, *Lloydia*, **41**, 348 (1978).
81. J. L. Hartwell, *Lloydia*, **33**, 97 (1970).

LIGUSTRUM

Source: *Ligustrum lucidum* Ait. (Family Oleaceae).

Common/vernacular names: Glossy privet, *nuzhenzi, dongqingzi.*

GENERAL DESCRIPTION

Erect large evergreen shrub or small tree, reaching over 10 m high; fruit a kidney shaped berry with one to two seeds, bluish black when ripe, about 1 cm long and 3–4 mm in diameter; native to eastern Asia; distributed throughout China; now cultivated as an ornamental plant in the United States. Part used is the ripe fruit, mostly wild crafted from October to December, rid of impurities and sun dried, or first briefly steamed and then sun dried; produced largely in southeastern provinces (ZHU).

Its Chinese synonym, *dongqingzi* can also be the fruit of *Ilex chinensis* Sims, another fairly commonly used drug, which is oval and slightly smaller, with four to five seeds. The two cannot be positively differentiated and identified in the literature unless Latin binomials or other collaborative information are also given (CMH).

CHEMICAL COMPOSITION

Contains triterpenoids, including oleanolic acid (ligustrin, 0.7–4.3%), acetyloleanolic acid, and ursolic acid;[1–5] secoiridoid glycosides (ligustroside, 10-hydroxyligustroside, lucidumosides A–C, oleuropein, 10-hydroxyoleuropein, nuezhenide, isonuezhenide, neonuezhenide, 4-hydroxy-β-phenylethyl-β-D-glucoside, etc.);[6,7] mannitol; fatty oil (10–15%), composed mainly of linoleic, linolenic, oleic, and palmitic acids, with presence of stearic acid not confirmed (HU, IMM-3);

Ligustrum also contains a unique volatile oil consisting primarily of esters and alcohols, with lesser amounts of thioketones, hydrocarbons, and traces of amines and aldehydes, but no terpene hydrocarbons.[8,9] Major components of the volatile oil include ethyl acetate (18.95%), thioketone (8.56%), α-butyl-benzenemethanol (5.6%), 4-acetyloxy-2-butanone (5.46%), 1-phenyl-1,2-butanediol (4.12%), 1,2-diphenyl-1,2-ethanediol (3.92%), hydrazine-methyl-oxalate (3.52%), α,α,4-trimethyl-3-cyclohexene-methanol (3.24%), 1-methyl-1-propyl-hydrazine (2.60%), and (Z)-1-(1-ethoxy-ethoxy)-3-hexene (1.89%).[9]

PHARMACOLOGY AND BIOLOGICAL ACTIVITIES

Ligustrum is one of the highly valued tonics in traditional Chinese medicine and has been shown to exhibit numerous biological effects in humans and experimental animals.

Ligustrum preparations (i.g. or i.m.) inhibited or prevented the leukopenia caused by chemotherapy (cyclophosphamide) and radiotherapy in mice and in cancer patients and are now clinically used for treating leukopenia in China. Extracts also have immunomodulating and antimutagenic effects both in humans (phase II clinical trial) and in experimental animals;[10–15] The acetone-soluble extractives had strong inhibitory effects on aflatoxin B1-induced mutation in *Salmonella typhimurium* strains TA_{98} and TA_{100}.[16]

The decoction of wine-cured ligustrum (steam-heated with wine, 20% w/w, until absorbed and dried) exhibited marked anti-inflammatory effects on different experimental inflammation models in animals.[17]

Water-soluble extractives increased blood flow volume in isolated rabbit coronary vessels; crude powder when fed to hyperlipemic rabbits (20 g/day, i.g., 30–60 days) markedly lowered blood lipids and prevented formation of atherosclerotic lesions;[16] decoction (i.g.) significantly decreased blood glucose levels in normal and alloxan diabetic mice (HU, WANG).[18]

Recent studies showed that ligustrum may be helpful in the treatment of PMS osteoporosis through its modulating effect on calcium balance and bone turnover.[19] Other recently reported activities include antibacterial against periodontic pathogens,[20] antiviral against HSV-1, Flu A, Para 31, and RSV viruses,[21] and antioxidant radical-scavenging activity.[6,21,22]

It is generally accepted that oleanolic acid (ligustrin) is responsible for most of the biological effects of ligustrum (e.g., immunomodulating, anti-inflammatory, and hypolipemic and antiatherosclerotic). In addition, it also has hepatoprotective effects (lowering SGPT, protecting from carbon tetrachloride damage),[23] as well as antiallergic, mild cardiotonic, diuretic, sedative, and antitumor (vs. S-180) effects.[2,11,24–26]

TOXICOLOGY

Oral toxicity is low: a single dose of 75 g of ripe fruit fed to rabbits did not elicit any toxic symptoms (WANG).

USES

Medicinal, Pharmaceutical, and Cosmetic. Powder and extracts of ligustrum are used in hair-growth products and in formulas for removing facial dark spots primarily intended for internal use for its traditionally alleged ability to prevent premature graying and its "beautifying" properties (per Li Shi-Zhen's *Ben Cao Gang Mu*, 1590).

Dietary Supplements/Health Foods. Powder and extracts are used as a tonic (usually combined with other herbs) in tablet, capsule, tea, or liquid (drink) form to improve body resistance to illnesses; also used in soup mixes.

Traditional Medicine. 7The first written record dates back to the *Shan Hai Jing* (ca. 800 BC); traditionally considered to be bitter and sweet tasting and neutral. It is one of the major *yin* tonics with vision-brightening and hair-darkening properties (*ming mu wu fa*). It invigorates the liver and kidney and is traditionally used for treating premature graying of hair, dizziness, and tinnitus, sore back and knees, and blurred vision; now also used to treat habitual constipation in the elderly as well as chronic benzene poisoning (NATIONAL).

Others. As it contains the highest concentration of oleanolic acid (4.33%) among 216 Chinese herbal drugs from 18 genera tested, ligustrum is a potential source of this compound.[5]

COMMERCIAL PREPARATIONS

Mainly crude (whole or powdered) and extracts.

Regulatory Status. Class 1 dietary supplement (can be safely consumed when used appropriately). It is an ethnic food.

REFERENCES

See the General References for BAILEY 1; CHP; CMH; FOSTER AND YUE; JIXIAN; HU; IMM-3; JIANGSU; HUANG; LU AND LI; MCGUFFIN 1 & 2; NATIONAL; WANG; ZHU.

1. H. Liu et al., *J. Pharm. Biomed. Anal.*, **32**, 479 (2003).
2. B.Wang and C. H. Jiang, *Zhongguo Yaoxue Zazhi*, **27**, 393 (1992).
3. Y. Rong and X. C. Ye, *Zhongguo Zhongyao Zazhi*, **14**, 41 (1989).
4. H. Y. Liu et al., *Zhongcaoyao*, **24**, 219 (1993).
5. N. J. Wu et al., *Zhongcaoyao*, **23**, 467 (1992).
6. Z. D. He et al., *Chem. Pharm. Bull. (Tokyo)*, **49**, 780 (2001).
7. Z. D. He et al., *Phytochemistry*, **56**, 327 (2001).
8. N. J. Wu et al., *Zhongcaoyao*, **24**, 4 (1993).
9. K. H. Li and C. S. Li, *Zhongchengyao*, **12**, 32 (1990).
10. D. Ma et al., *Arch. Anim. Nutr.*, **59**, 439 (2005).
11. Y. Sun et al., *Chin. J. Clin. Pharmacol.*, **6**, 72 (1990).
12. Y. Sun et al., *J. Biol. Response Mod.*, **2**, 227 (1983).
13. B. H. S. Lau et al., *Phytother. Res.*, **3**, 148 (1989).
14. M. Shoemaker et al., *Phytother. Res.*, **19**, 649 (2005).
15. M. Niikawa et al., *Mutat. Res.*, **319**, 1 (1993).
16. B. Zhou et al., *Shiyong Zhongxiyi Jiehe Zazhi (PJCM)*, **6**, 168 (1993).
17. Y. Dai et al., *Zhongguo Zhongyao Zazhi*, **14**, 47 (1989).
18. Z. Q. Hao et al., *Zhongguo Zhongyao Zazhi*, **17**, 429 (1992).
19. Y. Zhang et al., *Biol. Pharm. Bull.*, **29**, 291 (2006).
20. Q. Wang et al., *Zhonghua Kou Qiang Yi Xue Za Zhi*, **37**, 388 (2002).
21. S. C. Ma et al., *Chem. Pharm. Bull. (Tokyo)*, **49**, 1471 (2001).
22. H. Li and Q.Wang, *Anal. Bioanal. Chem.*, **378**, 1801 (2004).
23. T. K. Yim et al., *Phytother. Res.*, **15**, 589 (2001).
24. Y. Dai et al., *Zhongguo Yaoli Xuebao*, **10**, 381 (1989).
25. Y. Dai et al., *Zhongguo Yaoli Xuebao*, **9**, 562 (1988).
26. Y. Dai et al., *Chin. J. Pharmacol. Toxicol.*, **3**, 96 (1989).

LIME OIL

Source: *Citrus aurantifolia* (Christm.) Swingle (syn. *C. medica* L. var. *acida* Brandis) (Family Rutaceae).

Common/vernacular names: Bitter orange, Sour or Seville orange.

GENERAL DESCRIPTION

Evergreen tree with stiff sharp spines; up to about 4.5 m high; leaves medium sized, ovate, bluntly pointed, at tips, rounded to cunate at base; flowers large fragerant;[1] native to southern Asia; cultivated in south Florida, the West Indies (e.g., Cuba),[2] and Central America (e.g., Mexico). Part used is the fresh peel of the green unripe fruit from which lime oil is obtained by cold expression (expressed lime oil). Distilled lime oil is obtained by steam distillation of the whole crushed fruit or juice of the crushed fruit. Another essential oil, centrifuged lime oil, is obtained by centrifuging the pulp and oil mixture of the fruit in high-speed centrifuges, thus separating the oil from the pulp.

Distilled lime oil is of much more economic importance than the other two lime oils. Major producers of the distilled oil include Mexico and the West Indies. Terpeneless lime oil is produced from the distilled oil.

CHEMICAL COMPOSITION

Distilled lime oil contains a large amount (ca. 75%) of terpene hydrocarbons of which d-limonene is a chief component, with α- and β-pinenes, camphene, sabinene, terpinolene, α-phellandrene, γ-terpenene, p-cymene, and others also reported present. Other types of components include oxygenated compounds (citral, α-terpineol, 1,8-cineole, 1,4-cineole, linalool, fenchol, etc.) and sesquiterpenes (e.g., α-bergamotene, β-caryophyllene, and β-bisabolene).[3–5] Germacrene B (0.35%) has been found to be an important fragrance component (with a sweet, woody-spicy; geranium-like note) for distinguishing lime oil from lemon oil.[6,7]

Expressed (cold-pressed) lime oil contains similar constituents as the distilled oil, but with lesser amounts of reaction or degradation products such as p-cymene. It also contains anthranilates as well as large amounts (ca. 7%) of substituted coumarins (limettin, bergapten, imperatorin, isoimperatorin, bergamottin, isopimpinellin, phellopterin, 8-geranoxypsoralen, oxypeucedanin hydrate, 5-geranoxy-7-methoxycoumarin, and 6,7-dimethoxycoumarin, with limettin being the major coumarin present).[5,8–13]

2,4,6-Trichloroanisole, a compound with a strong musty naphthalenic odor, has been isolated in trace amounts from distilled Mexican lime oil as well as from French geranium, Italian lemon, and American peppermint oils. It is believed to be of microbial rather than pesticide origin.[14]

PHARMACOLOGY AND BIOLOGICAL ACTIVITIES

Lime oil (30% in mustard oil) has a mosquito repellent effect that lasts for up to 5 h.[15] The oil also has insecticidal activity against mosquitoes, cockroaches, and houseflies.[16]

TOXICOLOGY

The phototoxicity of certain substituted coumarins is well documented (see **bergamot**). Expressed lime oil contains more of such compounds than the other citrus oils (e.g., **bergamot** and **grapefruit**) and has been reported to be phototoxic to humans.[17]

Both expressed lime oil and distilled lime oil have been demonstrated to promote tumor formation on the skin and in the forestomach epithelium of mice treated with 9,10-dimethyl-1,2-benzanthracene and benzo-[a]-pyrene, respectively.[17–19]

Distilled lime oil has been reported to be nonirritating, nonsensitizing, and nonphototoxic to human skin, though it was slightly irritating to rabbit skin.[20]

USES

Medicinal, Pharmaceutical, and Cosmetic. Expressed lime oil and distilled lime oil are used as fragrance components and fixatives (expressed oil, due to its coumarins) in cosmetics, including soaps, detergents, creams, lotions, and perfumes, with maximum use level of 1.5% reported for both oils in perfumes.[17,20]

Food. Distilled lime oil and terpeneless lime oil are extensively used in carbonated beverages (especially the terpeneless type) and in hard candy (especially the distilled type) for the popular lemon-lime flavor. Other food products in which they are also used include alcoholic beverages, frozen dairy desserts, baked goods, gelatins and puddings, meat and meat products, and others. Highest average maximum use level reported is about 0.078% (783 ppm) for lime oil (distilled) in candy.

COMMERCIAL PREPARATIONS

Lime oil (distilled, expressed, and terpeneless). Distilled lime oil is official in F.C.C.

Regulatory Status. GRAS (§182.20).

REFERENCES

See the General References for ARCTANDER; BAILEY 2; FEMA; GUENTHER; BRUNETON; MCGUFFIN 1 & 2; MORTON 2.

1. W. Swingle, The Botany of Citrus and its Wild Relatives of the Orange Subfamily, 1943, p. 403.
2. J. A. Pino and R. Tapanes, *J. Food Technol.*, **18**, 523 (1983).
3. J. Perez Zayas and R. Tapanes, *Rev. CENIC, Cienc. Fis.*, **5**, 1 (1974).
4. C. A. Slater, *J. Sci. Food Agric.*, **12**, 732 (1961).
5. C. A. Slater, *Chem. Ind. (London)*, **833** (1961).
6. B. C. Clark et al., *J. Agric. Food Chem.*, **35**, 514 (1987).
7. B. M. Lawrence, *Perfum. Flav.*, **16**, 59 (1991).
8. W. L. Stanley and L. Jurd, *J. Agric. Food Chem.*, **19**, 1109 (1971).
9. J. H. Tatum and R. E. Berry, *Phytochemistry*, **16**, 1091 (1977).
10. W. L. Stanley and S. H. Vannier, *Phytochemistry*, **6**, 585 (1967).
11. L. Haro-Guzman and R. Heut, *Fruits*, **25**, 887 (1970).
12. G. L. K. Hunter and M. G. Moshonas, *J. Food Sci.*, **31**, 167 (1966).
13. S. Jiwajinda et al., *Biosci. Biotechnol. Biochem.*, **64**, 420 (2000).
14. J. Stoffelsma and K. B. de Roos, *J. Agric. Food Chem.*, **21**, 738 (1973).
15. N. G. Das et al., *J. Vector. Borne. Dis.*, **40**, 49 (2003).
16. F. C. Ezeonu et al., *Bioresour. Technol.*, **76**, 273 (2001).
17. D. L. J. Opdyke, *Food Cosmet. Toxicol.*, **12**, 731 (1974).
18. F. J. C. Roe and W. E. H. Field, *Food Cosmet. Toxicol.*, **3**, 311 (1965).
19. W. E. H. Field and F. J. C. Roe, *J. Natl. Cancer Inst.*, **35**, 771 (1966).
20. D. L. J. Opdyke, *Food Cosmet. Toxicol.*, **12**, 729 (1974).

LOBELIA

Source: *Lobelia inflata* L. (Family Campanulaceae).

Common/vernacular names: Indian tobacco, wild tobacco, asthma weed, gagroot, emetic herb, and vomit wort.

GENERAL DESCRIPTION

A hairy annual or biennial herb with alternate leaves with light blue flowers; up to about 1 m high; native to North America from Labrador to Georgia and west to Arkansas. Part used is the aboveground herb.

CHEMICAL COMPOSITION

Contains about 0.48% pyridine (piperidine) alkaloids composed mainly of lobeline, with lesser amounts of lobelanine and lobelanidine. Other alkaloids present include norlobelanine (isolobelanine), lelobanidine, norlelobanidine, norlobelanidine, lobinine, isolobinine, lobinanidine, isolobinanidine, 8-methyl-10-ethyllobelidiol, and 8-methyl-10-phenyllobelidiol, among others.[1,2]

It also contains resin, gum, fats, chelidonic acid, β-amyrin palmitate, and others (KARRER; LIST AND HÖRHAMMER; MARTINDALE).[3–5]

The formation of lobeline and related alkaloids by *L. inflata* in tissue culture has been reported.[6]

Apart from *L. inflata*, the pyrrolidine alkaloids radicamine A and B have been isolated from *L. chinensis*.[7] A bioactive (α-glucosidase inhibitor) homonojirimycin glycoside has been isolated from *L. sessilifolia*.[8]

PHARMACOLOGY AND BIOLOGICAL ACTIVITIES

Lobeline is reported to have many of the pharmacological properties of nicotine, although it is less potent. Like nicotine, its action is first CNS stimulation followed by severe depression (GOODMAN AND GILMAN).

Subarnas et al. reported that the methanolic extract of *L. inflata* and β-amyrin palmitate (10 mg/kg) isolated from the extract displayed sedative/antidepressant activity in the forced swimming test in mice. Activation of noradrenergic activity was suggested to be involved.[3,9,10]

Recent *in vitro* studies show that lobeline promotes the release of dopamine from presynaptic storage vesicles and inhibits its uptake. This effect is nonaddictive and makes lobeline a potential candidate for the treatment of psychostimulant abuse.[11,12] Lobeline (30–300 μM) was also found to dose-dependently inhibit the release of catecholamines from the rat adrenal gland *in vitro*.[13]

Lobeline also has expectorant activities, and in large doses is emetic, purgative, and diuretic.

TOXICOLOGY

Overdosage may lead to convulsions and collapse, with disputed fatal results (MARTINDALE).[14]

USES

Medicinal, Pharmaceutical, and Cosmetic. Lobeline salts are used as the active ingredient in several brands of antismoking (smoking deterrent) preparations. Lobelia and its extracts are used as ingredients in cough preparations and in counterirritant preparations.

Dietary Supplements/Health Foods. Dried herb decreasingly available in capsules, tinctures, etc.; rare in tea formulations. Concern over potential toxicity, most unsubstantiated, has resulted in most manufacturers dropping it from product lines. Self-administration in any form is not advised (TYLER 1).

Traditional Medicine. Used as an antispasmodic, antiasthmatic, diaphoretic, expectorant, emetic, and sedative. Conditions for which it is used include asthma, whooping cough, bruises, sprains, ringworm, insect bites, poison ivy symptoms, and others.

In Chinese medicine related *Lobelia* species (e.g., *L. chinensis* Lour. and *L. sessilifolia* Lamb.) are also used, the former primarily for jaundice and sores, wounds, ringworm, snakebite and insect bites, and malaria; while the latter is used as an expectorant and antitussive and in treating bronchitis, ulcerous sores, snakebites, and insect and dog bites. Both herbs also contain lobeline and related alkaloids.

Several *Lobelia* species including *L. inflata* and *L. chinensis* (syn. *L. radicans*, Thunb.) have been used in cancers.[15]

COMMERCIAL PREPARATIONS

Crude and extracts. Crude, fluid extract, and tincture were formerly official in N.F. Strengths (see **glossary**) of extracts are expressed in weight-to-weight ratios.

Regulatory Status. Undefined in the United States as food ingredient; as of 1993 lobeline is no longer allowed as an ingredient in antismoking preparations.[16]

REFERENCES

See the General References for APPLEQUIST; BARNES; FOSTER; FOSTER AND DUKE; GOSSELIN; JIANGSU; LEWIS AND ELVIN-LEWIS; MARTINDALE; MCGUFFIN 1 & 2; TYLER 1.

1. W. A. Ayer and T. E. Habgood in R. H. F. Manske, ed., *The Alkaloids*, Vol. 11, Academic Press, New York, 1968, p. 459.
2. D. Gross in W. Herz et al., eds., *Fortschritte der Chemie Organischer Naturstoffe*, Vol. 29, Springer-Verlag, Vienna, 1971, p. 1.
3. A. Subarnas et al., *Life Sci.*, **52**, 289(1993).
4. M. S. Karawya et al., *J. Assoc. Offic. Anal. Chem.*, **54**, 1423 (1971).
5. T. E. Wallis, *J. Pharma. Pharmacol.*, **9**, 663 (1957).
6. H. Wysokinska, *Farm. Pol.*, **33**, 725 (1977).
7. M. Shibano et al., *Chem. Pharm. Bull. (Tokyo)*, **49**, 1362 (2001).
8. K. Ikeda et al., *Carbohydr. Res.*, **323**, 73 (2000).
9. A. Subarnas et al., *J. Pharm. Pharmacol.*, **45**, 545 (1993).
10. A. Subarnas et al., *J. Pharm. Sci.*, **81**, 620 (1992).
11. L. P. Dwoskin and P. A. Crooks, *Biochem. Pharmacol.*, **63**, 89 (2002).
12. L. Teng et al., *J. Neurochem.*, **71**, 258 (1998).
13. D. Y. Lim et al., *Auton. Neurosci.*, **110**, 27 (2004).
14. E. G. C. Clarke in R. H. F. Manske, ed., *The Alkaloids*, Vol. 12, Academic Press, New York, 1970, p. 513.
15. J. L. Hartwell, *Lloydia*, **31**, 71 (1968).
16. L. Israelsen, Personal Communication, 1993.

LOCUST BEAN GUM

Source: *Ceratonia siliqua* L. (Family Leguminosae or Fabaceae).

Common/vernacular names: Carob bean gum, carob gum, locust gum, and carob seed gum.

Locust bean gum is derived from the seed of the carob, *Ceratonia siliqua* L. (see **carob**). The seed consists of 33–46% endosperm, 30–33% seed coat (husk), and 23–30% embryo (germ).

The endosperm is the source of locust bean gum; it is separated from the dark seed coat and yellow embryo by a series of milling operations. The endosperm is then ground and graded according to particle size, color, and other parameters. Major producing countries include Spain, Italy, and Greece.

Food-grade locust bean gum is a white to yellowish white, nearly odorless powder. Many of its physical properties are similar to those of guar gum, but unlike guar gum it is not completely dispersed in water at room temperature. Consequently, it is customary to hydrate locust bean gum in hot water and then let its solution (sol) cool to achieve maximum viscosity.

Solutions of locust bean gum have a pH between 5 and 7 and are highly viscous like those of guar gum. They behave similarly toward pH changes (stable over pH 3–11), borate ions, and organic solvents. Locust bean gum is compatible with gelatin, starch, and other plant gums (see **guar gum**).

Locust bean gum can improve the character and strength of carrageenan and agar gels as well as retard syneresis in these gels (see **agar**). This property is not shared by guar gum.

CHEMICAL COMPOSITION

Like guar gum, commercial locust bean gum is not a pure galactomannan. It contains substances other than galactomannan (protein, moisture, ash, etc.) comparable in quantities to those of guar gum. Its molecular weight has been reported to be 310,000. It has a structure similar to that of guar gum, except that its D-galactose side chain is not attached to every other D-mannose unit as in guar but is attached to every consecutive mannose unit in blocks of 25 units separated by unsubstituted blocks of 85 mannose units each. Thus, locust bean gum contains fewer galactose units than guar gum.[1] A single heptasaccharide, digalactosylmannopentaose, was identified after enzymatic hydrolysis of the gum.[2]

PHARMACOLOGY AND BIOLOGICAL ACTIVITIES

Like guar gum, locust bean gum does not seem to be digested by animals. Its growth-depressing effects in laboratory animals have been reported, with inconclusive results as in guar gum.[3]

In a pilot trial, inclusion of guar gum in semisolid nutrient products, such as rice pudding, delayed gastric emptying rate in healthy human subjects.[4]

The presence of locust bean gum in infant milk formulas was shown to increase their viscosity and decrease the bioavailability of their calcium content.[5]

USES

Locust bean gum has gum, many of which are interchangeable (see guar gum).

It is also used in place of tragacanth in some pharmaceutical and cosmetic applications (see tragacanth).

Food. Gum is used in ice cream, cheese, meat and fish sauces, pie filling, soups, bakery goods, and so on.[6]

Others. Gum used in manufacture of papers as a sizing, bonding fibers together more efficiently than starch.[6]

COMMERCIAL PREPARATIONS

Various grades with different viscosities and particle sizes. It is official in F.C.C.

Regulatory Status. Has been affirmed as GRAS (§184.1343).

REFERENCES

See the General References for FURIA; GLICKSMAN; GRIEVE; LAWRENCE; MARTINDALE; MERCK; WHISTLER AND BEMILLER.

1. C. W. Baker and R. L. Whistler, *Carbohydr. Res.*, **45**, 237 (1975).
2. A. L. Davis et al., *Carbohydr. Res.*, **271**, 43 (1995).
3. S. E. Davis and B. A. Lewis, *ACS Symp. Ser.*, **15**, 296 (1975).
4. G. Darwiche et al., *BMC. Gastroenterol.*, **3**, 12 (2003).
5. D. Bosscher et al., *Int. J. Food Sci. Nutr.*, **54**, 261 (2003).
6. R. L. Whistler, *Econ. Bot.*, **36**, 195 (1982).

LOVAGE ROOT

Source: *Levisticum officinale* W. D. J. Koch (syn. *Angelica levisticum* Baill.) (Family Umbelliferae or Apiaceae).

Common/vernacular names: Smellage, smallage, and maggi herb.

GENERAL DESCRIPTION

A large, nonhairy perennial herb with a stout hollow stem; up to about 2 m high; native to the mountains of southern Europe; naturalized in North America; cultivated in central and southern Europe (e.g., France, Belgium, Czech Republic, Hungary, Slovakia, and

Germany). Parts used are the dried rhizome and roots of 2- to 3-year-old plants, collected in the spring. Lovage oil is obtained by steam distillation of the fresh root.

CHEMICAL COMPOSITION

Contains up to 1.8% (usually 0.5–1.0%) volatile oil composed of 70% phthalides (butylidene, dihydrobutylidene, and butylphthalides; sedanonic anhydride; ligustilide; senkyunolide; validene-4,5-dihydrophthalide; etc.), with lesser amounts of terpenoids (α- and β-pinenes, α- and β-phellandrenes, γ-terpinene, carvacrol, and l-α-terpineol), volatile acids (butyric acid, isovaleric acid, maleic acid, angelic acid, etc.),[1–4] 0.1–4.3% coumarins (coumarin, umbelliferone, bergapten, psoralen, etc.);[5–8] β-sitosterol and its glucoside,[9] gum, and resin, among others (KARRER; LIST AND HÖRHAMMER; STAHL). Heptanal, *trans*-2-nonenal, and *cis*-3-hexenyl butyrate are obtained by direct solvent extraction of the root, but are not in the root oil.[10]

The leaf oil contains α-terpinyl acetate (29%), *cis*- and *trans*-ligustilides (18%), α-phellandrene (17%), and α-terpineol (5%).[10,11] β-Phellandrene was recently reported to be the major volatile oil component (36–79%) in all organs except the roots.[12]

PHARMACOLOGY AND BIOLOGICAL ACTIVITIES

Lovage extracts and oil have been reported to exhibit strong diuretic effects on rabbits and mice (LIST AND HÖRHAMMER); and also spasmolytic effects.[13] Lovage root has recently been included in a review about herbals used for urinary tract problems.[14]

Phthalides have been reported to have sedative activities on mice (see ***celery***).

TOXICOLOGY

Some of the coumarins are known to be phototoxic to humans as well as useful in treating psoriasis (see ***bergamot***). Interaction with warfarin and the possibility of increasing the risk of bleeding are also a concern.[15]

Available data from one source indicate lovage root oil to be nonirritating and nonsensitizing to human skin, though one case of sensitization has been reported from another source. The phototoxic effects of lovage oil on humans are not known.[16]

USES

Medicinal, Pharmaceutical, and Cosmetic. Lovage oil is used as a fragrance component in soaps, creams, lotions; and perfumes; with maximum use level of 0.2% reported in perfumes.[16]

Food. Lovage oil and lovage extracts are used as flavor components in major food products, including alcoholic (liqueurs, etc.) and nonalcoholic beverages, frozen dairy desserts, candy, baked goods, gelatins and puddings, meat and meat products, and sweet sauces, among others. Average maximum use levels are generally below 0.005%, with the exceptions of 0.017% and about 0.013% (125 ppm) reported for lovage extract (type not indicated) in sweet sauces and in frozen dairy desserts, respectively.

Lovage (crude) is also reported used in alcoholic beverages, frozen dairy desserts, candy, and baked goods. Highest average maximum use level is 0.015% in alcoholic beverages.

Dietary Supplements/Health Foods. Root occasionally used in digestive formulations in capsules, tablets, and also as tea ingredient (FOSTER).

Traditional Medicine. Used as a diuretic, stomachic, expectorant, and emmenagogue. Conditions for which it is used include digestive problems, flatulence, gastric catarrh, skin problems, and menstrual difficulties.

During China's Cultural Revolution, the root was used as a substitute for wild-harvested *danggui* (*Angelica sinensis*), until cultivated supplies of the drug were developed (FOSTER AND YUE).

COMMERCIAL PREPARATIONS

Crude, extracts, and oil. Lovage oil is official in F.C.C. Strengths (see *glossary*) of extracts are expressed in flavor intensities.

Regulatory Status. Has been approved for food use (§172.510). The root is the subject of a German therapeutic monograph; allowed in irrigation therapy for inflammation of the lower urinary tract and to prevent kidney stones.[13]

REFERENCES

See the General References for ARCTANDER; BAILEY 2; BLUMENTHAL 1; DER MARDEROSIAN AND BEUTLER; FEMA; FERNALD; FOSTER; FOSTER AND YUE; GRIEVE; GUENTHER; LIST AND HÖRHAMMER; LUST; MCGUFFIN 1 & 2; ROSE; STAHL.

1. M. Sekulic and M. Smodlaka, *Arh. Farm. (Belgrade)*, **11**, 177 (1961).
2. G. Tibori et al., *Rev. Med. (Tirgu-Mures, Rom.)*, **20**, 222 (1974).
3. G. Pattenden, *Fortschritte der Chemie Organischer Naturstoffe*, Springer-Verlag, Vienna, 1978, p. 133.
4. M. J. M. Gijbels et al., *Planta Med.*, **44**, 207 (1982).
5. D. Albulescu et al., *Farmacia (Bucharest)*, **23**, 159 (1975).
6. A. D. Dauksha, *Aktual. Vop. Farm.*, **23**, (1968).
7. J. Karlsen et al., *Medd. Nor. Farm. Selsk.*, **30**, 169 (1968).
8. F. C. Fischer and A. B. Svendsen, *Phytochemistry*, **15**, 1079 (1976).
9. B. E. Nielsen and H. Kofod, *Acta Chem. Scand.*, **17**, 1167 (1963).
10. B. M. Lawrence, *Perfum. Flav.*, **15**, 57 (1990).
11. Toulemonde et al., in B. M. Lawrence et al., eds., *Flavors and Fragrances: A World Perspective*, Elsevier Science Publishers B.V., Amsterdam, 1988, p. 641.
12. E. Bylaite et al., *J. Agric. Food Chem.*, **48**, 6183 (2000).
13. Monograph, *Levistici radix*, Bundesanzeiger, no. 101 (June 1, 1990).
14. E. Yarnell, *World J. Urol.*, **20**, 285 (2002).
15. A. M. Heck et al., *Am. J. Health Syst. Pharm.*, **57**, 1221 (2000).
16. D. L. J. Opdyke, *Food Cosmet. Toxicol.*, **16**(Suppl. 1), 813 (1978).

LYCIUM FRUIT

Source: *Lycium barbarum* L. (*L. halimifolium* Mill.) and *L. chinense* Mill. (Family Solanaceae).

Common/vernacular names: DUKE of Argyll's tea tree, Ningxia *gouqi* (*L. barbarum*); Chinese matrimony vine (*L. chinense*); boxthorn, wolfberry; *gouqizi*.

GENERAL DESCRIPTION

Lycium barbarum is a deciduous shrub, upright or spreading, reaching 2–3 m high; fruit (berry) subglobose to ovoid or short oblong, scarlet to orange red when ripe, 0.8–2 cm long and 0.5–1.0 cm in diameter; native to northern Asia, distributed and abundant in northern China, especially in the province of Ningxia; naturalized in the United States.

Lycium chinense is also a deciduous shrub, up to about 1 m high, with slightly smaller fruit; native to eastern Asia; now distributed throughout China; naturalized in the United States. Young shoots (leaves) are a popular vegetable in southern China and Hong Kong and are now also available in major Chinatowns in the United States.[1]

Part used is the ripe fruit collected in summer or fall from both wild and cultivated plants, rid of stalk, left in a shady and airy area until skin is wrinkled, and then sun dried or oven dried until skin is dried but the whole fruit is still soft to the touch. Lycium fruit comes in numerous grades; top grades consist of fruits that are large, bright red or purplish red, soft to the touch, and sweet in taste. Ningxia is the major producing province, which also produces the best grades (ZHU); most of the lycium fruit imported into the United States comes from Ningxia produced from *L. barbarum*.[2]

CHEMICAL COMPOSITION

Most of the chemical and biological studies on lycium fruit have been performed on *L. barbarum*. Hence unless otherwise stated, information reported in the following is for fruits from this species.

According to a report from Ningxia, lycium fruit (Ningxia *gouqi*) contains large amounts of amino acids (8–10%), about half of which in the free form: aspartic acid (1.2%), proline (0.65%), glutamic acid (0.63%), alanine (0.37%), arginine (0.19%), serine (0.14%), and nine others. Even higher amounts were found in the leaves (10–17%); freeze-drying and conventional drying did not affect the concentration and relative distribution of these amino acids either in the fruit, fruit stalk, or in the leaf.[3] The high concentrations of amino acids were confirmed by a more recent study on lycium fruits from four different sources, which were found to contain total amino acids ranging from 9.9% to 18.8%.[4] However, another study found a much lower concentration of free amino acids (ca. 0.47%) in lycium fruit of unspecified origin.[5] Also, lycium fruit (from *L. chinense*) produced in Shandong Province is reported to contain much lower amounts of total amino acids (ca. 5.3%).[6]

Other constituents reported include carotenoid pigments (zeaxanthin, physalien (zeaxanthin dipalmitate) and cryptoxanthin);[6–8] betaine;[9] β-carotene (7.38–8.88 mg/100 g); vitamins B_1, B_2, C (and a vitamin C glycoside);[10] nicotinic acid;[2,11,12] 58–64% carbohydrates (47–56% sugars, 5.4–8.2% polysaccharides);[2,13] scopoletin and low levels of atropine;[14,15] taurine and γ-aminobutyric acid;[4] monomethyl succinante,[16] cinnamic acid; fats (8–12%) glycolipids and sphingolipids (cerebrosides);[17,18] proteins (11–20%), glycoproteins and cyclic peptides;[19–21] trace minerals; and others (HU).[2,22,23]

Bioactive phenolic amides (caffeoyl tyramines and feruloyl octopamine) and pyrrole derivatives were recently isolated from the bark and fruit of *L. chinense*, respectively.[24,25]

PHARMACOLOGY AND BIOLOGICAL ACTIVITIES

A highly valued *yin* tonic in traditional Chinese medicine, lycium fruit is regarded as having antiaging, body-strengthening, and vision-brightening properties, all of which have at least some scientific basis as evidenced by the following modern findings: (1) Oral administration of 50 g/day of lycium fruit for 10 successive days to 43 patients aged 60 years and over significantly raised the values of lysozyme, IgG, IgA, lymphocyte blastogenesis (Stimulation Index), cyclic AMP, and testosterone.[26] (2) Oral administration of 50 g/day of lycium fruit for 10 successive days to 25 healthy subjects aged 64–80 years significantly raised their serum levels of superoxide dismutase (SOD) and hemoglobin but lowered the level of lipid peroxides.[27] (3) In a double-blind, placebo-controlled study, oral administration of lycium fruit extractives to normal subjects aged 56–81 years significantly improved their immune functions, reduced senility symptoms, and increased total peripheral white cell count and neutrophils.[28] (4) Oral administration of 50 g/day of lycium fruit (containing ca. 5 mg β-carotene) to healthy subjects aged 18–25 years significantly improved vision and dark adaptation, with serum vitamin A reaching saturation level at 1.16 ± 0.15 mol/L after 34 days of intake.[12]

Many studies indicate that the polysaccharides are responsible for most of the biological activities of lycium fruit, including antiperoxidative effects on cell membrane of *Xenopus* oocytes, with activity comparable to that of SOD;[29] preventing lipid peroxidation in liver, spleen, and brain tissues of rats and mice induced by physical stress and by carbon tetrachloride;[30] increasing the interleukin-2 (IL-2) activity in adult mice and restoring the level of IL-2 activity of aged mice to that of adult mice;[31] markedly elevating the cellular immune response in mice, raising level of splenic plaque forming cells in aged mice to that of adult mice and enhancing T-lymphocyte proliferation and cytotoxicity of cytotoxic T-lymphocytes and natural killer cells in normal mice as well as in immunosuppressed mice treated with cyclophosphamide;[32,33] reducing the dosage of *Corynebacterium parvum* in a synergistic effect on the tumoristatic activity of mouse peritoneal macrophages against P815 and P388 cells;[34] and protecting genetic material from genetic damage (antimutagenic) by mitomycin both *in vitro* and in healthy subjects over 60 years old.[35,36]

The above-mentioned findings are further supported by recent research that demonstrated that lyceum has antioxidant,[11,37,38] neuroprotective (retinal ganglia and brain cells),[39,40] cytoprotective,[37] hepatoprotective (extract, glycolipids, pyrroles, and physalien),[17,18,25,41,42] immunomodulatory (polysaccharides) and anticancer activities in various animal models.[43–47]

Other biological effects include hypoglycemic (possibly by decreasing insulin resistance),[11,48] hypolipemic, preventing fatty liver, hypotensive, and increasing weight gain in mice (when used with ginseng) among others (WANG); markedly increasing tolerance to anoxia as well as increasing dermal hydroxyproline level (by 15.5%) in mice indicating increased collagen synthesis, and promoting human fibroblast survival resulting in skin protection.[49,50] Antimicrobial (antibacterial/antifungal);[24,51] and aphrodisiac effects have also been reported for lyceum.[52,53]

Juice expressed from fresh young leaves of *L. barbarum* or *L. chinense* collected in spring and summer was used in topically treating 1853 cases of mosquito and insect stings/bites with great success: 1703 cases cured with one to three applications, leaving no scars or pigmentation; only 10 cases did not respond.[54]

Possible interaction between *L. barbarum* and warfarin has been reported. The authors advised against the concomitant use of these entities.[55]

USES

Medicinal, Pharmaceutical, and Cosmetic. Traditionally considered of benefit to the complexion and to prolong life (*mei rong yan nian*), lycium fruit has been consumed for 2000 years in China for these purposes; in recent years it is also used successfully in the topical treatment of burns, ulcers, bedsores, frostbite, canker sores, and furuncles.[56] Its high contents of free amino acids, β-carotene, and bioactive polysaccharides as well recent evidence of its ability to increase skin hydroxyproline levels and at the same time with no known human toxicity make it a potentially useful cosmetic ingredient.

Dietary Supplements/Health Foods. Powder and extracts (water and hydroalcoholic) are used in tonic formulas (especially for vision and male problems) in tablet, capsule, or liquid form; also sold as whole for making tea (use 1–2 tablespoons/cup).

Traditional Medicine. First described in the *Ming Yi Bie Lu* (ca. AD 200), lycium fruit is one of the most commonly used Chinese yin tonics. Traditionally regarded as sweet tasting and neutral, liver- and kidney-nourishing, replenishing vital essence (*yi jing*), and vision improving. Used in treating general debility and deficient energy (*xu lao jing kui*), aching back and knee, tinnitus, dizziness, diabetes, blurred vision, cough, and nocturnal emission. It is said to grow muscles and to cause weight gain but is also used to reduce weight;[57] and in recent years also used in treating the damaging side effects of chemotherapy and radiotherapy.[58]

COMMERCIAL PREPARATIONS

Crude (whole) and extracts (water, hydroalcoholic, and oil). Crude can be powdered after baked completely dry.

Regulatory Status. It is an ethnic food; United States regulatory status not known.

REFERENCES

See the General References for BAILEY 1; CHP; FERNALD; FOSTER AND YUE; JIXIAN; HU; JIANGSU; HUANG; LU AND LI; MCGUFFIN 1 & 2; WANG; ZHU.

1. S. Y. Zee and L. H. Hui, *Hong Kong Food Plants*, The Urban Council, Hong Kong, 1981, p. 27.
2. Z. S. Qi et al., *Zhongyao Tongbao*, **11**, 41 (1986).
3. X. Z. Meng et al., *Zhongyao Tongbao*, **12**, 42 (1987).
4. H. Q. Chen et al., *Zhongguo Yaole Daxue Xuebao*, **22**, 53 (1991).
5. Y. X. Gong, *Zhongcaoyao*, **18**, 37 (1987).
6. J. Wang et al., *Shandong Zhongyi Zazhi*, **10**, 42 (1991).
7. Y. Peng et al., *Plant Foods Hum. Nutr.*, **60**, 161 (2005).
8. P. Weller and D. E. Breithaupt, *J. Agric. Food Chem.*, **51**, 7044 (2003).
9. Y. G. Shin et al., *J. Chromatogr. A*, **857**, 331 (1999).
10. Y. Toyoda-Ono et al., *J. Agric. Food Chem.*, **52**, 2092 (2004).
11. Q. Luo et al., *Life Sci.*, **76**, 137 (2004).
12. Y. Shen et al., *Acta Nutrimenta Sinica*, **12**, 420 (1990).
13. Q. Wang et al., *Zhongcaoyao*, **22**, 67 (1991).
14. M. Adams et al., *Phytochem. Anal.*, **17**, 279 (2006).
15. C. S. Li et al., *Zhongguo Zhongyao Zazhi*, **15**, 43 (1990).
16. R. D. Hiserodt et al., *J. Agric. Food Chem.*, **52**, 3536 (2004).
17. S. Y. Kim et al., *J. Nat. Prod.*, **60**, 274 (1997).
18. K. Jung et al., *Arch. Pharm. Res.*, **28**, 1381 (2005).
19. S. Yahara et al., *Chem. Pharm. Bull. (Tokyo)*, **41**, 703 (1993).
20. X. Qin et al., *Carbohydr. Res.*, **333**, 79 (2001).
21. X. Peng and G. Tian, *Carbohydr. Res.*, **331**, 95 (2001).
22. Q. H. Yuan, *Zhongguo Zhongyao Zazhi*, **14**, 42 (1989).
23. Y. Q. Shao et al., *Shandong Zhongyi Zazhi*, **9**, 38 (1990).
24. D. G. Lee et al., *Biotechnol. Lett.*, **26**, 1125 (2004).
25. Y. W. Chin et al., *Bioorg. Med. Chem. Lett.*, **13**, 79 (2003).
26. P. G. Xiao and K. J. Chen, *Phytother. Res.*, **2**, 55 (1988).
27. W. Li et al., *Zhongcaoyao*, **22**, 251 (1991).
28. D. Y. Li et al., **20**, 10, (1989).
29. X. Zhang and X. C. Xie, *Zhongguo Zhongyao Zazhi*, **18**, 110 (1993).
30. H. Zhan et al., *Chin. J. Pharmacol. Toxicol.*, **3**, 163 (1989).

31. C. S. Geng et al., *Chin. J. Pharmacol. Toxicol.*, **3**, 163 (1989).
32. C. S. Geng et al., *Zhonghua Laonian Yixue Zazhi*, **8**, 236 (1989).
33. B. K. Wang et al., *Chin. J. Pharmacol. Toxicol.*, **4**, 39 (1990).
34. Y. X. Zhang et al., *Chin. J. Pharmacol. Toxicol.*, **3**, 169 (1989).
35. M. X. Tao and Z. L. Zhao, *Zhongcaoyao*, **23**, 474 (1992).
36. M. D. Wang et al., *Zhongcaoyao*, **23**, 251 (1992).
37. M. S. Yu et al., *Int. J. Mol. Med.*, **17**, 1157 (2006).
38. S. J. Wu et al., *Phytother. Res.*, **18**, 1008 (2004).
39. H. C. Chan et al., *Exp. Neurol.*, (2006).
40. M. S. Yu et al., *Exp. Gerontol.*, **40**, 716 (2005).
41. K. T. Ha et al., *J. Ethnopharmacol.*, **96**, 529 (2005).
42. H. P. Kim et al., *Biol. Pharm. Bull.*, **25**, 390 (2002).
43. H. Gong et al., *Cancer Biother. Radiopharm.*, **20**, 155 (2005).
44. M. Zhang et al., *Life Sci.*, **76**, 2115 (2005).
45. G. Du et al., *J. Huazhong. Univ Sci. Technolog. Med. Sci.*, **24**, 518 (2004).
46. L. Gan et al., *Int. Immunopharmacol.*, **4**, 563 (2004).
47. L. Gan et al., *Eur. J. Pharmacol.*, **471**, 217 (2003).
48. R. Zhao et al., *Yakugaku Zasshi*, **125**, 981 (2005).
49. S. Q. Chen et al., *Zhongguo Yaoli Yu Linchuang*, **6**, 28 (1990).
50. H. Zhao et al., *Phytomedicine*, **12**, 131 (2005).
51. D. G. Lee et al., *Arch. Pharm. Res.*, **28**, 1031 (2005).
52. Q. Luo et al., *Life Sci.*, **79**, 613 (2006).
53. Y. Wang et al., *J. Ethnopharmacol.*, **82**, 169 (2002).
54. P. X. Gao and X. L. Dang, *Shaanxi Zhongyi*, **9**, 320 (1988).
55. A. Y. Lam et al., *Ann. Pharmacother.*, **35**, 1199 (2001).
56. L. J. Jiang et al., *Zhongyi zazhi*, **26**, 69 (1985).
57. H. X. Jing, *Xinzhongyi* **37**(1988).
58. S. N. Gu et al., *Zhongguo Yaoli Yu Linchuang*, **6**, 38 (1990).

MAGNOLIA FLOWER

Source: *Magnolia biondii* Pamp. (syn. *M. fargesii* (Fin. et Gagnep.) Cheng), *M. denudata* Desr. (syn. *M. heptapeta*, (Buc'hoz) Dandy), *M. sprengeri* Pamp., *M. sargentiana* Rehd. Et Wils. (syri. *M. emargenata* Cheng), *M. wilsonii* (Fin. et Gagnep.) Rehd. et Wils., *M. salicifolia* Maxim., and other *Magnolia* species (Family Magnoliaceae).

Common/vernacular names: Magnolia flower bud, *xinyi*, *xinyihua*, *shin-i*, and flos magnoliae.

GENERAL DESCRIPTION

Small to large trees, mostly deciduous, widely distributed in China; some extensively cultivated. At least 10 species of *Magnolia* serve as source of magnolia flower. However, *M. biondii*, *M. denudata*, and *M. sprengeri* are currently the major sources; and *M. liliflora* Desr., although still listed in many major works as the major source, is not a current source.[1,2] Produced mainly in eastern and southern provinces in China.

Magnolia salicifolia is source of Japanese magnolia flower.[3]

Part used is the flower bud collected before opening in early spring and carefully sun or oven dried. The dried bud measures 1–5 cm long and has a diameter of 0.5–2 cm at midsection, depending on the source.[4] It is hairy and its overall appearance resembles that of a pussy willow bud, but it emits a strong characteristic eucalyptus-like odor when crushed.

CHEMICAL COMPOSITION

Contains 0.4–3.2% (v/w) volatile oil, with *M. denudata* yielding the highest and *M. sargentiana* the lowest;[5] neolignans and lignans; alkaloids; flavonoids (tiliroside and others); tannins; and others (HU; WANG).[3,6–9]

More than 70 components of the volatile oil have been identified. Major ones include α-pinene (0.63–5.67%), β-pinene (0.84–39.05%), camphene (0.39–5.54%), limonene (0.96–10.14%), 1,8-cineole (0.76–17.48%), *p*-cymene (0.50–25.69%), linalool (0.28–2.83%), camphor (0.14–44.20%), and α-terpineol (0.49–4.00%), which are present in flower buds of all eight *Magnolia* (including *M. biondii*, *M. denudata*, *M. sprengeri*, *M. sargentiana*, and *M. salicifolia*) and one *Michelia* species tested; others include myrcene (0.40–2.70%), sabinene (1.15–13.42%), fenchone (0.06–9.62%), terpinen-4-ol (0.90–10.00%), borneol (0.40–3.18%), bornyl acetate (0.42–8.40%), methyleugenol (0.22–19.45%), caryophyllene oxide (0.76–3.30%), and eudesmol (0.70–4.41%), present in the majority of species tested. The proportions of these constituents vary considerably among the species. Thus β-pinene is the major component in *M. denudata* (39.05%) and camphor, the major component in *M. biondii* (44.20%) and *M. sargentiana* (40.00%). Although safrole is not present in the others tested, it is present in *M. salicifolia* (Japanese *xinyi*) in major concentration (29.87%) along with methyleugenol (19.45%) and fenchone (9.62%), which are absent or present only in minor amounts in the other species (HU).[5]

Neolignans and lignans include fargesone A, B, and C; denudatin B; pinoresinol dimethyl ether; lirioresinol B dimethyl ether; magnolin; and fargesin (from flower buds of *M. biondii*) as well as magnoshinin and magnosalin (from Japanese *xinyi*).[6,7] Other recently isolated compounds include a bicycle and tricyclo octane- and nonane-type neolignans, magliflonenone, veraguensin, and others (from twigs, leaves, and flower buds of *M. denudata*).[10–12] The bioactive lignans magnone A & B and magnosalicin have also been isolated from the flower buds of *M. fargesii* and *M. salicifolia*, respectively.[13,14]

Alkaloids include *d*-coclaurine, *d*-reticuline, 1-*N*-methylcoclaurine, and yuzirine (all benzylisoquinolines), which are present in minor to trace amounts in flower buds of *M. salicifolia* and *M. biondii*.[3] Taspine (ethylamine chromene alkaloid) and (−)-asimilobine (aporphine alkaloid) have recently been identified in *M. x soulangiana*.[15]

PHARMACOLOGY AND BIOLOGICAL ACTIVITIES

Magnolia flower has numerous biological activities, including antihistaminic in guinea pig trachea (due to volatile oil, alcoholic, and water extracts);[16,17] protective against allergic asthma in guinea pig (volatile oil);[17] inhibition of histamine release from rat mast cells;[14] anti-inflammatory in mice (magnoshinin and magnosalin);[7,11] inhibitor of TNF-α in murine macrophages;[18] Ca^{2+}-antagonistic on the taenia coli of guinea pig (neolignans);[6] central dopaminergic modulating in mice (*d*-coclaurine and *d*-reticuline);[8] neuromuscular blocking in isolated frog skeletal muscle and nerve-muscle preparations (alkaloids);[3] hypotensive in several species of experimental animals (alcoholic and water extracts); uterus stimulating in animals (decoction and fluid extract); antifungal, antibacterial, and antiviral (decoction); frog skeletal muscle contracting (decoction) and relaxant (alkaloids); local anesthetic in animals (decoction and infusion); and others (JIANGSU; WANG).

Magnolia flower in various forms (decoction, alcoholic extract, volatile oil, etc.) and in combination with other herbal drugs has been reported highly effective both locally and internally in treating allergic rhinitis (e.g., hay fever), chronic rhinitis, and paranasal sinusitis (WANG).[19,20]

Magnolia extracts and/or their constituents have also been investigated in the following activities/conditions:

1. *Diabetic nephropathy in rats* (*magnolol*). Reduced fasting blood glucose and retarded the development of complications.[21]
2. *Acetylcholinesterase inhibition* (*taspine*). Dose-dependent inhibition (IC_{50} 0.33 μM).[15]
3. *Apoptosis/Angiogenesis*. Induction of mitochondrial- and caspase-dependent apoptosis in mast cells;[22] and antiangiogenic activity (magnosalin) in rat vascular endothelial cells.[23]
4. *Platelet activation and complement fixation.* Magnone A & B exhibited antiplatelet-activating-factor activities at IC_{50} 38 and 100 nM, respectively;[13] while tiliroside had potent anticomplement activity at IC_{50} 54 nM.[9]
5. *Transdermal absorption of drugs.* The essential oil of *M. fargesii* enhanced the *in vitro* skin permeation of theophylline and catechin.[24]

TOXICOLOGY

Magnolia flower has very low toxicity: i.v. injection of decoction in dogs (1.0 g/kg) and rabbits (4.75 g/kg) produced no fatalities. The LD_{50} (i.p.) of its tincture (after alcohol removal) in rats and mice were 22.5 ± 0.96 and 19.9 ± 0.25 g/kg, respectively, based on the crude drug (WANG).

USES

Medicinal, Pharmaceutical, and Cosmetic. Extracts used in skin care products to minimize or counteract undesirable irritant effects of other cosmetic ingredients; also used for its traditional skin-whitening properties.

Dietary Supplements/Health Foods. Used in allergy and cold preparations (JIANGSU).

Traditional Medicine. Traditionally considered acrid tasting and warming; disperses wind and cold (*san feng han*) and clears the nasal cavity. Chinese traditional use records date back 3000 years to the *Wu Shi Er Bing Fang* (*Prescriptions for 52 Diseases*; 1065-771 BC). Used both internally and externally to treat nasal congestion, running nose, the common cold, and headache as well as facial dark spots; also used topically to treat toothache.

COMMERCIAL PREPARATIONS

Crude and extracts (mainly hydroalcoholic and hydroglycolic).

Regulatory Status. United States regulatory status not determined.

REFERENCES

See General References for JIANGSU; JIXIAN; HU; MCGUFFIN 1 & 2; NATIONAL; WANG; HONGKUI;

1. Z. F. Wu et al., *Zhongguo Zhongyao Zazhi*, **16**, 13 (1991).
2. W. Z. Song, *Zhongcaoyao*, **15**, 26 (1984).
3. I. Kimura et al., *Planta Med.*, **48**, 43 (1983).
4. Y. Y. Tong and W. Z. Song, *Yaoxue Xuebao*, **20**, 22 (1985).
5. Z. L. Xu et al., *Zhongguo Zhongyao Zazhi*, **14**, 38 (1989).
6. C. C. Chen et al., *Planta Med.*, **54**, 438 (1988).
7. M. Kimura et al., *Planta Med.*, **51**, 291 (1985).
8. H. Watanabe et al., *Planta Med.*, **42**, 213 (1981).
9. K. Y. Jung et al., *Biol. Pharm. Bull.*, **21**, 1077 (1998).
10. J. Li et al., *Chem. Pharm. Bull. (Tokyo)*, **53**, 235 (2005).
11. J. Du et al., *J. Asian Nat. Prod. Res.*, **3**, 313 (2001).
12. M. Kuroyanagi et al., *Chem. Pharm. Bull. (Tokyo)*, **48**, 832 (2000).
13. K. Y. Jung et al., *J. Nat. Prod.*, **61**, 808 (1998).
14. T. Tsuruga et al., *Chem. Pharm. Bull. (Tokyo)*, **39**, 3265 (1991).
15. J. M. Rollinger et al., *J. Nat. Prod.*, **69**, 1341 (2006).
16. R. D. Xiang et al., *Zhongcaoyao*, **16**, 22 (1985).
17. D. Q. Zhouo et al., *Zhongcaoyao*, **22**, 81 (1991).
18. S. H. Chae et al., *Arch. Pharm. Res.*, **21**, 67 (1998).
19. T. C. Wang et al., *Chin. J. Integr. Trad. Western Med.*, **4**, 728 (1984).
20. Y. Ren, *Zhongyao Tongbao*, **10**, (1985).
21. E. J. Sohn et al., *Life Sci.*, (2006).
22. G. C. Kim et al., *Int. Arch. Allergy Immunol.*, **131**, 101 (2003).
23. S. Kobayashi et al., *Biol. Pharm. Bull.*, **19**, 1304 (1996).
24. J. Y. Fang et al., *J. Pharm. Pharmacol.*, **56**, 1493 (2004).

MARJORAM (SWEET, POT, AND WILD)

Source: *Sweet marjoram Origanum majorana* L. (syn. *Majorana hortensis* Moench); ***Pot marjoram*** *Origanum onites* L. (syn. *Majorana onites* (L.) Benth.); ***Wild marjoram*** *Origanum vulgare* L. (see **oregano**) (Family Labiatae or Lamiaceae).

Common/vernacular names: Knotted marjoram (*O. majorana*).

GENERAL DESCRIPTION

There is much confusion regarding the plant sources of marjoram (pot, sweet), oregano, and Spanish origanum (see ***origanum oil, Spanish***).

No discrepancies seem to exist with respect to sweet marjoram, as most authors agree to *Origanum majorana* L., also known as *Majorana hortensis* Moench, as its major botanical source (ARCTANDER; BAILEY 1; FURIA AND BELLANCA; ROSENGARTEN; TERRELL; UPHOF; §182.10). However, the sources of pot marjoram have been attributed by numerous authors to three plant names, *Origanum onites* L., *O. vulgare* L., and *Majorana onites* Benth. One source lists *M. onites* as a synonym of *O. vulgare*, both of which are entered under the heading of *pot marjoram*, and *O. vulgare* is also described as the source of origanum, which is a separate entry (FURIA

AND BELLANCA). According to ROSENGARTEN, *O. vulgare* has been known as wild marjoram (in agreement with ARCTANDER; BAILEY 1; and TERRELL) and is one of the two major sources of **oregano** (European oregano). The other major type of oregano, Mexican oregano, is mostly derived from Lippia species, especially *L. graveolens* H. B. K. (ROSENGARTEN; TERRELL) (See **oregano**). The source of pot marjoram is attributed to a single species, *M. onites*, also called *O. Onites* (BAILEY 1; ROSENGARTEN; TERRELL). The major source of Spanish origanum is not an *Origanum* species but is *Thymus capitatus* (L.) Hoffmgg. et Link. Nevertheless, Origanum species are used and thus add to the confusion (see **origanum oil**, **Spanish**).

Sweet marjoram is a tender, bushy perennial hairy herb, up to about 0.6 m; leaves woolly hairy ovate to ovate lanceolate, slightly toothed; flowers purple rarely white in compact heads, forming a terminal trichotomous panicle;[1] native to the Mediterranean region; cultivated as an annual in colder climates. Part used is the dried flowering herb from which sweet marjoram oil is obtained by steam distillation in 0.2–0.8% yield. Major oil-producing countries include France, Tunisia, Morocco, and Bulgaria.

Pot marjoram is a sturdy perennial herb with coarse hairy, erect stems and sessile, hairy (long and soft) leaves; native to the Mediterranean region. Parts used are the dried leaves. It is considered of low quality and not much used.

CHEMICAL COMPOSITION

Sweet marjoram contains up to 3% volatile oil (usually less than 1%), consisting primarily of α-terpinene, γ-terpinene, *p*-cymene, 4-terpineol, sabinene, linalool, borneol, carvacrol, *cis*-sabinene hydrate, and *trans*-sabinene hydrate with linalyl acetate, ocimene, cadinene, geranyl acetate, citral, estragole, eugenol, and 3-carene, totaling more than 50 compounds. *cis*-Sabinene hydrate is reported to be responsible for the typical aroma of sweet marjoram oil (GUENTHER; LIST AND HÖRHAMMER).[2–6]

Other compounds present in sweet marjoram include flavonoid glycosides (luteolin-7-diglucoside, apigenin-7-glucoside, and diosmetin-7-glucuronide), 6-hydroxyluteolin and 6-hydroxyapigenin glycosides, arbutin, methylarbutin,[7,8] tannins, caffeic acid, carnosic acid, carnosol,[9] labiatic acid, rosmarinic acid,[10] steroids (e.g., β-sitosterol), triterpenoids (oleanolic acid, ursolic acid, etc.), paraffins (e.g., *n*-triacontane), protein (ca. 13%), vitamins (especially A and C), and others (LIST AND HÖRHAMMER; MARSH).[9,11–13]

Pot marjoram is reported to contain 1.4–2.4% volatile oil, consisting mainly of carvacrol, linalool, and thymol (LIST AND HÖRHAMMER). A Turkish variety was recently reported to contain carvacrol, thymol, *p*-cymene, and γ-terpinene as the major constituents.[14] Fatty acids (e.g., linolenic, linoleic, and oleic) and tochopherols (α-, β-, γ- and δ-) are also present with γ-tochopherol being the major homologue.[15,16]

PHARMACOLOGY AND BIOLOGICAL ACTIVITIES

Extracts of sweet marjoram have antioxidant/free radical scavenging properties that are in part due to labiatic, ursolic and carnosic acids, and carnosol and phenolic compounds.[9,11,17–19] The antioxidant activity is reflected in the ability of the volatile oil and different marjoram extracts to act as liver and kidney chemopreventive agents against lead acetate toxicity in mice.[20]

The methanolic extract of sweet marjoram exhibited moderate antibacterial activity against *Helicobacter pylori*.[21] Its aqueous extract has also been reported to have antiviral activities against herpes simplex *in vitro*.[22]

Sweet marjoram is considered to have carminative, antispasmodic, diaphoretic, and diuretic properties. Ursolic acid has been reported to inhibit acetylcholineesterase and may thus have potential in the management of Alzheimer's disease.[23] Larvicidal activity of pot marjoram against *Culex* mosquito larvae has also been reported.[24]

TOXICOLOGY

Fresh sweet marjoram may cause inflammation of the skin and eyes (LIST AND HÖRHAMMER).

Sweet marjoram oil has been reported to be nonirritating and nonsensitizing to human skin.[25]

USES

Medicinal, Pharmaceutical, and Cosmetic. Sweet marioram oil is used as a fragrance component in soaps, detergents, creams, lotions, and perfumes, with maximum use level of 0.6% reported in perfumes.[25]

Food. Sweet marjoram oil and oleoresin are used as flavor ingredients in most food categories, including alcoholic (bitters, vermouths, etc.) and nonalcoholic beverages, frozen dairy desserts, candy, baked goods, gelatins and puddings, meat and meat products, condiments and relishes, and others. Average maximum use levels reported are generally below 0.004%.

Sweet marjoram is also used in baked goods, meat and meat products, condiments, and relishes, soups, snack foods, processed vegetables, and others, with highest average maximum use level of about 1% (9946 ppm) reported in baked goods.

Traditional Medicine. Sweet marjoram has been used for treating similar types of conditions as *Origanum vulgare* (see **oregano**). It has also been used in cancers.[26]

COMMERCIAL PREPARATION

Sweet marjoram crude, oil, and oleoresin. Sweet marjoram oil is official in F.C.C.

Regulatory Status. GRAS (Sweet marjoram, §182.10 and §182.20; pot marjoram, §182.10).

REFERENCES

See the General References for ARCTANDER; BAILEY 1; FEMA; GUENTHER; GRIEVE; MARTINDALE; MCGUFFIN 1 & 2; ROSE; ROSENGARTEN; TERRELL.

1. K. R. Kiritikar and B. D. Basu, *Indian Medicinal Plants*, International Book Distributors, 1999, p. 1985.
2. J. Novak et al., *Biochem. Syst. Ecol.*, **28**, 697 (2000).
3. R. Granger et al., *Riv. Ital. Essenze, Profumi, Piante Offic., Aromi, Saponi, Cosmet.*, **57**, 446 (1975).
4. J. Taskinen, *Acta Chem. Scand., Ser. B*, **28**, 1121 (1974).
5. G. Graner, *Präp. Pharm.*, **4**, 86 (1968).
6. E. Sarer et al., *Planta Med.*, **46**, 236 (1982).
7. J. Kawabata et al., *Biosci. Biotechnol. Biochem.*, **67**, 445 (2003).
8. M. H. Assaf et al., *Planta Med.*, **53**, 343 (1987).
9. E. Vagi et al., *J. Agric. Food Chem.*, **53**, 17 (2005).
10. U. Gerhardt and A. Schroeter, *Fleischwirtschaft*, **63**, 1628 (1983).
11. H. J. Heo et al., *Mol. Cells*, **13**, 5 (2002).
12. W. Olechnowicz-Stepien and E. Lamer-Zarawska, *Herba Pol.*, **21**, 347 (1975).
13. G. Lossner, *Planta Med.*, **16**, 54 (1968).
14. F. Demirci et al., *J. Agric. Food Chem.*, **52**, 251 (2004).
15. V. Lagouri and D. Boskou, *Int. J. Food Sci. Nutr.*, **47**, 493 (1996).
16. N. Azcan et al., *Lipids*, **39**, 487 (2004).

17. Y. Saito et al., *Eiyo To Shokuryo*, **29**, 505 (1976).
18. K. Herrmann, *Z. Lebensm. Unters. Forsch.*, **116**, 224 (1962).
19. S. Zalewski, *Przemysl Spozywczy*, **16**, 237 (1962).
20. I. M. el-Ashmawy et al., *Basic Clin. Pharmacol. Toxicol.*, **97**, 238 (2005).
21. G. B. Mahady et al., *Phytother. Res.*, **19**, 988 (2005).
22. E. C. Herrmann and L. S. Kucera, *Proc. Soc. Exp. Biol. Med.*, **124**, 874 (1967).
23. Y. K. Chung et al., *Mol. Cells*, **11**, 137 (2001).
24. Cetin H and A. Yanikoglu, *J. Vector Ecol.*, **31**, 118 (2006).
25. D. L. J. Opdyke, *Food Cosmet. Toxicol.*, **14**, 469 (1976).
26. J. L. Hartwell, *Lloydia*, **32**, 247 (1969).

MILK THISTLE

Source: *Silybum marianum* (L.) Gaertner. (syn. *Carduus marianum* L.) (Family Asteraceae or Compositae).

Common/vernacular names: Mary thistle, St. Mary thistle, Marian thistle, lady's thistle, holy thistle (not to be confused with blessed thistle *Cnicus benedictus*).

GENERAL DESCRIPTION

Stout, branching annual or biennial, 1–2 m high; leaves large, alternate, white mottled, glabrous, margins scalloped, with sharp spines on lobes; heads, solitary, up to 6 cm in diameter; purple, black shiny seeds (fruit) crowned with a spreading pappus; indigenous to Mediterranean region, southwest Europe; cultivated for centuries and naturalized in much of Europe; also naturalized in North America, especially California; South America from Uruguay to Chile, to Ecuador; weedy in Australia; common in abandoned fields, old pastures, and roadsides (FOSTER).

The part used is the seed (fruit).

CHEMICAL COMPOSITION

A flavanolignan complex, silymarin, was first isolated from the seeds in 1968.[1,2] Silymarin (4–6% in ripe fruits) consists primarily of three flavanolignans, silybin (silibinin), silychristin (silichristin) and sylichristin B, and silidianin.[3,4] Other flavanolignans include dehydrosilybin, 3-desoxysilichristin, deoxysilydianin (silymonin), siliandrin, silybinome, silyhermin, and neosilyhermin.[4] Other constituents include taxifolin, apigenin, silybonol; a fixed oil (16–18%), consisting largely of linoleic and oleic acids, plus myristic, palmitic, and stearic acids; betaine hydrochloride, triamine, histamine, and others.[1,2,4–6] The lanostane triterpene, marianine, and the triterepene glycosides, marianosides A and B were recently isolated from the whole plant.[7]

PHARMACOLOGY AND BIOLOGICAL ACTIVITIES

Few plant principles have been as extensively investigated in recent years as silymarin (WEISS). Primary activity is as a hepatoprotective and antioxidant.[1] Anticancer and cancer chemopreventive activities are also drawing interest in addition to other investigated activities.

The antioxidant activity has been demonstrated in several models of oxidative stress, such as H_2O_2-induced stress in human keratinocytes and mouse fibroblasts (silymarin and flavanolignans);[8] superoxide radical and lipid peroxidation in PMNs and hepatocytes (silymarin and flavanolignans);[9–11] high-sucrose diet in hypertriglyceridemic rats (silymarin);[12] UV-induced stress in rats (silymarin);[13] and copper-induced oxidation of human LDL (silymarin).[14]

Hepatoprotective activity of silymarin has been demonstrated in numerous experimental models of toxic liver damage, including carbon tetrachloride, alcohol, galactosamine, thioacetamide, hepatotoxic cold-blood frog virus (FV$_3$), lanthanides, and the toxins of *Amanita phalloides* (deathcap fungus) phalloidin and α-amanitin.[1,15,16] Efficacy is based on several separate mechanisms of action. Silymarin stimulates RNA polymerase A, enhancing ribosome protein synthesis and resulting in activating the regenerative capacity of the liver through cell development.[1,5,17] Silybin inhibits Kupffer cell functions responsible for the formation of ROSs and mediators of inflammation.[18] Silymarin interacts with hepatic cell membranes, blocking binding cites and hindering the uptake of toxins as demonstrated in rabbit liver microsomes[19] and mononuclear lipid layers.[20] Strong antioxidant (free radical scavenging activity 10-fold greater than vitamin E), blocking the release of malonyldialdehyde, and antiperoxidative activity has been reported.[9,11,21,22] Clinical studies have suggested that pretreatment with silymarin inhibits alcohol-, industrial- chemical-, and psychopharmaceutical-induced liver damage, accelerating normalization of impaired liver function.[1,17] Patients who received silymarin showed an accelerated improvement of increased serum levels of GOT, GPT, and γ-GT.[23] Silibinin administered in i.v. infusion has shown protective and curative effect on liver damage resulting from the highly toxic compounds phalloidin and α-amanitin (from *Amanita phalloides*). The antihepatotoxic effect of silymarin was found to depend on the time interval in which poisoning and therapy took place as well as the degree of liver damage. A multicenter trial involving 220 cases of *Amanita* poisoning treated in German, French, Swiss, and Austrian hospitals was carried out from 1979 to 1982, using silibinin in supportive treatment. Use of silibinin as an adjunct to current methods has lowered mortality rates below any levels that have previously been achieved.[24–26] Silymarin products were used with success in toxic-metabolic liver damage, ranging from fatty liver through fatty liver hepatitis to actual hepatic cirrhosis, caused by toxic substances, drugs, or exposure to irradiation.[6]

The chemopreventive effects of milk thistle have been investigated in various studies. Silybin was found to protect the following organs from chemical- or radiation-induced toxicities: rat cardiac muscle cells,[27,28] kidney cells of rats and monkeys,[29,30] skin cells of mice,[31–33] and rat gastric mucosa and pancreas.[34,35] Silymarin was also found to inhibit carcinogenesis in rat colon cells,[36] mice urinary bladder,[37] rat tongue cells,[38] and human breast cancer cells.[39]

The anticancer effect of milk thistle has equally been investigated in numerous animal models and in human patients. Silybin was found to arrest cell growth and progression of lung and prostate tumors in mice and rats,[40–42] and to inhibit human hepatocellular, prostate, and bladder carcinomas.[43–48] Mechanisms attributed to the anticancer effect include inactivation of MAPK and other kinase signaling pathways,[43,45,49] suppression of Topo-II gene expression,[44] reduced metastasis by inhibiting urokinase-plasminogen activator and matrix metalloproteinase,[50] inhibition of telomerase,[47] and caspase 3-controlled apoptosis.[41,46,51]

Milk thistle extract and its flavanolignans were recently found to display immunomodulatory/stimulatory,[13,52–54] antimicrobial (silybin against Gram-positive bacteria),[55] anti-inflammatory[56,57] (proinflammatory at high dose),[57] antidiabetic,[58,59] P-glycoprotein inhibitory,[60,61] and cholesterol-lowering effects.[62–64]

TOXICOLOGY

Even in large doses silymarin is devoid of toxic effects and in particular has no harmful action an the embryo.[1,6,17] In isolated cases, a mild laxative effect has been observed.[16] Also, in a recent study, dietary silymarin administration enhanced mammary carcinogenesis in rats and mice possibly due to an estrogen-like activity.[65]

Many reviews of the pharmacological activities, clinical efficacy, and new potentials of milk thistle have been published.[40,66–77]

USES

Medicinal, Pharmaceutical, and Cosmetic. In Europe, clinical use is widespread for toxic liver damage in supportive treatment of chronic inflammatory liver disorders and cirrhosis, including chronic hepatitis and fatty infiltration of the liver by alcohol and other chemicals. In infusion therapy, silibinin preparations used for supportive treatment of *Amanita* mushroom poisoning.[16]

Foods. Historically, grown in Europe as a vegetable, for edible peeled stalks; leaves as a salad green or potherb (with spines removed); roots eaten (soaked overnight in water to remove bitterness); flower receptacle eaten like artichoke; roasted seeds used as a coffee substitute.[1]

Dietary Supplements/Health Foods. Seeds or seed extract in teas, capsules, tablets, tinctures, or other preparations, primarily as a liver detoxicant (FOSTER).

Traditional Medicine. Milk thistle seeds have been used continuously for 2000 years for liver conditions (first mentioned by Pliny in the 1st century).[1] Historical references are particularly abundant in herbals of the Middle Ages, including the hepatoprotective activity.[6] Eclectic physicians in 19th-century America used seeds for liver congestion. Use of the seed for treatment of liver diseases was revitalized by the German physician Rademacher in the mid-19th century. Reinvestigation of the value of milk thistle in modern practice began with H. Schulz in 1929 and G. Madaus in 1938.[6]

COMMERCIAL PREPARATIONS

Crude, ethanolic extracts, tablets, or capsules (35–70 mg) standardized to 70% silymarin (calculated as silibinin) in average daily dose of 200–400 mg.

Regulatory Status. Class 1 dietary supplement (can be safely consumed when used appropriately). Milk thistle is the subject of a positive German therapeutic monograph indicated for toxic liver damage, liver cirrhosis, and supportive treatment of chronic inflammatory liver disease.[16]

REFERENCES

See the General References for APPLEQUIST; BARNES; BARRETT; BISSET ET AL.; BLUMENTHAL 1 & 2; BRUNETON; DER MARDEROSIAN AND BEUTLER; DER MARDEROSIAN AND LIBERTI; DUKE 4; FELTER AND LLOYD; FOSTER; FOSTER AND DUKE; GRIEVE; HUANG; MARTINDALE; MCGUFFIN 1 & 2; STEINMETZ; TYLER 1–3; UPHOF; WEISS; WREN.

1. S. Foster, *Milk Thistle-Silybum Marianum*, Botanical Series, no. 305, American Botanical Council, Austin, TX, 1991.
2. H. Wagner et al., *Arzeim.-Forsch.*, **18**, 688 (1968).
3. W. A. Smith et al., *Planta Med.*, **71**, 877 (2005).
4. H. Wagner and O. Seligmann, *Advances in Chinese Medicinal Materials Research*, World Scientific Pub. Co., Singapore, 1985, p. 247.
5. D. V. C. Awang, *Can. Pharm. J.*, 403 (1993).
6. G. Hahn and A. Meyer, *Dtsch. Apoth.*, **40**, 2 (1988).
7. E. Ahmed et al., *Chem. Pharm. Bull. (Tokyo)*, **54**, 103 (2006).
8. A. Svobodova et al., *Burns*, **32**, 973 (2006).

9. M. Zielinska-Przyjemska and K. Wiktorowicz, *Phytother. Res.*, **20**, 115 (2006).
10. Z. Varga et al., *Phytother. Res.*, **15**, 608 (2001).
11. E. Bosisio et al., *Pharmacol. Res.*, **25**, 147 (1992).
12. N. Skottova et al., *Pharmacol. Res.*, **50**, 123 (2004).
13. S. K. Katiyar, *Int. J. Oncol.*, **21**, 1213 (2002).
14. N. Skottova et al., *Phytother. Res.*, **13**, 535 (1999).
15. C. S. Lieber et al., *J. Clin. Gastroenterol.*, **37**, 336 (2003).
16. *Monograph Cardui Mariae Fructus, Bundesanzeiger*, no. 50 (March 13, 1986).
17. H. Hikino and Y. Kiso in H. Wagner, H. Hikino, and N.R. Farnsworth, eds., *Economic and Medicinal Plant Research*, Vol. 2, Academic Press, New York, 1968, p. 39.
18. C. Dehmlow et al., *Hepatology*, **23**, 749 (1996).
19. A. Wissemann in D. Lorenz et al., eds., *Aktuelle Hepatologie*, Hansiches Verlagskontor, Lübek, 1990, p. 55.
20. T. Parasassi et al., *Cell Biochem. Funct.*, **2**, 85 (1984).
21. A. Bindoli et al., *Biochem. Pharmacol.*, **26**, 2405 (1977).
22. A. Valenzuela and R. Guerra, *Experientia*, **42**, 139 (1986).
23. V. Fintelmann and A. Albert, *Therapiewoche*, **30**, 5589 (1080).
24. K. Hruby, *Forum*, **8**, 23 (1984).
25. K. Hruby, *Intensivmedizin*, **24**, 269 (1987).
26. K. Hruby et al., *Wien. Klin. Wochenschr*, **95**, 225 (1983).
27. B. Zhou et al., *Biol. Pharm. Bull.*, **29**, 1900 (2006).
28. S. Chlopcikova et al., *Phytother. Res.*, **18**, 107 (2004).
29. G. Karimi et al., *Evid. Based Complement Alternat. Med.*, **2**, 383 (2005).
30. J. Sonnenbichler et al., *J. Pharmacol. Exp. Ther.*, **290**, 1375 (1999).
31. S. Dhanalakshmi et al., *Carcinogenesis*, **25**, 1459 (2004).
32. M. Lahiri-Chatterjee et al., *Cancer Res.*, **59**, 622 (1999).
33. S. K. Katiyar et al., *J. Natl. Cancer Inst.*, **89**, 556 (1997).
34. A. de la Lastra AC et al., *Planta Med.*, **61**, 116 (1995).
35. S. J. von et al., *Cell. Mol. Life Sci.*, **53**, 917 (1997).
36. H. Kohno et al., *Int. J. Cancer*, **101**, 461 (2002).
37. P. Q. Vinh et al., *Jpn. J. Cancer Res.*, **93**, 42 (2002).
38. Y. Yanaida et al., *Carcinogenesis*, **23**, 787 (2002).
39. X. Zi et al., *Clin. Cancer Res.*, **4**, 1055 (1998).
40. R. P. Singh and R. Agarwal, *Eur. J. Cancer*, **41**, 1969 (2005).
41. A. Tyagi et al., *Prostate*, **53**, 211 (2002).
42. R. P. Singh et al., *Cancer Res.*, **62**, 3063 (2002).
43. L. Varghese et al., *Clin. Cancer Res.*, **11**, 8441 (2005).
44. P. R. vis-Searles et al., *Cancer Res.*, **65**, 4448 (2005).
45. R. P. Singh and R. Agarwal, *Mutat. Res.*, **555**, 21 (2004).
46. A. Tyagi et al., *Carcinogenesis*, **25**, 1711 (2004).
47. P. Thelen et al., *J. Urol.*, **171**, 1934 (2004).
48. N. Bhatia et al., *Cancer Lett.*, **147**, 77 (1999).

49. P. N. Chen et al., *Chem. Biol. Interact.*, **156**, 141 (2005).
50. S. C. Chu et al., *Mol. Carcinog.*, **40**, 143 (2004).
51. H. G. Yoo et al., *Int. J. Mol. Med.*, **13**, 81 (2004).
52. J. Schumann et al., *J. Hepatol.*, **39**, 333 (2003).
53. C. Wilasrusmee et al., *Med. Sci. Monit.*, **8**, BR439 (2002).
54. Z. Amirghofran et al., *J. Ethnopharmacol.*, **72**, 167 (2000).
55. D. G. Lee et al., *Arch. Pharm. Res.*, **26**, 597 (2003).
56. V. J. Johnson et al., *Planta Med.*, **69**, 44 (2003).
57. O. P. Gupta et al., *Phytomedicine*, **7**, 21 (2000).
58. H. F. Huseini et al., *Phytother. Res.*, (2006).
59. M. Maghrani et al., *J. Ethnopharmacol.*, **91**, 309 (2004).
60. S. Zhou et al., *Drug Metab Rev.*, **36**, 57 (2004).
61. M. Maitrejean et al., *Bioorg. Med. Chem. Lett.*, **10**, 157 (2000).
62. N. Skottova and V. Krecman, *Physiol Res.*, **47**, 1 (1998).
63. N. Skottova et al., *Acta Univ Palacki. Olomuc. Fac. Med.*, **141**, 87 (1998).
64. V. Krecman et al., *Planta Med.*, **64**, 138 (1998).
65. B. Malewicz et al., *Carcinogenesis*, **27**, 1739 (2006).
66. M. Torres et al., *P. R. Health Sci. J.*, **23**, 69 (2004).
67. F. A. Crocenzi and M. G. Roma, *Curr. Med. Chem.*, **13**, 1055 (2006).
68. A. Rambaldi et al., *Am. J. Gastroenterol.*, **100**, 2583 (2005).
69. V. Kren and D. Walterova, *Biomed. Pap. Med. Fac. Univ. Palacky. Olomouc. Czech. Repub.*, **149**, 29 (2005).
70. K. R. Ball and K. V. Kowdley, *J. Clin. Gastroenterol.*, **39**, 520 (2005).
71. S. K. Katiyar, *Int. J. Oncol.*, **26**, 169 (2005).
72. M. D. Tanamly et al., *Dig. Liver Dis.*, **36**, 752 (2004).
73. J. Boerth and K. M. Strong, *J. Herb. Pharmacother.*, **2**, 11 (2002).
74. R. P. Singh and R. Agarwal, *Curr. Cancer Drug Targets.*, **4**, 1 (2004).
75. B. P. Jacobs et al., *Am. J. Med.*, **113**, 506 (2002).
76. R. Saller et al., *Drugs*, **61**, 2035 (2001).
77. S. Foster, ed., *Milk Thistle Bibliography and Abstracts*, American Botanical Council, Austin, TX 1995.

MINTS

Source: Peppermint Mentha × piperita L. (hybrid of *M. spicata* L. and *M. aquatica* L.); **Spearmint** Mentha spicata L. (syn. *M. viridis* L.); **Cornmint** Mentha arvensis L. var. *piperascens* Malinvaud (Family Labiatae or Lamiaceae).

Common/vernacular names: Field mint, Japanese mint (*M. arvensis*).

GENERAL DESCRIPTION

Closely related perennial aromatic herbs with runners or stolons by which they are propagated; leaves of spearmint are sessile (no petioles), while those of peppermint and cornmint are petioled and short petioled, respectively; up to about 1 m high; cultivated worldwide. Each species has numerous varieties, strains, or chemotypes that produce essential oils with widely different chemical compositions.[1,2] There are 20 true species of *Mentha*, represented by as many as 2300 named variations, half of which are synonyms; half are legitimate infraspecific names.[3] Commercial varieties of mints produce oils that can be distinguished by their relative contents of menthol and carvone. Parts used are the dried leaves and the fresh or partially dried whole, aboveground flowering herb. The former furnishes the spice, while the latter is used for the production of the essential oil.

United States is the major producer of peppermint and spearmint and their essential oils, especially in Idaho, Indiana, Oregon, Washington, and Wisconsin (USDA). Major producers of cornmint and cornmint oil include Japan, Taiwan, and Brazil.

CHEMICAL COMPOSITION

Peppermint yields 0.1–1.0% (usually 0.3–0.4%) of volatile oil that is composed mainly of menthol (29–48%), menthone (20–31%), and menthyl acetate (ca. 310%), with smaller amounts of menthofuran (1–7%) and limonene.[1,4-6] Other constituents present in peppermint oil include viridiflorol,[7,8] pulegone (1–11%),[9] 1,8-cineole (6–7.5%), piperitone, caryophyllene, bisabolene, isomenthone, isomenthol, α- and β-pinenes,[10] neomenthol,[5] ledol, *d-trans*-sabinene hydrate,[11] and bicycloelemene, among others (REMINGTON).[12]

Other constituents present in peppermint include flavonoids (e.g., menthoside, isorhoifolin, hesperetin, eriodictyol-7-*O*-rutinoside, luteolin-7-*O*-rutinoside, rutin, narirutin, diosmin, and hesperidin),[13-15] phytol, tocopherols (α and γ), carotenoids (e.g., α- and β-carotenes), betaine, choline, azulenes, chromone glycosides,[14] caffeic and rosmarinic acids,[13,16] and tannin.[17-22] An *O*-acetylated xyloglucan polysaccharide was isolated from the cell suspension cultures of peppermint.[23]

Spearmint yields normally about 0.7% volatile oil, consisting of 50–70% carvone, with lesser amounts of dihydrocarvone, phellandrene, and limonene.[24] Other compounds reported to be present include 6-hydroxycarvone,[25] menthone, menthol, pulegone, piperitenone, piperitenone oxide,[26] *cis*-carveyl acetate,[27] carveol, myrcene, α-and β-pinenes, cineole, linalool, α-terpineol, terpinen-4-ol, terpinolene,[1] dihydrocarveol, dihydrocarveol acetate; caryophyllene, 3-octyl acetate, 3-octanol; menthofuran,[28] and *cis*-hexenyl isovalerate, among others (JIANGSU; REMINGTON).[29]

Other constituents present in spearmint include flavonoids (e.g., diosmin and diosmetin),[30] monoterpene glycosides (spicatoside A and B)[31] and probably similar compounds as those found in peppermint.

Monomenthyl succinate is a common constituent in peppermint and spearmint.[32]

Cornmint contains 1–2% volatile oil that consists of a high concentration of menthol (70–95%),[33-35] menthone (10–20%),[36] menthyl acetate, isomenthone, thujone, apinene, phellandrene, piperitone, menthofuran, and others (JIANGSU; LIST AND HÖRHAMMER; MORTON 3).[28]

A cultivar of *M. arvensis* has been reported to contain high concentrations of menthofuran (42%) and lesser amounts of menthol (21%), menthyl acetate (14%), and menthone (0.71%).[37]

Other compounds present in cornmint include oligosaccharides (raffinose and stachyose),[38] resin, tannin, and rosmarinic acid (JIANGSU).[22]

PHARMACOLOGY AND BIOLOGICAL ACTIVITIES

Volatile oils obtained from peppermint, cornmint, and other mint species have antimicrobial, antimalarial and antigiradial activities *in vitro*.[39-50] Peppermint extracts have been reported to have antiviral activities against Newcastle disease, herpes simplex, vaccinia, Semliki Forest, and West Nile viruses in egg and cell culture systems (see **balm**).[51,52] The ethyl acetate extract of Moroccan *M. longifolia* was also found to possess inhibitory activity against HIV-1 reverse transcriptase.[53]

Peppermint oil has been demonstrated to exhibit spasmolytic activity on smooth muscles of experimental animals.[54] It was also found to be effective in alleviating the symptoms of irritable bowel syndrome.[55] Other effects of mint on the GIT, for example, as a spasmolytic and in nonulcer dyspepsia, have been reviewed.[56–59]

The antioxidant activity[13,41,43,60] of mint oil and extracts has been demonstrated alongside of cancer chemopreventive,[61,62] radioprotective,[63–67] and antimutagenic[68] effects in a number of different *in vitro* assays. Peppermint and cornmint oils have also been reported to have cytotoxic properties.[69]

The flavonoids in peppermint leaves reportedly have choleretic activity in dogs.[19]

Azulene isolated from peppermint had antiinflammatory and antiulcer effects in experimental animals.[28,49,70]

Recently reported effects of cornmint include acetylcholinesterase inhibition (linarin) and reversible antifertility.[71–73]

TOXICOLOGY

Menthol, the major component of peppermint and cornmint oils, may cause allergic reactions (e.g., contact dermatitis, flushing, and headache) in certain individuals. Applying a menthol-containing ointment to the nostrils of infants for the treatment of cold symptoms may cause instant collapse (MARTINDALE). Two recent studies, however, showed that peppermint extracts and their flavonoid glycosides (luteolin rutinoside) exhibited antihistaminic activities in experimental models of allergic rhinitis.[14,71,74] Other adverse effects of peppermint tea after excessive oral administration in rat models include reduction of iron absorption and possible hepatic and reproductive toxicity.[75–77]

USES

Medicinal, Pharmaceutical, and Cosmetic. Peppermint oil is extensively used as a flavoring agent, carminative, antiseptic, and local anesthetic in cold, cough, and other preparations (lozenges, syrups, ointments, tablets, etc.).

Enteric-coated peppermint oil capsules have been examined as a useful treatment for irritable bowel syndrome; enteric coating allows oil to reach colon in an unmetabolized state; treatment is contraindicated with meals (dosage recommended between meals), and in achlorhydria.[78–80]

Peppermint oil has been recommended as an adjunct to colonoscopy; a diluted suspension of the oil is sprayed on the endoscope to reduce colonic spasm.[81,82]

Spearmint oil is used primarily as a flavoring agent; it is also used as a carminative.

The mint oils (especially spearmint oil) are commonly used as fragrance components in toothpastes, mouthwashes, gargles, soaps, detergents, creams, lotions, and perfumes. Maximum use levels reported for spearmint and cornmint oils are 0.4–0.8%, respectively, in perfumes.[29,34]

Food. Spearmint oil and peppermint oil (usually rectified) are extensively used in flavoring chewing gums, candies, and chocolates as well as in most other food products, including alcoholic (liqueurs, etc.) and nonalcoholic beverages, frozen dairy desserts, baked goods, gelatins and puddings, processed fruits, and sweet sauces. The highest average maximum use levels reported are 0.104% for peppermint oil in candy and about 0.132% (1318 ppm) for spearmint oil in baked goods.

Spearmint leaves and extracts are reportedly used in alcoholic and nonalcoholic beverages. Spearmint leaves are also used in baked goods, gelatins and puddings, and meat and meat products. Highest average maximum use level is about 0.597% (5967 ppm) for spearmint leaves in baked goods.

Dietary Supplements/Health Foods. Leaves (or oil) of peppermint and spearmint, widely used as primary or adjunct flavoring for herb teas; capsules, tablets, tincture, and so on, in formulations for digestion, colds, and fevers (FOSTER).

Traditional Medicine. Peppermint, spearmint, and their oils reportedly used in both western and eastern cultures as aromatic, stomachic, stimulant, antiseptic, local anesthetic, and antispasmodic in treating indigestion, nausea, sore throat, diarrhea, colds, headaches, toothaches, and cramps (LEUNG).

Cornmint is used in China for treating similar conditions; also used in relieving earache and treating tumors and sores.

All three mints have been reported to be used in cancers.[83]

Others. Peppermint oil and menthol are widely used in flavoring tobacco.

Cornmint oil is mainly used for the production of menthol. Due to the high concentrations of menthol in this oil, it will solidify at room temperature. Much of the menthol can be removed from the crude oil (which may contain about 90% menthol) by freezing. The "dementholized" oil still contains about 55% menthol; this oil is the commercial cornmint oil. It can be further used as a source of menthol.

COMMERCIAL PREPARATIONS

Crudes (peppermint and spearmint) and oils. Peppermint, spearmint, and peppermint oil and spearmint oil are official in N.F. Peppermint oil, spearmint oil, and dementholized cornmint oil are official in F.C.C.

Regulatory Status. Peppermint, spearmint, and their derivatives (e.g., oils) are GRAS (§182.10 and §182.20). Peppermint-derived entities (oil, extracts, and leaves) are also considered to be safe when used in cosmetic formulations.[84]

The European monograph suggests use of the oil, only under the direction of a physician, for bile duct and gallbladder inflammation and gallstones; also for spasms of the upper gastrointestinal tract, flatulence, symptomatic treatment of irritable bowel syndrome, and catarrh of the respiratory tract. External use includes oral mucosa inflammations, rheumatic conditions, and local muscle and nerve pain, as well as skin conditions such as pruritus and urticaria (ESCOP 3).

Peppermint leaf and oil are subjects of German therapeutic monographs; leaves in infusion, or extract for spastic complaints of the gastrointestinal tracts as well as gallbladder and bile ducts, at average daily dose of 3–6 g of the leaves; 5–15 g tincture.[85,86]

REFERENCES

See the General References for APhA; ARCTANDER; BAILEY 2; BLUMENTHAL 1 & 2; ESCOP 3; FARNSWORTH 1–4; FEMA; FOGARTY; FOSTER; GOSSELIN; GUENTHER; JIANGSU; LIST AND HÖRHAMMER; MARTINDALE; NANJING; REMINGTON; ROSENGARTEN; TERRELL; UPHOF; USD 26TH.

1. F. W. Hefendehl and M. J. Murray, *Planta Med.*, **23**, 101 (1973).
2. B. M. Lawrence and J. K. Morton, *An. Acad. Bras. Cienc.*, **44**(Suppl.) (1972).
3. A. O. Tuckler et al., *Taxon*, **29**, 233 (1980).
4. M. B. Embong et al., *Can. Inst. Food Sci. Technol. J.*, **10**, 247 (1977).
5. K. Belafi-Rethy et al., *Acta Chim. (Budapest)*, **76**, 167 (1973).
6. A. K. Saxena et al., *Indian Perfum.*, **20**, 1 (1976).
7. W. Rojahn et al., *Dragoco Rep. (Ger. Ed.)*, **24**, 230 (1977).
8. D. Karasawa and S. Shimizu, *Shinshu Daigaku Nogakubu Kiyo*, **13**, 89 (1976).
9. A. Morkunas et al., *Polez. Rast. Priblat. Respub. Beloruss., Mater Nauch. Knof.*, **2**, 225 (1973).

10. I. Calvarano, *Essenze Deriv. Agrum.*, **39**, 77 (1969).
11. Y. Ono et al., *Bunseki Kagaku*, **24**, 589 (1975).
12. B. M. Lawrence et al., *Flav. Ind.*, **3**, 467 (1972).
13. Z. Sroka et al., *Z. Naturforsch. C*, **60**, 826 (2005).
14. T. Inoue et al., *Biol. Pharm. Bull.*, **25**, 256 (2002).
15. B. G. Hoffmann and L. T. Lunder, *Planta Med.*, **50**, 361 (1984).
16. U. Gerhardt and A. Schroeter, *Fleischwirtschaft*, **63**, 1628 (1983).
17. E. V. Gella et al., *Farmatsevt. Zh.*, **21**, 58 (1966).
18. H. Wagner et al., *Chem. Ber.*, **102**, 2083 (1969).
19. I. K. Pasechnik, *Farmacol. Toksikol.*, **29**, 735 (1966).
20. M. F. Shakhova and L. O. Shnaidman, *Rast. Resur.*, **4**, 53 (1968).
21. N. M. Solodovnichenko and Y. G. Borisyuk, *Farmatsevt. Zh.*, **17**, 44 (1962).
22. B. E. Ellis and G. H. N. Towers, *Biochem. J.*, **118**, 291 (1970).
23. K. Maruyama et al., *Phytochemistry*, **41**, 1309 (1996).
24. M. J. Murray et al., *Crop Sci.*, **12**, 723 (1972).
25. T. Tsuneya et al., *Koryo*, **104**, 23 (1973).
26. S. Shimizu et al., *Int. Congr. Essent. Oils (Pap.)*, **6**, 13 (1974).
27. T. Nagasawa et al., *Koryo*, **108**, 45 (1974).
28. I. C. Nigam and L. Levi, *J. Pharm. Sci.*, **53**, 1008 (1964).
29. D. L. J. Opdyke, *Food Cosmet. Toxicol.*, **16**(Suppl. 1), 871 (1978).
30. S. S. Subramanian and A. G. R. Nair, *Phytochemistry*, **11**, 452 (1972).
31. J. Zheng et al., *J. Asian Nat. Prod. Res.*, **5**, 69 (2003).
32. C. Marin and C. Schippa, *J. Agric. Food Chem.*, **54**, 4814 (2006).
33. M. L. Sharma et al., *Indian Perfum.*, **16**, 27 (1972).
34. M. M. Chopra and K. L. Handa, *Indian Perfum.*, **16**, 15 (1972).
35. A. K. Srivastava et al., *Indian Perfum.*, **20**, 61 (1976).
36. M. J. Murray et al., *Crop Sci.*, **12**, 742 (1972).
37. N. G. R. Donalisio et al., *Int. Congr. Essent. Oils (Pap.)*, **6**, 10 (1974).
38. A. Lombard et al., *J. Chromatogr.*, **134**, 242 (1977).
39. H. Cetin et al., *Phytother. Res.* (2006).
40. F. Vidal et al., *Exp. Parasitol.* (2006).
41. D. Yadegarinia et al., *Phytochemistry*, **67**, 1249 (2006).
42. A. K. Tripathi et al., *J. Med. Entomol.*, **41**, 691 (2004).
43. N. Mimica-Dukic et al., *Planta Med.*, **69**, 413 (2003).
44. G. Iscan et al., *J. Agric. Food Chem.*, **50**, 3943 (2002).
45. H. Imai et al., *Microbios*, **106**(Suppl. 1), 31 (2001).
46. C. C. Tassou et al., *J. Appl. Bacteriol.*, **78**, 593 (1995).
47. A. Sanyal and K. C. Varma, *Indian J. Microbiol.*, **9**, 23 (1969).
48. A. C. Pizsolitto et al., *Rev. Fac. Farm. Odontol. Araraquara*, **9**, 55 (1975).
49. K. K. Abdullin, *Uch. Zap. Kazansk. Vet. Inst.*, **84**, 75 (1962).
50. F. M. Ramadan et al., *Chem. Mikrobiol. Technol. Lebensm.*, **1**, 96 (1972).
51. A. Schuhmacher et al., *Phytomedicine*, **10**, 504 (2003).
52. E. C. Herrmann Jr. and L. S. Kucera, *Proc. Soc. Exp. Biol. Med.*, **124**, 874 (1967).
53. S. Amzazi et al., *Therapie*, **58**, 531 (2003).

54. T. Shipochliev, *Vet. Med. Nauki*, **5**, 63 (1968).
55. H. G. Grigoleit and P. Grigoleit, *Phytomedicine*, **12**, 612 (2005).
56. H. G. Grigoleit and P. Grigoleit, *Phytomedicine*, **12**, 607 (2005).
57. H. G. Grigoleit and P. Grigoleit, *Phytomedicine*, **12**, 601 (2005).
58. M. H. Pittler and E. Ernst, *Am. J. Gastroenterol.*, **93**, 1131 (1998).
59. C. J. Thompson and E. Ernst, *Aliment. Pharmacol. Ther.*, **16**, 1689 (2002).
60. M. H. Ka et al., *J. Agric. Food Chem.*, **53**, 4124 (2005).
61. R. M. Samarth et al., *Int. J. Radiat. Biol.*, **82**, 331 (2006).
62. R. M. Samarth et al., *Mutagenesis*, **21**, 61 (2006).
63. R. M. Samarth et al., *Phytother. Res.*, **18**, 546 (2004).
64. R. M. Samarth and A. Kumar, *J. Radiat. Res. (Tokyo)*, **44**, 101 (2003).
65. G. C. Jagetia and M. S. Baliga, *Strahlenther. Onkol.*, **178**, 91 (2002).
66. R. M. Samarth et al., *Indian J. Exp. Biol.*, **40**, 1245 (2002).
67. R. M. Samarth et al., *Indian J. Exp. Biol.*, **39**, 479 (2001).
68. T. W. Yu et al., *Environ. Mol. Mutagen.*, **44**, 387 (2004).
69. K. Silyanovska et al., *Parfüm. Kosmet.*, **50**, 293 (1969).
70. G. N. Maksimenko, *Farmakol. Toksikol.*, **27**, 571 (1964).
71. P. P. Oinonen et al., *Fitoterapia*, **77**, 429 (2006).
72. N. Sharma and D. Jacob, *J. Ethnopharmacol.*, **80**, 9 (2002).
73. N. Sharma and D. Jacob, *J. Ethnopharmacol.*, **75**, 5 (2001).
74. T. Inoue et al., *Biol. Pharm. Bull.*, **24**, 92 (2001).
75. M. Akdogan et al., *Hum. Exp. Toxicol.*, **23**, 21 (2004).
76. M. Akdogan et al., *Toxicol. Ind. Health*, **20**, 119 (2004).
77. M. Akdogan et al., *Urology*, **64**, 394 (2004).
78. K. W. Somerville et al., *Br. J. Clin. Pharmacol.*, **18**, 638 (1984).
79. W. D. Rees et al., *Br. Med. J.*, 835 (1979).
80. S. Foster, *Peppermint—Mentha × Piperita*, Botanical Series, no. 301, American Botanical Council, Austin, TX, 1991.
81. R. J. Leicester and R. H. Hunt, *Lancet* (1982).
82. H. L. Duthie, *Br. J. Surg.*, **68**, 820 (1981).
83. J. L. Hartwell, *Lloydia*, **32**, 247 (1969).
84. B. Nair, *Int. J. Toxicol.*, **20**(Suppl. 3), 61 (2001).
85. Monograph *Menthae piperitae aetheroleum*, Bundesanzeiger, no. 50 (March 13, 1986).
86. Monograph *Menthae piperitae folium*, Bundesanzeiger, no. 223 (November 30, 1985); corrected (March 13, 1990).

MISTLETOE

Source: *Viscum album* L. (Family Loranthaceae or Viscaceae).

Common/vernacular names: European mistletoe.

GENERAL DESCRIPTION

Parasitic shrub, stems yellow-green, up to 100 cm long; forming tufts; leaves opposite, obovate–oblong, 2–8 cm long; flowers deciduous, in cymes, unisexual, tetramerous; berry white, translucent;[1] occurring on woody angiosperms and gymnosperms; most of Europe; naturalized in one California county.

The part used is the herb.

CHEMICAL COMPOSITION

All plant parts contain β-phenylethylamine, tyramine, and related compounds; polypeptides, including viscotoxins I, II, III, IVb (II, III, IVb identical to viscotoxins B, A-2, and A-3, respectively);[2-4] glycoprotein lectins, including viscumin and lectins 1, II, and III;[5-13] flavonoids/phenylpropanoids, including syringin, coniferin, syringenin apiosylglucoside and 4,4″-diglucoside (eleutheroside E), 5,7-dimethoxyflavanone apiosylglucoside, 5,7-dimethoxynaringenin, 2′-hydroxy-4′,6′-dimethoxychalcone glucoside, and others;[14-16] phenolic acids (nature depends on the plant on which mistletoe grows), for example, caffeic, gentisic, vanillic, digallic, and salicylic acids;[17] polysaccharides[18] rich in glucose, galactose, rhamnose, arabinose, xylose, and galacturonic acid (WREN).

Sterols and triterpenes (β-amyrin, β-sitosterol, lupeol, stigmasterol, viscin, betulinic acid, oleanolic acid, and ursolic acid),[19,20] fatty acids (oleic, linoleic, palmitic, and stearic acids),[19] and an acyclic monoterpene glycoside (2,6-dimethylocta-2,7-diene-1,6-diol 4-O-apiosylglucoside) have recently been isolated from mistletoe.[21]

PHARMACOLOGY AND BIOLOGICAL ACTIVITIES

Various pharmacological activities include immunostimulant, antineoplastic, hypotensive, cardiotonic, sedative, and antispasmodic (WREN).

Immunostimulatory activity has been confirmed in many *in vitro* and *in vivo* models in addition to human clinical studies.[22-29] Proposed mechanisms include stimulation of both the cellular (increased production of lymphocytes, leukocytes, and natural killer cells) and humoral (the production of lymphokines by lymphocytes) immune responses.[23,25-30]

A single study, however, reported that isolated polysaccharides from the stem (a galacturonan), and from the berries (an arabinogalactan), failed to increase phagocytosis of granulocytes and macrophages.[31]

The most studied activity for mistletoe over the last decade is its antitumor effect. Numerous reports have been published on its efficacy in various human cancer conditions, and against many tumors and cancer cell lines *in vivo* and *in vitro*, respectively.[2,32-41] Antitumor mechanisms involve the immunomodulatory effect mentioned above for lectins, as well as apoptosis, antiangiogenesis, cell cycle disruption, and direct cytotoxic effects on cancer cells.[19,42-51]

Isolated polypeptides, viscotoxins II, III, and IVb, have been associated with cardiotoxicity but have also been found to exhibit cytotoxic activity against human tumor cells of the KB and HeLa lines in tissue culture.[4] A peptide with a molecular weight of 5000 was found to be cytotoxic to Dalton's lymphoma ascites tumor cells *in vitro* in mice, without affecting normal lymphocytes, indicating a cell-dependent specificity;[52] also cyctotoxic to Ehrlich ascites cells, both prophylactically and after tumor development.[53] Commercial mistletoe products have been used to treat various cancers in Europe with clinical success. A group of 50 cases of carcinomatous pleural effusions were treated with a topical preparation for an average of 3.3 application over 18 days; exudation disappeared in 92% of the patients.[54] In postoperative ovarian cancer patients, a mistletoe preparation statistically increased survival.[55]

Anti-inflammatory activity has been reported for mistletoe extract, its agglutinin-I lectin and its flavonoid constituents in the carrageenan-induced rat paw edema, in chronic hepatitis and in the inflammatory response through neutrophils.[14,56,57]

The aqueous extract and phenolic fractions of mistletoe exhibited *in vitro* vasodilating effect on the coronary blood vessels, rat aortic rings, and perfused heart model. The effect may be mediated through modulation of nitric oxide production.[15,58] Lignans are believed to be responsible for cardiotonic activity, significantly inhibiting cAMP-phosphodiesterase activity.[59]

Other recently reported activities for mistletoe include antioxidant,[60,61] antidiabetic,[61,62] and antiviral effects (against human parainfluenza virus).[63] Many reviews about the pharmacology and therapeutic utility of mistletoe in oncology have been published.[41,64–70]

TOXICOLOGY

Often regarded as a poisonous plant, toxic effects are primarily associated with parenteral administration. Large quantities taken orally may be locally irritating and necrotizing. Given potential toxicity, self-medication is not advised (FROHNE AND PFANDER). One case of anaphylaxis after mistletoe administration has recently been reported in which viscotoxin-specific IgE could be detected.[71] An analysis of more than 300 reported mistletoe ingestion cases in the United States, however, found that a majority of patients were asymptomatic; no fatalities occurred; ingestion of fewer than three berries or two leaves will likely not produce serious toxicity.[13] Similar conclusions to the generally accepted safety of mistletoe, especially the standardized preparations, have appeared in more recent reports.[25,72–74]

USES

Medicinal, Pharmaceutical, and Cosmetic. In German phytomedicine, fresh plant, cut, or powdered herb in injectable solutions are used for supportive therapy of nonspecific irritation in malignant tumors; degenerative inflammatory joint diseases; contraindicated in protein hypersensitivity and chronic progressive infections. Reported side effects include fever, headaches, angina pectoris symptoms, and allergic reactions.[75]

Traditional Medicine. Mistletoe has been employed for the treatment of various disorders, including sores and abscesses, and nervous disorders; an abortifacient and antineoplastic for over two millenia.[76]

COMMERCIAL PREPARATIONS

At least five parenterally administered product forms available in Germany.[76]

Regulatory Status. Leaves subject of a German therapeutic monograph.[75] Fruit indicated for numerous conditions in combination with other herbs is not recommended due to lack of documentation of efficacy.[77]

REFERENCES

See the General References for APPLEQUIST; BARNES; BLUMENTHAL 1; DER MARDEROSIAN AND LIBERTI; FOSTERANDDUKE; GRIEVE; MARTINDALE; MCGUFFIN 1 & 2; STEINMETZ; TUTIN 1; TYLER 1; WEISS; WREN.

1. K. R. Kirtikar and B. D. Basu, *Indian Medicinal Plants*, International Book Distribution, Dehradun, India, 1999, p. 2183.
2. K. Urech et al., *Anticancer Res.*, **26**, 3049 (2006).
3. V. Mishra et al., *Arch. Biochem. Biophys.*, **423**, 288 (2004).
4. J. Konopa et al., *Hoppe Seylers Z. Physiol. Chem.*, **361**, 10 (1980).
5. K. Urech et al., *Arzneim.-Forsch.*, **56**, 428 (2006).
6. R. Wacker et al., *J. Pept. Sci.*, **11**, 289 (2005).
7. S. Y. Lyu et al., *Arch. Pharm. Res.*, **23**, 592 (2000).
8. G. M. Stein et al., *Anticancer Res.*, **19**, 1037 (1999).
9. H. S. Lee et al., *Cell Mol. Life Sci.*, **55**, 679 (1999).

10. R. Krauspenhaar et al., *Biochem. Biophys. Res. Commun.*, **257**, 418 (1999).
11. W. B. Park et al., *Arch. Pharm. Res.*, **21**, 429 (1998).
12. S. Eschenburg et al., *Biochem. Biophys. Res. Commun.*, **247**, 367 (1998).
13. B. Olin, Ed., *Lawrence Rev. Natl. Prod.* (Dec. 1992).
14. D. D. Orhan et al., *Z. Naturforsch. C*, **61**, 26 (2006).
15. D. Deliorman et al., *J. Ethnopharmacol.*, **72**, 323 (2000).
16. H. Wagner et al., *Oncology*, **43**, 16 (1986).
17. M. Luczkiewicz et al., *Acta Pol. Pharm.*, **58**, 373 (2001).
18. U. Edlund et al., *Arzneim.-Forsch.*, **50**, 645 (2000).
19. K. Urech et al., *J. Pharm. Pharmacol.*, **57**, 101 (2005).
20. M. J. Jung et al., *Arch. Pharm. Res.*, **27**, 840 (2004).
21. D. Deliorman et al., *Fitoterapia*, **72**, 101 (2001).
22. A. Bussing, *Arzneim.-Forsch.*, **56**, 508 (2006).
23. R. Huber et al., *J. Soc. Integr. Oncol.*, **4**, 3 (2006).
24. R. Klopp et al., *Anticancer Res.*, **25**, 601 (2005).
25. W. Dohmen et al., *Anticancer Res.*, **24**, 1231 (2004).
26. G. Stein et al., *Eur. J. Med. Res.*, **3**, 194 (1998).
27. T. Hajto et al., *Anticancer Drugs*, **8**(Suppl. 1), S43 (1997).
28. N. Bloksma et al., *Immunobiology*, **156**, 309 (1979).
29. E. G. Coeugniet and E. Elek, *Oncology*, **10**, 27 (1987).
30. K. Hostanska et al., *Nat. Immun.*, **15**, 295 (1996).
31. E. Jordan and H. Wagner, *Oncology*, **43**, 8 (1986).
32. T. Hajto et al., *Arzneim.-Forsch.*, **56**, 441 (2006).
33. G. Kelter and H. H. Fiebig, *Arzneim.-Forsch.*, **56**, 435 (2006).
34. G. Bar-Sela et al., *Anticancer Res.*, **26**, 709 (2006).
35. I. F. Pryme et al., *Histol. Histopathol.*, **21**, 285 (2006).
36. U. Elsasser-Beile et al., *Anticancer Res.*, **25**, 4733 (2005).
37. M. B. Enesel et al., *Anticancer Res.*, **25**, 4583 (2005).
38. M. Rostock et al., *Anticancer Res.*, **25**, 1969 (2005).
39. U. Elsasser-Beile et al., *J. Urol.*, **174**, 76 (2005).
40. A. Thies et al., *Toxicology*, **207**, 105 (2005).
41. M. Schink, *Anticancer Drugs*, **8**(Suppl. 1), S47 (1997).
42. M. Harmsma et al., *Arzneim.-Forsch.*, **56**, 474 (2006).
43. E. Kovacs et al., *Arzneim.-Forsch.*, **56**, 467 (2006).
44. M. D. Mossalayi et al., *Arzneim.-Forsch.*, **56**, 457 (2006).
45. J. P. Duong Van Huyen et al., *Cancer Lett.*, **243**, 32 (2006).
46. W. B. Park et al., *Cancer Biother. Radiopharm.*, **16**, 439 (2001).
47. M. S. Kim et al., *Gen. Pharmacol.*, **34**, 349 (2000).
48. A. Savoie et al., *J. Leukoc. Biol.*, **68**, 845 (2000).
49. H. Bantel et al., *Cancer Res.*, **59**, 2083 (1999).
50. T. J. Yoon et al., *Cancer Lett.*, **136**, 33 (1999).
51. H. Franz, *Oncology*, **43**, 23 (1984).
52. G. Kuttan et al., *Cancer Lett.*, **41**, 307 (1988).
53. G. Kuttan et al., *J. Ethnopharmacol.*, **29**, 35 (1990).

54. G. Slazer and H. Muller, *Praxis und Klinik der Pneumologia*, **32**, 721 (1978).
55. W. Hassauer et al., *Onkologie*, **2**, 28 (1979).
56. K. J. Tusenius et al., *Arzneim.-Forsch.*, **55**, 749 (2005).
57. V. Lavastre et al., *Clin. Exp. Immunol.*, **137**, 272 (2004).
58. F. A. Tenorio et al., *Fitoterapia*, **76**, 204 (2005).
59. H. Wagner et al., *Planta Med.*, **52**, 102 (1986).
60. U. E. Onay et al., *Fitoterapia* (2006).
61. D. D. Orhan et al., *J. Ethnopharmacol.*, **98**, 95 (2005).
62. A. M. Gray and P. R. Flatt, *J. Endocrinol.*, **160**, 409 (1999).
63. A. Karagoz et al., *Phytother. Res.*, **17**, 560 (2003).
64. S. Elluru et al., *Arzneim.-Forsch.*, **56**, 461 (2006).
65. E. Ernst et al., *Int. J. Cancer*, **107**, 262 (2003).
66. G. S. Kienle et al., *Eur. J. Med. Res.*, **8**, 109 (2003).
67. P. J. Mansky, *Semin. Oncol.*, **29**, 589 (2002).
68. H. Stauder and E. D. Kreuser, *Onkologie*, **25**, 374 (2002).
69. U. Mengs et al., *Anticancer Res.*, **22**, 1399 (2002).
70. N. Zarkovic et al., *Cancer Biother. Radiopharm.*, **16**, 55 (2001).
71. C. Bauer et al., *Ann. Allergy Asthma Immunol.*, **94**, 86 (2005).
72. M. Augustin et al., *Arzneim.-Forsch.*, **55**, 38 (2005).
73. A. M. Burger et al., *Anticancer Res.*, **23**, 3801 (2003).
74. W. M. van et al., *Am. J. Ther.*, **6**, 37 (1999).
75. Monograph Visci *albi herba*, Bundesanzeiger, no. 228 (December 5, 1984).
76. I. A. Bowman, *HerbalGram*, **26**, 16 (1992).
77. Monograph Visci *albi fructus*, Bundesanzeiger, no. 228 (July 14, 1993).

MONOSODIUM GLUTAMATE (MSG)

Source: Wheat gluten and other natural sources of glutamic acid are used for the manufacture of MSG.

Common/vernacular names: Sodium glutamate, monosodium L-glutamate.

GENERAL DESCRIPTION

Monosodium glutamate (MSG) is a monosodium salt of L-glutamic acid, which is a common amino acid present in high concentrations in brain tissues and is found in most proteins.[1] Certain proteins such as gluten (corn and wheat) are very rich in this amino acid. Wheat gluten contains about 38% glutamic acid and has been used as one of the major sources for the manufacture of MSG.[2–6]

Monosodium glutamate can generally be produced by three methods: (1) hydrolysis of proteins such as gluten or proteins present in sugar beet wastes, (2) synthesis, and (3) microbial fermentation.[5,6] In the hydrolysis method, the protein is hydrolyzed with a strong mineral acid to free amino acids, and the glutamic acid is then separated from the mixture, purified, and converted to its monosodium salt, MSG. This used to be the major method of MSG manufacture. Currently, most of the world production of MSG is by bacterial fermentation. In this method, bacteria (especially strains of *Micrococcus glutamicus*) are grown aerobically in a liquid nutrient medium containing a carbon source (e.g., dextrose or citrate), a nitrogen source such as ammonium ions or urea, and mineral ions and growth factors. The bacteria selected for this process have the ability to synthesize and excrete glutamic acid into the medium where it accumulates. The glutamic acid is separated from the fermentation broth by filtration, concentration, acidification, and crystallization, followed by conversion to its monosodium salt MSG.[5,6]

Monosodium glutamate is considered to be a flavor enhancer, which enhances or intensifies the flavor of other foods and salty taste.[7,8] Its flavor-enhancing property was discovered in the early 1900s by the Japanese who first started producing it in a commercial scale and have since been the major producer.[9] Monosodium glutamate has a sweetish, meaty taste.

PHARMACOLOGY AND BIOLOGICAL ACTIVITIES

Monosodium glutamate has been used in treating mental retardation and hepatic coma that is accompanied by a high blood level of ammonia. It has been reported to be effective in lowering the blood level of ammonia in many cases, though the mechanism of action is still unknown (MARTINDALE; USD 26TH).

TOXICOLOGY

Many scientific investigations have demonstrated MSG to cause brain damage in infant experimental animals (mice, rats, rhesus monkeys, etc.) and to produce the so-called Chinese restaurant syndrome (burning sensation, facial pressure, and chest pain) in humans.[10–23] However, as many research efforts have been spent by as many different groups of investigators resulting in negative findings.[24–35] It should be noted that only one laboratory has reported both positive and negative findings.[15,18,26] The other laboratories have observed consistently either one or the other but not both.[13,14,16,17,22–25,29] Three recent reports further support the lack of serious side effects for MSG in humans.[36–38] Other recent studies, however, showed that MSG can induce diabetes, hepatotoxicity, nephrotoxicity, and visual impairment in mice and rats.[39–41]

Monosodium glutamate has also been reported to have emetic properties in experimental animals (dogs, cats, and monkeys),[42] to induce clonic movements and tonic-clonic seizures in rats,[43,44] to cause obesity in mice,[45] to cause sterility in female mice, and other activities.[19,28,46]

USES

Medicinal, Pharmaceutical, and Cosmetic. MSG is used as flavor enhancer in certain liver and protein hydrolysate preparations.

Food. MSG is very popular in Japanese and Chinese cooking. It is liberally used in restaurants, especially in soups and vegetarian dishes.

MSG is used very extensively in processed foods, including nonalcoholic beverages, candy, baked goods, meat and meat products, condiments and relishes, breakfast cereals, milk products, cheeses, processed fruits, processed vegetables, soups, snack foods, nut products, gravies, seasonings and flavorings, and fish products, among others. Highest average maximum use levels are about 24.68% (246,785 ppm), 1.07%, and about 0.803% (8034 ppm) reported in seasonings and flavorings, breakfast cereals, and soups, respectively.

COMMERCIAL PREPARATIONS

Monosodium glutamate. It is official in N.F. and F.C.C.

Regulatory Status.   GRAS.

REFERENCES

See the General References for FEMA; FURIA; MARTINDALE; MERCK; USD 26TH.

1. A. White et al., *Principles of Biochemistry*, 2nd ed., McGraw-Hill, New York, 1959.
2. K. Hess and E. Hille, *Z. Lebensm. Unters. Forsch.*, **115**, 211 (1961).
3. H. Ludewig and T. Messing, *Ger. (East)*, 37074 (1965).
4. E. Magyar et al., *Staerke*, **15**, 12 (1963).
5. W. L. Faith et al., *Industrial Chemicals*, 3rd ed., John Wiley & Sons, New York, 1965, p. 521.
6. R. Powell, *Monosodium Glutamate and Glutamic Acid*, Noyes Development Corp., Park Ridge, NJ, 1968.
7. F. Bellisle, *Ann. N. Y. Acad. Sci.*, **855**, 438 (1998).
8. A. Okiyama and G. K. Beauchamp, *Physiol. Behav.*, **65**, 177 (1998).
9. H. J. Sanders, *Chemistry*, **40**, 23 (1967).
10. T. Uehara et al., *Okinawa-ken Kogai Eisei Kenkyusho Ho*, **10**, 34 (1976).
11. M. J. Kuhar, *Res. Commun. Chem. Pathol. Pharmacol.*, **2**, 95 (1971).
12. T. Fujiwara et al., *Jutsugo Taisha Kenkyu Kaishi*, **10**, 385 (1976).
13. E. A. Arees and J. Mayer, *Science*, **170**, 549 (1970).
14. P. E. Araujo and J. Mayer, *Am. J. Physiol.*, **225**, 764 (1973).
15. N. Lemkey-Johnston and W. A. Reynolds, *J. Neuropathol. Exp. Neurol.*, **33**, 74 (1974).
16. N. Snapir et al., *Pathol. Eur.*, **8**, 265 (1973).
17. B. Robinson et al., *Poult. Sci.*, **54**, 234 (1975).
18. N. Lemkey-Johnston et al., *J. Comp. Neurol.*, **167**, 481 (1976).
19. J. L. Everly, *Diss. Abstr. Int. B*, **33**, 1351 (1972).
20. S. Ungthavorn et al., *J. Fac. Med. Chulalongkorn Univ., Bangkok*, **16**, 265 (1971).
21. H. H. Schaumburg et al., *Science*, **163**, 826 (1969).
22. J. W. Olney and L. G. Sharpe, *Science*, **167**, 1017 (1970).
23. J. W. Olney, *Science*, **165**, 1029 (1969).
24. R. Heywood et al., *Toxicol. Lett.*, **1**, 151 (1977).
25. G. Owen et al., *Toxicol. Lett.*, **1**, 217 (1978).
26. W. A. Reynolds et al., *Science*, **172**, 1342 (1971).
27. P. Morselli and S. Garattini, *Nature (London)*, **227**, 611 (1970).
28. G. Bazzano et al., *Science*, **169**, 1208 (1970).
29. A. J. Newman et al., *Toxicology*, **1**, (1973).
30. M. E. Semprini et al., *Nutr. Metab.*, **16**, 276 (1974).
31. B. L. Oser et al., *Food Cosmet. Toxicol.*, **13**, 7 (1975).
32. N. J. Adamo and A. Ratner, *Science*, **169**, 673 (1970).
33. I. P. Barchenko and S. G. Vasiliu, *Vop. Ratsion Pitan.*, **6**, 21 (1970).
34. I. Rosenblum et al., *Toxicol. Appl. Pharmacol.*, **18**, 367 (1971).
35. S. Matsuyama et al., *Natl. Inst. Anim. Health Q.*, **13**, 91 (1973).
36. M. Freeman, *J. Am. Acad. Nurse Pract.*, **18**, 482 (2006).
37. K. Beyreuther et al., *Eur. J. Clin. Nutr.*, (2006).
38. R. Walker and J. R. Lupien, *J. Nutr.*, **130**, 1049S (2000).
39. M. Nagata et al., *Exp. Anim.*, **55**, 109 (2006).

40. G. G. Ortiz et al., *Biomed. Pharmacother.*, **60**, 86 (2006).
41. C. Praputpittaya and A. Wililak, *Nutr. Neurosci.*, **6**, 301 (2003).
42. W. P. Pi and M. T. Peng, *Taiwan I Hseuh Hui Tsa Chih*, **64**, 669 (1965).
43. C. N. Stewart et al., *Toxicol. Appl. Pharmacol.*, **23**, 635 (1972).
44. C. B. Nemeroff and F. D. Crisley, *Pharmacol. Biochem. Behav.*, **3**, 927 (1975).
45. J. Bunyan et al., *Br. J. Nutr.*, **35**, 25 (1976).
46. P. Cooper, *Food Cosmet. Toxicol.*, **15**, 347 (1977).

MUSK

Source: *Moschus moschiferus* L. (Family Moschidae).

Common/vernacular names: Tonquin musk, musk Tonquin, and deer musk.

GENERAL DESCRIPTION

The musk deer is a relatively small solitary animal standing about 0.5 m high at the shoulder and measuring 0.65–0.95 m long; both male and female are devoid of antlers. It is distributed in mountainous regions of Asia such as northern India, Tibet, and southern, western, northern, and northeastern China (JIANGSU).[1,2]

Musk is the highly odoriferous secretion derived from the musk gland present under the abdomen near the pubis of the male animal. There are two methods of obtaining musk. In the first method, the male musk deer is trapped and killed in late winter or early spring and the whole musk gland is immediately removed from the abdomen. After drying, the whole gland is known as a pod and the secretion inside in the form of granules is called grained musk, or musk grains. This used to be the only method of collecting musk. Nowadays musk is collected more and more from male musk deer raised in captivity. In this method, the male deer is tied to a special table on its back and the secretion is carefully removed from the musk gland with a special sterilized spoon. Healthy male deer at least 3 years old are used. Musk is collected once a year in late winter or early spring; sometimes it is also collected twice a year, in March or April and in July or August. The fresh secretion is a dark brown viscous semisolid, which turns into brownish yellow or purplish red granules when dried (JIANGSU; POUCHER).

Major musk producers include China and India.

It should be noted that the term musk is sometimes also used to describe substances or compounds that have an odor similar to that of true musk grains; these "musks" can be of synthetic, for example, musk xylene, nitromusks and polycyclic musks,[3,4] or natural origins (MERCK).[5]

CHEMICAL COMPOSITION

Musk normally contains as its odorous and active principles 0.3–2% muscone together with small quantities of normuscone (JIANGSU; LIST AND HÖRHAMMER).[6–8]

Other constituents present in musk include steroids (cholesterol and its esters, cholest-4-en-3-one, and possibly the bufadienolides resibufogenin and cinobufagin,[9] and others); paraffins (long chain and branched); triglycerides (oleodipalmitin, palmitodiolein, and triolein; methyl palmitate and methyl oleate; wax;[7,8] protein (musk protein MP-1, MW ca. 35,000 Da);[10] muscopyridine, hydroxymuscopyridine A and B, and other nitrogenous substances (e.g., urea and ammonium carbonate);[11] mineral salts; and fatty acids (JIANGSU; LIST AND HÖRHAMMER).

PHARMACOLOGY AND BIOLOGICAL ACTIVITIES

Musk has anti-inflammatory and antihistaminic activities on experimental animals.[12–14] Its anti-inflammatory activity was greater than that of phenylbutazone against arthritis in rats induced by injection of dead tubercle bacteria in liquid paraffin.[13] Its water-soluble fraction has the strongest anti-inflammatory activity, being 36 times that of hydrocortisone in mouse ear edema induced by croton oil. The active principle is a polypeptide with a molecular weight of about 10,000 whose structure has not been determined.[15,16]

Musk has also been reported to have spasmolytic, CNS-depressant, as well as stimulant, antibacterial, and other activities (JIANGSU).[14,17,18]

Muscone has recently been reported to exhibit weak estrogenic activity *in vitro*.[19]

USES

Medicinal, Pharmaceutical, and Cosmetic. The use of musk in cosmetics (e.g., fragrance and perfumes) dates back to at least 1300 years when it was widely used during the Tang and Sung dynasties (NANJING).

It is used as a fragrance component and fixative in perfumes (especially oriental and heavy floral types).

Food. Used for its "rounding off" effect in nut, caramel, and fruit-type flavors in major food products, including alcoholic and nonalcoholic beverages, frozen dairy desserts, candy, baked goods, and gelatins and puddings. Use levels are very low, generally much below 0.0001% (1 ppm).

Traditional Medicine. Musk has been used in Chinese medicine for thousands of years in treating stroke, coma, neurasthenia, convulsions, heart pains, ulcerous sores, and other conditions. It has been used in the clinical treatment of angina pectoris with results (ca. 74%) comparable or better than those of nitroglycerin (JIANGSU).

The musk used in Chinese medicine has been treated differently than the musk that is intended for export. For example, the former is dried only in the shade, while the latter is sun or heat dried (NANJING).

COMMERCIAL PREPARATIONS

Crude and extracts (e.g., tincture and absolute). Musk was formerly official in U.S.P.

Regulatory Status. GRAS (§182.50).

REFERENCES

See the General References for ARCTANDER; FEMA; JIANGSU; JIXIAN; MARTINDALE; NANJING; POUCHER.

1. *The Larousse Encyclopedia of Animal Life*, McGraw-Hill, New York, 1967, p. 596.
2. M. Burton, ed., *The World Encyclopedia of Animals*, Funk and Wagnalls, New York, 1972, p. 255.
3. H. U. Kafferlein et al., *Crit Rev. Toxicol.*, **28**, 431 (1998).
4. P. Steinberg et al., *Arch. Toxicol.*, **75**, 562 (2001).
5. W. E. Brugger and P. C. Jurs, *J. Agric. Food. Chem.*, **25**, 1158 (1977).
6. H. Y. Hsu et al., *T'ai-wan Yao Hsueh Tsa Chih*, **25**, 26 (1973).
7. J. C. Do, *Yongnam Taehakkyo Chonyonmul Hwahak Yonguso Yongu Pogo*, **3**, 19 (1976).
8. J. C. Do et al., *Chem. Pharm. Bull.*, **23**, 629 (1975).

9. H. Song et al., *Biomed. Chromatogr.*, **14**, 130 (2000).
10. K. S. Ahn et al., *Biol. Pharm. Bull.*, **25**, 418 (2002).
11. D. Q. Yu and B. C. Das, *Planta Med.*, **49**, 183 (1983).
12. R. K. Mishra et al., *J. Pharm. Pharmacol.*, **14**, 830 (1962).
13. H. H. Siddiqui, *Indian J. Pharm.*, **27**, 80 (1965).
14. S. D. S. Seth et al., *Jpn. J. Pharmacol.*, **23**, 673 (1973).
15. X. Y. Zhu et al., *Yaoxue Xuebao*, **23**, 406 (1988).
16. X. Y. Zhu et al., *Acta Acad. Med. Sin.*, **11**, 52 (1989).
17. M. Kimura et al., *Yakugaku Zasshi*, **88**, 130 (1968).
18. A. Mukhopadhyay et al., *Indian J. Pharm.*, **35**, 169 (1973).
19. N. Bitsch et al., *Arch. Environ. Contam. Toxicol.*, **43**, 257 (2002).

MUSTARD

Source: **Brown mustard** *Brassica juncea* (L.) Czern. et Coss. (syn. *Sinapis juncea* L.); **Black mustard** *Brassica nigra* (L.) Koch; **White mustard** *Sinapis alba* L. (syn. *B. alba* (L.) Rabenh.; *B. alba* (L.) Boiss.; *B. hirta* Moench) (Family Cruciferae or Brassicaceae).

Common/vernacular names: Indian mustard (*B. juncea*), yellow mustard (*S. alba*).

GENERAL DESCRIPTION

Annual or biennial herbs; brown and white mustards up to about 1 m, black mustard up to 3 m high; white mustard and black mustard are native to Eurasia (probably the Mediterranean region), and brown mustard is native to Asia. All three are cultivated worldwide. Parts used are their dried ripe seeds.

Mustard flour (ground mustard) is powdered mustard seeds with their seed coats removed. It often consists of a mixture of brown (or black) and white seeds, especially certain British and Chinese types. The more pungent ("hot") mustard flours are those with the fixed oil removed. The fixed oil constitutes more than one-third of the flour and does not contribute to the pungency or aroma of the mustard. Ground mustard does not have any pungent aroma when dry. This aroma is produced when the mustard comes in contact with water whereby the enzyme systems present in the mustard hydrolyze sinigrin (a glucoside of brown and black mustard), setting free allyl isothiocyanate, which is mainly responsible for the aroma.

Mustard oil (volatile) is prepared by steam distillation from brown or black mustard after expressing the fixed oil (expressed mustard oil) and macerating in warm water to allow the hydrolysis of sinigrin by the enzyme myrosin. The volatile oil consists mainly of allyl isothiocyanate.

White mustard does not produce a volatile oil by water and steam distillation.

CHEMICAL COMPOSITION

Brown mustard (*B. juncea*) contains the glucosinolate sinigrin (potassium myronate) and the enzyme myrosin (myrosinase); sinapic acid (3,5-dimethoxy-4-hydroxycinnamic acid); sinapine (sinapic acid choline ester); fixed oils (25–37%), consisting mainly of glycerides of erucic, eicosenoic, arachidic, nonadecanoic, behenic, oleic, and palmitic acids, among others;[1] proteins (e.g., globulins);[2] and mucilage (JIANGSU; NANJING).

Sinigrin on hydrolysis by myrosin (myrosinase) yields allyl isothiocyanate, glucose, and potassium bisulfate. Allyl isothiocyanate is volatile; its yield from *B. juncea* is 0.25–1.4% (usually ca. 0.9%).[3–8] Other minor volatile components that are also set free by enzymatic hydrolysis include methyl, isopropyl, *sec*-butyl, butyl, 3-butenyl, 4-pentenyl, phenyl, 3-methylthiopropyl, benzyl, and β-phenylethyl isothiocyanates.[7–9]

Black mustard (*B. nigra*) contains similar constituents as *B. juncea* (LIST AND HÖRHAMMER).

White, or yellow, mustard (*S. alba*) contains the glucosinolate sinalbin, which on hydrolysis by enzymes present (myrosin or glucosinolases) yields *p*-hydroxybenzyl isothiocyanate (a relatively nonvolatile compound),[3,5,6,9,10] *p*-hydroxybenzylamine,[11] and other similar components (proteins, fixed oils, sinapine, rhamnogalacturonan mucilage, etc.) as brown mustard (JIANGSU; MARSH).[1–3,6,12–14]

Apart from the seeds, the flavonoid glycosides isorhamnetin 7-*O*- and 3,7-di-*O*-glucoside and kaempferol 7-*O*-triglucoside were isolated from the leaves of *B. juncea*.[15–17] The roots and shoots of *B. alba*, on the other hand, yielded 3,5,6,7,8-pentahydroxy flavone, 3,5,6,7,8-pentahydroxy-4′-methoxy flavone, and 2′,3′,4′,5′,6′-pentahydroxy chalcone.[18]

PHARMACOLOGY AND BIOLOGICAL ACTIVITIES

Volatile mustard oil (or allyl isothiocyanate) is an extremely powerful irritant and produces blisters on the skin. It is also tear producing (lachrymatory) and has counterirritant properties when greatly diluted (e.g., 1 in 50). It should not be tasted or inhaled when undiluted. It is one of the most toxic essential oils (GOSSELIN; JIANGSU; MARTINDALE).[19]

p-Hydroxybenzyl isothiocyanate from white mustard does not have lachrymatory properties, but has a very pungent taste (ARCTANDER).

Isothiocyanates such as those present in mustard have been implicated in endemic goiter (hypothyroidism with thyroid enlargement). They have also been reported to produce goiter in experimental animals (LIENER).[20]

Volatile mustard oil has strong antimicrobial (bacteria and fungi) properties.[21,22]

Sinigrin has been reported to be toxic to certain insect larvae but harmless to others.[23]

One report suggests that glucosinolate products of the Brassicaceae may have protective effects against a variety of carcinogens.[24]

The hypoglycemic effect of *B. juncea* has been investigated by more than one group over the last decade. Different effects were observed in rat models of diabetes ranging from increased hepatic glycogen levels, reduced serum levels of glucose, to preventing the development of insulin resistance.[25–27] Antioxidant free radical scavenging activity and protection against diabetic oxidative stress were also displayed by some of the tested extracts and/or purified compounds, such as isorhamnetin mono- and diglucoside.[15–17,26,28,29]

Sinapic acid isolated from *B. juncea* has peroxynitrite scavenging activity in addition to a dose-dependent inhibition of nitration of BSA and LDL *in vitro*.[30]

USES

Medicinal, Pharmaceutical, and Cosmetic. The volatile oil (allyl isothiocyanate) is used in certain rubefacient and counterirritant liniments.

Food. Mustard (especially white) is extensively used in prepared mustards, where it is commonly used with vinegar and other spices; the bright yellow color is usually due to turmeric. Other food products in which mustard is used include baked goods, meat and meat products, processed vegetables, fats and oils, snack foods gravies, and nut products, among others. Highest average maximum use levels are about 12.4% (124,274 ppm) and 10.1% (101,503 ppm) reported for yellow and brown mustards, respectively, in condiments and relishes (e.g., prepared mustards).

Mustard oil (allyl isothiocyanate) is widely used as a flavor ingredient in nonalcoholic beverages, frozen dairy desserts, candy, baked goods, gelatins and puddings, meat and meat products, condiments and relishes (e.g., pickles), fats and oils, and gravies. Highest average maximum use level reported is about 0.02% (201 ppm) in gravies.

Dietary Supplements/Health Foods. Sometimes used in ointments or liniments intended to relieve symptoms of colds. Mustard oil, absorbed through the skin, is eliminated via the lungs, hence use in liniment preparations for relief of lung congestion (WEISS).

Traditional Medicine. Both brown (black) and white mustards are used as an appetizer, stimulant, emetic, diuretic, and rubefacient. They are more commonly used externally for treating rheumatism, arthritis, and lumbago. Black mustard is also used in footbaths for sore and aching feet.

Brown and white mustards are used in Chinese medicine for essentially the same types of illnesses (JIANGSU; NANJING).

Others. Mustard oil (allyl isothiocyanate) is used in cat and dog repellents.

Expressed mustard oil is used in soap making and as lubricant and illuminant.

COMMERCIAL PREPARATIONS

Crude (whole, ground) and volatile oil. Brown and black mustards were formerly official in N.F. and U.S.P. Mustard oil (volatile oil) formerly official in U.S.P. and currently official in F.C.C. where it is listed as equivalent to synthetic allyl isothiocyanate, except that its source (natural or synthetic) has to be indicated on the label. It is only required that volatile mustard oil or synthetic allyl isothiocyanate contain 93% (or more) allyl isothiocyanate (C_3H_5NCS); the remaining can be minor amounts of mosly known isothiocyanates (natural mustard oil) or impurities of unknown composition (synthetic allyl isothiocyanate) that can vary with the synthetic processes.

Regulatory Status. Has been affirmed as GRAS (§184.1527). White mustard seed is the subject of a German therapeutic monograph; allowed externally in poultice for upper respiratory tract congestion and supportive treatment of joint and soft tissue diseases.[31]

REFERENCES

See the General References for ARCTANDER; BAILEY 1; BLUMENTHAL 1; CLAUS; FEMA; GOSSELIN; JIANGSU; LIST AND HÖRHAMMER; LUST; NANJING; ROSE; ROSENGARTEN; TERRELL; UPHOF; WEISS.

1. J. Pokorny and I. Zeman, *Nahrung*, **15**, 35 (1971).
2. S. L. MacKenzie and J. A. Blakely, *Can. J. Bot.*, **50**, 1825 (1972).
3. M. S. Karawya et al., *Egypt J. Pharm. Sci.*, **16**, 113 (1975).
4. H. Nakabayashi et al., *Utsunomiya Daigaku Nogakubu Gakutsu Hokoku*, **8**, 1 (1972).
5. B. Reichert, *Dtsch. Apoth. Ztg.*, **20**, 234 (1968).
6. A. Kjär in L. Zechmeister, ed., *Fortschritte der Chemie organischer Naturstoffe*, Vol. 18, Sringer-Verlag, Vienna, 1960, p. 122.
7. B. E. Wallbank and G. A. Wheatley, *Phytochemistry*, **15**, 763 (1976).
8. M. Kojima et al., *Yakugaku Zasshi*, **93**, 453 (1973).
9. Anon., *Fed. Regist.*, **42**, 43092 (1977).
10. J. R. Vose, *Phytochemistry*, **11**, 1649 (1972).

11. P. O. Larsen, *Biochim. Biophys. Acta*, **107**, 134 (1965).
12. W. Cui et al., *Carbohydr. Res.*, **292**, 173 (1996).
13. S. A. Taille and F. E. Weber, U.S. 3,878,195 (1975).
14. V. K. Morozov, *Maslob. Zhir. Prom.*, **25**, 15 (1959).
15. J. S. Choi et al., *Arch. Pharm. Res.*, **25**, 625 (2002).
16. J. E. Kim et al., *Arch. Pharm. Res.*, **25**, 621 (2002).
17. T. Yokozawa et al., *J. Agric. Food Chem.*, **50**, 5490 (2002).
18. M. A. Ponce et al., *Phytochemistry*, **65**, 3131 (2004).
19. E. Skramlik, *Pharmazie*, **14**, 435 (1959).
20. P. Ahmad and A. J. Muztar, *Pak. J. Biochem.*, **4**, 72 (1971).
21. K. K. Abdullin, *Zap. Karansk. Vet. Inst.*, **84**, (1962).
22. I. Slavenas, *Tr. l-oi (Pervoi) Biokhim. Kon. Pribaltiisk Resp. i Belorussii Tartusk. Gos. Univ. Est. SSR, Estonsk, Biokhim. Obshchestvo, Tartu.*, 258 (1960).
23. P. A. Blau et al., *Science*, **200**, 1296 (1978).
24. R. McDannell et al., *Food Chem. Toxicol.*, **26**, 59 (1988).
25. S. P. Yadav et al., *J. Ethnopharmacol.*, **93**, 113 (2004).
26. H. Y. Kim et al., *Phytother. Res.*, **17**, 465 (2003).
27. B. A. Khan et al., *Indian J. Biochem. Biophys.*, **32**, 106 (1995).
28. T. Yokozawa et al., *J. Nutr. Sci. Vitaminol. (Tokyo)*, **49**, 87 (2003).
29. B. A. Khan et al., *Indian J. Exp. Biol.*, **35**, 148 (1997).
30. Y. Zou et al., *J. Agric. Food Chem.*, **50**, 5884 (2002).
31. Monograpph *Sinapis albae semen.*, *Bundesanzeiger*, no. 22 (February 1, 1990).

MYRRH

Source: *Commiphora myrrha* (Nees) Engl.; *C. molmol* Engl. ex Tschirch; *C. madagascariensis* Jacq. (syn. *C. abyssinica* (Berg) Engl.); *C. erythraea* (Ehrenb.) Engl.; and other *Commiphora* species (Family Burseraceae).

Common/vernacular names: African myrrh, Somali myrrh (*C. myrrha*); Arabian myrrh, Yemen myrrh (*C. madagascariensis*); myrrha, gum myrrh.

GENERAL DESCRIPTION

Commiphora species yielding myrrh are shrubs to small trees up to about 10 m high; native to northeastern Africa and southwestern Asia, especially the Red Sea region (e.g., Somalia, Yemen, and Ethiopia). The whitish gray bark has schizogenous gum-oil-resin cavities.[1] Part used is the exudation from the natural fissures in the bark or from man-made incisions. The exudation is a pale yellow liquid, which soon hardens to form yellowish red or reddish brown tears or masses that are then collected. Myrrh is an oleogum resin.

CHEMICAL COMPOSITION

Myrrh contains 1.5–17% (usually ca. 8%) volatile oil composed of heerabolene, limonene, dipentene, pinene, eugenol, cinnamaldehyde, cuminaldehyde, cumic alcohol, *m*-cresol, cadinene, curzerene (11.9%), curzerenone (11.7%), dihydropyrocurzerenone (1.1%), furanoeudesma-1,3-diene (12.5%), 1,10(15)-furanodiene-6-one (1.2%), lindestrene (3.5%), and furanogermacranes among others;[1–4] up to 40% (usually ca. 20%) resins consisting of α-, β-, and γ-commiphoric acids; commiphorinic acid, α- and β-heerabomyrrhols, heeraboresene, commiferin, campesterol, β-sitosterol, cholesterol, α-amyrone, 3-epi-α-amyrin, and others;[5,6] about 60% gum, which on hydrolysis yields arabinose, galactose, xylose, and 4-O-methylglucuronic acid; and others (GUENTHER; JIANGSU; LIST AND HÖRHAMMER).[4,6]

A mixture of furanoeudesma-1,3-diene and lindestrene possesses a typical myrrh fragrance, and the resinous myrrh fragrance is best represented by dihydropyrocurzerenone.[1]

Two furanosesquiterpenoids and a furanodienone have been isolated from a hexane extract of *C. erythraea*, traded as opopanax.[1]

PHARMACOLOGY AND BIOLOGICAL ACTIVITIES

Myrrh is reported to have astringent properties on mucous membranes as well as antimicrobial activities *in vitro* (JIANGSU; MARTINDALE). Sesquiterpene-rich fractions of *C. molmol* have antibacterial and antifungal activities against Gram-positive and Gram-negative bacteria and *Candida albicans* at MIC of 0.18–2.8 μg/mL.[7] Effect on mucous membranes may attribute to the gastric antiulcer and cytoprotective effects of an aqueous myrrh suspension observed in mice at 250–1000 mg/kg.[8]

Some reports have been published lately on the efficacy of myrrh as an antischistosomal[9,10] drug against infection with Egyptian *Schistosoma haematobium* and *S. mansoni* (causative agents of bilharzia disease) in experimental animals and in humans.[11–13] At least two reports, however, disputed that claim.[14,15] Molluscicidal effects against *Biomphlaria arabica* (*Schistosoma* vector) and *Bithynia connollyi* snails have also been reported.[9,10]

A preparation made from *C. molmol* oleogum resin was effective in the treatment of fascioliasis (liver fluke worm infection) in seven human patients.[13]

Furanosesquiterpenoids are ixodicidal/acaricidal (tick-killing) against the larvae of the African tick *Rhipicephalus appendiculatus*.[1] *C. molmol* extracts are also effective against the fowl tick *Argas persicus*.[16]

C. molmol oleogum resin has selective anticarcinogenic/antitumor activity (125–500 mg/kg dose) in mice bearing Ehrlich carcinoma cells.[17,18]

Z-Guggulsterone (4,17(20)-*trans*-pregnanediene-3,16-dione) isolated from the oleogum resin of *Commiphora mukul* (Hook. ex Stocks) Engl. has thyroid-stimulating activity in rats.[19,20]

TOXICOLOGY

Myrrh has been reported to be nonirritating, nonsensitizing, and nonphototoxic to human and animal skins.[21] However, the essential oil and some sesquiterpene constituents (curzerenone, furanodienone, and furanodesmadiene) were irritant in the open ear assay in mice.[22]

The chemistry, pharmacology, and uses of myrrh have recently been reviewed.[23,24]

USES

Medicinal, Pharmaceutical, and Cosmetic. Myrrh is used mainly as an astringent in certain mouthwashes and gargles.

Myrrh oil is used as a fragrance component or fixative in soaps, detergents, creams, lotions, and perfumes, with maximum use level of 0.8% reported in perfumes.[21] The sweet, warm balsamic notes of absolute, oil, or resinoid used in oriental spice fragrances, woody and forest notes; often blended with geranium, musk, patchouli, and other heavy floral bases.[1]

Food. Myrrh is reported to be used in nonalcoholic beverages with an average maximum use level of 0.025%.

Myrrh oil is used as a flavor component in major food products, including alcoholic and nonalcoholic beverages, frozen dairy desserts, candy, baked goods, gelatins and puddings, and meat and meat products. Highest average maximum use level is about 0.002% reported in alcoholic beverages (25 ppm), baked goods (23.5 ppm), and gelatins and puddings (20 ppm).

Dietary Supplements/Health Foods. Primarily in tinctures, and other formulations as astringent for mucous membranes of the throat and mouth, indigestion, bronchial congestion, and emmenagogue; also in lip balms for chapped lips; externally for wounds, hemorrhoids, sores, and so on. (LUST).

Traditional Medicine. Myrrh has been used since ancient times in Western culture as a stimulant, antiseptic, expectorant, antispasmodic, emmenagogue, and stomachic. Conditions for which it is used include cancers, leprosy, syphilitic ulcers, sores, sore throat, asthma, coughs, bad breath, weak gums, gingivitis, and loose teeth.[4,25]

Myrrh was introduced in Chinese medicine around the 7th century and has since been used mainly in treating conditions involving bleeding, pain, and wounds (e.g., bleeding hemorrhoids, menstrual difficulties, sores, tumors, and arthritic pain).[25]

COMMERCIAL PREPARATIONS

Crude, extracts (tincture, fluid extract, etc.), and oil. Crude and tincture were formerly official in N.F. Myrrh oil is official in F.C.C.

Regulatory Status. Has been approved for food use (§172.510). Myrrh powder and tincture subjects of a positive German therapeutic monograph for treatment of mild inflammations of the oral and pharyngeal mucosa.[26]

REFERENCES

See the General References for ARCTANDER; BLUMENTHAL 1; CLAUS; DUKE 4; FEMA; GRIEVE; HUANG; JIANGSU; LIST AND HÖRHAMMER; LUST; MCGUFFIN 1 & 2; NANJING; ROSE; TUCKER AND LAWRENCE; TYLER 1; UPHOF.

1. A. O. Tucker, *Econ. Bot.*, **40**, 425 (1986).
2. F. Ahmed et al., *Pharmazie*, **61**, 728 (2006).
3. N. Zhu et al., *J. Nat. Prod.*, **64**, 1460 (2001).
4. R. Pernet, *Lloydia*, **35**, 280 (1972).
5. E. Mincione and C. Iavarone, *Chim. Ind. (Milan)*, **54**, 525 (1972).
6. E. Mincione and C. Iavarone, *Chim. Ind. (Milan)*, **54**, 424 (1972).
7. P. Dolara et al., *Planta Med.*, **66**, 356 (2000).
8. M. M. al-Harbi et al., *J. Ethnopharmacol.*, **55**, 141 (1997).
9. N. M. Shoukry, *J. Egypt. Soc. Parasitol.*, **36**, 701 (2006).
10. E. M. Al-Mathal and M. A. Fouad, *J. Egypt. Soc. Parasitol.*, **36**, 305 (2006).
11. A. M. Massoud et al., *J. Egypt. Soc. Parasitol.*, **34**, 1051 (2004).
12. A. A. bo-Madyan et al., *J. Egypt. Soc. Parasitol.*, **34**, 423 (2004).
13. A. Massoud et al., *Am. J. Trop. Med. Hyg.*, **65**, 96 (2001).
14. R. Barakat et al., *Am. J. Trop. Med. Hyg.*, **73**, 365 (2005).
15. S. Botros et al., *Am. J. Trop. Med. Hyg.*, **71**, 206 (2004).
16. A. M. Massoud et al., *J. Egypt. Soc. Parasitol.*, **35**, 667 (2005).
17. M. M. al-Harbi et al., *Chemotherapy*, **40**, 337 (1994).

18. S. Qureshi et al., *Cancer Chemother. Pharmacol.*, **33**, 130 (1993).
19. Y. B. Tripathi et al., *Planta Med.*, **50**, 78 (1984).
20. Y. B. Tripathi et al., *Planta Med.*, **54**, 271 (1988).
21. D. L. J. Opdyke, *Food Chem. Toxicol.*, **14**, 621 (1976).
22. M. A. Saeed and A. W. Sabir, *Fitoterapia*, **75**, 81 (2004).
23. L. O. Hanus et al., *Biomed. Pap. Med. Fac. Univ. Palacky. Olomouc. Czech. Repub.*, **149**, 3 (2005).
24. E. S. El Ashry et al., *Pharmazie*, **58**, 163 (2003).
25. J. L. Hartwell, *Lloydia*, **31**, 71 (1968).
26. Monograph *Myrrha*, *Bundesanzeiger*, no. 193 (October 15, 1987).

NETTLE

Source: *Urtica dioica* L. (Family Urticaceae).

Common/vernacular names: Stinging nettle.

GENERAL DESCRIPTION

Herbaceous perennial up to 30–150 cm, monoecious or dioecious, stems four angled; leaves and stems with stinging hairs; leaves opposite, ovate to cordate oblong–lanceolate, serrate;[1] flowers green, in axillary panicles; June to September; found in waste places, moist thickets, fields, pastures; most of North America and Europe.[1]

American material is designated *U. dioica* subsp. *gracilis* (Ait.) Seland., consisting of six varieties. The American material differs from the typical European *Urtica dioica* subsp. *dioica* primarily in that it has male and female flowers on the same plant. Some botanists treat the varieties of *U. dioica* subsp. *gracilis* as separate species.

The parts used are the herb, leaves, and root.

CHEMICAL COMPOSITION

Herb contains amines, including histamine, serotonin (5-HT), acetylcholine; flavonoids, including isoquercitrin, rutin, kaempferol, and isorhamnetin (in flowers) and their glycosides;[2,3] a lectin, *Urtica dioica* agglutinin (made of seven isolectins);[4,5] ubiquitous compounds, including carotenoids, vitamin C, triterpenes, sterols, such as β-sitosterol, campeterol, and stigmasterol, formic, citric, oxalic, and tartaric acids, and relatively high amounts of calcium and potassium salts, and silicic acids (BRADLY; WREN).[6,7]

The lignans (+)-neoolivil, (−)-secoisolariciresinol, pinoresinols, and others were isolated from the roots.[8]

PHARMACOLOGY AND BIOLOGICAL ACTIVITIES

Until the early 1990s, the described activities included diuretic, hemostatic, CNS-depressant, antispasmodic,[9] and antiallergenic. Since then the list has expanded to include antioxidant, analgesic, anti-inflammatory, antimicrobial, antihyperglycemic, antiulcer, antiplatelet aggregation, immunomodulatory and cardiovascular activities, as well as potential for treatment of benign prostatic hyperplasia (BPH) symptoms.

Diuretic activity has been suggested in animal models;[10] nettle juice produced a diuretic effect in a clinical study with patients suffering from myocardial or chronic venous insufficiency.[11]

Hemostatic and mild hypotensive activities have been reported (BRADLY). In an *in vitro* study, the aqueous extract of *U. dioica* dose dependently inhibited thrombin-induced platelet aggregation. The observed effect has been attributed to the flavonoid constituents of the extract.[12] The hypotensive effects of *U. dioica* extracts (aqueous and methanolic, *in vivo* and in isolated organs) have also been reported by more than one group and the mechanism of action is suggested to be through vasodilation mediated by nitric oxide release, potassium channel opening, and negative inotropic action.[13–15]

Reported antimicrobial activity include antibacterial activity against *Staphylococcus aureus*, *Staphylacoccus albus*, and others.[16,17] *In vitro* antiviral activity against feline immunodeficiency virus (FIV) has also been reported.[18]

A freeze-dried nettle extract produced positive, though limited, results in a double-blind clinical study in the treatment of allergic rhinitis.[19]

Another study evaluated the effects of an ethanol extract of the roots of *U. dioica* and *U. urens* in 67 men over 60 years of age suffering from prostatic adenoma. Symptoms of nocturia were alleviated, especially in less severe cases.[20] A recent double-blind, placebo-controlled study in 620 patients reported significant improvements in urinary symptoms associated with BPH, such as urinary flow rate and postvoid residual urine volume, after 6 months of treatment. There was no effect on prostate volume, serum prostate-specific antigen (PSA), and testosterone levels.[21] Two studies that targeted the direct effect of *U. dioica* on the epithelial and stromal cells of the prostate showed a significant and concentration-dependent antiproliferative effect of the methanolic extract and its polysaccharide fraction on such cells both *in vitro* and *in vivo*.[22,23] A review about the utility of nettle in managing BPH symptoms is available.[24]

The anti-inflammatory activity of leaf extracts has been studied *in vitro* and dose-dependent inhibitions of primary T-cell responses, expression of T helper cytokines, and NF-κB have been reported.[25–27] Similarly, the antioxidant activity of the extract has been established in a number of models of hepatoprotection, free radical/superoxide/hydrogen peroxide scavenging, and lipid peroxidation.[16,28–30] Analgesic and antiulcer activity were also observed in the acetic acid-induced writhing and the ethanol-induced ulcerogenesis tests, respectively.[16]

The cholesterol lowering effect of an aqueous extract of *U. dioica* fed to rats for 30 days (150 mg/kg/day) was recently reported with no hepatotoxic effects observed.[31]

Flavonoid glycosides of the aerial parts displayed *in vitro* immunostimulatory effects at 4–16 μg/mL in the Boyden migration chamber and NBT reduction tests.[3]

The antihyperglycemic effect of *U. dioica* was demonstrated in rat models in two independent studies whereby the effect was attributed to reduced intestinal glucose absorption and/or increased insulin secretion.[32,33]

Fresh plants sting on contact, due to amines in bladder-like base of stinging hairs. Upon contact, the hair tip breaks off, injecting the amines and folic acid into the skin. Burning sensation may last for up to 1 h (FOSTER AND CARAS).

USES

Medicinal, Pharmaceutical, and Cosmetic. In German phytomedicine, an average daily dose of 8–10 g of the herb is used for supportive treatment of rheumatic complaints, inflammation of the lower urinary tract, and for treatment of renal gravel; also for benign prostatic hyperplasia.[34,35]

Nettle extract is reported to be used as a biological additive in shampoos, permanent waves; hair conditioners; skin fresheners, and miscellaneous skin care products.

Food. Boiled fresh greens eaten as a potherb.

Dietary Supplements/Health Foods. Dried leaf (or roots) in teas, capsules, tablets, tinctures, primarily as a nutritional supplement for vitamin and mineral content; chlorophyll source (FOSTER).

Traditional Medicine. In European and American folk traditions, leaf tea used as blood purifier, diuretic, astringent; for anemia, gout, glandular diseases, rheumatism, poor circulation, enlarged spleen, mucous discharges of lungs, internal bleeding, and diarrhea, dysentery (FOSTER AND DUKE).

Other. *Urtica dioica* agglutinin has been suggested as a useful probe for the analysis of T-cell activation by superantigens.[5]

COMMERCIAL PREPARATIONS

Crude dried or fresh, or freeze-dried; extract. Nettles are used as a commercial source of chlorophyll.

Regulatory Status. Undetermined in the United States. Nettle herb, leaves, and roots are subjects of positive German therapeutic monographs.[34,35] Herb and leaves indicated for supportive treatment of rheumatic complaints; internally for irrigation therapy in inflammatory conditions of the lower urinary tract, and prevention and treatment of gravel.[34] Root allowed for symptomatic relief of urinary difficulties associated with prostate enlargement (without decreasing enlargement).[35]

REFERENCES

See the General References for APPLEQUIST; BLUMENTHAL 1 & 2; BRADLY; BRUNETON; DER MARDEROSIAN AND BEUTLER; DUKE 4; FOSTER; FOSTER AND CARAS; FOSTER AND DUKE; GRIEVE; MARTINDALE; MCGUFFIN 1 & 2; NIKITAKIS; STEINMETZ; TYLER 1; WEISS; WREN.

1. K. R. Kirtikar and B. D. Basu, *Indian Medicinal Plants*, International book distribution, Dehradun, India, 1999, p. 2341.
2. N. Chaurasia and M. Wichtl, *Planta Med.*, **53**, 432 (1987).
3. P. Akbay et al., *Phytother. Res.*, **17**, 34 (2003).
4. K. Harata et al., *Acta Crystallogr. D. Biol. Crystallogr.*, **57**, 1513 (2001).
5. A. Galelli and P. Truffa-Bachi, *J. Immunol.*, **151**, 1821 (1993).
6. H. Y. Fu et al., *Ann. Bot. (Lond)*, **98**, 57 (2006).
7. T. Hirano et al., *Planta Med.*, **60**, 30 (1994).
8. M. Schottner et al., *Planta Med.*, **63**, 529 (1997).
9. E. Madrid et al., *Ann. R. Acad. Farm.*, **53**, 284 (1987).
10. H. Schilcher, *Dtsch. Apoth. Ztg.*, **124**, 2429 (1984).
11. H. W. Kirchhoff, *Z. Phytotherapie*, **4**, 621 (1984).
12. H. M. El et al., *Phytother. Res.*, **20**, 568 (2006).
13. A. Legssyer et al., *Phytother. Res.*, **16**, 503 (2002).
14. L. Testai et al., *J. Ethnopharmacol.*, **81**, 105 (2002).
15. A. Tahri et al., *J. Ethnopharmacol.*, **73**, 95 (2000).
16. I. Gulcin et al., *J. Ethnopharmacol.*, **90**, 205 (2004).
17. L. Lezhneva et al., *Rastit Resur.*, **22**, 255 (1986).
18. R. E. Uncini Manganelli et al., *J. Ethnopharmacol.*, **98**, 323 (2005).
19. P. Mittman et al., *Planta Med.*, **56**, 44 (1990).
20. P. Belachie and O. Lievoux, *Phytother. Res.*, **5**, 267 (1991).
21. M. R. Safarinejad, *J. Herb. Pharmacother.*, **5**, 1 (2005).
22. L. Konrad et al., *Planta Med.*, **66**, 44 (2000).
23. J. J. Lichius et al., *Pharmazie*, **54**, 768 (1999).
24. E. Koch, *Planta Med.*, **67**, 489 (2001).
25. J. Broer and B. Behnke, *J. Rheumatol.*, **29**, 659 (2002).
26. S. Klingelhoefer et al., *J. Rheumatol.*, **26**, 2517 (1999).
27. K. Riehemann et al., *FEBS Lett.*, **442**, 89 (1999).
28. M. Kanter et al., *World J. Gastroenterol.*, **11**, 6684 (2005).
29. M. Kanter et al., *J. Vet. Med. A Physiol. Pathol. Clin. Med.*, **50**, 264 (2003).
30. M. K. Turkdogan et al., *Phytother. Res.*, **17**, 942 (2003).
31. C. F. Daher et al., *Fitoterapia*, **77**, 183 (2006).
32. M. Bnouham et al., *Fitoterapia*, **74**, 677 (2003).
33. B. Farzami et al., *J. Ethnopharmacol.*, **89**, 47 (2003).
34. Monograph *Urticae herba and Urticae folium, Bundesanzeiger*, no. 76 (April 23, 1987).
35. Monograph *Urticae radix, Bundesanzeiger*, no. 173 (September 18, 1986); revised (March 6, 1990).

NUTMEG (AND MACE)

Source: *Myristica fragrans* Houtt. (syn. *M. officinalis* L.) (Family Myristicaceae).

Common/vernacular names: Myristica (nutmeg); macis (mace).

GENERAL DESCRIPTION

The nutmeg tree is an evergreen tree with spreading branches and dense foliage; up to about 20 m high; leaves coriaceous, elliptic–oblong, and at times oblanceolate, cordate at tip, and acute at base; flowers bracteolate; fruits ovoid, subglobose, or pyriform; native to the Moluccas and nearby islands; cultivated in Indonesia (Java, Moluccas, etc.), Sri Lanka, and the West Indies (especially Grenada). Its fruit is fleshy like an apricot and up to 6 cm long; on ripening it splits in half, exposing a bright red net-like aril wrapped around a dark reddish brown and brittle shell within which lies a single seed. The net-like aril is mace, which on drying turns from red to yellowish or orange brown. The dried brown seed, after shell is broken and discarded, is nutmeg.

Major producers of mace and nutmeg include Indonesia, Sri Lanka, and Grenada. The first two produce the East Indian nutmegs, while Grenada produces the West Indian nutmegs. East Indian nutmegs are considered superior in flavor to their West Indian counterparts. Nutmeg oil (myristica oil) is generally produced by steam distillation of worm-eaten nutmegs; these give a higher yield of essential oil, as the worms have eaten much of the starchy and fatty portions of the nutmegs, leaving behind portions that are rich in volatile oil (ARCTANDER; ROSENGARTEN).

CHEMICAL COMPOSITION

Nutmeg contains 2–16% (usually ca. 10%) volatile oil;[1] 25–40% fixed oils consisting of free myristic acid and triglycerides of lauric, tridecanoic, palmitic, stearic, and myristic acids as well as branched isomers of myristic and stearic acids, with the West Indian nutmeg having larger amounts than East Indian nutmeg;[1–4] starch (ca. 30%); protein (ca. 6%); an oleanolic acid glycoside (saponin);[2] sclareol;[5] diarylpropanoids (dimeric phenylpropanoids), such as macelignan, *meso*-dihydroguaiaretic acid, and otobaphenol;[6–12] catechins, proanthocyanidins;[13] and others (LIST AND HÖRHAMMER; MARSH; STAHL).[1,14,15] Two resorcinols (malabricone B and C) have been isolated from mace.[16]

Nutmeg oil contains large amounts of monoterpene hydrocarbons (ca. 88%) with camphene and pinene as the major components and dipentene, sabinene, cymene, α-thujene, and γ-terpinene, among others in lesser amounts; monoterpene alcohols (geraniol, *d*-borneol, linalool, terpineol, etc.);[1,3,14] and 4–8% myristicin and smaller amounts of safrole and elemicin, with the East Indian oil higher in myristicin content than the West Indian oil.[17] Other constituents reportedly present include copaene, *trans*- and *cis*-sabinene hydrate, *cis*-piperitol, eugenol, isoeugenol, methyleugenol, dehydrodiisoeugenol, and *cis-p*-menth-2-enol, among others (LIST AND HÖRHAMMER).[3,4,9,15,17]

Mace and mace oil contain similar constituents as nutmeg and nutmeg oil, except with less fixed oil and more myristicin content (LIST AND HÖRHAMMER; MARSH; STAHL).[6,10,14,15,18]

Nutmeg, mace, and their extracts have strong antioxidant activities that do not seem to be due to their essential oils.[19–21]

PHARMACOLOGY AND BIOLOGICAL ACTIVITIES

Nutmeg in sufficient dosage is reported to have psychotropic properties (hallucinations, feelings of unreality, euphoria, delusions, etc.); these have been proposed to be due to myristicin or to its metabolic conversion to amphetamine-like compounds. However, the psychotropic effects of myristicin need further investigation, and the amphetamine-like metabolite(s) theory has recently been questioned.[22] Other CNS activities reported for nutmeg include an antidepressant effect of the seed hexane extract in mice (10 mg/kg, 3 day, oral administration) using the forced swim test and the tail suspension test. Interaction of the extract with adrenergic, dpaminergic, and serotonergic receptors has been proposed.[23] Another group reported earlier that trimyristin and the acetone-insoluble component of the hexane extract of the seed (10–100 mg/kg and 30–300 mg/kg, respectively, i.p. administration) exhibited anxiogenic activity in mice using the elevated plus-maze, open-field, and hole-board tests. The authors suggested that the observed activity is linked to serotonin and GABA.[24]

The effect of nutmeg hexane extract on memory and learning has also been studied in mice (5–20 mg/kg, oral administration) whereby a significant enhancement has been observed in the elevated plus-maze and passive-avoidance tests.[25] This effect probably involves the deactivation of brain acetylcholinesterase as compared to the untreated control group.[26]

Extracts and pure compounds of nutmeg and mace have demonstrated strong antimicrobial activities. Myristicin, myristic acid, and trimyristin (myristic acid triglyceride), isolated from the chloroform extract of the seed, exhibited antibacterial effects against Gram-positive and Gram-negative bacteria.[27] Macelignan isolated from the seed possessed a strong inhibitory activity against *Streptococcus mutans* responsible for the development of dental caries. The same compound was also active against other oral microorganisms.[11] The malabricones isolated from mace exhibited strong antifungal and antibacterial activities against a panel of selected microorganisms.[16] Nutmeg oil and mace oil have been reported to have larvicidal properties.[28]

Nutmeg extracts have been demonstrated to inhibit the *in vitro* biosynthesis of prostaglandin by rat kidney tissue, and ground nutmeg fed orally to rats has decreased kidney prostaglandin levels, these effects being similar to those produced by indomethacin.[29]

The anti-inflammatory and analgesic activities of nutmeg and mace have been established in carrageenan-induced paw edema and acetic acid-induced writhing, respectively, in rats.[30] Vascular permeability in mice is attributed to myristicin.

Other activities recently reported for nutmeg and/or its constituents include antidiabetic (*in vitro*),[12] antiulcer (*in vivo*),[31] antidiarrheal (*in vivo*),[32] aphrodisiac (*in vivo*),[33,34] hepatoprotective (*in vivo*),[35] and hypolipidemic (*in vivo*)[36,37] effects.

TOXICOLOGY

East Indian nutmeg oil has been reported to be nonirritating and nonsensitizing to human skin, though it was moderately irritating to rabbit skin when applied undiluted for 24 h under occlusion.[38] Toxic symptoms due to the ingestion of large doses of nutmeg (>5 g) or nutmeg oil include nausea, vomiting, stupor, disorientation, flushing, tachycardia, and dryness of mouth. Lethal cases of nutmeg abuse have been reported (GOSSELIN; MARTINDALE; MERCK).[38,39]

USES

Medicinal, Pharmaceutical, and Cosmetic. Nutmeg oil is used in certain analgesic ointments and tonic preparations. It is also used as a flavoring agent in pharmaceuticals.

In European phytomedicine, nutmeg and/or mace have been used for gastrointestinal ailments, including diarrhea, gastric spasms, intestinal catarrh, and flatulence, though claimed efficacies are not sufficiently documented to warrant a positive regulatory status for claims.[40]

Nutmeg oil is used as a fragrance component in soaps, detergents, creams, lotions, and perfumes. Maximum use level reported is 0.3% for the East Indian oil in perfumes.[38]

Food. Nutmeg and mace are widely used as domestic spices in cooking, in cakes, drinks (e.g., eggnog and spiced hot wines), and other foods.

Nutmeg and mace are extensively used as flavor ingredients in many food products, including nonalcoholic beverages, baked goods, meat and meat products, condiments and relishes, processed vegetables, soups, snack foods, and gravies. Nutmeg is also used in alcoholic beverages, frozen dairy desserts, gelatins and puddings, breakfast cereals, and others. Highest average maximum use level is about 0.3% reported for nutmeg in gelatins and puddings (3125 ppm), sweet sauces (2600 ppm), and baked goods (2594 ppm).

Nutmeg oil, mace oil, and mace oleoresin are used in most above major food products, including alcoholic and nonalcoholic beverages, frozen dairy desserts, candy, baked goods, gelatin and puddings, meat and meat products, and condiments and relishes, among others. Highest average maximum use levels are about 0.078% (775 ppm) and 0.065% reported for mace oleoresin in alcoholic beverages and candy, respectively.

Dietary Supplements/Health Foods. Nutmeg is used in tea flavoring.

Traditional Medicine. Nutmeg and mace have been used for centuries in both Western and Eastern cultures mainly as carminative and stimulants in treating flatulence, indigestion, nausea, and other stomach as well as kidney problems. They have also been reportedly used in cancers.[41] In the Peruvian Andes, nutmeg chopped in pork fat has been used externally in massages for paralysis, rheumatism, and as an antiparasitic (for mange).[42] Mace has been used externally as a treatment for rheumatism.[43]

In European tradition, nutmeg and mace are reportedly used for gastric spasms, diarrhea, flatulence, and others.[40]

Others. The fixed oil of nutmeg (nutmeg butter) obtained by expression is used in soap making and in candles.

Due to the presence of sclareol in nutmeg oil distillation waste, it can be recovered as a by-product of the nutmeg oil industry (see *clary sage*).[5]

COMMERCIAL PREPARATION

Crude, extracts (e.g., oleoresin), and oils (mace, East and West Indian nutmeg). Mace and nutmeg were formerly official in N.F. Nutmeg oil is official in N.F. and F.C.C., with specifications that sources (East Indian or West Indian) be indicated on the label.

Regulatory Status. GRAS (§182.20). Subject of a German therapeutic monograph; claimed efficacies not sufficiently documented; allowed as flavor or fragrance corrigent.[40]

REFERENCES

See the General References for ARCTANDER; BARRETT; BLUMENTHAL 1; BRUNETON; CLAUS; DER MARDEROSIAN AND BEUTLER; FEMA; GOSSELIN; GRIEVE; GUENTHER; GUPTA; JIANGSU; JIXIAN; MARTINDALE; MCGUFFIN 1 & 2; ROSENGARTEN; TERRELL; UPHOF.

1. A. T. Shulgin et al., *Public Health Serv. Publ.*, **1645**, 202 (1967).
2. I. P. Varshney and S. C. Sharma, *Indian J. Chem.*, **6**, 474 (1968).
3. K. J. Sanford and D. E. Heinz, *Phytochemistry*, **10**, 1245 (1971).
4. Z. Mobarak et al., *Chemosphere*, **6** (1977).
5. N. F. Novotel'nova et al., *U. S. S. R.*, **161842** (1964).
6. J. E. Forrest et al., *J. Chem. Soc. Perkin Trans. I*, (2), 205 (1974).
7. A. Isogai et al., *Agric. Biol. Chem.*, **37**, 193 (1973).
8. A. Isogai et al., *Agric. Biol. Chem.*, **37**, 1479 (1973).

9. T. P. Forrest et al., *Naturwissenschaften*, **60**, 257 (1973).
10. D. J. Harvey, *J. Chromatogr.*, **110**, 91 (1975).
11. J. Y. Chung et al., *Phytomedicine*, **13**, 261 (2006).
12. S. Yang et al., *Phytother. Res.*, **20**, 680 (2006).
13. J. M. Schulz and K. Herrmann, *Z. Lebensm. Unters. Forsch.*, **171**, 278 (1980).
14. A. T. Weil, *Econ. Bot.*, **19**, 194 (1965).
15. J. E. Forrest and R. A. Heacock, *Lloydia*, **35**, 440 (1972).
16. K. Y. Orabi et al., *J. Nat. Prod.*, **54**, 856 (1991).
17. J. Baldry et al., *Int. Flav. Food Addit.*, **7**, 28 (1976).
18. J. E. Forrest et al., *Experientia*, **29**, 139 (1973).
19. A. Palitzsch et al., *Fleischwirtschaft*, **54**, 63 (1974).
20. Y. Kihara and T. Inoue, *Nippon Shokuhin Kogyo Gakkaishi*, **9**, 290 (1962).
21. Y. Saito et al., *Eiyo To Shokuryo*, **29**, 505 (1976).
22. J. Beyer et al., *Ther. Drug Monit.*, **28**, 568 (2006).
23. D. Dhingra and A. Sharma, *J. Med. Food*, **9**, 84 (2006).
24. G. S. Sonavane et al., *Pharmacol. Biochem. Behav.*, **71**, 239 (2002).
25. M. Parle et al., *J. Med. Food*, **7**, 157 (2004).
26. D. Dhingra et al., *J. Med. Food*, **9**, 281 (2006).
27. B. Narasimhan and A. S. Dhake, *J. Med. Food*, **9**, 395 (2006).
28. K. Oishi et al., *Nippon Suisan Gakkaishi*, **40**, 1241 (1974).
29. V. Misra, et al., *Indian J. Med. Res.*, **67**, 482 (1978).
30. O. A. Olajide et al., *Phytother. Res.*, **13**, 344 (1999).
31. M. Jan et al., *J. Ayub Med. Coll. Abbottabad*, **17**, 69 (2005).
32. J. K. Grover et al., *Methods Find. Exp. Clin. Pharmacol.*, **24**, 675 (2002).
33. Tajuddin et al., *BMC Complement. Altern. Med.*, **5**, 16 (2005).
34. Tajuddin et al., *BMC Complement. Altern. Med.*, **3**, 6 (2003).
35. T. Morita et al., *J. Agric. Food Chem.*, **51**, 1560 (2003).
36. A. Ram et al., *J. Ethnopharmacol.*, **55**, 49 (1996).
37. A. Sharma et al., *Indian J. Physiol. Pharmacol.*, **39**, 407 (1995).
38. D. L. J. Opdyke, *Food Chem. Toxicol.*, **14**, 631 (1976).
39. B. C. Sangalli and W. Chiang, *Clin. Toxicol.*, **38**, 671 (2000).
40. Monograph *Myristicae semen; Myristicae arillus*, *Bundesanzeiger*, no. 173, (September 18, 1986).
41. J. L. Hartwell, *Lloydia*, **33**, 288 (1970).
42. V. De Feo, *Fitoterapia*, **63**, 417 (1992).
43. S. S. Handa et al., *Fitoterapia*, **63**, 3 (1992).

OLIBANUM

Source: *Boswellia carterii* Birdw., *B. frereana* Birdw., *B. serrata* Roxb. ex Colebr., *B. bhau-dajiana* Birdw., and other *Boswellia* species (Family Burseraceae).

Common/vernacular names: Frankincense, olibanum gum, Bible incense, (*B. carterii*); African elemi (*B. frereana*); Indian olibanum, Indian frankincense (*B. serrata*).

GENERAL DESCRIPTION

Olibanum-yielding *Boswellia* species are shrubs to small trees belonging to the same family as myrrh-producing trees (see **myrrh**). They are native to the Red Sea region, growing wild throughout northeastern Africa. Olibanum is the exudation (an oleogum resin) from their bark. It is collected by making incisions in the bark; the milky liquid flows out and solidifies on the branches or on the ground as tears or masses. These are then sorted and graded. Major frankincense-producing countries include Somalia and Ethiopia.

CHEMICAL COMPOSITION

Contains 3–1% volatile oil;[1,2] 60–70% resins composed mainly of 3-acetyl-β-boswellic acid and α-boswellic acid;[1,3,4] about 20% gum made up of galactose, arabinose, and 4-O-methylglucuronic acid; 5–8% bassorin; and bitter substances (JIANGSU; LIST AND HÖRHAMMER).[1,5]

The volatile oil contains predominantly terpene hydrocarbons (especially pinene, dipentene, limonene, α-thujene, α- and β-phellandrenes, *p*-cymene, myrcene, and terpinene; also α-copaene, β-bourbonene, β-ylangene, β-guaiene, *trans*-bergamotene, β-cadinene, humulene, β-caryophyllene, farnesene, etc.), with lesser amounts of oxygenated compounds (farnesol, borneol, carvone, *trans*-pinocarveol, terpinen-4-ol, menthadien-7-ol, verbenone, *trans*-verbenol, bornyl acetate, terpinyl acetate, ethyl laurate, etc.). Other important compounds reportedly present in olibanum oil include octyl acetate, octanol, incensole, isoincensole, incensole oxide, isoincensole oxide, incensole acetate, and isoincensole acetate among others.[3,6–14] The relative proportions of the volatile compounds vary considerably according to the types and sources of olibanum (GUENTHER; JIANGSU; LIST AND HÖRHAMMER).[1,9] A study of the essential oil from an olibanum sample of Somalian origin revealed a high content (62.1%) of esters comprised mostly of 1-octyl acetate, while terpene hydrocarbons amounted to only 17% of the oil.[2]

In addition to the known boswellic acids mentioned above, more than 15 triterpene acids have recently been isolated from olibanum resin.[15] They belong to the lupane (e.g., lupeolic acid), ursane (e.g., 11-keto-β-boswellic acid), oleanane (e.g., 3-α-acetyl-11-keto-α-boswellic acid), and tirucallane (e.g., 3-oxo- and 3-hydroxy-tirucallic acids) classes of triterpenes.[16–19]

PHARMACOLOGY AND BIOLOGICAL ACTIVITIES

The nonphenolic fraction of Indian olibanum (from *B. serrata*) has been reported to have strong analgesic effects on rats; it also had depressant effects.[20]

Anti-inflammatory and antiarthritic activities have been established *in vitro* and in the carrageenan-induced paw edema test, in arthritic rats and in dogs with inflammatory joint and spinal disease.[15,21,22] The anti-inflammatory activity is apparently mediated through inhibition of TNF-α, IL-1β, NO, MAPK, and 5-LO, as well as P-selectin-mediated recruitment of inflammatory cells.[23–26] Earlier clinical trials in India have been conducted on arthritis patients with positive results, and boswellic acid and other related pentacyclic triterpene acids are marketed as antiarthritic drugs in India.[27] More recent trials have demonstrated the efficacy of olibanum preparations in various colitis conditions.[24,28,29]

Due to its high content of boswellic acid, olibanum has been demonstrated to have antioxidant properties on seed oils (cottonseed and sunflower); 0.1% olibanum has activity comparable to 0.02% butylated hydroxyanisole (BHA).[4] The water-soluble extract inhibits NO production in rat macrophages resulting in both hepato- and renal protection.[30] The hexane extract has also been reported to possess hepatoprotective activity.[31]

Olibanum has been reported to possess anticancer and cancer chemopreventive effects in different conditions. As such, it was successful in reversing breast cancer brain metastases in a patient who was not responsive to standard therapy.[32] Acetyl boswellic acids induce malignant cell apoptosis through caspase activation.[16,33,34] Terpene acids of olibanum are also cytotoxic against human neuroblastoma, meningioma, and leukemia cells at μM level IC_{50}'s.[17,35,36]

Other reported activities of olibanum resin and/or boswellic acids include immunomodulatory,[18,37–39] P-glycoprotein inhibition,[40] antidiarrheal,[41] and hypolipidemic[30,36] activities.

TOXICOLOGY

Olibanum and olibanum absolute have been reported to be nonirritating and nonsensitizing to human skin; olibanum absolute was also nonphototoxic. The phototoxicity of olibanum gum is not known. Olibanum gum was moderately irritating to rabbit skin when applied undiluted for 24 h under occlusion.[42,43]

The biological effects, molecular targets, and potential of olibanum and boswellic acids in chronic inflammation and other conditions have recently been reviewed.[44–46]

Medicinal, Pharmaceutical, and Cosmetic. Olibanum, oil, and extracts (absolute and resinoid) are used as fixative and/or fragrance components in soaps, detergents, creams, lotions, and perfumes (e.g., oriental types), with maximum use level of 0.8% reported for the gum and absolute in perfumes.[42,43] In perfumes, the absolute, oil, and resinoid are used for a fresh balsamic, dry, resinous, somewhat green note in oriental bases, ambers, florals, colognes, male fragrances, and so on.[47]

Food. Only olibanum oil is used in food products, including alcoholic and nonalcoholic beverages, frozen dairy desserts, candy, baked goods, gelatins and puddings, and meat and meat products. Highest average maximum use level is about 0.001% (11.2 ppm) in meat and meat products.

Dietary Supplements/Health Foods. A proprietary extract of *B. serrata* from India, containing a mixture of triterpene pentacyclic acid derivatives of boswellic acids, has appeared on the market in recent years, with claims for anti-inflammatory and antiarthritic activities.

Traditional Medicine. Olibanum (more popularly known as frankincense) has been used since antiquity as an incense in India, China, Egypt, and the Catholic Church. It was an ingredient of the embalming liquid ancient Egyptians used on their dead. It has been used as a stimulant, respiratory antiseptic, diuretic, and emmenagogue in both Western and Eastern cultures. Conditions for which it has been and still is used include syphilis, rheumatism, painful menstruation, pimples, sores, tumors, cancers, asthma, sore throat, abdominal pain, stomach troubles, and nervous problems, among others (JIANGSU; LIST AND HÖRHAMMER; ROSE).[1,48] In Ayurvedic medicine, the oleogum resin of *B. serrata*, known as "Salai guggal," has been used as a treatment for rheumatism, nervous diseases, and as a topical anti-inflammatory.[27]

COMMERCIAL PREPARATIONS

Crude, extracts (e.g., resinoid), and oil. Olibanum oil is official in F.C.C.

Regulatory Status. Has been approved for food use (§172.510).

REFERENCES

See the General References for ARCTANDER; FEMA; GUENTHER; JIANGSU; LIST AND HÖRHAMMER; MERCK; TUCKER AND LAWRENCE; UPHOF.

1. R. Pernet, *Lloydia*, **35**, 280 (1972).
2. S. M. Abdel Wahab et al., *Planta Med.*, **53**, 382 (1987).
3. G. Snatzke and L. Vertesy, *Monatsh. Chem.*, **98**, 121 (1967).
4. A. H. Y. Abdel-Rahman, *Grasas Aceites (Seville)*, **27**, 175 (1976).
5. A. K. Sen, Sr. et al., *Carbohydr. Res.*, **223**, 321 (1992).
6. S. Hamm et al., *Phytochemistry*, **66**, 1499 (2005).
7. S. A. Higazy et al., *Egypt. J. Food Sci.*, **1**, 203 (1973).
8. S. A. Higazy et al., *Egypt. J. Food Sci.*, **2**, 29 (1974).
9. H. Obermann, *Dragoco Rep. (Ger. Ed.)*, **24**, 260 (1977).
10. S. Corsano and R. Nicoletti, *Tetrahedron*, **23**, 1977 (1967).
11. R. Nicoletti and M. L. Forcellese, *Tetrahedron*, **24**, 6519 (1972).
12. M. L. Forcellese et al., *Tetrahedron*, **28**, 325 (1972).
13. L. Peyron et al., *Riv. Ital. Essenze, Profumi, Piante Offic., Aromi, Saponi, Cosmet., Aerosol*, **63**, 133 (1981).
14. P. Maupetit, *Perfum. Flav.*, **9**, 19 (1985).
15. N. Banno et al., *J. Ethnopharmacol.*, **107**, 249 (2006).
16. B. Buchele et al., *Planta Med.*, **72**, 1285 (2006).
17. T. Akihisa et al., *Biol. Pharm. Bull.*, **29**, 1976 (2006).
18. F. A. Badria et al., *Z. Naturforsch. C*, **58**, 505 (2003).
19. G. Culioli et al., *Phytochemistry*, **62**, 537 (2003).
20. A. Kar and M. K. Menon, *Life Sci.*, **8**, 1023 (1969).
21. J. Reichling et al., *Schweiz. Arch. Tierheilkd.*, **146**, 71 (2004).
22. M. Duwiejua et al., *Planta Med.*, **59**, 12 (1993).
23. B. Gayathri et al., *Int. Immunopharmacol.*, **7**, 473 (2007).
24. C. Anthoni et al., *Am. J. Physiol Gastrointest. Liver Physiol.*, **290**, G1131 (2006).
25. H. Safayhi et al., *Planta Med.*, **66**, 110 (2000).
26. H. Safayhi et al., *J. Pharmacol. Exp. Ther.*, **261**, 1143 (1992).
27. S. S. Handa et al., *Fitoterapia*, **63**, 3 (1992).
28. I. Gupta et al., *Planta Med.*, **67**, 391 (2001).
29. I. Gupta et al., *Eur. J. Med. Res.*, **2**, 37 (1997).
30. R. S. Pandey et al., *Indian J. Exp. Biol.*, **43**, 509 (2005).
31. Y. J. et al., *Pak. J. Pharm. Sci.*, **19**, 129 (2006).
32. D. F. Flavin, *J. Neurooncol.*, **82**, 91 (2007).
33. L. Xia et al., *Mol. Cancer Ther.*, **4**, 381 (2005).
34. K. Hostanska et al., *Anticancer Res.*, **22**, 2853 (2002).
35. Y. S. Park et al., *Planta Med.*, **68**, 397 (2002).
36. M. T. Huang et al., *Biofactors*, **13**, 225 (2000).
37. P. Pungle et al., *Indian J. Exp. Biol.*, **41**, 1460 (2003).
38. I. Gupta et al., *Eur. J. Med. Res.*, **3**, 511 (1998).
39. A. Wildfeuer et al., *Arzneim.-Forsch.*, **48**, 668 (1998).
40. C. C. Weber et al., *Planta Med.*, **72**, 507 (2006).

41. F. Borrelli et al., *Br. J. Pharmacol.*, **148**, 553 (2006).
42. D. L. J. Opdyke, *Food Cosmet. Toxicol.*, **16**(Suppl. 1), 837 (1978).
43. D. L. J. Opdyke, *Food Cosmet. Toxicol.*, **16**(Suppl. 1), 835 (1978).
44. D. Poeckel and O. Werz, *Curr. Med. Chem.*, **13**, 3359 (2006).
45. H. P. Ammon, *Planta Med.*, **72**, 1100 (2006).
46. E. Basch et al., *J. Herb. Pharmacother.*, **4**, 63 (2004).
47. A. O. Tucker, *Econ. Bot.*, **40**, 425 (1986).
48. J. L. Hartwell, *Lloydia*, **31**, 71 (1968).

ONION

Source: *Allium cepa* L. (Family Amaryllidaceae or Liliaceae).

GENERAL DESCRIPTION

A perennial or biennial herb with hollow leaves and a taller and thicker scape (flowering stem), also hollow; up to about 1.2 m high; generally considered to be native of western Asia; has long been cultivated worldwide and much varied. Part used is the fleshy bulb; onion oil is obtained by steam distillation.

There are numerous forms and varieties of onion, the most common ones being the white globe, yellow globe, and red globe onions. Due to the particularly high variability of onion as opposed to most other plants, results of chemical and biological studies from different countries should be evaluated with caution as they may be based on different varieties of *A. cepa* that are often not specified in the reports.

CHEMICAL COMPOSITION

Like garlic, onion contains many organic sulfur compounds, including *trans*-S-(1-propenyl) cysteine sulfoxide, S-methylcysteine sulfoxide, S-propylcysteine sulfoxide, and cycloalliin.[1,2] Except for cycloalliin, these sulfur compounds are converted to simpler sulfur compounds by the enzyme alliinase released when the onion is cut or crushed. These simpler compounds are unstable and undergo further decomposition to sulfides (di-, tri-, etc.) and other compounds that are responsible for the onion flavor (especially methylpropyl disulfide, methylpropyl trisulfide, and dipropyl trisulfide).[3,4]

The lachrymating (tear-producing) principle in crushed or cut onion is thiopropanal S-oxide (propanethial S-oxide) produced from its precursor, *trans*-S-(1-propenyl)-cysteine sulfoxide, by the action of alliinase.[5–8]

Onion also contains a trace of volatile oil composed mainly of sulfur compounds, with dipropyl disulfide as the main component (but not an important flavor contributor).[1,3] Others present include the three important flavor contributors methylpropyl disulfide, methylpropyl trisulfide, and dipropyl trisulfide, as well as allylpropyl disulfide, dimethyl disulfide, 3,4-dimethylthiophene, methyl-*cis*-propenyl disulfide, and others.[1,3,9]

Other constituents present in onion include phenolic acids (caffeic, sinapic, *p*-coumaric, protocatechuic acids, etc.), flavonoids (e.g., quercetin, isorhamnetin, taxifolin, and their glucosides),[10–13] anthocyanins (e.g., cyanidin, carboxypyranocyanidin, and peonidin glycosides),[12–16] sterols (cholesterol, stigmasterol, β-sitosterol, etc.),[17,18] saponins (e.g., tropeosdies and ascalonicosides),[11,19] sugars, vitamins (A, C, B, and B_2), pectin, and peptides (e.g., alliceptin),[20] among others (JIANGSU; LIST AND HÖRHAMMER).[21]

PHARMACOLOGY AND BIOLOGICAL ACTIVITIES

Onion is reported to have many similar pharmacological properties as garlic, including antihypercholesterolemic (in experimental animals), hypoglycemic (in humans and experimental animals),[22–26] and antifungal (see *garlic*).[27] The *S*-methyl cysteine sulfoxide component of onion reduced blood lipids and glucose when fed to rats at a dose of 200 mg/kg for 45 days.[28,29] Consumption of brown onions (a variety of *A. cepa*) also reduced plasma triacylglycerol in pigs. The levels of other plasma lipids, however, were not affected.[30] The hypoglycemic and antioxidant effects of dietary onion supplementation and *S*-methyl cysteine have been repeatedly demonstrated in diabetic rats.[29,31–34]

The antimicrobial and anthelmintic activities of onion have been reported. Aqueous extracts of onion were antifungal (different *Candida* and *Malassezia* strains) in a dose-dependent manner.[35] The peptide allicepin inhibited the growth of several fungal species including *Botrytis*, *Fusarium*, and others.[20] Onion extracts and onion oil have also demonstrated antibacterial and antidermatophytic activities against numerous Gram-positive and Gram-negative bacteria, as well as *Trichophyton mentagrophytes*, respectively.[36,37] Oral administration of onion oil (5 mg/kg/day for 2 weeks) to rats infected with *Trichinella spiralis* worms resulted in a significant decline in the number of adult worms and larvae.[38]

Other reported activities for onion include antiplatelet aggregation and an increase in fibrinolytic activity (with conflicting results).[39,40] The antiplatelet activity is believed to be sulfur dependent.[41] Saponins isolated from *A. cepa* had an antispasmodic activity on the isolated guinea pig ileum.[11] Onion juice was also effective in the treatment of alopecia areata, a condition characterized by patchy hair loss, as demonstrated by a pilot clinical trial in 23 patients.[42]

A preliminary study suggests that increased consumption of *Allium* vegetables (including onion) correlated to a significant reduction in gastric cancer risk.[43]

Reviews of the therapeutic value of onion have been published.[44]

USES

Food. Onion oil is extensively used as a flavor ingredient in most food products, including alcoholic and nonalcoholic beverages, frozen dairy desserts, candy, baked goods, gelatins and puddings, meat and meat products, condiments and relishes, fats and oils (e.g., salad dressings and salad oils), soups, snack foods, and gravies. Highest average maximum use level reported is about 0.083% (826 ppm) in fats and oils.

Dietary Supplements/Health Foods. Supplements of dehydrated onion (capsules and tablets); contradictory results have been reported for studies using fresh onions compared with supplements.[39]

Traditional Medicine. Although not as valued or widely used as garlic, onion is used as antispasmodic, carminative, diuretic, expectorant, stomachic, and antihelmintic for many of the same conditions as is garlic (see garlic). Its use in Chinese medicine is of only recent origin and is limited (NANJING).

COMMERCIAL PREPARATIONS

Oil; official in F.C.C.

Regulatory Status. GRAS (§182.20). Onion is the subject of a positive German therapeutic monograph, indicated for antibacterial, lipid lowering, blood pressure lowering, and antiplatelet aggregation activity.[45]

REFERENCES

See the General References for ARCTANDER; BAILEY 2; BLUMENTHAL 1; DER MARDEROSIAN AND BEUTLER; DUKE 4; FEMA; GRIEVE; GUENTHER; GUPTA; JIANGSU; LIST AND HÖRHAMMER; LUST; MCGUFFIN 1 & 2; TERRELL.

1. J. R. Whitaker, *Adv. Food Res.*, **22**, 73 (1976).
2. A. I. Virtanen and C. G. Spare, *Suomen Kemistilehti*, **34**, 72 (1961).
3. W. G. Galetto and A. A. Bednarczyk, *J. Food Sci.*, **40**, 1165 (1975).
4. M. Boelens et al., *J. Agric. Food Chem.*, **19**, 984 (1971).
5. C. G. Spare and A. I. Virtanen, *Acta Chem. Scand.*, **17**, 641 (1963).
6. W. F. Wilkens, *Cornell Univ. Agric. Expt. Sta. Mem.*, **385** (1964).
7. M. H. Brodnitz and J. V. Pascale, *J. Agric. Food Chem.*, **19**, 269 (1971).
8. T. H. Maugh, *Science*, **204**, 293 (1979).
9. M. H. Brodnitz et al., *J. Agric. Food Chem.*, **17**, 760 (1969).
10. P. Bonaccorsi et al., *J. Agric. Food Chem.*, **53**, 2733 (2005).
11. G. Corea et al., *J. Agric. Food Chem.*, **53**, 935 (2005).
12. J. P. Varshney and T. Ali, *Indian J. Appl. Chem.*, **34**, 142 (1971).
13. M. Leonte and A. Leonte, *Lucr. Stiint. Inst. Politeh. Galati*, **4**, 195 (1970).
14. T. Fossen et al., *Phytochemistry*, **64**, 1367 (2003).
15. T. Fossen and O. M. Andersen, *Phytochemistry*, **62**, 1217 (2003).
16. N. Terahara et al., *Biosci. Biotechnol. Biochem.*, **58**, 1324 (1994).
17. L. A. R. Sallam et al., *Qual. Plant Foods Hum. Nutr.*, **24**, 159 (1974).
18. T. Itoh et al., *Phytochemistry*, **16**, 140 (1977).
19. D. Nitschke and A. Smoczkiewiczowa, *Zesz. Nauk Akad. Ekon. Poznaniu, Ser. 1.*, **69**, 104 (1976).
20. H. X. Wang and T. B. Ng, *J. Pept. Sci.*, **10**, 173 (2004).
21. A. L. Feldman et al., *Konserv. Ovoshchesuch. Prom.*, **4**, 19 (1973).
22. R. C. Jain and C. R. Vyas, *Br. Med. J.*, **2**, 730 (1974).
23. R. K. Gupta et al., *Indian J. Exp. Biol.*, **15**, 313 (1977).
24. K. K. Sharma et al., *Indian J. Med. Res.*, **65**, 422 (1977).
25. K. T. Augusti, *Indian J. Exp. Biol.*, **14**, 110 (1976).
26. K. T. Augusti and M. E. Benaim, *Clin. Chim. Acta*, **60**, 121 (1975).
27. P. S. Shekhawat and R. Prasada, *Sci. Cult.*, **37**, 40 (1971).
28. K. Kumari and K. T. Augusti, *J. Ethnopharmacol.*, **109**, 367 (2007).
29. K. Kumari et al., *Indian J. Biochem. Biophys.*, **32**, 49 (1995).
30. E. Ostrowska et al., *Br. J. Nutr.*, **91**, 211 (2004).
31. K. E. Campos et al., *Int. J. Food Sci. Nutr.*, **54**, 241 (2003).
32. K. Kumari and K. T. Augusti, *Indian J. Exp. Biol.*, **40**, 1005 (2002).
33. S. M. Kelkar et al., *Indian J. Biochem. Biophys.*, **38**, 277 (2001).
34. A. Helen et al., *Toxicol. Lett.*, **116**, 61 (2000).
35. M. Shams-Ghahfarokhi et al., *Fitoterapia*, **77**, 321 (2006).
36. J. H. Kim, *J. Nihon Univ. Sch. Dent.*, **39**, 136 (1997).
37. A. N. Zohri et al., *Microbiol. Res.*, **150**, 167 (1995).
38. N. M. bu El Ezz, *J. Egypt. Soc. Parasitol.*, **35**, 511 (2005).
39. J. Kleijnen et al., *Br. J. Clin. Pharmacol.*, **28**, 535 (1989).
40. B. S. Kendler, *Prev. Med.*, **16**, 670 (1987).

41. I. L. Goldman et al., *Thromb. Haemost.*, **76**, 450 (1996).
42. K. E. Sharquie and H. K. Al-Obaidi, *J. Dermatol.*, **29**, 343 (2002).
43. W. C. You et al., *J. Natl. Cancer Inst.*, **81**, 162 (1989).
44. K. T. Augusti, *Indian J. Exp. Biol.*, **34**, 634 (1996).
45. Monograph *Allii cepae bulbus*, *Bundesanzeiger*, no. 50 (March 13, 1986).

ORANGE (BITTER AND SWEET)

Source: **Bitter orange** *Citrus aurantium* L. (syn. *C. aurantium* L. ssp. *amara* (L.) Engl.; *C. vulgaris* Risso; *C. bigaradia* Risso); **Sweet orange** *Citrus sinensis* (L.) Osb. (syn. *C. aurantium* L. var. *sinensis* L.; *C. aurantium* L. var. *dulcis* Pers.; *C. aurantium* L. ssp. *dulcis* L.; *C. aurantium Risso* ssp. *sinensis* Engl.) (Family Rutaceae).

GENERAL DESCRIPTION

Bitter orange is an evergreen glabrous (nonhairy) tree with long but not very sharp spines and very fragrant flowers; membranes and pulp of fruit very bitter and sour; up to about 10 m high; native to southern China and northeastern India; cultivated in China, southern Europe, and the United States. Its trunk is more resistant to plant diseases than those of other citrus trees and consequently serves as stock for the less resistant sweet orange. Parts used are the peel of the fruit, freshly picked flowers, and leaves and twigs.

Bitter orange oil is obtained from the fresh peel by cold expression in about 0.15% yield, by machine or hand. Its major producers include Guinea, southern European countries (especially Spain and Italy), Brazil, and the West Indies. The Guinean and Spanish oils are considered of the best quality.

Neroli oil (orange flower oil) is obtained from the freshly picked flowers of bitter orange by steam distillation in about 0.1% yield. Major producers include France, Italy, and Tunisia. The distillation water from which the oil layer has been removed is called orange flower water.

Petitgrain oil is obtained from the leaves (also twigs) of bitter orange by steam distillation in about 0.2% yield. Major producers include France, Haiti, Paraguay, and Guinea.

Sweet orange exists in numerous varieties (e.g., Navel, Jaffa, and Valencia). It is a smaller tree than the bitter orange tree, less hardy, and with few or no spines; fruits smaller, with sweet pulp and nonbitter membranes; generally believed to be native to China; extensively cultivated worldwide, especially in the United States (e.g., California and Florida) and Mediterranean countries. Part used is the peel of the partially or fully ripe fruit. The peel used for the preparation of Sweet Orange Peel Tincture N.F. is specified to be derived from the nonartificially colored ripe fruit and devoid of the albedo (inner white portion of the rind).

Sweet orange oil, commonly known simply as orange oil, is obtained by one of the three major methods: (1) cold expression (hand or machine) of the fresh peel, (2) steam distillation of the fresh peel (either already expressed or not yet processed), and (3) distillation of the essences recovered as a by-product in the manufacture of orange juice concentrates. Major producers of cold-expressed sweet orange oil include the United States, Cyprus, Guinea, Israel, and Brazil, while the United States is the major producer of the distilled sweet orange oil.

Terpeneless orange oils (both bitter and sweet) are obtained from the respective oils by vacuum distillation and/or extraction with dilute ethyl alcohol whereby all or most of the terpenes (e.g., limonene) are removed.

CHEMICAL COMPOSITION

In addition to a volatile oil (1–2.5%), bitter orange peel contains appreciable quantities of neohesperidin (up to ca. 14% in unripe peel, usually 2.4–2.8% in ripe peel), naringin (0.9–4%), rhoifolin, lonicerin, hesperidin, and other polymethoxyflavonoids (tangeretin, nobiletin, sinensetin, auranetin, rutin, etc.);[1,2] vitamins (A, B_1, and C); coumarins (e.g., 6,7-dimethoxycoumarin and umbelliferone);[3] carotenoid pigments (citraurin, violaxanthin, and cryptoxanthin); pectin, citrantin; and others (FARNSWORTH 1–4; JIANGSU; LIST AND HÖRHAMMER). High levels of the alkaloid synephrine (six isomers) are also present in bitter orange peel.[4,5]

The volatile oil (bitter orange oil) contains more than 90% monoterpenes (main d-limonene, also myrcene, campherr pinene, ocimene, p-cymene, etc.); small amounts of alcohols (linalool, terpinene nerol, farnesol, nerolidol, octanol, etc. usually 0.5–1% aldehydes (mainly decanal also nonanal, dodecanal, citronellal, neral acetaldehyde, formaldehyde, etc.), and ketones (carvone, α-ionone, and jasmine); free acids (octadecadienoic, pelargonic, cinnamic, acetic, etc.);[6] about 2.4% esters (linalyl acetate, decyl pelargonate, octyl acetate, geranyl acetate, etc.); coumarins (osthole and auraptenol);[7,8] and others (JIANGSU; LIST AND HÖRHAMMER).

Neroli oil is reported to contain linalool (ca. 34%); linalyl acetate (6–17%), limonene (15%), β-pinene (11%), nerolidol (6%), geraniol, nerol, methyl anthranilate; indole, and jasmone as well a small amounts of other compounds including citral, nonanal, cis-8-heptadecene, 2,5- dimethyl-2-vinyl-4-hexenal, neryl acetate, and valeric acid, among others (GUENTHER; JIANGSU; LIST AND HÖRHAMMER; MERCK; REMINGTON).[9–11]

Petitgrain oil contains large amounts of esters (40–80%, depending on sources; composed mainly of linalyl acetate, with lesser quantity of geranyl acetate). Other compounds present include linalool, nerol, α-terpineol, geraniol, nerolidol, farnesol, and limonene (GUENTHER; MERCK).[10,11]

Sweet orange peel contains 1.5–2% volatile oil, numerous flavonoids including polymethoxyflavones, O- and C-glycosylated flavones (neohesperidin, hesperidin, naringin, tangeretir auranetin, nobiletin, etc.),[12–17] vitamins (e.g., C and E), limonin,[18] coumarins (e.g., 6,7-dimethoxycoumarin),[3] phenolic acids, for example, hydroxycinnamic acid,[15] carotenoids, pectin, citrantin, and other similar constituents present in bitter orange peel (JIANGSU; LIST AND HÖRHAMMER). The juices of different varieties of Sicilian sweet orange were also found to contain many of the flavonoids, carotenoids, and vitamins present in the peel.[19]

Sweet orange oil is very similar to bitter orange oil in chemical composition, containing about 90% or more of d-limonene and normally 1.2–2.5% aldehydes (mostly decanal, octanal, etc.).[20–22] The cold-expressed oil also contains coumarins (e.g., bergapten and auraptenol), acids (octadecadienoic, etc.), valencene, α-ylangene,[23] and other compounds present in bitter orange oil (JIANGSU; LIST AND HÖRHAMMER; MERCK; REMINGTON).[6,22] A study on the use of absorbents for cold-pressed orange oil found that silica gel produced the best results, reducing monoterpene hydrocarbons, while maintaining high oxygenated component content, especially decanal.[24,25]

Sweet orange oil does not taste bitter as bitter orange oil. Important flavor contributors reported in sweet orange essence include octanol, ethyl butyrate, and acetaldehyde, while α-terpineol and trans-2-hexenal contribute to the off flavor of the oil.[21,22]

Certain flavonoid glycosides (e.g., naringin and neohesperidin) are intensely bitter; the specific linkages in the sugars present are reported to be the determining factors for their bitterness.[26]

PHARMACOLOGY AND BIOLOGICAL ACTIVITIES

Both bitter and sweet orange peels exhibit numerous distinct pharmacological activities, including anti-inflammatory (due to flavonoids, especially naringin), antibacterial and antifungal (flavonoids and pectin), antihypercholesterolemic in humans and laboratory animals due to the contained pectin (see ***pectin***), and choleretic; among others (FARNSWORTH 1–4; JIANGSU; LIST AND HÖRHAMMER). The ethnopharmacology and traditional uses of citrus products have recently been reviewed.[27]

Citrantin is reported to have antifertility properties in rabbits when fed orally at a dose of 0.75 mg/kg/day (FARNSWORTH 1–4; JIANGSU).

Bitter orange, sweet orange, and neroli oils have been reported to exhibit antifungal and antibacterial activities *in vitro* (see ***lemon oil*** and ***pine needle oil***).[26,28–31] An emulsion of bitter orange oil exhibited antifungal activity against topical dermatophyte infection in humans (*Tinea pedis*, *T. cruris*, and *T. corporis*) with an 80% cure rate after 1–2 weeks of treatment.[32] The oil of sweet orange, citral, and linalool were effective against some Gram-positive and Gram-negative bacteria *in vitro*.[33] 3-(4-Hydroxy-3-(3-methyl-2-butenyl)-phenyl)-2 (*E*)-propenal, a phenolic cinnamaldehyde found in wound gum from injured peels of *C. sinensis*, is antifungal against *Penicillium digitatum* and *Cladosporium cucumerinum*.[34]

The volatile oil of sweet orange peel exhibited insecticidal activity against mosquito, housefly, and cockroach after 60 min of its application as a room spray.[35]

The vitamins, flavonoids, and phenolic constituents of sweet orange peel and juice were found to possess antioxidant free radical scavenging activity.[15,19]

TOXICOLOGY

Sweet orange oil has been reported to promote tumor formation on mouse skin treated with a primary carcinogen (see ***lemon oil*** and ***lime oil***).[36,37] Its major component, *d*-limonene, is also reported to have anticarcinogenic activity.[36]

Ingestion of large amounts of orange peel (bitter or sweet) by children has been reported to cause intestinal colic, convulsions, and even death (JIANGSU).

Bitter orange oil is reported to have distinct phototoxic activity, while none is reported for expressed sweet orange oil even though both oils contain coumarins (see ***lime oil***).[36,38]

Expressed sweet orange oil, bitter orange oil, and neroli oil are generally reported to be nonirritating and nonsensitizing to humans.[31,36,38] However, limonene present in citrus oils has been known to cause contact dermatitis in humans (LEWIS AND ELVIN-LEWIS).

Many reports of adverse effects due to the use of bitter orange, as a replacement for ephedra, in weight-loss dietary supplements have recently emerged. Some of the reported effects include ischemic colitis, angina, elevated blood pressure, and other cardiovascular effects.[39–45] Synephrine and/or octopamine are speculated to play a role in the advent of such effects.[41–43]

USES

Medicinal, Pharmaceutical, and Cosmetic. Bitter orange peel is occasionally used as an ingredient in certain stomachic, carminative, and laxative preparations.

Neroli oil (orange flower oil), orange flower water, and sweet orange peel tincture are used in flavoring pharmaceuticals.

Bitter and sweet orange, neroli, and petitgrain oils are extensively used as fragrance components in soaps, detergents, creams, lotions, and cosmetics. Highest maximum use level is 1.0% reported for bitter orange oil in perfumes.[31,36,38]

Food. Bitter orange oil is extensively used as a flavor component in most major food products, including alcoholic (e.g., liqueurs, especially triple sec) and nonalcoholic beverages, frozen dairy desserts, candy, baked goods, gelatins and puddings, meat and meat products, and condiments and relishes. Highest average maximum use level reported is about 0.043% (424 ppm).

Neroli and petitgrain oils are also widely used in major food products, including alcoholic and nonalcoholic beverages, frozen dairy desserts, candy, baked goods, and gelatins and puddings, among others, with highest average maximum use levels of about 0.001% reported for neroli oil in alcoholic beverages (11.5 ppm) and baked goods (11.3 ppm) and about 0.004% (37.7 ppm) for petitgrain oil in condiments and relishes.

Sweet orange oil (produced by various methods) is by far the most widely used in food products, including all above categories in addition to jams and jellies, gravies, sweet sauces, breakfast cereals, and processed vegetables, among others. Highest average maximum use level is 0.75% reported for the distilled oil in sweet sauces. Highest average maximum use level reported for its terpeneless type is 0.01% in breakfast cereals.

Dietary Supplements/Health Foods. Dried bitter and sweet orange peels used in tea formulations and in digestive formulas. Due to the presence of synephrine, a sympathomimetic alkaloid, the use of bitter orange peel in "ephedra-free" weight-loss dietary supplements has significantly increased after the ban of ephedra in 2004.

Traditional Medicine. Dried bitter orange peel and, to a lesser extent, sweet orange peel are used as tonic and carminative in treating dyspepsia. In Chinese medicine, dried sweet orange peel is used to reduce phlegm and in treating coughs, colds, anorexia (lack of appetite), and malignant breast sores. Dried bitter orange and less commonly the dried peel are used in treating prolapse of the uterus and of the anus (also rectum), diarrhea, blood in feces, and others, in addition to above conditions (JIANGSU). They have also been reported to be used in cancers.[46]

Bitter orange flowers and oil have reportedly been used in European tradition as prophylactics for gastrointestinal complaints, nervous conditions, gout, sore throat, as a sedative, and for sleeplessness.

Others. Due to the high contents of d-limonene, in both bitter and sweet orange oils, the oils are also used as major sources of d-limonene, which in turn serves as a starting material for the synthesis of carvone, an important flavor chemical (see *caraway* and *spearmint*).

Neohesperidin and naringin from bitter orange peel could serve as starting materials for the production of neohesperidin dihydrochalcone (NHDHC) and naringin dihydrochalcone (NDHC); both are sweeteners, with NDHC being as sweet as saccharin and NHDHC 20 times more so.

COMMERCIAL PREPARATIONS

Dried peels (bitter and sweet), their extracts (solid, fluid, and tincture), and oils. Bitter and sweet orange peels were formerly official in N.F. and U.S.P., respectively. Bitter orange oil was formerly official in N.F., and sweet orange oil formerly in U.S.P. Both oils as well as petitgrain oil Paraguay are official in F.C.C., while neroli oil, orange flower water, and sweet orange peel tincture are official in N.F. Qualities of oils vary according to sources. Fluid extracts (see *glossary*) prepared by the direct extraction method have flavor qualities superior to those produced by the dilution method.

Regulatory Status. All GRAS (§182.20). Bitter orange flowers subject of a German therapeutic monograph; traditional uses are not recommended as efficacy has not been demonstrated; use is allowed as a flavor corrigent.[47] Sweet orange peels subject of a positive monograph for treatment of loss of appetite.[48]

REFERENCES

See the General References for ARCTANDER; BAILEY 1; BAILEY 2; BIANCHINI AND CORBETTA; BLUMENTHAL 1; FARNSWORTH 1–4; FEMA; GOSSELIN; GUENTHER; JIANGSU; LIST AND HÖRHAMMER; LUST; MARTINDALE; TERRELL; TUCKER AND LAWRENCE; UPHOF.

1. T. Nakabayashi, *Nippon Nogei Kagaku Kaishi*, **35**, 945 (1961).

2. S. Natarajan et al., *Econ. Bot.*, **30**, 38 (1976).

3. J. H. Tatum and R. E. Berry, *Phytochemistry*, **16**, 1091 (1977).
4. L. Mattoli et al., *J. Agric. Food Chem.*, **53**, 9860 (2005).
5. D. B. Allison et al., *Int. J. Obes. (Lond)*, **29**, 443 (2005).
6. G. Lamonica et al., *Essenze Deriv. Agrum.*, **45**, 115 (1973).
7. W. L. Stanley et al., *Tetrahedron*, **21**, 89 (1965).
8. W. L. Stanley and L. Jurd, *J. Agric. Food Chem.*, **19**, 1106 (1971).
9. B. Corbier and P. Teisseire, *Recherches*, **19**, 289 (1974).
10. W. D. Fordham in L. W. Codd et al., eds., *Chemical Technology: An Encyclopedic Treatment*, Vol. 5, Barnes & Noble, New York, 1972, p. 1.
11. M. Stoll in A. Standen, ed., *Kirk-Othmer Encyclopedia of Chemical Technology*, Vol. 14, 2nd ed., Wiley–Interscience, New York, 1967, p. 178.
12. C. O. Green et al., *Biomed. Chromatogr.*, **21**, 48 (2007).
13. S. Li et al., *J. Agric. Food Chem.*, **54**, 4176 (2006).
14. S. Li et al., *Biomed. Chromatogr.*, **20**, 133 (2006).
15. M. A. Anagnostopoulou et al., *Biomed. Chromatogr.*, **19**, 138 (2005).
16. C. W. Wilson and P. E. Shaw, *J. Agric. Food Chem.*, **25**, 221 (1977).
17. H. Wagner et al., *Chem. Ber.*, **102**, 2089 (1969).
18. W. C. Scott, *Proc. Fla. State Hort. Soc.*, **83**, 270 (1970).
19. A. R. Proteggente et al., *Free Radic. Res.*, **37**, 681 (2003).
20. M. G. Moshonas and E. D. Lund, *J. Food Sci.*, **34**, 502 (1969).
21. E. D. Lund and W. L. Bryan, *J. Food Sci.*, **42**, 385 (1977).
22. P. E. Shaw and R. L. Coleman, *J. Agric. Food Chem.*, **19**, 1276 (1971).
23. S. K. Ramaswami et al. in B. M. Lawrence et al., eds., *Flavors and Fragrances: A World Perspective*, Elsevier Science Publishers B.V., Amsterdam, 1988, p. 951.
24. O. J. Ferrer and R. F. Matthews, *J. Food Sci.*, **52**, 801 (1987).
25. B. M. Lawrence, *Perfum. Flav.*, **15**, 45 (1990).
26. S. Kamiya et al., *Agric. Biol. Chem.*, **38**, 1785 (1974).
27. B. A. Arias and L. Ramon-Laca, *J. Ethnopharmacol.*, **97**, 89 (2005).
28. M. S. Subba et al., *J. Food Sci.*, **32**, 225 (1967).
29. B. G. V. N. Rao and P. L. Joseph, *Riechst., Aromen, Korperpflegem.*, **21**, 405 (1971).
30. D. I. Murdock and W. E. Allen, *Food Technol.*, **14**, 441 (1960).
31. D. L. J. Opdyke, *Food Cosmet. Toxicol.*, **12**, 813 (1976).
32. W. Ramadan et al., *Int. J. Dermatol.*, **35**, 448 (1996).
33. K. Fisher and C. A. Phillips, *J. Appl. Microbiol.*, **101**, 1232 (2006).
34. R. R. Stange, Jr et al., *J. Nat. Prod.*, **56**, 1627 (1993).
35. F. C. Ezeonu et al., *Bioresour. Technol.*, **76**, 273 (2001).
36. D. L. J. Opdyke, *Food Cosmet. Toxicol.*, **12**, 733 (1974).
37. F. M. Nacino et al., *Kalikasan*, **4**, 240 (1975).
38. D. L. J. Opdyke, *Food Cosmet. Toxicol.*, **12**, 735 (1974).
39. S. Sultan et al., *Mayo Clin. Proc.*, **81**, 1630 (2006).
40. C. A. Gange et al., *Mayo Clin. Proc.*, **81**, 545 (2006).
41. L. T. Bui et al., *Ann. Pharmacother.*, **40**, 53 (2006).
42. B. Min et al., *Pharmacotherapy*, **25**, 1719 (2005).

43. C. A. Haller et al., *Am. J. Med.*, **118**, 998 (2005).
44. F. Firenzuoli et al., *Phytomedicine*, **12**, 247 (2005).
45. D. L. Nykamp et al., *Ann. Pharmacother.*, **38**, 812 (2004).
46. J. L. Hartwell, *Lloydia*, **34**, 103 (1971).
47. Monograph, *Aurantii flos*, Bundesanzeiger (July 14, 1993).
48. Monograph, *Citri sinensis pericarpium*, Bundesanzeiger, (February 1, 1990).

OREGANO

Source: *Origanum vulgare* L. (Family Labiatae or Lamiaceae), *Lippia graveolens* H. B. K. (syn. *L. berlandieri* Schauer), *L. palmeri* S. Wats. (Family Verbenaceae), and other *Origanum* and *Lippia* species.

Common/vernacular names: Wild marjoram, origanum, European oregano (*O. vulgare*); Mexican oregano, Mexican marjoram, and Mexican wild sage (*Lippia* species).

GENERAL DESCRIPTION

Oregano is derived from several genera of plants from mainly two families (e.g., *Origanum, Lippia, Coleus, Lantana,* and *Hyptis*). It is not one or two well-defined species but rather any one of more than two dozen known species that yield leaves or flowering tops having the flavor recognized as being oregano. The most commonly used plants are *Origanum vulgare, O. onites* (as Turkish oregano), *Lippia graveolens*, and *L. palmeri* (see **marjoram**).[1,2]

European oregano (*O. vulgare*) is a hardy perennial herb with erect, more or less hairy, branching stems, and hairy leaves; up to about 0.9 m high; native to Europe. Parts used are the dried herbs and leaves.

Mexican oregano is usually derived from *Lippia graveolens* and occasionally from *L. palmeri*. Both are aromatic shrubs native to Mexico; *L. graveolens* is larger and is sometimes a tree, up to 9 m high. Parts used are the dried leaves.[1]

CHEMICAL COMPOSITION

Origanum vulgare (also known as wild marjoram) contains 0.1–1.0% volatile oil composed of thymol, carvacrol, β-bisabolene, caryophyllene, *p*-cymene, borneol, linalool, linalyl acetate, geranyl acetate, α-pinene, β-pinene, α-terpinene, *cis*-sabinene, germacrene, ocimene, and others, with highly variable relative proportions, depending on sources; the phenols (thymol and carvacrol) content has been reported to vary from 0% to 90% in the oil (JIANGSU; LIST AND HÖRHAMMER; MARSH).[3–9] Other constituents include luteolin, apigenin, chrysoeriol, diosmetin, quercetin, eriodictyol, cosmoside, vicenin-2, thymoquinol glycosides, caffeic acid, ursolic acid, oleanolic acid, rosmarinic acid, jasmonic acid, lithospermic acid, aristolochic acids I and II, D-(+)-raffinose,[10] protein, and vitamins for example, tochopherols (LIST AND HÖRHAMMER; MARSH).[11–14]

The essential oil of Greek oregano (*O. vulgare* subsp. *viride* (Boiss) Hayek; syn. *O. heracleoticum* L.) has been reported to contain carvacrol, thymol, linalool, borneol, and other similar constituents found in *O. vulgare* as well as carvone, camphor, amyl alcohol, and 4,5-epoxy-*p*-menthlene.[15,16]

The essential oil of *O. onites* has been found to be highly variable in wild populations in Greece, ranging from 1.85% to 4.37%, with carvacrol as the major component (51–84.5%), plus borneol (1.3–8.2%), *p*-cymene (5.1–12.2%), γ-terpinene (2.3–13.6%), in addition to minor components including β-phellandrene, *trans*-thujanol, cis-thujanol, linalyl acetate, α-muurolene, and others.[2]

The essential oils of numerous *Lippia* species have been reported to have highly variable compositions.[17–22] Nevertheless, *Lippia* species (e.g., *L. graveolens* and *L. origanoides* H. B. K.) known to be used as sources of oregano contain *p*-cymene, 1,8-cineole, thymol, and/or carvacrol as major components.[21–23]

The flavanones pinocembrin and naringenin along with lapachenole have been isolated from *L. graveolens* (aerial parts and root).[24]

PHARMACOLOGY AND BIOLOGICAL ACTIVITIES

Thymol and carvacrol, usually the major phenols present in oregano, have strong fungicidal, anthelmintic, irritant, and other properties (see **thyme**).

Antibacterial effects have recently been reported for oregano against *Clostridium perfringens*.[25] The antifungal effects have also been reported against *Trichophyton mentagrophytes* and *Candida albicans*.[26,27] The oil also has antimalarial activity against Culex pipiens larvae;[28] and *in vitro* antiprotozoal activity against *Trypanosoma cruzi* and *Giardia intestinalis*.[29,30]

Aqueous oregano extract exhibited antihyperglycemic effect in diabetic rats.[31] The aqueous and methanolic extracts also displayed inhibitory effects against aldose reductase and soybean lipoxygenase *in vitro*.[32]

The volatile oil was reported to have nonspecific immunostimulatory effect *in vitro* and to enhance growth in pigs.[33]

The essential oil of European oregano is reported to have spasmolytic, expectorant, diuretic, and choleretic properties (JIANGSU; LIST AND HÖRHAMMER).

TOXICOLOGY

Lapachenole has antifertility as well as carcinogenic activities.[24]

Aristolochic acids and D-raffinose have antithrombin activity in leukemia cells.[10]

USES

Food. European oregano is extensively used as a major flavor ingredient in pizza. The more spicy Mexican oregano is widely used in Mexican dishes (chili, chili con carne, etc.); it is less preferred for use in pizza than the milder European type. Oregano is also widely used in other foods, including alcoholic beverages, baked goods, meat and meat products, condiments and relishes, milk products, processed vegetables, snack foods, fats and oils, and others. Highest average maximum use level reported is about 0.3% in condiments and relishes (3168 ppm) and milk products (2900 ppm).

Traditional Medicine. *Lippia graveolens* is used in Mexico as a stimulant emmenagogue and demulcent.[34]

Origanum vulgare has been used as a stimulant, carminative, diaphoretic, and nerve tonic and as a cure for asthma, coughs, indigestion, rheumatism, toothaches, headaches, spider bites, and coronary conditions (ROSENGARTEN).

In European phytomedicine, oregano and its preparations have been used for the treatment of respiratory ailments, coughing, bronchitis; antispasmodic and expectorant. Also used as an appetite stimulant, diuretic, and mild sedative.[35]

In China, in addition to some of the above uses, *O. vulgare* is used to treat fevers, vomiting, diarrhea, jaundice, and itchy skin conditions (JIANGSU).

COMMERCIAL PREPARATIONS

Mainly the spice.

Regulatory Status. GRAS (Mexican oregano, §182.20; Origanum, §182.20). *O. vulgare* is the subject of a German therapeutic monograph; use not recommended, since efficacy has not been confirmed.[35]

REFERENCES

See the General References for BAILEY 1; BLUMENTHAL 1; FEMA; GOSSELIN; JIANGSU; LIST AND HÖRHAMMER; LUST; MARTINDALE; ROSE; ROSENGARTEN; TERRELL; TUCKER 1–3.

1. L. Calpouzos, *Econ. Bot.*, **8**, 222 (1954).
2. D. Vokou et al., *Econ. Bot.*, **42**, 407 (1988).
3. M. Hazzit et al., *J. Agric. Food Chem.*, **54**, 6314 (2006).
4. M. R. Rodrigues et al., *J. Agric. Food Chem.*, **52**, 3042 (2004).
5. D. Mockute et al., *Phytochemistry*, **57**, 65 (2001).
6. S. Afsharypour et al., *Planta Med.*, **63**, 179 (1997).
7. U. Asllani, *Bull. Shkencave Nat., Univ. Shteteror Tiranes*, **28**, 61 (1974).
8. H. Maarse and F. H. L. Van Os, *Flav. Ind.*, **4**, 481 (1973).
9. C. H. Brieskorn and H. Brunner, *Planta Med.*, (Suppl.), 96 (1967).
10. E. Goun et al., *Fitoterapia*, **73**, 692 (2002).
11. C. Koukoulitsa et al., *J. Agric. Food Chem.*, **54**, 5388 (2006).
12. V. Lagouri and D. Boskou, *Int. J. Food Sci. Nutr.*, **47**, 493 (1996).
13. W. Olechnowicz-Stepien and E. Lamer-Zarawska, *Herba Pol.*, **21**, 347 (1975).
14. U. Gerhardt and A. Schroetu, *Fleischwirtschaft*, **63**, 1628 (1983).
15. B. M. Lawrence et al., *Phytochemistry*, **13**, 1012 (1974).
16. V. Staikov et al., *Soap, Perfum. Cosmet.*, **41**, 327 (1968).
17. P. Rovesti, *Riv. Ital. Essenze, Profumi, Piante Offic., Aromi, Saponi, Cosmet.*, **54**, 254 (1972).
18. C. A. N. Catalan et al., *Riv. Ital. Essenze, Profumi, Piante Offic., Aromi, Saponi, Cosmet.*, **59**, 513 (1977).
19. M. Leao da Silva et al., *Acta Amazonica*, **3**, 41 (1973).
20. J. A. Retamar et al., *Essenze Deriv. Agrum.*, **45**, 31 (1975).
21. A. Alpande de Morais et al., *An. Acad. Bras. Cienc.*, **44** (Suppl.), 315 (1972).
22. W. H. Stahl et al., *J. Assoc. Anal. Chem.*, **52**, 1184 (1969).
23. C. M. Compadre et al., *Planta Med.*, **53**, 495 (1987).
24. X. A. Dominguez et al., *Planta Med.*, **55**, 208 (1989).
25. V. K. Juneja et al., *J. Food Prot.*, **69**, 1546 (2006).
26. S. Inouye et al., *J. Infect. Chemother.*, **12**, 349 (2006).
27. V. Manohar et al., *Mol. Cell. Biochem.*, **228**, 111 (2001).
28. H. Cetin and A. Yanikoglu, *J. Vector. Ecol.*, **31**, 118 (2006).
29. G. F. Santoro et al., *Parasitol. Res.*, **100**, 783 (2007).
30. M. Ponce-Macotela et al., *Parasitol. Res.*, **98**, 557 (2006).
31. A. Lemhadri et al., *J. Ethnopharmacol.*, **92**, 251 (2004).
32. C. Koukoulitsa et al., *Phytother. Res.*, **20**, 605 (2006).
33. B. M. Walter and G. Bilkei, *Tijdschr. Diergeneeskd.*, **129**, 178 (2004).
34. P. C. Standley, *Trees and Shrubs of Mexico*, Vol. 23, Part 3 of *Contributions from the United States National Herbarium*, Smithsonian Press, Washington, DC, 1923, p. 1242.
35. Monograph, *Origani vulgaris herba*, *Bundesanzeiger*, no. 122 (July 6, 1988).

ORIGANUM OIL, SPANISH

Source: *Thymus capitatus* (L.) Hoffmanns. et Link (syn. *Coridothymus capitatus* Reichb. f.; *Satureja capitata* L.) and carvacrol-rich **Origanum** species (Family Labiatae or Lamiaceae).

Common/vernacular names: Origanum oil.

GENERAL DESCRIPTION

Thymus capitatus is a subshrub with stout, erect branches, and very short, stiff linear leaves; up to about 30 cm high; native to the Mediterranean region. Part used is the flowering top from which Spanish origanum oil is obtained by steam distillation. Spanish origanum oil is characterized by its content of carvacrol, which is its major phenolic component. *Origanum* species that are used for the production of Spanish origanum oil are the ones that yield essential oils with carvacrol as their major phenolic component (see **marjoram**). Major producing countries include Spain, Israel, Lebanon, and Turkey.

CHEMICAL COMPOSITION

Contains 60–75% phenols consisting mostly of carvacrol (13–74%) and thymol (139%),[1–3] plus α- and β-pinene, thujone, camphene, carene, myrcene, terpinene, limonene, α-phellandrene, *cis-O*-ocimene, γ-terpinene, and others.[1–6]

PHARMACOLOGY AND BIOLOGICAL ACTIVITIES

The oil has antimicrobial activities against bacteria and fungi, as well as mosquitocidal activity against *Culex pipiens* adults and larvae.[7–10] Carvacrol has antifungal and anthelmintic activities, though they are weaker than those of thymol (see **thyme**).

TOXICOLOGY

Origanum oil has been reported to be nonirritating, nonsensitizing, and nonphototoxic to human skin but is moderately to severely irritating to rabbit and mouse skin when applied undiluted.[11]

USES

Medicinal, Pharmaceutical, and Cosmetic. Origanum oil is used as a fragrance component in soaps, detergents, creams, lotions, and perfumes, with maximum use level of 0.2% reported in perfumes.[11]

Food. Used as a flavor component in most food products, including alcoholic and nonalcoholic beverages, frozen dairy desserts, candy, baked goods, gelatins and puddings, meat and meat products, condiments and relishes, soups, and gravies. Average maximum use levels reported are usually below 0.004%, except in condiments and relishes, soups, and gravies, which are about 0.007% (74.3 ppm), 0.062%, and 0.01% (99 ppm), respectively.

COMMERCIAL PREPARATIONS

Oil; it is official in F.C.C.

Regulatory Status. Has been approved for food use (§172.510); listed as *Thymus capitatus* (Spanish "origanum").

REFERENCES

See the General References for ARCTANDER; BAILEY 2; FEMA; GUENTHER; MARTINDALE; MERCK; POLUNIN AND SMYTHIES; TERRELL; TUCKER 1–3.

1. D. V. Zaitschek and S. Levontin, *Harokeach Haivri*, **14**, 284 (1971).
2. B. G. Skrubis, *Flav. Ind.*, **3**, 566 (1972).

3. J. M. Hagemann et al., *Lipidis*, **2**, 371 (1967).
4. M. De Vincenzi and M. R. Dessi, *Fitoterapia*, **62**, 39 (1991).
5. B. M. Lawrence, *Perfum. Flav.*, **9**, 41 (1984).
6. B. M. Lawrence, *Perfum. Flav.*, **13**, 69 (1988).
7. V. Manohar et al., *Mol. Cell Biochem.*, **228**, 111 (2001).
8. G. Arras and M. Usai, *J. Food Prot.*, **64**, 1025 (2001).
9. S. A. Mansour et al., *J. Nat. Toxins*, **9**, 49 (2000).
10. O. Kandil et al., *J. Ethnopharmacol.*, **44**, 19 (1994).
11. D. L. J. Opdyke, *Food Cosmet. Toxicol.*, **12**(Suppl.), 945 (1974).

PARSLEY

Source: *Petroselinum crispum* (Mill.) Nym. ex. A. W. Hill (syn. *P. sativum* Hoffm.; *P. hortense* Hoffm.; *Apium petroselinum* L.; *Carum petroselinum* Benth. et Hook. f.) (Family Umbelliferae or Apiaceae).

Common/vernacular names: Common parsley, garden parsley.

GENERAL DESCRIPTION

A nonhairy biennial or short-lived perennial with a much-branched stem; up to about 0.7 m high; often cultivated as an annual for its foliage; native to the Mediterranean region; extensively cultivated (especially in California, Germany, France, Belgium, and Hungary), with numerous varieties. Parts used are the ripe fruits (commonly called seeds), the aboveground herb, and the leaves. Parsley seed oil and parsley herb oil are obtained from the fruits and the aboveground herb, respectively, by steam distillation. The leaves (devoid of stems), after drying, furnish the familiar dehydrated parsley flakes; when fresh, they are also a familiar culinary herb carried in most supermarkets.

The major producers of parsley oils (seed and herb) include France, Hungary, Germany, and the Netherlands. Domestically used dehydrated parsley flakes are mostly produced in California.

Parsley herb oil has a flavor that resembles the fresh herb, but parsley seed oil has a distinctly different flavor.

Parsley oleoresin is obtained from the fruits by solvent extraction (see **glossary**).

CHEMICAL COMPOSITION

Parsley seed contains 2–7% volatile oil;[1] 13–22% fixed oil consisting mainly of petroselinic acid (*cis*-6-octadecenoic acid), with lesser amounts of palmitic, myristic, stearic, oleic, linoleic, and myristolic acids as well as 7-octadecenoic acid;[2–5] flavonoids (apiin (apigenin-7-apiosylglucoside) and luteolin-7-apiosylglucoside);[6] and traces of bergapten (LIST AND HÖRHAMMER).

Parsley seed oil contains mainly apiole, myristicin, tetramethoxyallybenzene, and α-pinene.[1,7–9] It also contains petroselinic acid and other volatile fatty acids (see also **celery seed oil**).[4]

Parsley oleoresin contains mainly apiole (CLAUS).

Parsley herb (leaf) contains 0.05–0.3% volatile oil;[1] furocoumarins, consisting mostly of bergapten (up to 0.02%) with smaller amounts of xanthotoxin (up to 0.003%) and isopimpinellin;[10,11] flavonoids (apiin, luteolin-7-apiosylglucoside, apigenin7-glucoside, and luteolin-7-diglucoside); 2–22% protein; fats (ca. 4%); vitamins (especially A and C); sugars; and others (LIST AND HÖRHAMMER; MARSH).[12] Oxypeucedanin has been reported as a major furocoumarin (though absent in some varieties).[13]

Parsley herb (leaf) oil contains myristicin (up to 85% in the curly moss variety), β-phellandrene, 1,3,8-*p*-menthatriene, myrcene, apiole, terpinolene, and 1-methyl-4-isopropenylbenzene as major components.[1,8,9,14] Other compounds present include α- and β-pinenes, *trans*-β-ocimene, γ-terpinene, methyl disulfide, α-terpineol, α-copaene, caryophyllene, and carotol, among others.[14–16] An assessment of 104 accessions found leaf oil content to range from 0.00% to 0.16% (fresh weight). Major constituents were 1,3,8-*p*-menthatriene (68%), myristicin (60%), β-phellandrene (33%), apiol (22%), myrcene (16%), plus terpinolene and 1-methyl-4-isopropenylbenzene at 13%. Thymol, first reported for the leaf oil, was found at 2% or less in seven samples. More than 45 components have been reported from parsley leaf oil.[17]

1,3,8-*p*-Menthatriene is reported to be one of the compounds that contribute to the aroma of parsley.[14]

β-Elemene, myristicin, and 1,3,8-*p*-menthatriene have been found to accumulate in detectable amounts in young undifferentiated cultures of *P. crispum* cv. "Paramount."[18]

PHARMACOLOGY AND BIOLOGICAL ACTIVITIES

Parsley (herb, seed, extracts, and oils) has been reported to have numerous pharmacological and biological properties, including laxative, hypotensive, antimicrobial, and tonic (on uterine muscles).[19–23] Recent reviews list parsley among other common vegetables that may be useful for reducing bone resorption and maintaining bone health as well as for cancer chemoprevention.[24–26] Polyacetylenes present in parsley and other members of the Apiaceae have been reported to posses antitumor, antimicrobial, and anti-inflammatory activities in addition to some undesirable activities such as neurotoxicity and allergic skin reactions.[27]

Myristicin is reported to have psychedelic activities as well as toxic properties (see **nutmeg**), and apiole is reported to have antipyretic properties (MERCK).

Furocoumarins are phototoxic and may cause skin inflammations or contact dermatitis (see **bergamot** and **lime**).[11]

Apiole is reported to be a spasmolytic, vasodilator, and emmenagogue.

TOXICOLOGY

Parsley seed oil, due to high concentrations of apiole, may cause vascular congestion and increase smooth muscle contractibility in the bladder, intestine, and uterus; damage to kidney epithelia and heart arrhythmia reported; may cause fatty liver, emaciation, and bleeding of mucous membranes of the gastrointestinal tract.[28]

USES

Medicinal, Pharmaceutical, and Cosmetic. In German phytomedicine the cut or ground herb is used; daily doses of 6.0 g in infusions or other galenic preparations for systemic irrigation for ailments of the lower urinary tract; also irrigation therapy for the prevention of renal gravel.[29] In France, topical preparations are used for dermatological conditions, such as cracks, grazes, chapped skin, and insect bites (BRADLY).

Parsley seed oil is used as a fragrance component in soaps, detergents, creams, lotions, and perfumes (especially Oriental types, men's fragrances and colognes), with maximum use level of 0.2% reported in perfumes.[19]

Food. Parsley (both fresh and dehydrated) is widely used in home cooking. Chinese parsley is not a variety of parsley but is the young leaf of coriander and has a chemical composition and aroma quite different from parsley (see **coriander**).

Parsley, parsley oils (herb and seed), and oleoresin are extensively used in flavoring meat sauces, sausages, canned meats, pickles, and spice blends.

Parsley is also used in baked goods, fats and oils, processed vegetables, soups, snack foods, gravies, and others. Highest average maximum use level reported is about 1.5% (14,963 ppm) in processed vegetables.

Other food products in which parsley oils and oleoresin are used include alcoholic and nonalcoholic (e.g., soft drinks) beverages, frozen dairy desserts, candy, baked goods, gelatins and puddings, and soups, among others. Highest average maximum use level reported is about 0.039% (392 ppm) for the oleoresin in condiments and relishes (e.g., pickles).

Dietary Supplements/Health Foods. Parsley herb or root sometimes used as a tea ingredient; also in diuretic formulations (FOSTER).

Traditional Medicine. Parsley herb and seed are reportedly used to treat jaundice, menstrual difficulties, asthma, coughs, indigestion, and dropsy, usually in the form of a tea. Parsley herb tea is used to treat gallstones, dyspepsia, dysuria, and rheumatic conditions. In combination with garlic and rue, the leaves are used as a vulnerary in Italy.[30] The herb is also used as a breath freshener when chewed (LUST; ROSENGARTEN). Parsley seed traditionally used as an abortifacient.

COMMERCIAL PREPARATIONS

Flakes, seed, oils (herb and seed), and extracts (e.g., oleoresin). Seed and oleoresin were formerly official in U.S.P. Both oils are official in F.C.C.

Regulatory Status. GRAS (§182.10 and §182.20) with botanical source listed as *Petroselinum crispum* (Mill.) Mansf. Parsley herb is the subject of a positive German therapeutic monograph. Use is contraindicated in pregnancy and inflammatory kidney diseases.[29] Parsley fruits are covered by a German monograph; however, use is not recommended because efficacy is not well documented and potential risks outweigh benefits.[28]

REFERENCES

See the General References for ARCTANDER; BAILEY 1; BARNES; BLUMENTHAL 1; BRADLY; DER MARDEROSIAN AND BEUTLER; FEMA; FOSTER; GUENTHER; MCGUFFIN 1 & 2; ROSENGARTEN; TERRELL.

1. C. Franz and H. Glasl, *Ind. Obst. Gemueseverwert*, **59**, 176 (1974).
2. S. I. Balbaa et al., *Egypt J. Pharm. Sci.*, **16**, 383 (1976).
3. E. Constantinescu et al., *Riv. Ital. Essenze, Profumi, Piante Offic., Aromi, Saponi, Cosmet., Aerosol*, **54**, 419 (1972).
4. A. R. S. Kartha and R. A. Khan, *Chem. Ind. (London)*, (52) 1869 (1969).
5. O. S. Privett et al., *J. Am. Oil Chem. Soc.*, **40**, 28 (1963).
6. J. B. Harborne and C. A. Williams, *Phytochemistry*, **11**, 1741 (1972).
7. H. Wagner and J. Hölzl, *Dtsch. Apoth. Ztg.*, **108**, 1620 (1968).
8. J. B. Harborne et al., *Phytochemistry*, **8**, 1729 (1969).
9. C. Franz and H. Glasl, *Qual. Plant. Plant Foods Hum. Nutr.*, **25**, 253 (1976).
10. G. Innocenti et al., *Planta Med.*, **29**, 165 (1976).
11. J. Palicska and B. Lengyel, *Borgyogy Venerol. Szemle*, **45**, 118 (1969).
12. R. A. Komarova, *Sb. Tr. Aspir. Molodykh Nauch. Sotrudnikov, Vses. Nauch. Issled Inst. Rastenievod.*, **8**, 276 (1967).
13. S. K. Chaudhary et al., *Planta Med.*, **52**, 462 (1986).
14. R. Kasting et al., *Phytochemistry*, **11**, 2277 (1972).
15. G. G. Freeman et al., *J. Sci. Food Agric.*, **26**, 465 (1975).
16. J. Garnero and Y. Chretien-Bessiere, *Fr. Ses Parfums*, **11**, 332 (1968).
17. J. E. Simon and J. Quinn, *J. Agric. Food Chem.*, **36**, 467 (1988).
18. A. A. Gdolade and G. B. Lockwood, *Fitoterapia*, **62**, 237 (1991).
19. D. L. J. Opdyke, *Food Cosmet. Toxicol.*, **13** (Suppl.) 897 (1975).
20. F. Kaczmarek, et al., *Biul. Inst. Roslin Leczniczych*, **8**, 111 (1962).
21. I. Tsonev et al., *Farmatsiya (Sofia)*, **17**, 39 (1967).
22. J. Kresanek and J. Vittek, *Farm. Obzor.*, **31**, 202 (1962).
23. A. Sharaf et al., *Qual. Plant. Mater. Veg.*, **17**, 337 (1969).
24. S. E. Putnam et al., *Phytother. Res.*, (2006).
25. S. Ren and E. J. Lien, *Prog. Drug Res.*, **48**, 147 (1997).

26. J. D. Potter and K.Steinmetz, *IARC Sci. Pub.*, **61** (1996).
27. L. P. Christensen and K. Brandt, *J. Pharm. Biomed. Anal.*, **41**, 683 (2006).
28. Monograph, *Petroselini fructus*, *Bundesanzeiger*, no. 43 (March 2, 1989).
29. Monograph, *Petroselini herba/radix*, *Bundesanzeiger*, no. 43 (March 2, 1989).
30. V. De Feo et al., *Fitoterapia*, **63**, 337 (1992).

PASSION FLOWER

Source: *Passiflora incarnata* L. (Family Passifloraceae).

Common/vernacular names: Maypop, maypop passion flower, passiflora, apricot vine, wild passion flower, and passion vine.

GENERAL DESCRIPTION

A perennial vine with deeply three- to five-lobed leaves and large showy flowers; climbing by axillary tendrils to about 9 m; native to the United States; distributed in southern states from Virginia to Florida and west to Texas; commercial cultivation in Florida and Guatemala. Part used is the dried flowering and fruiting top (FOSTER).

CHEMICAL COMPOSITION

Contains small and highly variable amounts (<0.01–0.09%) of indole alkaloids, consisting mainly of harman, with lesser amounts of harmol, harmaline, harmine, and harmalol.[1–6] Presence of the last four alkaloids has been disputed,[1,7] but their presence was later confirmed at a total concentration of 0.7 ppm for all five alkaloids.[8]

Other constituents present include flavonoids (isovitexin 2″-β-D-glucoside, isoorientin 2″-β-D-glucoside, apigenin, luteolin, quercetin, kaempferol, schaftoside, isoschaftoside, saponaretin, saponarin, vitexin, orientin, and rutin);[9–15] a cyanogenic glucoside, gynocardin (0.01%);[16,17] sugars (raffinose and sucrose predominant);[18] sterols (stigmasterol and sitosterol); *n*-nonacosane, and gum, among others (LIST AND HÖRHAMMER).

Maltol and ethyl maltol have been isolated from the plant.[19] The coumarins umbelliferone and scopoletin have been detected in the root.[20] Passion flower has recently been identified as a rich source of lycopene.[21]

PHARMACOLOGY AND BIOLOGICAL ACTIVITIES

Passion flower has been reported to have analgesic and sedative properties. It depresses motor activity, increases rate of respiration, and produces a transient reduction in blood pressure.

Antispasmodic, antiasthmatic, antitussive, anxiolytic, and hypotensive activities have been reported;[22–26] a synergistic association of sedative activity at high dosage and anxiolytic activity at low dosage, have been observed.[26] Animal studies have repeatedly reported a depression of motor reflexes (ESCOP 2).[27–29] The sedative and anxiolytic effects have also been demonstrated in humans.[30]

Oral administration and peritoneal injections in rats decreased brain stimulus in a number of pharmacological models; the latter significantly prolonged sleep time and produced protective anticonvulsive activity; locomotor activity was also affected. Active components ascribed to both hydrophilic and lipophilic fractions; however, activity cannot be attributed to specific alkaloids or flavonoids; neuropharmacological activity cannot be clearly attributed to a single chemical fraction. Fractions of a fluid extract have been shown to raise the nociceptive threshold of rats in the tail-flick and hot-plate tests and to prolong sleeping time as well as protect the animals from the convulsive effect of pentylenetetrazole. There are claims that the active compounds have not yet been characterized but are neither alkaloids nor flavonoids.[31] However, one research group emphasized the role of a new "trisubstituted benzoflavone" in the CNS effects of passion flower.[32] Another group earlier suggested that the flavonoids, alkaloids, and maltol were involved in the observed behavioral effects.[33] Also, the harman alkaloids and flavonoids were earlier reported to have tranquilizing effects.[34]

The efficacy of passion flower in the reduction of withdrawal symptoms associated with benzodiazepines, opiates, marijuana, alcohol, and nicotine dependence has been demonstrated.[35–39]

Passion flower was shown to be effective as an aphrodisiac, to enhance libido and virility and to prevent azospermia in mice and rats. These effects were most pronounced with reduced sexual function associated with delta-9-tetrahydrocannabinol (Δ^9-THC), alcohol, nicotine, and aging.[36,40–42]

TOXICOLOGY

Although it appears to be nontoxic (GOSSELIN; MERCK), one case possibly associated with passion flower has been reported for a patient who developed severe nausea, vomiting, drowsiness, and cardiovascular complications.[43]

USES

Medicinal, Pharmaceutical, and Cosmetic. Fresh or dried herb and extracts are used as ingredients in some sedative preparations for nervous anxiety. Preparations include tisanes (tea), tinctures, fluid extracts, solid extracts, and sedative chewing gums available in Europe; used in European proprietary tranquilizing phytomedicines in combination with valerian and hawthorn. Passionflower and hawthorn combinations are employed as antispasmodics for digestive spasms, such as in gastritis and colitis.[44]

A European monograph indicated use for nervous tension, especially in cases of sleep disturbance or exaggerated awareness of heart palpitations at doses ranging from 0.5 to 2 g of the herb and 2.5 g of the herb in infusion, and so on (ESCOP 2).

Food. Extract (type not specified) of passion flower is reported to be used as a flavor component in alcoholic beverages; nonalcoholic beverages, and frozen dairy desserts with average maximum use levels of 0.05%, 0.32%, and 0.05%, respectively.

Dietary Supplements/Health Foods. Herb in capsules, tablets, teas, and other product forms; primarily in sleep aid formulations (FOSTER).

Traditional Medicine. Used as a sedative in treating neuralgia, insomnia, restlessness, headache, hysteria, epilepsy, and other nervous conditions. Also used in bath mixtures for its allegedly calming and soothing effects.

Fruits were cultivated or managed for fruit production before arrival of Europeans in Algonkian settlements in Verginia.[45]

COMMERCIAL PREPARATIONS

Crude and extracts. Crude was formerly official in N.F. Strengths (see *glossary*) of extracts are expressed in weight-to-weight ratios. European monograph specifications comply with French, German, and Swiss pharmacopoeias, with crude drug to contain not less than 0.8% total flavonoids, calculated as vitexin. A German monograph specifies harmala alkaloids not to exceed 0.01%. Other *Passiflora* species are often distributed in commerce as passion flower (*P. incarnata*).

Regulatory Status. Has been approved for food use (§172.510). The dried flowering and fruiting tops of *P. incarnata* in the fourth (1916) and fifth (1926) N.F.; dropped in sixth

N.F. (1936): formerly an approved OTC sedative and sleep aid; removed in 1978. Herb subject of a positive German therapeutic monograph for the treatment of nervous anxiety, in daily dosages equivalent to 4–8 g.[46]

REFERENCES

See the General References for APPLEQUIST; BAILEY 1; BARNES; BLUMENTHAL 1; BRADLY; CLAUS; ESCOP 2; FEMA; FERNALD; FOSTER; GOSSELIN; GRIEVE; KROCHMAL AND KROCHMAL; LEWIS AND ELVIN-LEWIS; LIST AND HÖRHAMMER; LUST; ROSE; TERRELL; UPHOF.

1. W. Poethke et al., *Planta Med.*, **18**, 303 (1970).
2. E. Bennati and E. Fedeli, *Boll. Chim. Farm.*, **107**, 716 (1968).
3. E. Bernnati, *Boll. Chim. Farm.*, **110**, 664 (1971).
4. J. Lutomski et al., *Herbal Pol.*, **14**, 139 (1968).
5. N. Svanidze et al., *Rev. Cubana Farm.*, **8**, 309 (1974).
6. J. Lutomski and B. Malek, *Planta Med.*, **27**, 381 (1975).
7. J. Loehdefink and H. Kating, *Planta Med.*, **25**, 101 (1974).
8. E. A. Abourashed et al., *Pharm. Biol.*, **41**, 100 (2003).
9. E. A. Abourashed et al., *Pharm. Biol.*, **40**, 81 (2002).
10. N. M. Gavasheli et al., *Khim. Prir. Soedin.*, **10**, 95 (1974).
11. H. Schilcher, *Z. Naturforsch. B*, **23**, 1393 (1968).
12. N. M. Gavasheli, *Soobhch. Akad. Nauk Gruz. SSR*, **60**, 353 (1970).
13. L. Qimin et al., *J. Chromatogr.*, **562**, 435 (1991).
14. H. Geiger and K. R. Markham, *Z. Naturforsch.*, **41c** 949 (1986).
15. C. Congora, et al., *Helv. Chim. Acta*, **69**, 251 (1986).
16. K. C. Spencer and D. S. Seigler, *Planta Med.*, **50**, 356 (1984).
17. K. C. Spencer and D. S. Seigler, *Phytochemistry*, **24**, 2615 (1985).
18. N. M. Gavasheli et al., *Khim. Prir. Soedin.*, **11**, 84 (1975).
19. N. Aoyagi et al., *Chem. Pharm. Bull.*, **22**, 1008 (1974).
20. N. M. Gavasheli et al., *Khim. Prir. Soedin.*, **9**, 552 (1973).
21. E. Mourvaki et al., *J. Med. Food*, **8**, 104 (2005).
22. K. Dhawan and A. Sharma, *Fitoterapia*, **73**, 397 (2002).
23. K. Dhawan et al., *Phytother. Res.*, **17**, 821 (2003).
24. B. R. Olin, ed., *Lawrence Rev. Nat. Prod.*, May 1989.
25. R. Kimura et al., *Chem. Pharm. Bull.*, **28**, 2570 (1980).
26. R. Della Loggia et al., *Rivista di Neurologia*, **51**, 297 (1981).
27. K. Dhawan et al., *J. Altern. Complement. Med.*, **8**, 283 (2002).
28. K. Dhawan et al., *Fitoterapia*, **72**, 922 (2001).
29. K. Dhawan et al., *J. Ethnopharmacol.*, **78**, 165 (2001).
30. S. Akhondzadeh et al., *J. Clin. Pharm. Ther.*, **26**, 363 (2001).
31. E. Speroni and A. Minghetti, *Planta Med.*, **54**, 488 (1988).
32. K. Dhawan et al., *J. Ethnopharmacol.* **94**, 1 (2004).
33. R. Soulimani et al., *J. Ethnopharmacol.*, **57**, 11 (1997).
34. J. Lutomski et al., *Planta Med.*, **27**, 112 (1975).

35. K. Dhawan et al., *J. Pharm. Pharm. Sci.*, **6**, 215 (2003).
36. K. Dhawan and A. Sharma, *Br. J. Pharmacol.*, **138**, 117 (2003).
37. K. Dhawan et al., *J. Ethnopharmacol.*, **81**, 239 (2002).
38. K. Dhawan et al., *J. Pharm. Pharmacol.*, **54**, 875 (2002).
39. K. Dhawan et al., *Addict. Biol.*, **7**, 435 (2002).
40. K. Dhawan et al., *Phytother. Res.*, **17**, 401 (2003).
41. K. Dhawan et al., *J. Med. Food*, **5**, 43 (2002).
42. K. Dhawan and A. Sharma, *Life Sci.*, **71**, 3059 (2002).
43. A. A. Fisher et al., *J. Toxicol. Clin. Toxicol.*, **38**, 63 (2000).
44. Brasseur and L. Angenot, *J. Pharm. Belg.* **38**, 15 (1984).
45. K. J. Gremillion, *J. Ethnobiol.*, **9**, 135 (1989).
46. Monograph, *Passiflorae herba*, *Bundesanzeiger*, no. 223 (November 30, 1985).

PATCHOULY OIL

Source: *Pogostemon cablin* (Blanco) Benth. (syn. *P. heyneanus* Benth.; *P. patchouly* Pellet.) (Family Labiatae or Lamiaceae).

Common/vernacular names: Patchouli oil.

GENERAL DESCRIPTION

A perennial herb with a sturdy, hairy stem, much branched at the top; leaves opposite, with fragrant odor when rubbed; up to about 1 m high; native to tropical Asia (especially Indonesia and the Philippines) and extensively cultivated in the tropics (Indonesia, Philippines, Malaysia, India, southern China, Seychelles, Brazil, etc.). Parts used are the dried leaves from which patchouly oil is obtained by steam distillation. The leaves are usually subjected to some sort of fermentation or curing process before distillation so as to increase oil yield. Major oil-producing countries include Malaysia, Indonesia, the Seychelles, China, India, and the Philippines. The oil is also distilled in Europe and the United States.

CHEMICAL COMPOSITION

The leaves contain 1.5–4% volatile oil that is composed mainly of patchouly alcohol (ca. 32–40%) and other sesquiterpenes such as pogostol, bulnesol, norpatchoulenol (ca. 2.2%), α-guaiene, α-bulnesene, and β-patchoulene.[1–6] Other compounds found in the oil include cycloseychellene (a tetracyclic sesquiterpene);[1,7,8] patchoulipyridine, epiguaipyridine, and guaipyridine (sesquiterpene alkaloids);[9,10] eugenol, cinnamaldehyde, and benzaldehyde;[11] pogostone or dhelwangine (a lactone);[12] and oxygenated sesquiterpenes (e.g., 1α,5α-epoxy-α-guaiene and epoxycaryophyllene);[13] among others (LIST AND HÖRHAMMER).[14–16]

Patchouly alcohol and norpatchoulenol are mostly responsible for the odor of patchouly oil.[5,13]

Flavonoids (ombuine, pachypodol, kumatakenin, and 5,7-dihydroxy-3′,4′-diemthoxyflavanone) and flavonoid glycosides (di- and trimethyleriodictyol, and trimethylkaempferol) have been isolated, in addition to licochalcone A, from the aerial parts of *P. cablin*.[17,18]

PHARMACOLOGY AND BIOLOGICAL ACTIVITIES

Dhelwangine (pogostone) is reported to have antimicrobial activities (fungi and bacteria) and is responsible for the bactericidal properties of patchouly oil (LIST AND HÖRHAMMER).[8] The essential oil, patchoulyl alcohol, and pogostone are also active against five species of periodontopathic bacteria.[19]

Eugenol, cinnamaldehyde, and benzaldehyde isolated from *P. heyneanus* reportedly have insecticidal activity against insects in stored grain.[11] The same effect was displayed by the essential oil against *Sopdoptera littoralis* larvae when applied by fumigation.[20] The undiluted essential oil provided 2 h of complete repellency and protection from *Aedes aegypti* mosquito bites when applied to the skin of human volunteers.[21]

Sesquiterpene hydroperoxides isolated from the acetone extract possessed *in vitro* trypanocidal activity against epimastigotes of *Trypanosoma cruzi* at MLC 0.84–1.7 µM.[22]

α-Bulnesene was found to inhibit platelet aggregation through an antagonistic concentration-dependent *in vitro* binding to the PAF receptor and inhibition of arachidonic acid-induced platelet aggregation in rabbit.[23]

Constituents of the *n*-hexane extract (patchouli alcohol, pogostol, stigmast-4-en-3-one, retusin, and pachypodol) displayed antiemetic activities in an emesis model in chicks.[24]

Lichochalcone A, isolated via cytotoxicity-guided fractionation, displayed *in vitro* antitumor activity against HL-60 cells.[18]

TOXICOLOGY

Results from one short-term feeding study indicate patchouly oil to be nontoxic to rats.[25] The flavonoids isolated via bioassay-guided fractionation from patchouli displayed *in vitro* antimutagenic activity against furylfuramide and Trp-P-1 in the Ames test at less than 1 µM levels.[17]

USES

Medicinal, Pharmaceutical, and Cosmetic. Patchouly oil is extensively used as a fragrance component in cosmetic preparations. It is one of the most used ingredients in perfumes (especially Oriental types) and is also widely used in soaps and in depilatory creams (to mask the undesirable odor of the active hair-removal ingredients).

Patchouly resinoid (concrete) is used as a fixative.

Food. Patchouly oil is extensively used as a flavor ingredient in most major food products, including alcoholic and nonalcoholic beverages, frozen dairy desserts, candy, baked goods, gelatins and puddings, and meat and meat products. Use levels reported are generally very low, mostly below 0.0002% (2.21 ppm).

Dietary Supplements/Health Foods. Patchouly oil, widely available in health food stores, primarily used topically as a fragrance (ROSE).

Traditional Medicine. The herb is used in Chinese medicine to treat colds, headaches, nausea, vomiting, diarrhea, and abdominal pain, usually decocted with other drugs (JIANGSU).

It is also used to treat bad breath either used alone as a decoction for gargling or combined with cardamom and another herb (*Eupatorium japonicum* Thunb.), the latter especially for bad breath associated with drinking of alcohol.

The leaf, fruit, and flower of *P. heyneanus* have been used in tumors.[26]

COMMERCIAL PREPARATIONS

Mainly the oil. A number of adulterants have been reported in commercial supplies, including gurjun balsam oil, copaiba balsam oil, cedarwood oil, and others.[27]

Regulatory Status. Has been approved for food use (§172.510).

REFERENCES

See the General References for ARCTANDER; BAILEY 2; FEMA; GUENTHER; HONGKUI; JIXIAN; JIANGSU; MCGUFFIN 1 & 2; NANJING; TERRELL; UPHOF.

1. Y. C. Tsai et al., *Fitoterapia*, (2006).
2. W. D. Fordham, L. W. Codd, et al., *Chemical Technology: An Encyclopedic Treatment*, Barnes and Noble, New York, 1972, p. 1.
3. P. Teisseire et al., *Recherches*, **19**, 8 (1974).
4. H. Hikino et al., *Chem Pharm. Bull.*, **16**, 1608 (1968).
5. F. W. Hefendehl, *Seifen, Ole, Fette, Wachse*, **103**, 159 (1977).
6. A. Akhila and M. C. Nigam, *Fitoterapia*, **55**, 363 (1984).
7. V. V. Dhekne and S. K. Paknikar, *Indian J. Chem.*, **12**, 1016 (1974).
8. S. J. Terhune et al., *Int. Congr. Essent. Oils (Pap)*, **6**, 153 (1974).
9. G. Beuchi et al., *J. Am. Chem. Soc.*, **88**, 3109 (1966).
10. Van der Gen et al., *Recl. Trav. Chim. Pays Bas*, **91**, 1433 (1972).
11. R. S. Deshpande et al., *Bull. Grian Technol.*, **12**, 232 (1974).
12. *K'o Hsueh T'ung Pao*, **22**, 318 (1977).
13. P. Teisseire, *Riv. Ital. Essenze, Profumi, Piante Offic., Aromi, Saponi, Cosmet., Aerosol*, **55**, 572 (1973).
14. S. Nakahara et al., *Phytochemistry*, **2712**, (1975).
15. B. M. Lawrence, *Perfum. Flav.*, **6**, 73 (1981).
16. B. M. Lawrence, *Perfum. Flav.*, **15**, 75 (1990).
17. M. Miyazawa et al., *J. Agric. Food Chem.*, **48**, 642 (2000).
18. E. J. Park et al., *Planta Med.*, **64**, 464 (1998).
19. K. Osawa et al., *Bull. Tokyo Dent. Coll.*, **31**, 17 (1990).
20. R. Pavela, *Fitoterapia*, **76**, 691 (2005).
21. Y. Trongtokit et al., *Phytother. Res.*, **19**, 303 (2005).
22. F. Kiuchi et al., *Chem. Pharm. Bull. (Tokyo)*, **52**, 1495 (2004).
23. H. C. Hsu et al., *Biochem. Biophys. Res. Commun.*, **345**, 1033 (2006).
24. Y. Yang et al., *Phytomedicine*, **6**, 89 (1999).
25. B. L. Oser et al., *Food Cosmet. Toxicol.*, **3**, 563 (1965).
26. J. L. Hartwell, *Lloydia*, **32**, 247 (1969).
27. A. Akhila and R. Tewari, *Curr. Res. Aromat. Plants*, **6**, 38 (1984).

PECTIN

Source: *Pectin* Primarily from Lemon peel *Citrus limon* (L.) Burm. f. (syn. *C. limonum* Risso) (Family Rutaceae).

GENERAL DESCRIPTION

Pectin or pectic substances are complex polysaccharides universally present in the cell walls of plants, especially in the spaces between cell walls (middle lamellae) where they act as binders to hold adjacent cell walls together. The peel (especially albedo, the white portion) of citrus fruits (especially lemon, lime, orange, and grapefruit) and apple pomace is rich in pectin. Other pectin-rich plant materials are sugar beet pulp, sunflower heads, carrot, potato, and tomato.[1–5]

Commercial pectin is isolated mainly from lemon peel and to a lesser degree from apple pomace. Lemon peel contains 2–4% pectin when fresh and 20–40% when dried, while dried apple pomace contains 10–20% pectin. The manufacturing process generally involves extraction of the raw materials with water containing a mineral acid (e.g., nitric acid, hydrochloric acid, or sulfur dioxide) at pH 1.5–3.0 and at an elevated temperature of 60–100°C, centrifugation, and filtration. The filtrate can be concentrated and spray dried or roller dried to yield a low-grade pectin. However, most often the filtrate is treated either with alcohols (especially isopropyl) or with soluble salts of aluminum or, less often, copper to precipitate the pectin, which is then washed, dried, and milled (WHISTLER AND BEMILLER).[6]

Two types of pectins are produced: low-ester (low-methoxy) and high-ester (high-methoxy) pectins. To obtain low-ester pectin, extended hydrolysis or de-esterification is allowed to take place some time (step) during the manufacturing process. Pectins are usually standardized to uniform grades: "100 gel power" for low-ester pectin and "150 jelly grade" for high-ester pectin. Diluents used are dextrose or other sugars. The "150 jelly grade" or "150 grade" means 1 part of the pectin will set 100 parts of sugar in solution to a jelly of standard strength and firmness containing 65% of sugar (GLICKSMAN; MARTINDALE; USD 26th).

The major pectin producer is the United States (especially California). Other producing countries include the UK, France, and Germany.

Pectin dissolves in water to form a viscous sol with an acidic pH. Its solutions (sols) are most stable at a pH range of 3–4. Outside this range, the viscosity and gel strength of these solutions decrease due to decomposition of pectin. They are especially unstable under alkaline conditions. Pectin is also incompatible with heavy metals, tannin, and salicylic acid. It is insoluble in alcohol and other organic solvents.

The gelling time of pectin varies with type and ranges from 20–70 s (rapid set) to 180–250 s (slow set). The gel strength and gelling time are the two most important factors determining the quality of pectin (GLICKSMAN; WHISTLER AND BEMILLER).

CHEMICAL COMPOSITION

In addition to pectin itself, commercial pectin normally contains sugar (e.g., dextrose) and sodium citrate or other buffer salts (sodium and potassium carbonates, lactates, etc.).

The molecular weight of pectin has been reported to range from 150,000 to 400,000. It is a complex polysaccharide, with D-galacturonic acid as its major sugar component. Other sugars present include D-galactose, L-arabinose, and L-rhamnose. Its molecule consists mainly of a linear partially methylated galacturonoglycan of $\alpha(1 \rightarrow 4)$-linked D-galactopyranosyluronic acid. Pectins from different sources may vary widely in their methoxy content (0.2–12%) and in their relative proportions of sugars (GLICKSMAN; MERCK; WHISTLER AND BEMILLER).[1]

PHARMACOLOGY AND BIOLOGICAL ACTIVITIES

Pectin does not appear to be digested by animals or humans. However, on passage through their gastrointestinal tracts, much of the pectin is degraded by bacteria present in their large intestine.[4,7] There is evidence indicating that a small amount of the products of degradation is absorbed, resulting in a hemostatic effect manifested in a shorter coagulation time of drawn blood samples (WHISTLER AND BEMILLER).

As with other gums (see ***algin***, ***guar***, etc.), the cholesterol-lowering property of pectin has been extensively studied. Many studies have demonstrated pectin to have antihypercholesterolemic activities in laboratory animals (rats, fowl, dogs, etc.) and in humans, with the high molecular weight and high-methoxy pectin being the most active.[4,7–16] Protopectin and low-methoxy pectin however do not appear to be active.[4] One study on humans did not produce antihypercholesterolemic effects in patients with hyperlipoproteinemia.[17]

Pectin (from apple) has been reported to exhibit bactericidal effects on Gram-negative and nonspore-forming Gram-positive bacteria, including *Salmonella typhi* and *Escherichia coli* from the former group and *Staphylococcus aureus* from the latter. It had little or no activity on spore-forming bacilli, yeast, and fungi.[18,19]

Other biological and pharmacological properties of pectin include increasing the excretion of heavy metals (e.g., lead and mercury) in experimental animals (see also *algin*),[20,21] lowering fat absorption in rats,[15] and others (MARTINDALE).[4]

USES

Medicinal, Pharmaceutical, and Cosmetic. Pectin (150 grade) is used as an ingredient in numerous antidiarrheal preparations, often in combination with kaolin. It is used as an emulsifier and/or thickener in creams and lotions. It is currently being considered as a promising excipient in oral formulations intended for site-specific drug delivery to the colon.[22]

Food. By far the largest use of pectin is in jams, jellies, and preserves, with low-methoxy pectin being primarily used in low-sugar or sugar-free products.

Other food products in which pectin (both or either type) is used include candy, frozen dairy desserts, bakers' jellies, and nonalcoholic beverages (GLICKSMAN; WHISTLER AND BEMILLER).

Pectin is also used in edible films with antimicrobial activities that are intended to enhance safety, reduce spoilage, and extend the shelf life of ready-to-eat foods.[23]

The chemistry and uses of pectin have been the subject of a comprehensive review by Thakur et al.[24]

COMMERCIAL PREPARATIONS

Several types, including 100 and 150 grades. Pectin (150 grade) is official in U.S.P. and both 100 and 150 grades are official in F.C.C.

Regulatory Status. GRAS (§182.1775).

REFERENCES

See the General References for APhA; DER MARDEROSIAN AND BEUTLER; DER MARDEROSIAN AND LIBERTI; GLICKSMAN; LAWRENCE; MARTINDALE; USD 26th; WHISTLER AND BEMILLER.

1. R. L. Whistler in H. W. Schulz et al., eds., *Symposium on Foods: Carbohydrates and Their Roles*, AVI, Westport, CT, 1969, p. 73.
2. R. M. McCready and H. S. Owens, *Econ. Bot.*, **8**, 29 (1954).
3. J. W. Kesterson and R. Hendrickson, *Econ. Bot.*, **12**, 164 (1958).
4. W. L. Chenoweth and G. A. Leveille, *ACS Symp. Ser.*, **15**, 312 (1975).
5. D. B. Nelson et al., *Food Colloids*, AVI, Westport, CT, 1977, p. 418.
6. A. H. Rouse and P. G. Crandall, *J. Food Sci.*, **43**, 72 (1978).
7. T. A. Miettinen and S. Tarpila, *Clin. Chim. Acta*, **79**, 471 (1977).
8. P. A. Judd et al., *Nutr. Metab.*, **21**(Suppl. 1) 84 (1977).
9. S. Kiriyama et al., *J. Nutr.*, **97**, 382 (1969).
10. D. J. A. Jenkins et al., *Ann. Int. Med.*, **86**, 20 (1977).
11. T. A. Anderson and R. D. Bowman, *Proc. Soc. Exp. Biol. Med.*, **130**, 665 (1969).
12. H. Fisher et al., *J. Atheroscler. Res.*, **6**, 292 (1966).
13. H. Fisher et al., *Science*, **146**, 1063 (1964).
14. K. Tsuji et al., *Eiyogaku Zasshi*, **26**, 113 (1968).
15. M. L. W. Chang and M. A. Johnson, *J. Nutr.*, **106**, (1976).
16. D. Mathe et al., *J. Nutr.*, **107**, 466 (1977).
17. F. Delbarre et al., *Am. J. Clin. Nutr.*, **30**, 463 (1977).

18. M. A. El-Nakeeb and R. T. Yousef, *Planta Med.*, **18**, 201 (1970).
19. M. A. El-Nakeeb and R. T. Yousef, *Planta Med.*, **18**, 295 (1970).
20. O. D. Livshits, *Vopr. Pitan*, **28**, 76 (1969).
21. O. G. Arkhipova and L. A. Zorina, *Prof. Zabolevaniya u Khim. Prom*, **210** (1965).
22. S. A. Sande, *Expert Opin. Drug Deliv.*, **2**, 441 (2005).
23. A. Cagri et al., *J. Food Prot.*, **67**, 833 (2004).
24. B. R. Thakur et al., *Crit Rev. Food Sci. Nutr.*, **37**, 47 (1997).

PEPPER (BLACK AND WHITE)

Source: *Piper nigrum* L. (Family Piperaceae).

GENERAL DESCRIPTION

A perennial woody vine with many nodes, climbing to about 5 m, stem terete; leaves coriaceous, broadly ovate, acuminate, glabrous, 5–9 nerved with rounded base; flowers in spikes; fruits globose 6 mm in diameter, yellowish and then becoming black; native to southwestern India and widely cultivated in tropical countries.[1]

Black pepper is the dried full-grown but unripe fruit.

White pepper is the dried ripe fruit with the outer part of the pericarp removed by soaking in water, followed by rubbing. It is less aromatic than black pepper but has a more delicate flavor.

Pepper oil is the volatile oil obtained from black pepper by steam distillation; yield is usually 2–4%.

The oleoresin is obtained from black pepper by solvent extraction followed by removal of the solvent.

Major producers of black and white peppers include India, Indonesia, Malaysia, and China.

Black and white peppers should not be confused with red pepper or cayenne pepper, which is obtained from *Capsicum* species (see **capsicum**).

CHEMICAL COMPOSITION

Black pepper contains 2–4% volatile oil and 5–9% piperine, piperidine, piperettine, and a few other minor alkaloids (piperyline, piperolein A, piperolein B, piperanine, etc.).[2–4] Piperine and piperanine are the known pungent principles. Fruit maturity has been found to have little effect on piperine content.[5] Chavicine, formerly believed to be *cis,cis*-piperine, is reported to be a mixture of piperine and minor alkaloids.[6] Many other amide alkaloids have recently been isolated, for example, pipwaqarine, piptigrine, pipnoohine, pipyahyine, pellitorine, guineensine, pipercide, retrofractamide A, dipiperamides D and E (bisalkaloids), nigramides A–S (dimeric amides), and others.[7–14]

White pepper contains little volatile oil but has the same pungent principles and alkaloids as black pepper. Both also contain about 11% protein, 65% carbohydrates, lipids, crude fiber, and others (MARSH).

Black pepper is also reported to contain flavonol glycosides (especially those of kaempferol, rhamnetin, and quercetin) in considerable concentration, as well as sterols (stigmastane-3,6-dione and stigmast-4-ene-3,6-dione) and polysaccharides.[15–17]

Pepper oil contains a complex mixture of monoterpenes (70–80%), sesquiterpenes (20–30%), and small amounts of oxygenated compounds, with no pungent principles present. Concentration and composition vary, depending on sources. Major monoterpenes include α-thujene, α-pinene, camphene, sabinene, β-pinene, myrcene, 3-carene, limonene, and β-phellandrene. Sesquiterpenes include β-caryophyllene (major component), β-bisabolene, β-farnesene, *ar*-curcumene, humulene, β-selinene, α-selinene, β-elemene; α-cubebene, α-copaene, and sesquisabinene. Oxygenated components include linalool, l-terpinen-4-ol, myristicin, nerolidol, safrole, β-pinone, and *N*-formypiperidine, among others (JIANGSU).[4,18–26]

The oleoresin contains pungent principles and volatile oil.

PHARMACOLOGY AND BIOLOGICAL ACTIVITIES

Black pepper has diaphoretic, carminative, and diuretic properties as well as stimulating activities on the taste buds and olfactory system, producing a reflex increase in gastric secretion especially in elderly patients.[27] Ingestion of piperine has been reported to stimulate gastric secretion in mice.[28] It has also been reported to have strong lipolytic activity that resides in the outer layer of the fruit.[29]

A 0.1 g dose of pepper (type not specified) when held in the mouth without swallowing caused a temporary increase in blood pressure of all 24 subjects tested; pulse rate was not significantly affected (JIANGSU).

Aqueous and organic extracts and many alkamides of black pepper exhibited antibacterial activity against both Gram-positive and Gram-negative bacteria.[14,30] Larvicidal and insecticidal activities have also been reported.[7,9,10,13,31,32]

More recently black pepper has been reported to exhibit antioxidant/free radical scavenging activity[33–35] and to reduce carcinogenesis in various experimental models.[36–40] The polysaccharides were also reported to have immunomodulatory activity.[16]

TOXICOLOGY

Data from one source indicate black pepper oil to be nonirritating and nonsensitizing to human skin but moderately irritating to rabbit skin when applied undiluted under occlusion for 24 h. Its phototoxicity on humans is not known.[41]

USES

Medicinal, Pharmaceutical, and Cosmetic. Used in certain tonic and rubefacient preparations.

Food. Both black and white peppers are used extensively as domestic spices. They are also widely used as flavor ingredients in most major food products, including nonalcoholic beverages, candy, baked goods, meat and meat products, cheese, and condiments and relishes. Their oleoresins and oils are used in major categories of foods, including alcoholic beverages, frozen dairy desserts, and gelatins and puddings. Highest average maximum use level reported is 0.42% for white pepper in nut products and about 0.2% for oleoresins in baked goods.

Traditional Medicine. Used as a stimulant, carminative, and tonic. Both types of pepper have been reportedly used in cancers.[42]

In Chinese medicine, white pepper is used to treat stomachache, malaria, and cholera in addition to above uses.

Others. Along with clove, allspice, and ginger, black pepper has been reported to have antioxidant properties and can be a potential source of natural antioxidants.[43]

COMMERCIAL PREPARATIONS

Crude, oleoresin (black pepper), and oil (black pepper). Black pepper was formerly official in N.F. Black pepper oil is official in F.C.C.

Regulatory Status. Both are GRAS (§182.10 and §182.20).

REFERENCES

See the General References for ARCTANDER; BARRETT; BRUNETON; FEMA; GOSSELIN; GRIEVE; GUENTHER; GUPTA; HUANG; JIXIAN; JIANGSU; MARTINDALE; MASADA; MERCK; ROSENGARTEN; YOUNGKEN.

1. K. R. Kirtikar and B. D. Basu, *Indian Medicinal Plants*, International Book Distribution, Dehradun, India, 1999, p. 2135.
2. J. T. Traxler, *J. Agric. Food Chem.*, **19**, 1135 (1971).
3. M. L. Raina et al., *Planta Med.*, **30**, 198 (1976).
4. C. K. Atal et al., *Lloydia*, **38**, 256 (1975).
5. E. R. Jansz et al., *J. Sci. Food Agric.*, **35**, 41 (1984).
6. R. Grewe et al., *Chem. Ber.*, **103**, 3752 (1970).
7. B. S. Siddiqui et al., *Nat. Prod. Res.*, **19**, 143 (2005).
8. K. Wei et al., *J. Org. Chem.*, **70**, 1164 (2005).
9. B. S. Siddiqui et al., *Chem. Pharm. Bull. (Tokyo)*, **52**, 1349 (2004).
10. B. S. Siddiqui et al., *Nat. Prod. Res.*, **18**, 473 (2004).
11. K. Wei et al., *J. Nat. Prod.*, **67**, 1005 (2004).
12. S. Tsukamoto et al., *Bioorg. Med. Chem.*, **10**, 2981 (2002).
13. I. K. Park et al., *J. Agric. Food Chem.*, **50**, 1866 (2002).
14. S. V. Reddy et al., *Phytomedicine*, **11**, 697 (2004).
15. K. Wei et al., *Magn. Reson. Chem.*, **42**, 355 (2004).
16. H. Chun et al., *Biol. Pharm. Bull.*, **25**, 1203 (2002).
17. B. Voesgen and K. Herrmann, *Z. Lebensm. Unters. Forsch.*, **170**, 204 (1980).
18. A. Orav et al., *J. Agric. Food Chem.*, **52**, 2582 (2004).
19. J. Debrauwere and M. Verzele, *Bull. Soc. Chim. Belg.*, **84**, 167 (1975).
20. C. J. Muller and W. G. Jennings, *J. Agric. Food Chem.*, **15**, 762 (1967).
21. H. M. Richard and W. G. Jennings, *J. Food Sci.*, **36**, 584 (1971).
22. G. F. Russell and W. G. Jennings, *J. Agric. Food Chem.*, **17**, 1107 (1969).
23. S. J. Terhune et al., *Can. J. Chem.*, **53**, 3285 (1975).
24. V. S. Govindarajan in T. E. Furia, ed., *CRC Critical Reviews in Food Science and Nutrition*, Vol. 9, CRC Press, Cleveland, OH, 1977, p. 115.
25. B. M. Lawrence, *Perfum. Flav.*, **10**, 51 (1985).
26. B. M. Lawrence, *Major Tropical Spices—Pepper (Piper nigrum L.) Essential Oils 1979–1980*, Allured Publishing Corp., Weaton, IL, 1981.
27. T. Ebihara et al., *J. Am. Geriatr. Soc.*, **54**, 1401 (2006).
28. I. M. Ononiwu et al., *Afr. J. Med. Med. Sci.*, **31**, 293 (2002).
29. E. Halbert and D. G. Weeden, *Nature (London)*, **212**, 1603 (1966).
30. N. M. Chaudhry and P. Tariq, *Pak. J. Pharm. Sci.*, **19**, 214 (2006).
31. M. Rasheed et al., *Nat. Prod. Res.*, **19**, 703 (2005).
32. I. M. Scott et al., *J. Econ. Entomol.*, **97**, 1390 (2004).
33. I. Gulcin, *Int. J. Food Sci. Nutr.*, **56**, 491 (2005).
34. M. Kaleem et al., *Indian J. Physiol Pharmacol.*, **49**, 65 (2005).
35. R. S. Vijayakumar et al., *Redox. Rep.*, **9**, 105 (2004).
36. N. Nalini et al., *J. Med. Food*, **9**, 237 (2006).
37. K. Selvendiran et al., *Pulm. Pharmacol. Ther.*, **19**, 107 (2006).

38. K. Selvendiran et al., *Mol. Cell. Biochem.*, **268**, 141 (2005).
39. C. R. Pradeep and G. Kuttan, *Clin. Exp. Metastasis*, **19**, 703 (2002).
40. H. R.El et al., *Food Chem. Toxicol.*, **41**, 41 (2003).
41. D. L. J. Opdyke, *Food Cosmet. Toxicol.*, **16**(Suppl. 1) 651 (1978).
42. J. L. Hartwell, *Lloydia*, **33**, 288 (1970).
43. B. Al-Jalay et al., *J. Food Prod.*, **50**, 25 (1987).

PINE BARK, WHITE

Source: *Pinus strobus* L. (Family Pinaceae).

Common/vernacular names: Eastern white pine.

GENERAL DESCRIPTION

An evergreen tree much used for its timber, with leaves in five-leaved fascicles (clusters); branches horizontal and in regular whorls; bark on old trunks deeply fissured; up to about 45 m high, occasionally reaching 67 m; native to northeastern North America and distributed throughout eastern United States and Canada, from Newfoundland west to Iowa and south to Georgia; also cultivated in Europe. Part used is the dried inner bark after having removed the outer layer (cork).

CHEMICAL COMPOSITION

Reported to contain mucilage, coniferin, coniferyl alcohol, diterpenoids (strobol, strobal, *cis*- and *trans*-abienol, manoyl oxide, etc.), a triterpenoid (3β-methoxyserrat-l4-en-2l-one), a volatile oil, and others (LIST AND HÖRHAMMER).[1,2]

The bark of Scotch pine (*P. sylvestris*), a related species, has been reported to contain similar but more terpenic compounds, including monoterpenes, diterpenes, and triterpenes as well as sterols, many of which are present in rosin or turpentine (see **turpentine** and **rosin**).[3]

PHARMACOLOGY AND BIOLOGICAL ACTIVITIES

White pine bark has been reported to have expectorant, demulcent, and diuretic properties (CLAUS; WREN).

USES

Medicinal, Pharmaceutical, and Cosmetic. Used as a constituent in some cough syrups that are based on White Pine Compound or related formulations. More commonly used in Europe than the United States.

Dietary Supplements/Health Foods. Bark rarely used in tea formulations.

Traditional Medicine. Has been used by American Indians for centuries to treat coughs, colds, and congestion; also as a poultice to treat wounds, sores, abscesses, boils, rheumatism; bruises, felons, and inflammation (FOSTER AND DUKE). It is still used as a home remedy.

COMMERCIAL PREPARATIONS

White pine bark, White Pine Compound, and their extracts; both were formerly official in N.F: Strengths (see **glossary**) of extracts are expressed in weight-to-weight ratios.

Regulatory Status. White pine bark has been approved for use in alcoholic beverages only (§172.510). Oil of pine (*P. sylvestris* and other species) subject of a German therapeutic monograph; allowed internally and externally for congestions of the respiratory tract; externally for rheumatic and neuralgic complaints.[4]

REFERENCES

See the General References for APhA; BAILEY 1; FOSTER AND DUKE; GOSSELIN; GRIEVE; LIST AND HÖRHAMMER; LUST; MCGUFFIN 1 & 2; SARGENT; TERRELL; WREN.

1. D. F. Zinkel et al., *Phytochemistry*, **11**, 425 (1972).
2. D. F. Zinkel et al., *Phytochemistry*, **11**, 3387 (1972).
3. T. Norin and B. Winell, *Acta Chem. Scand.*, **26**, 2297 (1972).
4. Monograph, *Pini aetheroleum*, *Bundesanzeiger*, no. 154 (August 21, 1985); revised (March 13, 1990).

PINE NEEDLE OIL (DWARF AND SCOTCH)

Source: *Dwarf pine* Pinus mugo Turra (syn. *P. montana* Mill.) and *P. mugo* var. *pumilio* (Haenke) Zenari (syn. *P. pumilio* Haenke); Scotch pine *Pinus sylvestris* L. (Family Pinaceae).

Common/vernacular names: Swiss mountain pine oil (*P. mugo*).

GENERAL DESCRIPTION

Dwarf pine is a prostrate shrub or pyramidal tree (up to ca: 12 m high) with leaves in two-leaved fascicles (clusters), stiff and twisted; native to mountains of central and southern Europe.

Scotch pine is a tree also with stiff and twisted leaves in two-leaved fascicles; bark deeply fissured; up to about 40 m high; native to Eurasia and cultivated in eastern United States.

Parts used are the leaves (needles) and twigs from which the essential oils are obtained by steam distillation.

Dwarf pine needle oil (dwarf pine oil, pine needle oil N.F., Pinus pumilio oil, or pumilio pine oil) is produced mainly in Austria (Tirol) and Italy.

Scotch pine needle oil (Scotch pine oil or Pinus sylvestris oil) is mostly produced in Austria (Tirol), Russia, and Scandinavia.

CHEMICAL COMPOSITION

Dwarf pine needle oil has been reported to contain mostly monoterpene hydrocarbons (ca. 70%), including *d*-limonene, 3-carene, α- and β-pinenes, β-phellandrene, dipentene, camphene, and myrcene; 4–10% bornyl acetate and other esters; aldehydes (e.g., hexanal, cuminaldehyde, and anisaldehyde); *d*-cryptone; small amounts of sesquiterpenes (e.g., cadinene); and alcohols; among others (LIST AND HÖRHAMMER; MARTINDALE; MERCK; REMINGTON).[1–3]

The presence of the simple aldehydes is believed to be responsible for the characteristic sweet balsamic odor of dwarf pine oil.[1]

Scotch pine needle oil contains 50–97% monoterpene hydrocarbons composed mostly of α-pinene, with lesser amounts of 3-carene, dipentene, β-pinene, *d*-limonene, α-terpinene, γ-terpinene, *cis*-β-ocimene, myrcene, camphene, sabinene, terpinalene, and others.[2,4–9] Other compounds reported present include bornyl acetate (3–3.5%), borneol, 1,8-cineole, citral, terpineol, T-cadinol, T-muurolol, α-cadinol, caryophyllene, chamazulene, butyric acid, valeric acid, caproic acid, and iscaproic acid (MERCK; LIST AND HÖRHAMMER).[2,6,8,10,11]

PHARMACOLOGY AND BIOLOGICAL ACTIVITIES

Both dwarf pine oil and Scotch pine oil have varying degrees of antimicrobial activities. Limonene, dipentene, and bornyl acetate have been reported as the active principles responsible for the antiviral and antibacterial activities of some essential oils (also see ***turpentine***).[3,4,12]

TOXICOLOGY

Dwarf pine oil (but not Scotch pine oil) has been reported to be irritating to human skin. Both have also been demonstrated to be sensitizing to certain individuals; both were nonphototoxic.[3,4]

USES

Medicinal, Pharmaceutical, and Cosmetic. Dwarf pine needle oil is used as a fragrance and flavor component in pharmaceutical preparations, including cough and cold medicines, vaporizer fluids, nasal decongestants, and analgesic ointments.

Both dwarf pine oil and Scotch pine oil are used as fragrance ingredients in soaps, detergents, creams, lotions, and perfumes, with maximum use level of 1.2% reported for both oils in perfumes.[3,4]

Food. Dwarf pine needle oil and Scotch pine oil are used as flavor components in major food products, including alcoholic and nonalcoholic beverages, frozen dairy desserts, candy, baked goods, and gelatins and puddings. Use levels are generally low, with average maximum level usually below 0.001%.

COMMERCIAL PREPARATION

Both oils. Dwarf pine needle oil is official in N.F. Both oils are official in F.C.C. Only leaves are specified in N.F. for the production of dwarf pine oil.

Regulatory Status. Both oils have been approved for food use (§172.510).

REFERENCES

See the General References for ARCTANDER; BAILEY 1; FEMA; FERNALD; GOSSELIN; GUENTHER; LIST AND HÖRHAMMER; TERRELL; UPHOF.

1. W. D. Fordham in L. D. Coddet al., eds., *Chemical Technology: An Encyclopedic Treatment*, Vol. 5, Barnes & Noble, New York, 1972, p. 1.
2. R. M. Ikeda et al., *J. Food Sci.*, **27**, 455 (1962).
3. L. J. Opdyke, *Food Cosmet. Toxicol.*, **14**, 843 (1976).
4. D. L. J. Opdyke, *Food Cosmet. Toxicol.*, **14**, 845 (1976).
5. R. Hiltunen, *Ann. Acad. Sci. Fenn., Ser. A4*, **208**, 1.
6. Y. A. Poltavchenko and G. A. Rudakov, *Biol. Nauki*, **15**, 95 (1972).
7. J.-C. Chalchát et al., *Phytochemistry*, **24**, 2443 (1985).
8. J.-C. Chalchát et al., *Planta Med.*, **51**, 285 (1985).
9. B. M. Lawrence, *Perfum. Flav.*, **16**(2), **59** (1991).
10. S. Z. Ivanova et al., *Khim. Drev.*, **1**, 103 (1978).
11. R. D. Kolesnikova et al., *Rastit. Resur.*, **13**, 351 (1977).
12. L. Joubert and M. Gattefosse, *Mezhdunar. Kongr. Efirnym Maslam (Mater.), 4th*, **1**, 99 (1968).

PIPSISSEWA

Source: *Chimaphila umbellata* Nutt. (syn. *C. corymbosa* Pursh) (Family Pyrolaceae or Ericaceae).

Common/vernacular names. Chimaphila, prince's pine, bitter wintergreen, spotted wintergreen, and holly.

GENERAL DESCRIPTION

Perennial evergreen herb; leaves lanceolate, leathery, prominently toothed; with long creeping subterranean shoots; up to about 30 cm high; flowers pentamerous, nodding, petals waxy, native to Eurasia and northern North America (Quebec to Georgia and west to British Columbia, California); also in Central America. Part used is the dried leaf. Commercial supplies in the United States largely sourced from the Pacific Northwest in recent years.

CHEMICAL COMPOSITION

Contains arbutin (ca. 7.5% in aerial parts)[1] and isohomoarbutin, as well as other glycosides (e.g., reinfolin and glucosides of homogentisic acid and toluquinol) and flavonoids (hyperoside, avicularin, kaempferol, etc.);[1-6] about 0.2% chimaphilin (2,7-dimethyl-1,4-naphthoquinone);[6,7] ursolic acid; epicatechin gallate;[1] β-sitosterol; taraxasterol; nonacosane and hentriacontane; methyl salicylate; resins; tannins; gums; starch; sugar; and others (LIST AND HÖRHAMMER; MERCK).

PHARMACOLOGY AND BIOLOGICAL ACTIVITIES

Pipsissewa has been reported as one of numerous plants that have elicited hypoglycemic activity in experimental animals.[8]

Arbutin and its derivatives have urinary antiseptic properties due to formation of the hydroquinone aglycone after hydrolysis (see *uva ursi*).

Chimaphilin is reported to have urinary antiseptic as well as tonic and astringent activities; it also has bacteriostatic properties (LIST AND HÖRHAMMER; MERCK).

USES

Medicinal, Pharmaceutical, and Cosmetic. Has been used in urinary antiseptic preparations.

Food. Pipsissewa extracts (especially solid) are used as flavor components mainly in beverages (e.g., root beer, sarsaparilla, and other soft drinks) and candy. Other food products in which they are used include frozen dairy desserts, baked goods, gelatins and puddings, and sweet sauces. Highest average maximum use level is about 0.03% reported for the extract (type not specified) in sweet sauces (365 ppm) and baked goods (290 ppm).

Dietary Supplements/Health Foods. Herb used in teas as flavor ingredient; also in capsules and tablet formulations for traditional uses (below).

Traditional Medicine. Used as diuretic (also as antidiuretic), astringent, diaphoretic, and mild disinfectant and in treating bladder stones, usually in the form of a tea; also reported to be an antispasmodic for epilepsy and nervous disorders (FOSTER AND DUKE). Used externally in treating ulcerous sores, blisters, and others. It is also reportedly used in cancers.[9]

Others. Herb reportedly used in poison baits for rodents.[10]

COMMERCIAL PREPARATIONS

Crude and extracts (e.g., solid). Crude was formerly official in N.F. Strengths (see *glossary*) of extracts are either expressed in weight-to-weight ratios or in flavor intensities.

Regulatory Status. GRAS (§182.20).

REFERENCES

See the General References for BAILEY 1; FEMA; FOSTER AND DUKE; GRIEVE; LEWIS AND ELVIN-LEWIS; LUST; UPHOF; YOUNGKEN.

1. A. A. Trubachev, *Tr. Leningrad, Khim. Farm. Inst.*, **21**, 176 (1967).
2. E. Walewska and H. Thieme, *Pharmazie*, **24**, 423 (1969).
3. E. Walewska, *Herba Pol.*, **17**, 242 (1971).
4. A. A. Trubachev and V. S. Batyuk, *Farmatsiya (Moscow)*, **18**, 48 (1969).
5. A. A. Trubachev and V. S. Batyuk, *Khim. Prir. Soedin.*, **4**, 320 (1968).
6. K. H. Bolkhart and M. H. Zenk, *Z. pflanzenphysiol.*, **61**, 356 (1969).
7. K. H. Bolkhart et al., *Naturwissenschaften*, **55**, 445 (1968).
8. N. R. Farnsworth and S. B. Segelman, *Tile Till*, **57**, 52 (1971).
9. J. L. Hartwell, *Lloydia*, **34**, 103 (1971).
10. D. M. Secoy and A. E. Smith, *Econ. Bot.*, **37**(1), **28** (1983).

PODOPHYLLUM (PODOPHYLLIN)

Source: *Podophyllum peltatum* L. (Family Berberidaceae).

Common/vernacular names: Mayapple, mandrake, American mandrake, devil's apple, wild lemon, and vegetable mercury.

GENERAL DESCRIPTION

Perennial herb with an erect stem bearing at its apex one or two large peltate (shield-like) leaves that are five to nine lobed and measure up to about 33 cm across; a solitary flower borne on fork between the two leaves; up to about 45 cm high; native to eastern North America from Quebec to Florida and west to Minnesota and Texas. Parts used are the dried rhizome and roots collected early in the spring or in the fall after the aboveground parts have died down. The resin (podophyllin) is prepared from the root and rhizome by alcohol extraction followed by partial removal of the solvent and precipitating the resin with acidified water.

Indian podophyllin obtained similarly from Indian podophyllum (rhizome and roots of *Podophyllum hexandrum* Royle, also known as *P. emodi* Wall.) is now more widely used. In recent years, drastic decline from commercial harvest of wild populations in India and Pakistan has prompted listing of the species under the provisions of Appendix II of the Convention on International Trade of Endangered Species of Wild Fauna and Flora (CITES).[1]

CHEMICAL COMPOSITION

Podophyllum contains lignans (mainly podophyllotoxin, β-peltatin, and α-peltatin, with minor amounts of 4′-demethylpodophyllotoxin, dehydropodophyllotoxin, desoxypodophyllotoxin, etc.), lignan glucosides (e.g., those of podophyllotoxin, β-peltatin, α-peltatin, 4′-demethylpodophyllotoxin, and podorhizol), flavonoids (e.g., quercetin, quercetin-3-galactoside, kaempferol, kaempferol-3-glucoside or astragalin, and isorhamnetin), starch, gum, and others (LIST AND HÖRHAMMER; MORTON 3).[2-4]

Podophyllin or podophyllum resin is the alcohol-soluble and water-insoluble resinous material present in podophyllum in 3–6%. Due to the method of preparation, podophyllin generally contains little or no lignan glucosides. Most of the active principles present are the lignan aglycones consisting of about 20% podophyllotoxin, 13% β-peltatin,

7% α-peltatin, and small amounts of 4-demethylpodophyllotoxin and dehydropodophyllotoxin. Quercetin (ca. 5%) is also reportedly present.[5]

Indian podophyllin contains no peltatins; it contains much higher amounts of podophyllotoxin (ca. 40%), 4'-demethylpodophyllotoxin (ca. 2%), and dehydropodophyllotoxin (2–3%).[5]

PHARMACOLOGY AND BIOLOGICAL ACTIVITIES

Podophyllin has strong cathartic and antineoplastic properties. These properties are due to the lignans and their glucosides present, with the former being more potent.[6] The cathartic properties are reported to be due mainly to the peltatins, while the antitumor properties are attributed to podophyllotoxin and 4-demethylpodophyllotoxin as well as the peltatins.[5–8]

Podophyllin was successfully utilized in the treatment of penile warts and HIV-related hairy leukoplakia of the tongue in 244 and 10 patients, respectively.[9,10]

Immunotherapeutic features of podophyllotoxin have been investigated. It inhibits mitogen-induced human lymphocyte proliferation and macrophage growth factor-stimulated macrophage proliferation. Alone, it induces lymphocyte-activating factor/interleukin 1, and with mitogen, induced T-cell growth factor/interleukin 1. The compound directly stimulates macrophage proliferation and potentiates the effects of low doses of macrophage growth factor.[11]

Podophyllotoxin has been found to be a microtubule-disrupting agent with strong antimitotic activity, blocking cells in mitosis. Semisynthetic podophyllotoxin derivatives, VP 16213 (etoposide) and VM26 (teniposide), have no antimitotic activity but are highly active in causing breakage of DNA strands (topoisomerase inhibition), thus blocking cancer cells in a premitotic stage of the cell cycle.[11–14]

Indian podophyllum extracts and their quercetin 3-galactoside have antioxidant radioprotective effects on the hematopoietic system in mice.[15–17] Aqueous extracts also displayed anti-inflammatory effects against lipopolysaccharide-induced inflammation in mice.[18]

TOXICOLOGY

Podophyllin and podophyllotoxin have been demonstrated to be strongly embryocidal and growth retarding but not teratogenic in animals (e.g., mice and rats).[19–23]

Podophyllin is very irritating to mucous membranes (especially those of the eye) and to the skin. It is also highly toxic, and fatalities as a result of its use have been reported (MARTINDALE).

USES

Medicinal, Pharmaceutical, and Cosmetic. Podophyllin is used in treating venereal warts (condyloma acuminatum) and other papillomas, most commonly as a dispersion in Compound Benzoin Tincture or as a solution in alcohol. It is also rather extensively used in laxative preparations, usually along with other cathartic agents.

A semisynthetic podophyllotoxin derivative, VP 16213 (etoposide), has been approved in the United States as a chemotherapeutic agent (often in combination regimes with cisplatin, bleomycin, and others) for the treatment of refractory testicular tumors and for the treatment of small cell lung cancer. VM26 (teniposide) is being investigated for the treatment of acute lymphoblastic leukemia, various lymphomas, and other carcinomas. Semisynthetic derivatives are being used in Europe for the treatment of psoriasis and rheumatoid arthritis.[11–14]

Traditional Medicine. Used as a cathartic and in treating jaundice, fever, liver ailments, and syphilis; also used in cancers.[24]

COMMERCIAL PREPARATIONS

Crude (podophyllum) and resin (podophyllin); both are official in U.S.P. Current U.S.P. standards do not specify contents or identities of active principles, and strengths (see *glossary*) of podophyllin usually vary, depending on suppliers.

An adulterant of Indian podophyllum has been identified as *Ainsliaea latifolia* (D. Don) Sch. Bip. (Family Asteraceae). It appears that its rhizome has been used as adulterant for a long time.[25]

Regulatory Status. The once widespread use of podophyllum resins in OTC laxative preparations has been disallowed.[13] Subject of a German therapeutic monograph; external preparations for condyloma acuminatum.[26]

REFERENCES

See the General References for BAILEY 1; DER MARDEROSIAN AND BEUTLER; GOSSELIN; GRIEVE; KROCHMAL AND KROCHMAL; LIST AND HÖRHAMMER; LUST; MARTINDALE; MCGUFFIN 1 & 2; MORTON 3; STAHL; TYLER 3.

1. S. FOSTER, *HerbalGram*, **23**, 19 (1990).
2. J. L. Hartwell and A. W. Schrecker, *Fortschritte der Chemie Organischer Naturstoffe*, Springer-Verlag, Vienna, 1958, p. 83.
3. A. Stoll et al., *J. Am. Chem. Soc.*, **76**, 6413 (1954).
4. A. Wartburg, *Helv. Chim. Acta*, **40**, 1331 (1957).
5. H. Auterhoff and O. May, *Planta Med.*, **6**, 240 (1958).
6. I. H. Emmenegger et al., *Arzeim.-Forsch.*, **11**, 327 (1961).
7. H. Staehelin, *Planta Med.*, **22**, 337 (1972).
8. S. M. Kupchan et al., *J. Pharm. Sci.*, **54**, 659 (1965).
9. D. J. White et al., *Genitourin. Med.*, **73**, 184 (1997).
10. G. Gowdey et al., *Oral Surg. Oral Med. Oral Pathol. Oral Radiol. Endod.*, **79**, 64 (1995).
11. Q. Y. Zheng et al., *Int. J. Immunopharmacol.*, **9**, 539 (1987).
12. B. F. Issell et al., eds., *Etoposide (VP-16): Current Status and New Develo-pment*, Academic Press, Orlando, FL, 1984.
13. B. R. Olin, ed., *Lawrence Rev. Nat. Prod.* (1992).
14. Anon., *Am. Pharm.*, **NS24** (3) 31 (1984).
15. R. K. Sagar et al., *Planta Med.*, **72**, 114 (2006).
16. R. Chawla et al., *Z. Naturforsch. C*, **60**, 728 (2005).
17. R. Chawla et al., *Mol. Cell Biochem.*, **273**, 193 (2005).
18. H. Prakash et al., *J. Pharm. Pharm. Sci.*, **8**, 107 (2005).
19. J. B. Thiersch, Proceedings of Third International Congress on Chemotherapy, Stuttgart 2, 1741 (1963).
20. M. G. Joneja and W. C. LeLiever, *Toxicol. Appl. Pharmacol.*, **27**, 408 (1974).
21. J. B. Thiersch, *Proc. Soc. Exp. Biol. Med.*, **113**, 124 (1963).
22. M. G. Joneja and W. C. LeLiever, *Can. J. Genet. Cytol.*, **15**, 491 (1973).
23. Anon., *Fed. Regist.*, **40**, 12902 (1975).
24. J. L. Hartwell, *Lloydia*, **31**, 71 (1968).
25. H. S. Puri and S. P. Jain, *Planta Med.*, **54**, 269 (1988).
26. Monograph *Podophylli peltati rhizoma* and *Podophylli peltati resina*, Bundesanzeiger, no. 50 (March 13, 1986); revised (March 13, 1990).

PORIA

Source: *Poria cocos* (Schw.) Wolf. (syn. *Pachyma cocos* Franch.; *Pachyma hoelen* Sacc.; *Sclerotium cocos* Schw.) (Family Polyporaceae).

Common/vernacular names: Yunling (Yunnan *fuling*), *fulingge* (whole poria), *baifuling* (white poria), *chifuling* (red poria), *fulingpi* (poria skin), *fuling*, hoelen, Indian bread, and tuckahoe.

GENERAL DESCRIPTION

Poria cocos is a polypore fungus parasitic on roots or saprophytic on rotten stumps of certain pine trees growing in loose sandy soil, the most common including *Pinus massoniana* Lamb, *P. densiflora* Sieb. et Zucc., *P. yunnanensis* Franch., *P. taiwanensis* Hayata, and *P. thunbergii* Parl. Part used is its sclerotium (fungal tissue mass), which is irregularly shaped, ranging from fist size to large masses with diameter of 10–30 cm or larger, growing 20–30 cm deep in the ground. Its presence in the native state ready for harvesting is indicated by several signs, including cracks on ground around the tree trunk, which when struck with a shovel produce a hollow sound; white mycelia (fungal tissue strands) or grayish white powder on surface above root system around the tree; and little or no vegetative growth around the tree where water also drains quickly after rain. Cultivated *fuling is* harvested after 2 years, with best quality after 3–4 years. The sclerotia from wild plants are dug up between July and March, while those of cultivated plants are dug in autumn, rid of dirt, and allowed to "sweat" (ferment) a few times (5–8 days each) until most moisture has evaporated and then dried completely in an airy, shady place. The resulting product is wrinkled and brown or dark brown called *fulingge* (whole *fuling*). *Fulingge* is white in the middle (with brown root embedded) and light brown to pink toward the outside.

From different parts of *fulingge* after sweating and before final drying, numerous products of various grades are produced: *fulingpi* (*fuling* skin), *baifuling* (white *fuling*), *chifuling* (red *fuling*), *fushen* (slices of peeled *fulingge* containing pieces of pine root), *fushenmu* (pine root removed from *fuling*), and others (CMH; JIANGSU; ZHU).[1,2]

CHEMICAL COMPOSITION

Contains ca. 93% of a polysaccharide named β-pachyman composed of pachymaran in β-(1 → 3) linkages and side chains of β-(1 → 6) linkages. The fresh sclerotia contain heteropolysaccharides composed of D-glucose, D-galactose, D-mannose, D-fucose, and D-xylose.[3] A bioactive water-soluble β-glucan has also been isolated from the *P. cocos*.[4]

Other constituents include triterpene acids, including pachymic acid, eburicoic acid, tumulosic acid, dehydrotrametenonic acid, dehydroeburiconic acid, poricoic acid, pinicolic acid, and poriatin (mixture), and others;[5–8] adenine, ergosterol, choline, lecithins, histidine, sucrose, fructose, proteases, trace minerals, and others (ETIC; HU; JIANGSU; NATIONAL).[9–14]

PHARMACOLOGY AND BIOLOGICAL ACTIVITIES

Poriatin has been shown to potentiate the effects of antitumor agents against mouse sarcoma S_{180}, with tumor inhibition of 38.9% for dactinomycin (vs. 19.6% when used without poriatin), 48% for mitomycin (vs. 35% for mitomycin alone), 69.0% for cyclophosphamide (vs. 32.3% for cyclophosphamide alone), and 59.1% for fluorouracil (vs. 38.6% for fluorouracil alone); against murine leukemia (L1210), increasing survival time by 168.1% when cyclophosphamide was used with poriatin (vs. 70% when cyclophosphamide was used alone), with the optimal dose for poriatin at 1/40 of LD_{50}.[10] Poriatin also has immunoregulatory effects in mice.[9,14] Other triterepene acids were active against prostate cancer and colon carcinoma.[15,16] Inhibition of topoisomerase-II and tumor promotion have also been reported.[8,16–18]

Polysaccharides (especially pachymaran) have exhibited antitumor (mouse sarcoma, S_{180}) and immunomodulating effects in mice, enhancing phagocytosis of macrophages and activating T and B lymphocytes (HU).[19–21] Antitumor activities were also reported against human breast carcinoma MCF-7, leukemic U937, and HL-60 cells.[4,22] Acetone extractives of *fuling* had marked inhibitory effects on mutagenesis induced by aflatoxin B, per the Ames test.[23]

A *P. cocos* preparation, a hydroalcoholic extract, as well as some of the triterpene acids displayed anti-inflammatory activities in different *in vivo* models.[7,24–26] Polysaccharides isolated from *P. cocos* were found to induce NF-κB and nitric oxide production, as well as iNOS expression.[27,28]

Extracts of submerged cultures of *P. cocos* exhibited strong nematicidal activity against the nematodes *Pangrellus redivivue*, *Bursaphelenchus xylophilus*, and *Meloidogyne arenaria*. 2,4,6-Triacetylene octane diacid was found to be the most potent component of the extracts.[29]

Other reported activities of *P. cocos* include antioxidant activity of the aqueous extract in different *in vitro* assays, such as lipid peroxidation inhibition and superanion scavenging.[30] *P. cocos* triterepenes exhibited antiemetic activity in a leopard frog model of emesis.[31] The ethanolic extract of *P. cocos* had an immunosuppressant effect resulting in suppressed rejection of heart transplants and prolonged survival in rats.[32]

Decoctions of both *fuling* and *fushen* had sedative effects in mice, with those of *fushen* being stronger. The diuretic effects of *fuling* have been reported but are so far equivocal; however, at doses of 30–100 g/day, it had a strong diuretic effect in patients with cardiac edema.[33] Other biological effects include hypoglycemic in rabbits, relaxation of isolated rabbit intestine, inhibiting isolated frog heart, and antibacterial *in vitro* (alcohol but not water extract) (WANG).

USES

Medicinal, Pharmaceutical, and Cosmetic. *Fuling* (especially *baifuling*) is an ingredient in several well-known skin (facial) treatment formulas used by women of the Chinese imperial courts to maintain clean and "radiant" skin and to prevent pimples, dark spots, and wrinkles. Powder used in facial scrubs and extracts (hydroalcoholic) in moisturizing and nourishing creams and lotions for its traditional cleansing, whitening, soothing, and moisturizing properties (ETIC).[34]

Food. Used as an ingredient in *fuling* breads and *fuling* cakes that are traditionally eaten for their tonic properties.

Dietary Supplements/Health Foods. Used as an ingredient in various herbal formulas for its traditional diuretic, tonic, and calming properties, especially in weight-control and sedative products.

Traditional Medicine. First mentioned in the *Shi. Jing* (ca. 400–300 BC) and its medicinal properties later described for the first time in the *Shen Nong Ben Cao Jing* (ca. 200 BC–AD 100), *fuling* is a widely used tonic food in China. Traditionally considered as sweet and bland tasting, neutral, diuretic (li shui shen shi), kidney invigorating, and nerve calming, it is used in treating dysuria (urination difficulties), edema, cough due to phlegm retention (*tan yin ke sou*), diarrhea, nervousness, insomnia, spermatorrhea, forgetfulness, and so on. It is extensively used in diet therapy, especially for older people, in soups, cakes, breads, and wines, and so on.[35–37]

Fushen is considered superior to *fuling* in treating nervousness and insomnia.

COMMERCIAL PREPARATIONS

Crude and extracts (aqueous and hydroalcoholic). Crude normally comes as *baifuling* (white), *chifuling* (red), and *fushen* in thick slices and as a powder; identity of powder can be verified by microscopy (XU AND XU); extracts in liquid and powdered forms with no established standards.

Regulatory Status. Fuling is an ethnic food; its U.S. regulatory status not determined.

REFERENCES

See the General References for CHP; CMH; DER MARDEROSIAN AND BEUTLER; HONGKUI; HU; HUANG; IMM-CAMS; JIANGSU; LU AND LI; NATIONAL; WANG; XU AND XU; ZHU.

1. B. Liu, *Medicinal Fungi of China*, People's Press, Taiyuan, Shanxi, 1984, p. 89.
2. Z. G. Song, *Sichuan Zhongyi*, **9**, 51 (1990).
3. Y. Wang et al., *Carbohydr. Res.*, **339**, 327 (2004).
4. M. Zhang et al., *Oncol. Rep.*, **15**, 637 (2006).
5. T. Akihisa et al., *Biosci. Biotechnol. Biochem.*, **68**, 448 (2004).
6. Z. Song et al., *Anal. Sci.*, **18**, 529 (2002).
7. K. Yasukawa et al., *Phytochemistry*, **48**, 1357 (1998).
8. M. Ukiya et al., *J. Nat. Prod.*, **65**, 462 (2002).
9. J. Xu, et al., *Faming Zhuanli Gongbao*, **5**, 8 (1989).
10. D. D. Li et al., *Zhongguo Kangshengsu Zazhi*, **15**, 63 (1990).
11. C. Goro, *Nature*, **225**, 943 (1970).
12. J. Hamuro et al., *Nature*, **233**, 486 (1971).
13. H. Saito et al., *Agric. Biol. Chem.*, **32**, 1261 (1968).
14. G. J. Wang et al., *Zhongguo Kangshengsu Zazhi*, **17**, 42 (1992).
15. L. Gapter et al., *Biochem. Biophys. Res. Commun.*, **332**, 1153 (2005).
16. G. Li et al., *Arch. Pharm. Res.*, **27**, 829 (2004).
17. Y. Mizushina et al., *Cancer Sci.*, **95**, 354 (2004).
18. T. Kaminaga et al., *Oncology*, **53**, 382 (1996).
19. C. X. Chen, *Zhongcaoyao*, **16**, 40 (1985).
20. S. C. Lu et al., *Diyi Junyi Daxue Xuebao*, **10**, 267 (1990).
21. Y. Jin et al., *Carbohydr. Res.*, **338**, 1517 (2003).
22. Y. Y. Chen and H. M. Chang, *Food Chem. Toxicol.*, **42**, 759 (2004).
23. C. C. Ruan et al., *Chin. J. Cancer*, **8**, 29 (1989).
24. S. M. Fuchs et al., *Skin Res. Technol.*, **12**, 223 (2006).
25. M. J. Cuellar et al., *Chem. Pharm. Bull. (Tokyo)*, **45**, 492 (1997).
26. H. Nukaya et al., *Chem. Pharm. Bull. (Tokyo)*, **44**, 847 (1996).
27. K. Y. Lee et al., *Int. Immunopharmacol.*, **4**, 1029 (2004).
28. K. Y. Lee and Y. J. Jeon, *Int. Immunopharmacol.*, **3**, 1353 (2003).
29. G. H. Li et al., *J. Microbiol.*, **43**, 17 (2005).
30. S. J. Wu et al., *Phytother. Res.*, **18**, 1008 (2004).
31. T. Tai et al., *Planta Med.*, **61**, 527 (1995).
32. G. W. Zhang et al., *Chin Med. J. (Engl.)*, **117**, 932 (2004).
33. A. Q. Kang and Z. X. Zhang, *Tianjin Zhongyi*, **14** (1989).
34. X. F. Zhang, *Dazhong Zhongyiyao*, **37** (1991).
35. S. Y. Chen, *Edible Fungi China*, **9**, 24, 25 (1990).
36. S. Y. Chen, *Edible Fungi China*, **9**, 42 (1990).
37. S. Y. Chen, *Edible Fungi China*, **9**, 37 (1990).

PSYLLIUM

Source: *Plantago psyllium* L., *P. indica* L. (syn. *P. arenaria* Waldst. et Kit.), and *P. ovata* Forsk. (syn. *P. decumbens* Forsk. and *P. ispaghula* Roxb.) (Family Plantaginaceae).

Common/vernacular names: Black psyllium (*P. indica*); blond psyllium, ispaghula, ispagol, spogel, Indian plantago (*P. ovata*); brown psyllium, French psyllium, Spanish psyllium (*P. psyllium*, *P. indica*); plantago; plantain. *P. arenaria* is accepted in the taxonomic literature as the correct name for *P. psyllium* (ambiguous name) and *P. indica* (illegitimate name); however, both names are still used in commercial trade over *P. arenaria*.

GENERAL DESCRIPTION

All three are annual herbs, less than 0.5 m high. *Plantago ovata* is stemless (acaulescent) or nearly so; it is native to the Mediterranean, North Africa, and western Asia; extensively cultivated in India and Pakistan. *Plantago psyllium* and *P. indica* have erect and branched stems; native to the Mediterranean region; cultivated in Spain and southern France.

Parts used are the dried ripe seed and its husk, a thin membranous layer on the seed coat. When soaked in water, the seed increases in volume many fold but contracts to its original volume when excess alcohol is added.[1]

CHEMICAL COMPOSITION

Contains 10–30% mucilage; present mainly in the husk; it is composed of a mixture of polysaccharides with D-xylose as the major residue, and L-arabinose and an aldobiouronic acid also present.[2]

Other constituents present include monoterpene alkaloids such as (+)-boschniakine (indicaine), (+)-boschniakinic acid (plantagonine), and indicainine;[3–5] aucubin (a glucoside);[3,4] planteose (a trisaccharide), sucrose, glucose, and fructose;[3,6,7] sterols (β-sitosterol, stigmasterol, and campesterol) and triterpenes (α- and β-amyrins);[8] a fixed oil and fatty acids (e.g., linoleic, oleic, stearic, and palmitic acids); tannins; and others (JIANGSU; MORTON 3).

Acteoside and isoacteoside (phenylethanoid glycosides) have recently been isolated from psyllium.[9]

PHARMACOLOGY AND BIOLOGICAL ACTIVITIES

Psyllium seed has laxative properties due to the swelling of its husk in water to form a gelatinous mass, thus keeping the feces hydrated and soft. The resulting bulk promotes peristalsis and laxation and is thus useful in the treatment of chronic constipation (GOODMAN AND GILMAN).[10,11] Psyllium has recently been reported to reduce postoperative pain and tenesmus.[12]

An alcoholic extract of blond psyllium seed (*P. ovata*) has been reported to lower the blood pressure of anesthetized dogs and cats and to inhibit isolated rabbit, rat, and guinea pig ileum; it also has cholinergic activity (JIANGSU).[13]

Blond psyllium seed powder has been reported to counteract strongly the deleterious effects of feeding a supplement of 2% sodium cyclamate, 2% FD & C red no. 2, and 4% polyoxyethylene (20) sorbitan monostearate to rats.[14]

In many animal and human studies, hydrophilic muciloid psyllium preparations were found to lower serum cholesterol, LDL cholesterol, and LDL:HDL ratios,[15] and triglycerides,[16] by various mechanisms including binding bile and altering the metabolism of bile acids, hence increasing fecal excretion while increasing bile salt synthesis from cholesterol (ESCOP 2).[17–22]

Psyllium seed husk and aqueous extract have antihyperglycemic/antidiabetic effects that may be attributed to reduced glucose absorption.[19,23–26] There are indications that psyllium may have potential in the management of child and adolescent obesity through modulating glucose homeostasis and lipid/lipoprotein profile.[27]

Psyllium polysaccharides have been shown to promote wound healing and to reduce scar formation through various mechanisms in *in vivo* and *in vitro* models.[28]

Blond psyllium seeds are a partly fermentable dietary fiber supplement that increases stool bulk and has mucosa-protective effects. Rats fed 100–200 g of blond psyllium seeds/kg in a fiber-free elemental diet for 4 weeks were found to have an increase of fecal fresh weight up to 100%, fecal dry weight up to 50%, and fecal water content up to 50%. Length and weight of the large intestine (but not small intestines) increased significantly. The seeds and husks also increase total fecal bile secretion while reducing activity of β-glucuronidase.[29]

TOXICOLOGY

The husk has been reported to depress the growth of chickens by 15% when 2% of it was included in an otherwise balanced diet.[30]

USES

Medicinal, Pharmaceutical, and Cosmetic. The husks are widely used as an ingredient in bulk laxatives; seeds are also used, but less so.

When seeds are used, they should not be ground or chewed, as it has been reported that they release a pigment that deposits in renal tubules. (USD 26th).[2]

A European monograph indicates use for habitual constipation, in case where soft stool is desired as in cases of anal fissures, hemorrhoids, and post-rectal surgery; irritable bowel syndrome, diverticulosis; as a dietary supplement where increases in dietary fiber are required; and adjuvant therapy for diarrhea (ESCOP 2).

Food. The husk mucilage is used as a thickener or stabilizer in certain frozen dairy desserts.

Dietary Supplements/Health Foods. Ground seeds or husks used in various dietary supplement formulations for increased fiber, cholesterol reduction, laxative activity, for example, weight loss products.

In 1989, psyllium was introduced into processed foods, including General Foods' Benefits cereal and Kellogg's Heartwise cereal, with claims for cholesterol-reducing benefits and soluble fiber health benefits. Shortly after introduction, the FDA suggested the products were "misbranded drugs" due to insufficient evidence to support the labeling claims. Benefits was removed from the market, while Heartwise is still marketed.

The introduction of these products largely stimulated the regulatory and legislative debate on "acceptable health claims" for foods, indirectly leading to passage of the Nutrition and Labeling Act of 1990 and the Dietary Supplement Health and Education Act of 1994 (DSHEA).

One report linked a case of anaphylactic reaction to Heartwise; asthma and anaphylaxis had previously been associated with rare allergies often resulting from industrial or occupational sensitivity.[4] In response, Kellogg added a label, reading: "New Users: A very small percentage of individuals, particularly some nurses and health care providers who have been occupationally exposed to psyllium dust, may develop a sensitivity to psyllium. This sensitivity may result in an allergic reaction."[31]

Traditional Medicine. Used in the United States and Europe, primarily as a bulk laxative to treat chronic constipation; also used as an emollient and demulcent. Fresh leaves of *Plantago* species applied topically for poison ivy, insect bites and stings; an uncontrolled study reported that use of the fresh leaves prevented itching and spread of dermatitis in poison ivy-induced dermatitis.[32]

Used in India as a diuretic and to treat diarrhea, gonorrhea, urethritis, hemorrhoids, kidney, and bladder problems, among others (MORTON 3).

In China both the seeds and whole herbs of related *Plantago* species (e.g., *P. asiatica* L. and *P. depressa* Willd.) are used for similar purposes. In addition, the seeds are used to treat hematuria (bloody urine), coughing, high blood pressure (with ca. 50% success in clinical trials), and other ailments. The seeds used in Chinese medicine are often specially treated by frying with saltwater before drying; these cannot be directly compared with the psyllium seeds used in the United States (FARNSWORTH 1–4; JIANGSU; NANJING).

Seeds of psyllium and other *Plantago* species as well as their roots, juice, leaves, and whole herbs have been reported to be used in treating cancer.[33]

COMMERCIAL PREPARATIONS

Husks (blond) and whole seeds (black and blond); whole seeds contain both husk and kernels; husks are official in U.S.P. Blond psyllium husk subject of a positive German therapeutic monograph used for habitual constipation and supportive treatment of irritable bowel syndrome.[34]

REFERENCES

See the General References for BLUMENTHAL 1; ESCOP 2; GRIEVE; MARTINDALE; MERCK; MORTON 3; TERRELL; TUCKER 1–3; UPHOF; YOUNGKEN.

1. R. Wasicky, *Planta Med.*, **9**, 232 (1961).
2. J. N. BeMiller in R. L. Whistler, ed., Academic Press, New York, 1973, p. 339.
3. M. S. Karawya et al., *U. A. R. J. Pharm. Sci.*, **12**, 53 (1971).
4. S. I. Balbaa et al., *U. A. R. J. Pharm. Sci.*, **12**, 35 (1971).
5. G. A. Cordell in R. H. F. Manske, ed., *The Alkaloids*, Vol. 16, Academic Press, New York, 1977, p. 431.
6. M. S. Karawya, *Planta Med.*, **20**, 14 (1971).
7. D. French et al., *J. Am. Chem. Soc.*, **75**, 709 (1953).
8. Y. Nakamura et al., *Nutr. Cancer*, **51**, 218 (2005).
9. L. Li et al., *J. Chromatogr. A*, **1063**, 161 (2005).
10. J. A. Marlett et al., *Am. J. Clin. Nutr.*, **72**, 784 (2000).
11. J. W. McRorie et al., *Aliment. Pharmacol. Ther.*, **12**, 491 (1998).
12. D. M. Kecmanovic et al., *Phytother. Res.*, **20**, 655 (2006).
13. M. L. Khorana et al., *Indian J. Pharm.*, **20**, 3 (1958).
14. B. H. Ershoff, *J. Food Sci.*, **41**, 949 (1976).
15. J. W. Anderson et al., *Arch. Intern. Med.*, **148**, 292 (1988).
16. A. Danielsson et al., *Acta Hepatogastroenterol.*, **26**, 148 (1979).
17. Y. C. Liu et al., *Ann. Nutr. Metab.*, **48**, 374 (2004).
18. K. G. Allen et al., *J. Agric. Food Chem.*, **52**, 4998 (2004).
19. M. Sierra et al., *Eur. J. Clin. Nutr.*, **56**, 830 (2002).
20. A. L. Romero et al., *J. Nutr.*, **132**, 1194 (2002).
21. J. W. Anderson et al., *Am. J. Clin. Nutr.*, **71**, 1433 (2000).
22. J. W. Anderson et al., *Am. J. Clin. Nutr.*, **71**, 472 (2000).
23. J. M. Hannan et al., *Br. J. Nutr.*, **96**, 131 (2006).
24. S. A. Ziai et al., *J. Ethnopharmacol.*, **102**, 202 (2005).
25. M. Rodriguez-Moran et al., *J. Diabetes Complications*, **12**, 273 (1998).
26. A. C. Frati Munari et al., *Arch. Med. Res.*, **29**, 137 (1998).

27. L. A. Moreno et al., *J. Physiol Biochem.*, **59**, 235 (2003).
28. W. Westerhof et al., *Drugs Exp. Clin. Res.*, **27**, 165 (2001).
29. E. Leng-Peschlow, *Br. J. Nutr.*, **66**, 331 (1991).
30. P. Vohra and F. H. Kratzer, *Poult. Sci.*, **43**, 1164 (1964).
31. D. R. Schaller, *N. Engl. J. Med.*, **323**, 1073 (1990).
32. S. Duckeet, *N. Engl. J. Med.*, **303**, 583 (1980).
33. J. L. Hartwell, *Lloydia*, **33**, 288 (1970).
34. Monograph *Plantaginis ovatae testa and semen, Bundesanzeiger*, no. 22 (February 1, 1990).

QUASSIA

Source: *Picrasma excelsa* (Sw.) Planch. and *Quassia amara* L. (Family Simaroubaceae).

Common/vernacular names: Jamaican quassia (*P. excelsa*); Surinam quassia (*Q. amara*); quassia wood, bitterwood.

GENERAL DESCRIPTION

Picrasma excelsa is a tree with a trunk diameter of 0.5–1 m; up to about 25 m high; native to the West Indies and growing in Jamaica and other Caribbean Islands. *Quassia amara* is a shrub or small tree up to about 3 m high; native to northern South America and growing in Surinam, Brazil, Colombia, Venezuela, and other tropical American countries. Part used is the wood.

CHEMICAL COMPOSITION

Surinam quassia (*Q. amara*) contains quassin, quassinol, 18-hydroxyquassin, and neoquassin, while Jamaican quassia (*P. excelsa*) contains isoquassin (picrasmin), neoquasin, and 18-hydroxyquassin as their bitter principles (LIST AND HÖRHAMMER).[1,2] These bitter principles are reported to be about 50 times more bitter than quinine (STAHL).

Other constituents include two additional quassinoids, quassimarin and simalikalactone D, isolated from *Q. amara*,[3] β-sitosterol and β-sitostenone;[4] 1.8% thiamine (in *P. excelsa*); and alkaloids of the β-carboline type including canthin-6-one, 5-methoxycanthin-6-one, 4-methoxy-5-hydroxycanthin-6-one, and N-methyl-l-vinyl-β-carboline (LIST AND HÖRHAMMER; WILLAMAN AND SCHUBERT).[5]

Jamaican quassia is reported to contain no tannin (MARTINDALE).

The Asian *P. quassioides* D. Don. Bern is reported to contain the carboline alkaloids 1-carboxy-β-carboline and picrasidines A, B, L, M, and P. Quassinoids include kumijians A, B, C, and G; picrasins A, B, C, D, E, F, G, N, O, and Q; and kusulactone (GLASBY 2).

PHARMACOLOGY AND BIOLOGICAL ACTIVITIES

Quassia has bitter tonic properties but in large doses is reported to cause stomach irritation and to produce vomiting (LEWIS AND ELVIN-LEWIS).

Quassinoids are reported to be amebicidal *in vitro* and *in vivo* (WREN).

An earlier publication mentioned that the reported traditional antimalarial activity was not confirmed in quassinoids.[6] More recent reports, however, showed that the methanol and hexane extracts of *Q. amara* exhibited strong antimalarial activity in mice infected with *Plasmodium berghei*.[7] The activity was comparable to that of chloroquine and was later attributed to the presence of the quassinoid simalikalactone D.[8,9]

Q. amara leaf methanol extract has antimicrobial activities against Gram-positive, Gram-negative bacteria, and fungi,[10] while simalikalactone D is believed to be responsible for the strong antiviral activity of the chloroform and ethyl acetate extracts.[11]

The hexane extract of *Q. amara* bark possesses considerable antinociceptive and anti-inflammatory effects when administered i.p. in mice.[12] The same group also reported that different extracts of the bark exhibited antiulcerogenic effects with oral and i.p. administration in mice.[13]

Quassimarin has been reported to have antileukemic properties.[3]

TOXICOLOGY

Different extracts of *Q. amara* have been reported to possess antifertility activity supported by *in vivo* experiments in mice and rats. This activity affects sperm counts, motility, and viability and seems to be mediated through inhibition of sex hormone production.[14–16]

USES

Medicinal, Pharmaceutical, and Cosmetic. Occasionally used in certain laxative preparations.

Food. Extracts and purified mixtures of bitter principles (commercially known simply as "quassin") are used to impart a bitter flavor to various food products, especially alcoholic (e.g., liqueurs and bitters) and nonalcoholic beverages. Food products in which the extracts are used include alcoholic and nonalcoholic beverages, frozen dairy desserts, candy, baked goods, and gelatins and puddings. Highest average maximum use level reported is about 0.007% (71.8 ppm) in nonalcoholic beverages. Used as a flavoring substitute for quinine.

Traditional Medicine. Used as a bitter and anthelmintic (via enema). Also used to treat fevers.

Others. Bark reportedly used as an insecticide.

COMMERCIAL PREPARATIONS

Crude, extracts, and quassin; crude and fluid extract were formerly official in N.F. Strengths (see *glossary*) of extracts are usually expressed in weight-to-weight ratios.

Regulatory Status. Has been approved for food use (§172.510).

REFERENCES

See the General References for BARNES; BIANCHINI AND CORBETTA; CLAUS; DER MARDEROSIAN AND BEUTLER; GLASBY 2; GOSSELIN; GRIEVE; LIST AND HÖRHAMMER; MCGUFFIN 1 & 2; MERCK; TERRELL; UPHOF.

1. J. Polonsky, in W. Herz et al., eds., *Fortschritte der Chemie Organischer Naturstoffe*, Vol. 30, Springer-Verlag, Vienna, 1973, p. 101.
2. H. Wagner et al., *Planta Med.*, **38**, 204 (1980).
3. S. M. Kupchan and D. R. Streelman, *J. Org. Chem.*, **41**, 3481 (1976).
4. D. Lavie and I. A. Kaye, *J. Chem. Soc.*, 5001 (1963).
5. H. Wagner et al., *Planta Med.*, **36**, 113 (1979).
6. D. H. Bray et al., *Phytother. Res.*, **1**, 22 (1987).
7. E. O. Ajaiyeoba et al., *J. Ethnopharmacol.*, **67**, 321 (1999).
8. S. Bertani et al., *J. Ethnopharmacol.* (2006).
9. S. Bertani et al., *J. Ethnopharmacol.*, **108**, 155 (2006).
10. E. O. Ajaiyeoba and H. C. Krebs, *Afr. J. Med. Med. Sci.*, **32**, 353 (2003).
11. S. Apers et al., *Planta Med.*, 68, 20 (2002).
12. W. Toma et al., *J. Ethnopharmacol.*, **85**, 19 (2003).
13. W. Toma et al., *Biol. Pharm. Bull.*, **25**, 1151 (2002).
14. S. Parveen et al., *Reprod. Toxicol.*, **17**, 45 (2003).
15. Y. Raji and A. F. Bolarinwa, *Life Sci.*, **61**, 1067 (1997).
16. V. C. Njar et al., *Planta Med.*, **61**, 180 (1995).

QUEBRACHO

Source: *Aspidosperma quebracho-blanco* Schlecht. (Family Apocynaceae) and *Schinopsis quebracho-colorado* (Schlecht.) Barkl. et T. Meyer (syn. *S. lorentzii* (Griseb.) Engl. and *Quebrachia lorentzii* Griseb.) (Family Anacardiaceae).

Common/vernacular names: Quebracho blanco, white quebracho (*A. quebracho-blanco*); quebracho colorado, red quebracho (*S. quebracho-colorado*).

GENERAL DESCRIPTION

Large trees native to Argentina and neighboring countries. Part used is their dried bark.

Although red quebracho has been approved for food use, it is doubtful that it is actually used in food products. Its wood extracts, containing large amounts of tannins and an alkaloid, have been primarily used in tanning leather and in dyeing. There has been relatively little scientific work (chemical or pharmacology) done on red quebracho relating to its food or drug use (MERCK).

CHEMICAL COMPOSITION

White quebracho contains 0.3–1.5% indole alkaloids, including aspidospermine, aspidospermatine, aspidosamine, yohimbine (quebrachine), *l*-quebrachamine, eburnamenine, aspidospermidine, *l*-pyrifolidine, deacetylpyrifolidine, rhazidine, and akuammidine, among others (GLASBY 1; LIST AND HÖRHAMMER; RAFFAUF; WILLAMAN AND SCHUBERT).[1–4]

Numerous alkaloids, including rhazinilam (a lactam), have also been isolated from the leaves of *A. quebracho-blanco*.[2,5–9]

Other constituents reportedly present in white quebracho include 3–4% tannin, sugars, β-sitosterol (quebrachol), and the triterpenic alcohols lupeol and α-amyrin (LIST AND HÖRHAMMER).[10]

PHARMACOLOGY AND BIOLOGICAL ACTIVITIES

White quebracho alkaloids (especially aspidospermine, aspidosamine, yohimbine, quebrachamine, and akuammidine) have been reported to have numerous pharmacological properties, including hypotensive, spasmolytic, diuretic, peripheral vasoconstrictor, arterial hypertensive, respiratory stimulant, uterine sedative, and local anesthetic, among others (LIST AND HÖRHAMMER; MARTINDALE).[2]

In a recent study, the bark extract has been demonstrated to bind with high affinity to human penile alpha$_2$ adrenoreceptors and with less affinity to the alpha$_1$ receptor. The authors' results provide support for the use of quebracho in erectile dysfunction and they attributed the reported activity to yohimbine present in the extract.[11] Aspidospermine and quebrachamine were also found to be similar to yohimbine in possessing adrenergic blocking activities in various urogenital tissues.[1]

TOXICOLOGY

Large doses of white quebracho may cause nausea and vomiting (MARTINDALE).

USES

Medicinal, Pharmaceutical, and Cosmetic. Now rarely used in pharmaceutical preparations.

Food. Quebracho is primarily used in foods. Its extract (type not specified) is used as a flavor ingredient in major categories of foods, including alcoholic and nonalcoholic beverages, frozen dairy desserts, candy, baked goods, and gelatins and puddings. Highest average maximum use level is about 0.003% reported in candy (29.8 ppm) and baked goods (34.5 ppm).

Dietary Supplements/Health Foods. Sometimes used as a tea ingredient; seldom used in the United States; liquid extracts used in respiratory preparations in the United Kingdom (WREN).

Traditional Medicine. White quebracho is used as a febrifuge, antispasmodic, and respiratory stimulant. Also used as an aphrodisiac. The latex of *A. nitidum* is used in Columbia for controlling leprosy. *A. schultesii* latex is used topically to control sores of probable fungal origin.[12]

COMMERCIAL PREPARATIONS

Mainly crude (white quebracho); white quebracho was formerly official in U.S.P.

Regulatory Status. Has been approved for food use (§172.510).

REFERENCES

See the General References for CLAUS; FEMA; GLASBY 1; GOSSELIN; GRIEVE; LEWIS AND ELVIN-LEWIS; LIST AND HÖRHAMMER; MARTINDALE; MCGUFFIN 1 & 2; TERRELL; UPHOF; WREN; YOUNGKEN.

1. H. F. Deutsch et al., *J. Pharm. Biomed. Anal.*, **12**, 1283 (1994).
2. R. L. Lyon et al., *J. Pharm. Sci.*, **62**, 218 (1973).
3. P. Tunmann and D. Wolf, *Z. Naturforsch. B*, **24**, 1665 (1969).
4. S. Markey et al., *Tetrahedron Lett.*, 157 (1967).
5. P. S. Benoit et al., *J. Pharm. Sci.*, **62**, 1889 (1973).
6. R. L. Lyon et al., *J. Pharm. Sci.*, **62**, 833 (1973).
7. D. J. Abraham et al., *Tetrahedron Lett.*, 909 (1972).
8. H. K. Schnoes et al., *Tetrahedron Lett.*, 993 (1962).
9. R. L. Lyon, *Diss. Abstr. Int. B*, **35**, 2673 (1974).
10. P. Tunmann and G. Hermann, *Pharmazie*, **25**, 361 (1970).
11. H. Sperling et al., *J. Urol.*, **168**, 160 (2002).
12. R. E. Schultes and R. F. Raffauf, *The Healing Forest*, Dioscorides Press, Portland, OR, 1990.

QUILLAIA

Source: *Quillaja saponaria* Mol. (Family Rosaceae).

Common/vernacular names: Soapbark, soap tree bark, murillo bark, quillaja, Panama bark, Panama wood, and China bark.

GENERAL DESCRIPTION

A large evergreen tree with shiny coriaceous (leathery) leaves and thick bark; native to Chile and Peru; cultivated in southern California. Part used is the dried inner bark deprived of cork.

CHEMICAL COMPOSITION

Quillaia contains 9–10% triterpenoid saponins consisting of glycosides of quillaic acid (quillaja sapogenin);[1–3] tannin; 11% calcium oxalate; sugars; starch; and others (LIST AND HÖRHAMMER).

PHARMACOLOGY AND BIOLOGICAL ACTIVITIES

Saponins are generally reported to have widely different (both in kind and in intensity) pharmacological and biological activities. Some of the more important activities include hemolytic (strong *in vitro*, much weaker *in vivo*),[4] local irritant, inflammatory (e.g., on intestine),[5–7] anti-inflammatory; antimicrobial,[8] cytotoxic,[9] and antihypercholesterolemic in laboratory animals (also see ***alfalfa***).[10–13]

It was recently shown that an aqueous extract of quillaia possessed marked cytoprotective against cellular infection by vaccinia virus, HSV-1, varicella zoster, HIV-1 and -2, and reovirus. The demonstrated activity was achieved at 0.1 μg/mL, which was noncytotoxic to host cells and was sustained for at least 16 h after the extract had been removed from the cell culture medium.[14]

Powdered quillaja bark or saponin concentrate (saponin) has highly local irritant and sternutatory (causing sneezing) properties. It also has expectorant properties as well as depressant activities on the heart and respiration. Saponin is reported to be too strongly hemolytic and irritating in the gastrointestinal tract to be used internally. Severe toxic effects due to large doses include liver damage, gastric pain, diarrhea, hemolysis of red blood corpuscles, respiratory failure, convulsions, and coma (LIST AND HÖRHAMMER; MARTINDALE).

An immunostimulating complex formed from a semipurified quillaia saponin fraction by a protein antigen has proven useful as a protective vaccine for equine influenza virus; also the subject of HIV research in humans.[15]

TOXICOLOGY

The chronic effects in humans due to the ingestion of low levels of saponin (especially in root beer) are not known. However, two studies (one short-term in rats and the other long-term in mice) from one laboratory have indicated quillaia saponins to be nontoxic.[2,16]

USES

Medicinal, Pharmaceutical, and Cosmetic. Quillaia extracts are used in some dermatological creams and in hair tonic preparations and shampoos for treating dandruff.

Food. Quillaia extracts (especially saponins concentrate) are used quite extensively as a foaming agent in root beer and cocktail mixes (see *yucca*). Other food products in which quillaia (product form not specified) is also reportedly used include frozen dairy desserts, candy, baked goods, and gelatins and puddings. Highest average maximum use level is about 0.01% for alcoholic and nonalcoholic beverages.

Traditional Medicine. Used for the relief of coughs, chronic bronchitis, and other pulmonary ailments; a component in herbal vaginal douches. Also used to relieve itchy scalp or dandruff, skin sores, and athlete's foot.

Others. Saponin used as a foaming agent in fire extinguishers (MABBERLY).

COMMERCIAL PREPARATIONS

Crude, extracts, and saponins concentrates; crude was formerly official in N.F.

Regulatory Status. Has been approved for food use (§172.510). Quillaia extract was included in a recent FAO/WHO committee report on the toxicology of various food additives and their daily intake.[17]

REFERENCES

See the General References for BAILEY 2; CLAUS; DER MARDEROSIAN AND BEUTLER; FEMA; GOSSELIN; GRIEVE; LIST AND HÖRHAMMER; LUST; MABBERLY; MCGUFFIN 1 & 2; ROSE; STAHL; TERRELL; UPHOF; WREN; YOUNGKEN.

1. R. A. Labriola and V. Deulofeu, *Experientia*, **25**, 124 (1969).
2. I. F. Gaunt et al., *Food Cosmet. Toxicol.*, **12**, 641 (1974).
3. R. Higuchi et al., *Phytochemistry*, **26**, 229 (1987).
4. C. D. Thron, *J. Pharmacol. Exp. Ther.*, **145**, 194 (1964).
5. R. Richou et al., *Rev. Immunol. Ther. Antimicrob.*, **33**, 155 (1969).
6. R. Richou et al., *Compt. Rend.*, **260**, 3791 (1965).

7. P. Lallouette et al., *C. R. Acad. Sci. Paris, Ser. D*, 582 (1967).
8. B. Wolters, *Planta Med.*, **14**, 392 (1966).
9. C. D. Thron et al., *Toxicol. Appl. Pharmacol.*, **6**, 182 (1964).
10. G. Wulff, *Dtsch. Apoth. Ztg.*, **108**, 797 (1968).
11. G. Vogel, *Planta Med.*, **11**, 362 (1963).
12. A. J. George, *Food Cosmet. Toxicol.*, **3**, 85 (1965).
13. E. Heftmann, *Phytochemistry*, **14**, 891 (1975).
14. M. R. Roner et al., *J. Gen. Virol.*, **88**, 275 (2007).
15. P. Newmark, *Biotechnology*, **6**, 23 (1988).
16. J. C. Phillips et al., *Food Cosmet. Toxicol.*, **17**, 23 (1979).
17. World Health Organ Tech. Rep. Ser., **934**, 1 (2006).

REHMANNIA

Source: *Rehmannia glutinosa* (Gaertn.) Libosch. ex Fisch. et Mey. (syn. *Rehmannia glutinosa* Libosch.; *R. glutinosa* Libosch. form *hueichingensis* (Chao et Schih) Hsiao; *R. glutinosa* (GaertiL) Libosch.; *R. chinensis* Libasch. ex Fisch. et Mey.; and *Digitalis glutinosa* Gaertn.) (Family Scrophulariaceae or Gesneriaceae).

Common/vernacular names: Radix Rehmanniae, Rhizoma Rehmanniae, *di*HUANG ("earth yellow," general name for rehmannia), *xian di*HUANG (fresh rehmannia), *sheng di*HUANG (raw rehmannia), *gan di*HUANG (dried rehmannia), *shu di*HUANG (cooked or cured rehmannia), *huaiqing di*HUANG (*R. glutinosa* f. *hueichingensis*), and Chinese foxglove.

GENERAL DESCRIPTION

Hardy herbaceous perennial, 25–40 cm high; whole plant covered with long soft hairs (pubescent-hirsute); root/rhizome thick and fleshy, tuberous, cylindrical or spindle-shaped, which is the part used, being referred to as root by some authors (CHP; CMH; FOSTER AND YUE; HU; IMM-2; MA; ZHU) and as rhizome by others (IMM-CAMS; JIANGSU; NATIONAL). The plant is distributed throughout most of China, especially northern, northeastern, and eastern and central provinces; extensively cultivated (IMM-CAMS).

Three major types of rehmannia are generally used, namely, fresh rehmannia (*xian di*HUANG), raw or dried rehmannia (*sheng di*HUANG or *gan di*HUANG), and cured or cooked rehmannia (*shu di*HUANG). However, only the latter two are available in the United States. Most rehmannia is produced from cultivated plants. Henan province is the largest producer, which also produces the best grades (ZHU).

The tubers are carefully dug up in autumn (sometimes also in spring) to avoid bruises that would promote rotting and rid of rootlets and dirt; these constitute fresh rehmannia, which can be used immediately or laid on the ground and covered with dry sand and dirt for later use as needed for up to 3 months.

Methods for producing dried or raw rehmannia from fresh rehmannia vary but basically consist of baking the fresh tubers at carefully controlled temperatures until the inside turns black, followed by kneading to round masses that are further dried to completion or first sliced and then sun dried to completion.

Cured or cooked rehmannia is generally produced by two methods: wine curing and steaming. To produce wine-cured rehmannia, dried rehmannia is first partially rehydrated, followed by mixing with wine (30–50% w/w) in a sealed container and steaming with the container sealed until all wine is absorbed. The tubers are then removed, partially dried, cut into thick slices, and further dried to completion. To produce steam-cured rehmannia, dried rehmannia is first partially rehydrated followed by steaming for hours until the tuber turns black and moist, which is then dried or first sliced into thick pieces and then dried (CMH; FOSTER AND YUE; JIANGSU; MA).

CHEMICAL COMPOSITION

Raw and cured rehmannia contain similar types of chemical components. The main difference is their relative proportions as a result of processing. The chemical constituents present include at least 25 iridoid, naphthopyrone, and phenethyl alcohol glycosides (catalpol 0.08–0.5%, with highest concentration in fresh and lowest in cured rehmannia; dihydrocatalpol; leonuride; aucubin; monomelittoside; melittoside; rehmanniosides A, B, C, and D; acteoside; isoacteoside; purpureaside C; echinacoside; cistanosides A and F; jionosides A_1 and B_1; etc.);[1–9] at least 15 free amino acids (ca. 0.16–6.15%), with higher concentrations in raw than in cured rehmannia as well as a thaumatin-like protein;[3,4,10–12] fatty acid esters (ca. 0.01%) composed mainly of methyl linoleate, methyl palmitate, and methyl-*n*-octadecanoate;[13] β-sitosterol, daucosterol, palmitic acid, succinic acid, and (S_8) cyclic compounds;[14] also sugars, including stachyose (a tetrasaccharide, 32.1–48.3%), sucrose, and monosaccharides (>3 times more in cured than in raw rehmannia);[15] trace minerals;[16] 1-ethyl-β-D-galactoside; mannitol; campesterol; α-aminobutyric acid; and so on. (HU; IMM-2).[3,4,17]

A number of polysaccharides have been reported that include a water-soluble polysaccharide fraction (RPS-b) composed of galactose, glucose, xylose, mannose, and arabinose in a 12:6:2:2:1 ratio, with molecular weights of 162,000 (70%), 66,000 (15%), 37,000 (8%), and 3000 (7%),[9] and four acidic polysaccharides (rehmannan FS-I and II and rehmannan SA and SB) composed of arabinose, galactose, rhamnose, galacturonic acid and glucuronic acid in different molar ratios.[18–20]

Compounds recently isolated from rehmannia include new furan and bis-furan analogs,[21] a new dimethoxymethylpyranone,[22] together with iridoids and phenethyl alcohol glycosides.[22]

PHARMACOLOGY AND BIOLOGICAL ACTIVITIES

Almost all the pharmacologic studies on rehmannia were originally conducted in China and Japan and most of the earlier reports did not specify the types of rehmannia used.

Later more specific studies have shown that cured rehmannia has tranquilizing, hypotensive, and diuretic activities and that the methods of curing (wine or steam) do not significantly affect these activities or its chemistry and clinical efficacy.[23] While dried rehmannia has exhibited both immunoenhancing and immunosuppressive effects in mice, the immunoenhancing effects are reduced while its immunosuppressive activities still remain after curing; it is suggested that the ether-soluble β-sitosterol, daucosterol and 1-ethyl-β-D-galactoside play a role in the immunosuppression.[24] Cured rehmannia has "blood tonic" effects, normalizing red cell counts and hemoglobin values in experimental posthemorrhagic anemia in mice, as well as markedly enhancing hematopoiesis (production and differentiation of CFU-S and CFU-E);[25] it has also shown "*yin*-nourishing" effects, being able to normalize serum levels of aldosterone, T_3, T_4, and other parameters in a hyperthyroid *yin*-deficiency rat model.[26]

The polysaccharide fraction, PRS-b, has exhibited immunomodulating/antitumor effects. Thus, when administered i.p., it was found to inhibit the growth of various transplanted tumors (sarcoma 5180, Lewis lung carcinoma, melanoma B16, hepatoma H22, etc.) in mice. It was also effective when administered p.o. in experiments with S180 but was ineffective *in vitro* against 5180 and HL22 cells.[9] The acidic polysaccharides also exhibited an immunomodulatory activity in the carbon clearance assay mediated through potentiation of the reticuloendothelial system.[18–20]

Other effects of *di*HUANG include hypoglycemic,[27,28] hypotensive, hypertensive, vasoconstricting, vasodilating, shortening of rabbit blood coagulation time, cardiotonic, liver protective, antifungal, anti-inflammatory, and so on; also found effective in treating rheumatoid arthritis, eczema, urticaria, and neurodermatitis (raw rehmannia, p.o.) (IMM-2; JIANGSU; WANG). More recently reported activities include antioxidant,[29–31] neuroprotectant,[32–35] antiallergic,[36] antiosteoporotic,[37,38] antitumor,[39] and antinephrotoxic.[40]

TOXICOLOGY

Toxicity of *di*HUANG (type not specified) is low: decoction and alcoholic extract administered to mice p.o. 60 g/kg for 3 days caused no deaths or adverse reactions after observation for 1 week (WANG).

USES

Medicinal, Pharmaceutical, and Cosmetic. Extracts of *di*HUANG (especially raw rehmannia) are used in skin care (e.g., toilet water, cleanser, and bath preparations) and hair care products (e.g., shampoos) for its anti-inflammatory and antimicrobial effects, as well as its traditional detoxicant and nourishing properties (ZHOU).

Dietary Supplements/Health Foods. Used in soup mixes where both raw and cured rehmannia are frequently used together; powder and extracts used in general and blood tonic formulas (FOSTER AND YUE).

Traditional Medicine. Earliest records of *di*HUANG date back to the *Shen Nong Ben Cao Jing* (ca. 200 BC–AD 100) for dried rehmannia and to the *Ben Cao-Tu Jing* (AD 1061) for cured rehmannia. Both are traditionally considered sweet tasting; the former cold while the latter slightly warming. Although both are *yin* nourishing, dried rehmannia is normally used for its heat-dispersing and blood-cooling properties in treating conditions related to febrile diseases (e.g., restlessness and thirst, skin eruptions, measles, vomiting blood, nosebleed, diabetes, restless fetus, etc.), and cured rehmannia is commonly used for its blood-tonifying properties and its beneficial effects on the vital essence (*jing*) in such conditions as lumbago, weak knees, night sweat, spermatorrhea (nocturnal emission), hectic fever, diabetes, palpitation, irregular menses, metrorrhagia (uterine bleeding), vertigo, tinnitus, and premature graying of hair. The two types of rehmannia are frequently used together in formulations and are among the most commonly used Chinese herbs. *di*HUANG is also highly valued as a tonic food for disease prevention and for prolonging life.[41,42]

COMMERCIAL PREPARATIONS

Crude and extracts of *sheng di*HUANG or *gan di*HUANG (raw or dried rehmannia) and *shu di*HUANG (cured rehmannia).

Regulatory Status. Rehmannia is an ethnic food. Class 2d dietary supplement (contraindicated with diarrhea and lack of appetite). However, it is not on the United Kingdom's General Sale List, and is not covered by a Commission E monograph in Germany. The U.S. Food and Drug Administration (FDA) has not granted generally recognized as safe (GRAS) status to rehmannia.

REFERENCES

See the General References for CHP; CMH; FOSTER AND YUE; HU; HUANG; IMM-2; IMM-CAMS; JIANGSU; JIXIAN; LU AND LI; MA; MCGUFFIN 1 & 2; NATIONAL; WANG; XIAO; ZHOU.

1. S. M. Wong et al., *Planta Med.*, **54**, 566 (1988).
2. H. Sasaki et al., *Planta Med.*, **55**, 458 (1989).
3. M. Tomoda et al., *Chem Pharm. Bull.*, **19**, 1455 (1971).
4. M. Tomoda et al., *Chem. Pharm. Bull.*, **19**, 2411 (1971).
5. M. Yoshikawa et al., *Chem. Pharm. Bull.*, **34**, 1403 (1986).
6. M. Yoshikawa et al., *Chem. Pharm. Bull.*, **34**, 2294 (1986).
7. M. Y. Ni and B. L. Bian, *Zhongguo Zhongyao Zazhi*, **14**, 40 (1989).
8. Y. Shoyama et al., *Phytochemistry*, **25**, 1633 (1986).
9. L. Z. Chen et al., *Chin. J. Pharmacol. Toxicol.*, **7**, 153 (1993).
10. C. H. Pan et al., *Biosci. Biotechnol. Biochem.*, **63**, 1138 (1999).
11. Y. X. Gong, *Zhongcaoyao*, **18**, 37 (1987).
12. M. Y. Ni et al., *Zhongguo Zhongyao Zazhi*, **14**, 21 (1989).

13. B. L. Bian et al., *Zhongguo Zhongyao Zazhi*, **16**, 339 (1991).
14. M. Y. Ni et al., *Zhongguo Zhongyao Zazhi*, **17**, 297 (1992).
15. Z. Y. Liu, *Zhongyao Tongbao*, **9**, 17 (1984).
16. M. Y. Ni and B. L. Bian, *Zhongyao Tongbao*, **13**, 18 (1988).
17. S. J. Wu et al., *Zhongcaoyao*, **15**, 6 (1984).
18. M. Tomoda et al., *Biol. Pharm. Bull.*, **17**, 1456 (1994).
19. M. Tomoda et al., *Chem. Pharm. Bull. (Tokyo)*, **42**, 1666 (1994).
20. M. Tomoda et al., *Chem. Pharm. Bull. (Tokyo)*, **42**, 625 (1994).
21. Y. S. Li et al., *Nat. Prod. Res.*, **19**, 165 (2005).
22. N. T. Anh et al., *Pharmazie*, **58**, 593 (2003).
23. Z. L. Cao et al., *Henan Zhongyi*, 36 (1989).
24. Z. L. Cao et al., *Zhongyao Tongbao*, **13**, 22 (1988).
25. Y. Yuan et al., *Zhongguo Zhongyao Zazhi*, **17**, 366 (1992).
26. S. L. Hou and J. W. Sheng, *Zhongguo Zhongyao Zazhi*, **17**, 301 (1992).
27. R. X. Zhang et al., *Pharmazie*, **59**, 552 (2004).
28. R. Zhang et al., *J. Ethnopharmacol.*, **90**, 39 (2004).
29. H. H. Yu et al., *Am. J. Chin Med.*, **34**, 1083 (2006).
30. H. H. Yu et al., *J. Ethnopharmacol.*, **107**, 383 (2006).
31. S. S. Kim et al., *Redox. Rep.*, **10**, 311 (2005).
32. J. Liu et al., *Brain Res.*, **1123**, 68 (2006).
33. Y. Y. Tian et al., *Life Sci.*, **80**, 193 (2006).
34. H. Yu et al., *Pharmacol. Res.*, **54**, 39 (2006).
35. D. Q. Li et al., *Toxicon*, **46**, 845 (2005).
36. H. Kim et al., *Int. J. Immunopharmacol.*, **20**, 231 (1998).
37. K. O. Oh et al., *Clin. Chim. Acta*, **334**, 185 (2003).
38. M. Kubo et al., *Biol. Pharm. Bull.*, **17**, 1282 (1994).
39. L. Z. Chen et al., *Zhongguo Yao Li Xue Bao*, **16**, 337 (1995).
40. D. G. Kang et al., *Biol. Pharm. Bull.*, **28**, 1662 (2005).
41. M. Y. Chang, *Zhongguo Shipin*, 25 (1987).
42. X. P. Li, *Zhongchengyao*, **15**, 47 (1993).

RHUBARB

Source: *Chinese rhubarb* Rheum officinale Baill., *R. palmatum* L., *R. tanguticum* Maxim. ex Reg., and other *Rheum* species or hybrids grown in China; **Indian rhubarb** *Rheum australe* D. Don. (syn. *R. emodi* Wall.); **Garden rhubarb** *Rheum* × *cultorum* Hort. (syn. *R. rhabarbarum* L. erroneously attributed to *R. rhaponticum* L.) (Family Polygonaceae).

Common/vernacular names: Medicinal rhubarb (*R. officinale*); Himalayan rhubarb (*R. australe*); common rhubarb, pie plant (*R.* × *cultorum*).

GENERAL DESCRIPTION

Large and sturdy, perennial herbs with large leaves borne on thick petioles; stem up to 2–3 m high; native to Asia (e.g., China, India, and southern Siberia); widely cultivated. Parts used are the dried rhizome and roots deprived of periderm (corky layer). Only plants 3 y or older are used. Chinese rhubarb, especially those from *R. officinale* and *R. palmatum*, are considered to be of the best quality. Those species with palmate rather than undulate leaves are generally considered the official drug source species or substitutes in China.

The species cultivated as ornamental plants in the United States are generally *R. palmatum* and *R.* × *cultorum*; the latter is also grown for its edible stalks (petioles). *R.* × *cultorum*, a hybrid that evolved in the 18th century, probably involved the rare eastern European species *R. rhaponticum*, a binomial commonly associated with the garden rhubarb (FOSTER AND YUE; MABBERLY).[1,2]

CHEMICAL COMPOSITION

Chinese rhubarbs contain two major classes of active constituents, anthraglycosides and tannins.

The amounts of anthraglycosides and related free anthraquinones range from about 1% (in *R. tanguticum*) to more than 5% (in *R. palmatum* and *R. officinale*).[3,4] Anthraglycosides constitute 75% or more of the mixture, with the rest being free anthraquinones that are mainly chrysophanic acid (chrysophanol), emodin (rheum emodin), aloe-emodin, rhein, and physcion (parietin). The major anthraglycosides are *O*-glucosides (chrysophanol-1-monoglucoside, emodin-6-monoglucoside, aloe-emodin-8-monoglucoside, rhein-8-monoglucoside, physcione monoglucoside, etc.), with lesser amounts of the dianthrone glycosides sennosides A, B, C, D, E, and F (see *cascara*, *frangula*, and *senna*).[3–7] The new anthrone *C*-glucosides 10-hydroxycascaroside C and D and 10-*R*-chrysaloin 1-*O*-β-D-glucopyranoside were isolated from the roots of *R. emodi* together with cascaroside C and D and cassialoin.[8] Although the total anthracene derivatives remain relatively constant year-round, the relative concentrations of anthrone and anthraquinone forms of the glycosides vary with time of harvest.[9] A recent study found physcion-8-*O*-β-D-gentiobioside to be a major constituent of the hydroxyanthracene complex.[10] New oxanthrone esters (revandchinone-1 and -2), anthraquinone ether (revandchinone-3), and oxanthrone ether (revandchinone-4) were reported in *R. emodi* rhizome.[11] Other isolated constituents with laxative activity from Chinese rhubarb include rheinosides A, B, C, and D.[12]

Nonanthraquinone glycosides have also been isolated from Chinese rhubarb; they are stilbene glycosides (e.g., 3,5,4′-trihydroxystilbene-4′-*O*-β-D-glucopyranoside) related to rhaponticin (rhapontin) found in *R. rhaponticum* (JIANGSU).[11] 6-Acyl stilbene and anthraquinone glycosides were reported for the first time in Japanese *R. palmatum*. Also, a new 6-sulfated anthraquinone (emodin 8-*O*-β-D-glucopyranosyl-6-*O*-sulfate) was recently isolated from Nepalese *R. emodi* together with two rare auronols (carpusin and maesopsin).[13]

The tannins in Chinese rhubarb are of both the catechin and gallic acid types, including *d*-catechin, *d*-epicatechin gallate, glucogallin (galloylglucose), and others (LIST AND HÖRHAMMER).[14,15]

Chinese rhubarb also contains a trace amount of volatile oil consisting of 100 constituents, including chrysophanic acid and other anthraquinones, diisobutyl phthalate, cinnamic acid, phenylpropionic acid, and ferulic acid, but no *p*-hydroxycinnamic, caffeic, or quinic acid.[16,17] In a recent report, the stalks of uncooked *R. rhabarbarum* yielded 59 volatile monoterpene alcohols, aldehydes, and acids to which the characteristic odor of rhubarb is largely attributed.[18]

Other constituents present in Chinese rhubarbs include calcium oxalate (ca. 6%), fatty acids (oleic acid, palmitic acid, etc.), sugars (glucose, fructose, etc.), rutin and other flavonoids, starch (ca. 16%), resins, and others (BRADLY; JIANGSU; NANJING; STAHL).

Indian rhubarb and garden rhubarb have been much less extensively investigated than Chinese rhubarb. Indian rhubarb is reported to contain tannins and anthraglycosides, many of which are probably the same as those of Chinese rhubarb. Garden rhubarb has been reported to contain chrysophanic acid (probably also its glucosides) but not the other anthraquinones or sennosides found in Chinese rhubarb; it contains rhaponticin not present in Chinese or Indian rhubarb (MARTINDALE; NANJING; STAHL).[4]

PHARMACOLOGY AND BIOLOGICAL ACTIVITIES

Chinese rhubarb has both astringent and cathartic properties. In small doses or when used under certain conditions, it is astringent and is used to treat diarrhea, while in larger doses or under other conditions it is cathartic (JIANGSU; MARTINDALE; STAHL).[19]

The cathartic properties of Chinese rhubarb are due mainly to the sennosides and the other anthraglycosides (see cascara, frangula, and senna). Laxative action occurs as the result of inhibition of water and electrolyte absorption from the large intestine, influencing intestinal motility. Anthraquinone glycosides are bound to emodin aglycones by microbes, partially absorbed, then reduced to anthranols and anthrones, which are responsible for the laxative effect.[2,20] Experiments on isolated gastric muscle strips of guinea pigs showed that rhubarb extract has a direct excitatory action. The cholinergic M and N as well as the L-type calcium channel are believed to be partly involved in this effect on the smooth muscle.[21] The tannins, on the other hand, are responsible for the astringent properties of rhubarb (see *tannic acid*).

Rhubarb extract and its different phenolic glycosides have demonstrated antitumor activities in different assays for cell proliferation, apoptosis and metastasis prevention, and against a number of human tumor cell lines, for example, salivary gland and oral squamous cell carcinoma.[22–25] Those with weak antitumor activity may still have potential as cancer chemopreventive agents.[24]

Extracts of *R. emodi* and its anthraquinones possess antimicrobial properties against Gram-positive and Gram-negative bacteria and different types of fungi.[11,26] Marked activity of the extract was also demonstrated *in vitro* and *in vivo* against *H. pylori*.[27]

Other pharmacological properties of rhubarb and its extracts include a hypotensive vasorelaxant effect,[28,29] anti-inflammatory effects (involving NO, TNF-α signaling),[28,30–32] antioxidant,[33–35] and cholesterol lowering effects,[36,37] among numerous others (JIANGSU).[38]

There are recent indications that rhaponticin and its aglycone metabolite (rhapontigenin) may have potential in the management of Alzheimer's disease and diabetes through a cytoprotective effect and α-glucosidase inhibition, respectively.[39,40]

TOXICOLOGY

Leaf blades of rhubarb are poisonous, and accidental ingestion will cause severe vomiting and may also cause liver and kidney damage. The toxic effect was formerly attributed to oxalates present in the leaves, but later data indicate it to be due to monoanthrones (see *cascara*).[7] Many cases of renal failure in rats and chronic renal toxicity in humans have been reported.[41,42] However, a recent report claims that extracts from Indian *R. emodi* have nephroprotective and renal function improving effects.[43]

USES

Medicinal, Pharmaceutical, and Cosmetic. Rhubarb and its extracts are used in certain laxative preparations.

Food. Chinese rhubarb is used as a flavor component (bitter note) in major food products, including alcoholic (e.g., bitters) and nonalcoholic beverages, frozen dairy desserts, candy, baked goods, and gelatins and puddings. Highest average maximum average use level is 0.05% in alcoholic beverages and baked goods.

Dietary Supplements/Health Foods. Chinese rhubarb used in various laxative formulations; included as an ingredient in the unconventional cancer treatment formula Essiac; for which the garden rhubarb is sometimes erroneously substituted. Tonic rhubarb alcoholic beverages are made in China and Italy.[2]

Traditional Medicine. Chinese rhubarb is considered a very valuable and versatile drug in Chinese medicine having been used for thousands of years not only for treating constipation but also for other conditions such as chronic diarrhea, thermal burns, jaundice, sores, and cancers (WANG).[44] It has also been used in treating upper GI bleeding with considerable success.[45–47]

COMMERCIAL PREPARATIONS

Crudes (Chinese, numerous grades; Indian) and extracts (e.g., solid and fluid). Crude, tincture, and fluid extract were formerly official in N.F. Strengths (see *glossary*) of extracts are expressed in weight-to-weight ratios.

Regulatory Status. Chinese rhubarb and garden rhubarb have been approved for food use; the latter in alcoholic beverages only (§172.510); Indian rhubarb is not listed. Chinese rhubarb (*R. palmatum*) is the subject of a positive German therapeutic monograph.[20]

REFERENCES

See the General References for BAILEY 1; BAILEY 2; BLUMENTHAL 1; BRADLY; CLAUS; FEMA; FOSTER AND YUE; JIANGSU; LIST AND HÖRHAMMER; MARTINDALE; NANJING; STAHL; TERRELL; UPHOF.

1. D. E. Marshall, *A Bibliography of Rhubarb and Rheum Spices*, National Agricultural Library, Bibliography and Literature of Agriculture, no. 62, Beltsville, MD, 1988.
2. C. M. Foust, *Rhubarb: The Wondrous Drug*, Princeton University Press, Princeton, NJ, 1992.
3. J. H. Zwaving, *Planta Med.*, **21**, 254 (1972).
4. B. Klimek, *Ann. Acad. Med. Lodz.*, **14**, 133 (1973).
5. J. H. Zwaving, *Pharm. Weekbl.*, **109**, 1169 (1974).
6. H. Oshio et al., *Chem. Pharm. Bull.*, **22**, 823 (1974).
7. F. H. L. van Os, *Pharmacology*, **14** (Suppl. 1), 7 (1976).
8. L. Krenn et al., *Chem. Pharm. Bull. (Tokyo)*, **52**, 391 (2004).
9. E. H. C. Verhaeren et al., *Planta Med.*, **45**, 15 (1982).
10. L. Holzschuh et al., *Planta Med.*, **46**, 159 (1982).
11. K. S. Babu et al., *Phytochemistry*, **62**, 203 (2003).
12. T. Yamagishi et al., *Chem. Pharm. Bull.*, **35**, 3132 (1987).
13. L. Krenn et al., *J. Nat. Prod.*, **66**, 1107 (2003).
14. G. Nonaka et al., *Chem. Pharm. Bull.*, **25**, 2300 (1977).
15. H. Friedrich and J. Hohle, *Arch. Pharm.*, **299**, 857 (1966).
16. C. Frattini et al., *Riv. Ital. Essenze, Profumi, Piante Offic., Aromi, Saponi, Cosmet., Aerosol*, **58**, 132 (1976).
17. C. Frattini et al., *Riv. Ital. Essenze, Profumi, Piante Offic., Aromi, Saponi, Cosmet., Aerosol*, **56**, 597 (1974).
18. M. Dregus and K. H. Engel, *J. Agric. Food Chem.*, **51**, 6530 (2003).
19. J. W. Fairbairn, *Pharmacology*, **14** (Suppl. 1), 48 (1976).
20. Monograph *Rhei radix*, *Bundesanzeiger*, no. 228 (December 5, 1984); revised (April 27, 1989).
21. M. Yu et al., *World J. Gastroenterol.*, **11**, 2670 (2005).
22. Q. Huang et al., *Med. Res. Rev.* (2006).
23. X. Zhou et al., *Bioorg. Med. Chem. Lett.*, **16**, 563 (2006).
24. Y. Q. Shi et al., *Anticancer Res.*, **21**, 2847 (2001).
25. O. K. Kabiev and S. M. Vermenichev, *Vopr. Onkol.*, **12**, 61 (1966).
26. S. K. Agarwal et al., *J. Ethnopharmacol.*, **72**, 43 (2000).

27. M. Ibrahim et al., *World J. Gastroenterol.*, **12**, 7136 (2006).
28. M. K. Moon et al., *Life Sci.*, **78**, 1550 (2006).
29. M. Y. Yoo et al., *Phytother. Res.*, (2006).
30. Y. Q. Zhao et al., *World J. Gastroenterol.*, **10**, 1005 (2004).
31. C. C. Wang et al., *Planta Med.*, **68**, 869 (2002).
32. T. Kageura et al., *Bioorg. Med. Chem.*, **9**, 1887 (2001).
33. Y. Cai et al., *J. Agric. Food Chem.*, **52**, 7884 (2004).
34. A. Iizuka et al., *J. Ethnopharmacol.*, **91**, 89 (2004).
35. H. Matsuda et al., *Bioorg. Med. Chem.*, **9**, 41 (2001).
36. I. Abe et al., *Planta Med.*, **66**, 753 (2000).
37. V. Goel et al., *Br. J. Nutr.*, **81**, 65 (1999).
38. X. S. Gao et al., eds., Abstracts of the First International Symposium on Rhubarb, Chengdu, 29–31 May 1990, Institute of Chinese Materia Medica, China Academy of Traditional Chinese Medicine, Beijing, 1990.
39. F. Misiti et al., *Brain Res. Bull.*, **71**, 29 (2006).
40. B. K. Suresh et al., *Bioorg. Med. Chem. Lett.*, **14**, 3841 (2004).
41. Z. Kang et al., *J. Tradit. Chin Med.*, **13**, 249 (1993).
42. T. Yokozawa et al., *Nippon Jinzo Gakkai Shi*, **35**, 13 (1993).
43. M. M. Alam et al., *J. Ethnopharmacol.*, **96**, 121 (2005).
44. J. L. Hartwell, *Lloydia*, **33**, 288 (1970).
45. D. H. Jiao et al., *Pharmacology*, **20** (Suppl. 1), 128 (1980).
46. D. A. Sun et al., *Chin. J. Integr. Trad. West. Med.*, **6**, 4589 (1986).
47. D. H. Jiao et al., *Zhejiang Zhongyi Zazhi*, **23**, 179 (1988).

ROSE HIPS

Source: *Rosa canina* L., *R. gallica* L., *R. rugosa* Thunb., *R. villosa* L. (syn. *R. pomifera* Herrm.), and other *Rosa* species (Family Rosaceae).

Common/vernacular names: Hipberries.

GENERAL DESCRIPTION

Prickly bushes or shrubs; native to Europe and Asia; extensively cultivated. Part used is the ripe fruit. *R. canina* is the major commercial source of rose hips; with ellipsoid, globose, or ovoid fruits. The *R. canina* group is separated into 17 or more species; found in most of Europe (TUTIN 2).

CHEMICAL COMPOSITION

Rose hips contain high concentrations of vitamin C (ascorbic acid), ranging from 0.24% to 1.25% depending on sources, climate, degrees of ripeness, and other factors (compare with *acerola*).[1–5]

Other constituents present in rose hips include carotenoids (0.01–0.05%),[4–6] flavonoids (0.01–0.35%),[5,7] pectic substances (3.4–4.6%), polyphenols (2.02–2.64%), leucoanthocyanins (1.35–1.75%), catechins (0.8–0.91%),[5] riboflavin,[4] sugars (glucose, fructose, sucrose, etc.), and plant acids (e.g., citric and malic), among others (JIANGSU). Depending on the species, rose hips may also contain purgative glycosides (e.g., multiflorin A and multiflorin B in *Rosa multiflora* Thunb.),[8] saponins (ca. 17% in *R. laevigata*

Michx.), and other compounds that have widely different pharmacological properties (JIANGSU) (see **Cherokee rosehip**).

More recently identified constituents include condensed tannins and ellagitannins (e.g., rudosin E),[1,9,10] glycoproteins and glycolipids,[11–13] and ursane-type triterpenes (e.g., rosamultin).[14,15] A sesquiterpene, (+)-4-epi-α-bisabolol, has been isolated from the leaves of two *R. rugosa* hybrids.[16]

PHARMACOLOGY AND BIOLOGICAL ACTIVITIES

Vitamin C has antiscorbutic properties.

Mild laxative and diuretic effects reported, probably attributable to malic and citric acid content (TYLER 1).

Antioxidant activity has been displayed by rose hips extracts and its different constituents, such as glycoproteins, tannins, rosamultin, polyphenolics, and vitamin C.[9,12,17–20]

The triterpenes of *R. rugosa* have antinociceptive and anti-inflammatory properties in mice and rats.[15] Anti-inflammatory activity was also displayed by a galactolipid isolated from *R. canina*.[13] Preparations made from the powdered rose hips were effective in reducing the painful symptoms of knee and hip osteoarthritis in human patients.[21,22]

The glycopeptides of *R. rugosa* strongly inhibit HIV-1 reverse transcriptase *in vitro*.[11] Anti-HIV protease activity was also exhibited by *R. rugosa* extract and its rosamultin constituent.[14]

R. rugosa ellagitannin, rugosin E, induces rabbit and human platelet aggregation at EC_{50} of 1.5 and 3.2 mM, respectively. Rugosin E was suggested to be a platelet receptor agonist of ADP.[10]

USES

Food. Rose hips are mainly used as a source of natural vitamin C. However, as with acerola, much of the vitamin C in rose hips is destroyed during ordinary drying or extraction (MERCK).[23] For this reason, most rose hips products are supplemented with synthetic vitamin C. They are usually in the form of tablets or capsules.

Dietary Supplements/Health Foods. Rose hips are widely used as an herb tea ingredient; also in capsules, tablets, and so on as a questionable source of vitamin C.[24]

COMMERCIAL PREPARATIONS

Crude and extracts. Usually adjusted with vitamin C to a specified amount.

Regulatory Status. GRAS (§182.20); with only *Rosa alba* L., *R. centifolia* L., *R. damascena* Mill., *R. gallica* L., and their varieties listed. Subject of a German therapeutic monograph; claimed efficacy to prevent colds and flu (due to vitamin C) content are not sufficiently documented; corrigent for tea mixtures.[24]

REFERENCES

See the General References for BAILEY 1; BIANCHINI AND CORBETTA; BLUMENTHAL 1; BRUNETON; DER MARDEROSIAN AND BEUTLER; DUKE 4; JIANGSU; MCGUFFIN 1 & 2; NANJING; TUTIN 2; TYLER 1; UPHOF.

1. J. P. Salminen et al., *J. Chromatogr. A*, **1077**, 170 (2005).
2. Z. Butkiene, *Liet. TSR Mokslu Akad Darb., Ser. C*, **1**, 51 (1969).
3. A. N. Nizharadze et al., *Konservn. Ovoshchesuch. Prom-st.*, **4**, 36 (1977).
4. G. P. Shnyakina and E. P. Malygina, *Rastit Resur.*, **11**, 390 (1975).

5. S. G. Mel'yantseva, *Konservn. Ovoshchesush. Prom-st.*, **2**, 13 (1978).
6. T. Hodisan et al., *J. Pharm. Biomed. Anal.*, **16**, 521 (1997).
7. M. Retezeanu et al., *Farmacia (Bucharest)*, **20**, 167 (1972).
8. S. Tagaki et al., *Yakugaku Zasshi*, **96**, 1217 (1976).
9. T. B. Ng et al., *J. Pharm. Pharmacol.*, **58**, 529 (2006).
10. C. M. Teng et al., *Thromb. Haemost.*, **77**, 555 (1997).
11. M. Fu et al., *J. Pharm. Pharmacol.*, **58**, 1275 (2006).
12. T. B. Ng et al., *J. Pharm. Pharmacol.*, **56**, 537 (2004).
13. E. Larsen et al., *J. Nat. Prod.*, **66**, 994 (2003).
14. J. C. Park et al., *J. Med. Food*, **8**, 107 (2005).
15. H. J. Jung et al., *Biol. Pharm. Bull.*, **28**, 101 (2005).
16. Y. Hashidoko et al., *Biosci. Biotechnol. Biochem.*, **66**, 2474 (2002).
17. T. B. Ng et al., *Biochem. Cell Biol.*, **83**, 78 (2005).
18. P. J. Cheol et al., *J. Med. Food*, **7**, 436 (2004).
19. E. J. Cho et al., *Am. J. Chin. Med.*, **32**, 487 (2004).
20. D. A. els-Rakotoarison et al., *Phytother. Res.*, **16**, 157 (2002).
21. K. Winther et al., *Scand. J. Rheumatol.*, **34**, 302 (2005).
22. E. Rein et al., *Phytomedicine*, **11**, 383 (2004).
23. S. Mrozewski, *Przem. Spozyw.*, **22**, 294 (1968).
24. *Bundesanzeiger*, **164**, (1990).

ROSE OIL (AND ABSOLUTE)

Source: *Rosa alba* L., *R. centifolia* L., *R. damascena* Mill., *R. gallica* L., and their varieties (Family Rosaceae).

Common/vernacular names: Bulgarian otto of rose, Bulgarian rose oil, Bulgarian attar of rose (*R. damascena* var. *alba*); Moroccan otto of rose, Moroccan rose oil (*R. centifolia*); Turkish otto of rose, Turkish rose oil, Turkish attar of rose (*R. damascena*); French rose absolute, rose de mai absolute (*R. centifolia*).

GENERAL DESCRIPTION

Small prickly shrubs up to about 1.2–2.4 m high; generally considered to be natives of Europe and western Asia; widely cultivated. Parts used are the fresh flowers, from which rose oil is obtained normally in 0.02–0.03% yield by steam distillation and rose absolute by petroleum ether extraction followed by alcohol solubilization (see **concrete** and **absolute** in **glossary**).[1–3] Rose water (more specifically called stronger rose water) is the aqueous portion of the steam distillation after rose oil is removed.

Flowers are usually harvested early in the morning when their essential oil content is the highest.[1,4] Major rose oil and absolute producing countries include France, Bulgaria, Morocco, Turkey, Italy, and China.

It has been proposed that *R. damascena* evolved as an eastern Mediterranean hybrid between *R. gallica* and *R. Phoenicia*.[5]

Major commercial cultivars of *R. damascena* include "Trigintipetala" ("Kananlik rose"), which is confused in the horticultural trade with "Prof. Emile Perrot," "Alika," and "Bella Donna" (TUCKER 1–3).

Major commercial cultivars of *R. gallica* include "Conditorum" (Hungarian rose) and "Officinalis" (apothecary rose) (TUCKER 1–3).

CHEMICAL COMPOSITION

Rose oil contains usually as its major components geraniol, citronellol, nerol, β-phenethyl alcohol, geranic acid, and eugenol, which together make up 55–75% of the oil, with citronellol up to 60%.[4,6,7] These components are found in free and bound forms in ratios specific to individual species or cultivars.[8] Other components present include terpene hydrocarbons (especially α- and β-pinenes, myrcene, etc.), esters, C_{14} to C_{23} n-paraffins making up the so-called stearoptenes (accounting for 15–23% of the oil), nerol oxide, and others (GUENTHER; JIANGSU).[4,9–11]

Important fragrance components of rose oil include (4R)-cis-rose oxide, (4R)-trans-rose oxide, (±)-nerol oxide, 3-(4-methyl-3-pentenyl)-2-buten-4-olide, 3-methyl-4-(3-methyl-2-butenyl)-2-buten-4-olide, cis- and trans-2-(3-methyl-2-butenyl)-3-methyl tetrahydrofuran, β-damascenone, 3-hydroxy-β-damascenone, and β-damascone.[8]

Rose absolute contains mainly phenethyl alcohol, with lesser amounts of citronellol, geraniol, and nerol.[12] Other compounds present include eugenol esters, and others.[13] A major review of rose oil and extracts has been published.[8]

PHARMACOLOGY AND BIOLOGICAL ACTIVITIES

Rose oil when added to the food has been reported to have choleretic effects on cats.[14]

Bulgarian rose oil has been reported to decrease urinary corticosteroids and serum ceruloplasmin, as well as elicit other effects on laboratory animals when administered intraperitoneally or intravenously.[15]

Available data from one source indicate rose oil (Moroccan, Bulgarian, and Turkish) to be nonirritating, nonsensitizing, and nonphototoxic to human skin, but slightly to moderately irritating to rabbit skin when applied undiluted.[16–18] Rose absolute French was nonirritating and nonphototoxic, but it produced one sensitization reaction in 25 subjects tested.[19]

The anxiolytic activity of rose oil has recently been reported by two groups. In the first study, the oil was active in mice using the Geller and Vogel conflict tests. The authors concluded that the active constituents were 2-phenethyl alcohol and citronellol.[20] In the second study, inhalation of rose oil to adult male rats (1–5% w/w) produced an anxiolytic effect comparable to diazepam (1 and 2 mg/kg, i.p.) in the elevated plus-maze test.[21]

The essential oil of R. damascena displayed a strong relaxant effect on the precontracted guinea pig tracheal chains at 0.75% and 1.0%. The effect was comparable to that of theophylline at 0.75 and 1.0 mM.[22]

USES

Medicinal, Pharmaceutical, and Cosmetic. Rose oil and stronger rose water are used primarily as fragrance components and as astringent in pharmaceutical preparations (e.g., cold creams, ointments, and lotions).

Rose oil and rose absolute are extensively used as fragrance ingredients in perfumes, creams, lotions, soaps, and sometimes detergents. Maximum use level is 0.2% reported for French rose absolute and rose oils Moroccan, Bulgarian, and Turkish in perfumes.[15–18]

Food. Rose oil and absolute are used extensively as flavor ingredients (usually in very low use levels) in fruit-type flavors. Food products in which they are used include alcoholic and nonalcoholic beverages, frozen dairy desserts, candy, baked goods, and gelatins and puddings. Reported average maximum use levels are generally below 0.0002% (2 ppm).

Dietary Supplements/Health Foods. Rose oil used in aromatherapy for aphrodisiac and rejuvenating claims (ROSE).

Others. Rose oil and to a lesser extent rose absolute are used in flavoring tobacco.

COMMERCIAL PREPARATIONS

Rose oil (Moroccan, French, Turkish, etc.) and absolute (French).

Rose oil is official in N.F. and F.C.C. Stronger rose water is official in N.F.

Regulatory Status. GRAS (§182.20). *Rosa gallica* flowers are the subject of a German therapeutic monograph; allowed for mild inflammations of the oral and pharyngeal mucosa.[23]

REFERENCES

See the General References for ARCTANDER; BAILEY 1; BLUMENTHAL 1; FEMA; GUENTHER; JIANGSU; MARTINDALE; ROSE; TUCKER 1–3; YOUNGKEN.

1. S. Kapetanovic, *Kem. Ind.*, **21**, 355 (1972).
2. D. Ivanov, *Parfums, Cosmet., Savons Fr.*, **2**, 153 (1972).
3. S. Kapetanovic, *Kem. Ind.*, **23**, 629 (1974).
4. M. R. Narayana, *Indian Perfum.*, **13**, 46 (1969).
5. M. P. Widrlechner, *Econ. Bot.*, **35**, 42 (1981).
6. Y. Ohno and S. Tanaka, *Agric. Biol. Chem.*, **41**, 399 (1977).
7. Y. Mikhailova et al., *Dokl. Bolg. Akad. Nauk*, **30**, 89 (1977).
8. B. M. Lawrence, *Perfum. Flav.*, **16**, 43 (1991).
9. V. Staikov, *Dokl. Acad. Sel'skokhoz. Nauk Bolg.*, **4**, 199 (1971).
10. G. Igolen et al., *Riv. Ital. Essenze, Profumi, Piante offic., Aromi, Saponi, Cosmet.*, **50**, 352 (1968).
11. C. Ehret and P. Teisseire, *Recherches*, **19**, 287 (1974).
12. M. S. Karawya et al., *Bull. Fac. Pharm., Cairo Univ.*, **13**, 183 (1974).
13. Y. Ohno et al., *Kanzei Chuo Bunsekishoho*, **15**, 47 (1975).
14. E. N. Vasil'eva and V. P. Gruncharov, *Farmakol. Toksikol. (Moscow)*, **35**, 312 (1972).
15. A. Maleev et al., *Eksp. Med. Morfol.*, **10**, 149 (1971).
16. D. L. J. Opdyke, *Food Cosmet. Toxicol.*, **12**(Suppl.), 981 (1974).
17. D. L. J. Opdyke, *Food Cosmet. Toxicol.*, **12**(Suppl.), 979 (1974).
18. D. L. J. Opdyke, *Food Cosmet. Toxicol.*, **13**(Suppl.), 913 (1975).
19. D. L. J. Opdyke, *Food Cosmet. Toxicol.*, **13**(Suppl.), 911 (1975).
20. T. Umezu et al., *Life Sci.*, **72**, 91 (2002).
21. R. N. de Almeida et al., *Pharmacol. Biochem. Behav.*, **77**, 361 (2004).
22. M. H. Boskabady et al., *J. Ethnopharmacol.*, **106**, 377 (2006).
23. Monograph *Rosae flos*, *Bundesanzeiger*, no. 164 (September 1, 1990).

ROSELLE

Source: *Hibiscus sabdariffa* L. (Family Malvaceae).

Common/vernacular names: Hibiscus, Jamaica sorrel, and Guinea sorrel.

GENERAL DESCRIPTION

Large strong annual, often shrub-like; up to about 2.4 m high; native to the Old World tropics; widely cultivated in the tropics and subtropics (e.g., Sudan, China, Thailand, Egypt, Mexico, and the West Indies). Parts used are the dried red, fleshy sepals (calyx) together with the bracts. Major producing countries include Sudan, China, and Thailand.

CHEMICAL COMPOSITION

Roselle contains about 1.5% anthocyanins (see *grape skin extract*) consisting mainly of delphinidin-3-sambubioside (hibiscin or daphniphylline) and cyanidin-3-sambubioside, with cyanidin-3-glucoside, delphinidin-3-glucoside, delphinidin, and other pigments in lesser amounts;[1–7] large amounts (23%) of hibiscic acid;[7–9] fruit acids consisting mostly of citric acid (12–17%), protocatechuic acid,[10,11] with small amounts of malic, tartaric, and other acids also present;[9] resins; sugars and polysaccharides;[12] trace of an unidentified alkaloid; and others (LIST AND HÖRHAMMER).[13]

The pigments have been tested as food colors (red) and found to be stable in fruit jellies and jams but unstable (hence unsuitable) in carbonated beverages.[5,6]

PHARMACOLOGY AND BIOLOGICAL ACTIVITIES

The chemopreventive activity of hibiscus extracts has been extensively studied recently. The protective effect was demonstrated against human carcinomas,[14–18] chemically induced toxicity,[19,20] and hepatotoxicity.[11,21–27]. Antimutagenic activity has also been demonstrated both *in vitro* and *in vivo*.[28] The chemopreventive activity is attributed mainly to the antioxidant effect of the anthocyanins present in hibiscus extracts.[29–33]

Roselle decoction or infusion reportedly has hypotensive properties with no side effects.[9,13] This effect has been investigated in more depth in experimental animals and in humans over the past decade and the results seem to support earlier studies especially in mild to moderate hypertension.[34–37] Suggested mechanisms of action for the hypotensive activity include inhibition of Ca^{2+} influx into vascular smooth muscle, NO–cGMP-relaxant pathway, and possible acetylcholine- and histamine-like vasorelaxation.[38,39]

Roselle methanolic and aqueous extracts have been reported to have antispasmodic activity on intestinal and uteral muscles as well as anthelmintic (tapeworm) properties.[40,41] Extracts have been shown to both stimulate and inhibit isolated rabbit intestine. An aqueous extract of the calyces has been found to decrease intestinal motility in intact rats, as well as the oral-caecal transit time in dogs.[42]

The extracts and protocatechuic acid reportedly have bactericidal properties *in vitro* against Gram-positive bacteria (LIST AND HÖRHAMMER).[9,10,13,41]

Other activities reported for roselle include hypolipidemic,[31,43,44] analgesic/anti-inflammatory,[45] and α-amylase inhibitory activities.[8]

TOXICOLOGY

Adverse effects of roselle are rare. One study reported that sub-chronic administration of an aqueous extract of hibiscus induced testicular toxicity in rats.[46]

The phytochemistry, pharmacology, and toxicology of roselle have been recently reviewed.[47]

USES

Food. Roselle is reportedly used as a flavor ingredient in alcoholic and nonalcoholic beverages, frozen dairy desserts, candy, baked goods, and gelatins and puddings. Average maximum use level reported is 0.02% for all above categories.

Roselle in coarse ground form is used in herb teas, primarily to impart a deep reddish-brown color and a pleasant slightly acid taste.

Fresh roselle is used in jams, juices, and jellies with a cranberry-like flavor. Also used in wines.

Dietary Supplements/Health Foods. Dried flowers one of the most widely used herbal tea ingredients, imparting red color and tart flavor.

Traditional Medicine. Leaves are used in Egypt for treating heart and nerve diseases.[48]

Roselle (calyx) is used as a refrigerant. It has also been used in cancer.[49]

In European tradition, flowers reportedly used for loss of appetite, colds, upper respiratory tract congestion, circulatory impairment, diuresis, and mild laxation.[50]

COMMERCIAL PREPARATIONS

Mainly crude.

Regulatory Status. Has been approved for use in alcoholic beverages only (§172.510). Flowers subject of a German therapeutic monograph; claimed efficacies are not substantiated.

REFERENCES

See the General References for APPLEQUIST; BAILEY 1; BAILEY 2; BLUMENTHAL 1; BRUNETON; DER MARDEROSIAN AND BEUTLER; FEMA; LIST AND HÖRHAMMER; MCGUFFIN 1 & 2; MORTON 2; TERRELL; TYLER 1; UPHOF.

1. T. Frank et al., *J. Clin. Pharmacol.*, **45**, 203 (2005).
2. C. T. Du and F. J. Francis, *J. Food Sci.*, **38**, 810 (1974).
3. M. Shibata and M. Furukawa, *Shokubutsugaku Zasshi*, **82**, 341 (1969).
4. M. S. Karawya et al., *Egypt J. Pharm. Sci.*, **16**, 345 (1976).
5. W. B. Esselen and G. M. Sammy, *Food Prod. Dev.*, **9**, 37 (1975).
6. W. B. Esselen and G. M. Sammy, *Food Prod. Dev.*, **7**, 80 (1973).
7. H. Schilcher, *Dtsch. Apoth. Ztg.*, **116**, 1155 (1976).
8. C. Hansawasdi et al., *Biosci. Biotechnol. Biochem.*, **65**, 2087 (2001).
9. J. Kerharo, *Plant. Med. Phytother.*, **5**, 277 (1971).
10. K. S. Liu et al., *Phytother. Res.*, **19**, 942 (2005).
11. T. H. Tseng et al., *Chem. Biol. Interact.*, **101**, 137 (1996).
12. B. M. Muller and G. Franz, *Planta Med.*, **58**, 60 (1992).
13. A. Sharaf, *Planta Med.*, **10**, 48 (1962).
14. H. H. Lin et al., *Chem. Biol. Interact.*, **165**, 59 (2007).
15. Y. C. Chang et al., *Toxicol. Appl. Pharmacol.*, **205**, 201 (2005).
16. D. X. Hou et al., *Arch. Biochem. Biophys.*, **440**, 101 (2005).
17. T. H. Tseng et al., *Biochem. Pharmacol.*, **60**, 307 (2000).
18. T. H. Tseng et al., *Cancer Lett.*, **126**, 199 (1998).
19. A. Amin and A. A. Hamza, *Asian J. Androl.*, **8**, 607 (2006).
20. A. Adetutu et al., *Phytother. Res.*, **18**, 862 (2004).
21. J. Y. Liu et al., *Food Chem. Toxicol.*, **44**, 336 (2006).
22. A. Amin and A. A. Hamza, *Life Sci.*, **77**, 266 (2005).
23. B. H. Ali et al., *Phytother. Res.*, **17**, 56 (2003).
24. W. L. Lin et al., *Arch. Toxicol.*, **77**, 42 (2003).
25. C. L. Liu et al., *Food Chem. Toxicol.*, **40**, 635 (2002).
26. C. J. Wang et al., *Food Chem. Toxicol.*, **38**, 411 (2000).
27. T. H. Tseng et al., *Food Chem. Toxicol.*, **35**, 1159 (1997).
28. T. Chewonarin et al., *Food Chem. Toxicol.*, **37**, 591 (1999).
29. Y. C. Chang et al., *Food Chem. Toxicol.*, **44**, 1015 (2006).
30. E. O. Farombi and A. Fakoya, *Mol. Nutr. Food Res.*, **49**, 1120 (2005).

31. V. Hirunpanich et al., *J. Ethnopharmacol.*, **103**, 252 (2006).
32. V. Hirunpanich et al., *Biol. Pharm. Bull.*, **28**, 481 (2005).
33. S. M. Suboh et al., *Phytother. Res.*, **18**, 280 (2004).
34. A. Herrera-Arellano et al., *Phytomedicine*, **11**, 375 (2004).
35. I. P. Odigie et al., *J. Ethnopharmacol.*, **86**, 181 (2003).
36. P. C. Onyenekwe et al., *Cell Biochem. Funct.*, **17**, 199 (1999).
37. F. M. Haji and T. A. Haji, *J. Ethnopharmacol.*, **65**, 231 (1999).
38. M. Ajay et al., *J. Ethnopharmacol.* (2006).
39. B. J. Adegunloye et al., *Afr. J. Med. Med. Sci.*, **25**, 235 (1996).
40. A. M. Salah et al., *Phytother. Res.*, **16**, 283 (2002).
41. M. B. Ali et al., *J. Ethnopharmacol.*, **31**, 249 (1991).
42. M. B. Ali et al., *Fitoterapia*, **62**, 475 (1991).
43. O. Carvajal-Zarrabal et al., *Plant Foods Hum. Nutr.*, **60**, 153 (2005).
44. C. C. Chen et al., *J. Agric. Food Chem.*, **51**, 5472 (2003).
45. A. A. Dafallah and Z. al-Mustafa, *Am. J. Chin. Med.*, **24**, 263 (1996).
46. O. E. Orisakwe et al., *Reprod. Toxicol.*, **18**, 295 (2004).
47. B. H. Ali et al., *Phytother. Res.*, **19**, 369 (2005).
48. A. M. Osman et al., *Phytochemistry*, **14**, 829 (1975).
49. J. L. Hartwell, *Lloydia*, **33**, 97 (1970).
50. Monograph *Hibisci fios, Bundesanzeiger*, no. 22 (February 1, 1990).

ROSEMARY

Source: *Rosmarinus officinalis* L. (Family Labiatae or Lamiaceae).

GENERAL DESCRIPTION

A small evergreen shrub with thick aromatic, linear leaves; up to about 2 m high; native to the Mediterranean region, cultivated worldwide (California, England, France, Spain, Portugal, Morocco, China, etc.). Part used is the dried leaf, which supplies the spice. Rosemary oil is prepared by steam distillation of the fresh flowering tops. Major oil-producing countries include Spain, France, and Tunisia.

CHEMICAL COMPOSITION

Contains about 0.5% volatile oil; flavonoids (diosmetin, diosmin, genkwanin, genkwanin-4'-methyl ether, 6-methoxygenkwanin, luteolin, luteolin glucuronides, 6-methoxyluteolin, 6-methoxyluteolin-7-glucoside, 6-methyoxyluteolin-7-methyl ether, hispidulin, apigenin, etc.);[1–5] phenolic acids (rosmarinic, labiatic, chlorogenic, neochlorogenic, and caffeic acids);[3] carnosic acid;[6–8] rosmaricine and isorosmaricine (reaction products of carnosic acid);[6,7] triterpenic acids (mainly ursolic, oleanolic, and betulinic acids, with traces of 19-α-hydroxyursolic, 2-β-hydroxyoleanolic, and 3-β-hydroxyurea-12,20(30)dien-l7-oic acids);[9,10] the diterpenes rosmanol, 7-ethoxyrosmanol, carnosol, seco-hinokiol, and rosmaquinone A and B;[11–15] and others (JIANGSU; MARSH).

The essential oil contains mainly monoterpene hydrocarbons (α- and β-pinenes, camphene, limonene, etc.), cineole (eucalyptol), and borneol with camphor, linalool, verbenol, verbenone, terpineol, 3-octanone, and isobornyl acetate also present (GUENTHER; JIANGSU).[16–20]

PHARMACOLOGY AND BIOLOGICAL ACTIVITIES

Rosemary extracts have antioxidative properties comparable to those of butylated hydroxyanisole (BHA) and butylated hydroxytoluene (BHT); carnosic acid and labiatic acid are reported as active components (see *sage*).[13,14,21–25] As such, rosemary constituents and extracts show potential as chemopreventive, radioprotective, neuroprotective, antimutagenic, antihepatotoxic, and antiulcerogenic agents.[26–35]

Rosemary oil and extracts have antimicrobial (bacteria and fungi) activities.[16,17,21,36–39] Ursolic acid also has trypanocidal activity *in vitro*.[9]

CNS activities have been reported for rosemary. For example, it stimulates the locomotor activity of mice when administered orally or by inhalation, and cineole is believed to be the active principle.[40] Aqueous and ethanolic extracts of the aerial parts reduced morphine withdrawal in mice;[41] while the volatile oil aroma significantly enhanced the quality of memory and cognitive performance in clinically tested human subjects.[42]

Diosmin is reported to be stronger than rutin in decreasing capillary permeability and fragility as well as less toxic (JIANGSU). A rosmaricine derivative (*O,O,N*-trimethylrosmaricine) has been demonstrated to exhibit significant smooth-muscle stimulant effects *in vitro* as well as moderate analgesic activity.[7] Rosemary extracts and volatile oil were reported to decrease lung inflammation and to activate the peroxisome proliferator-activated receptor gamma (PPARgamma), which may contribute to the anti-inflammatory activity.[43,44] Rosemary extracts have also exhibited inhibitory activities against urease, and rosmanol was found to be the active agent.[15] The activation of PPARgamma may be responsible for the hypoglycemic effect of rosemary extracts.[44] Earlier testing in rabbits, however, showed that rosemary had a hyperglycemic and insulin-lowering effect.[33]

Rosemary galenicals are reported to be spasmolytic on gall passages and small intestines and to have a positive inotropic effect, increasing coronary flow-through.[45] Other recently reported activities include diuretic effect in rat models and P-glycoprotein inhibiting effect in tumor cells.[46,47] The pharmacology of rosemary has also been reviewed.[48]

TOXICOLOGY

Rosemary oil has been reported to be nonirritating and nonsensitizing to human skin but moderately irritating to rabbit skin when applied undiluted.[39] There is one recent report, however, of a case of severe rosemary-induced contact dermatitis in a 53-year-old European man.[49]

USES

Medicinal, Pharmaceutical, and Cosmetic. Rosemary leaves are used in European phytomedicine for dyspeptic complaints and as supportive therapy for rheumatic diseases; externally for circulatory problems; in baths, the herb is used as an external stimulant for increased blood supply to the skin.[45]

Rosemary oil is extensively used in cosmetics as a fragrance component and/or a masking agent. Products in which it is used include soaps, detergents, creams, lotions, and perfumes (especially colognes and toilet waters). Maximum use level reported is 1% in the last category.[39]

Food. Both the spice and oil are extensively used in foods.

The spice is reported used in alcoholic beverages, baked goods, meat and meat products, condiments and relishes, processed vegetables, snack foods, gravies, and others, with highest average maximum use level of about 0.41% (4098 ppm) in baked goods.

The oil is used in alcoholic and nonalcoholic beverages, frozen dairy desserts, candy, baked goods, gelatins and puddings, meat and meat products, and condiments and relishes, among others, with highest average maximum use level of about 0.003% (26.2 ppm) reported in meat and meat products.

Dietary Supplements/Health Foods. More widely used in Europe than the United States in infusions, powder, dry extracts, or other galenic preparations for internal and external use, primarily as a stomachic (FOSTER).[45]

Traditional Medicine. Has been used since ancient times in Europe as a tonic, stimulant, and carminative and in treating dyspepsia (indigestion), stomach pains, headaches, head colds, and nervous tension.

In China, rosemary herb (leaves and branches) has been used for centuries in treating singular conditions, particularly headaches. An infusion of a mixture of rosemary herb and borax is used for the prevention of baldness. A similar formula is also used in European folk medicine for the prevention of scurf and dandruff (GRIEVE).

Rosemary has been reported used in cancers.[50]

Others. Rosemary can serve as a source of natural antioxidants.[51,52]

COMMERCIAL PREPARATIONS

Crude and oil. Crude was formerly official in U.S.P., and oil was formerly in N.F. Oil is currently official in F.C.C.

Regulatory Status. GRAS (§182.10 and §182.20). Rosemary leaves are the subject of a positive German therapeutic monograph; allowed internally for dyspeptic complaints; externally for supportive therapy for rheumatic diseases and circulatory problems.[50]

REFERENCES

See the General References for APPLEQUIST; ARCTANDER; BAILEY 1; BARNES; BIANCHINI AND CORBETTA; BLUMENTHAL 1; BRUNETON; CLAUS; DER MARDEROSIAN AND BEUTLER; DER MARDEROSIAN AND LIBERTI; FEMA; FOSTER; GRIEVE; JIANGSU; LUST; MCGUFFIN 1 & 2; ROSENGARTEN; TERRELL; UPHOF.

1. N. Okamura et al., *Phytochemistry*, **37**, 1463 (1994).
2. C. H. Brieskorn et al., *Dtsch. Lebensm. Rundsch.*, **69**, 245 (1973).
3. V. I. Litvinnenko et al., *Planta Med.*, **18**, 243 (1970).
4. V. Plouvier, *C. R. Acad. Sci., Paris, Ser. D*, **269**, 646 (1969).
5. C. H. Brieskorn and H. Michel, *Tetrahedron Lett.*, **30**, 3447 (1968).
6. L. D. Yakhontova et al., *Khim. Prir. Soedin.*, **7**, 416 (1971).
7. A. Boido et al., *Studi Sassar., Sez. 2*, **53**, 383 (1975).
8. C. H. Brieskorn and H. J. Dömling, *Z. Lebensm. Unters. Forsch.*, **14**, 10 (1969).
9. F. Abe et al., *Biol. Pharm. Bull.*, **25**, 1485 (2002).
10. C. H. Brieskorn and G. Zweyrohn, *Pharmazie*, **25**, 488 (1970).
11. A. A. Mahmoud et al., *Phytochemistry*, **66**, 1685 (2005).
12. C. L. Cantrell et al., *J. Nat. Prod.*, **68**, 98 (2005).
13. H. H. Zeng et al., *Acta Pharmacol. Sin.*, **22**, 1094 (2001).
14. H. Haraguchi et al., *Planta Med.*, **61**, 333 (1995).
15. T. Hayashi et al., *Planta Med.*, **53**, 394 (1987).
16. S. Santoyo et al., *J. Food Prot.*, **68**, 790 (2005).
17. A. Angioni et al., *J. Agric. Food Chem.*, **52**, 3530 (2004).
18. J. Cabo Torres et al., *Boll. Chim. Farm.*, **111**, 573 (1972).
19. A. Koedam and M. J. M. Gijbels, *Z. Naturforsch., C. Biosci.*, **33**C, 144 (1978).
20. B. G. Skrubis, *Flav. Ind.*, **3**, 566 (1972).

21. S. Moreno et al., *Free Radic. Res.*, **40**, 223 (2006).
22. H. Schulze et al., *Fieischwirtschaft*, **51**, 303 (1971).
23. A. Palitzsch et al., *Fleischwirtschaft*, **54**, 63 (1974).
24. Y. Watanabe and Y. Ayano, *Eiyo To Shokuryo*, **27**, 181 (1974).
25. Y. Saito et al., *Eiyo To Shokuryo*, **29**, 505 (1976).
26. K. Alexandrov et al., *Cancer Res.*, **66**, 11938 (2006).
27. S. Costa et al., *J. Appl. Toxicol.*, (2006).
28. M. J. Del Bano et al., *J. Agric. Food Chem.*, **54**, 2064 (2006).
29. J. I. Sotelo-Felix et al., *J. Ethnopharmacol.*, **81**, 145 (2002).
30. F. A. Fahim et al., *Int. J. Food Sci. Nutr.*, **50**, 413 (1999).
31. P. C. Dias et al., *J. Ethnopharmacol.*, **69**, 57 (2000).
32. E. A. Offord et al., *Carcinogenesis*, **16**, 2057 (1995).
33. M. T. Huang et al., *Cancer Res.*, **54**, 701 (1994).
34. S. J. Kim et al., *Neuroreport*, **17**, 1729 (2006).
35. K. Kosaka and T. Yokoi, *Biol. Pharm. Bull.*, **26**, 1620 (2003).
36. C. J. Del et al., *J. Food Prot.*, **63**, 1359 (2000).
37. B. G. V. N. Rao and S. S. Nigam, *Indian J. Med. Res.*, **58**, 627 (1970).
38. B. G. V. N. Rao and P. L. Joseph, *Riechst., Aromen, Körperpflegem.*, **21**, 405 (1971).
39. D. L. J. Opdyke, *Food Cosmet. Toxicol.*, **12**(Suppl.), 977 (1974).
40. K. A. Kovar et al., *Planta Med.*, **53**, 315 (1987).
41. H. Hosseinzadeh and M. Nourbakhsh, *Phytother. Res.*, **17**, 938 (2003).
42. M. Moss et al., *Int. J. Neurosci.*, **113**, 15 (2003).
43. K. Inoue et al., *Basic Clin. Pharmacol. Toxicol.*, **99**, 52 (2006).
44. O. Rau et al., *Planta Med.*, **72**, 881 (2006).
45. Monograph, *Rosmarini folium, Bun desanzeiger*, no. 223 (November 30, 1985); revised (November 28, 1986; March 13, 1990).
46. M. Haloui et al., *J. Ethnopharmacol.*, **71**, 465 (2000).
47. C. A. Plouzek et al., *Eur. J. Cancer*, **35**, 1541 (1999).
48. M. R. al-Sereiti et al., *Indian J. Exp. Biol.*, **37**, 124 (1999).
49. I. Gonzalez-Mahave et al., *Contact Dermatitis*, **54**, 210 (2006).
50. J. L. Hartwell, *Lloydia*, **32**, 247 (1969).
51. S. S. Chang et al., *J. Food Sci.*, **42**, 1102 (1977).
52. A. G. Maggi, *Neth. Appl.* 6,600,754 (1966).

ROYAL JELLY

Source: Pharyngeal glands of the worker bee (honeybee).

Common/vernacular names: Queen bee jelly, apilak, Weiselfuttersaft, Gelee royale, and *feng* WANG *jiang*.

GENERAL DESCRIPTION

Royal jelly is a milky white, viscous substance secreted by the pharyngeal glands of the worker bee (honeybee), *Apis mellifera* L., an insect of the family Apidae. It is the food for all bee larvae for the first three days of life but is reserved as food for queen bees for the rest of their lives, hence the name "royal jelly." It is collected from bee hives by manual scraping using nonmetallic utensils or by suction through a glass tube, at the same time removing impurities such as wax and insect fragments. The crude royal jelly is placed in sterilized glass jars and refrigerated until enough is collected for processing or shipment to central processing facilities. Processing involves filtration to further remove impurities of smaller particle size; the resulting liquid is either frozen in special plastic containers or freeze-dried (lyophilized). Normally the fresher materials are used for direct freezing while the less fresh materials are freeze-dried. Most of the royal jelly used in the United States is imported from China; the best grades are produced in northeastern provinces.

CHEMICAL COMPOSITION

The chemistry of royal jelly has been extensively studied. Chemical constituents reported to be present: hydroxy fatty acids, including 10-hydroxy-*trans*-2-decenoic acid (royal jelly acid; 10-HDA), which ranges between 1.64% and 4.24% (usually 2.0–2.5%) in fresh royal jelly and up to 7.0% in the freeze-dried product (usual range: 4.5–5.5%),[1–5] 10-hydroxydecanoic acid (0.60–1.25%);[2,4,6] 10-acetoxydecanoic acid, 11-*S*-hydroxydodecanoic acid, hydroxy-2*E*-decenoic acid 10-phosphate, and others;[7–9] gluconic acid,[2] sebacic acid and 2-decenedioic acid,[3] *p*-hydroxybenzoic acid,[6] 3-hydroxydecanoic, 8-hydroxydecanoic, and 3,10-dihydroxydecanoic acids;[10] sterols including methylenecholesterol (ca. 0.3%), cholesterol, stigmastanol, stigmasterol, and testosterone;[6,11–14] acetylcholine (467–1113 µg/g);[15–17] free amino acids (total 1.59% according to one report),[18] including proline, lysine, glutamic acid, serine, alanine, arginine, aspartic acid, glycine, isoleucine, leucine, methionine, tyrosine, valine, glutamine, and taurine, with proline accounting for 54–60% of all free amino acids present, followed by lysine and glutamic acid;[18–20] peptides[21–23] and glycoproteins;[24–26] and biopterin, among others (JIANGSU; MERCK).[27]

Adenosine monophosphate and adenosine monophosphate N1-oxide were recently isolated from royal jelly.[28]

The following is its general composition: 24–70% water (usually ca. 65%);[14] to 31% crude protein (usually ca. 12%); 4.9–23% total lipids, including phospholipids, glycerides, waxes, and fatty acids; 8.5–16% carbohydrates; vitamins, including thiamine 1.2–7.4 mg%, riboflavine 5.2–10 mg%, niacin 60–150 mg%, ascorbic acid 12 mg%, pyridoxine 2.2–10.2 mg%, B_{12} 0.15 mg%, pantothenic acid 65–200 mg%, biotin 0.9–3.7 mg inositol 80–150 mg%, folic acid 0.2 mg%, and vitamin E <0.2 mg%; and trace elements (MERCK).[13,27,29]

Generally recognized as the major active principle of royal jelly, 10-HDA has been reported to be very stable to heat, remaining chemically intact despite deterioration of other components; royal jelly itself is also quite stable when refrigerated or in the frozen or dried state.[30–33] However, one study found that honey accelerated the decomposition of 10-HDA when the two were dissolved in water along with royal jelly.[4]

PHARMACOLOGY AND BIOLOGICAL ACTIVITIES

Many pharmacological effects have been attributed to royal jelly, such as antihypertensive, antimicrobial, estrogenic, immunomodulatory, and others.

Royal jelly exhibited a transient vasodilating effect on dog femoral artery, which was due to its acetylcholine constituent.[15] Peptides present in royal jelly were found to possess antihypertensive effect when orally administered in hypertensive rats for 28 days.[22] A similar effect was observed in rats administered a royal jelly protein hydrolysate, and the antihypertensive effect was attributed to the inhibition of ACE.[34]

Royal jelly has exhibited weak to strong antibacterial activities against several bacterial species (*Bacillus subtilis*, *Staphylococcus aureus*, *Escherichia coli*, *Streptococcus hemolyticus*, *Enteroeoccus*, etc.) *in vitro* and *in vivo* (experimentally infected rats), with

10-HDA being the major active agent.[35–38] The peptides, jelleine I-III, were recently shown to be active against Gram-positive and Gram-negative bacteria and yeast.[21]

Estrogenic activity, prevention of osteoporosis, and stimulation of bone marrow formation have recently been demonstrated in a number of animal (ovarectomized rats) and *in vitro* (estrogen receptor binding and collagen formation) models.[39–41]

Royal jelly inhibits the production of inflammatory mediators (proinflammatory cytokines) in mouse peritoneal macrophages.[42] It also inhibits the formation of dermatitis-like skin lesions in mice when administered orally.[43]

Other activities of 10-HDA and royal jelly include: strongly inhibiting growth of transplantable tumor and leukemia in mice, with enhanced phagocytosis of peritoneal macrophages;[44,45] protecting DNA of peripheral leukocytes from damage by 4-nitrosoquinoline-*N*-oxide (4-NQO);[46] immunopotentiating/immunomodulatory,[47–49] antioxidant,[50,51] liver-protectant, radiation-protectant, and so on.[27,52]

TOXICOLOGY

Although rare, royal jelly can cause allergic reactions, ranging from asthma to anaphylaxis, in some individuals (JIANGSU).[27,53–56] Possible interaction with warfarin has also been reported.[57]

USES

Medicinal, Pharmaceutical, and Cosmetic. Royal jelly (in liquid, powder, or extract form) is used in various types of skin care products (creams, lotions, soap, etc.) for its alleged antiwrinkle, skin-nourishing and whitening properties; also used in hair care and oral (e.g., toothpaste) products.

Dietary Supplements/Health Foods. Extensively used as a food supplement in tablet, capsule, or liquid (honey syrup) form, often in combination with Asian ginseng and other herbs for their "energizing" effects.

Traditional Medicine. Used as a general nutrient and tonic by different cultures. Although adopted only recently (20th century) into traditional Chinese medical practice, it is considered sweet and sour tasting, neutral and to have invigorating, nourishing, and strengthening properties; used in treating malnutrition in children, general debility in the elderly, chronic hepatitis, diabetes, rheumatism and arthritis, and hypertension (JIANGSU).

COMMERCIAL PREPARATIONS

Royal jelly normally is available in frozen or freeze-dried form (strength = 3.5×). Quality is determined by the amount of 10-HDA present: high-quality frozen royal jelly contains about 2% while the freeze-dried powder contains about 5% 10-HDA.

Regulatory Status. U.S. regulatory status not determined.

REFERENCES

See the General References for DER MARDEROSIAN AND BEUTLER; JIANGSU; MERCK; NATIONAL; TYLER.

1. V. E. Tyler, *The New Honest Herbal*, George F. Stickley Co., Philadelphia, PA, 1987, p. 197.
2. T. Echigo et al., *Tamagawa Daigaku Nogakubu Kenkyu Hokoku*, (22), 67 (1982).
3. W. H. Brown and R. J. Freure, *Can. J. Chem.*, **37**, 2042 (1959).
4. M. Matsui, *Shokuhin Eiseigaku Zasshi*, **29**, 297 (1988).
5. L. P. Zhang, *Zhongguo Zhongyao Zazhi*, **15**, 30 (1990).

6. W. H. Brown et al., *Can. J. Chem.*, **39**, 1086 (1961).
7. E. Melliou and I. Chinou, *J. Agric. Food Chem.*, **53**, 8987 (2005).
8. M. Genc and A. Aslan, *J. Chromatogr. A*, **839**, 265 (1999).
9. N. Noda et al., *Lipids*, **40**, 833 (2005).
10. N. Weaver et al., *Lipids*, **3**, 535 (1968).
11. M. Barbier and D. Bogdanovsky, *Compt. Rend.*, **252**, 3407 (1961).
12. J. Matsuyama et al., *Tamagawa Daigaku Nogakubu Kenkyu Hokoku*, 46 (1973).
13. Y. F. Yang, *Zhongcaoyao*, **19**, 33 (1988).
14. J. Vittek and B. L. Slomiany, *Experientia*, **40**, 104 (1984).
15. M. Shinoda et al., *Yagaku Zasshi*, **98**, 139 (1978).
16. S. M. Abdel Wahab et al., *Egypt J. Pharm. Sci.*, **20**, 353 (1982).
17. W. B. Rice and F. C. Lu, *Can. Pharm. J.*, **97**, 34 (1964).
18. K. Baek and B. Y. Cho, *Kangwon Tachak Yon'gu Nonmunjip*, **6**, 7 (1972).
19. J. J. Pratt Jr. and H. L. House, *Science*, **110**, 9 (1949).
20. T. Takenaka and T. Echigo, *Honeybee Sci.*, **5**, 7 (1984).
21. R. Fontana et al., *Peptides*, **25**, 919 (2004).
22. K. H. Tokunaga et al., *Biol. Pharm. Bull.*, **27**, 189 (2004).
23. K. Bilikova et al., *FEBS Lett.*, **528**, 125 (2002).
24. M. Kimura et al., *Biosci. Biotechnol. Biochem.*, **67**, 2055 (2003).
25. Y. Kimura et al., *Biosci. Biotechnol. Biochem.*, **67**, 1852 (2003).
26. M. Kimura et al., *Biosci. Biotechnol. Biochem.*, **66**, 1985 (2002).
27. B. X. Wang, *Therapeutic Efficacy of Honeybee Products*, Jilin People's Publishers, Changchun, China, 1981.
28. N. Hattori et al., *Biosci. Biotechnol. Biochem.*, **70**, 897 (2006).
29. S. R. Howe et al., *J. Apic. Res.*, **24**, 52 (1985).
30. L. SaitamaYoho Co., Ltd., *Honeybee Sci.*, **2**, 123 (1981).
31. T. Takenaka et al., *Nippon Shokuhin Kogyo Gakkaishi*, **33**, 1 (1986).
32. E. L. Lee et al., *Shih Pin K'o Hsueh (Taipei)*, **15**, 81 (1988).
33. A. Dietz and M. H. Haydak, *J. Georgia Entomol. Soc.*, **5**, 203 (1970).
34. T. Matsui et al., *J. Nutr. Biochem.*, **13**, 80 (2002).
35. M. S. Blum et al., *Science*, **130**, 452 (1959).
36. K. Yatsunami and T. Echigo, *Bull. Faculty Agric., Tamagawa Univ.*, **25**, 13 (1985).
37. Kh. N. Muratova et al., *Eksp. Khir. Anesteziol.*, **12**, 52 (1967).
38. Y. D. Xu, *Zhongguo Yangfeng*, 28 (1989).
39. Y. Narita et al., *Biosci. Biotechnol. Biochem.*, **70**, 2508 (2006).
40. S. Hidaka et al., *Evid. Based. Complement. Alternat. Med.*, **3**, 339 (2006).
41. S. Mishima et al., *J. Ethnopharmacol.*, **101**, 215 (2005).
42. K. Kohno et al., *Biosci. Biotechnol. Biochem.*, **68**, 138 (2004).
43. Y. Taniguchi et al., *Int. Immunopharmacol.*, **3**, 1313 (2003).
44. G. F. Townsend et al., *Nature*, **183**, 1270 (1959).
45. J. Z. Dai et al., *Yiyao Gongye*, **16**, 219 (1985).
46. X. L. Yang et al., *Beijing Yike Daxue Xuebao*, **22**, 75 (1990).
47. I. Okamoto et al., *Life Sci.*, **73**, 2029 (2003).
48. H. Oka et al., *Int. Immunopharmacol.*, **1**, 521 (2001).
49. L. Sver et al., *Comp. Immunol. Microbiol. Infect. Dis.*, **19**, 31 (1996).

50. T. Nagai et al., *J. Med. Food*, **9**, 363 (2006).
51. S. Inoue et al., *Exp. Gerontol.*, **38**, 965 (2003).
52. Z. Q. Xie et al., *Zhongguo Yike Daxue Xuebao*, **21**, 167 (1990).
53. R. Leung et al., *Clin. Exp. Allergy*, **27**, 333 (1997).
54. M. Harwood et al., *N. Z. Med. J.*, **109**, 325 (1996).
55. F. C. Thien et al., *Clin. Exp. Allergy*, **26**, 216 (1996).
56. M. Takahashi et al., *Contact Dermatitis*, **9**, 452 (1983).
57. N. J. Lee and J. D. Fermo, *Pharmacotherapy*, **26**, 583 (2006).

RUE

Source: *Ruta graveolens* L. (Family Rutaceae).

Common/vernacular names: Common rue, garden rue.

GENERAL DESCRIPTION

An erect glaucous and nonhairy perennial herb, with a strong disagreeable odor; up to about 1 m high; native to the Mediterranean region; cultivated worldwide (Europe, Africa, Asia, America, etc.). Part used is the dried herb. Rue oil is obtained by steam distillation of the fresh flowering plant. *Ruta montana* L. and *R. bracteosa* L. are also reported used; Algerian rue oil is derived from these species.

CHEMICAL COMPOSITION

Common rue (*R. graveolens*) contains a volatile oil (ca. 0.1%), rutin (ca. 2%),[1] numerous acridone and quinolone alkaloids (γ-fagarine, arborinine, kokusaginine, skimmianine, graveoline, graveolinine, 6-methoxydictamnine, rutacridone, etc.)[2–7] coumarin derivatives (bergapten, xanthotoxin, rutamarin, psoralens, isoimperatorin, pangelin, rutarin, chalepensin, etc.),[2,6,8–13] and others (KARRER; JIANGSU). An unusual coumarin, naphthoherniarin, was also isolated from root of common rue.[14]

Rue oil contains varying amounts of 2-nonanone, 2-decanone, and 2-undecanone (methyl-*n*-nonyl ketone) as major components, with 2-undecanone about 90% in Algerian rue oil.[15–18] Other components include 2-heptanone, 2-octanone, 2-nonanol, 2-undecanol, undecyl-2-acetate, anisic acid, phenol, guaiacol, small amounts of coumarins (bergapten, herniarin, and xanthotoxin), monoterpenes (cineole, α- and β-pinenes, limonene),[17,19–21] and others (KARRER; JIANGSU; MARTINDALE).

The water-soluble benzofuranoid glycosides cnidioside A, methylcnidioside, and methylpicraquassioside A; as well as 3′,6-disinapoylsucrose and 3′-sinapoyl-6-feruloylsucrose, have recently been isolated from the aerial parts of *R. graveolens* growing in China.[22]

PHARMACOLOGY AND BIOLOGICAL ACTIVITIES

Rue alkaloids (especially arborinine and γ-fagarine) as well as furocoumarins (e.g., bergapten and xanthotoxin) and rue oil are reported to have spasmolytic effects on smooth muscles (e.g., isolated rabbit ileum). The extract inhibited the inflammatory response (iNOS expression and COX-2 transcription) to LPS in murine macrophage cells.[23] Arborinine and furocoumarins also have anti-inflammatory and antihistaminic properties (JIANGSU).[2,3,9] Furocoumarins have phototoxic properties and are useful in treating psoriasis (see ***bergamot***).[18] Rutin, first isolated from common rue, has numerous pharmacological activities (see ***rutin***).

Antifertility and abortifacient activities have been reported for the extract in pregnant mice and rats;[24–26] and chalepensin has been shown to have (anti-implantation) antifertility effects in rats.[12] Rue oil has been reported to cause abortion in pregnant guinea pigs and in pregnant women.[21] The antiandrogenic activity of the extract was also exhibited in male rats.[27]

Rue oil has been reported to have anthelmintic (worm, leech, and nematode) activities *in vitro* that were attributed to its major component, 2-undecanone.[18]

Different extracts and alkaloids of *R. graveolens* were reported to posses antifungal activities against different species of *Colletotrichum*, *Botrytis cineara* and, *Fusarium oxysporum*;[28–30] as well as antitumor activity against Ehrlich ascites carcinoma, HeLa, MCF7, and A431 cell lines *in vitro* and in animal models.[29,31–33]

Other reported activities of rue include antioxidant (inhibition of pig liver aldehyde oxidase) and hypotensive (positive inotropic and chronotropic effects) activities.[34,35]

TOXICOLOGY

When applied to human skin, rue oil may produce a burning sensation, erythema (redness), and vesication (blisters). Taken internally, it causes severe stomach pain, vomiting, exhaustion, confusion, and convulsion. Large doses may be fatal (JIANGSU; MARTINDALE).

A single oral dose of 400 mg/kg given to guinea pigs has been reported to be fatal due to internal hemorrhages, particularly of the adrenal gland, liver, and kidney. However, an oral daily dose of 30 mg given to human subjects for three months did not cause abnormal liver functions (JIANGSU). Oral administration of 5 g/kg/d of *R. graveolens* leaves was toxic to Nubian goats and resulted in death after 1–7 days of administration. Toxicity symptoms included tremor, dyspnea, frequent urination, incoordinated movements, ataxia, and recumbency.[36]

USES

Medicinal, Pharmaceutical, and Cosmetic. Rue oil is used as a fragrance ingredient in soaps, detergents, creams, lotions, and perfumes, with maximum use level of 0.15% reported in perfumes.[18]

Food. Rue oil is used as a flavor component (e.g., coconut type) in most major food products, including alcoholic (vermouths, bitters, etc.) and nonalcoholic beverages, frozen dairy desserts, candy, baked goods, gelatins and puddings, among others. Average maximum use levels reported are below 0.001%.

Rue is also used in certain foods, including nonalcoholic beverages, frozen dairy desserts, candy, and baked goods. Average maximum use levels are below 0.0002% (2 ppm).

Traditional Medicine. Rue is used as an emmenagogue, intestinal antispasmodic, uterine stimulant, hemostatic, and vermifuge.

In Chinese medicine, rue is also used for essentially the same purposes. In addition, it is used to treat colds, fevers, toothache, and especially snake and insect bites.

Others. Rue oil can serve as a source of natural 2-undecanone, which is a starting material for the synthesis of methylnonyl acetaldehyde, a valuable perfume chemical (ARCTANDER).

COMMERCIAL PREPARATIONS

Crude and oil. Crude was formerly official in U.S.P., and oil in N.F. is official in F.C.C.

Regulatory Status. Has been affirmed as GRAS (rue, §184.1698; oil of rue, §184.1699).[37,38] Subject of a German therapeutic monograph; effectiveness of claimed application is not verified.[39]

REFERENCES

See the General References for ARCTANDER; BAILEY 1; BIANCHINI AND CORBETTA; BLUMENTHAL 1; DER MARDEROSIAN AND BEUTLER; GRIEVE; JIANGSU; KARRER; LUST; MARTINDALE; MCGUFFIN 1 & 2; MERCK; TERRELL; UPHOF.

1. F. R. Humphreys, *Econ. Bot.*, **18**, 195 (1964).
2. I. Novak et al., *Acta Pharm. Hung.*, **37**, 131 (1967).
3. I. Novak et al., *Planta Med.*, **15**, 132 (1967).
4. J. Reisch et al., *Phytochemistry*, **15**, 240 (1976).
5. O. Nieschulz, *Sci. Pharm. Proc., 25th*, **2**, 559 (1965).
6. A. Gonzalez et al., *An. Quim.*, **70**, 60 (1974).
7. K. Szendrei et al., *Herba Hung.*, **10**, 131 (1971).
8. A. L. Hale et al., *J. Agric. Food Chem.*, **52**, 3345 (2004).
9. I. Novak et al., *Pharmazie*, **20**, 738 (1965).
10. E. Varga et al., *Fitoterapia*, **47**, 107 (1976).
11. J. Reisch et al., *Magy. Kem. Foly.*, **78**, 6 (1972).
12. Y. C. Kong et al., *Planta Med.*, **55**, 176 (1989).
13. A. Nahrstedt et al., *Planta Med.*, **51**, 517 (1985).
14. Z. Rozsa et al., *Planta Med.*, **55**, 68 (1989).
15. V. De Feo et al., *Phytochemistry*, **61**, 573 (2002).
16. T. M. Andon and N. V. Belova, *Rastit Resur.*, **11**, 539 (1975).
17. D. H. E. Tattje, *Pharm. Weekbl.*, **105**, 1241 (1970).
18. D. L. J. Opdyke, *Food Cosmet. Toxicol.*, **13**, 455 (1975).
19. D. H. E. Tattje et al., *Pharm. Weekbl.*, **109**, 881 (1974).
20. K. H. Kubeczka, *Phytochemistry*, **13**, 2017 (1974).
21. Anon., *Fed. Regist.*, **39**, 34215 (1974).
22. C. C. Chen et al., *J. Nat. Prod.*, **64**, 990 (2001).
23. S. K. Raghav et al., *J. Ethnopharmacol.*, **104**, 234 (2006).
24. T. G. de Freitas et al., *Contraception*, **71**, 74 (2005).
25. J. L. Gutierrez-Pajares et al., *Reprod. Toxicol.*, **17**, 667 (2003).
26. M. Gandhi et al., *J. Ethnopharmacol.*, **34**, 49 (1991).
27. N. A. Khouri and Z. El-Akawi, *Neuro. Endocrinol. Lett.*, **26**, 823 (2005).
28. K. M. Meepagala et al., *Phytochemistry*, **66**, 2689 (2005).
29. A. Ivanova et al., *Fitoterapia*, **76**, 344 (2005).
30. A. Oliva et al., *J. Agric. Food Chem.*, **51**, 890 (2003).
31. B. Rethy et al., *Planta Med.* (2006).
32. K. C. Preethi et al., *Asian Pac. J. Cancer Prev.*, **7**, 439 (2006).
33. S. Pathak et al., *Int. J. Oncol.*, **23**, 975 (2003).
34. P. Saieed et al., *Chem. Pharm. Bull. (Tokyo)*, **54**, 9 (2006).
35. K. W. Chiu and A. Y. Fung, *Gen. Pharmacol.*, **29**, 859 (1997).
36. S. E. el Agraa et al., *Trop. Anim. Health Prod.*, **34**, 271 (2002).
37. Anon., *Fed. Regist.*, **41**, 236 (1976).
38. Anon., *Fed. Regist.*, **43**, 3704 (1978).
39. Monograph, *Ruta graveolens*, *Bundesanzeiger*, no. 43 (March 2, 1989).

RUTIN

Source: *Sophora japonica* L., *Eucalyptus macrorhyncha* F. v. M., and *Fagopyrum esculentum* Moench are the major sources of rutin.

Common/vernacular names: Quercetin-3-rutinoside, rutoside, eldrin, and sophorin.

GENERAL DESCRIPTION

Rutin is a glycoside containing quercetin as its aglycone and rutinose (rhamnose and glucose) as its sugar portion. It is widely distributed in the plant kingdom, being found in many families of higher plants as well as in ferns.[1–4]

It is present in high concentrations in leaves of *Eucalyptus macrorhyncha* (10–24%), flowers of *Viola tricolor* L. var. *maxima* (18–21%), flower buds of *Sophora japonica* (13–30%), and in buckwheat, *Fagopyrum esculentum* (0.1–6.4%).[1–4]

Currently, *S. japonica*, *E. macrorhyncha*, and *F. esculentum* are the major sources of rutin. Its production is relatively simple to rather complex, depending on the raw material and the process used.

PHARMACOLOGY AND BIOLOGICAL ACTIVITIES

Rutin has been reported to have many pharmacological properties. The most well-known is its ability to decrease capillary permeability and fragility, though evidence is inconclusive (CLAUS; MARTINDALE; REMINGTON; USD 26th).[1–5] It is considered a "vitamin P" or "permeability" vitamin.

Other pharmacological activities include antiedematous,[6] antiatherogenic, and antiadepogenic in chicks and mice on high-fat diets,[7,8] and increasing survival time of rats fed a thrombogenic diet,[9] cancer chemopreventive inhibiting tumor formation in mouse skin and colon by the carcinogens benzo(a)pyrene and azoxymethane, respectively.[10,11] Antispasmodic activity has been observed in isolated guinea-pig colon and rat duodenum.[12,13] Rutin has been found to induce a release of endogenous histamine and 5-hydroxytryptamine 2 h after administration; however, a predicted antispasmodic effect, as observed *in vitro*, was absent *in vivo*, suggesting a physiological antagonism mechanism contrary to previously reported *in vitro* results.[13] Also reported are hypotensive, antiplatelet,[14] and protective activities against X-ray irradiation in rats and mice,[15,16] among others[17,18] (JIANGSU).

Rutin is active in different models of acute and chronic inflammation and has been shown to attenuate the proinflammatory cytokine production.[19–22] Rutin is a noncompetitive inhibitor of angiotensin II and prostaglandin E_2 on the guinea-pig ileum.[23]

Antioxidant activities have also been reported for rutin, whereby protection against hepatotoxicity, myocardial infarction, diabetes complications, UV irradiation, and ethanol-induced gastric lesions have been observed.[24–28]

TOXICOLOGY

Rutin has generally been considered to lack toxicity (USD 26th). However, it has been reported by one group of investigators to cause concretion formation in human as well as in laboratory animals.[29,30]

USES

Rutin was formerly an official drug in the United States and has been used in treating capillary hemorrhage due to increased capillary fragility in degenerative vascular diseases (e.g., arteriosclerosis and hypertension), diabetes, and allergic manifestations.

Although no longer official in N.F., rutin still is quite widely used, both as a prescription drug and as a vitamin supplement.[31] The latter use is by far the more widespread. It is usually used in formulations with vitamin C or together with other bioflavonoids as well as rose hips, especially in the health food industry.

Flower buds of *Sophora japonica* containing high concentrations (usually ca. 20%) of rutin have been used for centuries in Chinese medicine for the treatment of internal bleeding (e.g., bloody urine, spitting blood, and intestinal bleeding) and bleeding hemorrhoids. It is also used for the prevention of strokes (FARNSWORTH 1–4; FOGARTY; JIANGSU).

COMMERCIAL PREPARATIONS

Pure rutin; often adulterated; it was formerly official in N.F.

REFERENCES

See the General References for APhA; BAILEY 2; JIANGSU.

1. F. R. Humphreys, *Econ. Bot.*, **18**, 195 (1964).
2. G. Berti and F. Bottari in L. Reinhold and Y. Liwschitz, eds., *Progress in Phytochemistry*, Vol. 1, Interscience, New York, 1968, p. 589.
3. J. Davidek, *Veda Vyzkum Prumyslu Potravinarskem*, **12**, 179 (1963).
4. V. A. Bandyukova and N. V. Sergeeva, *Khim. Prir. Soedin.*, **4**, 524 (1974).
5. K. Venkataraman in L. Zechmeister, ed., *Fortschritte der Chemie Organischer Naturstoffe*, Vol. 17, Springer-Verlag, Vienna, Austria, 1959, p. 1.
6. V. M. Samvelyan and L. G. Khlgayan, *Farmakol. Toksikol. (Moscow)*, **32**, 447 (1969).
7. I. Choi et al., *Biofactors*, **26**, 273 (2006).
8. A. N. Chernov et al., *Lipidy Tkanei Eksp. Giperkholesterinemin*, 115 (1967).
9. R. C. Robbins, *J. Atheroscler. Res.*, **7**, 3 (1967).
10. K. Yang et al., *Carcinogenesis*, **21**, 1655 (2000).
11. B. L. Van Duuren et al., *J. Natl. Cancer Inst.*, **46**, 1039 (1971).
12. O. Altinkurt and Y. Öztürk, *J. Fac. Pharm. Ankara*, **17**, 49 (1987).
13. N. Yildizoglu et al., *Phytother. Res.*, **5**, 19 (1991).
14. J. R. Sheu et al., *J. Agric. Food Chem.*, **52**, 4414 (2004).
15. U. Undeger et al., *Toxicol. Lett.*, **151**, 143 (2004).
16. R. P. Webster et al., *Cancer Lett.*, **109**, 185 (1996).
17. V. A. Baraboi, *Fenol'nye Soedin, Ikh Biol. Funkts., Mater Vses. Simp.*, **1**, 353 (1966).
18. G. Vogel and H. Ströcker, *Arzneim.-Forsch.*, **16**, 1630 (1966).
19. K. H. Kwon et al., *Biochem. Pharmacol.*, **69**, 395 (2005).
20. L. Selloum et al., *Exp. Toxicol. Pathol.*, **54**, 313 (2003).
21. T. Guardia et al., *Farmaco*, **56**, 683 (2001).
22. I. B. Afanas'eva et al., *Biochem. Pharmacol.*, **61**, 677 (2001).
23. O. Altinkurt and N. Abacioglu, *Arzneim.-Forsch.*, **30**, 610 (1980).
24. N. Kamalakkannan and P. S. Prince, *Mol. Cell. Biochem.*, **293**, 211 (2006).
25. M. Karthick and P. P. Stanely Mainzen, *J. Pharm. Pharmacol.*, **58**, 701 (2006).
26. M. Alia et al., *Eur. J. Nutr.*, **45**, 19 (2006).
27. H. Palmer et al., *J. Photochem. Photobiol. B*, **67**, 116 (2002).
28. C. C. La et al., *J. Ethnopharmacol.*, **71**, 45 (2000).
29. K. Pfeifer et al., *Dtsch. Gesundheitsw.*, **24**, 260 (1969).
30. K. Pfeifer et al., *Dtsch. Gesundheitsw.*, **25**, 386 (1970).
31. N. R. Farnsworth in L. P. Miller, ed., *Phytochemistry*, Vol. 3, Van Nostrand Rheinhold, New York, 1973, p. 351.

SAFFRON

Source: *Crocus sativus* L. (Family Iridaceae).

Common/vernacular names: Saffron crocus.

GENERAL DESCRIPTION

A perennial herb with a large fleshy corm from which leaves and flowers are produced in the fall. Flowers violet, throat of perianth bearded, anthers yellow, style armed exerted orange-red subulate tips; entire or lobed.[1] Native to the eastern Mediterranean region; cultivated as an annual or perennial worldwide (Spain, France, Italy, India, etc.). Part used is the dried stigma; 100,000–140,000 flowers are reportedly required to yield 1 kg saffron. Hence, it is the most expensive spice. Major producing countries include Spain, France, Turkey, and India.

CHEMICAL COMPOSITION

Contains about 2% crocin-1 (ester of crocetin with 2 molecules gentiobiose); about 2% picrocrocin (bitter principle); small amounts of crocin-2 (ester of crocetin with 1 molecule each of gentiobiose and glucose), crocin-3 (monogentiobiose ester of crocetin), and crocin-4 (ester of monomethylcrocetin with 1 molecule glucose); small amounts of free crocetin (α-crocetin), methylcrocetin (β-crocetin), *trans*-dimethylcrocetin (γ-crocetin), *cis*-dimethylcrocetin and other crocetin esters; kaempferol;[2] starch (ca. 13%); vitamins B and 132; fixed oils (8 to 13%); other carotenoids; and 0.4–1.3% of a volatile oil consisting of safranal, oxysafranal, pinene, cineole, isophorone, naphthalene, 2-butenoic acid lactone, 2-phenylethanol, 3,5,5-trimethyl-4-hydroxy-1-cyclohexanone-2-ene, 4-hydroxy-2,6,6-trimethyl-1-cyclohexene-1-carboxaldehyde, and others (JIANGSU; LIST AND HÖRHAMMER).[3–7] Crocin is mainly responsible for the color of saffron, while picrocrocin is responsible for its bitter taste and aroma (after hydrolysis to yield safranal).

At least 12 new monoterpenes have recently been isolated from saffron, including crocusatins A–L; as well as the new (3*S*)-4-dihydroxybutyric acid.[8–10]

PHARMACOLOGY AND BIOLOGICAL ACTIVITIES

Extracts of saffron have been demonstrated to have various pharmacological properties, including stimulation of the uteri of experimental animals, lowering the blood pressure of anesthetized dogs and cats as well as stimulating respiration, and strongly inhibiting the contraction of isolated toad and rat hearts and guinea pig tracheal chains, among others (JIANGSU).[11,12]

Many recent clinical trials from the same source suggest that saffron may be effective in the management of mild to moderate depression. In some of these trials, saffron was favorably comparable to fluoxetine.[13–17]

A number of studies suggest that saffron extracts have *in vitro* anticancer activity, limiting the growth of experimentally induced cancers by inhibiting cellular nucleic acid synthesis and by other DNA-mediated mechanisms.[18–23] The anticancer and cancer chemopreventive activities of saffron have been reviewed.[24–26]

Saffron extracts and their crocin compounds inhibit amyloid-beta protein deposition and improve learning and memory functions in mice which suggests a possible benefit in the management of Alzheimer's disease and other memory impairment conditions.[27–29] This effect may be attributed to the antioxidant/free radical scavenging activity frequently reported for saffron and its major constituents.[30–36] The antioxidant activity has also been implicated in the antigenotoxic/antimutagenic effects of saffron.[35,37–40]

Other recently reported activities of saffron include antitussive,[41] anticonvulsant,[34] antinociceptive/anti-inflammatory,[42] and antiplatelet effects.[43]

Crocetin has been found to increase oxygen diffusion in plasma by 80%. Crocetin binds to serum albumin and has been found to lower serum cholesterol levels in rabbits.[44,45] Crocetin has recently been shown to reduce the deleterious effects caused by insulin-resistance in rats.[46] The biological activities of crocetin have been reviewed.[47]

TOXICOLOGY

No risk associated with food or therapeutic use at 1.5 g or less. Toxic reactions have been reported at a dose of 5 g including necrosis of the nose, thrombocytopenia and uremia collapse. Associated symptoms include vomiting, bleeding of the uterus, bloody diarrhea, nose bleed, vertigo, dizziness, and others.[48]

USES

Medicinal, Pharmaceutical, and Cosmetic. Extracts (e.g., tincture) are used as fragrance components in perfumes (especially oriental types). An Australian patent has been issued for use of an aqueous extract of the corm (in combination with other ingredients for treatment of baldness).[45]

Food. Saffron is used both as a coloring (yellow) and as a flavoring agent. It is often used as a domestic spice, especially in Spanish and French cooking (e.g., arroz con pollo and bouillabaisse).

Saffron and saffron extract (type not specified) are used in alcoholic (e.g., bitters and vermouths) and nonalcoholic beverages, candy, baked izoods, and other food products. Highest average maximum use level is about 0.1% (969 ppm) reported for the crude in baked goods.

Traditional Medicine. Reportedly used as a sedative, diaphoretic, antispasmodic, emmenagogue, anodyne, and aphrodisiac. Conditions for which it is used include coughs, whooping cough, stomach gas, and insomnia. A tincture formerly used as a sedative in Germany (WEISS).

In Chinese medicine, it is traditionally used to treat conditions resulting from depression, fright, or shock; spitting blood; pain and difficulties in menstruation and after childbirth; and others (JIANGSU).

Saffron has been widely used in cancers, which has been experimentally confirmed *in vitro.*[18,19,24–26,49]

COMMERCIAL PREPARATIONS

Mainly crude (Spanish). it was formerly official in N.F.

Regulatory Status. GRAS (§182.10 and §182.20); also has been approved as a food color additive exempt from certification (§73.500). Subject of a German therapeutic monograph, indicated as a sedative for spasms and asthma, with the caveat that claimed efficacy is not documented.[48]

REFERENCES

See the General References for APPLEQUIST; ARCTANDER; BARRETT; BISSET; BLUMENTHAL 1; BRUNETON; CLAUS; DER MARDEROSIAN AND BEUTLER; DUKE 4; FEMA; GRIEVE; HUANG; JIXIAN; JIANGSU; LIST AND HÖRHAMMER; LUST; MCGUFFIN 1 & 2; ROSENGARTEN; TERRELL; UPHOF; WEISS.

1. K. R. Kirtikar and B. D. Basu, *Indian Medicinal Plants*, International Book Distributor, Dehradun, India, 1995, p. 246.
2. I. Kubo and I. Kinst-Hori, *J. Agric. Food Chem.*, **47**, 4121 (1999).
3. M. Carmona et al., *J. Agric. Food Chem.*, **54**, 973 (2006).
4. V. K. Dhingra et al., *Indian J. Chem.*, **13**, 339 (1975).
5. P. Duquenois, *Bull. Soc. Pharm. Strasbourg*, **15**, 149 (1972).
6. N. S. Zarghami, *Diss. Abstr. Int. B*, **31**, 5235 (1971).

7. C. D. Kanakis et al., *J. Agric. Food Chem.*, **52**, 4515 (2004).
8. C. Y. Li et al., *J. Nat. Prod.*, **67**, 437 (2004).
9. C. Y. Li and T. S. Wu, *J. Nat. Prod.*, **65**, 1452 (2002).
10. C. Y. Li and T. S. Wu, *Chem. Pharm. Bull. (Tokyo)*, **50**, 1305 (2002).
11. P. Y. Chang et al., *Yao Hsueh Hsueh Pao*, **11**, 94 (1964).
12. M. H. Boskabady and M. R. Aslani, *J. Pharm. Pharmacol.*, **58**, 1385 (2006).
13. B. A. Akhondzadeh et al., *Prog. Neuropsychopharmacol. Biol. Psychiatry*, (2006).
14. E. Moshiri et al., *Phytomedicine*, **13**, 607 (2006).
15. S. Akhondzadeh et al., *Phytother. Res.*, **19**, 148 (2005).
16. A. A. Noorbala et al., *J. Ethnopharmacol.*, **97**, 281 (2005).
17. S. Akhondzadeh et al., *BMC Complement. Altern. Med.*, **4**, 12 (2004).
18. F. I. Abdullaev et al., *Biofactors*, **3**, 201 (1992).
19. S. C. Nair et al., *Cancer Lett.*, **57**, 109 (1991).
20. M. Ashrafi et al., *Int. J. Biol. Macromol.*, **36**, 246 (2005).
21. D. C. Garcia-Olmo et al., *Nutr. Cancer*, **35**, 120 (1999).
22. J. Escribano et al., *Cancer Lett.*, **100**, 23 (1996).
23. P. A. Tarantilis et al., *J. Chromatogr. A*, **699**, 107 (1995).
24. F. I. Abdullaev and J. J. Espinosa-Aguirre, *Cancer Detect. Prev.*, **28**, 426 (2004).
25. F. I. Abdullaev, *Exp. Biol. Med. (Maywood.)*, **227**, 20 (2002).
26. S. C. Nair et al., *Cancer Biother.*, **10**, 257 (1995).
27. M. A. Papandreou et al., *J. Agric. Food Chem.*, **54**, 8762 (2006).
28. N. Pitsikas and N. Sakellaridis, *Behav. Brain Res.*, **173**, 112 (2006).
29. K. Abe and H. Saito, *Phytother. Res.*, **14**, 149 (2000).
30. S. Saleem et al., *J. Med. Food*, **9**, 246 (2006).
31. H. Hosseinzadeh and H. R. Sadeghnia, *J. Pharm. Pharm. Sci.*, **8**, 394 (2005).
32. H. Hosseinzadeh et al., *J. Pharm. Pharm. Sci.*, **8**, 387 (2005).
33. A. N. Assimopoulou et al., *Phytother. Res.*, **19**, 997 (2005).
34. H. Hosseinzadeh and F. Talebzadeh, *Fitoterapia*, **76**, 722 (2005).
35. K. Premkumar et al., *Phytother. Res.*, **17**, 614 (2003).
36. S. K. Verma and A. Bordia, *Indian J. Med. Sci.*, **52**, 205 (1998).
37. K. Premkumar et al., *Hum. Exp. Toxicol.*, **25**, 79 (2006).
38. I. Das et al., *Asian Pac. J. Cancer Prev.*, **5**, 70 (2004).
39. K. Premkumar et al., *Asia Pac. J. Clin. Nutr.*, **12**, 474 (2003).
40. K. Premkumar et al., *Drug Chem. Toxicol.*, **24**, 421 (2001).
41. H. Hosseinzadeh and J. Ghenaati, *Fitoterapia*, **77**, 446 (2006).
42. H. Hosseinzadeh and H. M. Younesi, *BMC Pharmacol.*, **2**, 7 (2002).
43. S. W. Jessie and T. P. Krishnakantha, *Mol. Cell. Biochem.*, **278**, 59 (2005).
44. B. R. Olin ed., *Lawrence Rev. Nat. Prod.*, April 1993.
45. T. L. Miller et al., *J. Pharm. Sci.*, **71**, 173 (1982).
46. L. Xi et al., *J. Nutr. Biochem.*, **18**, 64 (2007).
47. M. Giaccio, *Crit. Rev. Food Sci. Nutr.*, **44**, 155 (2004).
48. Monograph, *Croci stigma*, Bundesanzeiger, no. 76 (April 23, 1987).
49. J. L. Hartwell, *Lloydia*, **32**, 247 (1969).

SAGE

Source: *Sage Salvia officinalis* L.; **Spanish sage** *Salvia lavandulaefolia* Vahl (Family Labiatae or Lamiaceae).

Common/vernacular names: Garden sage, true sage, and Dalmation sage (*S. officinalis*).

GENERAL DESCRIPTION

Salvia officinalis is a small, evergreen shrubby perennial with woody stems near the base and herbaceous ones above, much branched; up to about 0.8 m high; native to the Mediterranean region; cultivated worldwide (Albania, Turkey, Greece, Italy, United States, etc.). Part used is the dried leaf from which sage oil is obtained by steam distillation.

Salvia lavandulaefolia is closely related to *S. officinalis*. It grows wild in Spain and southwestern France. Spanish sage oil is obtained by steam distillation of its leaves.

A study found that most commercial sage sold in the United States (from 50% to 95%) was represented by *S. fruticosa* Mill. (*S. triloba* L. f.), characterized by compound or simple leaves with 1–2 pairs of lateral segments and a large terminal segment, rather than *S. officinalis* as purported (TUCKER 1–3).[1]

CHEMICAL COMPOSITION

Sage (*S. officinalis*) contains 1.0–2.8% volatile oil;[2] quinone- and abietane-type diterpenes: carnosol, 12-*O*-methyl carnosol, carnosic acid, picrosalvin, salvin, salvin monomethyl ether, royleanonic acid, royleanone, rosmanol, isorosmanol, 7-methoxyrosmanol, horminone, acetyl horminone, and galdosol;[3–11] flavonoids including genkwanin, 6-methoxygenkwanin, apigenin, luteolin, luteolin-7-methyl ether, 6-methoxyluteolin, 6-methoxyluteolin-7-methyl ether, hispidulin, cirsimaritin, and salvigenin;[6,7,12–14] phenolic acids (rosmarinic, labiatic, caffeic, hydroxycinnamic) and phenolic glycosides of caffeic and benzoic acid, and trace amounts of chlorogenic acid;[5,15–21] anthraquinones;[7] salviatannin (a tannin of the condensed catechin type that on storage undergoes degradation to phlobaphenes);[22] polysaccharides;[23] and others (MARSH; MARTINDALE; STAHL).[21]

Sage oil contains α- and β-thujones (normally ca. 50%) as the main components. Other compounds present include cineole, borneol, viridiflorol, 2-methyl-3methylene-5-heptene, 1,8-cineole, camphor, and limonene and sesquiterpenes (STAHL).[2,24–27]

Spanish sage contains a volatile oil composed of highly variable amounts of camphor (11–34%), cineole (18–35%), limonene (1–41%), camphene (5–30%), α-pinene (4–20%), β-pinene (6–19%), linalool, linalyl acetate, borneol, and others (GUENTHER; STAHL).[28–31] It also contains numerous polyphenolic compounds, including luteolin-4'-*O*-glucuronide, rosmarinic acid, salvigenin, eupatorin, nepetin, and apigenin, among others.[32]

PHARMACOLOGY AND BIOLOGICAL ACTIVITIES

Sage reportedly has antibacterial, fungistatic, virustatic, astringent, secretion-stimulating, and perspiration-inhibiting effects.[33]

Phenolic acids (e.g., salvin and salvin monomethyl ether) isolated from sage have antimicrobial activities, especially against *Staphylococcus aureus*.[10,21] Antileishmanial activity has recently been reported for sage polyphenolics, which also exhibited immunomodulatory activity.[15] Spanish sage oil has also been reported to have antimicrobial properties.[29]

Sage oil has been reported to have neurotropic antispasmodic effects against acetylcholine spasms in laboratory animals.[34]

Sage extracts, like those of rosemary, have strong antioxidant activities *in vitro* and *in vivo*; labiatic acid, carnosic acid, and the phenolic acids are reported to be the active compounds (see **rosemary**).[5,7,20,35,36] Sage oil displayed chemopreventive activity against tumor progression and skin papillomas in mice.[37] Antimutagenic activities have also been reported for sage oil and extract both *in vitro* and *in vivo*.[24,38]

Sage leaf extracts and Spanish sage oil were found to enhance memory and to be beneficial in the management of Alzheimer's disease through interaction with the cholinergic system and protection against beta-amyloid protein neurotoxicity, respectively.[39–44]

Sage polysaccharides displayed immunomodulatory activity in the *in vitro* comitogenic thymocyte test.[23,45]

Other recently reported activities for sage include anti-inflammatory,[12,46] antidiabetic,[47,48] antitumor;[3] in addition to benzodiazepine receptor binding,[6] reduced lipid absorption (carnosic acid),[4] and activation of peroxisome proliferator-activated receptor gamma.[49] Infusions and suspensions of Spanish sage have hypoglycemic activity in rabbits.[50]

Sage is also reported to have fish odor-suppressant properties.[51]

TOXICOLOGY

Spanish sage oil was nonirritating and nonsensitizing to human skin and skin of laboratory animals; it was also nonphototoxic on mice and swine.[29]

Although sage oil contains more thujone than absinthium oil, it has not been reported to be toxic (see ***absinthium***). Two recent studies, however, reported that the oil may be hepatotoxic at high concentrations.[35,52] Dalmation sage oil has been reported to be nonirritating and nonsensitizing to human skin when tested in a diluted form. When applied undiluted, it produced one irritation reaction in 20 subjects and was moderately irritating to rabbits.[53]

USES

Medicinal, Pharmaceutical, and Cosmetic. The dried leaves, the essential oil, tincture, and fluid extract are used in European phytomedicine for dyspeptic symptoms and diaphoretic effects; external use (gargles and rinses) for inflamed mucous membranes of the oral mucosa and throat.[33]

Both sage oil and Spanish sage oil are used (the former much more extensively) as fragrance components in soaps, detergents, creams, lotions, and perfumes (e.g., colognes and after-shave lotions), with maximum use level of 0.8% reported for both oils in perfumes.[29,53] Spanish sage oil is generally more commonly used in soaps, detergents, and industrial fragrances.

Food. Sage is widely used as a flavor ingredient in baked goods, meat and meat products, condiments and relishes, processed vegetables, soups, gravies, fats and oils, and others. Highest average maximum use level reported is 0.477% in baked goods.

Sage oleoresin is also widely used in baked goods, meat and meat products, and condiments and relishes. Highest average maximum use level reported is about 0.014% (139 ppm) in meat and meat products.

Sage oil and Spanish sage oil are extensively used in most categories of food products, including alcoholic (e.g., vermouths and bitters) and nonalcoholic beverages, frozen dairy desserts, candy, baked goods, gelatins and puddings, meat and meat products, and condiments and relishes. Highest average maximum use levels reported are about 0.013% (126 ppm) and 0.004% (40.5 ppm) for sage oil and Spanish sage oil, respectively, in meat and meat products.

Dietary Supplements/Health Foods. Dried leaves used as a tea ingredient; occasionally tablets, capsules, tincture, and so on, for traditional indications (FOSTER).

Traditional Medicine. Sage is used as tonic, digestive, antiseptic, astringent, and antispasmodic. It is used to reduce perspiration (e.g., night sweats), to stop the flow of milk, to treat nervous conditions (e.g., trembling, depression, and vertigo dysmenorrhea, diarrhea, gastritis, sore throat, insect bites, and others, usually the form of a tea or infusion. Sage has been reported used cancers.[54]

Others. Like rosemary, sage can serve source of natural antioxidants (see *rosemary*).

COMMERCIAL PREPARATIONS

Sage, sage oleoresin, and Dalmation sage oil; Spanish sage oil. Sage was formerly official in N.F. Dalmation sage oil and Spanish sage oil are official in F.C.C.

Regulatory Status. GRAS (sage, §182.10 and §182.20; Spanish sage, §182.20); no thujone limit for sage oil is specified (see *absinthium*). Subject of a positive German therapeutic monograph; allowed for internal use in dyspeptic complaints, and as diaphoretic; internally for inflamed oral mucous membranes.[33]

REFERENCES

See the General References for APPLEQUIST; ARCTANDER; BAILEY 1; BARNES; BIANCHINI AND CORBETTA; BLUMENTHAL 1; DER MARDEROSIAN AND BEUTLER; FEMA; FOSTER; GUENTHER; KROCHMAL AND KROCHMAL; LUST; POLUNIN AND SMYTHIES; ROSENGARTEN; TUCKER 1–3.

1. A. O. Tucker et al., *Econ. Bot.*, **34**, 16 (1980).
2. A. Ceylan, *Ege Univ. Ziraat Fak., Derg., Ser A*, **13**, 283 (1976).
3. D. Slamenova et al., *Basic Clin. Pharmacol. Toxicol.*, **94**, 282 (2004).
4. K. Ninomiya et al., *Bioorg. Med. Chem. Lett.*, **14**, 1943 (2004).
5. T. C. Matsingou et al., *J. Agric. Food Chem.*, **51**, 6696 (2003).
6. D. Kavvadias et al., *Planta Med.*, **69**, 113 (2003).
7. K. Miura et al., *J. Agric. Food Chem.*, **50**, 1845 (2002).
8. K. Miura et al., *Phytochemistry*, **58**, 1171 (2001).
9. E. Ghigi et al., *Ann. Chim. (Rome)*, **59**, 510 (1969).
10. V. N. Dobrynin et al., *Khim. Prir. Soedin.*, **5**, 686 (1976).
11. C. H. Brieskorn and H. J. Dömling, *Z. Lebensm. Unters. Forsch.*, **141**, 10 (1969).
12. D. Baricevic et al., *J. Ethnopharmacol.*, **75**, 125 (2001).
13. C. H. Brieskorn and W. Biechele, *Arch. Pharm. (Weinheim)*, **304**, 557 (1971).
14. C. H. Brieskorn and W. Biechele, *Dtsch. Apoth. Ztg.*, **111**, 141 (1971).
15. O. A. Radtke et al., *Z. Naturforsch. C*, **58**, 395 (2003).
16. C. T. Ho et al., *Biofactors*, **13**, 161 (2000).
17. Y. Lu and L. Y. Foo, *Phytochemistry*, **55**, 263 (2000).
18. M. Wang et al., *J. Agric. Food Chem.*, **48**, 235 (2000).
19. M. Wang et al., *J. Nat. Prod.*, **62**, 454 (1999).
20. K. Herrmann, *Z. Lebensm. Unters. Forsch.*, **116**, 224 (1962).
21. N. Z. Alimkhodzhaeva and R. L. Khazanovich, *Mater. Yubileinoi Resp. Nauchn. Konf. Farm., Posvyashch. 50-Letiyu Obraz. SSSR*, **37**, (1972).
22. D. Murko et al., *Planta Med.*, **25**, 295 (1974).
23. P. Capek and V. Hribalova, *Phytochemistry*, **65**, 1983 (2004).
24. B. Vukovic-Gacic et al., *Food Chem. Toxicol.*, **44**, 1730 (2006).
25. C. H. Brieskorn and S. Dalferth, *Liebigs Ann. Chem.*, **676**, 171 (1964).
26. M. B. Embong et al., *Can. Inst. Food Sci. Technol. J.*, **10**, 201 (1977).

27. C. Karl et al., *Planta Med.*, **44**, 188 (1982).
28. B. M. Lawrence et al., *J. Chromatogr.*, **50**, 59 (1970).
29. D. L. J. Opdyke, JT Food Cosmet. Toxicol., **14**(Suppl.) 857 (1976).
30. C. H. Brieskorn and S. Dalferth, *Dtsch. Apoth. Ztg.*, **104**, 1388 (1964).
31. M. E. Crespo et al., *Planta Med.*, **52**, 366 (1986).
32. S. Cañigueral et al., *Planta Med.*, **55**, 92 (1989).
33. Monograph, *Salviae folium, Bundesan Zeiger*, no. 90 (May 15, 1985); revised (March 13, 1990).
34. A. M. Debelmas and J. Rochat, *Plant. Med. Phytother.*, **1**, 23 (1967).
35. C. F. Lima et al., *J. Ethnopharmacol.*, **97**, 383 (2005).
36. W. Bors et al., *Biol. Res.*, **37**, 301 (2004).
37. H. U. Gali-Muhtasib and N. I. Affara, *Phytomedicine*, **7**, 129 (2000).
38. M. Vujosevic and J. Blagojevic, *Acta Vet. Hung.*, **52**, 439 (2004).
39. M. Eidi et al., *Nutrition*, **22**, 321 (2006).
40. T. Iuvone et al., *J. Pharmacol. Exp. Ther.*, **317**, 1143 (2006).
41. D. O. Kennedy et al., *Neuropsychopharmacology*, **31**, 845 (2006).
42. N. T. Tildesley et al., *Pharmacol. Biochem. Behav.*, **75**, 669 (2003).
43. S. Akhondzadeh et al., *J. Clin. Pharm. Ther.*, **28**, 53 (2003).
44. N. S. Perry et al., *J. Pharm. Pharmacol.*, **53**, 1347 (2001).
45. P. Capek et al., *Int. J. Biol. Macromol.*, **33**, 113 (2003).
46. M. Hubbert et al., *Eur. J. Med. Res.*, **11**, 20 (2006).
47. C. F. Lima et al., *Br. J. Nutr.*, **96**, 326 (2006).
48. M. Eidi et al., *J. Ethnopharmacol.*, **100**, 310 (2005).
49. O. Rau et al., *Planta Med.*, **72**, 881 (2006).
50. J. Jimenez et al., *Planta Med.*, **52**, 260 (1986).
51. T. Kikuchi et al., *Eiyo To Shokuryo*, **21**, 253 (1968).
52. C. F. Lima et al., *Food Chem. Toxicol.*, (2006).
53. D. L. J. Opdyke, *Food Cosmet. Toxicol.*, **12**(Suppl.), 987 (1974).
54. J. L. Hartwell, *Lloydia*, **32**, 247 (1969).

SANDALWOOD OIL

Source: *Santalum album* L. (Family Santalaceae).

Common/vernacular names: Santal oil, East Indian sandalwood oil, white sandalwood oil, yellow sandalwood oil, and white saunders oil.

GENERAL DESCRIPTION

A small evergreen tree, heavily branched, up to about 9 m high; with opposite leathery leaves; elliptic–lanceolate, glabrous, entire, thin; flowers, brownish purple, in terminal and axillary paniculate cymes, shorter than leaves; sapwood odorless, heart wood yellowish brown strongly scented;[1] native to and cultivated in tropical Asia (especially India, Sri Lanka, Malaysia, Indonesia, and Taiwan). Part used is the heartwood. Sandalwood oil is obtained in 3–5% yield from the coarsely powdered dried heartwood by steam or water distillation.

India is the major producer of sandalwood oil (East Indian sandalwood oil). A closely related oil, Australian sandalwood oil, is derived from the wood of *Eucarya spicata* sprag.

et Summ. (syn. *Santalum spicatum* DC.); it has a different topnote than the East Indian oil but is similar to the East Indian oil in overall odor.[2]

CHEMICAL COMPOSITION

Sandalwood oil contains up to 90% or more of α- and β-santalols. Minor constituents present include 6% sesquiterpene hydrocarbons (mostly α- and, β-santalenes and epi-β-santalene with small amounts of α-and β-curcumenes, possibly β-farnesene, and dendrolasin), dihydro-β-agarofuran, santene, teresantol, borneol, teresantalic acid, tricycloekasantalal, santalone, and santanol, among others (JIANGSU).[2–7]

α-Santalol (ca. 46%) and β-santalol (ca. 20%) account for most of the odor of sandalwood oil.[3,8]

The heartwood of *S. album* contains terpenes, sesquiterpenes, neolignans, and aromatic esters.[9,10]

Australian sandalwood oil has recently been reported to contain α- and β-santalols, α-bisabolol, α-bergamotol, farnesol, nuciferol, and lanceol as the major components.[11]

PHARMACOLOGY AND BIOLOGICAL ACTIVITIES

Sandalwood oil is reported to have diuretic and urinary antiseptic properties. Santalol can reportedly cause contact dermatitis in sensitive individuals (CLAUS; LEWIS AND ELVIN-LEWIS).

Sandalwood oil has been reported to be nonirritating, nonsensitizing, and nonphototoxic to human skin, though it was slightly irritating to mouse skin and irritating to rabbit skin when applied undiluted.[12] Two reports described that sandalwood oil had a chemopreventive effect that prevented the incidence and multiplicity of skin papillomas in mice and may thus be useful against skin cancer.[13,14]

Antiviral activity has been reported against HSV-1 and 2 whereby the oil dose-dependently prevented viral replication and was more effective against HSV-1.[15] The oil also has antifungal activities.[8] Six sesquiterpenes isolated from the oil displayed antibacterial activity against *Helicobacter pylori*, and two of them were strongly active against a clarithromycin-resistant strain of *H. pylori*.[16]

The effects of transdermal absorption of sandalwood oil and α-santalol on various physiological parameters and mental/behavioral conditions in humans were investigated. α-Santalol had a general relaxing and sedative effect; while the oil had a relaxing effect on the physiological parameters, for example, blood pressure, pulse rate, and skin temperature, and a stimulant effect on behavior.[17]

USES

Medicinal, Pharmaceutical, and Cosmetic. Extensively used as a fragrance ingredient in soaps, detergents, creams, lotions, and perfumes (especially oriental types), with maximum use level of 1% reported in perfumes.[12] It is also commonly used in incenses.

Food. Used as a flavor component in major categories of food products, including alcoholic and nonalcoholic beverages, frozen dairy desserts, candy, baked goods, and gelatins and puddings, with reported average maximum use levels generally below 0.001%.

Dietary Supplements/Health Foods. Sandalwood oil is used in aromatherapy formulations (ROSE).

Traditional Medicine. In Chinese medicine, sandalwood oil is reportedly used to treat stomachache, vomiting, and gonorrhea (JIANGSU). Oil formerly used in Europe for pains, fevers, and "strengthening the heart."

COMMERCIAL PREPARATIONS

Oil. It was formerly official in N.F. and is official in F.C.C.

Regulatory Status. Has been approved for food use (§172.510). Wood is the subject of a German therapeutic monograph; allowed as an antibacterial and spasmolytic for supportive therapy for lower urinary tract infections.[18]

REFERENCES

See the General References for ARCTANDER; CLAUS; DER MARDEROSIAN AND BEUTLER; FEMA; GRIEVE; GUENTHER; JIXIAN; JIANGSU; MCGUFFIN 1 & 2; TERRELL; UPHOF.

1. K. R. Kirtikar and D. B. Basu, *Indian Medicinal Plants*, International Book Distribution, Dehradun, India, 1999, p. 2186.
2. W. D. Forham in L. W. Codd et al., eds., *Chemical Technology: An Encyclopedic Treatment*, Vol. 5, Barnes & Noble, New York, 1972, p. 1.
3. E. Demole et al., *Helv. Chim. Acta*, **59**, 73 (1976).
4. D. R. Adams et al., *Phytochemistry*, **14**, 1459 (1975).
5. S. K. Panitkar and C. G. Naik, *Tetrahedron Lett.*, **1293** (1975).
6. H. C. Kretschmar et al., *Tetrahedron Lett.*, **37** (1970).
7. B. M. Lawrence, *Perfum. Flav.*, **6**, 27 (1981).
8. A. Dikshit and A. Husain, *Fitoterapia*, **55**, 171 (1984).
9. T. H. Kim et al., *J. Nat. Prod.*, **68**, 1805 (2005).
10. T. H. Kim et al., *Chem. Pharm. Bull. (Tokyo)*, **53**, 641 (2005).
11. R. Shellie et al., *J. Chromatogr. Sci.*, **42**, 417 (2004).
12. D. L. J. Opdyke, *Food Cosmet. Toxicol.*, **12**(Suppl.), 989 (1974).
13. C. Dwivedi and Y. Zhang, *Eur. J. Cancer Prev.*, **8**, 449 (1999).
14. C. Dwivedi and A. bu-Ghazaleh, *Eur. J. Cancer Prev.*, **6**, 399 (1997).
15. F. Benencia and M. C. Courreges, *Phytomedicine*, **6**, 119 (1999).
16. T. Ochi et al., *J. Nat. Prod.*, **68**, 819 (2005).
17. T. Hongratanaworakit et al., *Planta Med.*, **70**, 3 (2004).
18. Monograph, *Santali albi lignum*, *Bundesanzeiger*, no. 43 (March 2, 1989).

SARSAPARILLA

Source: *Smilax medica* Schlecht. (syn. *S. aristolochiifolia* Mill.), *S. regelii* Killip et Morton, *S. officinalis* Kunth, *S. febrifuga* Kunth, *S. ornata* Lem. and other Smilax species (Family Liliaceae).

Common/vernacular names: Mexican sarsaparilla (*S. medica*), Honduras sarsaparilla (*S. regelii* and *S. officinalis*), Ecuadorian sarsaparilla (*S. febrifuga* and other Smilax species), Jamaican sarsaparilla (*S. regelii*).

GENERAL DESCRIPTION

Mostly climbing or trailing perennial vines with prickly stems, short and thick rhizomes, and very long slender roots; native to tropical America and the West Indies. Part used is the dried root.

CHEMICAL COMPOSITION

Contains steroids (sarsasapogenin, smilagenin, sitosterol, stigmasterol, and pollinastanol) and their glycosides (saponins) including sarsasaponin (parillin), smilasaponin (smilacin), sarsaparilloside, and sitosterol glucoside, among others (KARRER; REMINGTON).[1–5]

Other constituents present include starch, resin, ethyl alcohol, and a trace of a volatile oil (KARRER; MERCK).

PHARMACOLOGY AND BIOLOGICAL ACTIVITIES

Sarsaparilla saponins have been reported to facilitate the absorption of other drugs when coadministered with sarsaparilla.

Sarsaparilla has been used in the United States in treating syphilis and rheumatism, but its effectiveness has not been substantiated.

Sarsaparilla products, along with other sterol containing plants, have in recent years been touted as performance-enhancing or body-building substitutes for anabolic steroids sold primarily to athletes. No human or animal studies substantiate these claims. Plant sterols cannot be bio-chemically transformed *in vivo* into steroidal compounds, and have not been shown to promote anabolic effects in humans.[6]

Sarsaparilla preparations may cause gastric irritation or temporary kidney impairment; adverse drug interaction includes increased absorption of Digitalis glycosides, while accelerating elimination of hypnotic drugs.[7]

Sarsaparilla is reported to have hepatoprotective;[8] diuretic, and anti-inflammatory activity.[9]

Weak antifungal activity against *Candida albicans*, *C. glabrata* and *C. tropicalis* has been demonstrated by saponins isolated from the roots of *S. medica* (MIC 12.5–50 mg/mL).[2]

USES

Medicinal, Pharmaceutical, and Cosmetic. It is used in certain tonic preparations. In European tradition the root used for skin disease, particularly psoriasis, as well as rheumatic complaints, and kidney disease; diuretic and diaphoretic. Claims for efficacy have not been substantiated.[7]

Food. Sarsaparilla extracts are used extensively as flavor components in root beer even though they are essentially odorless and have hardly any taste; and it is doubtful that the reported average maximum use level of approximately 0.001% (12.9 ppm), even for the strongest commercial extract (solid extract), will contribute to the foaming properties of the root beer (see *quillaia* and *yucca*). Other food products in which they are used include frozen dairy desserts, candy, and baked goods, with the highest average maximum use level of 0.2% reported in baked goods.

Dietary Supplements/Health Foods. Root used as a flavoring ingredient in teas; numerous product forms, including tablets, capsules, and nutritional powders touted as an anabolic enhancing dietary supplement for athletes and bodybuilders. Claims that the ingredient contains testosterone are not substantiated (TYLER 1).

Traditional Medicine. Used generally as a tonic. Rhizome of Mexican sarsaparilla is reportedly used in Mexico in treating gonorrhea, skin diseases, rheumatism, fevers, and digestive disorders, usually as a decoction. Rhizome and root of Honduras sarsaparilla are used for similar purposes.

In Chinese medicine, the roots and/or rhizomes of several related *Smilax* species native to China are used. They include *S. sieboldi* Miq., *S. stans* Maxim., *S. scobinicaulis* C. H. Wright, and *S. glabra* Roxb., among others. As with sarsaparilla, they are used mainly in treating rheumatism, arthritis, sores, and skin problems. The rhizome of *S. glabra* is also used in treating mercury poisoning, syphilis, and acute bacterial dysentery, among others. In clinical observations, its effectiveness on primary syphilis has been reported to be about 90% (negative blood test). It is sometimes decocted with other plant drugs (JIANGSU). Numerous *Smilax* species have been used in cancers both in the Old World and the New World.[10]

COMMERCIAL PREPARATIONS

Crude and extracts; crude and fluid extract were formerly official in N.F. Strengths (see *glossary*) of extracts are expressed in weight-to-weight ratios.

Hemidesmus indicus R. Br. (Family Asclepidaceae) is reported to be a widespread adulterant to commercial supplies of sarsaparilla, which is easily distinguished organoleptically by its strong vanilla fragrance.[11]

Regulatory Status. Has been approved for food use (§172.510). Subject of a German therapeutic monograph; not recommended since claims for skin diseases and psoriasis have not been substantiated.[7]

REFERENCES

See the General References for ARCTANDER; BLUMENTHAL 1; BRADLY; CLAUS; FEMA; GOSSELIN; JIANGSU; LUST; STAHL; TERRELL; TUCKER AND LAWRENCE; TUCKER 1–3; TYLER 1; UPHOF; YOUNGKEN.

1. M. Sautour et al., *Planta Med.*, **72**, 667 (2006).
2. M. Sautour et al., *J. Nat. Prod.*, **68**, 1489 (2005).
3. R. R. Bernardo et al., *Phytochemistry*, **43**, 465 (1996).
4. M. Devys et al., *C. R. Acad. Sci., Ser. D*, **269**, 2033 (1969).
5. R. Tschesche et al., *Chem. Ber.*, **102**, 1253 (1969).
6. R. L. Barron and G. J. Vanscoy, *Ann. Pharmacother.*, **27**, 607 (1993).
7. Monograph, *Sarsaparillae radix. Bundesanzeiger*, 164 (9-1-1990).
8. S. Rafatullah et al., *Int. J. Pharmacogn.*, **29**, 296 (1991).
9. A. M. Ageel et al., *Drugs Exp. Clin. Res.*, **15**, 369 (1989).
10. J. L. Hartwell, *Lloydia*, **33**, 97 (1970).
11. C. Hobbs, *HerbalGram*, **17**, 10 (1988).

SASSAFRAS

Source: *Sassafras albidum* (Nutt.) Nees [syn. *S. officinale* Nees et Eberm.; *S. variifolium* (Salisb.) Kuntzel] (Family Lauraceae).

Common/vernacular names: Common sassafras.

GENERAL DESCRIPTION

An aromatic deciduous tree with leaves ranging in shape from three-lobed to two-lobed to unlobed; up to about 40 m high; native to eastern United States from Maine to Florida and west to Michigan and Texas. Part used is the dried root bark.

Safrole-free sassafras extract is obtained by dilute alcoholic extraction of the bark followed by concentrating under vacuum, diluting the concentrate with water, and separating and discarding the oily fraction.[1]

CHEMICAL COMPOSITION

Contains 5–9% volatile oil, about 0.02% alkaloids (boldine, norboldine, isoboldine, cinnamolaurine, norcinnamolaurine, and reticuline), two lignans (sesamin and desmethoxyaschantin), sitosterol, tannins, resin, and starch (CLAUS).[2]

The volatile oil contains safrole (80–90%) as the major component. Other compounds present include α-pinene, α- and β-phellandrenes, methyleugenol, 5-methoxyeugenol, asarone, piperonylacrolein, apiole, coniferaldehyde, camphor, myristicin, thujone; *l*-menthone, caryophyllene, elemicin, copaene, anethole, and eugenol, among others.[3–5]

PHARMACOLOGY AND BIOLOGICAL ACTIVITIES

Sassafras and its oil have been reported to have carminative and diaphoretic properties. The oil also reportedly has anti-infective and pediculicide (lice destroying) activities (MERCK).

Safrole is a hepatotoxin that has produced hepatomas (liver tumors) in laboratory animals (GOSSELIN; LEWIS AND ELVIN-LEWIS).[4–7]

Safrole is a strong inhibitor of human cytochrome P450 enzymes.[8]

TOXICOLOGY

Sassafras tea has been associated with clinical diaphoresis with hot flashes.[9]

USES

Food. Sassafras, its extracts, and oil were formerly extensively used in flavoring root beer; this use has been discontinued. Only safrole-free bark extract is reported used in nonalcoholic beverages and in candy, with average maximum use levels of 0.022% and 0.015%, respectively. As most of the flavor is removed along with safrole, these uses of the safrole-free extract are rather limited.

Dietary Supplements/Health Foods. Bulk sassafras is readily available; usually labeled "not for internal use". Popularity as a "spring tonic," continues in the Ozarks and Appalachians, where fresh root is available in the produce section of supermarkets in spring (FOSTER).

Traditional Medicine. Sassafras is traditionally used in treating bronchitis, high blood pressure of elderly people, rheumatism, gout, arthritis, skin problems, and kidney problems, among others, usually as a tea or infusion, used both internally and externally.

Sassafras has also been used in cancers.[10]

Others. Safrole present in sassafras oil is used as a starting material for the synthesis of heliotropin (piperonal), an important fragrance and flavor chemical.

COMMERCIAL PREPARATIONS

Crude, oil, and safrole-free extract. Crude and oil were formerly official in N.F.

Regulatory Status. Only safrole-free sassafras extract (§172.580) and safrole-free sassafras leaves and extracts (§172.510) have been approved for food use. Safrole, sassafras and sassafras oil are prohibited from use in foods (§189.180).[11]

REFERENCES

See the General References for BAILEY 1; BARNES; CLAUS; DER MARDEROSIAN AND BEUTLER; FEMA; FOSTER; FOSTER and DUKE; KROCHMAL AND KROCHMAL; LUST; MCGUFFIN 1 & 2; MORTON 1; SARGENT; TERRELL; TYLER 1–3; UPHOF.

1. Anon., *Fed. Regist.*, **27**, 9449 (1962).
2. B. K. Chowdhury et al., *Phytochemistry*, **15**, 1803 (1976).
3. D. P. Kamdem and D. A. Gage, *Planta Med.*, **61**, 574 (1995).
4. M. L. Sethi et al., *Phytochemistry*, **15**, 1773 (1976).

5. A. B. Segelman et al., *J. Am. Med. Assoc.*, **236**, 477 (1976).
6. F. Homburger et al., *Arch. Pathol.*, **73**, 118 (1962).
7. P. Borchert et al., *Cancer Res.*, **33**, 575 (1973).
8. Y. F. Ueng et al., *Food Chem. Toxicol.*, **43**, 707 (2005).
9. J. D. Haines, *Postgrad. Med.*, **90**, 75 (1991).
10. J. L. Hartwell, *Lloydia*, **32**, 247 (1969).
11. Anon., *Fed. Regist.*, **41**, 19207 (1976).

SAVORY (SUMMER AND WINTER)

Source: *Summer Savory Satureja hortensis* L. (syn. *Calamintha hortensis* Hort.); ***Winter savory*** *Satureja montana* L. (syn. *S. obovata* Lag.; *Calamintha montana* Lam.) (Family Labiatae or Lamiaceae).

GENERAL DESCRIPTION

Summer savory (*S. hortensis*) is an annual herb with oblong-linear leaves and hairy, erect branching stems; up to about 45 cm high; native to Europe and widely escaped from cultivation elsewhere (e.g., United States). Parts used are the dried leaves and tender stems; summer savory oil is obtained by steam distillation of the whole dried herb. Major producing countries include Spain, France, and the United States.

Winter savory (*S. montana*) is a bristly perennial subshrub with a woody base and oblong-linear leaves; up to about 38 cm high; native to the Mediterranean region and widely cultivated. Parts used are the leaves and tender stems.

The savory used in American households is normally summer savory.

CHEMICAL COMPOSITION

Summer savory (*S. hortensis*) contains a volatile oil (ca. 1%) consisting mostly of carvacrol and monoterpene hydrocarbons (β-pinene, *p*-cymene, β-phellandrene, limonene, camphene, etc.), with borneol, cineole, camphor, and others also present (GUENTHER; ROSENGARTEN).[1–4]

Other constituents present include labiatic acid, proteins (ca. 7%), vitamins (especially vitamin A), and minerals (especially Ca and K), among others (MARSH).[5]

Winter savory (*S. montana*) contains a volatile oil (1.6%) composed mainly of carvacrol, *p*-cymene, and thymol (total phenols ca. 50%), with lesser amounts of α-and β-pinenes, limonene, cineole, borneol, and α-terpineol.[6–8] It also contains triterpenic acids (ursolic and oleanolic acids).[9] Flavonoids include apigenin, apigenin-4-methyl ether, scutellarein-6,7dimethyl ether, and others.[10]

PHARMACOLOGY AND BIOLOGICAL ACTIVITIES

Summer savory oil has antimicrobial (fungi and bacteria) activities. The essential oil, ground material, and extract of Turkish *S. hortensis* were fungicidal against *Alternaria mali* and *Botrytis cinerea* at different concentrations.[11] Another Turkish group reported that the essential oil, methanol, and hexane extracts were potent against more than 100 laboratory strains of bacteria and fungi.[12,13] The essential oil of Greek savory (*S. spinosa*) exhibited antibacterial activity against a number of Gram-positive and Gram-negative bacteria.[14] It has recently been reported that the oil of winter savory has a destructive effect of the cell membranes of *Escherichia coli* and *Listeria mono- cytogenes*.[15]

Summer savory oil is also reported to have spasmolytic on rat-isolated ileum as well as *in vivo* antidiarrheal effects in mice.[16]

Antioxidant activities have been reported for *S. hortensis* (ethnolic extract and essential oil) and *S. montana* (essential oil) in different *in vitro* models of oxidative stress.[6,12,17,18]

The aqueous extract of summer savory was found to be anti-inflammatory against rhinosinusitis in rabbit at 250 mg/kg. A hydroalcoholic extract, polyphenolic fraction, and the essential oil also displayed antinociceptive and anti-inflammatory effects when administered orally (50–2000 mg/kg) to mice and rats in different models of pain and inflammation.[13,19]

Winter savory has been reported to exhibit diuretic activity in rats.[20]

TOXICOLOGY

When applied undiluted to the backs of hairless mice, summer savory was lethal to half of the animals in 48 h; it was also strongly irritating to rabbit and guinea pig skin, though it was nonphototoxic to hairless mice and swine. In diluted form, it was nonirritating and nonsensitizing to human skin.[2]

USES

Food. Summer savory is quite extensively used as a flavor component in baked goods, meat and meat products (e.g., canned meats), condiments and relishes (e.g., spice blends), processed vegetables, soups, and gravies, among others, with highest average maximum use level of 0.519% reported in condiments and relishes.

Summer savory oil and oleoresin are used in candy, baked goods, meat and meat products (e.g., canned meats), condiments and relishes, and others, with highest average maximum use level of about 0.036% reported for the oil in meat and meat products (358 ppm) and in condiments and relishes (373 ppm).

Winter savory oil and oleoresin are not widely used. Food products in which they are used include candy, baked goods, meat and meat products, and condiments and relishes. Highest average maximum use level reported is about 0.013% (127 ppm for the oleoresin in condiments and relishes).

Traditional Medicine. Both summer savory and winter savory are used as tonic, carminative, astringent, and expectorant in treating stomach and intestinal disorders (e.g., cramps, nausea, and indigestion), diarrhea, and sore throat, generally in the form of a tea. Fresh summer savory is also used for insect bites (e.g., bee sting); for this purpose it is rubbed on the affected area.

Savory has been used as an aphrodisiac.

COMMERCIAL PREPARATIONS

Crude, oleoresin, and oil (mostly summer savory). Summer savory oil is official in F.C.C.

Regulatory Status. GRAS (§182.10 and §182.20).

REFERENCES

See the General References for ARCTANDER; BAILEY 1; FEMA; FERNALD; GUENTHER; LUST; ROSE; ROSENGARTEN; UPHOF.

1. M. S. Karawya et al., *Am. Perfum. Cosmet.*, **85**, 23 (1970).
2. D. L. J. Opdyke, *Food Cosmet. Toxicol.*, **14**(Suppl.) 859 (1976).
3. A. Herisset et al., *Plant. Med. Phytother.*, **8**, 287 (1974).
4. B. M. Lawrence, *Perfum. Flav.*, **13**, 46 (1988).

5. K. Herrmann, *Z. Lebensm. Unters. Forsch.*, **116**, 224 (1962).
6. A. Radonic and M. Milos, *Free Radic. Res.*, **37**, 673 (2003).
7. B. Srepel, *Acta Pharm. Jugoslav.*, **4**, 167 (1974).
8. M. Paulet and D. Felisaz, *Riv. Ital. Essenze, Profumi, Piante Offic., Aromi, Saponi, Cosmet., Aerosol*, **53**, 618 (1971).
9. J. Susplugas et al., *Trav. Soc. Pharm. Montp.*, **29**, 129 (1969).
10. E. Wollenweber and K. M. Valant-Vetschera, *Fitoterapia*, **62**, 5, 462 (1991).
11. N. Boyraz and M. Ozcan, *Int. J. Food Microbiol.*, **107**, 238 (2006).
12. M. Gulluce et al., *J. Agric. Food Chem.*, **51**, 3958 (2003).
13. F. Sahin et al., *J. Ethnopharmacol.*, **87**, 61 (2003).
14. N. Chorianopoulos et al., *J. Agric. Food Chem.*, **52**, 8261 (2004).
15. M. Oussalah et al., *J. Food Prot.*, **69**, 1046 (2006).
16. V. Hajhashemi et al., *J. Ethnopharmacol.*, **71**, 187 (2000).
17. F. Mosaffa et al., *Arch. Pharm. Res.*, **29**, 159 (2006).
18. A. Radonic and M. Milos, *Nahrung*, **47**, 236 (2003).
19. C. Uslu et al., *J. Ethnopharmacol.*, **88**, 225 (2003).
20. G. Stanic and I. Samarzija, *Phytother. Res.*, **7**, 363 (1993).

SAW PALMETTO

Source: *Serenoa repens* (Bart.) Small. (syn. *Sabal serrulata* Schultes and Schultes) (Family Palmaceae).

Common/vernacular names: Sabal.

GENERAL DESCRIPTION

Shrub, 3–4 m tall; leaves palmate, without continuing rib, divided into lance-shaped linear-lanceolate leaflets, to 2.5 dm long; petioles armed with spiny teeth; inflorescence many branched, less than 1 m, subtending the leaves, with white flowers; fruit a prominent olive like mesocarp, 16–25 mm long, yellowish turning blue-black when ripe, with single large oblong seed to 18 mm long (17.7% of dry fruit weight),[1] ripening September through December; forming large colonies in southeastern United States. in coastal plain from South Carolina, to south Mississippi and throughout Florida (FOSTER AND DUKE).

The part used is the fruit.

CHEMICAL COMPOSITION

Fruits reported to contain 1–2% essential oil (formed by reaction of alcohols and acids during distillation); fixed oil with 75% free fatty acids and 25% neutral substances including free or esterified sterols or esters of fatty acids with alcohols; acids including caproic, caprylic (ca. 1.3%), caprinic (ca. 1.8%), lauric (ca. 24%), linoleic (ca. 3.6%), linolenic; myristic (ca. 11.6%), oleic (ca. 33.2%), palmitic (ca. 8.7%), myristoleic and stearic acids;[2,3] anthranilic acid in fruit alcohol extracts;[4,5] sterols including β-sitosterol, β-sitosterol 3-*O*-β-D-glucoside, campesterol, stigmasterol, lupeol, 24-methylene-cycloartanol;[6,7] alcohols including hexacosanol, 1-octacosanol, farnesol, and phytol;[3,8] acylglycerides (monolaurin and monomyristin);[9] polysaccharides S1, S2, S3, and S4 with varying ratios of glucose, galactose, mannose, fucose, arabinose, rhamnose, and glucuronic acid.[10–14]

PHARMACOLOGY AND BIOLOGICAL ACTIVITIES

Fruits are considered to have diuretic, sedative, antiandrogenic, anti-inflammatory, antiexudative, and both antiestrogenic and estrogenic effects.

In mice experiments intraperitoneal administration of partially purified β-sitosterol fractions of the fruit extract showed significant estrogenic activity;[7] an anti-estrogenic effect has also been demonstrated.[15]

A number of earlier double-blind, placebo controlled clinical studies (involving over 600 patients) suggest that lipoid hexane extracts or supercritical CO_2 extracts improve objective and subjective symptoms, including dysuria, nocturia, and frequent and poor urinary flow in benign prostatic hyperplasia (BPH).[16–22]

A fruit extract has been shown to selectively antagonize and prevent binding to 52% of dihydrotestosterone receptors in the prostate.[23] Oral administration of a fruit extract in mice and rats has demonstrated inhibition of up to 90% of the activity of prostate 5-α-reductase (that transforms testosterone into metabolites increasing the size of the prostate).[24]

The use of saw palmetto in BPH has continued as the main focus of numerous reports over the last decade. Such reports include *in vivo* experiments in animal models,[25–28] *in vitro* experiments in prostate cells, and human clinical trials.[28–47] The majority of these reports were in favor of saw palmetto's efficacy in improving lower urinary tract symptoms. Some studies showed that it favorably compared with finasteride and tamsulosin, the first-line treatments for BPH.[29,32,35,39] There is no consensus yet on the mechanism of action of saw palmetto, but one or more of the following may be involved: 5-α-reductase inhibition, indirect sympathomimetic effect ($α_1$-adrenergic receptor binding), anti-inflammatory and/or hormonal antiandrogenic effects.[25,26,30,31]

Saw palmetto extract and its myristoleic acid component were reported to possess antitumor activity against human urological and pancreatic cancer cell lines that gave an indication of potential use in prostate cancer.[2,48]

TOXICOLOGY

Saw palmetto is generally regarded as safe with high long-term tolerability.[36,41] One case, however, was recently reported in which saw palmetto appeared to be responsible for development of recurrent pancreatitis in a 55-year-old male.[49] A study to investigate its effect on liver functions in rat showed that saw palmetto did not result in any toxicity at 5× the maximum recommended human doses.[50]

The pharmacology and therapeutic significance of saw palmetto has been the subject of many reviews over the last decade.[34,51–56]

USES

Medicinal, Pharmaceutical, and Cosmetic. In European phytomedicine, cut-sifted or powdered crude fruit in galenical preparations; hexane, or ethanol extracts are used in antitestosterone and antiexudative preparations for treatment of difficulty in micturition in BPH, stages I and II. According to one source, the drug only relieves symptoms associated with prostate enlargement, without reducing enlargement.[57] Some recent studies, however, reported the reduction in prostate size.[36]

Food. Historically used as a food source by indigenous groups of Florida; survival food of early European settlers (FOSTER AND DUKE).

Dietary Supplements/Health Foods. Fruits in tea, capsules, tablets, tinctures, and other product forms as a dietary supplement, primarily as an endocrine and anabolic agent; also for prostate enlargement (FOSTER AND DUKE).

Traditional Medicine. Fruit traditionally considered expectorant, sedative, diuretic; used to treat prostate enlargement and inflammation; also used for colds, coughs, irritated mucous membranes, sore throat, asthma, chronic bronchitis, head colds, and migraine; suppository of powdered fruits used as a uterine and vaginal tonic; also a folk cancer remedy (FOSTER AND DUKE).

Others. Leaf wax investigated as a potential valuable material for wax-consuming industries, but was found to be of less value than the principal hard vegetable waxes currently available.[58]

COMMERCIAL PREPARATIONS

Crude, ethanol, hexane, and supercritical CO_2 extracts.

Regulatory Status. Class 1 dietary supplement (can be safely consumed when used appropriately). The fruit is the subject of a positive German therapeutic monograph, indicated for micturition trouble in BPH stages I and II.[57]

REFERENCES

See the General References for BRUNETON; FOSTER and DUKE; MARTINDALE; MCGUFFIN 1 & 2; STEINMETZ; TYLER 1–3; WEISS; WREN.

1. J. B. Hilmon, Autecology of Saw Pal Metto, Ph.D. Dissertation, DUKE University, 1968.
2. K. Iguchi et al., *Prostate*, **47**, 59 (2001).
3. W. Breu, *Arzneim.-Forsch.*, **42**, 547 (1992).
4. R. Hansel et al., *Planta Med.*, **12**, 169 (1964).
5. R. Hansel et al., *Planta Med.*, **14**, 261 (1966).
6. W. R. Sorenson and D. Sullivan, *J. AOAC Int.*, **89**, 22 (2006).
7. M. I. Elghamry and R. Hansel, *Experientia*, **25**, 8 (1969).
8. G. Jommi et al., *Gazzetta Chim. Ital.*, **118**, 823 (1988).
9. H. Shimada et al., *J. Nat. Prod.*, **60**, 417 (1997).
10. G. Harnischfeger, *Z. Phytother.*, **10**, 71 (1989).
11. P. Hatinguais et al., *Trav. Soc. Pharm. Montp.*, **41**, 253 (1981).
12. H. Wagner et al., *Planta Med.*, **41**, 244 (1981).
13. H. Wagner et al., *Arzneim.-Forsch.*, **34**, 659 (1984).
14. H. Wagner et al., *Arzneim.-Forsch.*, **35**, 1069 (1985).
15. F. Di Silverio et al., *Eur. Urol.*, **21**, 309 (1992).
16. E. Emili et al., *Urologia*, **50**, 1040 (1983).
17. C. Boccafoschi et al., *Urologia*, **50**, 1257 (1983).
18. G. Champault et al., *Br. J. Clin. Pharmacol.*, **18**, 461 (1984).
19. A. Tasca et al., *Minerva Urol. Nefrol.*, **37**, 87 (1985).
20. P. Cukier et al., *C. R. Ther. Pharmacol. Clin.*, **4**, 15 (1985).
21. C. Casarosa et al., *Clin. Ther.*, **10**, 585 (1988).
22. S. M. Adriazola et al., *Arch. Esp. Urol.*, **45**, 211 (1992).
23. C. Sultan et al., *J. Steroid Biochem.*, **20**, 515 (1984).
24. A. Stenger et al., *Gaz. Med. Fr.*, **89**, 2041 (1982).
25. N. Cao et al., *Prostate*, **66**, 115 (2006).
26. T. Oki et al., *J. Urol.*, **173**, 1395 (2005).

27. N. Talpur et al., *Mol. Cell. Biochem.*, **250**, 21 (2003).
28. M. F. Palin et al., *Endocrine*, **9**, 65 (1998).
29. F. Hizli and M. C. Uygur, *Int. Urol. Nephrol.*, (2007).
30. F. K. Habib et al., *Int. J. Cancer*, **114**, 190 (2005).
31. A. C. Buck, *J. Urol.*, **172**, 1792 (2004).
32. E. M. Gong and G. S. Gerber, *Am. J. Chin. Med.*, **32**, 331 (2004).
33. G. S. Gerber and J. M. Fitzpatrick, *BJU Int.*, **94**, 338 (2004).
34. P. Boyle et al., *BJU Int.*, **93**, 751 (2004).
35. S. A. Kaplan et al., *J. Urol.*, **171**, 284 (2004).
36. Y. A. Pytel et al., *Adv. Ther.*, **19**, 297 (2002).
37. X. Giannakopoulos et al., *Adv. Ther.*, **19**, 285 (2002).
38. G. S. Gerber et al., *Urology*, **58**, 960 (2001).
39. L. S. Marks et al., *Urology*, **57**, 999 (2001).
40. M. Goepel et al., *Prostate*, **46**, 226 (2001).
41. L. S. Marks et al., *J. Urol.*, **163**, 1451 (2000).
42. F. Vacherot et al., *Prostate*, **45**, 259 (2000).
43. C. W. Bayne et al., *J. Urol.*, **164**, 876 (2000).
44. C. W. Bayne et al., *Prostate*, **40**, 232 (1999).
45. M. Goepel et al., *Prostate*, **38**, 208 (1999).
46. S. F. Di et al., *Prostate*, **37**, 77 (1998).
47. M. Paubert-Braquet et al., *Pharmacol. Res.*, **34**, 171 (1996).
48. K. Ishii et al., *Biol. Pharm. Bull.*, **24**, 188 (2001).
49. I. Jibrin et al., *South. Med. J.*, **99**, 611 (2006).
50. Y. N. Singh et al., *Phytomedicine*, (2006).
51. C. Ulbricht et al., *J. Soc. Integr. Oncol.*, **4**, 170 (2006).
52. A. L. Avins and S. Bent, *Curr. Urol. Rep.*, **7**, 260 (2006).
53. T. J. Wilt et al., *JAMA*, **280**, 1604 (1998).
54. G. L. Plosker and R. N. Brogden, *Drugs Aging*, **9**, 379 (1996).
55. A. E. Gordon and A. F. Shaughnessy, *Am. Fam. Physician*, **67**, 1281 (2003).
56. T. Wilt et al., *Cochrane Database Syst. Rev.*, CD001423 (2002).
57. Monograph, *Sabal Fructus, Bundesanzeiger*, no. 43 (March 2, 1989); revised (February 1, 1990; January 17, 1991).
58. E. A. Wilder and E. D. Kitzke, *Science*, **120**, 108 (1954).

SCHISANDRA

Source: *Schisandra chinensis* (Turcz.) Baill., *S. sphenanthera* Reid. et Wils., and other **Schisandra** species (Family Schisandraceae).

Common/vernacular names: Wuweizi, meaning "five-flavor seed" (general term for all varieties); northern schisandra or *beiwuweizi* (*S. chinensis*); southern schisandra or *nanwuweizi*, western schisandra or *xiwuweizi* (*S. sphenanthera*); gomishi.

GENERAL DESCRIPTION

Schisandra chinensis is a deciduous woody vine, up to 8 m long; berries bright red when mature; native to northern and northeastern

China and adjacent regions of Russia and Korea. Part used is the fully ripe, sun-dried fruit that yields northern schisandra; it is oval and wrinkled, with a diameter of 5–8 mm, ranging from bright red, dull red, to purplish red; flesh is soft, with a weak characteristic odor and tastes primarily sour and sweet, with a salty note; its 1–2 yellowish brown, kidney-shaped seeds are fragrant when crushed and taste simultaneously pungent, bitter, and salty; these five flavor elements give schisandra its name, "five-flavor seed." Northern schisandra is mainly produced in northern and northeastern provinces, including Inner Mongolia, Ningxia, Shanxi, Hebei, Liaoning, Jilin, and Heilongjiang.

Schisandra sphenanthera is a climbing shrub similar to *S. chinensis*, up to about 5 m long; native to western, central, and southern China. Its fully ripe, sun-dried fruit yields southern or western schisandra, which is similar in properties to northern schisandra fruit, but is smaller, with thinner flesh and is reddish brown to dull brown. Southern schisandra is mainly produced in western, central, and southern provinces, including Gansu, Shaanxi, Henan, Hunan, Hubei, and Sichuan.

Although several other *Schisandra* species also serve as commercial or potential commercial sources of schisandra fruit, its current major sources are *S. chinensis* and *S. sphenanthera*.[1,2]

CHEMICAL COMPOSITION

Schisandra chinensis fruit contains roughly 1–3% volatile oil (composed mainly of citral, sesquicarene, β-2-bisabolene, aylangene, β-chamigrene, α-chamigrene, and chamigrenal), 12% citric acid, 10% malic acid, cinnamic acid, small amounts of tartaric acid, flavonoids (quercetin and kaempferol), monosaccharides, resins, pectin, vitamins A, C, and E, phospholipids, sterols, tannins, and so on.[3–6]

Seeds of *S. chinensis* contain about 19% lignans (those of *S. sphenanthera* about 10%) that are the major active principles.[1] Extensive chemical studies have been performed on these schisandra lignans, primarily by Russian, Japanese, and Chinese researchers.[3,7] Russian researchers named them schizandrins and schizandrols while Japanese researchers called them gomisins and the Chinese called them wuweizisus and wuweizi esters, among others.[7–12] Over 40 such lignans have been isolated from the nonsaponifiable fraction of the seed oil. The more important ones include: schizandrin (=schizandrol A), schizandrin A (=deoxyschizandrin), schizandrin B (=γ-schizandrin), schizandrin C, schizandrol B, schisantherin D, schisandrene, schizandrer A, schizandrer B (=schisantherin), pseudo-γ-schizandrin, wuweizisu A (=schizandrin A), wuweizisu B, wuweizisu C, gomisins A–R and derivatives, wuweizi ester A (=gomisin C, schisantherin A, schizandrer A), wuweizi ester B (=gomisin B, schisantherin B, schizandrer B), wuweichun A (=schizandrin), wuweichun B, and others.[1,3,7–14]

Other lignans isolated from seeds of other *Schisandra* species (e.g., *S. sphenanthera*, *S. henryi*, *S. Propinqua* and *S. rubriflora*) include: schisanhenol, schisandrone, epischisandrone, enshicine, epienshicine, wulignan A1, wulignan A2, epiwulignan A1, rubrisandrin A and B, (−)-rubschisandrin, rubschisantherin, schisanhenol acetate, schisanhenol B, propinquanins A–D, and others.[7,13,15–19]

Over the past 5 y, numerous triterpenes have been isolated from *S. Sphenanthera* and other species (*S. lancifolia*, *S. rubrifolia*, *S. propinqua*, *S. micrantha*, and *S. henryi*). Among these triterpenes are sphenadilactones A and B, lancifodilactones A–N, lancifoic acid A, nigranoic acid, rubriflordilactones A and B, schiprolactone A, schisanlactone B, schisandronic acid, micrandilactone A, micranoic acids A and B, and others.[16,20–29]

A new sesquiterpene, plenoxide, was recently isolated from *S. plena*.[30]

PHARMACOLOGY AND BIOLOGICAL ACTIVITIES

Schisandra fruit exhibits a wide variety of pharmacological activities in humans and in laboratory animals, including: "adaptogenic" properties (increasing nonspecific resistance) similar to those of ginseng and Siberian ginseng, but with weaker effects and lower toxicities;[31] stimulating the central nervous system and improving the reflexes, endurance, and work performance of healthy individuals;[32] tranquilizing and anticonvulsive effects in rodents, with schizandrol A as an active agent;[33] antidepressant effect in mice;[34] antifatigue effects in rodents and horses and improving markedly the performance of race horses;[35] heart stimulating, vasodilating and blood pressure normalizing, especially in circulatory failure;[36] promoting thymidine incorporation in DNA synthesis in human lymphocytes, and enhancement as well as suppression of immune functions;[36,37] stimulating rabbit uterus *in vivo* and *in vitro*;[36,38] stimulating respiration in animals, with lignans as active agents; antitussive and expectorant effects in mice; antibacterial *in vitro*;[38,39] and others (JIANGSU; JILIN; WANG).

The most extensively studied effect of schisandra fruit is its antioxidant activity[14,40–43] and ability to protect human and animal liver from toxins and diseases. Studies over the past 20 y have shown it to have the following antihepatotoxic effects: lowering serum alanine aminotransferase (SGPT) levels in humans and in experimental animals, and improving symptoms in patients with chronic hepatitis; protecting the liver from damage caused by carbon tetrachloride and other toxins; promoting the biosynthesis of serum and liver proteins; stimulating the formation of liver glycogen; induction of cytochrome P-450 of animal liver microsomes; inhibiting microsomal lipid peroxidation by liver toxins; inhibiting the covalent binding between liver toxins and microsomal lipids; lowering the mortality rate of animals poisoned by acetaminophen, digitoxin, and indomethacin; increasing serum and liver cyclic AMP levels in mice; and others.[7,8,10,11,39,40,44–48] These antihepatotoxic effects are due mainly to the lignans present in the seed, especially wuweizisu C, sehisantherin D, gomisin A, gomisin C, gomisin N, deoxygomisin A,[7,10] schizandrin A, schizandrin B, schizandrin C, schizandrol A, schizandrol B, schizandrer A, schizandrer B, and schisanhenol, among others.[8,11,45,47,48] Myocardial protecting effects have also been reported for *S. chinensis*.[49,50]

The acetone extract of schisandra fruit exhibited marked inhibition against the mutagenic effect of aflatoxin B_1 per the Ames test.[51]

Anti-HIV activity has been reported for many triterpenes (e.g., micrandilactones and lancifodilactones) and a few lignans (e.g., rubrisandrins A and B) from different *Schisandra* species.[15,21,23–25]

Other recently reported activities of schisandra fruit include reduced proteoglycan degradation (antihyaluronidase activity);[52] and antiplatelet activation (PAF antagonism).[53]

TOXICOLOGY

Adverse side effects of schisandra fruit are rare; they include stomach upset, decreased appetite, and urticaria (WANG).[54]

USES

Medicinal, Pharmaceutical, and Cosmetic. Biphenyldimethyldicarboxylate (BDD), an intermediate of schizandrin C synthesis, is now used in China to treat viral hepatitis with much higher efficacy than silymarin.[8]

Dietary Supplements/Health Foods. Extracts and crude powder are used in preparations marketed mainly for their antioxidant and energy enhancing as well as adaptogenic effects. Crude is also used in soup mixes (FOSTER AND YUE).

Traditional Medicine. In traditional Chinese medicine, schisandra fruit is considered a lung astringent, and kidney and male tonic. It is used in treating cough, asthma, insomnia, neurasthenia, chronic diarrhea, night sweat, spontaneous sweating, involuntary seminal discharge, thirst, impotence, physical exhaustion, and excessive urination. The official *Chinese Pharmacopoeia* lists its daily oral dose as 1.5–6 g.

The vines and roots of various *Schisandra* species (e.g., *S. sphenanthera* and *S. henryi*) are used in China to treat painful joints, rheumatism, and traumatic injuries.

COMMERCIAL PREPARATIONS

Crude and extracts. Seeds must be crushed before extraction. Most extracts are expressed in weight-to-weight ratios with no uniform quality standards; some extracts are standardized to schizandrin content.

Regulatory Status. Class 1 dietary supplement (can be safely consumed when used appropriately). Schisandra is an ethnic food.

REFERENCES

See the General References for CHP; DER MARDEROSIAN AND BEUTLER; FOSTER AND YUE; HONGKUI; HU; IMM-3; JIANGSU; JILIN; LU AND LI; MCGUFFIN 1 & 2; NATIONAL; WANG; ZHU.

1. W. Z. Song and Y. Y. Tong, *Yaoxue Xuebao*, **18**, 138 (1983).
2. W. Z. Song, *Zhongyao Tongbao*, **13**, 3 (1988).
3. Y. S. Huang et al., *Chin. J. Pharmacol. Toxicol.*, **4**, 275 (1990).
4. Y. Ohta and Y. Hirose, *Tetrahedron Lett.*, **20**, 2483 (1968).
5. R. Sladkovsky et al., *J. Pharm. Biomed. Anal.*, **24**, 1049 (2001).
6. X. N. Li et al., *Phytochem. Anal.*, **14**, 23 (2003).
7. H. Hikino et al., *Planta Med.*, **50**, 213 (1984).
8. P. G. Xiao and K. J. Chen, *Phytother. Res.*, **2**, 55 (1988).
9. R. Tan et al., *Planta Med.*, **52**, 49 (1986).
10. Y. Kiso, *Planta Med.*, **51**, 331 (1985).
11. G. T. Liu in H. M. Chang et al., eds., *Advances in Chinese Medicinal Materials Research*, World Scientific Publish, Singapore, 1985, p. 257.
12. Y. Ikeya et al., *Chem. Pharm. Bull.*, **30**, 3207 (1982).
13. H. B. Zhai and P. Z. Cong, *Yaoxue Xuebao*, **25**, 110 (1990).
14. Y. W. Choi et al., *J. Nat. Prod.*, **69**, 356 (2006).
15. M. Chen et al., *J. Nat. Prod.*, **69**, 1697 (2006).
16. L. J. Xu et al., *Planta Med.*, **72**, 169 (2006).
17. L. N. Li and H. Xue, *Planta Med.*, **51**, 217 (1985).
18. J. S. Liu et al., *Huaxue Xuebao*, **46**, 483 (1988).
19. H. J. Wang and Y. Y. Chen, *Yaoxue Xuebao*, **20**, 832 (1985).
20. W. L. Xiao et al., *Org. Lett.*, **8**, 1475 (2006).
21. W. L. Xiao et al., *J. Nat. Prod.*, **69**, 277 (2006).
22. W. L. Xiao et al., *Org. Lett.*, **8**, 991 (2006).
23. R. T. Li et al., *Chem. Commun. (Camb.)*, **2936** (2005).
24. W. L. Xiao et al., *Org. Lett.*, **7**, 2145 (2005).
25. W. L. Xiao et al., *Org. Lett.*, **7**, 1263 (2005).
26. Y. G. Chen et al., *Arch. Pharm. Res.*, **26**, 912 (2003).
27. R. T. Li et al., *Chem. Pharm. Bull. (Tokyo)*, **51**, 1174 (2003).
28. R. T. Li et al., *Org. Lett.*, **5**, 1023 (2003).

29. Y. G. Chen et al., *Fitoterapia*, **72**, 435 (2001).
30. R. T. Li et al., *J. Asian Nat. Prod. Res.*, **7**, 847 (2005).
31. I. I. Brekhman and I. V. Dardymov, *Ann. Rev. Pharmacol.*, **419**, (1969).
32. B. X. Wang, *Tianjin Yiyao Zazhi*, **7**, 338 (1965).
33. X. Y. Niu et al., *Yaoxue Xuebao*, **18**, 491 (1983).
34. J. L. Hancke et al., *Planta Med.*, **53**, (1987).
35. F. Ahumada et al., *Phytother. Res.*, **3**, 175 (1989).
36. K. Sun, *Yaoxue Xuebao*, **7**, 277 (1959).
37. M. Li and H. Liu, *Zhongcaoyao*, **14**, 31 (1983).
38. S. S. Liu, ed., *Zhongyao Yanjiu Wen Xian Zhaiyao (1820–1961)*, Science Press, Beijing, 1975, p. 85.
39. S. S. Liu, ed., *Zhongyao Yanjiu Wen Xian Zhaiyao (1962–1974)*, Science Press, Beijing, 1979, p. 108.
40. T. M. Zhang et al., *Acta Pharmacol. Sinica*, **10**, 353 (1989).
41. T. J. Lin et al., *Chin. J. Pharmacol. Toxicol.*, **3**, 153 (1989).
42. P. Y. Chiu et al., *Planta Med.*, **68**, 951 (2002).
43. S. P. Ip et al., *Pharmacol. Toxicol.*, **78**, 413 (1996).
44. M. Li, *Zhongyao Tongbao*, **9**, 41 (1984).
45. G. T. Liu et al., *Chin. J. Integr. Trad. West. Med.*, **3**, 182 (1983).
46. H. Nagai et al., *Planta Med.*, **55**, 13 (1989).
47. G. T. Liu and H. L. Wei, *Acta Pharmacol. Sin.*, **6**, 41 (1985).
48. G. T. Liu et al., *Yaoxue Xuebao*, **15**, 206 (1980).
49. J. S. You et al., *Chang Gung. Med. J.*, **29**, 63 (2006).
50. P. C. Li et al., *Am. J. Chin. Med.*, **24**, 255 (1996).
51. C. C. Ruan et al., *Chin. J. Cancer*, **8**, 29 (1989).
52. S. I. Choi et al., *J. Ethnopharmacol.*, **106**, 279 (2006).
53. I. S. Lee et al., *Biol. Pharm. Bull.*, **22**, 265 (1999).
54. H. W. Song and F. X. Wang, *Zhongguo Zhongyao Zazhi*, **15**, 51 (1990).

SENNA

Source: *Cassia senna* L. (syn. *C. acutifoli* Del.; *Senna alexandrina* Mill.) and *C. angustifolia* Vahl (syn. *Senna alexandrina* Mill.) (Family Leguminosae or Fabaceae).

Common/vernacular names: Alexandrian senna, Khartoum senna (*C. senna*); Tinnevelly senna, Indian senna (*C. angustifolia*).

GENERAL DESCRIPTION[1]

Alexandrian senna (*C. senna*) is an herbaceous subshrub up to about 1 m high; native to the Nile region of northern Africa (Egypt, Sudan, etc.) and cultivated there as well as in tropical Asia (e.g., southern China and India).

Indian or Tinnevelly senna (*C. angustifolia*) is also an herbaceous subshrub up to about 1 m high; native to India and northeastern Africa; cultivated mainly in southern and northwestern India and in Pakistan.

Parts used are the dried leaflets (commonly called leaves) and pods. Alexandrian senna leaves and pods are reportedly derived mostly from wild plants, while Indian senna leaves are collected at 3–5 months after planting and pods 1–2 months hence.

The two species of senna are very closely related and have formerly been recognized as a single species, though later research data have established them as distinct species.[2,3] Recent taxonomic treatment once again merges *C. senna*, *C. acutifolia*, and *C. angustifolia* into one taxonomic entity, *Senna alexandrina* Mill., which is not generally recognized in the trade.

Most of the senna currently used in the United States is Indian senna (Tinnevelly senna).

CHEMICAL COMPOSITION

Alexandrian and Indian senna leaves have similar chemical compositions, especially in their anthracene derivatives. They contain as their active constituents dianthrone glucosides (usually 1.5–3.0%), consisting mostly of sennosides A and B (rhein-dianthrone glucosides), with minor amounts of sennosides C and D (rhein-aloe-emodin-heterodianthrone glucosides) and aloe-emodin dianthrone glucoside also present.[4–10] These dianthrone glycosides are reportedly absent in fresh leaves, and it appears that they are formed during the drying process through enzymatic oxidation of monoanthrone glycosides that are present in fresh leaves but normally absent in dried leaves (also see *cascara*).[5,11]

There is also evidence of the existence of primary glycosides of the sennosides (with additional sugar molecules) that are more active than the sennosides.[6]

Senna leaves also contain small amounts of free anthraquinones (rhein, aloe-emodin, chrysophanol, etc.) and their *O*-glycosides and *C*-glycosides.[2,5–8,11,12]

Alexandrian senna leaves generally have a higher sennoside content than Indian senna leaves.

Two benzophenone glucosides have recently been isolated from Tinnevelly senna pods and were characterized as 6′-carboxy-2′,6-dihydroxy-2-β-glucopyranosyloxy-4′-hydroxymethyl benzophenone and 4′,6′-dicarboxy-2′,6-dihydroxy-2-β-glucopyranosyloxy-4′-hy-roxymethyl benzophenone (cassiaphenone A-2-glucoside and cassiaphenone B-2-glucoside, respectively). The naphthalene glycoside tinnevellin-8-glucoside and kaempferol were also isolated.[13]

Other constituents present in senna leaves include free sugars (glucose, fructose, sucrose, and pinitol), a mucilage (consisting of galactose, arabinose, rhamnose, and galacturonic acid), and polysaccharides (in *C. angustifolia*);[14,15] flavonoids (isorhamnetin, kaempferol, etc.); a trace of volatile oil; and resins; among others (LIST AND HÖRHAMMER).

Senna pods normally contain 2–5% sennosides, with Alexandrian pods having higher values than Indian pods. In addition to sennosides A and B, a closely related glucoside, named sennoside A_1, has been isolated from Alexandrian senna pods.[16]

A galactomannan consisting of D-glucose and D-mannose in a 3:2 molar ratio has recently been isolated from the seeds of Indian senna.[17]

PHARMACOLOGY AND BIOLOGICAL ACTIVITIES

Sennosides are cathartic, with a similar mode of action as cascarosides (see *cascara*).

Sennosides A and C have equal purgative potency in mice but sennoside C has potentiating effects on the activity of sennoside A, exerting a potentiating effect of about 1.6 when 20% of the dose of sennoside A is replaced by sennoside C.[18]

Senna products along with cascara products are generally considered the drugs of choice among anthraquinone cathartics, and are also generally considered safe (APhA).[19,20] Products containing purified sennosides (20 mg) reduced colonic transit time in healthy human volunteers.[21] Senna preparations are continuously being evaluated as alternative laxatives in bowel preparation for colonoscopy. In this respect, they have comparative efficacy as sodium phosphate and PEG-electrolyte lavage solutions.[22,23]

There are controversial data on the adverse effects of senna. On the one hand, it is reported that excessive or prolonged use of senna, as with laxatives in general, may lead to colon damage and other problems (APHA; MARTINDALE).[24] Stimulating laxatives should not be used more than 1–2 weeks without medical advice.[25] Daily treatment, as with any laxative, is not recommended. Chronic abuse can disturb electrolyte balance, leading to potassium deficiency, heart dysfunction, and muscular weakness, especially under concomitant use of heart-affecting glycosides, thiazide diuretics, corticoadrenal steroids, and licorice root.[25] A number of toxicity cases due to chronic use of senna have recently been reported and include skin breakdown and blisters leading to severe diaper rash in children,[26] acute liver failure with renal impairment,[27] and subacute cholestatic hepatitis.[28] On the other hand, it is reported that senna can be safely administered when given in doses sufficient to produce a motion of physiological water content, even over a long period of time. Senna does not induce specific lesions in the nerve plexus of the intestinal wall, and when used rationally, does not lead to electrolyte losses or habituation.[29] Nonsignificant side effects were observed in rats receiving 750–1500 mg/kg/day of senna for 13 weeks. The side effects completely disappeared after 8 weeks of recovery.[30] It was also found that daily administration of up to 300 mg/kg/day of senna for 2 years was not carcinogenic to mice.[31] Earlier studies showed that senna extracts do not promote malignant tumors in rat colons at laxative doses and that there was no genotoxic risk associated with the use of senna in animals and humans.[32,33]

One of the polysaccharides exhibited a significant inhibitory effect against solid sarcoma 180 in CD1 mice.[14]

An ethanolic extract of a preparation containing *C. senna* (Senokot®) inhibited the effects of different mutagenic agents (e.g., benzo[*a*]pyrene, aflatoxin B1) in the Ames test.[34]

USES

Medicinal, Pharmaceutical, and Cosmetic. Senna leaves, pods, their extracts, and sennosides are extensively used as active ingredients in laxative preparations (syrups, tablets, etc.). Often termed a "bowel irritant" or "stimulant," recent scientific consensus suggests such terms should be avoided, in favor of a more specific characterization as a prokinetic agent with a secretory component.[29]

Dietary Supplements/Health Foods. Senna leaves or pods used in laxative formulations, usually in tablet, capsule or tea bag form.

Traditional Medicine. Senna leaves have been used for centuries in both western and eastern cultures as a laxative, usually taken as a tea or swallowed in powdered form.

COMMERCIAL PREPARATIONS

Crudes (leaves and pods), extracts (solid, fluid, etc.), and sennosides A and B (15–100%). Senna leaf (both Indian and Alexandrian), its fluid extract, and sennosides A and B are official in U.S.P. Commercial grades of leaves are graded according to their size and shape (entire, broken, etc.), with whole leaves being the most expensive, yet they may often contain less sennosides than the lower grades (e.g., broken leaves). Strengths (see *glossary*) of extracts are based on weight-to-weight ratios, though more and more manufacturers are using sennoside contents as in house standards or guides.

Regulatory Status. Only Alexandrian senna (leaf and pod) has been approved for food used (§172.510). Senna leaves, pods, and their preparations are subjects of a positive German therapeutic monograph.[10,25,35]

REFERENCES

See the General References for BARNES; BLUMENTHAL 1; CLAUS; GUPTA; JIANGSU; LIST AND HÖRHAMMER; LUST; MORTON 3; NANJING; TUCKER 1–3; WREN; YOUNGKEN.

1. G. Franz, *Pharmacology*, **47**(Suppl. 1), 2 (1993).
2. J. W. Fairbairn and A. B. Shrestha, *Lloydia*, **30**, 67 (1967).
3. J. Lemli et al., *Planta Med.*, **49**, 36 (1983).
4. A. Stoll and B. Becker in L. Zechmeister, ed., *Fortschritte der Chemie organischer Naturstoffe*, Vol. 7, Springer-Verlag, Vienna, Austria, 1950, p. 248.
5. J. W. Fairbairn, *Lloydia*, **27**, 79 (1964).
6. F. H. L. Van Os, *Pharmacology*, **14** (Suppl. 1), 7 (1976).
7. H. Friedrich and S. Baier, *Planta Med.*, **23**, 74 (1973).
8. W. D. Brendel and D. Schneider, *Planta Med.*, **25**, 342 (1974).
9. R. Atzorn et al., *Planta Med.*, **41**, 1 (1981).
10. Monograph *Sennae*, *Bundesanzeiger*, no. 228 (December 5, 1984).
11. J. Lemli and J. Cuveele, *Plant. Med. Phytother.*, **10**, 175 (1976).
12. S. C. Y. Su and N. M. Ferguson, *J. Pharm. Sci.*, **62**, 899 (1973).
13. C. Terreaux et al., *Planta Med.*, **68**, 349 (2002).
14. B. M. Müller et al., *Planta Med.*, **55**, 536 (1989).
15. J. Lemli and J. Cuveele, *Plant. Med. Phytother.*, **10**, 175 (1976).
16. B. Christ et al., *Arzeim.-Forsch.*, **28**, 225 (1978).
17. N. Alam and C.Gupta, *Planta Med.*, **52**, 308 (1986).
18. K. Kisa et al., *Planta Med.*, **42**, 302 (1981).
19. Anon., *Fed. Regist.*, **40**, 10902 (1975).
20. J. W. Fairbairn, ed., *The Anthraquinone Laxatives*, S. Karger, Basel, Switzerland, 1976, in *Pharmacology*, **14** (Suppl. 1), 1, 1976.
21. K. Ewe et al., *Pharmacology*, **47** (Suppl. 1), 242 (1993).
22. S. Kositchaiwat et al., *World J. Gastroenterol.*, **12**, 5536 (2006).
23. F. Radaelli et al., *Am. J. Gastroenterol.*, **100**, 2674 (2005).
24. B. Smith, *The Neuropathology of the Alimentary Tract*, Arnold, London, 1972.
25. Monograph *Sennae folium*, *Bundesanzeiger* (July 21, 1993).
26. H. A. Spiller et al., *Ann. Pharmacother.*, **37**, 636 (2003).
27. B. Vanderperren et al., *Ann. Pharmacother.*, **39**, 1353 (2005).
28. A. Sonmez et al., *Acta Gastroenterol. Belg.*, **68**, 385 (2005).
29. E. Leng-Peschlow, *Intl. J. Exp. Clin. Pharmacol.*, **44**, S1 (1992).
30. U. Mengs et al., *Arch. Toxicol.*, **78**, 269 (2004).
31. J. M. Mitchell et al., *Arch. Toxicol.*, **80**, 34 (2006).
32. N. Mascolo et al., *Dig. Dis. Sci.*, **44**, 2226 (1999).
33. D. Brusick and U. Mengs, *Environ. Mol. Mutagen.*, **29**, 1 (1997).
34. A. A. al-Dakan et al., *Pharmacol. Toxicol.*, **77**, 288 (1995).
35. Monograph *Sennae fructus*, *Bundesanzeiger* (July 21, 1993).

SOUR JUJUBE KERNEL (SEE ALSO JUJUBE, COMMON)

Source: *Ziziphus spinosa* Hu and *Z. jujuba* Mill. var. *spinosa* (Bge.) Hu ex H.F. Chow (Family Rhamnaceae). Also listed as sources/synonyms: *Z. jujuba* Mill.; *Z. jujuba* Mill. var. *spinosus* Bge. [*Z. spinosa* (Bge.) Hu].

Common/vernacular names: Sour date kernel, spiny jujube kernel, *suan zao ren*, *zao ren*, and *shan zao ren*.

GENERAL DESCRIPTION

Deciduous shrub, 1–3 m high, bearing two types of spines, one sturdy and erect and the other small and curved; fruit (drupe) almost round, 1–1.4 cm in diameter, dull reddish brown when ripe, with thin flesh, and tastes sour; grows in northern and central China. Part used is the seed. Fruits are wildcrafted when they ripen and turn red in autumn, soaked overnight and rid of flesh; the pits are then crushed open and the seeds collected and dried. The better grades are full and deep red or purplish brown in appearance and contain less than 5% impurities (e.g., husks). Major producing provinces include Hebei, Shaanxi, Liaoning, and Henan.

CHEMICAL COMPOSITION

Contains flavonoids that include swertisin, spinosin (2″-β-O-glucopyranosyl swertisin), and zivulgarin (4″-β-O-glucopyranosyl swertisin); triterpenes and triterpene saponin glycosides, including betulin, betulinic acid (free and esterified),[1] jujubosides A, B, B$_1$ and jujubogenin, ceanothic acid, alphitolic acid; numerous alkaloids, including aporphines, benzylisoquinolines, and 14-membered cyclopeptides (e.g., zizyphusine, nuciferine, nornuciferine, coclaurine, norisocorydine, caaverine, *N*-methylasimilobine, sanjoinines A, B, D, F, G1, sanjoinenine, etc.) (STEINER); ferulic acid; β-carotene; sterols (daucosterol); fatty oil (ca. 32%, composed of 42% oleic and 46% linoleic acids, as well as linolenic, lauric, myristic, palmitic, palmitoleic, and stearic acids);[2–4] and cyclic AMP and cyclic GMP, among others (IMM-3).[5–9]

PHARMACOLOGY AND BIOLOGICAL ACTIVITIES

Suan zao ren is probably the best known and most frequently used Chinese herbal sedative. Modern studies have shown it to have strong sedative and hypnotic effects in humans and in experimental animals (mice, rats, guinea pigs, cats, rabbits, and dogs),[6] with the flavonoid glycosides (spinosin, swertisin, and zivulgarin), alkaloids, and saponins (jujubosides A and B) being the active principles (STEINER).[10–13] The sedative effect was reported to occur at higher doses (>1.0 g/kg), while lower doses (0.5–1.0 g/kg) resulted in an anxiolytic effect in behavioral evaluation models in rats and mice.[14] Further evaluation of jujuboside A showed that the reported CNS-depressant effects occur through an inhibitory effect on the excitatory, glutamate-mediated hippocampal functions and that an anticalmodulin action may be involved.[15–17]

An ethanolic extract of *suan zao ren* (per os) markedly increased feeding and weight gain as well as immune response in mice. Its polysaccharides (per os) also exhibited immunopotentiating and radiation-protective effects in mice.[18,19]

The water extractives of the seed had antiarrhythmic activities in experimental animals, and its total saponins had protective effects on cultured rat myocardial cells.[20,21]

Oral administration of the fatty oil to quails (2.5 mL/kg/day) for 53 days not only markedly reduced their serum lipids but also significantly inhibited blood platelet aggregation.[2]

Seed extracts rich in fatty acids and triglycerides displayed COX-2 inhibitory activity, which was attributed to the triglyceride 1,3-di-*O*-[9(Z)-octadecenoyl]-2-*O*-[9(Z),12(Z)-octadecadienoyl]glycerol and a fatty acid mixture of linoleic, oleic, and stearic acids.[1]

The triterpene glycoside ziziphin isolated from the leaves has sweetness inhibitory (antisweetness) activity in humans.[22,23]

Other pharmacologic effects of *suan zao ren* include analgesic, antipyretic, antispasmodic, hypotensive, uterine stimulant, and others.[6]

TOXICOLOGY

Toxicities are very low: 150 g/kg crude drug or 50 g/kg decoction fed to mice did not produce any toxic reactions, and no adverse side effects have been reported from clinical use of this drug. However, since it is known to be a uterine stimulant, caution is advised in pregnant women (JIANGSU; WANG).[6]

USES

Medicinal, Pharmaceutical, and Cosmetic. Its water and/or hydroalcoholic extracts are used along with Dahurian angelica in freckle removal creams and lotions; also used in other skin care products (sunscreen, acne, deodorant, antiperspirant, etc.) for its traditional skin-protectant, antiperspirant, calming, and nourishing properties (ZHOU).

Dietary Supplements/Health Foods. Powder and extract used in herbal sedative formulas in tea, capsule, or tablet form.

Traditional Medicine. Sour jujube was first recorded over 2000 years ago and the fruit was listed in the *Shen Nong Ben Cao Jing* (ca. 200 BC–AD 100) as a tonic in the superior category. Use of the seed, *suan zao ren*, however, was not described until the 3rd century AD.[24] Traditionally regarded as sweet and sour tasting, neutral, heart nourishing and tranquilizing (*yang xin an shen*), promoting the secretion of body fluids, and stopping excessive perspiration. It is one of the major traditional Chinese brain tonics for treating neurasthenia, insomnia, excessive dreams (nightmares), night sweat, forgetfulness, and palpitations. Traditional medical writings distinguish between raw and stir-fried *suan zao ren*: the former for treating sleepiness while the latter for treating insomnia (LiShi-Zhen's Ben Cao Gang Mu, ca. AD 1590). However, modern chemical studies have not confirmed the validity of this distinction.[12,25,26]

Suan zao ren is normally used in the crushed or powdered form.

COMMERCIAL PREPARATIONS

Crude (raw and stir-fried) in whole or powdered form; whole seeds must be crushed or powdered before extraction.

Regulatory Status. Class 2b dietary supplement (not to be used during pregnancy).

REFERENCES

See the General References for CHP; FOSTER AND YUE; IMM-3; JIANGSU; LU AND LI; NATIONAL; STEINER; XIAO 4,6; ZHOU; ZHU.

1. B. N. Su et al., *Planta Med.*, **68**, 1125 (2002).
2. S. X. Wu et al., *Zhongguo Zhongyao Zazhi*, **16**, 435 (1991).
3. J. Zhao et al., *J. Chromatogr. A*, **1108**, 188 (2006).
4. J. L. Guil-Guerrero et al., *Plant Foods Hum. Nutr.*, **59**, 23 (2004).
5. K. H. Shin et al., *Planta Med.*, **44**, 94 (1982).
6. G. X. Hong and B. Cao, *Zhongyao Tongbao*, **12**, 51 (1987).

7. L. Zeng et al., *Acta Bot. Sin.*, **28**, 517 (1986).
8. S. Y. Li and R. Y. Zhang, *Zhongyao Tongbao*, **11**, 43 (1986).
9. L. Zeng et al., *Yaoxue Xuebao*, **22**, 114 (1987).
10. W. S. Woo et al., *Phytochemistry*, **18**, 353 (1979).
11. I. Watanabe et al., *Jpn. J. Pharmacol.*, **23**, 563 (1973).
12. J. Wang, *Zhongchengyao*, **11**, 19 (1989).
13. C. L. Yuan et al., *Zhongyao Tongbao*, **12**, 34 (1987).
14. W. H. Peng et al., *J. Ethnopharmacol.*, **72**, 435 (2000).
15. M. Zhang et al., *Planta Med.*, **69**, 692 (2003).
16. C. Shou et al., *Planta Med.*, **68**, 799 (2002).
17. C. H. Shou et al., *Acta Pharmacol. Sin.*, **22**, 986 (2001).
18. X. C. Lang et al., *Zhongyao Tongbao*, **13**, 43 (1988).
19. X. C. Lang et al., *Zhongguo Zhongyao Zazhi*, **16**, 366 (1991).
20. S. Y. Xu et al., *Zhongcaoyao*, **18**, 18 (1987).
21. X. J. Chen et al., *Acta Pharmacol. Sin.*, **11**, 153 (1990).
22. R. Suttisri et al., *J. Ethnopharmacol.*, **47**, 9 (1995).
23. Y. Kurihara, *Crit. Rev. Food Sci. Nutr.*, **32**, 231 (1992).
24. F. X. Liu et al., *Zhongguo Zhongyao Zazhi*, **16**, 444 (1991).
25. F. X. Liu and J. F. Gao, *Zhongguo Zhongyao Zazhi*, **15**, 28 (1990).
26. H. Jin and Y. M. Xu, *Liaoning Zhongyi Zazhi*, **20**, 40 (1993).

SQUILL

Source: *Urginea maritima* (L.) Baker (syn. *U. Scilla* Steinh., *Scilla maritima* L., *Drimia maritima* (L.) Stearn) and ***U. indica*** (Roxb.) Kunth. (syn. *Scilla indica* L., *Drimia indica* (L.) Stearn) (Family Liliaceae).

Common/vernacular names: Scilla, sea onion, sea squill, Mediterranean squill, red squill, white squill (*U. maritima*); Indian squill, white squill (*U. indica*).

GENERAL DESCRIPTION

Urginea maritima is a bulbous perennial herb belonging to the lily family; native to the Mediterranean region. Parts used are the dried fleshy inner scales of the bulb. White squill is derived from the white variety, while red squill is from its red variety. *U. indica* is used as an acceptable substitute for *U. maritima*. Recent taxonomic treatments placed the taxa in the genus *Drimia*, however, the genus name *Urginea* persists in the trade and scientific literature (BRADLY).

Urginea maritima is a species complex, now consisting of six or more species, with both genetic and phytochemical differences. *U. maritima sensu strictu*, native to the Iberian peninsula, a hexaploid ($2n = 60$), has the highest bufadienolide content (BRADLY).[1]

CHEMICAL COMPOSITION

White squill contains as active constituents several steroid glycosides (bufadienolides), including scillaren A (scillarenin rhamnoglucoside), glucoscillaren A (scillaren A glucoside), proscillaridin A (scillarenin rhamnoside), scillaridin A, scillicyanoside, scilliglucoside, scilliphaeoside (12β-hydroxyproscillaridin A), glucoscilliphaeoside (12β-hydroxyscillaren), and others,[2–7] with the most important ones being scillaren A and proscillaridin A.[3,5] Scillaren B has been used to describe a mixture of squill glycosides as opposed to pure scillaren A. Two new bufadienolides were isolated from *U. maritima sensu strictu* and were identified as 11α-acetyl-γ-bufotalin glucorhamnoside and 11α-hydroxyscilliglucoside.[8]

Other constituents present in white squill include flavonoids (vitexin, isovitexin, orientin, isoorientin, scoparin, vicenin-2, quercetin, dihydroquercetin or taxifolin, dihydroqueretin-4'-monoglucoside, etc.), stigmasterol, scilliglaucosidin, mucilage (glucogalactans), lignans, and others (KARRER; MERCK).[4–6,9–11]

Red squill contains scilliroside and cardiac glycosides as white squill (KARRER; MARTINDALE);[12] it also contains sinistrin, a fructan polysaccharide, and other fructo-oligosaccharides.[13,14]

Indian squill has same types of active bufadienolides as the Mediterranean squill, but with contents of proscillaridin A and scillaren A in some cytotypes (diploid and tetraploid races) considerably higher than the latter.[15] A novel bioactive 29 kDa glycoprotein and an antifungal chitinase (also 29 kDa) were recently isolated from the bulb.[16–18]

PHARMACOLOGY AND BIOLOGICAL ACTIVITIES

The glycosides present in squill have digitalis-like cardiotonic properties, which are due to their aglycones (GOODMAN AND GILMAN).[19] Action is more rapid, but shorter in duration, than that of digitalis glycosides.[20,21]

White squill also reportedly has expectorant, emetic, and diuretic properties (MARTINDALE; MERCK).

The recently reported glycoprotein of white squill has a potent *in vitro* and *in vivo* antiangiogenic and proapoptotic activity against a mouse mammary carcinoma. The mechanism of actions appears to be mediated through NF-κB and caspase-activated DNase.[17]

The chitinase isolated from white squill displayed significant fungistatic activity against *Fusarium oxysporum* and *Rhizocotonia solani*.[16,18] The mechanism of antifungal action was proposed to be through a chitinolytic effect on fungal cell wall.[16]

Methanol extracts of red squill have been claimed to be effective as hair tonics in treating chronic seborrhea and dandruff, with the active principle being attributed to scilliroside.[22]

TOXICOLOGY

Red squill is very toxic to rats, the active principle being scilliroside. It is also extremely irritating to the skin, and handling with rubber gloves is strongly recommended (MARTINDALE; REMINGTON). Poisoning of livestock has been observed in North Africa. A case of fatal poisoning due to ingestion of two bulbs of *U. maritima* has also been reported in a human female. Death was due to cardiac glycoside toxicity.[23,24]

USES

Medicinal, Pharmaceutical, and Cosmetic. White squill is used mainly as an expectorant in some cough preparations. In German phytomedicine used for mild cardiac insufficiency and impaired kidney function.[20]

Traditional Medicine. Has been used for centuries in Europe as a diuretic, emetic, expectorant, and cardiac stimulant. It has also been used in cancers.[21]

Others. The primary use of red squill is as an effective rat poison. However, it is generally not effective with mice. It is quite safe with humans and animals such as pets (cats and dogs) and domestic animals (e.g., hogs) because it causes vomiting in these animals promptly, thus not allowing enough poison to be ingested (REMINGTON).

COMMERCIAL PREPARATIONS

Crude and extracts; crude, tincture, and fluid extract were formerly official in N.F. Strengths (see *glossary*) of extracts are expressed in weight-to-weight ratios.

Regulatory Status. Subject of a German therapeutic monograph, indicated for diminished kidney capacity and mild cardiac insufficiency (contraindicated with digitalis glycosides or potassium deficiency).[20]

REFERENCES

See the General References for BLUMENTHAL 1; BRADLY; BRUNETON; CLAUS; DER MARDEROSIAN AND BEUTLER; GOSSELIN; LEWIS AND ELVIN-LEWIS; MARTINDALE; MCGUFFIN 1 & 2; TERRELL; TUCKER 1–3; UPHOF; YOUNGKEN.

1. L. Krenn et al., *Planta Med.*, **57**, 560 (1991).
2. H. Lichti et al., *Helv. Chim. Acta*, **56**, 2088 (1973).
3. M. S. Karawya et al., *Planta Med.*, **23**, 213 (1973).
4. A. Von Wartburg et al., *Helv. Chim. Acta*, **51**, 1317 (1968).
5. P. Garcia Casado et al., *Pharm. Acta Helv.*, **52**, 218 (1977).
6. M. Iizuka et al., *Chem. Pharm. Bull. (Tokyo)*, **49**, 282 (2001).
7. B. Kopp et al., *Phytochemistry*, **42**, 513 (1996).
8. L. Krenn et al., *Fitoterapia*, **71**, 126 (2000).
9. F. S. Hakim and F. J.Evans, *Pharm. Acta Helv.*, **51**, 117 (1976).
10. M. Fernandez et al., *Phytochemistry*, **14**, 586 (1975).
11. M. Fernandez et al., *Galencia Acta*, **24**, 45 (1971).
12. A. Von Wartburg, *Helv. Chim. Acta*, **49**, 30 (1966).
13. T. Spies et al., *Carbohydr. Res.*, **235**, 221 (1992).
14. W. Praznik and T. Spies, *Carbohydr. Res.*, **243**, 91 (1993).
15. S. Jha and S. Sen, *Planta Med.*, **47**, 43 (1983).
16. S. R. Shenoy et al., *Biotechnol. Prog.*, **22**, 631 (2006).
17. A. V. Deepak and B. P. Salimath, *Biochimie*, **88**, 297 (2006).
18. A. V. Deepak et al., *Biochem. Biophys. Res. Commun.*, **311**, 735 (2003).
19. C. L. Gemmill, *Bull. N. Y. Acad. Med.*, **50**, 747 (1974).
20. Monograph *Scillae bulbus*, Bundesanzeiger, no. 154 (August 21, 1985); corrected (March 2, 1989).
21. W. E. Court, *Pharm. J.*, **235**, 194 (1985).
22. S. Giannopoulos, Ger. Offen., 2, 715,214 (1977), through *Chem. Abstr.*, **88**, 55079e (1978).
23. L. elBahri et al., *Vet. Hum. Toxicol.*, **42**, 108 (2000).
24. Y. Tuncok et al., *J. Toxicol. Clin. Toxicol.*, **33**, 83 (1995).

STEVIA

Source: *Stevia rebaudiana* (Bertoni) Bertoni (syn. *Eupatorium rebaudianum* Bertoni) (Family Compositae or Asteraceae).

Common/vernacular names: Sweet herb.

GENERAL DESCRIPTION

Stevia is an herbaceous perennial to 80 cm high, leaves opposite, toothed; indigenous to highlands in the Amambay and Iguaqu districts of the border area of Brazil and Paraguay;[1] commercially produced in Paraguay,

Brazil, Japan, Korea, Thailand, and China. Much of the Asian production is for export to Japan, the largest consumer.

CHEMICAL COMPOSITION

The main constituents of the leaves responsible for sweetness are the ent-kaurene diterpene glycosides stevioside (2.2–18.5%), rebaudiosides A and C, and dulcoside A; rebaudiosides B, D, and E and steviolbioside are of much less importance. Labdane diterpenes include jhanol, austroinulin, 6-O-acetylaustroinulin; triterpenes include β-amyrin acetate and lupeol; other constituents include β-sitosterol, stigmasterol, tannins, and a volatile oil (0.12–0.43%) from which at least 25 compounds have been identified.[2]

PHARMACOLOGY AND BIOLOGICAL ACTIVITIES

Stevioside is about 300 times sweeter than sucrose at 0.4% sucrose concentration, 150 times sweeter at 4% sucrose, and 100 times sweeter at 10% sucrose concentration.[2]

Other reported activities include claims of hypoglycemic activity of leaf extracts and stevioside in animals or humans; however, follow-up studies have failed to support a hypoglycemic effect. Weak antimicrobial activity of unidentified fractions has been reported against *Pseudomonas aeruginosa* and *Proteus vulgaris*.[2]

TOXICOLOGY

Given the interest in the plant and its glycosides as sweeteners, most of the biological testing has been to assess toxicity. Stevioside has been found to be nontoxic in acute toxicity tests with rabbits, guinea pigs, and fowl, being excreted without structural modification. A leaf extract (with 50% stevioside) administered i.p. to rats has an LD_{50} of 3.4 g/kg. Separate 2.0 g/kg doses of stevioside, rebaudiosides A–C, steviolbioside, or dulcoside A administered to mice by oral intubation showed no acute toxicity. Two weeks after administration, no significant differences in body or organ weights were reported.[2]

Subacute toxicity studies on rats over a 50-day period up to 7.0% concentration of stevioside in feed produced no remarkable toxic effects.[2]

Studies on the effect of stevioside on fertility found that it produced no abnormal mating performance or fertility, and produced no teratogenic effects. Various tests have not shown mutagenic or genotoxic activity; however, mutagenic activity of metabolized steviol has been reported.[2]

USES

Food. Stevia leaf is reported to have been used by Indian groups in Paraguay as a sweetener for mate (*Ilex paraguayensis*) as early as the 16th century.[2] Use as a sweetener by the Guaraní Paraguayan Indians, Mestizos, and others is documented back to 1887.[3] It was planted in England during World War II as a possible sugar substitute.[3] Stevioside and stevia leaves are used as nonfermentative, noncaloric commercial sweetening agent primarily in Japan, as well as Paraguay and Brazil. As much as 1700 metric tons dry leaf were used in Japan in 1987.[4] Given high stability to heat and acid, a sweetness similar to sucrose (with only a mild aftertaste), it has been found to be of use as a sweetener in many food categories in Japan. Reported Japanese use includes flavoring pickles, dried seafood, fish, and meat products; vegetables; confectionery; desserts; soy sauce; miso (and other products with high levels of sodium chloride); soft drinks; and others.[2]

Dietary Supplements/Health Foods. Used since the early 1980s; dried leaf a sweetening agent in herbal teas, as well as leaf powder or extracts as a table or cooking sweetener.[5]

Traditional Medicine. Stevia has been traditionally used in Paraguay to treat diabetes and as a contraceptive.[1]

COMMERCIAL PREPARATIONS

Crude leaf, extracts, and stevioside.

Regulatory Status. Current status: Class 1 dietary supplement (can be safely consumed when used appropriately).

Earlier status: Since May 17, 1991, the FDA has treated stevia leaf as an "unsafe food additive," banning importation into the United States.[6] A petition by the American Herbal Products Association to allow the herb to be sold as a food flavor, arguing that the ingredient is a food, commonly used before 1958 (rather than a food additive), and hence exempt from the import alert, has been rejected by FDA.[4,5,7] The agency claims that there are not sufficient data to establish common food use prior to 1958 and raised questions about possible toxic effects related to reduced fertility rates in mice.[7,8]

REFERENCES

See the General References for APPLEQUIST; BRUNETON; DER MARDEROSIAN AND BEUTLER; GLASBY 2; MCGUFFIN 1 & 2; UPHOF.

1. D. D. Soejarto et al., *Econ. Bot.*, **37** (1), 71 (1983).
2. A. D. Kinghorn and D. D. Soejarto in H. Wagner, H. Hikino, and N. R. Farnsworth, eds., *Economic and Medicinal Plant Research*, Vol. 1, Academic Press, Orlando, FL, 1985, p. 1.
3. W. H. Lewis, *Econ. Bot.*, **46**(3), 336 (1992).
4. M. Blumenthal, *Whole Foods*, **29** (1992).
5. M. Blumenthal, *HerbalGram*, **26**, 22 (1992).
6. FDA Import Alert No. 45-06 (May 17, 1991).
7. M. Blumenthal, *HerbalGram* (1994).
8. P. Nunes et al., *Brazilian Review of Pharmacy*, **69**, 1, 46 (1988) (in M. BLUMENTHAL, *HerbalGram*, **31**, 1994).

STORAX

Source: *American storax Liquidambar styraciflua* L.; ***Levant (Asiatic) storax*** *Liquidambar orientalis* Mill. (Family Hamamelidaceae or Altingiaceae).

Common/vernacular names: Liquid storax (*L. orientalis*), styrax.

GENERAL DESCRIPTION

Liquidambar styraciflua (sweet gum or red gum) is a deciduous tree with five to seven lobed leaves that resemble stars and are subtruncate at their base; trunk tall and straight, with furrowed bark; up to about 46 m high; native to North and Central America (Connecticut to Florida and west to Oklahoma and Texas; central Mexico to Nicaragua); cultivated in eastern United States as ornamental tree. Part used is the exudation (balsam) collected in natural pockets between the wood and the bark and may be located by excrescences on the trunk. The balsam flows readily into containers when the pockets are tapped, yielding the crude American storax, a solid or semisolid. Major producing countries include Honduras and Guatemala. *Liquidambar orientalis* (oriental sweet gum or Turkish sweet gum) is a deciduous tree smaller than American sweet gum; leaves usually deeply five lobed with irregular secondary lobes; up to about 15 m high; native to Asia Minor. Levant storax is obtained as a pathological secretion by pounding the tree bark, inducing the sapwood to secrete the balsam, which then accumulates in the bark. The balsam is collected by stripping the bark, followed by pressing it and boiling it with water. Levant storax is often packed in tin cans with a layer of water to prevent the evaporation of volatile ingredients; it is a viscous semi-liquid.

CHEMICAL COMPOSITION

American storax and Levant storax are reported to contain similar constituents in highly variable concentrations, including free cinnamic acid (5–15%), 5–10% styracin (cinnamyl cinnamate), about 10% phenylpropyl cinnamate, a resin (storesin) consisting of triterpenic acids (oleanolic and 3-epioleanolic acids) and their cinnamic acid esters, and a volatile oil.[1] The volatile oil present in Levant storax is usually less than 1%, but that in American storax has been reported to range from less than 7% to over 20% (ARCTANDER; LIST AND HÖRHAMMER; MERCK; NANJING; REMINGTON).

Compounds present in storax oil include styrene, phenylpropyl alcohol, cinnamic alcohol, benzyl alcohol, ethyl alcohol, and vanillin (KARRER; LIST AND HÖRHAMMER).

New lupane (lupenoic acids) and oleanane (oleanenoic acids) triterpenes have recently been isolated from the cones and stem bark.[2,3]

PHARMACOLOGY AND BIOLOGICAL ACTIVITIES

Storax has antiseptic and expectorant properties. It is also reported to have antimicrobial and anti-inflammatory properties (JIANGSU). An ethanolic solution of storax was effective against a broad spectrum of bacteria (Gram-positive, Gram-negative, and acid-fast) in the agar diffusion assay at 10% concentration.[4]

In a study to evaluate the antioxidant activity of a number of gums, resins, and pigments, storax was among those that protected human LDL against copper-induced oxidation *in vitro*.[5]

Inhalation of the essential oil of a Chinese antiepileptic storax pill (*SuHeXiang Wan*) delayed the onset of pentylenetetrazole-induced convulsions in mice. This is an indication of a CNS mechanism involving GABAergic neuromodulation, which is supported by the prolonged pentobarbital-induced sleeping time and inhibition of brain lipid peroxidation observed in the same study.[6]

Some of the recently isolated triterpenes possessed antitumor and cancer chemopreventive activities. The chemopreventive activity was displayed in a mouse skin carcinogenesis assay, a UV-induced photocarcinogenesis assay in mice, and a TPA-induced Epstein-Barr virus early antigen activation assay *in vitro*.[2,7] The antitumor activity was demonstrated *in vitro* in a panel of 39 human cancer cell lines.[3]

USES

Medicinal, Pharmaceutical, and Cosmetic. Storax is mainly used as an ingredient of Compound Benzoin Tincture (see ***aloe***, ***balsam tolu***, and ***benzoin***).

Storax oil and resinoid are used as fragrance components and/or fixatives in soaps and perfumes (especially oriental and floral types).

Food. Storax and storax extracts (e.g., resinoid and absolute) are used as flavor components or fixatives in major food products, including alcoholic and nonalcoholic beverages, frozen dairy desserts, candy, baked goods, and gelatins and puddings, among others. Use levels are usually quite low

(0.001%), with the highest average maximum levels being about 0.002% reported for storax in candy (15.3 ppm) and baked goods (25 ppm).

Traditional Medicine. American storax is used as antiseptic and expectorant in treating wounds and skin problems (e.g., scabies), as well as coughs and colds.

Levant storax is used in Chinese medicine for similar purposes as is American storax in the West. In addition, it is used in treating epilepsy. Both American storax and Levant storax have been used in cancers.[8]

COMMERCIAL PREPARATIONS

American and Levant storax, purified storax, oil, and extracts (e.g., resinoid). Storax is official in U.S.P.; purified storax was formerly official in U.S.P. According to ARCTANDER, storax oil and storax resinoid have been frequently adulterated.

Regulatory Status. Has been approved for food use (§172.510).

REFERENCES

See the General References for ARCTANDER; BAILEY 2; FEMA; GRIEVE; GUENTHER; JIANGSU; KROCHMAL AND KROCHMAL; LIST AND HÖRHAMMER; LUST; MORTON 3; NANJING; REMINGTON; SARGENT; TERRELL; TYLER 3; UPHOF.

1. S. Hunek, *Tetrahedron*, **19**, 479 (1963).
2. Y. Fukuda et al., *J. Nat. Prod.*, **69**, 142 (2006).
3. K. Sakai et al., *J. Nat. Prod.*, **67**, 1088 (2004).
4. O. Sagdic et al., *Phytother. Res.*, **19**, 549 (2005).
5. N. K. Andrikopoulos et al., *Phytother. Res.*, **17**, 501 (2003).
6. B. S. Koo et al., *Biol. Pharm. Bull.*, **27**, 515 (2004).
7. Y. Fukuda et al., *Cancer Lett.*, **240**, 94 (2006).
8. J. L. Hartwell, *Lloydia*, **32**, 247 (1969).

TAGETES

Source: *Tagetes erecta* L., *T. patula* L., and *T. minuta* L. (syn. *T. glandulifera* Schrank) (Family Compositae or Asteraceae).

Common/vernacular names: African marigold, Aztec marigold, big marigold (*T. erecta*); French marigold (*T. patula*); Mexican marigold (*T. minuta*); marigold.

GENERAL DESCRIPTION

Strong-scented annual herbs, usually 0.3–1 m high; leaves strong scented, pinnately divided leaves;[1] *T. erecta* bears the largest flower heads (ca. 5–10 cm across) among the three species; generally considered to be natives of Mexico (*T. erecta* and *T. patula*) and South America (*T. minuta*); cultivated or found growing wild worldwide, including Ethiopia, Kenya, Nigeria, and Australia (*T. minuta*) and Europe, India, and China (*T. erecta* and *T. patula*).

Tagetes oil is obtained by steam distillation of the aboveground parts of all the three species (especially *T. minuta*).

Tagetes meal is the dried ground flower petals of Aztec marigold (*T. erecta*); tagetes extract is the hexane extract of the flower petals of Aztec marigold.[2]

CHEMICAL COMPOSITION

Tagetes oil from *T. minuta* contains tagetones, ocimene, β-myrcene, linalool, limonene, α- and β-pinenes, carvone, citral, camphene, and salicylaldehyde as major components, with phenylethanol, valeric acid, ocimenones [e.g., (5E)-ocimenone], geraniol, p-cymene, sabinene, cineole, linalyl acetate, linalool monoxide, aromadendrene, and α-terpineol, among others.[3–8]

The volatile oils from *T. erecta* and *T. patula* have been reported to have compositions similar to that from *T. minuta*.[9,10] A recent report identified 30 compounds in *T. patula* volatile oil with piperitone, piperitenone, and terpinolene as the major components (>50%).[11]

The flower petals of Aztec marigold (*T. erecta*) is reported to contain mainly carotenoids, especially lutein and its esters (dipalmitate, dimyristate, and monomyristate) (JIANGSU) and zeaxanthin (*cis* and *trans* isomers) as the major pigments; α-terthienyl is also present.[12–15]

Different organs of *T. patula* were reported to contain triterpenes, steroids, and thiphenes.[16,17] The roots contain citric and malic acids, pyridine hydrochloride, and 2-hydroxy-5-hydroxymethylfuran.[18] Pyrethrins were isolated from callus tissue cultures of *T. erecta* and thiophenes were also isolated from callus cultures of *T. patula*.[19,20]

PHARMACOLOGY AND BIOLOGICAL ACTIVITIES

Tagetes oil (from *T. minuta*) has been reported to have tranquilizing, hypotensive, bronchodilatory, spasmolytic, and anti-inflammatory properties in experimental animals.[15,21] However, it was also reported that s.c. injection of the volatile oil of *T. minuta* (0.1–0.45 mg/kg) resulted in an anxiogenic effect in domestic chicks.[22] The methanolic extract of *T. patula* flowers inhibited acute and chronic inflammatory responses in mice and rats.[23]

Patulin, derived from *T. patula*, has been shown to reduce capillary permeability; is antispasmodic and hypertensive.[15] Hypotensive activity was demonstrated in rats by citric and malic acids isolated from the methanolic extract of *T. patula* root. Pyridine hydrochloride isolated from the same extract resulted in a hypertensive effect.[18]

α-Terthienyl from *T. erecta* is reportedly a nematocide and larvicide.[4] (5E)-Ocimenone and thiophenes reportedly have a toxic effect against malaria mosquito larvae.[4,17,19] The essential oil of *T. patula* also demonstrates the same larvicidal effect against three mosquito species as well as on adult ones.[16,24]

The essential oil of *T. patula* and its major components (piperitone, piperitenone, and terpinolene) have antifungal activity.[11,25] The total leaf extract of *T. minuta* and its major flavonoid (quercetagetin-7-arabinogalactoside) possess considerable antimicrobial activity against Gram-positive and Gram-negative bacteria in comparison to chloramphenicol.[26]

Carotenoids of *T. erecta* have an antimutagenic effect that may be mediated through complexation with mutagenic agents.[27]

TOXICOLOGY

Tagetes species have been reported to cause contact dermatitis.[28,29]

USES

Medicinal, Pharmaceutical, and Cosmetic. Tagetes oil (from *T. minuta*) is used as a fragrance component in perfumes.

Food. Tagetes oil is used as a flavor component in most major food products, including alcoholic and nonalcoholic beverages, frozen dairy desserts, candy, baked goods, gelatins and puddings, and condiments and relishes. Highest average maximum use level reported is 0.003% in condiments and relishes.

Traditional Medicine. Flower heads and foliage of *T. erecta* are used as anthelmintic and emmenagogue and in treating colic.

The herb of *T. minuta* is used as stomachic, carminative, diuretic, and diaphoretic.

In China, the flower heads of *T. erecta* are used in treating whooping cough, coughs, colds, mumps, mastitis, and sore eyes, usually as a decoction. The leaves are used in treating sores and ulcers (JIANGSU). The whole herb of *T. patula* is used in coughs and dysentery, taken internally in the form of a powder or a decoction (JIANGSU).

In India the juice of the leaves of *T. erecta* is used as a treatment for eczema.[30]

In Peru the aerial parts of *T. minuta* are used in decoction as a digestive, vermifuge, cholagogue, sedative in gastric pain, and antiabortifacient.[31]

Others. Tagetes meal and tagetes extract are extensively used in chicken feed to give the characteristic yellow color to chicken skin and egg yolk. *T. erecta* flowers are used as a mosquito repellent, *T. minuta* oil for treatment of wound maggots, *T. lucida* Cav. in fumigants and repellents for mosquitoes, and *T. terniflora* H. B. & K as an insecticide in South America.[32]

COMMERCIAL PREPARATIONS

Tagetes oil and tagetes (Aztec marigold) meal and extract.

Regulatory Status. Tagetes oil has been approved for food use (§172.510); tagetes (Aztec marigold) meal and extract have been approved for use in chicken feed to enhance the yellow color of chicken skin and eggs, exempt from certification (§73.295).

REFERENCES

See the General References for ARCTANDER; BAILEY 1; FEMA; FERNALD; JIANGSU; MCGUFFIN 1 & 2; TERRELL; UPHOF.

1. K. R. Kirtikar and B. D. Basu, *Indian Medicinal Plants*, Vol. 2, International book distribution, Dehradun, 1999, p. 1385.
2. Anon., *Fed. Regist.*, **31**, 5069 (1966).
3. P. C. Carro de la Torre and J. Retamar, *Arch. Bioquim., Quim. Farm.*, **18**, 39 (1973).
4. A. Maradufu et al., *Lloydia*, **41**, 181 (1978).

5. S. H. L. De Mucciarelli and A. L. Montes, *An. Soc. Cient. Argent.*, **190**, 145 (1970).
6. N. A. Kekelidze et al., *Aktual. Probl. Izuch. Efirnomaslich. Rast. Efirn. Masel*, **135**, (1970).
7. Y. N. Gupta and K. S. Bhandari, *Indian Perfum.*, **19**, 29 (1975).
8. M. Koketsu et al., *An. Acad. Bras. Cienc.*, **48**, 743 (1976).
9. Y. N. Gupta and K. S. Bhandari, *Indian Perfum.*, **18**, 29 (1974).
10. Y. N. Gupta et al., *Indian Perfum.*, **17**, 24 (1973).
11. C. Romagnoli et al., *Protoplasma*, **225**, 57 (2005).
12. R. Tsao et al., *J. Chromatogr. A*, **1045**, 65 (2004).
13. W. L. Hadden et al., *J. Agric. Food Chem.*, **47**, 4189 (1999).
14. T. Phillip and J. W. Berry, *J. Food Sci.*, **40**, 1089 (1975).
15. R. A. Bye Jr., *Econ. Bot.*, **40**, 103 (1986).
16. H. Bano et al., *Pak. J. Pharm. Sci.*, **15**, 1 (2002).
17. M. J. Perich et al., *J. Am. Mosq. Control Assoc.*, **11**, 307 (1995).
18. R. Saleem et al., *Arch. Pharm. Res.*, **27**, 1037 (2004).
19. T. Rajasekaran et al., *Indian J. Exp. Biol.*, **41**, 63 (2003).
20. R. Sarin, *Fitoterapia*, **75**, 62 (2004).
21. N. Chandhoke and B. J. R. Ghatak, *Indian J. Med. Res.*, **57**, 864 (1969).
22. R. H. Marin et al., *Fundam. Clin. Pharmacol.*, **12**, 426 (1998).
23. Y. Kasahara et al., *Phytother. Res.*, **16**, 217 (2002).
24. M. J. Perich et al., *J. Med. Entomol.*, **31**, 833 (1994).
25. D. Mares et al., *Microbiol. Res.*, **159**, 295 (2004).
26. M. L. Tereschuk et al., *J. Ethnopharmacol.*, **56**, 227 (1997).
27. M. E. Gonzalez de et al., *Mutat. Res.*, **389**, 219 (1997).
28. J. C. Mitchell in V. C. Runeckles, ed., *Recent Advances in Phytochemistry*, Vol. 9, Plenum Press, New York, 1975, p. 119.
29. E. Rodriguez et al., *Phytochemistry*, **15**, 1573 (1976).
30. M. B. Siddiqui et al., *Econ. Bot.*, **43**, 480 (1989).
31. V. De Feo, *Fitoterapia*, **63**, 417 (1992).
32. D. M. Secoy and A. E. Smith, *Econ. Bot.*, **37**, 28 (1983).

TAMARIND

Source: *Tamarindus indica* L. (Family Leguminosae or Fabaceae).

Common/vernacular names: Tamarindo.

GENERAL DESCRIPTION

Evergreen tree with large trunk and dark gray bark; up to about 20 m high; leaves 5–12.5 cm long, rachis slender, channeled, stipules linear caduceus; leaflets sessile 10–20 pairs, closely set on rachis;[1] native to tropical Asia and Africa; cultivated worldwide (e.g., China, India, South Florida, Africa, and the West Indies). Part used is the partially dried ripe fruit (pod). The fruit is a legume 6–15 cm long, with a thin brittle shell and a brown, sweet–sour, stringy pulp, enclosing up to 12 seeds. The brittle shell is removed and the fruit is preserved in syrup, or the whole fruit (shell, pulp, and seeds) is mixed with salt and pressed or made into cakes or balls. Producing countries include tropical American countries (e.g., Jamaica and the Antilles) and India. The West Indian countries produce the syrup-preserved product, while East Indian countries produce the salt-preserved product.

CHEMICAL COMPOSITION

Contains plant acids (16–18%) composed mainly of d-tartaric acid (up to ca. 18%), with minor amounts of l-malic acid.[2–4] Citric acid has also been reported as a major component in the old literature, though it has not been detected in Indian tamarind (YOUNGKEN).[2] Other constituents include polyphenolics (catechin, epicatechin, and procyanidin),[5] flavonoids (taxifolin, apigenin, eriodictyol, luteolin, and naringenin),[5] sugars (20–40%),[3,6] pectin, protein (2.8%), fat, vitamins (e.g., B_1 and C), minerals (Ca, K, P, etc.), and tartrate (MERCK; WATT AND MERRILL).[3,6,7]

It also contains a volatile fraction that consists of over 60 identified compounds, including limonene, terpinen-4-ol, neral, α-terpineol, geranial, and geraniol, which are responsible for its citrus note; methyl salicylate, safrole, ionones (β- and γ-), cinnamaldehyde, and ethyl cinnamate, which contribute to its warm spicy notes; piperitone; and several pyrazines and alkylthiazoles that normally occur only in roasted or fried foods such as roasted peanuts, coffee, and potato chips.[6]

β-Sitosterol and a bitter principle tamarindienal (5-hydroxy-2-oxo-hexa-3,5-dienal) are reported from the dried fruit pulp (ca. 0.67%).[4]

A bioactive protein (ca. 12–15 kDa) was recently isolated from the seeds.[8] A mucoadhesive polysaccharide was also isolated from the seeds,[9] while a galactose-specific lectin was isolated from the fruits.[10]

PHARMACOLOGY AND BIOLOGICAL ACTIVITIES

Tamarind has mild laxative properties, which are reportedly destroyed on cooking (JIANGSU).

Aqueous extract of the fruit pulp has been found to be highly toxic to all life cycle stages of the parasitic blood trematode, *Schistosoma mansoni*; molluscicidal against *Bulinus truncatus* vector snails.[4]

Tamarindienal is fungicidal against *Aspergillus niger* and *Candida albicans*; antibacterial against *Bacillus subtilis*, *Staphylococcus aureus*, *Escherichia coli*, and *Pseudomonas aeruginosa*.[4]

The crude extract of fruit pulp has a hypolipemic and antioxidant effect in hamster rats.[11] The phenolic constituents contribute to the antioxidant activity.[5] The hypolipemic effect was also demonstrated in laying hens whose serum cholesterol levels significantly decreased after dietary supplementation with 2% tamarind.[12]

Seed extract inhibited the hydrolytic enzymes and reversed the toxic effects of snake venom injected in test animals 10 min before administration of the extract. Among the inhibited enzymes are serine protease, hyaluronidase, and 5′-nucleotidase.[8,13] Another enzyme inhibited by the methanolic extract of the stem bark is *Clostridium chauvoei* neuraminidase (IC_{50} 100 μg/mL).[14]

Tamarind food supplementation enhances the excretion of excess fluoride and the retention of calcium in children living in fluoride endemic areas.[15]

The seed polysaccharide (xyloglucan) has a corneal wound healing effect in albino rabbits that is mediated through the integrin recognition system.[16]

TOXICOLOGY

There are indications that prolonged occupational exposure to tamarind powder may result in allergy and impairment of lung function.[17]

USES

Food. Widely used in Asia as an ingredient in chutneys and curries and in pickling fish; also extensively used in making a refreshing drink in the tropics where tamarind grows.

Extracts are widely used as a flavor ingredient in Worcestershire and other steak sauces.

Other food categories in which tamarind extracts are used include alcoholic and nonalcoholic beverages, frozen dairy desserts, candy, baked goods, gelatins and puddings, fats and oils, and gravies. Highest average maximum use level is about 0.81% (8072 ppm) reported in gravies.

Traditional Medicine. Used mainly as a refrigerant (in a refreshing drink to cool down fever) and a laxative. In China it is also used to treat nausea in pregnancy and as an anthelmintic for children. In Saudi Arabia the fruit pulp juice is used for stomach problems, colds, and fevers; a thick paste of ground seeds has been used as a cast for broken bones.[18]

COMMERCIAL PREPARATIONS

Crude (in cake form or preserved in syrup) and extracts; crude was formerly official in N.F. Extracts come in varying flavor strengths based primarily on total acid contents; tartaric and citric acids are most commonly used.

Regulatory Status. GRAS (§182.20).

REFERENCES

See the General References for ARCTANDER; BAILEY 2; FEMA; GRIEVE; JIANGSU; LUST; MARTINDALE; MCGUFFIN 1 & 2; MORTON 2; TERRELL; UPHOF; WHISTLER AND BEMILLER; YOUNGKEN.

1. K. R. Kirtikar and B. D. Basu, *Indian Medicinal Plants*, International Book Distribution, Dehradun, 1999, p. 887.
2. Y. S. Lewis and S. Neelakantan, *Food Sci. (Mysore)*, **9**, 405 (1960).
3. Y. S. Lewis et al., *Food Sci. (Mysore)*, **10**, 49 (1961).
4. E. S. Imbabi et al., *Fitoterapia*, **63**, 537 (1992).
5. Y. Sudjaroen et al., *Food Chem. Toxicol.*, **43**, 1673 (2005).
6. P. L. Lee, et al., *J. Agric. Food Chem.*, **23**, 1195 (1975).
7. A. Vialard-Goudou et al., *Qualitas Plant. et Materiae Vegetabiles*, **3–4** (1958).
8. J. M. Fook et al., *Life Sci.*, **76**, 2881 (2005).
9. E. Ghelardi et al., *Antimicrob. Agents Chemother.*, **48**, 3396 (2004).
10. R. Coutino-Rodriguez et al., *Arch. Med. Res.*, **32**, 251 (2001).
11. F. Martinello et al., *Food Chem. Toxicol.*, **44**, 810 (2006).
12. S. R. Chowdhury et al., *Poult. Sci.*, **84**, 56 (2005).
13. S. Ushanandini et al., *Phytother. Res.*, **20**, 851 (2006).
14. N. M. Useh et al., *J. Enzyme Inhib. Med. Chem.*, **19**, 339 (2004).
15. A. L. Khandare et al., *Nutrition*, **20**, 433 (2004).
16. S. Burgalassi et al., *Eur. J. Ophthalmol.*, **10**, 71 (2000).
17. A. Steger et al., *Allergy*, **55**, 376 (2000).
18. H. A. Abulafatih, *Econ. Bot.*, **41**, 354 (1987).

TANNIC ACID

Source: Nutgalls from oaks (*Quercus* spp.).

Common/vernacular names: Tannin, gallotannic acid, and gallotannin.

GENERAL DESCRIPTION

Tannins are complex polyphenolic substances isolated from nutgalls formed on young twigs of certain Middle Eastern oak trees *Quercus infectoria* Olivier and other related *Quercus* species (Family Fagaceae) by insects (e.g., *Cynips gallae-tinctoriae* Olivier). The nutgalls are collected preferably before the insect matures and leaves the gall through a hole bored in the wall, as galls with holes contain less tannin. Other sources of tannin include the seedpods of Tara [*Caesalpinia spinosa* (Mol.) O. Kuntze] of South America and Chinese nutgalls formed by insects on certain *Rhus* species growing in China [e.g., *R. chinensis* Mill. and *R. potaninii* Maxim. (Family Anacardiaceae)].

Tannic acid is generally extracted from the galls by a mixture of solvents involving water, alcohol, ether, and acetone. The complex tannic acid is separated from the simple acids such as gallic acid (3,4,5-trihydroxybenzoic acid) and is then purified.

Tannic acid is very soluble in water, alcohol, acetone, and glycerol; it is practically insoluble in carbon disulfide, chloroform, benzene, ether, hexane, and fixed oils. Its aqueous solution is acidic and decomposes on standing.

CHEMICAL COMPOSITION

Commercial tannic acid is not the flavanol (catechin) type. It is a mixture of glycosides of phenolic acids (mainly gallic acid). Its structure varies in complexity from a few to many molecules of gallic acid per molecule of sugar (especially glucose). Pharmaceutical-grade tannic acid is generally considered to be pentadigalloylglucose (consisting of 10 gallic acid molecules per glucose molecule) or its higher molecular weight derivatives. However, in reality, the pharmaceutical-grade (or food-grade) tannic acid in commerce can differ widely in molecular weight and structure. There are no official tests that distinguish these differences. It may also contain gallic and digallic acids as well as other impurities.

PHARMACOLOGY AND BIOLOGICAL ACTIVITIES

The most well-known property of tannins is their astringency, due to their ability to precipitate proteins.[1] Tannic acid lotions are thus used for skin conditions such as burns and hyperhidrosis.[2,3]

Tannins (both gallic acid and catechin types) have both carcinogenic and anticancer properties in experimental animals or systems.[4-8] They have also been implicated in human cancers (LEWIS AND ELVIN-LEWIS).[7] More recent research show that tannic acid has potential in the chemotherapy of cancer due to its growth inhibition of different tumor cell lines *in vitro*.[9-11] *In vivo* studies in mice bearing malignant tumors also show that tannic acid can act as a chemopreventive agent and increase the survival rate at low and high doses, respectively.[12,13]

The antioxidant effects of tannic acid have been recently demonstrated in a number of models of hepatotoxicity, free radical inhibition, NO production, and others.[14-20]

Tannins have been reported to have numerous other biological or pharmacological properties, including antimicrobial/antiviral (see ***balm***),[21-24] growth suppression (in rats),[4,25] nonspecific CNS depression (in mice),[26] cariostasis (in hamsters),[27] hypoglycemic (in humans),[28] antifertility (in males),[29] and, more recently, an antiamyloidogenic activity *in vitro* through destabilizing Alzheimer's β-amyloid fibrils[30] (see ***catechu***).[31,32] Earlier studies in mice also showed that tannic acid inhibits hyaluronidase enzyme and may have potential in managing subcutaneous poisoning resulting from snakebites.[33]

TOXICOLOGY

Toxic effects of tannic acid in humans include fatal liver damage, which may result from the use of tannic acid on burns or as an ingredient of enemas. There is evidence that the toxic agent may not be tannic acid but rather digallic acid, which is present as an impurity.[4] Ingestion of large doses of tannic acid may cause gastric irritation, nausea, and vomiting (MARTINDALE; USD 26th).

USES

Medicinal, Pharmaceutical, and Cosmetic. Mainly used as an ingredient in some ointments and suppositories for treating hemorrhoids and in some burn lotions.

Food. Tannic acid is used in clarifying alcoholic beverages (especially beer and wines). It is also used as a flavor ingredient in most major categories of foods, including alcoholic and nonalcoholic beverages, frozen dairy desserts, candy, baked goods, gelatins and puddings, and meat and meat products. Highest average maximum use level is about 0.018% (182 ppm) reported in frozen dairy desserts.

Traditional Medicine. Nutgalls (especially Chinese nutgalls) containing high concentrations of tannic acid (50–80%) have been used for centuries in China in treating numerous conditions, including bleeding, chronic diarrhea or dysentery, bloody urine, hard-to-heal sores, painful joints, and persistent cough, both internally and externally in the form of a powder or decoction.

Nutgalls have also been used in cancers.[34]

Others. Tannic acid is used extensively in tanning hide and in the manufacture of inks. For these purposes, industrial grades are usually used.

COMMERCIAL PREPARATIONS

Numerous grades (pharmaceutical, food, industrial, etc.). Tannic acid is official in U.S.P. and in F.C.C. Grades differ in their molecular complexities.

Regulatory Status. GRAS (§182.20); source not limited to *Quercus* species.

REFERENCES

See the General References for APhA; CLAUS; FEMA; GOSSELIN; HORTUS 3rd; JIANGSU; MARTINDALE; NANJING; SAX; YOUNGKEN.

1. E. C. Bate-Smith, *Phytochemistry*, **12**, 907 (1973).
2. C. L. Goh and K. Yoyong, *Singapore Med. J.*, **37**, 466 (1996).
3. P. Hupkens et al., *Burns*, **21**, 57 (1995).
4. Z. Glick, *Diss. Abstr. Int. B*, **29**, 1416 (1968).
5. A. Oler et al., *Food Cosmet. Toxicol.*, **14**, 565 (1976).
6. G. J. Kapadia et al., *J. Natl. Cancer Inst.*, **57**, 207 (1976).
7. J. F. Morton, *Quart. J. Crude Drug Res.*, **12**, 1829 (1972).
8. H. H. S. Fong et al., *J. Pharm. Sci.*, **61**, 1818 (1972).
9. P. J. Naus et al., *J. Hepatol.*, **46**, 222 (2007).
10. C. Marienfeld et al., *Hepatology*, **37**, 1097 (2003).
11. H. Kamei et al., *Cancer Biother. Radiopharm.*, **14**, 135 (1999).
12. T. Koide et al., *Cancer Biother. Radiopharm.*, **14**, 231 (1999).
13. C. Nepka et al., *Cancer Lett.*, **141**, 57 (1999).
14. L. R. Richards et al., *Biomed. Sci. Instrum.*, **42**, 357 (2006).
15. I. H. El-Sayed et al., *Toxicol. Ind. Health*, **22**, 157 (2006).
16. A. Sehrawat et al., *Redox. Rep.*, **11**, 85 (2006).
17. R. G. Andrade Jr. et al., *Biochimie*, **88**, 1287 (2006).
18. R. G. AndradeJr. et al., *Arch. Biochem. Biophys.*, **437**, 1 (2005).
19. R. C. Srivastava et al., *Cancer Lett.*, **153**, 1 (2000).

20. N. S. Khan et al., *Chem. Biol. Interact.*, **125**, 177 (2000).
21. T. Wauters et al., *Can. J. Microbiol.*, **47**, 290 (2001).
22. K. T. Chung et al., *Food Chem. Toxicol.*, **36**, 1053 (1998).
23. F. Uchiumi et al., *Biochem. Biophys. Res. Commun.*, **220**, 411 (1996).
24. V. E. Sokolova et al., *Prikl. Biokhim. Mikrobiol.*, **5**, 694 (1969).
25. M. A. Joslyn and Z. Glick, *J. Nutr.*, **98**, 119 (1969).
26. R. N. Takahashi et al., *Planta Med.*, **4**, 272 (1986).
27. A. Stralfors, *Arch. Oral Biol.*, **12**, 321 (1967).
28. H. Gin et al., *Metabolism*, **48**, 1179 (1999).
29. I. A. Taitzoglou et al., *Reproduction*, **121**, 131 (2001).
30. K. Ono et al., *Biochim. Biophys. Acta*, **1690**, 193 (2004).
31. B. Toth, *Kolor. Ert.*, **9**, 77 (1967).
32. G. V. N. Rayudu et al., *Poult. Sci.*, **49**, 957 (1970).
33. U. R. Kuppusamy and N. P. Das, *Pharmacol. Toxicol.*, **72**, 290 (1993).
34. J. L. Hartwell, *Lloydia*, **32**, 153 (1969).

TARRAGON

Source: *Artemisia dracunculus* L. (Family Compositae or Asteraceae).

Common/vernacular names: Estragon.

GENERAL DESCRIPTION

A green, nonhairy perennial herb with an erect branched stem; up to about 1.2 m high; native to Europe (southern Russia) and western Asia; cultivated in Europe (France, Germany, Italy, etc.), United States (California), Argentina, and other countries. Parts used are the leaves or aboveground herb; from the latter, tarragon oil is obtained by steam distillation. Two varieties traded: *A. dracunculus* cv. "Sativa," French tarragon, and *A. dracunculus*, Russian tarragon (syn. *A. redowskii*). Russian tarragon does not have the fine flavor of French tarragon. Major oil-producing countries include France and the United States.

CHEMICAL COMPOSITION

The aboveground herb usually contains 0.25–1% volatile oil; coumarins (coumarin, esculetin dimethyl ether, herniarin, scopoletin, etc.);[1–3] isocoumarins (e.g., artemidin and artemidinal);[4,5] flavonoids (e.g., rutin and quercetin);[6] sterols (β-sitosterol, stigmasterol, etc.);[1,2] a saturated hydrocarbon, $C_{29}H_{60}$ (m.p. 63–64°C);[1] tannin; protein; and others (LIST AND HÖRHAMMER; MARSH).

Tarragon oil consists mainly of estragole (methyl chavicol, 70–81%).[7–10] Other components present include capillene, ocimene, nerol, thujone, 1,8-cineole, 4-methoxycinnamaldehyde, α-pinene (0.89%), β-phellandrine (1.07%), limonene (2.68%), and γ-terpinene (10.40%), among others;[7,8,11,12] nerol has been reported to be the major component of an oil of British origin.[11] Elemicin, *trans*-isoelemicin, eugenol, methyl eugenol, and *trans*-methyl isoeugenol were found in the oil of Russian tarragon.[13] The composition of the essential oil of Iranian *A. dracunculus* was reported to include *trans*-anethole and α-*trans*-ocimene as the major constituents (21.1% and 20.6%, respectively) in addition to limonene, α- and β-pinene, *allo*-ocimene, methyleugenol, α-terpinolene, bornyl acetate, and bicyclogermacrene.[14] According to another report, Turkish *A. dracunculus* contained Z-anethole as the major constituent (81%) together with Z- and E-β-ocimene, limonene, and methyleugenol.[15]

The root has been reported to contain several oligosaccharides (including inulobiose), polyacetylenes, artemidiol [3-(1,2-dihydroxybutyl)-isocoumarin], and other isocoumarins.[16–19]

Another isocoumarin, artemidinol, has also been reportedly isolated from tarragon.[20]

Fractionation of cell tissue extract of *A. dracunculus* guided by human benzodiazepine receptor binding leads to the isolation of delorazepam and temazepam at concentrations ranging 100–200 ng/g. This is the first report on the presence of such compounds in plant cells.[21]

The aerial parts contain the alkamides, pellitorine, and neopellitorine A and B. The latter two were isolated as new compounds from *A. dracunculus*.[3]

4,5-Di-*O*-caffeoylquinic acid, davidigenin, 6-demethoxycapillarisin, and 2′,4′-dihydroxy-4-methoxydihydrochalcone have recently been reported in the ethanolic extract of tarragon.[22]

PHARMACOLOGY AND BIOLOGICAL ACTIVITIES

Tarralin, an ethanolic extract of *A. dracunculus*, administered to genetically diabetic mice by oral gavage resulted in a significant antihyperglycemic effect in comparison to troglitazone and metformin. The effect was attributed to the inhibition of aldose reductase by the four compounds mentioned above,[22] which was comparable to that produced by quercitrin as a reference control.[22,23]

The essential oil is strongly antibacterial (though estragole at 81% of oil content is not responsible for activity)[9,10] and antifungal against *Colletrotichum acutatum*, *C. fragariae*, and *C. gloeosporioides* with 5-phenyl-1,3-pentadiyne and capillarin as the active constituents.[24] The methanolic extract is also active against the Gram-negative bacteria *Escherichia coli*, *Shigella*, *Listeria monocytogenes*, and *Pseudomonas aeruginosa*.[25]

The essential oil of Iranian tarragon displayed a dose- and time-dependent anticonvulsant activity in mice at ED_{50} values of less than 1 mL/kg. Sedation and motor impairment were also produced at certain anticonvulsant doses.[14]

TOXICOLOGY

Estragole, the main constituent of tarragon oil, has been reported to produce tumors in mice (see **sweet basil** and **avocado**).

Although undiluted tarragon oil has been reported to be irritating to rabbit skin and the backs of hairless mice, it was found to be nonirritating and nonsensitizing to humans at a concentration of 4% in petrolatum; it was also not phototoxic.[26] The ethanolic extract (tarralin) has recently been shown to be devoid of acute and chronic toxicity in rats at 1000 mg/kg/day oral dose. Mutagenic activity in the Ames test was also negative.[27]

USES

Medicinal, Pharmaceutical, and Cosmetic. The oil is used as a fragrance component in soaps, detergents, creams, lotions, and perfumes, with maximum use level of 0.4% reported in perfumes.[26]

Food. The leaf is used commonly as a domestic herb. It is also used extensively as a flavor component in numerous food products, including nonalcoholic beverages, candy, meat and meat products, condiments and relishes (e.g., vinegar), fats and oils, and gravies. Highest average maximum used level reported is about 0.27% (2731 ppm) in condiments and relishes.

In addition to all the above food categories, the oil is used in alcoholic beverages (e.g., liqueurs), frozen dairy desserts, baked goods, and gelatins and puddings. Highest average maximum use level reported is about 0.04% (414 ppm) in baked goods.

Traditional Medicine. The herb is used as a stomachic, diuretic, hypnotic, emmenagogue, and in treating toothache. Also reported to be used for treating tumors.[28]

COMMERCIAL PREPARATIONS

Mainly the leaves and oil; oil is official in F.C.C. According to ARCTANDER, adulteration of tarragon oil has been common.

Regulatory Status. GRAS (§182.10 and §182.20).

REFERENCES

See the General References for ARCTANDER; BAILEY 1; FEMA; FOSTER; GRIEVE; GUENTHER; GUPTA; LIST AND HÖRHAMMER; LUST; ROSENGARTEN; TERRELL; UPHOF.

1. A. Mallabaev et al., *Khim. Prir. Soedin.*, **5**, 320 (1969).
2. P. Tunmann and E. Mann, *Z. Lebensm.-Unters. Forsch.*, **138**, 146 (1968).
3. B. Saadali et al., *Phytochemistry*, **58**, 1083 (2001).
4. A. Mallabaev et al., *Khim. Prir. Soedin.*, **7**, 120871 (1971).
5. A. Mallabaev et al., *Khim. Prir. Soedin.*, **6**, 531 (1971).
6. G. A. Lukonikova, *Prikl. Biokhim. Mikrobiol.*, **1**, 594 (1965).
7. H. Thieme and N. T. Tam, *Pharmazie*, **23**, 339 (1968).
8. G. M. Nano et al., *Riv. Ital. Essenze, Profumi, Piante offic., Aromi, Saponi, Cosmet., Aerosol*, **98**, 409 (1966).
9. S. G. Deans and K. P. Svoboda, *J. Hortic. Sci.*, **63**, 503 (1988).
10. B. M. Lawrence, *Perfum Flav.*, **15**, 75 (1990).
11. D. V. Banthrope et al., *Planta Med.*, **20**, 147 (1971).
12. S. R. Srinivas, *Atlas of Essential Oils*, Bronx, NY, 1986.
13. B. M. Lawrence, *Perfum. Flav.*, **13**, 44 (1988).
14. M. Sayyah et al., *J. Ethnopharmacol.*, **94**, 283 (2004).
15. S. Kordali et al., *J. Agric. Food Chem.*, **53**, 9452 (2005).
16. A. Lombard et al., *Atti Accad. Sci. Torino, Cl. Sci. Fis. Mat. Nat.*, **109**, 439 (1975).
17. A. Mallabaev and G. P. Sidyakin, *Khim. Prir. Soedin.*, **6**, 720 (1974).
18. F. Bohlmann and K. M. Kleine, *Chem. Ber.*, **95**, 39 (1961).
19. H. Greger et al., *Phytochemistry*, **16**, 795 (1977).
20. A. Mallabaev and G. P. Sidyakin, *Khim. Prir. Soedin.*, **6**, 811 (1976).
21. D. Kavvadias et al., *Biochem. Biophys. Res. Commun.*, **269**, 290 (2000).
22. S. Logendra et al., *Phytochemistry*, **67**, 1539 (2006).
23. D. M. Ribnicky et al., *Phytomedicine*, **13**, 550 (2006).
24. K. M. Meepagala et al., *J. Agric. Food Chem.*, **50**, 6989 (2002).
25. M. Benli et al., *Cell Biochem. Funct.* (2006).
26. D. L. J. Opdyke, *Food Cosmet. Toxicol.*, **12**, 709 (1974).
27. D. M. Ribnicky et al., *Food Chem. Toxicol.*, **42**, 585 (2004).
28. J. L. Hartwell, *Lloydia*, **31**, 71 (1968).

TEA

Source: *Camellia sinensis* (L.) Kuntze (syn. *C. thea* Link; *C. theifera* Griff.; *Thea sinensis* L.; *T. bohea* L.; *T. viridis* L.) and its varieties (Family Theaceae).

GENERAL DESCRIPTION

Tea originated in China; its use dates back to several thousand years. It is an evergreen shrub to occasionally a tree, much branched; young leaves hairy; up to about 9 m high if free growing, but usually maintained at 1–1.5 m high by regular pruning; native to the mountainous regions of southern China, Japan, and India; extensively cultivated in China, India, Japan, Sri Lanka, Indonesia, and other tropical and subtropical countries (Kenya, Uganda, Turkey, Argentina, etc.); cultivated in the United States in the Carolinas. Parts used are the dried, cured leaf bud and the two adjacent young leaves together with the stem, broken between the second and third leaf; older leaves are also used but are considered of inferior quality. The young leaves and leaf bud together are called "tea flush" and are collected from spring to fall (JIANGSU).[1,2]

There are two major kinds of tea, black tea and green tea. The main difference between black tea and green tea is that during tea manufacture the former undergoes a fermentation step whereby enzymes (polyphenol oxidase, peptidase, alcohol dehydrogenase, etc.) present in tea flush convert certain constituents present (proteins, amino acids, fatty acids, polyphenols, etc.) to compounds that are responsible for the characteristic aroma and flavor of black tea.[1–3] In green tea manufacture, this fermentation step is eliminated by initially subjecting the flush to steaming (Japanese process) or dry heating (Chinese process), whereby the enzymes are inactivated.[1,2]

There are many grades of black tea and green tea that differ enormously in price. India and Sri Lanka are the major black tea producers, and China and Japan are the major green tea producers.

Tea drinking has evolved into a very delicate art in some Asian countries (especially China, Japan, and India), where apart from the tea used, types of water (spring, well, etc.), brewing utensils, and brewing conditions, among others, are carefully controlled by some connoisseurs. Certain types of tea in China are served in thimble-sized cups and are so strong (astringent) that they would be unpalatable to most Americans.

CHEMICAL COMPOSITION

The chemistry of tea is extremely complicated.[1,2] Both black tea and green tea contain caffeine (1–5%), with small amounts of other xanthine alkaloids (theobromine, theophylline, dirnethylxanthine, xanthine, adenine, etc.) also present.[4] Part of the caffeine is in bound form (see **kola nut**). They also contain large amounts of tannins or phenolic substances (5–27%) consisting of both catechin (flavanol) and gallic acid units, with those in green tea being higher than those in black tea (JIANGSU; LIST AND HÖRHAMMER; STAHL).[1–3,5–8]

Other components present in tea include 4–16.5% fats; flavonoids (quercetin, quercitrin, rutin, etc.); anthocyanins; amino acids (higher in green tea);[2] triterpenoid saponin glycosides (theasaponin, isotheasaponins, and assamsaponins);[9–16] sterols; vitamin C; flavor and aroma chemicals including theaflavin, thearubigin, l-epicatechin gallate, theogallin, theaspirone, dihydroactinidiolide, dimethyl sulfide, ionones (α- and β-), damascones (α- and β-), jasmone, furfuryl alcohol, geranial, trans-hexen-2-al, and others, totaling over 300 compounds; proteins; polysaccharides;[17] pigments (carotenoids);[18] and others (JIANGSU; LIST AND HÖRHAMMER).[1–3]

Theaflavin, thearubigin, and l-epicatechin gallate are reported to be important taste components of black tea, while its important aroma compounds include ionones, damascones, theaspirone, trans-hexen-2-al, and dihydroactinidiolide.[1–3,19]

Dimethyl sulfide has been reported to be an important aroma compound of green tea,[2] along with benzylaldehyde, benzyl alcohol, cyclohexanones, dihydroactinodiolide, cis-hexen-3-ol, hexenyl hexanoate, cis-jasmone, linalool, linalool oxides, nerolidol, and phenylethanol.[20]

A cup of tea contains comparable amounts of caffeine (ca. 100 mg) as a cup of coffee (see **coffee**), but it has a much higher tannin content.

PHARMACOLOGY AND BIOLOGICAL ACTIVITIES

Caffeine, the most commonly known constituent of tea, has diuretic and CNS-stimulant activities, as well as numerous other kinds of pharmacological properties (see **coffee**).

All varieties of tea (black, green, Assam, etc.) have been extensively investigated over the past decade and numerous pharmacological effects and biological activities have been reported.

Tea has been reported effective in clinically treating bacterial dysentery (>95% effective in acute cases; 85% chronic), amebic dysentery (100%, 12 cases), acute gastroenteritis (100%, 20 cases), acute and chronic enteritis (>90% acute; >83% chronic), acute infectious hepatitis, and dermatitis due to handling of rice plants, among others (JIANGSU). Antidiarrheal effects of black and green tea extracts have been attributed to both an antimicrobial effect on *Salmonella typhimurium, S. typhi* and *Vibrio cholera* as well as a direct effect on the opioid system that controls gastric motility.[21,22] Antimicrobial effects against methicillin-resistant *Staphylococcus aureus* and *Yersinia enterocolitica* have also been reported.[23,24] Catechins isolated from an ethyl acetate fraction of green tea exhibited very strong trypanocidal effects against *Trypanosoma cruzi* at 50% inhibitory concentrations as low as 0.12 pM.[25] Various constituents of tea were reported to prevent the adhesion of pathogenic (e.g., *Helicobacter pylori* and *Propionibacterium acnes*) and cariogenic (e.g., *Streptococcus mutans* and *S. sobrinus*) bacteria to their respective targets.[17,26–28]

Tannins have anticancer, carcinogenic, and other activities (see **tannic acid**).[29] Cancer-inducing *N*-nitrosation by-products has been shown to be inhibited by green tea polyphenols. Expression of experimentally induced chromosome damage in bone marrow cells (*in vitro*) has been found to be suppressed by pretreatment with green tea polyphenols.[30] A recent study found that green tea polyphenols and epigallocatechin gallate arrested skin tumor growth, size, and number in mice.[31] Green tea catechins suppressed the growth of prostate cancer and epithelial ovarian cancer cell lines.[32] Nonphenolic fractions of green tea exhibited an anticarcinogenic effect when tested in a *Salmonella typhimurium* model of genotoxicity, while steroidal saponins isolated from a tea root extract were cytotoxic to two human leukemia cell lines.[33,34]

Certain constituents found in tea, especially the tannic substances catechin, epigallocatechin, and epigallocatechin gallate, have been reported to have antioxidative properties, with epigallocatechin being the strongest.[8,35–38] Following this line are many reports on the anti-inflammatory,[39–42] chemopreventive,[43,44] hepatoprotective,[45–47] and anticataract effects[48] of different tea extracts and their constituents in various *in vitro* and *in vivo* models. The chemopreventive, antioxidant, traditional, and future therapeutic uses of tea have been reviewed.

Tea has been reported to have antiatherosclerotic effects, "vitamin P" activities, and others.[2] More recently reported effects include gastroprotective/antiulcer,[11,12,14,49,50] estrogenic,[51,52] immunomodulatory,[53] antiallergic,[54] and neuromuscular-protecting effect against botulinum and tetanus toxins.[55,56] There also indications of possible benefits of tea in reducing the side effects of diabetes.[57,58]

TOXICOLOGY

An undesirable proconvulsive effect was observed in three models of experimentally induced convulsions in mice after acute and chronic administration of extracts of both black and green tea. Such effect was proposed to be mediated through Ca^{2+} channels.[59]

USES

Food. The major use of tea domestically is as a beverage usually in the form of tea bags (pure ground tea) or as instant teas (tea extracts mixed with other ingredients). The tea used is mostly black tea.

Tea extract (type not specified) is also reportedly used as a flavor component in most major food products, including alcoholic beverages, frozen dairy desserts, candy, baked goods, and gelatins and puddings. Highest average maximum use level is 3% in baked goods.

Dietary Supplements/Health Foods. Various products containing green tea extracts are appearing in this market, presumably in response to reports of antioxidant activity.

Traditional Medicine. The common tea bag is used as a wash for sunburn, as a poultice for baggy eyes, and as a compress for headache or tired eyes (ROSE); also used to stop bleeding of a tooth socket.

Tea has been used for millennia in Chinese medicine as a stimulant, diuretic, stomachic, expectorant, and antitoxic. Conditions for which it is traditionally used include headache, dysentery, and excess phlegm. In India the leaf juice is used as a topical hemostatic for cuts and injuries.[30]

Others. Tea (especially low grade) is a potential source of food colors (green, yellow, orange, black, etc.).[60,61]

COMMERCIAL PREPARATIONS

Crude and extracts.

Regulatory Status.   GRAS (§182.20)

REFERENCES

See the General References for BARRETT; BLUMENTHAL 2; DER MARDEROSIAN AND BEUTLER; GOODMAN AND GILMAN; HUANG; JIANGSU; LEWIS AND ELVIN-LEWIS; LIST AND HÖRHAMMER; MARTINDALE; MCGUFFIN 1 & 2; TERRELL.

1. G. W. Sanderson in V C. Runeckles and T. C. Tso, eds., *Recent Advances in Phytochemistry*, Vol. 5, Academic Press, New York, 1972, p. 247.
2. R. L. Wiskremasinghe in C. O. Chichester et al., eds., *Advances in Food Research*, Vol. 24, Academic Press, New York, 1978, p. 229.
3. D. Reymond, *Chemtech*, **7**, 664 (1977).
4. M. Mironescu, *Rev. Fiz. Chim. Ser. A*, **11**, 218 (1974).
5. C. T. Wu et al., *Chung Kuo Nung Yeh Hua Hseuh Hui Chih*, **13**, 159 (1975).
6. D. J. Cattell and H. E. Nursten, *Phytochemistry*, **15**, 1967 (1976).
7. T. Bryce, et al., *Tetrahedron Lett.*, 463 (1972).
8. G. Kajimoto et al., *Eiyo To Shokuryo*, **22**, 473 (1969).
9. M. Yoshikawa et al., *Chem. Pharm. Bull. (Tokyo)*, **55**, 57 (2007).
10. K. Kobayashi et al., *Phytochemistry*, **67**, 1385 (2006).
11. T. Morikawa et al., *J. Nat. Prod.*, **69**, 185 (2006).
12. M. Yoshikawa et al., *Chem. Pharm. Bull. (Tokyo)*, **53**, 1559 (2005).
13. Y. Lu et al., *Phytochemistry*, **53**, 941 (2000).
14. T. Murakami et al., *Chem. Pharm. Bull. (Tokyo)*, **47**, 1759 (1999).
15. I. Kitagawa et al., *Chem. Pharm. Bull. (Tokyo)*, **46**, 1901 (1998).
16. Y. M. Sagesaka et al., *Biosci. Biotechnol. Biochem.*, **58**, 2036 (1994).
17. J. H. Lee et al., *J. Agric. Food Chem.*, **54**, 8717 (2006).
18. K. Higashi-Okai et al., *J. UOEH*, **23**, 335 (2001).
19. P. Coggon et al., *J. Agric. Food Chem.*, **25**, 278 (1977).
20. H. N. Graham, *Prev. Med.*, **21**, 334 (1992).
21. S. E. Besra et al., *Phytother. Res.*, **17**, 380 (2003).
22. M. Shetty et al., *J. Commun. Dis.*, **26**, 147 (1994).
23. T. S. Yam et al., *J. Antimicrob. Chemother.*, **42**, 211 (1998).
24. T. S. Yam et al., *FEMS Microbiol. Lett.*, **152**, 169 (1997).

25. C. Paveto et al., *Antimicrob. Agents Chemother.*, **48**, 69 (2004).
26. L. Z. Touyz and R. Amsel, *Quintessence. Int.*, **32**, 647 (2001).
27. J. M. Hamilton-Miller, *J. Med. Microbiol.*, **50**, 299 (2001).
28. A. Rasheed and M. Haider, *Arch. Pharm. Res.*, **21**, 348 (1998).
29. H. E. Kaiser, *Cancer*, **20**, 614 (1967).
30. R. R. Rao and N. S. Jamir, *Econ. Bot.*, **36**, 176 (1982).
31. Z. Y. Wang et al., *Cancer Res.*, **52**, 6657 (1992).
32. M. H. Ravindranath et al., *Evid. Based Complement Alternat. Med.*, **3**, 237 (2006).
33. P. Ghosh et al., *Leuk. Res.*, **30**, 459 (2006).
34. Y. Okai and K. Higashi-Okai, *Teratog. Carcinog. Mutagen.*, **17**, 305 (1997).
35. Y. S. Lin et al., *J. Agric. Food Chem.*, **51**, 975 (2003).
36. M. Sano et al., *Biol. Pharm. Bull.*, **18**, 1006 (1995).
37. M. A. Bokuchava et al., *Maslo-Zhir. Prom-st.*, **2**, 15 (1975).
38. S. Uchida et al., *Life Sci.*, **50**, 147 (1992).
39. D. Das et al., *Life Sci.*, **78**, 2194 (2006).
40. P. Chattopadhyay et al., *Life Sci.*, **74**, 1839 (2004).
41. C. Adcocks et al., *J. Nutr.*, **132**, 341 (2002).
42. P. Sur et al., *Phytother. Res.*, **15**, 174 (2001).
43. A. Halder et al., *J. Environ. Pathol. Toxicol. Oncol.*, **24**, 141 (2005).
44. I. P. Lee et al., *J. Cell Biochem. Suppl.*, **27**, 68 (1997).
45. A. Singal et al., *Phytother. Res.*, **20**, 125 (2006).
46. H. A. El-Beshbishy, *J. Biochem. Mol. Biol.*, **38**, 563 (2005).
47. Y. C. Hung et al., *J. Agric. Food Chem.*, **52**, 5284 (2004).
48. S. K. Gupta et al., *Ophthalmic Res.*, **34**, 258 (2002).
49. S. Maity et al., *Jpn. J. Pharmacol.*, **78**, 285 (1998).
50. S. Maity et al., *J. Ethnopharmacol.*, **46**, 167 (1995).
51. A. S. Das et al., *Life Sci.*, **77**, 3049 (2005).
52. A. S. Das et al., *Asia Pac. J. Clin. Nutr.*, **13**, 210 (2004).
53. E. Zvetkova et al., *Int. Immunopharmacol.*, **1**, 2143 (2001).
54. M. Akagi et al., *Biol. Pharm. Bull.*, **20**, 565 (1997).
55. E. Satoh et al., *J. Toxicol. Sci.*, **27**, 441 (2002).
56. E. Satoh et al., *Pharmacol. Toxicol.*, **90**, 199 (2002).
57. G. T. Mustata et al., *Diabetes*, **54**, 517 (2005).
58. A. Gomes et al., *J. Ethnopharmacol.*, **45**, 223 (1995).
59. A. Gomes et al., *Phytother. Res.*, **13**, 376 (1999).
60. M. A. Bokuchava and G. N. Pruidze, *Br.* 1,181,079 (1970).
61. M. A. Bokuchava, et al., *Biokhim. Progr. Tekhnol. Chai. Proizvod., Akad. Nauk SSSR, Inst. Biokhim.*, 335 (1966).

THYME

Source: *Thymus vulgaris* L. (Family Labiatae or Lamiaceae).

Common/vernacular names: Common thyme, garden thyme, and French thyme.

GENERAL DESCRIPTION

There are many species and varieties of thyme whose classification is complicated. Estimates of legitimate species range from 100 to 400. The most commonly used species is *Thymus vulgaris*. It is an erect evergreen subshrub with numerous white hairy stems and a woody fibrous root; up to about 45 cm high; native to the Mediterranean region (Greece, Italy, Spain, etc.); extensively cultivated in France, Spain, Portugal, Greece, and the United States (California). Parts used are the dried or partially dried leaves and flowering tops from which thyme oil is produced by water and steam distillation.

Other thyme species used include: *T.* × *citriodorus* (Pers.) Schreb. (syn. *T. serpyllum* L. var. *vulgaris* Benth.), which is known as lemon thyme; *T. zygis* L.; and *T. serpyllum* L., known as creeping thyme, wild thyme, or mother of thyme.

Thyme oil is derived from *T. vulgaris* and *T. zygis* and its var. *gracilis* Boiss. Two types of thyme oil are produced, red thyme oil and white thyme oil. White thyme oil is obtained from red thyme oil by redistillation; it has been reported to be much adulterated (ARCTANDER). The major oil-producing country is Spain (see also *origanum*, *Spanish*).

CHEMICAL COMPOSITION

Common thyme contains 0.8–2.6% (usually ca. 1%) volatile oil consisting of highly variable amounts of phenols (20–80%);[1–5] monoterpene hydrocarbons (e.g., 51% according to one report) such as *p*-cymene and γ-terpinene;[4,6,7] and alcohols (e.g., linalool, α-terpineol, and thujan-4-ol each of which can be the major component and constitute up to 80% or more of the volatile oil).[2,3,5,6,8,9] Thymol is normally the major phenolic component in common thyme with carvacrol being only a minor component.[2–5,8–10] Thymol, carvacrol, and *p*-cymene occurring as glucosides or galactosides have been reported.[11–13]

Other constituents present include tannins; flavonoids (e.g., quercitin, eridictyol, cirsilineol);[14,15] caffeic, rosmarinic, labiatic, ursolic, and oleanolic acids;[16] hydroxyjasmone glucoside;[17] acetophenone glycosides;[18] and polysaccharides[19,20] (STAHL).

Wild thyme (*T. serpyllum*) contains 0.4–2.3% volatile oil consisting of highly variable amounts of phenols (e.g., 47–74%);[21,22] alcohols;[21] and monoterpene hydrocarbons. Either thymol or carvacrol can be the major phenol in wild thyme, depending on sources (see *origanum*, *Spanish*).[2,3,10]

PHARMACOLOGY AND BIOLOGICAL ACTIVITIES

Thyme oil is reported to have antispasmodic, expectorant, and carminative properties; it also has antimicrobial (bacteria and fungi) activities. These activities are mainly due to thymol and carvacrol, with the former being more potent.[5,23–25]

The most frequently studied activity of thyme is the antioxidant effect of its extracts, volatile oil, and main constituents. Free radical scavenging and protective effects against hepatotoxicity, DNA damage, brain phospholipid oxidation have been demonstrated.[14,16,26–31] Thyme oil and thymol have been demonstrated to have antioxidative activities on dehydrated pork.[32] The labiatic acid present in thyme as well as in marjoram, oregano, sage, and other plants of the mint family also has antioxidative properties (see *marjoram*, *rosemary*, and *sage*).

When administered to rabbits orally or per intramuscular injection, the oil caused arterial hypotension accompanied by increased rhythmic contraction of the heart, and in higher dosage also increased respiratory frequency.[33] When given intravenously to cats as a 5% emulsion in saline solution, thyme oil also increased respiratory volume and lowered blood pressure.[34]

Fluid extracts of *T. vulgaris* and *T. serpyllum* have been shown to have a relaxant activity on the smooth muscle of guinea-pig trachea (bronchodilation) and ileum (spasmolytic) *in vitro*. However, this activity is not due to thymol or carvacrol but is probably due to unidentified nonphenolic components.[35–37]

Thyme has been shown to have an antithyrotropic effect in rats;[38] while thymol has been reported to have a sedative effect (GABA modulator)[39] and to inhibit platelet aggregation *in vitro*.[40]

Thyme oil has been reported to be lethal to mosquito larvae.[41] Its major monoterpenes were also found to effectively repel mosquitoes, with α-terpinene being most active at 2% concentration.[6]

Thyme oil is reported to have strongly fungicidal (e.g., against *Aspergillus* and *Candida*), antibacterial (especially Gram-positive and *Helicobacter pylori*), antitrypanosomal, and anthelmintic (especially hookworms) as well as mildly local irritant properties.[42–46] However, it is considered to be very toxic. Toxic symptoms include nausea, vomiting, gastric pain, headache, dizziness, convulsions, coma, cardiac, and respiratory collapse (GOSSELIN; MERCK; USD 26th).

TOXICOLOGY

Red thyme oil has been reported to be nonirritating, nonsensitizing, and nonphototoxic to human skin but severely irritating to mouse and rabbit skin when applied undiluted.[47]

A comprehensive review on the history, traditional uses, chemistry, pharmacology, and toxicology of thyme has recently been published.[48]

USES

Medicinal, Pharmaceutical, and Cosmetic. Thyme oil is used as a flavor component, antispasmodic, carminative, counterirritant, or rubefacient in certain cough drops, antiseptic mouthwashes, and liniments. Thymol is similarly used; in addition, it is used in antifungal preparations (for fungal skin infections), dental formulations, and others.

In German phytomedicine, preparations (tea) prescribed at 1–2 g of dried herb (calculated to contain at least 0.5% phenols, calculated as thymol) are used for alleviating the symptoms of bronchitis, whooping cough, and catarrhs of the upper respiratory tract.[49]

Thyme oil is also used in toothpastes, soaps, detergents, creams, lotions, and perfumes, with maximum use level of 0.8% of the red type reported in perfumes.[47]

Food. Lemon thyme is used primarily in spice blends (especially for salads).

Thyme is widely used in baked goods, meat and meat products, condiments and relishes, processed vegetables, soups, gravies, and fats and oils, among others. Highest average maximum use level is about 0.172% (1716 ppm) reported in meat and meat products.

Thyme oil, white thyme oil, tincture, and fluid extract are used as flavor components in most major food products, including alcoholic (e.g., liqueurs) and nonalcoholic beverages, frozen dairy desserts, candy, baked goods, gelatins and puddings, meat and meat products, and condiments and relishes. Average maximum use levels are usually below 0.003%.

Dietary Supplements/Health Foods. Thyme is sometimes used as a flavoring ingredient in teas.

Traditional Medicine. Common thyme (both fresh and dried) is reportedly used as an anthelmintic, antispasmodic, bronchospasmolytic, carminative, sedative, diaphoretic, and expectorant, usually in the form of an infusion or tincture. Conditions for which it is used include acute bronchitis, laryngitis, whooping cough, chronic gastritis, diarrhea, and lack of appetite. It is also used externally in baths to help rheumatic and skin pro-blems (bruises, sprains, etc.). In Chinese medicine it is used to treat similar conditions (JIANGSU).

Creeping thyme is used for similar purposes as common thyme. Its infusion is also reputed to be useful in the treatment of alcoholism (LUST).

Several species of thyme (e.g., *T. vulgaris*, *T. serpyllum*, and *T. zygis*) have been used in cancers.[50]

COMMERCIAL PREPARATIONS

Crude and oils. Thyme and thyme oil were formerly official in N.F. Thyme oil is official in F.C.C.

Regulatory Status. GRAS: Common thyme and creeping thyme (§182.10 and §182.20); derivatives (e.g., essential oils) of *T. vulgaris*, *T. zygis* var. *gracilis*, and *T. serpyllum* (§182.20). *T. vulgaris* is the subject of a positive German therapeutic monograph, indicated for symptoms of bronchitis and congestion of the upper respiratory tract.[50]

REFERENCES

See the General References for ADA; APhA; APPLEQUIST; ARCTANDER; BAILEY 2; BARNES; BIANCHINI AND CORBETTA; BLUMENTHAL 1; BRUNETON; CLAUS; FEMA; FOSTER; GOSSELIN; GRIEVE; GUENTHER; JIANGSU; LUST; MARTINDALE; MCGUFFIN 1 & 2; MORTON 3; ROSE; ROSENGARTEN; STAHL; UPHOF; USD 26th.

1. M. C. az-Maroto et al., *J. Agric. Food Chem.*, **53**, 5385 (2005).
2. B. M. Lawrence, *Perfum. Flav.*, **2**, 3, (1977).
3. J. D. Miguel et al., *J. Agric. Food Chem.*, **24**, 833, (1976).
4. A. J. Poulose and R. Croteau, *Arch. Biochem. Biophys.*, **187**, 307, (1978).
5. M. Simeon de Bouchberg et al., *Riv. Ital. Essenze, Profumi, Piante Offic., Aromi, Saponi, Cosmet., Aerosol*, **58**, 527, (1976).
6. B. S. Park et al., *J. Am. Mosq. Control Assoc.*, **21**, 80, (2005).
7. M. Hudaib et al., *J. Pharm. Biomed. Anal.*, **29**, 691, (2002).
8. A. B. Svendsen and J. Karlsen, *Planta Med.*, **14**, 376, (1966).
9. M. S. Karawya and M. S. Hifnawy, *J. Assoc. Offic. Anal. Chem.*, **57**, 997, (1974).
10. M. Grims and R. Senjkovic, *Acta Pharm. Jugoslav.*, **17**, 3, (1967).
11. K. Skopp and H. Hoerster, *Planta Med.*, **29**, 208, (1976).
12. H. Takeuchi et al., *Biosci. Biotechnol. Biochem.*, **68**, 1131, (2004).
13. J. Kitajima et al., *Phytochemistry*, **65**, 3279, (2004).
14. H. Haraguchi et al., *Planta Med.*, **62**, 217, (1996).
15. Y. Morimitsu et al., *Biosci. Biotechnol. Biochem.*, **59**, 2018, (1995).
16. A. Dapkevicius et al., *J. Nat. Prod.*, **65**, 892, (2002).
17. J. Kitajima et al., *Chem. Pharm. Bull. (Tokyo)*, **52**, 1013, (2004).
18. M. Wang et al., *J. Agric. Food Chem.*, **47**, 1911, (1999).
19. H. Chun et al., *Biol. Pharm. Bull.*, **24**, 941, (2001).
20. H. Chun et al., *Chem. Pharm. Bull. (Tokyo)*, **49**, 762, (2001).
21. U. Asllani, *Bul. Shkencave Natyr., Univ. Shteteror Tiranes*, **27**, 111, (1973).
22. M. Mihajlov and J. Tucakov, *Bull. Acad. Serbe Sci. Arts, Cl. Sci. Med.*, **44**, 57, (1969).
23. A. C. Pizsolitto et al., *Rev. Fac. Farm. Odontol. Araraquara*, **9**, 55, (1975).
24. M. De Vincenzi and M. R. Dessi, *Fitoterapia*, **62**, 39, (1991).
25. D. Patakova and M. Chladek, *Pharmazie*, **29**, 140, (1974).
26. S. Aydin et al., *Mutat. Res.*, **581**, 43, (2005).
27. S. Aydin et al., *J. Agric. Food Chem.*, **53**, 1299, (2005).
28. K. Miura et al., *J. Agric. Food Chem.*, **50**, 1845, (2002).
29. K. A. Youdim and S. G. Deans, *Br. J. Nutr.*, **83**, 87, (2000).
30. K. A. Youdim and S. G. Deans, *Mech. Ageing Dev.*, **109**, 163, (1999).
31. K. A. Youdim and S. G. Deans, *Biochim. Biophys. Acta*, **1438**, 140, (1999).
32. H. Fujio et al., *Nippon Shokuhin Kogyo Gakkaishi*, **16**, 241, (1969).

33. K. M. Kagramanov et al., *Azerb. Med. Zh.*, **54**, 49, (1977).
34. T. Shipochliev, *Vet. Med. Nauki*, **5**, 63, (1968).
35. C. O. Van Den Broucke and J. A. Lemli, *Planta Med.*, **41**, 129, (1981).
36. M. H. Boskabady et al., *Phytother. Res.*, **20**, 28, (2006).
37. A. Meister et al., *Planta Med.*, **65**, 512, (1999).
38. H. Sourgens et al., *Planta Med.*, **45**, 78, (1982).
39. C. M. Priestley et al., *Br. J. Pharmacol.*, **140**, 1363, (2003).
40. K. Okazaki et al., *Phytother. Res.*, **16**, 398, (2002).
41. D. Novak, *Arch. Roum. Pathol. Exp. Microbiol.*, **27**, 721, (1968).
42. I. Rasooli and P. Owlia, *Phytochemistry*, **66**, 2851, (2005).
43. R. Giordani et al., *Phytother. Res.*, **18**, 990, (2004).
44. M. Marino et al., *J. Food Prot.*, **62**, 1017, (1999).
45. M. Tabak et al., *J. Appl. Bacteriol.*, **80**, 667, (1996).
46. G. F. Santoro et al., *Parasitol. Res.*, (2006).
47. D. L. J. Opdyke, *Food Cosmet. Toxicol.*, **12**(Suppl.), 1003 (1974).
48. E. Basch et al., *J. Herb. Pharmacother.*, **4**, 49, (2004).
49. Monograph Thymi *herba* (*T. vulgaris*), *Bundesanzeiger*, no. 228 (December 5, 1984).
50. J. L. Hartwell, *Lloydia*, **32**, 247, (1969).

TIENCHI GINSENG

Source: *Panax notoginseng* (Burk.) F. H. Chen [syn. *P. pseudo-ginseng* Wall. var. *notoginseng* (Burk.) Hoo et Tseng; *P. sanchi* Hoo] (Family Araliaceae). *P. pseudoginseng* Wall (also a synonym for ginseng) is sometimes listed as source; this has resulted in much confusion whenever herbal formulas are presented in English with no Chinese name given.

Common/vernacular names: Sanqi, sanchi ginseng, tianqi, tianchi, shen sanqi, tian sanqi, and *renshen sanqi*.

GENERAL DESCRIPTION

Perennial herb, 30–60 cm high, with spindle-shaped, fleshy main root; distributed in southern China; now mostly cultivated in the provinces of Yunnan, Guangxi, Guangdong, Jiangxi, and Hubei, especially at altitudes between 800 and 1000 m.

Part used is the tuberous root collected from 3- to 4-year-old plants before flowering or after fruits have ripened; spring harvest is of better quality and preferred. After being rid of root crown, lateral roots, rootlets, sand and dirt, the root is exposed to the hot sun for a day followed by gentle hand kneading and again sun-drying. This process is repeated until root is hard and completely dried, which is then placed in a vessel along with pieces of wax and shaken back and forth until its surface turns shiny and dark brown (MA). Yunnan, Guangdong, and Guangxi are the major producers.

CHEMICAL COMPOSITION

Sanqi contains 4.42–12% saponin glycosides (ginsenosides and protopanaxatriols) (WANG);[1–4] amino acids, including dencichine (β-*N*-oxalo-L-α,β-diaminopropionic acid); volatile oil; flavonoids; phytosterols (e.g., β-sitosterol, stigmasterol, and daucosterol); polysaccharides (e.g., an arabinogalactan named sanchinan-A);[5] trace minerals, and others (HU; WANG).[2]

Saponin glycosides are similar to those of ginseng (dammarane type) and include: ginsenosides R_{b1}, R_d, R_e, R_{g1}, R_{g2}, and R_{h1}, with R_{b1} and R_{g1} in predominant amounts; notoginsenosides R_1, R_2, R_3, R_4, and R_6 in minor amounts; and gypenoside XVII in trace amount. The sapogenins so far identified are 20(S)-protopanaxadiol and 20(S)-protopanaxatriol; no oleanolic acid is present, hence no ginsenoside R_0, (see **ginseng**).[1–9] Saponin glycoside contents in raw and, cured *sanqi* are similar, except that there is an increase of monodesmosides (ginsenosides R_{g2}, R_{g3} and notoginsenoside R_2) and a decrease of bidesmosides (ginsenosides R_{bi}, Rd, Re, R_{gi} and notoginsenosides R_1 and R_4) in the cured drug.[8] Also, compared to raw *sanqi*, steam-curing markedly increased amounts of total extractives and saponin glycosides while deep frying reduced amounts of both.[10]

PHARMACOLOGY AND BIOLOGICAL ACTIVITIES

Sanqi has broad biologic activities, including hematologic, cardiovascular, immunomodulating, anti-inflammatory, and effects on the nervous system (central stimulant, due to panaxatriol saponins), digestive system (antiulcer in rats), metabolism, and endocrine system.[11–13]

Sanqi is best known for its traditional use as a hemostatic, and modern studies have verified the rationale for this use by demonstrating its shortening of coagulation time as well as its hemostatic effects in experimental animals (dencichine being an active component).[11] On the other hand, *sanqi* has also exhibited anticoagulant and antiplatelet aggregation activities under other conditions. Its total saponins have hemolytic activity *in vitro*, though some of the saponins only hemolyzed red cells of certain animals (guinea pig and monkey) but not others (rabbit, sheep, and pigeon) (WANG).[14,15] *Sangi* has been reported to have both hypolipemic and hyperlipemic effects, and a recent study found raw *sanqi* to be hypolipemic but cured *sanqi* to be hyperlipemic.[12] Oral administration of total saponins in rabbits inhibited aortic atherosclerotic plaque formation; their oral administration in rats increased the contents of prostacyclin in carotid artery and decreased the amount of thromboxane A in blood platelets.[16]

Most of the cardiovascular studies on *sanqi* have been performed in China using the total saponins. Their effects include: antiarrhythmic in mice, rats, and rabbits, with the anaxatriol saponins being most active;[17,18] vasodilating and hypotensive in experimental animals;[19–22] protective against experimental hemorrhagic shock in rabbits;[23] protective against experimental myocardial injury in rats, inhibiting lipid peroxidation, and preventing reduction of superoxide dismutase activity, with ginsenosides R_{b1} and R_{g1} being the most active;[24,25] selective blocking of calcium channels;[26] and increasing coronary blood flow volume and reducing coronary arterial resistance as well as peripheral vascular resistance, among others (WANG).[22]

Sanqi decoction stimulated the activities of natural killer cells, macrophages, and plaque-forming cells while crude polysaccharides stimulated the activities of only macrophages and plaque-forming cells in mice;[27] sanchinan-A (a polysaccharide) has activating effects on the reticuloendothelial system as per the carbon clearance test in mice.[5] Similar immunomodulatory properties have recently been reported for the saponin and polysaccharide fractions.[14,15,28,29]

Sanqi total saponins and those from the rootlets all exhibited strong anti-inflammatory activities in several experimental models, with rootlet saponins (100 mg) stronger than cortisone (50 mg) in the ear edema (1.74 ± 0.41 vs. 5.54 ± 0.83 mg; control: 9.35 ± 0.72 mg) induced by croton seed oil in mice.[30,31] Oral administration of *sanqi* powder to rats markedly reduced lipid peroxide formation and greatly increased superoxide dismutase activity in brain tissue but not in other tissues (heart, liver, and lung).[31] Total *sanqi* saponins (i.p.) markedly decreased adrenal ascorbic acid in rats and increased plasma corticosteroids in guinea pigs;[32] rootlet total saponins also greatly increased plasma corticosterone in mice.[30] Other anti-inflammatory mechanisms mediated by *sanqi* include attenuation of pro-inflammatory mediators, such as COX-2, TNF-α and NO.[33–36]

The aqueous extract of *sanqi* and trilinolein have hepatoprotective and cardioprotective effects, respectively, that may be due to antioxidant activities.[37–39]

Rootlet total saponins have been shown to promote growth in mice and to have androgenic activities in rats.[30] Ginsenoside Rg_1, on the other hand, was found to possess *in vitro* estrogen-like activity in MCF-7 cells.[40]

TOXICOLOGY

Oral toxicity of *sanqi* powder is low: stomach feeding of 15 g/kg to mice did not cause any fatalities and examination of tissues (heart, liver, kidney, spleen, and gastrointestinal tract) did not reveal any abnormalities.[11] The pharmacology of tienchi ginseng has recently been reviewed.[41]

USES

Medicinal, Pharmaceutical, and Cosmetic. Powder and extracts used primarily in Asia as an ingredient in shampoos and in skin care products (e.g., acne creams and lotions) for its vasodilating effects and its traditional ability to remove dark spots (ZHOU).[42]

Dietary Supplements/Health Foods. Powder used in tonic formulas, usually in tablet or capsule form.

Traditional Medicine. First described in Li Shi-Zhen's *Ben Cao Gang Mu* (ca. 1590 A.D.), *sanqi* is traditionally regarded to be sweet and slightly bitter tasting, warming, and to have stasis-dispersing, hemostatic, anti-swelling, and analgesic properties. Used in hemorrhages of various kinds (e.g., coughing blood, vomiting blood, nose bleeds, hematochezia, and metrorrhagia), traumatic injuries with bleeding and pain, stabbing pain in chest and abdomen, coronary heart disease, and other conditions (CHP; JIANGSU; WANG).

Its most well-known use is as a major ingredient in *Yunnan Bao Yao*, which was carried by both Chinese and American airmen (the Flying Tigers) during World War II for stopping bleeding resulting from wounds and injuries.

COMMERCIAL PREPARATIONS

Crude comes in over 20 grades with top grades being heavy and hard, 3–6 cm long, and with a grayish brown or grayish green fracture; powdered crude is most likely produced from the crown, lateral roots, rootlets, or other traditionally considered low-grade materials.

Regulatory Status. Class 2d dietary supplement (contraindicated for hypertension).

REFERENCES

See the General References for BARNES; CHP; HUANG; IMM-1; IMM-CAMS; JIXIAN; JIANGSU; MA; NATIONAL; WANG; ZHOU; ZHU.

1. H. Sun et al., *Int. Immunopharmacol.*, **6**, 14 (2006).
2. F. Y. Gan and G. Z. Zheng, *Zhongguo Yaoxue Zazhi*, **27**, 138 (1992).
3. H. Matsuura et al., *Chem. Pharm. Bull.*, **31**, 2281 (1983).
4. C. R. Yang et al., *Yaoxue Tongbao*, **20**, 337 (1985).

5. K. Ohtani et al., *Planta Med.*, **53**, 166 (1987).
6. Q. Du et al., *J. Chromatogr. A*, **1008**, 173 (2003).
7. M. Yoshikawa et al., *J. Nat. Prod.*, **66**, 922 (2003).
8. C. R. Yang et al., *Zhongyao Tongbao*, **10**, 33 (1985).
9. J. B. Wan et al., *J. Sep. Sci.*, **29**, 2190 (2006).
10. S. Zhang, *Zhongchengyao*, **11**, 20 (1989).
11. B. H. Zhang, *Zhongcaoyao*, **15**, 34 (1984).
12. G. Z. Chen et al., *Chin. J. Integr. Trad. West. Med.*, **4**, 540 (1984).
13. H. X. Quan et al., *Henan Zhongyi*, **10**, 41 (1990).
14. H. X. Sun et al., *Vaccine*, **22**, 3882 (2004).
15. H. X. Sun et al., *Acta Pharmacol. Sin.*, **24**, 1150 (2003).
16. L. Shi et al., *Acta Pharmacol. Sin.*, **11**, 29 (1990).
17. S. Liu and J. X. Chen, *Acta Pharmacol. Sin.*, **5**, 100 (1984).
18. B. Y. Gao et al., *Yaoxue Xuebao*, **27**, 641 (1992).
19. K. Yanai et al., *Biosci. Biotechnol. Biochem.*, **70**, 2501 (2006).
20. J. D. Wang and J. X. Chen, *Acta Pharmacol. Sin.*, **5**, 181 (1984).
21. J. X. Wu and J. X. Chen, *Acta Pharmacol. Sin.*, **9**, 147 (1988).
22. Y. P. Zhou and W. H. Liu, *Zhongcaoyao*, **19**, 25 (1988).
23. L. X. Li et al., *Acta Pharmacol. Sin.*, **9**, 52 (1988).
24. X. Li et al., *Acta Pharmacol. Sin.*, **11**, 26 (1990).
25. T. B. Ng et al., *J. Ethnopharmacol.*, **93**, 285 (2004).
26. Z. G. Xiong et al., *Acta Pharmacol. Sin.*, **10**, 122 (1989).
27. J. M. Wang et al., *Zhongguo Yiyao Xuebao*, **4**, 29 (1989).
28. H. Sun et al., *Chem. Biodivers.*, **2**, 510 (2005).
29. Y. Zhu et al., *Planta Med.*, **72**, 1193 (2006).
30. Q. S. Chen et al., *Zhongyao Tongbao*, **12**, 45 (1987).
31. E. B. Dong et al., *Zhongcaoyao*, **21**, 26 (1990).
32. J. D. Wang and J. X. Chen, *Acta Pharmacol. Sin.*, **5**, 50 (1984).
33. U. H. Jin et al., *Phytother. Res.*, (2006).
34. Y. Wang et al., *Burns*, **32**, 846 (2006).
35. H. S. Zhang and S. Q. Wang, *Vascul. Pharmacol.*, **44**, 224 (2006).
36. A. Rhule et al., *J. Ethnopharmacol.*, **106**, 121 (2006).
37. W. H. Park et al., *Life Sci.*, **76**, 1675 (2005).
38. C. F. Lin et al., *Phytother. Res.*, **17**, 1119 (2003).
39. P. Chan et al., *Acta Pharmacol. Sin.*, **23**, 1157 (2002).
40. R. Y. Chan et al., *J. Clin. Endocrinol. Metab.*, **87**, 3691 (2002).
41. T. B. Ng, *J. Pharm. Pharmacol.*, **58**, 1007 (2006).
42. Y. X. Ling, *Shanghai Zhongyiyao Zazhi*, 22 (1991).

TRAGACANTH

Source: *Astragalus* spp. (especially *A. gummifer* Labill.) (Family Leguminosae or Fabaceae).

Common/vernacular names: Gum tragacanth.

GENERAL DESCRIPTION

Tragacanth is the dried gummy exudate from various species of *Astragalus* (e.g., *A. gummifer*). They are low thorny shrubs, up to about 1 m high, native to the mountainous regions of the Middle East (especially Iran,

Turkey, and Syria). Tragacanth is obtained by tapping the branches and tap roots (near the ground surface) whereby the gum exudes, which after drying becomes horny and is collected. The word tragacanth is derived from Greek, meaning goat's horn, probably describing the appearance and texture of the gum. Iran is the major producer; it also produces the best quality product. Tragacanth occurs in two main forms, ribbons and flakes; ribbons are considered to be of superior quality (FURIA; GLICKSMAN).[1]

Tragacanth partly dissolves and partly swells in water to give a viscous colloidal solution (sol); the maximum viscosity is only attained after 24 h at room temperature or after heating for 8 h at 40°C or 2 h at 50°C. The viscosity of its solutions is generally regarded as the highest among all the plant gums (see *guar*, *karaya*, and *locust bean*). A 1% solution has a slightly acidic pH (5.1–5.9). Its solutions are stable under acidic conditions (down to pH 2), with maximum viscosity at pH 5; lowering or increasing its pH beyond 5 would decrease its viscosity.[2] Solutions are also heat stable.

Like karaya gum, tragacanth forms heavy, thick gel-like pastes at high concentrations (e.g., 2–4%).

Solutions or gels of tragacanth are especially susceptible to microbial degradation, sometimes even in the presence of a preservative. The most effective preservative system for tragacanth gels or sols at neutral and acidic pHs has been reported to be a mixture of parabens (MARTINDALE; WHISTLER AND BEMILLER).

Tragacanth is compatible with other polysaccharides and with proteins.

CHEMICAL COMPOSITION

Tragacanth is reported to contain 20–30% of a water-soluble fraction called tragacanthin (consisting of tragacanthic acid and an arabinogalactan) and 60–70% of a water-insoluble fraction called bassorin as well as 1–3% starch, 1–4% cellulose; 3% ash, small amounts of invert sugar, 2–3% of a volatile acid (probably acetic acid), and about 15% water. Tragacanthic acid is composed of 40% D-galacturonic acid, 40% D-xylose, 10% L-fucose, 4% D-galactose, and three aldobiuronic acids. Arabinogalactan consists of 75% L-arabinose, 12% D-galactose, 3% D-galacturonic acid, and small amounts of L-rhamnose. Bassorin is reported to contain about 5% methoxyl groups (LIST AND HÖRHAMMER).[3,4]

The structures and molecular weights of bassorin and tragacanthic acid have not been reported, but the molecular weight of tragacanth itself has been reported to be 310,000 and 840,000 (FURIA; GLICKSMAN; WHISTLER AND BEMILLER). The structure of the arabinogalactan component has recently been determined by NMR and MS and was shown to contain maltose oligosaccharides in addition to the complex branched polymer of arabinose, glucose, galactose, and galacturonic acid.[5] A nitrogen content of about 0.17–0.58% has been attributed to the presence of a proteinaceous fraction rich in hydroxyproline, praline, serine, and valine.[6]

Tragacanthin dissolves in water to form a sol, while bassorin swells to form a thick gel.

PHARMACOLOGY AND BIOLOGICAL ACTIVITIES

Tragacanth, as a component (5%) of a cholesterol-rich diet fed to rats for 4 weeks, significantly lowered plasma cholesterol and LDL in comparison to a cholesterol-rich diet devoid of tragacanth.[7] A similar effect was observed in four out of five male volunteers who administered 9.9 g tragacanth per day for 21 days. Although their serum cholesterol levels did not change, their fecal fat concentration significantly increased.[8]

The application of tragacanth solution to hydroxyapatite *in vitro* and as a mouth wash to humans significantly reduced the adhesion and dental biofilm development of *Streptococcus mutans* to hydrxoyapatite and to teeth, respectively.[9]

Tragacanth has also been reported to have strong inhibitory activity on cancer cells (MORTON 3).

TOXICOLOGY

Tragacanth preparations contaminated with enterobacteria have been reported to cause death of fetuses when administered intraperitoneally to pregnant mice.[10] However, non-contaminated tragacanth has been found to be safe at doses up to ca. 10 g/day.[8,11]

USES

Tragacanth has been in use since ancient times. Most of its uses are based on its emulsifying, thickening, and suspending abilities as well as its stability to acid and heat.

Medicinal, Pharmaceutical, and Cosmetic. Extensively used in vaginal jellies and creams, in emulsions (e.g., cod liver oil), low-calorie syrups and elixirs, and as binding agent or demulcent in tablets and lozenges.

It is also used in toothpastes and hand lotions, among others.

Food. Tragacanth is used extensively in salad dressings (due to its acid stability), flavor formulations (emulsions), ice cream, and in confectionery. Other food products in which it is used include nonalcoholic beverages, baked goods, gelatins and puddings, meat and meat products, condiments and relishes, processed fruits, processed vegetables, gravies, and others. Highest average maximum use level reported is about 1.3% (13,015 ppm) in fats and oils (e.g., salad dressings).

COMMERCIAL PREPARATIONS

Ribbon, flake, and powder in numerous grades; tragacanth is official in N.F. and F.C.C. Grades meeting N.F. or F.C.C. standards vary considerably in the viscosity of their sols.

Regulatory Status. Has been affirmed as GRAS (§184.1351).

REFERENCES

See the General References for FURIA; GLICKSMAN; LAWRENCE; LIST AND HÖRHAMMER; MARTINDALE; MORTON 3; UPHOF; WHISTLER AND BEMILLER.

1. H. S. Gentry, *Econ. Bot.*, **11**, 40 (1957).
2. T. W. Schwarz et al., *J. Am. Pharm. Assoc., Sci. Ed.*, **47**, 695 (1958).
3. G. O. Aspinall and J. Baillie, *J. Chem. Soc.*, 1702 (1963).
4. G. O. Aspinall and J. Baillie, *J. Chem. Soc.*, 1086 (1967).
5. C. A. Tischer et al., *Carbohydr. Res.*, **337**, 1647 (2002).
6. D. M. Anderson et al., *Food Addit. Contam.*, **2**, 231 (1985).
7. S. Amer et al., *Pak. J. Pharm. Sci.*, **12**, 33 (1999).
8. M. A. Eastwood et al., *Toxicol. Lett.*, **21**, 73 (1984).
9. A. Shimotoyodome et al., *Biofouling*, **22**, 261 (2006).
10. H. Frohberg et al., *Arch. Toxikol.*, **25**, 268 (1969).
11. D. M. Anderson, *Food Addit. Contam.*, **6**, 1 (1989).

TURMERIC

Source: *Curcuma longa* L. (syn. *C. domestica* Val. and *C. domestica* Loir.) (Family Zingiberaceae).

Common/vernacular names: Curcuma; Indian saffron.

GENERAL DESCRIPTION

A perennial herb of the ginger family with a thick rhizome from which arise large, oblong, and long-petioled leaves; up to about 1 m high; flowers in spikes; peduncle ca. 15 cm long, concealed by the sheath, flowering bracts pale green;[1] native to southern Asia; extensively cultivated in India; China, Indonesia, and other tropical countries (e.g., Jamaica and Haiti). Part used is the cured (boiled, cleaned, and sun-dried) and polished rhizome. India is the major producer of turmeric (up to 94% of annual world production).[2,3]

CHEMICAL COMPOSITION

Contains 0.3–7.2% (usually 4–5%) of an orange-yellow volatile oil that is composed mainly of turmerone (ca. 60%), ar-turmerone, α-atlantone, γ-atlantone, and zingiberene (25%), with minor amounts of 1,8-cineole, α-phellandrene, d-sabinene, borneol, and dehydroturmerone, among others;[4-6] yellow coloring matter including 0.3–5.4% curcumin, monodesmethoxycurcumin, and didesmethoxycurcumin;[4-6] p-coumaroylferuloylmethane and di-p-coumaroylmethane;[7] sugars (28% glucose, 12% fructose, and ca. 1% arabinose); fixed oil; protein (ca. 8%); minerals (especially high in potassium); vitamins (especially C); resin; and others (JIANGSU; LIST AND HÖRHAMMER; MARSH).[4]

Turmeric and its water-, alcohol-, and ether-soluble fractions have been reported to have antioxidative activities;[8] curcumin is mostly responsible for these activities.[9-11]

PHARMACOLOGY AND BIOLOGICAL ACTIVITIES

Curcuma extract, volatile oil, and its curcumin components have *in vitro* and *in vivo* anti-inflammatory activity that may be due to inhibition of eicosanoid (leukotrienes/thromboxanes) biosynthesis.[12-16] A fraction of curcuma oil (b.p: 80–110 °C) has been demonstrated to have anti-inflammatory and antiarthritic activities in rats.[17] An essential oil-depleted extract is effective against experimental rheumatoid arthritis. Involved targets include NF-κB, chemokine, COX-2, and others.[18,19] Curcumin has also been reported to exhibit antiedemic effects in rats (MARTINDALE).

Antioxidant activity of curcuma, through free radical scavenging and inhibition of lipid peroxidation, has been demonstrated.[15,20] This may account for the efficacy of curcuma as a hepatoprotective, cardioprotective, and antigenotoxic agent.[21-24] *In vitro* protection of mouse liver and cultured rat hepatocytes from injury induced by carbon tetrachloride and galactosamine (curcumin, p-coumaroylferuloylmethane, and di-p-coumaroylmethane) has been reported earlier.[7]

Turmeric extracts have recently exhibited efficacy in managing type-2 diabetes hyperglycemia[25,26] and reducing the side effects (nephrotoxicity and cataract) of induced hyperglycemia[27,28] by targeting the peroxisome proliferator-activated receptor (PPAR) and due to their antioxidant activity, respectively. Curcumin was also able to reduce hyperlipidemia in diabetic rats by increasing cholesterol catabolism.[29]

Healing of skin excision wounds as well as peptic ulcers have recently been reported for turmeric and curcumin.[30-33]

Other activities of turmeric and its derivatives include choleretic in dogs (aqueous extracts); hypotensive in dogs (alcohol extract); antibacterial (curcumin, volatile oil, etc.);[34] insecticidal against houseflies (petroleum ether extracts);[35] antidepressant MAO inhibition (aqueous extract);[36] modulation of multidrug-resistant proteins (curcuminoids);[37,38] and others (JIANGSU; LIST AND HÖRHAMMER).

Choleretic action of the essential oil is attributed to tolmethyl carbinol.[3] Extracts have antispasmodic activity on isolated guinea pig ileum; lower serum cholesterol and triglyceride levels in mice; anticancer activity against Dalton's lymphoma cells in the Chinese hamster; plus anticoagulant, antifungal, and antimutagenic, among others.[12] The pharmacological activities of turmeric have been reviewed.[39,40]

TOXICOLOGY

One study found no visible signs of acute toxicity with extracts, though at 3 g/kg CNS stimulation was observed; in chronic toxicity feeding, an increase in weights of the heart and lung was observed, though general visceral condition was normal; increased sperm motility (without increase in sperm count) suggested a possible androgenic effect.[41]

USES

Medicinal, Pharmaceutical, and Cosmetic. Cut or ground drug (1.5–3 g) or other preparations are used in European phytomedicine for dyspeptic conditions. Use is contraindicated in obstruction of gall passages; used only under medical advice for gallstones.[42]

The essential oil (called curcuma oil) is used to a limited extent in certain perfumes (especially oriental types).

Food. Turmeric is a major ingredient of curry powder and is also used in prepared mustard. Turmeric and turmeric oleoresin are used extensively both for their color and flavor in many food products, including baked goods, meat and meat products, condiments and relishes (especially pickles), fats and oils, egg products, soups, and gravies, among others. Highest average maximum use levels are 22% and about 0.883% (8834 ppm) reported for turmeric in seasonings and flavorings and in condiments and relishes, respectively.

Dietary Supplements/Health Foods. Turmeric is used as an antioxidant in capsules, tablets; flavoring in tea (LEUNG).

Traditional Medicine. Reportedly used in Chinese medicine to treat numerous conditions including flatulence, liver problems, menstrual difficulties, bloody urine, hemorrhage, toothache, bruises and sores, chest pain, and colic, usually decocted with other drugs. It is also used as a poultice to relieve pain and itching of sores and ringworms (JIANGSU).

Root tuber of *Curcuma* spp. is used in China for treating epilepsy and unconsciousness due to febrile diseases (*re bing shen hun*) (CHP; JIANGSU).

In Polynesia, the root is used for asthma, skin diseases, constipation, and religious rituals.[43]

COMMERCIAL PREPARATIONS

Crude and oleoresin. Crude was formerly official in U.S:P. Strength (see **glossary**) of oleoresin is often expressed in terms of curcumin content.

Regulatory Status. GRAS (§182.10 and §182.20); turmeric and turmeric oleoresin have also been approved as food colorants exempt from certification (§73.600 and §73.615). The rhizome is subject of a positive German therapeutic monograph for treatment of dyspeptic conditions.[42]

REFERENCES

See the General References for ARCTANDER; BAILEY 1; BLUMENTHAL 1; CHP; DER MARDEROSIAN AND BEUTLER; FEMA; GUPTA;; JIXIAN; JIANGSU; LEUNG; LIST AND HÖRHAMMER; MARTINDALE; MCGUFFIN 1 & 2; ROSENGARTEN; STAHL; TERRELL; UPHOF.

1. K. R. Kirtikar and B. D. Basu, *Indian Medicinal Plants*, International book distributors, Dehradun, India, 1999, p. 2423.
2. M. Ilays, *Econ. Bot.*, **32**, 238 (1988).
3. G. S. Randhawa and R. K. Mahey in L. E. Craker and J. E. Simon, eds., *Herbs, Spices, and Medicinal Plants: Recent Advances in Botany, Horticulture, and Pharmacology*, Vol. 3, Oryx Press, Phoenix, 1988, p. 71.

4. A. Khalique and M. N. Amin, *Sci. Res. (Dacca, Pakistan)*, **4**, 193 (1967).
5. C. R. Mitra, *Riechst., Aromen, Körperpflegem.*, **25**, 15 (1975).
6. N. Krishnamurthy et al., *Trop. Sci.*, **18**, 37 (1976).
7. Y. Kiso et al., *Planta Med.*, **49**, 185 (1983).
8. F. Hirahara et al., *Eiyogaku Zasshi*, **32**, 1 (1974).
9. S. B. Xu et al., *Zhongcaoyao*, **22**, 140 (1991).
10. S. B. Xu et al., *Zhongcaoyao*, **22**, 264 (1991).
11. B. L. Zhao et al., *Cell Biophys.*, **14**, 175 (1989).
12. H. P. T. Ammon and M. A. Wahl, *Planta Med.*, **57**, 1 (1991).
13. H. P. T. Ammon et al., *Planta Med.*, **58**, 226 (1992).
14. N. Chainani-Wu, *J. Altern. Complement. Med.*, **9**, 161 (2003).
15. R. S. Ramsewak et al., *Phytomedicine*, **7**, 303 (2000).
16. K. C. Srivastava et al., *Prostaglandins Leukot. Essent. Fatty Acids*, **52**, 223 (1995).
17. D. Chandra and S. S. Gupta, *Indian J. Med. Res.*, **60**, 138 (1972).
18. J. L. Funk et al., *Arthritis Rheum.*, **54**, 3452 (2006).
19. J. L. Funk et al., *J. Nat. Prod.*, **69**, 351 (2006).
20. J. C. Tilak et al., *Phytother. Res.*, **18**, 798 (2004).
21. I. M. El-Ashmawy et al., *Basic Clin. Pharmacol. Toxicol.*, **98**, 32 (2006).
22. M. Miyakoshi et al., *Biofactors*, **21**, 167 (2004).
23. I. Mohanty et al., *Life Sci.*, **75**, 1701 (2004).
24. U. R. Deshpande et al., *Indian J. Exp. Biol.*, **36**, 573 (1998).
25. M. Kuroda et al., *Biol. Pharm. Bull.*, **28**, 937 (2005).
26. T. Nishiyama et al., *J. Agric. Food Chem.*, **53**, 959 (2005).
27. S. Sharma et al., *Clin. Exp. Pharmacol. Physiol.*, **33**, 940 (2006).
28. P. Suryanarayana et al., *Invest. Ophthalmol. Vis. Sci.*, **46**, 2092 (2005).
29. P. S. Babu and K. Srinivasan, *Mol. Cell. Biochem.*, **166**, 169 (1997).
30. D. C. Kim et al., *Biol. Pharm. Bull.*, **28**, 2220 (2005).
31. S. Kundu et al., *Int. J. Low. Extrem. Wounds*, **4**, 205 (2005).
32. G. C. Jagetia and G. K. Rajanikant, *J. Surg. Res.*, **120**, 127 (2004).
33. C. Prucksunand et al., *Southeast Asian J. Trop. Med. Public Health*, **32**, 208 (2001).
34. J. Lutomski et al., *Planta Med.*, **26**, 9 (1974).
35. R. S. Dixit and S. L. Perti, *Bull. Reg. Res. Lab. Jammu, India*, **1**, 169 (1963).
36. Z. F. Yu et al., *J. Ethnopharmacol.*, **83**, 161 (2002).
37. W. Chearwae et al., *Cancer Chemother. Pharmacol.*, **57**, 376 (2006).
38. W. Chearwae et al., *Biochem. Pharmacol.*, **68**, 2043 (2004).
39. C. C. Araujo and L. L. Leon, *Mem. Inst. Oswaldo Cruz*, **96**, 723 (2001).
40. *Altern. Med. Rev.*, **6**(Suppl.), S62 (2001).
41. S. Qureshi et al., *Planta Med.*, **58**, 124 (1992).
42. Monograph, *Curcumae longae rhizoma, Bundesanzeiger*, no. 223 (November 30, 1985); revised (September 1, 1990).
43. W. McClatchey, *Econ. Bot.*, **47**, 291 (1993).

TURPENTINE (AND ROSIN)

Source: *Longleaf pine* Pinus palustris Mill; *slash pine* P. elliottii Engelm.; and other *Pinus* spp.

Common/vernacular names: Gum terpentine, gum thus, turpentine oil, and turpentine balsam.

GENERAL DESCRIPTION

The term "turpentine" is rather loosely used to describe either the oleoresin obtained from the longleaf pine (*Pinus palustris*), slash pine (*P. elliottii*), and other *Pinus* species that yield exclusively terpene oils, or the essential oil obtained from the above oleoresin. The oleoresin is commonly called gum turpentine or turpentine balsam while the essential oil is called turpentine oil; both the oleoresin and the essential oil are also called simply turpentine. To avoid confusion, gum turpentine (though not a true gum; see **glossary**) is here reserved only for the oleoresin, while turpentine or turpentine oil (spirits of turpentine) is used for the essential oil. Thus, gum turpentine on steam distillation yields turpentine (turpentine oil) and rosin (a terpenic resin), also known as colophony.

Turpentine and rosin are also produced by solvent extraction of heartwood chips of pine stumps, which are by-products of the lumber industry, and as by-products of the paper (sulfate or kraft pulping) industry. The last source is reported to account for the largest volumes of turpentine and rosin produced in the United States. These products derived from pines and other resinous conifers are commonly called naval stores.

Turpentine is the largest (in volume) essential oil in the world. Its current biggest producer is the United States; other major producing countries include New Zealand, China, Mexico, Portugal, the former U.S.S.R., and the Scandinavian countries.

CHEMICAL COMPOSITION

On account of their widely different sources (plant, geography, method of manufacture, etc.) turpentine and rosin often vary considerably in their relative composition.

Turpentine contains mostly monoterpene hydrocarbons, the major ones being α-pinene (45–95%), β-pinene (0–35%), and 3-carene (20–60%).[1–4] Others present in lesser amounts include camphene, dipentene, terpinolene, β-myrcene, β-phellandrene, and *p*-cymene (LIST AND HÖRHAMMER).[1,5,6]

Rosin contains mainly diterpene resin acids including abietic acid (22–50%), dehydroabietic acid (6–30%), palustric acid (10–25%), neoabietic acid (4–20%), isopimaric acid (10–17%), and pimaric acid (4–6%).[1] Others present include dehydropimaric acid, levopimaric acid, and sandaracopimaric acid.[1,5,6] It also contains small amounts of diterpene alcohols and aldehydes, sterols (mainly sitosterol), and phenolic compounds.[1]

PHARMACOLOGY AND BIOLOGICAL ACTIVITIES

Turpentine has rubefacient and counterirritant properties.

The turpentine exudate of *P. nigra* has been shown to possess concentration-dependent antioxidant activity in a number of *in vitro* models. The activity was stronger than that of a-tocopherol. In the same study, the exudate demonstrated an analgesic effect comparable to that of the metamizol reference control.[7]

Turpentine has contact allergenic activities that are mainly due to the pinenes and 3-carene as well as dipentene (MARTINDALE; MORTON 3).[8–12] It also has antimicrobial activities that are reportedly due to α-pinene, 3-carene, and dipentene (LEWIS AND ELVIN-LEWIS).[13–15]

Turpentine oil has been reported to promote tumor development on rabbit but not mouse skin.[16,17] Inhalation and ingestion of turepentine oil resulted in a serious case of extensive lung tissue necrosis in a toddler.[18] The lung of a 20-year-old male developed chemical pneumonitis and a bronchopleural fistula after inhalation of turpentine oil.[19] Other toxic effects of turpentine include erythema, urticaria, headache, insomnia, coughing, vomiting, hematuria, albuminuria, and coma (MARTINDALE).[8] Turpentine has caused fatalities in children ingesting as little as 15 mL.[20]

Turpentine is used in experimental animal models to induce systemic inflammatory immune response.[21,22]

TOXICOLOGY

Rosin also has irritant properties on some individuals (MARTINDALE).

USES

Medicinal, Pharmaceutical, and Cosmetic. Turpentine is used as an ingredient in many ointments, liniments, and lotions for treating minor aches and pains as well as colds. Rosin is an ingredient in some soaps and ointments; it is also used as a fixative in perfumes.

Many recent studies have focused on the use of rosin-based polymers for drug delivery in the form of enteric coating, cream bases, and nanoparticles.[23–26]

Turpentine oil has been used in endodontic retreatment (repeated root canal procedure) to soften or dissolve the rubbery latex (gutta-percha) in the root canal cavity.[27]

Food. Steam-distilled turpentine oil is reportedly used as a flavor component in most major food products, including alcoholic and nonalcoholic beverages, frozen dairy desserts, candy, baked goods, gelatins and puddings, meat and meat products, and condiments and relishes. Highest average maximum use level is about 0.002% (20.6 ppm) in baked goods.

Traditional Medicine. Turpentine is mainly used as a counter-irritant and rubefacient in treating rheumatism and aching muscles.

In Chinese medicine, gum turpentine and rosin (mainly from *Pinus tabulaeformis* Carr., *P. massoniana* Lamb., and *P. yunnanensis* Franch.) have been used for centuries in treating rheumatism, stiff joints, toothache, boils, and sores. Furthermore rosin is used in treating ringworms, chronic bronchitis, and neurogenic dermatitis, among others. They are used both internally and externally.

Others. The major use of turpentine oil is as a solvent (e.g., paints) and as a starting material for the synthesis of useful chemicals such as camphor, menthol, terpin hydrate (an expectorant), α-terpineol and other fragrance compounds, and resins (adhesives, chewing gum, etc.), among others.[1]

COMMERCIAL PREPARATIONS

Rosin, turpentine oil, and rectified turpentine oil; all were formerly official in N.F.

Regulatory Status. Turpentine oil has been approved for food use, and rosin has been approved for use in alcoholic beverages only (§172.510).

REFERENCES

See the General References for ARCTANDER; CLAUS; FEMA; GUENTHER; JIANGSU; LIST AND HÖRHAMMER; MORTON 3; NANJING; USDA.

1. D. F. Zinkel, *Chemtech*, **5**, 235 (1975).
2. Y. G. Drochnev, *Lesokhim. Podsochka*, **3**, 7 (1977).
3. W. D. Fordham in L. W. Coddet al., eds., *Chemical Technology: An Encyclopedic Treatment*, Vol. **5**, Barnes & Noble, New York, 1972, p. 1.

4. J. Fousseteau et al., *Bull. Soc. Fr. Dermatol. Syphiligr.*, **77**, 415 (1970).
5. I. I. Bardyshev et al., *Isv. Vyssh. Ucheb. Zaved., Les Zh.*, **12**, 161 (1969).
6. I. I. Bardyshev et al., *Vestsi Akad. Nayuk Belarus. SSR, Ser Khim. Navuk*, **5**, 123 (1971).
7. I. Gulcin et al., *J. Ethnopharmacol.*, **86**, 51 (2003).
8. P. Mikhailov et al., *Allerg. Asthma*, **16**, 201 (1970).
9. P. Mikhailov and N. Berova, *Dermatol Veenerol. (Sofia)*, **19**, 20 (1970).
10. M. Pambor, *Dermatol Monats.*, **162**, 992 (1976).
11. D. L. J. Opdyke, *Food Cosmet. Toxicol.*, **12**, 703 (1974).
12. E. Rudzki et al., *Contact Dermatitis*, **24**, 317 (1991).
13. B. N. Uzdennikov, *Nauch. Tr. Tyumen. Sel. Khoz. Inst.*, **7**, 116 (1970).
14. S. V. Iliev and G. Y. Papanov, *Khim. Farm. Zh.*, **3**, 35 (1969).
15. B. N. Uzdennikov, *Tr. Tyumenskogo Sel'skokhoz. Inst.*, **7**, 120 (1970).
16. F. Homburger and E. Boger, *Cancer Res.*, **28**, 2372 (1968).
17. F. J. C. Roe and W. E. H. Field, *Food Cosmet. Toxicol.*, **3**, 311 (1965).
18. A. J. Khan et al., *Pediatr. Emerg. Care*, **22**, 355 (2006).
19. A. Rodricks et al., *J. Assoc. Physicians India*, **51**, 729 (2003).
20. B. R. Olin, *Lawrence Rev. Nat. Prod.*, (1993).
21. M. A. Elhija et al., *Am. J. Reprod. Immunol.*, **55**, 136 (2006).
22. C. Pous et al., *Inflammation*, **156**, 197 (1992).
23. C. M. Lee et al., *Biotechnol. Lett.*, **27**, 1487 (2005).
24. P. M. Satturwar et al., *Int. J. Pharm.*, **270**, 27 (2004).
25. S. V. Fulzele et al., *AAPS PharmSciTech.*, **3**, E31 (2002).
26. V. T. Dhanorkar et al., *J. Cosmet. Sci.*, **53**, 199 (2002).
27. G. J. Kaplowitz, *Int. Endod. J.*, **29**, 93 (1996).

UVA URSI

Source: *Arctostaphylos uva-ursi* (L.) Spreng. (syn. *Arbutus uva-ursi* L.) and its varieties *coactitis* and *adenotricha* Fern. et Macbr. (Family Ericaceae).

Common/vernacular names: Bearberry, common bearberry, beargrape, hogberry, and rockberry.

GENERAL DESCRIPTION

Trailing evergreen shrub rooting along the branches, often forming dense mats; with decumbent, much branched, irregular stems; small leathery, obovate to spatulate leaves; up to about 15 cm high; fruits berries, reddish; native to the temperate regions of the northern hemisphere (e.g., Europe, northern United States, Canada, and Asia). Part used is the dried leaf. The major producing country is Spain.

CHEMICAL COMPOSITION

Contains as its active principles 5–18% (usually 7–9%) arbutin (hydroquinone β-glucoside) and lesser amounts of methylarbutin, with concentrations varying according to ages of leaves, season, localities, and other factors;[1–3] a third glucoside, piceoside, has also been isolated.[4] Other constituents include flavonoids (quercetin mono- and diglucosides, myricetin),[5] allantoin,[6] tannins (6–27.5%) of the gallic and ellagic acid types,[2] ursolic acid (0.4–0.75%), phenolic acids (e.g., gallic, ellagic, and quinic acids), an iridoid glycoside (monotropein), uvaol, trace of volatile oil, resin, and others (ESCOP 3; LIST AND HÖRHAMMER; STAHL).[7]

PHARMACOLOGY AND BIOLOGICAL ACTIVITIES

Uva ursi is reported to have diuretic and astringent as well as urinary antiseptic properties (MARTINDALE).

Arbutin undergoes hydrolysis to yield hydroquinone (in the intestines), which has urinary disinfectant activities. *In vitro* results suggest that antibacterial activity is due either to hydroquinone sulfate ester or to the free hydroquinone. Arbutin is reported to be an effective urinary disinfectant if taken in large doses and if the urine is alkaline (LIST AND HÖRHAMMER).[8] The disinfectant effect may be useful in the treatment and prevention of kidney stone formation as was demonstrated in female rats.[9] The aqueous extract of bearberry also had a diuretic effect on rats.[10] However, caution should be exercised when using large doses of arbutin as hydroquinone is toxic. Toxic symptoms include tinnitus (ringing in the ear), vomiting, delirium, convulsions, and collapse; death may result (ESCOP 3; GOSSELIN; MERCK).

The 50% ethanolic leaf extracts of *A. uva-ursi* and other species inhibited melanin biosynthesis and displayed SOD-like activity, which may be useful in skin-whitening preparations.[11]

Bearberry extract exhibited strong antioxidant and genoprotective effects on U937 cells subjected to various oxidative stress conditions, such as exposure to hydrogen peroxide and etoposide.[12]

TOXICOLOGY

Use is contraindicated in kidney disorders and irritated digestive conditions. Crude drug preparations may induce nausea and vomiting (ESCOP 3). Ingestion of uva ursi for three years resulted in reduced visual acuity and retinal toxicity (bull's-eye maculopathy) in a 56-year-old woman possibly due to inhibition of melanin synthesis.[13]

Uva ursi has been included, among other herbs, in a recent review about urinary antiseptic herbs and botanical medicines for the urinary tract.[14]

USES

Medicinal, Pharmaceutical, and Cosmetic. Crude and extracts are used quite extensively as components in certain diuretic as well as laxative preparations. Also used as a urinary disinfectant, especially in Europe.[8] Reported dosage is 1.5–2.5 g of crude drug, infusion, or cold aqueous extract, containing not less than 6.0% of hydroquinone derivatives calculated as anhydrous arbutin. Treatment is limited to 7 days or less (ESCOP 3).

Dietary Supplements/Health Foods. Crude and extracts used in various capsule, tablet, and tea formulations with an intended diuretic or urinary antiseptic effect; also as tea flavoring due to tannin content (FOSTER AND DUKE).

Traditional Medicine. Reportedly used as a diuretic, astringent, and urinary antiseptic. Conditions for which it is used included chronic cystitis, nephritis, kidney stones, and bronchitis, usually in the form of a tea or tincture (LUST).

COMMERCIAL PREPARATIONS

Crude and extracts; crude and fluid extract were formerly official in N.F. and U.S.P. Strengths (see *glossary*) of extracts are generally expressed in weight-to-weight ratios and sometimes in arbutin contents.

Regulatory Status. Undetermined in the United States. Subject of a positive German therapeutic monograph for urinary tract infections.[15]

REFERENCES

See the General References for APhA; APPLEQUIST; BAILEY 1; BARNES; BIANCHINI AND CORBETTA; BLUMENTHAL 1; BRUNETON; CLAUS; DER MARDEROSIAN AND BEUTLER; ESCOP 3; FOSTER AND DUKE; GRIEVE; LEWIS AND ELVIN-LEWIS; LIST AND HÖRHAMMER; LUST; STAHL; TERRELL; UPHOF; YOUNGKEN.

1. V. Moretti, *Boll. Soc. Ital. Farm. Osp.*, **23**, 207 (1977).
2. A. A. Makarov, *Uch. Zap. Yakutsk. Gos. Univ.*, **18**, 41 (1971).
3. E. Rubine, *Nauch. Tr. Irkutsk. Gos. Med. Inst.*, **113**, 9 (1971).
4. G. A. Karikas, et al., *Planta Med.*, **53**, 307 (1987).
5. K. E. Denford, *Experientia*, **29**, 939 (1973).
6. E. Constantinescu et al., *Herba Hung.*, **8**, 101 (1969).
7. T. Kawai et al., *Osaka Kogyo Daigaku Kiyo, Rikohen*, **19**, 1 (1974).
8. D. Frohne, *Planta Med.*, **18**, 1 (1970).
9. F. Grases et al., *Int. Urol. Nephrol.*, **26**, 507 (1994).
10. D. Beaux et al., *Phytother. Res.*, **13**, 222 (1999).
11. H. Matsuda et al., *Biol. Pharm. Bull.*, **19**, 153 (1996).
12. N. M. O'Brien et al., *J. Med. Food*, **9**, 187 (2006).
13. L. Wang and L. V. Del Priore, *Am. J. Ophthalmol.*, **137**, 1135 (2004).
14. E. Yarnell, *World J. Urol.*, **20**, 285 (2002).
15. Monograph *Urvae ursi folium*, *Bundesanzeiger*, no. 228 (December 5, 1984).

VALERIAN ROOT

Source: *Valeriana officinalis* L. and *V. jatamansii* Jones (syn. *V. wallichii* DC.) (Family Valerianaceae).

Common/vernacular names: Common valerian, Belgian valerian, all heal, fragrant valerian, garden valerian (*V. officinalis*); Indian valerian (*V. jatamansii*).

GENERAL DESCRIPTION

Common valerian (*V. officinalis*) is a perennial herb growing up to about 1.5 m high; with deeply dissected leaves each bearing 7 to 10 pairs of lance-shaped leaflets; stems erect, longitudinally grooved, and hollow; native to Eurasia and naturalized in North America.

Indian valerian (*V. jatamansii*) is a perennial herb with basal (lower) leaves long-petioled, large, and heart-shaped; stems green or with a purplish tint and fine hairy; up to 0.7 m high; native to Asia, especially the Himalayan region (e.g., India and southwestern China).

The genus *Valeriana* includes about 250 northern temperate species; also from South Africa and the Andes; 20 species indigenous to Europe; 16 species are found in the United States and Canada.[1] Parts used are the dried rhizomes and roots, commonly called valerian roots. Major producers of common valerian root include Belgium, France, the former U.S.S.R, and China, while India is the major producer of Indian valerian.

CHEMICAL COMPOSITION

Common valerian contains as its primary active constituents several iridoid compounds called valepotriates including valtrates (valtrate, valtrate isovaleroxyhydrin, acevaltrate, valechlorine, etc.), didrovaltrates (didrovaltrate, homodidrovaltrate, deoxydodidrovaltrate, homodeoxydodidrovaltrate, isovaleroxyhydroxydidrovaltrate, etc.), and isovaltrates (isovaltrate, 7-epideacetylisovaltrate, etc.);[2–7] valtrate and didrovaltrate are the major valepotriates.[2] In addition, it contains valerosidatum (an iridoid ester glycoside)[8] and a volatile oil (0.5–2%) consisting of many components including bornyl acetate and isovalerate (major compounds), caryophyllene, α- and β-pinenes, the sesquiterpenes valerenal, valerenic acid, acetylvalerenolic acid, valeranone and valerenal, β-ionone, eugenyl isovalerate, isoeugenyl isovalerate, patchouli alcohol, valerianol, borneol, camphene, β-bisabolene, ledol, isovaleric acid, and terpinolene, among others (JIANGSU).[9–17] $(-)$-3β,4β-Epoxyvalerenic acid was isolated as a new compound from *V. officinalis*.[18]

Common valerian also contains several alkaloids including actinidine, valerianine, valerine, and chatinine.[19–27]

Other constituents present in common valerian include the flavone 6-methylapigenin and the flavonoid glycosides hesperidin and linarin;[28,29] lignans and lignan glycosides (pinoresinols, massoniresinol, and olivil);[30] choline (ca. 3%), methyl-2-pyrrolyl ketone, chlorogenic acid, and caffeic acid,[21] β-sitosterol and glycosides of clionasterol;[6,31] tannins; gums; and others (JIANGSU).

Indian valerian is generally reported to contain similar constituents as common valerian including valepotriates, valerosidatum, and volatile oil.[8,22,23] It also contains 2″-*O*- and 3″-*O*-2-methylbutyryl esters of acetylated linarin (JIANGSU).[24]

Valepotriates are also present in aerial parts of Indian valerian, but they are absent in those of common valerian.[3,25,26]

PHARMACOLOGY AND BIOLOGICAL ACTIVITIES

As with the chemical studies, most biological work on valerian has been performed on common valerian.

Valerian is reputed for its CNS-depressant and sleep enhancing activities, and it is reported to have antispasmodic and equalizing (sedative in states of agitation and stimulant in fatigue) effects (JIANGSU; STAHL).[27] Many studies have been performed in animal models and in humans to investigate this effect. For example, in a double-blind study, an aqueous extract of common valerian root was found to decrease sleep latency in eight human subjects who had problem falling asleep.[32] Valerian extract has been reported to affect the sleep–wake cycle in rats at doses of 1–3 g/kg.[33] Valerian was also shown to be effective in long-term treatment of sleep difficulties in children with intellectual deficits.[34] Many other clinical trials have been performed on preparations containing valerian with overall results supporting its sleep enhancing effects. Other trials, however, reported that valerian had no effect on insomnia and/or anxiety.[35–37]

The valepotriates have been reported to be responsible for the CNS-depressant and antispasmodic effects in laboratory animals.[6,38,39] Valerenal, valerenic acid, and, to a lesser extent, valeranone have central depressive effect in mice.[12,40] Another study, however, reported that valepotriates, valerenic acid, valeranone, and the volatile oil of valerian were ineffective as CNS-depressant agents when tested individually even though the whole extract was active.[41] The direction of recent studies shifted toward investigating the mechanisms involved in the CNS activity of valerian. In addition to the established involvement of the GABA receptor,[42–48] valerian extracts were found to bind to benzodiazepine,[30,46] melatonin,[49] serotonin,[49,50] and adenosine receptors.[30,51–53] Bioassay-guided fractionation of a polar extract showed that a glycoside of the lignan olivil was a partial agonist at the adenosine A_1 receptor, while fractionation of a nonpolar extract showed that isovaltrate was an antagonist/inverse agonist at the A_1 receptor.[30,51] Valerian extracts may have potential in the treatment of morphine withdrawal as demonstrated in a mouse model of morphine dependence.[54] The possible benefits of valerian in sleep and anxiety disorders have been the subject of many recent reviews.[55–59]

Other activities of valerian include hypotension in experimental animals; antibacterial, especially against Gram-positive bacteria (due to the alkaloids); antidiuretic; protective against experimental liver necrosis;[60] and others (JIANGSU). Valtrate and didrovaltrate as well as baldrinal (a valtrate degradation product) also have antitumor activities against experimental tumors.[23] An ethanol extract of valerian is said to have antidandruff properties.[61] Valerian displayed a neuroprotective effect against amyloid-β peptide that may be of benefit in cases of neurodegenerative disorders associated with old age.[62] Valerenic acid, acetylvalerenolic acid, and valerenal, isolated from the ethyl acetate extract of *V. officinalis*, were recently found to inhibit NF-κB, which provides support for its use as an anti-inflammatory remedy in European traditional medicine.[9]

TOXICOLOGY

Valerian oil has been reported to be the least toxic among numerous common volatile oils tested orally in rats.[63] Common valerian preparations are considered safe despite the known *in vitro* cytotoxic activity of valepotriates, as these compounds degrade easily, are absent from most products, and are poorly absorbed via oral administration. Acute side effects are not common (ESCOP 1), but certain germ cell toxicity in mice (testicular chromosomal aberrations and spermatozoan abnormalities) have recently been reported.[64]

USES

Medicinal, Pharmaceutical, and Cosmetic. Crude, extracts, tinctures are used in certain sedative preparations, especially in Europe, at doses equivalent to 2–3 g of the drug, 1–3 times per day.[41,65]

Food. Extracts and the essential oil (produced by steam distillation) of common valerian are used as flavor components in most major food products, including alcoholic (liqueur, beer, etc.) and nonalcoholic (e.g., root beer) beverages, frozen dairy desserts, candy, baked goods, gelatins and puddings, and meat and meat products, among others. Highest average maximum use levels are about 0.01% of the extract (no type given) reported in alcoholic beverages (96.1 ppm) and baked goods (94.3 ppm), and about 0.002% (16.7%) reported for the oil in baked goods.

Dietary Supplements/Health Foods. Widely used in sleep aid and sedative formulations, alone or in combination with other herbs, in various dosage forms (teas, tincture, capsule, tablet, etc.).

Traditional Medicine. Common valerian root (fresh or dried) is used as antispasmodic, carminative, stomachic, and sedative. Conditions for which it is used include migraine, insomnia, hysteria, neurasthenia, fatigue, stomach cramps that cause vomiting, and other nervous conditions, usually as a tea or an infusion. It is also used externally to treat sores and pimples.

In Chinese medicine, both common and Indian valerian roots as well as those of *V. coreana* Briq., *V. stubendorfi* Kreyer ex Kom., *V. amurensis* P. Smirn. ex Kom., and *V. hardwickii* Wall. are similarly used. In addition, they are used in treating chronic backache, numbness due to rheumatic conditions, colds, menstrual difficulties, and bruises and sores, among others, generally as a decoction or alcoholic infusion.

COMMERCIAL PREPARATIONS

Crude, extracts, and oil; crude and fluid extract were formerly official in N.F. Strengths (see *glossary*) of extracts are expressed in weight-to-weight ratios, although certain manufacturers have their own in house bioassays.

Regulatory Status. Has been approved for food use (§172.510); only *V. officinalis* is listed. Subject of a positive German therapeutic monograph.

REFERENCES

See the General References for APPLEQUIST; ARCTANDER; BAILEY 1; BARRETT; BIANCHINI AND CORBETTA; BLUMENTHAL 1; BRUNETON; DER MARDEROSIAN AND BEUTLER; ESCOP 1; FEMA; FOSTER; FOSTER AND DUKE; GOSSELIN; GRIEVE; JIANGSU; LUST; MARTINDALE; MCGUFFIN 1 & 2; STAHL; TYLER 1; UPHOF.

1. S. Foster, *Valerian*, Botanical Series No. 312, American Botanical Council, Austin, TX, 1991.
2. S. Popov et al., *Phytochemistry*, **13**, 2815 (1974).
3. E. D. Funke and H. Friedrich, *Planta Med.*, **28**, 215 (1975).
4. N. Marekov et al., *Izv. Khim.*, **8**, 672 (1975).
5. J. H. Van Meer et al., *Pharm. Weekbl.*, **112**, 20 (1977).
6. V. D. Petkov et al., *Dokl. Bolg. Akad. Nauk*, **27**, 1007 (1974).
7. G. Verzarne-Petri et al., *Herba Hung.*, **15**, 79 (1976).
8. P. W. Thies, *Tetrahedron Lett.*, 2471 (1970).
9. N. J. Jacobo-Herrera et al., *Phytother. Res.*, **20**, 917 (2006).
10. H. Wagner et al., *Arzneim.-Forsch.*, **22**, 1204 (1972).
11. E. Lemberkovics et al., *Sci. Pharm.*, **45**, 281 (1977).
12. H. Hendricks et al., *Phytochemistry*, **16**, 1853 (1977).
13. H. Hoerster et al., *Phytochemistry*, **16**, 1070 (1977).
14. G. Ruecker and J. Tautges, *Phytochemistry*, **15**, 824 (1976).
15. G. Jommi et al., *Collect. Czech. Chem. Commun.*, **34**, 593 (1969).
16. E. Pethes et al., *Sci. Pharm.*, **43**, 173 (1975).
17. H. Hendricks et al., *Planta Med.*, **42**, 62 (1981).
18. H. R. Dharmaratne et al., *Planta Med.*, **68**, 661 (2002).
19. K. Torssell and K. Wahlberg, *Acta Chem. Scand.*, **21**, 53 (1967).
20. R. D. Johnson and G. R. Waller, *Phytochemistry*, **10**, 3334 (1971).

21. G. R. Szentpetery et al., *Pharmazie*, **18**, 816 (1963).
22. G. D. Joshi et al., *Perfum. Essent. Oil Records*, **59**, 187 (1968).
23. C. Bounthanh et al., *Planta Med.*, **41**, 21 (1981).
24. V. M. Chari et al., *Phytochemistry*, **16**, 1110 (1977).
25. J. Hoelzl and K. Jurcic, *Planta Med.*, **27**, 133 (1975).
26. E. D. Funke and H. Friedrich, *Phytochemistry*, **13**, 2023 (1974).
27. E. Cionga, *Pharmazie*, **16**, 43 (1961).
28. S. Fernandez et al., *Pharmacol. Biochem. Behav.*, **77**, 399 (2004).
29. M. Marder et al., *Pharmacol. Biochem. Behav.*, **75**, 537 (2003).
30. B. Schumacher et al., *J. Nat. Prod.*, **65**, 1479 (2002).
31. S. V. Pullela et al., *Planta Med.*, **71**, 960 (2005).
32. P. D. Leathwood and F. Chauffard, *Planta Med.*, **51**, 144 (1985).
33. K. Shinomiya et al., *Acta Med. Okayama*, **59**, 89 (2005).
34. A. J. Francis and R. J. Dempster, *Phytomedicine*, **9**, 273 (2002).
35. B. P. Jacobs et al., *Medicine (Baltimore)*, **84**, 197 (2005).
36. K. T. Hallam et al., *Hum. Psychopharmacol.*, **18**, 619 (2003).
37. J. R. Glass et al., *J. Clin. Psychopharmacol.*, **23**, 260 (2003).
38. P. Manolov and V. Petkov, *Farmatsiya (Sofia)*, **26**, 29 (1976).
39. K. W. Von Eickstedt and S. Rahman, *Arzeim.-Forsch.*, **19**, 316 (1969).
40. H. Hendricks et al., *Planta Med.*, **51**, 28 (1985).
41. J. Krieglstein and D. Grusla, *Dtsch. Apoth. Ztg.*, **40**, 2041 (1988).
42. T. Komori et al., *Chem. Senses*, **31**, 731 (2006).
43. J. G. Ortiz et al., *Phytother. Res.*, **20**, 794 (2006).
44. C. S. Yuan et al., *Anesth. Analg.*, **98**, 353 (2004).
45. A. M. Fields et al., *J. Altern. Complement. Med.*, **9**, 909 (2003).
46. J. G. Ortiz et al., *Neurochem. Res.*, **24**, 1373 (1999).
47. M. S. Santos et al., *Arch. Int. Pharmacodyn. Ther.*, **327**, 220 (1994).
48. C. Cavadas et al., *Arzneimittelforschung*, **45**, 753 (1995).
49. E. A. Abourashed et al., *Phytomedicine*, **11**, 633 (2004).
50. B. M. Dietz et al., *Brain Res. Mol. Brain Res.*, **138**, 191 (2005).
51. S. K. Lacher et al., *Biochem. Pharmacol.*, **73**, 248 (2007).
52. Z. Vissiennon et al., *Planta Med.*, **72**, 579 (2006).
53. C. E. Muller et al., *Life Sci.*, **71**, 1939 (2002).
54. M. Sharifzadeh et al., *Addict. Biol.*, **11**, 145 (2006).
55. S. Bent et al., *Am. J. Med.*, **119**, 1005 (2006).
56. L. S. Miyasaka et al., *Cochrane Database Syst. Rev.*, CD004515 (2006).
57. C. Stevinson and E. Ernst, *Sleep Med.*, **1**, 91 (2000).
58. F. Donath et al., *Pharmacopsychiatry*, **33**, 47 (2000).
59. P. J. Houghton, *J. Pharm. Pharmacol.*, **51**, 505 (1999).
60. M. A. Farooki, *Pakistan Med. Forum.*, **1**, 19 (1966).
61. T. Abe, *Jpn Kokai* 72 47, 664 (1972).
62. J. O. Malva et al., *Neurotox. Res.*, **6**, 131 (2004).
63. E. Skramlik, *Pharmazie*, **14**, 435 (1959).
64. A. A. Al-Majed et al., *Food Chem. Toxicol.*, **44**, 1830 (2006).
65. C. Hobbs, *HerbalGram*, **21**, 19 (1989).

VANILLA

Source: *Vanilla planifolia* Andr. [syn. *V. fragrans* (Salisb.) Ames] and ***V. tahitensis*** J. W. Moore (Family Orchidaceae).

Common/vernacular names: Bourbon vanilla, Réunion vanilla, Mexican vanilla (*V. planifolia*); Tahiti vanilla (*V. tahitensis*); common vanilla.

GENERAL DESCRIPTION

Large green-stemmed perennial herbaceous vines, reaching a length of about 25 m or more in their wild state; native to tropical America (especially Mexico); cultivated in the tropics (Madagascar, Comoros Islands, French Polynesia, Tahiti, Indonesia, Reunion, Seychelles, Mexico, Tanzania, Uganda, etc.). Part used is the fully grown but unripe fruit (a capsule) commonly called pod or bean, collected 4–9 months after pollination; pollination is all done artificially, except in Mexico where it is partly performed artificially and partly by certain indigenous hummingbirds and butterflies not found elsewhere. The pods are then subjected to a complicated labor-intensive curing (fermentation) and drying process that requires 5–6 months to complete. During this period, vanillin is enzymatically produced and may accumulate as white crystals on the surface of the beans; the pods also turn brown and lose 80% of their weight. A much faster process that takes only a few days is reportedly used in Uganda where the beans are processed mechanically. By far the largest vanilla producer is Madagascar. The Mexican beans and the Bourbon beans, the latter grown in Madagascar, are considered of the best quality; they are larger than the Tahiti beans. The United States is the leading importer and consumer.[1]

The quality of vanilla beans does not depend on the vanillin content, even though vanillin is generally recognized as having the "vanilla" odor. Other constituents present together with vanillin as a whole are responsible for the flavor and quality of vanilla and its extracts. Value is determined by fragrance rather than vanillin content.[1]

The so-called "single-strength vanilla extract" is comparable to a 1 : 0.1 or 10% tincture (see ***glossary***) containing at least 35% alcohol. The beans used are required by federal regulations to contain no more than 25% moisture, otherwise proportionately more beans will have to be used. The so-called "10-fold extract" is comparable to a 1:1 fluid extract (see ***glossary***) in terms of extract to beans (crude) ratio and is 10 times stronger than the single-strength extract.

Due to the high price of vanilla and the low cost of vanillin, vanilla extracts have been extensively adulterated (ARCTANDER). There is still no simple method to detect with certainty whether or not a vanilla extract is authentic.[2,3]

CHEMICAL COMPOSITION

Vanilla contains vanillin (1.3–3.0%) as the major flavor component, with over 150 other aroma chemicals also present, most of which are present in traces, including *p*-hydroxybenzaldehyde, acetic acid, isobutyric acid, caproic acid, eugenol, furfural, *p*-hydroxybenzyl methyl ether, vanillyl ethyl ether, anisyl ethyl ether, and acetaldehyde.[2,4–8] The vanillin content differs in different varieties of vanilla, with Bourbon beans containing generally higher amounts than Mexican and Tahiti beans (MARTINDALE; ROSENGARTEN; STAHL).

GC–MS analysis of three vanilla extracts (Tahitian, Indonesian, and Bourbon) revealed the presence of ethyl hexanoate, ethyl nonanoate, ethyl decanoate, methyl *p*-methoxybenzoate, methyl 3-phenyl-2-propenoate, *p*-methoxybenzaldehyde, 5-propenyl-1,3-benzodioxole at various concentrations, in addition to vanillin.[9]

The epicuticular wax contains β-dicarbonyl compounds with long aliphatic chains and an isolated *cis* double bond. The hydrocarbon chain length is 16–24 carbons long, including the 2,4-dione moiety.[10]

The ethyl acetate fraction of the alcoholic extract of the leaves and stems contains 4-ethoxymethylphenol and 4-butoxymethylphenol, 4-hydroxy-2-methoxycinnamaldehyde, 3,4-dihydroxyphenylacetic acid, in addition to vanillin.[11]

The vanilla β-D-glucosidase has recently been purified and characterized as a homogenous tetramer (201 kDa) of four identical subunits.[12]

Other constituents present include resins, sugars, and fixed oil.

PHARMACOLOGY AND BIOLOGICAL ACTIVITIES

The ethyl acetate fraction of the leaves and stems was shown to be toxic against mosquito larvae with 4-butoxymethylphenol as the most active component.[11]

The odor of vanilla was shown to enhance the sweetness sensation of aspartame (odor-induced taste enhancement) in a group of human volunteers. The effect was mainly attributed to olfactory stimulation rather than taste bud stimulation.[13]

A 75% aqueous methanol extract of vanilla pods inhibited quorum sensing (method of chemical communication) in *Chromobacterium violaceum* via reduction of violacin production. The authors suggested that this effect may be useful in preventing bacterial pathogenesis.[14]

TOXICOLOGY

Vanilla has been reported to have allergenic properties in humans, but vanillin was found not to be the principal active agent.[15]

USES

Many of the uses of vanilla have been replaced by vanillin. However, vanillin cannot replace vanilla in many applications where a delicate natural vanilla flavor or fragrance is called for.

Medicinal, Pharmaceutical, and Cosmetic. Vanilla extracts (especially tincture N.F.) are used in pharmaceutical preparations such as syrups, primarily as a flavoring agent.

Vanilla extracts (tincture, absolute, etc.) are used as fragrance ingredient in perfumes.

Food. Vanilla, vanilla extract, and vanilla oleoresin are widely used as flavor ingredients in most food products, including alcoholic (e.g., liqueurs) and nonalcoholic beverages, frozen dairy desserts (especially ice cream and yogurt), candy, baked goods, gelatins and puddings, and others. Highest average maximum use level is about 0.964% (9642 ppm) reported for vanilla in baked goods.

COMMERCIAL PREPARATIONS

Crude and extracts. Crude and tincture are official in N. F.

Regulatory Status. GRAS (§182.10, §182.20, and §169.3).

REFERENCES

See the General References for ARCTANDER; BAILEY 2; CLAUS; DER MARDEROSIAN AND BEUTLER; FEMA; GUENTHER; KARRER; MARTINDALE; MCGUFFIN 1 & 2; MERCK; REMINGTON; ROSENGARTEN; STAHL; TERRELL; UPHOF.

1. E. Westpahl and P. C. M. Jansen, eds., *Plant Resources of South-East Asia: A Selection*, Pudoc, Wageningen, Netherlands, 1989.
2. G. E. Martin et al., *Food Technol.*, **29**, 54 (1975).
3. G. E. Martin et al., *Food Sci.*, **42**, 1580 (1977).

4. B. M. Lawrence, *Perfum Flav.*, **2**, 3 (1977).
5. H. Shiota and K. Itoga, *Korya*, **113**, 65 (1975).
6. I. Klimes and D. Lamparsky, *Int. Flav. Food Addit.*, **7**, 272 (1976).
7. H. Bohnsack, *Riechst., Aromen, Korperpflegem.*, **17**, 133 (1967).
8. H. Bohnsack, *Riechst., Aromen, Korperpflegem.*, **15**, 284 (1965).
9. T. Sostaric et al., *J. Agric. Food Chem.*, **48**, 5802 (2000).
10. B. Ramaroson-Raonizafinimanana et al., *J. Agric. Food Chem.*, **48**, 4739 (2000).
11. R. Sun et al., *J. Agric. Food Chem.*, **49**, 5161 (2001).
12. E. Odoux et al., *J. Agric. Food Chem.*, **51**, 3168 (2003).
13. N. Sakai et al., *Percept. Mot. Skills*, **92**, 1002 (2001).
14. J. H. Choo et al., *Lett. Appl. Microbiol.*, **42**, 637 (2006).
15. D. L. J. Opdyke, *Food Cosmet. Toxicol.*, **14**, 633 (1976).

WINTERGREEN OIL

Source: *Gaultheria procumbens* L. and other *Gaultheria* species (Family Ericaceae).

Common/vernacular names: Checkerberry, teaberry, and gaultheria oil.

GENERAL DESCRIPTION

An evergreen shrub with slender and extensively creeping stems from which arise erect branches bearing at the top oval, leathery leaves with toothed (often bristly) margins; up to about 15 cm high; native to North America, growing from Newfoundland to Manitoba and south to Georgia and Alabama. Part used is the leaf. Wintergreen oil is obtained by steam distillation of the warm water-macerated leaves whereby, gaultherin present in the leaves is enzymatically hydrolyzed to yield methyl salicylate (which is subsequently distilled with steam), D-glucose, and D-xylose (see *sweet birch oil*). The yield of oil is normally 0.5–0.8% from the leaves (LIST AND HÖRHAMMER).[1]

CHEMICAL COMPOSITION

Wintergreen oil contains almost exclusively methyl salicylate ($\geq$98%).[2]

PHARMACOLOGY AND BIOLOGICAL ACTIVITIES

Gaultherin has analgesic anti-inflammatory effects in mice at 200 mg/kg. The effect is comparable to that of aspirin but has the advantage of being devoid of gastric ulcerogenic effect.[3]

TOXICOLOGY

Adverse effects resulting from topical application and accidental ingestion of wintergreen oil have been reported. In the first case, symptoms of acute salicylism were observed (tinnitus, vomiting, and acid–base disturbance).[4,5] Laryngeal edema resulted after accidental ingestion.[6]

Also see *sweet birch oil*.

USES

Medicinal, Pharmaceutical, and Cosmetic. Wintergreen oil is used interchangeably with sweet birch oil or methyl salicylate (see *sweet birch oil*).

Food. Wintergreen oil is used for similar flavoring purposes as sweet birch oil, though its use levels reported are generally lower than those of sweet birch oil; the highest average maximum use level is about 0.04% (405 ppm) in candy (see *sweet birch oil*).

Dietary Supplements/Health Foods. The dried leaves are used as a tea flavoring ingredient.

Traditional Medicine. Leaf tea reportedly for colds, headache, stomachache, fevers, kidney ailments; externally, wash for rheumatism, sore muscles, and lumbago (FOSTER AND DUKE).

COMMERCIAL PREPARATIONS

Volatile oil; formerly official in U.S.P. and currently official in F.C.C. (see *sweet birch oil*). Wintergreen oil is slightly levorotatory, while sweet birch oil and synthetic methyl salicylate are optically inactive.

Regulatory Status. Not listed under §172.510, §182.10, or §182.20.

REFERENCES

See the General References for APhA; ARCTANDER; BAILEY 1; DER MARDEROSIAN AND BEUTLER; FERNALD; FEMA; FOSTER and DUKE; GRIEVE; GUENTHER; LIST AND HÖRHAMMER; MARTINDALE; MCGUFFIN 1 & 2; TERRELL; UPHOF; USD 26TH.

1. C. C. Mu and I. C. Yang, *Yao Hsueh Hsueh Pao*, **13**, 451 (1966).
2. B. R. Olin, ed., *Lawrence Rev. Nat. Prod.*, (1992).
3. B. Zhang et al., *Eur. J. Pharmacol.*, **530**, 166 (2006).
4. A. J. Bell and G. Duggin, *Emerg. Med. (Fremantle.)*, **14**, 188 (2002).
5. T. Y. Chan, *Hum. Exp. Toxicol.*, **15**, 747 (1996).
6. M. Botma et al., *Int. J. Pediatr. Otorhinolaryngol.*, **58**, 229 (2001).

WITCH HAZEL

Source: *Hamamelis virginiana* L. (Family Hamamelidaceae).

Common/vernacular names: Hamamelis.

GENERAL DESCRIPTION

A deciduous shrub or small tree flowering in the fall; up to about 7.5 m high; native to North America; distributed from Quebec to Georgia and west to Minnesota. Parts used are the dried leaves, bark, and partially dried dormant twigs. Witch hazel water (also known as hamamelis water and distilled witch hazel extract) is obtained from the recently cut and partially dried dormant twigs. *Hamamelis vernalis* Sarg. leaves and twigs vicariously enter commercial supplies from the Ozark plateau. The twigs are macerated for about 24 h in twice their weight of warm water followed by distilling and adding the required amount of alcohol to the distillate and thoroughly mixing, such that 1000 volumes witch hazel water are derived from 1000 parts crude.

CHEMICAL COMPOSITION

Witch hazel leaf contains 8–10% tannin that is composed of hamamelitannin or digallyhamamelose, gallotannins, and/or proanthocyanidins (LIST AND HÖRHAMMER).[1–4] Other constituents present include free gallic acid, free hamamelose, saponins, choline, resins, flavonoids [quercetin, kaempferol, astragalin (kaempferol-3-glucoside), quercitrin, afzelin, myricitrin, etc.], 0.5% volatile oil (*n*-hexen-2-al, hexenol, α- and β-ionones, safrole, sesquiterpenes, etc.), and others (HÖRHAMMER).[5–8]

The bark contains 1–7% hamamelitannin and smaller amounts of condensed tannins (e.g., *d*-gallocatechin, *l*-epicatechin gallate, and *l*-epigallocatechin); saponins; fixed oil (0.6%); wax; 0.5% volatile oil (sesquiterpenes, a phenol, etc.); polysaccharides and a resin (LIST AND HÖRHAMMER).[4,9–11]

Witch hazel water contains a trace of volatile oil consisting of eugenol, carvacrol, and probably similar compounds as the volatile oils of leaf and bark.[8] As it is a steam distillate, it does not contain tannins.

PHARMACOLOGY AND BIOLOGICAL ACTIVITIES

Witch hazel leaf, witch hazel bark, and witch hazel water have all been reported to have astringent and hemostatic properties. These properties can be attributed to the tannins contained in the leaves and bark, but it is not known what is responsible for these activities in hamamelis water (LIST AND HÖRHAMMER; MARTINDALE). On the other hand, results from two comparative trials utilizing hamamelis distillate creams showed that these preparations were not as effective as a low-dose hydrocortisone cream in the treatment of atopic eczema and UV-induced erythema.[12,13]

Topical preparations containing hamamelis extract are effective in various skin disorders in adults and in children (diaper dermatitis, minor injuries, and localized inflammation of the skin).[9,14–16] Anti-inflammatory activity (in paw edema model) has also been demonstrated when the hydroalcoholic extract was orally administered in rat.[17] Antioxidant activity as well as inhibition of 5-LO and TNF-α may be involved in the anti-inflammatory effects of witch hazel.[18–21]

Witch hazel was found to have cell protective/antimutagenic effects,[22–25] and to display antibacterial (against periodontopathic anerobes/facultative aerobes) and antiviral (against HSV-1) activities.[20,26]

USES

Medicinal, Pharmaceutical, and Cosmetic. Witch hazel leaf extract, witch hazel bark extract, and witch hazel water are all used as astringent and hemostatic in preparations (suppositories, ointments, lotions, cloth wipes, etc.) for use in treating hemorrhoids, itching, irritations, and minor pains, with witch hazel water the most commonly used. They are also used in eye drops, shaving lotions, and others.

In European phytomedicine, preparations are used as an astringent, anti-inflammatory, and local hemostyptic for mild skin injuries, hemorrhoids, varicose veins, and local inflammations of the skin and mucous membranes.[27]

The bottled "itch hazel" in most domestic medicine cabinets is simply witch hazel water.

Dietary Supplements/Health Foods. Leaves are sometimes used as tea ingredient.

Traditional Medicine. The leaf and bark are reportedly used internally to treat diarrhea and externally to treat mouth and throat irritations, hemorrhoids, eye inflammation, insect bites, minor burns, and other skin irritations, usually as a decoction, poultice, or ointment. They have also been used in cancers.[28]

COMMERCIAL PREPARATIONS

Crudes (leaf and bark), extracts (solid, fluid, etc.), and witch hazel water. Witch hazel leaf and its fluid extracts were formerly official in N.F.; witch hazel bark was formerly official in U.S.P.; and witch hazel water was formerly official in N.F. Strengths (see *glossary*) of extracts and distillate are expressed in weight-to-weight or volume-to-weight ratios.

Regulatory Status. Used in OTC preparations in the United States. The leaves and branches are the subject of a positive German therapeutic monograph, indicated for minor skin injuries, local skin and mucous membrane irritation, hemorrhoids, and varicose veins.[27]

REFERENCES

See the General References for APPLEQUIST; BAILEY 1; BARNES; BLUMENTHAL 1; CLAUS; DER MARDEROSIAN AND BEUTLER; FERNALD; FOSTER; GRIEVE; KROCHMAL AND KROCHMAL; LIST AND HÖRHAMMER; LUST; MARTINDALE; MCGUFFIN 1 & 2; TERRELL; USD 26TH.

1. A. Dauer et al., *Planta Med.*, **69**, 89 (2003).
2. G. Netien and R. Rochan, *Bull. Trav. Soc. Pharm. Lyon*, **12**, 121 (1968).
3. H. Friedrich and N. Krueger, *Planta Med.*, **26**, 327 (1974).
4. B. Vennat et al., *Planta Med.*, **54**, 454 (1988).
5. P. Bernard et al., *J. Pharm. Belg.*, **26**, 661 (1971).
6. W. Messerschmidt, *Arch. Pharm. (Weinheim)*, **300**, 550 (1967).
7. W. Messerschmidt, *Arzneim.-Forsch.*, **18**, 1618 (1968).
8. H. Janitstyn, *Parfüm Kosmet.*, **45**, 335 (1964).

9. A. Deters et al., *Phytochemistry*, **58**, 949 (2001).
10. H. Glick et al., *Carbohydr. Res.*, **39**, 160 (1975).
11. H. Friedrich and N. Krueger, *Planta Med.*, **25**, 138 (1974).
12. H. C. Korting et al., *Eur. J. Clin. Pharmacol.*, **48**, 461 (1995).
13. H. C. Korting et al., *Eur. J. Clin. Pharmacol.*, **44**, 315 (1993).
14. H. H. Wolff and M. Kieser, *Eur. J. Pediatr.*, (2006).
15. B. J. Hughes-Formella et al., *Skin Pharmacol. Appl. Skin Physiol.*, **15**, 125 (2002).
16. B. J. Hughes-Formella et al., *Dermatology*, **196**, 316 (1998).
17. M. Duwiejua et al., *J. Pharm. Pharmacol.*, **46**, 286 (1994).
18. S. Habtemariam, *Toxicon*, **40**, 83 (2002).
19. C. Hartisch et al., *Planta Med.*, **63**, 106 (1997).
20. C. A. Erdelmeier et al., *Planta Med.*, **62**, 241 (1996).
21. H. Masaki et al., *Free Radic. Res. Commun.*, **19**, 333 (1993).
22. A. Dauer et al., *Phytochemistry*, **63**, 199 (2003).
23. A. Dauer et al., *Planta Med.*, **64**, 324 (1998).
24. H. Masaki et al., *J. Dermatol. Sci.*, **10**, 25 (1995).
25. H. Masaki et al., *Biol. Pharm. Bull.*, **18**, 59 (1995).
26. L. Lauk et al., *Phytother. Res.*, **17**, 599 (2003).
27. Monograph, *Hamamelidis folium et cortex*, *Bundesanzeiger*, no. 154 (August 21, 1985); revised (March 13, 1990).
28. J. L. Hartwell, *Lloydia*, **32**, 247 (1969).

WOODRUFF, SWEET

Source: *Galium odoratum* (L.) Scop. (syn. *Asperula odorata* L.) (Family Rubiaceae).

Common/vernacular names: Woodruff, master of the wood, and woodward.

GENERAL DESCRIPTION

A small perennial herb with a creeping rhizome from which smooth erect stems arise (up to 30 cm high) bearing lance-shaped leaves in whorls of six to eight (usually eight); native to Eurasia and northern Africa; naturalized in North America. Part used is the dried whole flowering herb that does not have any odor when fresh but develops a new-mown hay (coumarin-like) odor on drying. European countries (especially Germany) are the major producers.

CHEMICAL COMPOSITION

Generally reported to contain coumarin in bound form (glycoside) that is set free by enzymatic action during wilting or drying. However, one study did not detect any coumarins in woodruff.[1]

Other constituents reported present include asperuloside (0.05%), monotropein, tannins, anthracene and naphthalene derivatives, traces of nicotinic acid, fixed oil, and bitter principle, among others (LIST AND HÖRHAMMER; MERCK).[2,3]

PHARMACOLOGY AND BIOLOGICAL ACTIVITIES

Asperuloside and the leaves are reported to have anti-inflammatory activity (LIST AND HÖRHAMMER).[4]

Asperuloside has been suggested as a starting material for prostaglandins.[5]

Coumarin is reported to be toxic (see ***deertongue***).

USES

Medicinal, Pharmaceutical, and Cosmetic. Used in sachets. Extracts (e.g., concrete and absolute) are used as fragrance components in perfumes, mostly in Europe.[6]

Food. It is reported used as a flavor component in various major food products, including alcoholic (May wines, vermouths, bitters, etc.) and nonalcoholic beverages, frozen dairy desserts, candy, baked goods, and gelatins and puddings. Highest average maximum use level is 0.04% in baked goods.

Traditional Medicine. The herb is traditionally used as a diaphoretic, antispasmodic, sedative (particularly for children and elderly people), and diuretic. Conditions for which it is used include restlessness, insomnia, stomachache, migraine, neuralgia, and bladder stones, usually as a tea (FOSTER). In European tradition, reportedly used for prophylaxis and therapy of respiratory conditions, as well as gallbladder, kidney, and circulatory disorders; topically for venous conditions, hemorrhoids; anti-inflammatory; claimed efficacy not documented.[7]

COMMERCIAL PREPARATIONS

Mainly crude.

Regulatory Status. Has been approved for use in alcoholic beverages only (§172.510). Subject of a German therapeutic monograph; use not recommended as efficacy is not documented.[7]

REFERENCES

See the General References for BAILEY 1; BIANCHINI AND CORBETTA; DER MARDEROSIAN AND BEUTLER; FEMA; FERNALD; FOSTER; LIST AND HÖRHAMMER; LUST; MCGUFFIN 1 & 2; TUCKER 1–3; UPHOF.

1. M. I. Borisov, *Khim. Prir. Soedin.*, **10**, 82 (1974).
2. A. Buckova et al., *Acta Fac. Pharm., Univ. Comeniana*, **19**, 7 (1970).
3. A. R. Burnett and R. H. Thomson, *J. Chem. Soc., C*, (7), 854 (1968).
4. N. Mascolo et al., *Phytother. Res.*, **1**(1), 28 (1987).
5. W. F. Berkowitz et al., *J. Org. Chem.*, **47**, 824 (1982).
6. L. Trabaud, *Perfum. Essent. Oil Records*, **54**, 382 (1963).
7. Monograph, *Galii odorati herba*, *Bundesanzeiger*, no. 193 (October 15, 1987).

XANTHAN GUM

Source: Derived from a pure-culture fermentation of glucose using the bacterium *Xanthomonas campestris*.

GENERAL DESCRIPTION

Xanthan gum is a polysaccharide secreted by certain species of bacteria. It is produced by a pure-culture fermentation of glucose using the bacterium *Xanthomonas campestris*, reported to be a minor plant pathogen, though the strain used is reportedly nonpathogenic and nontoxic to humans.[1–5]

During fermentation the bacteria utilize the glucose and other nutrients in the culture medium to produce a high molecular weight polysaccharide that they excrete into the culture broth (see also **monosodium glutamate**). The polysaccharide is recovered by precipitation and purification with isopropyl alcohol, followed by drying and milling to yield commercial xanthan gum.

Xanthan gum is a cream-colored powder that is readily soluble in cold or hot water to form neutral, viscous, and nonthixotropic (thinning when disturbed) solutions. They have relatively high viscosity that is unusually stable toward changes in temperatures, acidity, alkalinity, and salt content. They also have good freeze–thaw stability (GLICKSMAN).[1,3]

The production and properties of xanthan gum have been reviewed.[6]

CHEMICAL COMPOSITION

Xanthan gum is a very long, linear polymer. Its molecular weight has been reported to range from 1 million to 10 million. The molecule consists of a chain composed of D-glucose, D-mannose, and D-glucuronic acid, with short side chains. Pyruvic acid, present in side chains, accounts for 2.5–4.8 of the molecule (GLICKSMAN; WHISTLER AND BEMILLER).[5]

Xanthan gum samples with high pyruvate contents have been reported to yield solutions with higher viscosity than those of low pyruvate samples.[5]

PHARMACOLOGY AND BIOLOGICAL ACTIVITIES

Xanthan gum has been reported to be nontoxic to three studied animal species (rat, cat, and dog).[7] However no toxicity or safety data on humans are reported.

Xanthan gum has been reported to be active against Ehrlich ascites tumor and S-180 in mice, being synergistic with 5-fluorouracil or bleomycin in S-180.[8]

The inclusion of xanthan gum in black currant soft drinks reduced teeth enamel loss in contrast to regular black currant drinks. This enamel saving effect was comparable to that of a black currant drink supplemented with calcium.[9,10]

USES

Medicinal, Pharmaceutical, and Cosmetic. Xanthan gum is used as a stabilizer, thickener, and an emulsifying agent in water-based pharmaceutical and cosmetic preparations. It has been evaluated as a carrier for the controlled release of drugs and as a component of a transparent blood analog fluid for *in vitro* hemodynamic studies.[11,12]

Food. Xanthan gum was approved for food use in 1969 as a stabilizer, emulsifier, thickener, suspending agent, bodying agent, and foam enhancer.[2] It was also listed as an optional emulsifier for French dressing in 1971 under the Standard of Identity for French dressing.[10] Since then, it has been extensively used in many types of food products, including salad dressings (e.g., French), dairy products (chocolate milk drinks, puddings, cheese spread, etc.) canned products (meat, fish, and poultry), and others.

Other. Xanthan gum has recently been used with success as a constituent of microbial culture media alone and combined with agar.[13]

COMMERCIAL PREPARATIONS

Available in powdered forms. Xanthan gum is official in N.F. and F.C.C.

Regulatory Status. Has been approved for food use (§172.695).

REFERENCES

See the General References for FURIA; GLICKSMAN; LAWRENCE; MARTINDALE; WHISTLER AND BEMILLER.

1. C. T. Blood in L. W. Coddet al., eds., *Chemical Technology: An Encyclopedic Treatment*, Vol. 5, Barnes & Noble, New York, 1972, p. 27.
2. *Fed. Regist.*, **34**, 5376 (1969).
3. J. K. Rocks, *Food Technol.*, **25**, 476 (1971).
4. R. W. Silman and P. Rogovin, *Biotechnol. Bioeng.*, **14**, 23 (1972).
5. P. A. Sandford et al., *ACS Symp. Ser.*, **45**, 192 (1977).
6. F. Garcia-Ochoa et al., *Biotechnol. Adv.*, **18**, 549 (2000).
7. W. H. McNeely and P. Kovacs, *ACS Symp. Ser.*, **15**, 269 (1975).
8. M. Oda, *Yakuri to Chiryo*, **13**, 5743 (1985).
9. N. X. West et al., *Br. Dent. J.*, **196**, 478 (2004).
10. *Fed. Regist.*, **36**, 9010 (1971).
11. A. G. Andreopoulos and P. A. Tarantili, *J. Biomater. Appl.*, **16**, 34 (2001).
12. K. A. Brookshier and J. M. Tarbell, *Biorheology*, **30**, 107 (1993).
13. S. B. Babbar and R. Jain, *Curr. Microbiol.*, **52**, 287 (2006).

YARROW

Source: *Achillea millefolium* L. (Family Compositae or Asteraceae).

Common/vernacular names: Milfoil, common yarrow, nosebleed, and thousand leaf.

GENERAL DESCRIPTION

A perennial herb with a simple stem bearing aromatic bipinnately parted and dissected leaves, giving a lacy appearance; up to about 1 m high; native to Eurasia and naturalized in North America; found in most temperate zones of the world (e.g., United States, Canada, throughout Europe, and northern China). There are numerous varieties or forms. The taxon *A. millefolium* is a species complex represented by a number of other species.[1] Part used is the entire flowering aboveground herb.

CHEMICAL COMPOSITION

Much work has been performed on yarrow especially regarding its volatile oil composition, which is highly variable.

Yarrow contains about 0.1–1.4% volatile oil that is composed of azulene (0.51%), α- and β-pinenes (mainly β-), caryophyllene, borneol, terpineol, cineole, bornyl acetate, camphor, sabinene, isoartemisia ketone, and other compounds (including a trace of thujone).[2-11] The relative composition varies considerably (especially its azulene content) depending on sources (HÖRHAMMER).[5-8,11,12] *A. millefolium* L. sensu stricto (a hexaploid) contains no azeulene.[1] Its oil primarily contains monoterpenes including linalool (26%), camphor (18%), borneol, 1,8-cineole, and others.[13]

Other constituents reported include sesquiterpenes and sesquiterpene lactones (methyl esters of achimillic acids A–C, α-peroxyachifolid, achillin, millefin, desacetylmatricarine, costunolide, leucodin, etc.);[14-18] flavonoids (free and glycosides, e.g., rutin, apigenin, luteolin, and casticin);[2,18-20] tannins; resin; coumarins; saponins;[21] sterols (e.g., β-sitosterol and its acetate);[18] alkanes (mainly tricosane, heptadecane, and pentacosane); lignans;[2] fatty acids (linoleic, palmitic, oleic acids, etc.);[22] sugars (glucose, galactose, sucrose, arabinose, inositol, dulcitol, mannitol, etc.); alkaloids or bases (betaine, choline, trigonelline, betonicine, and stachydrine); amino acids (alanine, leucine, lysine, histidine, glutamic acid; etc.); and acids (succinic acid and salicylic acid),[2,23] among others (JIANGSU; LIST AND HÖRHAMMER).

PHARMACOLOGY AND BIOLOGICAL ACTIVITIES

An aqueous extract of yarrow flower heads has been reported to have anti-inflammatory activity in laboratory animals; this activity was due to a mixture of protein–carbohydrate complexes that had very low toxicity.[24,25]

Yarrow alkaloids (bases) are reported to have hypotensive and weakly antipyretic, astringent, antibacterial, and choleretic properties. Yarrow extracts also exhibited hemostatic properties, among others (JIANGSU).[26,27]

More recently investigated activities of yarrow include choleretic effects and stimulation of bile flow in isolated perfused rat liver;[28] antiulcer/protective effect on the gastric mucosa of rats;[29] antitumor activity (mainly due to the flavonoid casticin and the achimillic acid sesquiterpenes);[16,19] estrogenic activity (in recombinant MCF-7 cells);[2] spasmolytic effects (in isolated rabbit jejunum and guinea pig ileum);[30,31] antioxidant and/or hepatoprotective effects;[31,32] and antimicrobial effect of the extract and volatile oil against Gram-positive and Gram-negative bacteria and fungi.[32,33]

TOXICOLOGY

Yarrow may cause contact dermatitis in certain individuals.[17,34] Reproductive toxicity in males (antispermatogenic effect) has also been reported.[35,36]

USES

Medicinal, Pharmaceutical, and Cosmetic. In European phytomedicine, herb (at a daily dose of 4.5 g) is used for appetite loss, dyspeptic complaints, spasmodic gastrointestinal disturbances, and so on. Use is contraindicated in known hypersensitivity to other Compositae members.[27]

Extracts used in preparations (e.g., baths) for their alleged soothing and quieting effects on the skin (DE NAVARRE),[37] in hair tonic and antidandruff preparations.[38] However, one safety assessment report concluded that data were not sufficient to support the safe incorporation of yarrow in cosmetic products and recommended more tests to be conducted.[39]

Dietary Supplements/Health Foods. Used in fever and cold formulation, mostly in combinations, as minor ingredient. Tincture used as topical styptic (FOSTER).

Food. Used in bitters and vermouths, with average maximum use level of less than 0.001% (5 ppm) reported.

Yarrow flower is used in herb teas.

Traditional Medicine. Reportedly used as a tonic, carminative, febrifuge, antispasmodic, astringent, hemostatic, and others, usually in the form of an infusion or decoction, or the fresh juice. Conditions for which it is used include lack of appetite, stomach cramps, flatulence, gastritis, enteritis, internal and external bleeding of all kinds (coughing blood, nosebleed, hemorrhoidal bleeding, bloody urine, etc.), wounds, sores, and skin rash.

In China, both the fresh herb and the dried herb are used to treat similar conditions. The fresh herb (mashed as a poultice) is especially recommended for all sorts of sores and wounds as well as dog and snakebites, while the dried herb is recommended for internal bleeding (especially menstrual and hemorrhoidal).

It has been used in cancers.[40]

COMMERCIAL PREPARATIONS

Mainly crude; it was formerly official in U.S.P.

Regulatory Status. Has been approved for use in alcoholic beverages only; finished beverage must be thujone-free (§172.510). Yarrow oil normally contains little or no thujone, whereas sage oil contains normally about 50% thujone (see *sage*). The herb is the subject of a positive German therapeutic monograph.[27]

REFERENCES

See the General References for APPLEQUIST; BAILEY 1; BARNES; BIANCHINI AND CORBETTA; BLUMENTHAL 1; BRUNETON; DER MARDEROSIAN AND BEUTLER; FEMA; FERNALD; FOSTER; FOSTER AND DUKE; GRIEVE; GUPTA; JIANGSU; KROCHMAL AND KROCHMAL; LIST AND HÖRHAMMER; LUST; MCGUFFIN 1 & 2;

1. R. F. Chandler, et al., *Econ. Bot.*, **36**, 203 (1982).
2. G. Innocenti et al., *Phytomedicine* (2006).
3. J. Rohloff et al., *J. Agric. Food Chem.*, **48**, 6205 (2000).
4. G. Verzarne-Petri and A. S. Shalaby, *Acta Agron. Acad. Sci. Hung.*, **26**, 337 (1977).
5. J. Kozlowski and J. Lutomski, *Planta Med.*, **17**, 226 (1969).
6. H. Popescu and H. Winand, *Clujul Med.*, **50**, 78 (1977).
7. A. Ruminska, *Acta Agrobot.*, **23**, 53 (1970).
8. M. Y. Haggag et al., *Planta Med.*, **27**, 361 (1975).
9. R. B. Chelishvili and A. I. Tavberidze, *Maslo-Zhir. Prom.*, **2**, 24 (1974).
10. G. Verzarne-Petri and H. N. Cuong, *Acta Pharm. Hung.*, **47**, 134 (1977).

11. A. J. Falk et al., *Lloydia*, **37**, 598 (1974).
12. A. Orav et al., *Nat. Prod. Res.*, **20**, 1082 (2006).
13. L. Hofmann et al., *Phytochemistry*, **31**, 537 (1992).
14. S. J. Smolenski et al., *Lloydia*, **30**, 144 (1967).
15. S. Z. Kasymov and G. P. Sidyakin, *Khim. Prir. Soedin.*, **2**, 246 (1972).
16. T. Tozyo et al., *Chem. Pharm. Bull. (Tokyo)*, **42**, 1096 (1994).
17. B. M. Hausen et al., *Contact Dermatitis*, **24**, 274 (1991).
18. S. Glasl et al., *Z. Naturforsch. C*, **57**, 976 (2002).
19. K. Haidara et al., *Cancer Lett.*, **242**, 180 (2006).
20. I. D. Neshta et al., *Khim. Prir. Soedin.*, **5**, 676 (1972).
21. K. S. Tillyaev et al., *Rast. Resur.*, **9**, 58 (1973).
22. C. Ivanov and L. Yankov, *God. Vissh. Khimikotekhnol. Inst. Sofia*, **14**, 61, 73 (1970).
23. C. Ivanov and L. Yankov, *God. Vissh. Khimikotekhnol. Inst. Sofia*, **14**, 195, 223 (1971).
24. A. S. Goldberg and E. C. Mueller, *J. Pharm. Sci.*, **58**, 938 (1969).
25. A. Goldberg et al., U.S. Pat. 3,552,350 (1970).
26. F. W. Kudrzycka-Bieloszabska and K. Glowniak, *Diss. Pharm. Pharmacol.*, **18**, 449 (1966).
27. Monograph *Achillea millefolium*, *Bundesanzeiger*, no. 22 (February 1, 1990).
28. B. Benedek et al., *Phytomedicine*, **13**, 702 (2006).
29. A. M. Cavalcanti et al., *J. Ethnopharmacol.*, **107**, 277 (2006).
30. R. Lemmens-Gruber et al., *Arzneimittelforschung*, **56**, 582 (2006).
31. S. Yaeesh et al., *Phytother. Res.*, **20**, 546 (2006).
32. F. Candan et al., *J. Ethnopharmacol.*, **87**, 215 (2003).
33. G. Stojanovic et al., *J. Ethnopharmacol.*, **101**, 185 (2005).
34. J. C. Mitchell in V C. Runeckles, ed., *Recent Advances in Phytochemistry*, Vol. 9, Plenum Press, New York, 1975, p. 119.
35. P. R. Dalsenter et al., *Reprod. Toxicol.*, **18**, 819 (2004).
36. T. Montanari et al., *Contraception*, **58**, 309 (1998).
37. P. Alexander, *Cosmet. Perfum.*, **88**, 35 (1973).
38. H. Greger and O. Hofer, *Planta Med.*, **55**, 216 (1989).
39. *Int. J. Toxicol.*, **20**(Suppl. 2), 79 (2001).
40. J. L. Hartwell, *Lloydia*, **31**, 71 (1968).

YERBA SANTA

Source: *Eriodictyon californicum* (Hook. et Arn.) Torr. (syn. *E. glutinosum* Benth. and *Wigandia californicum* HOOK et Arn.) (Family Hydrophyllaceae).

Common/vernacular names: Eriodictyon, bear's weed, consumptives' weed, mountain balm, and tarweed.

GENERAL DESCRIPTION

An evergreen aromatic shrub with woody rhizomes from which arise glutinous nonhairy stems bearing lance-shaped leaves (5–15 cm long) that are glutinous and nonhairy above but hairy beneath; up to 2.2 m high; native to California, extending north to Oregon and south to northern Mexico. Part used is the dried leaf.

CHEMICAL COMPOSITION

Contains the flavanones eriodictyonine (homoeriodictyol, 3–6%); eriodictyol (5,7,4,3'-tetrahydroxyflavanone, 0.23–0.6%); other flavonoids including sterubin, 3'-methyl-4'-isobutyryleriodictyol, pinocembrin, sakuranetin, cirsimaritin, chrysoeriol, hispidulin, chrysin, among others; eriodictyonic acid (probably impure eriodictyonine); a resin consisting of triacontane, pentatriacontane, cerotic acid, chrysoeriodictyol, xanthoeriodictyol (xanthoeriodol), and eriodonol (eriodonal), among others; tannin; gum; sugars; fats; formic and acetic acids; a trace of volatile oil; and others (LIST AND HÖRHAMMER).[1]

PHARMACOLOGY AND BIOLOGICAL ACTIVITIES

Eriodictyol is reported to have expectorant properties (MERCK).

A recent sensory study of taste modification demonstrated that flavanones of yerba santa (eriodictyol, homoeriodictyol, and sterubin) had a significant masking effect on the bitter taste of caffeine and other bitter compounds, such as salicin, paracetamol, and quinine. The sodium salt of homoeriodictyol had a dose-dependent bitter masking effect against caffeine and amarogentin.[2]

USES

Medicinal, Pharmaceutical, and Cosmetic. Yerba Santa extracts are used mainly in flavoring pharmaceutical preparations, particularly those containing bitter drugs such as quinine whose bitter taste can be masked by the yerba santa extracts.

Food. The fluid extract is used as a flavor component in alcoholic and nonalcoholic beverages, frozen dairy desserts, and baked goods, with highest average maximum use level of 0.05% in baked goods.

Dietary Supplements/Health Foods. Tincture available, primarily from local California manufacturers, used for traditional indications. Traditional Medicine. Traditionally used as an expectorant, antispasmodic, febrifuge, and tonic. Conditions for which it is used include asthma and chronic bronchitis. Also used as a poultice to treat wounds, bruises, sprains, and insect bites (LUST; UPHOF).

COMMERCIAL PREPARATIONS

Crude and extracts. Crude, fluid extract, and aromatic syrup are official in N.F. Strengths (see glossary) of extracts are expressed in weight-to-weight ratios.

Regulatory Status. Has been approved for food use (§172.510).

REFERENCES

See the General References for DER MARDEROSIAN AND BEUTLER; FEMA; LIST AND HÖRHAMMER; LUST; MARTINDALE; MCGUFFIN 1 & 2; MUNZ AND KECK; REMINGTON; UPHOF.

1. Y. L. Liu et al., *J. Nat. Prod.*, **55**, 357 (1992).
2. J. P. Ley et al., *J. Agric. Food Chem.*, **53**, 6061 (2005).

YLANG YLANG OIL

Source: *Cananga odorata* (Lam.) HOOK. f. et Thoms. forma **genuina** (Family Annonaceae).

GENERAL DESCRIPTION

Both ylang ylang oil and cananga oil are derived from the same species, *Cananga odorata*, a tree native to tropical Asia (especially Indonesia and the Philippines). The species is reported to exist in different forms; ylang ylang is a different form than cananga (see *cananga oil*). Part used is the fresh flower picked early in the morning. Ylang ylang oil is obtained by steam distillation or water and steam distillation of the freshly picked flowers. The first distillate (usually ca. 40% of the total distillate) constitutes ylang ylang extra, which is considered to be the best grade. As distillation continues, other inferior grades (first, second, and third) are obtained in decreasing degrees of quality. Major producers include the Comoro Islands, Madagascar, and Réunion Island (ARCTANDER; GUENTHER).[1,2]

CHEMICAL COMPOSITION

Ylang ylang oil contains d-α-pinene, linalool, geraniol, sesquiterpenes (caryophyllene, γ-, σ- and ε-cadinenes, ylangene, farnesol, farnesyl acetate, γ-muurolene, etc.),[3] acids (acetic, valeric, isovaleric, benzoic, 2-methylbutyric, hexanoic, heptanoic, octanoic, nonanoic, and *trans*-geranic acids), phenols (eugenol and isoeugenol), benzyl acetate, methyl benzoate, p-cresol methyl ether, eugenol methyl ether (methyl eugenol), geranyl acetate, p-tolyl methyl ether, safrole, and isosafrole, among others (LIST AND HÖRHAMMER; MASADA).[4-6]

PHARMACOLOGY AND BIOLOGICAL ACTIVITIES

The oil has a smooth muscle relaxing effect in animals. *In vitro* and *in vivo* relaxing effects on the urinary bladder of experimental animals have been reported and are suggested to be mediated by c-AMP.[7] A CNS calming effect was also observed in humans after inhalation and transdermal absorption of ylang ylang oil. Signs of relaxation included a drop in blood pressure and pulse rate, an increase in skin temperature, and enhanced alertness.[8,9]

Ylang ylang oil was found to have a moderate acaricidal effect when tested among other Korean herbs against *Dermatophagoides farinae* and *D. pteronyssinus*.[10]

TOXICOLOGY

Data from one source indicate ylang ylang oil to be nonirritating to mouse skin, slightly irritating to rabbit skin, and nonirritating and nonsensitizing to human skin; no phototoxic effects were reported.[11]

USES

Medicinal, Pharmaceutical, and Cosmetic. Ylang ylang oil is extensively used as fragrance components in soaps, detergents, creams, lotions, and perfumes (especially floral and heavy oriental types), with maximum use level of 1% reported in perfumes. The higher grades are generally used in perfumes, while the lower grades are used in scenting soaps and detergents.

Food. Used as a flavor component (e.g., fruit flavors) in major food products, including alcoholic and nonalcoholic beverages, frozen dairy desserts, candy, baked goods, gelatins and puddings, and others. Use levels reported are generally below 0.001% (5.03 ppm).

COMMERCIAL PREPARATIONS

Available in several grades, especially extra and third. Ylang ylang oil and cananga oil have been extensively adulterated (ARCTANDER).

Regulatory Status. GRAS (§182.20).

REFERENCES

See the General References for ARCTANDER; FEMA; FURIA AND BELLANCA; GUENTHER; MASADA; MERCK; UPHOF.

1. W. D. Fordham in L. W. Codd et al., eds., *Chemical Technology: An Encyclopedic Treatment*, Vol. 5, Barnes and Noble, New York, 1972, p. 1.

2. M. Stoll in A. Standen, ed., *Kirk-Othmer Encyclopedia of Chemical Technology*, Vol. 14, 2nd ed., Interscience, New York, 1967, p. 178.
3. E. M. Gaydou et al., *J. Agric. Food Chem.*, **34**, 481 (1986).
4. D. B. Katague and E. R. Kirch, *J. Pharm. Sci.*, **52**, 252 (1963).
5. J. A. Wenninger et al., *Proc. Sci. Sect. Toilet Goods Assoc.*, **46**, 44 (1966).
6. R. Timmer et al., *Int. Flavors Food Addit.*, **6**, 189 (1975).
7. H. J. Kim et al., *J. Korean Med. Sci.*, **18**, 409 (2003).
8. T. Hongratanaworakit and G. Buchbauer, *Planta Med.*, **70**, 632 (2004).
9. T. Hongratanaworakit and G. Buchbauer, *Phytother. Res.*, **20**, 758 (2006).
10. I. S. Rim and C. H. Jee, *Korean J. Parasitol.*, **44**, 133 (2006).
11. D. L. J. Opdyke, *Food Cosmet. Toxicol.*, **12**(Suppl.) (1974).

YOHIMBE

Source: *Pausinystalia yohimba* Pierre ex Beille. (*P. johimbe* (Schumann) Beille, *Corynanthe johimbi* Schumann) (Family Rubiaceae).

GENERAL DESCRIPTION

Tall evergreen forest tree with large, glabrous, leathery leaves with upcurving lateral nerves fading out at margins; flowers white arranged in umbel-like clusters at the ends of the shoots. The range of this West African tree extends from southwestern Nigeria to Gabon.[1]

The part used is the bark, entering commerce in flattened or slightly quilled pieces.

CHEMICAL COMPOSITION

Bark contains a number of alkaloids, the most important of which is yohimbine (reportedly up to 6% in crude bark) (17α-hydroxy-yohimban-16α-carboxylic acid methyl ester); also yohimbinine, α-yohimbane, yohimbenine, isoyohimbine (mesoyohimbine), dihydroyohimbine, corynantheine, and others.[2]

PHARMACOLOGY AND BIOLOGICAL ACTIVITIES

Biological activity mainly attributed to the alkaloid yohimbine (rather than crude bark) includes aphrodisiac and CNS stimulant activity. Early studies suggest that the alkaloid stimulated the respiratory center in small doses, while depressed respiration in large doses (USD 23RD & 26TH). It readily penetrates the CNS causing a complex response including antidiuresis (due to release of antidiuretic hormone), general excitation, with elevated blood pressure, heart rate, and increased motor activity in both humans and animals. It is generally considered an α_2-adrenergic blocking agent.[3] According to Goodman and Gilman, the activity of yohimbine may result from its activity as a relatively selective inhibitor of α_2-adrenergic receptors, enhancing neural release of norepinephrine at concentrations less than those required to block postsynaptic α_1 receptor.[4] Yohimbine also blocks peripheral 5-HT receptors (GOODMAN AND GILMAN). Aphrodisiac activity has also been attributed to the enlargement of blood vessels in the genitalia, transmission of nerve impulses to genital tissue, and an increased reflex excitability in the sacral region of the spinal cord. Yohimbine is reportedly a monoamine oxidase inhibitor (DUKE 2; TYLER 1); it also possesses a weak calcium channel blocking effect.[5]

Results from a recent study indicated that an aqueous extract of *P. yohimbe* displays endothelin-like actions and affects nitric oxide production in renal circulation, thus augmenting the already established α-adrenergic antagonist effect of yohimbine.[6]

A widely reported clinical study at Queens University (Kingston, Ontario) found that in 23 impotent men, yohimbine improved measurements of nocturnal penile tumescence in 43% of those studied. Yohimbine was determined to be most effective in patients with vascular dysfunction.[7]

TOXICOLOGY

Yohimbine is reported to produce nausea, salivation, irritability, elevated heart rate, and blood pressure (USD 23RD & 26TH). A case of yohimbine-induced skin eruption, renal failure, and lupus-like syndrome has been reported.[8] There is a possible risk of hypertension when yohimbine is administered in combination with tricyclic antidepressants.[9] A recent study in mice showed that in spite of an enhancing effect of an aqueous extract of yohimbe on weight of seminal vesicles and count/motility of spermatozoa, there was an overall reduction in fertility resulting from the development of spermatozoan abnormalities and chromosomal aberrations.[10]

USES

Medicinal, Pharmaceutical, and Cosmetic. Yohimbine has been applied in prescription drugs for impotence of vascular, diabetic, or psychogenic origin; contraindicated when chronic inflammation of the prostate gland or related organs is present; as an MAO inhibitor, it is contraindicated with tyramine-containing foods, antidepressants, and other mood-changing drugs. Risks associated with use of the crude drug have included excitation, tremor, sleeplessness, anxiety, increase in blood pressure, tachycardia, nausea, and vomiting. Observations of interactions with psychopharmacological drugs have been reported but are not well documented.[11] Cardiovascular and behavioral effects of clonidine are antagonized by yohimbe.[12]

Dietary Supplements/Health Foods. Bark in teas, capsules, tablets, tinctures, and other product forms has been promoted as an aphrodisiac and to increase athletic performance (as an alternative to anabolic steroids) (TYLER 1).

Traditional Medicine. Bark reportedly used as an aphrodisiac for centuries, though little evidence support claims that it has an effect on sexual desire or performance. The majority of pharmacological data are on the alkaloid yohimbine, rather than the crude drug.

COMMERCIAL PREPARATIONS

Crude drug. yohimbine hydrochloride (prescription), often with other drugs such as strychnine, thyroid hormones, and/or methyltestosterone (TYLER 1).

Regulatory Status. Crude bark undetermined in the United States, though of substantial regulatory agency interest. The bark is the subject of a negative German monograph, as the efficacy of the bark and its preparations for the claimed applications (male impotence, aphrodisiac activity, stimulant) are not well documented.[11] The FDA placed yohimbe on the "unsafe herb" list as of March 1997.

REFERENCES

See the General References for BLUMENTHAL 1; BRUNETON; DER MARDEROSIAN AND BEUTLER; DUKE 2; GOODMAN AND GILMAN; LEWIS AND ELVIN-LEWIS; MARTINDALE; MCGUFFIN 1 & 2; TYLER 1; TYLER 3; USD 23RD & 26TH; WREN.

1. R. W. J. Keay, *Trees of Nigeria*, Clarendon Press, Oxford, 1989.
2. J. W. Schermerhorn and M. W. Quimby, eds., *The Lynn Index*, Vol. V, Massachusetts College of Pharmacy, Boston, 1962.
3. A. J. Riley, *Br. J. Clin. Pract.*, **48**, 133 (1994).
4. B. B. Hoffman and R. J. Lefkowitz, *New Engl. J. Med.*, **302**, 1390 (1980).
5. K. Watanabe et al., *J. Pharm. Pharmacol.*, **39**, 439 (1987).
6. A. A. Ajayi et al., *Methods Find. Exp. Clin. Pharmacol.*, **25**, 817 (2003).
7. A. Moreles et al., *J. Urol.*, **128**, 45 (1982).
8. B. Sandler and P. Aronson, *Urology*, **41**, 343 (1993).
9. A. Fugh-Berman, *Lancet*, **355**, 134 (2000).
10. A. A. Al-Majed et al., *Asian J. Androl.*, **8**, 469 (2006).
11. Monograph *Yohimbe Cortex*, *Bundesanzeiger*, no. 193 (October 15, 1987).
12. R. D. Robson et al., *Eur. J. Pharmacol.*, **47**, 431 (1978).

YUCCA

Source: *Mohave yucca Yucca schidigera* Roezl ex Ortgies (syn. *Y. mohavensis* Sarg.), *Joshua tree Yucca brevifolia* Engelm. (syn. *Y. arborescens* Trel.), and other *Yucca* spp. (Family Liliaceae or Agavaceae).

GENERAL DESCRIPTION

Mohave yucca (*Y. schidigera*) is a tree seldom exceeding 4.5 m in height with a simple or branched trunk that is 15–20 cm in diameter; leaves narrow, up to 1.5 m long, with few coarse marginal fibers; native to southwestern United States (especially southern Nevada, northwestern Arizona, the Mojave Desert, and southern California to northern Baja California).

Joshua tree (*Y. brevifolia*) is a tree up to about 20 m high, branched mostly at 1–3 m aboveground with lance-shaped leaves (blades 20–35 cm long) crowded in dense clusters near ends of branches; native to southwestern United States (especially the Mojave Desert in California, southwestern Utah, and western Arizona).

Parts used are the leaves from which a solid extract is prepared by hot water extraction.

CHEMICAL COMPOSITION

Yucca extracts contain steroidal saponins, with the major sapogenins being sarsasapogenin and tigogenin in Mohave yucca and Joshua trees, respectively (see **sarsaparilla** and **fenugreek**).[1] The concentration of saponins in these extracts has not been reported, except that Mohave yucca extracts contain about 60% solids.[2] Spirostanol saponins (schidigera-saponins A–F) and furostanol saponin glycosides have been isolated as new saponin compounds from Mohave yucca.[3,4]

Biologically active hydroxystilbenes, such as resveratrol, and other spirophenolic compounds (yuccaols A–C) were also isolated from the bark of Mohave yucca.[5,6]

PHARMACOLOGY AND BIOLOGICAL ACTIVITIES

Saponins generally have hemolytic properties *in vitro*, but when given orally or intravenously, these activities are much weaker. They also have numerous other pharmacological activities (see **quillaia**).

Water extracts of *Y. glauca* Nutt. in mice experiments have shown antitumor activity against B16 melanoma (FOSTER AND DUKE).

The phenolic constituents of *Y. schidigera* were found to possess antiproliferative activity against Kaposi's sarcoma cells and that, at 25 µM, the yuccaols were more active than resveratrol in inhibiting the mediators of proliferation in the investigated cells.[7]

The *Y. schidigera* phenolics were investigated *in vitro* for other activities by one group and were found to possess antioxidant (free radical scavenging), anti-inflammatory (inhibition of NF-κB and iNOS), and antiplatelet adhesion activities.[5,8–12]

The antigiardial effect of a Muhave yucca powder preparation was studied *in vitro* in a trophozoite adherence inhibition assay and *in vivo* in infected gerbils and lambs. In both cases, the saponin-rich butanol extract displayed the highest antigiardial activity.[13] Saponin-rich fractions and saponins from *Y. schidigera* possess antimicrobial activities against food-deteriorating fungi and ruminal bacteria,[4,14] and a protein isolated from the leaves of *Y. recurvifolia* has potent activity against HSV-1.[15]

TOXICOLOGY

The effects of yucca extracts (or saponins) in humans due to long-term ingestion of small amounts of these extracts (especially in root beer) are not known. However, one 12-week study indicated Mohave yucca extract (ca. 60% solids) to be nontoxic to rats; it also demonstrated that the extract was about half as hemolytic *in vitro* as commercial soapbark saponin.[16]

USES

Food. Extract (especially Mohave yucca) is extensively used as the foaming agent in root beer and other frothy drinks, with average maximum use level of about 0.062% (618 ppm) reported.

Dietary Supplements/Health Foods. Used in various preparations to treat arthritic conditions (TYLER 1).

Traditional Medicine. American Indian groups used root of various *Yucca* spp. in salves or poultices for sores, skin diseases, inflammations, and to stop bleeding; in steam baths for sprains and broken limbs; as hair wash for dandruff and baldness (FOSTER AND DUKE).

COMMERCIAL PREPARATIONS

Extracts (mainly Mohave yucca).

Regulatory Status. Both have been approved for food use (§172.510).

REFERENCES

See the General References for FEMA; FOSTER AND DUKE; HOCKING; MUNZ AND KECK; SARGENT; TYLER 1; UPHOF.

1. M. E. Wall and C. S. Fenske, *Econ. Bot.*, **15**, 131 (1961).
2. B. L. Oser, *Food Cosmet. Toxicol.*, **4**, 57 (1966).
3. W. Oleszek et al., *J. Agric. Food Chem.*, **49**, 4392 (2001).
4. M. Miyakoshi et al., *J. Nat. Prod.*, **63**, 332 (2000).
5. S. Piacente et al., *J. Nat. Prod.*, **67**, 882 (2004).
6. W. Oleszek et al., *J. Agric. Food Chem.*, **49**, 747 (2001).
7. C. Balestrieri et al., *Biochem. Pharmacol.*, **71**, 1479 (2006).
8. P. Cheeke et al., *J. Inflamm. (Lond)*, **3**, 6 (2006).

9. B. Olas et al., *Nutrition*, **21**, 199 (2005).
10. S. Marzocco et al., *Life Sci.*, **75**, 1491 (2004).
11. B. Olas et al., *Nutrition*, **19**, 633 (2003).
12. B. Olas et al., *Platelets*, **13**, 167 (2002).
13. T. A. McAllister et al., *Vet. Parasitol.*, **97**, 85 (2001).
14. Y. Wang et al., *J. Appl. Microbiol.*, **88**, 887 (2000).
15. K. Hayashi et al., *Antiviral Res.*, **17**, 323 (1992).
16. L. Liberti, *Lawrence Rev. Nat. Prod.* (1988).

Indian Traditional Medicine

AYURVEDA

The word *Ayurveda* is composed of two words "*ayus*" meaning "life" and "*veda*" meaning "knowledge," that is, "science of a long life." Ayurveda deals with healthy living, along with therapeutic measures that relate to physical, mental, social, and spiritual harmony and is among the few traditional systems of medicine that involve surgery.[1] Though documented references to the precise timing of its origin are not available, its age has been established on the basis of correlating evidence with other disciplines as well as circumstantial evidence to be between 2500 and 600 BC.[2] Ayurveda was first described by Agnivesha, in his book *Agnivesh Tantra*, written during the Vedic period. The book was later revised by Charaka and renamed as *Charaka Samhitā*.[1,2,4] Another early text of Ayurveda is the *Sushruta Samhitā*, which in addition to the *Charaka Samhitā* served as the textual material in the ancient Universities of Takshashila and Nalanda.[3,5] The Ayurvedic system of medicine was orally transferred via the Gurukul system until a script came into existence. The earliest scripts have been written on perishable materials such as *Taalpatra* (leaf surfaces). These scripts were later written on stone and copper sheets. Ayurvedic practices have evolved over time and flourished during the time of Buddha (around 520 BC).[2] During this period, mercury, sulfur, and other metals were used in conjunction with herbs to prepare medications. Also in this period, Ayurveda evolved and flourished with the invention of new drugs, new methodologies, and innovations. The practice of the accompanying surgery also subsided during that period.

During the rule of Chandragupta Maurya (AD 375–415), Ayurveda was part of mainstream Indian medical techniques, and it continued to be so until the British invasion. Chakrapani Dutta (DuttaSharma) who was a *rajabaidya* (Chief Doctor) of King Nayapala (AD 1038–1055) wrote books on Ayurveda such as *Chakradutta* and others. It is believed by some practitioners that Chakradutta is the essence of Ayurveda. Later, about 200 years ago, Pranacharya Shri Sadanand Sharma wrote the *Ras Tarangini*, a book with modernized Ayurveda practices. In this book, advances in chemistry were also included. The book describes the use of many chemical substances, such as sulfates and nitrates, in medicine.

Over the course of time, Ayurveda has evolved as a fully developed medical science with eight branches that are, more or less, parallels to the modern western system of medicine. Ayurveda deals elaborately with measures of healthy living during the entire span of life and its various phases. Besides dealing with principles for maintenance of health, it has also developed a wide range of therapeutic measures to combat illness. These principles of positive health and therapeutic measures related to physical, mental, social, and spiritual welfare of human beings. Ayurveda operates on the precept that all materials of vegetable, animal, or mineral origin have some medicinal value. Ayurvedic medicines are made from a single herb or mixture of herbs, alone or in combination with minerals and other ingredients of animal origin. Several classical treatises indicate the presence of two schools: the physicians (*Atreya Sampradaya*) and the surgeons (*Dhanvantri Sampradaya*), and eight disciplines (*Ashtanga Ayurveda*) classified as (1) *Kaya chikitsa* (Internal Medicine), (2) Kaumarabhrtya (Pediatric), (3) *Shalya* (Surgery), (4) *Shalakya* (Otorhinolaryngology and Ophthalmology), (5) *Agad Tantra* (Toxicology), (6) *Rasayana* (Geriatrics), (7) *Bhoot Vidya* (microorganism and spirits, or Psychiatry), and (8) *Vajikaran* (Eugenics and aphrodisiacs).[6–9] The most important and massive ancient compilation "Charka Samhita" contains several chapters dealing with therapeutic or internal medicine. About 600 drugs of plant, animal, and mineral origin are described in it. Besides, it also deals with other branches of Ayurveda such as anatomy, physiology, etiology, prognosis, pathology, treatment, and medicine.[3,4] An equally exhaustive ancient compilation, "Sushruta Samhita" relates more to the school of surgery. More than 100 kinds of surgical instruments including scalpels, scissors, forceps, specula, and so on are described along with their use. Dissection and operative procedures are also explained. In addition to the therapeutic uses of vegetables and dead animals, topics such as anatomy, embryology, and toxicology are also mentioned together with ca. 650 drugs.[5]

According to Ayurveda, all objects in the universe including the human body are composed of five basic elements (*Panchamahabhutas*), namely, earth, water, fire, air, and vacuum (ether). There is a balanced distribution of these elements in different proportions to suit the needs and requirements of different structures and functions of the body and its parts. It is believed that illness and disease result from an imbalance in the five elements. The "Charaka Samhita" also defines 32 causes of disease resulting in imbalance. Among them are pathogens such as parasites, amoebae, or bacteria (*krimija*); genetic factors (*janaja*), poisons, trauma, psychiatric factors, and so on. The most important category is *prajnaparadha* or intellectual insufficiency. Life according to Ayurveda is conceived as the union of body, senses, mind, and soul and is synonym of dhari, jivita, nityaga, and anubandha. The living man is a conglomeration of three doshas/humors (*Vata*, *Pitta*, and *Kapha*; dosha means "that which changes"), seven basic tissues (*Rasa*, *Rakta*, *Mansa*, *Meda*, *Asthi*, *Majja*, and *Shukra*), and the waste products of the body (feces, urine, and sweat).[5,8] The growth and decay of the body and its constituents revolve around food that gets processed into humors, tissues, and wastes. Ingestion, digestion, absorption, assimilation, and metabolism of food have interplay in health and disease, which are significantly affected by psychological mechanisms as well as by bio-fire (*Agni*).

In Ayurvedic philosophy, the five elements combine in pairs to form three dynamic forces or interactions called doshas or *prakruti*. The three active doshas are *Vata*, *Pitta*, and *Kapha*:[5,7–9]

VATA

Vata is made up of the two elements space and air. Charaka Samhita defines the characteristics of *Vata* dosha as dry and rough (*rookshaha*); cool (*sheetoha*); light-weightless (*laghuhu*); very tiny, penetrating molecules (*sookhshmaha*); always moving (*chalota*); broad, unlimited, unbounded akash means unbounded space (*vishadaha*); and rough (*kharaha*).

PITTA

Pitta is made up of the two elements fire and water. Charaka Samhita defines pitta dosha as hot and a little oily (*sahasnehamushnam*); sharp, burning (*tikshnam*); liquid and acidic (*dravamlam*); always flowing in an unbounded manner (*saram*); pungent and sharp (*katuhu*). *Pitta* contains fire, but it also contains water. It is the source of the flame, but not the flame itself. People with more *Pitta* in their constitutions tend to be of medium proportions, with a frame that is neither petite nor heavy, warm skin that is very fair or ruddy and may be sensitive, and fine hair that tends toward premature graying or thinning. They are sharp and determined in thought, speech, and action. There is an element of purpose to their step and intensity to their voice. Ambition is usually their second name. They are moderate sleepers and gravitate toward cooler environments.

KAPHA

Kapha is made up of the two elements water and earth. Charaka Samhita defines the characteristics of *Kapha* dosha as heavy, since both water and earth are heavy elements, (*guru*); cold (*sheetoha*); soft (*mridu*); oily, offering lubrication (*snigdha*); sweet (*madhura*); stable, offering immunity (*sthira*); and slippery (*tikshila*).

Normally, treatment measures involve use of medicines, specific diet, and prescribed activity routines. Use of these three measures is done in two ways. In one approach of treatment, the three measures antagonize the disease by counteracting the etiological factors and various manifestations of the disease (*Vipreeta* treatment). In the second approach, the same three measures of medicine, diet, and activity are targeted to exert effects similar to the etiological factors and manifestations of the disease process (*Vipreetarthkari* treatments). Falling under these two approaches, the treatment of disease can broadly be classified into

1. **Shodhana therapy (purification treatment)**
   Shodhana aims at removal of the causative factors of somatic and psychosomatic diseases. The process involves internal and external purification. The usual practices involved are *Panchkarma* (medically induced emesis, purgation, oil enema, decoction enema, and nasal administration of medicines) and pre-*panchkarma* procedures (external and internal oleation and induced sweating). *Panchkarma* treatment focuses on metabolic management. It provides needed purificatory effect, besides conferring therapeutic benefits. This treatment is especially helpful in neurological disorders, musculoskeletal disease conditions, and certain vascular or neurovascular states, and respiratory diseases and metabolic and degenerative disorders.

2. **Shamana therapy (palliative treatment)**
   Shamana therapy involves suppression of damaged humors (doshas). It is the process by which disturbed humor subsides or returns to normal without creating imbalance of other humors. This treatment is achieved by use of appetizers, digestives, exercise, and exposure to sun, fresh air, and so on. In this form of treatment, palliatives and sedatives are used.

3. **Pathya Vyavastha (prescription of diet and activity)**
   Pathya Vyavastha comprises indications and contraindications in respect of diet, activity, habits, and emotional status. This is done with a view to enhance the effects of therapeutic measures and to impede the pathogenetic processes. Emphasis on do's and don'ts of diet and so on is laid with the aim to stimulate *Agni* and optimize digestion and assimilation of food in order to ensure restrengthening of tissues.

4. **Nidan Parivarjan (avoidance of disease causing and aggravating factors)**
   Nidan Parivarjan is to avoid the known disease causing factors in the diet and lifestyle of the patient. It also encompasses the idea to refrain from precipitating or aggravating factors of the disease.

5. **Satvavajaya (psychotherapy)**
   Satvavajaya is concerned mainly with mental disturbances. This includes restraining the mind from desires for unwholesome objects and cultivation of courage, memory, and concentration. The study of psychology and psychiatry has been developed

extensively in Ayurveda and has wide range of approaches in the treatment of mental disorders.

6. **Rasayana therapy (use of immunomodulators and rejuvenation medicines)**

Rasayana therapy deals with promotion of strength and vitality. The integrity of body matrix, promotion of memory, intelligence, immunity against disease, the preservation of youth, luster and complexion, and maintenance of optimum strength of the body and senses are some of the positive benefits credited to this treatment. Prevention of premature wear and tear of body tissues and promotion of total health content of an individual are the roles that *Rasayana* therapy plays.

In Ayurveda, regulation of diet as therapy has great importance. This is because it considers human body as the product of food. An individual's mental and spiritual development as well as his temperament is influenced by the quality of food consumed by him. Food in human body is transformed first into chyle or *Rasa* and then successive processes involve its conversion into blood, muscle, fat, bone, bone marrow, and reproductive elements. Thus, food is basic to all the metabolic transformations and life activities. Lack of nutrients in food or improper transformation of food leads to diseases.

UNANI MEDICINE

The Unani system of medicine originated in Greece. The Greek philosopher physician Hippocrates (460–377 BC) is considered as the founder of Unani medicine.[10] Before Hippocrates, medicine was restricted only to the Aesclabius family. Later, a number of Greek scholars enriched the system considerably, but it was Galen (AD 131–210) who stabilized its foundation. Thus, it was earlier known as "Galenic" and later became "Unani" system of medicine when many Arab and Persian scholars further enriched this science (Arabic name for "Greek"). The Egyptians developed this system by preparing the medicine in different dosage forms such as alcohol, oils, powder and ointment, and so on, and the Persians encouraged and developed physicians and philosophers. On this foundations, Islamic physicians such as Al-Razi (Rhazes) (AD 850–925) and Ibn Sina (Avicenna) (AD 980–1037), Al-Zahrawi (Albucasis) the surgeon and Ibn-el-Nafis, to name only a few, constructed an imposing structure. Unani medicine got enriched by incorporating what was best in the contemporary systems of traditional medicine in Egypt, Syria, Iraq, Persia, India, China, and other Middle East and Far East countries. The Unani system received great impetus during the reign of the Abbasids and became a respectable and "rational" science. In India, the Unani system of medicine was introduced by the Arabs and soon it took firm roots in the soil.[11] When the Mongols ravaged Persian and Central Asian cities such as Shiraz, Tabrez, and Geelan, scholars and physicians of Unani medicine fled to India where, during the 13th and 17th centuries, this system reached its peak. The scholars and physicians of Unani medicine who settled in India subjected Indian drugs to clinical trials. In this period, they invented the process of distillation, sublimation, calcination, and fermentation to promote the efficacy of the medicine and to remove the impurities and toxic effects of the drug. As a result of their experimentation, numerous native drugs were added to their own system, further enriching its wealth. Among those who made valuable contributions to the Unani system were Abu Bakr Bin Ali Usman Ksahani, Sadruddin Damashqui, Bahwa bin Khwas Khan, Ali Geelani, Akbal Arzani, and Mohammad Hashim Alwi Khan. During the British rule, Unani medicine suffered a setback and its development was hampered due to withdrawal of governmental patronage. However, since the system enjoyed faith among the masses, it continued to be practiced. The very eminent scholar of this period was Ibn Sina (Avicenna AD 980–1037). He was the great philosopher physician who gave shape to the Unani medicine and redefined many of its concepts. His book *Alqanoon-fil-Tibb* (The Law of Medicine or The Canon in Medicine as commonly known in the West) was an internationally accepted book on medicine, which was taught in European countries until the 17th century.[12]

Of the famous Unani medicine practitioners, *Hakim* (Arabic for wise man or physician) Shareef Khan (1725–1807) wrote his book *Ilaj-ul-Amrad* (Arabic for Treatment of Diseases) in Delhi during the Mongol period. Another physician, Hakim Ajmal Khan (1864–1927), was the first person to introduce modern research in Indian systems of medicine and under his supervision, various alkaloids, for example, ajmaline and reserpine, were isolated from Asarol (*Rauwolfia serpentina*). He also translated 88 Unani books from Arabic and Persian to the Urdu language.

According to Unani medicine, it is established that disease is a natural process and that symptoms are the reactions of the body to a certain disease. It believes in the humoral theory that presupposes the presence of four fluids or humors: *damm* (Arabic for blood), *balgham* (phlegm), *safra* (yellow bile), and *sauda* (black bile) in the body. Each humor has its own temperament. So, blood is hot and moist, phlegm is cold and moist, yellow bile is hot and dry, and black bile is cold and dry. The temperaments of persons are expressed by the words sanguine, phlegmatic, choleric, and melancholic according to the preponderance of the respective humors blood, phlegm, yellow bile, and black bile. If the four main humors and the four primary qualities were all in a state of mutual equilibrium, one is considered healthy. It is believed that every person has a unique humoral constitution that represents his healthy state, and, to maintain the correct humoral balance, there is a power of self-preservation or adjustment called *al-Quwwat-ul-Mudabbira* (medicatrix naturae) in the body. If this power weakens, imbalance in the humoral composition occurs and causes disease. Great reliance is placed in this power. The medicines used in this system, in fact, help the body to regain this power to an optimum level and thereby restore humoral balance, thus retaining health. Also, correct diet and digestion are considered to maintain humoral balance.

The diagnosis of diseases in Unani medicine is through examination of pulse, urine, and stool. This system observes the influence of surroundings and ecological conditions such as air, food, drinks, body movement and repose, psychic movement and repose, sleep and wakefulness, and excretion and retention on the state of health. This influence causes a dominance of one of the four humors in every human body. Unani medicine believes that it is this dominance that gives a man his individual habit and complexion, that is, his temperament.

In short, Unanipathy aims at maintaining proper health by conserving symmetry in the different spheres of a man's life. Unani practitioners not only cure bodily diseases but also act as ethical instructors. When the equilibrium of the humors is disturbed and functions of the body are abnormal, in accordance to its own temperament and environment, that state is called disease. Unani medicine believes in promotion of health, prevention of disease, and cure. The health of a human being is based on six essentials that have to be maintained in order to prevent diseases. The six essentials are (1) atmospheric air, (2) food and drink, (3) sleep and wakefulness, (4) excretion and retention, (5) physical activity and rest, and (6) mental activity and mental relaxation. Another distinctive feature of the Unani system of medicine is its emphasis on diagnosing a disease through examination of *Nabd* (Arabic for pulse), *baul* (urine), and *boraz* (stool).

Disease treatment in Unani medicine can follow one or more of the following methods: (1) Ilaj-bil-Tadbeer (regimenal therapy), (2) Ilaj-bil-Ghiza (dietotherapy), (3) Ilaj-bil-Dava (pharmacotherapy), and (4) Ilaj-bil-Yad (surgery).[10]

1. **Ilaj-bil-Tadbeer (Arabic for regimenal therapy)**
It includes treatments of certain ailments with exercise, massage, *hammam* (Turkish Bath), douches (Cold and Hot), venesection,

cupping, diaphoresis, diuresis, cauterization, purging, emesis, exercise, and leeching.
2. **Ilaj-bil-Ghiza (dietotherapy)**
In this type of treatment, different diets are recommended for the patients depending on the kind of disease. Regulating the quantity and quality of food may also be employed.
3. **Ilaj-bil-Dawa (pharmacotherapy)**
The basic concept of treatment is to correct the disease that may be caused by abnormal humors, due to internal or external causes, through (a) drugs of opposite temperament to the temperament of the disease (*Ilaj-bil-didd*), or (b) drugs of similar temperament as of the temperament of the disease (*Ilaj-bil-mithl*). The drugs used are mostly of the plant origin, but some drugs of animal and mineral origin may also be used. Patients are treated either by single drugs (crude drugs) or by compound drugs (formulations of single drugs).
4. **Ilaj-bil-Yad (surgery)**
Surgery has also been in use in this system for quite long. In fact, the ancient physicians of Unani medicine were pioneers in this field and had developed their own instruments and techniques.

SIDDHA SYSTEM

The word *Siddha* comes from the word "*Siddhi*," which means an object with perfection or heavenly bliss. The principles of this system have close similarity to that of Ayurveda. According to Indian history, prior to Aryans migration, the Dravidians were the first inhabitants of India of whom the Tamilians were the most prominent. Thus, this system of medicine can also be called the Dravidian system of medicine, since it is unique only to the Dravian Tamils and was perfected by the Siddhars.[13] The Siddha system of medicine flourished in the south India, whereas the Ayurveda was prevalent in the north. According to tradition, it is believed that Lord Shiva extended the knowledge of Siddha system of medicine to his wife Parvati who handed it down to Nandi Deva and the Siddhars.[14] After that the Siddha principles were presented to the followers of Lord Siva and Sakti, starting with Siddhar Nantheesar, then Siddhar Thirumoolar, Agathiyar, and other disciples along with the 18 Siddhars and so on.[15] The origin of the documented Siddha system of medicine is generally attributed to the great Siddha Ayastiyar, whose writings are still standard books of medicine and surgery in daily use among the Siddha medical practitioners. According to this system, the human body is a replica of the universe as well as the food and drugs irrespective of their origin. Like Ayurveda, this system believes that all objects in the universe including human body are composed of five basic elements, namely, the earth, water, fire, air, and sky. The food that the human body takes and the drugs it uses are all made of these five elements. The proportion of these elements present in the drugs may vary and their preponderance or otherwise is responsible for certain actions and therapeutic results. As in Ayurveda, this system also considers the human body as a conglomeration of three humors, seven basic tissues, and the waste products of the body such as feces, urine, and sweat. The food is considered as the basic building material of human body, which gets processed into humors, body tissues, and waste products. An equilibrium of humors is considered healthy and its disturbance or imbalance leads to disease or sickness. This system also deals with the concept of salvation in life. The exponents of this system consider the achievement of this state to be possible by medicines and meditation. The system has a rich and unique wealth of drug knowledge in which use of metals and minerals is advocated. There are 25 varieties of water-soluble inorganic compounds called "*uppu*" and different types of alkalis and salts. There are 64 varieties of mineral drugs that do not dissolve in water but emit vapors when put in fire. There are seven drugs that do not dissolve in water but emit vapors on heating. The system has metals and alloys, which melt when heated and solidifies on cooling, classified together. These include items such as gold, silver, copper, tin, lead, and iron, which are incinerated by special processes and used in medicine. There is a group of drugs that exhibit sublimation on heating and includes mercury and its different forms such as red sulfide of mercury, mercuric chloride and red oxide of mercury, and so on. Sulfur, which is insoluble in water, finds a crucial place in Siddha materia medica along with mercury for use in therapeutics and in maintenance of health. In addition, there are drugs obtained from animal sources.[16,17] The Siddhars were also aware of several chemical procedures, such as calcination, sublimation, distillation, fusion, separation, conjunction or combination, congelation, cibation, fermentation, exaltation (the process of refining gold), fixation (bringing to the condition of being nonvolatile), purification, incineration of metals, liquefaction, extraction, and so on. They were also skilled pharmacists and as such were engaged in boiling, dissolving, precipitating, and coagulating chemical substances.[17]

The Siddha system claims to be capable of treating all types of diseases other than emergency cases. In general, this system is effective in treating all types of skin problems, particularly psoriasis, sexually transmitted diseases, urinary tract infections, diseases of liver and gastrointestinal tract, general debility, postpartum anemia, diarrhea, and general fevers in addition to arthritis and allergic disorders.

The diagnosis of diseases involves identifying its causes. Identification of causative factors is through the examination of pulse, urine, eyes, voice, body color, tongue, and the status of the digestive system. This system provides a detailed procedure of urine examination, which includes study of its color, smell, density, quantity, and oil drop spreading pattern. This treatment is individualistic, which ensures that mistakes in diagnosis or treatment are minimal. In general, single or compound medicines are advised for the patients based on *naadi* (pulse) diagnosis methods and the variations of their *vali* or *vaatham*, *azhal* or *pitham*, and *aiyyan* or *kapam* (similar to *vata*, *pita*, and *kapha* in Ayurveda). These three vital forces of cosmic elements are controlled by the functions of the *Punchaboothas*. The five major concepts of *Punchaboothas* are *nilam* or *prithivi* (earth), *neer* or *appu* (water), *kattru* or *vayu* (air), *neruppu* or *theyu* (fire), and *veli* or *akash* (space). According to this theory, all the substances in the universe are created under the actions or reactions of the *Punchaboothas* functions only. If the *Punchaboothas* ratio is disturbed, any disease may attack the body (human beings, animals, birds, flies, etc.) by the way of the deficiency of certain vitamins and minerals. Siddhars also cautioned the administration of certain *basmas* and *sinduras*, which are oxidized stages of metals and minerals, and advised for intaking periods and diet restrictions, according to age, climate, and location.[14,17]

Siddhars classified diseases into different categories that accounted for a total of 4448 human diseases. The fundamental subjects of Siddha methodology are (1) *Vadham* (Alchemy), (2) *Aithiyam* (Medicine), (3) *Yogam* (Yoga), and (4) *Gnanam* or *Thathuvam* (Philosophy). Siddhars explored and explained the reality of nature and its relationship to man by their yogic awareness and experimental findings. They postulated the concept of spiritualism for self-improvement and the practices propounded by them came to be known as the Siddha system. The spiritual scientists of Tamil Nadu introduced the eight mighty Siddhic processes, or octomiracle, "*Atta-Ma-Siddhi*," which could keep the body strong and perfect for external life, where there is no death or rebirth. The *Att-Ma-Siddhi* are (1) *ANIMA*—the faculty of reducing gross body to the size of an atom and to enable him to fly in space, (2) *MAHIMA*—power of expanding oneself without limit, (3) *KARIMA*—power of reducing the primordial elements within oneself to a point desired, (4) *LAHIMA*— power of becoming as light as a feather, (5) *PRAPTHI*—faculty of knowing everything, past, present, and future, and to secure everything as desired, (6) *PRAHAMIYAM*—power of penetration like rays by which one can attain immortality, (7) *ESATHWAM*—supreme power over animates and inanimate in the universe, and (8) *VASITHAM*—power of securing any object.[15,16,18]

REFERENCES

1. A. S. Chopra in Helaine Selin and H. Shapiore, eds., *"Ayurvedas"*, *Medicine Across Cultures*, Kluwer Academic Publisher, USA, 2003, p. 75–83.
2. G. Dwivedi and S. Divedi, History of Medicine: Sushrut—the Clinician—teacher par Excellence. *Indian J. Chest Dis. Allied Sci.*, **49**, 243–244 (2007).
3. P. V. Sharma, translator and editor *Sushrut Samhita*, with English translations of text and Dalhana's commentary along with critical notes, 3 Vols, Visvabharathi, Varanasi, 1983.
4. P. V. Sharma, translator and editor *Caraka-Samhita*, Text with English translation and commentaries, 4 Vols, Chaukhambha Orientalia, India, 1996.
5. E. A. Underwood and P. Rhodes, Encyclopedia Britannica, 2008.
6. http://indianmedicine.nic.in/ayurveda. asp. Introduction to Ayurveda. *Ministry of Health and Family Welfare, Govt. of India*. Retrieved on February 28, 2009.
7. C. Dwarakanath, *The Fundamentals Principles of Ayurveda*, Chowkhamba Krishnadas Academy, 2003.
8. Swami Sadashiva Tirtha, *The Ayurveda Encyclopedia: Natural Secrets to Healing, Prevention & Longevity*, Ayurvedic Holistic Center Press, Bayville, NY, 1998.
9. J. R. S. Murthy, *Astanga Hrayam*, Krishna Academy, Varanasi, 1994.
10. Bilal Ahmad and Jamal Akhtar, *Review Article Unani System of Medicine*, *Phcog. Rev.*, **1**, 210–214 (2007).
11. http://indianmedicine.nic.in/unani.asp. *Ministry of Health and Family Welfare, Govt. of India*. Retrieved on February 28, 2009.
12. O. C. Gruner, *The Canon of Medicine*, English Translation, Luzac & Co., London, 1930.
13. A. Neelavathi, A brief introduction of Siddha system of medicine. *Bull. Indian Inst. Hist. Med. (Hyderabad)*, **9**, 38–40 (1979).
14. http://indianmedicine.nic.in/siddha.asp. *Ministry of Health and Family Welfare, Govt. of India*. Retrieved on February 28, 2009.
15. T. G. Ramamurti Iyer, *The Handbook of Indian Medicine or The Gems of Siddha System*, Chaukhamba Sanskrit Pratishthan, 2005.
16. P.V. Sharma in P.V. Sharma, ed., Siddha medicine. *History of Medicine in India*, The Indian National Science Academy, New Delhi, 1992, p. 445–450.
17. B. V. Subbarayappa in D.M. Bose, S. N. Sen, and B. V. Subarayappa, eds., Chemical practices and alchemy. *A Concise History of Science in India*, Indian National Science Academy, New Delhi, 1971, p. 315–335.
18. K. H Krishnamurthy and G. C. Mouli, Siddha system of medicine: a historic appraisal. *Indian J. Hist. Sci.*, **19**(1), 43–53 (1984).

POPULAR INDIAN MEDICINAL HERBS

ASHWAGANDHA

Source: *Withania somnifera* (L.) Dunal. (syn. *Physalis somnifera* Linn.) (Family Solanaceae).

Common/vernacular names: Ashwagandha.

GENERAL DESCRIPTION

Branched undershrub, branches clothed with mealy stellate hoary tomentum. Leaves ovate, subacute, entire, minutely stellate pubescent with 6–8 pairs of veins. Flowers 5–6 together in sessile or sessile umbellate cymes, greenish to lurid yellow. Calyx with stellate tomentose, green. Fruits berry, red, smooth, enclosed in inflated calyx. Seeds yellowish scruffy.[1]

CHEMICAL COMPOSITION

Alkaloids, steroids, and saponins represent major groups of secondary constituents found in ashwagandha. The bioactive compounds include ashwagandhine, cuscohygrine, anahygrine, tropine, withaferin A, withanolides, withasomniferin, withasomidienone, withasomniferols, withanone, withaniol, sitoindosides, and acylsteryl glucosides.

PHARMACOLOGY AND BIOLOGICAL ACTIVITIES

Antioxidant, immunomodulatory, hematopoietic, anxiolytic, antidepressant, tonic, anti-inflammatory, antitumor, hypolipidemic, and antibacterial activities. The pharmacological activity of the roots is attributed to the presence of several alkaloids.[2,3]

USES

The roots are prescribed for hiccup, female disorders, cough, rheumatism, and dropsy and as a sedative in cases of senile debility. Aswagandha is useful in the treatment of inflammatory conditions, ulcers and scabies when applied locally, and rheumatoid arthritis. Internally, it is given in marasmus in children. The root is given for dyspepsia, and it is prescribed for lumbar pain and for abortifacient properties. The leaves are bitter and given in fever. They are bruised and applied to lesions, painful swellings, and sore eyes. A paste made from the leaves is prescribed for syphilitic sores. It is used in antispasmodic effects against several spasmogens on intestinal, uterine, bronchial, tracheal, and blood vascular muscles. The green berries are bruised and rubbed on ringworm in human beings and on animal sores and girth galls in horses. Ashwagandha is classified as tranquilizer, adaptogenic, and anti-inflammatory. The herbal drug was found to decrease the degree of anxiety and depression and can be used as antidepressant. Along with *Panax ginseng* and *Tribulus terrestris*, ashwagndha was found to give improvement over all psychomotor functions including adaptability of patients, to various stresses, and in the building of tissues. Ashwagandha is used in several Ayurvedic preparations along with other herbal drugs such as "Laksha Guggulu," "Raktawardak," "Abana," and BR-16A in the treatment of hypercholesterolemia, mental disturbances, and convulsions.

REFERENCES

1. G. L. Gupta and A. C. Rana, *Pharmacognosy Rev.*, **1**, 129 (2007).
2. K. R. Kirtikar and B. D. Basu, *Indian Medicinal Plants*, Vol. 1, Dehradun, India, 1999, p. 1774–1776.
3. N. P. Bector et al., *Indian J. Med. Res.*, **56**, 1581 (1968).

GOKSHURA

Source: *Tribulus terrestris* Linn. (Family Zygophyllaceae)

Common/vernacular names: Gokshura, Ikshugancdha.

GENERAL DESCRIPTION

A variable, prostrate annual herb, up to 90 cm in length. Stems and branches pilose. Roots slender, cylindrical, somewhat fibrous, 10–15 cm long, light brown, and faintly aromatic. Leaves opposite, paripinnate; leaflets 5–8 pairs, subequal, oblong to linear–oblong; mucronate, sericeous-villous with appressed hairs beneath more few on the upper surface. Flowers axillary, leaf-opposed, solitary, pale yellow to yellow; pedicel 1.2 cm long, slender, and hairy. Fruits globose, consisting of 5–12 woody cocci, each with two pairs of hard, sharp, divaricate spines, one pair longer than the other; seeds several in each coccus with transverse partitions between them.[1]

CHEMICAL COMPOSITION

Plant contains saponins, which, on hydrolysis, yield the steroidal sapogenins: diosgenin, gitogenin, chlorogenin, ruscogenin, and 25-D-spirosta-3,5-diene. Saponins having high hemolytic index are present in leaves and roots, but absent in stems or seeds; three saponins in leaves and two in roots have been identified. The flavonoids kaempferol, kaempferol-3-glucoside, kaempferol-3-rutinoside, and a new flavonoid tribuloside have also been isolated from the leaves and fruits;[2,3] alkaloids are present in the fresh aerial parts of the plants, the major ones being β-carboline indole-amines, namely, harmane and norharmane. Lignanamides and cinammic acid amides have also been isolated from *T. terrestris*. The presence of spirostanol and furostanol saponins is a characteristic feature of this plant. The primary bioactive components are tigogenin, neotigogenin, gitogenin, neogitogenin, terrestriamide, protodioscin, prototribestin, pseudoprotodioscin, dioscin, tribestin, tribulosin, hecogenin, neohecogenin, diosgenin, ruscogenin, chlorogenin, and sarsasapogenin.[4] The alkaloid, harman, has been reported from the herb and harmine from the seeds.[5] The fruits also contain alkaloids, resin, and a fixed oil (3.5–5.0%), consisting mainly of unsaturated acids, tannins, reducing sugars, sterols, an essential oil, nitrates, peroxidase, diastase, and traces of a glucoside. The fatty oil has excellent drying properties and the seed-cake is rich in phosphorus and nitrogen.

PHARMACOLOGY AND BIOLOGICAL ACTIVITIES

Antimicrobial tonic, immunomodulatory, cytotoxic, hypoglycemic, and anthelmintic activities.

USES

Leaves and fruits are used as a sweetish cooling tonic, aphrodisiac, and are useful in urinary discharge, to alleviate burning sensation, to reduce inflammation, cough, asthma, and pain, and to cure skin and heart diseases. The leaves are also used to purify blood. The seeds are a cooling diuretic that reduce inflammation and expel bladder stones. The leaves and tender shoots are used by economically disadvantaged classes as a potherb, either alone or mixed with other herbs. It is used to increase menstrual flow and to alleviate local symptoms of gonorrhea. A decoction is used as mouthwash to treat painful gums and to reduce inflammation. The fruits have diuretic and tonic properties and are used for the treatment of renal calculi and painful micturition. In China, the fruits are a reputed tonic and astringent; used for coughs, scabies, and ophthalmic disorders. In India, flour from the fruits is made into bread and eaten during the times of scarcity. The root is credited with laxative and tonic properties. It forms a constituent of the well-known Ayurvedic medicines Dashamoolarishtha and Amritha Prasa Ghritha, prescribed for several diseases.

REFERENCES

1. K. R. Kirtikar and B. D. Basu, *Indian Medicinal Plants*, Vol. 1, Dehradun, India, 1999, p. 419–423.
2. R. N. Chopra and S. Ghosh, *Indian J. Med. Res.*, **17**, 377 (1929).
3. B.C. Bose et al., *Indian J. Med. Res.*, **17**, 291 (1963).
4. I. Kostova and D. Dinchev, *Phytochemistry Rev.*, **4**, 111 (2005).

GREATER GALNANGA

Source: *Alpinia galanga* (L.) Willd. (Syn. *Amomum galanga* (Linn.) Lour.) (Family Zingiberaceae).

Common/vernacular names: Kulinjan, Greater Galangal.

GENERAL DESCRIPTION

An herb, 1.8–2.4 m in height, with tuberous aromatic rootstocks. Leaves oblong–lanceolate, acute, glabrous, 30–60 cm long, with rounded ligule. Flowers greenish white, streaked with red, in dense-flowered, 30 cm long panicles; capsules orange or red, globose. The dried rhizome provides the drug Greater Galangal. The rhizome is 2.5–10.0 cm thick and is reddish brown externally and light orange-brown inside. It has a tough and fibrous fracture, and a spicy pungent taste.[1]

CHEMICAL COMPOSITION

On steam distillation, fresh rhizomes yield an essential oil (0.04%) with a peculiar strong and spicy odor. The oil contains methyl cinnamate, cineol, camphor, and probably *d*-pipene, fenchyl acetate, β-caryophyllene, β-farnesene, caryophyllene oxide, 1,8-cineole, limonene, camphor, β-terpineol, and cubenol. (*S*)-1′-Acetoxychavicol acetate is a characteristic aroma compound.[3] A hydrocarbon ($C_{15}H_{30}$) is also reported to be present. The leaves yield an essential oil containing mostly methyl cinnamate. Other constituents include galanganal, galanganols, phenylpropanoids, *p*-hydroxybenzaldehyde, galangin, alpinin, kaempferide, and 3-dioxy-4-methoxy flavone.[2] Several volatile constituents have been identified in the essential oils obtained from the rhizome and leaves. The major constituents of the rhizome oil were myrcene (94.51%), (*Z*)-β-ocimene (2.05%), and α-pinene (1.16%) and that of the leaf oil were myrcene (52.34%), (*Z*)-β-ocimene (17.06%), α-pinene (9.00%), and borneol (4.13%). Ethyl-*trans*-cinnamate and ethyl-4-methoxy-*trans*-cinnamate were identified in the root oil.

PHARMACOLOGY AND BIOLOGICAL ACTIVITIES

Antifungal, antioxidant, antiulcer, antiinflammatory, hypoglycemic, immunostimulating, antitumor, and cytotoxic activities have been reported.[3] It has an antispasmodic effect that alleviates asthma and also exhibits antiamphetamine and diuretic properties.

The seeds contain two potent antiulcer principles, namely, 1′-acetoxychavicol acetate and 1′-acetoxyeugenol acetate. Caryophyllene oxide, caryophyllenol I, caryophyllenol II, pentadecane, 7-heptadecane, and fatty acid methyl esters have also been isolated.[1] The seeds contain diterpenes-I and -II (15α- and 15β-isomers), which have been patented in Japan as antitumor agents.[4] It also exerted anti-inflammatory activity in both acute and chronic inflammation. The ethanolic extract of the herb showed significant antiulcer activity in rats, which has been attributed to the antisecretory and cytoprotective activities of the plant.

USES

The oil is carminative and in moderate doses has an antispasmodic and bactericidal properties. The oil is also used in perfumery.

The drug is used in rheumatism and bronchial catarrh. It is considered a tonic, stomachic, carminative, and stimulant and is used as a fragrant adjunct to complex preparations and also in cough and digestive mixtures. Its chief use is for clearing the voice. The drug has an expectorant action and is useful in many respiratory ailments, especially of children.

The rhizomes are used as a condiment in Indonesia and India for seasoning fish and in pickling.

REFERENCES

1. K. R. Kirtikar and B. D. Basu, *Indian Medicinal Plants*, Vol. 4, Dehradun, India, 1999, p. 2445.
2. T. Morikawa et al., *Chem. Pharm. Bull.*, **53**, 625 (2005).
3. A. Tewari et al., *J. Med. Arom. Plant Sci.*, **21**, 1155 (1999).
4. G. Q. Zheng et al., *J. Agric. Food Chem.*, **41**, 153 (1993).

GURMAR

Source: *Gymnema sylvestre* (Retz.) R. Br. (syn. *Periploca sylvestris* Retz. Obs) (Family Asclepiadaceae).

Common/vernacular names: Kavali, Pitani.

GENERAL DESCRIPTION

Woody, branched climber, young stem and branches are pubescent, often dense, terete. Leaves ca. 5–6 cm long, ovate, elliptic to lanceolate, shortly acuminate, pubescent on both surfaces, dense on the abaxial side, rounded to cordate at base with pubescent petiole. Flowers in sessile to pedunculate cymes. Peduncles are densely pubescent. Calyx pubescent, flowers yellow, tube campanulate, lobes deltoid to ovate, thick, glabrous. Fruit is a follicle 4–6 cm long, terete, rigid, lanclolate, attenuate into a beak, and glabrous. Seeds with broad marginal wing, narrowly ovoid–oblong, glabrous, and brown.

CHEMICAL COMPOSITION

The leaves of *G. sylvestre* were found to have oleanane-type triterpenes and/or their glycosides gymnemic acid A,[1] 3β,16β,22β,28-tetrahydroxy-olean-12-en-30-oic acid,[2] sodium salt of alternoside II, 21β-O-benzoylsitakisogenin 3-O-β-D-glucopyranosyl (1 → 3)-β-D-glucuronopyranoside, potassium salt of longispinogenin 3-O-β-D-glucopyranosyl (1 → 3)-

β-D-glucuronopyranoside, potassium salt of 29-hydroxylongispinogenin 3-*O*-β-D-glucopyranosyl (1 → 3)-β-D-glucuronopyranoside,[3] longispinogenin 3-*O*-β-D-glucuronopyranoside, 21β-benzoylsitakisogenin 3-*O*-β-D-glucuronopyranoside, 3-*O*-β-D-glucopyranosyl (1 → 6)-β-D-glucopyranosyl oleanolic acid 28-*O*-β-D-glucopyranosyl ester, oleanolic acid 3-*O*-β-D-xylopyranosyl (1 → 6)-β-D-glucopyranosyl (1 → 6)-β-D-glucopyranoside, 3-*O*-β-D-xylopyranosyl (1 → 6)-β-D-glucopyranosyl (1 → 6)-β-D-glucopyranosyl oleanolic acid 28-*O*-β-D-glucopyranosyl ester, 3-*O*-β-D-glucopyranosyl (1 → 6)-β-D-glucopyranosyl oleanolic acid 28-β-D-glucopyranosyl (1 → 6)-β-D-glucopyranosyl ester, oleanolic acid 28-*O*-β-D-glucopyranosyl ester, oleanolic acid 3-*O*-β-D-glucopyranosyl (1 → 6)-β-D-glucopyranoside,[4] gymnemosides a–b,[5] gymnemic acids III, IV, V, VIII–XII, XV–XVIII,[6–8] and gymnemasins A–D.[9] The dammarane-type triterpene glycosides were also reported from the leaves (gymnemasides I–VII, gypenoside XXVIII, XXXVII, LV, LXII, and LXIII),[10] in addition to kaempferol and quercetin-type flavonol glycosides.[11]

PHARMACOLOGY AND BIOLOGICAL ACTIVITIES

Some oleanane-type saponins isolated from *G. sylvestre* exhibited antisweetness activity.[3,12] When chewed, the leaves produce a prolonged desensitization effect for sweet and bitter tastes. The sour taste is not affected, while the salty taste is very slightly affected. The saponin fraction,[13] and ethanolic extract of the leaves,[14,15] showed significant antibacterial and antifungal activities. The aqueous extract of *G. sylvestre* possessed hypoglycemic activity.[16] The leaf powder stimulates the heart and the circulatory system, increases the secretion of urine, and activates the uterus.

USES

The plant is stomachic, stimulant, laxative, and diuretic. The laxative property is attributed to the presence of anthraquinone derivatives. It is useful in cough, liver problems, and sore eyes. The leaves have sometimes been used as a remedy for diabetes. The drug has been used for dysgeusia (bad taste) and furunculosis.

REFERENCES

1. Y. Wang et al., *Huaxi Yaoxue Zazhi*, **19**, 336 (2004).
2. S. Peng et al., *Chin. Chem. Lett.*, **16**, 223 (2005).
3. W. Ye et al., *J. Nat. Prod.*, **64**, 232 (2001).
4. W. Ye et al., *Phytochemistry*, **53**, 893 (2000).
5. N. Murakami et al., *Chem. Pharm. Bull.*, **44**, 469 (1996).
6. H. Liu et al., *Chem. Pharm. Bull.*, **40**, 1366 (1992).
7. K. Yoshikawa et al., *Chem. Pharm. Bull.*, **40**, 1779 (1992).
8. K. Yoshikawa et al., *Chem. Pharm. Bull.*, **41**, 1730 (1993).
9. N. P. Sahu et al., *Phytochemistry*, **41**, 1181 (1996).
10. K. Yoshikawa et al., *Phytochemistry*, **31**, 237 (1992).
11. X. Liu et al., *Carbohydr. Res.*, **339**, 891 (2004).
12. K. Yoshikawa et al., *Chem. Pharm. Bull.*, **40**, 1779 (1992).
13. K. V. Gopiesh and K. Venkatesan, *World J. Microbiol. Biotech.*, **24**, 2737 (2008).
14. R. K. Satdive et al., *Fitoterapia*, **74**, 699 (2003).
15. K. Shankar et al., *Int. J. Chem. Sci.*, **5**, 993 (2007).
16. B. Khan et al., *Pak. J. Pharm. Sci.*, **18**, 62 (2005).

KALMEGH

Source: *Andrographis paniculata* (Burm. f.) Wall. ex Nees (Family Acanthaceae).

Common/vernacular names: Kreata, Chiretta, Chuan xin lian, Kalmegh, and Bhunimba/ Mahatikta.

GENERAL DESCRIPTION

Annual erect herb, branched, branches quadrangular, slightly winged at upper end. Leaves ca. 6 × 2 cm, lanceolate, acute at apex, glabrous, slightly undulate, pale beneath, base tapering, with 4–6 pairs of veins. Flowers small, solitary, in axillary or terminal racemes or panicle, with bracts and bracteolates. Flowering pedicle are glandular pubescent. Sepals linear–lanceolate, glandularly pubescent; corolla rose colored, slightly hairy outside, two lipped, upper lipped oblong two toothed at apes, lower deeply three cleft. Filaments hairy in upper part, anthers bearded at base. Fruit a capsule, linear–oblong, acute at ends. Seeds yellowish brown, rugosely pitted, subquadrangular, and glabrous.[1,2]

CHEMICAL COMPOSITION

Rich in diterpenoids and diterpenoid glycosides, including andrographolide, neoandrographolide, deoxyandrographolide, isoandrographolide, 14-deoxyandrographolide, 14-deoxy-11,12-didehydroandrographolide, andrographiside, deoxyandrographiside, homoandrographolide, andrographan, andrographon, andrographosterin, and stigmasterol. The primary bioactive component is probably andrographolide. Flavonoids have also been isolated.[3]

PHARMACOLOGY AND BIOLOGICAL ACTIVITIES

A broad range of pharmacological effects have been reported, including hepatoprotective, immunostimulanting, anti-inflammatory, antimalaria, antidiarrheal, hypoglycemic, and antifertility activities.

USES

Kalmegh has astringent, anodyne, tonic, and alexipharmic properties useful in dysentery, cholera, influenza, bronchitis, swelling and itching, pile, and gonorrhea.

Traditional Medicine. Juice extracted from the leaves is mixed with certain spices such as cardamoms, cloves, cinnamon, and so on and is sun dried and made into globules, which are given to infants to relieve griping, irregular stools, and loss of appetite. The roots and leaves are febrifuge, used as antihelmintic. An infusion of the whole leaf is also used against fever. A decoction or infusion of leaf is useful in general debility and dyspepsia.[1]

COMMERCIAL PREPARATIONS

Kalmegh forms a part of a number of drug formulations available in Indian market. It is a major constituent of an Ayurvedic drug (SG-1 Switradilepa), which is effective in treating vitiligo. It is also used in certain homoeopathic preparations.

REFERENCES

1. K. R. Kirtikar and B. D. Basu, *Indian Medicinal Plants*, Vol. 3, Dehradun, India, 1991, p. 1884.
2. T. Cook, *Flora of Presidency of Bombay*, Vol. 2, Botanical Survey of India, Calcutta, 1906, p. 451.
3. S. K. Mishra, et al., *Pharmacognosy Rev.*, **1**, 283 (2007).

MAHBALA

Source: *Sida rhombifolia* L. (Family Malvaceae).

Common/vernacular names: Mahabala.

GENERAL DESCRIPTION

A small undershrub, branches with stellate hairs. Leaves variable in shape ca. 5 cm long, rhomboidal–lanceolate to obovate truncate, cuneate at base, coarsely toothed, dark green glabrous above, tomentose beneath. Flowering pedicel axillary yellow or white, solitary or in pairs crowded toward the end of the branches. Calyx angular, lobes triangular, acute. Flowers yellow, up to 1.5 cm in diameter. Fruits globose, enclosed by calyx, carpel 7–10 with two short awns. Seed black, smooth.

CHEMICAL COMPOSITION

Three types of alkaloids, β-phenethylamines (β-phenethylamine, ephedrine, Ψ-ephedrine), quinazolines (vasicine, vasicinol, vasicinone), and carboxylated tryptamines (S-(+)-N_b-methyltryptophan methyl ester), in addition to choline and betaine, were found in *Sida rhombifolia*.[1] Recently, seven ecdysteroids and/or their glycosides 25-acetoxy-20-hydroxyecdysone-3-O-β-D-glucopyranoside, pterosterone-3-O-β-D-glucopyranoside, ecdysone-3-O-β-D-glucopyranoside, ecdysone, 20-hydroxyecdysone, 2-deoxy-20-hydroxyecdysone-3-O-β-D-glucopyranoside, and 20-hydroxyecdysone-3-O-β-D-glucopyranoside were isolated from *S. rhombifolia*.[2] Sterols (β-sitosterol, stigmasterol, campesterol) and alkanes (nonacosane, hentriacontane) were also identified.[3]

PHARMACOLOGY AND USES

The plant is useful in tuberculosis and rheumatism. The root and leaves are aphrodisiac, tonic, good in urinary disorders, burning sensations, piles, and all kinds of inflammation. Stem mucilage is employed as demulcent and emollient; it is also used internally in skin diseases and as a diuretic and febrifuge. Roots are used in the treatment of rheumatism and leucorrhoea (vaginal discharge). Ethanolic extract of the plant depresses the activity of the smooth muscles of the ileum of guinea pig. The yellow flower is eaten with wild ginger to ease labor pains. Leaf tips crushed in water are drunk to ease menstruation problems. Leaf tips are chewed with *Areca catechu* to induce abortion or to cause sterility. The flower is chewed to give temporary relief from toothache. The plant is a major ingredient in several Ayurvedic medicines. The alcoholic extract of the roots exhibited antibacterial and antipyretic activities. It also exhibited antimalarial activity *in vitro* against erythrocytic stages of *Plasmodium berghei*. It is a potential source of natural antioxidants.[4] Ethyl acetate extract showed potent cytotoxicity.[5]

REFERENCES

1. A. Prakash et al., *Planta Med.*, **43**, 384 (1981).
2. A. N. Jadhav et al., *Chem. Biodivers.*, **4**, 2225 (2007).
3. M. M. Goyal and K. K. Rani, *Fitoterapia*, **60**, 163 (1989).
4. K. Dhalwal et al., *J. Med. Food*, **10**, 683 (2007).
5. M. E. Islam et al., *Phytother. Res.*, **17**, 973 (2003).

NEEM

Source: *Azadirachta indica* A. Juss. (syn. *Melia azadirachta* L.) (Family Meliaceae).

Common/vernacular names: Neem/Margos tree.

GENERAL DESCRIPTION

Large tree, up to 50 ft tall. Leaves simple, pinnately compound 16–20 cm long, crowded near the end of branches, leaflets subopposite, obliquely lanceolate, sometimes falcate, acuminate, serrate, glabrous on both the surfaces, base unequal, acute. Flowers in branched glabrous panicles, with bract. Bracts minute, lanceolate, caduceus. Calyx pueblos divided at the base. Corolla oblong–obovate, slightly puberlous outside, white, slightly fragrant. Fruits a drupe, ca. 1.5 cm long, glabrous, one-seeded.[1]

CHEMICAL COMPOSITION

Hundreds of compounds have been isolated from neem. The main class includes phenolic diterpenoids (protomeliacins) and limonoids (*C*-secomeliacins).[2] The main bioactive compounds include azadirachtin, azadiradione, azadirone, nimbin, nimbinin, nimbolins, nimbolide, margolone, margolonone, isomargolonone, meliantriol, melianone, melianol, and salanin.[3]

PHARMACOLOGY AND BIOLOGICAL ACTIVITIES

Insecticidal, antifeedant, antibacterial, antifungal, antimalarial, antiviral, cytotoxic, antipyretic, anti-inflammatory, analgesic, antiulcerogenic, antihypertensive, antihyperglycaemic, hepatoprotective, and immunomodulatory activities.[2,3]

USES

The bark exudes a clear, bright, amber-colored gum, known as the East India gum. The gum is a stimulant, demulcent, and tonic and is useful in catarrhal and other infections. The oil is used in some chronic skin diseases and ulcers. It is a common external application for rheumatism, leprosy, and sprain. The warm oil relieves ear trouble; the oil also cures dental and gum troubles. A few drops of oil taken in betel leaf provide relief in asthma. The oil is reported to have antifertility properties. It possesses antiseptic and antifungal activity and is found to be active against both Gram-positive and Gram-negative organisms. Leaves are used for skin disorders. Nimbin has been found to be antipyretic and nonirritant.

Medicinal, Pharmaceutical, and Cosmetic. Oil from the seed kernel is employed in the cosmetic preparations such as creams, hair lotions, medicated soaps, washing soaps and toothpastes; it can be mixed with other oils also for soaps. Regular application of hair oil containing oil of margosa is reported to prevent baldness and graying of hair. The shell from the seeds can be used for the production of activated carbon and tooth powder. The major use of neem oil is in the soap industry. Soaps and shampoos prepared from neem oil control ticks and fleas. Neem shampoo is used for greasy scalp. A prickly heat powder is also made from neem. A neem face pack is prepared for oily and pimple-prone skin. Tooth powder and toothpaste prepared from neem are effective dentifrices. Patented extracts of neem bark Silvose T and Silvose TRS are used as toothpaste and mouth wash.

Traditional Medicine. Neem extracts have been reported to possess antidiabetic, antibacterial, and antiviral properties and are used in cases of stomach worms and ulcers. The stem and root bark and young fruits are reported to possess astringent, tonic, and antiperiodic properties. The root bark is reported to be more active than the stem bark and young fruits. The bark is reported to be beneficial in malarial fever and useful in cutaneous diseases. The leaf paste is useful in cowpox ulceration. Fresh mature leaves along with seeds of *Psoralea corylifolia* Linn. and *Cicer arietinum* Linn. are used to prepare a very effective medicine for leucoderma. The hot infusion of leaves is much used as anodyne for swollen glands, bruises, and sprains. The shade-dried leaves, on steam distillation, yield a golden yellow essential oil (0.13%). The phosphate buffer, ether, and alcoholic extracts of the leaves inhibit the activity of *Micrococcus pyogenes* var. *aureus*. The essential oil possesses marked antibacterial properties and inhibits the growth of *Mycobacterium tuberculosis*. The fruit is used as a tonic, antiperiodic, purgative, emollient, and as an antihelminthic. It is beneficial in urinary diseases and the powdered seeds mixed with honey are reported to be given in piles.

COMMERCIAL PREPARATIONS

Leaves are regularly fed to cattle and goats to increase the secretion of milk, immediately after parturition. They are carminative and aid digestion. An ointment is prepared from neem for dermatological use. Another neem cream is a fly and mosquito repellent. A wound dressing contains neem oil as one of its active ingredients. It is recommended only for animals. Neem leaf decoction is used as a galactogogue for initiating milk secretion in nursing mothers and also recommended for diabetes mellitus of adults, nonketonic diabetes, as well as in cases of insulin sensitivity. Tablets and injections are being formulated for chronic malaria. A neem leaf preparation is also recommended as a local sedative for external applications.

REFERENCES

1. K. R. Kirtikar and B. D. Basu, *Indian Medicinal Plants*, Vol. 1, International Book Distributors, Dehradun, India, 1999, p. 536–541.
2. A. K. Gupta and N. Tandon, *Indian Medicinal Plants*, Vol. 3, Indian Council of Medicinal Research, New Delhi, 2004, p. 316–453.
3. A. C. Vishnukanta, *Pharmacognosy Rev.*, **2**, 173 (2008).

TERMINALIA/BHIBITKI

Source: *Terminalia bellerica* (Gaertn.) Roxb (syn. *Myrobalanus bellirica* Gaertn.) (Family Combretaceae).

Common/vernacular names: Terminalia.

GENERAL DESCRIPTION

A large tree, up to 50 ft in height. Leaves alternate coriaceous, ca. 16 × 6–8 cm, broadly elliptic, or elliptic ovoid, rounded or rarely subacute or shortly acuminate at the apex, surface puberlous when young, glabrous, reticulate when old, entire along the margins, pellucid, narrow at base, with 6–8 pairs of veins, with a prominent midrib on both the surfaces. Petiole 4–8 cm long, with glands at the apex, flowers pale greenish yellow, with odors, in spikes longer than the petiole, but shorter than the leaves. Male flowers in upper spikes, lower flowers hermaphrodite, sessile. Calyx pubescent, outside, densely villous within, teeth broadly triangular, acute. Fruit a drupe ca. 2.5 cm long, narrowing into a short stalk, with minute pale tomentum, obscurely angled when dried.

CHEMICAL COMPOSITION

Phytochemical studies revealed the presence of cardenolides cannogenol 3-O-β-D-galactopyranosyl-(1 → 4)-O-L-rhamnopyranoside, sarmentogenin-3-O-β-D-glucopyranosyl (1 → 4)-(L-rhamnopyranoside),[1,2] lignans (termilignan, thannilignan, and anolignan B),[3] a flavan (7-hydroxy-3′,4′-methylenedioxyflavan),[3] and triterpenoids and/or their glycosides (2,3,7,23-tetrahydroxyolean-12-en-28-oic acid, 2,3,19,23,24-pentahydroxyolean-12-en-28-oic acid, β-D-glucopyranosyl 2,3,7,23-tetrahydroxyolean-12-en-28-oate, β-D-galactopyranosyl 2,3,19,23,24-pentahydroxyolean-12-en-28-oate, arjungenin, arjunglycoside, belleric acid, and bellericoside).[4,5] The kernels of *T. bellirica* contain 40% oil and 35% protein. The oil has palmitic (35%), oleic (24%), and linoleic (31%) acids as major fatty acids.[6]

PHARMACOLOGY/BIOLOGICAL ACTIVITIES AND USES

The kernels possess narcotic properties and are sometimes eaten with betel nut and leaf for the treatment of dyspepsia; the ripe fruit is used as an astringent. The pulp is employed in dropsy, piles, diarrhoea, and leprosy, also occasionally in fever. When half ripe, it is used as a purgative due to the presence of an oil that has properties similar to those of castor oil. The oil is considered a good application for hair; it is also employed as an application for rheumatism. The gum is believed to be demulcent and purgative. The bark is a mild diuretic.[7] The dried fruit constitutes the drug "Bahera." The fruit is one of the constituents of the popular Indian medicine "Triphala" or the three myrobalans; the other two being *T. chebula* and *Emblica officinalis*. Triphala possesses rejuvenating, astringent, cardioprotective, aperient, antiflatulant, antacid, anthelmintic, cardioprotective, laxative, and antibacterial properties.[8] Triphala has significant hypoglycemic effect and possesses antiinflammatory, antiarthritic, and analgesic properties. It is also used extensively as an adjunct to many other Unani and Ayurvedic medicines such as lohasava, given for anaemia, Triphala gugglu (TG3X) for obesity; Kachnar gugglu for glandular swelling; Majoon-e-fanjnosh for liver ailments, ascites, spermatorrhoea, and dropsy. Majoon-e-flasia given in arthritis, gout, rheumetic pains, and anorexia, and Hepatogard® a drug that exhibited hepatoprotective activity against paracetamol-induced liver injury in rats. The fruit is used against gastrointestinal disorders. The fruit is also an ingredient of "Livol," an herbal drug having antihepatotoxic activity. It is also given in viral hepatitis. Using herbal eye drops containing *T. bellirica*, encouraging results have been obtained in cases of myopia, corneal opacity, terigium, and immature cataract, chronic and acute infective conditions.[8] The fruit possesses myocardial depressive activity. The alcoholic extract of the fruit exerted a negative chronotropic and inotropic and hypotensive effects of varying magnitude in a dose-dependant fashion on isolated rat and frog atria and rabbit heart.[9]

Antiviral activity, including anti-HIV-1 activity, antibacterial and antifungal activities of crude extracts of *T. bellerica* fruits have been reported.[3] The lignans (termilignan, thannilignan, and anolignan B) and flavan (7-hydroxy-3′,4′-methylenedioxyflavan) isolated from it possessed demonstrable anti-HIV-1, antimalarial, and antifungal activities *in vitro*.[3] *T. bellerica* fruit possesses a combination of anticholinergic and Ca^{2+} antagonist effects, which explain its folkloric use in colic, diarrhea, and asthma.[10]

REFERENCES

1. R. N. Yadava and K. Rathore, *Fitoterapia*, **72**, 310 (2001).

2. R. N. Yadava and K. Rathore, *J. Inst. Chemists*, **72**, 154 (2000).

3. R. Valsaraj et al., *J. Nat. Prod.*, **60**, 739 (1997).
4. S. B. Mahato et al., *Tetrahedron*, **48**, 2483 (1992).
5. A. K. Nandy et al., *Phytochemistry*, **28**, 2769 (1989).
6. C. Rukmini et al., *J. Am. Oil Chem. Soc.*, **63**, 360 (1986).
7. K. R Kirtikar and B. D. Basu, *Indian Medicinal Plants*, Vol. 2, Dehradun, India, 1999, p. 1018.
8. M. A. Iyengar and S. Dwivedi, *Indian Drugs*, **26**, 655 (1989).
9. S. Srivastava et al., *Indian Drugs*, **29**, 144 (1992).
10. A. H. Gilani et al., *J. Ethnopharmacol.*, **116**, 528 (2008).

VASAKA

Source: *Justicia adhatoda* L. (syn. *Adhatoda vasica* Nees.; *Adhatoda zeylanica* Medic.) (Family Acanthaceae).

Common/vernacular names: Adulsa, Vasaka.

GENERAL DESCRIPTION

A dense erect much branched, gregarious shrub, up to 6–7 ft tall, stem quadrangular to nearly terete yellowish, glabrous. Leaves 15–16 cm long, elliptic–lanceolate, acuminate, minutely puberlous when young, glabrous when mature, entire, dark green, pale beneath, tapering at base, with 10–12 pairs of veins. Flowers white in dense axillary and terminal peduncluate spikes, ca. 6 cm long. Peduncles 4–6 cm long, stout. Flowers surrounded by bract and bracteoles. Calyx, puberlous, corolla white with a few irregular lines on the throat, pubescent outside. Fruit a capsule, ca. 2.5 cm long, clavate, subacute, bluntly acute, pointed four-seeded, pubescent. Seeds orbicular, oblong, tubercular verrucose, and glabrous.

CHEMICAL COMPOSITION

The pyrroloquinazoline-type alkaloids (peganidine, desmethoxyaniflorine, 7-methoxyvasicinone, L-vasicine, L-vasicinone, L-vasicol, anisotine, 3-hydroxyanisotine, vasnetine),[1–3] triterpenoids (3α-hydroxy-D-friedoolean-5-ene, epitaraxerol),[1] aliphatic long chains (29-methyltriacontan-1-ol, 37-hydroxyhexatetracont-1-en-15-one, and 37-hydroxyhentetracontan-19-one),[4,5] indole alkaloids (9-acetamido-3,4-dihydropyrido[3,4-*b*]indole),[6] β-sitosterol,[4] and β-sitosterol-β-D-glucoside[6] were isolated from *A. vasica*. The essential oils mainly α-phellandrene (56.23%), longifolene (25.10%), and α-cedrene (18.67%) from stems and longifolene (25.12%), α-cedrene (16.67%), and α-phellandrene (58.21%) were also identified from its roots.[7]

PHARMACOLOGY AND BIOLOGICAL ACTIVITIES

The essential oils possess antimicrobial and antihelminthic activities.[7] The methanolic extracts from the leaves and its alkaloids vasicine and vasicinol showed sucrase inhibitory activity with sucrose as a substrate.[8] The extract of *A. vasica* showed antioxidant and radical scavenging activities,[9] and anticestodal efficacy.[10] The alkaloid fractions of the methanolic extract showed anti-inflammatory[11] and antibacterial[12] activities. The leaf powder exhibited a considerable degree of antiulcer[13] and hepatoprotective activities.[14] *A. vasica* also showed antitussive properties.[15]

USES

Generally used in as a popular remedy for the treatment of different respiratory conditions, cold, wheezing, asthma, and tuberculosis.

Traditional Medicine. The shrub is the source of the drug, vasaka, which is well known in the traditional medicine for bronchitis. The drug comprises fresh or dried leaves, mixed with stems or other aerial parts, and is employed as fresh juice, decoction, and infusion and in powder form; also given as alcoholic extract, liquid extract, or syrup. The leaves, flowers, fruits, and roots are extensively used for treating cold, cough, whooping cough and chronic bronchitis and asthma, as a sedative expectorant, antispasmodic, and as antihelminthic. Dried leaves are also smoked as a cigarette. In chronic bronchitis, it is efficacious and affords relief, especially when the sputum is thick and tenacious. The leaf juice is used to cure diarrhea, dysentery, and glandular tumors. The powder is reported to be used as poultice on rheumatic joints, a counterirritant in inflammatory swellings, fresh wounds, and urticaria. An ointment is prepared from the alcoholic extract of leaves and is used for healing wounds in veterinary medicine.[16]

REFERENCES

1. A. Rehman et al., *Nat. Prod. Lett.*, **10**, 249 (1997).
2. R. Thappa et al., *Phytochemistry*, **42**, 1485 (1996).
3. B. Joshi et al., *J. Nat. Prod.*, **57**, 953 (1994).
4. R. S. Singh et al., *Fitoterapia*, **63**, 262 (1992).
5. R. S. Singh, et al., *Phytochemistry*, **30**, 3799 (1991).
6. M. P. Jain et al., *Phytochemistry*, **19**, 1880 (1980).
7. K. Sarada et al., *J. Microb. World*, **10**, 86 (2008).
8. H. Gao et al., *Food Chem.*, **108**, 965 (2008).
9. R. M. Samarth et al., *Food Chem.*, **106**, 868 (2008).
10. A. K. Yadav and V. J. Tangpu, *J. Ethnopharmacol.*, **119**, 322 (2008).
11. A. Chakraborty and A. H. Brantner, *Phytotherapy Res.*, **15**, 532 (2001).
12. A. H. Brantner and A. Chakraborty, *Pharm. Pharmacol. Lett.*, **8**, 137 (1998).
13. N. Shrivastava et al., *J. Herb Pharmacother.*, **6**, 43 (2006).
14. D. Bhattacharyya, et al., *Fitoterapia*, **76**, 223 (2005).
15. J. N. Dhuley, *J. Ethnopharmacol.*, **67**, 361 (1999).
16. Wealth of India, *A Dictionary of Indian Raw Materials*, First Supplement Series, Vol. 1, National Institute of Science Communication, New Delhi, 2002.

Chinese Cosmetic Ingredients

As in other ancient cultures, use of herbs by the Chinese to modify and improve physical appearance dates back thousands of years. During the course of using herbs for treating illnesses and undesirable physical conditions, they have accumulated considerable experience in cosmetic treatments and much of this experience has been documented, and this documentation dates back at least 2000 years. Common modern-day cosmetic problems such as facial dark spots, wrinkles, pimples, and dry skin were already dealt with in ancient medical treatises such as the HUANG *Di Nei Jing*, which was written 2000 years ago. However, unlike their Western counterparts, the Chinese have always considered external beauty as an extension of one's whole self. Thus, cosmetic conditions are often simply considered as manifestations of an unbalanced whole. Hence, treatments are often a combination of both internal and external applications.

Among the more than 6000 documented natural drugs used in traditional Chinese medicine, one can find a sizable number that are used in treating skin and oral problems that can be considered as "cosmetic" conditions.[1,2]

To take advantage of the vast resource of Chinese herbal medicine as a source of cosmetic ingredients, it is necessary to define in a broad sense the scope of "cosmetic" properties and "cosmetic" usages.

For practical purposes, any herb or herbal formula reported to have the following properties, either traditional or modern, can be considered a potential source of natural cosmetic ingredient(s).

TRADITIONAL PROPERTIES

alleviates depression,
benefits/improves complexion,
detoxifying,
invigorates/nourishes blood,
lightens skin,
moistens skin/removes dryness,
prevents scar formation,
promotes flesh growth,
promotes hair growth/prevents hair loss,
removes heat,
removes swelling,
and so on.

MODERN PROPERTIES

analgesic,
antiallergic,
anti-inflammatory,
antihistaminic,
antimicrobial,
antipruritic,
astringent,
local anesthetic,
vasodilator,
whitening,
wound healing,
and so on.

Alternatively, any herb or herbal formula that has been recorded as beneficial in external application in one or more of the following conditions can be considered within the scope of "cosmetic" usage:

dark spots on skin,
lacquer sores,
acne/pimples,

chapped skin,
dandruff,
burns,
dry skin,
itching,
hand/facial wrinkles,
skin rash,
wounds,
premature graying of hair,
hair loss,
insect bites,
snakebites,
urticaria,
vitiligo,
ringworm,
sore gums,
toothache,
eczema,
skin sores,
and so on.

In fact, among the more than 100,000 Chinese herbal formulas recorded over the past 3000 years, at least 10% of them are for the treatment of above conditions.

Thus, in the *Wu Shi Er Bing Fang* (*Prescriptions for Fifty-two Diseases*) compiled sometime between 1065 BC and 771 BC, 283 known prescriptions from 247 drugs were recorded. Among them, over one third are for treating diseases/conditions that are relevant to cosmetics:

Disease	No. of Prescriptions
Skin ulcers/carbuncles	42
Wounds/injuries	17
Frostbite	14
Snakebite	13
External hemorrhoids	4
Poison arrow wounds	7
Lacquer sores (dermatitis)	7

The most extensive collection of formulas ever published is the *Pu Ji Fang* (*Prescriptions for General Relief*), a formulary describing 61,739 prescriptions for treating diseases and human conditions of practically every kind.

It was compiled during the later part of the 14th century AD by ZHU Xiao and others of the Ming court. In this compilation, there are at least 5000 formulas for treating skin sores and ulcers, wounds and injuries, insect and snakebites, hemorrhoids, and ringworm; and at least 2000 for hair, facial, and oral problems, including 350 for facial dark spots; pimples, rash, chapped skin, scars and "lack of luster," 400 for hair, beard and eyebrow conditions (e.g., premature graying and hair loss), 80 for dryness in mouth, bad breath, and mouth sores, and 1000 for tooth and gum problems (e.g., toothache, swollen gums, tooth decay, yellow and stained teeth, etc.). Most of these formulas are for topical application and thus are potential source for cosmetic ingredients.

The most extensive work on natural products ever published is the *Zhong Yao Da Ci Dian* (*Encyclopedia of Chinese Materia Medica*). Compiled by the JIANGSU Institute of New Medicine and published in 1977, it describes 5767 natural drugs, many in great detail, with 4500 drawings. It contains both traditional (properties, uses, use history, etc.) and modern (pharmacology, chemistry, clinical reports, etc.) information, with a comprehensive index/appendix. The index/appendix lists hundreds of substances and chemical compounds that exhibit certain pharmacological properties, such as anti-inflammatory, antimicrobial, (including antifungal), analgesic, wound-healing, antiallergic, antihistaminic, vasodilating, and others, which can provide useful cosmetic ingredients.

Among the hundreds to thousands of single herbs and herbal formulas recorded to have "cosmetic" properties, numerous natural drugs have been repeatedly used. A few of them are already being used in Western cosmetics and may be familiar to some readers. And a few dozen more are now being used in certain Asian cosmetics, which are bound to find their way into cosmetic products in the United States in the foreseeable future.

Chinese herbal medicine as a potential source of modern cosmetic ingredients can be appreciated from a recent analysis of 85 beautifying formulas in the *Qian Jin Fang*

(*Precious Formulas*), a formulary compiled by the well-known physician, Sun Si-Miao, during the 7th century AD, which contains 8200 formulas. Among the 160 natural substances used in these beauty formulas, 105 were herbs, 32 were animal by-products, and 23 were minerals. At least two dozen of these substances repeatedly appeared in the formulas, some over 20 times. Most of them are still being used in Chinese cosmetic products.

Although most of the cosmetic usage of Chinese herbal drugs and formulas is based on traditional rationale, some of the uses do appear to have a scientific basis.

Thus, among the herbs used to remove dark spots on the skin, some (especially *chuanxiong*, *fangfeng*, *gaoben*, and *danggui*) have been shown to exhibit marked tyrosinase inhibitory effects *in vitro*.[3,4]

Among more than 100 Chinese herbal drugs screened for *in vitro* antihistaminic activities, numerous were demonstrated to exhibit strong effects. They include *wumei*, *xinyi*, *baizhi*, ginger, and star anise; all have traditionally been used to treat cosmetic problems.[5–7]

Also, many of the herbal drugs traditionally used for their healing and antiswelling properties (e.g., in wounds and burns) have been scientifically demonstrated to have anti-inflammatory effects. They include *xinyi*, *gaoben*, *huzhang*, *lianqiao*, *wumei*, *fangfeng*, *chishao*, *sanqi*, and *rehmannia* (see corresponding entries).

Based on above criteria, the following herbs and/or their derivatives (extracts, essential oils, special chemical fractions, etc.) can serve as potential sources of cosmetic ingredients.

ATRACTYLODES (*BAIZHU* AND *CANGZHU*)
BAIZHU (ATRACTYLODES)
BLETILLA TUBER (*BAIJI*)
CANGZHU (ATRACTYLODES)
DAHURIAN ANGELICA (*BAIZHI*)
DITTANY BARK (*BAIXIANPI*)
FORSYTHIA FRUIT (*LIANQIAO*)
GARDEN BURNET (*DIYU*)
HONEYSUCKLE FLOWER (*JINYINHUA*)
KNOTWEED, GIANT (*HUZHANG*)
LIGUSTICUM (*GAOBEN*)
LUFFA (*SIGUALUO*)
MUME (SMOKED PLUM OR *WUMEI*)
PEARL (*ZHENZHU* OR MARGARITA) AND MOTHER-OF-PEARL (*ZHENZHUMU*)
PEONY (PEONY BARK AND PEONY ROOT)
PEONY BARK (*MUDANPI*)
PEONY ROOT, RED AND WHITE (*SHAOYAO*: *CHISHAOYAO* AND *BAISHAOYAO*)
PHELLODENDRON BARK (HUANG*BAI*)
PURSLANE, COMMON (*MACHIXIAN*)
RED SAGE (*DANSHEN*)
SAFFLOWER (FALSE SAFFRON; *HONGHUA*)
SICHUAN LOVAGE (*CHUANXIONG*)
SKULLCAP, BAIKAL (HUANG*QIN*)

REFERENCES

1. A. Y. Leung, D & C I, April, 34 (1989).
2. A. Y. Leung, Chinese Medicinals in J. Janick and J. E. Simon, eds., *Advances in New Crops*, Timber Press, Portland, OR, 1990, p. 499; reprinted in *HerbalGram*, **23**, 21 (1990).
3. X. T. Liu, *Zhongchengyao*, **13**(3), 9 (1991).
4. Y. Masamoto et al., *Planta Med.*, **40**, 361 (1980).
5. R. D. Xiang et al., *Zhongcaoyao*, **15**(2), 22 (1985).
6. Z. X. Zhang and L. S. Liu, *Zhong Chengyao*, **14**(11), 30 (1992).
7. D. Q. Zhou et al., *Zhongcaoyao*, **22**(2), 81 (1991).

INDIVIDUAL HERBS

ATRACTYLODES (*BAIZHU* AND *CANGZHU*)

Atractylodes yields two widely used herbs, *baizhu* and *cangzhu* (see individual entries).

BAIZHU

Rhizome of *Atractylodes macrocephala* Koidz. (Family Compositae or Asteraceae), an erect perennial herb, up to ca. 80 cm high. Thick fist-shaped rhizome is collected from 2- to 3-year-old plants in early winter after lower leaves have withered and turned yellow, rid of sand, dirt, and rootlets, and oven or sun dried. Produced mainly in eastern and central provinces.

First recorded use dates back at least 2000 years when it was simply called ZHU, not distinguishable from **cangzhu** (see that entry); listed in the *Shen Nong Ben Cao Jircg* (ca. 200 BC–100 AD) under the first category of drugs to which most tonics belong. It was not until around AD 600 that *baizhu* became a distinct separate entity. Traditionally regarded as bitter and sweet tasting; warming; and to have spleen and qi-invigorating, as well as wetness-drying (*zao shi*), diuretic and fetus-calming properties. Used in treating spleen-deficient (*pi xu*) conditions, including dyspepsia (indigestion), flatulence, diarrhea and fluid retention; also used for spontaneous perspiration, restless fetus, and as an ingredient in numerous well-known tonic foods, including soups, cakes, and specialty rice (PENG).

In recent years also used in treating constipation, leukopenia, and toxic effects due to chemotherapy and radiotherapy (WANG).[1,2]

Contains 0.25–1.42% volatile oil (depending on geographic sources and processing methods), with atractylon (a furan derivative) as its major component (9.59–27.4%) (IMM-1); others include sesquiterpene lactones [e.g., selina-4(15),7(11)-dien-8-one; atractylenolide I; atxactylenolide II; atractylenolide III; and 8-β-ethoxyasterolid], scopoletin, and acetylenes.[3]

Also contains Atractylodes polysaccharides, AM-1 (β-D-mannan), AM-2 (β-D-fructan), and AM-3 (β-D-mannan), with molecular weight of 3.1×10^4, 1.1×10^4, and 1.2×10^4, respectively.[4,5]

Although high concentrations of vitamin A substances (0.2599%) were earlier reported to be present in *baizhu*, later findings could not substantiate them (WANG).

The atractytenolides have been reported to have anti-inflammatory activities while the polysaccharides were shown to have immunopotentiating effects.[4–6]

In endurance tests, decoctions of cured *baizhu* markedly prolonged the swimming time of mice while those of raw *baizhu* were not active. On the other hand, decoction of raw *baizhu* enhanced the phagocytosis activity of mouse phagocytes while that of cured *baizhu* had no such effect.[7]

Other documented pharmacologic effects of *baizhu* decoctions in animals (p.o.) include: diuretic; hypoglycemic; improving stamina; liver protectant (preventing liver glycogen reduction caused by carbon tetrachloride); and anticoagulant. The volatile oil was active against experimental tumors (JIANGSU; ZHOU AND WANG).[8]

Toxicity of *baizhu* is low: LD_{50}, of its decoction in mice was 13.3 g/kg (i.p.); and a daily oral dose of 0.5 g/kg decoction in rats moderately lowered white blood cells in 14 days, which was contrary to clinical results (WANG).

Used in many traditional topical formulas for wrinkles and dark spots on the skin, especially the face and hand.

REFERENCES

See the General References for CHP; HONGKUI; IMM-1; JIXIAN; JIANGSU; MCGUFFIN 1 & 2; PENG; WANG; ZHOU AND WANG.

1. C. X. Zou et al., *Zhongyiyao Xinxi*, (3), 34 (1992).
2. C. Q. Ling, *Henan Zhongyi*, **13**, 94 (1993).
3. Z. L. Chen, *Planta Med.*, **53**, 493 (1987).
4. Y. C. Gu et al., *Zhongcaoyao*, **23**, 507 (1992).
5. Y. C. Gu et al., *Zhongguo Yaoxue Zazhi*, **28**, 275 (1993).
6. K. Endo et al., *Chem. Pharm. Bull.*, **27**, 2954 (1979).
7. B. Sun et al., *Shandong Zhongyi Xueyuan Xuebao*, **17**, 51 (1993).
8. Z. H. Li et al., *Zhongcaoyao*, **17**(10), 37 (1986).

BLETILLA TUBER (*BAIJI*)

Rhizome of *Bletilla striata* (Thunb.) Reichb. f. (Family Orchidaceae), a perennial herb, 30–70 cm high, with a thick and fleshy rhizome; widely distributed in central, eastern, and southern China; also cultivated. Rhizome is collected in autumn and early winter and cured by steaming or boiling in water until thoroughly cooked and then sun- or oven-dried.

Has been traditionally used for over 2000 years to treat and heal sores, burns, wounds, and chapped skin as well as to stop bleeding (both internal and external); also has astringent and antiswelling properties.

Bletilla tuber contains large amounts of starch (30.48% in fresh tuber) and mucilage (Bletilla mannan), which is a polysaccharide consisting of mannose and glucose in a 4 : 1 ratio.

Its methanol extractives (phenolic fraction) have been shown to have antimicrobial activities (esp. against *Staphylococcus aureus*).[1] Decoction has exhibited protective effects on HCl-induced gastric mucosa damage in rats.[2]

Used in skin creams and lotions for its healing and whitening effects (per Li Shi-Zhen's *Ben Cao Gang Mu*). Its mucilage powder (water extracted, alcohol precipitated) is used as an adhesive and molding material, especially in dentistry.[3]

REFERENCES

See the General References for FOSTER AND YUE; HU; IMM-1; JIXIAN; JIANGSU; MCGUFFIN 1 & 2.

1. S. Takagi et al., *Phytochemistry*, **22**, 1011 (1983).
2. Z. G. Geng et al., *Zhongcaoyao*, **21**(2), 24 (1990).
3. J. D. Zhang and F. H. Tao, *Zhongguo Zhongyao Zazhi*, **14**(4), 34 (1989).

CANGZHU

Rhizomes of *Atractylodes lancea* (Thunb.) DC., *Atractylodes chinensis* Koidz., and *Atractylodes japonica* Koidz. ex Kitam. (Family Compositae or Asteraceae), all perennial herbs, 30–80 cm high, with thick rhizomes. They are also known as *nan cangzhu* (southern), *bei cangzhu* (northern), and *dong cangzhu* (eastern). Rhizomes normally collected in spring or autumn (preferably autumn) from 1- to 2-year-old plants (if from cultivation), rid of residual stems, rootlets, and dirt, and sun dried. Southern *cangzhu* is produced mainly in JIANGSU, Hubei, Hunan and other relatively southern provinces; northern *cangzhu* is produced primarily in northern provinces, including Inner Mongolia, Hebei,

Shanxi and Shaanxi; and eastern *cangzhu* is produced mainly in northeastern provinces, including Heilongjiang, Jilin, and Liaoning (IMM-1; JIANGSU). Eastern *cangzhu* (*A. japonica*) is considered equivalent to **baizhu** (see that entry) in Japan (HU).[1] The *cangzhu* available in the United States is mostly southern or northern *cangzh*.

First recorded use dates back at least 2000 years when it was simply referred to as ZHU in the *Shen Nong Ben Cao Jing* (ca. 200 BC–AD100) without differentiation from **baizhu**. Traditionally considered as acrid and bitter tasting and warming; spleen-invigorating; wetness-drying (*zaoshi*); anti-inflammatory (*qu feng*); alleviating depression (*jie yu*); cold-dispersing; and eye-brightening. Used in treating various conditions, including abdominal distention; diarrhea; edema; lack of appetite; indigestion; tiredness and sleepiness; rheumatism and arthritic pain; common cold; eczema; night blindness; and warts, among others.[2]

Also, a traditional practice during an epidemic was to burn *cangzhu* to drive off the "evil" that was believed to cause the epidemic. Recently, this practice has been adapted for the routine sterilization of operating rooms, using 1 g/m^3 of *cangzhou* with considerable success and safety. The fumes from burning *cangzhu* not only killed bacteria (esp. *Staphylococcus aureus*) but also killed viruses and fungi;[3] however, they had no effect on mosquitoes and other insects (WANG).

Contains 1.1–9.0% (w/w) volatile oil, depending on botanical, geographic, and reporting sources, with southern *cangzhu* in the general range of 5–9%, northern *cangzhu* 3–5%, and eastern *cangzhu* 1–3% (IMM-1; JIANGSU; NATIONAL). The major volatile components present in southern *cangzhu* include atractylol (mixture of β-eudesmol and hinesol), elemol, and atractylodin (an acetylenic furan);[4] minor components present include atractylon, vitamin A substances, vitamin B, and inulin. Major compounds present in northern *cangzhu* include atractylodin (12.50–20.93%), atractylon, and atractylol. The volatile oil in eastern *cangzhu* contains little or no atractylodin; its major components being atractylol and atractylon (IMM-1; JIANGSU); also contains hypoglycemic glycans (atractans A, B, and C).[1]

Although fumes from burning *cangzhu* have strong disinfectant effects, its decoction does not. The fumes are reported to be nonirritating and to have very tow toxicity: mice and rats exposed to burning *cangzhu* and *aiye* (*Artemisia argyi* leaves) for 0.5–2 h exhibited no abnormalities on external and pathological examination; and over 4000 normal subjects exposed nightly to one round of burning *cangzhu-aiye* incense for 30 days did not show any adverse reactions (WANG).

Used as a component in hair tonic liniments and in skin care formulas (e.g., acne creams) for its antimicrobial activity as well as its whitening properties.[5]

REFERENCES

See the General References for CHP; HONGKUI; HU; IMM-1; IMM-CAMS; JIANGSU; JIXIAN; MCGUFFIN 1 & 2; NATIONAL; WANG; ZHOU.

1. C. Konno et al., *Planta Med.*, **51**, 102 (1985).
2. C. F. Zhang, *Zhongyiyao Xuebao*, (4), 32 (1992).
3. L. E. Jiang and Z. X. Zhu, *Chin. J. Integr. Trad. West. Med.*, **9**, 245 (1989).
4. H. Y. Sun and A. W. Wang, *Zhongcaoyao*, **23**, 298 (1992).
5. Li Shi-Zhen's, Ben Cao Gang Mu, 16th century, reprinted.

DAHURIAN ANGELICA (*BAIZHI*)

Root of *Angelica dahurica* (Fisch. ex Hoffm.) Benth. et Hook. f. and *Angelica dahurica* (Fisch. ex Hoffm.) Benth. et. Hook. f. var. *formosana* (Boiss.) Shan et Yuan. (Family Umbelliferae or Apiaceae), both perennial herbs, 1–2.5 m high; the former grows throughout northeastern China while the latter is distributed in southeastern China and Taiwan. Both are now extensively cultivated; and Dahurian angelica produced from different regions provides the various commercial types available, especially *Chuan* (Sichuan) *baizhi* (from *A. dahurica* var. *formosana* grown in Sichuan), *Yu baizhi* (from *A. dahurica* grown in Yu County in Henan), and *Hang baizhi* (from *A. dahurica* var. *formosana* grown in Hangzhou, Zhejiang Province) (IMM-1).[1] The types imported into the United States are *Chuan baizhi* and *Hang baizhi*. *Dian* (Yunnan) *baizhi* (from *Heracleum scabridum* Franch. grown in Yunnan) is used as "*baizhi*" in Yunnan and has also been the subject of studies reported in the Chinese literature.

The root is harvested in late summer to early autumn from plants planted the year before, rid of stem remnants and leaves, rootlets and dirt and sun-dried; sulfur fumigation is often used. In the production of *Hang baizhi*, the cleaned root is mixed with lime and kept for a week before drying. Thick, starchy, hard, and heavy roots with strong aroma are considered of best quality (XU AND XU; ZHU).

Traditionally considered acrid tasting and warming; removing rheumatic and arthritic pain; sinus-clearing, pain relieving, antiswelling, and pus expelling (*pai nong*); and used to treat the common cold and headache, nasal congestion, sinusitis, toothache, leucorrhea, and others (CHP; JIANGSU).

Used externally since ancient times in treating various skin conditions such as acne, dark spots, freckles, carbuncles, ringworm, scabies, and itching, among others. Now also used in treating psoriasis, eczema, and vitiligo. First known use was recorded in the *Wu Shi Er Bing Fang* (*Prescriptions for Fifty-two Diseases*, ca. 1065–771 BC) in a formula together with ginger and magnolia flower (see *magnolia*) for the topical treatment of carbuncles. *Baizhi* has since been extensively used in many beauty formulas, including numerous skin and hair formulas of the imperial court, such as Empress Dowager Cixi's "Fragrant Hair Powder" for reportedly keeping her hair full and dark.[2] It is the single most frequently used herbal ingredient in traditional beauty formulas.[3]

Baizhi contains numerous furocoumarins (see also *angelica*), including: byakangelicin (ca. 2%), anhydrobyakangelicin, byakangelicol (ca. 0.2%), neobyakangelicol, oxypeucedanin (2.56%), oxypeucedanin hydrate, imperatorin (1.86%), isoimperatorin (1.96%), alloisoimperatorin, phellopterin, xanthotoxin, scopoletin, 5-(2-hydroxy-3-methoxy-3-methylbutoxy) psoralen, knidilin, demethylsuberosin, cedrelopsin, bergapten, and others (HU; IMM-1).[4–10] Also reported present are sterols (sitosterol and daucosterol), stearic acid, and (S)-2-hydroxy-3,4-dimethyl-2-buten-4-olide that gives *baizhi* its characteristic odor.[6,7,11] Sulfur fumigation during processing was shown to markedly reduce the concentrations of furocoumarins in *baizhi*.[12]

Decoction has antibacterial and antifungal activities *in vitro* as well as anti-inflammatory, analgesic, and antipyretic effects in experimental animals, with effects comparable among the four types of *baizhi* tested (*Yu, Chuan, Hang, and Dian*).[1] Alcohol extractives exhibited antihistaminic effects on isolated guinea pig trachea.[13]

Among the numerous furocoumarins tested, several (oxypeucedanin hydrate, imperatorin, phellopterin, etc.) activated adrenaline- and ACTH-induced lipolysis while others (byakangelicin, neobyakangelicol, isopimpinellin, etc.) strongly inhibited insulin-stimulated lipogenesis in fat cells.[14]

Although furocoumarins have photosensitizing effects, their presence in *baizhi* has not prevented its extensive use in China for over 3000 years, both internally and externally, without any major toxic side effects.

Powder and extracts used in skin care products (especially acne creams, freckle removing and antifungal creams) and in hair care products (antidandruff shampoos, hair-growth tonics, etc.) for its antimicrobial and anti-inflammatory effects and its traditional anti-itching, wound-healing, and skin-whitening (tyrosinase inhibitory) properties (ZHOU).[15]

REFERENCES

See the General References for CHP; HONGKUI; HU; HUANG; IMM-1; JIANGSU; JIXIAN; MCGUFFIN 1 & 2; ZHOU; ZHU.

1. H. Y. Li et al., *Zhongguo Zhongyao Zazhi*, **16**, 560 (1991).
2. A. Y. Leung, *Drug Cosmet. Ind.*, 34 (1989); (4), *Chem. Abstr.*, 95, 209449j (1981).
3. W. X. Hong, *Fujian Zhongyiyao*, **18**(1), 39 (1987).
4. J. Han and Q. Yang, *Beijing Yike Daxue Xuebao*, **21**, 186 (1989).
5. L. R. Wang et al., *Yaoxue Xuebao*, **131**, (1990)
6. R. Y. Zhang et al., *Beijing Yixueyuan Xuebao*, **17**, 104 (1985)
7. H. Q. Zhang et al., *Yaoxue Tongbao*, **15**(9), 2 (1980).
8. J. Han and H. L. Zhang, *Zhongcaoyao*, **17**(8), 0 (1986).
9. M. Kozawa et al., *Shoyakugaku Zasshi*, **35**(2), 90 (1981); through *Chem. Abstr.*, 95, 209449j (1981).
10. J. M. Zhou et al., *Zhongcaoyao*, **18**(6), 2 (1987).
11. K. Baba et al., *Planta Med.*, **51**, 64 (1985).
12. H. Y. Li et al., *Zhongguo Zhongycao Zazhi*, **16**, 27 (1991). (1990).
13. R. D. Xiang et al., *Zhongcaoyao*, **16**(2), 22 (1985)
14. Y. Kimura et al., *Planta Med.*, **45**, 183 (1982).
15. Y. Masamoto et al., *Planta Med.*, **40**, 361 (1980).

DITTANY BARK (*BAIXIANPI*)

Root bark of *Dictamnus albus* L. var. *dasycarpus* (Turcz.) T. N. Liou et Y. H. Chang (syn. *Dictamnus dasycarpus* Turcz.) (Family Rutaceae), a strong scented perennial herb, up to 1 m high, with fleshy root; widely distributed in northern, northeastern and central China as well as Mongolia and Siberia. Root is dug up in spring and autumn, rid of rootlets, and washed free of dirt; bark is then removed, cut in sections, and normally sun-dried.

First described in the *Shennong Ben Cao Jing* (ca. 200 BC–AD 100) as cold-natured and bitter-tasting. Traditionally used to treat pruritus (itching), eczema, urticaria, ringworm, arthritis, and rheumatism; also used to treat acute and chronic hepatitis as well as traumatic injuries and external bleeding. Its use in pruritus is common.

Root contains 0.19–0.39% alkaloids, mainly quinoline type (dictamnine, skimmianine, preskimmianine, isodictamnine, dasycarpamine, γ-fagarine (8-methoxydictamnine), isomaculosindine, trigonelline, choline, etc.); lactones (dictamnolactone, obaculactone or limonin, rutaevin,[1] fraxinellone, etc.); sterols (sitosterol and campesterol); saponins and volatile oil. Fraxinellone has antifertility (anti-implantation) activities.[2]

Its decoction exhibited antifungal activities *in vitro* and lowered experimental fever in rabbits.

Extracts used in cosmetic creams and bath preparations for their antifungal and allegedly whitening effects.

REFERENCES

See the General References for ETIC; HUANG; JIANGSU; JILIN; JIXIAN; MCGUFFIN 1 & 2; NATIONAL.

1. Z. C. Wang et al., *Zhongguo Zhongyao Zazhi*, **17**, 551 (1992).
2. W. S. Woo et al., *Planta Med.*, **53**, 399 (1987).

FORSYTHIA FRUIT (*LIANQIAO*)

Fruit of *Forsythia suspensa* (Thunb.) Vahl. (Family Oleaceae), a deciduous shrub, 2–4 m high, bearing bright golden yellow flowers early in spring; native to China, distributed mostly in northern and northeastern provinces; cultivated in China; also cultivated as an ornamental plant in the United States. Fruit is a dehiscent capsule; both the newly ripened (not yet opened) and the fully ripened (opened) fruits are used, with the latter more common, which is also the one imported into the United States. This is mainly wild-crafted in October when the capsule has turned yellow and opened, rid of impurities and sun dried. This *lianqiao* does not have seeds; major producing provinces include Shanxi, Henan, Shaanxi, and Shandong (ZHU).

One of the most commonly used herbal drugs in China, *lianqiao* has enjoyed a continuous documentation of at least 3000 years. Traditionally considered bitter tasting and cooling and to have the following properties: heat dissipating, detoxifying, breaking up nodules, and removing swelling. Widely used in treating "toxic" and "hot" conditions that correlate to inflammatory and infectious diseases, such as erysipelas, carbuncles, inflammation of the lymph nodes (*luo li*), influenza, the common cold with its associated symptoms, fevers, skin eruption (rash), and urination difficulties (e.g., frequent, slow, or painful urination). Now widely used in commercial herbal formulas for treating the common cold and influenza as well as allergies.

Lianqiao contains triterpenoids, including oleanolic acid (0.73–2.28%),[1–3] ursolic acid, betulinic acid, amyrin acetate, isobauerenyl acetate, and 20(*S*)-dammar-24-ene-3β,20-diol-3-acetate;[4] forsythol,[5] forsythosides A, C, and D (1.46–1.63%), forsythin (phillyrin, phillyroside), forsythigenin (phillygenol), caffeic acid, sterols, and rutin (IMM-3).[5–7] High concentrations (ca. 4%) of an antiviral and antibacterial volatile oil have also been reported in the seeds,[5] which is probably also present in minor amounts in the dried ripe fruit.

Decoctions have exhibited broad biological activities: strong antibacterial *in vitro* (especially against *Staphylococcus aureus* and *Shigella dysenteriae*), with forsythol being one of the active components; antiviral *in vitro*; antipyretic in rabbits; and protective against liver damage by carbon tetrachloride in rats. Other activities include diuretic, anti-inflammatory, hypotensive, antioxidant, and others (WANG).[8]

Oleanolic acid has antiallergic, liver protective, and weak cardiotonic activities, among others (see also **ligustrum**). Acute toxicity is low: LD_{50} of decoction (1:1) in mice was 29.37 g/kg (s.c.) (WANG).

Used in hair care (hair-growth liniments, antidandruff shampoos, etc.), foot care (e.g., athlete's foot) and skin care products (moisturizing, nourishing, and acne creams) for its antibacterial and antifungal activities and for its traditional skin-protective, wrinkle-removal, and moisture-preserving properties (ETIC; ZHOU).

REFERENCES

See the General References for BAILEY 2; BRUNETON; CHP; DER MARDEROSIAN AND BEUTLER; HONGKUI; HU; HUANG; IMM-3; JIANGSU; JIXIAN; MCGUFFIN 1 & 2; WANG; ZHOU; ZHU.

1. H. B. Li et al., *Zhongchengyao*, **11**(6), 33 (1989).
2. N. J. Wu et al., *Zhongcaoyao*, **23**, 467 (1992).
3. B. Wang and C. H. Jiang, *Zhongguo Yaoxue Zazhi*, **27**, 393 (1992).
4. W. Y. Hu and S. D. Luo, *Zhongcaoyao*, **22**, 147 (1991).
5. L. Q. Fang and J.X. Guo, *Zhongchengyao*, **12**(9), **35** (1990).
6. Y. Y. Cui et al., *Yaoxue Xuebao*, **27**, 603 (1992).
7. J. X. Guo et al., *Zhongchengyao*, **12**(8), 29 (1990).
8. Y. L. Zhou and R. X. Xu, *Zhongguo Zhongyao Zazhi*, **17**, 368 (1992).

GARDEN BURNET (*DIYU*)

Root and rhizome of *Sanguisorba officinalis* L. or its variety, *S. officinalis* L. var. *longifolia* (Bert.) Yu et Li (Family Rosaceae), a sturdy perennial with deep red or purplish spikes, also known as salad burnet or burnet-bloodwort, native to Eurasia, widely distributed in China, and can now also be found in the United States from Maine to Minnesota. Root with rhizome is collected in spring before budding or in autumn after aboveground parts have withered, rid of rootlets, washed and dried; or first sliced and then dried. Major production in China is along eastern coastal and adjacent provinces.

First recorded use dates back 2000 years in the *Shen Nong Ben Cao Jing*. Traditionally regarded as bitter- and sour-tasting, cold-natured, and to have blood-cooling, hemostatic, heat-clearing, and detoxifying (*qing re jie du*) properties. Used in treating nosebleeds, vomiting blood, hematochezia (bloody stools), bleeding hemorrhoids, bloody diarrhea, metrorrhagia (*beng lou*), burns and scalds, eczema, skin sores, and swelling; also in dog and snakebites; among others (CHP; JIANGSU). Some of the uses have been well documented in recent years, especially for burns,[1–4] metrorrhagia,[5,6] acne;[7,8] and eczema (JIANGSU; WANG).

Garden burnet contains 14.0–40.4% phenolic substances and 2.5–4.0% triterpene glycosides as its major constituents. Phenolic substances include: (+)-catechin; (+)-gallocatechin; (+)-catechol; (+)-gallocatechol; ellagic acid; and leucoanthocyanins. The triterpene glycosides include: Ziyu-glycoside I [aglycone = pomolic acid (19-α-hydroxyursolic acid); glycone = arabinose and glucose]; Ziyu glycoside II (pomolic acid and arabinose); sanguisorbins A, B, C, D, and E (aglycone = ursolic acid; glycone = arabinose and glucose); sauvissimoside R_1 and pomolic acid 28-O-β-D-glucopyranoside (JIANGSU; JILIN) [9–11]

Decoction had antimicrobial activities *in vitro* and antiemetic effects in pigeons. Crude powder fed to mice and rabbits markedly shortened bleeding and coagulation times; effects were only partially due to tannic substances present; when applied to experimental burns in rabbits and dogs, it reduced inflammation and promoted healing, with effects superior to those of tannins.

Toxicity of garden burnet is low: daily feeding of water extract (1:3) 20 mL/kg to rats for 10 days did not produce obvious toxic symptoms (JIANGSU).

Extracts used in acne creams, bath preparations, and toilet waters for their antimicrobial effects as well as their alleged properties in preventing eczema and contact dermatitis, especially in children and infants (ZHOU).

REFERENCES

See the General References for BARNES; CHP; FERNALD; FOSTER AND DUKE; GRIEVE; HUANG; IMM-2; JIANGSU; JILIN; MCGUFFIN 1 & 2; WANG; ZHOU.

1. P. J. Li and C. Z. Li, *Hebei Zhongyi*, **11**(6), 16 (1989).
2. W. B. Lu and Y. Q. Zhao, *Henan Zhongyi*, **8**(5), 30 (1988).
3. Y. L. Chen and H. Y. Zhou, *Zhongguo Yiyuan Yaoxue Zazhi*, **9**, 277 (1989).
4. T. S. Wang and B. X. Jiang, *Zhongyi*, **2**(1), 41 (1989).
5. G. S. Xu, *Shaanxi Zhongyi*, **10**, 204 (1989).
6. Z. Y. Zhang, *Xinzhongyi*, **23**(4), 18 (1991).
7. G. P. Zhou et al., *Chin. J. Dermatol.*, **24**, 192 (1991).
8. M. X. Liu, *Faming Zhuanli Gongbao*, **7**(21), 11 (1991).
9. G. W. Qin et al., *Zhongcaoyao*, **22**, 483 (1991).
10. F. Abe et al., *Chem. Pharm. Bull.*, **35**, 1148 (1987).
11. I. Yosioka et al., *Chem. Pharm. Bull.*, **19**, 1700 (1971).

HONEYSUCKLE FLOWER (*JINYINHUA*)

Flower buds of *Lonicera japonica* Thunb. and numerous other *Lonicera* species (*L. confusa* DC., *L. hypoglauca* Miq., *L. dasysryla* Rehd., etc.) collectively known as honeysuckle (Family Caprifoliaceae). They are erect or climbing shrubs with opposite leaves and mildly fragrant to very fragrant flowers, with *L. japonica* (Japanese honeysuckle) being the major source. Native to Asia, Japanese honeysuckle now runs wild in many parts of North America, especially eastern United States; its climbing or twining stem reaches 9 m long and its flowers are very fragrant.

The flower buds are collected from both wild and cultivated plants in late spring to early summer. Traditionally, they are picked in the morning after the dew has evaporated and sun-dried or air-dried in the shade, avoiding harsh midday and early afternoon sun. The resulting dried flower buds have a characteristic aroma but not that of fresh honeysuckle flower. Produced mainly in Henan and Shandong from cultivated plants and in other provinces (Guangxi, Zhejiang, Sichuan, etc.) from wild plants. Honeysuckle flower comes in numerous grades; Henan produces the best grade while Shandong produces the largest quantity. Top grades consist of minimal amounts of opened flowers, leaves, and twigs (ZHU).

Also known in Chinese as *rendong* ("winter-resistant") and sHUANG*hua* ("double flower"), honeysuckle has a long recorded history, dating back to the *Ming Yi Bie Lu* (ca. AD 200) according to some records (JIANGSU PROVINCIAL 3).[1] Traditionally considered sweet tasting, cold, fever-relieving, and detoxicant (*qing re jie du*), it is extensively used in treating "heat"-related conditions such as fevers, inflammations, and infections (especially viral and bacterial). Also, together with other detoxifying herbs (especially dandelion, licorice, chrysanthemum, mung bean, soybean, etc.), honeysuckle flower is often used for what Chinese medicine calls "toxic" conditions, such as swellings, sores, and boils as well as in food, drug, and industrial (pesticide, heavy metal, etc.) poisoning.

Flower (*jinyinhua*) contains 0.089–12.00% chlorogenic and isochlorogenic acids (amounts varying greatly depending on botanical sources and methods of processing),[2–5] a volatile oil composed predominantly of linalool, aromadendrene and geraniol;[6] saponins (ca. 1% in *L. fulvotomentosa* Hsu et S. C. Cheng);[7,8] flavonoids, including luteolin and luteolin-7-glucoside; inositol (ca. 1%); and tannins. Leaves and stems contain flavonoids, loganin, and secologanin.[9]

Honeysuckle flower is active against various bacteria (*Staphylococcus aureus*, *Salmonella typhi*, *Mycobacterium tuberculosis*, dysentery bacilli, etc.) and viruses (e.g., HIV and influenza viruses) *in vitro* and/or *in vivo*; also active against dermatophytes but to a less extent. It has anti-inflammatory effects on several experimental inflammation models. Other effects include: strengthening body resistance and activating phagocytosis of leukocytes in mice; lowering absorption of cholesterol in the intestinal tract of rabbits; and others (NATIONAL; WANG).[10,11]

Some of the biological activities are due to chlorogenic acid and saponins. In addition, chlorogenic acid inhibited nitrosation *in vitro* and *in vivo* in rats;[12] inhibited tumor formation in experimental animals;[13] and also has central stimulant effects in mice and rats when administered per os, effect being 1/6 that of caffeine (WANG) (see also *coffee*).

Toxicity of *jinyinhua* is low: LD_{50} in mice was 53 g/kg (s.c.). *Jinyinhua* is a major ingredient in some well-known Chinese cold remedies such as *Yinqiao Jiedu San*, which is also sometimes used externally to treat itching and inflammatory conditions. Aqueous and hydroalcoholic extracts of *jinyinhua* are used in skincare products (creams, lotions, cleansers) for its antimicrobial and astringent properties (ETIC).

REFERENCES

See the General References for CHP; FERNALD; FOSTER AND YUE; HU; HUANG; JIANGSU; JIANGSU PROVINCIAL 3; JIXIAN; LEUNG; MCGUFFIN 1 & 2; NATIONAL; WANG; ZHU.

1. G. M. Ding and L. C. Sun, *Jiangxi Zhongyiyao*, (5), 45 (1988).
2. X. G. Dong et al., *Zhongyao Tongbao*, **10**(5), 31 (1985).
3. H. R. Li et al., *Jilin Zhongyiyao*, (2), 39 (1989).
4. H. B. Lin et al., *Shandong Zhongyi Zazhi*, **9**(4), 34 (1990).
5. B. Q. Liu et al., *Jilin Zhongyiyao*, (3), 41 (1992).
6. G. L. Wang et al., *Zhongguo Zhongyao Zazhi*, **17**, 268 (1992).
7. J. Liu et al., *Acta Pharmacol. Sin.*, **9**, 395 (1988).
8. Q. Mao and X. S. Jia, *Yaoxue Xuebao*, **24**, 269 (1989).
9. A. Hermans-Lokkerbol and R. Verpoorte, *Planta Med.*, **53**, 546 (1987).
10. G. Y. Song et al., *Zhongcaoyao*, **16**(5), 37 (1985).
11. Y. S. Nan et al., *Zhongchengyao*, **11**(8), 17 (1989).
12. B. Pignatelli et al., *Carcinogenesis*, **3**, 1045 (1982)
13. Anon., *C&EN*, **69**(37), 27 (1991).

KNOTWEED, GIANT (*HUZHANG*)

Rhizome and root of *Polygonum cuspidatum* Sieb. et Zucc. (Family Polygonaceae), a stout perennial with mottled stems, 1–2.5 m high, also known as Japanese knotweed and "*huzhang* = tiger cane" in Chinese; the plant is native to eastern Asia, has escaped in North America, and is now a weed throughout New England and neighboring states and Canada; young shoots edible.[1] Rhizome and root are dug up in spring or autumn, cut into sections, and sun dried.

Recorded use in China dates back at least 2000 years. Traditionally considered slightly bitter and cold; used to treat arthritic pain, jaundice (*shi re* HUANG *dan*), amenorrhea, abdominal mass (*zheng jia*), cough with excessive phlegm, traumatic injuries, skin sores and boils, and burns and scalds (CHP); also used in Japan to treat suppurative dermatitis, gonorrhea, favus, athlete's foot, and hyperlipemia.[2]

Recently used in treating burns, acute viral hepatitis, leukocytopenia due to chemotherapy and radiotherapy, and acute infections (lung infection, appendicitis, etc.) with considerable success (WANG).[3]

Contains anthraquinones and their glycosides (chrysophanol, physcione, emodin, emodin-8-O-D-glucoside (polygonin), physcione-8-O-D-glucoside, etc.); stilbenes (resveratrol (3,5,4'-trihydroxy stilbene)) and piceid (polydatin; resveratrol-3-O-D-glucoside); 2-methoxy-6-acetyl-7-methyljuglone (a naphthoquinone); and others, including fallacinol, citreorosein, questin, questinal, protocatechuic acid, (+)-catechin, 2,5-dimethyl-7-hydroxychromone, torachrysone-8-O-D-glucose, 7-hydroxy-4-methoxy-5-methylcoumarin, condensed tannin, and polysaccharides (IMM-1; JIANGSU).[2,4–7]

Aqueous extracts have antibacterial (polygonin and polydatin active principles) and strong antiviral activities *in vitro*, especially against influenza virus Asian strain 68-1, $ECHO_{11}$, HSV, and Coxsackie enteroviruses A and B as well as hepatitis B virus (WANG).[8]

Huzhang decoction exhibited antitussive (polydatin an active principle) and antihistaminic effects in experimental animals; decoction and polydatin also had hypotensive and vasodilating actions; polydatin (not decoction; per os) markedly lowered serum lipid levels in rats; an alcoholic extract was antioxidant *in vitro*;[9] resveratrol and polydatin (i.p., or per os) reduced triglyceride synthesis from palmitate in mouse liver;[4] polydatin also had liver-protectant and antioxidant (vs. lipid peroxidation) effects as well as inhibited platelet aggregation of rabbits both *in vitro* and *in vivo*.[4,10]

The anthraquinones and their glycosides (esp. emodin) exhibited cytotoxic effects on HL-60 cells.[6] Emodin and the stilbenes have recently been shown to be inhibitors of a protein-tyrosine kinase partially purified from bovine thymus.[11]

Extracts used in skin lotions and creams (e.g., antifatigue, massage, and cleansing creams) for their antimicrobial and astringent properties (ZHOU); their intensely yellow color may limit their scope of application (see **phellodendron bark**).

REFERENCES

See the General References for CHP; FERNALD; HUANG; JIANGSU; JIXIAN; MCGUFFIN 1 & 2; WANG; ZHOU; ZHU.

1. J. Richardson, *Wild Edible Plants of New England*, Delorme Publishing Co., Yarmouth, ME, 1981.
2. Y. Kimura et al., *Planta Med.*, **48**, 164 (1983).
3. W. Z. Huang, *Fujian Zhongyiyao*, **18**(4), 27 (1987).
4. C. W. Shan, *Yaoxue Xuebao*, **23**, 394 (1988).
5. H. Arichi et al., *Chem. Pharm. Bull.*, **30**, 1766 (1982).
6. S. F. Yeh et al., *Planta Med.*, 413 (1988).
7. C. G. Ouyang, *Zhongcaoyao*, **18**(8), 45 (1987).
8. J. Y. Yang et al., *Chin. J. Integr. Trad. West. Med.*, **9**, 494 (1989).
9. Y. L. Zhou and R. X. Xu, *Zhongguo Zhongyao Zazhi*, **17**, 368 (1992).
10. P. Tong and Z. T. Zhang, *Zhongguo Yaoxue Zazhi*, **26**, 363 (1991).
11. G. S. Jayatilake et al., *J. Nat. Prod.*, **56**, 1805 (1993).

LIGUSTICUM (*GAOBEN*)

Roots and rhizomes of several *Ligusticum* species, especially *Ligusticum sinense* Oliv. and *L. jeholense* Nakai et Kitag. (Family Umbelliferae or Apiaceae), which are aromatic perennial herbs, erect, up to 1 m high, the former with irregular cylindrical to round rhizome bearing many thin roots while the latter has a short rhizome. *Ligusticum sinense* is distributed and produced in central China, including the provinces of Hubei, Shaanxi, and Sichuan while *L. jeholense* (syn. Liaoning *gaoben*) is distributed and produced in northeastern provinces, including Liaoning, Jilin, Inner Mongolia, Hebei, and Shandong (HU; JIANGSU). Roots and rhizomes are mostly wild crafted, in spring and autumn, and after being rid of dirt, stem, and shoots, they are sun-dried or dried by artificial heat (ZHU).

Although *Gaoben* is closely related to Sichuan lovage (chuanxiong) and is used interchangeably in some areas, they are distinctly different drugs, with some distinctly different uses. They can be differentiated by their microscopic features (XU AND XU) as well as chemical compositions.[1]

Earliest written record dates back to the *Shan Hai Jing* (ca. 800 BC); and later also described in the *Shen Nong Ben Cao Jing* (ca. 200 BC–AD 100). Traditionally regarded as acrid tasting and warming; removing rheumatic and arthritic pain (*qu feng*); dispersing cold; eliminating wetness; and stopping pain. Used to treat the common cold and headache associated with it, headache on top of head (*dian ding tong*), migraine, rheumatic and arthritic pain, acne, acne rosacea, skin blemishes (freckles and dark spots), and abdominal pain and diarrhea; also used in treating ringworm, scabies, and dandruff (CHP; JIANGSU; NATIONAL). In Jiangxi, *gaoben* (*chaxiong, L. sinense*) is often brewed with tea for the prevention of diseases.[2] Like Sichuan lovage, it is one of the most commonly used ingredients in traditional Chinese beauty formulas (see **sichuan lovage**).[3]

Contains 0.3–1.8% volatile oil;[1–4] β-sitosterol, ferulic acid, and others. Amounts of volatile oil and its individual components vary with the geographic and botanical sources; major components present include: neocnidilide (0–25.57%), cnidilide (2.93–10.78%), myristicin (1.63–9.08%), ligustilide (0–6.23%), butylidene phthalide (0–2.01%), β-phellandrene (0–33.32%), 4-terpinyl acetate (3.59–13.82%), limonene (0–14.44%), terpineol-4 (2.7–8.0%), and terpinolene (2.67–3.24%).[4] Other compounds present include methyleugenol, butyl phthalide, 3-butylidene-4,5-dihydrophthalide, and senkyunolides A, G, H, and I (HU; IMM-2).[2,5]

Decoction (15–30%) has exhibited antifungal activities against dermatophytes *in vitro* (JIANGSU). Fat- and water-soluble extractives reduced inflammation (croton seed oil induced otitis in mice) by 75.3% and 72.9%, respectively; water extract also active against experimental edema (egg white induced).[5]

The neutral fraction of the volatile oil (*L. sinense*) has been shown to have numerous biological activities, including: sedative, analgesic, antipyretic, and anti-inflammatory in experimental animals;[6,7] antispasmodic and antihistaminic;[8] markedly decreasing oxygen consumption, prolonging survival time, increasing ability of tissue to tolerate anoxia and extending survival time in mice under cerebral ischemic anoxia.[9]

Ferulic acid and ligustilides are some of the active principles of *gaoben* (see also **Sichuan lovage**).

Powder and extracts used in hair-care and skin-care products, (especially acne and whitening creams), often together with *Dahurian angelica*, for many of the same functions (e.g., antiallergic, anti-inflammatory, and tyrosinase inhibitory) as Sichuan lovage;[3–10,11] also imparts special aroma to products.

REFERENCES

See the General References for CHP; HONGKUI; HU; IMM-2; JIANGSU; JIXIAN; MCGUFFIN 1 & 2; XU AND XU; ZCYX.

1. B.C. Zhang et al., *Zhongcaoyao*, **17**(8), 34 (1986).
2. Q. S. Li et al., *Zhongcaoyao*, **24**(4), 180 (1993).

3. W. X. Hong, *Fujian Zhongyiyao*, **18**(1), 39 (1987).
4. B. Dai *Yaoxue Xuebao*, **23**, 361 (1988).
5. Y. G. Xi et al., *Zhongcaoyao*, **18**(2), 6 (1987).
6. Y. Q. Shen et al., *Chin. J. Integr. Trad. West. Med.*, **7**, 738 (1987).
7. Y. Q. Shen et al., *Zhongcaoyao*, **20**(6), 22 (1989).
8. G. J. Chen et al., *Zhongyao Tongbao*, **12**(4), 48 (1987).
9. C. K. Tang and Q. Y. Xu, *Zhongguo*, **17**, 745 (1992).
10. X.T. Liu, *Zhongchengyao*, **13**(3), 9 (1991).
11. Y. Masamoto et al., *Planta Med.*, **40**, 361 (1980).

LUFFA (*SIGUALUO*)

Luffa is also called loofah, vegetable sponge, or dishcloth gourd. It is the fibrous remains of the old mature fruit of either *Luffa cylindrica* (L.) Roem. (water gourd, smooth loofah) or *Luffa acutangula* Roxb. (silky gourd, angled loofah) (Family Cucurbitaceae). Both are annual vines, native to tropical Asia. The former is cultivated throughout China while the latter mainly in the southern provinces of Guangdong and Guangxi. The fruit of *L. cylindrica* is elongated and smooth while that of *L. acutangula* is elongated but bears ten prominent longitudinal ridges. Young fruits are eaten as vegetables.[1] For medicinal and cosmetic uses, the fruit of *L. cylindrica* is allowed to grow old and is harvested in autumn, usually after the first frost. The pulp, skin, and seeds are then removed by rubbing or the fruit is soaked in water until the skin and pulp disintegrate, which are then washed off along with the seeds; the resulting sponge-like luffa is then sun dried. This is the form most familiar to Westerners. The mature fruit of *L. acutangula* is also collected in the fall but is dried without removing skin and seeds.

Luffa has been used in Chinese medicine since the 16th century. Traditionally considered sweet tasting and neutral and to promote blood circulation and facilitate energy flow in the body (*huoxue tongluo*) as well as having anti-inflammatory, fever-reducing, and detoxifying properties, among others (LEUNG). Used in treating numerous conditions, especially rheumatism, arthritic pain, muscle pain, chest pain, amenorrhea, and lack of milk flow in nursing mothers. Luffa charcoal has recently been reported effective in the topical treatment of shingles (herpes zoster) in the face and eye region (LEUNG).[2]

Apart from the presence of polysaccharides (including cellulose, xylan, and mannangalactan), the chemistry of luffa sponge is basically unknown.

Decoctions of luffa sponge (i.p. or s.c.) exhibited marked anti-inflammatory, analgesic, and tranquilizing effects in mice.[3,4]

Toxicity of luffa is very low: LD_{50} (i.p.) of decoction in mice was 137.40 ± 16.71 g/kg.[4]

Luffa sponge is used to remove dead skin tissue and to stimulate the skin; powdered luffa and extracts used in facial scrubs, skin cleansers, and other skin-care products for their anti-inflammatory and traditional detoxicant properties.

REFERENCES

See the General References for CHEUNG AND LI; CHP; IMM-3; JIANGSU; LEUNG; MCGUFFIN 1 & 2; NATIONAL.

1. S. Y. Zee and L. H. Hui, *Hong Kong Food Plants*, The Urban Council of Hong Kong, 1981, p. 42.
2. X. L. Gu, *Zhejiang Zhongyi Zazhi*, **23**, 88 (1988).
3. B. Kang et al., *Shiyong Zhongxiyi Zazhi*, **6**, 227 (1993).
4. B. Kang et al., *Zhongcaoyao*, **24**, 248 (1993).

MUME (SMOKED PLUM OR *WUMEI*)

It is the dried unripe fruit of *Prunus mume* (Sieb.) Sieb. et Zucc., a deciduous tree up to 10 m high, also known as Japanese apricot (Family Rosaceae). The green, about-to-ripen fruit is collected in May and oven-dried at ca. 40°C for 2–3 days followed by leaving in the closed oven for 2–3 more days until it turns black. Produced mostly in southern provinces, especially Sichuan, which is the largest producer; Zhejiang produces the best *wumei* that is large and jet black, with thick meat and small pit, and tastes sour (ZHU).

Its earliest record dates back to the *Shen Nong Ben Cao Jing* (ca. 200 BC–AD 100). Traditionally regarded as sour tasting, astringent, neutral, and to promote secretion of body fluids as well as expel parasites; used to treat chronic cough, chronic diarrhea, diabetes, ascariasis (roundworm infection) and hookworm infection, neurodermatitis, eczema, and hard-to-heal sores, among other conditions (CHP; IMM-3; JIANGSU). In recent years, often used as an ingredient with other detoxicant herbs (e.g., schisandra, licorice, and *fangfeng*) both internally and externally in the treatment of allergic conditions such as asthma, urticaria, allergic rhinitis, and pruritus;[1,2] also to treat polyps, tumors, and capillary hemangioma (ZHOU AND WANG).[3] It is also boiled in water and sweetened with sugar to make *suan mei tang* (sour plum decoction), a refreshing drink very popular in southern China and Taiwan.

Wumei contains various plant acids, especially citric acid (19%) and malic acid (15%); oleanolic acid; β-sitosterol; amino acids; carbohydrates; wax; and others (HU; JIANGSU). Seed contains amygdalin (see **almond**).

Decoctions and alcoholic extracts of *wumei* have exhibited marked *in vitro* antibacterial effects against numerous bacteria (both Gram-positive and Gram-negative); and its decoction strongly active against pathogenic fungi *in vitro*. Its alcoholic extract also exhibited strong antihistaminic effects on isolated guinea pig trachea;[4] and acetone-soluble extractives were strongly active against aflatoxin B_1 mutagenesis.[5] Oleanolic acid has been reported to have various biological activities, including antiallergic (see ***ligustrum***). It appears that the use of *wumei* in treating allergic conditions has some scientific basis.

Extracts (water and hydroalcoholic) are used in antiallergic ointments and in skin creams and lotions for their antimicrobial and antiallergic effects; also used in hair care products for their traditional hair darkening and growth-stimulating properties.[6]

REFERENCES

See the General References for BAILEY 1; CHP; HU; HUANG; IMM-3; JIANGSU; JIXIAN; MCGUFFIN 1 & 2; WANG; ZHOU AND WANG; ZHU.

1. Z. H. Dong, *Zhongchengyao*, **11**(11), 37 (1989).
2. L. H. Liu and Z. X. Tang, *Zhongyi Zazhi*, (1), 15 (1989).

3. J. P. Yang, *Jiangsu Zhongyi*, **14**(1), 28 (1993).
4. R. D. Xiang et al., *Zhongcaoyao*, **15**(2), 22 (1985).
5. C. C. Ruan et al., *Chin. J. Cancer*, **8**(1), 29 (1989).
6. J. H. Chen, *Shandong Zhongyi Xueyuan Xuebao*, **15**(1), 58 (1991).

PEARL (*ZHENZHU* OR *MARGARITA*) AND MOTHER-OF-PEARL (*ZHENZHUMU*)

Pearl is found in certain mollusks, including oysters, clams, and mussels. It is composed of concentric layers of nacre secreted by these animals in response to irritation, especially that caused by a foreign substance. Nacre is also the substance that makes up mother-of-pearl, the shiny layer that forms the inner lining of the shells. The most common source species of Chinese pearl and/or mother-of-pearl are *Pteria martensii* (Dunker) (syn. *Pinctada martensii*), *Pteria margaritifera* (L.) (syn. *Pinctada margaritifera*) (Family Pteriidae); *Hyriopsis cumingii* (Lea) and *Cristaria plicata* (Leach) (Family Unionidae). The first two are marine while the latter two are freshwater species.

Pearl is collected from both natural and artificially implanted animals, washed free of mucilage and towel dried. The better grades are large, round, white, and lustrous and, when broken, show distinct layers of nacre deposits and no hard nucleus. Major producers are coastal provinces in China, especially Guangxi, Guangdong, Hainan, and Zhejiang (CMH; ZCYX; ZHU).

Mother-of-pearl is produced primarily from the shells of the freshwater species, which are boiled in alkaline water followed by soaking in fresh water. The dark surface layer is then scraped off and the shells are baked until crisp. Better grades are white, come in large pieces, and are crisp. Much of the commercial mother-of-pearl is produced from shells that are by-products of buttons production. Major producers are the same as those of pearl (NATIONAL; ZCYX).

For medicinal or cosmetic use, pearl must be ground to an extremely fine powder while mother-of-pearl can be used as a coarse powder or simply broken up in pieces. A typical traditional method of producing pearl powder is to wrap the clean pearls in cheesecloth or muslin, place them between two pieces of tofu (bean curd) so that the pearls are completely embedded in the tofu, and cook them for 2 h. After cooking, the pearls are washed with clean water, placed in a mortar with a suitable amount of water, and ground to an extremely fine powder (when no more sound is produced by the pearl particles rubbing against the mortar) and dried. The tofu treatment makes the pearl easier to grind and at the same time preserving many of its active components that would otherwise be destroyed if it were to be subjected to high baking temperatures to render it crisp (JIANGSU; MA).

Earliest recorded use of pearl in traditional Chinese medicine dates back to the *Ben Cao Jing Ji*ZHU (ca. AD 500). Traditionally regarded as sweet and salty tasting, cold, and to have tranquilizing, vision-brightening, detoxifying, and healing properties, it is used in treating anxiety, infantile convulsions, insomnia, epilepsy, nebula or opacity of the cornea (*yunyi*), sore throat, mouth sores, and difficult-to-heal sores and ulcers (CHP; JIANGSU; NATIONAL).

Pearl powder has long been regarded as of special benefit to the skin, having been recorded in the *Hai Yao Ben Cao* (ca. AD 907–925) and the *Ben Cao Gang MU* (AD 1593) as good for removing facial dark spots and making one's skin smooth and young-looking. For this, it can be applied directly to the skin or taken internally once every 10–15 days, up to 3 g each time (CHP; JIANGSU).[1]

Mother-of-pearl is traditionally considered salty tasting, cold and to have some of the properties and uses of pearl (e.g., vision-brightening, calming, and treating insomnia). However, it is traditionally not known to have healing properties nor special benefits to the skin (CHP; JIANGSU; NATIONAL).

Pearl contains mostly calcium carbonate, ranging from ca. 81–95%, depending on source species, natural or cultured. It also contains ca. 6–13% organic substances that include 16 amino acids (leucine, methionine, alanine, phenylalanine, glycine, aspartic acid, glutamic acid, proline, serine, etc.) and a small amount of taurine; numerous trace minerals, and so on. (HU; JIANGSU; NATIONAL).[2]

Mother-of-pearl contains similar constituents as pearl but with lesser amount of organic substances (JIANGSU) that include conchiolin (a protein).

At 3% level in an ointment, pearl has exhibited strong wound-healing activities, ranking second among 39 Chinese traditional drugs tested.[3] Treatment of 20 cases of canker sore due to adverse reactions to chemotherapy by topical application of pearl powder (4 × daily; 3 g/time) resulted in complete resolution of the condition in 3–7 days.[4]

Studies on mother-of-pearl have shown it to have numerous biological effects, including antianoxia effects in mice (water extract; not conchiolin);[5] cardiovascular effects in patients with coronary heart disease, markedly decreasing their serum lipid peroxide levels;[6] antiulcer effects in humans;[7] antihistaminic; inhibiting contraction of isolated uterus and intestine of guinea pig, and so on (JIANGSU).[8] Its acute toxicity is low: LD_{50} in rats were >21,500 mg/kg (p.o.) and >31,600 mg/kg (s.c.).[8]

It appears that both pearl and mother-of-pearl are used in cosmetics, even though the latter lacks prior use documentation.

Pearl powder is extensively used in skincare products (e.g., acne and freckle creams and lotions, nourishing creams, etc.) for its healing and traditional skin-lightening, smoothing, and antiwrinkle properties. However, as pearl is a very expensive ingredient and there is no meaningful assays to determine its identity and quality, it is prone to adulteration, especially with mother-of-pearl and other shell products.

REFERENCES

See the General References for CHP; CMH; HU; JIANGSU; JIXIAN; MA; NATIONAL; ZCYX; ZHU.

1. Y. B. Xie, *Dazhong Zhongyiyao*, (4), 39 (1990).
2. L. F. Wen, *Zhongguo Yaoxue Zazhi*, **24**, 276 (1989).
3. S. X. Sun, *Chin. J. Integr, Trad. West. Med.*, **6**, 408 (1986).
4. H. J. Yang et al., *Sichuan Zhongyi*, (4), 57 (1988).
5. W. Y. Yang et al., *Haiyang Yaowu Zazhi*, (1), 29 (1986).
6. Y. W. Huang et al., *Chin. J. Integr., Trad. West. Med.*, **7**, 596 (1987).
7. Z. H. Zhu et al., *Zhonghua Xiaolzua Zazhi*, **2**, 167 (1982).
8. Y. N. Yang et al., *Haiyang Yaowu Zazhi*, (1), 16 (1986).

PEONY (PEONY BARK AND PEONY ROOT)

Peony yields three commonly used herbs (see individual entries): **peony bark**(*mudanpi*) and **peony root**, **red** and **white** (*chishaoyao* and *baishaoyao*).

PEONY BARK (*MUDANPI*)

It is the root bark of tree peony, *Paeonia suffruticosa* Andr. (syn. *P. moutan* Sims and *P. arborea* Donn) (Family Paeoniaceae or Ranunculaceae), a small perennial deciduous shrub, 1–1.5 m high, with short robust stems and thick roots; native to China, now extensively cultivated, also as ornamental in the United States (HAY AND SYNGE). The roots from 3- to 5-year-old plants are dug up in autumn or early spring, rid of dirt and rootlets; the bark is removed and sun dried (*yuan danpi* or original bark) or the outer bark is first scraped off with a bamboo knife or broken porcelain and then dried (*gua danpi* or scraped bark). Produced mainly in central and eastern provinces. There are numerous grades between the two types of peony bark.

Earliest recorded medicinal use of peony bark in China dates back at least 2000 years to the *Shen Nong Ben Cao Jing* (ca. 200 BC–AD 100). Traditionally considered pungent and bitter tasting, cooling and to have heat-dispersing, blood-cooling, and blood-activating as well as stasis-removing properties; it is used in treating skin rashes, nosebleed, and vomiting blood due to "heat and toxins" (*wen du fa ban*) (e.g., viral or bacterial infections such as flu, measles, acute appendicitis, etc.), neurodermatitis, carbuncles, amenorrhea and dysmenorrhea, abdominal pain, hypertension, allergic sinusitis, urticaria, and traumatic injuries and contusions, and so on.

Peony bark contains paeonol, paeonoside (paeonol glucoside), paeonolide (paeonoside-arabinoside), and the monoterpene glycosides paeoniflorin, benzoylpaeoniflorin, and oxypaeoniflorin (JIANGSU PROVINCIAL 1);[1] 1,2,3,4,6-pentagalloylglucose;[2] a volatile oil (0.15–0.4%) and phytosterols (NATIONAL). The amounts of paeonol vary greatly depending on geographic and reporting sources, ranging from a low of 0.19–0.54%[1,3] to a high of 3.5%,[4] with a more common range of 1.08–2.51% (HU).[5]

Many of the biological activities of peony bark can be attributed to paeonol, which include: anti-inflammatory, analgesic, antipyretic, central depressant, antibacterial and antifungal, diuretic, and antiatherosclerotic and antiplatelet aggregation, among others.[3,6–10] The toxicities of paeonol are low, its LD_{50} in mice being 196 mg/kg (i.v.), 781 mg/kg (i.p.) and 3430 mg/k (p.o.).[8] 1,2,3,4,6-Pentagalloylglucose had antiviral activities.[2]

Extracts (water and hydroalcoholic) and paeonol are used in dental products (e.g., toothpaste for inflamed and sore gums), hair-care and skin-care products (e.g., antiallergy creams and lotions) for their antibacterial, anti-inflammatory, and traditional skin-soothing and skin-protectant properties (ETIC).

REFERENCES

See the General References for BAILEY 1; CHP; FOSTER AND YUE; HONGKUI; HU; HUANG; JIANGSU; JIANGSU PROVINCIAL 1; JIXIAN; MCGUFFIN 1 & 2; NATIONAL; ZHU.

1. J. Yu et al., *Yaoxue Xuebao*, **20**, 229 (1985).
2. M. Takechi and Y. Tanaka, *Planta Med.*, **45**, 252 (1982).
3. K. Kawashima et al., *Planta Med.*, **187**, (1985).
4. T. Tani et al., *J. Ethnopharmacol.*, **21**, 37 (1987).
5. Y. S. Zhou et al., *Zhongchengyao*, **14**(7), 23 (1992).
6. A. B. Wang and X. C. Tang, *Zhongcaoyao*, **14**(10), 26 (1983).
7. L. Shi et al., *Acta Pharmacol. Sin.*, **9**, 555 (1988).
8. Q. A. Li, *Zhongcaoyao*, **19**(6), 36 (1988).
9. T. Ohta et al., *Yakugaku Zasshi*, **81**, 100 (1961).
10. M. Harada et al., *Yakugaku Zasshi*, **92**, 750 (1972).

PEONY ROOT, RED AND WHITE (*SHAOYAO: CHISHAOYAO* AND *BAISHAOYAO*)

Peony root is known as *shaoyao* in Chinese; they are of two types: *chishaoyao* or *chishao* (red peony root) and *baishaoyao* or *baishao* (white peony root). Both are the roots of *Paeonia lactiflora* Pall. (syn. *P. albiflora* Pall., *P. edulis* Salisb., and *P. fragrans* Redoute) (Family Paeoniaceae or Ranunculaceae), a perennial herb, 60–80 cm high, with glabrous erect stems and thick cylindrical to spindle-shaped root; native to Siberia and China and distributed throughout northern, northeastern, eastern, and central China; extensively cultivated, also as ornamental in the United States (BAILEY 1; HAY AND SYNGE).

Red peony root is collected from wild plants of *P. lactiflora* (*P. obvata* Maxim. and *P. veitchii* Lynch are also used) in spring and autumn (autumn is preferred as better grades are obtained), rid of rhizomes and rootlets, washed free of dirt, and sun dried or dried in the shade; no cooking is involved. Inner Mongolia produces the best grade that is thick and long, fracture white, and starchy.

White peony root is collected from 3- to 4-year-old cultivated plants of *P. lactiflora* in summer and autumn, rid of rhizomes and rootlets, washed free of dirt, scraped off outer bark, and boiled in water for 5–15 min until soft; it is then sun dried. Zhejiang produces the best grade that is thick, tough, starchy, and with no white center or cracks.

Topical use of red peony root in treating carbuncles (*ju*) was first mentioned in the *Wu Shi Er Bing Fang* (1065–771 BC), and its medicinal properties and uses were first described in the *Shen Nong Ben Cao Jing* (ca. 200 BC–100 AD) while white peony root was not described medicinally until the mid-11th century in the *Tu Jing Ben Cao*.[1] However, according to another report, sauce made with white peony root is said to be a favorite of Confucius (ca. 500 BC); also, it was cooked with animal organs to prevent food poisoning.[2] Hence, there is still much confusion regarding the history and identity of the two herbal food/drugs. The major difference between the two is that red peony root, such as peony bark, is heat-dispersing and blood-cooling while white peony root is a liver and blood tonic. The following properties and uses are described in the current Chinese Pharmacopeia: red peony root is bitter tasting and slightly cold, with analgesic and stasis dispersing properties; it is used in treating tight chest and abdominal pain due to stagnation of Liver *qi* (*gan yu xie tong*) abdominal mass and "heat and toxins" conditions (see **peony bark**). White peony root is bitter and sour tasting, slightly cold with liver-calming (*ping gan*), analgesic blood-nourishing, and menstrual regulating properties. It is used in treating headache, dizziness, abdominal pain, tight chest, stiff and painful joints (limbs), pale complexion due to blood deficiency, irregular menses, spontaneous perspiration and night sweat, and so on (CHP). Whenever peony root (*shaoyao*) is prescribed without specifics, white peony root is normally used. White peony root is also more frequently used for general tonic purposes, and there are now numerous *baishao*-based food and drink products commercially available in China, including wines, fruit juices, and soft drinks.[2]

Both red and white peony roots contain very similar chemical components, especially the monoterpene glycosides paeoniflorin (3–5%), benzoylpaeoniflorin, and oxypaeoniflorin but little or no paeonol, paeonoside, or paeonolide, the latter only present in shrubby peonies (see peony bark).[3–5] Other constituents present include: albiflorin, lactiflorin, and (Z)-(1S,5R)-β-pinen-10-yl-vicianoside (monoterpene glycosides), β-sitosterol, β-sitosterol-α-glucoside, benzoic acid (ca. 1%), palmitic acid, *cis*-9,12-octadecadienoic acid, alkanes (C_{24}–C_{26}), daucosterol, gallic acid, methyl gallate, *d*-catechin, myoinositol, sucrose, and glucogallin, among others (JIANGSU).[5–9]

The glycosides (especially paeoniflorin) of peony root (red and white) are responsible for many of their biological activities, which include antibacterial, antifungal, and antiviral;[10,11] anti-inflammatory and immunomodulating;[10,12–14] analgesic; sedative; antispasmodic; antiplatelet aggregation; antifatigue, prolonging survival, and improving memory;[15] antitumor, enhancing phagocytosis of macrophages and elevating cyclic AMP levels;[16] antimutagenic;[17] and others in humans and experimental animals (HU; JIANGSU).

Acute toxicities of peony root and paeoniflorin are low: LD_{50} (p.o.) of white peony root in rats was 81 g/kg; LD_{50} (i.v.), and LD_{50} (i.p.) of paeoniflorin in mice were 3530 mg/kg and 9530 mg/kg, respectively (HU).

Aqueous and hydroalcoholic extracts of both red and white peony roots are used in skin care products for their antimicrobial (acne creams, etc.), anti-inflammatory, and astringent properties; used with Dahurian angelica (see that entry) in freckle-removal creams and lotions for their traditional ability to remove blood stasis and activate blood circulation; also used as part of a *n* natural preservative system in cosmetic products due to its relatively high content of benzoic acid (ETIC; ZHOU).

REFERENCES

See the General References for BAILEY 1; CHP; ETIC; FOSTER AND YUE; HONGKUI; HU; HUANG; JIANGSU; JIXIAN; MCGUFFIN 1 & 2; NATIONAL; ZHOU; ZHU.

1. R. J. Chai, *Beijing Zhongyi Xueyuan Xuebao*, **14**(1), 49 (1991).
2. J. C. Li, *Zhongyao Tongbao*, **12**(8), 54 (1987).
3. J. Yu et al., *Yaoxue Xuebao*, **20**, 229 (1985).
4. C. D. Jin et al., *Zhongcaoyao*, **24**, 183 (1993).
5. C. F. Wu, *Zhongyao Tongbao*, **10**(6), 43 (1985).
6. H. Y. Lang et al., *Planta Med.*, **50**, 501 (1984).
7. H. S. Chen et al., *Zhongguo Yaoxue Zazhi*, **28**, 137 (1993).
8. M. Kaneda et al., *Tetrahedron*, **28**, 4309 (1972).
9. H. Y. Lang et al., *Yaoxue Xuebao*, **18**, 551 (1983).
10. M. Y. Wang et al., *Liaoning Zhongyi Zazhi*, (9), 43 (1992).
11. B. Y. Liang et al., *Shanghai Zhongyiyao Zazhi*, (6), **4** (1989).
12. J. S. Liang et al., *Chin. J. Pharmacol. Toxicol.*, **4**, 258 (1990).
13. M. R. Liang et al., *Xinzhongyi*, (3), 51 (1989).
14. H. Zhang et al., *Chin. J. Pharmacol. Toxicol.*, **4**, 190 (1990).
15. D. Zhou et al., *Jilin Zhongyiyao*, (2), 38 (1993).
16. K. W. Huang et al., *Chin. J. Oncol.*, **6**, 319 (1984).
17. X. B. Ni, *Zhongcaoyao*, **22**, 429 (1991).

PHELLODENDRON BARK (HUANG*bai*)

Stem bark of *Phellodendron amurense* Rapt. (Amur corktree) or *P. chinense* Schneid. (Chinese corktree) (Family Rutaceae); the former (called HUANG*bai*) from northern and northeastern China while the latter (*chuan*HUANG*bai* or HUANG*pishu*) from central and southern China. Both deciduous trees, with *P. amurense* up to 25 m and *P. chinense* up to 12 m high. Bark is collected between March and

June from trees at least 10 years old, rid of outer cork layer, cut into small sections, and sun dried.

Recorded use of HUANG*bai* dates back at least 2000 years. Traditionally considered to taste bitter and cold, with heat-clearing, wetness-drying (*zaa shi*), fire-purging (*xie huo*), and detoxifying properties. Used in treating numerous heat (*re*) conditions, including acute bacterial dysentery, acute enteritis, acute icterohepatitis, jaundice, urinary infections, night sweating, wet dreams, leukorrhagia, and oral sores. Externally used in treating eczema, pruritus, and skin sores, among others.

In recent years, it has been extensively and successfully used in China as an ingredient in numerous formulas (in extract or powder form) for treating burns, often together with giant knotweed, garden burnet, baikal scullcap (see individual entries), and *zicao*;[1-7] also used in wounds and injuries and to treat acne and facial dark spots.[8,9]

Amur corktree bark contains alkaloids composed mainly of berberine (0.6–2.5%), phellodendrine, magnoflorine, jatrorrhizine, candicine, and palmatine; limonin (bitter principle); obakunone, dictamnolide, γ-sitosterol, β-sitosterol, 7-dehydrostigmasterol, stigmasterol, and mucilage, and so on.

Chinese corktree bark contains similar constituents, but with higher berberine content (4–8%). Aqueous extracts of Amur corktree bark have strong antioxidant[10] as well as strong and broad antibacterial and antifungal activities (berberine is one of the active components); the alkaloids (esp. berberine, phellodendrine, and palmatine) also have hypotensive action in animals; other activities include hypoglycemic, hypocholesterolemic, and blood platelet protective effects (JILIN; WANG).

Extracts used in baby powder and acne creams and as natural preservatives in products that are not incompatible with a yellow tone imparted by these extracts (ETIC; ZHOU).

REFERENCES

See the General References for CHP; ETIC; HU; JIANGSU; JILIN; HUANG; MCGUFFIN 1 & 2; NATIONAL; WANG; ZHOU.

1. Y. D. Ge, *Xinzhongyi*, (1), 25 (1986).
2. H. T. Mao et al., *Chin. J. Integr. Trad. West. Med.*, **7**, 532 (1987).
3. Y. H. Dong, *Zhejiang Zhongyi Zazhi*, **23**, 495 (1988).
4. Z. L. Li and Z. P. Xie, *Sichuan Zhongyi*, (2), 43 (1990).
5. H. Q. Song and F. X. Wang, *Yunnan Zhongyi Zazhi*, **12**(3), 34 (1991).
6. D. C. Chen, *Shiyong Zhongxiyi Jiehe Zazhi (PJCM)*, **4**, 414 (1991).
7. X. W. Sun et al., *Xinzhongyi*, (12), 30 (1992).
8. X. D. Zhou, *Chin. J. Integr. Trad. West. Med.*, **5**, 726 (1985).
9. L. T. Songand and H. J. Jiang, *Jilin Zhongyiyo*, (3), 36 (1992).
10. L. C. Song et al., *ShaanxiZhongyi*, **14**, 185 (1993).

PURSLANE, COMMON (*MACHIXIAN*)

Aboveground parts of *Portulaca oleracea* L. (Family Portulacaceae), a prostrate, smooth herbaceous annual with succulent spatula-shaped leaves, and tiny yellow flowers, up to 30 cm high; probably native to Eurasia but is now found worldwide and considered a weed throughout most of the United States and southern Canada; distributed in most of China; used both fresh and dried. To prepare the dried herb, aerial parts are collected in summer and early autumn when stems and leaves are in their fullest, washed free of dirt, briefly treated with boiling water, and sun dried. Produced throughout China.

Also known as garden purslane, green purslane, and pigweed, it is eaten as a salad and vegetable by people around the world; and is used medicinally for various conditions, including headache, stomachache, painful urination, dysentery, enteritis, mastitis, lack of milk flow in nursing mothers, and in postpartum bleeding; and externally in treating burns, earache, insect stings, inflammations, skin sores, ulcers, pruritus, eczema, and abscesses, for which the fresh herb is normally used as poultice or expressed juice (FOSTER AND DUKE; GRIEVE).[1]

Purslane's earliest recorded use in China dates back to about 500 AD in the *Ben Cao Jing Ji* ZHU. Traditionally considered sour tasting and cold, with heat-relieving and detoxicant (*qing re jie du*) as well as blood-cooling and hemostatic properties; used internally in treating bacillary dysentery, hematochezia (bloody stool), bleeding hemorrhoids and metrorrhagia; and externally to treat the same conditions listed above except in addition to using the fresh herb, the Chinese also use decoctions and powder of the dried herb for topical application. In recent years, it has also been used to treat colitis, acute appendicitis, diabetes, dermatitis, and shingles (IMM-4; JIANGSU).

Purslane contains large amounts of *l*-norepinephrine (*l*-noradrenaline; 0.25% in fresh herb), a neurohormone that has vasopressor and antihypotensive activities and reduces hemorrhage at the tissue level (JIANGSU; MARTINDALE).

It also contains numerous common nutrients (varying from low to high concentrations depending on report), including: vitamins (A, B_1, B_2, C), niacinamide, nicotinic acid, α-tocopherol, (β-carotene, etc.); minerals (especially potassium); fatty acids, especially omega-3 acids whose concentration in purslane is the highest found in leafy vegetables;[2] glutathione; glutamic acid; and aspartic acid. Other constituents include a mucilage composed of an acidic and a neutral fraction with structure determined,[3] calcium oxalate, malic and citric acids, dopamine and dopa, coumarins, flavonoids, alkaloids, saponins, and urea, among others (JIANGSU; WATT AND MERRILL).[1–7]

An aqueous extract of purslane exhibited skeletal muscle relaxant effects both *in vitro* and *in vivo*; it also relaxed guinea pig gastric fundus, taenia coli, and rabbit jejunum as well as contracted the rabbit aorta and raised blood pressure.[8–10] Topical application of the aqueous extract onto the skin was effective in relieving muscle spasms.[9]

Other biological effects include: antibacterial and antifungal; wound healing; anti-inflammatory; uterine stimulant; and diuretic in rabbits (JIANGSU; NATIONAL).[11,12]

Although norepinephrine may account for some pharmacologic activities, the active principles for most of the biological activities and medicinal properties of purslane are still unidentified.

Due to its high content of nutrients, especially antioxidants (vitamins A and C, α-tocopherol, β-carotene, glutathione) and omega-3 fatty acids, and its wound-healing and antimicrobial effects as well as its traditional use in the topical treatment of inflammatory conditions, purslane is a highly likely candidate as a useful cosmetic ingredient. Since most of the reported effects of purslane are due to its fresh juice or to its decoction, water extractives would be most suitable.

REFERENCES

See the General References for CHP; FERNALD; FOSTER and DUKE; FUJIAN; GRIEVE; HUANG; IMM-4; JIANGSU; JIXIAN; MARTINDALE; NATIONAL.

1. C. Whiteman, *Aust. J. Med. Herbalism*, **5**(2), 29 (1993).
2. A. P. Simopoulos et al., *J. Am. Coll. Nutr.*, **11**, 374 (1992).
3. E. S. Amin and S. M. El-Deeb, *Carbohydr. Res.*, **56**(1), 123 (1977).
4. F. R. Bharucha and G. V. Joshi, *Naturwissenschaften*, **44**, 263 (1957).
5. Zh. Stefanov et al., *Farmatsiya (Sofia)*, **16**(3), 27 (1966) (Bulg.).
6. J. Gillaspy, *The UWPT Newsletter*, **1**(2), 2 (1993).
7. T. M. Zennie and C. D. Ogzewalla, *Econ. Bot.*, **31**(1), 76 (1977).
8. F. Okwuasaba et al., *J. Ethnopharmacol.*, **17**, 139 (1986).
9. O. Parry et al., *J. Ethnopharmacol.*, **19**, 247 (1987).
10. O. Parry et al., *J. Ethnopharmacol.*, **22**, 33 (1988).
11. Q. G. Hao and C. L. Wang, *Shandong Zhongyi Zazhi*, **10**(3), 39 (1991).
12. S. J. Xu and L. N. Liu, *Shandong Zhongyi Zazhi*, **10**(3), 52 (1991).

RED SAGE (*DANSHEN*)

Root and rhizome of *Salvia miltiorrhiza* Bge. (Family Labiatae or Lamiaceae), a hairy perennial herb, 30–80 cm high; native to China and widely distributed there. Root (with rhizome) is dug up in spring and autumn, rid of rootlets, sand and dirt, and sun dried. Other species also used as source of *danshen* include *Salvia przewalskii* Maxim. (*gansu danshen*), *S. przewalskii* Maxim. var. *mandarinorum* (Diels) Stib., *S. bowleyana* Dunn (southern *danshen*) and *S. yunnanensis* C. H. Wright; they all have similar chemistry.

Recorded use dates back 2000 years. Considered one of the major *huo xue hua yu* (activating blood circulation to dissipate stasis) herbs, *danshen* is traditionally used in blood and blood circulation problems, including angina pectoris (chest pain), irregular menses, menstrual pain, amenorrhea, metrorrhagia (*xue beng*), leukorrhagia (*dai xia*), abdominal masses (*zheng jia ji ju*), abdominal pain, and insomnia due to palpitations and tight chest, among others.

Contains several phenanthrene diketones and derivatives: tanshinones I, IIA, IIB, V, and VI; isotanshinones I and II, cryptotanshinone, isocryptotanshinone, and dihydrotanshinone; hydroxytanshinone IIA and methyltanshinonate; tanshinol I and tanshinol II;[1] tanshindiols A, B, and C, nortanshinone and 3-β-hydroxytanshinone IIA.[2] Also contains miltirone, Ro-090680 and salvinone (diterpenoids);[2,3] salvianolic acids A and B, rosmarinic acid;[4,5] danshensu [D(+)-(β-(3,4-dihydroxyphenyl)-lactic acid).;[6,7] protocatechuic aldehyde, protocatechuic acid, oleanolic acid, ferruginol, dehydromiltirone, β-sitosterol, vitamin E, and others.[8–10]

Modern scientific studies have confirmed many of red sage's traditional properties and uses. Thus, its extracts and chemical components have been shown to have the following activities: anticoagulant or antiplatelet aggregation (decoction, injection, tanshinones, miltirone, ferruginol, Ro-090680, danshensu, protocatechuic aldehyde, salvinone, salvianolic acid A and rosmarinic acid);[3,4,6,8,11–13] antibacterial (decoction, alcoholic extract, tanshinones, methyltanshinonate); anti-inflammatory (tanshinones);[14] estrogenic and antiandrogenic (tanshinones); retardation of cholesterol biosynthesis in cells and inhibition of lipoprotein oxidation (danshensu);[7] antioxidant *in vitro* (strong: salvianolic acids A and B, rosmarinic acid; fair to weak: danshensu, alcoholic extract);[5,7,15] antimutagenic (acetone extractives);[16] as well as positive cardiovascular and other effects (WANG).[1,17,18]

Toxicity is low: 43 g/kg of decoction (i.p.) in mice caused no fatality within 48 h while 64 g/kg only resulted in 2 deaths in 10 mice; gastric feeding of 2% tanshinones emulsion to mice (0.5 ml) for 14 days and to rats (2.5 mL) for 10 days produced no obvious toxic reactions (WANG).

Tanshinone IIA sulfonate (a major active principle) did not promote growth or metastasis of Lewis carcinoma transplanted in mice.[19]

In addition to cardiovascular diseases, *danshen* extracts (per os) have been successfully used in treating acne, psoriasis, eczema and other skin diseases (WANG).[12,20]

Tanshinones can be extracted with ether, acetone, or related solvents while danshensu is extracted with water.

Extracts used in hair liniments and shampoos for their alleged ability to prevent hair loss and maintain hair color (see safflower); also used in skin creams and lotions for their alleged whitening effects (ZHOU).

REFERENCES

See the General References for BRUNETON; CHP; DER MARDEROSIAN AND BEUTLER; FOSTER AND YUE; HONGKUI; HU; HUANG; JIANGSU; JIXIAN; MCGUFFIN 1 & 2; WANG; ZHOU.

1. A. Yagi et al., *Planta Med.*, **55**, 51 (1989).
2. H. W. Luo et al., *Phytochemistry*, **24**, 815 (1985).
3. N. Wang et al., *Planta Med.*, **55**, 390 (1989).
4. L. N. Li et al., *Planta Med.*, 227 (1984).
5. Y. S. Huang and J. T. Zhang, *Yaoxue Xuebao*, **27**, 96 (1992).
6. C. Z. Li et al., *Chin. J. Integr. Trad. West. Med.*, **3**, 297 (1983).
7. X. M. Sun et al., *Zhongcaoyao*, **22**, 20 (1991).
8. H. W. Luo et al., *Yaoxue Xuebao*, **23**, 830 (1988).
9. X. M. Xu and Z. Y. Xiao, *Zhongcaoyao*, **15**(1), 1 (1984).
10. N. Wang and H. W. Luo, *Zhongcaoyao*, **20**(4), 7 (1989).
11. Y. D. Shi et al., *Zhongyao Tongbao*, **11**(7), 48 (1986).
12. D. B. Wang, *J. Trad. Chin. Med.*, **3**, 227 (1983).
13. Z. W. Zou et al., *Yaoxue Xuebao*, **28**, 241 (1993).
14. Y. G. Gao et al., *Chin. J. Integr. Trad. West. Med.*, **3**, 300 (1983).
15. Y. L. Zhou and R. X. Xu, *Zhongguo Zhongyao Zazhi*, **17**, 368 (1992).
16. C. C. Ruan et al., *Chin. J. Cancer*, **8**, 29 (1989).
17. P. G. Xiao and K. J. Chen, *Phytother. Res.*, **1**(2), 53 (1987).
18. H. J. Deng et al., *Zhongguo Zhongyao Zazhi*, **17**, 233 (1992).
19. M. Z. Liu et al., *Acta Pharmacol. Sin.*, **12**, 534 (1991).
20. D. B. Wang et al., *Chin. J. Dermatol.*, **21**, 167 (1988).

SAFFLOWER (FALSE SAFFRON; *HONGHUA*)

Flowers of *Carthamus tinctorius* L. (Family Compositae), a glabrous annual herb, up to about 1 m high, flowering from May to July; extensively cultivated worldwide, especially for its seeds. When the tubular florets turn from yellow to red, they are collected early in the morning and dried (sun, oven, or in the shade). Xinjiang Province is the largest producer.

First described in the *Shan Han Lu* (circa 3rd century AD), safflower is one of the commonly used *huo xue hua yu* (activating circulation to dissipate blood stasis) herbs. Traditionally used to invigorate blood, break up stasis, facilitate menstruation, and relieve pain; used in cardiovascular conditions (e.g., amenorrhea, menorrhalgia coronary heart disease, chest pain, and traumatic injuries). Although also used to "calm" live fetus and abort dead fetus, caution in pregnancy is normally advised.

Contains a complex mixture of red and yellow pigments including 20–30% safflower yellow (safflower yellow; SY) which is composed of safflomin A (75%), SY-2 (15%), SY-3, and SY-4, all chalcones;[1] glycosides of chalcone (e.g., carthamin; yellow) and quinone (e.g., carthamone; red), with the latter predominant in the commercial product; also colorless flavonoids and flavonoid glycosides (carthamidin, isocarthamidin, neocarthamin); safflower polysaccharide (glucose, xylose, arabinose, and galactose in β-linkages);[2] lignans, fatty acids; and others.[3]

Safflower yellow has immunosuppressive and strong anticoagulant activities;[1,4] and safflower polysaccharide has immunopotentiating effects.[2] Other activities of safflower extracts include: cardiac stimulant, vasodilating, hypolipemic, hypotensive, uterine stimulant, and so on.[5] Toxicities are low: i.v. LD_{50} of carthamin in mice is 2.35 ± 0.14 g/kg while safe oral dose is >8 g/kg; i.p. MLD of decoction in mice is 1.2 g/kg; and a 50% injection when dropped in rabbit eye produced no irritation to its conjunctiva.

Extract used in most Chinese hair growth liniments along with *danshen* and tonic herbs for its blood-invigorating and vasodilating effects that are thought to facilitate transportation of other tonic ingredients to nourish the hair follicles; in facial and body massage preparations and bath preparations for the same effects; also can be used as a coloring agent.

REFERENCES

See the General References for BRUNETON; DERMARDEROSIAN AND BEUTLER; DUKE 4; GRIEVE; HONGKUI; HU; HUANG; JIANGSU; JIXIAN; MCGUFFIN 1 & 2; WANG; ZHOU.

1. Z. W. Lu et al., *Acta Pharmacol. Sin.*, **12**, 537 (1991).
2. H. Huang et al., *Zhongcaoyao*, **15**(5), 21 (1984).
3. X. Q. An et al., *Zhongcaoyao*, **21**(4), 44 (1990).
4. Z. L. Huang et al., *Zhongcaoyao*, **18**(4), 22 (1987).
5. C. Z. Li et al., *Zhongcaoyao*, **14**(7), 27 (1983).

SICHUAN LOVAGE (*CHUANXIONG*)

Rhizome of *Ligusticum chuanxiong* Hort. (syn. *L. wallichii* auct. sin. non-Franch.) (Family Umbelliferae or Apiaceae), an aromatic perennial herb, up to 1 m high, with erect stem and irregularly knobby fist-like rhizome; all cultivated, mainly in southern China, including the provinces of Sichuan, Guizhou, and Yunnan, with Sichuan as the primary producer. Rhizome is collected in late May to early June from plants planted the year before; after being rid of stem, leaves, rootlets, and dirt, it is sun-dried or dried by artificial heat, followed by further removing of all rootlets and dirt by stirring in bamboo baskets (IMM-CAMS).

First recorded use dates back to the *Shen Nong Ben Cao Jing* (circa 200 BC–AD 100). Traditionally considered to be acrid tasting and warming; promoting blood circulation, and activating vital energy (*huo xue xing qi*); alleviating mental depression (*kai yu*); and removing rheumatic and arthritic pain (*qu feng*). Traditionally used in treating irregular menses, amenorrhea, dysmenorrhea, abdominal mass (*zheng jia*), abdominal pain, chest pain, swelling, and pain due to traumatic injuries, headache, rheumatism, and arthritic pain. Extensively used over the centuries as an ingredient in many famous beauty formulas (both internal and topical), ranking third or fourth among the most frequently used herbs in traditional Chinese beauty formulas (cosmetics).[1-3]

Contains alkaloids (0.15% in raw and 0.20% after wine-curing),[4] including tetramethylpyrazine (ligustrazine, chuanxiongzine), L-isobutyl-L-valine anhydride, and perlolyrine;[5] phthalides, including butylphthalide, 4-hydroxy-3-butylphthalide (chuanxiongol), butylidene phthalide, hydroxybutylidene phthalide, dihydroxybutylidene phthalide, ligustilide, sedanenolide, cnidilide, neocnidilide, and so on;[5-10] phenols, including ferulic acid, caffeic acid, chrysophanol, and vanillic acid;[5,11,12] volatile oil (1.0–1.6%, depending on sources) containing the phthalides;[13] adenine and adenosine;[14] spathulenol;[15] sedanonic acid, and others (HU; IMM-2).[11]

The chemical composition of *chuanxiong* is similar but distinctly different from that of *gaoben* (Chinese ligusticum).[13]

Chuanxiong has exhibited various biological activities, including: cardiovascular (coronary dilatation, increasing coronary, cerebral and renal blood flow, reducing vascular resistance, hypotensive, calcium antagonistic, and others, with tetramethylpyrazine being a major active component and perlolyrine a minor active component);[9,16-20] antispasmodic (ligustilide: cnidilide, ferulic acid, and alkaloids);[9,11] antiplatelet aggregation (tetramethylpyrazine and ferulic acid);[9,11] Sedative (volatile oil); antibacterial, antifungal, and antiviral; radiation protective and other effects (WANG; ZHOU AND WANG).

In addition to its cardiovascular and hematologic activities, tetramethylpyrazine also has antimetastatic effects in animals.[21] It is used in China to treat ischemic cerebrovascular diseases, coronary heart disease, and angina pectoris (WANG; ZHOU);[11,22] with one case of drug-induced skin rash reported.[23]

Adenosine has exhibited antiplatelet aggregation and central inhibitory activities (analgesic, reducing spontaneous motor activity, and prolonging death time induced by caffeine); also has weak muscle relaxant effects (see *ganoderma*).[24]

Ferulic acid has antiallergic, broad cardiovascular, and hematologic effects (antiplatelet aggregation, antithrombic, etc.) as well as inhibitory effects on lipid peroxidation (see *angelica*).[25,26]

Toxicities of ferulic acid and tetramethylpyrazine were both fairly low (LD_{50} (i.v.) = 866 ± 29 mg/kg and 416 ± 17 mg/kg, respectively, in mice), but their toxicities were much lower when used together, indicating the traditional rationale of using both *danggui* (ferulic acid) and *chuanxiorcg* (tetramethylpyrazine) in the same formula to obtain the desired effects with lower toxicity.[27]

Powder and extracts (aqueous, oil, and hydroalcoholic) are used in hair care products for their vasodilating and traditional hair-nourishing properties to prevent hair loss and premature graying; and in skin care products (e.g., cleansing creams, nourishing creams, acne creams and lotions) for their vasodilating, antimicrobial, antiallergic effects as well as traditional skin-whitening (tyrosinase inhibitory),[28,29] anti-swelling, and antiwrinkle properties (ETIC; ZHOU).[1-3] Tetramethylpyrazine hydrochloride eye drops are used in the prevention and treatment of near-sightedness.[30]

REFERENCES

See the General References for CHP; ETIC; HONGKUI; IMM-2; IMM-CAMS; JIANGSU; WANG; ZHOU; ZHOU AND WANG; ZHU.

1. W. X. Hong, *Fujian Zhongyiyao*, **18**(1), 39 (1987).
2. Y. B. Xie, *Jiangsu Zhongyi*, (5), 38 (1989).
3. J. H. Chen, *Shandong Zhongyi Xueyuan Xuebao*, **15**(1), 58 (1991).
4. Y. Q. Ou, *Zhongchengyao*, **11**(9), 18 (1989).
5. F. Y. Cao et al., *Zhongcaoyao*, **14**(6), 1 (1983).
6. M. Puech-Baronnat et al., *Planta Med.*, **50**, 105 (1984).
7. M. Kaouadji, *Plantes Med. Phytother.*, **17**, 147 (1983).
8. P. S. Wang et al., *Zhongcaoyao*, **16**(3), 41 (1985).
9. J. M. Xu, *Zhongchengyao*, **11**(1), 37 (1989).
10. Y. S. Wen et al., *Zhongcaoyao*, **17**(3), 26 (1986).
11. Y. J. Ma and S. S. Zhu, *Chin. J. Integr. Trad. West. Med.*, **4**, 574 (1984).
12. P. S. Wang et al., *Zhongcaoyao*, **16**(5), 45 (1985).
13. B. C. Zhang et al., *Zhongcaoyao*, **17**(8), 34 (1986).
14. Y. X. Wang et al., *Zhongcaoyao*, **16**(11), 17 (1985).
15. P. S. Wang et al., *Zhongcaoyao*, **16**(4), 30 (1985).
16. L. B. Hou et al., *Zhongcaoyao*, **23**(11), 583 (1992).
17. P. G. Xiao and K. J. Chen, *Phytother. Res.*, **1**(2), 53 (1987).
18. J. A. O. Ojewole, *Planta Med.*, **42**, 223 (1981).
19. M. G. Feng et al., *Acta Pharmacol. Sin.*, **9**, 548 (1988).
20. Y. L. Wang and Y. K. Ba, *Chin. J. Integr. Trad. West. Med.*, **5**, 291 (1985).
21. J. R. Liu and S. B. Ye, *Chin. J. Pharmacol. Toxicol.*, **7**, 149 (1993).
22. J. W. Yang, *Liaoning Zhongyi Zazhi*, **12**(5), 26 (1988).
23. K. L. Zhang et al., *Chin. J. Integr. Trad. West. Med.*, **6**, 375 (1986).
24. Y. Kasahara and H. Hikino, *Phytother. Res.*, **1**, 173 (1987).
25. J. L. Wu and D. Y. Wang, *Zhongguo Yaoxue Zazhi*, **28**, 267 (1993).
26. H. J. Hu and B. Q. Hang, *Acta Pharmacol. Sin.*, **12**, 426 (1991).
27. J. Xu et al., *Zhongguo Zhongguo Zazhi*, **17**, 680 (1992).
28. X. T. Liu, *Zhongchengyao*, **13**(3), 9 (1991).
29. Y. Masamoto et al., *Planta Med.*, **40**, 361 (1980).
30. X. L. Hu and X.Q. Hu, *Zhongcaoyao*, **14**(5), 16 (1983).

SKULLCAP, BAIKAL (HUANG*QIN*)

Root of *Scutellaria baicalensis* Georgi (Family Labiatae or Lamiaceae), a perennial herb up to 60 cm high, with a large and long taproot; plant native to eastern Asia. Root is collected in spring and autumn, usually from 3- to 4-year-old plants. After having rootlets and dirt removed, the root is partially dried, rid of root bark, and then further dried to completion. Produced mainly in northern China. Several other species of *Scutellaria* from other regions of China are also used, with specific names indicating their geographical or botanical origins (JIANGSU).[1]

Earliest use (in ointment form for treating wounds and cramps) was described in the *Wu Shi Er Bing Fang* (ca. 1065 BC to 771 BC).[2] It was later listed under the middle category of drugs in the *Shen Nong Ben Cao Jing* (ca. 200 BC–AD 100). Traditionally considered to taste bitter and cold and to have heat-clearing, wetness-drying (*zao shi*), fire-purging (*xie huo*), detoxifying, hemostatic, and fetus-calming properties; used to treat heat (*re*) related conditions, including restlessness and thirst (*fan ke*), cough, diarrhea, tight chest and abdominal distention (*pi man*), jaundice, fever, vomiting of blood, sores, carbuncles, furuncles, red eye with swelling and pain, and threatened abortion (restless fetus).

In recent years, Baikal skullcap is also used in treating burns, often combined with phellodendron, giant knotweed, and garden burnet (see individual entries), as well as different kinds of infections (bacterial, viral, etc.) and hypertension, among others (JIANGSU).

Baikal skullcap contains flavonoids and their glycosides, including baicalein (5,6,7-trihydroxyflavone), wogonin (5,7-dihydroxy-8-methoxyflavone), skullcapflavone I (5,6′-dihydroxy-7,8-dimethoxyflavone), skullcapflavone II (2′,5-dihydroxy-6,6′,7,8-tetramethoxyflavone), oroxylin A (5,7-dihydroxy-6-methoxyflavone), koganebanain (5,7-dihydroxy-6, 8,2′,3′-tetramethoxyflavone), (2S)-2′, 5,6′,7-tetrahydroxyflavanone, (2R, 3R)-2′,3,5, 6′,7-pentahydroxyflavanone, 2′,5,5′,7-tetrahydroxy-6′,8-dimethoxyflavone, baicalin (5,6,7-trihydroxyflavone-7-O-D-glucuronide) wogonoside (5,7dihydroxy-8-methoxyflavone-7-O-glucuronide), oroxylin A glucuronide, and others, with baicalin in major concentration (3.6–6.2%); benzoic acid, β-sitosterol, and so on (IMM-1; JIANGSU).[1] Roots of other *Scutellaria* species have similar chemistry as Baikal skullcap.[1]

Decoction and alcohol extractives have exhibited broad antibacterial and antifungal effects *in vitro*; also active against several viruses, including influenza strains PR_8 and Asian A; baicalein being one of the major active principles. (2S),2′,5,6′,7-Tetrahydroxyflavanone is active against Gram-negative bacteria (WANG).[3]

Baicalein, baicalin, wogonin, skullcapflavone II, and 2′,5,5′,7-tetrahydroxy-6′,8-dimethoxyflavone have anti-inflammatory and antiallergic effects in animals; skullcapflavone II also exhibited cytotoxic effects on L1210 cells (JIANGSU).[4–6]

Other biological effects of HUANG*qin* include: sedative; antipyretic; hypotensive, diuretic; antiarthritic; hypolipemic; cholagogic; antispasmodic; and detoxicant (WANG);[7] also strongly antioxidant (due to flavonoids).[8–10]

Baikal skullcap has very low toxicity: 10 g/kg of decoction p.o. and 2 g/kg i.v. of alcohol extractives produced only sedation but no deaths in rabbits; and LD_{50} of baicalin in mice is 3.081 g/kg (i.v.). However, 15 mg/kg of baicalin administered i.v. to rabbits was fatal within 48 h (WANG).

Extracts used in skin care products (esp. skin freshener, acne creams and lotions, etc.) for their astringent, anti-inflammatory and antimicrobial effects (ETIC; ZHOU); also used in toothpaste.[11]

REFERENCES

See the General References for BAILEY 1; CHP; ETIC; FOSTER AND YUE; HONGKUI; HU; HUANG; IMM-1; JIANGSU; JIXIAN; MCGUFFIN 1 & 2; WANG; ZHOU.

1. W. Z. Song, *Yaoxue Xuebao*, **16**, 139 (1981).
2. Q. M. Sun, *Yaoxue Tongbao*, **17**(5), 33 (1982).
3. M. Kubo et al., *Planta Med.*, **43**, 194 (1981).
4. Y. Kimura et al., *Planta Med.*, **51**, 132 (1985).
5. H. Otsuka et al., *J. Nat. Prod.*, **51**, 74 (1988).
6. S. H. Ryu et al., *Planta Med.*, **51**, 462 (1985).
7. Z. Li and X. Y. Guo, *Chin. J. Integr. Trad. West. Med.*, **9**, 698 (1989).
8. Y. Kimura et al., *Planta Med.*, **50**, 290 (1984).
9. Y. Kimura et al., *Chem. Pharm. Bull.*, **29**, 2610 (1981).
10. L. C. Song et al., *Shaanxi Zhongyi*, **14**, 185 (1993).
11. Z. Lin, *Dazhong Zhongyiyao*, (3), 47 (1990).

Appendix A
General References

American Dental Association (ADA). 1973. *Accepted Dental Therapeutics*, 35th ed. ADA, Chicago, IL.

American Herbal Products Association (AHPA). 1992. *Herbs of Commerce*. American Herbal Products Association, Austin, TX.

American Pharmaceutical Association (APhA). 1990. *Handbook of Nonprescription Drugs*, 10th ed. APhA, Washington, DC.

Applequist, W., illustrated by B. Alongi. 2006. *The Identification of Medicinal Plants. A Handbook of the Morphology of Botanicals in Commerce*. American Botanical Council, Austin, TX.

Arctander, S. 1960. *Perfumes and Flavor Materials of Natural Origin*. Published by the author, Elizabeth, NJ.

Ayensu, E. S. 1981. *Medicinal Plants of the West Indies*. Reference Publications, Algonac, MI.

Bailey, L. H. 1949. *Manual of Cultivated Plants*. MacMillan, New York.

Bailey, L. H. 1942. *The Standard Cyclopedia of Horticulture*, 3 Vols. MacMillan, New York.

Balsam, M. S. and E. Sagarin (eds.) 1972. *Cosmetics Science and Technology*, 2nd ed., 2 Vols. Wiley–Interscience, New York.

Barnes, J., L. A. Anderson, and J. D. Phillipson. 1996. *Herbal Medicines. A Guide for Health Care Professionals*. The Natural History Museum, London.

Barnes, J., L. A. Anderson, and J. D. Phillipson, 2002. *Herbal Medicines: A Guide For Healthcare Professionals*, 2nd ed. Pharmaceutical Press, London.

Barrett, M. 2004. *The Handbook of Clinically Tested Herbal Remedies*, 2 Vols. Haworth Press, Binghamton, NY.

Bauer, K., D. Garbe, and H. Surburg. 1990. *Common Fragrance and Flavor Materials*, 2nd ed. VCH Publishers, New York.

Bensky, D. and A. Gamble. 1986. *Chinese Herbal Medicine: Materia Medica*. Eastland Press, Inc., Seattle, WA.

Bianchini, F. and F. Corbetta. 1977. *Health Plants of the World-Atlas of Medicinal Plants*. Newsweek Books, New York.

Bisset, N. G. (ed. and transl.). 1994. *Herbal Drugs and Phytopharmaceuticals. A Handbook for Practice on a Scientific Basis*. CRC Press, Boca Raton, FL.

Blumenthal, M. (ed.), S. Klein (transl.). 1995. *German Bundesgesuntheitsamt (BGA) Commission E Therapeutic Monographs on Medicinal Products for Human Use*. American Botanical Council, Austin, TX (English translation).

Blumenthal, M. 2003. *The ABC Clinical Guide to Herbs*. American Botanical Council, Austin, TX.

Boulos, L. 1983. *Medicinal Plants of North Africa (Medicinal Plants of the World)*. Reference Publications, Algonac, MI.

Bown, D. 2003. *RHS Encyclopedia of Herbs and their Uses*. 3rd ed. Dorling Kindersley. London.

Boyle, W. 1991. *Official Herbs: Botanical Substances in the United States Pharmacopoeias 1820-1990*. Buckeye Naturopathic. East Palestine, Ohio.

Bradly, P. R. (ed.). 1992. *British Herbal Compendium*, Vol. 1. British Herbal Medicine Association, Dorset, UK.

Brinker, F. J. 2000. *The Toxicology of Botanical Medicines*. 3rd ed. Eclectic Medical. Sandy, Oregon.

Brouk, B. 1975. *Plants Consumed by Man*. Academic Press, London.

Bruneton, J. 1995. *Pharmacognosy: Phytochemistry of Medicinal Plants*, 2nd ed. Lavoisier, France.

Burkill, I. H. 1966. *A Dictionary of the Economic Products of the Malay Peninsula*, 2 Vols. Ministry of Agriculture and Cooperatives, Kuala Lumpur.

Cheung, S. C. and N. H. Li (eds.). 1978, 1981, 1983, 1985, 1986. *Chinese Medicinal Herbs of Hong Kong*. 5 Vols. Commercial Press, Hong Kong (in Chinese and English).

Chinese Ministry of Health Bureau of Drug Administration (CMH). 1990. *Manual of Chinese Drugs*, 2nd ed. People's Health Publications, Beijing (in Chinese).

Chinese Pharmacopeia Committee (CHP), Chinese Ministry of Health. *Chinese Pharmacopeia* (1990). People's Health Publications, Beijing (in Chinese).

Chittendon, F. J. 1956. *Dictionary of Gardening*, 4 Vols. Oxford University Press, London.

Claus, E. P. 1961. *Pharmacognosy*, 4th ed. Lea & Febiger, Philadelphia, PA.

Coon, N. 1974. *The Dictionary of Useful Plants*. Rodale, Emmaus, PA.

Council of Scientific and Industrial Research (CSIR). 1948–1985. *The Wealth of India*, 11 Vols. Publications & Information Directorate, Council of Scientific & Industrial Research, New Delhi.

Crellin, J. K. and J. Philpott. 1989. *Herbal Medicine Past and Present*, 2 Vols. Duke University Press, Durham, NC.

Cupp, M. J. and T. S. Tracy. 2003. *Dietary Supplements. Toxicology and Clinical Pharmacology*. Humana Press, Totowa, NJ.

Davidson, R. L. (ed.). 1980. *Handbook of Water Soluble Gums and Resins*. McGraw-Hill, New York.

Deng, W. L. (ed.). 1990. *Pharmacology and Applications of Traditional Chinese Formulas*. Chongqing Publications, Chongqing, China (in Chinese).

Der Marderosian, A. and L. Liberti. 1988. *Natural Product Medicine: A Scientific Guide to Foods, Drugs, Cosmetics*. George F. Stickley Co., Philadelphia, PA.

Der Marderosian, A. and J. Beutler (eds.). 2002. *The Review of Natural Products*, 2nd ed. Facts and Comparisons, St. Louis, MO.

Devon, T. K. and A. L. Scott. 1975. *Handbook of Naturally Occurring Compounds. Vol. 1. Acetogenins, Shikimates, and Carbohydrates.* Academic Press, New York.

Devon, T. K. and A. L. Scott. 1972. *Handbook of Naturally Occurring Compounds. Vol. 2. Terpenes.* Academic Press, New York.

Duke, J. A. 1981. *Handbook of Legumes of World Economic Importance*. Plenum Press, New York.

Duke, J. A. 1985. *CRC Handbook of Medicinal Herbs*. CRC Press, Boca Raton, FL.

Duke, J. A. 1989. *CRC Handbook of Nuts*. CRC Press, Boca Raton, FL.

Duke, J. A. et al. 2008. *Dukes Handbook of Medicinal Plants of the Bible*. CRC Press, Boca Raton, FL.

Economic and Technical Information Center (ETIC), Ministry of Light Industries. 1987. *Utilization of Natural Additives in Cosmetics*. ETIC, People's Republic of China (in Chinese).

Erichsen-Brown, C. 1989. *Medicinal and Other Uses of North American Plants*. Dover Publications, New York.

ESCOP, 1990. *Proposal for European Monographs*, Vol. 1. ESCOP Secretariat, Bevrijdingslaan, The Netherlands.

ESCOP, 1992. *Proposal for European Monographs*, Vol. 2. ESCOP Secretariat, Bevrijdingslaan, The Netherlands.

ESCOP, 1992. *Proposal for European Monographs*, Vol. 3. ESCOP Secretariat, Bevrijdingslaan, The Netherlands.

Evans, W. C. 2002. *Trease and Evans Pharmacognosy*, 15th ed. Bailliere Tindall, London.

Farnsworth, N. R. 1969. *The Lynn Index: A Bibliography of Phytochemistry*, Vol. 6. Norman R. Farnsworth, Pittsburgh, PA.

Farnsworth, N. R. 1974. *The Lynn Index: A Bibliography of Phytochemistry*, Vol. 8. Norman R. Farnsworth, Chicago, IL.

Farnsworth, N. R. 1975. *An Evaluation of Atlas of Common Chinese Drugs, a compilation by the*

Chinese College of Medical Sciences, Peking. University of Illinois Medical Center, Chicago, IL. See also *Herbal Pharmacology in the People's Republic of China.*

Farnsworth, N. R. et al. 1971. *The Lynn Index: A Bibliography of Phytochemistry*, Vol. 7. Norman R. Farnsworth, Chicago, IL.

Felter, H. W. and J. U. Lloyd. 1906. *King's American Dispensatory*, 18th ed., 2 Vols. Reprinted. 1983. Eclectic Medical Publications, Portland, OR.

Fernald, M. L. 1950. *Gray's Manual of Botany.* American Book, New York.

Flavor and Extract Manufacturer's Association of the United States (FEMA). 1975. *Results of Second FEMA Survey of Flavoring Ingredients: Average Maximum Use Levels.* FEMA, Washington, DC.

J. E. Fogarty International Center for Advanced Study in the Health Sciences (FOGARTY). 1974. *A Barefoot Doctor's Manual.* Department of Health Education, and Welfare Publication No. (NIH) 75-695. National Institutes of Health, Washington, DC (translation of Chinese text).

Food Chemicals Codex, 2nd ed. 1972. National Academy of Sciences, Washington, DC.

Foster, S. 1993. *Herbal Renaissance.* Gibbs Smith Publisher, Layton, UT.

Foster, S. and R. Caras. 1994. *A Field Guide to Venomous Animals and Poisonous Plants of North America (North of Mexico).* Houghton Mifflin Co., Boston, MA.

Foster, S. and J. A. Duke. 1990. *A Field Guide to Medicinal Plants: Eastern and Central North America.* Houghton Mifflin Co., Boston, MA.

Foster, S. and C. X. Yue. 1992. *Herbal Emissaries: Bringing Chinese Herbs to the West.* Healing Arts Press, Rochester, VT.

Frohne, D. and H. J. Pfander. 1983. *A Colour Atlas of Poisonous Plants.* Wolfe Publishing, Ltd., London.

Fugh-Berman, A. (ed.). 2002. *Five Minute Herb and Dietary Supplement Clinical Consult.* Lippincott Williams & Wilkins, Philadelphia, PA.

Fujian Provincial Institute of Medical and Pharmaceutical Research. 1979, 1982. *Records of Fujian Materia Medica*, 2 Vols. Fujian People's Press and Fujian Scientific and Technical Publications, Fuzhou, China (in Chinese).

Furia, T. E. (ed.). 1975. *Handbook of Food Additives*, 2nd ed. CRC Press, Cleveland, OH.

Furia, T. E. and N. Bellanca (eds.). 1975. *Fenaroli's Handbook of Flavor Ingredients*, 2nd ed., Vol. 1. CRC Press, Cleveland, OH.

Ghazanfar, S. A. 1994. *Handbook of Arabian Medicinal Plants.* CRC. Boca Raton, FL.

Glasby, J. S. 1976. *Encyclopedia of Antibiotics.* Wiley–Interscience, London.

Glasby, J. S. 1991. *Dictionary of Plants Containing Secondary Metabolites.* Taylor & Francis, New York.

Gleason, H. A. and A. Cronquist. 1991. *Manual of Vascular Plants*, 2nd ed. New York Botanical Garden, New York.

Glicksman, M. 1969. *Gum Technology in the Food Industry.* Academic Press, New York.

Goodman, L. S. and A. Gilman (eds.). 2005. *The Pharmacological Basis of Therapeutics*, 11th ed. MacMillan, New York.

Gosselin, R. E. et al. 1976. *Clinical Toxicology of Commercial Products: Acute Poisoning*, 4th ed. Williams & Wilkins, Baltimore, MD.

Grieve, M. 1992. *Modern Herbal: The Medicinal, Culinary, Cosmetic, and Economic Properties, Cultivation, and Folklore of Herbs, Grasses, Fungi, Shrubs, and Trees with all their Modern Scientific Uses.* Dorset Press, New York.

Guenther, E. 1948. *The Essential Oils*, 6 Vols. Van Nostrand, New York.

Gupta, A. K. 2003. *Quality Standards of Indian Medicinal Plants*, Indian Council of Medical Research, New Delhi.

Harborne, J. B. 1973. *Phytochemical Methods.* Chapman & Hall, London.

Harbourne, J. B. and H. Baxter. 1993. *Phytochemical Dictionary: A Handbook of Bioactive Compounds from Plants.* Taylor & Francis, London.

Hardin, J. W. and J. M. Arena. 1974. *Human Poisoning from Native and Cultivated Plants*, 2nd ed. Duke University Press, Durham, NC.

Harris, R. S. and E. Karmas (eds.). 1975. *Nutritional Evaluation of Food Processing*, 2nd ed. AVI, Westport, CT.

Hay, R. and P. M. Synge. 1975. *The Color Dictionary of Flowers and Plants for Home and Garden.* Crown, New York.

Herbal Pharmacology in the People's Republic of China: A Trip Report of the American Herbal Pharmacology Delegation. 1975. National Academy of Sciences, Washington, DC.

Hickman, J. C. (ed.). 1993. *The Jepsor Manual: Higher Plants of California.* University of California Press, Berkeley.

Higgins, S. T. and J. L. Katz (eds.). 1998. *Cocaine Abuse: Behavior, Pharmacology, and Clinical Applications.* Academic Press, London.

Hocking, G. M. 1955. *A Dictionary of Terms in Pharmacognosy and Economic Botany.* Thomas, Springfield, IL.

Hoffman, D. 1987. *The Herbal Handbook: A User's Guide to Medical Herbalism.* Healing Arts Press, Rochester, VT.

Hongkui, Z. 1995. *The Resource Atlas of Chinese Herbal Medicine.* Science Press, Bejing, China.

Hortus Third: A Concise Dictionary of Plants Cultivated in the United States and Canada. L. H. Bailey Hortorium Staff Cornell University. 1976. MacMillan, New York.

Hsu, H. H. et al. 1989. *Oriental Materia Medica: A Concise Guide.* Oriental Healing Arts Institute, Long Beach, CA.

Hu, S. L. 1989. *Indigenous Chinese Drugs.* Heilongjiang Scientific and Technical Publishers, Harbin, China (in Chinese).

Huang, K. C. 1999. *The Pharmacology of Chinese Herbs*, 2nd ed. CRC Press, Boca Raton, FL.

Institute of Materia Medica (IMM-1), Chinese Academy of Medical Sciences. 1982. *Records of Chinese Materia Medica*, Vol. 1. People's Health Publications. Beijing. (in Chinese)

Institute of Materia Medica (IMM-2), Chinese Academy of Medical Sciences. 1982. *Records of Chinese Materia Medica*, Vol. 2. People's Health Publications, Beijing, China (in Chinese).

Institute of Materia Medica (IMM-3), Chinese Academy of Medical Sciences. 1981. *Records of Chinese Materia Medica*, Vol. 3. People's Health Publications, Beijing, China (in Chinese).

Institute of Materia Medica (IMM-4) and Institute of Medicinal Plant Development, Chinese Academy of Medical Sciences. 1988. *Records of Chinese Materia Medica*, Vol. 4. People's Health Publications, Beijing, China (in Chinese).

Institute of Materia Medica, Chinese Academy of Medical Sciences (IMM-CAMS). 1979. *Techniques for Cultivating Chinese Herbs.* People's Health Publications, Beijing, China (in Chinese).

Isler, O. et al. (eds.). 1971. *Carotenoids.* Halsted (Wiley), New York.

Jiangsu Institute of Modern Medicine. 1977. *Encyclopedia of Chinese Drugs*, 3 Vols. Shanghai Scientific and Technical Publications, Shanghai, China (in Chinese).

Jiangsu Provincial Institute of Botany (Jiangsu Provincial 1). 1988. *Essentials of Medicinal Plants of New China*, Vol. 1. Shanghai Scientific and Technical Publications, Shanghai, China (in Chinese).

Jiangsu Provincial Institute of Botany (Jiangsu Provincial 3). 1990. *Essentials of Medicinal Plants of New China*, Vol. 3. Shanghai Scientific and Technical Publications, Shanghai, China (in Chinese).

Jilin Provincial Institute of Traditional Chinese Medicine and Materia Medica. 1982. *Records of Plant Drugs of Changbai Mountain.* Jilin People's Press, Changchun, China (in Chinese).

Jixian, G. 1997. *Pharmacopoeia of the Peoples Republic of China.* Chemical Industry Press, Beijing, China.

Johnson, A. H. and M. S. Peterson. 1974. *Encyclopedia of Food Technology.* AVI, Westport, CT.

Karrer, W. 1958. *Konstitution und Vorkommen der organischen Pflanzenstoffe (exclusive Alkaloide).* Birkhauser Verlag, Basel, Switzerland (in German).

Kartesz, J. T. and R. Kartesz. 1980. *A Synonymized Checklist of the Vascular Flora of the*

United States, Canada, and Greenland. University of North Carolina Press, Chapel Hill, NC.

Kartesz, J. T. 1994. *A Synonymized Checklist of the Vascular Flora of the United States, Canada, and Greenland*, 2nd ed., 2 Vols. Timber Press, Portland, OR.

Keay, R. W. J. 1989. *Trees of Nigeria.* Clarendon Press, Oxford, UK.

Kennedy, J. F. (ed.). 1988. *Carbohydrate Chemistry.* Oxford University Press, London.

Kindscher, K. 1992. *Medicinal Wild Plants of the Prairie.* University Press of Kansas, Lawrence, KS.

Kiple, K. F. and K. C. Ornelas. 2000. *The Cambridge World History of Food.* Cambridge University. New York.

Kirtikar, K. R. and B. D. Basu. 1999. *Indian Medicinal Plants.* International Book Distribution. Dehradun, India.

Kreig, M. B. 1964. *Green Medicine. The Search for Plants that Heal.* Rand McNally, Skokie, IL.

Krochmal, A. and C. Krochmal. 1975. *A Guide to the Medicinal Plants of the United States.* Quadrangle/The New York Times Book Co., New York.

Kutsky, R. J. 1973. *Handbook of Vitamins and Hormones.* Van Nostrand Reinhold, New York.

Lawrence, A. A. 1976. *Natural Gums for Edible Purposes.* Noyes, Park Ridge, NJ.

Leung, A. Y. 1984. *Chinese Herbal Remedies.* Universe Books, New York. (Republished as: *Chinese Healing Foods and Herbs.* 1993. AYSL Corp, Glen Rock, NJ.)

Lewis, W. H. and M. P. H. Elvin-Lewis. 1977. *Medical Botany. Plants Affecting Man's Health.* Wiley–Interscience, New York.

Liener, I. E. (ed.). 1969. *Toxic Constituents of Plant Foodstuffs.* Academic Press, New York.

List, P. H. and L. Hörhammer. 1969–1976. *Hagers Handbuch der Pharmazeutischen Praxis*, Vols. 2–5. Springer-Verlag, Berlin, Germany (in German).

Lloyd, J. U. 1921. *Origin and History of all the Pharmacopeial Vegetable Drugs, Chemicals, and Preparations, with Bibliography*, Vol. 1. The Caxton Press, Cincinnati, OH.

Lu, Q. Y. and M. Li. 1987. *Chinese Herbs for Life Extension.* People's Health Publications, Beijing, China (in Chinese).

Lucas, R. 1966. Nature's Medicines: *The Folklore, Romance and Value of Herbal Remedies.* Parker, West Nyack, NY.

Lust, J. B. 1974. *The Herb Book.* Benedict Lust, Simi Valley, CA.

Ma, X. M. 1980. *Processing Methods for Chinese Materia Medica. Revised.* Shaanxi Scientific and Technical Publishers, Xi'an, China (in Chinese).

Mabberly, D. J. 1997. *The Plant Book: A Portable Dictionary of the Higher Plants.* Cambridge University Press, New York.

Manniche, L. 2006. *An Ancient Egyptian Herbal.* British Museum, London.

Marsh, A. C. et al. 1977. *Composition of Foods, Spices and Herbs. Raw, Processed, Prepared.* Agriculture Handbook No. 8-2. Agricultural Research Service, U.S. Department of Agriculture, Washington, DC.

Martindale: The Extra Pharmacopoeia. 1982. The Pharmaceutical Press, London.

Masada, Y. 1976. *Analysis of Essential Oils by Gas Chromatography and Mass Spectrometry.* Halsted (Wiley), New York.

The Merck Index. An Encyclopedia of Chemicals and Drugs, 10th ed. 1983. Merck, Rahway, NJ.

McGuffin, M., J. I. Kartesz, A. Y. Leung, and A. O. Tucker. 2000. *Herbs of Commerce*, 2nd ed., American Herbal Products Association and Michael McGuffin, USA.

McGuffin, M., C. Hobbs, R. Upton, and A. Goldberg. 1997. *Botanical Safety Handbook.* The American Herbal Product Association, Boca Raton, FL.

McKenna, D. J., K. Jones and K. Hughes. 2002. *Botanical Medicines: The Desk Reference for Major Herbal Supplements.* 2nd ed. Haworth Herbal. Binghamton, New York.

Merory, J. 1968. *Food Flavorings, Composition, Manufacture and Use.* AVI, Westport, CT.

Moerman, D. E. 1986. *Medicinal Plants of Native America*, 2 Vols. Technical Reports,

No. 19, Research Reports in Ethnobotany, Contribution 2. University of Michigan Museum of Anthropology, Ann Arbor, MI.

Moore, M. 1989. *Medicinal Plants of the Desert and Canyon West*. Museum of New Mexico Press, Santa Fe, NM.

Moore, M. 1993. *Medicinal Plants of the Pacific West*. Red Crane Books, Santa Fe, NM.

Morton, J. F. 1974. *Folk Remedies of the Low Countries*. Seemann, Miami, FL.

Morton, J. F. 1976. *500 Plants of South Florida*. Seemann, Miami, FL.

Morton, J. F. 1977. *Major Medicinal Plants: Botany, Culture, and Uses*. Thomas, Springfield, IL.

Morton, J. F. 1981. *Atlas of Medicinal Plants of Middle America*. Charles C. Thomas, Springfield, IL.

Morton, J. F. 1987. *Fruits of Warm Climates*. Creative Resources Systems, Inc., Winterville, SC.

Munz, P. A. and D. D. Keck. 1968. *A California Flora*. University of California Press, Berkeley, CA.

Nadkarni, A. K. 2005. *Dr. Nadkarni's Indian Materia Medica*. Popular Prakashan, India.

Nakanishi, K. et al. (eds.). 1975. *Natural Products Chemistry*, 2 Vols. Kodansha, Tokyo and Academic, New York.

Nanjing Pharmaceutical Institute. 1960. *Materia Medica*. Shao Hwa Society for Cultural Services, Hong Kong (in Chinese).

National Collection of Chinese Herbal Drugs Editorial Committee. 1983. *National Collection of Chinese Herbal Drugs*, 2 Vols. People's Health Publications, Beijing, China (in Chinese).

National Formulary, 16th ed. (NF XVI) 1985. American Pharmaceutical Association, Washington, DC.

de Navarre, M. G. 1975. *The Chemistry and Manufacture of Cosmetics*, 2nd ed., Vols. 3 and 4. Continental, Orlando, FL.

Newall, C. A. et al. *Herbal Medicines A Guide for Healthcare Professionals*. 1996. The Pharmaceutical Press, London.

Nikitakis, J. M. 1988. *CTFA Cosmetic Ingredient Handbook*. The Cosmetic, Toiletry and Fragrance Association, Inc., Washington, DC.

Peng, M. Q. 1987. *Comprehensive Treatise on Chinese Medicinal Foods*. Sichuan Scientific and Technical Publications, Chengdu, China (in Chinese).

Peterson, M. S. and A. H. Johnson. 1978. *Encyclopedia of Food Science*. AVI, Westport, CT.

Petrides, G. A. 1988. *Peterson Field Guide to Eastern Trees*. Houghton Mifflin Co., Boston, MA.

Phillips, R. W. 1973. *Science of Dental Materials*, 7th ed. Saunders, Philadelphia, PA.

Polunin, O. and B. E. Smythies. 1973. *Flowers of Southwest Europe*. Oxford University Press, London.

Poucher, W. A. 1974. *Perfumes, Cosmetics, and Soaps. Vol. 1. The Law Materials of Perfumery*, 7th ed. Halsted (Wiley), New York.

Raffauf, R. F. 1970. *A Handbook of Alkaloids and Alkaloid-Containing Plants*. Wiley–Interscience, New York.

Ravindran, P. N., K. Nirmal-Babu and M. Shylaja. 2003. *Cinnamon and Cassia: The Genus Cinnamomum*. CRC, Boca Raton, Florida.

Remington's Pharmaceutical Sciences, 15th ed. 1975. Mack, Easton, PA.

Rose, J. 1976. *The Herbal Body Book*. Grosset & Dunlap, New York.

Rosengarten, F., Jr. 1969. *The Book of Spices*. Livingston, Wynnewood, PA.

Sargent, C. S. 1965. *Manual of Trees of North America*, 2 Vols. Dover, New York.

Sax, N. L. 1975. *Dangerous Properties of Industrial Materials*, 4th ed. Van Nostrand Reinhold, New York.

Simon, J. E., A. F. Chadwick, and L. E. Craker. 1984. *Herbs: An Indexed Bibliography, 1971–1980*. Archon Books, Hamden, CT.

De Smet, P. A. G. M., et al. 1997. *Adverse Effects of Herbal Drugs*. 3rd. Springer-Verlag, Berlin.

Stahl, E. (ed.). 1973. *Drug Analysis by Chromatography and Microscopy*. Ann Arbor Science, Ann Arbor, MI.

Steiner, R. P. (ed.). 1986. *Folk Medicine: The Art and the Science*. American Chemical Society, Washington, DC.

Steinmetz, E. F. 1957. *Codex Vegetabilis*. E. F. Steinmetz, Amsterdam, The Netherlands.

Tang, W. and G. Eisenbrand (eds.). 1992. *Chinese Drugs of Plant Origin*. Springer-Verlag, Berlin, Germany.

Terrell, E. E. 1977. *A Checklist of Names for 3,000 Vascular Plants of Economic Importance*. Agriculture Handbook No. 505. Agricultural Research Service. U.S. Department of Agriculture, Washington, DC.

Tisserand, R. and Balazs, T. 1995. *Essential Oil Safety: A Guide for Health Care Professionals*. Churchill Livingstone, New York.

Tu, G. S. 1988. *Pharmacopoeia of the People's Republic of China* (English Translation of 1985 *Pharmacopeia of the People's Republic of China*). China Pharmaceutical Books Co., Hong Kong, China.

Tucker, A. O. 1986. *Botanical Nomenclature of Culinary Herbs and Potherbs*. In L. E. Craker and J. E. Simon (eds.). *Herbs, Spices, and Medicinal Plants: Recent Advances in Botany, Horticulture, and Pharmacology*, Vol. 1. Oryx Press, Phoenix, AZ, pp. 33–80.

Tucker, A. O., J. A. Duke and S. Foster. 1989. *Botanical Nomenclature of Medicinal Plants*. In L. E. Craker and J. E. Simon (eds.) *Herbs, Spices, and Medicinal Plants: Recent Advances in Botany, Horticulture, and Pharmacology*, Vol. 4. Oryx Press, Phoenix, AZ, pp. 169–242.

Tucker, A. O. and B. M. Lawrence. 1987. *Botanical Nomenclature of Commercial Sources of Essential Oils, Concretes, and Absolutes*. In L. E. Craker and J. E. Simon (eds.). *Herbs, Spices, and Medicinal Plants: Recent Advances in Botany, Horticulture, and Pharmacology*, Vol. 2. Oryx Press, Phoenix, AZ, pp. 183–220.

Tutin, T. G. et al. (eds.). 1964. *Flora Europaea*, Vol. 1. Cambridge University Press, Cambridge, MA.

Tutin, T. G. et al. (eds.). 1968. *Flora Europaea*, Vol. 2. Cambridge University Press, Cambridge, MA.

Tutin, T. G. et al. (eds.). 1972. *Flora Europaea*, Vol. 3. Cambridge University Press, Cambridge, MA.

Tutin, T. G. et al. (eds.). 1976. *Flora Europaea*, Vol. 4. Cambridge University Press, Cambridge.

Tutin, T. G. et al. (eds.). 1980. *Flora Europaea*, Vol. 5. Cambridge University Press, Cambridge.

Tyler, V. E. 1993. *The Honest Herbal*, 3rd ed. Pharmaceutical Products Press, Binghamton, NY.

Tyler, V. E. 1994. *Herbs of Choice: The Therapeutic Use of Phytomedicinals*. Pharmaceutical Products Press, Binghamton, NY.

Tyler, V. E., L. R. Brady, and J. E. Robbers. 1988. *Pharmacognosy*, 9th ed. Lea & Febiger, Philadelphia, PA.

United States Department of Agriculture (USDA). 1978. *Agricultural Statistics*. U.S. Government Printing Office, Washington, DC.

The Dispensatory of the United States of America, 23rd ed. (USD 23rd). 1943. Lippincott, Philadelphia, PA.

The United States Dispensatory and Physicians Pharmacology, 26th ed. (USD 26th). 1967. Lippincott, Philadelphia, PA.

University of South Florida Institute for Systematic Botany (USFISB), *Atlas of Florida Vascular Plants*, http://www.plantatlas.usf.edu.

The United States Pharmacopeia, 21st rev. (USP XXI). 1985. U.S.P. Convention, Rockville, MD.

Uphof, J. C. T. 1968. *Dictionary of Economic Plants*. J. Cramer. Stechert-Hafner, New York.

Wallis, T. E. 1967. *Textbook of Pharmacognosy*, 5th ed. Churchill, London.

Wang, Y. S. et al. (eds.). 1983. *Pharmacology and Applications of Chinese Materia Medica*. People's Health Publications, Beijing, China (in Chinese) (English translation: Chang, H. M. and P. P. But. 1986. *Pharmacology and Applications of Chinese Materia Medica*. World Scientific, Hong Kong, China).

Watt, B. K. and A. L. Merrill. 1975. *Composition of Foods, Raw, Processed, Prepared*. Agriculture Handbook No. 8. Agricultural Research Service, U.S. Department of Agriculture, Washington, DC.

Watt, J. M. and M. G. Breyer-Brandwijk. 1962. *The Medicinal and Poisonous Plants of Southern and Eastern Africa*, 2nd ed. E. & S. Livington Ltd., Edinburgh and London

Weiss, R. F. 1988. *Herbal Medicine* (translated from German by A. R. Meuss). Beaconsfield Publishers Ltd., Beaconsfield, UK.

Weniger, B. and L. Robineau. 1988. *Elements for a Caribbean Pharmacopeia*. Tramil 3 Workshop, Havana, Cuba. November 1988.

Whistler, R. L. and J. N. BeMiller (eds.). 1973. *Industrial Gums*. Academic Press, New York.

Wichtl, M. 2004. *Herbal Drugs and Phytopharmaceuticals*. 3rd ed. Medpharm Scientific. Stuttgart.

Willaman, J. J. and B. G. Schubert. 1961. *Alkaloid-Bearing Plants and Their Contained Alkaloids*. U.S. Department of Agriculture Technical Bulletin 1234, Washington, DC.

Williamson, E. M. 2002. *Major Herbs of Ayurveda*. Churchill Livingstone. London.

Willoughby, M. J. and S. Y. Mills. 1996. *British Herbal Pharmacopoeia*. British Herbal Medical Association.

Wren, R. C., Revised by E. M. Williamson and F. J. Evans. 1988. *Potter's New Cyclopedia of Botanical Drugs and Preparations*, 8th ed. C. W. Daniel Co. Essex, UK.

Xiao, P. G. (ed.). 1988, 1988, 1989, 1989, 1989, 1989, 1990, 1990, 1990, 1990, 1991. *Pictorial Records of Chinese Herbal Drugs*, 10 Vols. plus index. People's Health Publications and Commercial Press, Beijing and Hong Kong, China.

Xu, G. J. and L. S. Xu (eds.). 1986. *Microscopy of Powdered Chinese Drugs*. People's Health Publications, Beijing, China (in Chinese).

Youngken, H. W. 1943. *Textbook of Phamacognosy*, 5th ed. Blakiston, Philadelphia, PA.

Zhong Cao Yao Xue Editorial Committee (ZCYX), Nanjing College of Pharmacy. 1987, 1976, 1980. *Chinese Herbal Drugs*, 3 Vols. Jiangsu People's Press and Jian Scientific and Technical Publications, Nanjing, China (in Chinese).

Zhou, J. H. and J. M. Wang (eds.). 1986. *Pharmacology of Chinese Materia Medica*. Shanghai Scientific and Technical Publicrtions, Shanghai, China (in Chinese).

Zhou, X. C. 1989. *Chinese Herbs and Cosmetics*. Vacation Publishers, Ltd., Taipei, China (in Chinese).

Zhu, S. H. 1990. *Chinese Materia Medica of Commerce*. People's Health Publications, Beijing, China (in Chinese).

Appendix B

Glossary/Abbreviations

ABSOLUTES. Entirely alcohol-soluble extracts prepared by alcohol extraction of concretes or related fat-soluble or waxy materials, alcohol-insoluble substances being removed before evaporation of the solvent

ABTS. 2,2′-Azino-bis-(3-ethyl-1,2-dihydrobenzothiazoline 6-sulfonate)

ALKALOIDS. Natural amines (nitrogen-containing compounds) that have pharmacological properties and that are generally of plant origin. They are widely distributed throughout the plant kingdom. They usually exhibit basic properties, though there are exceptions. Most alkaloids are insoluble or only slightly soluble in water, but their salts are water soluble. Many naturally derived drugs are alkaloids; well-known examples are morphine, codeine, cocaine, caffeine, nicotine, emetine, atropine, and quinine

ALT AND AST. Alanine aminotransferase and aspartate aminotransferase are enzymes located in liver cells that leak out into the general circulation when liver cells are injured. Previously known as the SGPT (serum glutamic-pyruvic transaminase) and the SGOT (serum glutaic-oxaloacetic transaminase), respectively

ATPASE. Adenosine triphosphatase

BALSAMS. Mixtures of resins that contain relatively large amounts of cinnamic or benzoic acid or their esters. Typical balsams are balsam Peru, balsam Tolu, styrax, and benzoin. Canada balsam, Oregon balsam, and copaiba balsam are not true balsams since they do not contain benzoic or cinnamic acid or their esters. Balsams are insoluble in water but soluble in alcohol

BHT. Butylated hydroxytoluene

BPH. Benign prostatic hyperplasia

BSA. Bovine serum albumin

CONCRETES. Water-insoluble but hydrocarbon-soluble extracts prepared from natural materials by using hydrocarbon-type solvents. They are primarily used in perfumery and in the preparation of absolutes

COX. Cyclooxygenase

COX-2. Cyclooxygenase-2

CRF. Corticotrophin-releasing factor

DA. Daltons

DECOCTIONS. Dilute aqueous extracts prepared by boiling the botanicals with water for a specific period of time, followed by straining or filtering. These are normally not commercially available in the United States

DPPH. 2,2-Diphenyl-1-picrylhydrazyl

DRY EXTRACTS. Same as powdered extracts. This term is used mainly in the United Kingdom and in Commonwealth countries

EBV. Epstein-Barr virus

EDHF. Endothelium-derived hyperpolarizing factor

ELIXIRS. Clear, sweetened, hydroalcoholic liquids intended for oral use. They contain flavoring substances and, in the case of medicated elixirs, active medicinal agents. Their primary solvents are alcohol and water, with glycerin, sorbitol, and syrup sometimes used as additional solvents and/or sweetening agents. They are prepared by simple solution or admixture of the several ingredients

ENZYMES. Proteins produced by living organisms that can bring about specific changes in other compounds (called substrates). Enzymes are also called organic catalysts; they are not consumed in the reactions they catalyze but are regenerated at the end of such reactions. The most commonly used enzymes in the food and drug industries are proteases, amylases, lipases, and pectinases

ESSENTIAL OILS. Also known as volatile oils, ethereal oils, or essences. When exposed to the air they evaporate at room temperature. They are usually complex mixtures of a wide variety of organic compounds (e.g., hydrocarbons, alcohols, ketones, phenols, acids, ethers, aldehydes, esters, oxides, sulfur compounds, etc.). They generally represent the odoriferous principles of the plants from which they are obtained. Most of these compounds are derived from isoprene and are terpenes at different stages of oxidation. Essential oils are generally isolated by distillation (most commonly steam distillation), solvent extraction, or expression. Clove, cinnamon, and peppermint oils are obtained by steam distillation, whereas bergamot and lemon oils and sweet and bitter orange oils are obtained by expression

EXTRACTS. Generally but not necessarily concentrated forms of natural substances obtained by treating crude materials containing these substances with a solvent and then removing the solvent completely or partially from the preparations. Most commonly used extracts are fluid extracts (liquid extracts), solid extracts, powdered extracts (dry extracts), tinctures, and native extracts

F.C.C. Food Chemicals Codex

FATS. Glycerol esters (glycerides) of fatty acids. They are semisolids or solids at room temperature and are generally produced from botanicals by expression and from animal materials by extraction or rendering

FATTY ACIDS. Carboxylic acids obtained from natural sources, mostly from fats. They can be both saturated and unsaturated. Examples of saturated fatty acids are palmitic and stearic acids; unsaturated ones are oleic, linoleic, and linolenic acids

FIXATIVES. Materials, usually high boiling and of high molecular weight, that retard the evaporation of the more volatile components in perfume formulations

FIXED (FATTY) OILS. Chemically the same as fats. They differ only physically from fats in that they are generally liquids at room temperature

FLU A. Influenza A virus

FLUID EXTRACTS. These extracts are commonly hydroalcoholic solutions with strengths of 1:1. The alcohol content varies with each product. Fluid extracts are prepared either from native extracts or solid extracts by adjusting to the prescribed strength with alcohol and water or by direct extraction of the botanicals with alcohol–water mixtures as directed in the official compendia. The latter method usually produces more desirable products due to the fewer steps involved in processing. Fluid extracts are also known as liquid extracts

GABA. γ-Aminobutyric acid

GLYCOSIDES. Sugar-containing compounds that on hydrolysis yield one or more sugars. They contain two components in their molecules, glycone, and aglycone. The glycone is the sugar component, which can be glucose, rhamnose, xylose, arabinose, or other sugars. When the glycone is glucose, the glycoside is commonly known as a glucoside. The aglycone is the nonsugar component of the glycoside; it can be any type of compound such as sterols, triterpenes, anthraquinones, hydroquinones, tannins, carotenoids, and anthocyanidins. They are a very important group of natural products, are widely present in plants, and constitute major classes of drugs. Well-known drug examples are digitalis glycosides, sennosides, cascarosides, ginseng glycosides, rutin, and arbutin. Glycoside-containing materials that are used in foods include grape skin color (betanin), soapbark, fenugreek, alfalfa (saponins), and licorice (glycyrrhizin)

GRANULAR EXTRACTS. Produced in the same way as powdered extracts. They are also of the same potency as powdered extracts. The only difference between them is that granular extracts have larger particle sizes

GRAS. Generally regarded as safe

GUMS. Hydrocolloids. They are polysaccharides of high molecular weight and can be dissolved or dispersed in water to form a viscous colloidal solution. The most commonly used natural gums are seaweed extracts (agar, algin, carrageenan, furcellaran), tree exudates (acacia, ghatti, karaya, tragacanth), tree extracts (larch gum), seed gums (guar, locust bean, quince seed), and microbial gums (dextran, xanthan)

GUM RESINS. Resins occurring admixed with gums. They usually also contain small amounts of volatile oils and sometimes are also called oleogum resins. Common examples are myrrh, gamboge, asafetida, galbanum, and olibanum

HDL. High-density lipoprotein

HIV. Human immunodeficiency virus

HSV-1. Herpes simplex virus type 1

5-HT. 5-Hydroxytryptamine

I.M. Intramuscular

I.P. Intraperitoneal

I.V. Intravenous

IC_{50}. 50% inhibitory concentration

IGE. Immunoglobulin E

IL-6. Interleukin-6

INFUSIONS. Infusions are sometimes the same as decoctions. They are generally dilute aqueous extracts containing the water-soluble ingredients of the botanicals. They are prepared by extracting the botanicals with boiling water. The resulting extracts are not concentrated further. Because of the dilute and aqueous nature of infusions and decoctions, they are very susceptible to microbial deterioration. Infusions are not normally available commercially in the United States

INOS. Inducible nitric oxide synthase

LD_{50}. Lethal dose that kills 50% of a population

LDL. Low-density lipoprotein

LIPASES. Lipolytic enzymes that hydrolyze fats or fixed oils into their glycerol and fatty acid components. They are chiefly used in the dairy industry as flavor producers or modifiers and in medicine as digestive aids

LIPIDS. Fatty materials that are soluble in fat solvents (ether, chloroform, alcohol, etc.). They include fatty acids, fats, waxes, fixed oils, steroids, lecithins, and fat-soluble vitamins (vitamins A, D, and K)

LIQUID EXTRACTS. British equivalents of fluid extracts. They are used in the United Kingdom and the Commonwealth countries

M.W. Molecular weight

MAO. Monoamine oxidase

MENSTRUUMS (MENSTRUA). Solvents used for extraction, for example, alcohol, acetone, and water

MIC. Minimum inhibitory concentration. Minimum concentration that inhibits the growth of a microorganism

NATIVE EXTRACTS. In the commercial manufacture of extracts, a botanical is first extracted with an appropriate solvent such as denatured alcohol, alcohol, methanol, water, or mixtures of these solvents. The extract is then concentrated under reduced pressure at low temperatures until all solvent is removed. The viscous, semisolid concentrated extract at this state is called a native extract by some manufacturers. The native extracts are usually of high potency from which solid, fluid, and powdered extracts of various strengths can be prepared by diluting with suitable diluents. If the botanical has resins and volatile oils as its active principles and the solvent used is a fat solvent, the resulting native extract is equivalent to a prepared oleoresin

N.F. National Formulary

NF-κB. Nuclear factor kappa B

NO. Nitric oxide

OLEOGUM RESINS. See gum resin

OLEORESINS. Mixtures of mostly resins and volatile oils. They either occur naturally or are prepared by solvent extraction of botanicals. Prepared oleoresins are made by extracting the oily and resinous materials from botanicals with fat solvents (hexane, acetone, ether, alcohol). The solvent is then removed under vacuum, leaving behind a viscous, semisolid extract that is an oleoresin. Examples of prepared oleoresins are paprika, ginger, and capsicum (see also native extracts). Examples of natural oleoresins are gum turpentine, Oregon balsam, and Canada balsam

PAF.   Platelet-activating factor

POWDERED EXTRACTS.   Prepared from native extracts by diluting the native extracts to the specified strengths with appropriate diluents (lactose, dextrose, sucrose, starch, etc.) and/or anticaking agents (calcium phosphate, magnesium carbonate, magnesium oxide, etc.), followed by drying, usually under vacuum, to yield dry solids. These are then ground into fine powders to form powdered extracts or into coarse granules to produce granular extracts

PROTEASES   (PROTEINASES).   Proteolytic enzymes that act on proteins by attacking specific peptide linkages in the proteins and hydrolyzing them. Depending on their specific applications, commonly used proteases can be of plant, animal, or microbial origin. They find uses in tenderizing meat, modifying dough in baking, chillproofing beer, cheese making, in wound debridement, as digestive aids, in relieving inflammations, bruises, and blood clots, as well as in other industries (leather, textile, dry cleaning, waste control). Examples of widely used plant proteases are bromelain, ficin, and papain; common animal proteases are pepsin and rennin. Proteases are usually divided into two types. Endopeptidases break up internal peptide bonds of the protein chain, producing peptides. Exopeptidases, on the other hand, cleave terminal peptide linkages, producing amino acids. Most commercial proteases are mixtures of different protease fractions and usually have both endopeptidase and exopeptidase activities. Commercial proteases come in many different grades that vary widely in proteolytic strengths. Few published studies on proteases, particularly commercial plant proteases, specify activity of enzymes used, and hence results are generally quantitatively irreproducible

ROS.   Reactive oxygen species

RESINS.   Natural products that either occur naturally as plant exudates or are prepared by alcohol extraction of botanicals that contain resinous principles. Naturally occurring resins are solids or semisolids at room temperature. They are soluble in alcohol and alkali solutions but are insoluble in water. They are usually noncrystalline, transparent or translucent, and soften or melt on heating. Chemically they are complex oxidation products of terpenes. They rarely occur in nature without being mixed with gums and/or volatile oils, forming gum resins, oleoresins, and oleogum resins. Hence in commerce the term "resins" is often used to include all above resinous materials. During preparation of a resin, the alcoholic extract is poured into an excess of water or acidified water and the precipitated resin is collected, washed, and dried. Typical examples of prepared resins are podophyllum and jalap resins. A prepared resin may also be derived from a natural oleoresin by removing the volatile oil by heat or from a natural gum resin by extracting its resin with alcohol followed by removal of the solvent. Resins prepared by alcohol extraction of natural resinous materials are sometimes referred to as resinoids. Resinoids may be considered as purified forms of certain resins; they are usually prepared from resins by extraction with hydrocarbons

SAPONINS.   Glycosides generally with sterols or triterpenes as their aglycones (nonsugar component), although there are exceptions. They have the ability of forming foams when their aqueous solutions are shaken. The aglycone portions are called sapogenins. Many saponins are hemolytic. However their foam-forming properties are utilized in beverages. Common examples are extracts of soapbark, yucca, and sarsaparilla. Other saponin-containing natural ingredients include alfalfa, fenugreek, senega, ginseng, and licorice

SOD.   Superoxide dismutase

SOFT EXTRACTS.   A British term that is equivalent to our native or solid extracts. It is used in the United Kingdom and the Commonwealth countries

SOLID EXTRACTS.   Also known as pilular extracts. They are usually thin to thick, viscous liquids or semisolids prepared from native extracts by adjusting the latter to the correct strength with suitable diluents (liquid glucose, corn syrup, glycerol, propylene glycol, etc.). Solid extracts are generally of the same strength as their corresponding powdered extracts

STRENGTH OF EXTRACTS. The potencies or strengths of botanical drug extracts are generally expressed in two ways. If they contain known active principles, for example, alkaloids in belladonna and ipecac, their strengths are commonly expressed in terms of their content of active compounds. Otherwise they are expressed in terms of their total extractives in relation to the crude drug. Thus, a strength of 1:4 means one part of extract is equivalent to or derived from four parts of crude drug. This method of expressing drug strength is not accurate since-in commerce an extract of a certain strength, for example, 1:3, may be different in actual potency depending on the manufacturer and the process or equipment used in the production. Examples of such variations can be found in extracts of senna and cascara, where the active principles are well known; yet, the strengths are sometimes expressed in terms of total extractives. Consequently, a solid extract of cascara (1:3) or a fluid extract of senna (1:1) from one supplier may contain several times more anthraglycosides than officially equivalent products from another supplier. Even though all these products may pass current or past pharmacopoeial specifications, some of them may actually contain little or no active ingredients. Furthermore, in recent years, strengths of extracts are also expressed in reverse numerical order (e.g., 4:1 instead of 1:4), causing much confusion. Hence one should not rely solely on numerical designations in reporting or interpreting strengths

TINCTURES. Alcoholic or hydroalcoholic solutions usually containing the active principles of botanicals in comparatively low concentrations. They are generally prepared either by maceration or percolation or by dilution of their corresponding fluid extracts or native extracts. The strengths of tinctures are generally 1:0.1 or 1:0.2

TNF-α.   Tumor necrosis factor alpha

TPA.   12-O-Tetradecanoylphorbol-13-acetate

TRP-P-1.   3-Amino-1,4-dimethyl-5H-pyrido[4,3-b]indole

U.S.P.   United States Pharmacopoea

VOLATILE OILS.   See essential oils

VSV.   Vesicular stomatitis virus

WAXES. Esters of fatty acids with alcohols, both of high molecular weight and straight-chained. But in reality waxes also contain free fatty acids, free fatty alcohols, and hydrocarbons. They are extensively used in pharmaceutical and cosmetic ointments, creams, and lotions

Appendix C
Botanical Terms

ABAXIAL. Side of an organ away from the axis or center of the axis; dorsal

ACUMINATE. An acute apex whose sides are somewhat concave and taper to a protracted point

ACUTE. Sharp, ending in a point, the sides of the tapered apex essentially straight or slightly convex

ADAXIAL. Side toward the axis

ADVENTITIOUS. On occasion, rather than resident in regular places and order, as those arising about wounds

ALTERNATE. Arrangement of leaves or other parts not opposite or whorled; placed singly at different heights on the axis or stem

ANNUAL. Of one season's duration from seed to maturity and death

ANTHER. Pollen-bearing part of the stamen, borne at the top of the filament or sometimes sessile

APEX. Tip or distal end

APICULATE. Terminated by an apicula, a short, sharp, flexible point

APPRESSED. Closely and flatly pressed against; adpressed

ARBORESCENT. Tree-like habit

ARIL (ARILLUS). Appendage or an outer covering of a seed, growing out from the hilum or funiculus; sometimes it appears as a pulpy covering

AXIL. Upper angle that a petiole or peduncle makes with the stem that bears it

AXILLARY. In an axil

BERRY. Pulpy, indehiscent, few- or many-seeded fruit; technically the pulpy fruit resulting from a single pistil, containing one or more seeds but no true stone, as the tomato or grape

BIENNIAL. Of two seasons duration, from seeds to maturity and death

BILABIATE. Two-lipped, often applied to a corolla or calyx; each lip may or may not be lobed or toothed

BLADE. Expanded part of a leaf or a petal

BRACT. Much-reduced leaf, particularly the small or scale-like leaves in a flower cluster or associated with the flowers; morphologically a foliar organ

BRACTEATE. Bearing bracts

BRACTEOLE. Secondary bract; a bractlet

BRISTLY. Bearing stiff strong hairs or bristles

BULB. Thickened part in a resting state and made up of scales or plates on a much shortened axis

BULBLET. Little bulb produced in a leaf axil, inflorescence, or other unusual area

BUSH. A low and thick shrub, without distinct trunk

CALYX. Outer whorl of floral envelopes, composed of the sepals; the latter may be distinct, or connate in a single structure, sometimes petaloid as in some ranunculaceous flowers

CAP. A convex removable covering of a part, as of a capsule; in the grape, the cohering petals falling off as a cap

CAPITATE. Headed; in heads; formed like a head; aggregated into a very dense or compact cluster

CAPSULE. Dry fruit resulting from the maturing of a compound ovary (of more than one carpel), usually opening at maturity by one or more lines of dehiscence

CARPEL. One of the foliar units of a compound pistil or ovary; a simple pistil has one carpel. A foliar, usually ovule-bearing unit of a simple ovary, two or more combined by connation in the origin or development of a compound ovary; a female or mega sporophyll of an angiosperm flower

CATKIN. Scaly-bracted, usually flexuous spike or spike-like inflorescence of cymules; ament; prominent in willows, birches, oaks

CAULINE. Pertaining or belonging to an obvious stem or axis, as opposed to basal or rosulate

CLEFT. Divided to or about the middle into divisions, as a palmately or pinnately cleft leaf

CLONE. Group of individuals resulting from vegetative multiplication; any plant propagated vegetatively and therefore presumably a duplicate of its parent. Originally spelled clon, but changed to clone by its coiner (H. J. Webber) for reasons of philology and euphony; proposed in application to horticultural varieties

COMPOUND LEAF. Leaf of two or more leaflets; in some cases (Citrus) the lateral leaflets may have been lost and only the terminal leaflets remain. Ternately compound when the leaflets are in 3's; *palmately compound* when 3 or more leaflets arise from a common point to be palmate (if only 3 are present they may be sessile); *pinnately compound* when arranged along a common rachis (or if only 3 are present at least the terminal leaflet is stalked); *odd-pinnate* if a terminal leaflet is present, and the total number of leaflets for the leaf is an odd number; *even-pinnate* if no terminal leaflet is present and the total is an even number

CONE. Dense and usually elongated collection of flowers or fruits comprising usually sporophylls and bracts on a central axis, the whole forming a detachable homogeneous fruit-like body; some cones are of short duration, as the staminate cones of pines, and others become dry and woody persistent parts

CORDATE. Heart-shaped; with a sinus and rounded lobes at the base, and ovate in general outline; often restricted to the basal portion rather than to the outline of the entire organ

CORIACEOUS. Of leathery texture as a leaf of *Buxus*

CORM. Solid bulb-like part of the stem, usually subterranean, as the "bulb" of *Crocus* and *Gladiolus*

COROLLA. Inner circle or second whorl of floral envelopes; if the parts are separate, they are petals and the corolla is said to be polypetalous; if not separate, they are teeth, lobes, divisions, or undifferentiated, and the corolla is said to be gamopetalous or sympetalous

CORYMB. Short and broad, more or less flat-topped indeterminate inflorescence, the outer flowers opening first

CREEPER. Trailing shoot that takes root mostly throughout its length; sometimes applied to a tight-clinging vine

CRENATE. Shallowly round-toothed or obtusely toothed, scalloped

CROWN. Corona; also that part of the stem at the surface of the ground; also a part of the rhizome with a large bud, suitable for the use in propagation

CULM. Stem of grasses and bamboos, usually hollow except at the swollen nodes

CULTIGEN. Plant or group known only in cultivation; presumably originating under domestication; contrast with indigen

CUNEATE. Wedge shaped; triangular, with the narrow end at point of attachment, as the bases of leaves or petals

CUSPIDATE. With an apex somewhat abruptly and sharply concavely constricted into an elongated, sharp-pointed tip

CYME. A broad, more or less flat-topped, determinate flower cluster, with central flowers opening first

DECIDUOUS. Falling at the end of one season of growth of life, as the leaves of nonevergreen trees

DECUMBENT. Reclining or lying on the ground, but with the end ascending

DEHISCENCE. Method or process of opening of a seed pod or anther: *loculicidally* dehiscent when the split opens into a cavity or locule, *septicidally* when opening at point of union of septum or partition to the side wall, *circumscissile* when the top valve comes off as a lid, *poricidal* when opening by means of pores whose valves are often flap-like

DENTATE. With sharp, spreading, rather coarse indentations, or teeth that are perpendicular to the margin

DIGITATE. Hand-like; compound with the members arising from one point, as the leaflets of horse chestnut

DIOECIOUS. Having staminate and pistillate flowers on different plants; a term properly applied to a taxonomic unit, not to flowers

DISSECTED. Divided into many slender segments

DORSAL. Back; relating to the back or outer surface of a part or organ, as the lower side of a leaf; the opposite of ventral

DRUPE. Fleshy one-seeded indehiscent fruit with seed enclosed in a stony endocarp (a pyrene); stone fruit

ELLIPTIC. Oval in outline, being narrowed to rounded ends and widest at or about the middle

ELONGATE. Lengthened; stretched out

EMBRYO. The rudimentary plant in the seed, usually developing from the zygote

ENDOCARP. Inner layer of the pericarp or fruit wall

ENDOSPERM. The starch- and oil-containing tissue of many seeds; often referred to as the albumen

ENTIRE. With a continuous margin; not in any way intended; whole (may or may not be hairy or ciliate)

EVERGREEN. Remaining green in its dormant season; sometimes applied to plants that are green throughout the year, properly applied to plants and not to leaves, but due to the persistence of leaves

EXOCARP. Outer layer of the pericarp or fruit wall

FASCICLE. Condensed or closer cluster, as of flowers, or of most pine leaves

FLORETS. Individual flowers, especially of composites and grasses; also other very small flowers that make up a very dense form of inflorescence

FLOWER. An axis bearing one or more pistils or one or more stamens or both: when only the former, it is a *pistillate flower*; when only the latter, a *staminate flower*; when both are present, it is a *perfect flower* (i.e., bisexual or hemaphroditic). When this perfect flower is surrounded by a perianth representing two floral envelopes (the inner envelope the corolla, the outer the calyx), it is a *complete flower*

FOLLICLE. Dry dehiscent fruit opening only on the dorsal (front) suture and the product of a simple pistil

FROND. Leaf of a fern; sometimes used in the sense of foliage; especially of palms of other compound leaves. Used by Linnaeus for the leaves of palms

FUSIFORM. Spindle-shaped; narrowed both ways from a swollen middle

GLABROUS. Not hairy; often incorrectly used in the sense of smooth

GLAND. Properly, a secreting part or prominence or appendage, but often used in the sense of a gland-like body

GLANDULAR. Having or bearing secreting organs, or glands

GLAUCOUS. Covered with a bloom, or whitish substance that rubs off

GLUTINOUS. Sticky

HASTATE. Shape of an arrowhead, but with the basal lobes pointed or narrow and standing nearly or quite at right angles; halberd shaped

HEAD. Short dense spike; capitulum

HEMI-. In Greek compounds, signifying half

HERB. Plant naturally dying to ground; without persistent stem aboveground; lacking definite woody firm structure

HERBACEOUS. Not Woody; dying down each year; said also of soft branches before they become woody

HERBAGE. Vegetative parts of plant

HIRSUTE. With rather rough or coarse hairs

HISPID. Provided with stiff or bristly hairs

HUSK. Outer covering of some fruits (as Physalis or Juglans), usually derived from the perianth or involucre

HYBRID. Plant resulting from a cross between parents that are genetically unlike; more commonly, in descriptive taxonomy, the offspring of two different species or their infraspecific units

INCISED. Cut; slashed irregularly, more or less deeply and sharply; an intermediate condition between toothed and lobed

INDEHISCENT. Not irregularly opening, as a seed pod or anther

INDIGEN. Indigenous inhabitat; a native

INFERIOR. Beneath, lower, below; as an inferior ovary, one that seemingly is below the calyx leaves

INFLORESCENCE. Mode of flower bearing; technically less correct but much more common in the sense of a flower cluster

INTERNODE. Part of an axis between two nodes

INVOLUCRE. One or more whorls of small leaves or bracts (phyllaries) standing close underneath a flower or flower cluster

LABIATE. Lipped; or, a member of the labiatae

LAMELLATE. Provided with many fin-like blades or cross partitions

LANCEOLATE. Lance-shaped; much longer than broad; widening above the base and tapering to the apex

LATERAL. On or at the side

LATEX. Milky sap

LEAF STALK. Stalk of a leaf; petiole

LEAFLET. One part of a compound leaf; secondary leaf

LEGUME. Simple fruit dehiscing on both sutures, and the product of a simple unicarpellate ovary

LENTICULAR. Lens-shaped

LIANOUS. Vine-like

LIUGULATE. Strap-shaped, as a leaf, petal, or corolla

LIGULE. Strap-shaped organ or body; particularly, a strap-shaped corolla, as in the ray flowers of composites; also a projection from the top of the sheath in grasses and similar plants

LINEAR. Long and narrow, the sides parallel or nearly so, as blades of most grasses

LOBE. Any part or segment of an organ; specifically, a part of petal or calyx or leaf that represents a division to about the middle

MEMBRANOUS. Membranaceous. Of parchment-like texture

5-MEROUS. In composition, referring to the numbers of parts; as flowers 5-merous, in which the parts of each kind or series are 5 or in 5s

MIDRIB. The main rib of a leaf or leaf-like part, a continuation of the petiole

MONOECIOUS. With staminate and pistillate flowers on the same plant, as in corn

MONOTYPIC. In reference to a genus, composed of a single species

NODE. Joint where a leaf is borne or may be borne; also, incorrectly, the space between two joints, which is properly an internode

NUT. An indehiscent one-celled and one-seeded hard and bony fruit, even if resulting from a compound ovary

NUTLET. Small or diminutive nut; nucule

OBLANCEOLATE. Reverse of lanceolate, as a leaf boarder at the distal third than at the middle and tapering toward the base

OBLONG. Longer than broad, and with the sides nearly or quite parallel most of their length

OBOVATE. Reverse of ovate, the terminal half broader than the basal

OBTUSE. Blunt, rounded

OPPOSITE. Two at a node, on opposing sides of an axis

ORBICULATE. Circular or disc-shaped, as leaf or Nelumbo

OVATE. With an outline like that of hen's egg, the broader and below the middle

OVOID. Solid that is oval (less correctly ovate) in flat outline

PALMATE. Lobed or divided or ribbed in a palm-like or hand-like fashion; digitate, although this word is usually restricted to leaves compound rather than to merely ribbed on lobed

PANICLE. An indeterminate branching raceme; an inflorescence in which the branches of the primary axis are racemose and the flowers pedicellate

PAPPUS. Peculiar modified outer perianth series of composites, borne on the ovary (persisting in fruit), being plumose, bristle-like, scales, or otherwise; once commonly accepted as a modified calyx, but now believed in some genera of compositae to represent a modification of the corolla

PEDICLE. Stalk of one flower in a cluster

PEDUNCLE. Stalk of a flower cluster or of a solitary flower when that flower is the remaining member of an inflorescence

PELTATE. Attached to its stalk inside the margin; peltate leaves are usually shield shaped

PENDULOUS. Drooping, hanging downward

PERENNIAL. Of three or more season's duration

PERICARP. Wall of a ripened ovary, that is, the wall of a fruit; sometimes used loosely to designate a fruit

PERSISTENT. Remaining attached; not falling off

PETAL. One unit of the inner floral envelope or corolla of a polypetalous flower, usually colored and more or less showy

PETALOID. Petal-like; in color and shape resembling a petal

PETIOLE. Leaf stalk

PINNATISECT. Cut down to the midrib in a pinnate way

PISTILLATE. Having pistils and no functional stamens; female

POD. Dehiscent dry fruit; rather general uncritical term, sometimes used when no other more specific term is applicable, as for the fruit of Nelumbo

POLLEN. Spores or grains borne by the anther, containing the male element (gametophyte)

PRICKLE. Small, weak, spine-like body borne irregularly on the bark or epidermis

PROLIFEROUS. Bearing offshoots or redundant parts; bearing other similar structures on itself

PROSTRATE. General term for lying flat on the ground

PUBESCENT. Covered with short soft hairs: downy

PUNGENT. Ending in a stiff sharp point or tip; also, acrid (to the taste)

PYRIFORM. Pear shaped

RACEME. Simple, elongated, indeterminate inflorescences with pedicelled or stalked flowers

RACHIS. Plural, rachides, or rachises. Axis bearing flowers or leaflets; petiole of a fern frond

RECEPTACLE. Torus; the more or less enlarged or elongated end of the stem or flower axis on which some or all of the flower parts are borne; sometimes the receptacle is greatly expanded, as in the compositae

RECURVED. Bent or curved downward or backward

RENIFORM. Kidney-shaped

RESINOUS. Containing or producing resin, said of bud scales when coated with a sticky exudate of resin (as in *Aesculus* spp.)

RHIZOME. Underground stem; rootstock: distinguished from a root by presence of nodes, buds, or scale-like leaves

ROOTSTOCK. Subterranean stem; rhizome

ROSETTE. Arrangement of leaves radiating from a crown or center and usually at or close to the earth, as in Taraxacum (dandelion)

ROTUND. Nearly circular; orbicular inclining to be oblong

RUNNER. Slender trailing shoot taking root at the nodes

SERRATE. Said of a margin when saw-toothed with the teeth pointing forward

SERRULATE. Minutely serrate

SESSILE. Not stalked; sitting

SHEATH. Any long or more or less tubular structure surrounding or organ or part

SHRUB. A woody plant that remains low and produces shoots or trunks from the base, not tree-like nor with a single bole; a descriptive term not subject to strict circumscription

SOLITARY.   Borne singly or alone

SPATULATE.   Spoon-shaped

SPIKE.   Usually unbranched, elongated, simple, indeterminate inflorescence whose flowers are sessile, the flowers either congested or remote; a seemingly simple inflorescence whose flowers may actually be composite heads (Liatris), or other inflorescence types (*Phleum*)

SPINE.   A strong and sharp pointed woody body mostly arising from the wood of the stem

SPORE.   A simple reproductive body, usually composed of a single detached cell and containing a nucleated mass of protoplasm (but not embryo) and capable of developing into a new individual; used particularly in reference to the pteridophytes and lower plants

STALK.   Stem of any organ, as the petiole peduncle, pedicel, filament, stipe

STEM.   Main axis of a plant, leaf-bearing, and flower-bearing as distinguished from the root-bearing axis

STIGMA.   Part of the pistil that receives the pollen

STIPE.   Stalk of a pistil or other small organ when axile in origin; also the petiole of the fern leaf

STOLON.   Shoot that bends to the ground and takes root; more commonly, a horizontal stem at or below surface of the ground that gives rise to a new plant at its tip

STROBILE.   Cone

SUB-.   As a prefix, usually signifying somewhat, slightly, or rather

SUBSHRUB.   A suffrutescent perennial (the stems basally woody), or a very low shrub often loosely treated as a perennial

SUCCULENT.   Juicy; fleshy; soft, and thickened in texture

TENDRIL.   Rotating or twisting thread-like process of extension by which a plant grasps an object and clings to it for support; morphologically, it may be a stem or a leaf

TERMINAL.   At the tip, apical, or distal end

TORMENTOSE.   With tomentum; densely woolly or pubescent; with matted soft wool-like hairiness

TUBER.   Short congested part; usually defined as subterranean (as of a root-stock) although this is not essential

TUBEROUS.   Bearing or producing tubers

TWIG.   Young woody stem; more precisely the shoot of a woody plant representing the growth of the current season and terminated basally by circumferential terminal bud-scar

UMBEL.   An indeterminate, often flat-topped inflorescence whose pedicles and peduncles (rays) arise from a common point resembling the stays of an umbrella; umbels are characteristic of the umbelliferae and are there usually compound, each primary ray terminated by a secondary umbel (umbellet)

UNDULATE.   Wavy (up and down, not in and out), as some leaf or petal margins

UNISEXUAL.   Of one sex; staminate only or pistillate only

VESICLE.   A small bladdery sac or cavity filled with air or fluid

VISCID.   Sticky, or with appreciable viscosity

WHORL.   Three or more leaves or flowers at one node, in a circle

WING.   Thin, dry, or membranaceous expansion or flat extension or appendage of an organ; also the lateral petals of a papilionaceous flower

WOOLY.   Provided with a long, soft, and more or less matted hairs; like wool; lanate

Appendix D
Morphological Description of Plant Organs

LEAF SHAPE

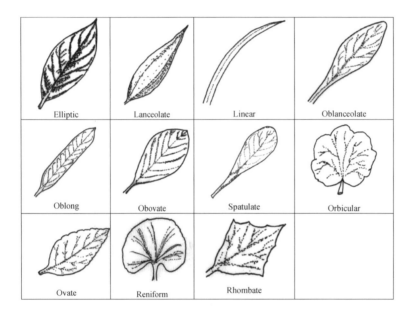

LEAF MARGIN

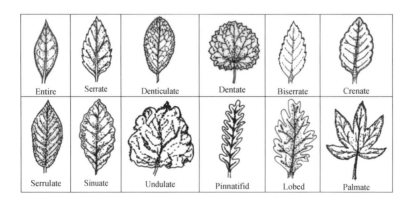

Appendix D

LEAF APEX

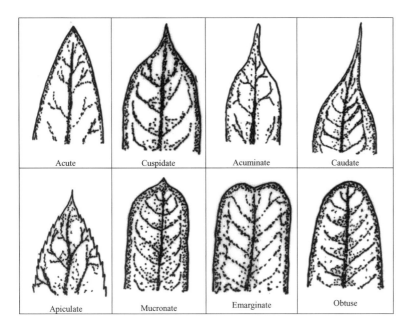

LEAF BASE

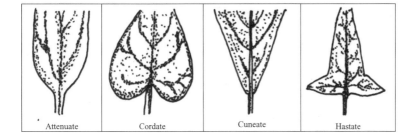

FLOWERS

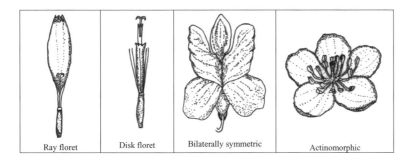

FLOWER CLUSTER (INFLORESCENCE)

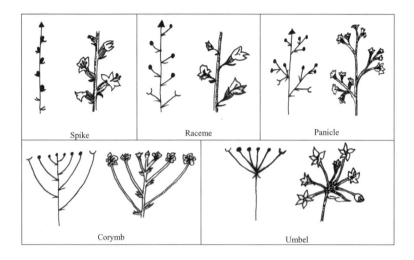

FRUIT

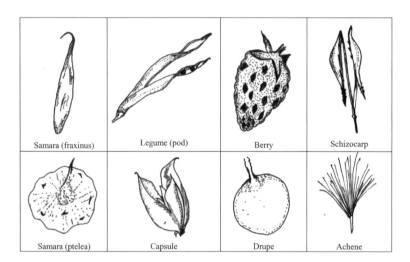

Appendix D

UNDERGROUND ORGANS

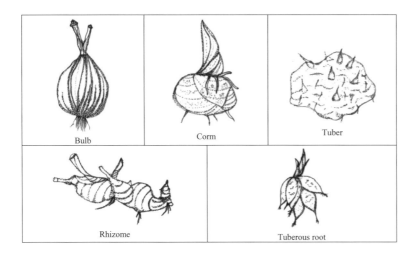

General Index

Abdominal distention, remedies, cangzhu, 662
Abelmoschus moschatus, 30
Abies balsamea, 64, 65
Abortifacient, parsley, 489
Abortion:
    mahabala, threatened, remedies, Baikal 651
    rue oil causing, 543
    skullcap, 685
Abrasions, remedies, *hypericum*, 373
Abscesses, remedies:
    alkanet, 19
    astragalus, 57
    broom tops, 116
    burdock, 121
    castor oil, 155
    chervil, 185
    kava, 397
    licorice root, 417
    purslane, 679
    white pine bark, 501
    see also entries under skin
Absidia species, inhibition by henna, 355
ABSINTHIUM, 1–3
Absolute cire d'abeille, 84
ACACIA, 152, *see also* Cassie absolute
Acacia catechu, 157
Acacia farnesiana, 152
Acacia senegal, 4
Acacia species, 4
Acanthopanax senticosus, *see* Eleuthero
Acanthopeltis species, 9
ACEROLA, 6–7
Aches, *see* Pain
Achillea millefolium, 626
Achiote, *see* Annatto
Achiotillo, *see* Annatto
Achras sapota, 188
Achras zapotilla, 188

Acid-base indicators (pH) indicators, 356
Acid fast beverage color, 210
Acne, remedies,
    asparagus, 52–53
    Baikal skullcap, 684–686
    balm of Gilead buds, 64–65
    chaparral, 174–176
    Dahurian angelica, 663–664, 670, 677
    garden burnet, 666, 678, 685
    goldenseal, 336–338
    lavender, 406–408
    ligusticum, 669–670
    olibanum, 471–472
    phellodendron bark, 677–678
    red sage, 680–681
    sour jujube kernel, 572–573
    tienchi ginseng, 598–600
    valerian root, 612–614
    see also Skin disorders
ACONITE, 7–9
Aconitum carmichaelii, 8
Aconitum chinese, 7, 8
Aconitum kusnezoffi, 8
Aconitum napellus, 7, 8
Aconitum species, 8
***Acorus calamus*, 127, 128**
Actaea racemosa, 97
Adaptogenic:
    eleuthero, 263
    gotu kola, 339
    schisandra, 566
Addictive properties, coca, 213
Adhatoda vasica 655
Adhatoda zeylanica Medic, 655
Adhesives:
    algin, 16–17
    bletilla tuber, 661
    α-Adrenergic blocking agent, 631
    yohimbe, 631–632
Adrenolytic, codonopsis, 220
Adulsa, 655

Aesculus hippocastanum, **364**
Agathosma betulina, **117**
Agathosma crenulata, 117
Agavaceae, 633
Agnivesha, 637
Agropyron repens, 247
Agropyron, *see* Doggrass
Ague grass, *see* Aletris
Agueweed, *see* Boneset
Ahnfeltia species, 9
Ahuacate, 59
AIDS, improved helper: suppressor ratio, garlic, 309
Aiyyan, 643
Akbal Arzani 640
Alant, *see* Elecampane
Albuminuria:
    as toxic symptom, turpentine, 608
    see also Kidney damage
Alcohol:
    abuse, remedies:
        borage, 110–112
        evening primrose, 276–278
        kudzu root, 399–402
        thyme, 595
    antagonist, ginseng, 330–333
    intoxication, remedies, bee pollen, 81–83
    see also CHEMICAL INDEX
Alcoholic beverages, *see* Beverages, alcoholic and nonalcoholic
Alcoholic bitters, *see* Bitters, alcoholic
Alder buckthorn, *see* Buckthorn, alder
ALETRIS, 11–12
Aletris farinosa, 11
Aletris foliata, 12
Aletris formosana, 12
Aletris species, 12
Aletris spicata, 24
ALFALFA, 12–15

Algae:
  brown, as algin source, 16, 17
  red, 9–11
ALGIN, 16–17
Alkaloid, *see* CHEMICAL INDEX
ALKANET, 18–19
Alkanna red, *see* CHEMICAL INDEX
Alkanna tinctoria, 18
Allergens:
  balsam tolu, 71–72
  bee pollen, 81–83
  beeswax, 84–85
  boronia absolute, 112–113
Allergens:
  bromelain, 113–114
  cade oil, 123
  castor oil, 154–155
  cedarwood oil, 162–164
  celery seed, 165–166
  chlorogenic acid, 223
  costus oil, 230–231
  fennel, 283–285
  ficin, 291–293
  fo-ti, 293–297
  grapefruit oil, 343–345
  hops, 359–360
  ipecac, 379–381
  kava, 395–397
  lemon oil, 410–411
  mints, 443–446
  mistletoe, 448–450
  quillaia, 518–519
  psyllium, 511–513
  royal jelly, 538–540
  turpentine, 607–608
  vanilla, 616–617
  see also Antiallergic
Allergic contact dermatities, *see* Dermatitis, contact
All heal, 612
Alligator pear, *see* Avocado
***Allium cepa*, 474, 475**
***Allium sativum*, 308**
Allium, *see* Garlic
ALMONDS, 21–23
Alpinin, 647
Aloe africana, 24
Aloe barbadensis, 24
Aloe ferox, 24, 27
Aloe species, 24, 25
Aloe spicata, 24

ALOE (AND ALOE VERA), 24–27
***Alpinia galanga*, 647**
Alternaria species, inhibition by henna, 355
Althea officinalis, 29
ALTHEA ROOT, 29–30
Altinigiaceae, *see* Hamamelidaceae
Altitude sickness, remedies, ganoderma, 300–304
Al-Razi, 640
Al Zahrawi, 640
Alqanoon, 641
Amaryllidaceae, 308, 474
Ambergris substitute
Ambreine, *see* Labdanum
AMBRETTE SEED, 30–32
Amebic dysentery, *see* Dysentery
Amebicidal properties, 309, 380, 515
Amenorrhea, remedies:
  bee pollen, 81–83
  castoreum, 156–157
  cinnamon and cassia, 196–200
  giant knotweed, 678, 685, 687
  gotu kola, 339–341
  hawthorn, 352–354
  Luffa, 671
  peony bark, 674–675
  red sage, 680–681
  safflower, 681–682
  Sichuan lovage, 682–683
American arborvitae, *see* Cedar leaf oil
American mandrake, *see* Podophyllum (podophyllin)
American wormseed oil, *see* Chenopodium oil
Amine, *see* CHEMICAL INDEX
Ammonia, *see* CHEMICAL INDEX
Ammonium chloride preparations, masking undesirable tastes, 417
Amnesia, remedies, ginseng, 333
Amomum cardamomum L., 142
Amomum galanga, 647
Amygalus communis, 22
Amygdalus dulcis, 22
Anabolic, ganoderma, 300–304

Anacardiaceae, 516, 586
Analeptic, castoreum, 156–157
Analgesic:
  allspice, 19–21
  aloe, 24–27
  balm of Gilead buds, 64–65
  Capsicum, 132–135
  cedar leaf oil, 161–162
  chamomile, 169–173
  Chaparral, 174–176
  cinchona, 194–195
  coffee, 222–224
  Dahurian angelica, 663–664
  dogwood, 249
  devil's claw, 242–243
  evening primrose, 276–278
  fangfeng, 281–282
  feverfew, 289–291
  ganoderma, 300–304
  gelsemium, 312–313
  gotu kola, 339–341
  henna, 355–356
  hops, 359–360
  horsetail, 367–369
  Kava, 395–397
  lemongrass, 412–414
  ligusticum, 669–670
  neem, 652–653
  nutmeg, 467–469
  olibanum, 471–472
  passion flower, 490–492
  peony bark, 674–675
  peony root, 676–677
  rosemary 535–537
  saffron, 547–548
  sour jujube kernel, 572–573
  sweet birch oil, 95–96
  wintergreen oil, 619
  see also Pain
Anal prolapse, remedies:
  horsetail, 367–369
  orange, 477–480
Ananas comosus, 113
Ananas sativus, *see Ananas comosus*
Anaphrodisiac, chaste-tree, 177–179
Anaphylactic shock:
  agents causing, psyllium, 511–513
  antianaphylactic properties, chamomile, 169–173
  as toxic Symptom,

General Index

Anchusa, *see* Alkanet
Andrographis paniculata, 192, 650
Andropogon citratus, 412
Andropogon flexuosus, 412
Andropogon nardus, 201
Anemia, remedies, 131, 158, 193, 389, 465
Anesthetics:
   belladonna, 88–89
   jasmine, 382–383
   kava, 395–397
   local:
      allspice, 19–21
      almonds, 21–23
      bloodroot, 102–103
      cinchona, 194–195
      coca, 212–214
      eugenol, 20
      magnolia flower, 434–435
      mints, 443–446
      quebracho, 516–518
Anethum, foeniculum, 283
Anethum graveolens, 244
Anethum sowa, **244**
ANGELICA, 32–34
Angelica anomala, 33
Angelica archangelica, 32, 33
Angelica dahurica, 663
Angelica formosana, 12, 33
Angelica levisticum, 427
Angelica sinensis, 33, 171
Angelica species, 33
Angelica tree, *see* Ash, prickly
Angina pectoris:
   as toxic symptom, mistletoe, 450
   remedies:
      bilberry, 94
      ginkgo, 327
      hawthorn, 353
      kudzu root, 401
      musk, 456
      Sichuan lovage, 683
   see also Pain, chest
Angostura bark, 34–35
"Angostura bitters", 35, 317, *see also* Bitter principles
Aniba rosaeodora, 106
Animal feed:
   alfalfa, 14
   fenugreek, 288

Animal repellents, mustard, 459
ANISE (AND STAR ANISE), 36–38
Anisette, *see* Beverages, alcoholic and nonalcoholic; Liquers
Anisum officinarum, 36
Anisum vulgare, 36
ANNATTO, 40–41
Annonaceae, 132, 629
Annatto, 40, 41
Anodyne, *see* Analgesic; Pain
Anorexia:
   remedies:
      blessed thistle, 101
      hypericum, 372
      as toxic symptom, chaparral, 175
Anoxia:
   anti-anoxia effects, mother of pearl, 674
   tolerance to, 220, 221, 431
Antacid preparations, ginger, 322
Anthelmintic:
   absinthium, 2
   aspidium, 54
   bromelain, 114
   carrot seed oil, 148
   cedar leaf oil, 161
   chenopodium oil, 180
   chirata, 193
   citronella oil, 202
   cloves, 210
   Dahurian angelica, 663
   dittany bark, 664
   ficin, 291
   filicin, 54
   garlic, 308
   gokshura, 646
   hawthorn, 353
   hypericum, 373
   Juniper berries, 391
   ligusticum, 670
   neem, 653
   mume, 672
   onion, 475
   oregano, 483
   origanum oil, 485
   papain, 114, 292
   quassia, 516
   roselle, 533

   rue, 543
   sage, 551
   sweet basil, 75
   tagetes, 582
   tamarind, 585
   thyme, 596
   Vasaka, 655
Anthemis nobilis, 169
Anthriscus cerefolium, 185
Anthriscus longirostris, 185
Antiadhesion, cranberry, 232
Antiaging, 26, 295, 430
Antiallergenic:
   Baikal skullcap, 685
   belladonna, 89
   bromelain, 114
   coffee, 223
   devil's claw, 243
   echinacea, 254
   ephedra, 266
   eyebright, 280
   Fangfen, 282
   forsythia fruit, 665
   ganoderma, 302
   ginseng, 332
   hydrangea, 370
   immortelle, 379
   licorice root, 416
   ligusticum, 670
   mume, 672
   nettle, 464
   papain, 114, 292
   rutin, 545
   Sichuan lovage, 683
Antiandrogenic:
   red sage, 680
   saw palmetto, 562
Antibacterial:
   absinthium, 2
   aconite, 8
   almonds, 21
   angelica, 33
   arnica, 43
   astragalus, 56
   avocado, 60
   Baikal skullcap, 685
   balm, 62
   balsam copaiba, 67
   balsam Peru, 70
   barberry, 73
   bayberry bark, 80
   black and pale catechu, 158
   blessed thistle, 101

Antibacterial (*Continued*)
  calendula, 130
  cangzhu, 662
  caraway, 138
  chaste-tree, 178
  cherokee rosehip, 181
  chicory root, 190
  cinchona, 195
  cinnamon, 197
  citronella oil, 202
  cloves, 210
  coriander, 228
  costus oil, 231
  cranberry, 232
  cubebs, 234
  cumin, 236
  Dahurian angelica, 663
  dill, 245
  doggrass, 248
  echinacea, 254
  elecampane, 260
  epimedium, 269
  eucalyptus, 272
  fennel, 284
  forsythia fruit, 665
  garlic, 308
  geranium oil, 319
  German chamomile, 171
  giant knotweed, 669
  goldenseal, 337
  grapefruit oil, 344
  Gurmar, 649
  henna, 356
  honey, 358
  honeysuckle flower, 667
  horsetail, 368
  hypericum, 372
  lemongrass, 413
  licorice root, 416
  magnolia flower, 435
  mahabala, 651
  mume, 672
  musk, 456
  mustard, 458
  neem, 652
  nettle, 464
  onion, 475
  orange, 478
  Patchouly oil, 493
  pectin, 496
  peony bark, 675
  peony root, 676
  phellodendron bark, 678
  pine needle oil, 503
  poria, 509
  purslane, 679
  red sage, 680
  balsam copaiba, 67
  rosemary, 536
  royal jelly, 539
  sage, 550
  sassafras, 558
  savory, 559
  Sichuan lovage, 683
  sweet bay, 77
  triphala, 654
  turmeric, 604
  uva ursi, 610
  valerian root, 613
  vasaka, 655
  yarrow, 626
  see also Antimicrobial; Antiseptic
Antibiotic, *see* Antimicrobial
Anticaking agent, castor oil, 154–155
Anticancer agents:
  aconite, 8
  aloe, 25
  asparagus, 52
  avocado, 60
  black berry bark, 97
  blessed thistle, 101
  bloodroot, 103
  boneset, 109–110
  bromelain, 114
  buckthorn, 119
  burdock, 121
  cade oil, 123
  cajeput oil, 125
  calendula, 130
  caraway, 139
  cascara sagrada, 150
  chaparral, 175
  codonopsis, 221
  cranberry, 233
  devil's claw, 243
  echinacea, 254
  euphorbia, 275
  ganoderma, 302
  greater galnanga, 647
  garlic, 310
  henna, 356
  honeysuckle flower, 667
  horsechestnut, 365
  Job's tears, 384
  licorice root, 416
  ligustrum, 421
  mistletoe, 449
  mother of pearl, 674
  onion, 475
  orange, 480
  papain, 114, 292
  pipsissewa, 504
  podophyllum, 506
  poria, 509
  rhubarb, 526
  royal jelly, 540
  rutin, 545
  saffron, 547
  sweet basil, 75
  tannic acid, 586
  tea, 592
  tragacanth, 602
  turmeric, 604
  wild cherry bark, 183
  yucca, 634
Anticholinergic:
  belladonna, 88
  ganoderma, 302
  see also Cholinergic activities
Anticoagulant:
  asafetida, 49
  baizhu, 660
  carrageenan, 146
  red sage, 680
  safflower, 682
  tienchi ginseng, 599
  turmeric, 604
Anticonvulsants:
  asafetida, 49
  barberry, 73
  bois de rose oil, 106
  calamus, 128
  celery seed, 165
  clary sage, 205
  ganoderma, 302
  ginseng, 333
  goldenseal, 337
  gotu kola, 340
  kava, 395
  lavender, 407
  licorice root, 416
  musk, 456
  passion flower, 490
  schisandra, 566

Antidandruff:
  balsam Peru, 70
  Cade oil, 123
  calamus oil, 128
  Dahurian angelica, 664
  garlic, 308
  goldenseal, 337
  Ligusticum, 670
  quillaia, 519
  rosemary, 536
  squill, 575
  sweet bay, 77
  valerian root, 613
  yucca, 634
  see also hair entries
Antidepressant:
  cangzhu, 662
  ginkgo, 326
  gotu kola, 340
  hypericum, 372
  saffron, 548
  sage, 551
  schisandra, 566
  Sichuan lovage, 683
Antidiarrheal, see Diarrhea, remedies
Antidiuretic:
  pipsissewa, 504
  valerian root, 613
  see also Diuretics
Antienletic:
  cloves, 210
  see also Emetics
Antiexudative:
  horsechestnut, 364
  saw palmetto, 562
Antifertility agents:
  asparagus, 53
  castor oil, 155
  chaparral, 175
  chirata, 193
  dittany bark, 664
  orange, 479
  oregano, 483
  neem, 652
  rue, 543
  stevia, 577
Antifibrillatory, barberry, 73
Antifoaming agent, Job's tears, 385
Antifungal:
  aconite, 8
  algin, 17

angelica, 33
arnica, 43
Baikal skullcap, 685
bergamot oil, 91
calendula, 130
cangzhu, 662
cedar leaf oil, 162
chaste-tree, 178
cherry laurel leaves, 184
cinnamon, 197
cinnamon and cassia, 196–200
citronella oil, 201–202
coriander, 227–228
Dahurian angelica, 663
dill, 245
dittany bark, 664
echinacea, 253
elecampane, 260
forsythia fruit, 665
garlic, 308
geranium oil, 319
German chamomile, 171
gotu kola, 340
grape skin extract, 342
greater galnanga, 648
Gurmar, 649
henna, 355
lemongrass, 413
ligusticum, 670
magnolia flower, 435
mustard, 458
neem, 652
onion, 475
orange, 478
oregano, 483
origanum oil, 485
patchouly oil, 493
peony bark, 675
peony root, 676
phellodendron bark, 678
purslane, 679
quebracho, 517
rehmannia, 522
rosemary, 536
sandalwood oil, 554
savory, 559
Sichuan lovage, 683
sweet bay, 77
tamarind, 584
thyme, 595
turmeric, 604
see also Antimicrobial

Antigalactogogue, chaparral, 174–176
Antigonadotropic, 225, 398
Antihemorrhagics, see Hemorrhage
Anlihemorrhagic vitamin, see Vitamin K
Antihistaminic:
  balm, 62
  caraway, 138
  Chinese herbals, 657–659
  cloves, 210
  codonopsis, 221
  Dahurian angelica, 663
  fangfeng, 281
  ganoderma, 302
  giant knotweed, 669
  hydrangea, 370
  Job's tears, 384
  lemon oil, 411
  ligusticum, 670
  magnolia flower, 435
  mother of pearl, 674
  mume, 672
  musk, 456
  rue, 542
Anti-hyaluronidase, henna, 356
Antihypercholesterolemics, see Cholesterol, blood
Antihypertensive, see Hypertension
Antihypoxic, ginkgo, 326
Antithyrotropic, balm, 62
Antiimplantation agent, chaparral, 175
Antiinfective activities, see Antimicrobial
Antiinflammatory:
  absinthium, 1
  acacia, 8
  aloe, 25
  arnica, 43
  ashwagandha, 645
  astragalus, 56
  Baikal skullcap, 685
  baizhu, 660
  balsam Canada, 66
  bee pollen, 82
  black cohosh, 97
  blue cohosh, 104
  boneset, 109
  borage, 111

Antiinflammatory (*Continued*)
  bromelain, 114
  calendula, 130
  cangzhu, 662
  capsicum, 133
  celery seed, 165
  chamomile, 171
  chaparral, 174
  chaste-tree, 177
  chickweed, 187
  chicory root, 190
  chirata, 193
  comfrey, 225–226
  Dahurian angelica, 663
  devil's claw, 242
  doggrass, 248
  dogwood, 249
  eleuthero, 263
  ephedra, 266
  epimedium, 269
  evening primrose, 277
  eyebright, 280
  fangfeng, 281
  feverfew, 290
  ficin, 292
  forsythia fruit, 665
  ganoderma, 303
  gentian, 316
  ginkgo, 326
  goldenseal, 337
  gotu kola, 340
  greater galnanga, 647
  guaiac wood oil, 346
  henna, 356
  honeysuckle flower, 667
  horsechestnut, 364
  horsetail, 368
  hypericum, 372
  immortelle, 378
  ipecac, 380
  Job's tears, 384
  jujube, 388
  lemon oil, 411
  licorice root, 416
  ligusticum, 670
  ligustrum, 421
  lovage root, 428
  luffa, 671
  mace, 467–469
  magnolia flower, 435
  margaspidin, 54
  mints, 445
  mistletoe, 449
  musk, 456
  neem, 652
  orange, 478
  papain, 114, 292
  peony bark, 675
  peony root, 677
  purslane, 679
  quillaia, 518
  red sage, 680
  rehmannia, 522
  rhubarb, 526
  royal jelly, 539
  rue, 542
  rutin, 545
  sarsaparilla, 556
  saw palmetto, 562
  storax, 579
  sweet birch oil, 96
  sweet woodruff, 622
  tagetes, 581
  tienchi ginseng, 599
  triphala, 654
  turmeric, 604
  vasaka, 655
  white pine bark, 501
  witch hazel, 620
  yarrow, 626
  yucca, 634
Antilipemic, Cherokee roschip, 181
Antimalarial:
  absinthium, 2
  black and white pepper, 498
  centaury, 168
  chirata, 192
  cinchona, 194
  deertongue, 241
  ginger, 321
  horsechestnut, 365
  horsetail, 368
  kalmegh, 650
  licorice root, 417
  lobelia, 425
  mahabala, 651
  neem, 652
  quassia, 515
Antimicrobial:
  alkanet, 18
  anise, 37
  bayberry bark, 80–81
  bletilla tuber, 661
  bloodroot, 102
  bois de rose oil, 106
  burdock, 120
  cade oil, 123
  cajeput oil, 125
  cangzhu, 662
  cascarilla bark, 151
  chamomile, 171
  chaparral, 175
  cinnamon, 196–200
  cloves, 210
  enhancement by bromelain, 114
  epimedium, 269
  fo-ti, 294
  galbanum, 299
  garden burnet, 666
  giant knotweed, 668
  gokshura, 646
  honey, 355–356
  honeysuckle flower, 667
  hops, 359
  immortelle, 378
  labdanum, 404
  lavender, 407
  lemongrass, 413
  lemon oil, 410
  mints, 444
  mume, 672
  mustard, 458
  myrrh, 461
  parsley, 488
  patchouly oil, 493
  peony root, 676
  pine needle oil, 503
  purslane, 679
  quillaia, 518
  rhubarb, 526
  rosemary, 536
  sage, 550
  sassafras, 558
  savory, 559
  Sichuan lovage, 682
  stevia, 577
  storax, 579
  tannic acid, 586
  tea, 592
  tragacanth, 602
  turmeric, 604
  wild cherry bark, 183

yucca, 634
see also Antiseptic
Antimitotic, podophyllum, 506
Antimutagenic:
  chaparral, 175
  Cherokee rosehip, 181
  fo-ti, 293–297
  Job's tears, 384
  lycium fruit, 430
  mume, 672
  poria, 509
  red sage, 680
  turmeric, 604
Antimycotic, 395
Antineoplastic, see Anticancer agents
Antiobesity preparations, see Obesity
Antioxidant:
  allspice, 20
  astragalus, 56
  Baikal skullcap, 685
  bee pollen, 82
  benzoin, 90
  black and white pepper, 499
  celery seed, 165
  chaparral, 175
  cloves, 210
  cocoa, 218
  as color stabilizers, beet, 87
  eleuthero, 263
  in foods, chaparral, 175
  forsythia fruit, 665
  ganoderma, 302
  ginger, 321
  greater galnanga, 647
  horehound, 363
  hyssop, 376
Job's tears, 385
  kudzu root, 401
  lemongrass, 413
  lemon oil, 410
  licorice, 416
  marjoram, 437
  milk thistle, 439
  nutmeg and mace, 467
  olibanum, 472
  red sage, 680
  rosemary, 536
  sage, 550
  schisandra, 566

tea, 592
turmeric, 604
vasaka, 655
Antipeptic:
  agar, 10
  almonds, 22
  carrageenan, 146
  chamomile, 171
Antiperiodics, 337, 652, 653
Antiperoxidative, lycium fruit, 431
Antiperspirant:
  astragalus, 57
  Cherokee rosehip, 181
  ephedra, 266
  peony root, 676
  sage, 550
  schisandra, 566
  sour jujube kernel, 573
Antiproteolytic activities:
  carrageenan, 146
  see also Bromelain; Ficin; Papain
Antiprurilics:
  balsam Peru, 70
  Cade oil, 123
  chaste-tree, 178
  chickweed, 187
  fangfeng, 282
  hawthorn, 353
  honeysuckle flower, 667
  jujube, 388
  quillaia, 519
  turmeric, 605
  witch hazel, 621
  see also skin entries
Antipyretic:
  absinthium, 1
  aconite, 8
  allspice, 20
  angostura bark, 35
  Baikal skullcap, 685
  balm of Gilead buds, 65
  barberry, 73
  bayberry bark, 80
  black cohosh, 98
  blessed thistle, 101
  boneset, 110
  borage, 111
  burdock, 120
  calamus, 128
  catnip, 160

cedar leaf oil, 162
centaury, 168
chirata, 193
cinchona, 194
cinnamon and cassia, 198
citronella oil, 202
coriander, 228
cranberry, 233
Dahurian angelica, 663
deertongue, 241
devil's claw, 243
dogwood, 249
ephedra, 267
eucalyptus, 273
fangfeng, 281
fenugreek, 287
feverfew, 290
forsythia fruit, 665
garlic, 310
gelsemium, 313
ginseng, 332
goldenseal, 337
golu kola, 339–341
guarana, 351
honeysuckel flower, 667
ipecac, 380
kudzu root, 401
lemongrass, 413
ligusticum, 670
luffa, 671
marjoram, 438
neem, 652
oregano, 483
parsley, 488
peony bark, 675
podophyllum, 506
quassia, 516
quebracho, 517
rue, 543
sandalwood oil, 554
sarsaparilla, 556
sour jujube kernel, 573
sweet birch oil, 96
tamarind, 584
wintergreen oil, 619
yarrow, 626
yerba santa, 629
Antirheumatic, see Arthritis; Inflammation; Rheumatism
Antiscorbutic, 529, see also Vitamin C

Antiseptic:
  balsam Canada, 66
  balsam copaiba, 68
  balsam Peru, 70
  balsam tolu, 71
  benzoin, 90
  bilberry, 94
  boneset, 110
  cade oil, 123
  cajeput oil, 126
  calendula, 131
  cangzhu, 662
  chamomile, 172
  cinchona, 195
  cinnamon and cassia, 199
  cloves, 210
  cranberry, 232
  echinacea, 255
  eucalyptus, 272
  eugenol, 20
  goldenseal, 337
  hypericum, 373
  juniper berries, 390
  kava, 395
  mints, 445
  myrrh, 462
  neem, 652
  olibanum, 472
  pipsissewa, 504
  sage, 551
  storax, 579
  sweet birch oil, 96
  west Indian bay, 79
  wintergreen oil, 619
  see also Antimicrobial
Antismoking preparations:
  dandelion root, 239
  gentian, 317
  licorice root, 417
  lobelia, 425
Antispasmodic:
  almonds, 22
  ambrette seed, 31
  angostura bark, 35
  anise, 38
  Baikal skullcap, 685
  balm, 62
  belladonna, 89
  black haw bark, 99
  blue cohosh, 105
  Bois de rose oil, 106
  calamus, 128
  calendula, 131

caraway, 138
cardamom, 141
cassie absolute, 153
catnip, 160
celery seed, 166
chamomile, 171
cherry laurel leaves, 184
cinchona, 195
citronella oil, 202
cloves, 211
coriander, 228
cumin, 236
dill, 246
dogwood, 249
euphorbia, 275
evening primrose, 278
fennel, 284
feverfew, 290
galbanum, 299
gelsemium, 313
greater galnanga, 647
hawthorn, 353
henna, 356
hops, 360
juniper berries, 390
kava, 395
lavender, 407
ligusticum, 670
lobelia, 425
lovage root, 428
marjoram, 437
mints, 446
mistletoe, 449
musk, 456
nettle, 464
onion, 475
oregano, 483
parsley, 488
passion flower, 490
peony root, 677
pipsissewa, 504
purslane, 679
quebracho, 517
red clover tops, 208
roselle, 533
rosemary, 537
rue, 543
rutin, 545
saffron, 548
sage, 550
savory, 560
Sichuan lovage, 683
sour jujube kernel, 573

sweet woodruff, 623
tagetes, 581
tagetes oil, 582
thyme, 596
turmeric, 604
valerian root, 612
vasaka, 656
yarrow, 627
yerba santa, 629
Antiswelling, see Edema
Antitestosterone, remedies, saw
    palmetto, 562
Antithrombotic, garlic, 309
Antithromboxane, aloe, 43
Antithyrotropic, 62, 595
Antitoxic, 294, 416, 593
Antitubercular:
  carrageenan, 146
  cassie absolute, 153
  chirata, 192
  eucalyptus, 273
  garlic, 310
  hops, 360
  kava, 397
  licorice root, 417
  prickly ash, 51
  see also Antibacterial;
      Anti microbia
Antitussive:
  apricot kernels, 23, 183
  chestnut leaves, 186
  dogwood, 249
  epimedium, 269
  ganoderma, 303
  giant knotweed, 669
  licorice root, 416, 417
  schisandra, 565, 566
  wild cherry bark, 182
  see also Cough
Antiviral:
  aconite, 8
  aloe, 25
  astragalus, 56
  Baikal skullcap, 685
  balm, 62
  black catechu, 158
  calendula, 130
  cangzhu, 662
  cedar leaf oil, 161
  Cherokee rosehip, 181
  cinnamomum, 199
  cloves, 210
  cubebs, 234

General Index 719

echinacea, 253
epimedium, 269
eucalyptus, 273
forsythia fruit, 665
garlic, 308
giant knotweed, 669
honeysuckle flower, 667
hypericum, 372
hyssop, 376
magnolia flower, 435
marjoram, 437
mints, 444
neem, 652
peony bark, 675
peony root, 677
pine needle oil, 503
sage, 550
Sichuan lovage, 683
tannic acid, 586
tea, 591
see also Antimicrobial
Anxiety:
remedies:
ashwagandha, 645
gotu kola, 340
hypericum, 372
passion flower, 491
pearl, 673
as toxic symptom, yohimbe, 632
Aphrodisiac:
avocado, 60
cassie absolute, 153
cumin, 236
damiana, 237
epimedium, 270
garlic, 310
gokshura, 646
jasmine, 383
karaya gum, 394
mahabala, 651
quebracho, 517
saffron, 548
savory, 560
yohimbe, 632
Apiaceae, see Umbelliferae
Apilak, 538
Apis cerana, **84**
Apis mellifera; Apis species, **84, 357, 538**
Apis species, **84, 357**
Apium carvi, 138

Apium graveolens, **165**
Apium petroselinum, 487
Apocynaceae, 516
Appendicitis, remedies:
giant knotweed, 669
Job's tears, 385
peony bark, 675
purslane, 679
Appetite depressant:
bromelain, 114
coca, 214
coffee, 223
ephedra, 267
guar gum, 348
see also Obesity
Appetite stimulant:
absinthium, 2
alfalfa, 14
allspice, 20
astragalus, 57
calamus, 128
cangzhu, 662
celery seed, 166
centaury, 168
chicory root, 190
cinchona, 195
codonopsis, 221
dandelion root, 239
eleuthero, 264
ganoderma, 303
gentian, 317
ginger, 322
ginseng, 333
hops, 360
ipecac, 380
jujube, 389
kalmegh, 650
mustard, 459
orange, 480
oregano, 483
roselle, 534
schisandra, 566
sweet basil, 75
thyme, 596
yarrow, 627
Appu, 643
Apricot kernals
Chinese medicine, 23
poisoning, remedies, wild cherry bark, 183
Apricot vine, see Passion flower
Arabic, gum, see Acacia

Araceae, 127
Araliaceae, 262, 330, 598
Arbutus ura-ursi, 610
Archangelica officinalis, 29
Arctium lappa, **120, 121**
Arctium majus, 120
Arctium minus, 120
Arctostaphylos uva-ursi,
Armoise, see Absinthium
ARNICA, 42–44
Arnica cordifolia, 42
Arnica fulgens, 42
***Arnica montana*, 42–44**
Arnica sororia, 42
Arnica species, 42
Arniotta, see Annatto
Arrhythmia:
protectants, devil's claw, 243
remedies:
chicory root, 191
cinchona, 195
ganoderma, 303
horehound, 363
parsley, 488
sour jujube kernel, 573
tienchi ginseng, 600
as toxic symptom, horehound, 363
see also cardiac entries
Arrow wood, see Buckthorn, alder
Arsenic poisoning, remedies, gotu kola, 340
***Artemisia absinthium*, 1–3**
Artemisia dracunculus, **588, 589**
Artemisia species, 2
Arteries, animal, preparation for human implantation, licin, 293
Arteriosclerosis, remedies:
garlic, 310
ginkgo, 327
rutin, 545
Arthritis, remedies:
aconite, 8
alfalfa, 14
angelica, 33
avocado, 60

Arthritis, remedies (*Continued*)
  Baikal skullcap, 685
  bayberry bark, 81
  bilberry, 94
  calamus, 128
  cangzhu, 662
  capsicum, 133
  cassie absolute, 153
  celery seed, 166
  chaparral, 176
  chirata, 193
  Dahurian angelica, 663
  devil's claw, 243
  dittany bark, 664
  echinacea, 255
  elder flowers, 258
  eleuthero, 264
  epimedium, 270
  fangfeng, 282
  feverfew, 290
  gentian, 317
  giant knotweed, 669
  ligusticum, 670
  luffa, 671
  mustard, 459
  myrrh, 462
  podophyllum, 506
  rehmannia, 522
  royal jelly, 540
  sarsaparilla, 556
  sassafras, 558
  Sichuan lovage, 683
  sweet birch oil, 96
  triphala, 654
  turmeric, 604
  wintergreen oil, 619
  yucca, 634
  see also Inflammation;
    Rheumatism
ARTICHOKE, 45–47
Aryans, **642**
ASAFETIDA, 48–49
Asarol, **641**
Ascaris, remedies, *see*
    Anthelmintics
Ascites, remedies, euphorbia,
ASH PRICKLY, 50–51
Ashtanga Ayurveda, 638
Ashwagandha, 645
***Ascophyllum nodosum*, 16**
Ascophyllum species,
ASPARAGUS, 52–53
  urine odor after ingestion, 52

Asparagus officinalis, 52
Aspergillus species, inhibition:
  grape skin extract, 342
  henna, 355
Asperula odorata, 622
Aspic, *see* Lavender
ASPIDIUM, 54–55
Aspidosperma
    quebracho-blanco, 516
Aspirin, 313, 396, 619
Asarol, 641
Asteraceae, 1, 42, 45, 100, 109,
    120, 129, 169, 189, 230,
    238, 240, 251, 259, 289,
    378, 439, 507, 576, 581,
    588, 626, 660
Asteraceae, *see* Compositae *Aster helenium*,
Aster officinalis, 259
Asthi, 638
Asthma:
  remedies:
    alfalfa, 14
    apricot kernels, 23, 183
    asafetida, 49
    belladonna, 89
    black haw bark, 100
    coca, 214
    cocoa, 218
    codonopsis, 221
    costus oil, 231
    dill, 246
    dogwood, 249
    elecampane, 260
    ephedra, 267
    euphorbia, 275
    feverfew, 290
    ganoderma, 303
    ginkgo, 327
    immortelle, 379
    jujube, 389
    licorice root, 417
    lobelia, 425
    marjoram, 438
    mume, 672
    myrrh, 462
    neem, 652
    olibanum, 472
    oregano, 483
    parsley, 488
    red clover tops, 208
    saw palmetto, 563
    schisandra, 566

    triphala, 654
    turmeric, 605
    vasaka, 656
    yerba santa, 629
  as toxic symptom:
    acacia, 5
    castor oil, 154
    gaur gum, 348
    ipecac, 380
    magnolia flower, 435
    psyllium, 512
    see also Bronchial ailments;
      Respiratory ailments
Asthma weed, *see* Lobelia
ASTRAGALUS, 55–57
***Astragalus gummifer*, 601**
***Astragalus membranaceus*,
    55, 56**
***Astragalus mongholicus*, 55, 56**
***Atragalus propinguus*, 55**
***Astragalus* species, 55, 56**
Astringent:
  alkanet, 19
  allspice, 20
  Baikal skullcap, 685
  bayberry bark, 81
  bee pollen, 83
  benzoin, 91
  bilberry, 94
  black and pale catechu, 158
  blackberry bark, 97
  bletilla tuber, 661
  cassie absolute, 153
  centaury, 168
  chestnut leaves, 186
  cinchona, 195
  cinnamon and cassia, 199
  comfrey, 226
  elder flowers, 258
  evening primrose, 278
  eyebright, 280
  geranium oil, 319
  giant knotweed, 669
  guarana, 351
  hawthorn, 353
  honeysuckle flower, 668
  horsechestnut, 365
  hypericum, 373
  kalmegh, 650
  kola nut, 399
  labdanum, 404
  lemongrass, 414
  myrrh, 461

General Index

neem, 652
nettle, 465
peony root, 677
pipsissewa, 504
rhubarb, 526
sage, 551
savory, 560
schisandra, 566
tannic acid, 586
tea, 591
terminali, 654
uva ursi, 611
West Indian bay, 79
wild cherry bark, 182
witch hazel, 621
yarrow, 627
Atherosclerosis,
    antiatherosclerotic
    properties:
  bee pollen, 82
  bilberry, 94
  Cherokee rosehip, 181
  foti, 295
  ligustrum, 421
  peony bark, 675
  tea, 592
  see also Cholesterol
Atharvaveda
Athlete's foot, remedies:
  clove tincture,
  forsythia fruit, 665
  giant knotweed, 668
  quillaia, 519
  see also Fungi; Ringworm;
    Skin conditions
Atony, chronic, gotu kola, 339
Atractylodes, 660
Atractylodes chinensis, **661**
Atractylodes japonica,
  661, 662
Atractylodes lancea, **661**
Atractylodes macrocephala, 660
Atreya Sampradaya, 637
Atropa belladonna, **88**
Attar of rose, see Rose oil (and
  absolute)
Aucklandia costus, 230
Axillary, 316, 350, 362, 464,
  490, 646, 650, 651, 655
AVOCADO, 59–60
Ayurveda, 637, 638, 640, 642,
  643
Azadirachta indica, **652**

Azhal, 643
Aztec marigold, see Tagetes

Bacillus subtilis, inhibition, 178
Back, weak, remedies,
  epimedium, 268–270
Backache, remedies:
  fennel, 285
  kava, 397
  ligustrunt, 421
  lycium fruit, 431
  valerian root, 614
Bacteria, see Antibacterial;
  Antimicrobial;
  Antitubercular;
  Mahabala
Bad breath, remedies:
  cloves, 211
  myrrh, 462
  parsley, 494
  patchouly oil, 494
Bahera, 654
Baidang, 220
Baikal skullcap, 684–685
Baishaoyao, 676–677
Baixianpi, 664
Baizhi, 663–664
BAIZHU, 660
Baked goods:
  colorants:
    annatto, 41
    capsicum, 134
    caramel color, 137
    turmeric, 605
  flavoring ingredients:
    absinthitim, 2
    agar, 10
    alfalfa, 14
    allspice, 20
    almonds, 23
    althea root, 30
    ambrette seed, 33
    angelica, 31
    angostura bark, 35
    anise, 38
    arnica, 44
    asafetida, 49
    balm, 63
    balsam copaiba, 68
    balsam Peru, 70
    balsam tolu, 72
    beeswax, 85
    benzoin, 91

    bergamot oil, 91
    black and pale catechu, 157
    black and white pepper, 499
    blackberry bark, 96
    bois de rose oil, 106
    boronia absolute, 113
    buchu, 117
    cajeput oil, 126
    cananga oil, 132
    cananga oil, 132
    capsicum, 134
    caraway, 139
    cardamom, 141
    carob, 144
    carrot seed oil, 148
    cascara sagrada, 150
    cascarilla bark, 152
    cassie absolute, 153
    castor oil, 155
    cedar leaf oil, 162
    celery seed, 166
    chamomile, 172
    cherry laurel leaves, 184
    Chervil, 185
chicory root, 190
cinchona, 195
cinnamon and cassia, 199
citronella oil, 202
civet, 204
clary sage, 205
cloves, 211
cocoa, 218
coffee, 224
coriander, 228
corn silk, 230
costus oil, 231
cubebs, 234
cumin, 236
damiana, 237
dandelion root, 239
dill, 246
doggrass, 248
elder flowers, 258
elecampane, 260
elemi gum, 262
eucalyptus, 273
fennel, 284
fenugreek, 287
galbanum, 300
garlic, 309
genet, 315
gentian, 317
geranium oil, 319

Baked goods:
    ginger, 322
    grapefruit oil, 344
    guaiac wood oil, 316
    hops, 360
    horehound, 363
    hyssop, 376
    immortelle, 379
    jasmine, 383
    juniper berries, 391
    kola nut, 399
    labdanum, 405
    lavender, 408
    lemongrass, 414
    lemon oil, 411
    licorice root, 417
    lime oil, 423
    lovage root, 428
    marjoram, 438
    mints, 445
    musk, 456
    mustard, 458
    myrrh, 462
    nutmeg and mace, 468
    olibanum, 472
    onion, 475
    orange, 479
    oregano, 483
    origanum oil, 485
    parsley, 488
    patchouly oil, 494
    pine needle oil, 503
    prickly ash, 51
    quassia, 516
    quebracho, 517
    quillaia, 519
    red clover tops, 208
    rhubarb, 526
    roselle, 533
    rosemary, 536
    rose oil, 106
    rue, 543
    saffron, 548
    sage, 551
    sandalwood oil, 554
    sarsaparilla, 556
    savory, 560
    storax, 579
    sweet bay, 77
    sweet birch oil, 96
    sweet woodruff, 623
    tagetes, 582
    tamarinds, 585
    tannic acid, 587
    tarragon, 589
    tea, 592
    thyme, 596
    turmeric, 605
    turpentine, 608
    valerian root, 613
    vanilla, 617
    West Indian bay, 79
    wild cherry bark, 183
    wintergreen oil, 623
    yerba santa, 629
    ylang ylang oil, 630
    fragrance ingredients:
        kudzu root, 399
    gums:
        algin, 17
        carrageenan, 148
        guar gum, 348
        locust bean gum, 428
        tragacanth, 603
    MSG, 453
    other ingredients:
        beeswax, 85
        bromelain, 113
        papain, 114, 292
        poria, 509
Bakers' color, 137
Baldness, *see* Hair loss
BALM, LEMON, 62–63,
    see also Yerba Santa
**BALM OF GILEAD BUDS,
    64–65** *see also* Balsam
    Canada
Balmony, East Indian, *see* Chirata
BALSAM CANADA, 65–66
BALSAM COPAIBA, 67–68
**BALSAM FIR OREGON,
    68–69**
Balsamodendron opobalsamum,
    64
BALSAM PERU, 69–70
Balsam poplar buds, *see* Balm of
    giliad buds
BALSAM TOLU, 71–72
Bannal, *see* Broom tops
Barbados aloe, 24
Barbadoes cherry, *see* Acerola
BARBERRY, 72
Barbiturates, ginseng, 332
Bardana, *see* Burdock
Bark, apricot, treating apricot
    kernel poisoning, 23
Barosma betulina, 117
Barosma crenultaa, 117

Barosnta serratifolia, 117
BASIL, SWEET, 74
Basmas, 643
Bath preparations, ingredients:
    chamomile, 172
    fangfeng, 282
    ginseng, 333
    hyssop, 376
    thyme, 596
    yarrow, 627
BAY:
    SWEET, 76–77
    WEST INDIAN, 78–79
BAYBERRY BARK, 80–81
Bayberry wax, 81
Bay oil, *see* Myrcia oil
Bay rum, 78, 79
Bearberry, *see* Uva ursi
Beargrape, *see* Yerba santa
Bears weed, *see* Yerba santa
Beaver, *see* Castoreum
Bedsores, remedies, 70–71, 155,
    431
Bedwetting, remedies, fennel,
    285
Bee balm, *see* Balm
BEE POLLEN, 81–83
Beer:
    chillproofing:
        bromelain, 114
        ficin, 293
        papain, 114, 292
    hops, 360
    licorice, 417
    tannic acid, 587
    valerian root, 413
BEESWAX, 84–85
**BEET COLOR, RED,
    86–87**
Beggar's buttons, *see* Burdock
Bei cangzhu, 661
BELLADONNA, 88–89
Bell pepper, 133
Benedictine liqueur, flavor
        ingredient, chamomile,
        172
Benjamin, gum, 90
BENZOIN, 90–91
Benzoylaconine, *see* Picraconiiin
Berberidaceae, 72, 104, 268, 505
Berberis aquifolium, 72
Berberis, *see* Barberry
Berberis species, **72**
Berberis vulgaris, **72**

General Index

BERGAMOT OIL, 91–92
Beriberi, remedies, fenugreek, 288
Beta vulgaris, 86
Betula alba, 95
Betula carpinefolia, 95
Betulaceae, 95
Betula lenta, 95
Betula pendula, 95–96
Beverages, alcoholic and nonalcoholic:
  colorants:
    annatto, 41
    capsicum, 134
    caramel color, 137
    grape skin extract, 343
  flavoring ingredients:
    absinthium, 2
    alfalfa, 14
    allspice, 20
    almonds, 23
    aloe, 27
    althea root, 30
    ambrette seed, 31
    angelica, 33
    angostura bark, 35
    anise, 38
    arnica, 44
    artichoke, 47
    asafetida, 49
    asparagus, 53
    balm, 63
    balsam Canada, 66
    balsam copaiba, 68
    balsam Peru, 70
    balsam tofu, 72
    beeswax, 85
    benzoin, 91
    bergamot oil, 92
    bilberry, 94
    black and pale catechu, 158
    black and white pepper, 499
    blackberry bark, 97
    black haw bark, 100
    blessed thistle, 101
    bois de rose oil, 106
    boldo leaves, 108
    boronia absolute, 113
    buchu, 117
    cajeput oil, 126
    cananga oil, 132
    capsicum, 134
    caraway, 139
    cardamom, 141

carob, 144
carrot seed oil, 148
cascara sagrada, 150
cascarilla bark, 152
cassie absolute, 153
castoreum, 157
castor oil, 155
cedar leaf oil, 162
celery seed, 166
centaury, 168
chamomile, 172
cherry laurel leaves, 184
chervil, 185
chestnut leaves, 186
chicory root, 190
chirata, 193
cinchona, 195
cinnamon and cassia, 199
citronella oil, 202
civet, 204
clary sage, 205
cloves, 211
coca, 214
cocoa, 218
coffee, 224
coriander, 228
corn silk, 230
costus oil, 231
cubebs, 234
cumin, 236
damiana, 237
dandelion root, 239
dill, 246
doggrass, 248
elder flowers, 258
elecampane, 260
elemi gum, 262
eucalyptus, 273
fennel, 284–285
fenugreek, 287
galbanum, 300
garlic, 309
genet, 315
gentian, 317
geranium oil, 319
ginger, 322
ginseng, 333
grapefruit oil, 344
guaiac wood oil, 346
guarana, 350
hawthorn, 353
hops, 360
horehound, 363
hyssop, 376

immortelle, 379
jasmine, 383
juniper berries, 391
kola nut, 399
labdanum, 405
lavender, 408
lemongrass, 414
lemon oil, 411
licorice root, 417
lime oil, 423
lovage root, 428
marjoram, 438
mints, 445
musk, 456
mustard, 459
myrrh, 461–462
nutmeg and mace, 469
olibanum, 472
onion, 475
orange, 479
oregano, 483
origanum oil, 485
parsley, 488
passion flower, 491
patchouly oil, 494
pine needle oil, 503
prickly ash, 51
quassia, 516
quebracho, 517
red clover tops, 208
rhubarb, 526–527
roselle, 533–534
rosemary, 536
rose oil, 531
rue, 543
saffron, 548
sage, 551
sandalwood oil, 554
sarsaparilla, 556
sassafras, 558
storax, 579
sweet basil, 75
sweet bay, 77
sweet birch oil, 96
sweet woodruff, 623
tagetes, 582
tamarind, 585
tannic acid, 587
tarragon, 589
tea, 592
thyme, 596
turpentine, 608
valerian root, 613
vanilla, 617

West Indian bay, 79
  white pine bark, 501
  wild cherry bark, 183
  wintergreen oil, 619
  yarrow, 627
  yerba santa, 629
  ylang ylang oil, 630
foaming agents, 519
fragrance ingredients, 375, 504
gums:
  algin, 17
  guar gum, 348
  karaya gum, 394
  tragacanth, 603
other ingredients:
  beeswax, 85
  kava, 396
  MSG, 453
  mume, 672
  pectin, 497
  peony root, 677
  sweeteners, stevia, 577
Bhibitki, 653–654
Bhoot Vidya, 638
Big marigold, *see* Tagetes
Bignonia sempervirens, 312
BILBERRY, 93–94
Bile secretions, *see* Choleretic; Chologogic
Binding agents:
  agar, 10
  algin, 17
  carrageenan, 146
  guar gum, 348
  origanum oil, 485
  traeacanth, 603
"Biological scalpel", papain, 114, 292
BIRCH OIL, SWEET, 95–96
Bitter herb, *see* Centaury
Bitter principles, *see* CHEMICAL INDEX
Bitter root, *see* Gentian (and stemless)
Bitters alcoholic, flavor ingredients:
  absinthium, 2
  ambrette seed, 31
  angelica, 33
  angostura bark, 35
  anise, 37
  artichoke, 47
  balm, 63

blessed thistle, 101
boldo leaves, 108
cascarilla bark, 151–152
centaury, 168
chamomile, 172
cinchona, 195
cinnamon and cassia, 199
cloves, 211
coriander, 228
dandelion root, 239
elder flowers, 258
elecampane, 260
gentian, 317
ginger, 322
hyssop, 376
juniper berries, 391
lemon oil, 411
marjoram, 438
quassia, 516
rhubarb, 526
rue, 543
saffron, 548
sage, 551
sweet woodruff, 623
yarrow, 627
see also Beverages, alcoholic and nonalcoholic
Bitter stick, *see* Chirata
Bitter tonic:
  absinthium, 2
  angostura bark, 35
  barberry, 73
  cascara sagrada, 150
  cascarilla bark, 151
  centaury, 168
  chicory root, 191
  chirata, 193
  cinchona, 194
  goldenseal, 337
  horehound, 363
  quassia, 515
  see also CHEMICAL INDEX
Bitter wintergreen, see pipsissewa
Bitterwood, see Quassia
Bixaceae, 40
Bixa orellana, **40**
Black beans, Chinese medicine, 8
BLACKBERRY BARK, 96–97
BLACK COHOSH, 97–99
Black currant flavor, buchu, 117

Black cutch, *see* Catechu (black and pale)
Black dogwood, *see* Buckthorn, alder
BLACK HAW BARK, 99–100
Black snakeroot, *see* Black cohosh
Black tea, 591
Blackwort, *see* Comfrey
Bladder:
  problems, remedies:
    celery seed, 166
    horsetail, 368
    hypericum, 373
    psyllium, 512
    sweet birch oil, 96
    see also Cystitis; Urinary disorders
  stones, remedies:
    hydrangea, 370
    pipsissewa, 504
    sweet woodruff, 623
  see also Cancer
Blazing star, *see* Aletris
Bleeding, *see* Hemorrhage
Blepharitis, remedies, eyebright, 280
BLESSED THISTLE, 100–101
Bletilla striata, 530
BLETILLA TUBER (BAIJI), 661
Blindness, as toxic symptom:
  aspidium, 54
  cinchona, 195
Blisters:
  remedies, pipsissewa, 504
  as toxic symptoms:
    anise, 37
    mustard, 458
    rue, 543
  see also Dermatitis
Blood:
  clots, *see* Bromelain; Ficin; Papain
  disease, remedies, devil's claw, 243
  platelet, protective effect, phellodendron bark, 678
  poisoning, remedies, echinacea, 255

General Index

purifier:
  chaparral, 176
  gotu kola, 340
  nettle, 465
vessel, see Vasodilation; Vasoconstriction
viscosity, agents that lower, garlic, 309
see also Circulation; Platelet aggregation
Blood pressure:
  normalizer, schisandra, 566
  see also Hypotension; Hypertension
BLOODROOT, 65, 102–103
Blood urea, see Urea, blood concentration
Bloody urine, see Hematuria
BLUE COHOSH, 104–105
Blue gum, see Eucalyptus
Blue sailors, see Chicory root
Boils, remedies:
  black and pale catechu, 158
  blessed thistle, 101
  chickweed, 187
  devil's claw, 243
  giant knotweed, 668
  hawthorn, 353
  honeysuckle flower, 667
  turpentine, 608
  white pine bark, 501
see also skin entries
BOIS DE ROSE OIL, 106
BOLDO LEAVES, 107–108
Boldu boldus, 107
Boldus, 107
Bones, broken, remedies:
  tamarind, 585
  yucca, 634
BONESET, 109–110
Bookoo, see Buchu
BORAGE, 110–112
Boraginaceae, 18, 110, 225
Borago officinalis, 110
Borneol, 32, 74, 132, 141, 151, 161, 175, 197, 227, 251, 299, 321, 378, 404, 407, 413, 434, 437, 467, 471, 482, 502, 535, 550, 554, 559, 604, 612, 626, 647
BORONIA ABSOLUTE, 112–113
Boronia megastigma, 112

Boswellia carteri, **471**
Boswellia serrata, **471**
Boswellia species, **471**
Botulism, honey, 358
Bourbon geranium oil, 318–319
Bowel disorders:
  cramps, remedies, chaparral, 176
  remedies, gotu kola, 340
Brain damage, by MSG, 453
Brassica alba, 457–458
Brassicaceae, see Cruciferae
Brassica hirta, 457
Brassica juncea, 457
Brassica nigra, 457
Brazil root, see Ipecac
Breakfast cereals:
  flavoring ingredients:
    citronella oil, 202
    lemon oil, 411
    nutmeg and mace, 469
    orange, 480
  other additives:
    bromelain, 114
    guar gum, 348
    MSG, 453
Breast problems, remedies:
  dandelion root, 239
  evening primrose, 277
  hyssop, 376
  orange, 480
  purslane, 679
  tagetes, 581
Breath:
  difficulties, as toxic symptom, ginkgo, 327
  shortness of, remedies, 221, 310
see also Bad breath
Brewers' color, 292
BROMELAIN, 113–114, see also Ficin; Papain
Bromeliaceae, 113
Bronchial ailments, remedies:
  angelica, 33
  asafetida, 49
  ashwagandha, 645
  balsam copaiba, 68
  belladonna, 89
  black cohosh, 98
  carrageenan, 146
  cedarwood oil, 164
  chaparral, 175
  cocoa, 218
  comfrey, 226

  costus oil, 231
  dandelion, 239
  elecampane, 260
  epimedium, 270
  euphorbia, 275
  fenugreek, 287
  ganoderma, 303
  garlic, 310
  ginkgo, 326–327
  immortelle, 379
  juniper berries, 391
  kava, 397
  lobelia, 425
  oregano, 483
  quillaia, 519
  red clover tops, 208
  sassafras, 558
  saw palmetto, 563
  thyme, 596–597
  turpentine, 608
  uva ursi, 611
  vasaka, 656
  yerba santa, 629
see also Asthma; Respiratory ailments
Bronchitis, see Bronchial ailments
Bronchodilator, 37, 326, 581
BROOM TOPS, 115–116
Brown algae, as algin source, 16
Brown bark, see Cinchona (red and yellow)
Brown mustard, 457–458
Brucella species, inhibition by henna, 356
Bruises, remedies:
  arnica, 44
  balm of Gilead buds, 65
  calendula, 131
  chaparral, 176
  comfrey, 226
  dandelion root, 239
  hypericum, 373
  hyssop, 376
  lobelia, 425
  thyme, 596
  turmeric, 605
  valerian root, 614
  white pine bark, 501
  yerba Santa, 629
see also skin entries
Bucco, see Buchu
BUCHU, 117–118

BUCKTHORN, ALDER, 26, 118–119
Bucku, *see* Buchu
Bugloss, Spanish, *see* Alkanet
Buku, *see* Buchu
Bulnesia sarmienti, 345
Bur, 120
BURDOCK, 120–121
Burnet-bloodwort, 666
Burns:
   debridement, bromelain, 114
   remedies:
      aloe, 27
      Baikal skullcap, 685
      balsam Canada, 66
      bletilla tuber, 661
      bloodroot, 103
      calendula, 131
      chaparral, 176
      echinacea, 255
      eucalyptus, 273
      garden burnet, 666
      giant knotweed, 668
      ginger, 322
      hydrangea, 370
      hypericum, 373
      hyssop, 376
      immortelle, 379
      lycium fruit, 431
      phellodendron bark, 678
      purslane, 679
      red clover tops, 208
      rhubarb, 527
      tannic acid, 587
      witch hazel, 621
Burnt sugar coloring,
         see Caramel color
Burseraceae, 261, 460, 471
Buxus chinensis, 386
Byttneriaceae, *see* Sterculiaceae

Cacao, *see* Cocoa (cacao)
Cachou, *see* Catechu
      (black and pale)
CADE OIL, 123
Caesalpinia spinosa, 586
CAJEPUT OIL, 124–126
Cajuput, *see* Cajeput oil
Calamintha hortensis, *see*
      Satureja hortensis
Calamintha montana, *see*
      Satureja montana
CALAMUS, 127–128
Calcinations, 640, 643

Calcium antagonist, angelica, 33
Calcium channels, blocker,
      tienchi ginseng, 599
CALENDULA, 129–131
Calendula officinalis, 129
California bay, 76
Calisaya bark, *see* Cinchona (red
      and yellow)
Calmative, *see* Sedative
Camellia sinensis, **590**
Camellia thea, 590
Camellia theifera, 590
cAMP, agents affecting,
      astragalus, 56
Campanulaceae, 220, 424
Canada turpentine, 65
Canadian balsam, 65
Canadian beaver, *see* Castoreum
Cananga odorata, **132**, 629
CANANGA OIL, 132
Canaries, poisoning, avocado, 60
Canarium commune, **261**
Canarium luzonicum, **261**
Cancer:
   antileukemic:
      buckthorn, 119
      quassia, 515
   carcinogens:
      bayberry bark, 81
      bois de rose oil, 106
      calamus, 128
      capsicum, 134
      carrageenan, 146
      cedarwood oil, 162
      chervil, 185
      coffee, 223
      estragole, 284
      eucalyptus, 273
      eugenol, 199
      euphorbia, 275
      grapefruit oil, 344
      lemon oil, 410
      lime oil, 423
      orange, 479
      oregano, 483
      papain, 114
      sassafras, 558
      sweet basil, 75
      tannic acid, 586
      tea, 592
      turpentine, 608
   remedies:
      almond oil, 23
      aspidium, 55

balm, 63
balsam Canada, 66
balsam Peru, 70
balsam tolu, 72
black and white pepper,
   499
black catechu, 158
borage, 112
broom tops, 116
burdock, 121
cassie absolute, 153
cedar leaf oil, 162
centaury, 168
chamomile, 172
chaparral, 175
chenopodium oil, 180
cherry laurel leaves, 183
chicory root, 191
chirata, 193
cinchona, 195
cinnamon and cassia, 199
clary sage, 205
cocillana bark, 216
comfrey, 226
cranberry, 233
cubebs, 234
dandelion root, 239
elder flowers, 258
elecampane, 260
eleuthero, 264
eucalyptus, 273
eyebright, 280
fenugreek, 287
gelsemium, 313
gentian, 317
ginseng, 333
goldenseal, 338
honey, 358
hops, 360
horehound, 363
hydrangea, 370
hypericum, 373
hyssop, 376
ipecac, 380
jasmine, 383
jojoba, 387
juniper berries, 391
lavender, 408
licorice root, 418
lobelia, 425
marjoram, 438
mints, 446
mume, 672
myrrh, 462

General Index

nutmeg and mace, 469
olibanum, 472
papain, 114
patchouly oil, 494
podophyllum, 506
prickly ash, 51
psyllium, 513
red clover tops, 208
roselle, 534
rosemary, 537
saffron, 548
sage, 552
sarsaparilla, 557
sassafras, 558
saw palmetto, 563
squill, 575
storax, 580
sweet bay, 77
tarragon, 589
thyme, 596
West Indian bay, 79
witch hazel, 621
yarrow, 627

Candida albicans, inhibition:
chamomile, 171
cinnamon, 197
echinacea, 255
garlic, 309
immortelle, 378
labdanum, 404

Candles, ingredients, 81, 469

Candy:
colorants, 134
flavoring ingredients, 38, 166
absinthium, 2
alfalfa, 14
allspice, 20
almonds, 23
aloe, 27
althea root, 30
angelica, 33
angostura bark, 35
anise, 38
arnica, 44
asafetida, 49
balm, 63
balsam Canada, 66
balsam copaiba, 68
balsam Peru, 70
balsam tolu, 72
beeswax, 85
benzoin, 91
bergamot oil, 92
black and white pepper, 499

blackberry bark, 97
black catechu, 158
bois de rose oil, 106
boronia absolute, 113
buchu, 117
cajeput oil, 126
cananga oil, 132
capsicum, 134
caraway, 139
cardamon, 141
carob, 144
carrot seed oil, 148
cascara sagrada, 150
cascarilla bark, 152
cassie absolute, 153
castoreum, 157
castor oil, 155
cedar leaf oil, 162
celery seed, 166
chamomile, 172
cherry laurel leaves, 184
chervil, 185
chicory root, 190
cinchona, 195
citronella oil, 202
civet, 204
clary sage, 205
cloves, 211
coca, 214
cocoa, 218
coffee, 224
coriander, 228
corn silk, 230
costus oil, 231
cubebs, 234
cumin, 236
damiana, 237
dandelion root, 239
doggrass, 248
elder flowers, 258
elecampane, 260
elemi gum, 262
eucalyptus, 273
fennel, 284, 285
fenugreek, 287
galbanum, 300
garlic, 309
genet, 315
gentian, 317
geranium oil, 319
ginger, 322
grapefruit oil, 344–345
guaiac wood oil, 346
guarana, 350

hops, 360
horehound, 363
hyssop, 376
immortelle, 379
jasmine, 383
juniper berries, 391
kola nut, 399
Labdanum, 405
lavender, 408
lemongrass, 414
lemon oil, 411
licorice root, 417
lime oil, 423
lovage root, 428
marjoram, 438
mints, 445
musk, 456
mustard, 459
myrrh, 462
nutmeg and mace, 469
olibanum, 472
onion, 475
orange, 479
origanum oil, 485
parsley, 488
patchouly oil, 494
pine needle oil, 503
prickly ash, 51
quassia, 516
quebracho, 517
red clover tops, 208
rhubarb, 526
roselle, 533
rosemary, 536
rose oil, 531
rue, 543
saffron, 548
sage, 551
sandalwood oil, 554
sarsaparilla, 556
sassafras, 558
savory, 560
storax, 579–580
sweet birch oil, 96
sweet woodruff, 623
tagetes, 582
tamarind, 585
tannic acid, 587
tarragon, 589
tea, 592
thyme, 596
turpentine, 608
valerian root, 613
vanilla, 617

Candy (*Continued*)
  West Indian bay, 79
  wild cherry bark, 183
  wintergreen oil, 619
  ylang ylang oil, 630
  fragrance ingredients,
      Pipsissewa, 504
  gums:
    algin, 17
    karaya gum, 394
    tragacanth, 608
  other ingredients:
    beeswax, 85
    MSG, 453
    pectin, 497
    quillaia, 519
CANGZHU, 661–662
Canker sores, lycium fruit, 431
Cannabaceae, 359
Cape aloe, 24, 26, 27
Capillaries, substances that strengthen:
  grape skin extract, 342
  rosemary, 536
  rutin, 545
Caprifoliaceae, 99, 257, 667
CAPSICUM, 132–135
Capsicum annuum, **133**
Capsicum baccatum var. pendulum, **132**
Capsicum chinense, **132**
Capsicum frutescens, **132**
Capsicum pendulum, 132
Capsicum pubescens, **132**
CARAMEL COLOR, 136–137
"Carat", 143
CARAWAY, 138–139
Carbohydrate metabolism, ginseng effect, 332
Carbuncles, remedies:
  Baikal skullcap, 686
  Dahurian angelica, 663
  fangfeng, 282
  forsythia fruit, 665
  fo-ti, 297
  peony bark, 675
  peony root, 676
CARDAMON, 141
Cardiac conditions:
  remedies:
    bilberry, 94
    chicory root, 191
    gelsemium, 313
    ginkgo, 327
    guarana, 351
    hawthorn, 353
    marjoram, 438
    oregano, 483
    red sage, 681
    roselle, 534
    tienchi ginseng, 600
  rhythm, horehound effects, 362
  as toxic symptoms:
    aconite, 8
    aspidium, 54
    coffee, 223
    grape skin extract, 342
    thyme, 596
  see also Arrhythmia; Tachycardia
Cardiac depressants:
  aconite, 8
  belladonna, 89
  broom tops, 116
  carrot seed oil, 148
  chicory root, 190
  cinchona, 194
  costus oil, 231
  dill, 245
  pale catechu, 158
  quillaia, 519
  saffron, 547
  sweet bay, 77
Cardiac pain, *see* Pain
Cardiac stimulant:
  belladonna, 88
  cocoa, 218
  coffee, 223
  ephedra, 266
  fenugreek, 287
  ginger, 321
  pale catechu, 158
  safflower, 682
  schisandra, 566
  squill, 575
  thyme, 596
Cardioglycoside, dose resistance, adaptogenic activity, eleuthero, 263
Cardiotonic:
  hawthorn, 353
  ligustrum, 421
  mistletoe, 449–450
  rehmannia, 522
  squill, 575
Cardiovascular effects:
  angelica, 33
  ganoderma, 302
  kudzu root, 400
  mother of pearl, 674
  Sichuan lovage, 682
Carduus marianum L., 439
Cariostatic, tannins, 586
Carminative:
  allspice, 20
  anise, 37–38
  asafetida, 48
  balm, 63
  bean dish flavoring, chenopodium oil, 180
  black and white pepper, 499
  buchu, 117
  cajeput oil, 124
  calamus, 128
  capsicum, 134
  caraway, 138–139
  cardamon, 141–142
  celery seed, 166
  chamomile, 172
  chaparral, 175
  chastetree, 178
  cinnamon and cassia, 199
  cloves, 211
  coriander, 228
  costus oil, 231
  cubebs, 234
  cumin, 236
  dill, 246
  fennel, 283–285
  feverfew, 290
  galbanum, 300
  ginger, 321–322
  greater galnanga, 648
  juniper berries, 391
  lavender, 408
  marjoram, 437
  mints, 445
  neem, 653
  nutmeg and mace, 469
  onion, 475
  orange, 479–480
  oregano, 483
  prickly ash, 51
  rosemary, 537
  sassafras, 558
  savory, 560
  sweet basil, 75

General Index

sweet bay, 77
tagetes, 582
thyme, 595–596
valerian root, 614
yarrow, 627
CAROB, 142–144
Carob bean gun, *see* Locust bean gum
Carolina vanilla, *see* Deertongue
Carolina yellow jessamine, *see* Gelsemium
Carony bark, *see* Angostura bark
Carphephorus odoratissimus, 240
CARRAGEENAN, 145–147
Carragheenan, *see* Carrageenan
CARROT OILS, 147–148
Carthamus tinctorius **L., 681**
Carum, *see* Caraway
Carum carvi, **138**
Carum petroselinum, 487
Caryophyllaceae, 187
CASCARA SAGRADA, 149–150
    fluid extract, aromatic, flavor ingredients, coriander, 228
CASCARILLA BARK, 151–152
Casein, as enzyme substrate, 292
Cashou, *see* Catechu (black and pale)
Cassia, *see* Cinnamon (and cassia)
Cassia angustifolia, **568–569**
CASSIE ABSOLUTE, 152–153
Castanea americana, 186
Castanea dentata, **186**
Castanea sativa, 186
Castanea species, 186
Castor bean, *see* Castor oil
Castor canadensis, 156
CASTOREUM, 156–157
Castor fiber, **156**
Castoridae, 156
CASTOR OIL, 54, 154–155
Catarrh, *see* Mucous membrane inflammation
CATECHU (BLACK AND PALE), 157–158
Cathartic, *see* Laxative
CATNIP, 159–160
Cat repellents, ingredients, mustard, 459

Cattle, poisoning, avocado, 60
Caulophylfum, *see* Blue cohosh
Caulophyllum thalictroides, **104**
Cauterization, 642
Cayenne pepper, *see* Capsicum
Cayenne rosewood oil, *see* Bois de rose oil
Cedar:
    eastern red, 163
    red, 163
CEDAR LEAF OIL, 161–162
CEDARWOOD OIL, 162–164
Cedro oil, *see* Lemon oil
Cedrus atlantica, **162**
Celeriac, 165
CELERY SEED, 165–166
Cement, for lenses and microscopic slides, balsam Canada, 66
Centarium minus, 167
Centariurn umbellatum, 167
Centaurium erythraea, 167
CENTAURY, 167–168
Centella asiatica, 339
Centella coriacea, 339
Central nervous system, *see* CNS
Cephaelis acuminata, **380**
Cephaelis ipecacuanha, **379**
Ceratonia siliqua, **142, 426**
Cereals, *see* Breakfast cereals
Cerebral insufficiency, ginkgo, 326
Ceruloplasmin, serum levels, reduction by Bulgarian rose oil, 531
cGMP, agents affecting, astragalus, 56
Chakradutta, 637
Chamaemelum nobile, **169**
CHAMOMILE (GERMAN AND ROMAN), 169–173
Champaca wood oil, *see* Guaiac wood oil
Chapping, remedies, *see also* skin entries
CHASTE-TREE, 177–179
Checkerberry, *see* Wintergreen Oil
Cheeses:
    flavoring agents:

729

        black and white pepper, 499
        dandelion root, 239
        dill, 246
    other ingredients:
        algin, 17
        annatto, 41
        carrageenan, 146
        ficin, 293
        guar gum, 348
        karaya gum, 394
        MSG, 453
        xanthan gum, 624
    see also Dairy products
Chemotherapy, side effects, remedies:
    baizhu, 660
    lycium fruit, 432
    pearl, 674
Chenopodiaceae, 86, 179
Chenopodium ambrosioides, **179**
CHENOPODIUM OIL, 179–180
CHEROKEE ROSEHIP, 181–182
Cherry, *see* Acerola; Cherry bark, wild; Cherry laurel leaves
CHERRY BARK, WILD, 182–183
Cherry birch, *see* Birch oil, sweet
CHERRY LAUREL LEAVES, 183–184
CHERVIL, 185
Chest:
    pain, *see* Pain
    tight, remedies, Baikal skullcap, 685, 686
CHESTNUT LEAVES, 186
Chewing gum, ingredients:
    balsam tolu, 72
    chicle, 189
    guarana, 351
    mints, 445
    sweet birch oil, 96
    wintergreen oil, 619
Chicken:
    feed, tagetes, 582
    skin, yellow color, tagetes, 582
Chicken pox, remedies, Chaparral, 176
CHICKWEED, 187–188

CHICLE, 188–189
CHICORY ROOT, 189–191
Childbirth difficulties, remedies, 105, 333, 548
Chili pepper, see Capsicum
Chili powder, cumin, 236
Chilblains, remedies, 30, 134, 327
Chills, remedies, 81, 267, 397
Chimaphila corymbosa, 504
Chimaphila, see Pipsissewa
***Chimaphila umbellata*, 504**
China bark, see Cinchona (red and yellow); Quillaia
Chinese anise, 36–38
Chinese cosmetic ingredients, 657–659
  modern properties, 657
  traditional properties, 657
  see also specific ingredients
Chinese gelatin, see Agar
Chinese licorice, 415
Chinese medicine:
  aconite, 8
  almonds, 23
  ambrette seed, 31
  angelica, 33
  anise, 38
  asafetida, 49
  asparagus, 53
  astragalus, 57
  barberry, 73
  bee pollen, 83
  beeswax, 85
  belladonna, 89
  black and white pepper, 499
  black catechu, 158
  black cohosh, 98
  buchu, 117
  cardamon, 142
  carrot seed, 148
  cassie absolute, 153
  castor oil, 155
  chenopodium oil, 180
  Cherokee rosehip, 181
  cinchona, 195
  civet, 204
  cloves, 211
  codonopsis, 221
  coriander, 228
  corn silk, 230
  costus oil, 231
  dill, 246
  elder flowers, 258
  elecampane, 260
  eleuthero, 262
  epimedium, 270
  eucalyptus, 273
  euphorbia, 276
  fangfeng, 282
  fennel, 285
  fenugreek, 288
  fo-ti, 296
  ganoderma, 303
  garlic, 310
  gelsemium, 313
  gentian, 317
  ginger, 322
  ginkgo, 327
  ginseng, 333
  gotu kola, 340
  hawthorn, 353
  honey, 358
  hops, 360
  horsetail, 368
  hydrangea, 370
  jasmine, 383
  Job's tears, 385
  jujube, 388–389
  kudzu root, 401
  lemongrass, 414
  licorice root, 417
  ligustrum, 420
  lobelia, 425
  lycium fruit, 430
  magnolia flower, 435
  mints, 445
  musk, 456
  mustard, 459
  onion, 475
  orange, 480
  oregano, 483
  patchouly oil, 494
  poria, 509
  psyllium, 513
  rehmannia, 523
  rhubarb, 526
  rosemary, 536
  royal jelly, 540
  rue, 543
  rutin, 546
  saffron, 548
  sandalwood oil, 554
  sarsaparilla, 556
  schisandra, 566
  skin conditions, 586
  sour jujube kernel, 572
  storax, 580
  sweet basil, 75
  tagetes, 582
  tamarind, 585
  tannic acid, 587
  tea, 593
  thyme, 596
  tienchi ginseng, 599
  turmeric, 605
  turpentine, 608
  valerian root, 614
  wild cherry bark, 183
  yarrow, 627
Chinese parsley, 227, 228, 488
Chinese restaurant syndrome, MSG, 453
CHIRATA, 192–193
Chirayta, see Chirata
Chiretta, see Chirata
Chishaoyao, 676–677
Chittem bark, see Cascara sagrada
Chloral hydrate, antagonists, ginseng, 332
Chlorophyll, source for, alfalfa, 14
Chocolate:
  imitation flavor, carob, 144
  ingredients:
    cocoa, 214
    mints, 445
    milk, carrageenan, 146
Cholagogue, 582
Cholera, remedies:
  black and white pepper, 499
  costus oil, 231
  echinacea, 255
  fennel, 285
  ginger, 322
  kalmegh, 650
Choleretic:
  absinthium, 2
  artichoke, 47
  bayberry bark, 80
  boldo leaves, 107
  calendula, 130
  chlorogenic acid, 223
  dandelion root, 238
  gentian, 316
  goldenseal, 337
  horehound, 362

immortelle, 379
mints, 445
orange, 479
oregano, 483
rose oil, 531
turmeric, 604
yarrow, 626
see also Chologogic
Cholesterol:
agents that elevate, 5, 10
agents that lower:
acacia, 5
alfalfa, 14
algin, 17
artichoke, 46–47
bilberry, 94
carrageenan, 146
doggrass, 248
fo-ti, 295
garlic, 309
ginger, 321
ginkgo, 326–327
guar gum, 347–348
honeysuckle flower, 668
onion, 475
orange, 478
pectin, 496
phellodendron bark, 678
psyllium, 511–512
quillaia, 518
saffron, 548
turmeric, 604
yucca, 634
retardation of biosynthesis, red sage, 680
see also CHEMICAL INDEX
Cholethiasis, see Gall stones
Cholinergic:
psyllium, 511
see also Anticholinergic activities
Chologogic:
boldo leaves, 108
sweet bay, 77
see also Choleretic
Chondrus crispus, **145**
Chondrus extract, see Carrageenan
Chorea, remedies, black cohosh, 98
Chrysanthemum parthenium, 289
Chuan baizhi, 663

Chuandang, 185
Chuanxiong, 682–683
Chutney, ingredients, tamarind, 584
Chymopapain, in disk disorders,
Cichorium intybus, **189**
Cilantro, 227, 228
Cimicifuga dahurica,
Cimicifuga foetida,
Cimicifuga racemosa, **97**
Cimicifuga, see Black cohosh
Cimicifuga simplex, 97
Cimicifuga species, 97
Cinchona calisaya, **194**
Cinchona flava, 194
Cinchona ledgeriana, **194**
Cinchona pubescens, 194
CINCHONA (RED AND YELLOW), 194–195
Cinchona rubra, 194
Cinchona species, 194
Cinchona succirubra, **194**
Cinchonism, 195
Cinnamomum aromaticum, **196**
Cinnamomum burmanii, **196–197**
Cinnamomum cassia, 196–200
Cinnamomum loureirii, **196**
Cinnamomum obtusifolium var. *loureirii*, 196
Cinnamomum verum, **196**
Cinnamomum zeylanicum, **196**
CINNAMON (AND CASSIA), 196–200
Circulation:
activation:
hawthorn, 353
nettle, 465
peony root, 677
roselle, 534
rosemary, 536
Sichuan lovage, 682, 683
sweet basil, 75
collapse, as toxic symptom, chenopodium oil, 180
disorders, remedies:
bilberry, 94
calendula, 131
ginkgo, 327
sweet woodruff, 623
impaired cerebral, as toxic symptom, ephedra, 267

Cirrhosis:
liver, remedies, jasmine, 383
milk thistle, 440–441
Cistaceae, 404
Ciste, see Labdanum
Cistus ladaniferus, **404**
Cistus species, 404
CITRONELLA OIL (CEYLON AND JAVA), 201–202
Citrus aurantifolia, **422**
Citrus aurantifolia, **422**
Citrus bergamia, **91–92**
Citrus bigaradia, 477
Citrus decumana, var. *racemose*, 343
Citrus limon, **410, 495**
Citrus limonum, 410, 495
Citrus medica var. *acida*, 422
Cirrus paradisi, **343**
Citrus racemosa, 343
Citrus sinensis, **477**
Citrus vulgaris, see *Viverra Civetta*
CIVET, 203–204
Ciwujia, 264
CLARY SAGE, 204–206
Clearing agent, in microscopy, cedarwood oil, 162–164
Clonic movements, caused by MSG, 453
Clotbur, see Burdock
CLOVER TOPS, RED, 206–208
CLOVES, 209–211
Cnicus benedictus, **100, 439**
CNS:
poison, aconite, 8
reaction to absinthe liqueur, 2
CNS depressant:
belladonna, 88
bitter almond oil, 22–23
calamus, 128
gotu kola, 340
lavender, 407
lemongrass, 413
lobelia, 425
musk, 456
peony bark, 675
tannic acid, 586
valerian root, 612–613

CNS stimulant:
  belladonna, 88
  coca, 213
  cocoa, 218
  codonopsis, 220
  coffee, 223
  ephedra, 266–267
  gelsemium, 313
  kola nut, 398
  lobelia, 425
  tea, 591
  tienchi ginseng, 599
  yohimbe, 604
COCA, 212–214
Cocaine plant, 212
COCILLANA BARK, 215–216
Cocktail mixes:
  foaming agent, quillaia, 519
  see also Beverages, alcoholic and nonalcoholic
COCOA (CACAO), 216–219
  substitute, carob, 142–144
Coconut flavor, ingredient, rue, 543
CODONOPSIS, 220–221
Codonopsis pilosula, **220**
Codonopsis species, 220
Codonopsis tangshen, **220**
Codonopsis tubulosa, **220**
Coffea arabica, **222**
Coffea canephora, **222**
Coffea robusta, 222
Coffea species, **222**
COFFEE, 222–224
  chicory root, 191
  substitute:
    asparagus, 53
    dandelion root, 239
    Job's tears, 385
    milk thistle, 441
Cohosh, see Black cohosh; Blue cohosh
Coix lachryma-jobi, **384**
Coix ma-yuen, 384
Cola acuminate,
Cola drinks:
  caffeine, 214
  flavor ingredients:
    cinnamon and cassia, 199
    coca, 214
    ginger, 322
    guarana, 350
    kola nut, 398

Cola nitida, **214, 398**
Cola nut, see Kola nut
Cola species, 398
Colds:
  prevention:
    fo-ti, 296
    rose hips, 529
  remedies:
    angelica, 33
    astragalus, 57
    balm of Gilead buds, 103
    bayberry bark, 81
    belladonna, 89
    blessed thistle, 101
    borage, 110
    burdock, 121
    cajeput oil, 126
    cangzhu, 662
    catnip, 160
    chaparral, 176
    chaste-tree, 178
    cinchona, 195
    cinnamon and cassia, 199
    coca, 214
    coffee, 223
    Dahurian angelica, 663
    echinacea, 255
    ephedra, 267
    eucalyptus, 273
    fangfeng, 282
    forsythia fruit, 665
    garlic, 310
    ginger, 322
    goldenseal, 337
    honey, 358
    honeysuckle flower, 667–668
    horehound, 363
    hyssop, 376
    ipecac, 380
    Kava, 397
    kudzu root, 401
    lemongrass, 414
    ligusticum, 670
    magnolia flower, 437
    mints, 445–446
    mustard, 459
    orange, 480
    patchouly oil, 494
    pine needle oil, 503
    roselle, 534
    rosemary, 537
    rue, 543

    saw palmetto, 563
    storax, 580
    sweet basil, 75
    tagetes, 582
    tamarind, 585
    turpentine, 608
    valerian root, 614
    vasaka, 656
    white pine bark, 504
    wild cherry bark, 182–183
    wintergreen oil, 619
    yarrow, 627
  see also Cough
Coleus species, 482
Colic, remedies:
  catnip, 160
  feverfew, 290
  hawthorn, 353
  juniper berries, 391
  lavender, 408
  tagetes, 582
  triphala, 654
  turmeric, 605
Colic root, see Aletris
Colitis, remedies, 491, 679
Collagen hydrolysis, decreased, bilberry, 94
Collagen films:
  ficin, 293
  papain, 293
Collapse, as toxic symptom, 43, 180, 194, 425, 596, 610
Colle du japon, see Agar
Color, see food colors
Colpidium colpoda, inhibition by barberry, 73
Coma:
  remedies, musk, 456
  as toxic symptom:
    aspidium, 54
    quillaia, 519
    thyme, 596
    turpentine, 608
Comedogenic properties, cocoa butter, 218
COMFREY, 225–226
Commiphora erythraea, **460**
Commiphora madagascariensis, **460**
Commiphora molmol, **460–461**
Commiphora myrrha, **460**
Commiphora species, **460**

Compositae, 1, 42, 45, 100–101, 109, 120, 129, 167, 169, 189–190, 230, 238, 240, 251, 259, 289, 378, 439, 576, 581, 588, 626–627, 660–661, 681
Composition powder, bayberry bark, 81
Compound benzoin tincture:
  contact dermatitis from, 90
  ingredients, 90
    aloe, 26
    balsam tolu, 71
    podophyllum, 506
    storax, 579
Compound cardamom spirit, 141
Compound white pine syrup, ingredients, 103, 183, 501
Concretion formation, as toxic symptom, rutin, 545
Condiments and relishes:
  colorants:
    annatto, 41
    capsicum, 134
    caramel color, 137
    greater galangal, 648
    turmeric, 605
  flavor ingredients:
    allspice, 20
    asafetida, 49
    black and white pepper, 499
    boronia absolute, 113
    buchu, 117
    cajeput oil, 126
    capsicum, 134
    caraway, 139
    cardamon, 141
    carob, 144
    carrot seed oil, 148
    cascarilla bark, 152
    cedar leaf oil, 162
    celery seed, 166
    chervil, 185
    cinchona, 195
    cinnamon and cassia, 199
    clary sage, 205
    cloves, 211
    coriander, 228
    cubebs, 234
    cumin, 236
    dill, 246
    elemi gum, 262

fennel, 284–285
galbanum, 300
garlic, 309
ginger, 322
labdanum, 405
marjoram, 438
mustard, 458–459
nutmeg and mace, 469
onion, 475
orange, 479–480
oregano, 483
origanum oil, 485
parsley, 488
rosemary, 536
sage, 551
savory, 560
sweet bay, 77
tagetes, 582
tarragon, 589
thyme, 596
turmeric, 605
turpentine, 608
West Indian bay, 79
gums:
  algin, 17
  guar gum, 348
  tragacanth, 603
MSG, 453
Condyloma, see Warts
Confectionery, see Candy
Confusion, as toxic symptom, 89, 543
Congestion, see Colds; Cough
Conjunctivitis, remedies, 155, 280, 383
Connective tissue injuries, remedies, horsetail, 368
Constipation:
  remedies:
    baizhu, 660
    bee pollen, 83
    fo-ti, 297
    honey, 358
    horsetail, 368
    ligustrum, 421
    psyllium, 512
    turmeric, 605
  as toxic symptom, belladonna, 89
see also Cathartic; Laxative
Consumptives weed, see Yerba Santa

Contraceptives, see Antifertility agents
Convulsions:
  adaptogenic activity, eleuthero, 263
  infantile, remedies, 172, 673
  as toxic symptom:
    absinthium, 2
    aspidium, 54
    cedar leaf oil, 161
    cinchona, 195
    ginkgo, 327
    lobelia, 425
    orange, 479
    quillaia, 519
    rue, 543
    thyme, 596
    uva ursi, 610
  see also Anticonvulsant
Copaiba, 67
Copaifera officinalis, **67**
Copaifera species, **67**
Copaiva, 67
Copper, see CHEMICAL INDEX
CORIANDER, 227–228
Coriandrum sativum, **227**
Coridothymus capitatus, 485
Cornaceae, 249
Cornea, opacity, remedies, pearl, 673
Cornmint, 443–446
Corns, remedies, papain,
CORN SILK, 229–230
Corn syrup, caramel color, 136
Cornus florida, 249
Coronary dilation, see Vasodilation, coronary
Coronary heart disease:
  preventative, evening primrose, 277
  remedies:
    bilberry, 94
    borage, 111
    epimedium, 270
    ganoderma, 303
    ginkgo, 327
    Goldenseal, 338
    Gokshura, 646
    safflower, 682
    Sichuan lovage, 683
    tienchi ginseng, 600
  see also cardiac entries

Cortex chinae, *see* Cinchona
  (red and yellow)
Corticosteroids, urinary,
    reduction by Bulgarian
    rose oil, 531
Corynanthe johimbi, 631
Cosmetics:
  colorants, annatto, 41
  fragrance ingredient,
    allspice, 20
  ingredients:
    alkanet, 18–19
    almond oil, 23
    arnica, 43–44
    balsam copaiba, 67
    benzoin, 91
    bloodroot, 103
    castor oil, 155
    chamomile, 172
    echinacea, 255
    gentian, 317
    jojoba, 386–387
    locust bean guns, 427
  see also Chinese cosmetic
      ingredients; Creams
      and lotions; Ointments;
      Perfumes; Soaps and
      detergents
COSTUS OIL, 230–231
Couch grass, *see* Doggrass
Cough:
  remedies:
    althea root, 30
    angelica, 33
    anise, 38
    asafetida, 49
    ashwagandha, 645
    Baikal skullcap, 685–686
    balm of Gilead buds, 65
    balsam copaiba, 67
    balsam tolu, 71
    bloodroot, 103
    burdock, 121
    carrageenan, 146
    cedar leaf oil, 162
    cedarwood oil, 164
    chaparral, 176
    Cherokee rosehip, 181
    cherry laurel leaves, 184
    chestnut leaves, 186
    cocillana bark, 216
    cocoa, 218
    codonopsis, 221

comfrey, 226
costus oil, 231
damiana, 237
ephedra, 267
eucalyptus, 273
euphorbia, 275
evening primrose, 278
eyebright, 280
fennel, 284
fenugreek, 287
ganoderma, 303
garlic, 310
giant knotweed, 668
ginger, 322
ginkgo, 327
Gokshura, 646
greater galnanga, 648
honey, 358
horehound, 363
hydrangea, 370
hyssop, 376
immortelle, 379
ipecac, 380
kava, 397
licorice root, 417
lobelia, 425
lycium fruit, 432
marjoram, 438
mints, 445
myrrh, 462
orange, 480
oregano, 483
parsley, 488
pine needle oil, 503
poria, 509
psyllium, 513
quillaia, 519
red clover tops, 208
saffron, 548
saw palmetto, 563
schisandra, 566
squill, 575
storax, 580
tagetes, 582
tannic acid, 587
thyme, 596
tienchi ginseng, 600
white pine bark, 501
wild cherry bark, 182–183
as toxic symptom, turpentine,
    608
see also Antitussive;
    Whooping cough

Coumarins, psoriasis and vitiligo
    remedies, 33, 93
  see also CHEMICAL INDEX
Counterirritant:
  capsicum, 134
  cedar leaf oil, 161
  cinnamon and cassia,
      199
  cloves, 211
  lobelia, 425
  magnolia flower, 435
  mustard, 458
  sweet birch oil, 96
  thyme, 596
  turpentine, 607
  wintergreen oil, 619
  see also Rubefacient
Cramps, remedies, Baikal
    skullcap, 684, 686
CRANBERRY, 232–233
Crataegus cuneata, **352**
Crataegus laevigata, **352**
Crataegus monogyna, **352**
Crataegus oxyacantha, 352
Crataegus pinnatifida, **352**
Crataegus **species, 352**
Creams and lotions:
  antiseptics:
    balsam Canada, 66
    cade oil, 123
    chamomile, 172
  emollients:
    almond oil, 23
    aloe, 27
    castor oil, 155
    cocoa, 218
    comfrey, 226
  fixatives:
    balsam Canada, 66
    balsam copaiba, 67
    balsam tolu, 71
    benzoin, 91
    castoreum, 157
    cedarwood oil, 164
    civet, 204
    costus oil, 231
    deertongue, 241
    elemi gum, 261
    galbanum, 300
    lime oil, 423
    myrrh, 461
  fragrance ingredients:
    absinthium, 2

General Index

ambrette seed, 31
angelica, 33
anise, 38
balsam Canada, 66
balsam copaiba, 67
balsam Peru, 70
balsam tolu, 71
bayberry bark, 81
beeswax, 85
bois de rose oil, 106
cede oil, 123
cajeput oil, 126
calamus, 128
cananga oil, 132
caraway, 139
cardamon, 141
carrot seed oil, 148
cascarilla bark, 151–152
castoreum, 157
cedar leaf oil, 162
cedarwood oil, 164
celery seed, 166
centaury, 168
chamomile, 172
chenopodium oil, 180
cinnamon and cassia, 199
civet, 204
clary sage, 205
cloves, 210
coriander, 223
costus oil, 231
cubebs, 234
cumin, 236
deertongue, 241
dill, 246
elecampane, 260
elemi gum, 261
eucalyptus, 273
fennel, 284
fenugreek, 287
galbanum, 299–300
genet, 315
geranium oil, 319
ginger, 322
grapefruit oil, 344
guaiac wood oil, 346
honey, 358
hyssop, 376
jasmine, 383
juniper berries, 390
labdanum, 405
lavender, 408
lemongrass, 413

lemon oil, 411
lime oil, 423
lovage root, 428
marjoram, 438
mints, 445
myrrh, 461
nutmeg, 468
olibanum, 472
orange, 479
origanum oil, 485
parsley, 488
patchouly oil, 494
pine needle oil, 503
rosemary, 536
rose oil, 531
rue, 543
sage, 551
sandalwood oil, 554
sweet bay, 77
sweet birch oil, 96
tarragon, 589
thyme, 596
West Indian bay, 79
ylang ylang oil, 630
gums and wax:
  algin, 17
  Beeswax, 85
  carrageenan, 146
  guar gum, 348
  karaya gum, 394
  tragacanth, 603
  xanthan gum, 624
other ingredients:
  astragalus, 57
  avocado, 60
Baikal scullcap, 685, 686
  bee pollen, 83
  bergamot oil, 92
  bletilla tuber, 661
  borage, 111
  bromelain, 114
  calendula, 130
  chamomile, 172
  dittany bark, 664–665
  elder flowers, 258
  eleuthero, 263
  evening primrose, 278
  fangfeng, 282
  fo-ti, 296
  forsythia fruit, 665
  ganoderma, 303
  garden burnet, 666
  giant knotweed, 669

  ginkgo, 327
  ginseng, 333
  gotu kola, 340
  honeysuckle flower, 668
  hops, 360
  horse chestnut, 365
  horsetail, 368
  hypericum, 373
  jojoba, 386–387
  ligusticum, 670
  luffa, 671
  magnolia flower, 435
  mume, 672
  nettle, 465
  quillaia, 519
  papain,
  pearl, 674
  pectin, 497
  peony bark, 675
  peony root, 677
  phellodendron bark, 678
  poria, 509
  red sage, 681
  rehmannia, 523
  royal jelly, 540
  safflower, 682
  Sichuan lovage, 683
  sour jujube kernel, 573
  tienchi ginseng, 600
  turpentine, 608
  witch hazel, 621
  see also Ointments
Creme de cacao:
  flavor ingredients, cocoa, 218
  see also Liqueurs
Criollo cocoa, 216–217
Cristaria *plicata*, 673
***Crocus sativus*, 547**
***Croton eluteria*, 151**
Croup, see Laryngitis
Cruciferae, 457
Cubeba officinalis, 233
Cubeba, 233
CUBEBS, 233–234
Cucumber, increased yield by triacontanol, 85
Cucurbitaceae, 671
Culture media, agar, 11
CUMIN, 235–236, see also Anise (and star anise)
***Cuminum cyminum*, 235**
Cuminium odorum, 235

Cupressaceae, 161–162, 389
Curaçao aloe, 24
Curcuma domestica, 603
***Curcuma longa*, 603**
Curcuma, *see* Turmeric
Curry powder, spices:
  cardamon, 141
  cumin, 236
  fenugreek, 287
  tamarind, 584
  turmeric, 605
Cusparia febrifuga, 34
Cusparia, *see* Angostura bark
Cusparia trifoliata, 34
Cutch, *see* Catechu (black and pale)
Cuts, remedies:
  aloe, 27
  balm of Gilead buds, 65
  balsam Canada, 66
  balsam tolu, 71
  benzoin, 91
  see also skin entries
Cyamopsis psoralioides, 347
***Cyamopsis tetragonoloba*, 347**
***Cymbopogon citratus*, 412**
***Cymbopogon flexuosus*, 412**
***Cymbopogon nardus*, 201**
***Cymbopogon winterianus*, 201**
***Cynara scolymus*, 45**
Cynips gallae-tinctoriae, 586
Cyste, *see* Labdanum
Cystitis:
  remedies:
    balsam copaiba, 68
    buchu, 118
    corn silk, 230
    hops, 360
    kava, 396
    uva ursi, 611
  as toxic symptom, fo-ti, 296
  see also Bladder; Urinary disorders
Cytisus scoparius, **115**, 315
Cytotoxicity:
  Baikal skullcap, 685–686
  boneset, 109
  chaparral, 175
  coriander, 228
  eugenol, 197
  fennel, 284
  ganoderma, 303

greater galnanga, 647
mahabala, 651
mints, 445
mistletoe, 449
neem, 652
quillaia, 518–519
sweet basil, 75
yucca, 634

DAHURIAN ANGELICA (BAIZHI), 663–664
Dairy products:
  colorants, annatto, 41
  evaporated milk, carrageenan, 146
  flavoring ingredients:
    agar, 10
    coffee, 224
    grapefruit oil, 344–345
    oregano, 483
  gums:
    algin, 17
    carrageenan, 146
    guar gum, 348
    karaya gum, 394
    locust bean gum, 426
    xanthan gum, 624
  MSG, 453
  see also Desserts, frozen dairy
***DAMIANA*, 237**
DANDELION ROOT, 238–240
Dandruff, *see* Antidandruff
"Danggui", 33
Danshen, 680
Datura species, 89
***Daucus carota*, 147**
Deadly nightshade, *see* Belladonna
Deafness:
  remedies, kudzu root, 401
  temporary, as toxic symptom, 195, 303, 401
Death, *see* Poisoning, fatal
Debility, remedies, goldenseal, 338
Decaffeinated coffee, 222
Deer musk, *see* Musk
DEERTONGUE, 240–241
Delirium, as toxic symptom:
  aspidium, 54
  belladonna, 89
  Cinchona, 195

  coffee, 223
  uva ursi, 610
Delusions, nutmeg causing, 468
Dementia, remedies, 327
Demulcent:
  acacia, 5
  algin, 17
  almonds, 22
  althea root, 29
  borage, 112
  carrageenan, 145–147
  cassie absolute, 152–153
  chickweed, 187
  comfrey, 225
  corn silk,
  deertongue, 241
  honey, 358
  licorice root, 417
  oregano, 483
  Psyllium, 512
  tragacanth, 603
  white pine bark, 501
  see also Emollient
Dental materials and preparations, ingredients:
  agar, 10
  algin, 17
  balsam Canada, 66
  balsam Peru, 70
  benzoin, 91
  bletilla tuber, 661
  bloodroot, 103
  cajeput oil, 126
  cloves, 210
  eucalyptus, 273
  karaya gum, 394
  neem, 652
  peony bark, 675
  thyme, 596
  see also Gum problems; Toothache; Toothpastes
Dentin, hypersensitive, garlic, 310
Deodorants, ingredients, sour jujube kernel, 573
Depressant:
  antagonism by ginseng, 332
  olibanum, 471
  see also Cardiac depressant; CNS depressant; Respiratory depressant

Depression:
- ashwagandha, 645
- kava contraindication, 396
- remedies, *see* Antidepressant,
- as toxic symptom, horse chestnut, 365

Dermatitis:
- contact:
  - benzoin, 90
  - black and white pepper, 498–499
  - lavender, 408
  - origanum oil, 485
  - prevention, garden burnet, 666
  - rose oil, 530–532
  - tea, 592
- neurogenic, remedies, turpentine, 608
- remedies, 130, 512, 608
- suppurative, remedies, giant knotweed, 668
- as toxic symptom:
  - arnica,
  - artichoke, 46
  - balsam Peru, 70
  - bromelain, 113–114
  - celery seed, 166
  - chamomile, 171–172
  - chicory root, 190
  - cinchona, 195
  - cinnamon and cassia, 198–199
  - citronella oil, 202
  - costus oil, 231
  - dandelion root, 238
  - ficin, 291–293
  - garlic, 309
  - geranium oil, 319
  - ginkgo, 327
  - hops, 360
  - juniper berries, 390
  - licorice root, 417
  - marjoram, 438
  - mints, 445
  - orange, 479
  - papain,
  - parsley, 488
  - sandalwood oil, 554
  - sweet bay, 77
  - tagetes, 582
  - yarrow, 626

see also Allergy; skin entries

Desserts, frozen dairy:
- colorants, 605
- flavor ingredients:
  - absinthium, 2
  - alfalfa, 14
  - allspice, 20
  - almonds, 23
  - althea root, 30
  - ambrette seed, 31
  - angelica, 33
  - angostura bark, 35
  - anise, 38
  - arnica, 44
  - asafetida, 49
  - balm, 63
  - balsam Canada, 66
  - balsam copaiba, 68
  - balsam Peru, 70
  - balsam tolu, 72
  - beeswax, 85
  - benzoin, 91
  - bergamot oil, 92
  - black and pale catechu, 158
  - black and white pepper, 499
  - blackberry bark, 96
  - bois de rose oil, 106
  - boronia absolute, 113
  - buchu, 117
  - cajeput oil, 126
  - cananga oil, 132
  - capsicum, 134
  - caraway, 139
  - cardamom, 141
  - carob, 144
  - carrot seed oil, 148
  - cascara sagrada, 150
  - cascarilla bark, 152
  - cassie absolute, 153
  - castoreum, 157
  - castor oil, 155
  - cedar leaf oil, 162
  - celery seed, 166
  - chamomile, 172
  - cherry laurel leaves, 184
  - chervil, 185
  - chicory root, 190–191
  - cinchona, 195
  - cinnamon and cassia, 199
  - citronella oil, 202
  - civet, 204
  - clary sage, 205
  - cloves, 211
  - coca, 214
  - cocoa, 218
  - coffee, 224
  - coriander, 228
  - corn silk, 230
  - costus oil, 231
  - cubebs, 234
  - cumin, 236
  - dandelion root, 239
  - dill, 246
  - doggrass, 248
  - elder flowers, 258
  - elecampane, 260
  - elemi gum, 262
  - eucalyptus, 273
  - fennel, 285
  - fenugreek, 287
  - galbanum, 300
  - garlic, 309
  - genet, 315
  - gentian, 317
  - geranium oil, 319
  - ginger, 322
  - grapefruit oil, 345
  - hops, 360
  - horehound, 363
  - immortelle, 379
  - jasmine, 383
  - juniper berries, 391
  - kola nut, 399
  - labdanum, 405
  - lavender, 408
  - lemongrass, 414
  - lemon oil, 411
  - licorice root, 417
  - lime oil, 423
  - lovage root, 428
  - marjoram, 438
  - mints, 445
  - musk, 456
  - mustard, 459
  - myrrh, 462
  - nutmeg and mace, 469
  - olibanum, 472
  - onion, 475
  - orange, 479
  - origanum oil, 485
  - parsley, 488
  - passion flower, 491
  - patchouly oil, 494
  - pine needle oil, 503
  - prickly ash,
  - quassia, 516

Desserts, frozen dairy
(*Continued*)
    quebracho, 517
    red clover tops, 208
    rhubarb, 526
    roselle, 533
    rosemary, 536
    rose oil,106
    rue, 543
    sage, 551
    sandalwood oil, 554
    sarsaparilla, 556
    storax, 579
    sweet bay, 77
    sweet birch oil, 96
    sweet woodruff, 623
    tagetes, 582
    tamarind, 585
    tannic acid, 587
    tarragon, 589
    tea, 592
    thyme, 596
    turpentine, 608
    valerian root, 613
    vanilla, 617
    West Indian bay, 79
    wild cherry bark, 183
    wintergreen oil, 619
    yerba santa, 629
    ylang ylang ail, 630
fragrance ingredients,
    pipsissewa, 504
guaiac wood oil, 346
gums, wax and other
    ingredients:
    algin, 17
    beeswax, 85
    carrageenan, 146
    guar gum, 348
    karaya gum, 394
    locust bean gum, 427
    pectin, 497
    psyllium, 512
    quillaia, 519
    tragacanth, 603
sweeteners, stevia, 577
see also Dairy products
Detergents, *see* Soaps and
    detergents
Detoxicant:
    aconitine, honey, 358
    Baikal skullcap, 685
    fo-ti, 297

forsythia fruit, 665
garden burnet, 666
honeysuckle flower, 667
luffa, 671
mume, 672
pearl, 673
phellodendron bark, 678
purslane, 679
rehmannia, 523
Devil's apple, *see* Asafetida;
    Podophyllim
    (podophyllin)
DEVIL'S CLAW, 242–243
Dewberry bark, *see* Blackberry
    bark
Dhanvantri Sampradaya, 637
Dhari, 638
Diabetes, remedies:
    astragalus, 57
    bilberry, 94
    borage, 111
    codonopsis, 221
    corn silk, 230
    doggrass, 248
    evening primrose, 277
    ginseng, 332
    kudzu root, 401
    licorice, 417
    ligustrum, 421
    lycium fruit, 432
    mume, 672
    purslane, 679
    rehmannia, 523
    royal jelly, 540
    rutin, 545
    stevia, 578
Diagnostic reagents, ingredients,
    346
Diaper rash, remedies, 70, 83
Diaphoretic:
    aconite, 8
    angelica, 33
    arnica, 44
    balm, 62–63
    black and white pepper, 499
    boneset, 109
    burdock, 120
    cajeput oil, 124–126
    cassie absolute, 152–153
    catnip, 160
    citronella oil, 202
    deertongue, 241
    elder flowers, 258

ephedra, 266–267
fangfeng, 282
garlic, 309
ginger, 322
horehound, 363
hydrangea, 370
hyssop, 376
ipecac, 380
kava, 397
kudzu root, 401
lobelia, 425
marjoram, 437
oregano, 483
pipsissewa, 504
prickly ash, 51
saffron, 548
sage, 551
sarsaparilla, 556
sassafras, 558
sweet bay, 77
sweet woodruff, 623
tagetes, 582
thyme, 596
Diarrhea:
    remedies:
    acerola, 6
    aletris, 11–12
    alkanet, 19
    angostura bark, 35
    astragalus, 57
    avocado, 60
    Baikal skullcap, 85–86
    baizhu, 660
    bayberry bark, 81
    bee pollen, 83
    beeswax, 85
    bilberry, 94
    black and pale catechu, 158
    blackberry bark, 96
    black haw bark, 100
    cangzhu, 662
    capsicum, 134
    carob, 143–144
    catnip, 160
    chamomile, 172
    chaparral, 174–176
    Cherokee rosehip, 181
    cinnamon and cassia, 199
    cloves, 211
    cocoa, 218
    codonopsis, 221
    comfrey, 223
    elecampane, 260

fo-ti, 296
garden burnet, 266
garlic, 309–310
gentian, 317
geranium oil, 319
ginger, 322
ginkgo, 327
goldenseal, 338
guarana, 351
hawthorn, 353
hops, 360
horse chestnut, 365
hypcricum, 373
Job's tears, 385
jujube, 389
kola nut, 399
kudzu root, 401
labdanum, 405
ligusticum, 670
marjoram, 436–438
mints, 446
mume, 472
nettle, 465
nutmeg and mace, 468–469
orange, 480
oregano, 483
papain,
patchouly oil, 494
pectin, 497
poria, 509
prickly ash, 54
psyllium, 512
rhubarb, 526–527
sage, 551
savory, 560
schisandra, 566
tannic acid, 587
triphala, 654
thyme, 596
wild cherry bark, 183
witch hazel, 621
as toxic symptom:
aloe, 26
arnica, 43
aspidium, 54
balsam copaiba, 67
bromelain, 114
cinchona, 195
eucalyptus, 273
ipecac, 380
quillaia, 519
saffron, 548
vasaka, 656

Dichroa. fegrifirga, 73
Dicramnus albus L.var. dasycarpus, 664
Dictamnus dasycarpus, 664
Digestive aid:
blessed thistle, 101
bromelain, 114
Cherokee rosehip, 181
dill, 246
fenugreek, 287
feverfew, 290
ficin, 292
gentian, 317
ginger, 322
hawthorn, 353
papain,
Sage, 551
tagetes, 582
see also Indigestion
Digestive disorders, remedies:
chicory root, 191
cinchona, 195
clary sage, 205
Euphorbia, 275
hyssop, 376
lovage root, 428
sarsaparilla, 556
see also Heartburn; Indigestion
Digestive enzymes, inhibition, carob, 142–144
Digestive tonic, *see* Tonic
Digitalis glutinosa, 521
DILL AND INDIAN DILL, 244–246
Diluents, 292, 333, 342, 496, 697–699
Diosma, *see* Buchu
Diphtheria, remedies, 255, 310
Diplopia, as toxic symptom, kava, 396
Disinfectant, *see* Antiseptic
Disintegrating agents, in tablets, 17, 348
Disk disorders, degenerative, chymopapain in treating,
Disorientation, nutmeg causing, 468
DITTANY BARK (BAIXIANPI), 664
Diuretic:
aletris, 12
alfalfa, 14

arnica, 44
artichoke, 47
Asparagus, 53
astragalus, 57
Baikal skullcap, 685
baizhu, 660
balsam copaiba, 67
bee pollen, 83
bilberry, 93
black and white pepper, 499
black haw bark, 100
blessed thistle, 101
blue cohosh, 105
boldo leaves, 108
borage, 112
broom tops, 116
buchu, 117, 118
burdock, 121
carrot seed, 148
cedar leaf oil, 162
celery seed, 166
chaparral, 176
chervil, 185
chicory root, 191
citronella oil, 202
cocoa, 218
coffee, 223
corn silk, 229, 230
cranberry, 233
cubebs, 234
cumin, 236
damiana, 237
dandelion root, 238–240
dill, 246
doggrass, 248
elder flowers, 258
elecampane, 260
eleuthero, 263, 264
ephedra, 266, 267
forsythia fruit, 665
ganoderma, 303
Garlic, 309
Genet, 315
goldenseal, 337
gokshura, 646
guarana, 351
hawthorn, 353
Hops, 360
horehound, 363
horsetail, 367–369
hydrangea, 370
immortelle, 379
Job's tears, 385

Diuretic (*Continued*)
juniper berries, 390, 391
kava, 395, 397
kola nut, 399
lavender, 408
lemon oil, 411
ligustrum, 421
lobelia, 425
lovage root, 428
marjoram, 437
mustard, 459
nettle, 464, 465
olibanum, 472
onion, 475
oregano, 483
papain,
peony bark, 675
pipsissewa, 504
poria, 509
psyllium, 512
purslane, 679
quebracho, 517
red clover tops, 208
rehmannia, 522
rose hips, 529
roselle, 533
sandalwood oil, 554
sarsaparilla, 556
savory, 560
saw palmetto, 562
senna, 570
squill, 575
sweet basil, 75
sweet woodruff, 623
tagetes, 582
tarragon, 589
tea, 591, 593
uva ursi, 610, 611
white pine bark, 501
see also Antidiuretics
Diverticulosis, remedies, psyllium, 512
Diyu, 666
Dizziness:
remedies:
fo-ti, 297
ginseng, 333
ligustrum, 421
lycium fruit, 432
peony root, 676
as toxic symptom:
aspidium, 54
ganoderma, 303

ipecac, 380
saffron, 548
thyme, 596
DNA, ginseng, biosynthesis, 332
Dog bites, remedies, 425
DOGGRASS, 247–248
Dog repellents, ingredients, mustard, 459
DOGWOOD, JAMAICAN, 249, *see also* Buckthorn, alder
Dolichos lobatus, 399
Dong cangzhu, 661
Dongdang, 220
Doshas, 638, 639
Doshic imbalance
Double vision, *see* Visual disturbances
Dough, ingredients,
Dropsy, remedies, 98, 102, 103, 166, 216, 488, 645, 654
ashwagandha, 645
Drug absorption, sarsaparilla, 556
Drug centaurium, *see* Centaury
***Dryopteris felix-mas*, 54**
***Dryopteris marginalis*, 54**
Dry process coffee, 222
Dry skin, *see* Moisturizer; skin entries
Dry socket, remedies, 70, 126, 210
Dutch process cocoa, 217
Dwarf pine needle oil, 502, 503
Dyes, henna, 356
Dysentery, remedies:
acerola, 7
angostura bark, 35
avocado, 60
bayberry bark, 81
bilberry, 94
carrot seed, 148
chaparral, 174–176
Comfrey, 226
coriander, 228
costus oil, 231
cubebs, 234
eucalyptus, 273
euphorbia, 275
garlic, 310
geranium oil, 319
ginger, 321
ginkgo, 327
ginseng, 333
gotu kola, 340

hawthorn, 353
henna, 360
hops, 360
horsechestnut, 365
hypericum, 373
ipecac, 380
jasmine, 383
kudzu root, 401
nettle, 465
phellodendron bark, 678
purslane, 678
sarsaparilla, 556
tagetes, 582
tannic acid, 587
tea, 592
Dysmenorrhea remedies:
aletris, 12
allspice, 19–21
bee pollen, 83
black cohosh, 98
castoreum, 157
dill, 246
goldenseal, 336–338
horse chestnut, 364–366
olibanum, 471–472
peony bark, 675
red sage, 680–681
safflower, 681–682
saffron, 547–548
Sichuanlovage, 683
Dyspepsia, *see* Indigestion
Dysphoric mood, remedies, hypericum, 372
Dyspnea, as toxic symptom, ipecac, 380
Dysrythmia, hawthorn, 353
Dysuria, remedies:
ashwagandha, 645
chaparral, 176
dill, 246
forsythia fruit, 665
parsley, 488
poria, 509
saw palmetto, 562

Earache, remedies:
eyebright, 280
feverfew, 289, 290
garlic, 310
mints, 310
purslane, 679
East African cedarwood oil, 162–164

Eastern red cedar, 163
Eastern white cedar, 161
Eastern White pine, *see* Pine bark, White
East Indian balmony, *see* Chirata
East Indian geranium oil, 318–320
ECHINACEA, 251–256
Echinacea angustifolia, 251–256
Echinacea pallida, **251–256**
Echinacea purpurea, **251–256**
Eczema:
  prevention, garden burnet, 666
  remedies:
    bee pollen, 81
    borage, 111
    burdock, 120, 121
    Cade oil, 123
    calendula, 131
    cangzhu, 662
    chervil, 185
    chickweed, 187
    Dahurian angelica, 663
    dandelion root, 239
    dittany bark, 664
    echinacea, 255
    evening primrose, 277
    garden burnet, 666
    horsechestnut, 365
    Job's tears, 385
    mume, 672
    phellodendron bark, 678
    purslane, 678
    red sage, 681
    rehmannia, 522
    tagetes, 582
  see also skin entries
Edema, remedies:
  astragalus, 57
  bilberry, 94
  bletilla tuber, 662
  bromelain, 114
  cangzhu, 662
  chickweed, 192
  Dahurian angelica, 663–664
  deertongue, 241
  eleuthero, 263
  ephedra, 266
  forsythia fruit, 670
  garden brunet, 666
  ginkgo, 326
  gotu kola, 339–341

horsechestnut, 364
horsetail, 368
Job's tears, 385
papain,
poria, 509
rutin, 545
tienchi ginseng, 599
turmeric, 603–605
Egg production, depression of, alfalfa, 13
Egg products:
  additive, algin, 17
  colorants, 605
Egyptian medicine:
  castor oil, 155
  centaury, 168
  fenugreek, 287
  roselle, 532
Egyptian privet, *see* Henna
ELDER FLOWERS (AMERICAN AND EUROPEAN), 257–258
Eldrin, *see* Rutin
ELECAMPANE, 259–260
***ELEMI GUM*, 261–262**
Elephantiasis, remedies, gotu kola, 340
Elettaria cardamomum, **141, 142**
ELEUTHERO, 262–264
Eleutherococcus senticosus, **262**
Elytrigia repens, **247**
Embalming liquid, olibanum, 472
Embryotoxic, 384
Emetic herb, *see* Lobelia
Emetics:
  bayberry bark, 80
  boneset, 109
  broom tops, 116
  chamomile, 169–173
  chaparral, 176
  cocillana bark, 215
  honey, 358
  hydrangea, 370
  ipecac, 380
  lobelia, 424
  MSG, 453
  mustard, 459
  squill, 575
  see also Antiemetic
Emmenagogues:
  absinthium, 2

  avocado, 60
  balm, 62–63
  blue cohosh, 105
  carrot seed, 148
  cedar leaf oil, 161
  celery seed, 166
  chaste-tree, 178
  citronella oil, 202
  cumin, 236
  eucalyptus, 271–273
  feverfew, 290
  lovage root, 428
  myrrh, 462
  olibanum, 472
  oregano, 483
  parsley, 488
  rue, 543
  Saffron, 548
  tagetes, 582
  tarragon, 589
  see also Menstruation
Emollient:
  almonds, 22
  aloe, 27
  avocado, 59–60
  borage, 112
  castor oil, 154
  chickweed, 187
  cocoa, 218
  comfrey, 225
  honey, 358
  jojoba, 386
  mahabala, 651
  psyllium, 512
  see also Demulcent
Emulsifying agents:
  agar, 9–11
  algin, 16–17
  beeswax, 85
  carrageenan, 146
  tragacanth, 603
  xanthan gum, 624
Enema, honey in, 358
Enocianina, *see* Grape skin extract (enocianina)
Enteritis, remedies:
  chaparral, 174–176
  Cherokee rosehip, 181
  hawthorn, 353
  horsechestnut, 365
  Job's tears, 385
  phellodendron bark, 678
  purslane, 678

Enrerococcus, inhibition by royal jelly, 538–540
Enuresis, remedies, 181, 327
Enzyme:
  cleaner for soft lenses, 66
  see also CHEMICAL INDEX
EPHEDRA, 265–267
Ephedraceae, 265
***Ephedra equisetina*, 265**
***Ephedra intermedia*, 265**
***Ephedra sinica*, 265**
***Ephedra species*, 265**
Epidermophytom interdigitale, inhibition by echinacea, 253
Epilepsy, remedies:
  calamus, 128
  passion flower, 491
  pearl, 673
  Pipsissewa, 504
  storax, 580
  turmeric, 605
EPIMEDIUM, 268–270
***Epimedium brevicornum*, 268**
***Epimedium sagittatum*, 268, 269**
***Epimedium species*, 269**
***Epimedium wushanense*, 268**
Equalizing activities, valerian root, 612
***Equisetum* arvense, 367–369**
***Equisetum hymale*, 367, 368**
Ericaceae, 93, 232
***Eriodictyon californicum*, 628**
Eriodictyon glutinosum, 628
Eriodictyon, *see* Yerba santa
Erysipelas, remedies, forsythia fruit, 665
Erythema, as toxic symptom:
  anise, 37
  rue, 543
  turpentine, 608
  see also Dermatitis
Erythraea centaurium, 167
Erythroxylaceae, 212
Erythroxylon coca, 212
***Erythroxylum coca*, 212**
***Erythroxylum novogranatense*, 212, 213**

Escherichia coli:
  antiadhesion activity, cranberry, 232
  inhibition:
    barberry, 73
    cinnamon, 197
    echinacea, 253
    honey, 358
    immortelle, 378
    labdanum, 404
    pectin, 496
    royal jelly, 539
Esophageal obstruction, 348
Estragon oil, 588
Estrogenic:
  aletris, 12
  alfalfa, 12
  anethole, 37, 284
  eleuthero, 263
  Fennel, 284
  gelsemium, 312–313
  licorice root, 416
  red clover tops, 207
  red sage, 680
Eubatus,
Eucalyptol, *see* CHEMICAL INDEX
EUCALYPTUS, 271–273
***Eucalyptus globulus*, 271**
Eucalyptus macrorhyncha, 545
Eucarya spicata, 553
***Euchema* species**,
Eugenia aromatica, 209
Eugenia caryophyllata, 209
Eugenia pimenta, 19
Eupatorium japonicum, 494
***Eupatoriun perfoliatum*, 109**
Eupatorium, *see* Boneset
Eupatorium rebaudianum, 576
Eupatorium species, 109
Euphrasia officinalis, 279
Euphrasia rostkoviana, 279, 280
EUPHORBIA, 274–276
Euphorbia capitata, 274
Euphorbiaceae, 151, 154, 274
Euphorbia hirta, 274
Euphorbia humifusa, 276
Euphorbia lathyris, 276
Etrphorbia lunulata, 276
Euphorbia pekinensis, 276
***Euphrorbia pilulifera*, 274**
Euphorbia sieboldiana, 276
Euphorbia species, 276

Euphoria, nutmeg causing, 467
Euphrasia officinalis, 279
***Euphrasia rostkoviana*, 279**
***Euphrasia species*, 279**
European beaver. *see* Castoreurn
Evaporated milk, *see* Dairy products
EVENING PRIMROSE, 276–278
Evening trumpet flower, *see* Gelsemium
Exanthema, as toxic symptom, kava, 396
Excipient, acacia, 5
Excitement, substances causing, 223
Exhaustion, *see* Fatigue
Expectorant:
  angelica, 33
  anise, 37
  asafetida, 49
  balm of Gilead buds, 65
  balsam copaiba, 67-68
  balsam tolu, 71
  benzoin, 90
  borage, 112
  cajeput oil, 126
  caraway, 139
  cedar leaf oil, 161
  chaparral, 174–176
  chervil, 185
  cocillana bark, 215
  cocoa, 218
  comfrey, 225
  cubebs, 234
  doggrass, 248
  elemi gum, 262
  epimedium, 269
  eucalyptus, 272
  fennel, 284
  galbanum, 300
  garlic, 309
  horehound, 362
  immortelle, 379
  ipecac, 379
  kava, 395
  labdanum, 405
  licorice root, 417
  lobelia, 425
  lovage root, 428
  myrrh, 462
  onion, 475

oregano, 483
quillaia, 519
red clover tops, 207
saffron, 547–548
savory, 560
saw palmetto, 562
schisandra, 564–567
squill, 575
storax, 579
tea, 593
thyme, 594
while pine bark, 501
yerba Santa, 629
Expressed bergamot oil, 92
Extrasystoles, caffeine causing, 223
Eye balm, see Goldenseal
Eye-brightening, mother of pearl, 673–674
EYEBRIGHT, 204, 279–280, *see* also Clary sage
Eye disorders:
conjunctivitis, remedies:
eyebright, 280
jasmine, 382–383
degenerative retinal conditions, remedies, bilberry, 94
irritation, inflammation, tiredness:
remedies:
Clary sage, 204-206
eyebright, 280
goldenseal, 336
red clover tops, 207
tagetes, 581–582
tea, 590–593
witch hazel, 621
as toxic symptom:
capsicum, 134
castor oil, 154
marjoram, 438
podophyllum, 506
Eye products, *see* Ophthalmic product
Eye root, *see* Goldenseal

Fabaceae, *see* Leguminosae
Fabrics, desizing, proteases used, 114
"Face-lift" formulations, proteases used,
Face powder, corn silk, 230

Facial dark spots, remedies, magnolia flower, 435
Facial mask, peelable, ingredients, 5, 14, 17
Fagaceae, 186
Fagopyrum esculentum, 82, 545
Faintness, as toxic symptom, ipecac, 379–381
FANGFENG, 281–282
Fatigue:
remedies:
astragalus, 57
cangzhu, 662
coca, 214
ganoderma, 303
ginseng, 333
gotu kola, 339
jujube, 389
mental, epimedium, 270
peony root, 677
schisandra, 566
valerian root, 614
as toxic symptom, rue, 543
Fats and oils:
colorants:
alkanet, 19
annatto, 41
caramel color, 136
turmeric, 605
excretion, increase with bromelain, 114
flavor ingredients:
dill, 246
eucalyptus, 273
fennel, 284
garlic, 309
lemongrass, 414
lemon oil, 411
mustard, 459
onion, 475
oregano, 483
parsley, 488
sage, 551
tamarind, 584
tarragon, 589
thyme, 596
turmeric, 605
gums algin, 17
Fatty acids, *see* CHEMICAL INDEX
Febrifuge, see Antipyrctics
Felons, remedies, white pine bark, 501

Female disorders, remedies:
balm, 63
cinnamon and cassia, 199
dandelion root, 239
dogwood, 249
goldenseal, 337
see also Menstruation
Feminine hygiene sprays, ingredients, balsam Peru, 70
Fang wang jiang,
FENNEL, 283–285
FENUGREEK, 286–288
Fern, *see* Aspidium
***Ferula assa- foetida*, 48, 49**
Ferula galbanifua, see Ferula gummosa
Ferula gummosa, 299
***Ferula* species, 299**
Fetus, restless, remedies, rehmannia, 523
Fever:
as toxic symptom, 290–291
see also Antipyretics
FEVERFEW, 289–291
Fever tree, *see* Cinchona (red and yellow) Eucalyptus
Feverwort, *see* Boneset
Free from prussic acid (FFPA), 22, 23, 184
Fiber supplement, psyllium, 512
Fibrillations, *see* Cardiac
FICIN, 291–293, *see* also Bromelain and Papain
Ficus anthelmintica, 291
Ficus glabrata, 291
***Ficus insipida*, 291**
Ficus laurifolia, 291
Filariasis, remedies, ginkgo, 327
Filmforming agents, 5, 17, 69, 293, 497, 602
***Finocchio, see* Fennel**
Fir, balsam, *see* Balsam Canada
Fish, poisoning, avocado, 60
Fish fuddle, *see* Dogwood
Fish products:
gums:
agar, 10
algin, 17
xanthan gum, 624
MSG, 453
protein, liquefying with bromelain, 114

Fixatives:
  asafetida, 49
  balsam Canada, 66
  balsam copaiba, 67
  balsam Peru, 70
  benzoin, 91
  castoreum, 157
  cedarwood oil, 164
  civet, 204
  costus oil, 231
  deertongue, 241
  elemi gum, 261
  galbanum, 300
  guaiac wood oil, 346
  immortelle, 378
  labdanum, 405
  lime oil, 423
  musk, 456
  myrrh, 461
  olibanum, 472
  orange, 479
  patchouly oil, 494
  storax, 579
  turpentine, 608
Fixed oils, *see* CHEMICAL INDEX
Flatulence:
  remedies:
    aletris, 12
    asafetida, 49
    baizhu, 660
    blessed thistle, 101
    carrot seed, 148
    dandelion root, 239
    dill, 246
    fennel, 285
    gentian, 317
    goldenseal, 338
    juniper- berries, 391
    lavender, 408
    lovage root, 428
    nutmeg and mace, 468, 469
    saffron, 548
    turmeric, 605
    yarrow, 627
  side effects, guar gum, 348
Flavor enhancer, mints, 445, 446
Flavors, see individual flavoring ingredients
Flos daturae, 89
Flowering dogwood, 249
Fluid retention, remedies, baizhu, 660

Flushing, as toxic symptom, 89, 445, 468
Foaming agents in beverages, 519, 634
Foaming beverage color, 519, 634
Foams, castor oil in synthesis of, 155
Foeniculum capillaceum, 283
***Foeniculum officinale*, 283**
***Foeniculum vulgare*, 283**
Foenugreek, see Fenugreek
Food absorption, reduction by carrageenan, 146
Food colors:
  alkanet, 19
  annatto, 41
  calendula, 131
  capsicum, 134
  caramel color, 137
  carrot seed oil, 148
  cocoa, 218
  cranberry, 233
  grape skin extract, 343
  red beet color, 87
  roselle, 533
  saffron, 548
  tea, 593
  turmeric, 603
  see also CHEMICAL INDEX
Food of the gods, *see* Asafetida
Food poisoning:
  prevention, peony root, 676
  remedies, 155, 417, 676
Footsore, remedies, mustard, 459
Forastero cacao, 216
Forgetfulness, *see* Memory
FORSYTHIA FRUIT (LIANQIAO), 665
***Forsythia suspensa*, 665**
FO-TI (RAW AND CURED), 293–297
Fractures, remedies, 44, 365, 368, 600, 647, 676
Fragrances, *see* individual
Frangula alnus, 118
Frangula, *see* Buckthorn, alder
Frankincense, *see* Olibanum
Freckles, skin, remedies, centaury, 168
French marigold, *see Tagetes*
French psyllium, *see* Psyllium
Fright, saffron, 548
Frost bite, remedies:

  balm of Gilead buds, 65
  hawthorn, 353
  hyssop, 376
  lycium fruit, 431
Fructose, see CHEMICAL INDEX
Fruit bromelain, 113
Fruit flavors:
  cassie absolute, 152
  immortelle, 379
  lavender, 408
  musk, 456
  rose oil, 531
  ylang ylang oil, 629
Fruit juices, proteases used, 114
Fruits, processed:
  flavor ingredients, 228
  gums, 445
  MSG, 453
Fumigant, labdanum, 405
Fungi, *see* Antifungal activities
Furocoutnarins, *see* Coumarins
Furuncles, remedies, 431, 685

Gagroot, *see* Lobelia
Galactagogue, 178, 397
GALBANUM, 299–300
Galipea officinalis, 34–35
***Galium odoratum*, 622**
Gall bladder disorders:
  remedies:
    absinthium, 2
    blessed thistle, 101
    sweet woodruff, 623
  see also Chologogic; Gall Stones
Gallotannic acid, *see* Tannic acid
Gallotannin, *see* Tannic acid
Gall stones, remedies:
  boldo leaves, 108
  cascara sagrada, 150
  chicory root, 191
  parsley, 488
Gambier, *see* Catechu (black and pale)
Gambir, *see* Catechu (black and pale)
Gangrene, remedies, echinacea, 255
GANODERMA, 300–304
***Ganoderma japonicum*, 300–302**
***Ganoderma lucidum*, 300**
Ganoderma species, 301

Gaoben, 669–670
GARDEN BURNET (DIYU), 666
Garden purslane, 679
GARLIC, 308–310
Gas, *see* Carminative
Gastric disorders:
  gastritis:
    remedies:
      acacia, 5
      bayberry bark, 81
      calamus, 128
      calendula, 131
      gentian, 317
      horse chestnut, 365
      horsetail, 368
      passion flower, 491
      sage, 551
      sweet basil, 75
      thyme, 596
      yarrow, 627
    as toxic symptom, 26
  remedies:
    althea root, 30
    angelica, 33
    burdock, 121
    dandelion root, 239
    fennel, 285
    lovage root, 428
    nutmeg and mace, 469
    olibanum, 472
    savory, 560
    valerian root, 613
    yarrow, 626
Gastric irritation, as toxic symptom, 67, 586
Gastric pain, *see* Stomachache
Gastric secretions:
  agents that reduce, carrageenan, 146
  agents that stimulate, 195, 223, 499
Gastroenteritis, remedies, tea, 592
Gastrointestinal disorders:
  cramps, remedies:
    balm, 63
    capsicum, 134
    hops, 360
    mints, 446
    savory, 560
    valerian root, 614
    yarrow, 627

gastroenteritis, remedies, tea, 592
remedies:
  anise, 38
  bilberry, 94
  comfrey, 226
  evening primrose, 278
  gentian, 317
  juniper berries, 391
  nutmeg and mace, 468
  papain,
  savory, 560
  see also Gastric disorders; Intestinal disorders; Ulcers
Gastrointestinal irritation:
  substances that cause, 380, 408
  see also Gastric irritation
Gaultheria oil,
  see Wintergreen oil
***Gaultheria procumbens*, 619**
***Gaultheria* species, 619**
Gelatin:
  agar effect on gel strength, 9–10
  as enzyme substrate, 417
  see also Agar
Gelatins and puddings:
  flavoring ingredients:
    alfalfa, 14
    allspice, 20
    almonds, 23
    Althea root, 30
    ambrette seed, 31
    angelica, 33
    angostura bark, 35
    anise, 38
    arnica, 44
    asafetida, 49
    balm, 63
    balsam Canada, 66
    balsam copaiba, 68
    balsam Peru, 70
    balsam tolu, 72
    beeswax, 85
    benzoin, 91
    bergamot oil, 92
    black and pale catechu, 158
    black and white pepper, 499
    blackberry bark, 97
    bois de rose oil, 106
    boronia absolute, 113
    buchu, 117

cananga oil, 132
capsicum, 134
caramel, 137
caraway, 139
cardamom, 141
carob, 144
carrot seed oil, 148
cassie absolute, 153
castoreum, 157
cedar leaf oil, 162
celery seed, 166
chamomile, 172
chicory root, 190
cinnamon and cassia, 199
citronella oil, 202
civet, 204
clary sage, 205
cloves, 211
coffee, 224
coriander, 228
costus oil, 231
cubebs, 234
cumin, 236
damiana, 237
dandelion root, 239
doggrass, 248
elder flowers, 258
elecampane, 260
elemi gum, 262
eucalyptus, 273
fennel, 285
fenugreek, 287
galbanum, 300
garlic, 309
genet, 315
gentian, 317
geranium oil, 319
ginger, 322
grapefruit oil, 344–345
guaiac wood oil, 346
hops, 360
horehound, 363
hyssop, 376
immortelle, 379
jasmine, 383
juniper berries, 391
kola nut, 399
labdanum, 405
lavender, 408
lemongrass, 414
lemon oil, 411
licorice root, 417
lime oil, 423

Gelatins and puddings
    (*Continued*)
  lovaee root, 428
  marjoram, 438
  mints, 445
  musk, 456
  mustard, 459
  myrrh, 462
  nutmeg and mace, 469
  olibanum, 472
  onion, 475
  orange, 479
  origanum oil, 485
  parsley, 488
  patchouly oil, 494
  pine needle oil, 503
  prickly ash, 51
  quassia, 516
  quebracho, 517
  rhubarb, 526
  roselle, 533
  rosemary, 536
  rose oil, 531
  rue, 543
  sage, 551
  sandalwood oil, 554
  storax, 579
  sweet birch oil, 96
  sweet woodruff, 623
  tagetes, 582
  tamarind, 585
  tannic acid, 587
  tarragon, 589
  tea, 592
  thyme, 596
  turpentine, 608
  valerian root, 613
  vanilla, 617
  West Indian bay, 79
  wild cherry bark, 183
  wintergreen oil, 619
  ylang ylang oil, 630
  fragrance ingredients:
    kudzu root, 401
    pipsissewa, 504
  other ingredients:
    algin, 17
    beeswax, 85
    carrageenan, 146
    guar gum, 348
    quillaia, 519
    tragacanth, 603
    xanthan gum, 624

Gelidium amansii, 9
Gelidium cartilagineunt, 9
Gelidium species, 9
Gelling agents:
  agar, 10
  algin, 17
  carrageenan, 146
Gelose, *see* Agar
GELSEMIUM, 312–313
Gelsemium elegans, 313
Gelsemium nitidum, 312
Gelsemium sempervirens, **312**
Gel strength, agar, 10
GENET, 314–315
Genista juncea, 314
Genotoxicity, stevia, 577
Gentiana acaulis, **316**
Gentianaceae, 167, 192, 316
Gentian lutea, **316**
GENTIAN (AND STEMLESS), 316–317
Gentiana rigescens, 317
Gentiana scabra, 317
Gentiana species, 317
Gentiana triflora, 317
Gentian root, 35
GERANIUM OIL, ROSE, 318–320
Giddiness:
  remedies, lavender, 408
  as toxic symptom, 43, 313
Gigartina species, **145**
Gin, flavor ingredients, 194, 228
  see also Beverages, alcoholic and nonalcoholic
GINGER, 320–323
GINKGO, 325–327
Ginkgoaceae, 325
Ginkgo biloba, **325**
GINSENG (ORIENTAL AND AMERICAN), 330–333
  adulterants, 264
  substitute, codonopsis, 220
Glandular diseases, remedies:
  nettle, 465
  triphala, 654
Glaucoma, as toxic symptom, 102, 267
Glucoside, *see* CHEMICAL INDEX
Gluten, wheat, as source of MSG, 452

Glycogen, liver, ginseng effect, 332
Glycoside, *see* CHEMICAL INDEX
Glycyrrhiza glabra, **415**
Glycyrrhiza, *see* Licorice root
Glycyrrhiza uralensis, **415**
Goats, poisoning, avocado, 60
Goiter, as toxic symptom, mustard, 458
GOKSHURA, 646–647
GOLDENSEAL, 336–338
Gomme arabiyue, *see* Acacia
Gonadotropic, eleuthero, 263
Gonorrhea, remedies:
  chaparral, 176
  cubebs, 234
  euphorbia, 275
  giant knotweed, 668
  horsetail, 368
  kava, 396–397
  psyllium, 512
  sandalwood oil, 554
  sarsaparilla, 556
GOTU KOLA, 339–341
Gout, remedies:
  bilberry, 94
  burdock, 121
  cedar leaf oil, 162
  celery seed, 166
  chamomile, 172
  chervil, 185
  dandelion root, 239
  garlic, 310
  guaiac wood oil, 346
  horsetail, 368
  kava, 397
  nettle, 465
  orange, 480
  sassafras, 558
  sweet birch oil, 96
Gracilaria confervoides, **9**
Gracilaria species, **9**
Gramineae, 229, 247, 384, 412
Grape, Oregon, *see* Barberry
Grape bark, *see* Cocillana bark
GRAPEFRUIT OIL, 343–345
Grape skin color, 342
GRAPE SKIN EXTRACT (ENOCIANINA), 342–343
GREATER GALNANGA, 647–648

General Index

Gravies:
  colorants:
    annatto, 41
    caramel color, 137
    turmeric, 605
  flavoring ingredients,
    angostura bark, 35
    Bois de rose oil, 106
    cardamon, 141
    carob, 144
    castoreum, 157
    cinnamon and cassia, 199
    cloves, 211
    cumin, 236
    dill, 246
    fennel, 284
    galbanum, 300
    garlic, 309
    mustard, 458–459
    nutmeg and mace, 469
    onion, 475
    orange, 480
    origanum oil, 485
    parsley, 488
    red clover tops, 208
    rosemary, 536
    sage, 551
    savory, 560
    tamarind, 585
    tarragon, 589
    thyme, 596
    turmeric, 605
  gurmar, 649
  gums:
    carrageenan, 146
    guar gum, 348
    tragacanth, 603
  MSG, 453
Graying, premature, remedies, 297, 421, 523
Great bur, see Burdock
Grecian laurel, see Bay, sweet
Green purslane, 679
Griping:
  cause of, aloe, 25
  prevention, coriander, 228
  reduction:
    aloe, 26
    belladonna, 89
Growth:
  assents inhibiting:
    guar gum, 348
    locust bean gum, 427

  podophyllum, 506
  psyllium, 512
  tannic acid, 586
  regulator, plant, alfalfa, 13
Gua danpi, 675
Guaiacum officinale, 346
Guaiacuno sanctum, 346
GUAIAC WOOD OIL, 345–346
Guanhua dangshen, 220
Guapi, *see* Cocillana bar
GUARANA, 350–351
***Guarea rusbyi*, 215**
Guarea species, 215
Guarea spiciflora, 216
Guarea trichilioides, 216
GUAR GUM, 347–349
Guinea sorrel, *see* Roselle
Gum:
  blue, *see* Eucalyptus
  carob, *see* Locust bean gtrnn
  elemi gum, 261–262
  guar, 347–349
  seaweed, 16, 145
  sterculia, 393
  xanthan gum, 624–625
  see also CHEMICAL INDEX
Gum arabic, *see* Acacia
Gum benjamin, *see* Benzoin
Gum cistus, *see* Labdanum
Gummae mimosas, *see* Acacia
Gum problems, remedies:
  alfalfa, 14
  benzoin, 91
  chamomile, 172
  comfrey, 226
  goldenseal, 337
  myrrh, 462
  see also Toothache
Gum Senegal, *see* Acacia
Gum thus, *see* Turpentine (and rosin)
Gum tree, *see* Eucalyptus
Gum turpentine, 607608
Guru nut, *see* Kola nut
Gut, *see* gastrointestinal entries; intestinal entries
***Gymnema sylvestre*, 648–649**
Gynecological problems, remedies, feverfew, 290

Hair, discoloration, as toxic symptom, 396

Hair growth, stimulants, 53, 60, 195
Hair loss:
  prevention, rosemary, 536
  remedies:
    cade oil, 123
    fenugreek, 288
    ginger, 322
    saffron, 548
    yucca, 634
Hair products and preparations, ingredients:
    antimicrobial, cangzhu, 662
    astragalus, 57
    avocado, 60
    balsam Canada, 66
    balsam Peru, 70
    beeswax, 85
    broom tops, 16
    cade oil, 123
    calendula, 129
    castor oil, 155
    chamomile, 169
    chicle, 189
    cinchona, 195
    citronella oil, 202
    comfrey, 226
    Dahurian angelica, 664
    echinacca, 255
    evening primrose, 278
    fangfeng, 284
    forsythia fruit, 665
    fo-ti, 297
    gotu kola, 340
    henna, 356
    honey, 358
    horse chestnut, 365
    horsetail, 368
    jojoba, 386
    Job's tears, 385
    karaya gum, 394
    ligusticum, 670
    ligustrum, 421
    mume, 672
    nettle, 465
    peony bark, 675
    quillaia, 519
    red sage, 681
    rehmannia, 523
    royal jelly, 540
    safflower, 682
    Sichuan lovage, 683
    squill, 575

Hair products (*Continued*)
  sweet basil, 75
  tienchi ginseng, 600
  vasaka, 656
  West Indian bay, 77
  see also Cosmetics; Creams and lotions
Hakim, 641
Hakim Ajmal Khan, 641
Hallucinations, as toxic symptoms, 2, 89
Hamam, 620
Hang baizhi, 663
Hangover, remedies, 195, 214, 333, 401
Hardening of arteries, *see* Arteriosclerosis
***Harpagophytum procumbens*, 242**
Haw, southern black, *see* Black haw bark
HAWTHORN, 352–354
Hay fever:
  remedies, 14, 89, 134, 275
  as toxic symptom, castor oil, 154
  see also Respiratory ailments
Headache:
  remedies:
    ambrette seed, 31
    balm, 63
    cajeput oil, 126
    catnip, 160
    Dahurian angelica, 663
    devils claw, 243
    ephedra, 267
    eyebright, 280
    fangfeng, 282
    garlic, 310
    gentian, 317
    ginseng, 333
    guarana, 351
    henna, 356
    immortelle, 379
    jasmine, 383
    kudzu root, 401
    lavender, 408
    lemongrass, 414
    ligusticum, 670
    magnolia flower, 435
    marjoram, 438
    mints, 446
    oregano, 483
    passion flower, 491
    patchouly oil, 494
    peony root, 676
    purslane, 679
    rosemary, 537
    Sichuan lovage, 683
    tea, 593
    wintergreen oil, 619
  as toxic symptom:
    chenopodium oil, 180
    cinchona, 195
    coffee, 223
    mints, 446
    mistletoe, 450
    thyme, 596
    turpentine, 608
Healing:
  alkanet, 18
  aloe, 27
  althea root, 30
  arnica, 43
  astragalus, 57
  avocado, 60
  Baikal skullcap, 684, 686
  balsam Peru, 70
  blessed thistle, 101
  bletilla tuber, 661
  bloodroot, 103
  cade oil, 123
  calendula, 131
  chamomile, 172
  chaparral, 176
  comfrey, 225
  Dahurian angelica, 664
  echinacea, 253
  eucalyptus, 273
  evening primrose, 278
  galbanum, 300
  gelsemium, 313
  Gentian, 317
  goldenseal, 338
  gotu kola, 340
  horsetail, 368
  hydrangea, 370
  hypericum, 373
  jujube, 388
  lobelia, 425
  myrrh, 461
  pearl, 673–674
  purslane, 679
  storax, 580
  white pine bark, 501
  yarrow, 627
  yerba santa, 629
Heart, *see* Coronary heart disease; cardiac entries
Heart attack, *see* Myocardial
Heartburn, remedies, 239, 317
Heart pain, *see* Pain
Heart rate, elevated, as toxic symptom, yohimbe, 632
Heat:
  increased tolerance, codonopsis, 220–221
  stress, remedies, 351
Heavy metals, *see* CHEMICAL INDEX
Hectic fever, remedies, rehmannia, 523
Helenium grandiflorum, 259
Helianthus tuberosus, 45
***Helichrysum angustifolium*, 378**
Helichrysum arenarium, 378
Helichrysun italicum, 378
Helichrysum orientale, 378
Helichrysum species, 378
Helichrysum *stoechas*, 378
Heliotropin, sarsaparilla as starting material, 558
Hematologic, 258, 599, 683
Hematomas, remedies, horsechestnut, 365
Hematuria:
  remedies:
    comfrey, 225–226
    garlic, 308–310
    psyllium, 513
    rutin, 545–546
    tannic acid, 585–587
    turmeric, 608
    yarrow, 626–627
  as toxic symptom, turpentine, 608
  see also Bladder entries; kidney
Hemolytic:
  alfalfa, 13
  quillaia, 518
  tienchi ginseng, 599
  yucca, 633
Hemorrhage:
  remedies
    algin, 16–17
    calendula, 131

comfrey, 225–226
euphorbia, 274–276
ginger, 320–323
ginseng, 333
goldenseal, 337
myrrh, 460–462
nettle, 464
rue, 543
rutin, 545
tannic acid, 585–587
turmeric, 605
witch hazel, 620–621
yarrow, 626–627
as toxic symptoms:
  deertongue, 241
  ficin, 283–285
  rue, 543
see also Capillaries; Vitamin K
Hemorrhagic shock, protectants, tienchi ginseng, 599
Hemorrhoids, remedies:
  balm of Gilead buds, 65
  balsam Canada, 66
  balsam copaiba, 68
  balsam Peru, 70
  belladonna, 89
  bilberry, 94
  black and, pale catechu, 158
  cinchona, 195
  coriander, 228
  dill, 246
  garden burnet, 666
  goldenseal, 337
  horse chestnut, 365
  myrrh, 462
  psyllium, 512
  purslane, 679
  Rutin, 545
  sweet woodruff, 415, 423
  tannic acid, 587
  witch hazel, 621
  yarrow, 627
Hemostat:
  Baikal skullcap, 684
  bee pollen, 83
  bletilla tuber, 661
  calendula, 129–131
  dittany bark, 664
  garden burnet, 666
  horsetail, 368
  labdanum, 405
  nettle, 464
  purslane, 679

rhubarb, 524–527
tienchi ginseng, 599
witch hazel, 620
yarrow, 626
yucca, 633–634
Hemostyptic, horsetail, 367
HENNA, 355–356
Hepatic, *see* Liver entries
Hepatitis, remedies:
  chicory root, 191
  dandelion root, 239
  dittany bark, 664
  epimedium, 270
  ganoderma, 302
  garlic, 309
  giant knotweed, 668
  gotu kola, 340
  jasmine, 383
  licorice root, 416
  milk thistle, 539
  royal jelly, 540
  schisandra, 566
  tea, 592
see also Liver disorders
Hepatoprotective:
  baizhu, 660
  black and pale catechu, 157–158
  fo-ti, 296
  garlic, 308–310
  hypericin, 372
  jujube, 401
  Kalmegh, 650
  ligustrum, 421
  milk thistle, 439
  neem, 652
  rehmannia, 521–523
  royal jelly, 538–540
  schisandra, 564–567
  vasaka, 655
Herbal teas, 130, 373, 577
Herbicide, catnip, 159–160
Herculis' club, *see* Ash, prickly
Hernia, remedies, I, 211, 228
Herpes:
  antiviral, echinacea, 253
  genital, remedies, garlic, 309
Herpes simplex, remedies:
  balm, 62
  echinacea, 253
  giant knotweed, 668–669
  hyssop, 376
  marjoram, 437

mints, 444
see also Antiviral
Herpes zoster, remedies, gotu kola, 340
Hibiscus abelmoscres, see Abelmoschus moschatus
Hibiscus sabdariffa, 532
Hiccups, remedies, beeswax, 85
Hide, bating of, *see* Leather industry
Hippocrate, 640
Hoarhound, *see* Horehound
Hogberry, *see* Broom tops
Hogberry, *see* Uva ursi
Holly, *see* Pipsissewa
HONEY, 357–358
HONEYSUCKLE FLOWER (JINYINHUM), 667–668
Honghua, 681–682
Hookworms, *see* Antltelmintic
HOPS, 359–360
HOREHOUND, 362–363
Hormone:
  plant growth, *see* Growth regulator, plants
  steroid, potential source, fenugreek, 288
HORSE CHESTNUT, 364–366
Horseheal, *see* Elecampane
Horses, poisoning, avocado, 59–60
HORSETAIL, 367–369
Hot pepper, *see* Capsicum
Hound's tongue, *see* Deertongue
Hoxsey cancer cure, ingredients, 51, 119, 208
Huangbai, 659, 677–678
Huangqin, 684–686
Huanuco coca, *see* Coca
Huesito, *see* Acerola
Huisache, *see* Cassie absolute
Humoral, 263, 449, 641
Humours
  potentiating, euphorbia, 274–276
HYDRANGEA, 369–370
Hvdrangea arborescens, 370
Hydrangeaceac, *see* Saxifragaceae
Hydrangea macrophylla, 370

Hydrangea paniculata, 370
Hydrangea species, 370
Hydrangea strigosa, 370
Hydrangea umbellata, 370
Hydrastine cydrochloride, antagonist, goldenseal, 336
Hydrastis canadensus, 336
Hydrocolloids, see Gums
Hydrocotyle asiatica, 339
Hydrophyllaceae, 628
Hypercholeste-rolemics, *see* Cholesterol, blood
Hypcremesis gravidarum, ginger, 321
Hyperglycemia, inhibition, astragalus, 56
Hyperglycemic, 260, 266, 272, 384, 390, 536
Hypericaceac, 371
HYPERICUM, 371–374
Hypericum perforatum, 371
Hyperkinetic children, caffeine effect on, 223
Hyperlipemia, remedies:
  fo-ti, 295
  ganoderma, 303
  giant knotweed, 685
  hawthorn, 353
  tienchi ginseng, 599
Hypersensitive dentin, garlic, 310
Hypertension:
  remedies:
    algin, 16–17
    centaury, 168
    cherokee rosehip, 181
    chervil, 185
    cinnamon and cassia, 199
    corn silk, 229
    devil's claw, 246
    epimedium, 269
    evening primrose, 276–278
    ganoderma, 302
    garlic, 308
    guar gum, 348
    hawthorn, 353
    jujube, 389
    kudzu root, 399
    peony bark, 674
    psyllium, 511
    purslane, 678
    royal jelly, 539
    rutin, 545
    saffron, 682
    sassafras, 557–558
  as toxic symptom, yohimbe, 632
  see also Hypotension
Hypertensive:
  black and white pepper, 498–499
  ephedra, 266
  ganoderma, 302
  quebracho, 517
  rehmannia, 522
  tagetes, 581
Hyperthermia, adaptogenic activity, eleuthero, 263
Hypnotic:
  belladonna, 88
  chamomile, 169–173
  eucalyptus, 271–273
  hops, 360
  kava, 396
  tarragon, 589
Hypocholesterolemics, *see* Cholesterol, blood
Hypoglycemic:
  astragalus, 56
  baizhu, 660
  bilberry, 94
  black cohosh, 97–99
  burdock, 121
  carob, 142–144
  coriander, 227
  corn silk, 229
  damiana, 237
  elecampane, 260
  eleuthero, 263
  ephedra, 266
  eucalyptus, 271–273
  fenugreek, 287
  ganoderma, 303
  garlic, 309
  ginseng, 332
  guar gum, 347–349
  Job's tears, 384
  lycium fruit, 431
  onion, 475
  phellodendron bark, 678
  pipsissewa, 504
  poria, 509
  rehmannia, 522
  sage, 551
  stevia, 577
Hypoinsulinemic, carob, 142–144
Hypolipemic:
  Baikal skullcap, 685, 686
  bee pollen, 82
  fo-ti, 295
  ganoderma, 303
  ligustrum, 421
  lycium fruit, 431
  safflower, 682
  tienchi ginseng, 599
Hypotension:
  remedies:
    eleuthero, 262–264
    purslane, 678
  as toxic symptom:
    cedar leaf oil, 161
    ipecac, 380
  see also Hypertension
Hypotensive:
  asafetida, 49
  asparagus, 52
  astragalus, 56
  Baikal skullcap, 686
  black cohosh, 97–99
  calamus, 128
  carrageenan, 146
  celery seed, 165–166
  codonopsis, 220
  corn silk, 229
  costus oil, 231
  dill, 244–246
  elecampane, 260
  ephedra, 266
  epimedium, 269
  forsythia fruit, 665
  ganoderma, 302
  garlic, 308
  Giant knotweed, 669
  goldenseal, 336–338
  hawthorn, 353
  kudzu root, 400
  lycium fruit, 431
  magnolia flower, 435
  mistletoe, 449
  nettle, 464
  pale catechu, 158
  papain, 146
  parsley, 488
  passion flower, 490
  psyllium, 511–513
  quebracho, 517
  rehmannia, 522

rhubarb, 526
roselle, 533
rutin, 545
safflower, 682
saffron, 547–548
sour jujube kernel, 573
sweet bay, 78
tagetes, 581
thyme, 595
tienchi ginseng, 599
turmeric, 604
valerian root, 613
yarrow, 626
Hypothermic:
  cinnamon and cassia, 196–200
  see also Antipyretic
Hypothrobinemia, as toxic symptoms, saffron, 547–548
Hyptis species, 482
Hyriopsis cumingii, 673
HYSSOP, 375–377
Hyssopus officinalis, 375
Hysteresis, 10
Hysteria, remedies:
  asafetida, 48–49
  castoreum, 157
  garlic, 310
  jujube, 389
  passion flower, 491
  valerian root, 614

Ibn Sina, 640
Ilaj ul Amrad, 641
Ilaj-bil Ghiza, 642
Ilaj-bil Tadbeer, 641–642
Ilaj-bil-mithl, 642
Ilaj-bil-didd, 642
Ilaj-bit-Tadbeer, 641
Illicium lanceolatum, 37
Illicium religiosum, 37
Illicium, see Anise
Illicium verum, 36, 37
Immersion oil, 164
IMMORTELLE, 378–379
Immunoenhancing, rehmannia, 522
Immunomodulating:
  calendula, 130
  codonopsis, 221
  epimedium, 269
  ganoderma, 302
  ligustrum, 421

peony root, 677
tienchi ginseng, 599
Immunoreactivity, aloe, 25
Immunoregulating, 82
Immunostimulant:
  arnica, 43
  astragalus, 56
  barberry, 73
  boneset, 109
  calendula, 130
  cedar leaf oil, 162
  echinacea, 253
  elecampane, 260
  eleuthero, 263
  fangfeng, 281
  fo-ti, 294
  mistletoe, 449
  royal jelly, 539
  safflower, 682
  sour jujube kernel, 572
  tragacanth, 601–603
Immunosuppressive:
  carrageenan, 146
  rehmannia, 522
  safflower, 682
Impetigo, remedies, chaparral, 176
Impotence, remedies
  chaste-tree, 177–179
  Cherokee rosehip, 181
  epimediunm, 268
  fenugreek, 288
  ginseng, 333
  schisandra, 567
  yohimbe, 632
Incense, fragrance ingredients:
  costus oil, 231
  olibanum, 471
  sandalwood oil, 554
Incontinence, remedies, caraway, 139
Indian balsam, see Balsam Peru
Indian Medicine:
Indian Medicine:
  East:
    cardamon, 140
    castor oil, 154
    chirata, 192
    costus oil, 230
    euphorbia, 274
    fenugreek, 286
    gotu kola, 339
    honey, 358

psyllium, 511
North American:
  aletris, 11
  balsam Canada, 66
  black haw bark, 99
  bloodroot, 103
  blue cohosh, 104
  evening primrose, horsetail, 277, 367
  hydrangea, 369–370
  jojoba, 386–387
  sweet birch oil, 95
  white pine bark, 501
Indian plantago, see *Psyllium*
Indian red paini, see Bloodroot
Indian saffron, see Turmeric
Indian tobacco, see Lobelia
Indian tragacanth, see Kazaya gum Indian turmeric, see Goldenseal
Indigestion, remedies:
  angelica, 33
  baizhu, 660
  bilberry, 93–94
  black Catechu, 158
  blessed thistle, 101
  cangzhu, 660
  chamomile, 172
  chicory root, 191
  cinchona, 195
  dandelion root, 238–240
  devil's claw, 243
  fennel, 287
  ganoderma, 303
  goldenseal, 336–338
  hawthorn, 353
  majorana, 436
  mints, 445
  nutmeg and mace, 469
  orange, 480
  oregano, 483
  parsley, 488
  rosemary, 537
  sage, 551
  savory, 560
  yarrow, 626–627
  see also Digestive; Gastric; Gastrointestinal; Stomachic; Tonic
Indigo, 356
Indolent ulcers, see Skin
Industrial lubricants, ingredients, 155

Infant formulas, carrageenan, 146
Infections:
  remedies, dandelion, 239
  see also Antibacterial; Antifungal
Inflammation:
  agents that cause:
    Carrageenan, 146
    quillaia, 518
  see also Antiinflammatories; Arthritis; Mucous membrane; Rhenmatisin; skin entries
Influenza:
  prevention, rose hips, 537
  remedies:
    allspice, 20
    astragalus, 57
    catnip, 159–160
    echinacea, 253
    ephedra, 266
    eucalyptus, 272
    fangfeng, 282
    forsythia fruit, 665
    garlic, 310
    ginger, 322
    kalmegh, 650
    kudzu root, 401
  see also Antiviral
Ink:
  for food products, alkanet, 19
  Karaya gum, 394
  tannic acid in manufacture of, 587
Inotropic, rosemary, 536
Insect bite, remedies:
  balm, 62–65
  chaparral, 174–176
  comfrey, 226
  echinacea, 251–256
  lobelia, 425
  lycium fruit, 431
  psyllium, 512
  purslane, 679
  rue, 543
  sage, 551
  savory, 560
  sweet basil, 75
  witch hazel, 621
  yerba santa, 629

Insecticide:
  anise, 37
  Blessed thistle, 101
  cassie absolute, 152–153
  catnip, 160
  chaparral, 197
  cinnamon, 197
  fennel, 284
  garlic, 309
  gotu kola, 341
  myrrh, 479
  patchouly oil, 494
  quassia, 516
  sweet basil, 75
  tagetes, 582
  turmeric, 604
Insect repellent:
  cedar leaf oil, 162
  cedarwood oil, 162
  citronella oil, 202
  lavender, 408
Insomnia:
  remedies
    balm, 62–65
    chamomile, 172
    cherry laurel leaves, 184
    fo-ti, 297
    ganoderma, 303
    hawthorn, 352–354
    hops, 360
    hypericum, 372
    jasmine, 383
    lavender, 406–408
    licorice root, 417
    mother of pearl, 673
    passion flower, 491
    pearl, 673
    poria, 509
    red sage, 680
    saffron, 548
    schisandra, 566
    sour jujube kernel, 573
    sweet woodruff, 623
    valerian root, 613
  as toxic symptom:
    absinthium, 2
    coffee, 223
    ephedra, 266
    turpentine, 608
  see also Restless sleep
Instant coffee, 222
Intelligence quotient, increased, gotu kola, 339

Intertrigo, remedies, see also Intestinal tone and motility, deceased, ephedra,
Intestinal disorders:
  colon damage,as toxic symptom, senna, 569
  enteritis, remedies, 181, 276, 365, 385
  remedies:
    carrageenan, 145–147
    cocoa, 218
    hops, 360
    hyssop, 376
    kava, 397
    savory, 560
  see also Gastrointestinal disorders
Intestinal worms, see Anthhelentic
Intraocular pressure, as toxic symptoms, 223
Inula helenium, 259
Inula, see Elecampane
Invert sugar, caramel color, 136
IPECAC, 379–381
Iridaceae, 547
Irish broom, see Broom tops
Irish moss extact, see Carrageenan
Iron, see CHEMICAL INDEX
Irritability, as toxic symptom, yohimbe, 631–632
Irritable bladder, syndrome, kava, 396
Irritable bowel syndrome, remedies, 512
Irritant:
  eugenol, 20, 494, 630
  local cedarwood oil, 163
  quillaia, 518
  thyme, 596
  yucca, 633–634
  mustard, 458
  oregano, 483
  prickly ash, 50–51
  see also Counterirritant; Mucous membrane; Skin irritation
Ischemic crebrovascular diseases, remedies, Sichuan lovage, 683

Isinglass, Japanese, *see* Agar
Ispaghula, *see* Psyllium
Ispagol, *see* Psyllium
Itching:
    as toxic symptom, kava, 396
    see also Antipruritics

Jaguar gum, *see* Guar gum
Jamaica pepper, *see* Allspice
Jamaica sorrel, *see* Roselle
Jams and jellies, ingredients
    carrageenan, 146
    orange, 480
    red clover tops, 208
    roselle, 533
Japanese cedarwood oil, 163
Japanese gelatin , *see* Agar
Japanese isinglass, *see* Agar
Japanese star anise, 37, 38
JASMINE, 382–383, *see also* Gelsemium
Jasminum grandiflorum, 382
Jasminum lanceolarium, 383
Jasminum officinale, 382, 383
Jasminum sambac, 382, 383
Jasminum species, 382
Jaundice, remedies:
    Baikal skullcap, 686
    bayberry bark, 81
    Calendula, 131
    eyebright, 280
    gentian, 317
    giant knotweed, 668, 678, 685
    goldenseal, 336
    gotu kola, 340
    henna, 356
    lobelia, 425
    marjoram, 436–438
    oregano, 483
    parsley, 488
    phellodendron bark, 678
    podophyllum, 506
    rhubarb, 527
Jaundice root, *see* Goldenseal
Jerusalem artichoke, 45
Jessamine, *see* Gelsemium copaiba
Jesuits balsam, *see* Balsam copaiba
Jesuit's bark, *see* Cinchone (red and yellow)
Jinyinhua, 667–668
JOB'S TEARS, 384–385

Joints:
    aching and stiff, remedies:
        dandelion root, 239
        ephedra, 267
        eucalyptus, 273
        jasmine, 383
        Job's tears, 385
        ligustrum, 420–421
        lycium fruit, 429–432
        schisandra, 567
        tannic acid, 587
        turpentine, 608
    inflammatory diseases, mistletoe, 450
    weak knees, remedies:
        epimedium, 270
        rehmannia, 523
    see also Arthritis: Rhumatism
JOJOBA, 386–387
Joshua tree, 633
JUJUBE, COMMON, 388–389
Juices, cranberry, 232
JUNIPER BERRIES, 389–391
Juniper tar, *see* Cade oil
Juniperus communis, 389
Junipers mexicana, 162
Juniperus oxycedrus, **123**
Juniperus virginiana, 162
Justicia adhatoda L., **655**

Kadaya, see Karaya guns
KALMEGH, 650
KARAYA GUM, 347, 393–394
KAVA, 395–397
KAVALI, 648
Kelp, giant, 16
Kerololytic, Cade oil, 123
Keshan disease, Remedies; ganoderma, 303
Kapha, 638–644
Kher*, see* Acacia
Kidney:
    substances that cause damage:
        chenopodium oil, 181
        rhubarb, 526
    impairment, as toxic symptom, sarsaparilla, 556
    kidney ailments, remedies:
        bilberry, 94
        chaparral, 175
        cinnamon and cassia, 199

        clary sage, 205
        corn silk, 229
        fenugreek, 288
        horsetail, 368
        nutmeg and mace, 468
        psyllium, 512
        sarsaparilla, 556
        sassafras, 558
        squill, 575
        sweet basil, 75
        sweet woodruff, 623
        wintergreen oil, 619
    kidney stones, remedies:
        celery seed, 166
        centaury, 168
        doggrass, 248
        hydrangea, 370
        uva ursi, 610
    see also Bladder; Nephritis; Pyelitis; Urethritis
Knees, *see* Joints
Knotted marjoram, *see* Marjoram (Sweet, pot and wild)
KNOTWEED, GIANT (HUZUANG), 668, 669
KOLA NUT, 398–399
Kreata, 650
Kretak cigarettes, 211
KUDZU ROOT, 399–402
Kulinjan, 647

LABDANUM, 404–405
Labiatae, 74, 204, 362, 375, 406, 436, 443, 482, 485, 493, 535, 550, 559, 594, 680, 684, 686
Labor, bromelain effect on, 114
Lachrymatory, mustard, 458
Lactation, *see* Milk secretion
Lactobacillus acidophilus, inhibition by grape skin extract, 342
Laetrile, 23
Laksha guggulu, 645
Lamiaceae, *see* Lattatae
Laminaria digitata, **16**
Laminaria species, 16
Lantana species, 482
Lappa, *see* Burdock
Lard, chaparral as antioxidant, 175
Larrea tridentate, 175

Larvicidal:
  allspice, 20
  caraway, 138
  Chaparral, 174–176
  cinnamon, 196–200
  cloves, 210
  coriander, 228
  cumin, 236
  garlic, 309
  gentian, 316–317
  mustard, 457–459
  nutmeg and mace, 468, 469
  tagetes, 581
  thyme, 594–597
Laryngitis, remedies:
  balsam tolu, 71
  thyme, 596
Lauraceae, 59, 76, 106, 196, 557
Laurel, *see* Bay Sweet; Cherry laurel leaves
Laurocerasus leaves, *see* Cherry laurel leaves
Laurocerasus officinalis, 183
Laurus nobilis, 76, 78
Laurus persea, 59
Lavandin, lavender, 406
Lavandula angustifolia, 406
***Lavandula latifolia*, 406**
LAVENDER, 406–408
Lawsonia alba, 355
Lawsonia inertnis, 355
Laxative:
  agar, 10
  aletris, 9–11
  almond oil, 22
  aloe, 26
  asafetida, 49
  asparagus, 53
  bee pollen, 82
  belladonna, 89
  blue cohosh, 105
  boneset, 109–110
  broom tops, 116
  buchu, 117
  buckthorn, 119
  burdock, 121
  caraway, 139
  cardamon, 141
  carob, 142–144
  cascara sagrada, 150
  castor oil, 163
  celery seed, 166
  chicory root, 189–191
  Chirata, 192–193
  coriander, 227–228
  damiana, 237
  dandelion root, 238–240
  elder flowers, 257–258
  fennel, 283, 284
  ficin, 291–293
  fo-ti, 295
  genet, 315
  ginger, 322
  goldenseal, 337
  Gurmar, 649
  honey, 358
  hydrangea, 369–370
  juniper berries, 390
  karaya gum, 394
  licorice root, 417
  orange, 478
  parsley, 488
  podophyllum, 506
  psyllium, 511
  quassia, 515
  rhubarb, 525
  rose hips, 529
  roselle, 533
  senna, 569
  tamarind, 584
  triphala, 654
  uva ursi, 611
  see also Cathartic
Layor caring, *see* Agar
Learning, increasing, astragalus, 56
Leather industry:
  proteases used, 698
  tannic acid used, 585–587
Leprosy, remedies, quebracho, 517
Leche de higueron, *see* Ficin
Leche de oje, *see* Ficin
Ledebouriella divaricata, 281
Ledebouriella seseloides, 281
Ledger bark, *see* Cinchona (red and yellow)
Leguminosae, 4, 12, 55, 67, 69, 71, 115, 142, 152, 157, 206, 249, 286, 314, 347, 399, 415, 426, 583, 601
Lemon, wild, *see* Podophyllum
LEMONGRASS, 412–414
LEMON OIL (AND LEMON PETITGRAIN OIL), 410–411
Lemon thyme, 595, 596
Lenses, cement for, balsam Canada, 66
Leopard's bane, *see* Arnica
Leprosy remedies:
  gotu kola, 340
  hops, 360
  kava, 397
  myrrh, 462
  neem, 652
  see also Anlitubercular
Leptospermacm scoparium, inhibition by honey, 358
Lettuce:
  growth inhibition by asparagusic acid, 52
  increased yield by triacontanol, 85
Leucanthemum parthenium, 289
Leucorrhea, remedies, 68, 81
Leukemia, *see* Cancer
Leukocytopenia, remedies, 303
Leukopenia, remedies, 270
Leukorrhea, remedies, Cherokee rosehips, 181
Leukorrhagia, remedies, 678
Levant storax, 579, 580
Levisticum officinale, 427
Lianqiao, 659, 665
Liatris odoratissima, 240
Liatris, *see* Deertongue
Lice, destruction of, sassafras, 558
Licorice candy, ingredients, 417
LICORICE ROOT, 415–418
Ligusticum chuanxiong, 682
LIGUSTICUM (GAOBEM), 669–670
Ligusrictma jeholense, 669, 670
Ligusticum sinense, 669, 670
LIGUSTRUM, 420–421
Ligustrum lucidum, **420**
Liliaceae, 11, 24, 52, 308, 474, 555, 574, 633
LIME OIL, 422–423
Lion's tooth, *see* Dandelion root
Lipids:
  garlic effect on, 308
  ginseng effect on, 332
  see also CHEMICAL INDEX
Lipogenesis, inhibition, Dahurian angelica, 663

Lipolysis, ACTH-induced, Dahurian angelica, 663
Lippia berlandieri, 482
Lippia graveolens, 437, 482, 483
Lippia palmeri, 482
Lippia species, 437, 482, 483
Lippia vulgare, 482
Lips, cracked, remedies, balsam tolu, 71
Lipstick, ingredients, 23, 4,1 85, 130, 155, 218, 386, 387
Liqueurs, flavor ingredients:
 angelica, 33*****
 anise, 38
 artichoke, 47
 boldo leaves, 108
 carrot seed oil, 147, 148
 chamomile, 172
 cherry laurel leaves, 184
 cinchona, 195
 clary sage, 205
 cocoa, 218
 coffee, 224
 cubebs, 234
 elecampane, 259–260
 fennel, 284
 ginger, 322
 guarana, 350
 lemon oil, 411
 mint, 445
 orange, 479
 quassia, 516
 sweet basil, 75
 thyme, 596
 Valerian root, 613
 vanilla, 617
 see also Beverages, alcoholic and nonalcoholic
Liquidambar oriontalis, 579
Liquidambar styraciflua, **578**
Liquid extracts, *see* Fluid extracts
Listeria monocytogenes, 43, 589
Liver:
 cirrhosis, remedies, jasmine, 383
 damage, as toxic symptoml:
  chenopodium oil, 180
  deertongue, 241
  grape skin extract, 342–343
  quillaia, 519
  rhubarb, 526
  tannic acid, 586

detoxifying function, increased, immortelle, 378
disorders:
 absinthium, 2
 acerola, 7
 boldo leaves, 108
 cascara sagrada, 149–150
 chicory root, 190
 dandelion root, 239
 doggrass, 247–248
 immortelle, 378-379
 milk thistle, 441
 podophyllum, 505–507
 turmeric, 603–605
 see also Hepatitis
hydrolysates, MSG, 453
necrosis, protective against valerian root, 613
preventing fatty, lycium fruit, 429
protectant, *see* Hepatoprotective
Livestock feed, *see* Animal feed
LOBELIA, 424–425
Lobelia chinensis, 425
Lobelia inflata, 424
Lobelia sessilifolia, 425
Lobelia species, 425
LOCUST BEAN GUNI, 426–427
Loganiaceae, 312
Logwood, 356
Lonicera japonica, 667
Lonicera species, 667
Loranthaceae, 448
Lotions, *see* Creams and lotions
Lovage, *see* Sichuan lovage
LOVAGE ROOT, 427–429
Lucerne, *see* Alfalfa
Lucium barbarum, 429–431
Ludang, 220
Luffa acutangula, 671
Luffa cylindrica, 671
LUFFA (SIGUALUO), 671
Lumbago, remedies:
 black cohosh, 97–99
 capsicum, 134
 chamomile, 172
 devil's claw, 243
 guarana, 351
 mustard, 459

rehmannia, 523
wintergreen oil, 619
Lung problems:
 abscess, remedies, Job's tears, 385
 congestion, remedies, cocoa, 218
 infection, remedies, giant knotweed, 669
 mucous discharges, remedies, nettle, 465
 remedies:
  borage, 112
  hyssop, 376
  Wild cherry bark, 182–183
 see also Respiratory ailments
Lupulin, 359, 360
Lupus, remedies, 14, 340, 365, 632
European oregano, 482
Lycium chinense, 429
LYCIUM FRUIT, 429–432
Lycium halimifolium, 429
Lymphatic stress, adaptogenic activity, eleuthero, 263
Lymphedema, coumarins in treating, 241
Lymphocytopenia, as toxic symptom, caramel color, 136–137

Mansa, 638
Machixian, 678–679
Macis, *see* Nutmeg and mace
Macrocystis pyrifera, 16
Macrocystis species, 16
Macrophages, stimulation, tienchi ginseng, 599
Maggi herb, *see* Lovage root
Magnesium, *see* CHEMICAL INDEX
Magnolia, 434–435
Magnolia biondii, 434
Magnolia denudata, 434
Magnolia emargenata, 434
Magnolia fargesii, 434
MAGNOLIA FLOWER, 434–435
Magnolia heptapeta, 434
Magnolia salicifolia, 434
Magnolia sargentiana, 434
Magnolia sprengeri, 434

Magnolia wilsonii, 434
Mahabala, 651
Mahonia aquifolium, 72
Mahonia, trailing, *see* Barberry
Mahuang, 265
Majoon-e-flasia, 654
Majorana hortensis, 436
Majorana onites, 436
Malaria, *see* Antimalarial
Male fern, *see* Aspidium
Male sexual function, improved, epimedium, 269
Malnutrition, remedies, royal jelly, 540
Malphighia punicifolia, 6
Malpighia glabra, 6
Malt syrup, caramel color, 136
Malvaceae, 30
Mammary glands, inflammation, *see* Breast problems
Mandrake, *see* Podophyllurn
Manganese, *see* CHEMICAL INDEX
Manikara achras, 188
***Manilkara zapota*, 188**
Manilkara zapolilla, 188
Malt syrup, imitation, flavoring ingredient fenugreek, 286–288
Margarine, *see* Oleomargarine
Marginal fern, *see* Aspidium
Marigold, *see* Tagetes
Marijuana, absinthium, 2
MARJORAM (SWEET, POT, AND WILD), 436–438, *see also* Oregano
Marrubium, see Horehound
Marrubium vulgare, 362
Marshmallow root, *see* Althea root
Masking agents:
  acacia, 5
  anise, 38
  licorice root, 417
  rosemary, 536
  yerba same, 629
Massage cream, ingredients, 60
Master of the wood, *see* Woodruff, sweet
Mastitis, *see* Breast problems
Matricaria chamomilla, 169
Matricaria, *see* Chamomile

Mattress filler, genet, 315
Mayapple, *see* Podophyllum (podophyllin)
Maypop, *see* Passion flower
Meadow clover, *see* Clover tops, red
Measles, remedies:
  burdock, 121
  coriander, 228
  kudzu root, 401
  peony bark, 675
  rehmannia, 523
Meat and meat products:
  colorants:
    alkanet, 19
    annatto, 41
    capsicum, 134
    caramel color, 137
    turmeric, 605
  flavoring ingredients:
    alfalfa, 14
    allspice, 19
    anise, 36
    asafetida, 48
    balsam copaiba, 67
    bergamot oil, 92
    black and white pepper, 498
    bois de rose oil, 106
    boronia absolute, 113
    cajeput oil, 126
    capsicum, 135
    caraway, 139
    cardamon, 141
    carrot seed oil, 147, 148
    castor oil, 155
    cedar leaf oil, 162
    celery seed, 165
    chervil, 185
    cinnamon and cassia, 196
    cloves, 210
    coriander, 228
    cubebs, 234
    cumin, 236
    dill, 244–246
    elemi gum, 262
    eucalyptus, 273
    fennel, 284
    fenugreek, 287
    galbanum, 300
    garlic, 309
Meat and meat products, flavoring
    ginger, 323

    guaiac wood oil, 346
    hyssop, 376
    juniper berries, 390
    labdanum, 405
    lemongrass, 414
    lemon oil, 411
    licorice root, 416
    lime oil, 423
    lovage root, 428
    marjoram, 438
    mints, 445
    mustard, 459
    myrrh, 462
    nutmeg and mace, 469
    olibanum, 472
    onion, 475
    orange, 479
    oregano, 483
    parsley, 488
    patchouly oil, 494
    rosemary, 536
    sage, 551
    savory, 560
    sweet bay, 77
    tannic acid, 587
    tarragon, 589
    thyme, 596
    turmeric, 603
    turpentine, 608
    valerian root, 613
  gums:
    agar, 10
    algin, 16–17
    carrageenan, 145
    guar gum, 348
    karaya gum, 394
    neem, 652
    tragacanth, 601
    xanthan gum, 624
  other ingredients:
    locust bean gum, 427
    MSG, 452–453
    West Indian bay, 77, 78
  sweeteners, stevia, 577
  tenderizers:
    bromelain, 114
    ficin, 292
    papain, 292
Meda, 638
Mecca balsam, 64
***Medicago sativa*, 12**
***Melaleuca alternifora*, 124**
***Melaleuca leucadendron*, 125**

General Index

Melaleuca quinquenervia, 124
Melaleuca species, 124
Melia azadirachta, 652
Meliaceae, 215, 652
Melissa officinalis, 62
Melissa, *see* Balm
Memory, increasing:
 astragalus, 56
 epimedium, 270
 ginkgo, 326
 gotu kola, 339
 peony root, 677
 poria, 508–510
 sour jujube kernel, 572–573
Menopause, chaste-tree, 178
Menorrhagia:
 remedies, 199
 as toxic sympton, bromelin, 114
Menstrua, *see* Menstruums
Menstruation:
 complaints, remedies, hyssop, 376
 discomforts, remedies:
  aletris, 12
  balm, 62–65
  caraway, 139
  cinnamon and cassia, 200
  excessive, *see* Menorrhagia
  parsley, 488
  saffron, 548
  turmeric, 605
 irregular:
  catnip, 159–160
  rehmannia, 523
 other disorders, remedies:
  black cohosh, 98
  black haw bark, 99–100
  blue cohosh, 104–105
  chaste-tree, 178
  damiana, 237
  lavender, 408
  lovage root, 427–429
  myrrh, 460–462
  valerian root, 613
 regulation, remedies, peony root, 676
 see also Emmenagogue; Female disorders
Mentha anvensis var. *piperascens*, 443
Mentha spicata, 443

Mentha viridis, 443
Mentha X piperita, 443
Mercury, 292, 497, 505, 556, 637, 643
Mercury poisoning, remedies, sarsaparilla, 556
Metabolism, ginseng effect on, 332
Metabolizable energy, reduction, 348
Metals, 5, 87, 292, 496, 497, 637, 642, 643
Metrorrhagia, remedies:
 bee pollen, 83
 Cherokee rosehip, 181–182
 garden burnet, 666
 purslane, 679
 red sage, 680
 rehmannia, 523
 tienchi ginseng, 600
Mexican avocado, 59
Mexican marigold, *see* Tagetes
Mexican marjoram, *see* Oregano
Mexican tea, *see* Chenopodiuat oil
Mexican wild sage, *see* Oregano
Microcirculation diseases, bilberry, 93–94
Micrococcus glutamics MSG production, 452–453
Microscopy:
 cement, balsam Canada, CO, 65–66
 clearing agent, cedarwood oil, 162–164
Migraine, remedies:
 fangfeng, 282
 feverfew, 290
 gelsemium, 313
 guarana, 351
 immortelle, 379
 kava, 397
 kola nut, 399
 kudzu root, 399
 lavender, 408
 ligusticum, 670
 saw palmetto, 563
 sweet woodruff, 623
 valerian root, 614
Milfoil, 626
Milk products, *see* Dairy products

Milk secretion:
 cessation, remedies, sage, 551
 kava contraindication, 396
 stimulant:
  caraway, 139
  chaste-tree, 178
  dandelion root, 239
  dill, 244–246
  luffa, 671
  neem, 653
  purslane, 679
MILK THISTLE, 439–441
Mineralocorticoid activity, 80
Minerals, *see* CHEMICAL INDEX
MINTS, 443–446
MISTLETOE, 448–450
Mohave yucca, 633–634
Moisturizer:
 cassie absolute, 152
 cosmetic preparations, aloe, 27
 see also Skin protestant; Skin softner
Molasses, caramel color, 136
Molding material, bletilla tuber, 661
Monimiaceae, 107
Monkshood, *see* Aconite
Monoamine oxidase inhibitor, 194, 631
MONOSODIUM GLUTAMATC MSG, 452–453
Moraceae, 291, 359
Morning sickness, remedies, feverfew, 290
Moroccan geranium oil, 139, 319
Moroccan otto of rose, *see* Rose oil and absolute
Moroccan rose oil, *see* Rose oil
Moschidae, 455
Moshas moschiferus, 455
MOTHER-OF-PEARL (ZHENZHUMU), 673–674
Motor activity, suppression, lavender, 407
Motor disturbances, as toxic symptom ephedera, 266

Mountain balm, *see* Yerba Santa
Mountain tobacco, *see* Arnica
Mouth irritations, *see* Mouth problems
Mouth problems, remedies, 23, 75, 85, 89, 94, 97, 658
Mouth washes, ingredients:
 black and pale catechu, 158
 caraway, 139
 chamomile, 172
 cinnamon and cassia, 198
 comfrey, 226
 eucalyptus, 273
 gokshura, 646
 mints, 445
 myrrh, 461
 neem, 652
 sweet basil, 75
 thyme, 596
MSG, *see* Monosodium glutamate
Mucars, *see* Karaya gum
Mucilage, *see* CHEMICAL INDEX
Mucous membrane:
 anti adhesion activity, cranberry, 232
 inflammation, remedies:
  Bilberry, 94
  calendula, 130
  goldenseal, 337
  gokshura, 646
  sage, 551
 substances acting on:
  althea root, 29
  cubebs, 233, 234
  myrrh, 460–462
 substances irritant to:
  arnica, 42–44
  capsicum, 132–135
  chenopodium oil, 179–180
  podophyllum, 505–507
 remedies:
  chamomile, 172
  saw palmetto, 562, 563
 see also Demulcent; Emollient; Irritant: Skin irritation: Throat inflammation
Mudanpi, 674–675
Mongol, 640, 641

Multiple sclerosis, evening primrose, 277
MUME (SMOKED PLUM OR WUMEI), 672
Mumps virus, remedies, 62
 see also Antiviral
Murillo bark, *see* Quillaia
Muscatel flavor, flavoring ingredients, clary sage, 204, 205
Muscatel sage, *see* Clary sage
Muscle contractant, magnolia flower, 435
Muscle relaxant:
 angelica, 33
 kava, 396
 magnolia flower, 435
 purslane, 679
 see also Antispasmodic; Smooth muscles
Muscles:
 toxic effects on:
  coffee, 223
 remedies, sore and aching:
  broom tops, 116
  cajeput oil, 125
  hydrangea, 370
  turpentine, 608
  wintergreen oil,
 strain, remedies, horsechestnut, 365
 twitching, as toxic symptom, 365
Muscle stimulant, skeletal, 218, 223
Mushroom poisoning, remedim, 303, 441
MUSK, 455–456
Muskmallow, *see* Ambrette seed
Musk seed, *see* Ambrette seed
MUSTARD, 457–459
 prepared, turmeric, 605
Mutagenicity:
 caffeine, 223
 caramel color, 136–137
 cinnamon and cassia, 199
 stevia, 577
Myalgia, remedies, 373, 397
Mydriasis, ephedra, 266
Mycobacterium species inhibition by balm, 62

Mycobacterium species, inhibition, 378
Mycoses, *see* Antlfungal
Myocardial infarction, coffee linked to, 223
Myrcia oil, 78, 79
Myrcia, *see* Bay, West Indian
Myrica cerifera, 80
Myristicaceae, 467
Myristica fragrans, 467
Myristica officinalis, 467
Myristica, *see* Nutmeg (and Mace)
Myroxylon balsamum, 69, 71
Myroxylon pereirae, 69
Myraxylon toluiferum, 71
MYRRH, 460–462
Myrtaceae, 19, 78, 124, 209, 271
Myrobalanus bellirica Gaertn, 653
Myrtle, *see* Bayberry bark; Calamus

Nabd, 641
Nails:
 biting, remedies, capsicum, 134, 135
 discoloration, as toxic symptom, 356, 396
 injuries, remedies, horsetail, 368
Nan cangzhu, 661
Nalanda, 637
Narcotic:
 almonds, 22
 dogwood, 249
 genet, 315
 kava, 396
 lavender, 407
 potentiating activities:
  clary sage, 205
  lavender, 407
Naringin dihydrochalcone (NDHC), 480
Naringin extract, 344, 345
Nasal congestion, *see* Colds
Nasal sprays, ingredients, 199
Natural killer cells, stimulation, tichencing ginseng, 599
Nausea:
 remedies:
  cloves, 211

General Index

coriander, 228
elecampane, 260
ginger, 321
lavender, 408
mints, 446
nutmeg and mace, 468
patchouly oil, 494
savory, 560
tamarind, 585
as toxic symptom:
angostura bark, 35
aspidium, 54
bromelain, 114
castor oil, 154
chaparral, 175–176
coffee, 223
eucalyptus, 273
fo-ti, 296
ginkgo, 327
ipecac, 380
nutmeg, 468
quebracho, 517
tannic acid, 585
yohimbe, 632
uva ursi, 610
see also Gastic: Gastrointestinal
Naval stores, 607
Neisseria meningitidis, inhibition, 73
Neisseria species, inhibition by henna, 356
Neohesperidin dihydrochalcone (NHDHC), 480
Nepeta cataria, 159
Nephritis:
remedies:
aconite, 8
damiana, 237
euphorbia, 276
uva ursi, 611
urinary protein reduction, 57
as toxic symptom, aloe, 26
see also Kidney
Nephrolithiasis, see Kidney stones problems
Neer, 642, 643
Neruppu, 643
Nerve tonic, see Nervous problems
Nervine, see Nervous problems
Nervous problems, remedies:
asafetida, 49

balm, 63
black cohosh, 98
castoreum, 156
celery seed, 166
damiana, 237
hops, 360
kava, 396
labdanum, 405
marjoram, 436–438
olibanum, 472
orange, 480
oregano, 483
passion flower, 491
pipsissewa, 504
poria, 509
roselle, 534
rosemary, 537
sage, 551
valerian root, 614
see also Anxiety: Neuralgia;
NETTLE, 464–466
Neuralgia, remedies:
aconite, 8
belladonna, 89
capsicum, 133
devil's claw, 243
dill, 246
gelsemium, 313
Kola nut, 399
lavender, 408
passion flower, 491
sweet birch oil, 96
sweet woodruff, 623
wintergreen oil, 623
see also Nervous problems
Neurasthenia, remedies:
Cherokee rosehip, 181
epimedium, 270
ganoderma, 303
musk, 456
schisandra, 566
sour jujube kernel, 573
valerian root, 614
see also Nervous problems
Neuritis, remedies, 53
Neurodermatitis, remedies:
astragalus, 57
horsetail, 368
mume, 672
peony bark, 675
rehmannia, 522
Neurosis, experimental, ginseng effect, 332

Neurotoxicity, remedies, ginkgo, 326
Newcastle disease virus, see Antiviral
New-mown hay fragrance, 241
Niacin, see CHEMICAL INDEX
Nidan Parivarjan, 639
Night blindness, remedies, 662
Nilam, 643
Nightmares:
remedies:
ephedra, 267
ganoderma, 303
as toxic symptom, absinthium, 2
Night sweat, remedies:
rehmannia, 523
schisandra, 567
sour jujube kernel, 572
Nightshade, deadly, see Belladonna
Nipples, cracked, remedies, 71, 172
Nocturnal emission, see Spermatorrhea
Nonalcoholic beverages, see Beverages, alcoholic and nonalcoholic
Nosebleed:
remedies:
bee pollen, 83
black and pale catechu, 158
garden burnet, 666
ginseng, 333
peony bail, 675
rehmannia, 523
tienchi ginseng, 600
yarrow, 626
as toxic symptom, 627
see also Hemorrhage; Yarrow
Numbness of limbs, remedies, fo-ti, 296
Nut flavors, castor oil, 155
Nutgalls, tannic acid, 585
NUTMEG AND MACE, 468, 469
Nut products:
flavoring ingredients, mustard, 458
MSG, 453
Nutritive, honey, 358

Obesity, agent causing, MSG, 453

Occult blood, guaiac gum in, 346
Ocimum basilicun, **74**
Oculomotor paralysis, as toxic symptom, kava, 396
Oenothera biennis, **111, 276**
Oil of cade, *see* Cade oil
Oil of juniper tar, *see* Cade oil
Oil, *see* Fats and oils
Ointments:
  antiseptics:
    balsam Canada, 66
    cade oil, 123
    chamomile, 171
    neem, 653
    mints, 445
  base:
    almond oil, 22, 23
    castor oil, 155
    cocoa butter, 218
    beeswax, 84–85
  fragrance ingredients:
    lavender, 408
    rose oil, 532
  other ingredients:
    Baikal skullcap, 685, 686
    calendula, 129–131
    cedar leaf oil, 161–162
    chickweed, 187–188
    echinacea, 251–256
    fangfeng, 281–282
    gotu kola, 339–341
    thyme, 594–597
Oleaceae, 382, 420, 665
Oleogum resins, *see* Gum resins
Oleomargarine, colorants, 19. *see also* Fats and oils
OLIBANUM, 471–472
Onagraceae, 276
ONION, 474–475. *see also* Squill
Ophthalmic products, ingredients:
  belladonna, 88–89
  castor oil, 154–155
  cherry laurel leaves, 183–184
  cinchona, 194–195
  comfrey, 225–226
  elder flowers, 257–258
  eyebright, 279–280
  ginkgo, 325–327
  goldensea, 336–338

Sichuan lovage, 682–683
witch hazel, 620–621
Opobalsam, *see* Balsam tolu
Oral, *see* Mouth
ORANGE (BITTER AND SWEET), 477–480
  acerola, 6–7
Orange flower water, 477, 479, 480
Orange root, *see* Goldenseal
Orchanette, *see* Alkanet
Orchidaceae, 616, 661
Orchitis, remedies, ginger, 321
OREGANO, 436–438, 482–483
Oregon grape, *see* Barberry
Origanum majorana, **436**
ORIGANUM OIL, SPANISH, 485
Origanum onites, **436**
Origanum, *see* Oregano
Origanum species, **485**
Origanum vulgare, **436, 482, 483**
Osteoporosis, remedies, horsetail, 368
Otitis, remedies, kava, 397
Otto of rose, *see* Rose oil and
Ovarian cancer, coffee linked to, 449, 592
Oxytocic, 116, 315

Panchkarma, 639
Pachyma cocos, 508
Pachymma hoelen, 508
Padang-cassia, *see* Cinoamon
Paeonia suffruticosa, **674**
Paeonia arborea, **674**
Paeoniaceae, 674, 676
Paeonia lactiflora, **676**
Paeonia moutan, **674**
Pain:
  abdominal:
    remedies:
      aletris,11–12
      bee pollen, 81–83
      belladonna, 88–89
      capsicum, 132–135
      chamomile, 169–173
      cinnamon and cassia, 199
      jasmine, 382–383
      juniper berries, 389–391
      lavender, 406–408
      lemongrass, 412–414
      licorice root, 415–418

      ligusticum, 669–670
      olibanum, 471–472
      patchouly oil, 493–494
      peony bark, 674–675
      peony root, 676–677
      tagetes, 581–582
      tienchi ginseng, 598–600
      turmeric, 603–605
    as toxic symptom:
      arnica, 42–44
      castor oil, 154–155
      cinchona, 194–195
      fo-ti, 293–297
    see also Colic
  arthritic, *see* Arthritis
  chest, remedies:
    balsam Canada, 65–66
    cinnamon and cassia, 196–200
    luffa, 671
    musk, 30–31, 455–456
    peony root, 674–675
    red sage, 680-681
    safflower, 681–682
    Sichuan lovage, 682–683
    tienchi ginseng, 598–600
    turmeric, 603–605
  general, remedies:
    aconite, 7–9
    allspice, 19–21
    beeswax, 84–85
    cherry laurel leaves, 183–184
    civet, 203–204
    eleuthero, 262–264
    gelsemium, 312–313
    ipecac, 379–381
    Jasmine, 382–383
    Job's tears, 384–385
    myrrh, 460–462
    pine needle oil, 502–503
    saffron, 547-548
    sandalwood oil, 553–555
    turmeric, 603–605
    turpentine, 607–608
    wild cherry bark, 182
    wintergreen oil, 619
  kidney, remedies, feverfew, 289-291
  sternum, as toxic symptom, chaparral, 174 176
  see also Analgesic; Headache; Stomachache

Paints, ingredients, 155, 608
Palmaceae, 561
Palma christi, *see* Castor oil
Palmarosa oil, 319
Palpitation, remedies, 89, 221, 491, 523, 573
Panama bark, *see* Quillaia
Panama wood, *see* Quillaia
***Panax ginseng,* 330, 645**
***Panax notoginseng,* 598**
Panax pseudo-ginseng, 598
Panax quinquefolius, 330
Panax sanchi, 598
Panax schinseng, 330
Pancytopenia, alfalfa, 14
Pansinusitis, remedies, garlic, 309
PAPAIN, 114, 146. *see also* Bromelain and ficin
Papaveraceae, 102
Paper industry:
    genet, 314–315
    karaya gum, 393–394
Paperbark tree, *see* Cajeput oil
Papilloma, *see* Cancer
Papoose root, *see* Blue cohosh
Paprika, 132–135
Paralysis, as toxic symptom:
    cinchona, 194–195
    horsechestnut, 364–366
Paramecicidal, bayberry bark, 80
Parasitic disease, remedies:
    asparagus, 52–53
    balsam peru, 69–70
    cade oil, 123
    calendula, 129–131
    chaparral, 174–176
Parkinson's disease, remmedies, 89, 178
PARSLEY, 487–488
Passifloraceae, 490
***Passiflora incarnata,* 490**
Passiflora, *see* Passion flower
PASSION FLOWER, 490–492
Passion vine, *see* Passion flower
PATCHOULY OIL, 493–494
Pathya Vyavastha, 639
***Paulinia cupana,* 350**
Paullinia sorbilis, 350
Pausinystalia johimbe, 631
***Pausinystalia yohimba,* 631**
Pear, alligator, *see* Avocado

PEARL (ZHENZHU OR MARGARITA) 673–674
PECTIN, 495–497
Pedaliaceae, 242
Pediculicide, sassafras, 558
Pegu catchu, *see* Catechu (black and pale)
Pelargonium capitatum, 318
Pelargonium crispum, 318, 487
***Pelargonium graveolens,* 318**
Pelargonium odoratissimum, 318
Pelargonium radula, 318
Pelargonium roseum, 318
Pelargonium species, 318, 319
Penicillium notatum, inhibition by grape skin extract, 342
Penicillium species, inhibition by henna, 202, 355, 413, 479
Peony, 674–675
PEONY BARK (MUDANPI), 674–675
PEONY ROOT, RED AND WHITE (*SHAOYAO: CHISHAOYAO AND BAISHAOYAO*), 676–677
Pepper:
    bell, 133
    green, 133
    oil, stability, 498
    sweet, 133
    see also Capsicum; Cubebs
PEPPER (BLACK AND WHITE), 498–499
    Allspice, 499
Peppermint, 139, 443–446
Pepper- wood, *see* Ash, prickly
Pepsin, vegetable, *see* Papain
Perfumes:
    fixatives:
        asafetida, 48–49
        balsam Canada, 65–66
        balsam copaiba, 67–68
        balsam Peru, 69–70
        balsam tolu, 71–72
        benzoin, 90–91
        castoreum, 156–158
        cedarwood oil, 162–164
        civet, 203–204
        costus oil, 230–231

        deertongue, 241–242
        elemi gum, 261–262
        galbanum, 299–300
        immortelle, 378-379
        lime oil, 422–423
        musk, 455–456
        myrrh, 460–462
        storax, 578-580
        turpentine, 607–608
    fragrance Ingredients:
        ambrette seed, 30–31
        angelica, 32–34
        anise, 36–38
        asafetida, 48–49
        balsam Canada, 65–66
        balsam copaiba, 67–68
        balsam Peru, 69–70
        balsam tolu, 71–72
        beeswax, 84–85
        bois de rose oil, 106
        boronia absolute, 112–113
        cade oil, 123
        cajeput oil, 124–126
        calamus oil, 128
        cananga oil, 132
        caraway, 138–139
        cardamon, 141
        carrot seed oil, 147–148
        cascarilla bark, 151–152
        cassie absolute, 152–153
        castoreum, 156–157
        cedar leaf oil, 161–162
        cedarwood oil, 162–164
        celery *seed*, 165–166
        chamomile, 169–173
        chenopodium oil, 179–180
        cinnamon and cassia, 199
        citronella oil, 201–202
        civet, 203–204
        clary sage, 204–206
        cloves, 209–211
        coriander, 227–228
        costus oil, 230–231
        cubebs, 233–234
        cumin, 235–236
        deertongue, 240–241
        dill, 244–246
        elder flowers, 257–258
        elecampane, 259–260
        eucalyptus, 271–273
        fennel, 283–285
        fenugreek, 286–288
        galbanum, 299–300

Perfumes (*Continued*)
   genet, 314–315
   geranium oil, 318–320
   ginger, 320-323
   grapefruit oil, 343–345
   greater galangal, 647
   guaiac wood oil, 345–346
   hyssop, 375–377
   immortelle, 378–379
   jasmine, 382–383
   labdanum, 404-405
   lavender, 406–408
   lemongrass, 412–414
   lemon oil, 410–411
   lime oil, 422–423
   lovage root, 427–429
   marjoram, 436–438
   mints, 443–446
   musk, 455–456
   myrrh, 460–462
   olibanum, 471–472
   orange, 477–480
   origanum oil, 485
   parsley, 487–489
   patchouly oil, 493–494
   pine needle oil, 502–503
   rosemary, 535–537
   rose oil, 530-532
   rue, 542–543
   saffron, 547–548
   sage, 550–552
   sandalwood oil, 553–555
   sweet basil, 74–75
   sweet bay, 76–77
   sweet birch oil, 95–96
   sweet woodruff, 622–623
   tagetes, 581–582
   tarragon, 588–590
   thyme, 594–597
   turmeric, 603–605
   vanilla, 616-617
   West Indian bay, 78–79
   ylang ylang oil, 629–630
   other ingredients:
      arnica, 42–44
      balm, 62–63
      bergamot oil, 91–92
      ginseng, 330–333
Peristalsis, promotion, *see* Laxatives
"Permeability" vitamin, *see* Vitamin P

Persea americana, **59**
Persea gratissima, 59
Persian, 299, 415
Perspiration:
   enhancement, *see* Diaphoretic
   see also Antiperspirant
Peruvian balsam, 69
Peruvian bark, *see* Cinchona (red and yellow)
Pesticide, lemongrass, 412–414
Petitgrain oil:
   bitter orange, 477–480
   lemon oil, 422–423
Petroselinum crispum, **487, 489**
Petroselinum hortense, 487
Petroselinun sativum, 487
Peumus boldus, **107**
Phaeophyceae, 16
Phagocytosis, stimulation:
   arnica, 42–44
   calendula, 129–131
Pharyngitis, remedies, chestnut leaves, 186
Phellodendron amurense, **677**
PHELLODENDRON BARK (HUANGBAI), 677–678
Phenacetin, 313
pH indicators, 342
Phlebitis, remedies, horse chestnut, 365
Phlegm, remedies to reduce, 480
Phosphorus, *see* CHEMICAL INDEX
Photophobia, as toxic symptom, 89, 397
Photosensitization, 373
Phototoxicity:
   angelica, 32–34
   bergamot oil, 91–92
   cumin, 235–236
   fangfeng, 281–282
   ginger, 320–323
   grapefruit oil, 343–345
   lemon oil, 410–411
   lime oil, 422–423
   lovage root, 427–429
   orange, 477–480
   parsley, 487–489
Physalis somnifera, 645

Phytotoxins, absinthium, 1–3
Picrasma excelsa, **515**
Pie-plant, *see* Rhubarb
Pigments, plant, *see* Food colors; CHEMICAL INDEX
Pigweed, 679
Pill-bearing spurge, *see* Euhporbia
Pile:
   mahabala, 651
   Terminalia, 654

Pillows, genet in, 315
Pimenta acris, 78
Pimenta dioica, **19**
Pimenta leaf oil, 20
Pimenta officinalis, 19
Pimenta racemosa, 76
Pimenta racemosa, **76, 78**
Pimenta, *see* Allspice
Pimento, *see* Allspice
Pimpinella anisum, **36**
Pimples, *see* Acne
Pinaceae, 65, 68, 123, 162, 501, 502
Pineapple, 113
Pine, prince's, *see* Pipsissewa
PINE BARK, WHITE, 501–502
PINE NEEDLE OIL (DWARF AND SCOTCH), 502–503
Pines elliottii, **607**
Pinus massoniana, 508
Pinus mugo, **502**
Pinus palustris, **607**
Pinus pumilio, 502
Pinus pumilio oil, 502
Pinus species, 82, 607
Pinus strobus, 501
Pinus sylvestris, **502**
Pinus sylvestris oil, 501, 502
Pinus tabulaeformis, 608
Pinus yunnanensis, 608
Piperaceae, 233, 395, 498
Piper cubeba, **233**
Piper methysticum, **395**
Piper nigrum, **498**
PIPSISSEWA, 504
Piscidia communis, 21, 249
Piscidia erythrina, 249
Piscidia piscipula, **249**
Plantaginaceae, 511

Plantago arenaria, 511
Plantago asiatica, 513
Plantago decumbens, 511
Plantago depressa, 513
Plantago indica, 511
Plantago ispaghula, 511
Plantago ovata, 511
Plantago psyllium, 511
Plantago, see Psyllium
Plantago species, 513
Plantain, *see* Psyllium
Plant growth regulator, alfalfa, 13
Plant protease concentrate, *see* Bromelain
Plastics, synthesis, castor oil, 155
Platelet aggregation:
   activities, rehmannia, 521–523
   inhibition:
      bilberry, 94
      bromelain, 113, 114
      evening primrose, 277
      feverfew, 289
      ganoderma, 303
      ginkgo, 326
      kudzu root, 401
      onion, 475
      peony bark, 675
      peony root, 677
      red sage, 680
      Sichuan lovage, 683
      sour jujube kernel, 572
      tienchi ginseng, 599
   stimulation, epimedium, 269
Pleurisy, remedies, gotu, 340
Pneumonia, remedies, 49, 68, 96, 134
Poaceae, *see* Graminae
Podophyllin, 505–507
Podophyllum emodi, 505
Podophyllum hexandrum, 505
Podophyllum peltatum, 505
PODOPHYLLUM (PODOPHYLLIN), 505–507
Poet's jessamine, *see* Jasmine
***Pogostemon cablin*, 493**
Pogostemon heyneanus, 493
Pogostemon patchouly, 493
Poisoning:
   aconite, 8
   almonds, 22
   anise, 38

aspidium, 54
benzene, remedies, ligustrum, 421
cedar leaf oil, 161
chenopodium oil, 179–180
cinchona, 195
eucalyptus, 271–273
food, *see* Food poisoning
gelsemium, 313
horse chestnut, 365
horsetail, 368
lobelia, 425
mushroom, remedies:
   ganoderma, 303
   milk thistle, 440
nutmeg, 467–469
orange, 477–480
podophyllum, 505–507
remedies, honeysuckle flower, 667
rue, 542–543
sweet birch oil, 96
tamarind, 583–585
tannic acid, 586
turpentine, 607–608
uva ursi, 610–611
wild cherry bark, 182
wintergreen oil, 619
Poison ivy, remedies, 425, 512
Poliomyelitis, remedies, 270
Polygonaceae, 293, 524, 668
***Polygonum cuspidatum*, 668**
***Polygonum multiflorum*, 293**
Polypodiaceae, 54
Polyporaceae, 300, 508
Popinac, *see* Cassie absolute
Poplar buds, *see* Balm of gilead buds
Populus balsamifera, 64
***Populus candicans*, 64**
Populus species, 65
***Populus tacamahacca*, 64**
***Poria cocos*, 508**
PORIA (FULING), 508–510
Portulacaceae, 678
***Portulaca oleracea*, 678**
Post extraction alveolitis, *see* Dry socket
Postthrombotic syndrome ginkgo, 325–327
Potassium, *see* CHEMICAL INDEX
Prakruti, 638

Pregnancy:
   cautions, sour jujube kernel, 573
   difficulties, remedies, 333
   kava contraindication, 396
Premenstrual syndrome, remedies:
   black cohosh, 98
   borage, 111
   chaste tree, 178
   evening primrose, 277
Prepared mustard, 458
Preservative:
   benzoin, 91
   phellodendron bark, 677
   see also Antioxidant
Prithivi, 643
Prince's pine, *see* Pipsissewa
Privet, Egyptian, *see* Henna
Processed foods, *see* Baked goods; nonalcoholic; Candy; Condiments and relishes; Desserts, frozen dairy; Fats and oils; Gelatins and puddings; Meat and meat products; Vegetables, processed
Prostate:
   cancer, coffee linked to, 223
   hyperplasia, remedies:
      saw palmetto, 562
   hypertrophy, inhibition, bee pollen, 82
   prostatitis, remedies, echinacea, 255
Protease, *see* CHEMICAL INDEX
Protease concentrate, plant, *see* Bromelain
Protein, *see* CHEMICAL INDEX
Protein hydrolyzates:
   MSG, 452
   proteases in production, 698
***Prunus amygdalus*, 21**
Prunus armeniaca, 22
Prunus communis, 21
Prunus domestica, 22
Prunus dulcis, 21
***Prunus laurocerasus*, 183**
***Prunus mume*, 672**
Prunus persica, 23

Prunus serotina, 182
Prunus species, 22, 183
Pruritus:
  remedies:
    balsam Peru, 70
    dittany bark, 664
    fangfeng, 282
    fo-ti, 297
    kudzu root, 401
    mints, 446
    mume, 672
    phellodendron bark, 678
    purslane, 679
  as toxic syndrome, ganoderma, 303
Pseudomonas aeruginosa, inhibition, 275
Pseudotsuga douglasii, 68
Pseudotsuga mucronata, 68
***Pseudotsuga taxifolia*, 68**
Psoralens, *see* Conmarins
Psoriasis, remedies:
  angelica, 32–34
  bergamot oil, 92
  chickweed, 187
  Dahurian angelica, 663
  echinacea, 255
  immortelle, 379
  kudzu root, 401
  lemon oil, 411
  lovage root, 428
  papain, 114
  podophyllum, 506
  red sage, 681
  rue, 542
  see also Skin
Psychedelic activities, 488
Psychomotor retardation, remedies, 372
Psychotropic ipecacuanha, 379
Psychotropic, 467
PSYLLIUM, 511–513
***Pteria margaritifera*, 673**
***Pteria martensii*, 673**
Pteriidae, 673
Pterocladia species, 9
Ptosis, as toxic symptom, 313
Puccoon, red, *see* Bloodroot
Puddings, *see* Gelatins and puddings
***Pueraria lobata*, 399**

Puerariu montana, 399, 502
Puerariu pseudohirsuta, 399
Puerariu thomsonii, 399, 400
Puerariu thunbergiana, 399
Puerto Rican cherry, *see* Acerola
Pulse rate, arnica effect, 42–44
Pulse, weak, as toxic symptom, 327
Pumilio pine oil, 502
Punk tree, *see* Cajeput oil
Pupil, substances causing dilation:
  belladonna, 88
  gelsemium, 313
Pupils, dilated, as toxic symptom, 313
Purgative, *see* Laxative
Purple clover, *see* Clover tops, red
Purpura, remedies, jujube, 388–389
PURSLANE, COMMON (MACHIXIAN), 678–679
Pyelitis:
  remedies, corn silk, 230
  see also Kidney
Pyelonephritis, remedies, ganoderma, 303
Pyorrhea, remedies, 60
Pyrethrum parthenium, 289
Pyricularia orysae, 17
Pyrolaceae, 504
Pyrolaceae, *see* Ericaceae

Qinghai, 400
Quack grass, *see* Doggrass
QUASSIA, 515–516
Quassia amara, 515
Quebrachia lorentzii, 516
QUEBRACHO, 516–518
Queen Anne's lace, *see* Carrot oils
Quercetiin-rutinoside, *see* Rutin
Quercus infectoria, 585
Quercus species, 585
Quick grass, *see* Doggrass
QUILLAIA, 518–519
Quillaja, 518–519
***Quillaja saponaria*, 518**
Quitch grass, *see* Doggrass

Rabbits, poisoning, avocado, 60
Radiation, x-ray, adaptogenic activity, eleuthero, 263

Radiation -protectant:
  sour jujube kernel, 572–573
Radiotherapy, side effects, 432
  alfalfa, 14
  baizhu, 660
  ginseng, 332
  lycium fruit, 432
Ranunculaceae, 7, 97, 336, 674, 676
Rajabaidya, 637
Rash, *see* Skin rash
Raktawardak, 645
Ras Tarangini, 637
Rasa, 638, 640
Rasayana, 638, 640
Red algae, 9, 11, 145
Red bark, *see* Cinchona (reds
Red blood cells, hemolysis, hemolysis, *see* Hemolytic
Red cedar, 162, 163
Red pepper, *see* Capsicum
Red Peruvian bark, *see* Cinchona
Red puccoon, *see* Bloodroot
Red root, *see* Bloodroot
RED SAGE (DANSHEN), 680–681
Red thyme oil, thyme, 595, 596
Refrigerant:
  Baikal skullcap, 685, 686
  borne, 133, 140, 142, 216, 238
  codonopsis, 220–221
  cranberry, 232–233
  ginseng, 330–333
  kudzu root, 399–402
  rehmannia, 521–523
  roselle, 532–534
  schisandra, 564–567
  tamarind, 585
REHMANNIA, 521–523
Rehmannia glutinosa, 521
Rehmannia chinensis, 521
Release agent, candy, castor oil, 155
Religious rituals, turmeric, 605
Relishes, *see* Condiments and relishes
Renal failure, as toxic symptom, 526, 632
Renal gravel, 53, 465, 488
Rendong, 667
Resin tole, *see* Balsam loin

General Index

Respiratory ailments, remedies:
  benzoin, 91
  blue cohosh, 104
  chamomile, 171
  chestnut leaves, 186
  dandelion root, 239
  echinacea, 254
  eucalyptus, 272
  fennel, 284
  ginkgo, 326
  horehound, 363
  kava, 397
  roselle, 534
  sweet birch oil, 95–96
  sweet woodruff, 623
  see also Asthma: Bronchial
Respiratory center, paralysis, 632
Respiratory depression, as toxic, 313
Respiratory enzymes, inhibition, chaparral, 174–176
Respiratory failure, as toxic symptom:
  almonds, 23
  aspidium, 54
  belladonna, 89
  quillaia, 519
  thyme, 595
Respiratory stimulant:
  coffee, 223
  cubebs, 234
  dogwood, 249
  eucalyptus, 272
  fennel, 284
  ginger, 321
  labdanum, 405
  passion flower, 502
  quebracho, 517
  saffron, 547
  schisandra, 566
  vasaka, 656
Respiratory volume, increased, dill, 245
Restlessness:
  remedies:
    Baikal skullcap, 685, 686
    hops, 359–360
    lavender, 408
    passion flower, 491
    sweet woodruff, 623
  as toxic symptoms:
    belladonna, 89

coffee, 223
ginkgo, 327
Restless sleep, see Insomnia
Retinal conditions, degenerative, 94
"Reunion" basil oil, 74
Reynaud's disease, ginkgo, 326
Rhamnaceae, 118, 149, 388, 572
Rhamnus frangula, 118
Rhamnus purshiana, 149
Rheumatism:
  remedies:
    ashwagandha, 645
    aconite, 7–9
    aletris, 12
    ashwagandha, 645
    asparagus, 53
    balm of Gilead buds, 65
    bee pollen, 83
    belladonna, 89
    black cohosh, 98
    bloodroot, 103
    burdock, 121
    cajeput oil, 128
    cangzhu, 662
    capsicum, 134
    cassic absolute, 340
    cedar leaf oil, 164
    celery seed, 166
    Chaparral, 176
    chaste-tree, 180
    chenopodium oil, 180
    chestnut leaves, 186
    chickweed, 187
    cinnamon and cassia, 199
    Dahurian angelica, 663
    dandelion root, 239
    devil's claw, 242
    dittany bark, 664
    elder flowers, 258
    epimedium, 270
    fangfeng, 282
    feverfew, 290
    garlic, 310
    gelsemium, 313
    ginger, 321
    ginseng, 333
    gotu kola, 340
    greater galanga, 648
    guaiac wood oil, guarana, 346
    immortelle, 379

jasmine, 383
Job's tears, 385
kava, 397
lavender, 408
lemongrass, 414
ligusticum, 670
luffa, 671
marjoram, 446
mustard, 459
neem, 652
nettle, 465
olibanum, 472
oregano, 483
parsley, 488
prickly ash, 51
rosemary, 536
royal jelly, 540
sarsaparilla, 556
sassafras, 558
schisandra, 567
Sichuan lovage, 683
sweet birch oil, 96
Terminalia, 654
triphala, 654
thyme, 596
turpentine, 608
valerian root, 614
white pine bark, 501
wintergreen oil, 619
  see also Arthritis: Inflammation
Rheum emodi, 525
Rheum officinale, 524
Rheum palmatum, 524
Rheum rhaponticum, 524, 525
Rheum species, 524
Rheum tanguticum, 524, 525
Rh factor, ficin, 293
Rhinitis, remedies:
  eyebright, 280
  ganoderma, 303
  magnolia flower, 435
  mume, 672
  nettle, 464
Rhodophyceae, 9, 145
RHUBARB, 524–527
Rhus chinensis, 586
Rhus potaninii, 586
Rhus species, 586
Riboflavin, see CHEMICAL INDEX
Rice fungal disease, sodium alginate in remedies, 17

Ricinus communis, 154
Ricinus, see Castor Oil
Ringworms, remedies:
  ashwagandha, 645
  clove oil, 210
  eucalyptus, 273
  garlic, 310
  goldenseal, 338
  lobelia, 425
  turmeric, 605
  turpentine, 608
  see also Antifungal activities; Athlete's foot; Skin conditions
RNA, ginseng effect, 332
Rockberry, see Uva ursi
Rockrose, see Labdanum
Roman chamomile oil, 170
Rookshaha, 638
Root beer
  fragrance ingredients, 408, 503
    licorice root, 415
    sarsaparilla, 555
    sweet birch oil, 96
    valerian root, 612
    wintergreen oil, 619
  foaming agents, 519
  fragrance ingredients, pipsissewa, 504
Root canal fillings and sealers, see Dental material and preprations
Ropes, genet in, 315
Rosa alba, 530
Rosa Camellia, 181
Rosa canina, 528
Rosa centifolia, 529, 530
Rosa cherokensis, 181
Rosa damascena, 529–531
Rosa gallica, 532
Rosa laevigata, 181
Rosa nivea, 181
Rosa pomifer, see Rosa villosa
Rosa rugosa, 528, 529
Rosa sinica, 181
Rosa species, 528
Rosa ternata, 181
Rosa villosa, 528
ROSEHIP, CHEROKEE, 181–182
ROSE HIPS, 528–529
ROSELLE, 532–534

ROSEMARY, 535–537
ROSE OIL (AND ABSOLUTE), 530–532
Rosewood oil, see Bois de rose oil
Rosin, See Turpentine (and rosin)
Rosamarinus officinalis, 535
Roundworms, see Anthchnintic
ROYAL JELLY, 538–540
Ruhefacient:
  aconite, 7–9
  black and white pepper, 498, 499
  mustard, 457–459
  terpentine, 607
  see also Counterirritant
Rubiaceae, 157, 194, 222, 379
Rubus fructicosus, 96
RUE, 542–543
Rum cherry bark, see Cherry bark, wild
Ruminant bloat, alfalfa, 14
Ruta bracteosa, 542
Rutaceae, 477, 495, 542
Ruta graveolens, 542
Ruta montana, 542
RUTIN, 545–546

Sabal serrulata, 561
Sacred bark, see Cascara sagrada
SAFFLOWER (FALSE SAFFRON: HONGHUA), 681–682
Safra, 641
SAFFRON, 547–548
  substitute, calendula, 131
  see also Turmeric
Salad, dandelion leaves, 240
Salad burnet, 666
Salad dressings, gums, tragacanth, 602
  salanin, 652
  saram, 638
  yohimbe, 631–632
Salmonella species, inhibition by henna, 356
Salmonella typhi, inhibition by pectin, 496
Salmonella typhimurium, inhibition, 421
Salvia bowleyana, 680

Salvia lavandulaefolia, 550
Salvia miltiorrhiza, 680
Salvia officinalis, 550
Salvia przewalskii, 680
Salvia sclarea, 204
Salvia yunnanensis, 680
Sambucus canadensis, 257, 258
Sambucus formosana, 258
Sambucus nigra, 257, 258
Sambucus, see Elder flowers (American and European)
Sambucus williamsii, 258
SANDALWOOD OIL, 553–555
Sang, see Ginseng (Oriental and American)
Sanguinaria canadensis, 102
Sanguinaria, see Bloodroot
Sanguisorba officinalis, 666
Sanke weed, see Euphorbia
Sanqi, 598–600
Santalaceae, 553
Santal oil, see Sandalwood oil
Santalum album, 553
Santalum spicatum, 554
Sapindaceae, 350
Sapodilla, see Chicle
Saponin, see CHEMICAL INDEX
Saposhnikovia divaricata, 281
Sapota achras, 188
Sapotaceae, 188
Sarothamnus scoparius, 115
Sarothamnus scoparius, 115
Sarothamnus vulgaris, 115
SARSAPARILLA, 555–557
  fragrance ingredients, pipsissewa, 504
SASSAFRAS, 557–558
Sassafras albidum, 557
Sassafras officinale, 225, 226
Sassafras variifolium, 557
Satureja hortensis, 559
Satureja montana, 559
Satureja obovata, 559
Satvavajay, 639
Satvavajaya, 639
Sauda, 641
Shukra, 638
  mints, 443–446
  nutmeg and mace, 468, 469

orange, 477–480
fragrance ingredients
  kudzu root, 399–402
  pipsissewa, 504
gums:
  agar, 10
  carrageenan, 145–147
  guarana, 350–351
Sausage casings:
  colorants, annatto, 41
  proteases in manufacture, 698
Saussnrea lappa, **230**
SAVORY (SUMMER AND WINTER), 559–560
SAW PALMETTO, 561–563
Saxifragaceae, 369
Scabies, remedies, 2, 70, 126, 155, 272, 327, 340, 580, 645
Scabwort, *see* Elecampane
Scaling, as toxic symptom, anise, 37–38, *see also* Dermatitis
Scalp problems, remedies, cade oil, 123, *see also* antidandruff; Hair; Skin disorders
Scar tissue, inhibition, remedies, gotu kola, 340
Schinopsis quebracho-colorado, 516
Schinopsis lorentzii, 516
SCHISANDRA, 564–567
Schisandraceae, 564
Schisandra chinensis, **564, 565**
Schisandra species, 566, 567
Schisandra spenanthera, **564–567**
Sciatica, remedies:
  aconite, 8
  belladonna, 89
  chamomile, 172
  chaparral, 174–176
  gelsemium, 313
Scilla, *see* Squill
Scleroderma, remedies, gotu kola, 340
Sclerosis, remedies, 60, 277, *see also* Skin
Sclerotium cocos, 508
Scoparium, *see* Broom tops
Scotch broom, *see* Broom tops

Scrofula, remedies:
  coca, 212–214
  fo-ti, 297
  gotu kola, 340
  sweet birch oil, 96
  see also Antitubercular
Scrophulariaceac, 279, 521
***Scutellaria baicalensis*, 684, 686**
Sea ash, *see* Ash, prickly
Sea onion, *see* Squill
Seaweed extracts, 697
Seborrhea, remedies, 155, 575, *see also* Skin
Sedative, 23, 63, 182, 184
  aletris, 11–12
  ashwagandha, 645
  Baikal skullcap, 685, 685
  barberry, 73
  belladonna, 89
  black cohosh, 105
  calamus, 128
  castoreum, 156
  catnip, 160
  celery seed, 165
  centaury, 168
  cherry laurel leaves, 184
  chestnut leaves, 186
  chicory root, 189–191
  cinnamon and cassia, 196–200
  civet, 204
  coca, 212–214
  devil's claw, 242
  dill, 246
  doggrass, 248
  dogwood, 249
  elecampane, 259
  evening primrose, 278
  ganoderma, 302
  gelsemium, 313
  goldenseal, 337
  hops, 360
  jasmine, 395
  kava, 395
  lavender, 408
  ligusticum, 670
  ligustrum, 421
  lobelia, 425
  lovage root, 428
  mistletoe, 449
  orange, 480
  oregano, 483
  papain, 114
  passion flower, 490

peony root, 677
poria, 509
red clover tops, 208
saffron, 548
saw palmetto, 562
Sichuan lovage, 683
sour jujube kernel, 572
sweet bay, 76
sweet woodruff, 623
thyme, 595
valerian root, 612
vasaka, 656
wild cherry bark, 182
Sedge, sweet, *see* Calamus
See bright, *see* Clary sage
Seeds, 15, 22, 23, 30–34, 38, 40, 53, 59, 60, 111, 112, 119, 139–142, 144, 148, 154, 165, 166, 235, 244–246, 277, 278, 283, 286–288, 325, 327, 350, 398, 426, 467, 468, 511, 561, 572, 672
Seed gums, 697
Semliki Forest virus, *see* Antiviral
Seng, *see* Ginseng (Oriental and American)
SENNA, 568–570
Sequestrants:
  as color stabilizer, 87
  as protease activator, bromelain, 113
Serenoa repens, 561
Serology, ficin, cin, 293
Serotonin:
  horehound effect on, 363
  inhibition, feverfew, 289
Seven barks, *see* Hydrangea
Shaddock oil, *see* Grapefruit oil
Shalya, 638
Shaman, 639
Shampoo, *see* Hair products and prepration
Shaoyao, 674, 676
Sheetoha, 638, 639
Sherbets, *see* Desserts, frozen dairy
Shingles, remedies, 671, 679
Shock, remedies, saffron, 548
Shodhana, 639
Shortenings, chaparral as antioxidant, 175
Sushruta Samhita, 638

Shuanghua, 667
Siberian beaver, *see* Castoreum
SICHUAN LOVAGE (CHUANXIONG), 670, 682–683
Sida rhombifolia, **651**
Siddha, 642–644
Siddhar Thirumoolar, 642
Siddhar-Nantheesar, 642
Siddhars, 642, 643
Simmondsiaceae, 386
Simmondsia chinensis, 386
Sinapis alba, 457
Sinapis juncea, 457
Sinduras, 643
Sinusitis, remedies:
  bromelain, 114
  Dahurian angelica, 663
  garlic, 309
  kudzu root, 401
  magnolia flower, 435
  peony bark, 675
Skin:
  cangzhu, 662
  chapped remedies, bletilla tuber, 661
  conditions, remedies, 658
  dark spots, remedies, baizhu, 660
  discoloration, as toxic symptom, 396
  dry, remedies, cassie absolute, 153
  lesions:
    remedies, devil's claw, 243
    as toxic symptom, kava, 396
Skin blemishes, remedies, 670, *see also* Acne
Skin disorders:
  remedies:
    alfalfa, 14
    angelica, 33
    burdock, 121
    cade oil, 123
    cajeput oil, 125
    calamus, 128
    cedarwood oil, 162
    chamomile, 171
    chirata, 193
    cocillana bark, 215–216
    dandelion root, 239
    doggrass, 248
    eyebright, 280

gotu kola, 340
henna, 356
kava, 396
lovage root, 428
marjoram, 438
oregano, 483
parsley, 488
sarsaparilla, 556
sassafras, 558
storax, 579
thyme, 596
turmeric, 604
yarrow, 627
yucca, 634
as toxic symptom:
  acacia, 5
  cocoa, 218
  elecampane, 260
  ficin, 292
  kava, 396
  turpentine, 607
see also Allergy; Dermatitis, contact
Skin inflamation, *see* Dermatitis, contact
Skin irritation:
  remedies:
    chamomile, 172
    hyssop, 376
    witch hazel, 620
  as toxic symptoms:
    cananga oil, 132
    capsicum, 133
    chenopodium oil, 180
    clary sage, 205
    cloves, 210
    lemongrass, 413
    lime oil, 423
    nutmeg, 468
    olibanum, 472
    pine needle oil, 503
    podophyllum, 506
    rosemary, 536
    rue, 543
    sage, 551
    sandalwood oil, 554
    savory, 560
    squill, 575
    thyme, 596
    turpentine, 607
    ylang ylang oil, 630
  see also Allergy; Dermatitis, contact; Irritant

Skin protectant, 91, 218, 675, *see also* Demulcent; Emollient
Skin rash:
  remedies:
    boneset, 109–110
    calendula, 129–131
    cedar wood oil, 164
    dittany bark, 659
    fangfeng, 282
    fo-ti, 296
    kudzu root, 399–402
    mints, 446
    mume, 672
    peony bark, 675
    rehmannia, 523
    yarrow, 627
  as toxic symptom:
    balsam copaiba, 68
    bromelain, 121
    castor oil, 154
    chaste-tree, 177–179
    cinchona, 194–195
    fo-ti, 296
    kava, 396
    schisandra, 564–567
    turpentine, 607
  see also Dermatitis, contact
Skin sensitization, as toxic symptoms:
  cloves, 210
  lavender, 407
  lemongrass, 413
  lovage root, 428
  pine needle oil, 503
  see also Allergy; Dermatitis, contact Irritant
Skin softener:
  avocado, 59
  cocoa, 218
  comfrey, 226
  hops, 360
Skin sores, *see* Sores
Skin ulcers, remedies:
  gotu kola, 340
  horsechestnut, 365
  jasmine, 383
  lobelia, 424–425
  musk, 458
  myrrh, 461
  red clover tops, 207
  tagetes, 582

General Index

SKULLCAP, BAIKAL (HUANGQIN), 684–685
Sleep disorders, remedies:
  castoreum, 156–157
  chamomile, 172
  dill, 246
  dogwood, 249
  kava, 396
  passion flower, 490
  see also Insomnia
Sleepiness, remedies, 573
Smallage, see Lovage root
Smellage, see Lovage root
Smilax aristolochiifolia, 555
Smilax febrifuga, 555
Smilax glabra, 556
Smilax medica, 555
Smilax officinalis, 555
Smilax regelii, 555
Smilax scobinicaulis, 556
Smilax sieboldi, 556
Smilax species, 555–557
Smilax Ssans, 555–557
Smoking, see Antismoking prepration
Smooth hydrangea, see Hydrangea
Smooth muscles:
  contracting, 56
  relaxant:
    bromelain, 114
    carrot seed oil, 148
    cocoa, 218
    coffee, 223
    euphorbia, 275
    ganoderma, 303
    garlic, 309
    hops, 359–360
    kudzu root, 401
    lavender, 406–408
    mints, 443–446
    roselle, 533
    rue, 542–543
    savory, 559–560
  stimulant:
    euphorbia, 274–276
    fennel, 283–285
    fenugreek, 286
    psyllium, 511–513
    rosemary, 536
Snack foods:
  colorants, 605

flavoring ingredients:
  celery seed, 166
  cumin, 236
  dill, 246
  fennel, 284
  galbanum, 300
  garlic, 309
  marjoram, 438
  mustard, 458
  nutmeg and mace, 469
  onion, 475
  oregano, 483
  parsley, 488
  rosemary, 536
gums:
  algin, 16–17
  guar gum, 347–349
  MSG, 453
Snakebite, remedies:
  sblack cohosh, 97
  chaparral, 174–176
  echinacea, 255
  fennel, 285
  garden burnet, 666
  garlic, 310
  ginger, 322
  gotu kola, 340
  juniper berries, 391
  lobelia, 425
  rue, 543
  sweet basil, 59, 74, 75, 76, 77, 95
  yarrow, 627
Snakeroot, black, see Black cohosh
Sneezing, agents that cause, 519
Soapbark, see Quillaia
Soaps and detergents:
  emollients, 27
  fixatives:
    balsam Canada, 66
    balsam Peru, 70
    balsam tolu, 71
    benzoin, 90–91
    castoreum, 157
    cedarwood oil, 162
    civet, 204
    deertoneue, 241
    elemi gum, 261
    galbanum, 300
    lime oil, 423
    myrrh, 461

neem, 652
orange, 479
storax, 579
fragrance ingredients:
  absinthium, 2,
  ambrette seed, 31
  angelica, 33
  anise, 38
  balsam Canada, 66
  balsam Peru, 70
  balsam tolu, 71
  beeswax, 85
  bergamot oil, 91
  bois de rose oil, 106
  cade oil, 123
  cajeput oil, 126
  calamus oil, 128
  cananga oil, 132
  caraway, 13–139
  cardamon, 141
  carrot seed oil, 148
  cascarilla bark, 151
  castoreum, 156–157
  cedar leaf oil, 162
  cedarwood oil, 162
  celery *see*d, 166
  chamomile, 172
  chenopodium oil, 180
  cloves, 210
  cinnamon and cassia, 199
  citronella oil, 204
  civet, 204
  clary sage, 204
  coriander, 227–228
  cubebs, 233–234
  deertongue, 241
  dill, 246
  elecampane, 260
  eucalyptus, 273
  fennel, 284
  fenugreek, 287
  galbanum, 300
  genet, 315
  geranium oil, 319
  ginger, 322
  grapefruit oil, 344
  guaiac wood oil, 346
  hyssop, 376
  juniper berries, 390
  labdanum, 405
  lavender, 408
  lemongrass, 413
  lemon oil, 411

Soaps and detergents
(*Continued*)
  lime oil, 423
  lovage root, 428
  marjoram, 438
  mints, 445
  myrrh, 461
  nutmeg, 468
  olibanum, 472
  orange, 479
  origanum oil, 485
  parsley, 488
  patchouly oil, 494
  pine needle oil, 503
  rosemary, 536
  rose oil, 106
  rue, 543
  sage, 551
  sandalwood oil, 554
  storax, 579
  sweet basil, 75
  sweet bay, 77
  tarragon, 589
  thyme, 598
  West Indian bay, 79
  ylang ylang oil, 630
  other ingredients:
    bergamot oil, 91–92
    chamomile, 172
    ginseng, 330–333
    mustard, 457–459
    turpentine, 608
Soap tree bark, 518
Sodium glutamate, *see*
  Monosodium
  glutamate (MSG)
Sol, agar, 9–11
Solanaceae, 88
Sophora japonica, 545, 546
Sophorin, *see* Rutin
Sores, remedies:
  asparagus, 53
  astragalus, 57
  Baikal skullcap, 686
  balm, 65
  balm of Gilead buds, 65
  balsam Canada, 65
  balsam Peru, 70
  bee pollen, 83
  black and pale catcehu, 158

  bletilla tuber, 661
  bromelain, 113–114
  burdock, 121
  chaparral, 176
  chickweed, 187
  comfrey, 226
  devil's claw, 243
  echinaceae, 255
  fo-ti, 297
  garden burnet, 666
  gelsemium, 313
  gentian, 317
  giant knotweed, 678
  goldenseal, 338
  hawthorn, 353
  honeysuckle flower, 667
  hypericum, 373
  jasmine, 382–383
  jojoba, 387
  lavender, 408
  licorice root, 417
  lobelia, 425
  mints, 446
  mume, 672
  musk, 456
  myrrh, 462
  olibanum, 472
  orange, 480
  pearl, 673
  phellodendron bark, 678
  pipsissewa, 504
  prickly ash, 51
  purslane, 678
  quillaia, 519
  red clover tops, 207
  rhubarb, 527
  sarsaparilla, 556
  tagetes, 582
  tannic acid, 587
  turmeric, 605
  turpentine, 608
  valerian root, 614
  white pine bark, 501
  yarrow, 627
  yucca, 634
  see also Skin entries
Sore throat, *see* Throat, sore
Sorrel, *see* Roselle
Soups:
  colorants, 605
  flavoring ingredients:
    carrot seed oil, 148

    catnip, 160
    celery seed, 166
    cinnamon and cassia, 199
    cumin, 236
    ganoderma, 303
    marjoram, 438
    nutmeg and mace, 469
    onion, 475
    origanum oil, 485
    parsley, 488
    sage, 551
    savory, 560
    thyme, 596
    turmeric, 605
  fragrance ingredients, kudzu root, 503
  gum
    algin, 17
    guar gum, 348
    locust bean gum, 427
    MSG, 453
  other ingredients:
    Cherokee rosehip, 181
    codonopsis, 221
SOUR JUJUBE KERNEL
  (SUAN ZAO REN),
  572–573, *see also*
  Jujube, common
Southern wax myrtle, *see*
  Bayberry bark
Soy sauce, sweeteners, stevia, 577
Spadic, *see* Coca
Spanish broom, *see* Genet
Spartium junceum, 314
Spartiun scoparium, 115
Spasmolytics, *see*
  Antispasmodic
Spasms:
  as toxic symptom, MSG, 452–453
  see also Antispasmodic
Spearmint, 443–446, 480
Sperm, increased motility, turmeric, 605
Spermatocida, bayberry bark, 81
Spermatorrhea, remedies:
  Cherokee rosehip, 181–182
  epimedium, 270
  ginkgo, 325–327
  lycium fruit, 429–432

General Index

phellodendron bark, 677–678
poria, 509
rehmannia, 523
schisandra, 564–567
Spices:
　garlic, 309
　parsley, 488
　saffron, 547
　sweet basil, 75
　sweet bay, 77
　tarragon, 588–590
Spider bites, remedies, 483
Spigeliaceae, *see* Loganiaceac
Spikenard root, 65, 103
Spinal cord injury, remedies, 372
　saffron, 547–548
Spleen ailments, remedies, 239
Spogel, *see* Psyllium
Spotted wintergreen, *see* Pipsissewa
Sprains, remedies:
　arnica, 44
　horse chestnut, 365
　hydrangea, 370
　lavender, 408
　lobelia, 425
　thyme, 596
　yerba santa, 629
　yucca, 634
Spurge, pill-hearing, *see* Euphorhia
Squaw root, *see* Blue cohosh
SQUILL, 574–576
Stabilizing agents:
　agar, 9–11
　algin, 17
　carrageenan, 145
　guar gum, 348
　psyllium, 512
　xanthan guns, 624
Stag-bush, *see* Black saw bark
Stamina, improving, astragalus, 57
Staphylococcus, inhibition by hypericum, 372
Staphylococcus albus, inhibition by nettle, 464
Staphylococcus aureus, inhibition:
　avocado, 60
　chamomile, 171

cinnamon, 198
echinacea, 253
galbanum, 299
immortelle, 378
labdanum, 404
nettle, 464
pectin, 496
royal jelly, 539
sage, 550
Staphylococcus epidermidis, inhibition by barberry, 73
Staphylococcus species, inhibition by henna, 356
Star anise, *see* Anise and star anise
Starch, 8, 10, 27, 29, 32, 57, 80, 136, 141, 143, 151, 181, *see also* CHEMICAL INDEX
Stargrass, *see* Aletris
Starwort, yellow, *see* Elecampane
Stasis-dispersing, bee pollen, Stelaria media
Stem bromelain, 113, 114
Sterculia acuminata, 398
Sterculiaceae, 216, 393, 398
Sterculia gum, *see* Karaya gum Sterculia urens
Sterility, as toxic symptom in mice, MSG, 453
Sternutatory, *see* Allergy
Steroids, anabolic, sarsaparilla as substitutute, 556
STEVIA, 576–578
***Stevia rebaudiana*, 577**
Stiffening agent, *see* Gums
Stigmata maydis, *see* Corn silk
Stimulant:
　ambrette seed, 31
　anise, 38
　arnica, 43
　artichoke, 45–47
　asafetida, 49
　balm of Gilead buds, 65
　balsam copaiba, 67–68
　black and white pepper, 499
　boneset, 109
　buchu, 118
　cajeput oil, 124–126
　capsicum, 134

cardamon, 141
cardiac, *see* Cardiac stimulant
cascarilla bark, 151
cassie absolute, 152
cedar leaf oil, 161
citronella oil, 202
cocoa, 218
coffee, 223
costus oil, 231
cubebs, 234
cumin, 236
damiana, 237
dill, 245
eleuthero, 263
eucalyptus, 273
fennel, 283
feverfew, 290
galbanum, 300
ginger, 321
ginseng, 332
guarana, 350
honeysuckle flower, 668
hops, 360
ipecac, 380
kava, 395
kola nut, 398
labdanum, 405
lavender, 408
local, 94
　capsicum, 134
　elemi gum, 261
marjoram, 436–438
mints, 446
musk, 456
mustard, 459
myrrh, 461
nutmeg and mace, 469
olibanum, 472
oregano, 483
prickly ash, 51
respiratoiy, *see* Respiratory stimulant
roselle, 533
rosemary, 536
rue, 543
sweet bay, 77
tea, 591
uterine, *see* Uterine stimulant
valerian root, 612
St, John's bread, *see* Carob

Stomachache:
  remedies:
    aletris, 12
    black and white pepper, 499
    coriander, 228
    feverfew, 290
    garlic, 309
    gentian, 317
    ginger, 322
    lemongrass, 414
    purslane, 679
    rosemary, 537
    sandalwood oil, 554
    sweet birch oil, 619
    sweet woodruff, 623
    wintergreen oil, 96
    yarrow, 627
  as toxic symptoms:
    ganoderma, 303
    ginkgo, 327
    quillaia, 518–519
    rue, 543
    schisandra, 566
    thyme, 594–597
Stomach disorders, *see* Gastric disorder
Stomachic:
  blessed thistle, 103
  boldo leaves, 107
  buchu, 117
  calamus, 128
  capsicum, 132–135
  caraway, 138
  cardamon, 141
  celery seed, 166
  centaury, 168
  cinnamon and cassia, 199
  citronella oil, 202
  clary sage, 205
  coca, 214
  coriander, 228
  costus oil, 231
  elecampane, 260
  elemi gum, 262
  eyebright, 280
  fennel, 285
  lavender, 408
  lemongrass, 414
  lovage root, 428
  mints, 446
  myrrh, 462
  onion, 475
  orange, 479

rosemary, 537
tagetes, 582
tarragon, 589
tea, 593
valerian root, 614
see also Bitter tonic
Stomach irritation, *see* Gastric
Stomach pain, *see* Pain
Stomach secretions, *see* Gastic Secretions, 88, 146, 168, 195, 284, 358
Stones, *see* Bladder: Gall stones:
Stools, blood:
  remedies:
    garden burner, 666
    horsetail, 368
    orange, 477–480
    tienchi ginseng, 598–600
  as toxic syndrome, ganoderma, 303
STORAX, 578–580
Streptococcus hemolytica, inhibition by balm, 62
Streptococcus hemolyticus, inhibition, 539
Streptococcus pyogenes, inhibition, 272, 358
Streptococcus species, inhibition, 272, 356
Stress:
  adaplogenic activity, eleuthero, 263
  codonopsis, 221
  ginseng, 332
  gotu kola, 339
  kava, 396
Strokes:
  prevention:
    borage, 111
    evening primrose, 277
    rutin, 546
  remedies, 128, 267, 340
Stronger rose water, 530–532
Stupor, as toxic symptom, 365, 468
Styptic, *see* Hemorrhage
Styraceae, 90
Styrax, *see* Storax
Styrax benzoin, 90
Styrax paralleloneurus, **90**
Styrax species, 90
***Styrax tonkinensis*, 90**
Suann zao ren, 572, 573

Succory, *see* Chicory root
Sugar diabetes, *see* Diabetes mellitus
Suhria, species, 9
Sunburn, remedies:
  aloe, 27
  balm of Gilead buds, 65
  calendula, 130
  jujube, 388
  tea, 593
  see also Burns
Sunscreen and suntan preparations, ingredients:
  aloe, 27
  beeswax, 84
  buckthorn, 119
  carrot seed oil, 148
  cascara sagrada, 150
  celery seed, 165–166
  chamomile, 171
  cinnamon and cassia, 199
  immortelle, 379
  Job's tears, 384
  sour jujube kernel, 573
Sunstroke, remedies, gotu kola, 340
Suppositories, ingredients:
  agar, 10
  beeswax, 85
  belladonna, 89
  castor oil, 155
  cocoa butter, 218
  sassafras, 557–558
Suppuration, remedies, astragalus, 57
Surgery, proteases used, 293
Suspending agent:
  agar, 10
  algin, 17
  carrageenan, 146
  guar gum, 348
  karaya gum, 394
  ragacanth, 603
  xanthan gum, 624
Swallowing difficulty, as toxic symptoms, belladonna, 89
Sweating:
  lack of remedies, ephedra, 267
  see also Antiperspirant; Night sweats
Sweet acacia, *see* Cassie absolute

General Index

Sweet bark, *see* Cascarilla bark
Sweet cinnamon, *see* Calanlus
Sweet cumin, *see* Anise
Sweeteners, natural, sources:
   artichoke, 47
   honey, 358
   hydrangea, 370
   orange, 477
   stevia, 577
Sweet flag, *see* Calamus
Sweet myrtle, *see* Calamus
Sweet Orange Peel Tincture N.F., 477
Sweet pepper, 133
Sweet root, *see* Calamus
Sweet sauces, *see* Sauces
Sweet sedge, *see* Calamus
Sweetwood bark, *see* Cascarilla
Sweet wood, *see* Licorice root
Swelling, *see* Edema
***Swertia chirata*, 192**
Swiss mountain pine, 502
Sycocarpus rusbyi, *See* Guarea rusbyi, 215
***Symphyratm officinale*, 557**
Syneresis:
   agar, 10
   xanthan gum, 624
Syphilis, remedies:
   echinacea, 255
   gotu kola, 340
   myrrh, 462
   olibanum, 472
   podophyllum, 506
   sarsaparilla, 556
Syrups, flavor ingredients, bilberry, 94
***Syzyguiun aromaticum*, 209**

Tabasco pepper, *see* Capsicum
Tablets:
   binding and disintegrating agents:
      algin, 17
      carrageenan, 146
      honey, 358
      tragacanth, 603
   polishing agents, beeswax, 85
   protective coating component, castor oil, 155
Taalpatra, 637
Tachycardia:
   agents that cause:

      coffee, 223
      ephedra, 266
      ipecac, 380
      nutmeg, 468
   remedies, chicory root, 190
   as toxic symptom, yohimbe, 632
   see also Cardiac rhythm
TAGETES, 581–582
***Tagetes erecta*, 581**
Tagetes glandulifera, 581
***Tagetes minuta*, 581**
***Tagetes patula*, 581**
***Tagetes* species, 582**
Tailed pepper, *see* Cubebs
Takshashila, 637
TANNIC ACID, 585–587
Tanning, *see* Leather industry; tannin entries in CHEMICAL INDEX
Tannin, 7, 18, 29, 43, 62, 97, 99, 115, 120, 128, 143, 151, 158, 184, 215, 235, 237, 249, 265, 269, 355, 362, 376, 398, 408, 444, 496, 515, 517, 518, 550, 585, 586, 588, 591, 611, 620, 629, 669
Tapeworms, *see* Anthehnintic
Taraxacum mongolicum, 239
***Taraxacum officinale*, 238**
Taraxacum, *see* Dandelion root
***Taraxacum* species, 238–239**
Taraxacum vulgare, 238
TARRAGON, *see also* Estragon
Tarweed, *see* Yerba santa
Tasmanian blue gum, *see* Eucalyptus
TEA, 590–593
   chamomile, 171
   dandelion root, 239
   substitute, cranberry, 233
   see also Chenopodium oil
Teaberry, *see* Wintergreen oil
Tear producing, *see* Lachrymatory properties
Teeth, loose, *see* Gum problems; Pyorrhea
Tenderizers, meat, *see* Meat tenderizers
Tendonitis, remedies, horse chestnut, 365

Teratogens, 94, 103, 104, 166, 172, 223, 348, 386, 506, 577
Terigium, 654
Terminalia, 653
***Terminalia bellerica*, 653**
termilignan, 654
Textiles:
   karaya gum, 394
   sizing agents, 17
Thea bohea, 590
Theaceae, 590
Thea sinensis, 590
Thea viridis, 590
Theobroma, *see* Cocoa (cacao)
***Theobroma cacao* subsp. cacao, 216**
Thermal injury, remedies, aloe, 26
Thickening agents:
   algin, 17
   beeswax, 85
   carrageenan, 146
   guar gum, 348
   karaya gum, 394
   kudzu root, 401
   psyllium, 512
   tragacanth, 602
   xanthan gum, 624
Thirst quenching, *see* Refrigerant
Thoroughwort, *see* Boneset
Thousand leaf, *see* Yarrow
Throat inflammation, remedies:
   acacia, 5
   cinchona, 195
   comfrey, 226
   see also Demulcent
Throat, sore:
   remedies:
      althea root, 30
      black cohosh, 98
      bloodroot, 103
      burdock, 121
      eyebright, 280
      gentian, 317
      honey, 358
      horehound, 363
      horsetail, 368
      hyssop, 376
      licorice root, 417
      mints, 446
      myrrh, 462
      olibanum, 472

Throat, sore (*Continued*)
    orange, 480
    pearl, 673
    sage, 551
    savory, 560
    saw palmetto, 563
    as toxic symptom, belladonna, 89
    see also Demulcent
Thrombocytopenia, as toxic symptom, saffron, 548
Thrombophlebitis, remedies, horsechestnut, 365
***Thuja occidentalis*, 161**
Thuja oils, *see* Cedar leaf oil
Thumb sucking, remedies, capsicum, 134
THYME, 594–597
***Thymus captatus*, 485**
Thymus citriodorus, 595
Thymus serpyllum, 595–597
***Thymus vulgaris*, 594**
Thymus zygis, 595–597
Thyroid:
    enlargement, *see* Goiter
    regulation, cocoa, 218
    stimulants, myrrh, 461
Tianqi, 598
Tiaodang, 220
TIENCHI GINSENG (*SANQI*), 598–600
Tinnitus:
    remedies:
        fo-ti, 297
        ginkgo, 327
        ligustrum, 421
        lycium fruit, 432
        rehmannia, 523
    as toxic symptom, uva ursi, 610
Tiredness, *see* Fatigue
Tobacco, flavoring ingredients:
    cascarilla bark, 152
    clary sage, 206
    cloves, 211
    coriander, 228
    cubebs, 234
    deertongue, 241
    immortelle, 379
    licorice root, 418
    mints, 446
    rose oil, 532
    see also Arnica: Lobelia
Tofu remedies, pearl, 673

Toiletries, *see* Cosmetics
Toluiferum balsamum, 71
Tolu, *see* Balsam tote
Tomato, increased yield by triacontanol, 85
Tonic:
    aletris, 12
    astragalus, 56
    baizhu, 660
    barberry, 73
    bayberry bark, 81
    black and white pepper, 499
    black cohosh, 98
    bloodroot, 103
    boldo leaves, 108
    boneset, 110
    broom tops, 116
    buchu, 118
    cajeput oil, 126
    celery seed, 166
    chamomile, 172
    chaparral, 176
    Cherokee rosehip, 181
    chestnut leaves, 186
    chicory root, 191
    cinnamon and cassia, 199
    codonopsis, 221
    costus oil, 231
    damiana, 237
    dandelion root, 238, 239
    devil's claw, 243
    dogwood, 249
    eleuthero, 264
    epimedium, 269, 270
    feverfew, 290
    fo-ti, 294, 296, 297
    ganoderma, 303
    gentian, 317
    ginseng, 333
    goldenseal, 337
    gokshura, 646–647
    greater galanga, 648
    hydrangea, 370
    Job's tears, 384, 385
    jujube, 388
    kalmegh, 650
    kava, 395
    kola nut, 399
    lavender, 408
    licorice root, 417
    ligustrum, 421
    lycium fruit, 431

    mahabala, 651
    neem, 653
    nutmeg, 468
    orange, 480
    parsley, 488
    peony root, 676
    pipsissewa, 504
    poria, 509
    prickly ash, 51
    quillaia, 519
    rehmannia, 523
    rosemary, 537
    royal jelly, 540
    safflower, 682
    sage, 551
    sarsaparilla, 556
    savory, 560
    saw palmetto, 563
    schisandra, 566
    sour jujube kernel, 573
    yarrow, 627
    yerba santa, 629
    see also Bitter tonic
Tonic-clonic seizures, *see* Spasms
Tonic water, bitter, ingredients, cinchona, 195
Tonquim musk, *see* Musk
Tonsillitis, remedies, 121, 340
Tooth, bleeding socket, remedies, tea, 593
Toothache, remedies:
    allspice, 20
    avocado, 60
    cajeput oil, 126
    capsicum, 134
    chamomile, 172
    cloves, 210
    coriander, 228
    Dahurian angelica, 663
    dogwood, 249
    eucalyptus, 273
    garlic, 310
    ginger, 322
    lavender, 408
    magnolia flower, 435
    mahabala, 651
    marjoram, 438
    mints, 446
    oregano, 483
    prickly ash, 51
    rue, 543
    tarragon, 589

General Index

turmeric, 605
turpentine, 608
Toothache tree, *see* Ash, prickly
Tooth decay inhibition, *see* Cariostatic
Tooth injuries, remedies, horsetail, 368
Toothpastes:
  flavoring ingredients:
    anise, 38
    caraway, 139
    cloves, 210
    cinnamon and cassia, eucalyptus, 199
    mints, 445
    thyme, 596
  other ingredients:
    Baikal skullcap, 685
    carrageenan, 146
    echinacea, 255
    horse chestnut, 365
    neem, 652
    papain, 146
    royal jelly, 540
    Tragacanth, 603
Trachoma, remedies, garlic, 310
TRAGACANTH, *see also* Karaya gum
Trailing mahonia, *see* Barberry
Tranquilizers:
  balm, 63
  gotu kola, 339
  Job's tears, 384
  passion flower, 491
  rehmannia, 522
  schisandra, 566
  tagetes, 581
Traumatic injuries, remedies:
  bee pollen, 83
  bromelain, 114
  kudzu root, 401
  schisandra, 567
  tienchi ginseng, 600
Tremor:
  remedies, sage, 551
  as toxic symptom:
    absinthium, 2
    aspidium, 54
    yohimbe, 632
***Tribulus terrestris*, 646**
Trichomonas vaginalis, inhibition by Echinacea, 253

Trichomoniasis, remedies:
  garlic, 310
  ginger, 321
  licorice root, 416
Tricium, *see* Dograss
***Trifolium pratense*, 206**
Trifolium, *see* Clover tops, red
Triglycerides, blood, agents that lower:
  bilberry, 94
  garlic, 308
  turmeric, 604
  see also CHEMICAL INDEX
***Trigonella foenum-graectum*, 286**
***Trilisa odoratissima*, 240**
Trinitario cacao, 216
Triphala:
  Terminalia, 654
  Terminelia chebula, 654
  Emblica officinalis, 654
Triple sec, *see* Liqueurs
Triticm repens, *see Agropyron repens*
Trompillo, *see* Cocillana bark
True sweet basil oil, 74
True unicorn root, *see* Aletris
Trumpet flower, evening, *see* Gelsemium
Trypsin activity enhancement:
  allspice, 20
  cardamon, 141
  cloves, 211
  eugenol, 20
Tuberculosis, *see* Antitubercular
Tumors, *see* Cancer
Turkish attar of rose, *see* Rose oil (and absolute)
Turkish rose oil, *see* Rose oil (and absolute)
TURMERIC, 603–605
  Indian, *see* Goldenseal
  mustard, 605
Turnera aphrodisiaca, 237
Turneraceae, 237
***Turnera diffusa*, 237**
TURPENTINE (AND ROSIN), 607–608
  Canada, *see* Balsam, Canada
  spirits of, *see* Turpentine oil
Typhoid, remedies, 110, 255, 310

Ulcers:
  adaptogenic activity, eleuthero, 263
  agents causing, carrageenan, 146
  contraindication, coffee, 223
  genital, remedies, dill, 246
  inhibition, codonopsis, 221
  remedies:
    aloe, 25
    astragalus, 57
    balsam Peru, 70
    bayberry bark, 81
    bee pollen, 82
    black and pale catechu, 158
    blessed thistle, 101
    bromelain, 114
    calendula, 131
    carrageenan, 146
    chamomile, 172
    chaparral, 176
    chickweed, 187
    cinnamon, 198
    comfrey, 226
    devil's claw, 243
    eucalyptus, 273
    fenugreek, 287
    garlic, 310
    greater galnanga, 647
    gotu kola, 340
    hawthorn, 353
    honey, 358
    horsechesnut, 365
    hypericum, 373
    jujube, 389
    licorice root, 417
    lycium fruit, 431
    mints, 445
    mother of pearl, 674
    neem, 652, 653
    papain, 114
    pearl, 673
    prickly ash, 51
    purslane, 679
    tagetes, 582
    tienchi ginseng, 599
  see also Gastrointestinal disorders; skin entries
Umbelliferae, 32, 36, 48, 138, 147, 165, 185, 227, 235, 244, 281, 283, 299, 399, 427, 487, 663, 669, 682, 705

Umbellularia californica, 76
Uncaria gambir, 157
Unconsciousness, remedies, turmeric, 605
Unicorn root, *see* Aletris
Unionidae, 673
Upas, *see* Cocillana bark
Uppu, 642
Uragoga granatensis, 379
Uragoga ipecacuanha, 379
Urea, blood concentration, reduction by chamomile, 171
Uremia collapse, as toxic symptom, saffron, 548
Urethanes, castor oil in synthesis of, 155
Urethritis, remedies, 118, 512, *see also* Urinary disorders
Urginea maritima, 574
Urginea scilla, 574
Urinary antiseptic:
 boneset, 110
 buchu, 117
 cubebs, 234
 sandalwood oil, 554
 turpentine, 608
 uva ursi, 610
Urinary disorders:
 agents preventing infections, cranberry, 232, 233
 remedies:
  bilberry, 94
  blue cohosh, 105
  buchu, 117
  cardamom, 142
  cranberry, 233
  gotu kola, 340
  kava, 396
  nettle, 465
  phellodendron bark, 678
 as toxic symptom, ephedra, 267
 see also Bladder disorders; Kidney disorders
Urination:
 bloody, remedies, horsetail, 368
 difficulties, *see* Dysuria
 excessive, remedies, 567
 frequent, remedies, ginkgo, 327
 nocturia, remedies, saw palmetto, 562

painful, remedies, purslane, 679
Urticaceae, 464
***Urtica dioica*, 464**
Urticaria, *see* Skin rash
Uterine bleeding:
 remedies, 57
 as toxic symptom, saffron, 548
Uterine disorders:
 remedies;
  black cohosh, 98
  orange, 480
 see also Female disorders; Menstruation
Uterine inflammation, *see* Female disorders
Uterine prolapse remedies, orange, 480
Uterine sedative, 517
Uterine stimulant:
 barberry, 73
 bee pollen, 82
 blue cohosh, 104
 cedar leaf oil, 161
 corn silk, 229
 fenugreek, 287
 magnolia flower, 435
 purslane, 679
 rue, 543
 safflower, 682
 saffron, 547
 sour jujube kernel, 573
Uterine tonic, saw palmetto, 563
UVA URSI, 610–611

Vaccinia virus, *see* Antiviral
***Vaccinium macrocarpon*, 232**
***Vaccinium myrtillus*, 93**
Vaginal discharges, remedies, ginkgo, 327
Vaginal douche, quillaia, 519
Vaginal inflammation, *see* Female disorders
Vaginal jellies and creams, ingredients, tragacanth, 603
Vaginal prolapse, remedies, kava, 397
Vaginal tonic, saw palmetto, 563
Valeriana amurenis, 614
Valerianaceae, 612
Valeriana coreana, 614
Valeriana hardwickii, 614

***Valeriana jatamansii*, 612**
***Valeriana officinalis*, 612**
Valeriana stubendorfi, 614
VALERIAN ROOT, 612–614
***Vanilla planifolia*, 616**
***Vanilla tahitensis*, 616**
Vaporizer fluids, ingredients, 91, 273, 503
Varicose veins, remedies:
 bilberry, 94
 ginkgo, 326
 horsechestnut, 365
 quinine, 195
Varnishes, castor oil in production, 155
Vasoconstrictor:
 ephedra, 266
 pale catechu, 158
 quebracho, 517
 rehmannia, 522
Vasodilator:
 astragalus, 56
 bilberry, 94
 black cohosh, 98
 carrot seed oil, 148
 cocoa, 218
 codonopsis, 220
 coffee, 223
 costus oil, 231
 dill, 245
 ephedra, 266
 epimedium, 269
 ganoderma, 303
 giant knotweed, 669
 ginkgo, 326
 horehound, 363
 parsley, 488
 rehmannia, 522
 royal jelly, 539
 safflower, 682
 schisandra, 566
 Sichuan lovage, 682
 tienchi ginseng, 599
Vasomotor center:
 ginger effect on, 321
 see also Vasoconstriction
Vasoprotective, bilberry, 94
Vata, 638
Vata dosha, 638
Veda, 637
Vegetable mercury, *see* Podophyllum (podophyllin)

Vegetable pepsin, *see* Papain
Vegetables, processed:
  colorants caramel color, 137
  favouring ingredients
    cumin, 236
    fennel, 284
    marjoram, 438
    mustard, 458
    nutmeg and mace, 469
    orange, 480
    oregano, 483
    parsley, 488
    rosemary, 536
    sage, 551
    savory, 560
    thyme, 596
  gum:
    agar, 10
    algin, 17
    guar gum, 348
    tragacanth, 603
  MSG, 453
  sweeteners, stevia, 577
Venereal disease, remedies, cassie absolute, 153
Venereal warts, *see* Warts, venereal
Venotonic, horsechestnut, 364
Veratrum album, 317
Verbenaceae, 177, 482
Vermifuge, *see* Anthelmiones
Vermouth, ingredients:
  absinthium, 2
  ambrette seed, 31
  angelica, 33
  balm, 63
  centaury, 168
  chamomile, 172
  cinnamon and cassia, 199
  clary sage, 205
  cloves, 211
  coriander, 228
  elder flowers, 258
  elecampane, 260
  gentian, 317
  lemon oil, 411
  marjoram, 438
  rue, 543
  saffron, 548
  sage, 551
  sweet woodruff, 623
  yarrow, 627

  see also Beverages, alcoholic and nonalcoholic
Vertigo:
  remedies, 327, 523, 551
Vesication, *see* Blisters
***Viburnum prunifolium*, 99**
Viburum, *see* Black haw bark
Viola tricolor var, *maxima*, 545
Vipreeta, 639
Vipreetarthkari, 639
visamine
Vision:
  blurred, remedies, 421, 431
  disturbances, as toxic symptom, 89, 313
  improved, lycium fruit, 431
Vitaceae, 342
Vitality, increased, 14, 640
***Vitex agnus-castes*, 177**
Vitamin C, sources, 6–7, 528
"Vitamin P" activity, 545, 592
Vitamins, *see* CHEMICAL INDEX
Vitiligo, remedies, 92, 326, 650, 663, *see also* Skin disorders
***Vitis vinifera*, 342**
***Viverra civerta*, 203**
Viverra species, 203
***Viverra zibetha*, 203**
Viverridae, 203
Vomiting:
  remedies:
    Baikal skullcap, 685, 686
    bee pollen, 83
    cherry laurel leaves, 184
    garden burnet, 666
    gentian, 317
    ginseng, 333
    lavender, 408
    marjoram, 438
    oregano, 483
    patchouly oil, 494
    peony bark, 675
    rehmannia, 523
    sandalwood oil, 554
    tienchi gingseng, 600
    valerian root, 614
  as toxic symptom:
    absinthium, 2
    arnica, 43
    aspidium, 54
    balsam copaiba, 67

    bromelain, 114
    castor oil, 155
    chenopodium oil, 180
    eucalyptus, 273
    fo-ti, 296
    horsetnut, 365
    ipecac, 380
    nutmeg, 468
    quassia, 515
    quebracho, 517
    rhubarb, 526
    rue, 543
    saffron, 548
    tannic acid, 586
    thyme, 596
    turpentine, 608
    yohimbe, 632
  see also Nausea
Vomit wort, *see* Lobelia
Vulnerarv, *see* Healing

Warts:
  remedies:
    cangzhu, 662
    cedar leaf oil, 162
    chaparral, 176
    Jobs tears, 385
    sweet basil, 75
    venereal, remedies, 164, 506
Wax, *see* Beeswax; CHEMICAL INDEX
Wax myrtle, southern, *see* Barberry
Wax myrtle bark, 80
Weakness:
  remedies:
    astragalus, 57
    codonopsis, 221
    fo-ti, 297
    ganoderma, 303
  as toxic symptom, 313, 365
Weavers broom, *see* Genet
Weight control, *see* Appetite depressant; Obesity
Weight gain, lycium fruit, 431
Weight loss products, horsetail, 368
Wermutkraut, *see* Absinthium
Wermut, *see* Absintltiunt
West Indian avocado, 59
West lndian cherry, *see* Acerola
Wet process coffee, 222

Wheezing, remedies, ephedra, 267
White Pine compound, *see* Compound white pin syrup
White Saunders oil, *see* Sandalwood oil
White thyme oil, 595–596
Whitetube stargrass, *see* Aletris
White wax, *see* Beeswax
Whooping cough, remedies:
  belladonna, 89
  chestnut leaves, 186
  dogwood, 249
  elecampane, 260
  evening primrose, 278
  garlic, 310
  goldenseal, 338
  immortelle, 379
  lobelia, 425
  red clover tops, 208
  saffron, 548
  tagetes, 582
  thyme, 596
  see also Cough
Wigandia californicum, 628
Wild cherry bark, 65, 182
Wild jessamine, *see* Gelsemium
Wild lemon, *see* Podophyllum (podophyllin)
Wild majoram, *see* Oregano
Wild tobacco, *see* Lobelia
Wild vanialla, *see* Deertongue
Wine, tannic acid in, 587, *see also* Nonalcoholic
Wintergreen, flavoring, sweet birch oil, 619, *see also* Pipsissewa
WINTERGREEN OIL, 619
Witchgrass, *see* Doggrass

WITCH HAZEL, 620–621
Withania somnifera, 645
Wolfs banae (wolfsbatte), *see* Arnica; Aconite
Wood, pepper or yellow, *see* Ash, prickly
Woodbine, *see* Gelsemium
Wood oil, Champaca, *see* Gelsenium
WOODRUFF, SWEET, 622–623
Woodward, *see* Woodruff, Sweet
Worcestershire sauce, flavoring ingredient, 49, 584, *see also* Sauces
Work capacity, ginseng effect, 333
Wormseed oils, American, *see* Chenopodium oil
Worms, remedies for, *see* Chenopodium oil
Worm wood, *see* Absinthium
Wound healing, *see* Healing
Wrinkles, remedies:
  baizhu, 660
  cocoa, 218
  jujube, 388
  mother of pearl, 674
  poria, 509
  royal jelly, 540
  Sichuan lovage, 683
Wunuei, 672

Xanthomonas campestris, 624
XANTHAN GUM, 624–625
Xanthoxylum, *see* Ash, prickly
Xidang, 220
X-ray irradiation, rutin effect, 545

Yarrow, 626–627

Yellow bark, *see* Cinchona (red and yellow)
Yellow jasmine, *see* Gelsemium
Yellow root, *see* Goldenseal
Yellow starwort, *see* Elecampane
Yellow wax, *see* Beeswax
Yellow wood, *see* Ash, prickly
YERBA SANTA, 628–629
YLANG YLANG OIL, 37, 132, 629–630
Yogurt, carrageenan, 146
YOHIMBE, 631–632
Yuan danpi, 675
Yu baizhi, 663
YUCCA, 633–634
Yucca arborescens, 633
Yucca brerifolia, **633**
Yucca mohavensis, 633
Yucca schidigera, **633**
Yunnan Bao Yao, ingredients, tienchi ginseng, 600

Zanthoxylum americanum, **50**
*Zanthoxylum clava-herculi*s, **50**
Zea, *see* Corn silk
Zea mays subsp, **mays, 229**
Zhenzhu, 673–674
Zhenzhmnu, 673–674
Zhu, 660, 662
Zibeth, *see* Civet
Zicao, 678
Zingiberaceae, 140, 320, 603, 647
Zingiber officinale, **320**
Ziziphus jujuba, **388, 572**
Ziziphus sativa, 388
Ziziphus spinos, **572**
Ziziphus vulgaris, 388
Zygophyllaceae, 174, 345, 346, 646

Chemical Index

Abienol, 501
Abietic acid, 66, 79, 607
Aborigine, 126
Absinthin, 1
Acantolic acid, 352
Acetaldehyde, 32, 36, 251, 478, 543, 616
9-acetamido-3,4-dihydropyrido-[3,4-b]indole, 655
Acetic acid:
   burdock, 120
   geranium oil, rose, 319
   karaya gum, 394
   orange, 478
   tragacanth, 602
   vanilla, 616
   yerba santa, 629
   ylang ylang oil, 630
Acetone, 7, 40, 71, 113, 151, 163, 205, 299, 321, 367, 384, 494, 509, 566, 586, 672, 680, 681, 697
Acetophenone, 65, 156, 190, 404, 595
(S)-1′-Acetoxychavicol acetate, 647, 648
N-Acetylanonaine, 50
Acetylastragaloside I, 56
3-Acetyl-β-boswellic acid, 471, 472
Acetylcholine, 62, 98, 231, 352, 405, 464, 533, 539, 550
Acetylenes, 170, 660
2-Acetylnortracheloside, 101
Acetyloleanolic acid, 420
8 Acetylthio-p-menthan-3-one, 117
Acids:
   diterpene, juniper berries, 390
   mineral, reaction with algin, 16
   honey, 358
   nonhydroxy, burdock, 120

plant, 4, 13, 170, 181, 225, 229, 249, 342, 388, 528, 584, 672
   as acidulants in henna preparations, 355
   antigonadotropic activity of, 225, 398
   volatile, 59, 120, 412, 428, 602
Aconitine (acetyl benzoylaconine), 7, 8, 358, 401
α-Acoradiene (acorene), 163
Actein, 97
Acteoside, 362, 511, 521
Actinidine, 612
Adenine, 302, 350, 352, 508, 591, 683
Adenosine, 301–303, 352, 372, 384, 539, 613, 683
Adhumulone, 359
Adipic acid, as acidulant in henna preparations, 356
Adlupulone, 359
Aescin, 364–366
Aesculin (esculin), 364
Afzelin, 620
Agaropectin, 10
Agarose, 10
Agropyrene (1-phenyl-2,4-hexadiyne), 248
Ajoenes, 308, 309
Akuammidine, 517
Alanine, 56, 143, 339, 430, 539, 566, 626, 674
Alantic acid, 259
Alantolactone (helenin), 230, 231, 259, 260
Albaspidin, 54
Albiflorin, 676
Alcohols:
   bergamot oil, 92
   castoreum, 156
   cyclic, citronella oil, 202

diterpene, horehound, 362
ethyl:
   antagonistic effect of ginseng, 332
   effect on acacia, 4
   ficin inhibition by, 291–293
   storax, 579
   genet, 314
   geranium oil, 319
   ginger, 321
high molecular weight:
   alfalfa, 13
   chenopodium oil, 180
   esters, alkanet, 18
   jojoba, 386
West Indian bay, 76, 78
   lemongrass, 412–414
   lemon oil, 410–411
   ligustrum, 420–421
   orange, 477–480
   patchouly oil, 493–494
   pine needle oil, 502–503
   polyhydric, avocado, 59
   terpene, 13, 32, 42, 62, 71, 92, 138, 165, 234, 241, 262, 284, 339, 423, 471, 472, 531
   thyme, 594–597
Aldehydes:
   chicory root, 190
   cumin, 235–236
   deertongue, 241
   eucalyptus, 272
   ginger, 321
   grapefruit oil, 344
   orange, 478
   pine needle oil, 502
Aldobiouronic acid, 511
Alginates (algin), 16, 17
Alginic acid, 16, 17
Aliphatic, 59, 60
Aliphatic esters, 6, 217

Alkalis, use in cocoa processsing, 217, 218
Alkaloids:
  aconite, 7, 8
  alfalfa, 13
  angostura bark, 35
  ashwagandha, 645
  biological activities, 35, 88, 116, 194, 213, 337
  bloodroot, 102, 103
  boldo leaves, 107
  broom tops, 115, 116
  carboline, 190, 220, 515, 646
  centaury, 167
  coca, 212, 213
  cocillana bark, 215
  cocoa, 217, 218
  comfrey, 225, 226
  corn silk, 229
  cranberry, 232
  destruction by acacia, 5
  ephedra, 265–267
  ganoderma, 302
  gokshura, 646
  mahabala, 651
  vasaka, 655
  gelsemium, 313
  genet, 314, 315
  gentian, 316
  goldenseal, 336–338
  guarana, 350
  horehound, 362
  hydrangea, 370
  Indole:
    biological activities, 158, 490
    chicory root, 190
    gelsemium, 313
    pale catechu, 158
    quebracho, 517
  ipecac, 380
  isoquinoline:
    barberry, 72, 73
    biological activities, 72, 73, 336, 388
    goldenseal, 336
    jujube, 388
  lobelia, 424, 425
  lupine, blue cohosh, 104
  magnolia flower, 434–435
  monoterpene, 192, 390, 511
  pepper, 133, 135, 177, 499
  phellodendron bark, 678

  piperidine, lobelia, 424
  prickly ash, 50–51
  purslane, 679
  pyrrolizidine, 18, 109, 111, 225, 226, 252
  quassia, 515
  quinoline:
    angostura bark, 35
    biological activities, 35, 194
    cinchona, 194
    dittany bark, 664
  roselle, 532–534
  rue, 542–543
  sassafras, 557
  sesquiterpene, patchouly oil, 493–494
  Sichuan lovage, 683
  sour jujube kernel, 572
  tropane, 88, 89
  uses, 89
  valerian root, 612, 613
  xanthine, 350, 352, 591
  yarrow, 626
  yohimbe, 631, 632
Alkanes:
  bergamot oil, 92
  centaury, 168
  elder flowers, 257
  euphorbia, 274
  fenugreek, 287
  ganoderma, 301
  high-molecular weight, 85
  mahabala, 651
  yarrow, 626
Alkannin (anchusin, anchusic acid, or alkanna red), 18, 19
Alkylamide, 251–254
Alkylthiazoles, 584
Allantoin, 111, 225, 400, 610
Allicin, 308, 309
Alliinase, 308, 309, 474
Alliin (S-allyl-L-cyteine sulfoxide), 308
Alloxanthoxyletin, 50
1-Allyl-2, 4-dimethoxybenzene, 185
Allyl isothiocyanate, 457–459
Allylpropyl disulfide, 308, 474
Aloctin A,
Aloe-emodin:
  aloe, 24, 25

  biological activities, 525, 569
  cascara sagrada, 149–150
  rhubarb, 525
  senna, 569
Aloe-emodin glucoside, 150
Aloesin, 25, 26
Aloesone, 25
Aloins, 24–27, 150
Alphitolic acid, 572
Alpinin, 647
Aluminum, 292, 367, 395, 496
  ficin inactivation by, 292
Amarogentin, 192, 316, 629
Amarogentin (chiratin), 192
Ambrettolic acid, 31
Ambrettolide [(Z)-7-hexadecen-l6-olide], 31
Amides, 50, 114, 292, 430, 498
Amines:
  cocoa, 217
  hawthorn, 352
  licorice root, 416
  nettle, 464, 465
  reaction with acacia, 4
Amino acids:
  agar, 10
  alfalfa, 13
  almonds, 22
  aloe, 25
  asparagus, 52
  avocado, 59
  bee pollen, 82
  chamomile, 170
  cocoa, 217
  codonopsis, 220
  ficin, 292
  free:
    astragalus, 56
    coffee, 223
    cumin, 235
    fenugreek, 287
    as flavor precursors, coffee, 223
    jujube, 388
    lycium fruit, 431
    rehmannia, 521
    royal jelly, 539
  ganoderma, 301
  ginger, 321
  gotu kola, 339
  grape skin extract, 342
  Job's tears, 384

kava, 395
kudzu root, 400
licorice root, 416
lycium fruit, 430
mother-of-pearl, 674
mume, 672
papain,
pearl, 674
tea, 591
tienchi ginseng, 598
yarrow, 626
Aminobutyric acids, 56, 430, 522, 696
Ammonia, blood level, MSG ingestion and, 453
Ammonium alginate, 17
Ammonium bicarbonate, in cocoa processing, 217
Ammonium carbonate, 217
Amygdalin, 22, 23, 182, 672
Amyl alcohol, 482
2-n-Amylquinoline, 35
α-Amyrin, 101, 109, 241, 257, 274, 378, 461, 517
β-Amyrin:
  anise, 36
  calendula, 129
  centaury, 167
  chicle, 188
  dandelion root, 238
  elder flowers, 257, 258
  gentian, 316
  euphorbia, 274
  psyllium, 511
β-Amyrin acetate, 577
Amyrin palmitates, 257, 258, 424, 425
α-Amyrone, 461
Anabsin, 1
Anagyrine, 104, 314
Anahygrine, 645
Anaferine
Anchusic acid, see Alkannin
Anchusin, see Alkannin
Andro-stenedione, 177
Andrographolide, 650
Andrographosterin, 650
Anethole:
  anise, 36, 37
  biological activities, 37
  chervil, 185
  coriander, 227
  doggrass, 248

fennel, 283, 284
sassafras, 558
Angelic acid, 428
Angelicin, 32, 227
Angostura bitters 1 and 2, 35, 317
3,6-Anhydro-D-galatose, 145
3,6-Anhydro-α-L-, galactopyranose, 10
Anisaldehyde, 36, 37, 283, 284, 502
Anise ketone (*p*-methoxy-phenylacetone), 36
Anisic acid, 36, 404, 542
Anisotine, 655
Anisyl ethyl ether, 616
Anthocyanidins, 342, 696
Anthocyanins, 13, 93, 94, 182, 229, 232, 233, 258, 342, 474, 533, 591
Anthocyanosides, 93, 94
Anthracene derivatives, 150, 525, 569
Anthraglycosides:
  aloe, 24, 25, 27
  biological activities, 119, 526
  buckthorn, 119
  cascara sagrada, 149, 150
  rhubarb, 525, 526
  senna, 25, 119, 568–570
  see also Anthraquinones
Anthranilates, 382, 423, 478
Anthranilic acid, 561
Anthraquinones:
  biological activities, 150, 526, 569
  cascara sagrada, 149, 150
  cocillana bark, 215
  fo-ti, 294–296
  giant knotweed, 669
  rhubarb, 525, 526
Apigenin, 45, 143,
Apigenin-7- glucuronosylglucoside, 235
Apigetrin (apigenin-7-glucoside), 36, 169, 170, 235
Apiin (apigenin-7- apiosyl-glucoside), 165, 170, 185, 487
Apiole, 245, 283, 487, 488, 558
Arabinogalactan, 5, 130, 252, 253, 281, 449, 598, 602

Arabinose, 13, 48, 56, 111, 225, 263, 299, 339, 449, 461, 471, 496, 511, 522, 561, 569, 602, 604, 626, 666, 682, 696
Arabsin, 1
Arachidic acid, 400
Arborinine, 542
Arbutin, 93, 99, 237, 437, 504, 610, 611, 696
Arctic acid, 120
Arctigenin, 101, 120, 121
Arctiin, 101, 120, 121
Arctiol (8α-hydroxyeudesmol), 120
Arginine, 10, 13, 22, 52, 56, 212, 287, 400, 430, 539
Arnicin, 43
Arnicolides, 43
Arnidiol, 129, 238
Arnifolin, 43
Arnisterin, 43
Aromadendrene, 272, 581, 667
Aromoline, 73
Artabin, 1
Artabsin, 1
Artemidinol, 589
Artemisetin, 2
Asaresinotannols, 48
Asarinin, 50, 51
Asarone, 127, 148, 558
Asarylaldehyde, 127
Ascaridole, 107, 108, 180
Ascorbic acid, see Vitamin C
Ashwagandhine, 645
Asiaticoside, 339, 340
Asparagine, 13, 29, 52, 56, 222, 416
Asparagosides, 52
Asparagusic acid, 52
Aspartic acid, 10, 22, 52, 56, 222, 416, 430, 539, 674, 679
Asperuloside, 622, 623
Aspidinol, 54
Aspidosamine, 517
Aspidospermatine, 517
Aspidospermine, 517
Astragalans, 56
Astragalin (kaempferol-3-β-glucoside), 43, 93, 109, 359, 505, 620
Astragalosides, 56
Astrasieversianins, 56

Atlantones, 163, 604
Atractans, 662
Atractylenolides, 220, 660
Atractylodin, 662
Atractylon, 660, 662
Atractylon, 660, 662
Atropine (*dl*-hyoscyamine), 5, 88, 362, 430, 695
Aucubin, 177, 178, 279, 280, 511, 521
Auranetin, 478
Auraptenol, 478
Avicularin, 504
Avocatins, 59
Azadirachtin, 652
Azadiradione, 652
Azadirone, 652
Azulene, 1, 127, 170, 259, 299, 404, 444, 445, 626
Azulenes, 1, 170, 299, 444

Baicalein, 685–686
Baicalin, 685–686
Balsamic acids, *see* Benzoic acid and Cinnamic acid
Baptifoline, 104
Barbaloin, 24–25, 27, 149–150
Barbiturates, antagonistic effects of ginseng, 332
Barium, sodium alginate effect on retention of, 17
Bassorin, 471, 602
Behenic acid, 32, 252, 301, 404
Benzaldehyde:
 almonds, 22–23
 benzoin, 90
 biological activities, 22, 148, 184, 404, 494
 cananga oil, 132
 cherry bark, wild, 182
 cherry laurel leaves, 184
 cinnamon, 197
 labdanum, 404
 patchouly oil, 493–494
 tea, 591
2,3-Benzofuran, 241
Benzoic acid:
 Baikal skullcap, 685, 686
 balsam Peru, 70
 balsam tolu, 71
 benzoin, 90
 biological activities, 232
 castoreum, 156
 cranberry, 232

 jasmine, 382
 peony root, 676–677
Benzothiazoles, 190
Benzoylpaeoniflorin, 675, 676
Benzyl acetate, 132, 382, 630
Benzyl alcohol, 62, 70
 cananga oil, 132
 cassie absolute, 153
 jasmine, 382
 storax, 579
 tea, 591
Benzyl benzoate, 70, 90, 197, 284
Benzyl cinnamate, 70–71, 90
Benzylisothiocyanate, 458
Berbamine, 72–73
Berberine, 50, 72–73, 102, 336–337, 678
Bergamotenes, 37, 92, 148, 410, 423, 471
Bergamottin, 92, 344, 410, 423
Bergapten (5-methoxypsoralen):
 angelica, 32–33
 anise, 36
 bergamot oil, 92
 biological activities, 92, 542
 dill, 245
 grapefruit oil, 344
 lime oil, 423
 lovage root, 428
 orange, 478
 parsley, 487
 phototoxicity caused by, 33, 281, 344
 psoriasis treatment, 663
 rue, 542
Bergaptol, 92, 344
Bervulcine, 73
Betacyanins, 86–87
Betaine:
 ganoderma, 301–302
 kola nut, 398
 licorice root, 416
 lycium fruit, 430
 mahbala, 651
 mints, 444
Betalains, 86–87
Betanidin, 86
Betanin, 86–87, 696,
Betaxanthins, 86–87
Betonicine, 363, 626
Betulin, 230, 316, 388, 572
Betulinic acid, 388, 416, 449, 572, 665

Betulonic acid, 388
Bicycloelemene, 444
Biflavones, ginkgo, 325
Bilberry, 93–94
Biochanin A, 13, 207
Biochanin B (formononetin), 13, 56, 207, 400, 416
Biopterin, 539
Biotin, 13, 59, 331, 539
Bisabolone, 169
β-Bisabolene, 35, 37, 67, 92, 148, 235, 482, 612
Bisabolol oxides, 169–171
α-Bisabolone oxide A, 169
4,8″-Biscatechin, 59, 60
Bisulfites, enzyme activation by, 223, 292
Bitter acids, 359
Bitter principles:
 asparagus, 52
 avocado, 59
 cascara sagrada, 150
 centaury, 167
 chervil, 185
 chicory root, 190
 corn silk, 229
 damiana, 237
 gentian, 316
 horehound, 362
 hyssop, 376
 olibanum, 471
 quassia, 515–516
 saffron, 547
 sage, 551
 sweet woodruff, 622
Bixin, annatto, 40–41
trans-Bixin, *see* Isobixin
Boldine, 76, 107, 557
Borax (borates), reaction with acacia, 5
Boric acid, as acidulant in henna preparations, 356
Borneol:
 angelica, 32
 basil, 74
 cananga oil, 132
 catnip, 159
 cedar leaf oil, 161
 chaparral, 175
 coriander, 227
 echinacea, 251
 ginger, 321
 immortelle, 378
 labdanum, 404

lemongrass, 413
magnolia flower, 434
marjoram, 437
oregano, 482
pine needle oil, 502
rosemary, 535
sage, 550
sandalwood oil, 554
savory, 559
turmeric, 604
yarrow, 626
Bornyl acetate:
   biological activities, 503
   catnip, 159
   echinacea, 251
   hyssop, 376
   labdanum, 404
   magnolia flower, 434
   pine needle oil, 502–503
   valerian root, 612
   yarrow, 626
(+)-Boschniakine (indicaine), 511
(+)-Boschniakinic acid (plantagonine), 511
α-Boswellic acid, 471
Bourbonene, 62, 202, 319, 471
Brein, calendula, 129
Brein, 129, 261
Bulnesenes, 346, 493–494
Bulnesol, 299, 346, 493
Butylidene phthalide, 670, 683
Butylphthalide, 428, 683
Byakangelicin, 344, 663
Byakangelicol, 32, 663

Cadinene, 1, 37, 170, 234, 299, 344, 376, 390, 407, 437, 460, 502, 630
β-Cadinene:
   cade oil, 123
   cubebs, 234
   hyssop, 376
   marjoram, 437
   myrrh, 460
   olibanum, 471
   pine needle oil, 502
δ-Cadinene, 237, 319
γ-Cadinene, 630
Cadinol, 20, 35, 70, 234, 502
Cadmium, sodium alginate effect on retention of, 17
Caffeic acid:
   artichoke, 45–46

balm, 62
balm of Gilead bud, 65
bilberry, 93
coriander, 227
dandelion root, 238
dill, 245
echinacea, 252, 254
eleuthero, 263
eucalyptus, 272
eyebright, 279
forsythia fruit, 665
hawthorn, 352
horsetail, 367
immortelle, 378
marjoram, 437
mistletoe, 449
onion, 474
oregano, 482
sage, 550
Sichuan lovage, 683
thyme, 595
valerian root, 612
Caffeine:
   biological activities, 218, 223, 350, 398, 591–592
   cocoa, 217
   coffee, 222–224
   guarana, 350
   kola nut, 398
   tea, 591
   uses, 191, 223
Caffeoylquinic acids, 46, 121
Calamene, 127, 170, 237
Calamol, 127
Calcium:
   acacia, 4
   acerola, 6
   alfalfa, 13
   almonds, 23
   borage, 111
   chervil, 185
   coca, 212
   fennel, 283
   fenugreek, 287
   fo-ti, 295
   ganoderma, 302
   in gel formation, 17, 146
   ginseng, 331
   hawthorn, 352
   kava, 395
   nettle, 464
   red clover tops, 207
   savory, 559
   tamarind, 584

Calcium alginate, 16–17
Calcium carbonate, 674
Calcium oxalate:
   almonds, 22
   blackberry bark, 97
   cinnamon and cassia, 197
   purslane, 679
   quillaia, 518
   rhubarb, 525
Calcium phosphate, 156, 698
Calendula, 129–131
Calenduladiol:
   calendula, 129
Calendulosides:
   calendula, 129
β-Calycanthoside, see Eleutherosides
Campesterol:
   alfalfa, 13
   avocado, 59
   cloves, 210
   cocoa, 218
   dittany bark, 664
   elder flowers, 257
   euphorbia, 274
   juniper berries, 390
   mahbala, 651
   myrrh, 461
   psyllium, 511
   saw palmetto, 561
Camphene:
   absinthium, 1
   bay, 76
   chaparral, 175
   citronella oil, 202
   coriander, 227
   eucalyptus, 272
   ginger, 321
   horehound, 362
   hyssop, 376
   juniper berries, 390
   labdanum, 404
   lime oil, 423
   magnolia flower, 434
   nutmeg and mace, 467
   orange, 477
   origanum oil, 485
   pepper, 498
   pine needle oil, 502
   rosemary, 535
   sage, 550
   savory, 559
   tagetes, 581
   turpentine, 607

Camphor:
    basil, 74
    calamus, 127
    cedar leaf oil, 161
    chaparral, 175
    coriander, 227
    gotu kola, 339
    kulinjan, 647
    lavender, 407
    magnolia flower, 434
    oregano, 482
    rosemary, 535
    sage, 550
    sassafras, 558
    savory, 559
    turpentine as starting material, 608
    yarrow, 626
Canadine (tetrahydroberberine), 337
Canavanine, 14, 56
Candaline, 337
Capillene, tarragon, 588
Caprinic acid, 561
Caprylic acid, 561
Capsaicin, 133–135
Carbohydrates:
    allspice, 20
    anise, 36
    avocado, 59
    basil, 75
    bee pollen, 82
    bromelain, 114
    caramel color, 137
    cloves, 210
    coca, 212
    coffee, 222
    cranberry, 232
    ginseng effect on metabolism, 332
    lycium fruit, 430
    mume, 672
    pepper, 498
    royal jelly, 539
    see also Polysaccharides
Carbonyls, aromatic, 217
Carboxypeptidase, 25
Carboxylated tryptamines, 651
cardenolides, 654
3-Carene:
    biological activities, 607
    anise, 37
    fennel, 283

galbanum, 299
marjoram, 437
origanum oil, 485
pepper, 499
pine needle oil, 502
turpentine, 607
Carnosic acid, see also Antioxidants (in GENERAL INDEX)
Carnosol, 437, 535, 550
Carotene, 147, 148, 225, 340
α-Carotene, 43, 147, 444
β-Carotene:
    alfalfa, 13
    annatto, 40
    arnica, 43
    calendula, 130
    carrot root oil, 148
    eleuthero, 263
    gotu kola, 340
    jujube, 388
    lycium fruit, 430–431
    mints, 444
    prostate cancer risk, 130
    purslane, 679
    sour jujube kernel, 572
Carotenoids:
    annatto, 40
    arnica, 43
    capsicum, 133
    dandelion root, 238
    gotu kola, 339
    nettle, 464
    orange, 478
    rosehips, 528
    saffron, 547
    tagetes, 581–582
Carotol, 148, 487
Carpaine, 286
Carthamin, 682
Carvacrol, 483, 485, 595
    biological activities, 483, 485, 595
    catnip, 159
    doggrass, 248
    marjoram, 437
    oregano, 482–483
    origanum oil, 485
    savory, 559
    thyme, 595
    witch hazel, 620
Carvone:
    anise, 37

caraway, 138–139
coriander, 227
dill, 245–246
doggrass, 248
mints, 444
orange, 478, 480
tagetes, 581
Caryophyllene, 1, 32
    anise, 37
    artichoke, 46
    balsam copaiba, 67
    cade oil, 123
    cananga oil, 132
    carrot seed oil, 148
    catnip, 159
    cedarwood oil, 163
    cloves, 210
    cubebs, 234
    echinacea, 251
    ephedra, 265–266
    gotu kola, 339
    hops, 359
    hypericum, 372
    juniper berries, 390
    mints, 444
    olibanum, 471
    oregano, 482
    pepper, 499
    yarrow, 626
α-Caryophyllene, see Humulene
β-Caryophyllene, see Caryophyllene
    kulinjan, 647
Caryophyllene epoxide, 251
Caryophyllene oxide, 62, 67, 124, 148, 205, 227, 230, 237, 434, 647, 648
    kulinjan, 647
Cascarosides, 149–150, 525, 569, 696
Casein, as substrate, ficin, 292
Castoramine, 156
Castorin, 156
Catalpol, 521
Catapol, 279
Catechins, 197, 325, 592
    biological activities, 87, 107, 158, 232, 353, 435, 591–592
    cinnamon, 197
    cranberry, 232
    ephedra, 265
    fo-ti, 295

garden burnet, 666
ginkgo, 325
guarana, 350
hawthorn, 352
juniper berries, 390
kola nut, 398
nutmeg and mace, 467
peony root, 676
rosehips, 528
tea, 591–592
Catechol, 158, 666
Catecholamine, inhibition, epimedium, 269
Catechutannic acid, 157–158
Caulosides, 104
Ceanothic acid, 572
Cedrene, 163–164, 655
α-Cedrene, 655
Cedrenol, 163
Cedrol, 163
Cephaeline, 380
Cerolein, 85
Cetyl alcohol,
Chalcones, 65, 317, 359, 416, 458, 682
Chalepensin, 542–543
Chamazulene, 1, 170–171, 502
Chamigrenal, 565
Chamigrenes, 565
Chatinine, 612
Chavicine, 498
Chavicol, 78, 156, 197, 210
Chelerythrine, 50–51, 102
Chelidonic acid, 424
Chicoric acid, 190, 253
Chimaphilin, 504
Chiratin, 192
Chlorogenic acid, 120
  artichoke, 45–47
  biological activities, 46, 223, 253
  burdock, 120
  calendula, 130
  coffee, 223
  coriander, 227
  dill, 245
  echinacea, 252
  elder flowers, 257
  goldenseal, 337
  hawthorn, 352

honeysuckle flower, 667–668
rosemary, 535
sage, 550
valerian root, 612
Chlorogenin, 646
Cholesterol, 7, 10, 14, 17, 22, 26, 31, 46–47, 60, 82, 94, 107, 133, 143, 146, 156, 170, 198, 207, 242, 245, 248, 295, 303, 308, 321, 326, 327, 332, 347–348, 353, 368, 372, 390, 401, 440, 455, 461, 465, 474, 496, 511–512, 526, 539, 548, 584, 602, 604, 668, 680
Choline:
  biological activities, 275
  borage, 111
  chamomile, 170
  chicory root, 190
  dandelion root, 238
  elder flowers, 257
  euphorbia, 274
  fenugreek, 286
  ginseng, 331
  guarana, 350
  horehound, 362
  ipecac, 380
  licorice root, 416
  mahbala, 651
  mints, 444
  poria, 508
  valerian root, 612
  witch hazel, 620
  see also Lecithins
Chromium, 56
Chromones, 25, 281, 372, 444, 669
Chrysaloin, 149–150, 525
Chrysophanic acid (chrysophanol), 150, 295, 525
Chymopapain,
Cichoric acid, 251–254
Cichoriin (esculetin-7-glucoside), 190
Cimicifugoside, 97, 98
Cinchonidine, 194
Cinchonine, 194
Cineole, *see* 1,8-Cineole

1,4-Cineole, 36, 141, 234, 390, 423
1,8-Cineole:
  basil, 74
  bay, 76–77
  bois de rose oil, 106
  cardamon, 141
  chamomile, 170
  chaste-tree, 177
  damiana, 237
  eucalyptus, 272
  gotu kola, 339
  hyssop, 376
  immortelle, 378
  kulinjan, 647
  labdanum, 404
  lavender, 407
  lime oil, 423
  magnolia flower, 434
  mints, 444
  oregano, 483
  pine needle oil, 502
  rosemary, 535
  saffron, 547
  savory, 559
  tarragon, 588
  turmeric, 404
  yarrow, 626
Cinnamaldehyde:
  biological activities, 479, 494
  cinnamon, 197–199
  myrrh, 460
  patchouly oil, 493–494
  tamarind, 584
Cinnamein, 70
Cinnamic acid:
  balsam tolu, 71
  benzoin, 90
  castoreum, 156
  cinnamon, 197
  esters, 66
  gokshura, 646
  lycium fruit, 430
  rhubarb, 525
  storax, 579
Cinnamic alcohol, 197, 579
Cinnamolaurine, 557
Cinnamyl acetate, 197
Cinnamyl cinnamate, *see* Sytracin
Cistanosides, 521

Citral (a and b):
  balm, 62–63
  bay, 79
  biological activities, 413
  garlic, 308
  geranium oil, 319
  labdanum, 404
  lemongrass, 412–413
  lemon oil, 410
  lime oil, 423
  marjorain, 437
  orange, 478–479
  pine needle oil, 502
  schisandra, 565
  tagetes, 581
Citral a, *see* Geranial
Citral b, *see* Neral
Citrantin, 478–479
Citraurin, 478
Citric acid:
  as acidulant in henna preparations, 356
  biological activities, 232, 529
  bromelain inhibition, 114
  Cherokee rosehip, 181
  as color stabilizer, 87
  cranberry, 232
  ephedra, 265
  mume, 672
  nettle, 464
  purslane, 679
  rosehips, 529
  roselle, 533
  schisandra, 565
  tamarind, 584–585
Citronellal, 62, 106, 201–202, 319, 344, 478
Citronellol, 62, 159, 177, 201–202, 319, 412, 531
Citronellyl acetate, 159, 319, 344
Citronellyl formate, 319
Citropten, *see* Limettin
Cnicin, 101
Cnidilide, 670, 683
Cocaine, 213–214
Cohumulone, 359
Coixans, 384
Coixenolide, 384
Coixol, 384
Columbamine, 72–73
Colupulone, 359
Commiferin, 461
Commiphorinic acid, 461

Conchiolin, 674
Coniferaldehyde, 558
Coniferin, 138, 449, 501
Coniferyl acetate, 382
Coniferyl alcohol, 384, 501
Coniferyl benzoate, 90, 382
Coniferyl cinnamate, 90
Copaene, 37, 62, 67, 234, 237, 359, 461, 471, 487, 499, 558
Copper:
  alfalfa, 13
  astragalus, 56
  caramel color, 137
  enzyme inhibition or inactivation by, 292
  ganoderma, 302
  ginseng, 331
  in red beet color breakdown, 87
  red clover tops, 207
Corynantheine, 631
Costols, 230
Costunolides, 76, 230–231, 626
p-Coumaric acid, 182, 407
Coumarins:
  alfalfa, 13
  angelica, 32
  anise, 36
  astragalus, 56
  belladonna, 88
  bergamot oil, 92
  biological activities, 33, 51, 198, 241, 281, 410, 488, 542, 623
  black haw bark, 99
  celery seed, 165
  chamomile, 169
  chickweed, 187
  cinnamon and cassia, 197
  coriander, 227
  Dahurian angelica, 663
  deertongue, 241
  dill, 245
  fangfeng, 281
  formation, 241
  gelsemium, 313
  grapefruit oil, 344
  henna, 355
  jujube, 388
  kudzu root, 400
  lavender, 406–407
  lemon oil, 410–411
  licorice root, 416

  lime oil, 423
  lovage root, 428
  orange, 478
  passion flower, 490
  potential sources, 33, 92
  prickly ash, 50
  purslane, 679
  red clover tops, 207
  rue, 542
  sweet woodruff, 622
  tarragon, 588
  yarrow, 626
Coumestrol, 13, 207
Crategolic acid, 210
m-Cresol, 460
p-Cresol, 36, 123, 630
Crocetin (α-crocetin), 547–548
Crocins, saffron, 547
Cryptone, 235, 502
Cryptotanshinone, 680
Cryptoxanthin, 43, 52, 229, 238, 430, 478
Cubebenes, 62, 234, 499
Cubebic acid, 234
Cubebin, 234
Cubebol, 234
  kulinjan, 667
Cubenene,
Cumic alcohol (cuminyl alcohol), 460
Cuminaldehyde:
  biological activities, 236
  cinnamon and cassia, 197
  cumin, 235
  myrrh, 460
  pine needle oil, 502
Cuparene, 163
αr-Curcumene (α-curcumene), 36, 230, 319, 321, 499
β-Curcumene, 554
Curcumin, 604–605
Curzerene, 460
Curzerenone, 460–461
Cuscohygrine, 88, 212, 645
  ashwagandha, 645
Cuspareine, 35
Cusparine, 35
Cyanides, enzyme activation by, 292
Cyanidins, 93, 258, 295, 370, 474, 533
Cyanidin-3-sambubioside, 533

Cyclic AMP, 101, 388, 430, 566, 572, 677
Cyclic GMP, 572
Cycloalliin, 474
Cyclocostunolide, 230
Cyclohexanones, 591
Cyclopeptides, 572
Cycloseychellene, 493
Cymbopogonol, 412
Cymene, 197, 245, 467
p-Cymene:
   bay, 76
   bergamot oil, 92
   boldo leaves, 107
   chenopodium oil, 180
   coriander, 227
   cumin, 235
   doggrass, 248
   horehound, 362
   juniper berries, 390
   labdanum, 404
   lime oil, 423
   magnolia flower, 434
   marjoram, 437
   olibanum, 471
   orange, 478
   oregano, 482–483
   savory, 559
   thyme, 595
   turpentine, 607
Cynarin (1,5-di-O-caffeoylquinic acid), 45–47, 251, 253, 254
Cynaropicrin, 46, 230
Cysteine, enzyme activation by, 292
Cytisine, 314

Daidzein, 13, 207, 400, 401
Daidzein-4′,7-diglucoside, 400
Daidzin, 400, 401
Damaradienol, 259
Damascones, tea, 591
Damianin, 237
Danshensu, 680, 681
Daucol, 148
Daucosterol (eleutheroside A), 22, 262, 269, 522, 572, 598, 663, 676
   biological activities, 522
   Dahurian agelica, 663
   eleuthero, 262
   peony root, 676

   rehmannia, 522
   tienchi ginseng, 598
Decanal (decyl aldehyde), 46, 84, 227, 319, 413, 478
Decanol, 84, 410
γ-Decanolactone, 148
2-Decenedioic acid, 539
Decussatin (1- hydroxyl-2,6,8-Trimethoxyxanthone), 168, 192
Decyl acetate, 344
Dehydroabietic acid, 607
Dehydroascorbic acid, 6, 187
Dehydrocitronellol, 319
Dehydrodiisoeugenol, 467
5,9-Dehydronepetalactone, 159
3-Dehydronobilin, 170, 171
Dehydropimaric acid, 607
7-Dehydrostigmasterol, 678
Dehydroturmerone, 604
Delphinidin-3-glucoside, 533
Demethoxyisoguaiacins, 175
4′-Demethylpodophyllotoxin, 505, 506
Dencichine, 598, 599
Dendrolasin, 554
27-Deoxyactein, 97
14-Deoxy-11, 650
Deoxyandrographolide, 650
14-Deoxyandrographolide, 650
Deoxygomisin A, 566
Desaspidin, 54
Desmethoxyaschantin, 557
Desmethoxyaniflorine, 655
Dextrin, 265, 358
Dextrose, honey, 358
Dhelwangine, 493
Diallyl disulfide, 308, 309
Diallyl trisulfide, 308, 309
1,3-Di-O-caffeoylquinic acid, 46
Diarylpropanoids, 467
1,5-Di-O-caffeoylquinic acid, see Cyndrin
Dictamnolide, 678
12-Didehydroandrographolide, 650
Didrovaltrates, 612, 613
Diethyl phthalate, 299, 315
Diflavonols, ephedra, 266
Digallyhamamelose, 620
Dihydroactinidiolide, tea, 591
Dihydro-β-agarofuran, 554
Dihydroalantolactone, 259

Dihydrobutylidenephthalide, 428
Dihydrocapsaicin, 133
Dihydrocarvone, 138, 165, 245, 444
Dihydrocatalpol, 521
Dihydrochamazulenes, 1
Dihydrocinnamic acid,
Dihydrocoumarin, 241, 407
Dihydrocurcumene,
Dihydrodehydrocostus lactone, 230
Dihydrogambirtannine, 158
Dihydroguaiaretic acid, 174, 467
Dihydroyohimbine, 631
Dihydroisoalantolactone, 259
7,8-Dihydrokavain, 395, 396
Dihydromethysticin, 395, 396
Dihydronepetalactone, 159
Dihydropyrocurzerenone, 460, 461
Dihydroquercetin-4′-monoglucoside, 575
Dihydrotanshinone, 680
Dihydroxybutylidene phthalide, 683
3,10-Dihydroxydecanoic acid, 539
1,8-Dihydroxy- 3,7-dimethoxyxanthone, 192
4′,7-Dihydroxyflavone, 416
Dihydroxyfumaric acid, papain activation by,
Dihydroxymaleic acid, papain inhibition by,
Dihydroxystearic acid, 154
3,4′-Dihydroxystilbene, 370
Dihydroxytartaric acid, papain inhibition by,
Diisobutylphthalate, 525
β-Diketones, 378
Diketopiperazine, 217
Dillanoside, 245
Dillapiole, 245
1,2-Dimercaptopropanol, ficin activation by, 292
6,7-Dimethoxycoumarin, 400, 423, 478
γ-γ-Dimethylallyl ether, 50
9,10-Dimethyl-1,2-benzanthracene, agents promoting tumor formation by, 272, 344, 410, 423

cis-Dimethylcrocetin, 547
1,4-Dimethyl-3-cyclohexen-1-yl methyl ketone, 390
Dimethyldisulfide, 308, 474
5-[(3,6-Dimethyl-6-formyl-2-heptenyl)oxy] psoralen, 344
2,4-Dimethylheptane-3,5-dione, occurrence and aroma properties, 378
3,5-Dimethyloctane-4,6-dione, occurrence and aroma properties, 378
4,7-Dimethyl-6-octen-3-one, 378
Dimethyl sulfide, 251, 308, 319, 591
3,4-Dimethylthiophene, 474
Dimethyl trisulfide, 308
2,5-Dimethyl-2-vinyl-4-hexenal, 478
Dimethylxanthine,
6,3'-Di-O-demethylisoguaiacin, 175
Dioleolinolein, 22
Dioscin, 646
Diosgenin, gokshura, 646
Diosmetin, 437, 444, 482, 535
Diosmin, 117, 376, 411, 444, 535, 536
Diosphenol, 117
Dipentene (dl-limonene):
  biological activities, 503, 607
  cascarilla bark, 151
  cedar leaf oil, 161
  elemi gum, 261
  lemongrass, 412, 413
  myrrh, 460
  nutmeg, 467
  olibanum, 471
  pine, needle oil, 502, 503
  turpentine, 607
Dipropyl trisulfide, 474
Diterpene, labdane, 577
Diterpenoids, 66, 501, 650, 652, 680
  Kalmegh, 650
Diterpenoid glycosides, 650
Docoscnoic acid, 386
Docosenol, 386
Dopa, 207, 679

Dopamine, 217, 266, 295, 340, 425, 679
Dulcoside A, 577

Echinacoside, 251, 253–255, 521
Eicosenoic acid, 386
Eicosenol, 386
Elastin, as substrate for ficin, 292
Elema-1,3,11(13)- trien-12-ol, 230
Elemenes:
  bois de rose oil, 106
  carrot seed oil, 148
  cedarwood oil, 163
  geranium oil, 319
  juniper berries, 390
  parsley, 487
β-Elemen-7α-ol, 390
Elemicin, 261, 467, 558, 588
Elemol, 202, 227, 230, 261, 662
Eleutherosides, 262, 263, 449
Ellagic acid, 143, 158, 272, 610, 666
Emetine, 380, 381, 695
Emodin, 295, 525, 669
Emodin-1-glucoside, 119
Emodin-8-glucoside, 119
Emodin-8-O-β-gentiobioside, 119
Emulsin, 22, 182, 184
Ephedradines, 266
Ephedrans, 265, 266
Ephedrine, 266–267, 651
Ephedroxane, 265, 266
Epicatechin, 59, 93, 143, 277, 325, 352, 353, 359, 371, 504, 584, 620
Epicatechin gallate, 145, 504, 525, 591, 620
Epigallocatechin, 325, 592, 620
Epiguaipyridine, 493
10-Epijunenol, 299
Epimedosides, 269
Epimyrtine, 93
Epinepetalactone,
Epitaraxerol, 655
Epitestosterone, 177
1,10-Epoxynobilin, 170, 171
Epoxycaryophyllene, 493
Epoxyocimenes, 1
Ergosterol, 31, 301, 302, 508
Eriodictyol, 482, 584, 629
Eriodictyonine, 629

Euostoside, 177
Erucic acid, glycerol esters, 696
Erythrodiol, 167
Esculetin, 190, 245
Esculetin dimethyl ether, 588
Esculetiln-7-glucoside-see Cichoriin
Esculin (esculetin-6-glucoside), 190, 364
Estragole (methyl chavicol):
  anise, 36, 37
  avocado, 59
  basil, 74, 75
  bay, 79
  biological activities, 75, 589
  chervil, 185
  fennel, 283, 284
  marjoram, 437
  tarragon, 588, 589
Ethyl butyrate, 478
Ethyl cinnamate, 197, 584
1-Ethyl-β-D-galactoside, 522
Ethyl jasmonates, 382
N-Ethylmaleimide, protease inhibition or inactivation by,
Ethyl maltol, 490
Ethyl oleate, genet, 314
Ethyl palmitate, 314
Ethyl stearate, 314
Eucalyptol, see 1,8-Cineole
Eudesmanolides, 289
Eudesmols, 120, 165, 197, 434, 662
Eufoliatin, 109
Eugeniin, 210
Eugenol:
  allspice, 20
  artichoke, 46
  basil, 74
  bay, 76
  biological activities, 20, 76, 210, 378, 404
  boronia absolute, 113
  calamus oil, 127
  cananga oil, 132
  carrot seed oil, 148
  cinnamon and cassia, 197, 199
  cloves, 210, 211
  as flavoring ingredient, 211
  immortelle, 378
  labdanum, 404
  marjoram, 437

myrrh, 460
nutmeg and mace, 467
patchouly oil, 493, 494
rose oil, 531
sassafras, 558
vanilla, 616
witch hazel, 620
Eugenol acetate, 62, 197, 210
Eukovoside, 279
Eupatorin, 109, 550

γ-Fagarine, 50, 542, 664
Faradiol, 46, 129, 130
Farnesol, 31, 132, 153, 170, 202, 266, 382, 412, 413, 471, 478, 554, 561, 630
Farnesene, 65, 170, 177, 359, 471
β-Farnesene, 36, 37, 92, 148, 235, 331, 339, 499, 554, 647
Farnesiferols, 48
Farnesol:
  ambrette seed, 31
  cananga oil, 132
  cassie absolute, 153
  orange, 478
  jasmine, 382
  saw palmetto, 561
Fats and fixed oils, 22, 120, 190, 287, 457
  biological activities, 22
Fatty acids,
  essential, evening primrose, 278
  esters, 130, 257, 386, 521
  free, saw palmetto, 561
  high molecular weight, jojoba, 386
Febrifugine, 370
Fenchol (fenchyl alcohol), 283, 299, 423
Fenchone, 283, 284, 404, 416, 434
Fenchyl acetate, 647
Ferruginol, 205, 680
Ferulic acid, 48, 97, 187, 384
  asafetida, 48
  biological activities, 187
  eyebright, 279
  ligusticum, 670
  rhubarb, 525
  Sichuan lovage, 683
  sour jujube kernel, 572

Feruloylhistamine, biological activities, 266
Fiber, 13, 31, 52, 133, 143, 191, 315, 325, 347, 395, 427, 498, 512, 633
FICIN, 291–293
Filicin, 54
Filixic acids, 54
Fisetin, 157
Flavaspidic acids, 54
Flavokavins, 395
Flavonoids:
  alfalfa, 13
  alkanet, 18
  anise, 36
  arnica, 43
  asparagus, 52
  astragalus, 56, 57
  Baikal skullcap, 685
  balm, 62
  bee pollen, 82
  belladonna, 88
  bilberry, 93
  biological activities, 178, 368
  boneset, 109
  buchu, 117
  calendula, 130
  caraway, 138
  chamomile, 169, 170
  chaparral, 175
  chaste-tree, 177, 178
  chickweed, 187
  clover tops, 207
  cocillana bark, 215
  coriander, 227
  cumin, 235
  damiana, 237
  Devils claw, 242
  dill, 246
  echinacea, 251, 252
  elder flowers, 257
  ephedra, 265, 266
  epimedium, 269, 270
  eucalyptus, 272
  euphorbia, 274
  fennel, 283
  fenugreek, 287
  ginkgo, 325
  gokshura, 646
  gotu kola, 339
  hawthorn, 352, 353
  honeysuckle flower, 667
  hops, 359, 360

horsechestnut, 364
horsetail, 367, 368
hydrangea, 370
hypericum, 371, 372
hyssop, 376
immortelle, 378, 379
jujube, 388
Juniper berries, 390
kalmegh, 650
kudzu root, 400, 401
lavender, 407
licorice root, 416
magnolia flower, 434
marjoram, 437
mints, 444, 445
nettle, 464, 465
onion, 474
orange, 478, 479
oregano, 482
parsley, 487
passion flower, 490, 491
pipsissewa, 504
podophyllum, 505
purslane, 679
rhubarb, 525
rosehips, 181
rosemary, 535
safflower, 682
sage, 550
savory, 559
senna, 569
skullcap, Baikal, 685
sour jujube kernel, 572
squill, 575
tarragon, 588
tea, 591
thyme, 595
tienchi ginseng, 598
uva ursi, 610
witch hazel, 620
yarrow, 626
yerba santa, 629
Flavanolignans, 439, 440
Folic acid:
  alfalfa, 13
  astragalus, 56
  bee pollen, 82
  royal jelly, 539
Formic acid:
  geranium oil, 319
  honey, 358
  nettle, 464
  yerba santa, 629

Formononetin, *see* Biochanin B
Forsythigenin, 665
Forsythin, 665
Forsythol, 665
Forsythosides, 362, 665
Frangulin A and B, 119, 149
Fraxinellone, 664
Friedelin, 220, 231, 259, 274
Fructose, 6, 8, 13, 32, 59, 143, 190, 220, 225, 233, 238, 239, 252, 274, 316, 325, 331, 333, 339, 352, 388, 390, 395, 508, 511, 525, 528, 569, 604
Fucose, 508, 561, 602
Fukinanolide, 120
Fukinone, 120
Fumaric acid, papain activation by,
1, 10 (15)-Furanodiene-6-one, 461
Furanoeudesma-1,3-diene, 461
Furans, 190, 223, 302, 522
Furfural (furfurol), 6, 84, 190, 207, 378, 616
Furfuryl alcohol, 416, 591
Furocoumarins, *see* Coumarins
Furostanol Saponins, 633, 646

Galactomannan, 25, 143, 144, 222, 287, 347, 426
Galactose, 10, 13, 48, 111, 220, 263, 299, 393, 427, 449, 461, 471, 522, 561, 569, 602, 626, 682
Galacturonic acid:
 althea root, 29
 chamomile, 169
 galbanum, 299
 karaya gum, 393
 pectin, 496
 tragacanth, 602
Galanganal, 647
Galanganols, 647
Galangin, 128, 647
Galbanol, 299
Galipidine, 35
Galipine, 35
Gallic acid:
 biological activities, 586
 blackberry bark, 97
 catechu, 158
 ephedra, 265

eucalyptus, 272
henna, 355, 356
peony root, 676
rhubarb, 525
tannic acid, 586
tea, 591
uva ursi, 610
witch hazel, 620
Gallocatechin, 325, 620, 666
(+)-Gallo-catechol, 666
Gallotannins, 143, 279, 390, 585, 620
Gambirdine, 158
Gambirine, 158
Ganoderans, 302, 303
Ganoderic acids, 301–303
Ganoderols, 302
Geijerone, 390
Gelatin, as substrate for, ficin, 292
Gelsedine, 313
Gelsemicine, 313
Gelsemidine, 313
Gelsemine, 313
Gelsevirine, 313
Geniposide, 279
Genistein, 13, 115, 143, 187, 207
Genisteine, 115
Genkwanin, 535, 550
Genkwanin-4′-methyl ether, 535
Gentiacauloside, 316
Gentianine, 167, 286, 316
Gentianose, 316
Gentiobiose, 316, 547
Gentioflavoside, 167
Gentiopicrin (gentiopicroside), 167, 168, 316, 317
Gentisein, 316
Gentisic acid, 504
Gentisin, 316
Gentrogenin, 12
Geranial (citral a), 62, 63, 159, 321, 344, 410, 412, 584, 591
Geranic acid, 62, 531, 630
Geraniol:
 balm, 62
 basil, 74
 bay, 76, 77
 biological activities, 77, 148, 378
 carrot seed oil, 148
 cassie absolute, 153

catnip, 159
citronella oil, 201, 202
coriander, 227
garlic, 308
geranium oil, 319
honeysuckle flower, 667
immortelle, 378
labdanum, 404
lemongrass, 412, 413
orange, 478
rose oil, 531
tamarind, 584
tea, 591
ylang ylang oil, 630
7-Geranoxycoumarin, 344
8-Geranoxypsoralen, 410, 423
Geranyl acetate
 cananga oil, 132
 carrot seed oil, 148
 coriander, 227
 geranium oil, 319
 grapefruit oil, 344
 labdanum, 404
 marjoram, 437
 oragne, 478
 oregano, 482
 ylang ylang oil, 630
Geranyl tiglate, 319
Germacranolides, 1, 76, 120, 170, 171, 289
Germacratriene, 319, 359
Germacrene alcohol, 252
Germacrenes, 1, 35, 205, 251, 339, 423, 482
Germanium,
Gingediacetate, 321
Gingerols, 321, 322
Ginkgolides, 325–327
Ginsenosides, 331–333, 598–600
Gitogenin, 646
Glabranine, 416
Glabrol, 416
Glabrone, 416
Glycofrangulins (A and B),119
Glucogallin, 525, 676
Glucomannan, 25, 27
Gluconic acid, 10, 539
Glucoscillaren A, 574
Glucoscilliphaeoside (12β-hydroxyscillaren), 574
Glucose (dextrose), 10

Glucosides:
  alfalfa, 13
  aloe, 24
  anise, 36
  artichoke, 45
  balm, 62
  beet color, 86
  boneset, 109
  buchu, 117
  buckthorn, 119
  burdock, 120
  cascara sagrada, 149, 150
  centaury, 167
  chicory root, 190
  cocoa, 218
  coriander, 227
  cumin, 235
  dandelion root, 238, 239
  deertongue, 241
  dosgrass, 248
  fo-ti, 295, 296
  immortelle, 378
  kola nut, 398
  onion, 474
  passion flower, 490
  podophyllum, 505, 506
  red beet color, 86
  rhubarb, 525
  senna, 569
  see also Glycosides
Glucuronic acid, 13, 29, 48, 56, 228
  cranberry, 232
  juniper berries, 390
  karaya gum, 393
  xanthan gum, 624
Glutamic acid:
  astragalus, 56
  coffee, 222
  lycium fruit, 430
  purslane, 679
  royal jelly, 539
Glutathione, purslane, 679
Gluten, as source of MSG, 452
Glycerides, see Glycerol esters
Glycerol esters, 696
  see also Triglycerides
Glycine, as carboxy terminal residue,
Glycine betaine, 252
Glycolipids, 384, 430, 431, 529
Glycones, see Glucosides; Glycosides

Glycoprotein lectins, 449
Glycoproteins, 114, 130, 198, 252, 380, 398, 430, 529, 539, 575
Glycosides:
  aletris, 12
  anise, 36
  artichoke, 46
  asparagus, 52
  bayberry bark, 80
  biological activities, 119, 150
  black cohosh, 97, 98
  blue cohosh, 104
  buckthorn, 119
  calendula, 129, 130
  chickweed, 187
  comfrey, 225
  coumarin, celery seed, 165
  cyanogenic, damiana, 237
  diterpene, stevia, 577
  elder flowers, 257, 258
  eleuthero, 263
  fenugreek, 287
  terminalia, 654
  flavonoid, see Flavonoids
  flavonol:
    biological activities, 43, 232
    cranberry, 232
  gymnema sylvestre, 648
  giant knotweed, 669
  ginseng, 330
  grape skin extract, 342
  hops, 359
  horsechestnut, 364
  hydrangea, 370
  hyssop, 376
  iridoid:
    biological activities, 178
    chaste-tree, 177, 178
    Devil's claw, 242
    eyebright, 279, 280
    rehmannia, 521
    valerian root, 612
  isoflavone, kudzu root, 400, 401
  juniper berries, 390
  licorice root, 416, 417
  ligustrum, 420
  marjoram, 437
  monoterpene:
    peony bark, 675
    peony root, 676

  naplithopyrone rehmannia, 521, 522
  pepper, 498
  phenethyl alcohol, rehmannia, 521, 522
  β-phenethylamines, 651
  phenolic, see Phenols
  pipsissewa, 504
  rhubarb, 525, 526
  rose hips, 528
  rutin, 545
  saponin:
    Cherokee rosehip, 181
    eleuthero, 263
    tienchi ginseng, 598–600
  sarsaparilla, 556
  senna, 569, 570
  squill, 574–576
  steroid, see Steroids
  tannic acid, 586
  triterpene:
    astragalus, 56
    chickweed, 187
    garden burnet, 666
    jujube, 388
    sour jujube kernel, 572, 573
    woodruff, 622
  see also Glucosides
Glycyridione, 416
Glycyrol, 416
Glycyrrhetinic acid (glycyrrhetic acid), 416, 417
Glycyrrhizin (glycyrrhizic acid; glycyrrhizinic acid), ammoniated, 415–418, 696
Glyzarin, 416
Gomisins, 565, 566
Gonzalitosin I (5-hydroxy-7,3′,4′-trimethoxyflavone), 237
Graveoline, rue, 542
Guaianolides, 1, 46, 170, 289
Guaiazulene, 170, 299, 346
α-Guaiene, 346, 493
β-Guaiene, 471
δ-Guaiene, see Bulnesenes
Guaiol, 299, 346
Guaioxide, 346
Guaipyridine, 493
Guanine, 350, 352
Guaran,
  see also Galactomannan

Z-Guggulsterone, 461
L-Guluronic acid, 16
Gums:
   asafetida, 48, 49
   carob, 142–144
   chicle, 188–189
   comfrey, 225–226
   corn silk, 229
   cubebs, 234
   cumin, 235
   dandelion root, 238
   karaya gum, 393–394
   licorice root, 416
   lobelia, 424
   lovage root, 428
   myrrh, 460–462
   olibanum, 471–472
   passion flower, 490, 491
   valerian root, 612
   Yerba santa, 629
   see also Mucilage; Polysaccharides
γ-Gurjunene,
Gymnemic acid A, 648, 649
Gynosaponins, 332

Hamamelitannin, 620
Harmaline, 490
Harmalol, 490
Harman, 190, 490, 491, 646
Harpagide, 242
Harpagoside, 242–243
10-HAD, see 10-Hydroxy-trans-decenoic acid
Heavy metals:
   enzyme inactivation by, 114, 292
   pectin effect on, 497
   reaction with acacia, 5
   reaction with pectin, 496
Hecogenin, 646
Hederagenin, 13, 104
Heerabomyrrhols, 461
Heeraboresene, 461
Helichrysin, 378
Helinatriol C and F, 129
Helipyrone, 378
Hennosides, 355
Heptadecane, 626, 648
cis-8-Heptadecene, 478
Herculin, 50

Herniarin (umbelliferone methylether), 170, 407, 416, 542, 588
Hesperidin, 117, 376, 444, 478, 612
1,26-Hexacosanediol, 314
Hexacosanol-1, 237
Hexanol, 416
trans-2-H exenal, 212, 478
n-Hexen-2-al, 620
Hexenol, 620
cis-3-Hexenyl acetate, 382
Hexylmethylphthalate, 33
Hibiscic acid, 533
Hibiscin (daphniphylline delphinidin-3-sambubioside), 533
Higenamine, cardiac effects, 8
Hippoaesculin, 364–365
Hippuric acid, 232
Histamine, 101, 125, 151, 171, 231, 302, 435, 439, 464, 533, 545
Hormones, steroid:
   manufacture, 12
   potential sources, fenugreek, 287
Humic acid, papain activation by,
γ-Humulene, 67
Humulene (α-caryophyllene):
   balm of gilead buds, 65
   bergamot oil, 92
   cedarwood oil, 163
   cinnamon, 197
   geranium oil, 319
   ginseng, 331
   hops, 359
   hypericum, 372
   olibanum, 471
   patchouly oil, 493
Humulone, 359–360
Hydragenol,
Hydrangeic acid, 370
Hydrangenol, 370
Hydrangetin (7-hydroxy-8-methoxycoumarin), 370
Hydrangin, hydrangea, 370
l-α-Hydrastine, 337
Hydrocarbons:
   aromatic, chicory root, 190
   avocado, 59
   boronia absolute, 113
   cassie absolute, 153

   chenopodium oil, 180
   chervil, 185
   chicle, 188
   cocoa, 217
   saturated, tarragon, 588
Hydrocyanic acid (HCN: prussic acid), 22, 182, 184, 258
Hydrogen peroxide, ficin inhibition or inactivation by, 292
Hydroquinone, biological activities, 504, 610
p-Hydroxybenzaldehyde, 616, 647
   kulinjan, 647
Hydroxybenzoic acid, 93, 187, 539
37-hydroxyhexatetracont-1-en-15-one, 655
37-hydroxyhentetracontan-19-one, 655
p-Hydroxybenzyl isothiocyanate, 458
p-Hydroxybenzyl methyl ether, 616
Hydroxybutylidene phthalide, 683
Hydroxy carboxylic acids,
p-Hydroxycinnamic acid methyl ester, 252
Hydroxycinnamic derivatives, 376
3-Hydroxyanisotine, 655
7-Hydroxycoumarins, see Umbelliferone
Hydroxydecanoic acids, 539
10-Hydroxy-trans-2-decenoic acid, 539
1-Hydroxy-3,7-dimethoxyxanthone, 316
(S)-2-Hydroxy-3,4-dimethyl-2-buten-4-olide, 663
14-Hydroxygelsemicine, 313
7-Hydroxy-Hardwick acid, 67
(2S,3R,4R)-4-Hydroxyisoleucine, 287
d-2-Hydroxy-isopinocamphone, 376
Hydroxymuscopyridines, 455
2-α-Hydroxyoleanolic acid (crataegolic acid), 352
15-Hydroxypalmitic acid esters, 85

p-Hydroxyphenylacetic acid, 238
17α-Hydroxyprogesterone, 177
18-Hydroxyquassin, 515
Hydroxytanshinone, 680
3-β-Hydroxytanshinone IIA, 680
5-Hydroxytetrahydroberberine, 337
4-Hydroxythymol dimethyl ether, 42
10-Hydroxy-*trans*-2-decenoic acid, 42
1-Hydroxy-3,5,8-trimethoxyxanthone, 192
30β-Hydroxyursolic acid, 257
l-Hyoscine *N*-oxide, 88
Hyoscine (scopolamine), 8, 88–89
Hyoscyamine, 5
l-Hyoscyamine, 88
l-Hyoscyamine *N*-oxides, 88
Hypoxanthine, 350
Hyperforin, 371–373
Hypericin, 371–373
Hyperin, *see* Hyperoside
Hyperoside (quercetin-3-β-galactoside):
  bilberry, 93
  biological activities, 272, 353
  boneset, 109
  epimedium, 269
  eucalyptus, 272
  hawthorn, 352
  hypericum, 371
  pipsissewa, 504
Hyssopin, 376

Icariin, 269–270
Imidazoles, 137, 266
Imperatorin, 32, 281, 410, 423, 542, 663
Incensole, 471
Incensole acetate, 471
Indicainine, 511
Indole, 158, 478, 490, 517, 646, 655, 699
  vasaka, 655
Inositol, 248, 274, 386, 539, 626, 667
Intermedine, 111, 225
Inulin:
  artichoke, 46
  asparagus, 52
  burdock, 120

cangzhu, 662
chicory root, 190
costus oil, 230
dandelion root, 238
dill, 245
echinacea, 252
elecampane, 259
ephedra, 265
Iodoacetamide, enzyme inactivation or inhibition by, 292
Iodoacetates, enzyme inactivation or inhibition by, 292
Ionones, 163
  biological activities, 113
  boronia absolute, 112–113
  orange, 478
  source, lemongrass, 412
  starting materials, lemongrass, 414
  tamarind, 584
  tea, 591
  witch hazel, 620
Ipecacuanhin, 380
Ipecoside, 380
Iron:
  acerola, 6
  alfalfa, 13
  astragalus, 56
  caramel color, 137
  cardamom, 141
  coca, 212
  doggrass, 248
  fenugreek, 287
  ficin inactivation or inhibition by, 292
  fo-ti, 295
  ganoderma, 302
  ginseng, 331
  hawthorn, 352
  kava, 395
  in red beet color breakdown, 87
Isoacteoside, 511, 521
Isoalantolactone, 230, 259, 260
Isoandrographolide, 650
Isoartemisia ketone, 626
Isoastragalosides I and II, 56
Isobellidifolin (1,6,8-trihydroxy-4-methoxyxanthone), 92
Isobetanin, 86
Isobixin (*trans*- bixin), 40
Isoboldine, 107, 557

Isobornyl acetate, 535
L-Isobutyl-L-valine anhydride, 683
Isobutyric acid, 407, 616
Isobutyric acid thymyl ether, 42–43
Isochlorogenic acids, 252, 667
Isocorydine, 107
Isocorydine-*N*-oxide, 107
Isocoumarins, 370, 588–589
Isocryptotanshinone, 680
Isodihydronepetalactone, 159
Isoemodin, 150
Isoeugenol, 79, 127, 132, 467, 588, 630
Isoflavonoids, 56–57, 400, 416
Isofraxidin-7-*O*-α-L-glucoside (eleutheroside B$_1$), 262
Isogentisin, 316
Isohelichrysin, 378
Isohomoarbutin, 504
Isoimperatorin, 410, 423, 542, 663
Isoincensole, 471
Isoliquiritigenin, 416
Isomargolonone, 652
Isomenthol, 444
Isomenthone, 117, 319, 404, 444
Isoorientin, 36, 177, 412, 490, 575
Isopelletierine
Isophorone, 547
Isopimaric acid, 607
Isopimpinellin, 92, 423, 487, 663
Isopinocamphone, 376
S-Isopropyl-3-methylbutanethioate, 299
Isopulegone, 117
Isoquassin (picrasmin), 515
Isoquercitrin (trifoliin):
  bilberry, 93
  caraway, 138
  clover tops, 207
  coca, 212
  coriander, 227
  elder flowers, 257
  fennel, 283
  hops, 359
  juniper berries, 390
Isorhamnetin, 56, 82, 107, 143, 207, 245, 252, 458, 464, 474, 505, 569
Isosafrole, 630
Isotanshinones, 680

Isotetrandine (berbamine
    methylether), 72–73
Isotheyone,
Isothiocyanates, 457–459
Isotussilagine, 252
Isovaleric aldehyde, 378
Isovaltrates, 612–613
Isovitexin, 36, 177, 287, 490, 575
Isoyohimbine, 631

Jasmolactone, 382
Jasmone, 382, 478, 591
δ-Jasmonic acid lactone, 382
Jatrorrhizine, 72, 678
Jionosides, 521
Jujubogenin, 572
Jujubosides, 388, 572

Kaempferol:
    asparagus, 52
    biological activities, 378
    boneset, 109
    cloves, 210
    dill, 245
    elder flowers, 257
    Gymnema sylvestre, 648
    hydrangea, 370
    immortelle, 378
    passion flower, 490
    pepper, 498
    podophyllum, 505
    senna, 569
    witch hazel, 620
Kaempferol-3-glucuronide, 283
Kaempferol-3-rutinoside, 109,
    252, 325, 359, 646
Kaempferide, 647
Kavalactones, 395–396
Ketoalkenes, 252
Ketoalkynes, 252
Ketones:
    castoreum, 156
    deertongue, 241
    eucalyptus, 272
    juniper berries, 390
    kava, 395
    orange, 478
Ketosteroids, 177
Koganebanain, 685
Kokusaginine, 542
Kumatakenin, 56, 416, 493
Kumijians, 515
Kusulactone, 515

Labdanolic acid, 404
Labiatic acid:
    biological acitivities, 536, 550,
        595
    marjoram, 437
    rosemary, 536
    sage, 550
    savory, 559
    thyme, 595
Lacoumarin (5-allyloxy-7-
    hydroxycoumarin), 355
Lactates, 94, 496
Lactiflora, 676
Lactones:
    absinthium, 1–2
    angelica, 32
    chamomile, 170–171
    cocoa, 217
    dittany bark, 664
    lavender, 407
    yarrow, 626
    see also Sesquiterpene lactones
Lactucin, 190
Lactucopicrin (intybin), 190
Laetrile, 23
cis-Lanceol,
Lapachenole, 483
Lappaols A and B, 120
Lauric acid, 13, 22, 32, 120, 238,
    314, 321, 382, 390, 467,
    561, 572
Laurifoline, 50
Laurolitsine, 107
Lavandulyl acetate, 1, 407
Lawsone (2-hydroxy-1,4-
    naphthoquinone),
    355–356
Laxanthones I and II, 355
Lead:
    enzyme activation and
        stabilization by,
    pectin effect on, 497
Lecithins, 295, 321, 508, 697,
    see also Choline:
Lectins, 25, 154, 248, 258, 302,
    314, 449, 464, 584
Ledol, 272, 404, 444, 612
Leonuride, 521
Leucoanthocyanins, 528, 666
Leucocyanidin, 274, 342, 371
Levopimaric acid, 607
Levulose, 358
Licochalcones A and B, 416

Licoflavones, 416
Licoflavonol, 416
Licoisoflavones A and B, 416
Licoricone, 416
Lignanamides, 646
Lignans:
    ash, 50
    biological activities, 51, 101,
        121, 175, 435, 506, 654
    blessed thistle, 101
    burdock, 120
    chaparral, 174
    magnolia flower, 434
    podophyllum, 505
    safflower, 682
    sassafras, 557
    schisandra, 565–566
    terminalia, 654
Lignins, 54, 120, 325, 416
Lignoceric acid, 281
Ligustilide, 428
    biological activities, 670, 683
    ligusticum, 670
    Sichuan lovage, 683
Limettin (citropten), 344, 410,
    423
Limonene, 32
    angelica, 32
    anise, 36
    bay, 76
    bergamot oil, 92
    biological activities, 138, 503
    bois de rose oil, 106
    buchu, 117
    cananga oil, 132
    caraway, 138
    carot seed oil, 148
    for carvone synthesis, 480
    catnip, 159
    celery seed, 165
    chaparral, 175
    chaste-tree, 177
    chenopodium oil, 180
    citronella oil, 202
    coriander, 227
    cubebs, 234
    dill, 245
    ephedra, 265–266
    eucalyptus, 272
    fennel, 283
    grapefruit oil, 344
    horehound, 362
    hypericum, 372

Chemical Index

immortelle, 378
juniper berries, 390
kulinjan, 647
labdanum, 404
lavender, 407
lemon oil, 410
ligusticum, 670
lime oil, 423
magnolia flower, 434
myrrh, 460
olibanum, 471
orange, 477–478
origanum oil, 485
pepper, 499
pine needle oil, 502
rosemary, 535
sage, 550
savory, 559
tagetes, 581
tamarind, 584
tarragon, 588
dl-Limonene, *see* Dipentene
Limonin:
  orange, 478
  phellodendron bark, 678
Linalool:
  bergamot oil, 92
  boil de rose oil, 106
  boldo leaves, 107
  calamus oil, 127
  cananga oil, 132
  chaparral, 175
  cinnamon, 197
  citronella oil, 202
  clary sage, 205
  ephedra, 265–266
  garlic, 308
  genet, 314
  ginger, 321
  honeysuckle flower, 667
  hyssop, 376
  immortelle, 378
  jasmine, 382
  labdanum, 404
  lavender, 407–408
  lemon oil, 410
  lime oil, 423
  majoram, 437
  natural sources, 251
  orange, 478–479
  oregano, 482
  rosemary, 535
  sage, 550

sweet basil, 74–75
sweet bay, 76
tagetes, 581
tea, 591
thyme, 595
ylang ylang oil, 630
Linalool acetate, *see* Linalyl acetate
Linalool oxide, 106, 382, 591
Linalyl acetate:
  bergamot oil, 92
  clary sage, 205
  lavender, 407
  majoram, 437
  natural sources of, 251
  orange, 478
  oregano, 482–483
  sage, 550
Linalyl oxide, 416
Linarin derivative, 612
Lindestrene, 461
Linoleic acid:
  almonds, 22
  arnica, 42
  carob, 143
  castor oil, 154
  chervil, 185
  cocoa, 218
  coffee, 222
  elder flowers, 258
  eyebright, 279
  fennel, 283
  gelsemium, 313
  ginger, 321
  juniper berries, 390
  ligustrum, 420
  milk thistle, 439
  parsley, 487
  psyllium, 511
  saw palmetto, 561
  sour jujube kernel, 572–573
  terminalia, 654
Linolenic acid:
  arnica, 42
  chickweed, 187
  coriander, 227
  echinacea, 251–252
  eyebright, 279
  horse chestnut, 364
  jasmine, 382
  ligustrum, 420
  sour jujube kernel, 572
α-Linolenic acid, 111

γ-Linolenic acid, 18, 111, 187, 277
Lipids, 20, 36, 40, 82, 138, 187, 308, 310, 321, 421, 475, 572
  see also Fats and fixed oils
Liquiritigenin, 416
Liquiritin, 416
Lithospermic acid, 101, 225, 485
Lobelanidine, 424
Lobelanine, 424
Lobeline, 424–425
Loganin, 667
Longifolene, 655
Longispoinogenine, 129
Lonicerin, 478
Lunularic acid, 370
Lunularin, 370
Lupeol, 79, 130, 151, 170, 188, 192, 241, 314, 316, 449, 471, 517, 561, 577
Lupine alkaloids,
Luproside, 279
Lupulone, 359
Lutein (xanthophyll), 43, 52, 130, 133, 238, 581
Luteolin, 378
  ginkgo, 325
  honeysuckle flower, 667
  hops, 359
  horsetail, 367
  lavender, 407
  oregano, 482
  sage, 550
Luteolin-7-apiosylglucoside, 487
Luteolin-7-glucoside, 36, 177, 185, 235, 238–239, 252, 378, 535, 667
Luteolin-7-glucuronosyl glucoside, 235
Luteolin-β-rutinoside, 46
Lycium fruit, 429–432
Lycopene, 130
Lycopsamine, 111, 225
Lysine, 13, 287, 400, 539, 626
Lysozyme, fungal, 301

Magnesium:
  astragalus, 56
  bromelain activation by, 114
  chervil, 185
  ganoderma, 302
  kava, 395
  red clover tops, 207

Magnesium bicarbonate, use in cocoa processing, 217
Magnesium carbonate, use in cocoa processing, 217
Magnesium hydroxide, use in cocoa processing, 217
Magnoflorine, 50, 73, 104, 678
Magnolia flower, 434–435
Malic acid:
  biological activities, 232, 529
  Cherokee rosehip, 181
  corn silk, 229
  cranberry, 232
  ephedra, 265
  jujube, 388
  mume, 672
  purslane, 679
  rosehips, 528
  schisandra, 565
  tamarind, 584
Maltol, 490, 491
Malvidin, 93, 342
Manganese:
  alfalfa, 13
  astragalus, 56
  cardamom, 141
  fo-ti, 295
  ganoderma, 302
  ginseng, 331
  horsetail, 367
  in red beet color breakdown, 87
  sodium alginate effect on retention of, 17
Mangiferin, 192
β-D-Mannan, 660
Mannitol:
  doggrass, 248
  fangfeng, 281
  ganoderma, 301
  ligustrum, 420
Mannose:
  bletilla tuber, 661
  bromelain, 114
  xanthan gum, 624
D-Mannuronic acid, 16
Manoyl oxide, 501
Margolone, 652
Margolonone, 652
Marmin, 344
Marrubenol, 363
Marrubiin, 363, 376
Marrubiol, 362
Maslinic acid, 388

Matricin, 170–171
Matricarin, 170, 626
Medicagenic acid, 13–14
Medicagol, antifungal properties, 13
Medicarpin-β-D-glucoside, 13
Melanoidins, 137
Meliantriol, 652
Melianone, 652
Melianol, 652
Melittoside, 521
p-Menthadien-7-als, 235
1,3,8-p-Menthatriene, 487
3-p-Menthen-7-al, 235
Menthofuran, 444
Menthol:
  biological activities, 445
  doggrass, 248
  geranium oil, 319
  mints, 444
  turpentine as starting material, 608
Menthone, 117, 248, 407, 444, 558
Menthyl acetate, 444
Mercury, enzyme inhibition or inactivation by, 497
Mesodihydroguaiaretic acid, 174
2-Methoxy-6-acetyl-7-methyljuglone, 669
29-methyltriacontan-1-ol, 655
4-Methoxycinnamaldehyde, 588
12-Methoxydihydrocostunolide, 231
Methoxyeugenol, 558
1-Methoxygelsemine, 313
7-Methoxyvasicinone, 655
7-Methoxy-5-geranoxy-coumarin, 92, 410
3′-Methoxyisoguaiacin, 174
5-Methoxypsoralen, see Bergapten
Methyl allyl sulfides, 309
Methyl anthranilate, 382, 478
Methylarbutin, 437, 610
2-Methyl-3-butene-2-ol, 360
Methyl chavicol, see Estragole
24-Methylcholesta-7, 22-dien-3-β-ol, 301
Methyl cinnamate, 74–75, 197, 647
  kulinjan, 647
Methylcrocetin (β-crocetin), 547

S-Methylcysteine sulfoxide, 474
Methylcytisine, 104–105, 314
Methyl dehydrojasmonate, 382
Methyl eugenol:
  biological activities, 76, 79, 210
  lemongrass, 413
  magnolia flower, 434
  nutmeg and mace, 467
  tarragon, 588
  West Indian bay, 79
Methyl gallate, 676
4-Methylglucuronic acid, 299
Methylglyoxal, 223
Methylheptenol, 412–413
Methyl heptenone, 382
1-Methyl-4-isopropenylbenzene, 487
Methyl jasmonate, 382
N-Methyllaurotetanine, 107
Methyl linoleate, 314, 521
Methyl linolenate, 314
Methylmercaptan, 52
Methyl myrtenate, 376
Methyl n-octadecanoate, 522
Methyl oleate, 455
Methyl palmitate, 220, 455, 521
Methyl-cis-propenyl disulfide, 474
Methylpropyl disulfide, 474
Methylpropyl trisulfide, 474
Methyl salicylate:
  cananga oil, 132
  cassie absolute, 153
  sweet birch oil, 95–96
  tamarind, 584
  wintergreen oil, 619
Methyltanshinonate, 680
Millefin, 626
Miltirone, 686
Minerals:
  allspice, 20
  fenugreek, 287
  garlic, 308
  ginger, 321
  ginseng, 331
  grape skin extract, 342
  honey, 358
  kava, 395
  purslane, 679
  savory, 559
  tamarind, 584
  turmeric, 604

Chemical Index

Mistletoe, 448–450
Molybdenum, 56
Monomelittoside, 521
Monosaccharides, 522, 565
Monoterpenes, 371
Mucilage:
 althea root, 29
 borage, 111
 chestnut leaves, 186
 cinnamon and cassia, 197
 guarana, 350
 mustard, 457–458
 phellodendron bark, 678
 purslane, 679
 squill, 575
 white pine bark, 501
Multiflorins, 528
Muscone, 455–456
Muscopyridine, 455
γ_1-Muurolene, 319, 630
Muurolene, 287, 319, 483, 630
Myoinositol, 59, 676
Myrcene:
 catnip, 159
 chenopodium oil, 180
 clary sage, 205
 hops, 359
 hypericum, 371
 immortelle, 378
 juniper berries, 390
 labdanum, 404
 lemongrass, 412–413
 lemon oil, 410
 magnolia flower, 434
 olibanum, 471
 orange, 478
 origanum oil, 485
 parsley, 487
 pepper, 499
 pine needle oil, 502
 rose oil, 531
 tagetes, 581
 turpentine, 607
Myricadiol, 80–81
Myricin, 85
Myristic acid:
 ambrette seed, 31
 dandelion root, 238
 echinacea, 251
 jasmine, 382
 nutmeg and mace, 467–468
 parsley, 487
 saw palmetto, 561

Myristicin:
 angostura bark, 35
 biological activities, 467–468
 ligsusticum, 670
 nutmeg and mace, 467
 parsley, 487
Myristolic acid, 487
Myrosinase (myrosinase), 308, 457–458
Myrtenic acid, 376
Myrtenol methyl ether, 376
Myrtine, 93

Napelline, 7
Naphthalene, 210, 547, 569, 622
Naphthodianthrones, 371
1,7-Naphthoquinone
Narigenin, 43
Naringenin, 143, 378, 483, 584
Naringenin-5-glucoside, 378
Naringin, 344–345, 478, 480
Neoabietic acid, 66, 607
Neoandrographolide, 650
Neocnidilide, 670, 683
Neohecogenin, 646
Neoherculin, 50
Neohesperidin, 478, 480
Neolignans, 390, 434–435, 554
Neoliquiritin, 416
Neomenthol, 444
Neonepetalactone, 159
Neoquassin, 515
Neotegolic acid, 352
Neotigogenin, 287, 646
Neogitogenin, 646
Nepetalactone, 13, 159–160, 207
Nepetalic acid, 159
Nepetalic anhydride, 159
Neral, 62–63, 321, 344, 410, 412, 478, 584
Neral acetate, 62
Nerol:
 biological activities, 378
 catnip, 159
 immortelle, 378
 labdanum, 404
 lemon oil, 410
 orange, 478
 rose oil, 531
 tarragon, 588
Nerolidol:
 jasmine, 382
 orange, 478

Roman chamomile, 170
 tea, 591
Neryl acetate, 1, 205, 344, 378, 410, 478
Nettle, 464–466
Niacin, *see* Vitamin B$_3$
Niacinamide, 679
Nicotine, 212, 367–368, 425, 491
Nicotinic acid:
 astragalus, 56
 bee pollen, 82
 ganoderma, 302
 ginseng, 331
 honey, 358
 hypericum, 372
 lycium fruit, 430
 purslane, 679
 sweet woodruff, 622
Nimbin, 652
Nimbinin, 652
Nimbolins, 652
Nimbolide, 652
Nitidine, 50
n-Nonacosane, 222, 257, 314, 490
Nobiletin, 478
Nobilin, 170, 171
Nonadecanonic acid, glycerol esters, 696
Nonanal, 478
Nonanol, 410, 542
2-Nonanone, 542
Nonyl aldehydes, 227, 321
Nootkatone, 344
Norbixin, 40–41
Norboldine, 557
Norcinnamolaurine, 557
Nordihydrocapsaicin, 133
Nordihydroguaiaretic acid, 174
Norepinephrine, 213, 295, 340, 631, 679
l-Norepinephrine, 679
Norharmon,
Norisocorydine, 107, 572
Normuscone, 455
Norpatchoulenol, 493
d-Norpseudoephedrine, 265–266
Norsolorinic acid, 194
Nortanshinone, 680
Nortracheloside, 101
Notoginsenosides, 331, 599
cis-Nuciferol, 554

Obakunone, 678
Obamegine, 73
Ocimene, 74
    marjoram, 437
    orange, 478
    origanum oil, 485
    pine needle oil, 502
    tagetes, 581
    tarragon, 588
Ocimenone, 581
1,18-Octadecanediol, 314
cis-6,9,12-Octadecatrienoic acid, 277
Octanol, 410, 444, 471, 478
3-Octanone, 74, 535
Octyl acetate, 344, 444, 471, 478
1-Octylacetate, 471
Officinalisnin-I and -II, 52
Oleanic acid,
Oleanolic acid:
    biological activities, 388, 421
    calendula, 128
    cloves, 210
    eleuthero, 263
    forsythia fruit, 665
    gymnema sylvestre, 648
    hawthorn, 352
    hyssop, 376
    jujube, 388
    lavender, 407
    ligustrum, 420
    mume, 672
    oregano, 482
    rosemary, 535
    thyme, 595
Oleanonic acid, 388
Oleic acid:
    borage, 111
    castor oil, 154
    coriander, 227
    eyebright, 279
    fennel, 283
    gelsemium, 313
    ginger, 321
    glycerol esters, 456
    jojoba, 386
    juniper berries, 390
    ligustrum, 420
    milk thistle, 439
    parsley, 487
    psyllium, 511

saw palmetto, 561
sour jujube kernel, 572–573
terminalia, 654
Oligosaccharides, 82, 137, 188, 339, 386, 444, 575, 589, 602
Omega-3-acids, 679
Orientin, 177–178, 287, 412, 490, 575
Oroxylin A, 220–221, 685–686
Osthol, 32, 344, 478
Oxalic acid, 187, 404, see also Calcium oxalate
Oxazoles, as flavor components, coffee, 223
Oxidants, enzyme inactivation or inhibition by, 114
Oxidizing agents, see Oxidants
6-Oxo-cativic acid, 404
Oxogambirtannine, 158
21-Oxo-gelsemine, 31
Oxyberberine, 73
Oxypaeoniflorin, 675, 676
Oxypeucedanin, 32, 245, 423, 487, 663
Oxysafranal, 547

β-Pachyman, 508
Pachymaran, 508–509
Paeoniflorin, 675–677
Paeonol, 675–676
Paeonolide, 675, 676
Paeonoside, 675, 676
Palmatine, 72, 678
Palmitate, 36, 85, 187, 220, 257, 258, 314, 424–425, 430
Palmitic acid:
    almonds, 22
    ambrette seed, 31
    beeswax, 85
    borage, 111
    chervil, 185
    cocoa, 218
    codonopsis, 220
    coffee, 222
    echinacea, 251–252
    elder flowers, 257
    eyebright, 280
    gelsemium, 313
    ginger, 321
    glycerol esters, 456
    horsechestnut, 364

jasmine, 382
juniper berries, 390
ligustrum, 420
parsley, 487
peony root, 676
psyllium, 511
rehmannia, 522
saw palmetto, 561
Palmitoleic acid, 22, 386
Palustric acid, 607
Panacene, 331
Panasinsenes, 331
Panaxadiol, 331
Panaxans, 331–332
Panaxatriol, 331, 599
Panaxosides, 331
Panaxynol, 331
Pangelin, 542
Paniculatan, 370
Pantothenic acid, 13, 59, 331, 539
Papain, 114, 292, 698
Paraaspadin, 54
Paradisiol (intermedcol), 344
Paraffins, 13, 404, 437, 455–456, 531, see also Hydrocarbons
Parietin, 525
Parillin, 556
Parthenolide, 289–291
β-Patchoulene, 346
Patchoulipyridine, 493
Patulin, 581
Pectin, 495–497
    biological activities, 478–479
    dandelion root, 238
    ephedra, 265
    ginseng, 331
    hypericum, 372
    onion, 475
    orange, 478
    rosehips, 528
    tamrind, 584
    see also Pectin (GENERAL INDEX)
Pectin methylesterase, 13
Peganidine, 655
Peltatins, 505–506
Pentacosane, 82, 626
Pentadeca-8-en-2-one, 251
n-Pentatriacontane, 313
Pentosans, 40, 325

5-Pentyl-5-pentanolide, 407
Peonidin, 342, 474
Peptides, low molecular weight, almonds, 23
  see also Proteins
Peregrinol, 362
Perillaldehyde, 235, 344, 407
Perillyl acetate, 344
Perlolyrine, 220, 683
Peroxidase, 4, 114, 292, 308, 646
Petroselinic acid (cis-6-octadecenoic acid), 32, 165, 185, 227, 245, 283, 487
Petunidin, 342
Phellandrene, 32, 404
  balsam Canada, 66
  cinnamon, 197
  clary sage, 205
  cubebs, 234
  dill, 245
  eucalyptus, 272
  garlic, 308
  ginger, 321
  labdanum, 404
  lime oil, 423
  lovage root, 428
  olibanum, 471
  origanum oil, 485
  parsley, 487
  sassafras, 558
  turmeric, 604
α-Phellandrene, 20, 32, 66, 234, 245, 251, 272, 283, 423, 428, 485, 655
β-Phellandrene
  balsam Canada, 66
  coriander, 227
  cubebs, 234
  garlic, 308
  lavender, 407
  ligusticum, 670
  olibanum, 471
  oregano, 482
  parsley, 487
  pepper, 499
  pine needle oil, 502
  sassafras, 558
  savory, 559
  turpentine, 607
Phellodendrine, 678

Phellopterin, 32, 281, 410, 423, 663
Phenanthrene diketonesred sage, 680
β-Phenethyl alcohol (2-phenyl-ethanol), 65, 531
Phenylpropyl cinnamate, 579
Phenolic acids:
  arnica, 43
  astragalus, 56
  balm of Gilead buds, 65
  bilberry, 93
  biological activities, 168, 368, 550
  centaury, 167
  dandelion root, 238
  dill, 245
  elder flowers, 257
  euphorbia, 274
  hops, 359
  horsetail, 367
  onion, 474
  rosemary, 535
  sage, 550
  uva ursi, 610
Phenols:
  biological activities, 483
  castoreum, 156
  chicory root, 190
  citronella oil, 202
  eucalyptus, 272
  genet, 314
  oregano, 482
  origanum oil, 485
  reaction with acacia, 5
  savory, 559
  Sichuan lovage, 683
  thyme, 595
  West Indian bay, 79
  witch hazel, 620
  ylang ylang oil, 630
Phenylacetic acid, 382
Phenylethanol, 197, 547, 581, 591
β-Phenylethylamine, 352, 449
Phenylhydrazine, 77
Phenylpropanoids, 20–21, 36, 262–263, 283–284, 362, 449, 467, 647
  kulinjan, 647

Phenylpropyl alcohol, 20–21, 36, 262–263, 283–284, 362, 449, 467, 647
Phenylpropyl cinnamate, 579
Phosphatidic acid, 321
Phospholipids, 31, 41, 46, 217, 295, 384, 539, 565, 595
Phosphorus:
  Acerola, 6
  Alfalfa, 13
  Chervil, 185
  Coca, 212
  fo-ti, 295
  hawthorn, 352
  tamarind, 584
Phthalides:
  biological activities, 428
  celery seed, 165
  immortelle, 378
  lovage root, 428
  Sichuan lovage, 683
Phyllodulcin, 370
Physalien, 430, 431
Physcione (parietin), 525, 669
Phytoalexins, 13, 207
Phytol, 362, 372, 444, 561
Phytosterols:
  Devils claw, 242
  Epimedium, 269
  peony bark, 675
  tienchi ginseng, 598
Piceid, 669
Piceoside, 610
Picraconitine, 7, 8
Picrasidines, 515
Picrasins, 515
Picrocrocin, 547
Picrosalvin, 550
Pigments, 13, 18, 27, 87, 115, 130, 157, 158, 181, 218, 233, 342, 350, 387, 398, 430, 431, 478, 512, 533, 579, 581, 591, 682
Pimaric acid, 607
Pinene:
  Catnip, 159
  Cinnamon, 197
  citronella oil, 202
  myrrh, 460
  nutmeg and mace, 467
  olibanum, 471
  orange, 478
  saffron, 547

α-Pinene
  balsam Canada, 66
  biological activities, 125
  carrot seed oil, 148
  chamomile, 170
  chaparral, 175
  chaste-tree, 177
  coriander, 227
  cubebs, 234
  damiana, 237
  eucalyptus, 272
  galbanum, 299
  horehound, 362
  hypericum, 371
  immortelle, 378
  juniper berries, 390
  labdanum, 404
  lemon oil, 410
  lime oil, 423
  magnolia flower, 434
  mints, 444
  oregano, 482
  origanum oil, 485
  parsley, 487
  pepper, 498
  pine needle oil, 502
  rosemary, 535
  rose oil, 531
  sage, 550
  sassafras, 558
  storax, 578
  sweet bay, 76
  tagetes, 581
  tarragon, 588
  turpentine, 607
  yarrow, 626
  ylang ylang oil, 630
β-Pinene:
  balsam Canada, 66
  biological activities, 299
  carrot seed oil, 148
  chaste-tree, 177
  clary sage, 205
  coriander, 227
  cubebs, 234
  cumin, 235
  damiana, 237
  galbanum, 299
  hypericum, 371
  hyssop, 376
  immortelle, 378
  juniper berries, 390
  labdanum, 404
  lemon oil, 410
  lime oil, 423
  magnolia flower, 434
  mints, 444
  orange, 478
  oregano, 482
  origanum oil, 485
  parsley, 487
  pepper, 498
  pine needle oil, 502
  rosemary, 535
  rose oil, 531
  sage, 550
  savory, 559
  tagetes, 581
  turpentine, 607
  yarrow, 626
cis-Pinic acid, 376
cis-Pinonic acid, 376
Pinocampheol, 376
Pinocembrin, 483, 629
Piperidine, 76, 77, 395, 424, 498, 499
Piperitone epoxide, 117
Piperitol, 467
Piperitone, 117, 235, 444, 581, 582, 584
Pipermethystin, 395
Piperonylacrolein, 558
Pipsissewa, 504
Planteose, 511
Plant pigments, 13
Pluviatilol, 50
Poaesculin
Podophyllotoxin, 505, 506
Pogostol, 493, 494
Pogostone, 493
Pollinastanol, 556
Polyacetylenes, 101, 120, 251, 253, 281, 283, 331, 488, 589
Polyamines, 222, 223
Polydatin, 669
Polygonin, 669
Polyisoprenes, 188
Polypeptides:
  biological activities, 449
  ganoderma, 301
  mistletoe, 449
Polyphenols:
  biological activities, 62, 592
  cocoa, 217
  gotu kola, 339
  rosehips, 181
  sage, 550
Polysaccharides:
  Aloe, 25
  Asparagus, 52
  Astragalus, 56, 57
  Baizhu, 660
  biological activities, 43, 130, 221, 263, 332
  bletilla tuber, 661
  calendula, 130
  codonopsis, 220, 221
  Echinacea, 252–254
  Eleuthero, 263
  Epimedium, 269, 270
  Ganoderma, 302, 303
  Hypericum, 371
  karaya gum, 393, 394
  luffa, 671
  lycium fruit, 430, 431
  mistletoe, 449
  poria, 508, 509
  psyllium, 511, 512
  safflower, 682
  saw palmetto, 561
  synthesis, promotion by epimedium, 269
  tienchi ginseng, 598, 599
  tragacanth, 602
Polysulfanes, 299
Populin, 65
Poria, 508, 509
Poriatin, 508
Posthumulone, 359
Potassium:
  Alfalfa, 13
  Avocado, 59
  Borage, 111
  Chervil, 185
  Cocoa, 217
  dandelion root, 239
  fennel, 283, 284
  in gel carrageenan formation, 146
  horsetail, 367
  kava, 395
  nettle, 464
  purslane, 679
  savory, 559, 560
  tamarind, 583, 584
  turmeric, 604
Potassium alginate, 17

Chemical Index

Potassium carbonate pectin
  use in cocoa processing, 217
Potassium hydroxide, use in
    cocoa processing, 217
Potassium nitrate, 111
Pratensein, 207
Prehumulone, 359
Proanthocyanidene
Proanthocyanidins:
  Cinnamon, 197
  Ginkgo, 325
  Hawthorn, 352
  juniper berries, 390
  witch hazel, 620
Proanthocyanins
Procumbide, 242
Procyanidins, 93, 197, 198, 207, 213, 325, 353, 354, 371, 584
Progesterone, 177, 178, 207, 228, 384
Proline, 56, 143, 416, 430, 539, 674
Propanol, ficin inhibition by, 292
trans-S- (1- Propenyl) cysteine sulfoxide, 474
Propionic acid, 407, 416
S-Propylcysteine sulfoxide, 474
Propylene glycol alginate, physicochemical properties, 16
Proscillaridin A, 574, 575
Prostaglandins, 20, 33, 94, 107, 111, 193, 277, 308, 309, 468, 545, 623
Proteases:
  activators, 292
  bromelain, 113, 114
  ficin, 292, 293
  ganoderma, 301
  ginger, 321, 322
  papain, 114
  poria, 508
  specificity, 114, 292
Proteins:
  alfalfa, 13, 14
  allspice, 20
  almonds, 22, 23
  angostura bark, 35
  annatto, 40, 41
  avocado, 59
  bee pollen, 82
  capsicum, 133

caraway, 138
cardamom, 141
cloves, 210
coca, 212
cocoa, 217
coffee, 222, 223
coriander, 227
cranberry, 232
cumin, 235
dill, 245
epimedium, synthesis promotion, 269
fennel, 283
fenugreek, 287
as flavor precursors, coffee, 223
ganoderma, 301, 302
garlic, 308
ginger, 321
ginkgo, 325, 326
ginseng effects on biosynthesis, 332, 333
gotu kola, 340
guar gum, 347
honey, 358
Job's tears, 384
Jujube, 388
Kava, 395
kola nut, 398
leaf protein source, 14
lycium fruit, 430
marjoram, 437
MSG source, 452, 453
Mustard, 457, 458
nutmeg and mace, 467
oregano, 482
parsley, 487
pepper, 498
reaction with tannins
red clover tops, 207
royal jelly, 539
savory, 559
sweet basil, 75
tamarind, 584
tarragon, 588
tea, 591
turmeric, 604
Protocatechuic acid, 62, 197, 272, 474, 533, 669, 680
Protocatechuic aldehyde, 680
Protodioscin, 52, 646
Protopine, 102
Prototribestin, 646

Prulaurasin (*dl*-mandelonitrile glucoside), 184
Prunasin (*d*-mandelonitrile glucoside), 22, 182–184
Prussic acid, *see* Hydrocyanic acid
Pseudoephedrine, 265–267
Pseudoephedroxane, 265, 266
Pseudohypericin:
  biological activities, 372
  hypericum, 371, 372
Pseudoprotodioscin, 646
Pseudotropine
Pseudo-γ-schizandrin, 565
Psoralen, 50, 51, 227, 281, 344, 428, 542, 663
Pterocarpans, 207, 416
Puerarin, 400, 401
Pulegone, 117, 159, 407, 444
Pungent principles, 133, 321, 498, 499
Purpureaside C, 521
Pyrazines, as flavor components, 217
Pyridines, as flavor components, 287
Pyridoxine- *see* Vitamin B_6
α-Pyrones, 395
Pyrroles, as flavor components, 217
Pyrroles, cocoa, 217
Pyrroloquinazoline, 655
Pyruvic acid, 10, 103, 624

Quassimarin, 515
Quassin, 515, 516
Quassinoids, 515
Quassinol, 515
Quebrachamine, 517
Quercetin:
  black and pale catechu, 157–158
  boneset, 109
  dill, 245
  elder flowers, 257
  fenugreek, 287
  hydrangea, 370
  onion, 474
  pepper, 498
  podophyllum, 505, 506
  tea, 591
Quercetin-3-arabinoside, 283

Quercetin-3-β-galactoside,
see Hyperoside
Quercetin-3-glucuronide, 36, 138, 227, 283
Quercitrin:
  bilberry, 93
  biological activities, 272, 275, 378, 589
  epimedium, 269
  eucalyptus, 272
  hops, 359
  immortelle, 378
  tea, 591
Quilliac acid
Quinazolines
  mahbala, 651
Quinic acids, 232, 325, 610
Quinidine, 194, 195
Quinine, 194, 195, 417, 515, 516, 629, 695
Quinones, 364

Racemoside, 97
Radioisotopes, absorption, reduction of by algin, 16–17
Raffinose, 22, 339, 444, 490
Rebaudiosides, 577
Reducing agents, enzyme activation by
  bromolain
  ficin, 292
Rehmannia, 521–523
Rehmanniosides, 521
Resins:
  asafetida, 48
  balm of Gilead buds, 65
  balsam Canada, 65
  balsam tolu, 71–72
  benzoin, 90–91
  castoreum, 156–157
  Cherokee rosehip, 182
  chestnut leaves, 186
  chicle, 189
  chicory root, 189–191
  cinchona, 197
  cinnamon and cassia, 197
  corn silk, 229–230
  costus oil, 230
  cubebs, 234
  cumin, 235–236
  damiana, 237
  dandelion root, 238

  dill, 244–246
  echinacea, 251–256
  elecampane, 259–260
  euphorbia, 274
  galbanum, 299
  ginger, 321
  goldenseal, 337
  guarana, 350
  hops, 359
  horehound, 362
  hydrangea, 370
  hyssop, 376
  ipecac, 380
  Juniper berries, 390
  labdanum, 404
  lobelia, 424
  lovage root, 428
  myrrh, 460
  olibanum, 471
  red clover tops, 207
  Roselle, 533
  sarsaparilla, 556
  sassafras, 557
  senna, 569
  storax, 579
  turmeric, 604
  turpentine, 607
  witch hazel, 620
  yarrow, 626
  yerba santa, 629
Resveratrol, 93, 633, 634, 669
Reticuline, 76, 107, 434, 435, 557
Reynosin, 76, 289
Rhamnetin (7-methylquercetin), 88, 175, 210, 498
Rhamnoliquiritin, 416
Rhamnose, 13, 29, 48, 56, 220, 225, 339, 393, 449, 496, 522, 545, 561, 569, 602, 696
Rhamnosides, 269
Rhaponticin, 525, 526
Rhapontin, 295, 525
Rhazinilam, 517
Rhein, 295, 525, 569
Rheinosides, 525
Rhoifolin, 444, 478
Riboflavin, see Vitamin B_2
Ricin, 154
Ricinine, 154
Ricinoleic acid, 154
Rosmanol, 535, 536, 550
Rosmarinic acid

  biological activities, 225, 376
  hyssop, 376
  marjoram, 438
  mints, 444
  oregano, 487
  red sage, 680
  sage, 550
  sweet basil, 75
Royal jelly acid, see 10-Hydroxy-trans-2-decenoic acid
Rubidium, 56
Rusbyine, 215
Ruscogenin, 646
Rutacridone, 542
Rutamarin, 542
Rutarin, 542
Rutin (quercetin-3-rutinoside),
  boneset, 109
  calendula, 130
  chaparral, 175
  elder flowers, 257
  fennel, 283
  forsythia fruit, 665
  hawthorn, 352
  hops, 359
  hydrangea, 370
  juniper berries, 390
  onion, 474–475
  orange, 478
  passion flower, 490
  rhubarb, 525
  rue, 542
  tea, 591
Rutinose, 545
Rutoside, 252, 545

Sabinene:
  Catnip, 159–160
  cedarwood oil, 163
  cubebs, 234
  horehound, 362
  juniper berries, 390
  labdanum, 404
  lemon oil, 410
  lime oil, 423
  magnolia flower, 434
  marjoram, 437
  pepper, 499
  pine needle oil, 502
  storax, 578–580
  turmeric, 604
  yarrow, 626

Sabinene hydrate, 202, 235, 321, 437, 444, 467
Sabinyl acetate, 1, 163
Safflomin A, 682
Safflower yellow, 682
Safranal, 547
Safrole:
 biological activities, 558
 cananga oil, 132
 cinnamon, 137
 nutmeg and mace, 467
 in production of heliotropin, 558
 sassafras, 557
 tamarind, 584
 witch hazel, 620
 ylang ylang oil, 630
Sagittatins, 269
Salanin, 652
Salicin, 65, 629
Salicylaldehyde, 581
Salicylic acid (salicylates), 97, 325, 449, 496, 626
Salonitenolide, 101
Salvianolic acids, 680
Salviatannin, 550
Salvin, 550, 680
Salvin monomethyl ether, 550
Salvinone, 680
Sambunigrin (*l*-mandelonitrile glucoside), 184
Sanchinan-A, 599
Sandaracopimaric acid, 390, 607
Sanguidimerine
Sanguinarine, 102, 103
Sanguisorbins, 666
Santalenes, 554
Santalols, 554
Santalone, 554
Santamarin, 76, 289
Santanol, 554
Santene, 554
Sapogenins:
 fenugreek, 287
 ginseng, 331
 gotu kola, 339
 as potential source of steroid hormones, fenugreek, 287
 tienchi ginseng, 599
 yucca, 633

Saponins:
 aletris, 12
 alfalfa, 12
 ashwagandha, 645
 astragalus, 56
 biological activities, 52, 332, 518, 556, 572, 592, 599, 633, 649
 calendula, 129
 chickweed, 187
 codonopsis, 220
 comfrey, 225
 corn silk, 229
 dittany bark, 664
 fenugreek, 287
 gokshura, 646
 ginseng, 331
 guarana, 350
 honeysuckle flower, 667
 horehound, 362
 hydrangea, 370
 ipecac, 388
 lemongrass, 412
 nutmeg and mace, 467
 onion, 474
 purslane, 679
 quillaia, 518
 rosehips, 181
 sarsaparilla, 556
 triterpene, horse chestnut, 364
 triterpenoid, gotu kola, 339
 witch hazel, 620
 yarrow, 626
 yucca, 633
Sarothamnine, 115
Sarsaparilloside, 556
Sarsasapogenin
 gokshura, 646
Sarsasaponin (parillin), 556
Saussurine, 230
Sauvissimoside R_1, 666
Saw palmetto, 561–563
Schisandra, 564–567
Schisanhenol, 565, 566
Schisantherin D, 565
Schizandrers, 565
Schizandrins, 565
Schizandrols, 565
Scillaren A and B, 575
Scilliglaucosidin, 575
Scilliphaeoside (12β-hydroxy-proscillariden A), 574
Scilliroside, 575

Sclareol, 205, 206, 467, 469
Scolymoside (luteolin-7-β-rutinoside), 46
Scoparin (scoparoside), 115
Scopolamine, 5, 13, 88, 207
Scopoletin (7-hydroxy-6-methoxy-coumarin):
 angostura bark, 34–35
 baizhu, 660
 black haw bark, 99
 dill, 245
 gelsemium, 313
 lycium fruit, 430
 passion flower, 490
 tarragon, 588
Sebacic acid, 539
Secologanin, 667
Sedanenolide (3-*n*-butyl-4, 5-dihydrophthalide), 165
Sedanolide, 165
Sedanonic acid, 683
Sedanonic anhydride, 165, 428
Selina-3,7(11)-diene, 319, 359
Selinenes, 331, 359
Sempervirine, 313
Senkyunolide, 165, 428, 670
Sennosides, 525, 526, 569, 570, 696
Senticosides, 263
Serine, 25, 339, 416, 430, 539, 584, 602, 674
Serotonin
 antagonistic effects of horehound, 363
 inhibition of release, feverfew, 289
Sesamin (eleutheroside B_4), 262
Sesquicarene, 565
β-Sesquiphellandrene, 74, 321
β-Sesquiphellandrol, 321
Sesquiterpene esters, 251
Sesquiterpene Lactones:
 baizhu, 660
 biological activities, 43
 blessed thistle, 101
 boneset, 109
 chamomile, 170
 costus oil, 230
 dill, 244–246
 elecampane, 259

Sesquiterpenes:
  Arnica, 46
  balm, 62–63
  balsam copaiba, 67–68
  balsam tolu, 74
  bergamot oil, 92
  biological activities, 626
  calamus oil, 128
  cedarwood oil, 163
  citronella oil, 202
  Costus oil, 230
  cubebs, 234
  damiana, 237
  dandelion root, 238
  deertongue, 241
  eucalyptus, 272
  fenugreek, 287
  feverfew, 289
  galbanum, 299
  geranium oil, 319
  ginger, 321
  ginseng, 331
  grapefruit oil, 344
  hops, 359–360
  horehound, 362
  hypericum, 372
  hyssop, 375–377
  immortelle, 378
  juniper berries, 390
  labdanum, 404, 405
  lavender, 407
  lemon oil, 410
  lime oil, 423
  myrrh, 461
  olibanum, 471–472
  patchouly oil, 473
  pepper, 498
  pine needle oil, 502
  Roman chamomile, 170
  Sage, 550
  sandalwood oil, 554
  sassafras, 557–558
  storax, 578–580
  witch hazel, 620
  ylang ylang oil, 630
Shikimic acid, 274, 275
Shikonin, 18
Shogaols, 321
Siaresinolic acid, 90
Silibinin, 439–441
Silica, 368, 478
Silicates, 367
Silicic acid, 111, 367, 464

Silicon, 248
Silidianin, 439
Silybin, 439, 440
Silychristin, 439
Silymarin, 439–441, 556
Simalikalactone D, 515
Simmondsin, 386
Sinalbin, 458
Sinapic acid, 457, 458
Sinapyl alcohol, 263
Sinensals, 344
Sinensetin, 478
Sinigrin (potassium myronate),
    457, 458
β-Sitostenone, 515
Sitoindosides, 645
Sitosterol:
  centaury, 168
  cloves, 210
  corn silk, 229
  Dahurian agelica, 663
  dittany bark, 664
  elder flowers, 257
  genet, 314
  ginkgo, 325
  mahabala, 651
  passion flower, 490
  sarsaparilla, 556
  sassafras, 557
  turpentine, 607
β-Sitosterol:
  asparagus, 52
  avocado, 59
  Baikal skullcap, 686
  biological activities, 356,
    522, 562
  cinchona, 195
  cocillana bark, 215
  damiana, 237
  dill, 245
  echinacea, 252
  elecampane, 259
  fangfeng, 281
  fo-ti, 295
  ganoderma, 301
  ginkgo, 325
  ginseng, 331
  hawthorn, 352
  horehound, 362
  horse chestnut, 364
  hypericum, 372
  hyssop, 376
  immortelle, 378

  juniper berries, 390
  lemongrass, 412
  ligusticum, 670
  lovage root, 428
  mahbala, 651
  marjoram, 437
  mume, 672
  myrrh, 461
  nettle, 464
  onion, 474
  peony root, 676
  phellodendron bark, 678
  psyllium, 511
  quassia, 515
  quebracho, 517
  rehmannia, 522
  saw palmetto, 561
  stevia, 577
  tamarind, 584
  tienchi ginseng, 598
  valerian root, 612
  vasaka, 655
  yarrow, 626
γ-Sitosterol, 259
β-Sitosterol-α-glucoside, 676
β-sitosterol-β-D-glucoside,
    655
Skimmianine, 50, 542, 664
Skullcapflavones, 685, 686
Smilagenin, 556
Smilasaponin (similacin), 556
Sodium:
  Acacia, 4–5
  Cocoa, 217
  Ganoderma, 300–304
  Kava, 395
  retention due to glycyrrhizin,
    417
Sodium alginate, 16, 17
Sodium benzoate, in crude ficin,
    292
Sodium bicarbonate, use in cocoa
    processing, 216–219
Sodium carbonate
  use in cocoa processing, 217
Sodium citrate, 496
Sodium hydroxide, use in cocoa
    processing, 217
Sodium sulfite, inactivation of
    mutagenic activity of
    coffee, 223
Sorbic acid, as color stabilizer,
    87

Soyasapogenols, 13
Soyasaponins, 416
Sparteine, 104, 115, 116, 314, 315
Spathulenol, 124, 170, 197, 205, 683
Spermine, 222, 266
α-Spinasterol, 13, 180, 187, 189, 220, 364
Spinosin, 572
Spirostanol, 633, 646
Stachyose, 22, 444, 522
Starch:
  barberry, 72–73
  blackberry bark, 96–97
  bletilla tuber, 661
  cardamom, 141
  Cherokee rosehip, 181
  cinchona, 194
  coriander, 227
  effect on agar gel strength, 10
  ephedra, 265
  fo-ti, 295
  ginger, 321
  ginkgo, 325
  ginseng, 331
  goldenseal, 337
  guarana, 350
  hydrangea, 370
  ipecac, 380
  Job's tears, 384
  kava, 395
  kola nut, 398
  kudzu root, 400
  licorice root, 416
  nutmeg and mace, 467
  quillaia, 518
  saffron, 547
  sarsaparilla, 556
  sassafras, 557
  tragacanth, 602
Stearic acid:
  borage, 111
  Dahurian agelica, 663
  eyebright, 280
  horsechestnut, 364
  jasmine, 382
Stearoptenes, 531
Steric acid:
  castor oil, 154–155
  cocoa, 216–219
  gelsemium, 312–313
  ginger, 320–323
  parsley, 487–489

psyllium, 511–513
Steroids:
  aletris, 11–12
  alfalfa, 12–15
  almonds, 21–23
  aswagandha, 645
  asparagus, 52
  avocado, 59–60
  bee pollen, 81–83
  boneset, 109–110
  centaurs, 167–168
  chickweed, 187
  cloves, 209–211
  codonopsis, 220–221
  coffee, 222–224
  Dahurian angelica, 663–664
  dandelion root, 238–240
  dittany bark, 664
  elder flowers, 257–258
  euphorbia, 274–276
  evening primrose, 277
  fenugreek, 287
  forsythia fruit, 665
  ganoderma, 302
  ginkgo, 325
  gotu kola, 339
  horse chestnut, 364–366
  horsetail, 367
  immortelle, 378–379
  Job's tears, 384–385
  Jujube, 388–389
  juniper berries, 389–391
  kudzu root, 399–402
  licorice root, 416
  marjoram, 437
  musk, 455–456
  myrrh, 460–462
  onion, 474–475
  passion flower, 490–492
  psyllium, 511–513
  royal jelly, 538–540
  sarsaparilla, 556
  saw palmetto, 561–563
  sour jujube kernel, 572–573
  squill, 574
  tea, 592
  tarragon, 588–590
  turpentine, 607–608
  white pine bark, 501–502
  yarrow, 626–627
  yucca, 633
Steviolbioside, 577
Stevioside, 577, 578

Stigmasterol:
  cloves, 210
  corn silk, 229
  costus oil, 229
  dandelion root, 238
  dill, 244–246
  Echinacea, 252
  elecampane, 259
  fennel, 283
  licorice root, 416
  onion, 474
  kalmegh, 650
  passion flower, 490
  phellodendron bark, 678
  psyllium, 511
  sarsaparilla, 556
  horse chestnut, 364
  mahbala, 651
  saw palmetto, 561
  stevia, 577
  tienchi ginseng, 598
Stilbene glucosides, 295–296
Stilbene glycoside gallates, 295, 525
Stilbenes, 669
Stimast-7-enol, 187
Strobal, 501
Strobol, 501
Styracin (cinnamyl cinnamate), 70, 90, 579
Styrene, 70, 90, 579
Succinic acid, 8, 52, 325, 522, 626
Sucrose:
  Fangfeng, 281
  Ginkgo, 325
  honey, 358
  passion flower, 490
  peony root, 676
  psyllium, 511
Sugars:
  in acacia, 5
  acerola, 6–7
  alfalfa, 13
  almonds, 22
  angelica, 32
  asparagus, 52
  avocado, 59
  bee pollen, 82
  Cherokee rosehip, 181
  Chicle, 188
  cinnamon and cassia, 197
  codonopsis, 220
  comfrey, 225

Sugars: (*Continued*)
  coriander, 227
  dandelion root, 238
  effect on agar gel strength, 10
  elder flowers, 257
  ephedra, 265
  euphorbia, 274
  fennel, 283
  as flavor precursors, coffee, 223
  fo-ti, 295
  ganoderma, 301
  gentian, 316
  ginseng, 331
  goldenseal, 337
  gotu kola, 339
  grape skin extract, 342
  henna, 355
  jujube, 388
  juniper berries, 390
  kava, 395
  kola nut, 398
  licorice root, 415
  onion, 474
  parsley, 487
  pectin, 495
  quebracho, 517
  quillaia, 518
  red clover tops, 207
  rehmannia, 522
  rhubarb, 525
  rosehips, 181
  Roselle, 533
  Rutin, 545
  Tamarind, 584
  Turmeric, 604
  Vanilla, 617
  Yarrow, 626
  yerba santa, 629
Sulfates, 637
Sulfhydryl proteases, 113
Sulfides:
  enzyme activation by, 107
  onion, 474
  organic, flavor properties, 474
Sulfur compounds
  asafetida, 48
  asparagus, 52
  garlic, 308
  hops, 359
  mustard, 457–459
  onion, 474
Sumaresinolic acid, 71, 90

Superoxide dismutase, increased levels, fo-ti, 295
Supinidine, 111
Supinine, 111
Swerchirin (1,8-dihydroxy-3,5-dimethoxyxanthone), 192
Sweroside, 167, 192, 316, 317
Swertiamarin, 167, 316, 317
Swertianin (1,7,8-trihydroxy-3-methoxyxanthone), 192
Swertinin (7,8-dihydroxy-1,3-dimethoxyxanthone), 192
Syringin (eleutheroside B), 262, 449

Tagetones, 581
Tamarindienal, 584
Tangeretin, 478
Tangshenosides, 220
Tannins:
  acerola, 6–7
  annatto, 40
  balm, 62
  barberry, 72–73
  bayberry bark, 80
  biological activities, 529
  black and pale cateclau, 158
  blackberry bark, 97
  catnip, 159
  Cherokee rosehip, 181
  cherry laurel leaves, 184
  chestnut leaves, 186
  chicory root, 190
  cinchona, 194
  cinnamon and cassia, 196
  cocillana bark, 215
  cocoa, 217
  coffee, 222
  coriander, 227
  cumin, 235
  damiana, 237
  elder flowers, 258
  ephedra, 265
  epimedium, 269
  eucalyptus, 272
  as flavor precursors, coffee, 222
  gelsemium, 313
  gotu kola, 339
  grape skin extract, 342
  guarana, 350

  henna, 355–356
  honeysuckle flower, 667
  hops, 359
  horehound, 362–363
  hydrangea, 370
  hypericum, 372
  hyssop, 376
  ipecac, 380
  kola nut, 398
  lavender, 406–408
  magnolia flower, 434
  marjoram, 437
  mints, 443–446
  psyllium, 511
  pyrocatechol, comfrey, 225
  quebracho, 517
  quillaia, 518–519
  rhubarb, 525
  sage, 550–552
  sassafras, 557
  stevia, 577
  sweet woodruff, 622
  tannic acid, 585
  tarragon, 588–590
  tea, 591
  thyme, 595
  uva ursi, 610
  valerian root, 612
  witch hazel, 620
  yarrow, 626
  yerba santa, 628–629
Tanshindiols, 680
Tanshinols, 680
Tanshinones, 680, 681
ψ-Taraxasterol, 46, 130
Taraxasterol (α-lactucerol), 46, 120, 129, 130, 170, 188, 190, 238, 504
Taraxerol, 80, 170, 192, 220, 238, 274
Taraxerone, 80, 274
Taraxol, 238
Tarragon, 588–590
Tartaric acid:
  corn silk, 229
  grape skin extract, 342
  jujube, 388
  as papain inhibitor
  schisandra, 565
  tamarind, 584
Taurine, 430, 539, 674
Taxifolin (dihydroquercetin), 439, 474, 575, 584

Teresantalic acid, 554
Teresantol, 554
Terpenes:
  Angelica, 32
  Arnica, 42
  bergamot oil, 91–92
  cubebs, 234
  doggrass, 248
  genet, 314–315
  grapefruit oil, 344
  horehound, 362
  nutmeg and mace, 467–469
  orange, 477
  rose oil, 531
  savory, 559
  thyme, 595
Terpenoids:
  Boneset, 109
  ginkgo, 325
  juniper berries, 390
  lovage root, 428
Terpinene, 148, 245, 471, 478, 485
α-Terpinene
  chenopodium oil, 179–180
  cubebs, 234
  hyssop, 376
  marjoram, 437
  oregano, 482
  origanum oil, 485
  pine needle oil, 502
  storax, 578–580
β-Terpinene, 283, 404
γ-Terpinene:
  coriander, 227
  cubebs, 234
  cumin, 235
  juniper berries, 390
  lemon oil, 410
  lime oil, 423
  marjoram, 437
  oregano, 482
  origanum oil, 485
  parsley, 487
  pine needle oil, 502
  savory, 559
  storax, 578–580
  tarragon, 588
  thyme, 595
Terpinen-4-ol:
  catnip, 159–160
  hyssop, 376
  labdanum, 404

ligusticum, 669–670
magnolia flower, 434
marjoram, 437
rosemary, 535–537
tamarind, 584
yarrow, 626–627
α-Terpineol:
  cananga oil, 132
  cubebs, 234
  ephedra, 265
  lemon oil, 410
  lime oil, 423
  lovage root, 428
  magnolia flower, 434
  orange, 478
  parsley, 487
  sweet basil, 74
  tamarind, 584
  thyme, 595
β-Terpineol
  kulinjan, 647
Terpin hydrate, turpentine as starting material, 608
Terpinolene:
  lavender, 407
  ligusticum, 670
  lime oil, 423
  parsley, 487
  pine needle oil, 502–503
  turpentine, 607
Terpinyl acetate, 76, 141, 376, 428, 471, 670
α-Terthienyl, 581
Terrestriamide, 646
Testosterone, 175, 177, 430, 465, 539, 556, 562
2,3′,4,6-Tetrahydroxybenzophenone, 316
2′,5,5′,7-Tetrahydroxy-6′,8-dimethoxyflavone, 685, 686
5,7,3′,4′-Tetrahydroxyflavan-3,4-diol, 629, 685, 686
1,3,7,8-Tetrahydroxyxanthone, 192
22β,28-Tetrahydroxy-olean-12-en-30-oic acid[2], 648
Tetramethoxyallybenzene, 487
Tetramethylpyrazine, 265, 266, 299, 683
Theaflavin, 591
Thearubigin, 591
Theaspirone, 591

Theobromine:
  biological activities, 218
  cocoa, 217
  guarana, 350
  kola nut, 398
  tea, 591
Theogallin, 591
Theophylline, 350, 531, 591
Thermopsine, 314
Thiamine, see Vitamin B_1
Thioglycolate, enzyme activation by
Thiol esters, 299
Thiopropanal S-oxide (propanethial S oxide), 474
Thiosulfate, protease activation by, 308
Thiosulfinates, 308, 309
Thujan-4-ol, 595
Thujanols, 482
Thujone (α-thujone), biological activities, 1
  catnip, 159–160
  cubebs, 233–234
  juniper berries, 389–391
  labdanum, 404
  lavender, 406–408
  olibanum, 471–472
  pepper, 498–499
  sage, 550
  sassafras, 558
  tarragon, 588
β-Thujone (isothujone), 1, 2, 550
Thujopsene, 163
Thymol:
  arnica, 42
  biological activities, 483, 595
  catnip, 159
  doggrass, 248
  marjoram, 437
  oregano, 482
  origanum oil, 485
  parsley, 487
  savory, 559
  sweet basil, 75
  thyme, 595
Thymol methyl ether, 42
Tienchi ginseng, 598–600
Tiglic acid, 170, 319
Tigogenin
  gokshura, 646
Tolmethyl carbinol, 604

p-Tolyl methyl ether, 630
Trace elements, 13, 82, 388, 539
Trace minerals:
  lycium fruit, 430
  mother-of-pearl, 674
  pearl, 674
  poria, 508
  rehmannia, 522
  tienchi ginseng, 599
Trachelogenin, 101
Tracheloside, 101
Tragacanthic acid, 602
Tragacanthin, 602
Trehalose, 301, 302
Triacontane, 13, 227, 237, 437, 629
Triacontanol (myricyl alcohol):
  biological activities, 85
  chenopodium oil, 179–180
  coriander, 227–228
  echinacea, 251–256
  lemongrass, 412
  West Indian bay, 78–79
Tribestin, 646
Tribulosin, 646
Tricetin, 82, 325
2,4,6-Trichloroanisole, 423
Tricosane, 626
Tricycloekasanthial, 554
1,11-Tridecadiene-3,5,7,9-tetrayne, 120
1,3,11-Tridecatriene-5,7,9-triyne, 120
trans- Tridecene-(2) al-(1), 227
Trifoliin, *see* Isoquercitrin
Trifoside (5-hydroxy-7-methoxy-isoflavone-4′-O-β-D-glucopyranoside), 207
Triglycerides:
  Almonds, 22
  in blood, reduction, artichoke, 46
  cocoa, 217
  dill, 245
  ginger, 321
  musk, 455
  nutmeg and mace, 468
Trigonelline:
  biological activities, 217, 222
  coffee, 222
  fenugreek, 286
  as flavor precursor, coffee, 222

1,2,4-Trihydroxyheptadeca-16-ene, 59, 60
2,2,6-Trimethylcyclohexanone, 404
O,O,N- Trimethylrosmaricine, 536
Triolein, 22, 218, 455
Trisaccharide, 331, 347, 511
Triterpene acids
  balsam tolu, 71
  marjoram, 436–438
  poria, 508
  rosemary, 535–537
  savory, 559–560
  storax, 579
  thyme, 594–597
Triterpene alcohols, 170, 188
Triterpene lactones, 378
Triterpenes:
  Aspidium, 54
  Balm, 71
  balsam tolu, 71
  bayberry bark, 80
  black cohosh, 97
  blue cohosh, 104
  boneset, 109
  centaury, 167
  chaparral, 175
  chirata, 192
  codonopsis, 220
  comfrey, 225–226
  dandelion root, 238
  deertongue, 241
  Devil's claw, 242–243
  elder flowers, 257
  eleuthero, 262
  euphorbia, 274
  fennel, 283–285
  ganoderma, 301
  gentian, 316
  ginseng, 331
  gymnema sylvestre, 648
  immortelle, 378
  jujube, 388
  lavender, 407
  lemongrass, 412–414
  licorice root, 416
  lime oil, 422–423
  nettle, 464
  pentacyclic, 352, 471, 472
  psyllium, 511
  sour jujube kernel, 572
  stevia, 577

  tea, 590–593
  white pine bark, 501
Triterpenic alcohols:
  Quebracho, 517
Triterpenoids: (3α-hydroxy-D-friedoolean-5-ene)
  biological activities, 302, 655
  calendula, 129
  cranberry, 232
  forsythia fruit, 665
  iridoid, 178
  licorice root, 416
  ligustrum, 420
  vasaka, 655
Triticin, 248
Tropine, 212, 645
Trypsin, 20, 114, 141, 210, 211
Tryptamine, foods containing, yohimbe contraindication, 631–632
Turicine, 362
Turmerone, 604
Tussilagine, 252
Tyramine, 115, 116, 217, 352, 430, 449, 632
Tyrosinase, inhibition, fangfeng, 281
l-Tyrosine, 281
Tyrosine betaine, 266

Umbelliferone (7-hydroxy-coumarin):
  angostura bark, 34–35
  asafetida, 48
  biological activities, 283
  galbanum, 299
  German chamomile, 170
  Hydrangea, 370
  Lavender, 406
  passion flower, 490
Umbelliprenine, 32, 36, 245
2-Undecanone (methyl-*n*-nonyl ketone), 542, 543
1,3,5-Undecatrienes, 299
Urea, 5, 107, 536, 679
Utsadiol
12-Ursene-3,16,21-triol, 129
Ursolic acid:
  biological activities, 388, 437, 536
  cherry laurel leaves, 184
  elder flowers, 257

hawthorn, 352
hyssop, 376
jujube, 388
lactone, 378
lavender, 407
ligustrum, 420
oregano, 482
rosemary, 535
thyme, 595
uva ursi, 610
Uvaol, 378, 610

Valepotriates, 612, 613
Valeranone, 612, 613
Valerenal, 612, 613
Valerianine, 612
Valeric acid, 48, 99, 428, 478, 502, 581
Valerine, 212, 612
Valerosidatum, 612
Validene-4,5-dihydrophthalide, 428
Valine, as amino terminal residue, bromelain, 113–114
Valtra tes, 612, 613
Vanillic acid, 187, 197, 683
Vanillin:
  balsam Peru, 70
  balsam tolu, 71
  benzoin, 90
  biological activities, 617
  carrot seed oil, 148
  echinacea, 252
  storax, 579
  vanilla, 616
Vanillin glucoside, 248
Vanillyl ethyl ether, 616
L-vasicine, 655
    mahbala, 651
L-vasicinone, 655
    mahbala, 651
L-vasicol, 655
    mahbala, 651
Vasnetine, 655
Verbenol, 471, 535
Vicenin-2, 187, 287, 482, 587
Vicenin (6,8-di-C-glucosyl-5,7,3′-trihydroxyfl avone), 245
Villosin, 97
Vinyl dithiins, 308
Violaxanthin, 52, 130, 238, 478

Viridiflorol, 124, 272, 404, 444, 550
Viscotoxins, 449, 450
Viscumin, 449
Vitamin A:
  acerola, 6
  alfalfa, 13
  allspice, 20
  baizhu, 660
  bee pollen, 82
  cangzhu, 660
  capsicum, 133
  coca, 212
  dandelion root, 238
  doggrass, 248
  fenugreek, 287
  ginger, 321
  hypericum, 372
  jujube, 388
  marjoram, 437
  natural source, carrot seed oil, 148
  onion, 475
  orange, 478
  parsley, 487
  purslane, 679
  savory, 559
  schisandra, 565
  Starting materials, lemongrass, 414
  sweet basil, 75
  synthetic source, lemongrass, 414
Vitamin B
  almonds, 22
  cangzhu, 662
  dandelion root, 238
  doggrass, 248
Vitamin B_1(thiamine)
  allspice, 20
  acerola, 6
  bee pollen, 82
  chickweed, 187
  fenugreek, 287
  garlic, 308
  ginseng, 331
  honey, 358
  hawthorn, 352–354
  lycium fruit, 430
  Job's tears, 388
  onion, 475
  orange, 478
  purslane, 679

quassia, 515
royal jelly, 535
saffron, 547
tamarind, 584
Vitamin B_2 (riboflavin):
  acerola, 6–7
  allspice, 19–21
  bee pollen, 82
  chickweed, 187–188
  coca, 212
  fenugreek, 287
  garlic, 308
  ginseng, 331
  hawthorn, 352
  honey, 358
  jujube, 388
  lycium fruit, 430
  onion, 475
  purslane, 679
  rosehips, 181–182
  saffron, 547–548
Vitamin B_3 (niacin):
  acerola, 6
  alfalfa, 13
  allspice, 20
  chickweed, 187
  coffee, 222
  garlic, 308
  ginger, 321
  royal jelly, 539
Vitamin B_6 (pyridoxine), 13, 539
Vitamin B_{12} (cyanocobalamin), 13, 539
Vitamin C:
  Absinthium, 1–3
  acerola, 6
  alfalfa, 13
  allspice, 20
  almonds, 22
  bee pollen, 82
  bilberry, 93
  biological activities, 181, 529
  capsicum, 133
  Cherokee rosehip, 181
  Chickweed, 187
  as color stabilizer, 86
  coriander, 227
  corn silk, 229
  cranberry, 232
  dandelion root, 238
  elder flowers, 258

Vitamin C: (*Continued*)
  fenugreek, 287
  gotu kola, 340
  hawthorn, 352
  honey, 358
  hypericum, 372
  jujube, 388
  Juniper berries, 389–391
  lycium fruit, 430
  marjoram, 437
  nettle, 464
  orange, 478
  as papain inhibitor
  parsley, 487
  purslane, 679
  rosehips, 528
  royal jelly, 539
  schisandra, 565
  sweet basil, 75
  tamarind, 584
  tea, 591
  turmeric, 604
Vitamin D, 59
Vitamin E (tocopherols):
  alfalfa, 13
  almonds, 22
  bee pollen, 82
  coffee, 222
  eleuthero, 263
  fennel, 283
  mints, 444
  orange, 478
  purslane, 679
  red sage, 680
  royal jelly, 539
  schisandra, 565
  utilization interference of, 14
Vitamin K:
  alfalfa, 13
  bee pollen, 82
  corn Silk, 229
  henna, 356
Vitamins:
  absinthium, 1–3
  alfalfa, 13
  allspice, 20
  bee pollen, 82
  capsicum, 133
  cloves, 210
  coffee, 222
  dandelion root, 238
  doggrass, 248

fennel, 283
fenugreek, 287
garlic, 308
ginger, 321
ginseng, 331
jujube, 388
lycium fruit, 430
marjoram, 437
onion, 475
orange, 478
oregano, 482
parsley, 487
purslane, 679
red clover tops, 207
rosehips, 181
royal jelly, 539
saffron, 547
savory, 559
schisandra, 565
tamarind, 584
turmeric, 604
Vitexin:
  biological activities, 178
  fenugreek, 287
  hawthorn, 352
Vitexin-7-glucoside, 287
Vulgarol, 362
Vulgaxanthin-I and –II, 86

Wax:
  Alkanet, 18
  cherry laurel leaves, 184
  cinchona, 194
  eucalyptus, 272
  honey, 358
  immortelle, 378
  jojoba, 386
  lemon oil, 410
  licorice root, 416
  mume, 672
  musk, 455
  royal jelly, 539
Withaferin, 645
Withanolides, 645
Withasomniferin, 645
Withasomidienone, 645
Withasomniferols, 645
Withasomine,
Withanone, 645
Withaniol, 645
Wogonin, 187, 685, 686
Wogonoside, 685, 686

Wushanicariin, 269
Wuweizisu C, 565, 566

Xanthine, 350, 352, 591
Xanthohumol, 359, 360
Xanthomicrol, 75
Xanthones:
  biological activities, 192
  chirata, 192
  gentian, 316
  henna, 355
  hypericum, 372
Xanthophylls, 130
Xanthotoxin
      (8-methoxypsoralen):
  Angelica, 32, 33
  biological activities, 92, 442
  parsley, 487
  potential source, 33
  rue, 542
Xanthoxyletin, 50
Xanthyletin, 50
Xylan, 188, 671
Xylose, 10, 13, 143, 220, 225,
      449, 461, 508, 511, 522,
      602, 619, 682, 696

Yamogenin tetrosides B and C,
      287
α-Ylangene, 67, 197, 210, 478
β-Ylangene, 471
α-Yohimbane, 631
Yohimbine, 517, 631, 632
Yohimbinine, 631

Zeaxanthin, 52, 133, 154, 430,
      581
Zinc:
  Alfalfa, 13
  almonds, 23
  astragalus, 56
  enzyme inhibition or
      inactivation by, 385
  fo-ti, 295
  ganoderma, 300–304
  ginseng, 331
  in red beet color breakdown, 86
Zingerone, 321
Zingiberene, 321, 604
Zivulgarin, 572
Ziyu-glycosides, 666
Zizyphus saponins, 388